T0367923

Modeling Materials

Material properties emerge from phenomena on scales ranging from ångstroms to millimeters, and only a multiscale treatment can provide a complete understanding. Materials researchers must therefore understand fundamental concepts and techniques from different fields, and these are presented in a comprehensive and integrated fashion for the first time in this book.

Incorporating continuum mechanics, quantum mechanics, statistical mechanics, atomistic simulations, and multiscale techniques, the book explains many of the key theoretical ideas behind multiscale modeling. Classical topics are blended with new techniques to demonstrate the connections between different fields and highlight current research trends. Example applications drawn from modern research on the thermomechanical properties of crystalline solids are used as a unifying focus throughout the text.

Together with its companion book, *Continuum Mechanics and Thermodynamics* (Cambridge University Press, 2012), this work presents the complete fundamentals of materials modeling for graduate students and researchers in physics, materials science, chemistry, and engineering.

Ellad B. Tadmor is Professor of Aerospace Engineering and Mechanics, University of Minnesota. His research focuses on the development of multiscale theories and computational methods for predicting the behavior of materials directly from the interaction of the atoms.

Ronald E. Miller is Professor of Mechanical and Aerospace Engineering, Carleton University. He has worked in the area of multiscale materials modeling for over 15 years and has published more than 40 scientific articles in the area.

Modeling Materials

Continuum, Atomistic and Multiscale Techniques

ELLAD B. TADMOR
University of Minnesota, USA

RONALD E. MILLER
Carleton University, Canada

CAMBRIDGE
UNIVERSITY PRESS

CAMBRIDGE
UNIVERSITY PRESS

Shaftesbury Road, Cambridge CB2 8EA, United Kingdom

One Liberty Plaza, 20th Floor, New York, NY 10006, USA

477 Williamstown Road, Port Melbourne, VIC 3207, Australia

314–321, 3rd Floor, Plot 3, Splendor Forum, Jasola District Centre, New Delhi – 110025, India

103 Penang Road, #05–06/07, Visioncrest Commercial, Singapore 238467

Cambridge University Press is part of Cambridge University Press & Assessment,
a department of the University of Cambridge.

We share the University's mission to contribute to society through the pursuit of education, learning and research at the highest international levels of excellence.

www.cambridge.org
Information on this title: www.cambridge.org/9780521856980

© E. Tadmor and R. Miller 2011

This publication is in copyright. Subject to statutory exception and to the provisions of relevant collective licensing agreements, no reproduction of any part may take place without the written permission of Cambridge University Press & Assessment.

First published 2011
5th printing 2020

A catalogue record for this publication is available from the British Library

Library of Congress Cataloging-in-Publication data
Tadmor, Ellad B., 1965–
Modeling materials : continuum, atomistic, and multiscale techniques /
Ellad B. Tadmord, Ronald E. Miller.
p. cm.
Includes bibliographical references and index.
ISBN 978-0-521-85698-0 (hardback)
1. Materials – Mathematical models. I. Miller, Ronald E. II. Title.
TA404.23.T33 2011
620.1'10113 – dc23 2011025635

ISBN 978-0-521-85698-0 Hardback

Additional resources for this publication at www.cambridge.org/9780521856980

Cambridge University Press & Assessment has no responsibility for the persistence or accuracy of URLs for external or third-party internet websites referred to in this publication and does not guarantee that any content on such websites is, or will remain, accurate or appropriate.

Contents

Preface

Studying *materials* can mean studying almost anything, since all of the physical, tangible world is necessarily made of *something*. Normally, we think of studying materials in the sense of materials science and engineering – an endeavor to understand the properties of natural and man-made materials and to improve or exploit them in some way – but even this includes broad and disparate goals. One can spend a lifetime studying the strength and toughness of steel, for example, and never once concern oneself with its magnetic or electric properties. At the same time, *modeling* in science can mean many things to many people, ranging from computer simulation to analytical effective theories to abstract mathematics. To combine these two terms "modeling materials" as the title of a single book, then, is surely to invite disaster. How could it be possible to cover all the topics that the product *modeling* × *materials* implies? Although this book remains true to its title, it will be necessary to pick and choose our topics so as to have a manageable scope. To start with, then, we have to decide: what models and what materials do we want to discuss?

As far as *modeling* goes, we must first recognize the fact that materials exhibit phenomena on a broad range of spatial and temporal scales that combine together to dictate the response of a material. These phenomena range from the bonding of individual atoms governed by quantum mechanics to macroscopic deformation processes described by continuum mechanics. Various aspects of materials behavior and modeling, which tend to focus on specific phenomena at a given scale, have traditionally been treated by different disciplines in science and engineering. The great disparity in scales and the interdisciplinary nature of the field are what makes modeling materials both challenging and exciting. It is unlikely that any one researcher has sufficient background in engineering, physics, materials science and mathematics to understand materials modeling at every length and time scale. Furthermore, there is increased awareness that materials must be understood, not *only* by rigorous treatment of phenomena at each of these scales alone, but rather through consideration of the *interactions* between these scales. This is the paradigm of *multiscale modeling* that will also be a persistent theme throughout the book.

Recognizing the need to integrate models from different disciplines creates problems of nomenclature, notation and jargon. While we all strive to make our research specialties clear and accessible, it is a necessary part of scientific discourse to create and use specific terms and notation. An unintended consequence of this is the creation of barriers to interdisciplinary understanding. One of our goals is to try to facilitate this understanding by providing a unified presentation of the fundamentals, using a common nomenclature, notation and language that will be accessible to people across disciplines. The result is this book on *Modeling Materials* (MM) and our companion book on *Continuum Mechanics and Thermodynamics* (CMT) [TME12].

The subject matter in MM is divided into four parts. **Part I** covers continuum mechanics and thermodynamics concepts that serve as the basis for the rest of the book. The description of continuum mechanics and thermodynamics is brief and is only meant to make MM a stand-alone book. The reader is referred to CMT for a far deeper view of these subjects consistent with the rest of MM. **Part II** covers atomistics, discussing the basic structure and symmetries of crystalline materials, quantum mechanics and more approximate empirical models for describing bonding in materials, and molecular statics – a computational approach for studying static properties of materials at the atomic scale. **Part III** focuses on the atomistic foundations of continuum concepts. Here, using ideas from statistical mechanics, connections are forged between the discrete world of atoms – described by atomic positions, velocities and forces – and continuum concepts such as fields of stress and temperature. Finally, the subject of molecular dynamics (MD) is presented. We treat MD as a computational method for studying dynamical properties of materials at the atomic scale subject to continuum-level constraints, so it is yet another unique connection between the atomistic and continuum views. **Part IV** on multiscale methods describes a class of computational methods that attempt to model material response by simultaneously describing its behavior on multiple spatial and temporal scales. This final part of the book draws together and unifies many of the concepts presented earlier and shows how these can be integrated into a single modeling paradigm.

By bringing together this unusual combination of topics, we provide a treatment that is uniquely different from other books in the field. First, our focus is on a critical analysis and understanding of the fundamental *assumptions* that underpin these methods and that are often taken for granted in other treatments. We believe that this focus on fundamentals is essential for anyone seeking to combine different theories in a multiscale setting. Secondly, some of the topics herein are often treated from the perspective of the gaseous or liquid states. Here, our emphasis is on solids, and this changes the presentation in important ways. For example, in statistical mechanics we comprehensively discuss the subject of the stress tensor (not just pressure) and the concept of a restricted ensemble for metastable systems. Similarly, we talk at length about constant stress simulations in MD and how to correctly interpret them in a setting of finite deformation beyond that of simple hydrostatic compression. Third, while covering this broad range of topics we strive to regularly make connections between the atomistic, statistical and continuum worldviews. Finally, we have tried to create a healthy balance between fundamental theory and practical "how to." For example, we present, at length, the practical implementation of such topics as density functional theory, empirical atomistic potentials, molecular statics and dynamics, and multiscale partitioned-domain methods. It is our hope that someone with basic computer programming skills will be able to use this book to implement any of these methods, or at least to better understand an implementation in a pre-existing code.

Although the modeling methods we describe are, in principle, applicable to any material, we focus our scope of materials and properties on those that we, the authors, know best. The answer to "What materials?" then is crystalline solids and their thermomechanical (as opposed to electrical, optical or chemical) properties. For the most part, these serve as examples to illustrate the application and usefulness of the modeling methods that we describe, but we hope that the reader will also learn something new regarding the materials themselves along the way.

Even starting from this narrow mandate we have already failed, to some degree, in our goal of putting all the fundamentals in one place. This is because the binding of these subjects into a single volume becomes unwieldy if we wish to maintain the level of detail that we feel is necessary. To make room in this book, we sacrificed coverage of continuum mechanics, leaving only a concise summary in Chapter 2 of the key results needed to make contact with the rest of the topics. CMT [TME12], the companion volume to this one, provides the full details of the continuum mechanics and thermodynamics that we believe to be fundamental to materials modeling.

Both books, MM and CMT, are addressed to graduate students and researchers in chemistry, engineering, materials science, mathematics, mechanics and physics. The interdisciplinary nature of materials modeling means that researchers from all of these fields have contributed to and continue to be engaged in this field. The motivation for these books came from our own frustration, and that of ours students, as we tried to acquire the breadth of knowledge necessary to do research in this highly interdisciplinary field. We have made every effort to eliminate this frustration in the future by making our writing accessible to all readers with an undergraduate education in an engineering or scientific discipline. The writing is self-contained, introducing all of the necessary basic concepts and building up from there. Of course, by necessity that means that our coverage of the different topics is limited and skewed to our primary focus on materials modeling. At the end of each chapter, we recommend sources for further reading for readers interested in expanding their understanding in a particular direction.

Acknowledgments

One of our favorite teachers when we were graduate students at Brown University, Ben Freund, has said that writing a book is like giving birth. The process is long and painful, it involves a lot of screaming, but in the end something has to come out. We find this analogy so apt that we feel compelled to extend it: in some cases, you are blessed with twins.[1] As we initially conceived it, our goal was to have everything in a single volume. But as time went on, and what we were "carrying" grew bigger and bigger, it became clear that it really needed to be two separate books.

Since the book has been split in two, we choose to express our gratitude twice, once in each book, to everyone who has helped us with the project as a whole. Surely, thanking everyone twice is the least we can do. Some people helped in multiple ways, and so their names appear even more often. Our greatest debt goes to our wives, Jennifer and Granda, and to our children: Maya, Lea, Persephone and Max. They have suffered more than anyone during the long course of this project, as their preoccupied husbands and fathers stole too much time from so many other things. They need to be thanked for such a long list of reasons that we would likely have to split these two books into three if we were thorough with the details. Thanks, all of you, for your patience and support. We must also thank our own parents Zehev and Ciporah and Don and Linda for giving us the impression – perhaps mistaken – that everybody will appreciate what we have to say as much as they do.

The writing of a book as diverse as this one is really a collaborative effort with so many people whose names do not appear on the cover. These include students in courses, colleagues in the corridors and offices of our universities and unlucky friends cornered at conferences. The list of people that offered a little piece of advice here, a correction there, or a word of encouragement somewhere else is indeed too long to include, but there are a few people in particular that deserve special mention.

Some colleagues generously did calculations for us, verified results, or provided other contributions from their own work. We thank Quiying Chen at the NRC Institute for Aerospace Research in Ottawa for his time in calculating UBER curves with DFT. Tsveta Sendova, a postdoctoral fellow at the University of Minnesota, coded and ran the simulations for the two-dimensional NEB example we present. Another postdoctoral fellow at the University of Minnesota, Woo Kyun Kim, performed the indentation and thermal expansion simulations used to illustrate the hot-QC method. We thank Yuri Mishin (George Mason

[1] This analogy is made with the utmost respect for our wives, and anyone else who *actually has* given birth. Assuming labor has units of power, then we feel that the integral of this power over the very different timescales of the two processes should yield quantities of work that are on the same order of magnitude. Our wives disagree, no doubt in part because some of the power consumed by book-writing indirectly comes from them, whereas our contribution to childbirth amounts mainly to sending around e-mail photos of the newborn children.

University) for providing figures, and Christoph Ortner (Oxford University) for providing many insights into the problem of full versus sequential minimization of multivariate functions, including the example we provide in the book. The hot-QC project has greatly benefited from the work of Laurent Dupuy (SEA Saclay) and Frederic Legoll (École Nationale des Ponts et Chaussées). Their help in preparing a journal paper on the subject has also proven extremely useful in preparing the chapter on dynamic multiscale methods. Furio Ercolessi must be thanked in general for his fantastic web-based notes on so many important subjects discussed herein, and specifically for providing us with his molecular dynamics code as a teaching tool to provide with this book.

Other colleagues patiently taught us the many subjects in this book about which we are decidedly *not* experts. Dong Qian at the University of Cincinnati and Michael Parks at Sandia National Laboratories very patiently and repeatedly explained the nuances of various multiscale methods to us. Similarly, we would like to thank Catalin Picu at the Rensselaer Polytechnic Institute for explaining CACM, and Leo Shilkrot for his frank conversations about CADD and the BSM. Noam Bernstein at the Navy Research Laboratories was invaluable in explaining DFT in a way that an engineer could understand, and Peter Watson at Carleton University was instrumental in our eventual understanding of quantum mechanics. Roger Fosdick University of Minnesota discussed, at length, many topics related to continuum mechanics including tensor notation, material frame-indifference, Reynolds transport theorem and the principle of action and reaction. He also took the time to read and comment on our take on material frame-indifference.

We are especially indebted to those colleagues that were willing to take the time to carefully read and comment on drafts of various sections of the book – a thankless and delicate task. James Sethna (Cornell University) and Dionisios Margetis (University of Maryland) read and commented on the statistical mechanics chapter. Noam Bernstein (Navy Research Laboratories) must be thanked more than once for reading and commenting on both the quantum mechanics chapter and the sections on cluster expansions. Nikhil Admal, a graduate student working with Ellad at the University of Minnesota, contributed significantly to our understanding of stress and read and commented on the continuum mechanics chapter, Marcel Arndt helped by translating an important paper on stress by Walter Noll from German to English and worked with Ellad on developing algorithms for lattice calculations, while Gang Lu at the California State University (Northridge) set us straight on several points about density functional theory. Ryan Elliott, our coauthor of the companion book to this one, must also be thanked countless times for his careful reading of quantum mechanics and his many helpful suggestions and discussions. Other patient readers to whom we say "thank you" include Mitch Luskin from the University of Minnesota (numerical analysis of multiscale methods and quantum mechanics), Bill Curtin from Brown University (static multiscale methods), Dick James from the University of Minnesota (restricted ensembles and the definition of stress) and Leonid Berlyand from Pennsylvania State University (thermodynamics).

There are a great many colleagues that were willing to talk to us at length about various subjects in this book. We hope that we did not overstay our welcome in their offices too often, and that they do not sigh too deeply anymore when they see a message from us in their inbox. Most importantly, we thank them very much for their time. In addition to those already mentioned above, we thank David Rodney (Institut National Polytechnique Grenoble), Perry Leo and Tom Shield (University of Minnesota), Miles Rubin and Eli Altus (Technion), Joel Lebowitz, Sheldon Goldstein and Michael Kiessling (Rutgers),[2] and Andy Ruina (Cornell University). We would also be remiss if we did not take the time to thank Art Voter (Los Alamos National Laboratory), John Moriarty (Lawrence Livermore National Laboratory) and Mike Baskes (Sandia National Laboratory) for many insightful discussions and suggestions of valuable references.

There are some things in these books that are so far outside our area of expertise that we have even had to look beyond the offices of professors and researchers. Elissa Gutterman, an expert in linguistics, provided phonetic pronunciation of French and German names. As neither of us is an experimentalist, our brief foray into pocket watch "testing" would not have been very successful without the help of Steve Truttman and Stan Conley in the structures laboratories at Carleton University. The story of our cover images involves so many people, it deserves its own paragraph.

As the reader will see in the introduction to both books, we are fond of the symbolic connection between pocket watches and the topics we discuss herein. There are many beautiful images of pocket watches out there, but obtaining one of sufficient resolution, and getting permission to *use* it, is surprisingly difficult. As such, we owe a great debt to Mr. Hans Holzach, a watchmaker and amateur photographer at Beyer Chronometrie AG in Zurich. Not only did he generously agree to let us use his images, he took over the entire enterprise of retaking the photographs when we found out that his images did not have sufficient resolution! This required Hans to coordinate with many people that we also thank for helping make the strikingly beautiful cover images possible. These include the photographer, Dany Schulthess (www.fotos.ch), Mr. René Beyer, the owner of Beyer Chronometrie AG in Zurich, who compensated the photographer and allowed photographs to be taken at his shop, and also to Dr. Randall E. Morris, the owner of the pocket watch, who escorted it from California to Switzerland (!) in time for the photo shoot. The fact that total strangers would go to such lengths in response to an unsolicited e-mail contact is a testament to their kind spirits and, no doubt, to their proud love of the beauty of pocket watches.

We cannot forget our students. Many continue to teach us things every day just by bringing us their questions and ideas. Others were directly used as guinea pigs with early drafts of parts of this book.[3] Ellad would like to thank his graduate students and post-doctoral fellows over the last five years who have been fighting with this book for attention; specifically Nikhil Admal, Yera Hakobian, Hezi Hizkiahu, Dan Karls, Woo Kyun Kim, Leonid Kucherov, Amit Singh, Tsvetanka Sendova, Valeriu Smiricinschi, Slava Sorkin and Steve Whalen. Ron would likewise like to thank Ishraq Shabib, Behrouz Shiari and Denis Saraev, whose work helped shape his ideas about atomistic modeling. Harley Johnson and his 2008–2009 graduate class at the University of Illinois (Urbana-Champaign) used the book extensively and provided great feedback to improve the manuscript, as did Bill

[2] Ellad would particularly like to thank the Rutgers trio for letting him join them on one of their lunches to discuss the foundations of statistical mechanics – a topic which is apparently standard lunch fare for them along with the foundations of quantum mechanics.

[3] Test subjects were always treated humanely and no students were harmed during the preparation of this book.

Curtin's class at Brown University in 2009–2010. The 2009 and 2010 classes of Ron's "Microstructure and properties of engineering materials" class caught many initial errors in the chapters on crystal structures and molecular statics and dynamics. Some students of Ellad's continuum mechanics course are especially noted for their significant contributions: Yilmaz Bayazit (2008), Pietro Ferrero (2009), Zhuang Houlong (2008), Jenny Hwang (2009), Karl Johnson (2008), Dan Karls (2008), Minsu Kim (2009), Nathan Nasgovitz (2008), Yintao Song (2008) and Chonglin Zhang (2008).

Of course, we should also thank our own teachers. Course notes from Michael Ortiz, Janet Blume, Jerry Weiner and Tom Shield were invaluable to us in preparing our own notes and this book. Thanks also to our former advisors at Brown University, Michael Ortiz and Rob Phillips (both currently at Caltech), whose irresistible enthusiasm, curiosity and encouragement pulled us down this most rewarding of scientific paths.

We note that many figures in this book were prepared with the drawing package Asymptote (see http://asymptote.sourceforge.net/), an open-source effort that we think deserves to be promoted here. Finally, we thank our editor Simon Capelin and the entire team at Cambridge, for their advice, assistance and truly astounding patience.

Notation

In a book covering such a broad range of topics, notation is a nightmare. We have attempted, as much as possible, to use the most common and familiar notation from within each field as long as this did not lead to confusion. However, this does mean that the occasional symbol will serve multiple purposes, as the tables below will help to clarify. To keep the amount of notation to a minimum, we generally prefer to append qualifiers to symbols rather than introducing new symbols. For example, $\boldsymbol{f}$ is force, which if relevant can be divided into internal, $\boldsymbol{f}^{\text{int}}$, and external, $\boldsymbol{f}^{\text{ext}}$, parts.

We use the following general conventions:

- Descriptive qualifiers generally appear as superscripts and are typeset using a Roman (as opposed to Greek) nonitalic font.
- The weight and style of the font used to render a variable indicates its type. Scalar variables are denoted using an italic font. For example, T is temperature. Array variables are denoted using a sans serif font, such as A for the matrix A. Vectors and tensors (in the technical sense of the word) are rendered in a boldface font. For example, $\boldsymbol{\sigma}$ is the stress tensor.
- Variables often have subscript and superscript indices. Indices referring to the components of a matrix, vector or tensor appear as subscripts in italic Roman font. For example, v_i is the ith component of the velocity vector. Superscripts are used as counters of variables. For example, $\boldsymbol{F}^e$ is the deformation gradient in element e. Superscripts referring to atoms are distinguished by using a Greek letter. For example, the velocity of atom α is denoted $\boldsymbol{v}^\alpha$. Iteration counters appear in parentheses, for example $\boldsymbol{f}^{(i)}$ is the force in iteration i.
- The Einstein summation convention is followed on repeated indices (e.g. $v_i v_i = v_1^2 + v_2^2 + v_3^2$), unless otherwise clear from the context. (See Section 2.1.1 for more details.)
- One special type of superscript concerns the denotation of Bravais lattices and crystals. For example, the position vector $\boldsymbol{R}^{[\boldsymbol{\ell}\lambda]}$ denotes the λth basis atom associated with Bravais lattice site $\boldsymbol{\ell}$. (See Section 3.6 for details.)
- A subscript is used to refer to multiple equations on a single line, for example "Eqn. $(2.66)_2$" refers to the second equation in Eqn. (2.66) ("$a_i(\boldsymbol{x}, t) \equiv \dots$").
- Important equations are emphasized by shading.

Below, we describe the main notation and symbols used in the book, and indicate the page on which each is first defined. We also include a handy list of fundamental constants and unit conversions at the end of this section.

Mathematical notation

Notation	Description	Page
$\equiv$	equal to by definition	28
$:=$	variable on the left is assigned the value on the right	24
$\forall$	for all	28
$\in$	contained in	28
iff	if and only if	28
$O(f)$	terms proportional to order f	188
$O(n)$	orthogonal group of degree n	31
$SO(n)$	proper orthogonal (special orthogonal) group of degree n	31
$\mathbb{R}$	set of all real numbers	26
$\mathbb{R}^n$	real coordinate space (n-tuples of real numbers)	27
$\lvert \bullet \rvert$	absolute value of a real number	28
$\lVert \bullet \rVert$	norm of a vector	28
$\langle \bullet, \bullet \rangle$	inner product of two vectors	28
$\langle \bullet \vert \bullet \rangle$	inner product of two vectors (bra-ket notation)	161
$\langle \bullet \vert \bullet \vert \bullet \rangle$	bra-operator-ket inner product	163
$[uvw]$	direction in a crystal ($u\boldsymbol{a} + v\boldsymbol{b} + w\boldsymbol{c}$)	125
$\langle uvw \rangle$	family of crystal directions	125
(hkl)	Miller indices denoting a crystallographic plane	150
$\{hkl\}$	family of crystallographic planes	151
$\{M \mid C\}$	a set of members M such that conditions C are satisfied	245
$\overline{\bullet}$	time average of a quantity	388
$\langle \bullet \rangle$	phase average of a quantity	391
$\langle \bullet; f \rangle$	phase average of a quantity relative to distribution function f	391
$\Pr(O)$	probability of outcome O	395
$\mathrm{Var}(A)$	variance of A: $\mathrm{Var}(A) = \langle A^2 \rangle - (\langle A \rangle)^2$	396
$\mathrm{Cov}(A, B)$	covariance of A and B: $\mathrm{Cov}(A, B) = \langle AB \rangle - \langle A \rangle \langle B \rangle$	461
$\mathrm{Cov}_\chi(A, B)$	covariance of A and B in a restricted ensemble	577
$\widehat{f}(\boldsymbol{k})$	Fourier transform of $f(\boldsymbol{x})$	166
$\underset{\sim}{f}(s)$	Laplace transform of a $f(t)$	684
$\bullet^*$	complex conjugate	161
$\boldsymbol{A}^T$	transpose of a matrix or second-order tensor: $[\boldsymbol{A}^T]_{ij} = A_{ji}$	25
$\boldsymbol{A}^{-T}$	transpose of the inverse of$\boldsymbol{A}$: $\boldsymbol{A}^{-T} \equiv (\boldsymbol{A}^{-1})^T$	35
$\boldsymbol{a} \cdot \boldsymbol{b}$	dot product (vectors): $\boldsymbol{a} \cdot \boldsymbol{b} = a_i b_i$	28
$\boldsymbol{a} \times \boldsymbol{b}$	cross product (vectors): $[\boldsymbol{a} \times \boldsymbol{b}]_k = \epsilon_{ijk} a_i b_j$	30
$\boldsymbol{a} \otimes \boldsymbol{b}$	tensor product (vectors): $[\boldsymbol{a} \otimes \boldsymbol{b}]_{ij} = a_i b_j$	33
$\boldsymbol{A} : \boldsymbol{B}$	contraction (second-order tensors): $\boldsymbol{A} : \boldsymbol{B} = A_{ij} B_{ij}$	36
$\boldsymbol{A} \cdot\cdot \boldsymbol{B}$	transposed contraction (second-order tensors): $\boldsymbol{A} \cdot\cdot \boldsymbol{B} = A_{ij} B_{ji}$	36
$\lambda_\alpha^{\boldsymbol{A}}$, $\boldsymbol{\Lambda}_\alpha^{\boldsymbol{A}}$	αth eigenvalue and eigenvector of the second-order tensor $\boldsymbol{A}$	38
$I_k^{\boldsymbol{A}}$	kth principal invariant of the second-order tensor $\boldsymbol{A}$	38
$\det \boldsymbol{A}$	determinant of a matrix or a second-order tensor	26
$\mathrm{tr}\, \boldsymbol{A}$	trace of a matrix or a second-order tensor: $\mathrm{tr}\, \boldsymbol{A} = A_{ii}$	25

$\nabla\bullet$, grad $\bullet$	gradient of a tensor (deformed configuration)	40
$\nabla_0\bullet$, Grad $\bullet$	gradient of a tensor (reference configuration)	45
curl $\bullet$	curl of a tensor (deformed configuration)	41
Curl $\bullet$	curl of a tensor (reference configuration)	45
div $\bullet$	divergence of a tensor (deformed configuration)	41
Div $\bullet$	divergence of a tensor (reference configuration)	45
$\nabla^2\bullet$	Laplacian of a tensor (deformed configuration)	41
$đ$	inexact differential	81
$\boldsymbol{r}^{\alpha\hat{\beta}}$	position vector to closest periodic image of β to atom α	326
$\overset{\diamond}{\alpha}$, $\overset{\#}{\alpha}$	unit cell and sublattice of atom α in a multilattice crystal	564

General symbols – Greek

Symbol	Description	Page
Γ	phase space	382
$\boldsymbol{\Gamma}$, Γ_i	set of extensive kinematic state variables	63
$\boldsymbol{\Gamma}_i$	wave vector of the ith DFT plane wave basis function	211
$\boldsymbol{\gamma}$, γ_i	set of intensive state variables work conjugate with $\boldsymbol{\Gamma}$	78
γ	damping coefficient	511
γ_s	surface energy	340
γ_{GB}	grain boundary energy	346
γ_{SF}	stacking fault energy	357
δ_{ij}	Kronecker delta	25
ϵ	energy of an electron	164
$\boldsymbol{\epsilon}$, ϵ_{ij}	small strain tensor	49
ϵ_{ijk}	permutation symbol	26
ζ_i^α	fractional coordinates of basis atom α	142
$\kappa^{\alpha\beta\gamma\delta}$	scalar atomistic stiffness term relating bonds α–β and γ–δ	297
Λ	de Broglie thermal wavelength	241
Λ_i	projection operator	162
λ	Lamé constant	105
λ	plane wave wavelength	164
μ	shear modulus (solid)	105
$\mu(m)$	mth moment of a function	230
ν	Poisson's ratio	105
ν_e	number of atoms associated with element e in QC	612
Π	total potential energy of a system and the applied loads	107
$\boldsymbol{\Pi}^\alpha$, Π_i^α	pull-back momentum of atom α	452
ρ	mass density (deformed configuration)	52
ρ	electron density	188
ρ_0	mass density (reference configuration)	52
ρ^{pt}	pointwise (microscopic) mass density field	468

ρ^α	total electron density at atom α in a pair functional	263
$\Sigma(E;\Delta E)$	hypershell in phase space with energy E and thickness ΔE	404
$\boldsymbol{\sigma}$, σ_{ij}	Cauchy stress tensor	56
$\boldsymbol{\sigma}^{\text{inst}}$, $\sigma_{ij}^{\text{inst}}$	instantaneous atomic-level stress	457
$\boldsymbol{\sigma}^{\text{pt}}$, σ_{ij}^{pt}	pointwise (microscopic) Cauchy stress tensor	470
$\boldsymbol{\sigma}^{\text{pt,K}}$, $\sigma_{ij}^{\text{pt,K}}$	kinetic part of the pointwise (microscopic) Cauchy stress	471
$\boldsymbol{\sigma}^{\text{pt,V}}$, $\sigma_{ij}^{\text{pt,V}}$	potential part of the pointwise (microscopic) Cauchy stress	471
$\boldsymbol{\varphi}$, φ_i	deformation mapping	43
$\phi(r)$	pair potential as a function of distance r	251
φ	electron wave basis function	173
$\varphi^{\alpha\beta}$	scalar magnitude of force on atom α due to presence of atom β	291
χ	general, time-dependent electronic wave function	163
χ	characteristic function in restricted ensemble	554
ψ	specific Helmholtz free energy	95
ψ	general, time-independent electronic wave function	165
ψ^{sp}	single-particle, time-independent electronic wave function	194
Ω	volume of a periodic simulation cell in a DFT simulation	210
Ω_0	nonprimitive unit cell volume in reference configuration	124
$\widehat{\Omega}$	volume of the first Brillouin zone	208
$\widehat{\Omega}_0$	primitive unit cell volume in reference configuration	122
$\Omega(E;\Delta E)$	volume of hypershell $\Sigma(E;\Delta E)$ in phase space	404
ω	plane wave frequency	164

General symbols – Roman

Symbol	Description	Page
$\mathcal{A}$	macroscopic observable associated with phase function $A(\boldsymbol{q},\boldsymbol{p})$	387
$A(\boldsymbol{q},\boldsymbol{p})$	phase function associated with macroscopic observable $\mathcal{A}$	387
$\boldsymbol{A}_1$, $\boldsymbol{A}_2$, $\boldsymbol{A}_3$	reference *nonprimitive* lattice vectors	123
$\hat{\boldsymbol{A}}_1$, $\hat{\boldsymbol{A}}_2$, $\hat{\boldsymbol{A}}_3$	reference *primitive* lattice vectors	120
$\boldsymbol{a}$, a_i	acceleration vector	50
$\boldsymbol{a}_1$,$\boldsymbol{a}_2$,$\boldsymbol{a}_3$	nonprimitive lattice vector (deformed configuration)	561
$\mathcal{B}$	the first Brillouin zone	212
B	bulk modulus	112
$B(\boldsymbol{x};\boldsymbol{u},\boldsymbol{v})$	bond function at $\boldsymbol{x}$ due to the spatially averaged bond $\boldsymbol{u}$–$\boldsymbol{v}$	479
$\boldsymbol{B}_1$, $\boldsymbol{B}_2$, $\boldsymbol{B}_3$	reciprocal reference lattice vectors	147
$\boldsymbol{B}$, B_{ij}	left Cauchy–Green deformation tensor	47
BO	bond order	272
$\boldsymbol{b}$, b_i	body force (spatial description)	55
$\boldsymbol{b}$, b_i	Burgers vector	351
$\boldsymbol{b}^{\text{pt}}$, b_i^{pt}	pointwise (microscopic) body force field	470
$\mathcal{C}$	the DFT simulation cell	210

C_v	molar heat capacity at constant volume	69
$\boldsymbol{C}, C_{IJ}$	right Cauchy–Green deformation tensor	47
$\boldsymbol{C}, C_{IJKL}$	referential elasticity tensor	101
c_v	specific heat capacity at constant volume	70
$c_I, c_{Ij}, c^{\alpha}_{iI}$	Ith eigenvector solution and its components (associated with plane wave j or orbital i on atom α)	176
c, c_{ijkl}	spatial (or small strain) elasticity tensor	102
$\mathsf{c}, \mathsf{c}_{mn}$	elasticity matrix (in Voigt notation)	104
$D(E)$	density of states (statistical mechanics)	405
$D(\epsilon)$	electronic density of states	230
D^{α}_i	electronic density of states for orbital i on atom α	230
$\boldsymbol{D}, D_{iJkL}$	mixed elasticity tensor	102
$\boldsymbol{d}, d_{ij}$	rate of deformation tensor	50
$\mathcal{E}$	total energy of a thermodynamic system	68
E	Young's modulus	105
$E_{\text{free}}(Z)$	energy of a free (isolated) atom with atomic number Z	247
$E_{\text{coh}}, E^0_{\text{coh}}$	cohesive energy and equilibrium cohesive energy	332
$\boldsymbol{E}, E_{IJ}$	Lagrangian strain tensor	48
$\boldsymbol{e}_i$	orthonormal basis vectors	27
$\boldsymbol{F}^{\text{ext}}$	total external force acting on a system	54
$\boldsymbol{F}, F_{iJ}$	deformation gradient	46
f	occupancy of an electronic orbital	208
$f(\boldsymbol{q}, \boldsymbol{p}; t)$	distribution function at point $(\boldsymbol{q}, \boldsymbol{p})$ in phase space at time t	391
$f_{\text{mc}}(\boldsymbol{q}, \boldsymbol{p}; E)$	microcanonical (NVE) distribution function	407
$f_{\text{c}}(\boldsymbol{q}, \boldsymbol{p}; T)$	canonical (NVT) distribution function	427
$\boldsymbol{f}^{\alpha}, f^{\alpha}_i$	force on atom α	54
$\boldsymbol{f}^{\alpha\beta}, f^{\alpha\beta}_i$	force on atom α due to the presence of atom β	289
$\boldsymbol{f}^{\text{int},\alpha}, f^{\text{int},\alpha}_i$	internal force on atom α	289
$\boldsymbol{f}^{\text{ext},\alpha}, f^{\text{ext},\alpha}_i$	external force on atom α	289
f	column matrix of finite element nodal forces	603
$\boldsymbol{G}^{\alpha}, G^{\alpha}_i$	stochastic force on atom α	511
g	specific Gibbs free energy	96
$g(r)$	electron density function in a pair functional	264
$\mathcal{H}$	Hamiltonian of a system	159
$\boldsymbol{H}, H_i$	angular momentum	58
H_0	matrix of periodic cell vectors (reference configuration)	326
H	matrix of periodic cell vectors (deformed configuration)	326
$\hat{\mathsf{H}}$	matrix of reference primitive lattice vectors	120
$\boldsymbol{I}$	identity tensor	34
I	identity matrix	25
J	Jacobian of the deformation gradient	46
$\mathcal{K}$	macroscopic (continuum) kinetic energy	68
K	stiffness matrix or Hessian	312
$\boldsymbol{k}$	wave vector and Fourier space variable	146

$\mathcal{L}$	Lagrangian function	158
$\boldsymbol{L}$, L_i	linear momentum	54
$\boldsymbol{L}$, $\boldsymbol{L}_i$	vectors defining a periodic simulation cell (reference)	325
$\boldsymbol{l}$, l_i	vectors defining a periodic simulation cell (deformed)	563
$\boldsymbol{l}$, l_{ij}	spatial gradient of the velocity field	50
M	total mass of a system of particles	380
M_{cell}	mass of a unit cell	567
$\boldsymbol{M}^{\text{ext}}$	total external moment acting on a system	58
M	finite element mass matrix	660
m, m^α	mass, mass of atom α	54
$\mathcal{N}^\alpha$	set of atoms forming the neighbor list to atom α	324
N	number of particles/atoms	54
N_{B}	number of basis atoms	142
$\hat{N}_{\text{B}}$	number of basis atoms in the primitive unit cell	140
N_{lat}	number of lattice sites	563
n_{d}	dimensionality of space	22
$\mathcal{P}^{\text{def}}$	deformation power	86
$\mathcal{P}^{\text{ext}}$	external power	85
$\boldsymbol{P}$, P_{iJ}	first Piola–Kirchhoff stress tensor	59
$\mathfrak{P}^\alpha$, $\mathfrak{P}^\alpha_i$	reference momentum of atom α	452
p	pressure (or hydrostatic stress)	57
$\boldsymbol{p}^\alpha$, p^α_i	momentum of atom α	54
$\boldsymbol{p}^\alpha_{\text{rel}}$	center-of-mass momentum of atom α	380
p, p_i	momentum of an electron or atom	164
p, p_i	generalized momenta in statistical mechanics	382
$\Delta\mathcal{Q}$	heat transferred to a system during a process	68
$\mathbf{Q}$,$\mathsf{Q}_{\alpha i}$	orthogonal transformation matrix	31
$\boldsymbol{q}$, q_i	spatial heat flux vector	87
$\boldsymbol{q}_0$, q_{0I}	reference heat flux vector	88
q, q_i	generalized positions in statistical mechanics	382
$\bar{\mathsf{q}}$, $\bar{\mathsf{q}}_i$	generalized mean positions in restricted ensemble	556
$\mathcal{R}$	rate of heat transfer	85
$\boldsymbol{R}$, R_{iJ}	finite rotation (polar decomposition)	47
$\boldsymbol{R}^{[\boldsymbol{\ell}\lambda]}$	reference position of the λth basis atom of lattice site $\boldsymbol{\ell}$	141
$\boldsymbol{R}$, R_i	center of mass of a system of particles	380
$\boldsymbol{R}^\alpha$, R^α_i	reference position of atom α	242
r	spatial strength of a distributed heat source	87
r_0	reference strength of a distributed heat source	88
$\boldsymbol{r}^\alpha$, r^α_i	spatial position of atom α	54
$\bar{\boldsymbol{r}}^\alpha$, $\bar{r}^\alpha_i$	mean position of atom α in restricted ensemble	556
$\boldsymbol{r}^\alpha_{\text{rel}}$	center-of-mass coordinates of atom α	380
$\mathcal{S}$	electronic orbital overlap	225
$\mathcal{S}$	entropy	73
$\mathcal{S}^\lambda$	set of all atoms belonging to sublattice λ in a multilattice	564

S^I	shape function for finite element node I	602
S_E	hypersurface of constant energy E in phase space	382
$\boldsymbol{S}$, S_{IJ}	second Piola–Kirchhoff stress tensor	60
s	specific entropy	88
$\boldsymbol{s}$, s_{ijkl}	spatial (or small strain) compliance tensor	103
$\bar{\boldsymbol{s}}^\lambda$, $\bar{s}_i^\lambda$	shift vector of basis atom λ	560
$\mathcal{T}$	instantaneous microscopic kinetic energy	158
$\mathcal{T}^{\text{vib}}$	microscopic (vibrational) kinetic energy	379
$\mathcal{T}^{\text{el}}$	instantaneous kinetic energy of the electrons	190
$\mathcal{T}^{\text{s}}$	instantaneous kinetic energy of the noninteracting electrons	192
T	temperature	65
$\boldsymbol{T}$, T_i	nominal traction (stress vector)	60
$\boldsymbol{t}$, t_i	true traction (stress vector)	56
$\bar{\boldsymbol{t}}$, $\bar{t}_i$	true external traction (stress vector)	55
$\mathcal{U}$	internal energy	68
$\mathcal{U}$	potential energy of a quantum mechanical system	169
$U(\rho)$	embedding energy term in a pair functional	263
$U(z)$	unit step function (Heaviside function)	404
$\boldsymbol{U}$, U_{IJ}	right stretch tensor	47
u	spatial specific internal energy	85
u_0	reference specific internal energy	88
$\boldsymbol{u}$, u_i	displacement vector	48
$\widetilde{\boldsymbol{u}}$, $\widetilde{u}_i$	finite element approximation to the displacement field	602
u	column matrix of finite element nodal displacements	601
$\mathcal{V}$	potential energy of a classical system of particles	158
$\mathcal{V}^{\text{int}}$	internal (interatomic) part of the potential energy	240
$\mathcal{V}^{\text{ext}}$	total external part of the potential energy	240
$\mathcal{V}^{\text{ext}}_{\text{fld}}$, $\mathcal{V}^{\text{ext}}_{\text{con}}$	potential energy due to external fields and external contact	240
V_0	volume (reference configuration)	46
V	volume (deformed configuration)	46
V_0^α	volume of atom α (reference configuration)	457
V^α	volume of atom α (deformed configuration)	457
V_R	volume of region R in phase space	384
$\boldsymbol{V}$, V_{ij}	left stretch tensor	47
$\boldsymbol{v}$, v_i	velocity vector	50
$\boldsymbol{v}^{\text{pt}}$, v_i^{pt}	pointwise (microscopic) velocity field	468
$\boldsymbol{v}^\alpha$, v_i^α	velocity of atom α	54
$\boldsymbol{v}^\alpha_{\text{rel}}$, $v^\alpha_{\text{rel},i}$	velocity of atom α relative to center of mass	471
v	column matrix of finite element nodal velocities	660
$\Delta\mathcal{W}$	work performed on a system during a process	67
$\mathcal{W}$	virial of a system of particles	422
W	strain energy density function	96
$w(\boldsymbol{r})$, $\hat{w}(r)$	spatial averaging weighting function (general and spherical)	476
$\boldsymbol{w}$, w_{ij}	spin tensor	50

$\boldsymbol{w}^{\alpha}$, w_i^{α}	displacement of atom α relative to its mean position	556
$\boldsymbol{X}$, X_I	position of a point in a continuum (reference configuration)	43
$\mathbf{X}$	column matrix of finite element nodal coordinates	601
$\boldsymbol{x}$, x_i	position of a point in a continuum (deformed configuration)	43
$\boldsymbol{x}$, x_i	position of an electron	156
Z	atomic number	176
Z	partition function	426
Z^{K},Z^{V}	kinetic and potential parts of the partition function	427
$\hat{\boldsymbol{Z}}^{\alpha}$	position of basis atom α relative to the Bravais site	141
z	valence of an atom (or charge on an ion)	198

Fundamental constants

Avogadro's constant (N_{A})	6.0221×10^{23} mol^{-1}
Bohr radius (r_0)	0.52918 Å
Boltzmann's constant (k_{B})	1.3807×10^{-23} J/K
	8.6173×10^{-5} eV/K
charge of an electron ($\tilde{e}$)	1.6022×10^{-19} C
charge-squared per Coulomb constant ($\tilde{e}^2/4\pi\epsilon_0 \equiv e^2$)	14.4 eV · Å
mass of an electron (m^{el})	9.1094×10^{-31} kg
permittivity of free space (ϵ_0)	8.8542×10^{-12} C^2/(J · m)
Planck's constant (h)	6.6261×10^{-34} J · s
	4.1357×10^{-15} eV · s
Planck's constant, reduced ($\hbar = h/2\pi$)	1.0546×10^{-34} J · s
	6.5821×10^{-16} eV · s
universal gas constant (R_{g})	8.3145 J/(K · mol)

Unit conversion

1 fs	=	10^{-15} s (femto)
1 ps	=	10^{-12} s (pico)
1 ns	=	10^{-9} s (nano)
1 µs	=	10^{-6} s (micro)
1 ms	=	10^{-3} s (milli)
1 Å	=	10^{-10} m = 0.1 nm (ångstrom)
1 eV	=	1.60212×10^{-19} J
1 eV/Å	=	1.60212×10^{-9} N = 1.60212 nN
1 eV/Å^2	=	16.0212 J/m^2 = 16.0212 N/m
1 eV/Å$^{2.5}$	=	1.60212×10^{6} N/m$^{1.5}$ = 1.60212 MPa · $\sqrt{\mathrm{m}}$
1 eV/Å^3	=	1.60212×10^{11} N/m^2 = 160.212 GPa
1 amu	=	1.66054×10^{-27} kg = 1.03646×10^{-4}eV · ps^2/Å^2

1 Introduction

As we explained in the preface, modeling materials is to a large extent an exercise in multiscale modeling. To set the stage for the discussion of the various theories and methods used in the study of materials behavior, it is helpful to start with a brief tour of the structure of materials – and in particular *crystalline materials* – which are the focus of this book. In a somewhat selective way, we will discuss the phenomena that give rise to the form and properties of crystalline materials like copper, aluminum and steel, with the goal of highlighting the range of time and length scales that our modeling efforts need to address.

1.1 Multiple scales in crystalline materials

1.1.1 Orowan's pocket watch

The canonical probe of mechanical properties is the *tensile test*, whereby a standard specimen is pulled apart in uniaxial tension. The force and displacement are recorded during the test, and usually normalized by the specimen geometry to provide a plot of stress versus strain. In the discussion of an article by a different author on "the significance of tensile and other mechanical test properties of metals," Egon Orowan states: [Oro44]

> The tensile test is very easily and quickly performed, but it is not possible to do much with its results, because one does not know what they really mean. They are the outcome of a number of very complicated physical processes taking place during the extension of the specimen. The extension of a piece of metal is, in a sense, more complicated than the working of a pocket watch, and to hope to derive information about its mechanism from two or three data derived from measurements during the tensile test is perhaps as optimistic as would be an attempt to learn about the working of a pocket watch by determining its compressive strength.

It is straightforward to determine the compressive strength of a pocket watch (see Fig. 1.1). The maximum load required to crush it can be read from the graph of Fig. 1.2(a) and is 1.8 kN. But this number tells us neither anything of how a watch works under normal service conditions, nor its mechanisms of failure under compressive loads. By examining the internal structures and mechanisms, and by observing their response during the test we can start to learn, for example, how the gears interact or how the winding energy is stored. We might even be able to develop some hypotheses for how the various parts contribute to the peaks and valleys of the load versus displacement response. However, it is only

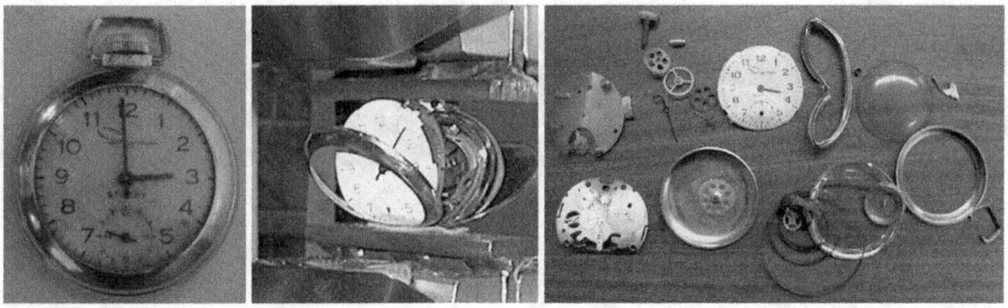

Fig. 1.1 "Orowan's pocket watch". An Ingraham pocket watch, circa 1960, was used to examine Orowan's claim. A twin-column test frame (manufactured by MTS Systems Corporation), instrumented with a 100 kN load cell was used to crush the watch between two flat plates, into each of which was machined a small notch to accept the shape of the watch. The test was displacement-controlled, with a constant rate of crushing that took about 2 minutes to complete.

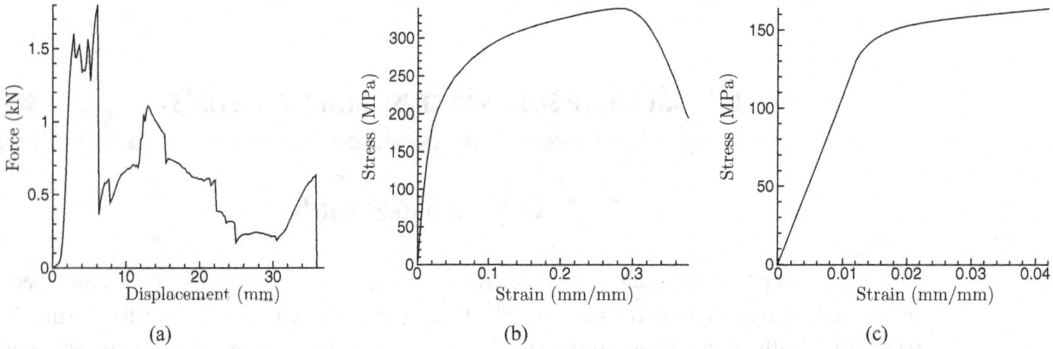

Fig. 1.2 A compressive test on a pocket watch in (a) is compared to a tensile test result for an annealed Cu–10%Ni alloy in (b) (adapted from [Cop10]) and a tensile test for compact bovine bone in (c) (adapted from [CP74, Fig. 5]).

through this combined approach of "macroscopic" testing (the curve of Fig. 1.2(a)) and "microscopic" observation and modeling (the analysis of the revealed springs and gears) that we can fully understand the pocket watch.

As Orowan suggests, the tensile test results of Figs. 1.2(b) and 1.2(c) for copper and bovine bone are not much more helpful than the pocket watch experiment in elucidating the internal microstructure[1] of these materials or how these microstructures respond to loading. Indeed, the two curves are strikingly similar aside from the differences of scale, but surely the mechanisms of failure in a metal alloy are profoundly different than those in a biological material like bone. And unlike the pocket watch, understanding the behavior of these materials requires a truly *microscopic* approach to reveal the complicated deformation mechanisms taking place in these materials as they are stretched to failure.

[1] The term "microstructure" refers to the internal structure of a material ranging from atomic-scale defects to larger-scale defect structures. In contrast, the "macrostructure" of the material (if such a term were used) would be the shape that the material is made to adopt as part of its engineering function.

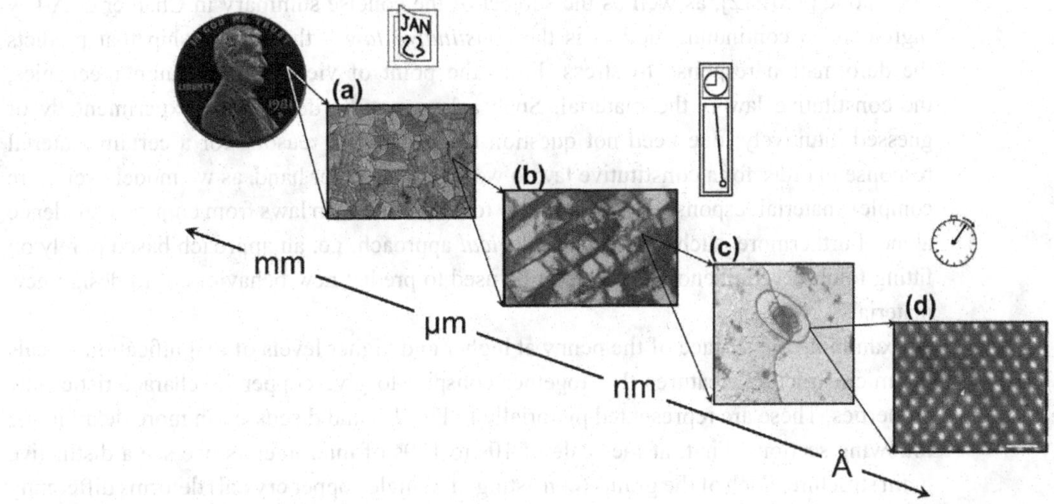

Fig. 1.3 Length and time scales in a copper penny. The macroscopically uniform copper has: (a) a grain structure on the scale of 10s to 100s of micrometers, (b) a dislocation cell structure on the scale of micrometers and (c) individual dislocations and precipitates on the nanometer scale. In (d), high-resolution transmission electron microscopy resolves individual columns of atoms in the dislocation core. This core structure has features on the ångstrom scale that affect the macroscopic plastic response. (Reprinted from: (a) [Wik08] in the public domain, (b) [GFS89] with permission of Elsevier, (c) [HH70] with permission of Royal Society Publishing and (d) [MDF94] with permission of Elsevier.)

Orowan's words sum up neatly the challenge of modeling materials. The macroscopic behavior we observe is built up of the intricate, complex interactions between mechanisms operating on a wide range of length and time scales. Studying a material from only the largest of scales is like studying a pocket watch with only a hammer; neither method will likely show us why things behave as they do. Instead, we need to approach the problem from a variety of observational and modeling perspectives and scales. Let us focus on just the question of deformation in crystalline materials, and look more closely at the operative length and time scales in the tensile stretching of a ductile metal like copper.

1.1.2 Mechanisms of plasticity

Whether we are considering the common tensile test or the complex minting of a coin, the same processes control the flow of deformation in crystalline materials like copper. The minting of the penny in Fig. 1.3 is a problem best studied with continuum mechanics, whereby the deformation can be predicted by a flow model driven by the stresses introduced by the die.[2] Such continuum modeling is the detailed subject of the companion volume

[2] The fact that the penny shown in Fig. 1.3 was minted in 1981 is no accident. Most pennies produced before 1982 were made of bronze (a copper alloy), and therefore the final microstructure (microscale arrangement of structures and defects in the material) is primarily that of cold-worked copper. Modern pennies, however, are actually composed of a zinc core that is forged and later plated with a thin layer of copper. As such, only 2.5% of the weight of a modern penny is copper, with a microstructure characteristic of plating, not cold-working.

to this one [TME12], as well as the subject of the concise summary in Chapter 2. A key ingredient to continuum models is the *constitutive law* – the relationship that predicts the deformation response to stress. From the point of view of continuum mechanics, the constitutive law *is* the material. Such a law may be determined experimentally or guessed intuitively; one need not question the underlying reasons for a certain material response in order for a constitutive law to work. On the other hand, as we model ever more complex material response, we are unlikely to determine such laws from empirical evidence alone. Furthermore, such a *phenomenological* approach, i.e. an approach based purely on fitting to observed phenomena, cannot be used to predict new behaviors or to design new materials.

Examining the surface of the penny at higher and higher levels of magnification reveals the microstructural features that together conspire to give copper its characteristic flow properties. These are represented pictorially in Fig. 1.3 and discussed in more detail in the following sections. First, at the scale of 10s to 100s of micrometers, we see a distinctive grain structure. Each of the grains (consisting of a single copper crystal) deforms differently depending on its orientation relative to the loading and local constraints. Within each grain, we see patterns of dislocations on the scale of a micrometer, resulting from the interactions between dislocations and the grain structure (dislocations are the subject of Section 6.5.5). At still smaller scales, we can see individual dislocations and their interaction with other microstructure features. Finally, at the smallest scales of atoms, we see that each grain is actually a single crystal, with individual dislocations being simple defects in the crystal packing. A daunting range of time scales is also at play. Although the minting of a penny may only take a few seconds, deformation processes such as creep and fatigue can span years. At the other extreme, vibrations of atoms on a femtosecond scale (1 fs = 0.000 000 000 000 001 s) contribute to the processes of solid-state diffusion that participate in these mechanisms of slow failure. Materials modeling is, at its core, an endeavor to develop constitutive laws through a detailed understanding of these microstructural features, and this requires the observation and modeling of the material at each of these different scales. In essence, this book is about the fundamental science behind such microstructural modeling enterprises.

1.1.3 Perfect crystals

It is likely that the copper in a penny started life by solidifying from the molten state. In going from the liquid to the solid state, copper atoms arrange themselves into the face-centered cubic (fcc) crystal structure shown in Figs. 1.4(a) and 1.4(b) (crystal structures are the subject of Chapter 3). While the lowest-energy arrangement of copper atoms is a single, perfect crystal of this type, typical solidification processes do not usually permit this to happen for large specimens. Instead, multiple crystals start to form simultaneously throughout the cooling liquid, randomly distributed and oriented, so that the final microstructure is *poly-granular*, i.e. it comprises the grains shown in Fig. 1.3(a), each of which is a single fcc crystal. Typical grains are 10–100 μm across and contain 10^{15} or more atoms. As such, they still represent an impressive extent of long-range order at the atomic scale, and the fcc crystal remains, by and large, the defining fine-scale structure of copper. This structure

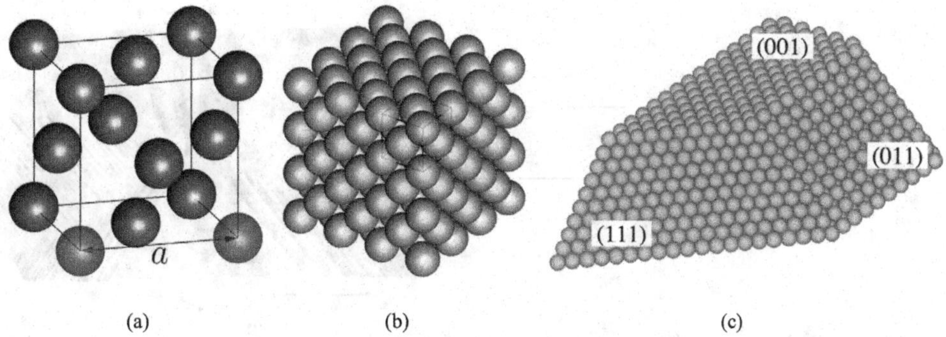

Fig. 1.4 The fcc unit cell in (a) is periodically copied through space to form a copper crystal in (b). In (c), we show three different types of free surface, indexed by the exposed atomic plane.

helps to explain the elastic properties of bulk copper, and also provides a rationale for the relatively soft, ductile nature of copper compared to other crystals. Why does copper prefer this particular crystal structure, while other elements or compounds adopt very different atomic arrangements? Why do other elements spontaneously change their crystal structure under conditions of changing temperature or stress? To understand these questions first requires modeling at the level of quantum mechanics, in order to characterize the features of bonding that make certain structures energetically favorable over others. The quantum mechanics of bonding is the subject of Chapter 4, while modeling techniques to investigate crystal structures and energetics are described in Chapter 6. Much can be gleaned about a material's properties by analyzing its overall crystal symmetry and examining the structure of its crystallographic planes. For example, the field of *crystal plasticity* is dedicated to the development of constitutive laws that predict the plastic deformation of single crystals. Some of the key physical inputs to such models are related to the so-called *slip systems*. These are the crystallographic planes and the directions on these planes along which the material can plastically deform through the passage of dislocations (a process referred to as *slip*). The number of available slip systems, their relative orientations and their respective resistances to slip deformation are determined by the structure of the underlying crystal. Thus, knowledge of the crystal structure guides us in the development of constitutive laws, dictating the appropriate symmetries and anisotropy in elastic and plastic response. Crystal structures and symmetry are the subjects of Chapter 3, while material symmetries are briefly reviewed in Section 2.5.4 (this topic is also covered in detail in the companion volume to this book [TME12]).

Figure 1.5 shows a striking manifestation of the effect of crystal structure on plastic flow. In this experiment, a large single crystal is specially prepared and notched, and the flow around the notch tip is imaged after a four-point bending test (Fig. 1.5(a)). The streaks shown in Fig. 1.5(b) are *slip lines* that form on the surface of the crystal during plastic deformation due to the motion of dislocations. As we move around the notch, we see that there are clearly different sectors that correspond to changes in the maximum stresses with respect to the orientation of the preferred slip directions in the crystal – the

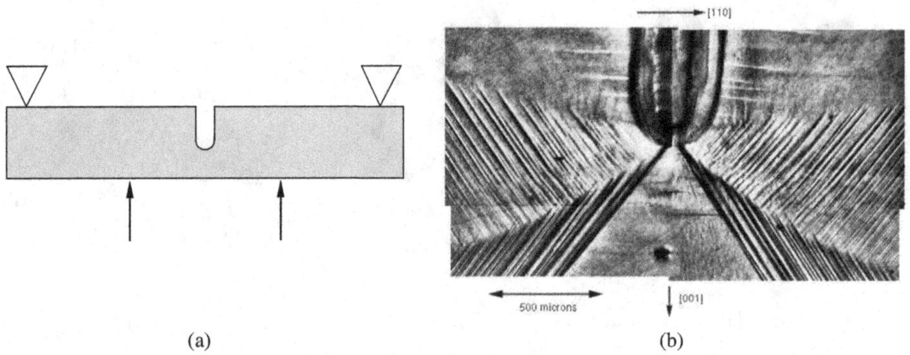

Fig. 1.5 (a) Schematic of a notched beam in four-point bending. (b) Slips lines formed during plastic flow around a notch (top of the picture) in a copper single crystal. The distinct sectors correspond to the activation of different slip systems as the stress changes around the notch. (Reproduced from [Shi96], with permission of Elsevier.)

orientation of these lines and the boundaries between the sectors is determined by the crystal structure.

The fcc crystal structure determines the shortest length scale of copper, which has a lattice constant of about $a = 3.6$ Å (see Fig. 1.4(a)). The lattice also determines the fastest time scales, as atomic vibrations due to thermal energy occur on a scale that is set by the bond stiffness and atomic spacing. The so-called Debye frequency, ω_D, provides an upper bound on the typical frequencies of atomic vibrations in a crystal, given by

$$\omega_D = v_s \left(\frac{3N}{4\pi V} \right)^{1/3},$$

where N/V is the number density of atoms and v_s is the mean speed of sound in the crystal. For copper, in which the speed of sound is about 3900 m/s, this corresponds to a frequency on the order of 10 cycles per picosecond (ps).[3] In other words, each oscillation of an atom about its equilibrium position takes a mere 0.1 ps. This number is important because it puts limits on the types of processes we can model using molecular dynamics (MD) simulations – any discrete time integration must take timesteps that are no longer than about a tenth of this oscillation time (MD is the subject of Chapter 9). On the other hand, these rapid oscillations mean that over time scales on the order of seconds, the number of vibrations is huge. It is the largeness of these numbers – the number of atoms and the number of oscillations – that accounts for the accuracy of statistical mechanics approaches (see Chapter 7) and leads to the microscopic origins of stress which is discussed in Chapter 8.

Although the perfect fcc structure explains many properties, it is far from the only factor in determining the behavior of copper. Much of the story comes not from the perfect crystal, but from *crystal defects*, such as free surfaces, grain boundaries, dislocations and vacancies. Let us consider each of these defects in turn.

[3] A picosecond is 10^{-12} seconds.

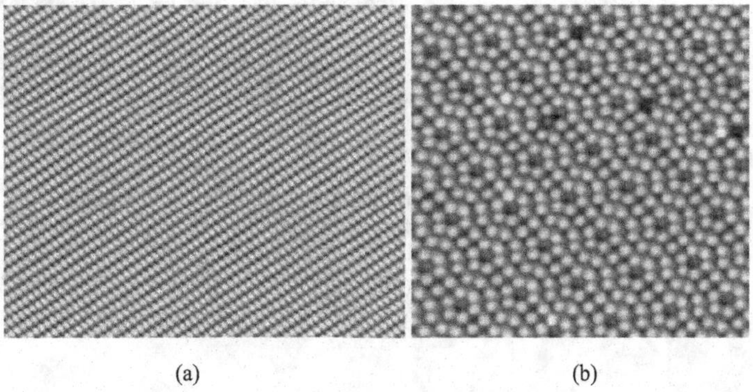

Fig. 1.6 STM atomic scale images of: (a) the (110) surface of Ag and (b) the (111) surface of Si. The width of view in each image is about 15 nm. (Reprinted from [Kah10] with the kind permission Professor Antoine Kahn.)

1.1.4 Planar defects: surfaces

The fact that even a defect-free crystal is finite means that free surfaces are ever-present in crystalline solids. We can imagine cleaving the crystal of Fig. 1.4(b) along any one of an infinite number of possible planes to create different arrangements of atoms at the surface. Three examples are shown in Fig. 1.4(c). Looking carefully at this figure reveals the distinct arrangements of the atoms on each of the three planes.

Structurally speaking, however, surfaces are more complex than merely the result of cleaving a perfect crystal. Instead, the undercoordination of surface atoms can lead to changes ranging from slight surface relaxation to dramatic reconstructions, where atoms are significantly rearranged from their bulk crystal positions. These changes in the surface structure affect such things as the reactivity of the surface with its environment, the rate at which other species of atoms may diffuse through the surface into the bulk and the mobility of atoms that are adsorbed on the surface. Materials modeling allows the direct calculation of the structure and energy of any one of these surfaces, and model calculations of surface relaxation and reconstruction have been performed for decades. More recently, experimental techniques have reached the point where the atomic surface structure can be observed directly with techniques such as scanning tunneling microscopy (STM). Some examples of STM images of surfaces (not of copper) are shown in Fig. 1.6. The images give subnanometric resolution of the surface elevation. In these examples, each hill corresponds to a single atom on an atomically flat planar surface. In Fig. 1.6(a), the (110) surface of silver shows very little difference from the ideal surface of a cleaved crystal (see Fig. 1.4(c)). On the other hand, the (111) surface of silicon shows a dramatic reconstruction that we will discuss further in Section 6.5.4. In [SK08], Salomon and Kahn used the silver surface as a template on which to grow patterns of silver nanowires.

Accurate atomic-scale calculations of the surface energy let us understand the driving forces for various materials behaviors. In Tab. 1.1, we report the surface energy values computed by Vitos *et al.* [VRSK98] for common surfaces in copper. These calculations were performed more than a decade ago using density functional theory (DFT), the subject

Table 1.1. Surface energies in copper, computed using density functional theory by Vitos *et al.* [VRSK98].

	Energy/area	
Surface	(J/m^2)	($meV/Å^2$)
(111)	1.952	121.8
(011)	2.237	139.6
(001)	2.166	135.2

Fig. 1.7 Dendritic growth forms: (a) 90° angles in cobalt and (b) 60° angles in ice (the snowflake shown is about 3 mm, tip-to-tip). ((a) Reproduced from [CGL+07], with permission from Elsevier. (b) Reprinted from [Lib05] with permission from IOP Publishing.)

of Section 4.4. Since systems of atoms tend to want to arrange themselves into low-energy configurations, the differences in energy between surfaces favor certain morphologies over others. For instance, free surface energetics plays an important role in determining the fracture response of crystals, since fracture is, almost by definition, the creation of new surfaces. In crystals with large differences between the energy cost of different types of surface, there will be a strong anisotropy in the fracture resistance of the material. In other cases, where the energy cost to create new surfaces is high compared with other competing deformation mechanisms, a crystal may not fracture at all. Indeed, this inherent toughness is one of the desirable properties of many fcc metals.

We have already mentioned how crystals usually solidify from the molten state. Surface energetics also plays a key role in establishing the final structure of the solidified state, as grain shapes vary from being more-or-less equiaxed as in Fig. 1.3(a) to the striking dendritic examples of Fig. 1.7. Although the same material can often form either dendrites or more regular grains depending on the cooling conditions, the details of these shapes are driven by the surface energetics. For example, the normals to the orthogonally arranged (100) planes are the preferred growth directions in cubic crystals like the cobalt in Fig. 1.7(a). As a result, the dendrites form at right angles. By way of contrast, the well-known hexagonal shape of the dendrites in a snowflake (see Fig. 1.7(b)) highlights the hexagonal (instead of cubic) symmetry of the preferred growth directions in crystalline ice.

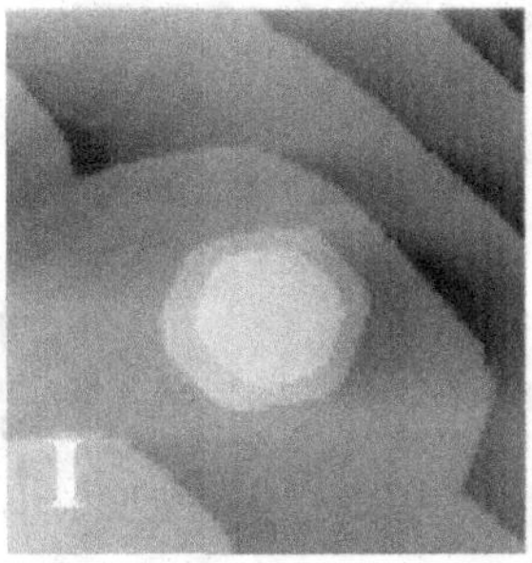

Fig. 1.8 An STM image of an island of Cu formed by vapor deposition on an atomically flat Cu (111) surface. The field of view is about 800 Å wide. (Reprinted with permission from [GIKI98], copyright 1998 by the American Physical Society.)

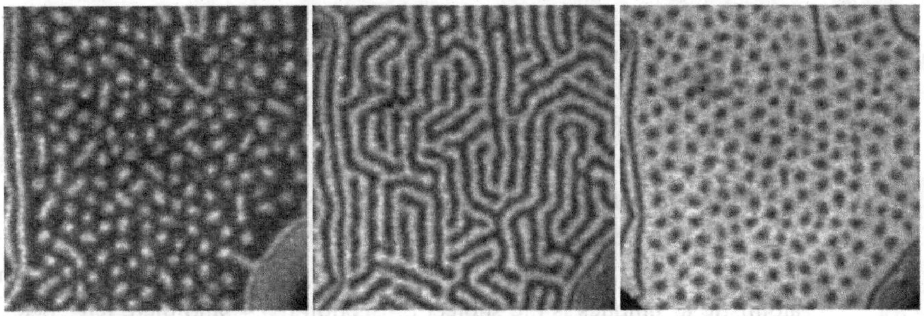

(a) (b) (c)

Fig. 1.9 Lead (light phase) deposited on copper (dark phase) can form: (a) isolated islands, (b) channels or (c) a cratered surface, depending on the amount of deposited lead. Each field of view is about 1.75 μm wide. (Reprinted from [PLBK01] with permission from Macmillan Publishers Ltd.)

As snowflakes can form directly from water vapor, so too can crystalline metals be made by depositing a vapor, atomic layer by atomic layer – a process referred to as *chemical vapor deposition*. Once again, surface energetics determines the morphology of the resulting structures and drives microstructural evolution. Figure 1.8 shows an example of the surface morphology during deposition of copper vapor onto an initially flat copper (111) surface. The steps and islands shown are each 2–4 atomic layers high, and adopt a morphology that is clearly influenced by the underlying surface structure (see the (111) face in Fig. 1.4(c)). Isolated islands like the one shown in the figure appear at a density of about 1 every 320 000 Å^2 (i.e. with an average spacing of about 600 Å). The islands then shrink or grow as atoms diffuse over the surface to lower the system energy. The rate of this process depends on the temperature and the deposition rates. For example, the white central island in Fig. 1.8 persisted for about 12 hours, slowly shrinking and changing shape due to diffusion. When this shape change led to the edge of the island contacting the edge of the larger island on which it rested, the white topmost island rapidly disappeared in less than an hour.

By depositing different atomic species, we can exploit surface energetics and morphology to cause the self-assembly of nano-scale patterns. For example, in Fig. 1.9, we show the

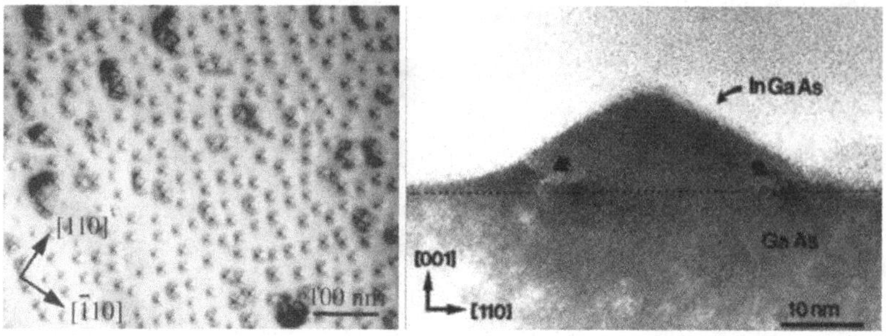

(a) (b)

Fig. 1.10 InGaAs deposited on GaAs forms isolated islands (shown in (a) top view and (b) side view) that can be utilized as quantum dots. (Reprinted from [TF00] with permission of Elsevier.)

deposition of lead atoms on a (111) copper surface. As the amount of deposited lead increases, the pattern changes from isolated lead islands (the light phase in the images), to interconnected channels, to a regular pattern of craters in a continuous lead surface layer.[4] Scientists are trying to exploit such patterns as templates to grow nano-scale devices. In other systems, such as the growth of InGaAs on GaAs in Fig. 1.10, the resulting pattern of islands can be used as so-called quantum dots, which researchers are currently trying to exploit for the development of solid-state quantum computers. Modeling surfaces using atomistic techniques is the subject of Section 6.5.4, and there have been many studies of vapor deposition processes using MD methods (see Chapter 9).

1.1.5 Planar defects: grain boundaries

The solidification of a solid from a melt naturally leads to the formation of a microstructure consisting of single-crystal grains separated by boundaries like those in the penny in Fig. 1.3(a). At the atomic level, these *grain boundaries* are simply the junctions between two crystals – a place where the atoms must strike a structural compromise between two different orientations (see Fig. 1.11). (Grain boundary modeling is also discussed in Section 6.5.4.) Although a typical grain boundary may be only a few ångstroms across in terms of the width of the region where the atomic structure is different from the bulk, the length of a single grain boundary can extend 100s of micrometers. The distances between grain boundaries, which is of course set by the average grain size, can vary from a few nanometers to several centimeters.

Grain boundaries play a key role in many processes. For example, they act as barriers to the motion of dislocations (another type of defect discussed shortly). Because grain boundaries tend to be more open structures than the surrounding bulk crystal, they are natural sites to which impurity atoms migrate. This makes the grain boundary the formation site of second phase precipitates that can, depending on the nature of the phases, either strengthen or weaken a solid (see Fig. 1.12). Indeed, such boundary phases are so common in complex

[4] Between each image, approximately one lead atom to every four copper surface atoms is deposited over a time span of about 400 s. The experimental temperature was 673 K.

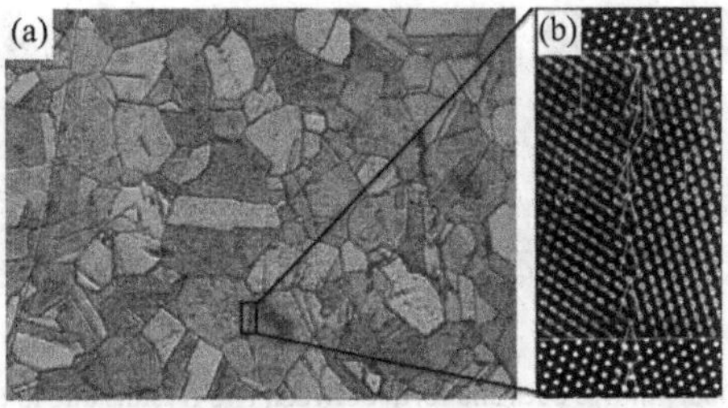

Fig. 1.11 Grain boundaries in Cu. The width of the view in (a) is on the order of a millimeter, while the close-up in (b) is only a few nanometers, and reveals the positions of individual atoms in the vicinity of the grain boundary. ((a) Reprinted from [Wik08] in the public domain, (b) Reprinted from [HCCK02], with permission of Elsevier.)

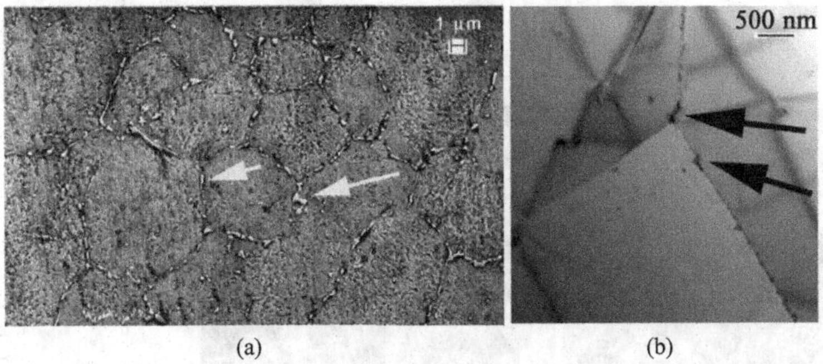

Fig. 1.12 Precipitation of second phase particles at grain boundaries in a stainless steel alloy at (a) low and (b) high magnification. In this case, the precipitates are chromium carbides and/or chromium nitrides that deplete Cr from the surrounding Fe matrix, making it susceptible to corrosion. (Reprinted from [HXW$^+$09] with permission of Elsevier.)

alloys that metallurgists often think of the properties of the grain boundary itself in terms of the effect of these precipitates. Once again, this is an example of the importance of many scales. At the atomic scale, a pure, clean boundary like the one shown in Fig. 1.11(b) might be inherently quite strong, but on the macroscale it can be significantly weakened by precipitates.

Just like the energetics of free surfaces and surface–liquid interfaces during solidification, the grain boundary energetics plays a key role in solid-state structural evolution driven by grain growth. Long-time phenomena like creep are partially governed by the abilities of grains to reshape, which is in turn dictated by grain boundary diffusion, sliding and structural energetics. The *mobility* of a grain boundary determines the grain boundary velocity as a function of a driving force (mechanical or thermal), and of course depends strongly on the type of grain boundary and the elements present in the neighboring grains. In Fig. 1.13 we show an example of in situ observations of a grain boundary in gold, moving in response to thermal vibrations of the atoms. This shows that some grain boundary motions are stochastic, thermally-driven processes, the frequency of which can be predicted by statistical mechanics models. The motion in this example involves the collective rearrangement of hundreds of atoms and takes place over time periods on the order of 10^{-2} s. During the experiments, the grain boundaries often moved back and forth several times between the two

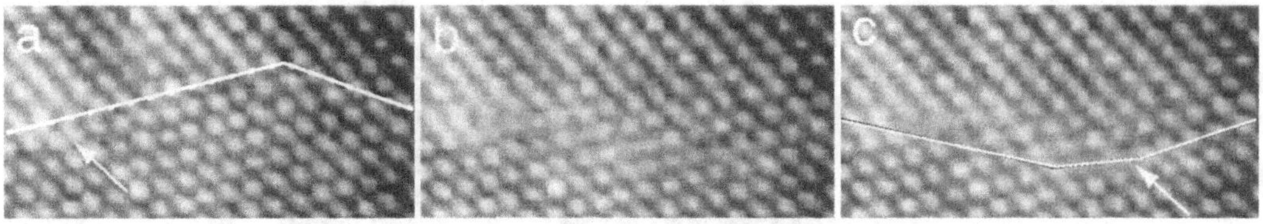

(a) (b) (c)

Fig. 1.13 A grain boundary moves back and forth between two crystals due to thermal fluctuations: (a) $t = 0.0$ s, (b) $t = 0.03$ s, (c) $t = 0.07$ s. The width of each image is approximately $4\ \mathrm{nm}$. (Reprinted with permission from [MTP02], © 2002 by the American Physical Society.)

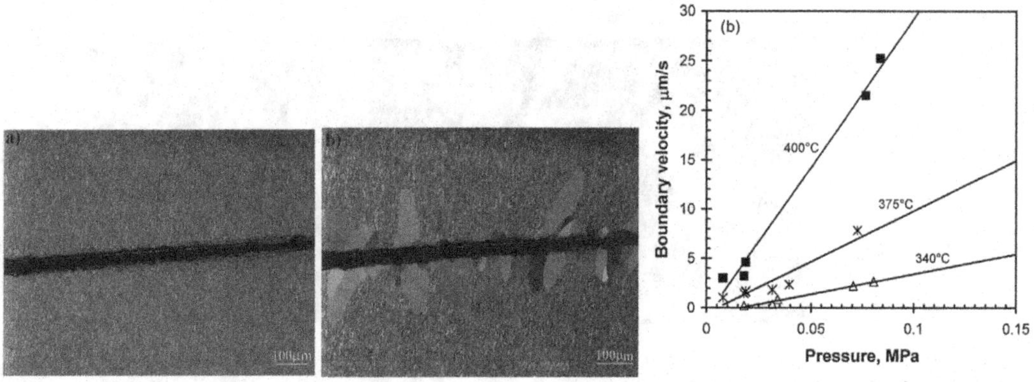

(a) (b) (c)

Fig. 1.14 By introducing a scratch on the surfaces of a heavily deformed single crystal as in (a), and annealing at various temperatures, Huang and Humphreys captured the motion of grain boundaries as new grains formed and grew (b). The grain boundary velocity as a function of temperature and pressure is shown in (c). (Reprinted from [HH99] with permission of Elsevier.)

end structures. Understanding this motion through modeling helps us unravel the complex mechanisms of creep and plastic flow.

An example where the mobility of grain boundaries was measured in response to applied pressure is the work of Huang and Humphreys [HH99]. We reproduce some of their results in Fig. 1.14, where we see that typical grain boundary velocities in this system are on the order of 5–25 $\mu\mathrm{m/s}$, eight orders of magnitude slower than the speed of sound. Relative to the previous example, this "macroscopic" motion of the boundaries is inhibited by interactions with defects in the original single crystal. Models designed to understand this motion must take into account the next crystalline defect we wish to discuss – the dislocation.

1.1.6 Line defects: dislocations

From the perspective of mechanical properties, the *dislocation* is perhaps the most important crystalline defect (dislocations are discussed in more detail in Section 6.5.5). These are essentially lines through a crystal along which there is a systematic error in the way the atoms are arranged. Transmission electron microscopy at the micrometer scale shows dislocations as tangled lines through the crystal, as seen in the example of Fig. 1.3(c). At the nanometer scale the perturbations of the perfect crystalline order inside the dislocation core appear quite subtle (see Fig. 1.3(d)), but their consequences are profound. Through progressive rearrangement of bonds in the dislocation core, this defect provides a low-stress mechanism by which crystal planes can easily slide over each other, without destroying the essential underlying crystal structure. The result is the highly ductile nature we observe in metals like aluminum, copper and iron. In broad terms, materials containing lots of dislocations that can move easily will be soft and deformable, while materials in which dislocation motion is inhibited will be hard and brittle. Striking a balance between these two extremes, by controlling and modifying the nucleation, multiplication and mobility of dislocations, is one of the primary disciplines of materials science. Indeed, a typical mechanical engineering program will include at least one course that is almost exclusively dedicated to "heating and beating" metallic alloys in order to achieve the right balance of these properties for a given application.

Dislocations are challenging to model because they exhibit important properties at many scales. The nature of the dislocation is such that it imparts a long-range elastic field on a crystal, inducing significant stresses over distances of many microns. At the same time, details of the atomic structure right inside the dislocation core (a region only a few nanometers wide) can profoundly change a dislocation's behavior. For example, two dislocations that produce identical elastic fields (determined by the dislocation's *Burgers vector*, see Section 6.5.5) can be completely different in terms of their tendency to move under applied stress, which depends in part on the crystallographic planes on which they travel. Between these two extremes, the interplay between the long-range stress fields and the dislocation cores leads to the formation of dislocation cell structures on a subgranular scale of a few microns. Under heavy plastic deformation or fatigue loading, dislocation patterns like those shown in Fig. 1.3(b) are often observed. These are due to complex elastic interactions between defects and multiple creation and annihilation processes during the deformation.

Like grain boundaries, dislocations have characteristic mobilities that depend on the elements in the crystal, the temperature, the applied stress and the type of dislocation. As one example of direct measurement of dislocation velocities, we show the results of Kruml *et al.* [KCDM02] in Fig. 1.15, who measured dislocation velocities in the range of 0.1–0.2 μm/s at stresses in the range of 10–50 MPa. While these are high velocities compared to atomic dimensions, they are relatively slow for dislocations, as these experiments include a number of velocity-inhibiting features such as point defects in the crystal. At the other extreme, Shilo and Zolotoyabko [SZ03] have made velocity measurements by exciting dislocations in $LiNbO_3$ with high-frequency sound waves. In this case, the dislocation is forced to vibrate around its initial position instead of traveling atomically long distances through the crystal, and as such the velocity is likely closer to the "true" unimpeded velocity of a dislocation through a perfect crystal. These measurements showed dislocation velocities as high as 3200 m/s, or about 80% of the speed of sound in the

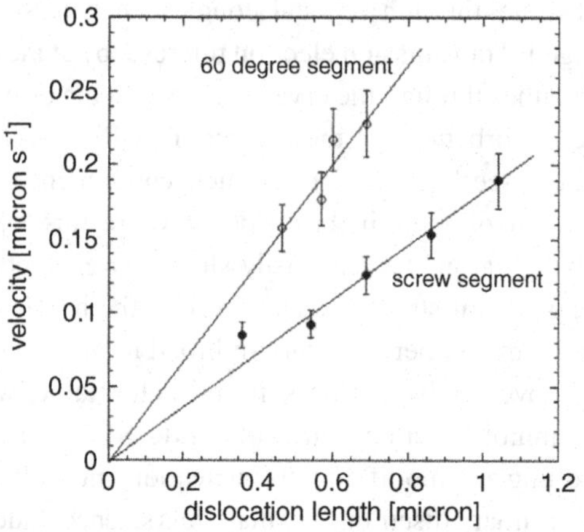

Fig. 1.15 Velocity of dislocations in Ge measured as a function of length at an applied shear stress of about 50 MPa. (Reprinted from [KCDM02] with permission from IOP Publishing.)

crystal.[5] This extreme range of velocities highlights, once again, the challenge of modeling materials: phenomena on many different scales matter. In this case, drastically different scales show drastically different dislocation mobilities.

In light of the above, it is no surprise that the modeling of dislocations also takes place on many different scales. At the atomic scale, DFT and molecular statics (MS) are used to study dislocation core structures. At slightly larger scales, MD is used to study interactions between different dislocations as well as between dislocations and other defects. These studies tell us something about strengthening processes and allow us to build a catalog of basic mechanisms that are used in larger-scale models. Understanding sources for new dislocations in materials is also important to the complete picture of how these defects multiply during deformation, and MD has been used to study these as well. The next larger scale used for dislocation simulations is associated with the formation of dislocation cell structures that exist on a micrometer scale. These structures are smaller than a typical grain, but still too large for fully atomistic simulations. So-called "discrete dislocation dynamics (DDD)" models have been developed to model dislocations on this scale.[6] A key input to DDD models is the elastic stress field around a dislocation – a fundamental result of continuum elasticity theory and one of the subjects of Section 6.5.5. Finally, at the largest scale, continuum models of plasticity treat dislocations as a continuous field with local "dislocation density." Constitutive laws, inspired by our understanding at smaller scales, are developed that prescribe the creation and interaction between densities of different types of dislocation. Such models make it possible to accurately predict large-scale plastic flow.

[5] This measurement compares well with atomistic models of isolated dislocations in Al under ideal conditions, which have predicted velocities on the order of 2000 m/s or about half the speed of sound at comparable stress levels [FOH08].

[6] In this book, we discuss specifics of dislocation modeling only briefly. However, an entire book has been written on the subject [BC06].

1.1.7 Point defects

Some of the slowest processes of interest to materials engineers are those that involve solid-state diffusion. Processes such as surface carburization of steel, recrystallization during annealing or precipitation hardening may take seconds at elevated temperatures or hours at lower temperatures, while the creep of materials under sustained temperature and stress may take years. All of these processes are extreme examples of the multiscale nature of materials, as they are on the one hand almost imperceivably slow and on the other hand utterly dependent on the behavior of the smallest and fastest moving of microstructural features: point defects.

Point defects include vacancies (the absence of a single atom from its lattice site in the perfect crystal) and interstitial atoms (an extra atom crammed into an interstice between perfect lattice sites). As atoms vibrate randomly due to thermal fluctuations, these point defects can occasionally jump from one lattice site or interstice to another. For large numbers of atoms under equilibrium conditions, these random jumps produce no net effect, as there are as many jumps in one direction as in any other. However, under any driving force, such as an applied stress or temperature gradient, more of these jumps will occur in a preferred direction. For example, interstitials may move preferentially from regions of high stress to regions of low stress. The collective action of large numbers of these "random but biased" events gives rise to macroscopic diffusion-driven processes.

Diffusion of one atomic species through the lattice of another is facilitated by the presence of vacancies, as otherwise the energy cost to exchange two atoms between lattice sites (or to squeeze an atom between interstices) can be too high. For many diffusion processes, the assumptions of *transition state theory* (TST) are valid[7] and the Arrhenius form prevails, whereby

$$\text{rate} \propto e^{-\Delta E / k_B T}.$$

This says that the rate of a particular process is proportional to the exponent of the activation energy (ΔE) over temperature (where k_B is Boltzmann's constant – a constant of nature that sets the energy scale). For vacancy-diffusion, there are two activation energies of interest: the energy required to create the vacancy (the so-called "vacancy formation energy" discussed in Section 6.5.3) and the energy to move a vacancy between sites (the "vacancy migration energy," ΔE_m). These quantities can be accurately computed using atomic-scale models. The great challenge here for modeling is that, on the one hand, the details of individual events matter (we must compute ΔE), but on the other hand these events occur only rarely relative to "atomistic" time scales (the scale of atomic vibrations). Thus, we need full atomistic simulations to get accurate estimates of ΔE, but we cannot run such simulations for long enough times to model the actual diffusion process. In Section 6.3, we discuss ways to compute the energetics of processes like individual diffusion steps, but these can

[7] The most important assumptions of TST for our purposes are that the motion between lattice sites that gives rise to diffusion is an equilibrium property of the system that can be computed from knowledge of the potential energy surface, and that motion between successive lattice sites is uncorrelated. See [Zwa01] for a discussion of TST.

only be reconciled with long-time processes within the context of statistical mechanics and using the probabilities of rare events that the Arrhenius form describes.

For example, let us set aside the issue of vacancy formation and consider the rate at which vacancies move between neighboring lattice sites. The *vacancy migration rate* can be expressed as

$$\Gamma = \nu_0 e^{-\Delta E_{\rm m}/k_{\rm B}T},$$

where ν_0 is a jump frequency that can be determined from the vibrational frequencies of the atomic normal modes. Atomistic simulations by Sorensen *et al.* [SMV00] found $\Delta E_{\rm m} = 0.69$ eV and $\nu_0 = 5.3$ THz for bulk fcc Cu. This sets the rate of vacancy migration at about $\Gamma = 5.5$ Hz at room temperature. Compared with the rate of atomic vibration (on the order of 10^{13} Hz), this gives a sense of what is meant by "rare" diffusion events. At the time of writing, practical MD simulations cannot even reach 1 μs in duration, but the above estimate means that it may take over 180 000 μs for for a single vacancy jump to occur.

The presence of dislocations and grain boundaries changes the rate of vacancy diffusion, since the more open structures of these defects permit vacancies to move more freely. For example, Huang *et al.* [HMP91] computed the migration energy of a vacancy through a dislocation in Cu to be about 7% less than in the bulk. Although this does not sound like much, the exponential dependence on the activation energy makes it significant. For example, using the data of Sorensen *et al.* cited above and assuming room temperature, this increases the rate of migration to nearly 38 Hz. Grain boundaries have an even more pronounced effect on diffusion rates, as shown in several studies (see, for example, [SMV00, SM03]).

1.1.8 Large-scale defects: cracks, voids and inclusions

The hot, violent and complex processing that many materials undergo means that they typically contain many more defects than those we have just described. Heat treatments and deformation processes typically introduce cracks and voids that can span scales from nanometers to meters. At the tip of any such crack we find the multiscale challenge of materials modeling once again, as large-scale crack geometry and loading conditions interact with micro- and nano-scale processes of plastic flow, tearing and atomic bond breaking. Fatigue processes, whereby a crack gradually grows due to cyclic loading and unloading, often take place on a time scale of days, months or years, even though each loading cycle may cause millions or billions of tiny processes of bond breaking and formation on the scale of picoseconds. Traditionally, fracture mechanics has been a part of continuum mechanics modeling, but it is increasingly the case that we turn to smaller scale models – crystal plasticity, discrete dislocation dynamics and atomistic simulations – to understand the detailed mechanisms of fracture.

Finally, most materials are more complex than a simple, single element such as pure Cu. Engineering materials are almost always combinations of elements, processed in such a way that multiple phases (different crystal structures with different compositions) coexist. We have already seen examples of this in Fig. 1.12. In Fig. 1.16, we see more examples that highlight the range of scales of such precipitates, often within the same sample. Through

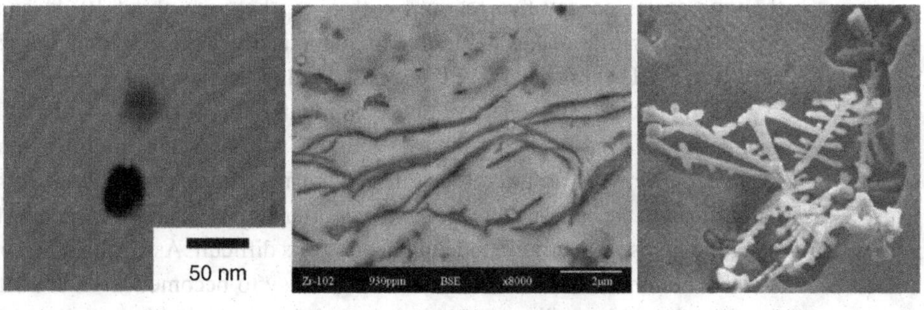

Fig. 1.16 Precipitates in engineering materials take on a wide range of sizes and shapes: (a) TiC precipitates in steel on the nanometer scale, (b) narrow, flat hydrides in a zirconium alloy and (c) a relatively large alumina precipitate in a steel alloyed with aluminum (the width of view in (c) is about 80 μm). ((a) Reproduced from [SSH+03], with permission from the IUCr (http://journals.iucr.org), (b) reproduced from figure 3a of [VAR+03], with the kind permission of Springer Science+Business Media, (c) reproduced from figure 2 of [RSF70], with the kind permission of Springer Science+Business Media.)

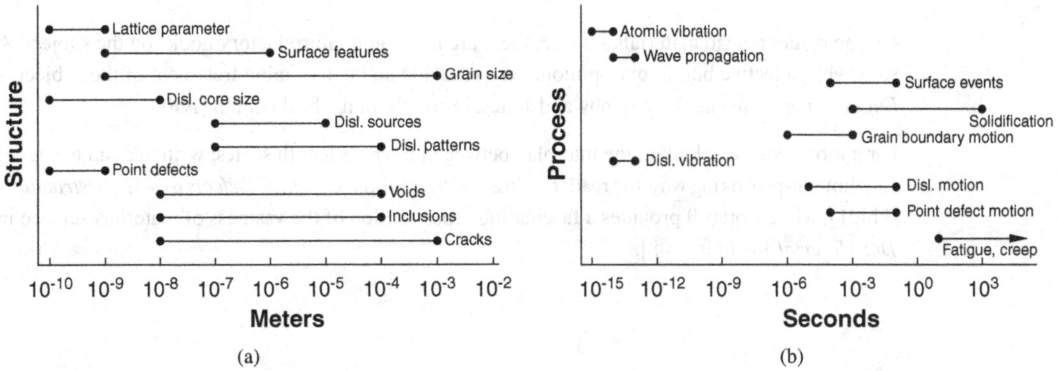

Fig. 1.17 (a) Length scales and (b) time scales associated with the material structures discussed in this chapter.

heat treatment and work, the sizes and distributions of these phases are controlled to optimize properties. Often, these optimization processes were determined through experimental trial and error, and it is only more recently that materials modeling has been used to understand the myriad of processes at many scales that gives rise to material properties.

1.2 Materials scales: taking stock

Figures 1.17(a) and 1.17(b) summarize the length and time scales that need to be addressed in the examples of materials phenomena presented here. The range of both length and time scales is daunting, as phenomena from each scale collectively contribute to the larger-scale behavior. It is interesting that while length scales are spread more or less across all scales, the time scales occupy two regimes with a gap between about 10^{-13} and 10^{-6} s. This *separation of scales* makes it possible to dramatically accelerate temporal processes using accelerated MD approaches like those described in Section 10.4.

With this admittedly selective survey of materials processes, we have laid down the gauntlet: with what modeling theories must we arm ourselves in order to understand the relationship between the microstructure and the mechanical properties of crystalline solids? Unfortunately, we believe there are *many* important topics, and that they come from disparate scientific fields. To be trite: modeling materials is difficult. As we dig down into each of the fundamental methods in this book, it will be easy to become lost in the details. We hope that you, the reader, will try to remind yourself regularly of this overview chapter and view each method within this context. By the end, we hope to convince you that the methods described in this book serve as the foundations for modeling materials, upon which your own materials science studies can be soundly built.

Further reading

- For the reader new to materials science, there are many good introductory books on the subject. A somewhat selective but, in our opinion, very readable and entertaining treatment of the subject is *Engineering Materials* 1 by Ashby and Jones, currently in its third edition [AJ05].
- For a more extensive look at the interplay between materials length scales, written in an engaging and thought-provoking way, the reader is directed to Phillips' *Crystals, Defects and Microstructures* [Phi01], while Cotterill provides a fascinating cross-section of the vastness of materials science in *The Material World* [Cot08].

PART I

CONTINUUM MECHANICS AND THERMODYNAMICS

2 Essential continuum mechanics and thermodynamics

A solid material subjected to mechanical and thermal loading will change shape and develop internal stresses. What is the best way to describe this? In principle, the behavior of a material (neglecting relativistic effects) is dictated by that of its atoms, which are governed by quantum mechanics. Therefore, if we could solve Schrödinger's equation (see Chapter 4) for 10^{23} atoms and evolve the dynamics of the electrons and nuclei over "macroscopic times" (i.e. seconds, hours and days) we would be able to predict material behavior. Of course when we say "material," we are already referring to a very complex system as demonstrated in the previous chapter. In order to predict the response of the material we would first have to construct its structure in the computer, which would require us to use Schrödinger's equation to simulate the process by which it was manufactured. Conceptually, it is useful to think of materials in this way, but we can quickly see the futility of the approach; state-of-the-art quantum calculations involve mere hundreds of atoms over a time of nanoseconds.

At the other extreme to quantum mechanics lie continuum mechanics and thermodynamics. These disciplines completely ignore the discreteness of the world, treating it in terms of "macroscopic observables," time and space averages over the underlying swirling masses of atoms. This leads to a theory couched in terms of continuously varying fields. Using clear thinking inspired by experiments it is possible to construct a remarkably coherent and predictive framework for material behavior. In fact, continuum mechanics and thermodynamics have been so successful that with the exception of electromagnetic phenomena, almost all of the courses in an engineering curriculum, from solid mechanics to aerodynamics, are simply an application of simplified versions of the general theory to situations of special interest. Clearly there is something to this macroscopically averaged view of the world. Of course, the continuum picture becomes fuzzy and eventually breaks down when we attempt to apply it to phenomena governed by small length and time scales. Those are exactly the "multiscale" situations that are of interest to us later in this book. But first we need to understand the limiting case where all is well with the macroscopic view.

The aim of this chapter is to provide a concise introduction to continuum mechanics, accessible to readers from different backgrounds, that covers the main issues and concepts that we will revisit later in the book. This chapter is a summary of the far more comprehensive book *Continuum Mechanics and Thermodynamics* [TME12] written by the authors as a companion to this book. Because this chapter is written as a compressed summary, its style and depth of coverage are very different from the rest of the book. (We feel that this is important to point out since you, our reader, will get the wrong impression of the nature of this book if you consider only this chapter.) Experts in continuum mechanics and thermodynamics can skip this chapter entirely for now, and merely use it as a handy reference to consult as necessary when reading the rest of the book. Readers who are new to the subject will benefit from the concise introduction presented here, but they are strongly encouraged to read the two books together. Unlike a typical continuum mechanics textbook, our companion book [TME12] places more emphasis on the basic assumptions and approximations inherent to the theory. This is vital background for a book like *Modeling Materials* that endeavors ultimately to couple continuum and atomistic representations and identify the connections between these parallel worldviews. Finally, [TME12] has been written with a broad audience in mind, making it accessible to readers without an engineering background, while also using the same notation and terminology as this book.

We begin this chapter with a discussion of the notation, definitions and properties of scalars, vectors and tensors. All physical variables must belong to this class and therefore this subject is a prerequisite, not only of continuum mechanics and thermodynamics, but of every other chapter in this book as well.

2.1 Scalars, vectors, and tensors

Continuum mechanics seeks to provide a fundamental model for material response. It is sensible to require that the predictions of such a theory should not depend on the irrelevant details of a particular coordinate system. The key is to write the theory in terms of variables that are unaffected by such changes; *tensors* (or *tensor fields*) are measures that have this property. Vectors and scalar invariants are special cases of tensors.

2.1.1 Tensor notation

Before discussing the definition and properties of tensors it is helpful to discuss the nuts and bolts of the notation of tensor algebra. In the process of doing so we will introduce important operations between tensors.

Tensors come in different flavors depending on the number of spatial directions that they couple. The simplest tensor has no directional dependence and is called a *scalar invariant* to distinguish it from a simple scalar. A *vector* has one direction. For two directions and higher the general term *tensor* is used. The number of spatial directions associated with a tensor is called its *rank* or *order*. We will use these terms interchangeably.

Indicial versus direct notation Tensors can be denoted using either *indicial notation* or *direct notation*. In both cases, tensors are represented by a symbol, e.g. m for mass, v for velocity and σ for stress. In indicial notation, the spatial directions associated with a tensor are denoted by indices attached to the symbol. Mass has no direction, so it has no indices, velocity has one index, stress two, and so on: m, v_i, σ_{ij}. The number of indices is equal to the rank of the tensor. Since tensor indices refer to spatial directions, the range of an index $[1, 2, \ldots, n_d]$ is determined by the dimensionality of space. We will be dealing mostly with three-dimensional space ($n_d = 3$); however, the notation we develop applies to any value of n_d. The tensor symbol with its indices represents the components of the tensor, for example

v_1, v_2 and v_3 are the components of the velocity vector. A set of simple rules for the interaction of indices provides a mechanism for describing all of the tensor operations that we will require. In fact, what makes this notation particularly useful is that *any operation defined by indicial notation has the property that if its arguments are tensors the result will also be a tensor.*

In direct notation, no indices are attached to the tensor symbol. The rank of the tensor is represented by the typeface used to display the symbol. Scalar invariants are displayed in a regular font while first-order tensors and higher are displayed in a bold font (or with an underline when written by hand): m, $\boldsymbol{v}$, $\boldsymbol{\sigma}$ (or m, $\underline{v}$, $\underline{\sigma}$ by hand). The advantage of direct notation is that it emphasizes the fact that tensors are independent of the choice of a coordinate system basis (whereas indices are always tied to a particular basis). Direct notation is also more compact and therefore easier to read. However, the lack of indices means that special notation must be introduced for different operations between tensors. Many symbols in this notation are not universally accepted and direct notation is not available for all operations. We will return to direct notation in Section 2.1.3 when we discuss tensor operations.

In some cases, the operations defined by indicial notation can also be written using the matrix notation familiar from linear algebra. Here vectors and second-order tensors are represented as column and rectangular matrices of their components, for example

$$[\boldsymbol{v}] = \begin{bmatrix} v_1 \\ v_2 \\ v_3 \end{bmatrix}, \quad [\boldsymbol{\sigma}] = \begin{bmatrix} \sigma_{11} & \sigma_{12} & \sigma_{13} \\ \sigma_{21} & \sigma_{22} & \sigma_{23} \\ \sigma_{31} & \sigma_{32} & \sigma_{33} \end{bmatrix}.$$

The notation $[\boldsymbol{v}]$ and $[\boldsymbol{\sigma}]$ is a shorthand representation for the column and rectangular matrices, respectively, formed by the components of the vector $\boldsymbol{v}$ and the second-order tensor $\boldsymbol{\sigma}$. This notation will sometimes be used when tensor operations can be represented by matrix multiplication and other matrix operations on tensor components.

Before proceeding to the definition of tensors, we begin by introducing the basic rules of indicial notation, starting with the most basic rule: the summation convention.

Summation and dummy indices Consider the following sum:

$$S = a_1 x_1 + a_2 x_2 + \cdots + a_{n_d} x_{n_d}.$$

We can write this expression using the summation symbol Σ:

$$S = \sum_{i=1}^{n_d} a_i x_i = \sum_{j=1}^{n_d} a_j x_j = \sum_{m=1}^{n_d} a_m x_m.$$

Clearly, the particular choice for the letter we use for the summation, i, j or m, is irrelevant since the sum is independent of the choice. Indices with this property are called *dummy indices*. Because summation of products, such as $a_i x_i$, appears frequently in tensor operations, a simplified notation is adopted where the Σ symbol is dropped and any index appearing twice in a product of variables is taken to be a dummy index and summed over. For example,

$$S = a_i x_i = a_j x_j = a_m x_m = a_1 x_1 + a_2 x_2 + \cdots + a_{n_d} x_{n_d}.$$

This convention was introduced by Albert Einstein in the famous 1916 paper where he outlined the principles of general relativity [Ein16]. It is therefore called *Einstein's summation convention* or just the *summation convention* for short. Some examples for $n_d = 3$ are:

$$a_i x_i = a_1 x_1 + a_2 x_2 + a_3 x_3, \qquad a_i a_i = a_1^2 + a_2^2 + a_3^2, \qquad \sigma_{ii} = \sigma_{11} + \sigma_{22} + \sigma_{33}.$$

It is important to point out that the summation convention only applies to indices that appear twice in a product of variables. A product containing more than two occurrences of a dummy index, such as $a_i b_i x_i$, is meaningless. If the objective here is to sum over index i, this would have to be written as $\sum_{i=1}^{n_d} a_i b_i x_i$. The summation convention does, however, generalize to the case where there are multiple dummy indices in a product. For example, a double sum over dummy indices i and j is

$$\begin{aligned} A_{ij} x_i y_j = A_{11} x_1 y_1 + A_{12} x_1 y_2 + A_{13} x_1 y_3 \\ + A_{21} x_2 y_1 + A_{22} x_2 y_2 + A_{23} x_2 y_3 \\ + A_{31} x_3 y_1 + A_{32} x_3 y_2 + A_{33} x_3 y_3. \end{aligned}$$

We see how the summation convention provides a very efficient shorthand notation for writing complex expressions. Finally, there may be situations where although an index appears twice in a product, we do *not* wish to sum over it. For example say we wish to set the diagonal components of a second-order tensor to zero: $A_{11} = A_{22} = A_{33} = 0$. In order to temporarily "deactivate" the summation convention we write:

$$A_{ii} := 0 \quad \text{(no sum)} \qquad \text{or} \qquad A_{\underline{i}\,\underline{i}} := 0,$$

where ":=" is the assignment operation setting the variable on the left to the value on the right (a notational convention we adopt throughout the book).

Free indices An index that appears only once in each product term of an equation is referred to as a *free index*. A free index takes on the values $1, 2, \ldots, n_d$, one at a time. For example,

$$A_{ij} x_j = b_i.$$

Here i is a free index and j is a dummy index. Since i can take on n_d separate values, the above expression represents the following system of n_d equations:

$$\begin{aligned} A_{11} x_1 + A_{12} x_2 + \cdots + A_{1 n_d} x_{n_d} &= b_1, \\ A_{21} x_1 + A_{22} x_2 + \cdots + A_{2 n_d} x_{n_d} &= b_2, \\ \vdots \qquad\qquad & \quad \vdots \\ A_{n_d 1} x_1 + A_{n_d 2} x_2 + \cdots + A_{n_d n_d} x_{n_d} &= b_{n_d}. \end{aligned}$$

Naturally, all terms in an expression must have the same free indices (or no indices at all). The expression $A_{ij} x_j = b_k$ is meaningless. However, $A_{ij} x_j = c$ (where c is a scalar) is fine. There can be as many free indices as necessary. For example the expression $D_{ijk} x_k = A_{ij}$ contains the two free indices i and j and therefore represents n_d^2 equations.

Matrix notation Indicial operations involving tensors of rank 2 or less can be represented as matrix operations. For example the product $A_{ij}x_j$ can be expressed as a matrix multiplication. For $n_d = 3$ we have

$$A_{ij}x_j = \mathsf{A}\mathsf{x} = \begin{bmatrix} A_{11} & A_{12} & A_{13} \\ A_{21} & A_{22} & A_{23} \\ A_{31} & A_{32} & A_{33} \end{bmatrix} \begin{bmatrix} x_1 \\ x_2 \\ x_3 \end{bmatrix} .$$

We use a sans serif font to denote matrices to distinguish them from tensors. Thus, A is a rectangular table of numbers. The entries of A are equal to the components of the tensor A, i.e. $\mathsf{A} = [A]$, so that $\mathsf{A}_{ij} = A_{ij}$. Column matrices are denoted by lower-case letters and rectangular matrices by upper-case letters.

The expression $A_{ji}x_j$ can be computed in a similar manner, but the entries of A must be *transposed* before performing the matrix multiplication, i.e. its rows and columns must be swapped. Thus,

$$A_{ji}x_j = \mathsf{A}^T\mathsf{x} = \begin{bmatrix} A_{11} & A_{21} & A_{31} \\ A_{12} & A_{22} & A_{32} \\ A_{13} & A_{23} & A_{33} \end{bmatrix} \begin{bmatrix} x_1 \\ x_2 \\ x_3 \end{bmatrix} ,$$

where the superscript T denotes the transpose operation. Similarly, the sum $a_i x_i$ can be written

$$a_i x_i = \mathsf{a}^T\mathsf{x} = \begin{bmatrix} a_1 & a_2 & a_3 \end{bmatrix} \begin{bmatrix} x_1 \\ x_2 \\ x_3 \end{bmatrix} .$$

The transpose operation has the important property that

$$(\mathsf{AB})^T = \mathsf{B}^T\mathsf{A}^T .$$

This implies that $(\mathsf{ABC})^T = \mathsf{C}^T\mathsf{B}^T\mathsf{A}^T$, and so on. Another example of a matrix operation is the expression, $A_{ii} = A_{11} + A_{22} + \cdots + A_{22}$, which is defined as the *trace* of the matrix A. In matrix notation this is denoted as $\operatorname{tr}\mathsf{A}$.

Kronecker delta The Kronecker delta is defined as follows:

$$\delta_{ij} = \begin{cases} 1 & \text{if } i = j, \\ 0 & \text{if } i \neq j. \end{cases} \tag{2.1}$$

In matrix form, δ_{ij} are the entries of the *identity matrix* I (for $n_d = 3$):

$$\mathsf{I} = \begin{bmatrix} 1 & 0 & 0 \\ 0 & 1 & 0 \\ 0 & 0 & 1 \end{bmatrix} . \tag{2.2}$$

Most often the Kronecker delta appears in expressions as a result of a differentiation of a tensor with respect to its components. For example, $\partial x_i / \partial x_j = \delta_{ij}$. This is correct as long as the components of the tensor are independent. An important property of δ_{ij} is *index substitution*: $a_i\delta_{ij} = a_j$, which can be easily demonstrated:

$$a_i\delta_{ij} = a_1\delta_{1j} + a_2\delta_{2j} + a_3\delta_{3j} = \left\{ \begin{array}{ll} a_1 & \text{if } j = 1 \\ a_2 & \text{if } j = 2 \\ a_3 & \text{if } j = 3 \end{array} \right\} = a_j.$$

Permutation symbol The permutation symbol ϵ_{ijk} for $n_{\mathrm{d}} = 3$ is defined as follows:

$$\epsilon_{ijk} = \begin{cases} 1 & \text{if } i, j, k \text{ form an even permutation of } 1, 2, 3, \\ -1 & \text{if } i, j, k \text{ form an odd permutation of } 1, 2, 3, \\ 0 & \text{if } i, j, k \text{ do not form a permutation of } 1, 2, 3. \end{cases} \tag{2.3}$$

Thus, $\epsilon_{123} = \epsilon_{231} = \epsilon_{312} = 1$, $\epsilon_{321} = \epsilon_{213} = \epsilon_{132} = -1$, and $\epsilon_{111} = \epsilon_{112} = \epsilon_{113} = \cdots = \epsilon_{333} = 0$. The permutation symbol has a number of important properties that are described in Section 2.2.6 in [TME12]. Here we focus on just one that we will need later in the book. The permutation symbol provides an expression for the determinant of a matrix:

$$\det \mathsf{A} = \epsilon_{ijk}\mathsf{A}_{1i}\mathsf{A}_{2j}\mathsf{A}_{3k}. \tag{2.4}$$

We will also need the derivative of the determinant of a matrix with respect to the matrix entries. This can be obtained from the above relation after some algebra (see Section 2.2.6 in [TME12]). The result is

$$\frac{\partial(\det \mathsf{A})}{\partial \mathsf{A}} = \mathsf{A}^{-T} \det \mathsf{A}. \tag{2.5}$$

The permutation symbol plays an important role in vector cross products. We will see this in Section 2.1.2.

Now that we have explained the rules for tensor component interactions, we turn to the matter of the definition of a tensor beginning from the special case of vectors.

2.1.2 Vectors and higher-order tensors

The typical high-school definition of a vector is "an entity with a magnitude and a direction," often stressed by the teacher by drawing an arrow on the board. This is clearly only a partial definition, since many things that are not vectors have a magnitude and a direction. This book, for example, has a magnitude (the number of pages in it) and a direction (front to back), yet it is not what we would normally consider a vector. It turns out that an indispensable part of the definition is the parallelogram law that defines how vectors are added together. This suggests that an *operational* approach must be taken to define vectors. However, if this is the case, then vectors can only be defined as a group and not individually. This leads to the idea of a *vector space*.

Vector spaces and the inner product and norm A real vector space V is a set, defined over the field of real numbers $\mathbb{R}$, where the following two operations have been defined:

1. *vector addition*: for any two vectors $\boldsymbol{a}, \boldsymbol{b} \in V$, we have $\boldsymbol{a} + \boldsymbol{b} = \boldsymbol{c} \in V$;
2. *scalar multiplication*: for any scalar $\lambda \in \mathbb{R}$ and vector $\boldsymbol{a} \in V$, we have $\lambda\boldsymbol{a} = \boldsymbol{c} \in V$,

with certain properties defined in Section 2.3.1 in [TME12]. At this point the definition is completely general and abstract. It is possible to invent many vector objects and definitions for addition and multiplication that satisfy these rules. The vectors that are familiar to us from the physical world have additional properties associated with the geometry of finite-dimensional space, such as distances and angles. The definition of the vector space must be extended to include these concepts. The result is the *Euclidean space* named after the Greek mathematician Euclid who laid down the foundations of "Euclidean geometry." We define these properties separately beginning with the concept of a finite-dimensional space.

Finite-dimensional spaces and basis vectors The dimensionality of a space is related to the concept of linear dependence. The m vectors $\boldsymbol{a}_1, \dots, \boldsymbol{a}_m \in V$ are *linearly dependent* if and only if there exist $\lambda_1, \dots, \lambda_m \in \mathbb{R}$ not all equal to zero, such that

$$\lambda_1 \boldsymbol{a}_1 + \cdots + \lambda_m \boldsymbol{a}_m = \boldsymbol{0}.$$

Otherwise, the vectors are *linearly independent*. The largest possible number of linearly-independent vectors is the dimensionality of the vector space. (For example in a three-dimensional vector space there can be at most three linearly independent vectors.) This is denoted by dim V. We limit ourselves to vector spaces for which dim V is finite.

Consider an n_d-dimensional vector space V^{n_d}. Any set of n_d linearly-independent vectors can be selected as a *basis* of V^{n_d}. The basis vectors are commonly denoted by $\boldsymbol{e}_i$, $i = 1, \dots, n_d$. Any other vector $\boldsymbol{a} \in V^{n_d}$ can be expressed as:

$$\boldsymbol{a} = a_1 \boldsymbol{e}_1 + \cdots + a_{n_d} \boldsymbol{e}_{n_d} = a_i \boldsymbol{e}_i, \tag{2.6}$$

where a_i are called the *components* of vector $\boldsymbol{a}$ with respect to the basis $\boldsymbol{e}_i$. The choice of basis vectors is not unique. However, the components of a vector in a particular basis are unique. This is easy to show by assuming the contrary and using the linear dependence of the basis vectors.

Euclidean space The real coordinate space $\mathbb{R}^{n_d}$ is an n_d-dimensional vector space defined over the field of real numbers. A vector in $\mathbb{R}^{n_d}$ is represented by a set of n_d real components relative to a given basis. Thus for $\boldsymbol{a} \in \mathbb{R}^{n_d}$ we have $\boldsymbol{a} = (a_1, \dots, a_{n_d})$, where $a_i \in \mathbb{R}$. Addition and multiplication are defined for $\mathbb{R}^{n_d}$ in terms of the corresponding operations familiar to us from the algebra of real numbers:

1. *Addition:* $\boldsymbol{a} + \boldsymbol{b} = (a_1, \dots, a_{n_d}) + (b_1, \dots, b_{n_d}) = (a_1 + b_1, \dots, a_{n_d} + b_{n_d})$.
2. *Multiplication:* $\lambda \boldsymbol{a} = \lambda(a_1, \dots, a_{n_d}) = (\lambda a_1, \dots, \lambda a_{n_d})$.

In order for $\mathbb{R}^{n_d}$ to be a Euclidean space it must possess an *inner product*, which is related to angles between vectors, and it must possess a *norm*, which provides a measure for the length of a vector. In this book we will be concerned primarily with three-dimensional Euclidean space for which $n_d = 3$.

Inner product and norm An inner product is a real-valued bilinear mapping. The inner product of two vectors $\boldsymbol{a}$ and $\boldsymbol{b}$ is denoted by $\langle \boldsymbol{a}, \boldsymbol{b} \rangle$. An inner product function must satisfy the following properties $\forall\, \boldsymbol{a}, \boldsymbol{b}, \boldsymbol{c} \in V$ and $\forall\, \lambda, \mu \in \mathbb{R}$:[1]

1. $\langle \lambda \boldsymbol{a} + \mu \boldsymbol{b}, \boldsymbol{c} \rangle = \lambda \langle \boldsymbol{a}, \boldsymbol{c} \rangle + \mu \langle \boldsymbol{b}, \boldsymbol{c} \rangle$ linearity with respect to first argument
2. $\langle \boldsymbol{a}, \boldsymbol{b} \rangle = \langle \boldsymbol{b}, \boldsymbol{a} \rangle$ symmetry
3. $\langle \boldsymbol{a}, \boldsymbol{a} \rangle \geq 0$ and $\langle \boldsymbol{a}, \boldsymbol{a} \rangle = 0$ iff $\boldsymbol{a} = \boldsymbol{0}$ positivity

For $\mathbb{R}^{n_d}$ the standard choice for an inner product is the *dot product*, $\langle \boldsymbol{a}, \boldsymbol{b} \rangle = \boldsymbol{a} \cdot \boldsymbol{b}$, from which the Euclidean norm follows as $\|\boldsymbol{a}\| = \sqrt{\boldsymbol{a} \cdot \boldsymbol{a}}$. This notation distinguishes the norm from the absolute value of a scalar, $|s| = \sqrt{s^2}$. A shorthand notation denoting $\boldsymbol{a}^2 \equiv \boldsymbol{a} \cdot \boldsymbol{a}$ is sometimes adopted. A vector $\boldsymbol{a}$ satisfying $\|\boldsymbol{a}\| = 1$ is called a *unit vector*. Geometrically, the dot product is $\boldsymbol{a} \cdot \boldsymbol{b} = \|\boldsymbol{a}\| \, \|\boldsymbol{b}\| \cos \theta(\boldsymbol{a}, \boldsymbol{b})$, where $\theta(\boldsymbol{a}, \boldsymbol{b})$ is the angle between vectors $\boldsymbol{a}$ and $\boldsymbol{b}$.

Frames of reference and position vectors The location of points in space and measurement of times requires the definition of a frame of reference.[2] We define a *frame of reference* as a rigid physical object, such as the earth, the laboratory or the "fixed stars," relative to which positions are measured, and a clock with which times are measured. Mathematically, the space associated with a frame of reference can be regarded as a set E of *points*, which are defined through their relation with a Euclidean vector space $\mathbb{R}^{n_d}$ (called the *translation space* of E). For every pair of points x, y in E, there exists a vector $\boldsymbol{v}(x, y)$ in $\mathbb{R}^{n_d}$ that satisfies the following conditions:

$$\boldsymbol{v}(x, y) = \boldsymbol{v}(x, z) + \boldsymbol{v}(z, y) \qquad \forall x, y, z \in E, \tag{2.7}$$

$$\boldsymbol{v}(x, y) = \boldsymbol{v}(x, z) \qquad \text{if and only if } y = z. \tag{2.8}$$

A set satisfying these conditions is called a *Euclidean point space*. A *position vector* $\boldsymbol{x}$ for a point x is defined by singling out one of the points as the *origin* o and writing: $\boldsymbol{x} \equiv \boldsymbol{v}(x, o)$. Equations (2.7) and (2.8) imply that every point x in E is uniquely associated with a vector $\boldsymbol{x}$ in $\mathbb{R}^{n_d}$. The vector connecting two points is given by $\boldsymbol{x} - \boldsymbol{y} = \boldsymbol{v}(x, o) - \boldsymbol{v}(y, o)$. The distance between two points and the angles formed by three points can be computed using the norm and inner product of the corresponding Euclidean vector space.

Orthonormal basis and the Cartesian coordinate system With the introduction of an origin o and position vectors above, a coordinate system is defined once a set of basis vectors is selected. An *orthogonal* basis is one satisfying the condition that all basis vectors are perpendicular to each other. If in addition the basis vectors have magnitude unity, the basis is called *orthonormal*. The requirements for an orthonormal basis are expressed mathematically by the condition

$$\boldsymbol{e}_i \cdot \boldsymbol{e}_j = \delta_{ij}, \tag{2.9}$$

[1] We use (but try not to overuse) the standard mathematical notation. $\forall$ should be read "for all" or "for every," $\in$ should be read "in," iff should be read "if and only if." The symbol "$\equiv$" means "equal by definition."

[2] The idea of a "frame of reference" and the underlying concepts of space and time have always been and remain controversial to this day. See Section 2.1 in [TME12] for more on this.

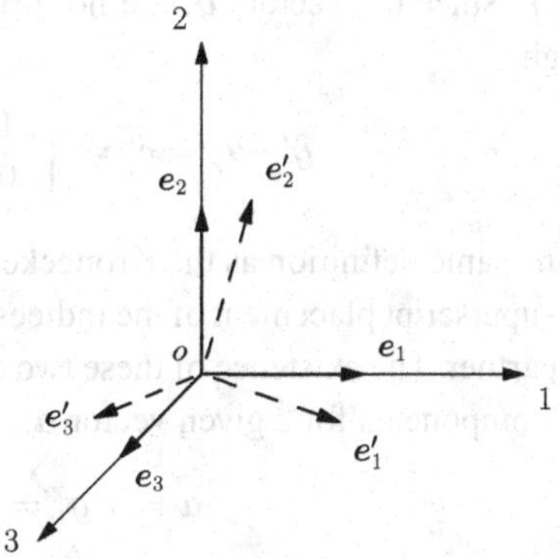

Fig. 2.1 The Cartesian coordinate system. The three axes and basis vectors e_i are shown along with an alternative rotated set of basis vectors e_i'. The origin of the coordinate system is o.

where δ_{ij} is the Kronecker delta defined in Eqn. (2.1). The orthonormal basis leads to the familiar *Cartesian coordinate system*, where e_i are unit vectors along the axis directions with an origin located at point o (see Fig. 2.1). By convention, we choose basis vectors that form a right-handed triad (this means that if we curl the fingers of the right hand, rotating them from e_1 towards e_2, the thumb will point in the positive direction of e_3). In an orthonormal basis, the indicial expression for the dot product is

$$a \cdot b = a_i b_i. \tag{2.10}$$

The components of a vector in an orthonormal basis are obtained by dotting the vector with the basis vector, which gives

$$a_i = a \cdot e_i. \tag{2.11}$$

Nonorthogonal bases and covariant and contravariant components The definitions given above for an orthonormal basis can be extended to nonorthogonal bases. We present a brief summary here for the sake of completeness and because nonorthogonal bases are closely related to the subject of crystal structures in Chapter 3. However, we will adopt an orthonormal description in the remainder of the book.

In $\mathbb{R}^3$, any set of three noncollinear, nonplanar and nonzero vectors forms a basis. There are no other constraints on the magnitude of the basis vectors or the angles between them. A general basis consisting of vectors that are not perpendicular to each other and may have magnitudes different from 1 is called a *nonorthogonal basis*. An example of such a basis is the set of lattice vectors that define the structure of a crystal (see Section 3.3). To distinguish such a basis from an orthonormal basis, we denote its basis vectors with $\{g_i\}$

instead of $\{e_i\}$. Since the vectors g_i are not orthogonal, a *reciprocal*[3] basis $\{g^i\}$ can be defined through

$$\hat{g}^i \cdot \hat{g}_j = \delta^i_j = \begin{cases} 1 & \text{if } i = j, \\ 0 & \text{if } i \neq j. \end{cases} \quad (2.12)$$

Thus, δ^i_j has the same definition as the Kronecker delta defined in Eqn. (2.1). Note that the subscript and superscript placement of the indices is used to distinguish between a basis and its reciprocal partner. The existence of these two closely related bases leads to the existence of two sets of components for a given vector $\boldsymbol{a}$:

$$\boldsymbol{a} = a^i \boldsymbol{g}_i = a_j \boldsymbol{g}^j, \quad (2.13)$$

where a^i are the *contravariant* components of $\boldsymbol{a}$ and a_i are the *covariant* components of $\boldsymbol{a}$. The connections between covariant and contravariant components are obtained by dotting Eqn. (2.13) with either $\boldsymbol{g}^k$ or $\boldsymbol{g}_k$, which gives

$$a^k = g^{jk} a_j \quad \text{and} \quad a_k = g_{ik} a^i, \quad (2.14)$$

where $g_{ij} = \boldsymbol{g}_i \cdot \boldsymbol{g}_j$ and $g^{ij} = \boldsymbol{g}^i \cdot \boldsymbol{g}^j$. The processes in Eqn. (2.14) are called *raising* or *lowering* an index.

Continuum mechanics can be phrased entirely in terms of nonorthogonal bases (and more generally in terms of curvilinear coordinate systems). However, the general derivation leads to notational complexity that can obscure the main physical concepts underlying the theory. We therefore limit ourselves to orthonormal bases in the remainder of the book.

Cross product We have already encountered the dot product that maps two vectors to a scalar. The *cross product* is a binary operation that maps two vectors to a new vector that is orthogonal to both with magnitude equal to the area of the parallelogram spanned by the two original vectors. The cross product is denoted by the $\times$ symbol, so that $\boldsymbol{c} = \boldsymbol{a} \times \boldsymbol{b} = A(\boldsymbol{a}, \boldsymbol{b})\boldsymbol{n}$, where $A(\boldsymbol{a}, \boldsymbol{b}) = \|\boldsymbol{a}\| \, \|\boldsymbol{b}\| \sin\theta(\boldsymbol{a}, \boldsymbol{b})$ is the area spanned by $\boldsymbol{a}$ and $\boldsymbol{b}$ and $\boldsymbol{n}$ is the unit vector normal to the plane defined by them. This definition is not complete since there are two possible opposite directions for the normal. The solution is to append to the definition the requirement that $(\boldsymbol{a}, \boldsymbol{b}, \boldsymbol{a} \times \boldsymbol{b})$ form a right-handed set. The indicial form of the cross product is

$$\boldsymbol{a} \times \boldsymbol{b} = \epsilon_{ijk} a_i b_j \boldsymbol{e}_k. \quad (2.15)$$

Change of basis We noted earlier that the choice of basis vectors $\boldsymbol{e}_i$ is not unique. There are in fact an infinite number of equivalent basis sets. Consider two orthonormal bases $\boldsymbol{e}_\alpha$ and

[3] The reciprocal basis vectors of continuum mechanics are closely related to the reciprocal lattice vectors of solid-state physics discussed in Section 3.7.1. The only difference is a 2π factor introduced in the physics definition to simplify the form of plane wave expressions.

$\mathbf{e}'_i$ as shown in Fig. 2.1. The two bases are related through the linear *transformation matrix* **Q**:

$$\mathbf{e}'_i = Q_{\alpha i}\mathbf{e}_\alpha \quad \Leftrightarrow \quad \begin{bmatrix} \mathbf{e}'_1 \\ \mathbf{e}'_2 \\ \mathbf{e}'_3 \end{bmatrix} = \begin{bmatrix} Q_{11} & Q_{12} & Q_{13} \\ Q_{21} & Q_{22} & Q_{23} \\ Q_{31} & Q_{32} & Q_{33} \end{bmatrix}^T \begin{bmatrix} \mathbf{e}_1 \\ \mathbf{e}_2 \\ \mathbf{e}_3 \end{bmatrix}, \tag{2.16}$$

where $Q_{\alpha i} = \mathbf{e}_\alpha \cdot \mathbf{e}'_i$. Note the transpose operation on the matrix **Q** in Eqn. (2.16). Since the basis vectors are unit vectors, the entries of **Q** are directional cosines, $Q_{\alpha i} = \cos\theta(\mathbf{e}'_i, \mathbf{e}_\alpha)$. The columns of **Q** are the coordinates of the new basis $\mathbf{e}'_i$ with respect to the original basis $\mathbf{e}_\alpha$. Note that **Q** is not symmetric since the representation of $\mathbf{e}'_i$ in basis $\mathbf{e}_\alpha$ is not the same as the representation of $\mathbf{e}_\alpha$ in $\mathbf{e}'_i$. As an example, consider a rotation by angle θ about the 3-axis. The new basis vectors are given by: $\mathbf{e}'_1 = \cos\theta\mathbf{e}_1 + \cos(90-\theta)\mathbf{e}_2$; $\mathbf{e}'_2 = \cos(90+\theta)\mathbf{e}_1 + \cos\theta\mathbf{e}_2$; $\mathbf{e}'_3 = \mathbf{e}_3$. The corresponding transformation matrix is

$$\mathbf{Q} = \begin{bmatrix} \cos\theta & -\sin\theta & 0 \\ \sin\theta & \cos\theta & 0 \\ 0 & 0 & 1 \end{bmatrix},$$

where we have used some elementary trigonometry.

Properties of Q The transformation matrix has special properties due to the orthonormality of the basis vectors that it relates.[4] The transformation matrix satisfies, $\mathbf{Q}^T\mathbf{Q} = \mathbf{I}$ and $\mathbf{Q}\mathbf{Q}^T = \mathbf{I}$, which implies that

$$\mathbf{Q}^T = \mathbf{Q}^{-1}. \tag{2.17}$$

In addition, it can shown that the determinant of **Q** equals only ± 1. Based on the sign of its determinant, **Q** can have two different physical significances. If $\det \mathbf{Q} = +1$, then the transformation defined by **Q** corresponds to a rotation, otherwise it corresponds to a rotation plus a reflection.[5] Only a rotation satisfies the requirement that the handedness of the basis is retained following the transformation; transformation matrices are therefore normally limited to this case.

Matrices satisfying Eqn. (2.17) are called *orthogonal* matrices. Orthogonal matrices with a positive determinant (i.e. rotations) are called *proper orthogonal*. The set of all 3×3 orthogonal matrices $O(3)$ forms a group under matrix multiplication called the *orthogonal group*. Similarly, the set of 3×3 proper orthogonal matrices form a group under matrix multiplication called the *special orthogonal group*, which is denoted $SO(3)$.

[4] See Section 2.3.4 in [TME12].

[5] A discussion of rotations and reflections in the context of crystal symmetries appears in Section 3.4.1.

Vector component transformation The transformation rule for vector components under a change of basis is obtained from the condition that a vector be *invariant* with respect to component transformation. Thus, for vector $\boldsymbol{a}$ we require $\boldsymbol{a} = a_\alpha \boldsymbol{e}_\alpha = a'_i \boldsymbol{e}'_i$, where a_α are the components of $\boldsymbol{a}$ in basis $\boldsymbol{e}_\alpha$ and a'_i are the components in $\boldsymbol{e}'_i$. After some minor algebra, the following transformation rule is obtained:[6]

$$a_\alpha = Q_{\alpha i} a'_i \quad \Leftrightarrow \quad [\boldsymbol{a}] = \mathbf{Q}\,[\boldsymbol{a}]'. \tag{2.18}$$

The prime on $[\boldsymbol{a}]'$ means that the components of $\boldsymbol{a}$ in the matrix representation are given in the basis $\{\boldsymbol{e}'_i\}$. The inverse relation is

$$a'_i = Q_{\alpha i} a_\alpha \quad \Leftrightarrow \quad [\boldsymbol{a}]' = \mathbf{Q}^T\,[\boldsymbol{a}]. \tag{2.19}$$

Higher-order tensors We now have a clear definition for vectors which we would like to generalize to higher-order tensors. A very general discussion of this subject, grounded in the ideas of linear algebra, which provides more physical insight is presented in Section 2.3.6 of [TME12]. Here we take a more limited view which will suffice for our needs in this book. We define a *second-order tensor* in the following way:

a second-order tensor $\boldsymbol{T}$ is a linear mapping transforming a vector $\boldsymbol{v}$ into a vector $\boldsymbol{w} = \boldsymbol{T}\boldsymbol{v}$.

The indicial form of the operation $\boldsymbol{T}\boldsymbol{v}$ will be considered in Section 2.1.3 where we discuss contracted multiplication. In similar fashion, we define a fourth-order tensor as a linear mapping transforming a second-order tensor to a second-order tensor.

Tensor component transformation We have stressed the point that tensors are objects that are invariant with respect to the choice of coordinate system. However, at a practical level, when performing calculations with tensors it is necessary to select a particular coordinate system and to represent the tensor in terms of its components in the corresponding basis. The invariance of the tensor manifests itself in the fact that the components of the tensor with respect to different bases cannot be chosen arbitrarily, but must satisfy certain transformation relations. We have already obtained these relations for vectors in Eqns. (2.18) and (2.19). Similar relations can be obtained for second-order tensors:

$$A'_{ij} = Q_{\alpha i} Q_{\beta j} A_{\alpha\beta} \quad \Leftrightarrow \quad [\boldsymbol{A}]' = \mathbf{Q}^T\,[\boldsymbol{A}]\,\mathbf{Q}. \tag{2.20}$$

[6] See Section 2.3.5 in [TME12].

Similarly for an nth-order tensor

$$B'_{i_1 i_2 \ldots i_n} = Q_{\alpha_1 i_1} Q_{\alpha_2 i_2} \cdots Q_{\alpha_n i_n} B_{\alpha_1 \alpha_2 \ldots \alpha_n}. \tag{2.21}$$

For the general case, there is no direct notation equivalent to the matrix multiplication form of the first- and second-order tensors.

2.1.3 Tensor operations

We now turn to the description and classification of tensor operations. Tensor operations can be divided into categories: (1) *addition* of two tensors; (2) *magnification* of a tensor; (3) *product* of two or more tensors to form a higher-order tensor; and (4) *contraction* of a tensor to form a lower-order tensor. Together, tensor products and tensor contraction lead to the idea of a *tensor basis*.

Addition Addition is defined for tensors of the same rank. For second-order tensors:

$$C_{ij} = A_{ij} + B_{ij} \quad \Leftrightarrow \quad \boldsymbol{C} = \boldsymbol{A} + \boldsymbol{B}.$$

The expression on the right is the direct notation for the addition operation. Indices i and j are free indices using the terminology of Section 2.1.1.

Magnification Magnification corresponds to a rescaling of a tensor by scalar multiplication. For a second-order tensor $\boldsymbol{A}$ and a scalar $\lambda \in \mathbb{R}$, a new second-order tensor $\boldsymbol{B}$ is defined by

$$B_{ij} = \lambda A_{ij} \quad \Leftrightarrow \quad \boldsymbol{B} = \lambda \boldsymbol{A}.$$

Tensor products Tensor products refer to the formation of a higher-order tensor by combining two or more tensors. For example, here we combine a second-order tensor $\boldsymbol{A}$ with a vector $\boldsymbol{v}$:

$$D_{ijk} = A_{ij} v_k \quad \Leftrightarrow \quad \boldsymbol{D} = \boldsymbol{A} \otimes \boldsymbol{v}. \tag{2.22}$$

Products of the form $A_{ij} v_k$ are called *tensor products*. In direct notation, this operation is denoted $\boldsymbol{A} \otimes \boldsymbol{v}$, where $\otimes$ is the tensor product symbol. The rank of the resulting tensor is equal to the sum of the ranks of the combined tensors. In this case, a third-order tensor is formed by combining a first- and a second-order tensor. We will be particularly interested in the formation of a second-order tensor from two vectors:

$$A_{ij} = a_i b_j \quad \Leftrightarrow \quad \boldsymbol{A} = \boldsymbol{a} \otimes \boldsymbol{b}. \tag{2.23}$$

The second-order tensor $\boldsymbol{A}$ is called the *dyad* of the vectors $\boldsymbol{a}$ and $\boldsymbol{b}$. Note that the order of the vectors in a dyad is important, i.e. $\boldsymbol{a} \otimes \boldsymbol{b} \neq \boldsymbol{b} \otimes \boldsymbol{a}$. In matrix notation the dyad is

$$[\boldsymbol{a} \otimes \boldsymbol{b}] = \begin{bmatrix} a_1 b_1 & a_1 b_2 & a_1 b_3 \\ a_2 b_1 & a_2 b_2 & a_2 b_3 \\ a_3 b_1 & a_3 b_2 & a_3 b_3 \end{bmatrix}.$$

Dyads lead to the important concept of a tensor basis which is discussed later.

Contraction Contraction corresponds to the formation of a lower-order tensor from a given tensor by summing over two of its components.[7] For example, if $\boldsymbol{D}$ is a third-order tensor, the three possible contraction operations give

$$u_i = D_{ijj}, \qquad v_j = D_{iji}, \qquad w_k = D_{iik},$$

where $\boldsymbol{u}$, $\boldsymbol{v}$ and $\boldsymbol{w}$ are vectors. We see that in indicial notation, contraction corresponds to a summation over dummy indices. Thus, any operation that includes dummy indices involves a contraction. Each contraction over a pair of dummy indices results in a reduction in the rank of the tensor by two orders. There is no general direct notation for tensor contraction. The exception is contraction operations that lead to scalar invariants. These are discussed at the end of this section.

Contracted multiplication Contraction operations can be applied to tensor products, leading to familiar multiplication operations from matrix algebra. For example, a contraction of the third-order tensor formed by the tensor product in Eqn. (2.22) gives

$$u_i = A_{ij} v_j \quad \Leftrightarrow \quad \boldsymbol{u} = \boldsymbol{A}\boldsymbol{v}. \tag{2.24}$$

The indicial expression can be written in matrix form as $[\boldsymbol{u}] = [\boldsymbol{A}]\,[\boldsymbol{v}]$. The direct notation appearing on the right of the above equation is adopted in analogy to the matrix operation. The matrix operation also lends to this operation its name of *contracted multiplication*. An important special case of Eqn. (2.24) follows when $\boldsymbol{A}$ is a dyad. In this case, the contracted multiplication satisfies the following relation:

$$(a_i b_j) v_j = a_i (b_j v_j) \quad \Leftrightarrow \quad (\boldsymbol{a} \otimes \boldsymbol{b})\boldsymbol{v} = \boldsymbol{a}(\boldsymbol{b} \cdot \boldsymbol{v}). \tag{2.25}$$

This identity can be viewed as a definition for the dyad as an operation that linearly transforms a vector $\boldsymbol{v}$ into a vector parallel to $\boldsymbol{a}$ with magnitude $\|\boldsymbol{a}\|\,|\boldsymbol{b} \cdot \boldsymbol{v}|$.

We use Eqn. (2.24) to define the *identity tensor* $\boldsymbol{I}$, as the second-order tensor that leaves any vector $\boldsymbol{v}$ unchanged when it operates on it: $\boldsymbol{I}\boldsymbol{v} = \boldsymbol{v}$. The components of the identity tensor (with respect to an orthonormal basis) are equal to the entries of the identity matrix introduced in Eqn. (2.2), $[\boldsymbol{I}] = \mathbf{I}$. Equation (2.24) can also be used to define the

[7] For a more general definition that is not phrased in terms of components, see Section 2.4.5 in [TME12].

transpose operation. The transpose of a second-order tensor $\boldsymbol{A}$, denoted $\boldsymbol{A}^T$, is defined by the condition:[8]

$$\boldsymbol{A}\boldsymbol{u} \cdot \boldsymbol{v} = \boldsymbol{u} \cdot \boldsymbol{A}^T \boldsymbol{v} \qquad \text{for all vectors } \boldsymbol{u} \text{ and } \boldsymbol{v}.$$

This implies that the components of $\boldsymbol{A}$ and $\boldsymbol{A}^T$ are related by $[\boldsymbol{A}^T]_{ij} = [\boldsymbol{A}]_{ji}$. The direct notation is adopted in analogy to the matrix notation.

So far we have considered the contraction of the third-order tensor obtained from the tensor product of a second-order tensor and a vector. If the tensor product involves two second-order tensors there are four possible contractions depending on which indices are contracted:

$$C_{ij} = A_{ik} B_{kj} \quad \Leftrightarrow \quad \boldsymbol{C} = \boldsymbol{A}\boldsymbol{B}, \qquad C_{ij} = A_{ik} B_{jk} \quad \Leftrightarrow \quad \boldsymbol{C} = \boldsymbol{A}\boldsymbol{B}^T,$$
$$C_{ij} = A_{ki} B_{kj} \quad \Leftrightarrow \quad \boldsymbol{C} = \boldsymbol{A}^T\boldsymbol{B}, \qquad C_{ij} = A_{ki} B_{jk} \quad \Leftrightarrow \quad \boldsymbol{C} = \boldsymbol{A}^T\boldsymbol{B}^T,$$

where the superscript T corresponds to the transpose operation defined above. A series of multiplications by the same tensor is denoted by an exponent:

$$\boldsymbol{A}^2 = \boldsymbol{A}\boldsymbol{A}, \qquad \boldsymbol{A}^3 = (\boldsymbol{A}^2)\boldsymbol{A} = \boldsymbol{A}\boldsymbol{A}\boldsymbol{A}, \quad \text{etc.}$$

The definition of tensor contraction allows us to define the inverse $\boldsymbol{A}^{-1}$ of a second-order tensor $\boldsymbol{A}$ through the relation

$$\boldsymbol{A}^{-1}\boldsymbol{A} = \boldsymbol{A}\boldsymbol{A}^{-1} = \boldsymbol{I}, \tag{2.26}$$

where $\boldsymbol{I}$ is the identity tensor defined above. In indicial form this is $A^{-1}_{ij} A_{jk} = A_{ij} A^{-1}_{jk} = \delta_{ik}$, and in matrix form $[\boldsymbol{A}^{-1}][\boldsymbol{A}] = [\boldsymbol{A}][\boldsymbol{A}^{-1}] = [\boldsymbol{I}]$. Comparing the last expression with Eqn. (2.26), we see that $[\boldsymbol{A}^{-1}] = [\boldsymbol{A}]^{-1}$. Consistent with this, the determinant of a second-order tensor is defined as the determinant of its components matrix:

$$\det \boldsymbol{A} \equiv \det [\boldsymbol{A}].$$

We will see later that $\det \boldsymbol{A}$ is invariant with respect to the coordinate system basis.

Given the above definitions, the expression in Eqn. (2.5) for the derivative of the determinant of a square matrix can be rewritten for a tensor as

$$\frac{\partial(\det \boldsymbol{A})}{\partial \boldsymbol{A}} = \boldsymbol{A}^{-T} \det \boldsymbol{A}, \tag{2.27}$$

where $\boldsymbol{A}^{-T} = (\boldsymbol{A}^{-1})^T$.

[8] See Section 2.4.3 in [TME12] for an alternative definition.

Scalar contraction Of particular interest are contraction operations that result in the formation of a zeroth-order tensor (i.e. a scalar invariant). Any tensor of even order can be reduced to a scalar by repeated contraction. For a second-order tensor $\boldsymbol{A}$, one contraction operation leads to a scalar. This is defined as the *trace* of $\boldsymbol{A}$:

$$\operatorname{tr} \boldsymbol{A} = \operatorname{tr}[\boldsymbol{A}] = A_{ii}. \tag{2.28}$$

Scalar contraction can also be applied to contracted multiplication. We have already seen an example of this in the dot product of two vectors, $\boldsymbol{a} \cdot \boldsymbol{b} = a_i b_i$. The dot product was defined in Section 2.1.2 as part of the definition of vector spaces. Other important examples of contractions leading to scalar invariants are the double contraction of two second-order tensors, $\boldsymbol{A}$ and $\boldsymbol{B}$, which can take two forms:

$$\boldsymbol{A} : \boldsymbol{B} = \operatorname{tr}[\boldsymbol{A}^T \boldsymbol{B}] = \operatorname{tr}[\boldsymbol{B}^T \boldsymbol{A}] = \operatorname{tr}[\boldsymbol{A}\boldsymbol{B}^T] = \operatorname{tr}[\boldsymbol{B}\boldsymbol{A}^T] = A_{ij} B_{ij}, \tag{2.29}$$

$$\boldsymbol{A} \cdot\cdot \boldsymbol{B} = \operatorname{tr}[\boldsymbol{A}\boldsymbol{B}] = \operatorname{tr}[\boldsymbol{B}^T \boldsymbol{A}^T] = \operatorname{tr}[\boldsymbol{B}\boldsymbol{A}] = \operatorname{tr}[\boldsymbol{A}^T \boldsymbol{B}^T] = A_{ij} B_{ji}. \tag{2.30}$$

The symbols $\cdot$, $:$ and $\cdot\cdot$ are the direct notation for the contraction operations.[9] It is worth pointing out that the double contraction $\boldsymbol{A} : \boldsymbol{B}$ is the inner product in the space of second-order tensors. The corresponding norm is $\|\boldsymbol{A}\| = (\boldsymbol{A} : \boldsymbol{A})^{1/2}$. The definition of the double-contraction operation is also extended to describe contraction of a fourth-order tensor $\boldsymbol{E}$ with a second-order tensor $\boldsymbol{A}$:

$$[\boldsymbol{E} : \boldsymbol{A}]_{ij} = E_{ijkl} A_{kl}, \qquad [\boldsymbol{E} \cdot\cdot \boldsymbol{A}]_{ij} = E_{ijkl} A_{lk}. \tag{2.31}$$

Finally, we note that when scalar contraction is applied to a contracted multiplication of the *same* vectors ($\boldsymbol{a} = \boldsymbol{b}$) or the *same* tensors ($\boldsymbol{A} = \boldsymbol{B}$) the results are scalar invariants of the tensors themselves. From the dot product we obtain the length squared of the vector $a_i a_i$ and from the tensor contractions, $A_{ij} A_{ij}$ and $A_{ij} A_{ji}$.

Tensor basis Using the definition of a dyad given above, it is straightforward to show that a second-order tensor can be represented in the following form:[10]

$$\boldsymbol{A} = A_{ij} (\boldsymbol{e}_i \otimes \boldsymbol{e}_j). \tag{2.32}$$

9 Note that this convention is not universally adopted. Some authors reverse the meaning of $:$ and $\cdot\cdot$. Others do not use the double dot notation at all and use $\cdot$ to denote scalar contraction for both vectors and second-order tensors.

10 See Section 2.4.6 in [TME12].

The dyads $e_i \otimes e_j$ can be thought of as the "basis tensors" relative to which the components of $\boldsymbol{A}$ are given in the same way that a vector can be represented as $\boldsymbol{a} = a_i \boldsymbol{e}_i$. It is straightforward to show that $\boldsymbol{e}_i \otimes \boldsymbol{e}_j$ form a linearly independent basis. The basis description can be used to obtain an expression for the components of $\boldsymbol{A}$:

$$A_{ij} = \boldsymbol{e}_i \cdot \boldsymbol{A}\boldsymbol{e}_j. \tag{2.33}$$

2.1.4 Properties of second-order tensors

Most of the tensors that we will be dealing with are second-order tensors. It is therefore worthwhile to review the properties of such tensors that we will need later in the book.

Orthogonal tensors A second-order tensor $\boldsymbol{Q}$ is called *orthogonal* if for every pair of vectors $\boldsymbol{a}$ and $\boldsymbol{b}$, we have

$$(\boldsymbol{Q}\boldsymbol{a}) \cdot (\boldsymbol{Q}\boldsymbol{b}) = \boldsymbol{a} \cdot \boldsymbol{b}. \tag{2.34}$$

Geometrically, this means that $\boldsymbol{Q}$ preserves the angles between, and the magnitudes of, the vectors on which it operates. A necessary and sufficient condition for this is $\boldsymbol{Q}^T\boldsymbol{Q} = \boldsymbol{Q}\boldsymbol{Q}^T = \boldsymbol{I}$, or equivalently

$$\boldsymbol{Q}^T = \boldsymbol{Q}^{-1}. \tag{2.35}$$

This condition is completely analogous to the one given for orthogonal matrices in Eqn. (2.17). As in that case, it can be shown that $\det \boldsymbol{Q} = \pm 1$. An orthogonal tensor $\boldsymbol{Q}$ is called *proper orthogonal* if $\det \boldsymbol{Q} = 1$, and *improper orthogonal* otherwise. A proper orthogonal transformation corresponds to a rotation. An improper orthogonal transformation involves a rotation and a reflection. The groups $O(3)$ and $SO(3)$ defined for orthogonal matrices in Section 2.1.2 also exist for orthogonal tensors.

Symmetric and antisymmetric tensors A *symmetric* second-order tensor $\boldsymbol{S}$ satisfies the condition: $\boldsymbol{S} = \boldsymbol{S}^T$ $(S_{ij} = S_{ji})$. An *antisymmetric* (also called *skew-symmetric*) tensor $\boldsymbol{A}$ satisfies the condition: $\boldsymbol{A} = -\boldsymbol{A}^T$ $(A_{ij} = -A_{ji})$. From this definition it is clear that $A_{11} = A_{22} = A_{33} = 0$. An important property related to the above definitions is that the contraction of any symmetric tensor $\boldsymbol{S}$ with an antisymmetric tensor $\boldsymbol{A}$ is zero, i.e. $\boldsymbol{S} : \boldsymbol{A} = S_{ij}A_{ij} = 0$.

Principal values and directions A second-order tensor $\boldsymbol{G}$ maps a vector $\boldsymbol{v}$ to a new vector $\boldsymbol{w} = \boldsymbol{G}\boldsymbol{v}$. We now ask whether there are special directions, $\boldsymbol{v} = \boldsymbol{\Lambda}$, for which

$$\boldsymbol{w} = \boldsymbol{G}\boldsymbol{\Lambda} = \lambda\boldsymbol{\Lambda}, \quad \lambda \in \mathbb{R},$$

i.e. directions that are not changed (only magnified) by the operation of $\boldsymbol{G}$. Thus we seek solutions to the following equation:

$$G_{ij}\Lambda_j = \lambda\Lambda_i \quad \Leftrightarrow \quad \boldsymbol{G}\boldsymbol{\Lambda} = \lambda\boldsymbol{\Lambda}, \tag{2.36}$$

or equivalently,

$$(Gij - \lambda\delta_{ij})\Lambda_j = 0 \quad \Leftrightarrow \quad (\boldsymbol{G} - \lambda\boldsymbol{I})\boldsymbol{\Lambda} = \boldsymbol{0}. \tag{2.37}$$

A vector $\boldsymbol{\Lambda}^{\boldsymbol{G}}$ satisfying this requirement is called an *eigenvector* (principal direction) of $\boldsymbol{G}$ with $\lambda^{\boldsymbol{G}}$ being the corresponding *eigenvalue* (principal value). The superscript "$\boldsymbol{G}$" denotes that these are the eigenvectors and eigenvalues of the tensor $\boldsymbol{G}$. Nontrivial solutions to Eqn. (2.37) require

$$\det(\boldsymbol{G} - \lambda\boldsymbol{I}) = 0.$$

For $n_{\mathrm{d}} = 3$, this is a cubic equation in λ that is called the *characteristic equation* of $\boldsymbol{G}$:

$$-\lambda^3 + I_1^{\boldsymbol{G}}\lambda^2 - I_2^{\boldsymbol{G}}\lambda + I_3^{\boldsymbol{G}} = 0, \tag{2.38}$$

where $I_1^{\boldsymbol{G}}$, $I_2^{\boldsymbol{G}}$, $I_3^{\boldsymbol{G}}$ are the *principal invariants* of $\boldsymbol{G}$:

$$I_1^{\boldsymbol{G}} = G_{ii} = \operatorname{tr}\boldsymbol{G}, \tag{2.39}$$

$$I_2^{\boldsymbol{G}} = \tfrac{1}{2}(G_{ii}G_{jj} - G_{ij}G_{ji}) = \frac{1}{2}\left[(\operatorname{tr}\boldsymbol{G})^2 - \operatorname{tr}\boldsymbol{G}^2\right] = \operatorname{tr}\boldsymbol{G}^{-1}\det\boldsymbol{G}, \tag{2.40}$$

$$I_3^{\boldsymbol{G}} = \epsilon_{ijk}G_{1i}G_{2j}G_{3k} = \det\boldsymbol{G}. \tag{2.41}$$

The characteristic equation (Eqn. (2.38)) has three solutions: $\lambda_\alpha^{\boldsymbol{G}}$ $(\alpha = 1,2,3)$. Since the equation is cubic and has real coefficients, in general it has one real root and two complex conjugate roots. However, in the special case where $\boldsymbol{G}$ is symmetric ($\boldsymbol{G} = \boldsymbol{G}^T$), all three eigenvalues are real.[11] Each eigenvalue $\lambda_\alpha^{\boldsymbol{G}}$ has an eigenvector $\boldsymbol{\Lambda}_\alpha^{\boldsymbol{G}}$ that is obtained by solving Eqn. (2.37) after substituting in $\lambda = \lambda_\alpha^{\boldsymbol{G}}$ together with the normalization condition $\left\|\boldsymbol{\Lambda}_\alpha^{\boldsymbol{G}}\right\| = 1$.

An important theorem states that the eigenvectors corresponding to distinct eigenvalues of a symmetric tensor $\boldsymbol{S}$ are orthogonal.[12] This together with the normalization condition means that

$$\boldsymbol{\Lambda}_\alpha^{\boldsymbol{S}} \cdot \boldsymbol{\Lambda}_\beta^{\boldsymbol{S}} = \delta_{\alpha\beta}. \tag{2.42}$$

[11] It is also common to encounter eigenvalue equations on an infinite-dimensional vector space over the field of complex numbers (see Chapter 4). For example, in quantum mechanics, the tensor operator is not symmetric but *Hermitian*, which means that $\boldsymbol{H} = (\boldsymbol{H}^*)^T$, where $*$ represents the complex conjugate. Hermitian tensors are a generalization of symmetric tensors, and it can be shown that Hermitian tensors have real eigenvalues and orthogonal eigenvectors, just like symmetric tensors.

[12] In situations where some eigenvalues are repeated, i.e. not all eigenvalues are distinct, it is still possible to generate a set of three mutually orthogonal vectors although the choice is not unique in this case. See Section 2.5.3 in [TME12] for more on this and a proof of Eqn. (2.42).

The fact that it is always possible to construct a set of three mutually orthonormal eigenvectors for a symmetric second-order tensor $\boldsymbol{S}$ suggests using these eigenvectors as a basis for a Cartesian coordinate system. This is referred to as the *principal coordinate system* of the tensor for which the eigenvectors form the *principal basis*. An important property of the eigenvectors that follows from this is the *completeness relation*:

$$\sum_{\alpha=1}^{3} \boldsymbol{\Lambda}_{\alpha}^{\boldsymbol{S}} \otimes \boldsymbol{\Lambda}_{\alpha}^{\boldsymbol{S}} = \boldsymbol{I}, \tag{2.43}$$

where $\boldsymbol{I}$ is the identity tensor. It is straightforward to show that in its principal coordinate system $\boldsymbol{S}$ is *diagonal* with components equal to its principal values, i.e. $S_{\underline{i}\,\underline{i}} = \lambda_i^{\boldsymbol{S}}$ and $S_{ij} = 0$ for $i \neq j$. This means that any symmetric tensor $\boldsymbol{S}$ may be represented as

$$\boldsymbol{S} = \sum_{\alpha=1}^{3} \lambda_{\alpha}^{\boldsymbol{S}} \boldsymbol{\Lambda}_{\alpha}^{\boldsymbol{S}} \otimes \boldsymbol{\Lambda}_{\alpha}^{\boldsymbol{S}}. \tag{2.44}$$

This is called the *spectral decomposition* of $\boldsymbol{S}$.

The quadratic form of symmetric second-order tensors A scalar functional form that often comes up with the application of tensors is the *quadratic form* $Q(\boldsymbol{x})$ associated with symmetric second-order tensors:

$$Q(\boldsymbol{x}) \equiv S_{ij} x_i x_j.$$

Special terminology is used to describe $\boldsymbol{S}$ if something definitive can be said about the sign of $Q(\boldsymbol{x})$, regardless of the choice of $\boldsymbol{x}$. Focusing on the positive sign definitions, we have

$$Q(\boldsymbol{x}) \begin{cases} > 0 & \forall \boldsymbol{x} \in \mathbb{R}^{n_d}, \boldsymbol{x} \neq \boldsymbol{0} \quad \boldsymbol{S} \text{ is } \textit{positive definite}, \\ \geq 0 & \forall \boldsymbol{x} \in \mathbb{R}^{n_d}, \boldsymbol{x} \neq \boldsymbol{0} \quad \boldsymbol{S} \text{ is } \textit{positive semi-definite}. \end{cases}$$

A useful theorem states that $\boldsymbol{S}$ is positive definite if and only if all of its eigenvalues are positive (i.e. $\lambda_{\alpha}^{\boldsymbol{S}} > 0, \forall \alpha$). Using the spectral decomposition in Eqn. (2.44), the square root, $\sqrt{\boldsymbol{S}}$, of a positive definite tensor $\boldsymbol{S}$ can be defined as

$$\sqrt{\boldsymbol{S}} \equiv \sum_{\alpha=1}^{3} \sqrt{\lambda_{\alpha}^{\boldsymbol{S}}} \left(\boldsymbol{\Lambda}_{\alpha}^{\boldsymbol{S}} \otimes \boldsymbol{\Lambda}_{\alpha}^{\boldsymbol{S}} \right). \tag{2.45}$$

2.1.5 Tensor fields

The previous sections have discussed the definition and properties of tensors as discrete entities. In continuum mechanics, we most often encounter tensors as spatially and temporally varying fields over a given domain. For three-dimensional objects, a tensor field $\boldsymbol{T}$

defined over a domain Ω is a function of the position vector $\boldsymbol{x} = x_i \boldsymbol{e}_i$ of points inside Ω:

$$\boldsymbol{T} = \boldsymbol{T}(\boldsymbol{x}, t) = \boldsymbol{T}(x_1, x_2, x_3, t), \quad \boldsymbol{x} \in \Omega(t).$$

Given the field concept, we can consider differentiation and integration of tensors.

Partial differentiation of a tensor field The partial differentiation of tensor fields with respect to their spatial arguments is readily expressed in component form:[13]

$$\frac{\partial s(\boldsymbol{x})}{\partial x_i}, \quad \frac{\partial v_i(\boldsymbol{x})}{\partial x_j}, \quad \frac{\partial T_{ij}(\boldsymbol{x})}{\partial x_k},$$

for a scalar s, vector $\boldsymbol{v}$ and second-order tensor $\boldsymbol{T}$. To simplify this notation and make it compatible with indicial notation, we introduce the *comma notation* for differentiation with respect to x_i:

$$(\cdot)_{,i} \equiv \frac{\partial(\cdot)}{\partial x_i}.$$

In this notation, the three expressions above are $s_{,i}$, $v_{i,j}$ and $T_{ij,k}$. Higher-order differentiation follows as expected: $\partial^2 s/(\partial x_i \partial x_j) = s_{,ij}$. The comma notation works in concert with the summation convention, e.g. $s_{,ii} = s_{,11} + s_{,22} + s_{,33}$ and $v_{i,i} = v_{1,1} + v_{2,2} + v_{3,3}$.

Differential operators Four important differential operators are the *gradient*, *curl*, *divergence* and *Laplacian*. These operators involve derivatives of a tensor field with respect to its vector argument. We define these operators below in terms of the components of the tensors relative to an orthonormal basis (i.e. for a Cartesian coordinate system).

The *gradients* of a scalar field $s(\boldsymbol{x})$, vector field $\boldsymbol{v}(\boldsymbol{x})$, and second-order tensor field $\boldsymbol{T}(\boldsymbol{x})$ are respectively[14]

$$\nabla s = \frac{\partial s}{\partial x_i} \boldsymbol{e}_i, \qquad \nabla \boldsymbol{v} = \frac{\partial v_i}{\partial x_j} (\boldsymbol{e}_i \otimes \boldsymbol{e}_j), \qquad \nabla \boldsymbol{T} = \frac{\partial T_{ij}}{\partial x_k} (\boldsymbol{e}_i \otimes \boldsymbol{e}_j \otimes \boldsymbol{e}_k). \tag{2.46}$$

[13] Differentiation with respect to time is more subtle and will be discussed in Section 2.2.5.

[14] It is important to point out that a great deal of confusion exists in the continuum mechanics literature regarding the direct notation for differential operators. The notation we introduce here for the grad, curl and div operations is based on a linear algebraic view of tensor analysis. The same operations are often defined differently in other books. The confusion arises when the operations are applied to tensors of rank one and higher, where different definitions lead to different components being involved in the operation. For example, another popular notation for tensor calculus is based on the del differential operator, $\boldsymbol{\nabla} \equiv \boldsymbol{e}_i \partial/\partial x_i$. In this notation, the gradient, curl and divergence are denoted by $\boldsymbol{\nabla}\square$, $\boldsymbol{\nabla} \times \square$ and $\boldsymbol{\nabla} \cdot \square$. This notation is self-consistent; however, it is not equivalent to the grad, curl and div notation used here. For example according to the del notation the gradient of a vector $\boldsymbol{v}$ is $\boldsymbol{\nabla v} = v_{j,i} \boldsymbol{e}_i \otimes \boldsymbol{e}_j$, which is the transpose of our definition. In our notation we retain an unbolded ∇ symbol for the gradient, but do not view it as a differential operator. Instead, we adopt the definition in the text which leads to the untransposed expression, $\nabla \boldsymbol{v} = v_{i,j} \boldsymbol{e}_i \otimes \boldsymbol{e}_j$. We will use the notation introduced here consistently throughout the book with the exception of Chapter 4, where the del notation is used in developing the Schrödinger equation in quantum mechanics. However, there it is applied to scalar functions where there is no confusion.

We see that the gradient operation increases the rank of the tensor by one: $[\nabla \boldsymbol{v}]_{ij} = v_{i,j}$ are the components of a second-order tensor, and $[\nabla \boldsymbol{T}]_{ijk} = T_{ij,k}$ are the components of a third-order tensor. For a scalar field, the gradient ∇s is the direction and magnitude of the maximum rate of increase of $s(\boldsymbol{x})$.

The *curl* of a vector field $\boldsymbol{v}(\boldsymbol{x})$ is a vector denoted curl $\boldsymbol{v}$, which is given by

$$\operatorname{curl} \boldsymbol{v} = -\epsilon_{ijk} \frac{\partial v_i}{\partial x_j} \boldsymbol{e}_k. \tag{2.47}$$

The curl is related to the local rate of rotation of the field. It plays an important role in fluid dynamics where it characterizes the vorticity or spin of the flow. The definition of a curl can be extended to higher-order tensors; see, for example, [CG01].

The *divergences* of a vector field $\boldsymbol{v}(\boldsymbol{x})$ and a second-order tensor field $\boldsymbol{T}(\boldsymbol{x})$ are respectively

$$\operatorname{div} \boldsymbol{v} = \frac{\partial v_i}{\partial x_i}, \qquad \operatorname{div} \boldsymbol{T} = \frac{\partial T_{ij}}{\partial x_j} \boldsymbol{e}_i. \tag{2.48}$$

We see that the divergence of a vector is a scalar invariant, $\operatorname{div} \boldsymbol{v} = v_{i,i}$, and the divergence of a second-order tensor is a vector, $[\operatorname{div} \boldsymbol{T}]_i = T_{ij,j}$. In instances where the divergence is taken with respect to an argument other than $\boldsymbol{x}$ it will be denoted by a subscript. For example, the divergence with respect to $\boldsymbol{y}$ of a tensor $\boldsymbol{T}$ is denoted $\operatorname{div}_y \boldsymbol{T}$. The divergence of a tensor field is related to the net flow of the field per unit volume at a given point.

The *Laplacian* of a scalar field $s(\boldsymbol{x})$ is a scalar denoted by $\nabla^2 s$. The Laplacian is defined by the following relation:

$$\nabla^2 s \equiv \operatorname{div} \nabla s = \frac{\partial^2 s}{\partial x_i \partial x_i} = s_{,ii}. \tag{2.49}$$

Divergence theorem The divergence theorem relates surface and volume integrals. Consider a closed volume Ω bounded by the surface $\partial\Omega$ with outward unit normal $\boldsymbol{n}(\boldsymbol{x})$ together with a smooth spatially-varying tensor field $\boldsymbol{B}(\boldsymbol{x})$ of any rank defined everywhere in Ω and on $\partial\Omega$. The divergence theorem for the tensor field $\boldsymbol{B}$ states

$$\int_{\partial\Omega} B_{ijk\ldots p} n_p \, dA = \int_\Omega B_{ijk\ldots p,p} \, dV \quad \Leftrightarrow \quad \int_{\partial\Omega} \boldsymbol{B}\boldsymbol{n} \, dA = \int_\Omega \operatorname{div} \boldsymbol{B} \, dV, \tag{2.50}$$

where the integral over $\partial\Omega$ is a surface integral (dA is an infinitesimal surface element) and the integral over Ω is a volume integral (dV is an infinitesimal volume element). Physically,

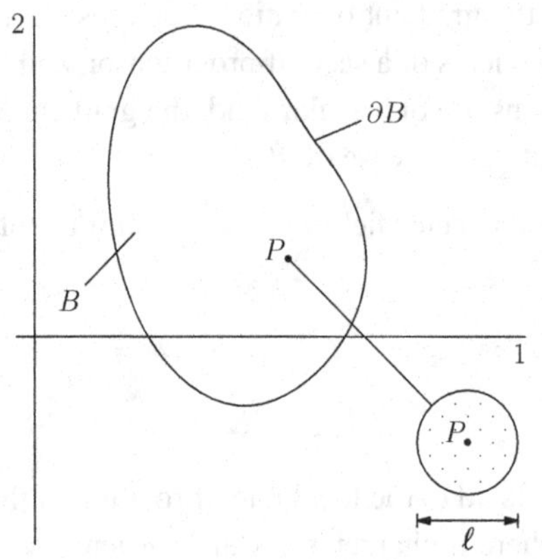

Fig. 2.2 A material body B with surface ∂B. A continuum particle P is shown together with a schematic representation of the atomic structure underlying the particle with length scale ℓ. The dots in the atomic structure represent the atoms.

the surface term measures the flux of $\boldsymbol{B}$ out of Ω, while the volume term is a measure of sinks and sources of $\boldsymbol{B}$ inside Ω. The divergence theorem is therefore a conservation law for $\boldsymbol{B}$. For the special case of first- and second-order tensors, Eqn. (2.50) gives

$$\int_{\partial\Omega} v_i n_i \, dA = \int_{\Omega} v_{i,i} \, dV, \qquad \int_{\partial\Omega} T_{ij} n_j \, dA = \int_{\Omega} T_{ij,j} \, dV. \tag{2.51}$$

2.2 Kinematics of deformation

Continuum mechanics deals with the change of shape (deformation) of bodies subjected to external loads that can be either forces or displacements. However, before we can discuss the physical laws governing deformation, we must develop measures that characterize and quantify it. This is the subject described by the *kinematics of deformation*.[15] Kinematics does not deal with predicting the deformation resulting from a given loading, but rather with the machinery for describing all possible deformations a body can undergo.

2.2.1 The continuum particle

A material body B bounded by a surface ∂B is represented by a continuous distribution of an infinite number of *continuum particles*. On the macroscopic scale, each particle is a point of zero extent much like a point in a geometrical space. It should therefore not be thought of as a small piece of material. At the same time, it has to be realized that a continuum particle derives its properties from a finite-sized region ℓ on the micro-scale (see Fig. 2.2). One can think of the properties of the particle as an average over the atomic behavior

[15] Webster's *New World College Dictionary* defines kinematics as "the branch of mechanics that deals with motion in the abstract, without reference to the force or mass."

within this domain. As one moves from one particle to its neighbor the microscopic domain moves over, largely overlapping the previous domain. In this way the smooth, field-like behavior we expect in a continuum is obtained.[16] A fundamental assumption of continuum mechanics is that it is possible to define a length ℓ that is large relative to atomic length scales and at the same time much smaller than the length scale associated with variations in the continuum fields.[17] This issue and the limitations that it imposes on the validity of continuum theory are discussed in Section 6.6 of [TME12].

2.2.2 The deformation mapping

A body B can take on many different shapes or *configurations* depending on the loading applied to it. We choose one of these to be the *reference configuration* and label it B_0. The reference configuration provides a convenient fixed state of the body to which other configurations can be compared to gauge their deformation. Any possible configuration can be taken as the reference. Typically the choice is dictated by convenience to the analysis. Often, it corresponds to the state where no external loading is applied to the body.

We denote the position of a particle P in the reference configuration by $\boldsymbol{X} = \boldsymbol{X}(P)$. Since particles cannot be formed or destroyed, we can use the coordinates of a particle in the reference configuration as a label distinguishing this particle from all others. Once we have defined the reference configuration, the *deformed configuration* occupied by the body is described in terms of a *deformation mapping* function $\boldsymbol{\varphi}$ that maps the reference position of every particle $\boldsymbol{X}$ to its deformed position $\boldsymbol{x}$:

$$x_i = \varphi_i(X_1, X_2, X_3) \quad \Leftrightarrow \quad \boldsymbol{x} = \boldsymbol{\varphi}(\boldsymbol{X}). \tag{2.52}$$

Here $\boldsymbol{X} \in B_0$ is a point in the reference configuration. In the deformed configuration the body occupies a domain B, which is the union of all positions $\boldsymbol{x}$ (see Fig. 2.3). In

[16] This is the approach taken in Section 8.2 where statistical mechanics ideas are used to obtain microscopic expressions for the continuum fields. See also footnote 31 on page 476 in that section.

[17] This microscopically-based view of continuum mechanics is not mandatory. Clifford Truesdell, one of the major figures in continuum mechanics who together with Walter Noll codified it and gave it its modern mathematical form, was a strong proponent of continuum mechanics as an independent theory eschewing perceived connections with other theories. For example in his manuscript with Richard Toupin, "The classical field theories" [TT60], it states: "The corpuscular theories and field theories are mutually contradictory as direct models of nature. The field is indefinitely divisible; the corpuscle is not. To mingle the terms and concepts appropriate to these two distinct representations of nature, while unfortunately a common practice, leads to confusion if not to error. For example, to speak of an element of volume in a gas as 'a region large enough to contain many molecules but small enough to be used as a element of integration' is not only loose but also needless and bootless." This is certainly true as long as continuum mechanics is studied as an independent theory. However, when attempts are made to connect it with phenomena occurring on smaller scales, as in this book, it leads to a dead end. Truesdell and Toupin even acknowledge this fact in the text immediately following the above quote where they discuss Noll's work on a microscopic definition of the stress tensor [Nol55]. Noll, following the work of Irving and Kirkwood [IK50], demonstrated that by defining continuum field variables as particular phase averages over the atomistic phase space, the continuum balance laws were exactly satisfied. Truesdell and Toupin consequently (and perhaps grudgingly) conclude that "those who prefer to regard classical statistical mechanics as fundamental may nevertheless employ the field concept as exact in terms of *expected values*" [TT60]. Irving and Kirkwood's and Noll's approach is discussed in Section 8.2.

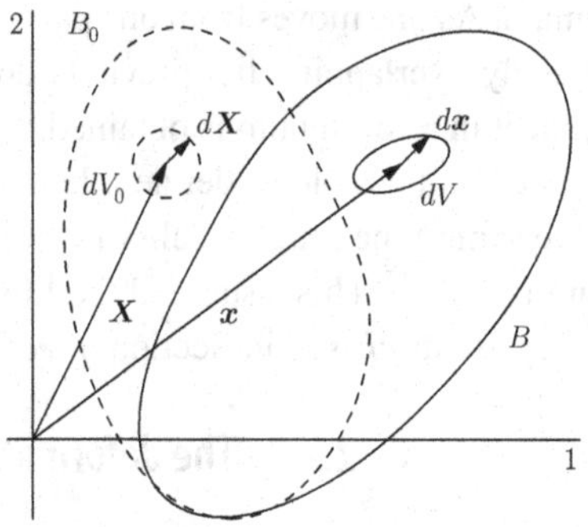

Fig. 2.3 The reference configuration B_0 of a body (dashed) and deformed configuration B.

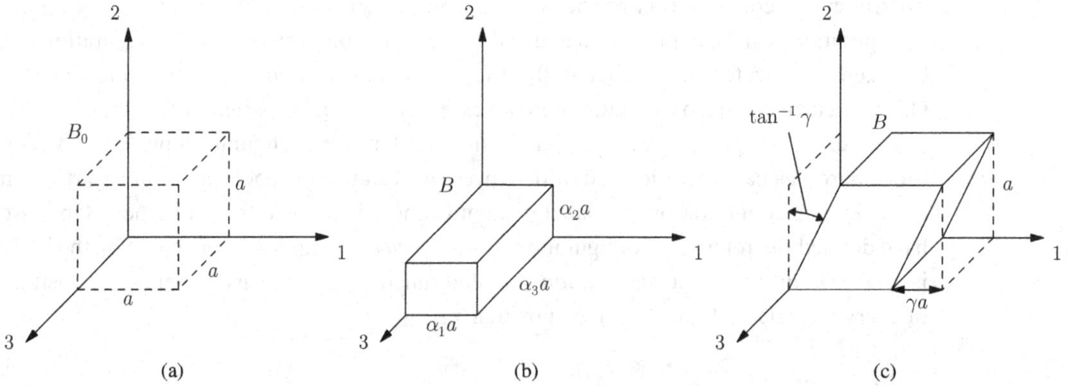

Fig. 2.4 Examples of deformation mappings: (a) reference configuration where the body is a cube (dashed); (b) uniform stretching deformation defined by $x_1 = \alpha_1 X_1, x_2 = \alpha_2 X_2, x_3 = \alpha_3 X_3$, where α_i are the stretch parameters; (c) simple shear defined by $x_1 = X_1 + \gamma X_2, x_2 = X_2, x_3 = X_3$, where γ is the shear parameter.

the above, we have adopted the standard continuum mechanics convention of denoting all things associated with the reference configuration with upper-case letters (as in $\boldsymbol{X}$) or with a subscript 0 (as in B_0) and all things associated with the deformed configuration in lower-case (as in $\boldsymbol{x}$) or without a subscript (as in B). Examples of uniform stretching and simple shear deformations are shown in Fig. 2.4.

A time-dependent deformation mapping, $\varphi(\boldsymbol{X}, t)$, is called a *motion*. In this case the reference configuration is often associated with the motion at time $t = 0$, so that $\varphi(\boldsymbol{X}, 0) = \boldsymbol{X}$, and the deformed configuration is associated with the motion at the "current" time t. For this reason the deformed configuration is also referred to as the *current configuration*.

2.2.3 Material and spatial descriptions

A scalar invariant field $\mathfrak{g}$, say the temperature, can be written as a function over the deformed or the reference configuration:

$$\mathfrak{g} = g(\boldsymbol{x}, t) \quad \boldsymbol{x} \in B \qquad \text{or} \qquad \mathfrak{g} = \breve{g}(\boldsymbol{X}, t) \quad \boldsymbol{X} \in B_0.$$

Table 2.1. The direct notation for the gradient, curl and divergence operators with respect to the material and spatial coordinates

Operator	Material coordinates	Spatial coordinates
gradient	$\nabla_0\square$ or Grad $\square$	$\nabla\square$ or grad $\square$
curl	Curl $\square$	curl $\square$
divergence	Div $\square$	div $\square$

The two descriptions are linked by the deformation mapping: $\breve{g}(\boldsymbol{X},t) \equiv g(\varphi(\boldsymbol{X},t),t)$. The first is the *spatial* or *Eulerian* description in which $g(\boldsymbol{x},t)$ provides the temperature at a particular position in space regardless of which particle is occupying it at time t. The second is the *material* or *referential* description, where $\breve{g}(\boldsymbol{X},t)$ gives the temperature of a particle $\boldsymbol{X}$ at time t regardless of where the particle is located in space. If the body occupies the reference state at $t = 0$, then instead of "referential" the term *Lagrangian* is used.

For obvious reasons the coordinates of a particle in the reference configuration $\boldsymbol{X}$ with components X_I are referred to as *material coordinates* and the coordinates of a spatial position $\boldsymbol{x}$ with components x_i are referred to as *spatial coordinates*. Here we have extended to the indices of coordinates and tensor components the convention of using upper-case and lower-case letters for the reference and deformed configurations, respectively. The introduction of upper-case and lower-case indices means that the summation convention introduced earlier now becomes case sensitive. Thus, $A_I A_I$ will be summed, but $b_i A_I$ will not.

Tensors are referred to as *material*, *spatial* or *mixed* depending on the configuration with which they are associated.[18] The distinction is made clear by the notation. Material tensors are associated with the reference configuration and are therefore denoted with upper-case letters and indices, e.g. A_I, B_{IJ}, etc. Spatial tensors are associated with the deformed configuration and are denoted with lower-case letters and indices, e.g. a_i, b_{ij}, etc. Second-order tensors and higher can also be mixed (also called *two-point tensors*) when some indices are material and some spatial. The case of the index indicates the configuration with which it is associated, e.g. F_{iJ}. Mixed tensors are denoted with an upper-case letter to indicate that they have at least one material index.

The introduction of referential and spatial descriptions for tensor fields means that the indicial and direct notation introduced earlier for differentiation (see Section 2.1.5) must be suitably amended. When taking derivatives with respect to positions it is necessary to indicate whether the derivative is with respect to $\boldsymbol{X}$ or $\boldsymbol{x}$. In indicial notation, the comma notation refers to the index of the coordinate. Again, we find that the case convention for indices is necessary. Thus, differentiation with respect to the material and spatial coordinates can be unambiguously indicated using the comma notation already introduced as $\square_{,I}$ or $\square_{,i}$, where $\square$ represents the tensor field being differentiated. The direct notation for the gradient, curl and divergence operators with respect to the material and spatial coordinates are given in Tab. 2.1. For example, $\nabla_0 \breve{g} = (\partial \breve{g}/\partial X_I)\boldsymbol{e}_I$ and $\nabla g = (\partial g/\partial x_i)\boldsymbol{e}_i$.

We defer the discussion of differentiation with respect to time to Section 2.2.5 where the time rate of change of kinematic variables is introduced.

[18] A more precise explanation of material, spatial and mixed tensors requires a deeper discussion of tensors than we have provided here. See Section 3.3 in [TME12].

2.2.4 Description of local deformation

The deformation mapping $\varphi(\boldsymbol{X})$ tells us how particles move, but it does not directly provide information on the change of shape of particles, i.e. strains in the material. This is important because materials resist changes to their shape and this information must be included in a physical model of deformation. To capture particle shape change, it is necessary to characterize the deformation in the infinitesimal neighborhood of a particle.

Deformation gradient The infinitesimal environment or *neighborhood* of a particle in the reference configuration is the sphere of volume dV_0 mapped out by $\boldsymbol{X} + d\boldsymbol{X}$, where $d\boldsymbol{X}$ is the differential of $\boldsymbol{X}$. The neighborhood dV_0 is transformed by the deformation mapping to an ellipsoid in the deformed configuration mapped out by $\boldsymbol{x} + d\boldsymbol{x}$ with volume dV as shown in Fig. 2.3. The material and spatial differentials are related by

$$dx_i = F_{iJ}\, dX_J \quad \Leftrightarrow \quad d\boldsymbol{x} = \boldsymbol{F}\, d\boldsymbol{X}, \tag{2.53}$$

where $\boldsymbol{F}$ is called the *deformation gradient* and is given by

$$F_{iJ} = \frac{\partial \varphi_i}{\partial X_J} = \frac{\partial x_i}{\partial X_J} = x_{i,J} \quad \Leftrightarrow \quad \boldsymbol{F} = \frac{\partial \boldsymbol{\varphi}}{\partial \boldsymbol{X}} = \frac{\partial \boldsymbol{x}}{\partial \boldsymbol{X}} = \nabla_0 \boldsymbol{x}. \tag{2.54}$$

The deformation gradient is a second-order two-point tensor and in general is *not* symmetric. It plays a key role in describing the *local* deformation in the vicinity of a particle.

Volume changes The local ratio of deformed-to-reference volume is given by

$$\frac{dV}{dV_0} = \det \boldsymbol{F} = J, \tag{2.55}$$

where $J = \det \boldsymbol{F}$ is the *Jacobian* of the deformation mapping. A volume preserving deformation satisfies $J = 1$ at all points. In order for the deformation mapping to be locally invertible, we must have $J \neq 0$ at all particles.

Area changes The cross product of two infinitesimal material vectors $d\boldsymbol{X}$ and $d\boldsymbol{Y}$ defines an *element of oriented area* in the reference configuration: $d\boldsymbol{A}_0 = d\boldsymbol{X} \times d\boldsymbol{Y} = \boldsymbol{N}\, dA_0$, where $\boldsymbol{N}$ is the normal in the reference configuration and dA_0 is the differential area there. The corresponding element of oriented area in the deformed configuration follows from *Nanson's formula*:

$$n_i\, dA = J F_{Ii}^{-1} N_I\, dA_0 \quad \Leftrightarrow \quad \boldsymbol{n}\, dA = J \boldsymbol{F}^{-T} \boldsymbol{N}\, dA_0. \tag{2.56}$$

Here $\boldsymbol{n}$ and dA are the normal and the differential area in the deformed configuration. Equation (2.56) plays a key role in the derivation of material and mixed stress measures.

Polar decomposition theorem The deformation gradient $\boldsymbol{F}$ represents an affine mapping[19] of the neighborhood of a material particle from the reference to deformed configuration. We state above that $\boldsymbol{F}$ provides a measure for the deformation of the neighborhood. This is a true statement but it is not precise. When we say "deformation" we are implicitly referring to changes in the shape of the neighborhood. This includes changes in lengths or *stretching* and changes in angles or *shearing* (see Fig. 2.4). However, the deformation gradient may also include a part that is simply a rotation of the neighborhood. Since rotation does not play a role in shape change, it would be useful to decompose $\boldsymbol{F}$ into its rotation and "shape-change" parts. It turns out that such a decomposition exists and is unique:

Polar decomposition theorem Any tensor $\boldsymbol{F}$ with positive determinant ($\det \boldsymbol{F} > 0$) can be uniquely expressed as

$$F_{iJ} = R_{iI}U_{IJ} = V_{ij}R_{jJ} \quad \Leftrightarrow \quad \boldsymbol{F} = \boldsymbol{R}\boldsymbol{U} = \boldsymbol{V}\boldsymbol{R}, \tag{2.57}$$

called the *right and left polar decompositions* of $\boldsymbol{F}$. Here $\boldsymbol{R}$ is a proper orthogonal transformation (finite rotation) and $\boldsymbol{U} = \sqrt{\boldsymbol{F}^T\boldsymbol{F}}$ and $\boldsymbol{V} = \sqrt{\boldsymbol{F}\boldsymbol{F}^T}$ are symmetric positive-definite tensors called, respectively, the *right and left stretch tensors*.[20] As indicated by their indices, $\boldsymbol{R}$ is a two-point tensor and $\boldsymbol{U}$ and $\boldsymbol{V}$ are material and spatial tensors, respectively. The left and right stretch tensors are related through the *congruence relation*,

$$\boldsymbol{V} = \boldsymbol{R}\boldsymbol{U}\boldsymbol{R}^T. \tag{2.58}$$

Deformation measures and their physical significance The right and left stretch tensors $\boldsymbol{U}$ and $\boldsymbol{V}$ characterize the shape change of a particle neighborhood, but they are inconvenient to work with because their components are irrational functions of $\boldsymbol{F}$ that are difficult to obtain. Instead we define the *right* and *left Cauchy–Green deformation tensors*, $\boldsymbol{C} = \boldsymbol{U}^2$ and $\boldsymbol{B} = \boldsymbol{V}^2$, which given the definitions of $\boldsymbol{U}$ and $\boldsymbol{V}$ are

$$C_{IJ} = F_{kI}F_{kJ} \quad \Leftrightarrow \quad \boldsymbol{C} = \boldsymbol{F}^T\boldsymbol{F} \quad \text{and} \quad B_{ij} = F_{iK}F_{jK} \quad \Leftrightarrow \quad \boldsymbol{B} = \boldsymbol{F}\boldsymbol{F}^T. \tag{2.59}$$

$\boldsymbol{C}$ and $\boldsymbol{B}$ are symmetric, material and spatial, respectively, second-order tensors.

For solids, $\boldsymbol{C}$ is most convenient. The physical significance of $\boldsymbol{C}$ is explained by considering Fig. 2.3, which shows the mapping of the infinitesimal material vector $d\boldsymbol{X}$ in the reference configuration to the spatial vector $d\boldsymbol{x}$. Define the lengths of these differential

[19] An affine mapping is a transformation that preserves collinearity, i.e. points that were originally on a straight line remain on a straight line. Strictly speaking, this includes rigid-body translation; however, the deformation gradient is insensitive to translation.

[20] See Eqn. (2.45) for the definition of the square root of a tensor.

vectors as $dS = \|d\boldsymbol{X}\|$ and $ds = \|d\boldsymbol{x}\|$. The diagonal elements of $\boldsymbol{C}$ are related to the stretches of differential material vectors originally oriented along the axis directions:

$$\alpha^{(1)} = \sqrt{C_{11}}, \qquad \alpha^{(2)} = \sqrt{C_{22}}, \qquad \alpha^{(3)} = \sqrt{C_{33}}.$$

For example, $\alpha^{(1)} = ds/dS$ for the case where $[d\boldsymbol{X}] = [dX_1, 0, 0]^T$, and similarly for the other stretches. The off-diagonal elements of $\boldsymbol{C}$ are related to the change in angle between differential material vectors originally oriented along the axis directions:

$$\cos\theta_{12} = \frac{C_{12}}{\sqrt{C_{11}}\sqrt{C_{22}}}, \qquad \cos\theta_{13} = \frac{C_{13}}{\sqrt{C_{11}}\sqrt{C_{33}}}, \qquad \cos\theta_{23} = \frac{C_{23}}{\sqrt{C_{22}}\sqrt{C_{33}}}.$$

For example, θ_{12} is the angle in the deformed configuration between two infinitesimal vectors originally oriented along the 1 and 2 directions in the reference configuration.

In its principal coordinate system, $\boldsymbol{C}$ is diagonal, i.e. $C_{\underline{I}\,\underline{I}} = \lambda_I^{\boldsymbol{C}}$ and $C_{IJ} = 0$ for $I \neq J$, where $\lambda_I^{\boldsymbol{C}}$ are the eigenvalues of $\boldsymbol{C}$. Given the physical significance of the components of $\boldsymbol{C}$, we see that $\lambda_I^{\boldsymbol{C}}$ are the squares of the stretches in the principal coordinate system, i.e. the squares of the *principal stretches*. In the principal coordinate system, the deformation corresponds to uniform stretching along the principal directions. The eigenvalues of the right stretch tensor are the square roots of the eigenvalues of $\boldsymbol{C}$. Therefore, the eigenvalues of $\boldsymbol{U}$ are the principal stretches. This is the reason for the term "stretch tensor."

An important deformation measure related to $\boldsymbol{C}$ is the *Lagrangian strain tensor* $\boldsymbol{E}$:

$$E_{IJ} = \frac{1}{2}(C_{IJ} - \delta_{IJ}) = \frac{1}{2}(F_{iI}F_{iJ} - \delta_{IJ}) \quad \Leftrightarrow \quad \boldsymbol{E} = \frac{1}{2}(\boldsymbol{C} - \boldsymbol{I}) = \frac{1}{2}(\boldsymbol{F}^T\boldsymbol{F} - \boldsymbol{I}). \tag{2.60}$$

The differential lengths in Fig. 2.3, $dS = \|d\boldsymbol{X}\|$ and $ds = \|d\boldsymbol{x}\|$, are related by

$$ds^2 - dS^2 = 2E_{IJ}dX_I dX_J.$$

The $\frac{1}{2}$ factor in the definition of the Lagrangian strain (which leads to the factor of 2 above) is introduced to agree with the infinitesimal definition of strain familiar from elasticity theory. To see this, introduce the displacement field $\boldsymbol{u}(\boldsymbol{X}, t)$ through the following relation:

$$\boldsymbol{\varphi}(\boldsymbol{X}, t) = \boldsymbol{X} + \boldsymbol{u}(\boldsymbol{X}, t). \tag{2.61}$$

The deformation gradient is then

$$\boldsymbol{F} = \frac{\partial\boldsymbol{\varphi}}{\partial\boldsymbol{X}} = \boldsymbol{I} + \nabla_0\boldsymbol{u}, \tag{2.62}$$

and the Lagrangian strain is

$$\boldsymbol{E} = \frac{1}{2}(\boldsymbol{F}^T\boldsymbol{F} - \boldsymbol{I}) = \frac{1}{2}\left[\nabla_0\boldsymbol{u} + (\nabla_0\boldsymbol{u})^T\right] + \frac{1}{2}(\nabla_0\boldsymbol{u})^T\,\nabla_0\boldsymbol{u}.$$

Neglecting the nonlinear part $\frac{1}{2}(\nabla_0\boldsymbol{u})^T\,\nabla_0\boldsymbol{u}$ and evaluating the expression at the reference configuration where $\nabla_0 = \nabla$ and the distinction between reference and deformed coordinates disappears, we obtain the *small strain tensor* ϵ:

$$\epsilon_{ij} = \frac{1}{2}(u_{i,j} + u_{j,i}) \quad \Leftrightarrow \quad \epsilon = \frac{1}{2}\left[\nabla u + (\nabla u)^T\right]. \tag{2.63}$$

In contrast to the Lagrangian strain tensor, the small strain tensor is *not* invariant with respect to finite rotations (see Exercise 2.7).

2.2.5 Kinematic rates

In order to study the dynamical behavior of materials, it is necessary to establish the time rate of change of the kinematic fields introduced so far in this chapter. To do so, we must first discuss time differentiation in the context of the referential and spatial descriptions.

Material time derivative The difference between the referential and spatial descriptions of a continuous medium becomes particularly apparent when considering the time derivative of tensor fields. Consider the field $\mathfrak{g}$, which can be written within the referential or spatial descriptions (see Section 2.2.3), $\mathfrak{g} = g(\boldsymbol{x}, t) = \breve{g}(\boldsymbol{X}(\boldsymbol{x}, t), t)$. Here $\mathfrak{g}$ represents the value of the field variable, while g and $\breve{g}$ represent the functional dependence of $\mathfrak{g}$ on specific arguments. There are two possibilities for taking a time derivative,

$$\left.\frac{\partial \breve{g}(\boldsymbol{X}, t)}{\partial t}\right|_{\boldsymbol{X}} \quad \text{or} \quad \left.\frac{\partial g(\boldsymbol{x}, t)}{\partial t}\right|_{\boldsymbol{x}},$$

where the notation $\square|_{\boldsymbol{X}}$ and $\square|_{\boldsymbol{x}}$ is used (this one time) to place special emphasis on the fact that $\boldsymbol{X}$ and $\boldsymbol{x}$, respectively, are held fixed during the partial differentiation. The first is called the *material time derivative* of $\mathfrak{g}$, since it corresponds to the rate of change of $\mathfrak{g}$ while following a particular material particle $\boldsymbol{X}$. The second derivative is called the *local rate of change* of $\mathfrak{g}$. This is the rate of change of $\mathfrak{g}$ at a fixed spatial position $\boldsymbol{x}$. The material time derivative is the appropriate derivative to use in considerations of the time rate of change of properties tied to the material itself, such as the rate of change of strain at a material particle. It is denoted by a superposed dot, $\dot{\square}$, or by $D\square/Dt$,

$$\dot{\mathfrak{g}} = \frac{D\mathfrak{g}}{Dt} = \frac{\partial \breve{g}(\boldsymbol{X}, t)}{\partial t}. \tag{2.64}$$

In the case where $\mathfrak{g}$ is the motion $\boldsymbol{x} = \varphi(\boldsymbol{X}, t)$, the first and second material time derivatives of $\boldsymbol{x}$ are the velocity and acceleration of a continuum particle $\boldsymbol{X}$:

$$\breve{v}_i(\boldsymbol{X}, t) = \dot{x} = \frac{\partial \varphi_i(\boldsymbol{X}, t)}{\partial t}, \qquad \breve{a}_i(\boldsymbol{X}, t) = \ddot{x} = \frac{\partial^2 \varphi_i(\boldsymbol{X}, t)}{\partial t^2}. \tag{2.65}$$

Although these fields are given as functions over the reference body B_0, they are spatial vector fields and therefore lower-case symbols are appropriate. Expressed in the spatial description, these fields are

$$v_i(\boldsymbol{x},t) \equiv \breve{v}_i(\boldsymbol{X}(\boldsymbol{x},t),t), \quad a_i(\boldsymbol{x},t) \equiv \breve{a}_i(\boldsymbol{X}(\boldsymbol{x},t),t). \tag{2.66}$$

In some cases, it may be necessary to compute the material time derivative within a spatial description. This can be readily done by using the chain rule,

$$\dot{\mathfrak{g}} = \frac{Dg(\boldsymbol{x},t)}{Dt} = \frac{\partial g(\boldsymbol{x},t)}{\partial t} + \frac{\partial g(\boldsymbol{x},t)}{\partial x_j}\frac{\partial x_j(\boldsymbol{X},t)}{\partial t} = \frac{\partial g(\boldsymbol{x},t)}{\partial t} + \frac{\partial g(\boldsymbol{x},t)}{\partial x_j} v_j(\boldsymbol{x},t), \tag{2.67}$$

where we have used Eqns. $(2.65)_1$ and $(2.66)_1$. The acceleration of a particle computed within the spatial description is

$$a_i = \frac{\partial v_i}{\partial t} + l_{ij} v_j \quad \Leftrightarrow \quad \boldsymbol{a} = \frac{\partial \boldsymbol{v}}{\partial t} + \boldsymbol{l}\boldsymbol{v}, \tag{2.68}$$

where we have defined $\boldsymbol{l}$ as the spatial gradient of the velocity field:

$$l_{ij} = v_{i,j} \quad \Leftrightarrow \quad \boldsymbol{l} = \nabla \boldsymbol{v}. \tag{2.69}$$

The velocity gradient can be divided into a symmetric part $\boldsymbol{d}$ called the *rate of deformation tensor* and an antisymmetric part $\boldsymbol{w}$ called the *spin tensor*:

$$d_{ij} \equiv \frac{1}{2}(l_{ij} + l_{ji}) = \frac{1}{2}(v_{i,j} + v_{j,i}), \qquad w_{ij} \equiv \frac{1}{2}(l_{ij} - l_{ji}) = \frac{1}{2}(v_{i,j} - v_{j,i}). \tag{2.70}$$

Note that the rate of deformation tensor is the material time derivative of the small strain tensor defined in Eqn. (2.63), $\boldsymbol{d} = \dot{\boldsymbol{\epsilon}}$, since $\boldsymbol{v} = \dot{\boldsymbol{u}}$.

Rate of change of local deformation measures The rate of change of the deformation gradient is defined as the material time derivative of $\boldsymbol{F}$ which gives[21]

$$\dot{F}_{iJ} = l_{ij} F_{jJ} \quad \Leftrightarrow \quad \dot{\boldsymbol{F}} = \boldsymbol{l}\boldsymbol{F}. \tag{2.71}$$

The rate of change of local deformation follows as

$$\dot{\overline{dx_i}} = \dot{\overline{F_{iJ} dX_J}} = l_{ij} F_{jJ}\, dX_J = l_{ij}\, dx_j. \tag{2.72}$$

We see that in a dynamical spatial setting, the velocity gradient plays a role similar to $\boldsymbol{F}$. The material time derivative of the Jacobian is

$$\dot{J} = J \operatorname{div} \boldsymbol{v} = J \operatorname{tr} \boldsymbol{l} = J \operatorname{tr} \boldsymbol{d}. \tag{2.73}$$

A motion that preserves volume, i.e. $\dot{J} = 0$, is called an *isochoric motion*.

[21] See Section 3.6.2 in [TME12] for the details of how this relation is obtained.

Reynolds transport theorem Consider an integral of the field $\mathfrak{g} = g(\boldsymbol{x}, t)$ over a sub-body E of the body B,

$$I = \int_E g(\boldsymbol{x}, t)\, dV,$$

where $dV = dx_1 dx_2 dx_3$. The material time derivative of I is

$$\dot{I} = \frac{D}{Dt}\int_E g(\boldsymbol{x}, t)\, dV = \int_E [\dot{g} + g(\operatorname{div} \boldsymbol{v})]\, dV. \tag{2.74}$$

This equation is called the *Reynolds transport theorem*. A useful corollary to this theorem for extensive properties, i.e. properties that are proportional to mass, is given in Section 2.3.1.

2.3 Mechanical conservation and balance laws

The kinematic fields given in the previous section describe the possible deformed configurations of a continuous medium. These fields on their own cannot predict the configuration a body will adopt as a result of a given applied loading. To do so requires a generalization of the laws of mechanics (originally developed for collections of particles) to a continuous medium, together with an application of the laws of thermodynamics. The result is a set of *universal* conservation and balance laws that apply to all bodies:

1. conservation of mass;
2. balance of linear and angular momentum;[22]
3. thermal equilibrium (zeroth law of thermodynamics);
4. conservation of energy (first law of thermodynamics);
5. second law of thermodynamics.

These equations introduce four new important quantities to continuum mechanics. The concept of *stress* makes its appearance in the derivation of the momentum balance equations. *Temperature*, *internal energy* and *entropy* star in the zeroth, first and second laws, respectively. In this section we focus on the mechanical conservation laws (mass and momentum) leaving the thermodynamic laws to Section 2.4.

2.3.1 Conservation of mass

A basic principle of classical mechanics is that mass is a fixed quantity that can be neither formed nor destroyed, but only deformed as a result of applied loads. In other words, the

[22] The balance of angular momentum is taken to be a basic principle in continuum mechanics. This is at odds with some physics textbooks that view the balance of angular momentum as a property of systems of particles in which the internal forces are central. Truesdell discusses this in his article "Whence the law of moment and momentum?" in [Tru68, p. 239]. He states: "Few if any specialists in mechanics think of their subject in this way. By them, classical mechanics is based on three fundamental laws, asserting the conservation or balance of *force, torque,* and *work,* or in other terms, of *linear momentum, moment of momentum,* and *energy.*"

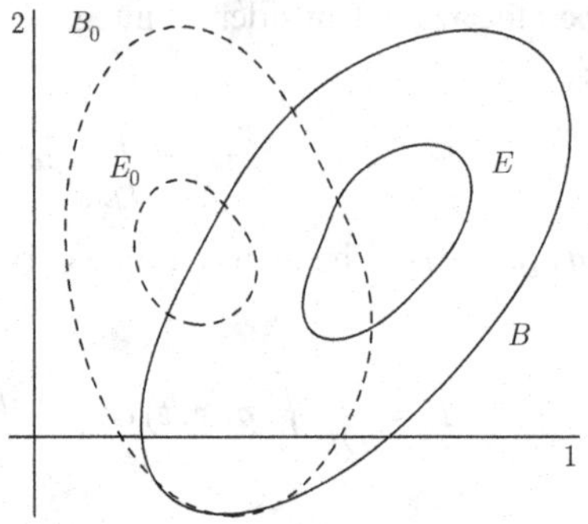

Fig. 2.5 A body B_0 and arbitrary sub-body E_0 are deformed to B and E in the deformed configuration.

total amount of mass in a closed system is conserved. For a system of particles this is a trivial statement that requires no further clarification. However, for a continuous medium it must be recast in terms of the mass density ρ, which is a measure of the distribution of mass in space.

Consider the body B_0 in Fig. 2.5, which is undergoing a deformation. In the absence of diffusion, the principle of conservation of mass requires that the mass of any subbody E_0 of the body remains unchanged by the deformation,

$$m_0(E_0) = m(E) \quad \forall E_0 \subset B_0,$$

where $m_0(\cdot)$ and $m(\cdot)$ represent the mass of a domain in the reference configuration and deformed configuration, respectively. Define $\rho_0 \equiv dm_0/dV_0$ as the *reference* mass density, so that $m_0(E_0) = \int_{E_0} \rho_0 \, dV_0$. Similarly $\rho \equiv dm/dV$ is the mass density in the deformed configuration,[23] so that $m(E) = \int_E \rho \, dV$. The conservation of mass is then

$$\int_{E_0} \rho_0 \, dV_0 = \int_E \rho \, dV \quad \forall E_0 \subset B_0.$$

Changing variables from dV to dV_0 ($dV = J dV_0$) and rearranging gives

$$\int_{E_0} (J\breve{\rho} - \rho_0) \, dV_0 = 0 \quad \forall E_0 \subset B_0,$$

where $\breve{\rho}(\boldsymbol{X}) = \rho(\boldsymbol{\varphi}(\boldsymbol{X}))$ is the material description of the mass density. From this point onward, when the description (material or spatial) is clear from the context we will suppress the $\breve{\square}$ notation. Thus, in the above equation $\breve{\rho}$ becomes simply ρ. Now, in order for this equation to be satisfied for all E_0 it must be satisfied pointwise, therefore

$$J\rho = \rho_0, \tag{2.75}$$

[23] The spatial mass density ρ is the standard definition of mass density – the one that would be measured in an experiment. The reference mass density (mass per unit original volume) ρ_0 is a mathematical convenience.

which is referred to as the material (referential) form[24] of the conservation of mass field equation. This relation makes physical sense. Since the total mass is conserved, the density of the material must change in correspondence with the local changes in volume. The spatial form of the conservation of mass is obtained from the condition $\dot{m}(E) = (D/Dt)\int_E \rho \, dV = 0$ and applying the Reynolds transport theorem in Eqn. (2.74). The result is

$$\dot{\rho} + \rho v_{k,k} = 0 \quad \Leftrightarrow \quad \dot{\rho} + \rho(\operatorname{div} \boldsymbol{v}) = 0. \tag{2.76}$$

An equivalent expression for Eqn. (2.76) is obtained by substituting Eqn. (2.67) for the material time derivative:

$$\frac{\partial \rho}{\partial t} + (\rho v_k)_{,k} = 0 \quad \Leftrightarrow \quad \frac{\partial \rho}{\partial t} + \operatorname{div}(\rho \boldsymbol{v}) = 0. \tag{2.77}$$

This is the common form of the *continuity equation*. However, Eqn. (2.76) is also referred to by that name. Finally, the continuity equation can be combined with the expression for material acceleration to form the following new relation:[25]

$$\rho a_i = \frac{\partial}{\partial t}(\rho v_i) + (\rho v_i v_j)_{,j} \quad \Leftrightarrow \quad \rho \boldsymbol{a} = \frac{\partial}{\partial t}(\rho \boldsymbol{v}) + \operatorname{div}(\rho \boldsymbol{v} \otimes \boldsymbol{v}). \tag{2.78}$$

This relation is used below in Section 2.3.2 and plays an important role in the definition of the microscopic stress tensor in Section 8.2.

Reynolds transport theorem for extensive properties Conservation of mass can be used to obtain a useful corollary to the Reynolds transport theorem in Eqn. (2.74) for the special case where the function g is an *extensive property*, i.e. a property that is proportional to mass. This means that $g = \rho\psi$, where ψ is a density field (g per unit mass). In this case,

$$\frac{D}{Dt}\int_E \rho\psi \, dV = \int_E \rho\dot{\psi} \, dV, \tag{2.79}$$

which is the Reynolds transport theorem for extensive properties.

2.3.2 Balance of linear momentum

The balance of linear momentum for a continuous medium is based on earlier ideas on the mechanics of particles which can be traced back to the work of Newton.

[24] The term *material form* indicates that the corresponding partial differential equation is defined with respect to material coordinates. For example here we have $J(\boldsymbol{X})\rho(\boldsymbol{X}) = \rho_0(\boldsymbol{X})$ for $\boldsymbol{X} \in \mathcal{B}_0$.

[25] See Section 4.1 in [TME12] for details.

Newton's laws for a system of particles In 1687, Isaac Newton published his *Philosophiae Naturalis Principia Mathematica* or simply *Principia*, where a unified theory of mechanics was presented for the first time. According to this theory, the motion of material objects is governed by three laws. Translated from the Latin, these laws state [Mar90]:

I *Every body remains in a state, resting or moving uniformly in a straight line, except insofar as forces on it compel it to change its state.*
II *The [rate of] change of momentum is proportional to the motive force impressed, and is made in the direction of the straight line in which the force is impressed.*
III *To every action there is always opposed an equal reaction.*

Mathematically, Newton's second law, also called the *balance of linear momentum*, is

$$\frac{D}{Dt}\boldsymbol{L}(t) = \boldsymbol{F}^{\text{ext}}(t), \tag{2.80}$$

where $\boldsymbol{F}^{\text{ext}}(t)$ is the total external force acting on a system at time t and $\boldsymbol{L}(t)$ is its linear momentum. (Note the use of the material time derivative here.) For a single particle with position $\boldsymbol{r}(t)$ and mass m,

$$\boldsymbol{L}(t) = m\dot{\boldsymbol{r}}(t), \quad \boldsymbol{F}^{\text{ext}}(t) = \boldsymbol{f}(t).$$

Here $\dot{\boldsymbol{r}}(t)$ is the velocity of the particle and $\boldsymbol{f}(t)$ is the force acting on it. Assuming m is constant, Eqn. (2.80) reduces to the more familiar form of Newton's second law:

$$m\ddot{\boldsymbol{r}}(t) = \boldsymbol{f}(t),$$

where $\ddot{\boldsymbol{r}}(t)$ is the acceleration of the particle.

Next, consider a system consisting of N particles with masses m^α $(\alpha = 1, 2, \dots, N)$ whose behavior is governed by classical mechanics. Let

$$\boldsymbol{r}^\alpha(t), \qquad \boldsymbol{v}^\alpha(t) = \dot{\boldsymbol{r}}^\alpha(t), \qquad \boldsymbol{p}^\alpha(t) = m^\alpha \boldsymbol{v}^\alpha(t), \tag{2.81}$$

denote the position, velocity and momentum of particle α as a function of time. Newton's second law holds individually for each particle:

$$\frac{d\boldsymbol{p}^\alpha(t)}{dt} = \boldsymbol{f}^\alpha(t), \tag{2.82}$$

where $\boldsymbol{f}^\alpha(t)$ is the force on particle α. It also holds for the entire system of particles with

$$\boldsymbol{L}(t) = \sum_{\alpha=1}^{N} \boldsymbol{p}^\alpha(t), \quad \boldsymbol{F}^{\text{ext}}(t) = \sum_{\alpha=1}^{N} \boldsymbol{f}^\alpha(t), \tag{2.83}$$

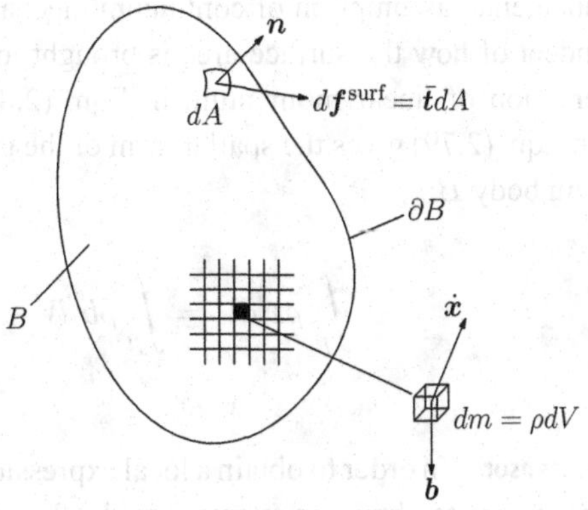

Fig. 2.6 A continuous body B with surface ∂B is divided into an infinite number of infinitesimal volume elements with mass dm and velocity $\dot{\boldsymbol{x}}$. Each volume element experiences a body force $\boldsymbol{b}$ per unit mass. Surface elements on ∂B, denoted dA, with normal $\boldsymbol{n}$ experience a force $d\boldsymbol{f}^{\text{surf}}$. See text for details.

together with Eqn. (2.80). This formulation applies to a system of interacting atoms which is considered extensively later in the book. For that case the more general Hamiltonian formulation of the balance of linear momentum described in Section 4.3 becomes advantageous. For now we continue with the balance laws for continuum systems.

Balance of linear momentum for a continuum system The extension of Newton's laws of motion from a system of particles to the differential equations for a continuous medium involved the work of many researchers over a 100 year period following the publication of Newton's *Principia* (see Section 4.2.1 of [TME12] for more details). The theory resulting from these efforts is described next.

Consider a continuous body B divided into infinitesimal volume elements as shown schematically in Fig. 2.6. The total linear momentum of the body, $\boldsymbol{L}(B)$, and the total external force acting on it, $\boldsymbol{F}^{\text{ext}}(B)$, are

$$\boldsymbol{L}(B) = \int_B \dot{\boldsymbol{x}}\rho \, dV, \qquad \boldsymbol{F}^{\text{ext}}(B) = \int_B \rho \boldsymbol{b} \, dV + \int_{\partial B} \bar{\boldsymbol{t}} \, dA. \tag{2.84}$$

The external force acting on the system has two contributions: (a) *body forces* resulting from long-range fields like gravity and electromagnetic interactions, and (b) *surface forces* across the boundary ∂B resulting from the short-range interaction of B with its surroundings.[26] Body forces are given in terms of a density field, $\boldsymbol{b}(\boldsymbol{x})$, of body force per unit mass. Surface forces (also called *contact forces*) are defined in terms of a surface density field of force per unit *spatial* area called the external *traction* or *stress vector* $\bar{\boldsymbol{t}}$ (see Fig. 2.6):

$$\bar{\boldsymbol{t}} \equiv \lim_{\Delta A \to 0} \frac{\Delta \boldsymbol{f}^{\text{surf}}}{\Delta A} = \frac{d\boldsymbol{f}^{\text{surf}}}{dA}. \tag{2.85}$$

[26] In reality, surface forces are also forces at a distance resulting from the interaction of atoms from the bodies coming into "contact." However, since the range of interactions is vastly smaller than typical macroscopic length scales it is more convenient to treat these separately as surface forces rather than as short-range body forces.

It is a fundamental assumption of continuum mechanics that this limit exists, is finite and is independent of how the surface area is brought to zero.[27] Substituting Eqn. (2.84) into the conservation of linear momentum in Eqn. (2.80) and using the Reynolds transport theorem in Eqn. (2.79) gives the spatial form of the global balance of linear momentum for a continuum body B:

$$\int_B \rho \ddot{\boldsymbol{x}}\, dV = \int_B \rho \boldsymbol{b}\, dV + \int_{\partial B} \bar{\boldsymbol{t}}\, dA. \tag{2.86}$$

Cauchy stress tensor In order to obtain a local expression for the balance of linear momentum it is first necessary to obtain an expression like that in Eqn. (2.86) for an arbitrary *internal* subbody E. This is not a problem for the body force term, but the external traction $\bar{\boldsymbol{t}}$ is defined explicitly in terms of the external forces acting on B across its outer surfaces. This dilemma was addressed by Cauchy in 1822 through his famous *stress principle* that lies at the heart of continuum mechanics. Cauchy's realization was that there is no inherent difference between external forces acting on the physical surfaces of a body and internal forces acting across virtual surfaces within the body. In both cases these can be described in terms of traction distributions. This makes sense since in the end external tractions characterize the interaction of a body with its surroundings (other material bodies) just like internal tractions characterize the interactions of two parts of a material body across an internal surface. A concise statement of Cauchy's stress principle is:

Cauchy's stress principle Material interactions across an internal surface in a body can be described as a distribution of tractions in the same way that the effect of external forces on physical surfaces of the body is described.

This may appear to be a very simple, almost trivial, observation, however, it cleared up the confusion resulting from nearly 100 years of failed and partly failed attempts to understand internal forces that preceded Cauchy. Cauchy's principle paved the way for the continuum theories of solids and fluids. Using special choices for internal subbodies (a "pillbox" and a tetrahedron), Cauchy was able to prove that the traction was related to a second-order stress tensor $\boldsymbol{\sigma}$, which we now call the *Cauchy stress tensor*, via the relation

$$t_i(\boldsymbol{n}) = \sigma_{ij} n_j \quad \Leftrightarrow \quad \boldsymbol{t}(\boldsymbol{n}) = \boldsymbol{\sigma}\boldsymbol{n}. \tag{2.87}$$

[27] From a microscopic perspective, the force $\Delta \boldsymbol{f}^{\rm surf}$ is taken to be the force resultant of all atomic interactions across ΔA. Notice that a term $\Delta \boldsymbol{m}^{\rm surf}$ accounting for the moment resultant of this microscopic distribution has not been included. This is correct as long as electrical and magnetic effects are neglected (we see this in Section 8.2 where we derive the microscopic stress tensor for a system of atoms interacting classically). If $\Delta \boldsymbol{m}^{\rm surf}$ is included in the formulation it leads to the presence of *couple stresses*, i.e. a field of distributed moments per unit area across surfaces. Theories that include this effect are called *multipolar*. Couple stresses can be important for magnetic materials in a magnetic field or polarized materials in an electric field. See for example [Jau67] or [Mal69] for more information on multipolar theories.

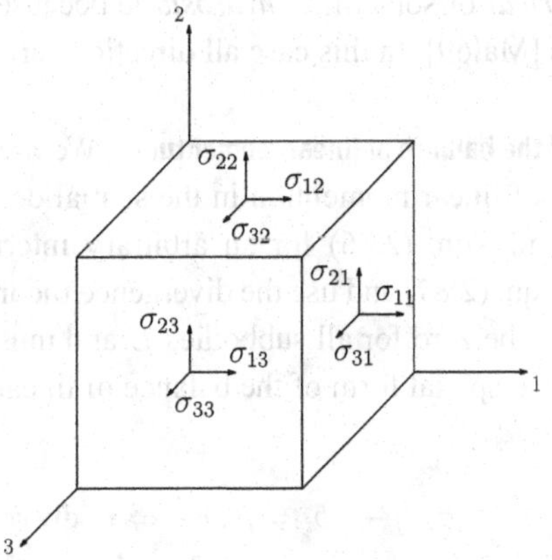

Fig. 2.7 Components of the Cauchy stress tensor. The components on the faces not shown are oriented in the reverse directions to those shown.

This important equation is referred to as *Cauchy's relation*. Note the absence of the bar over the traction; $\boldsymbol{t}$ is now the *internal* traction.

The physical significance of the components of $\boldsymbol{\sigma}$ becomes apparent when considering a cube of material oriented along the basis vectors (Fig. 2.7). The component σ_{ij} is the stress in the direction $\boldsymbol{e}_i$ on the face normal to $\boldsymbol{e}_j$. The diagonal components $\sigma_{11}, \sigma_{22}, \sigma_{33}$ are normal (tensile/compressive) stresses. The off-diagonal components $\sigma_{12}, \sigma_{13}, \sigma_{23}, \ldots$, are shear stresses.[28]

A commonly employed additive decomposition of the Cauchy stress tensor is

$$\sigma_{ij} = s_{ij} - p\delta_{ij} \quad \Leftrightarrow \quad \boldsymbol{\sigma} = \boldsymbol{s} - p\boldsymbol{I}, \tag{2.88}$$

where p is the *hydrostatic stress* or *pressure* and $\boldsymbol{s}$ is the *deviatoric* part of the Cauchy stress tensor:

$$p = -\frac{1}{3}\sigma_{kk} \quad \Leftrightarrow \quad p = -\frac{1}{3}\operatorname{tr}\boldsymbol{\sigma}, \qquad s_{ij} = \sigma_{ij} + p\delta_{ij} \quad \Leftrightarrow \quad \boldsymbol{s} = \boldsymbol{\sigma} + p\boldsymbol{I}. \tag{2.89}$$

Note that $\operatorname{tr}\boldsymbol{s} = \operatorname{tr}\boldsymbol{\sigma} + (\operatorname{tr}\boldsymbol{I})p = \operatorname{tr}\boldsymbol{\sigma} + 3p = 0$, thus $\boldsymbol{s}$ only includes information about shear stress. Consequently any material phenomenon that is insensitive to hydrostatic pressure, such as plastic flow in metals, will depend only on $\boldsymbol{s}$. A stress state with $\boldsymbol{s} = \boldsymbol{0}$ is

28 We note that in some books the stress tensor is defined as the transpose of the definition given here. Thus they define $\tilde{\boldsymbol{\sigma}} = \boldsymbol{\sigma}^T$. Both definitions are equally valid as long as they are used consistently. We prefer our definition of $\boldsymbol{\sigma}$, since it leads to the Cauchy relation in Eqn. (2.87), which is consistent with the linear algebra idea that the stress tensor operates on the normal to give the traction. With the transposed definition of the stress, the Cauchy relation would be $\tilde{\boldsymbol{\sigma}}^T\boldsymbol{n}$, which is less transparent. Of course, this distinction becomes moot if the stress tensor is symmetric, which as we will see later is the case for nonpolar continua.

called *spherical* or sometimes *hydrostatic* because this is the only possible stress state for static fluids [Mal69]. In this case all directions are principal directions (see page 37).

Local form of the balance of linear momentum We are now ready to derive the local form of the balance of linear momentum in the spatial description. We write the balance of linear momentum in Eqn. (2.86) for an arbitrary internal subbody E, substitute in Cauchy's relation in Eqn. (2.87) and use the divergence theorem in Eqn. (2.50). The resulting volume integral must be zero for all subbodies E and must therefore be satisfied pointwise which gives the local spatial form of the balance of linear momentum:[29]

$$\sigma_{ij,j} + \rho b_i = \rho a_i \quad \Leftrightarrow \quad \operatorname{div} \boldsymbol{\sigma} + \rho \boldsymbol{b} = \rho \boldsymbol{a} \quad \boldsymbol{x} \in \mathcal{B}. \tag{2.90}$$

An alternative form is obtained by using Eqn. (2.78) in the right-hand side of Eqn. (2.90):

$$\sigma_{ij,j} + \rho b_i = \frac{\partial(\rho v_i)}{\partial t} + (\rho v_i v_j)_{,j} \quad \Leftrightarrow \quad \operatorname{div} \boldsymbol{\sigma} + \rho \boldsymbol{b} = \frac{\partial(\rho \boldsymbol{v})}{\partial t} + \operatorname{div}(\rho \boldsymbol{v} \otimes \boldsymbol{v}). \tag{2.91}$$

Equation (2.91) is correct only if the continuity equation is satisfied (since it is used in the derivation of Eqn. (2.78)). It is therefore called the *continuity momentum equation*. It plays an important role in the statistical mechanics derivation of the microscopic stress tensor of Section 8.2. Finally, for static problems the balance of linear momentum simplifies to

$$\sigma_{ij,j} + \rho b_i = 0 \quad \Leftrightarrow \quad \operatorname{div} \boldsymbol{\sigma} + \rho \boldsymbol{b} = \boldsymbol{0} \quad \boldsymbol{x} \in \mathcal{B}, \tag{2.92}$$

which are referred to as the *stress equilibrium equations*.

2.3.3 Balance of angular momentum

In addition to requiring a balance of linear momentum, we must also require that the system be balanced with respect to angular momentum. The *balance of angular momentum* states that the change in angular momentum of a system is equal to the resultant moment applied to it. This is also called the *moment of momentum principle*. In mathematical form this is

$$\frac{D}{Dt}\boldsymbol{H}_0 = \boldsymbol{M}_0^{\text{ext}}, \tag{2.93}$$

where $\boldsymbol{H}_0$ is the angular momentum or moment of momentum of the system about the origin and $\boldsymbol{M}_0^{\text{ext}}$ is the total external moment about the origin. For a subbody E of a

[29] See Section 4.2.7 in [TME12] for details.

continuum system,[30]

$$\boldsymbol{H}_0(E) = \int_E \boldsymbol{x} \times (\rho \dot{\boldsymbol{x}})\, dV, \qquad \boldsymbol{M}_0^{\text{ext}}(E) = \int_E \boldsymbol{x} \times (\rho \boldsymbol{b})\, dV + \int_{\partial E} \boldsymbol{x} \times \boldsymbol{t}\, dA.$$

Substituting $\boldsymbol{H}_0(E)$ and $\boldsymbol{M}_0^{\text{ext}}(E)$ into Eqn. (2.93) and following a similar procedure to that in the previous section leads after some algebra to the result that

$$\sigma_{ij} = \sigma_{ji} \quad \Leftrightarrow \quad \boldsymbol{\sigma} = \boldsymbol{\sigma}^T. \tag{2.94}$$

Thus the balance of angular momentum implies that the Cauchy stress is symmetric.[31]

2.3.4 Material form of the momentum balance equations

The derivation of the balance equations in the previous sections is complete. However, for solids, it is often computationally more convenient[32] to solve the balance equations in a Lagrangian description. This is because in reference coordinates, the boundary of the body ∂B_0 is a constant whereas in the spatial coordinates the boundary ∂B depends on the motion, which is usually what we are trying to solve. Thus, we must obtain the material form (or referential form) of the balance of linear and angular momentum. In the process of doing so we will identify the first and second Piola–Kirchhoff stress tensors (and the related Kirchhoff stress tensor) that play important roles in the material description formulation.

Material form of the balance of linear momentum The derivation begins with the spatial form of the balance of linear momentum for an arbitrary subbody. All integrals are then mapped back to the reference configuration using a change of variables. This requires the application of Nanson's formula in Eqn. (2.56) for the surface integrals. Then following the procedure outlined in the previous section, the following relation is obtained:

$$P_{iJ,J} + \rho_0 b_i = \rho_0 a_i \quad \Leftrightarrow \quad \operatorname{Div} \boldsymbol{P} + \rho_0 \boldsymbol{b} = \rho_0 \boldsymbol{a}, \quad \boldsymbol{X} \in B_0, \tag{2.95}$$

which is the local *material* form of the balance of linear momentum. In this relation, $\boldsymbol{P}$ is the *first Piola–Kirchhoff stress tensor* which is related to the Cauchy stress tensor by

$$P_{iJ} = J \sigma_{ij} F_{Jj}^{-1} \quad \Leftrightarrow \quad \boldsymbol{P} = J \boldsymbol{\sigma} \boldsymbol{F}^{-T}, \qquad \sigma_{ij} = \frac{1}{J} P_{iJ} F_{jJ} \quad \Leftrightarrow \quad \boldsymbol{\sigma} = \frac{1}{J} \boldsymbol{P} \boldsymbol{F}^T. \tag{2.96}$$

[30] For an application of the balance of angular momentum to a system of particles, see Section 5.8.1.

[31] Note that in a multipolar theory with couple stresses $\boldsymbol{M}_0^{\text{ext}}(E)$ would also include contributions from distributed body couples and corresponding hypertractions. As a result $\boldsymbol{\sigma}$ would not be symmetric. Instead the balance of angular momentum would supply a set of three equations relating the Cauchy stress tensor to the *couple stress tensor*.

[32] See, for example, Chapter 9 in [TME12].

The first Piola–Kirchhoff stress tensor is a two-point tensor which is also called the *engineering stress* or *nominal stress* because it is the force per unit area in the *reference* configuration. The Cauchy stress, on the other hand, is the *true stress* because it is the force per unit area in the *deformed* configuration. The fact that there are different stress measures with different meanings is often something that is not appreciated by nonexperts in mechanics, especially since these differences vanish if the deformation is small (i.e. when $\boldsymbol{F} \approx \boldsymbol{I}$).

The material form of Cauchy's relation is

$$T_i = P_{iJ} N_J \quad \Leftrightarrow \quad \boldsymbol{T} = \boldsymbol{P}\boldsymbol{N}, \tag{2.97}$$

where $\boldsymbol{T}_i \equiv d\boldsymbol{f}/dA_0$ is the *nominal traction* as opposed to $\boldsymbol{t}$ defined in Eqn. (2.85) which can be called the *true traction*.

Material form of the balance of angular momentum The material form of the balance of angular momentum is

$$P_{kM} F_{jM} = P_{jM} F_{kM} \quad \Leftrightarrow \quad \boldsymbol{P}\boldsymbol{F}^T = \boldsymbol{F}\boldsymbol{P}^T. \tag{2.98}$$

Note that the first Piola–Kirchhoff stress tensor is in general *not* symmetric (i.e. $\boldsymbol{P} \neq \boldsymbol{P}^T$).

Second Piola–Kirchhoff stress tensor The first Piola–Kirchhoff stress tensor is a mixed tensor spanning the reference and deformed configurations. It is mathematically advantageous to define another stress tensor entirely in the reference configuration by pulling the force back to the reference configuration as if it were a kinematic quantity. This is the *second Piola–Kirchhoff stress tensor* $\boldsymbol{S}$, which is defined by

$$S_{IJ} = F^{-1}_{Ii} P_{iJ} \quad \Leftrightarrow \quad \boldsymbol{S} = \boldsymbol{F}^{-1}\boldsymbol{P}. \tag{2.99}$$

The relation between $\boldsymbol{\sigma}$ and $\boldsymbol{S}$ is obtained by using Eqn. $(2.96)_1$:

$$\sigma_{ij} = \frac{1}{J} F_{iI} S_{IJ} F_{jJ} \quad \Leftrightarrow \quad \boldsymbol{\sigma} = \frac{1}{J}\boldsymbol{F}\boldsymbol{S}\boldsymbol{F}^T. \tag{2.100}$$

Inverting this relation gives $\boldsymbol{S} = J\boldsymbol{F}^{-1}\boldsymbol{\sigma}\boldsymbol{F}^{-T}$, from which it is clear that $\boldsymbol{S}$ is symmetric since $\boldsymbol{\sigma}$ is symmetric. The second Piola–Kirchhoff stress tensor, $\boldsymbol{S}$, has no direct physical significance, but since it is symmetric it can be more convenient to work with than $\boldsymbol{P}$. The balance of linear momentum in terms of $\boldsymbol{S}$ follows from Eqn. (2.95) as

$$(F_{iI} S_{IJ})_{,J} + \rho_0 b_i = \rho_0 a_i \quad \Leftrightarrow \quad \operatorname{Div}(\boldsymbol{F}\boldsymbol{S}) + \rho_0 \boldsymbol{b} = \rho_0 \boldsymbol{a} \quad \boldsymbol{X} \in \mathcal{B}_0. \tag{2.101}$$

Our discussion so far has been entirely mechanical in nature and has not considered temperature or heat transfer. To account for thermal effects and to obtain insight into the form of the constitutive relations, we next turn to thermodynamics.

2.4 Thermodynamics

Thermodynamics is the theory dealing with the flow of heat and energy between material systems. This theory boils down to three fundamental laws based on empirical observation that all physical systems are assumed to obey. The *zeroth law of thermodynamics* is related to the concept of thermal equilibrium. The *first law of thermodynamics* is a statement of conservation of energy. The *second law of thermodynamics* is related to the directionality of thermodynamic processes. We will discuss each of these laws, but first we must describe the fundamental concepts in which thermodynamics is phrased beginning with the ideas of macroscopic observables and state variables.

2.4.1 Macroscopic observables, thermodynamic equilibrium and state variables

We know that all systems are composed of discrete particles that (to a good approximation, see Section 5.2) satisfy Newton's laws of motion. Thus, to have a complete understanding of a system it is necessary to determine the positions and momenta of the particles that make it up. However, as described further below, this is a hopeless task and we must make do with a much smaller set of variables.

Macroscopically observable quantities Fundamentally, a thermodynamic system is composed of a vast number of particles N, where N is huge (on the order of 10^{23} for a cubic centimeter of material). The *microscopic kinematics*[33] of such a system are described by a (time-dependent) vector in a $6N$-dimensional vector space, called *phase space*, corresponding to the complete list of particle positions and momenta,

$$\boldsymbol{y} = (\boldsymbol{r}^1, \dots, \boldsymbol{r}^N, m^1\dot{\boldsymbol{r}}^1, \dots, m^N\dot{\boldsymbol{r}}^N),$$

where $(m^1, \dots, m^N)$ are the masses of the particles.[34]

Although it is possible to image individual atoms, we can certainly never hope (nor wish) to record the time-dependent positions and momenta of *all* atoms in a macroscopic thermodynamic system. This would seem to suggest that there is no hope of obtaining a deep understanding of the behavior of such systems. However, for hundreds of years mankind has studied these systems using only a relatively crude set of tools and nevertheless has been able to develop a sophisticated theory of their behavior. The tools that were used for measuring kinematic quantities initially involved things like measuring sticks and lengths of string.

[33] See the definition of "kinematics" in footnote 15 on page 42.

[34] Depending on the nature of the material there may be additional quantities that have to be known, such as the charges of the particles or their magnetic moments. Here we focus on purely thermomechanical systems for which the positions and momenta are sufficient. We revisit the idea of a phase space in Section 7.1.

Later the advent of interferometry and lasers gave rise to laser extensometers and laser interferometers. All of these devices have two important characteristics in common. First, they have very limited spatial resolution relative to typical interparticle distances which are on the order of 10^{-10} m. Indeed, the spatial resolution of measuring sticks is typically on the order of 10^{-4} m and that of interferometry is on the order of 10^{-6} m (a micron). Second, these devices have very limited temporal resolution relative to characteristic atomic time scales, which are on the order of 10^{-13} s (for the oscillation period of an atom in a crystal). The temporal resolution of measuring sticks and interferometers relies on the device used to record measurements. The human eye is capable of resolving events spaced no less than 10^{-2} s apart. If a camera is used, then the shutter speed – typically on the order of 10^{-3} or 10^{-4} s – sets the temporal resolution. Clearly, these tools provide only very coarse measurements that correspond to some type of temporal and spatial averaging of the positions of the particles in the system.[35] Accordingly, the only quantities these devices are capable of measuring are those that are essentially uniform in space (over lengths up to their spatial resolution) and nearly constant in time (over spans of time up to their temporal resolution). We say that these quantities are *macroscopically observable*. The fact that such quantities exist is a deep truth of our universe, the discussion of which is outside the scope of our book.[36]

The measurement process described above replaces the $6N$ microscopic kinematic quantities with a dramatically smaller number of *macroscopic kinematic quantities*, such as the volume of the system and its shape relative to a given reference configuration as characterized by the strain tensor. In addition there are also nonkinematic quantities that are macroscopically observable, such as the number of particles in the system and its mass.

Thermodynamic equilibrium When a thermodynamic system experiences an external perturbation[37] it undergoes a *dynamical process* in which its microscopic kinematic vector and, in general, its macroscopic observables evolve as a function of time. It is empirically observed that all systems tend to evolve to a *quiescent and spatially homogeneous* (at the macroscopic length scale) terminal state where the system's macroscopic observables have reached constant limiting values. Also any fields, like density or strain, must be constant since the terminal state is spatially homogeneous. Once the system reaches this terminal condition it is said to be in a state of *thermodynamic equilibrium*. In general, even in this state, a system's microscopic kinematic quantities continue to change with time. However, these quantities are of no explicit concern to thermodynamic theory.

As you might imagine, thermodynamic equilibrium can be very difficult to achieve. We may need to wait an infinite amount of time for the dynamical process to obtain the limiting equilibrium values of the macroscopic observables. Thus except for gases, most systems

[35] See Section 1.1 for a discussion of spatial and temporal scales in materials.

[36] We encourage the reader to refer to Chapter 1 of Callen's book [Cal85] for a a more extensive introduction, similar to the above, and to Chapter 21 of [Cal85] for a discussion of the deep fundamental reason for the existence of macroscopically observable quantities (i.e. broken symmetry and Goldstone's theorem).

[37] An *external perturbation* is defined as a change in the system's surroundings which results in the surroundings doing work on, transferring heat to or transferring particles to the system.

never reach a true state of thermodynamic equilibrium. Those that do not, however, do exhibit a characteristic "two-stage dynamical process" in which the macroscopic observables first evolve at a high rate during and immediately after an external perturbation. These values then further evolve at a rate that is many orders of magnitude smaller than in the first stage. This type of system is said to be in a state of *metastable thermodynamic equilibrium* once the first part of its two-stage dynamical process is complete.[38]

State variables The behavior of a thermodynamic system does not depend on all of its macroscopic observables. Those observables that affect the behavior of the system are referred to as *state variables*:

> The macroscopic observables that are well defined and single-valued when the system is in a state of thermodynamic equilibrium are called *state variables*. Those state variables which are related to the kinematics of the system (volume, strain, etc.) are called *kinematic state variables*.

State variables can be divided into two categories: *extensive* and *intensive*. An extensive variable is one whose value depends on amount. Suppose we have two identical systems which have the same values for their state variables. The extensive variables are those whose values are exactly doubled when we treat the two systems as a single composite system. Kinematic variables like volume are naturally extensive. For example, if the initial systems both have volume V, then the composite system has total volume $2V$. Strain, which is also a kinematic variable, is not extensive; however, we can define a new quantity, "volume strain" as $V_0\boldsymbol{E}$, where V_0 is the reference volume, which is extensive.

> In general, we write that a system in thermodynamic (or metastable) equilibrium is characterized by a set of n^Γ independent extensive kinematic state variables, which we denote generically as $\boldsymbol{\Gamma} = (\Gamma_1, \ldots, \Gamma_{n^\Gamma})$. For a gas, $n^\Gamma = 1$ and $\Gamma_1 = V$. For a metastable solid, $n^\Gamma = 6$ and $\boldsymbol{\Gamma}$ contains the six independent components of the Lagrangian strain tensor multiplied by the reference volume.

Other important extensive variables include the number of particles making up the system and its mass. A special extensive quantity which we have not encountered yet is the total internal energy of the system $\mathcal{U}$. Later we will find that most extensive state variables are associated with corresponding *intensive* quantities which play an equally important role. Intensive state variables are quantities whose values are independent of amount. Examples include the temperature and pressure (or stress) of a thermodynamic system. Table 2.2 presents a list of the extensive and intensive state variables that we will encounter (not all of which have been mentioned yet) indicating the pairings between them.

Independent state variables and equations of state A system in thermodynamic equilibrium can have many state variables but not all can be specified independently. Consider the case

[38] The issue of metastable equilibrium within the context of statistical mechanics is discussed in Section 11.1.

Table 2.2. Extensive and intensive state variables. Kinematic state variables are indicated with a $*$

Extensive	Intensive
internal energy ($\mathcal{U}$)	
mass (m)	
number of particles (N)	chemical potential (μ)
volume (V)*	pressure (p)
(Lagrangian) volume strain ($V_0 \boldsymbol{E}$)*	elastic part of the (second Piola–Kirchhoff) stress ($\boldsymbol{S}^{(e)}$)
entropy (S)	temperature (T)

of an ideal gas[39] enclosed in a rigid, thermally-insulated box. This system is characterized by four state variables: N, V, p and T. However, based on empirical observation, we know that not all four of these state variables are thermodynamically independent. Any three will determine the fourth. We will see this later when we derive the ideal gas law in Eqn. (2.129). In fact, it turns out that *any* system in *true*[40] thermodynamic equilibrium is fully characterized by a set of three independent state variables. This is because all systems are fluidlike on the infinite time scale of thermodynamic equilibrium.[41] For a system in metastable equilibrium the number of thermodynamically independent state variables is equal to $n^\Gamma + 2$, where n^Γ is the number of independent kinematic state variables characterizing the system, as described above. (See Section 5.1.3 in [TME12] for an explanation how the independent state variables of a system can be identified.)

We adopt the following notation. Let B be a system in thermodynamic (or metastable) equilibrium and $\mathcal{B} = (\mathcal{B}_1, \mathcal{B}_2, \dots, \mathcal{B}_{\nu_B}, \mathcal{B}_{\nu_B+1}, \dots)$ be the set of all state variables, where $\nu^{\mathsf{B}} = n^\Gamma + 2$ is the number of independent properties characterizing system B. The nonindependent properties are related to the independent properties through *equations of state*[42]

$$\mathcal{B}_{\nu^{\mathsf{B}}+j} = f_j(\mathcal{B}_1, \dots, \mathcal{B}_{\nu_B}), \quad j = 1, 2, \dots. \tag{2.102}$$

We will see examples of equations of state later in Eqns. (2.110) and (2.128) (and following) where we discuss ideal gases in more detail.

[39] See page 69 for the definition of an ideal gas.

[40] We use the term "true thermodynamic equilibrium" for systems that strictly satisfy the definition of thermodynamic equilibrium as opposed to those that are in a state of metastable equilibrium.

[41] Consider, for example, a solid placed inside of a container in the presence of gravity. Given an infinite amount of time the atoms will eventually diffuse down filling the bottom of the container like a liquid. See Section 7.4.5 for a proof, based on statistical mechanics theory, that in the limit of an infinite number of particles (keeping the density fixed) the equilibrium properties of a system do not depend on any kinematic state variables other than the system's volume. Also see Chapter 11 for a detailed discussion of the metastable nature of solids.

[42] Equations of state are closely related to *constitutive relations*, which are described in Section 2.5. Typically, the term "equation of state" refers to a relationship between state variables that characterize the entire thermodynamic system, whereas "constitutive relations" relate density variables (per unit mass) defined locally at continuum points.

2.4.2 Thermal equilibrium and the zeroth law of thermodynamics

Up to this point we have referred to temperature without defining it, relying on you, our reader, for an intuitive sense of this concept. We now see how temperature can be defined in a more rigorous fashion.

Thermal equilibrium Our sense of touch provides us with the feeling that an object is "hotter than" or "colder than" our bodies, and thus, with an intuitive sense of temperature. This concept can be made more explicit by defining the notion of thermal equilibrium between two systems.

> Two systems A and B in thermodynamic equilibrium are said to be in *thermal equilibrium* with each other, denoted A $\sim$ B, if they remain in thermodynamic equilibrium after being brought into thermal contact while keeping their kinematic state variables and their particle numbers fixed.

Thus, heat is allowed to flow between the two systems but they are not allowed to transfer particles or perform work. Here, heat is taken as a primitive concept similar to force. Later, when we discuss the first law of thermodynamics, we will discover that heat is simply a form of energy.

A practical test for determining whether two systems, already in thermodynamic equilibrium, are in thermal equilibrium, can be performed as follows: (1) thermally isolate both systems from their common surroundings; (2) for each system, fix its number of particles and all but one of its kinematic state variables and arrange for the system's surroundings to remain constant; (3) bring the two systems into thermal contact; (4) wait until the two systems are again in thermodynamic equilibrium. If the free kinematic state variable in each system remains unchanged in stage (4), then the two systems were, in fact, in thermal equilibrium when they were brought into contact.[43]

As an example, consider the two cylinders of compressed gas with frictionless movable pistons shown in Fig. 2.8. In Fig. 2.8(a) the cylinders are separated and thermally isolated from their surroundings. The forces F^{A} and F^{B} are mechanical boundary conditions applied by the surroundings to the system. Both systems are in a state of thermodynamic equilibrium. Since the systems are already thermally isolated and the only extensive kinematic quantity for a gas is its volume, steps (1)–(3) of the procedure are achieved by arranging for F^{A} and F^{B} to remain constant and bringing the two systems into thermal contact. Thus, in Fig. 2.8(b) the systems are shown in thermal contact through a *diathermal partition*, which is a partition that allows only thermal interactions (heat flow) across it but is otherwise impermeable and rigid. If the volumes remain unchanged, $V^{\mathrm{A}'} = V^{\mathrm{A}}$ and $V^{\mathrm{B}'} = V^{\mathrm{B}}$, then A and B are in thermal equilibrium.

[43] Of course at the end of stage (4) the systems will be in thermal equilibrium regardless of whether or not they were so in the beginning. However, the purpose of the test is to determine whether the systems were in thermal equilibrium when first brought into contact.

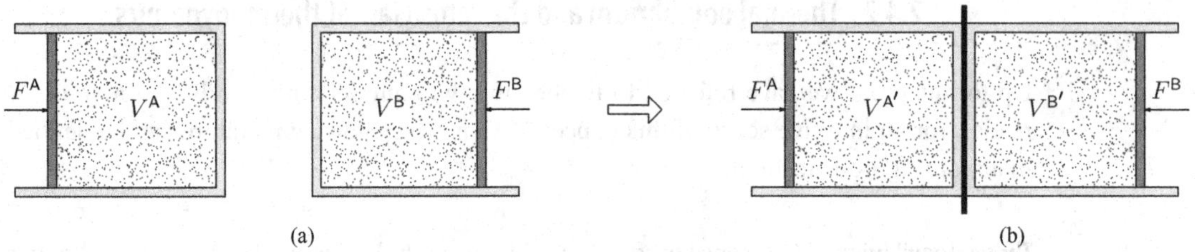

Fig. 2.8 Two cylinders of compressed gas A and B with movable frictionless pistons. (a) The cylinders are separated; each is in thermodynamic equilibrium. (b) The cylinders are brought into contact via a diathermal partition. If they remain in thermodynamic equilibrium they are said to be in thermal equilibrium with each other, $A \sim B$.

The *zeroth law of thermodynamics* is a statement about the relationship between bodies in thermal equilibrium.

> *Zeroth law of thermodynamics* Given three thermodynamic systems, A, B and C, each in thermodynamic equilibrium, then if $A \sim B$ and $B \sim C$ it follows that $A \sim C$.

Empirical temperature scales The concept of thermal equilibrium also suggests an empirical approach for defining temperature scales. The idea is to calibrate temperature using a thermodynamic system that has only one independent kinematic state variable. Thus, its temperature is in one-to-one correspondence with the value of its kinematic state variable. For example, the old-fashioned mercury-filled glass thermometer is characterized by the height (volume) of the liquid mercury in the thermometer. Denote the calibrating system as Θ and its single kinematic state variable as θ. Now consider two systems, A and B. For each of these systems, there will be values θ^A and θ^B for which $\Theta \sim A$ and $\Theta \sim B$, respectively. Then, according to the zeroth law, $A \sim B$, if and only if $\theta^A = \theta^B$. This introduces an *empirical temperature scale*. Different temperature scales can be defined by setting, $T = f(\theta)$, where $f(\theta)$ is a monotonic function. In our example of the mercury-filled glass thermometer, the function $f(\theta)$ corresponds to the markings on the side of the thermometer that identify the spacing between specified values of the temperature T. The condition for thermal equilibrium between two systems A and B is then

$$T^A = T^B. \tag{2.103}$$

In fact, we see below that there exists a uniquely defined, fundamental temperature scale called the *thermodynamic temperature* (or *absolute temperature*).[44] The thermodynamic temperature scale is defined for nonnegative values only, $T \geq 0$, and the state of *zero temperature* (which can be approached but never actually obtained by any real system) is uniquely defined by the general theory. Thus, the only unambiguous part of the scale is the unit of measure for temperature. In 1954, following a procedure originally suggested by Lord Kelvin, this ambiguity was removed by the international community's establishment

[44] The theoretical foundation for the absolute temperature scale and its connection to the behavior of ideal gases is discussed beginning on page 79.

of the *kelvin* temperature unit K at the Tenth General Conference of Weights and Measures. The kelvin unit is defined by setting the temperature at the triple point of water (the point at which ice, water and water vapor coexist) to 273.16 K. See [Adk83, Section 2] for a detailed explanation of empirical temperature scales.

2.4.3 Energy and the first law of thermodynamics

The zeroth law introduced the concepts of thermal equilibrium and temperature. The first law establishes the fact that heat is actually just a form of energy and leads to the idea of internal energy.

First law of thermodynamics Consider a thermodynamic system that is in a state of thermodynamic equilibrium (characterized by its temperature, the kinematic state variables, and a *fixed* number of particles); call it state 1. Now imagine that the system is perturbed by mechanical and thermal interactions with its environment. Mechanical interaction results from tractions applied to its surfaces and body forces applied to the bulk. Thermal interactions result from heat flux in and out of the system through its surfaces and internal heat sources distributed through the body. Due to this perturbation, the system undergoes a dynamical process and eventually reaches a new state of thermodynamic equilibrium; call it state 2. During this process mechanical work $\Delta\mathcal{W}_{12}^{\mathrm{ext}}$ is performed *on the system* and heat $\Delta\mathfrak{Q}_{12}$ is transferred into the system. Next consider a second perturbation that takes the system from state 2 to a third state; state 3. This perturbation is characterized by the total work $\Delta\mathcal{W}_{23}^{\mathrm{ext}}$ done on the system and heat $\Delta\mathfrak{Q}_{23}$ transferred to the system. Now suppose we have the special case where state 3 coincides with state 1. In other words, the second perturbation returns the system to its original state (original values of temperature and kinematic state variables) and also to the original values of total linear and angular momentum. In this case the total external work is called the *work of deformation*, $\Delta\mathcal{W}^{\mathrm{def}} = \Delta\mathcal{W}_{12}^{\mathrm{ext}} + \Delta\mathcal{W}_{21}^{\mathrm{ext}}$. This set of processes is called a thermodynamic cycle, since the system is returned to its original state. Through a series of exhaustive experiments in the nineteenth century, culminating with the work of James Prescott Joule, it was observed that in any thermodynamic cycle the amount of mechanical work performed on the system is always in constant proportion to the amount of heat *expelled by the system*:

$$\Delta\mathcal{W}^{\mathrm{def}} = -\mathcal{J}\Delta\mathfrak{Q},$$

where $\Delta\mathcal{W}^{\mathrm{def}}$ is the work (of deformation) *performed on the system* during the cycle, $\Delta\mathfrak{Q}$ is the external heat *supplied to the system* during the cycle, and $\mathcal{J}$ is *Joule's mechanical equivalent of heat*, which expresses the constant of proportionality between work and heat.[45] Accordingly, we can define a new heat quantity that has the same units as work $\mathcal{Q} = \mathcal{J}\mathfrak{Q}$, and then Joule's observation can be rearranged to express a conservation principle for any

45 Due to the success of Joule's discovery that heat and work are just different forms of energy, the constant bearing his name has fallen into disuse because independent units for heat (such as the calorie) are no longer part of the standard unit systems used by scientists.

thermodynamic system subjected to a cyclic process:

$$\Delta\mathcal{W}^{\mathrm{def}} + \Delta\mathcal{Q} = 0 \qquad \text{for any thermodynamic cycle.}$$

This implies the existence of a function that we call the *internal energy* $\mathcal{U}$ of a system in thermodynamic equilibrium.[46] The change of internal energy in going from one equilibrium state to another is, therefore, given by

$$\Delta\mathcal{U} = \Delta\mathcal{W}^{\mathrm{def}} + \Delta\mathcal{Q}. \tag{2.104}$$

If we consider the possibility of changes in the total linear and angular momentum of the system, we need to account for changes in the associated macroscopic kinetic energy $\mathcal{K}$. This is accomplished by the introduction of the *total energy* $\mathcal{E} \equiv \mathcal{K} + \mathcal{U}$. Then the total external work performed on a system consists of two parts, one that goes toward a change in macroscopic kinetic energy and the work of deformation that goes toward a change in internal energy: $\Delta\mathcal{W}^{\mathrm{ext}} = \Delta\mathcal{K} + \Delta\mathcal{W}^{\mathrm{def}}$. With these definitions, Eqn. (2.104) may, alternatively, be given as

$$\Delta\mathcal{E} = \Delta\mathcal{W}^{\mathrm{ext}} + \Delta\mathcal{Q}. \tag{2.105}$$

Equation (2.105) (or equivalently Eqn. (2.104)) is called the *first law of thermodynamics*. It shows that the total energy of a thermodynamic system and its surroundings is conserved. Mechanical and thermal energy transferred to the system (and lost by the surroundings) is retained in the system as part of its total energy. This consists of *kinetic energy* associated with motion of the system's particles (which also includes the system's gross motion) and *potential energy* associated with deformation. In other words, energy can change form, but its amount is conserved. Two conclusions can be drawn from the above discussion:

1. Equation (2.104) implies that the value of $\mathcal{U}$ depends only on the state of thermodynamic equilibrium. This means that it does not depend on the details of how the system arrived at any given state, but only on the values of the independent state variables that characterize the system. It is therefore a state variable itself. For example, taking the independent state variables to be the number of particles, the values of the kinematic state variables and the temperature, we have that[47] $\mathcal{U} = \widehat{\mathcal{U}}(N, \mathbf{\Gamma}, T)$. Further, we note that the internal energy is extensive.

[46] To see this consider any two thermodynamic equilibrium states, 1 and 2. Suppose $\Delta\mathcal{U}_{1\to2} = \Delta\mathcal{W} + \Delta\mathcal{Q}$ for one given process taking the system from 1 to 2. Now, let $\Delta\mathcal{U}_{2\to1}$ be the corresponding quantity for a process that takes the system from 2 to 1. The conservation principle requires that $\Delta\mathcal{U}_{2\to1} = -\Delta\mathcal{U}_{1\to2}$. In fact, this must be true for *all* processes that take the system from 2 to 1. The argument may be reversed to show that all processes that take the system from 1 to 2 must have the same value for $\Delta\mathcal{U}_{1\to2}$. We have found that the change in internal energy for any process depends only on the beginning and ending states of thermodynamic equilibrium. Thus, we can write $\Delta\mathcal{U}_{1\to2} = \mathcal{U}_2 - \mathcal{U}_1$, where $\mathcal{U}_1$ is the internal energy of state 1 and $\mathcal{U}_2$ is the internal energy of state 2.

[47] The symbol $\widehat{\mathcal{U}}$ is used to indicate the particular functional form where the energy is determined by the values of the number of particles, kinematic state variables, and temperature.

2. Joule's relation between work and heat implies that, although the internal energy is a state function, the work of deformation and heat transfer are not. Their values depend on the process that occurs during a change of state. In other words, $\Delta\mathcal{W}^{\text{def}}$ and $\Delta\mathcal{Q}$ are measures of energy transfer, but associated functions $\mathcal{W}^{\text{def}}$ and $\mathcal{Q}$ (similar to the internal energy $\mathcal{U}$) do not exist. Once heat and work are absorbed into the energy of the system they are no longer separately identifiable. (See Section 5.3 in [TME12] for further discussion of this point.)

Internal energy of an ideal gas It is instructive to demonstrate the laws of thermodynamics with a simple material model. Perhaps the simplest model is the *ideal gas*, where the atoms are treated as particles of negligible radius which do not interact except when they elastically bounce off each other. This idealization becomes more and more accurate as the pressure of a gas is reduced.[48] The reason for this is that the density of a gas goes to zero along with its pressure. At very low densities the size of an atom relative to the volume it occupies becomes negligible. Since the atoms in the gas are far apart most of the time, the interaction forces between them also become negligible.

Insight into the internal energy of an ideal gas was also gained from Joule's experiments mentioned earlier. Joule studied the free expansion of a thermally-isolated gas (also called "Joule expansion") from an initial volume to a larger volume and measured the temperature change. The experiment is performed by rapidly removing a partition that confines the gas to the smaller volume and allowing it to expand. Since no mechanical work is performed on the gas ($\Delta\mathcal{W}^{\text{def}} = 0$) and no heat is transferred to it ($\Delta\mathcal{Q} = 0$), the first law (Eqn. (2.104)) is simply $\Delta\mathcal{U} = 0$, i.e. the internal energy is constant in any such experiment.

Now, recall that volume is the only kinematic state variable for a gas, the total differential of internal energy associated with an infinitesimal change of state is thus[49]

$$d\mathcal{U} = \left.\frac{\partial\widehat{\mathcal{U}}}{\partial N}\right|_{V,T} dN + \left.\frac{\partial\widehat{\mathcal{U}}}{\partial V}\right|_{N,T} dV + nC_v dT, \tag{2.106}$$

where $n = N/N_{\text{A}}$ is the number of moles of gas (with Avogadro's constant $N_{\text{A}} = 6.022 \times 10^{23}\ \text{mol}^{-1}$), and

$$C_v = \frac{1}{n}\left.\frac{\partial\widehat{\mathcal{U}}}{\partial T}\right|_{N,V} \tag{2.107}$$

48 The formal definition of pressure is given in Eqn. (2.126).

49 The "vertical bar" notation $(\partial\square/\partial T)|_X$ is common in treatments of thermodynamics. It is meant to explicitly indicate which state variables (X) are to be held constant when determining the value of the partial derivative. For example, $(\partial\mathcal{U}/\partial T)|_{N,V} \equiv (\partial\widehat{\mathcal{U}}(N,V,T)/\partial T)$. However, $(\partial\mathcal{U}/\partial T)|_{N,p}$ is completely different. It is the partial derivative of the internal energy as a function of the number of particles, the pressure and temperature: $\mathcal{U} = \widetilde{\mathcal{U}}(N,p,T)$. That is, $(\partial\mathcal{U}/\partial T)|_{N,p} \equiv (\partial\widetilde{\mathcal{U}}(N,p,T)/\partial T)$. The main advantage of the notation is that it allows for the use of a single symbol ($\mathcal{U}$) to represent the value of a state variable. Thus, it avoids the use of individual symbols to indicate the particular functional form used to obtain the quantity's value: $\mathcal{U} = \widehat{\mathcal{U}}(N,V,T) = \widetilde{\mathcal{U}}(N,p,T)$. However, we believe this leads to a great deal of confusion, obscures the mathematical structure of the theory, and often results in errors by students and researchers who are not vigilant in keeping track of which particular functional form they are using. In this book, we have decided to keep the traditional notation while also using distinct symbols to explicitly indicate the functional form being used. Thus, the vertical bar notation is, strictly, redundant and can be ignored if so desired.

is the *molar heat capacity at constant volume*.[50] The molar heat capacity of an ideal gas is a universal constant. For a *monoatomic* ideal gas it is $C_v = \frac{3}{2}N_A k_B = 12.472\,\text{J}\cdot\text{K}^{-1}\cdot\text{mol}^{-1}$, where $k_B = 1.3807 \times 10^{-23}$ J/K is Boltzmann's constant (see Exercise 7.8 for a derivation based on statistical mechanics). For a real gas C_v is a material constant which can depend on the equilibrium state. A closely related property to C_v is the *specific heat capacity at constant volume*, which is the amount of heat required to change the temperature of a unit mass of material by 1 degree. The specific heat capacity c_v is related to the molar heat capacity C_v, through

$$c_v = \frac{C_v}{M}, \tag{2.108}$$

where M is the molar mass (the mass of one mole of the substance).

For a Joule expansion corresponding to an infinitesimal increase of volume dV at constant mole number, the first law requires $d\mathcal{U} = 0$. Joule's experiments showed that the temperature of the gas remained constant as it expanded ($dT = 0$), therefore the first and third terms of the differential in Eqn. (2.106) drop out and we have[51]

$$\left.\frac{\partial \widehat{\mathcal{U}}}{\partial V}\right|_{N,T} = 0. \tag{2.109}$$

This is an important result, since it indicates that the internal energy of an ideal gas does not depend on volume:

$$\mathcal{U} = \widehat{\mathcal{U}}(n, V, T) = n\mathcal{U}_0 + nC_v T. \tag{2.110}$$

Here the number of moles n has been used to specify the amount of gas (instead of the number of particles N) and $\mathcal{U}_0$ is the molar internal energy of an ideal gas at zero temperature. This is called *Joule's law*. It is exact for ideal gases, by definition, and provides a good approximation for real gases at low pressures.

Joule's law is an example of an equation of state as defined in Eqn. (2.102). Of course, other choices for the independent state variables could be made. For example instead of n, V and T, we can choose to work with n, V and $\mathcal{U}$, as the independent variables, in which case the equation of state for the ideal gas would be

$$T = \widehat{T}(n, V, \mathcal{U}) = (\mathcal{U} - n\mathcal{U}_0)/nC_v.$$

[50] Formally, the molar heat capacity of a gas at constant volume is defined as $C_v = (1/n)(\Delta Q_V/\Delta T)$, where ΔQ_V is the heat transferred under conditions of constant volume and n is the constant number of moles of gas. This is the amount of heat required to change the temperature of 1 mole of material by 1 degree. For a fixed amount of gas at constant volume, the first law reduces to $\Delta\mathcal{U} = \Delta Q$ (since no mechanical work is done on the gas), therefore the molar heat capacity is also $C_v = (1/n)(\partial\widehat{\mathcal{U}}/\partial T)|_{N,V}$. Similar properties can be defined for a change due to temperature at constant pressure and changes due to other state variables. See [Adk83, Section 3.6] for a full discussion.

[51] Actually, the temperature of a real gas does change in free expansion. However, the effect is weak and Joule's experiments lacked the precision to detect it. For an ideal gas, the change in temperature is identically zero. See Exercise 7.9 for proof of this using statistical mechanics.

Another possibility is to use n, p and T as the independent state variables, where p is the pressure – the thermodynamic tension associated with the volume as defined in Eqn. (2.126). In this case the internal energy would be expressed as $\mathcal{U} = \widetilde{\mathcal{U}}(n,p,T)$. It is important to understand that in this case the internal energy would *not* be given by Eqn. (2.110). It would depend explicitly on the pressure. See Section 7.3.5 for a derivation of the equations of state for an ideal gas using statistical mechanics.

2.4.4 Thermodynamic processes

States of thermodynamic equilibrium are of great interest, but the true power of the theory of thermodynamics is its ability to predict the state to which a system will transition when it is perturbed from a known state of equilibrium. In fact, it is often of interest to predict an entire series of equilibrium states that will occur when a system is subjected to a given series of perturbations.

General thermodynamic process We define a *thermodynamic process* as an ordered set or sequence of equilibrium states. This set need not correspond to any actual series followed by a real system. It is simply a string of possible equilibrium states for a system. For example for system B with independent state variables $\mathcal{B} = (\mathcal{B}_1, \mathcal{B}_2, \dots, \mathcal{B}_{\nu^{\mathrm{B}}})$, a thermodynamic process containing M states is denoted by

$$\mathfrak{B} = (\mathcal{B}^{(1)}, \mathcal{B}^{(2)}, \dots, \mathcal{B}^{(M)}),$$

where $\mathcal{B}^{(i)} = (\mathcal{B}_1^{(i)}, \mathcal{B}_2^{(i)}, \dots, \mathcal{B}_{\nu^{\mathrm{B}}}^{(i)})$ is the ith state in the thermodynamic process. The behavior of the dependent state variables follows through the appropriate equations of state. A general thermodynamic process can have any number of states M and there is no requirement that consecutive states in the process are close to each other. That is, the values of the independent state variables for stages i and $i+1$, $\mathcal{B}_\alpha^{(i)}$ and $\mathcal{B}_\alpha^{(i+1)}$, respectively, of a thermodynamic process need not be related in any way.

Quasistatic process Although the laws of thermodynamics apply equally to all thermodynamic processes, those processes that correspond to a sequence of successive small increments to the independent state variables are of particular interest. In the limit, as the increments become infinitesimal, the process becomes a continuous path in the *thermodynamic state space* (the ν^{B}-dimensional space of independent state variables):

$$\mathfrak{B} = \mathcal{B}(s), \qquad s \in [0,1].$$

Here functional notation is used to indicate the continuous variation of the independent state variables and s is used as a convenient variable to measure the "location" along the process.[52] Such a process is called *quasistatic*.

Quasistatic processes are singularly useful within the theory of thermodynamics for two reasons. First, such processes can be associated with phenomena in the real world, where small perturbations applied to a system occur on a time scale that is significantly slower

[52] The choice of domain for s is arbitrary and the unit interval, used here, bears no special significance.

than that required for the system to reach equilibrium. In the limit as the perturbation rate becomes infinitely slower than the equilibration rate, the thermodynamic process becomes quasistatic. Technically, no real phenomena are quasistatic since the time required for a system to reach true equilibrium is infinite. However, in many cases the dynamical processes that lead to equilibrium are sufficiently fast for the thermodynamic process to be approximately quasistatic. This is particularly the case if we relax the condition for thermodynamic equilibrium and accept metastable equilibrium instead. Indeed, nature is replete with examples of physical phenomena that can be accurately analyzed within thermodynamic theory when they are approximated as quasistatic processes.

Second, general results of the theory are best expressed in terms of infinitesimal changes of state. These results may then be integrated along any quasistatic process in order to obtain predictions of the theory for finite changes of state. The expressions associated with such finite changes of state are almost always considerably more complex than their infinitesimal counterparts and often are only obtainable in explicit form once the equations of state for a particular material are introduced.

The first law of thermodynamics speaks of the conservation of energy during thermodynamic processes, but it tells us nothing about the *direction* of such processes. How is it that if we watch a movie of a shattered glass leaping onto a table and reassembling, we immediately know that it is being played in reverse? The first law provides no answer – it can be satisfied for any process. Enter the second law of thermodynamics.

2.4.5 The second law of thermodynamics and the direction of time

The first law of thermodynamics tells us that the total energy of a system and the rest of the world is conserved. The flux of mechanical and thermal energy into and out of the system is balanced by the change in its energy. As far as we know, this is a basic property of the universe we inhabit, but it does not tell the whole story. Consider the following scenario:

1. A rigid hollow sphere filled with an ideal gas is placed inside of a larger, otherwise empty, sealed box that is thermally isolated from its surroundings.
2. A hole is opened in the sphere.
3. The gas quickly expands to fill the box.
4. After some time, the gas spontaneously returns, through the hole, to occupy only its original volume within the sphere.

This scenario is perfectly legal from the perspective of the first law. In fact, we showed in our discussion of Joule's experiments in Section 2.4.3 that the internal energy of an ideal gas remains unchanged by the free expansion in step 3. It is therefore clearly *not* a violation of the first law for the gas to return to its initial state. However, our instincts, based on our familiarity with the world, tell us that this process of "reverse-expansion" will never happen – the process has a unique *direction*. In fact, we can relate this directionality of thermodynamic processes to our concept of time and why we perceive that time always evolves from the "present" to the "future" and never from the "present" to the "past."

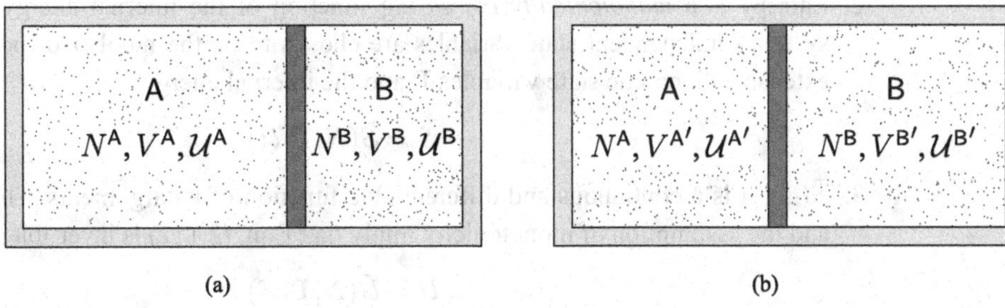

Fig. 2.9 An isolated system consisting of a rigid, sealed and thermally-isolated cylinder of total volume V, an internal frictionless, impermeable piston and two subsystems A and B containing ideal gases. (a) Initially the piston is fixed and thermally insulating and the gases are in thermodynamic equilibrium. (b) The new states of thermodynamic equilibrium obtained following a dynamical process once the piston becomes diathermal and is allowed to move.

Clearly, something in addition to the first law is necessary to describe the directionality of thermodynamic processes.

Entropy First let us set the scene for a discussion of the directionality of thermodynamic processes. Suppose we have a rigid, sealed and thermally isolated cylinder of volume V with a frictionless and impermeable internal piston that divides it into two compartments, A and B of initial volume V^{A} and $V^{\mathsf{B}} = V - V^{\mathsf{A}}$, respectively as shown in Fig. 2.9(a). Initially, the piston is fixed in place and thermally isolating. Compartment A is filled with N^{A} particles of an ideal gas with internal energy $\mathcal{U}^{\mathsf{A}}$ and compartment B is filled with N^{B} particles of another ideal gas with internal energy $\mathcal{U}^{\mathsf{B}}$. As long as the piston remains fixed and thermally insulating, A and B are isolated systems. If we consider the entire cylinder as a single isolated thermodynamic system consisting of two subsystems, the piston represents a set of *internal constraints*. We are interested in answering the following questions: If we release the constraints by allowing the piston to move and to transmit heat, in what direction will the piston move? How far will it move? And, why is the reverse process never observed, i.e. why does the piston never return to its original position?

Since nothing in our theory so far is able to provide the answers to these questions, we postulate the existence of a new state variable, related to the direction of thermodynamic processes, which we call *entropy*.[53] We will show below that requiring this variable to satisfy a simple extremum principle (the second law of thermodynamics) is sufficient to endow the theory with enough structure to answer all of the above questions. We denote entropy by the symbol $\mathcal{S}$ and assume that it has the following properties [Cal85]:

1. Entropy is extensive, therefore the entropy of a collection of systems is equal to the sum of their entropies,

$$\mathcal{S}^{\mathsf{A+B+C+\cdots}} = \mathcal{S}^{\mathsf{A}} + \mathcal{S}^{\mathsf{B}} + \mathcal{S}^{\mathsf{C}} + \cdots. \tag{2.111}$$

[53] The word *entropy* was coined in 1865 by the German physicist Rudolf Clausius as a combination of the Greek words *en* meaning *in* and *tropē* meaning *change or turn*.

2. Entropy is a *monotonically increasing* function of the internal energy $\mathcal{U}$, when the system's independent state variables are chosen to be the number of particles N, the extensive kinematic state variables $\mathbf{\Gamma}$ and the internal energy $\mathcal{U}$:

$$\mathcal{S} = \bar{\mathcal{S}}(N, \mathbf{\Gamma}, \mathcal{U}). \tag{2.112}$$

3. $\bar{\mathcal{S}}(\cdot,\cdot,\cdot)$ is a continuous and differentiable function of its arguments. This assumption and the assumption of monotonicity imply that Eqn. (2.112) is invertible, i.e.

$$\mathcal{U} = \bar{\mathcal{U}}(N, \mathbf{\Gamma}, \mathcal{S}). \tag{2.113}$$

Second law of thermodynamics The direction of a process can be expressed as a constraint on the way entropy can change during it. This is what the *second law of thermodynamics* is about. There are many equivalent ways that this law can be stated. We choose the statement attributed to Rudolf Clausius, which we find to be physically most transparent.

> *Second law of thermodynamics* An isolated system in thermodynamic equilibrium adopts the state that has the maximum entropy of all states consistent with the imposed kinematic constraints.

Let us see how the second law would apply for the cylinder with an internal piston shown in Fig. 2.9. The second law tells us that once the internal constraints are removed and the piston is allowed to move and to transmit heat the system will evolve to a new state (Fig. 2.9(b)) that maximizes its entropy. That is, we must consider changes to the system's unconstrained state variables, say ΔV^{A} and $\Delta\mathcal{U}^{\mathsf{A}}$ (which will imply the corresponding changes to V^{B} and $\mathcal{U}^{\mathsf{B}}$ that are imposed by the conservation constraints for the composite system), in order to find the maximum possible value for the entropy of the isolated composite system. Thus, the equilibrium value of entropy $\mathcal{S}'$ for the isolated composite system will be (assuming the piston is impermeable)[54]

$$\mathcal{S}' = \max_{0 \le V^{\mathsf{A}'} \le V} \left(\max_{\mathcal{U}^{\mathsf{A}'} \in \mathbb{R}} \left[\bar{\mathcal{S}}^{\mathsf{A}}(N^{\mathsf{A}}, V^{\mathsf{A}'}, \mathcal{U}^{\mathsf{A}'}) + \bar{\mathcal{S}}^{\mathsf{B}}(N^{\mathsf{B}}, V - V^{\mathsf{A}'}, \mathcal{U}^{\mathsf{A}} + \mathcal{U}^{\mathsf{B}} - \mathcal{U}^{\mathsf{A}'}) \right] \right),$$

where $\bar{\mathcal{S}}^{\mathsf{A}}(\cdot,\cdot,\cdot)$ and $\bar{\mathcal{S}}^{\mathsf{B}}(\cdot,\cdot,\cdot)$ are the entropy functions for the ideal gases of A and B, respectively.[55] The value of $V^{\mathsf{A}'}$ obtained from the above maximization problem determines the final position of the piston, and thus, provides the answers to the questions posed earlier in this section. In particular, we see that any change of the volume of A away from the equilibrium value $V^{\mathsf{A}'}$ must necessarily result in a decrease of the total entropy. As we will see next, this would violate the second law of thermodynamics. This is why real thermodynamic processes have a unique direction and are never observed to occur in reverse.

[54] Because the total volume and energy are conserved, we have $V^{\mathsf{B}'} = V - V^{\mathsf{A}'}$ and $\mathcal{U}^{\mathsf{B}'} = \mathcal{U} - \mathcal{U}^{\mathsf{A}'} = \mathcal{U}^{\mathsf{A}} + \mathcal{U}^{\mathsf{B}} - \mathcal{U}^{\mathsf{A}'}$.

[55] Note that, although the internal energy is extensive, it is not required to be positive. In fact, in principle $\mathcal{U}^{\mathsf{A}'}$ may take on any value as long as $\mathcal{U}^{\mathsf{B}'}$ is then chosen to ensure conservation of energy. Thus, the maximization with respect to energy considers all possible values of $\mathcal{U}^{\mathsf{A}'}$.

It is useful to rephrase the second law in an alternative manner.

Second law of thermodynamics (alternative statement) The entropy of an isolated system can never decrease in any process. It can only increase or stay the same.

Mathematically this statement is

$$\Delta\mathcal{S} \geq 0, \tag{2.114}$$

for any isolated system that transitions from one equilibrium state to another in response to the release of an internal constraint. It is trivial to show that the Clausius statement of the second law leads to this conclusion. Consider a process that begins in state 1 and ends in state 2. The Clausius statement of the second law tells us that $\mathcal{S}^{(2)} \geq \mathcal{S}^{(1)}$, therefore $\Delta\mathcal{S} = \mathcal{S}^{(2)} - \mathcal{S}^{(1)} \geq 0$, which is exactly Eqn. (2.114).

Note that the statements of the second law given above have been careful to stress that the law only holds for *isolated* systems. The entropy of a system that is not isolated can and often does decrease in a process. We will see this later.

Thermal equilibrium from an entropy perspective In order to see the connection between entropy and the other thermodynamic state variables whose physical significance is more clear to us (e.g. temperature, volume, and internal energy), we revisit the problem of thermal equilibrium discussed earlier in Section 2.4.2.

Let C be an isolated thermodynamic system made up of two subsystems, A and B, that are composed of (possibly different) materials. We take the independent state variables for each system to be the number of particles N, extensive kinematic state variables $\boldsymbol{\Gamma}$ and the internal energy $\mathcal{U}$. Since C is isolated, according to the first law its internal energy is conserved, i.e. $\mathcal{U}^{\mathsf{C}} = \mathcal{U}^{\mathsf{A}} + \mathcal{U}^{\mathsf{B}} =$ constant. This means that any change in internal energy of subsystem A must be matched by an equal and opposite change in B:

$$\Delta\mathcal{U}^{\mathsf{A}} + \Delta\mathcal{U}^{\mathsf{B}} = 0. \tag{2.115}$$

Like the internal energy, entropy is also extensive and therefore the total entropy of the composite system is $\mathcal{S}^{\mathsf{C}} = \mathcal{S}^{\mathsf{A}} + \mathcal{S}^{\mathsf{B}}$. However, entropy is generally not constant in a change of state of an isolated system. The total entropy is a function of the two subsystems' state variables $N^{\mathsf{A}}, \boldsymbol{\Gamma}^{\mathsf{A}}, \mathcal{U}^{\mathsf{A}}, N^{\mathsf{B}}, \boldsymbol{\Gamma}^{\mathsf{B}}$ and $\mathcal{U}^{\mathsf{B}}$. The first differential of the total entropy is then[56]

$$\begin{aligned} d\mathcal{S}^{\mathsf{C}} = {} & \left.\frac{\partial\bar{\mathcal{S}}^{\mathsf{A}}}{\partial N^{\mathsf{A}}}\right|_{\boldsymbol{\Gamma}^{\mathsf{A}},\mathcal{U}^{\mathsf{A}}} dN^{\mathsf{A}} + \sum_{\alpha} \left.\frac{\partial\bar{\mathcal{S}}^{\mathsf{A}}}{\partial\Gamma^{\mathsf{A}}_{\alpha}}\right|_{N^{\mathsf{A}},\mathcal{U}^{\mathsf{A}}} d\Gamma^{\mathsf{A}}_{\alpha} + \left.\frac{\partial\bar{\mathcal{S}}^{\mathsf{A}}}{\partial\mathcal{U}^{\mathsf{A}}}\right|_{N^{\mathsf{A}},\boldsymbol{\Gamma}^{\mathsf{A}}} d\mathcal{U}^{\mathsf{A}} \\ & + \left.\frac{\partial\bar{\mathcal{S}}^{\mathsf{B}}}{\partial N^{\mathsf{B}}}\right|_{\boldsymbol{\Gamma}^{\mathsf{B}},\mathcal{U}^{\mathsf{B}}} dN^{\mathsf{B}} + \sum_{\beta} \left.\frac{\partial\bar{\mathcal{S}}^{\mathsf{B}}}{\partial\Gamma^{\mathsf{B}}_{\beta}}\right|_{N^{\mathsf{B}},\mathcal{U}^{\mathsf{B}}} d\Gamma^{\mathsf{B}}_{\beta} + \left.\frac{\partial\bar{\mathcal{S}}^{\mathsf{B}}}{\partial\mathcal{U}^{\mathsf{B}}}\right|_{N^{\mathsf{B}},\boldsymbol{\Gamma}^{\mathsf{B}}} d\mathcal{U}^{\mathsf{B}}. \end{aligned}$$

Suppose we fix the values of A's kinematic state variables $\boldsymbol{\Gamma}^{\mathsf{A}}$ and its number of particles N^{A} (then the corresponding values for B are determined by constraints imposed by C's

[56] The notation $(\partial\bar{\mathcal{S}}/\partial\Gamma_\alpha)|_{N,\mathcal{U}}$ refers to the partial derivative of the function $\bar{\mathcal{S}}(N, \boldsymbol{\Gamma}, \mathcal{U})$ with respect to the αth component of $\boldsymbol{\Gamma}$ (while holding all other components of $\boldsymbol{\Gamma}$, N, and $\mathcal{U}$ fixed). We leave out the remaining components of $\boldsymbol{\Gamma}$ from the list at the bottom of the bar in order to avoid extreme notational clutter.

isolation), but allow energy (heat) transfer between A and B. Then the terms involving the increments of the extensive kinematic state variables and the increments of the particle numbers drop out. Further, since C is isolated the internal energy increments must satisfy Eqn. (2.115), so likewise $d\mathcal{U}^{\mathrm{A}} = -d\mathcal{U}^{\mathrm{B}}$. All of these considerations lead to the following expression for the differential of the entropy of system C:

$$d\mathcal{S}^{\mathrm{C}} = \left[\left.\frac{\partial \bar{\mathcal{S}}^{\mathrm{A}}}{\partial \mathcal{U}^{\mathrm{A}}}\right|_{N^{\mathrm{A}},\boldsymbol{\Gamma}^{\mathrm{A}}} - \left.\frac{\partial \bar{\mathcal{S}}^{\mathrm{B}}}{\partial \mathcal{U}^{\mathrm{B}}}\right|_{N^{\mathrm{B}},\boldsymbol{\Gamma}^{\mathrm{B}}} \right] d\mathcal{U}^{\mathrm{A}}. \tag{2.116}$$

Now, according to our definition in Section 2.4.2, A and B are in thermal equilibrium if they remain in equilibrium when brought into thermal contact. This implies that the composite system C, subject to the above conditions, is in thermodynamic equilibrium when A and B are in thermal equilibrium. Thus, according to the second law of thermodynamics, the first differential of the entropy, Eqn. (2.116), must be zero for all $d\mathcal{U}^{\mathrm{A}}$ in this case (since the entropy is at a maximum). This leads to

$$\left.\frac{\partial \bar{\mathcal{S}}^{\mathrm{A}}}{\partial \mathcal{U}^{\mathrm{A}}}\right|_{N^{\mathrm{A}},\boldsymbol{\Gamma}^{\mathrm{A}}} = \left.\frac{\partial \bar{\mathcal{S}}^{\mathrm{B}}}{\partial \mathcal{U}^{\mathrm{B}}}\right|_{N^{\mathrm{B}},\boldsymbol{\Gamma}^{\mathrm{B}}} \tag{2.117}$$

as the condition for thermal equilibrium between A and B in terms of their entropy functions. Now recall from Eqn. (2.103) that thermal equilibrium requires $T^{\mathrm{A}} = T^{\mathrm{B}}$ or equivalently $1/T^{\mathrm{A}} = 1/T^{\mathrm{B}}$. Here, we are referring explicitly to the thermodynamic temperature scale. Comparing these with the equation above it is clear that $\partial\bar{\mathcal{S}}/\partial\mathcal{U}$ is either[57] T or $1/T$. To decide which is the correct definition, we recall that the concept of temperature also included the idea of "hotter than." Thus, we must test which of the above options is consistent with our definition that if $T^{\mathrm{A}} > T^{\mathrm{B}}$, then heat (energy) will spontaneously flow from A to B when they are put into thermal contact.

To do this, consider the same combination of systems as before, and now assume that initially A has a higher temperature than B, i.e. $T^{\mathrm{A}} > T^{\mathrm{B}}$. Since the composite system is isolated, our definition of temperature and the first law of thermodynamics imply that heat will flow from A to B which will result in a decrease of $\mathcal{U}^{\mathrm{A}}$ and a correspondingly equal increase of $\mathcal{U}^{\mathrm{B}}$. However, the second law says that such a change of state can only occur if it increases the total entropy of the isolated composite system. Thus, we must have that

$$d\mathcal{S}^{\mathrm{C}} = \left[\left.\frac{\partial \bar{\mathcal{S}}^{\mathrm{A}}}{\partial \mathcal{U}^{\mathrm{A}}}\right|_{N^{\mathrm{A}},\boldsymbol{\Gamma}^{\mathrm{A}}} - \left.\frac{\partial \bar{\mathcal{S}}^{\mathrm{B}}}{\partial \mathcal{U}^{\mathrm{B}}}\right|_{N^{\mathrm{B}},\boldsymbol{\Gamma}^{\mathrm{B}}} \right] d\mathcal{U}^{\mathrm{A}} > 0.$$

Since we expect $d\mathcal{U}^{\mathrm{A}} < 0$, this implies that

$$\left.\frac{\partial \bar{\mathcal{S}}^{\mathrm{A}}}{\partial \mathcal{U}^{\mathrm{A}}}\right|_{N^{\mathrm{A}},\boldsymbol{\Gamma}^{\mathrm{A}}} < \left.\frac{\partial \bar{\mathcal{S}}^{\mathrm{B}}}{\partial \mathcal{U}^{\mathrm{B}}}\right|_{N^{\mathrm{B}},\boldsymbol{\Gamma}^{\mathrm{B}}}. \tag{2.118}$$

[57] Instead of T or $1/T$ any monotonically increasing or decreasing functions of T would do. We discuss this further below.

The derivatives in Eqn. (2.118) are required to be nonnegative by the monotonically increasing nature of the entropy (see property 2 on page 74). Therefore since $T^{\mathsf{A}} > T^{\mathsf{B}}$, the definition that satisfies Eqn. (2.118) is[58]

$$\left.\frac{\partial \overline{\mathcal{S}}}{\partial \mathcal{U}}\right|_{N,\Gamma} = \frac{1}{T}, \tag{2.119}$$

where $\mathcal{S}, \mathcal{U}$ and T refer to either system A or system B. The inverse relation is

$$\left.\frac{\partial \overline{\mathcal{U}}}{\partial \mathcal{S}}\right|_{N,\Gamma} = T. \tag{2.120}$$

Equations (2.119) and (2.120) provide the key link between entropy, temperature and the internal energy.

To ensure that the extremum point at which $d\mathcal{S}^{\mathsf{C}} = 0$ is a maximum, we must also require $d^2\mathcal{S}^{\mathsf{C}} \leq 0$. Physically, this means that the system is in a state of stable equilibrium. Let us explore the physical restrictions imposed by this requirement. The second differential follows from Eqn. (2.116) as

$$d^2\mathcal{S}^{\mathsf{C}} = \left[\left.\frac{\partial^2 \overline{\mathcal{S}}^{\mathsf{A}}}{\partial(\mathcal{U}^{\mathsf{A}})^2}\right|_{N^{\mathsf{A}},\Gamma^{\mathsf{A}}} + \left.\frac{\partial^2 \overline{\mathcal{S}}^{\mathsf{B}}}{\partial(\mathcal{U}^{\mathsf{B}})^2}\right|_{N^{\mathsf{B}},\Gamma^{\mathsf{B}}}\right](d\mathcal{U}^{\mathsf{A}})^2. \tag{2.121}$$

Now note that the following expression holds for systems A and B:

$$\left.\frac{\partial^2 \overline{\mathcal{S}}}{\partial \mathcal{U}^2}\right|_{N,\Gamma} = \left.\frac{\partial}{\partial \mathcal{U}}\right|_{N,\Gamma}\left.\frac{\partial \overline{\mathcal{S}}}{\partial \mathcal{U}}\right|_{N,\Gamma} = \left.\frac{\partial}{\partial \mathcal{U}}\right|_{N,\Gamma}\left(\frac{1}{T}\right) = -\frac{1}{T^2}\left.\frac{\partial \widehat{T}}{\partial \mathcal{U}}\right|_{N,\Gamma} = -\frac{1}{T^2 n C_v}, \tag{2.122}$$

where we have used Eqn. (2.119), C_v is the molar heat capacity at constant volume defined in Eqn. (2.107) and n is the number of moles. Substituting Eqn. (2.122) into Eqn. (2.121) and noting that at equilibrium both systems are at the same temperature T, we have

$$d^2\mathcal{S}^{\mathsf{C}} = -\frac{1}{T^2}\left[\frac{1}{n^{\mathsf{A}}C_v^{\mathsf{A}}} + \frac{1}{n^{\mathsf{B}}C_v^{\mathsf{B}}}\right](d\mathcal{U}^{\mathsf{A}})^2. \tag{2.123}$$

Since n^{A} and n^{B} are positive and arbitrary, the condition $d^2\mathcal{S}^{\mathsf{C}} \leq 0$ is satisfied provided that $C_v^{\mathsf{A}} \geq 0$ and $C_v^{\mathsf{B}} \geq 0$. We see that the heat capacity at constant volume of a material

58 As noted above, any monotonically decreasing function would do here, i.e. $\partial\overline{\mathcal{S}}/\partial\mathcal{U} = f_-(T)$. The choice of a particular function can be interpreted in many ways. From the above point of view the choice defines what entropy is in terms of the temperature. From another point of view, where we apply the inverse function to obtain $f_-^{-1}(\partial\overline{\mathcal{S}}/\partial\mathcal{U}) = T$, it defines the temperature scale in terms of the entropy. It turns out that the definition selected here provides a clear physical significance to both the thermodynamic temperature and the entropy. This is discussed further from a microscopic perspective in Section 7.3.4. When viewed from the macroscopic perspective the thermodynamic temperature scale is naturally related to the behavior of ideal gases as shown below starting on page 79.

must be a positive number if that material is to have the ability to reach thermal equilibrium with another system. Given Eqn. (2.122), this implies that $\partial^2\overline{\mathcal{S}}/\partial\mathcal{U}^2 \leq 0$, i.e. $\overline{\mathcal{S}}(N,\mathbf{\Gamma},\mathcal{U})$ must be a locally *concave* function of $\mathcal{U}$. Using similar reasoning it can be shown that it must also be locally concave with respect to its other variables. An even stronger conclusion can be reached that for stable materials the entropy function must be *globally* concave with respect to its arguments. This means that the concavity condition is satisfied not just for infinitesimal changes to the arguments but arbitrary finite changes as well. See Section 5.5.3 in [TME12] for details.

The introduction of entropy almost seems like the sleight of hand of a talented magician. This variable was introduced without any physical indication of what it could be. It was then tied to the internal energy and temperature through the thought experiment described above. However, this does not really provide a greater sense of what entropy actually is. An answer to that question will have to wait until the discussion of statistical mechanics in Chapter 7, where we make the connection between the dynamics of the atoms making up a physical system and the thermodynamic variables introduced here. In particular, in Section 7.3.4, we show that entropy has a very clear and, in retrospect, almost obvious significance. It is a measure of the number of microscopic kinematic vectors (microscopic states) that are consistent with a given set of macroscopic state variables. Equilibrium is thus simply the macroscopic state that has the most microscopic states associated with it and is therefore most likely to be observed. This is what entropy is measuring.

Internal energy and entropy as fundamental thermodynamic relations The entropy function $\overline{\mathcal{S}}(N,\mathbf{\Gamma},\mathcal{U})$ and the closely related internal energy function $\overline{\mathcal{U}}(N,\mathbf{\Gamma},\mathcal{S})$ are known as *fundamental relations* for a thermodynamic system. From them we can obtain all possible information about the system when it is in any state of thermodynamic equilibrium. In particular, we can obtain all of the equations of state for a system from the internal energy fundamental relation. As we saw in the previous section, the temperature is given by the derivative of the internal energy with respect to the entropy. This can, in fact, be viewed as the definition of the temperature, and in a similar manner we can *define* a state variable associated with each argument of the internal energy function. These are the intensive state variables that were introduced on page 63. So, we have the following definitions.

1. Temperature

$$T = \overline{T}(N,\mathbf{\Gamma},\mathcal{S}) \equiv \left.\frac{\partial\overline{\mathcal{U}}}{\partial\mathcal{S}}\right|_{N,\mathbf{\Gamma}}. \tag{2.124}$$

2. Thermodynamic tensions

$$\gamma_\alpha = \overline{\gamma}_\alpha(N,\mathbf{\Gamma},\mathcal{S}) \equiv \left.\frac{\partial\overline{\mathcal{U}}}{\partial\Gamma_\alpha}\right|_{N,\mathcal{S}}, \quad \alpha = 1,2,\ldots,n^\Gamma. \tag{2.125}$$

A special case is where the volume is the kinematic state variable of interest, say $\Gamma_1 = V$. In this case we introduce a negative sign and give the special name, *pressure*, and symbol, $p \equiv -\gamma_1$, to the associated thermodynamic tension. The negative sign is introduced so that, in accordance with our intuitive understanding of the concept, the pressure is positive and increases with decreasing volume. Thus, the definition of the pressure is

$$p = \bar{p}(N, \mathbf{\Gamma}, \mathcal{S}) \equiv -\left.\frac{\partial \bar{\mathcal{U}}}{\partial V}\right|_{N,\mathcal{S}}. \tag{2.126}$$

In general we will refer to the entire set of thermodynamic tensions with the symbol γ.

3. Chemical potential

$$\mu = \bar{\mu}(N, \mathbf{\Gamma}, \mathcal{S}) \equiv \left.\frac{\partial \bar{\mathcal{U}}}{\partial N}\right|_{\mathbf{\Gamma},\mathcal{S}}. \tag{2.127}$$

It is clear that each of the above defined quantities is intensive because each is given by the ratio of two extensive quantities. Thus, the dependence on amount cancels and we obtain a quantity that is independent of amount.

Fundamental relation for an ideal gas and the ideal gas law Recall that in Section 2.4.3 we found the internal energy of an ideal gas as a function of the mole number, the volume and the temperature:

$$\mathcal{U} = \widehat{\mathcal{U}}(n, V, T) = n\mathcal{U}_0 + nC_v T,$$

where $\mathcal{U}_0$ is the energy per mole of the gas at zero temperature. However, this equation is not a fundamental relation because it is not given in terms of the correct set of independent state variables. It is easy to see this. For instance, the derivative of this function with respect to the volume is zero. Clearly the pressure is not zero for all equilibrium states of an ideal gas. In order to obtain all thermodynamic information about an ideal gas we need the internal energy expressed as a function of the number of particles (or equivalently the mole number), the volume and the entropy. This functional form can be obtained from the statistical mechanics derivation in Section 7.3.5 or the classic thermodynamic approach in [Cal85, Section 3.4]. Taking the arbitrary datum of energy to be $\mathcal{U}_0 = 0$, we obtain

$$\mathcal{U} = \bar{\mathcal{U}}(n, V, \mathcal{S}) = Kn \exp\left(\frac{\mathcal{S}}{nC_v}\right)\left(\frac{V}{n}\right)^{-R_g/C_v}, \tag{2.128}$$

where K is a constant and $R_g = k_B N_A$ is the universal gas constant. Here, k_B is Boltzmann's constant and N_A is Avogadro's constant.[59] From this fundamental relation we can obtain all of the equations of state for the intensive state variables.

1. Chemical potential

$$\mu = \bar{\mu}(n, V, \mathcal{S}) = \frac{\partial \bar{\mathcal{U}}}{\partial n} = K \exp\left(\frac{\mathcal{S}}{nC_v}\right)\left(\frac{V}{n}\right)^{-R_g/C_v}\left[1 + \frac{R_g}{C_v} - \frac{\mathcal{S}}{nC_v}\right].$$

2. Pressure

$$p = \bar{p}(n, V, \mathcal{S}) = -\frac{\partial \bar{\mathcal{U}}}{\partial V} = K \exp\left(\frac{\mathcal{S}}{nC_v}\right)\left(\frac{V}{n}\right)^{-(\frac{R_g}{C_v}+1)}\frac{R_g}{C_v}.$$

[59] Recall that $k_B = 8.617 \times 10^{-5}$ eV/K $= 1.3807 \times 10^{-23}$ J/K and $N_A = 6.022 \times 10^{23}$ mol^{-1}.

3. Temperature

$$T = \bar{T}(n, V, \mathcal{S}) = \frac{\partial \bar{\mathcal{U}}}{\partial \mathcal{S}} = Kn \exp\left(\frac{\mathcal{S}}{nC_v}\right)\left(\frac{V}{n}\right)^{-R_g/C_v} \frac{1}{nC_v}.$$

We may now recover from these functions the original internal energy function and the *ideal gas law* by eliminating the entropy from the equation for the pressure and the temperature. First, notice that the temperature contains a factor which is equal to the internal energy Eqn. (2.128), giving $T = \mathcal{U}/nC_v$. From this we may solve for the internal energy and immediately obtain Eqn. (2.110) (where we recall that we have chosen $\mathcal{U}_0 = 0$ as the energy datum). Next we recognize that the equation for the pressure can be written $p = \mathcal{U}(1/V)(R_g/C_v)$. Substituting the relation we just obtained for the internal energy in terms of the temperature, we find that $p = nR_gT/V$ or

$$pV = nR_gT, \tag{2.129}$$

which is the *ideal gas law* that is familiar from introductory physics and chemistry courses.

From the ideal gas law we can obtain a physical interpretation of the thermodynamic temperature scale referred to after Eqn. (2.103) and defined in Eqn. (2.124). Since all gases behave like ideal gases as the pressure goes to zero,[60] gas thermometers provide a unique temperature scale at low pressure [Adk83]:

$$T = \lim_{p \to 0} \frac{pV}{nR_g}.$$

The value of the ideal gas constant R_g appearing in this relation (and by extension, the value of Boltzmann's constant k_B) is set by defining the thermodynamic temperature $T = 273.16$ K to be the triple point of water (see page 66).

Entropy form of the first law The above definitions for the intensive state variables allow us to obtain a very useful interpretation of the first law of thermodynamics in the context of a quasistatic process. Consider the first differential of internal energy:

$$d\mathcal{U} = \left.\frac{\partial \bar{\mathcal{U}}}{\partial N}\right|_{\Gamma,\mathcal{S}} dN + \sum_\alpha \left.\frac{\partial \bar{\mathcal{U}}}{\partial \Gamma_\alpha}\right|_{N,\mathcal{S}} d\Gamma_\alpha + \left.\frac{\partial \bar{\mathcal{U}}}{\partial \mathcal{S}}\right|_{N,\Gamma} d\mathcal{S}.$$

Substituting in Eqns. (2.124), (2.125) and (2.127), we obtain the result

$$d\mathcal{U} = \mu dN + \sum_\alpha \gamma_\alpha d\Gamma_\alpha + T d\mathcal{S}. \tag{2.130}$$

[60] See also Section 2.4.3 where ideal gases are defined and discussed.

Restricting our attention to the case where the number of particles is fixed, the first term in the differential drops out and we find

$$d\mathcal{U} = \sum_{\alpha} \gamma_\alpha d\Gamma_\alpha + TdS. \tag{2.131}$$

Equations (2.130) and (2.131) are called the *entropy form of the first law* of thermodynamics. If we compare the second of these equations with the first law in Eqn. (2.104), it is natural to make the identifications[61]

$$đ\mathcal{W}^{\text{def}} = \sum_{\alpha} \gamma_\alpha d\Gamma_\alpha, \qquad đ\mathcal{Q} = TdS, \tag{2.132}$$

which are increments of *quasistatic work* and *quasistatic heat*, respectively. These variables identify the work performed on the system and the heat transferred to the system as it undergoes a quasistatic process in which the number of particles remain constant, the kinematic state variables experience increments $d\mathbf{\Gamma}$ and the entropy is incremented by dS. In fact, we will take Eqn. (2.132) as an additional defining property of quasistatic processes.

An important special case is that of a thermally-isolated system undergoing a quasistatic process. In this situation there is no heat transferred to the system, $đ\mathcal{Q} = 0$. Since the temperature will generally not be zero, the only way that this can be true is if $dS = 0$ for the system. Thus, we have found that when a thermally-isolated system undergoes a quasistatic process its entropy remains constant, and we say the process is *adiabatic*.

Based on the identification of work as the product of a thermodynamic tension with its associated kinematic state variable, it is common to refer to these quantities as *work conjugate* or simply *conjugate* pairs. Thus, we say that γ_α and Γ_α are work conjugate, or that the pressure is conjugate to the volume.

Reversible and irreversible processes According to the statement of the second law in Eqn. (2.114), the entropy of an isolated system cannot decrease in any process, rather it must remain constant or else increase. Clearly, if an isolated system undergoes a process in which its entropy increases, then the reverse process can never occur. We say that such a process is *irreversible*. However, if the process leaves the system's entropy unchanged, then the reverse process is also possible and we say that the process is *reversible*.

In reality almost all processes are irreversible; however, it is possible to construct idealized reversible systems for transmitting work and heat to a system. A *reversible work source* supplies (or accepts) work from another system while keeping its own entropy constant. A *reversible heat source* accepts heat from another system by undergoing a quasistatic process at constant values of its particle number and kinematic state variables. (For a detailed discussion of reversible and irreversible processes and a description of reversible work and heat sources, see Section 5.5.7 in [TME12].)

[61] Here we use the notation $đ$ in $đ\mathcal{W}^{\text{def}}$ and $đ\mathcal{Q}$ to explicitly indicate that these quantities are not the differential of functions $\mathcal{W}^{\text{def}}$ and $\mathcal{Q}$. This will serve to remind us that the heat and work transferred to a system generally depend on the process being considered.

The idealized reversible work and heat sources are useful because they can be used to construct reversible processes. Indeed, for any system A and for any two of its equilibrium states $\mathcal{A}$ and $\mathcal{A}'$, we can always construct an isolated composite system – consisting of a reversible heat source, a reversible work source and A as subsystems – for which there exists a reversible process starting in state $\mathcal{A}$ and ending in state $\mathcal{A}'$. There are many such processes each of which results in the same amount of energy being transferred from A to the rest of the system. The distinguishing factor between the processes is exactly how this total energy transfer is partitioned between the reversible work and heat sources. Since the reversible work source does not change its entropy during any of these processes, the second law tells us that the total entropy change must satisfy

$$\Delta \mathcal{S} = \Delta \mathcal{S}^{\mathsf{A}} + \Delta \mathcal{S}^{\mathrm{RHS}} \geq 0,$$

where $\Delta \mathcal{S}^{\mathrm{RHS}}$ is the reversible heat source's entropy change and the equality holds only for reversible processes. Thus, we find that $\Delta \mathcal{S}^{\mathsf{A}} \geq -\Delta \mathcal{S}^{\mathrm{RHS}}$. If we consider an infinitesimal change of A's state, then this becomes $d\mathcal{S}^{\mathsf{A}} \geq -d\mathcal{S}^{\mathrm{RHS}} = -đ\mathcal{Q}^{\mathrm{RHS}}/T^{\mathrm{RHS}}$, since the reversible heat source supplies heat quasistatically. Finally, if the reversible heat source accepts an amount of heat $đ\mathcal{Q}^{\mathrm{RHS}}$, then the heat transferred to A is $đ\mathcal{Q}^{\mathsf{A}} = -đ\mathcal{Q}^{\mathrm{RHS}}$. Using this relation, we find that the minus signs cancel and we obtain (dropping the subscript A to indicate that this relation is true for any system)

$$d\mathcal{S} \geq \frac{đ\mathcal{Q}}{T^{\mathrm{RHS}}}. \tag{2.133}$$

This is called the *Clausius–Planck inequality*, which is an alternative statement of the second law. We emphasize that T^{RHS} is not, generally, equal to the system's temperature T. Thus, T^{RHS} is the "temperature at which heat is supplied to the system."

If we define the *external entropy input* as

$$d\mathcal{S}^{\mathrm{ext}} \equiv \frac{đ\mathcal{Q}}{T^{\mathrm{RHS}}},$$

then the difference between the actual change in the system's entropy and the external entropy input is called the *internal entropy production* and is defined as

$$d\mathcal{S}^{\mathrm{int}} \equiv d\mathcal{S} - d\mathcal{S}^{\mathrm{ext}}.$$

Then, according to the Clausius–Planck inequality, $d\mathcal{S}^{\mathrm{int}} \geq 0$. We can convert this into a statement about the work performed on the system by noting that the change of internal energy is by definition, $d\mathcal{U} = Td\mathcal{S} + \sum_\alpha \gamma_\alpha d\Gamma_\alpha$ and that the first law requires $d\mathcal{U} = đ\mathcal{Q} + đ\mathcal{W}^{\mathrm{def}}$ for all processes. Equating these two expressions for $d\mathcal{U}$, solving for $d\mathcal{S}$, and substituting the result and the definition of $d\mathcal{S}^{\mathrm{ext}}$ into the internal entropy production gives

$$d\mathcal{S}^{\mathrm{int}} = đ\mathcal{Q}\left(\frac{1}{T} - \frac{1}{T^{\mathrm{RHS}}}\right) + \frac{1}{T}\left(đ\mathcal{W}^{\mathrm{def}} - \sum_\alpha \gamma_\alpha d\Gamma_\alpha\right) \geq 0. \tag{2.134}$$

The equality holds only for reversible processes, in which case it is then necessary that $T = T^{\text{RHS}}$. We can further note that if $đQ > 0$ then $T < T^{\text{RHS}}$ and if $đQ < 0$, then $T > T^{\text{RHS}}$. Either way, the first term on the right-hand side of the inequality in Eqn. (2.134) is positive. This allows us to conclude that for any irreversible process

$$đW^{\text{def}} > \sum_{\alpha} \gamma_\alpha d\Gamma_\alpha.$$

Thus in an irreversible process, the work of deformation performed on a system is greater than it would be in a quasistatic process. The difference goes towards increasing the entropy.

So far, our discussion of thermodynamics has been limited to homogeneous thermodynamic systems. We now make the assumption of *local thermodynamic equilibrium* and derive the continuum counterparts to the first and second laws.

2.4.6 Continuum thermodynamics

Our discussion of thermodynamics has led us to definitions for familiar quantities such as the pressure p and temperature T as derivatives of a system's fundamental relation. This relation describes the system only for states of thermodynamic equilibrium, which by definition are homogeneous, i.e. without spatial and temporal variation. Accordingly, it makes sense to talk about the temperature and pressure of the gas inside the rigid sphere discussed at the start of Section 2.4.5 *before* the hole is opened. However, the temperature and pressure are not defined for the system *while the gas expands* after the hole is opened. This may seem reasonable to you because the expansion process is so fast (relative to the rate of processes we encounter on a day-to-day basis) that it seems impossible to measure the temperature of the gas at any given spatial position. However, consider the case of a large swimming pool, into which hot water pours from a garden hose. In this case your intuition and experience would lead you to argue that it is certainly possible to identify locations within the pool that are hotter than others. That is, we believe we can identify a spatially varying temperature field. The question we are exploring is: *is it possible to describe real processes using a continuum theory where we replace p, V and T with fields of pressure $p(\boldsymbol{x})$, density $\rho(\boldsymbol{x})$ and temperature $T(\boldsymbol{x})$*? As the above examples suggest, the answer depends on the conditions of the experiment.

It is correct to represent state variables as spatial fields provided that the length scale over which the continuum fields vary appreciably is much larger than the microscopic length scale. In fluids, this is measured by the *Knudsen number* $Kn = \lambda/L$, where λ is the mean free path (the average distance between collisions of gas atoms) and L is a characteristic macroscopic length (such as the diameter of the rigid sphere from Section 2.4.5). The continuum approximation is valid as long as $Kn \ll 1$. For an ideal gas where the velocities of the atoms are distributed according to the Maxwell–Boltzmann distribution (see Section 9.3.3), the mean free path is [TM04, Section 17.5]

$$\lambda = \frac{k_B T}{\sqrt{2}\pi\delta^2 p},$$

where δ is the atom diameter. For a gas at room temperature and atmospheric pressure, $\lambda \approx 70$ nm. That means that, for the gas in the rigid sphere, the continuum assumption is valid as long as the diameter of the sphere is much larger than 70 nm. However, if the sphere is filled with a rarefied gas ($p \approx 1$ torr), then $\lambda \approx 0.1$ mm. This is still small relative to, say, a typical pressure gauge, but we see that we are beginning to approach the length scale where the continuum model breaks down.[62]

By accepting the "continuum assumptions" and the existence of state variable fields, we are in fact accepting the *postulate of local thermodynamic equilibrium*. This postulate states that *the local and instantaneous relations between thermodynamic quantities in a system out of equilibrium are the same as for a uniform system in equilibrium.*[63] Thus, although the system as a whole is not in equilibrium, the laws of thermodynamics and the equations of state developed for uniform systems in thermodynamic equilibrium are applied locally. For example, for the expanding gas, the relation between pressure, density and temperature at a point,

$$p(\boldsymbol{x}) = \frac{k_{\mathrm{B}}}{m}\rho(\boldsymbol{x})T(\boldsymbol{x}),$$

follows from the ideal gas law in Eqn. (2.129) by setting $\rho = Nm/V$, where m is the mass of one atom of the gas.

In addition to the spatial dependence of continuum fields, a temporal dependence is also possible. Certainly the expansion of a gas is a time-dependent phenomenon. Again, the definitions of equilibrium thermodynamics can be stretched to accommodate this requirement provided that the rate of change of continuum field variables is slow compared with the atomistic equilibration time scale. This means that change occurs sufficiently slowly on the macroscopic scale so that all heat transfers can be approximated as quasistatic and that at each instant the thermodynamic system underlying each continuum particle has sufficient time to reach a close approximation to thermodynamic (or at least metastable) equilibrium.

Since the thermodynamic system associated with each continuum particle is not exactly in equilibrium, there is some error in the quasistatic heat transfer assumption and the use of the equilibrium fundamental relations to describe a nonequilibrium process. However, this error is small enough that it can be accurately compensated for by introducing an irreversible viscous, or dissipative, contribution to the stress. Thus, the total stress has an elastic contribution (corresponding to the thermodynamic tensions and determined by the equilibrium fundamental relation) and a viscous contribution.

Integrated form of the first law We now turn to the derivation of the local forms of the first and second laws of thermodynamics. The total energy of a continuum body B is

$$\mathcal{E} = \mathcal{K} + \mathcal{U}, \tag{2.135}$$

[62] See [Moo90], for an interesting comparison between the continuum case ($Kn \to 0$) and the free-molecular case ($Kn \to \infty$) for the expansion of a gas in vacuum.

[63] This is the particular form of the postulate of local thermodynamic equilibrium given by [LJCV08]. There are no clear quantitative measures that determine when this condition is satisfied, but experience has shown that it holds for a broad range of systems over a broad range of conditions [EM90]. When it fails, there is no recourse but to turn to a more general theory of *nonequilibrium statistical mechanics* that is valid far from equilibrium. This is a very difficult subject that remains an area of active research (see, for example, [Rue99] for a review). In this book we will restrict ourselves to nonequilibrium processes that are at least locally in thermodynamic equilibrium.

where $\mathcal{K}$ is the total kinetic energy and $\mathcal{U}$ is the total internal energy,

$$\mathcal{K} = \int_B \frac{1}{2}\rho \|\boldsymbol{v}\|^2 \, dV, \qquad \mathcal{U} = \int_B \rho u \, dV. \tag{2.136}$$

Here $\boldsymbol{v}$ is the velocity and u is called the *specific internal energy* (i.e. the internal energy per unit mass).[64] The first law of thermodynamics in Eqn. (2.105) can then be written as

$$\dot{\mathcal{E}} = \dot{\mathcal{K}} + \dot{\mathcal{U}} = \mathcal{P}^{\text{ext}} + \mathcal{R}, \tag{2.137}$$

where $\mathcal{P}^{\text{ext}} \equiv đ\mathcal{W}^{\text{ext}}/dt$ is the rate of external work (also called the external power) and $\mathcal{R} \equiv đ\mathcal{Q}/dt$ is the rate of heat supply. The rates of change of kinetic and internal energy are

$$\dot{\mathcal{K}} = \frac{D}{Dt}\int_B \frac{1}{2}\rho v_i v_i \, dV = \int_B \frac{1}{2}\rho(a_i v_i + v_i a_i) \, dV = \int_B \rho a_i v_i \, dV, \tag{2.138}$$

$$\dot{\mathcal{U}} = \frac{D}{Dt}\int_B \rho u = \int_B \rho \dot{u} \, dV, \tag{2.139}$$

where we have used the Reynolds transport theorem (Eqn. (2.79)).

External power $\mathcal{P}^{\text{ext}}$ A continuum body may be subjected to distributed body forces and surface tractions as shown in Fig. 2.6. The work per unit time transferred to the continuum by these fields is the external power,

$$\mathcal{P}^{\text{ext}} = \int_B \rho b_i v_i \, dV + \int_{\partial B} \bar{t}_i v_i \, dA, \tag{2.140}$$

where $\bar{\boldsymbol{t}}$ is the external traction acting on the surfaces of the body. Applying Cauchy's relation (Eqn. (2.87)) followed by the divergence theorem (Eqn. (2.50)) it can be shown that (see Section 5.6.1 of [TME12]):

$$\mathcal{P}^{\text{ext}} = \dot{\mathcal{K}} + \mathcal{P}^{\text{def}}, \tag{2.141}$$

where

$$\mathcal{P}^{\text{def}} = \int_B \sigma_{ij} d_{ij} \, dV \quad \Leftrightarrow \quad \mathcal{P}^{\text{def}} = \int_B \boldsymbol{\sigma} : \boldsymbol{d} \, dV, \tag{2.142}$$

64 In Eqn. (2.136), $\frac{1}{2}\rho \|\boldsymbol{v}\|^2$ is the macroscopic kinetic energy associated with the gross motion of a continuum particle, while u includes the strain energy due to deformation of the particle, the microscopic kinetic energy associated with vibrations of the atoms making up the particle and any other energy not explicitly accounted for in the system. This is stated here without proof. See Appendix A of [TME12] for a heuristic microscopic derivation. For a more rigorous proof based on nonequilibrium statistical mechanics, see [AT11].

is the deformation power (corresponding to the rate of the work of deformation $đ\mathcal{W}^{\text{def}}$ we encountered in Section 2.4.3). The deformation power is the portion of the external power contributing to the deformation of the body with the remainder going towards kinetic energy. We note that since $\boldsymbol{d} = \dot{\boldsymbol{\epsilon}}$ (see Eqn. (2.63)), Eqn. (2.142) can also be written

$$\mathcal{P}^{\text{def}} = \int_B \sigma_{ij}\dot{\epsilon}_{ij}\, dV \quad \Leftrightarrow \quad \mathcal{P}^{\text{def}} = \int_B \boldsymbol{\sigma} : \dot{\boldsymbol{\epsilon}}\, dV. \tag{2.143}$$

Returning now to the representation of the first law in Eqn. (2.137) and substituting in Eqn. (2.141), we see that the first law can be written more concisely as

$$\dot{\mathcal{U}} = \mathcal{P}^{\text{def}} + \mathcal{R}, \tag{2.144}$$

which is similar to the form obtained previously in Eqn. (2.104).

Alternative forms for the deformation power It is also possible to obtain expressions for the deformation power in terms of other stress variables that are often useful (see Section 5.6.1 of [TME12]). For the first Piola–Kirchhoff stress $\boldsymbol{P}$ we have

$$\mathcal{P}^{\text{def}} = \int_{B_0} P_{iJ}\dot{F}_{iJ}\, dV_0 \quad \Leftrightarrow \quad \mathcal{P}^{\text{def}} = \int_{B_0} \boldsymbol{P} : \dot{\boldsymbol{F}}\, dV_0, \tag{2.145}$$

where $\boldsymbol{F}$ is the deformation gradient. For the second Piola–Kirchhoff stress $\boldsymbol{S}$ we have

$$\mathcal{P}^{\text{def}} = \int_{B_0} S_{IJ}\dot{E}_{IJ}\, dV_0 \quad \Leftrightarrow \quad \mathcal{P}^{\text{def}} = \int_{B_0} \boldsymbol{S} : \dot{\boldsymbol{E}}\, dV_0, \tag{2.146}$$

where $\boldsymbol{E}$ is the Lagrangian strain tensor.

Elastic and viscous parts of the stress As indicated at the beginning of this section, a continuum particle will not generally be in a perfect state of thermodynamic equilibrium. Therefore, the stress will not usually be equal to the thermodynamic tensions that are work conjugate to the strain, i.e. the stress is *not* a state variable. To correct for this, continuum thermodynamic theory introduces the ideas of the *elastic part of the stress* $\boldsymbol{\sigma}^{(e)}$ and the *viscous part of the stress* [ZM67]:[65]

$$\boldsymbol{\sigma} = \boldsymbol{\sigma}^{(e)} + \boldsymbol{\sigma}^{(v)}. \tag{2.147}$$

By definition, the elastic part of the stress is given by the material's fundamental relation, and therefore it *is* a state variable. The viscous part of the stress is the part which is not associated with an equilibrium state of the material, and is therefore not a state variable.

[65] It is not definite that an additive partitioning can always be made. In plasticity theory, for example, it is common to partition the deformation gradient into a plastic and an elastic part, instead of the stress. See [Mal69, p. 267] for a discussion of this issue.

Table 2.3. Work conjugate pairs and viscous stress for a continuum system under finite strain. Representation in Voigt notation

α	$\Gamma^{\mathrm{i}}_\alpha$	γ_α	$\gamma^{(v)}_\alpha$
1	E_{11}	$S^{(e)}_{11}$	$S^{(v)}_{11}$
2	E_{22}	$S^{(e)}_{22}$	$S^{(v)}_{22}$
3	E_{33}	$S^{(e)}_{33}$	$S^{(v)}_{33}$
4	$2E_{23}$	$S^{(e)}_{23}$	$S^{(v)}_{23}$
5	$2E_{13}$	$S^{(e)}_{13}$	$S^{(v)}_{13}$
6	$2E_{12}$	$S^{(e)}_{12}$	$S^{(v)}_{12}$

Substituting Eqn. (2.147) into the definitions for the first and second Piola–Kirchhoff stresses, we can similarly obtain the elastic and viscous parts of these stress measures.

Work conjugate variables The three equations (2.143), (2.145) and (2.146) for the deformation power provide three pairs of variables whose product yields an internal energy density: $(\boldsymbol{\sigma}, \boldsymbol{\epsilon})$, $(\boldsymbol{P}, \boldsymbol{F})$ and $(\boldsymbol{S}, \boldsymbol{E})$. These work conjugate variables fit the general form given in Eqn. (2.132) except that for the continuum formulation the kinematic state variables are intensive. This allows us to use the general and convenient notation we introduced in Section 2.4.1. Thus in general, the deformation power is written

$$\mathcal{P}^{\mathrm{def}} = \int_{B_0} \sum_\alpha (\gamma_\alpha + \gamma^{(v)}_\alpha)\dot{\Gamma}^{\mathrm{i}}_\alpha \, dV_0, \tag{2.148}$$

where $\boldsymbol{\Gamma}^{\mathrm{i}} = (\Gamma^{\mathrm{i}}_1, \dots, \Gamma^{\mathrm{i}}_{n^\Gamma})$ is a relevant set of n^Γ intensive state variables that describe the local kinematics of the continuum, and $\boldsymbol{\gamma} = (\gamma_1, \dots, \gamma_{n^\Gamma})$ and $\boldsymbol{\gamma} = (\gamma^{(v)}_1, \dots \gamma^{(v)}_{n^\Gamma})$ are the thermodynamic tensions and their viscous counterparts, respectively, which when added together are work conjugate to the strain. For example, for Eqn. (2.146) we can make the assignment in Tab. 2.3 which is called *Voigt notation*.[66] For notational simplicity we will drop the superscript "i" on $\boldsymbol{\Gamma}^{\mathrm{i}}$ in subsequent chapters.

Heat transfer rate $\mathcal{R}$ The heat transfer rate $\mathcal{R}$ can be divided into two parts:

$$\mathcal{R} = \int_B \rho r \, dV - \int_{\partial B} h \, dA. \tag{2.149}$$

Here, $r = r(\boldsymbol{x}, t)$ is the strength of a distributed heat source per unit mass, and h is the *outward* heat flux across an element of the surface of the body with normal $\boldsymbol{n}$:

$$h = h(\boldsymbol{x}, t, \boldsymbol{n}) = \boldsymbol{q}(\boldsymbol{x}, t) \cdot \boldsymbol{n}, \tag{2.150}$$

where $\boldsymbol{q}$ is called the *heat flux vector*.[67]

[66] Voigt notation is a concatenated notation used for symmetric stress and strain tensors. The two coordinate indices of the tensor are replaced with a single index ranging from 1 to 6.

[67] The proof that h has the form in Eqn. (2.150) is similar to the one used by Cauchy for the traction vector. See Section 5.6.1 of [TME12] for details.

Local form of the first law (energy equation) Substituting Eqns. (2.139), (2.142), (2.149) and (2.150) into Eqn. (2.144), applying the divergence theorem and combining terms, gives

$$\int_B [\sigma_{ij} d_{ij} + \rho r - \rho \dot{u} - q_{i,i}] \, dV = 0.$$

This can be rewritten for any arbitrary subbody E, so it must be satisfied pointwise:

$$\sigma_{ij} d_{ij} + \rho r - q_{i,i} = \rho \dot{u} \quad \Leftrightarrow \quad \boldsymbol{\sigma} : \boldsymbol{d} + \rho r - \operatorname{div} \boldsymbol{q} = \rho \dot{u}. \tag{2.151}$$

This equation, called the *energy equation*, is the local spatial form of the first law of thermodynamics. It can be thought of as a statement of conservation of energy for an infinitesimal continuum particle. The first term in the equation ($\boldsymbol{\sigma} : \boldsymbol{d}$) is the portion of the mechanical power going towards deformation of the particle, the second term (ρr) is the internal source of heat,[68] the third term ($-\operatorname{div} \boldsymbol{q}$) is the inflow of heat through the boundaries of the particle and the term on the right-hand side ($\rho \dot{u}$) is the rate of change of internal energy. The energy equation can also be written in the reference configuration:

$$P_{iJ} \dot{F}_{iJ} + \rho_0 r_0 - q_{0I,I} = \rho_0 \dot{u}_0 \quad \Leftrightarrow \quad \boldsymbol{P} : \dot{\boldsymbol{F}} + \rho_0 r_0 - \operatorname{Div} \boldsymbol{q}_0 = \rho_0 \dot{u}_0, \tag{2.152}$$

where r_0, $\boldsymbol{q}_0$ and u_0 are respectively the specific heat source, heat flux vector and specific internal energy defined in the reference configuration.

Local form of the second law (Clausius–Duhem inequality) Having established the local form of the first law, we now turn to the second law of thermodynamics. Our objective is to obtain a local form of the second law. We begin with the Clausius–Planck inequality (Eqn. (2.133)) in its rate form,

$$\dot{\mathcal{S}} \geq \dot{\mathcal{S}}^{\text{ext}} = \frac{\mathcal{R}}{T^{\text{RHS}}}, \tag{2.153}$$

where T^{RHS} is the temperature of the reversible heat source from which the heat is *quasi-statically* transferred to the body. We now introduce continuum variables. The entropy $\mathcal{S}$ is an extensive variable, we therefore define the entropy content of an arbitrary subbody E as a volume integral over the *specific entropy* s:

$$\mathcal{S}(E) = \int_E \rho s \, dV. \tag{2.154}$$

The rate of heat transfer to E is

$$\mathcal{R}(E) = \int_E \rho r \, dV - \int_{\partial E} \boldsymbol{q} \cdot \boldsymbol{n} \, dA. \tag{2.155}$$

[68] The idea of an internal heat source is used to model interactions of the material with the external world that are like body forces but are otherwise not accounted for in the thermomechanical formulation. For example, electromagnetic interactions may cause a current to flow in the material and its natural electrical resistance will then generate heat in the material.

This can be substituted into Eqn. (2.153), but to progress further we must address an important subtlety. In principle there can be a reversible heat source associated with every point on the boundary of the body and the temperature of these sources is not necessarily equal to the temperature of the material point at the boundary. However, in continuum thermodynamics theory it is assumed that the boundary points are always in thermal equilibrium with their reversible heat sources. The argument is that even if the boundary of the body starts a process at a different temperature, a thin layer at the boundary heats (or cools) nearly instantaneously to the source's temperature. Also, it is assumed that the internal heat sources are always in thermal equilibrium with their material point.[69] Accordingly, we can substitute Eqn. (2.154) into Eqn. (2.153) and take the factor of $1/T$ inside the integrals where it is treated as a function of position and obtained from the material's fundamental relation. This means that the external entropy input rate is

$$\dot{S}^{\text{ext}}(E) = \int_E \frac{\rho r}{T}\, dV - \int_{\partial E} \frac{\boldsymbol{q} \cdot \boldsymbol{n}}{T}\, dA. \tag{2.156}$$

Substituting Eqns. (2.154) and (2.156) into Eqn. (2.153), applying the Reynolds transport theorem (Eqn. (2.79)) and the divergence theorem (Eqn. (2.50)) and using the arbitrariness of E gives the pointwise relation

$$\dot{s} \geq \dot{s}^{\text{ext}} = \frac{r}{T} - \frac{1}{\rho} \operatorname{div} \frac{\boldsymbol{q}}{T}, \tag{2.157}$$

where $\dot{s}^{\text{ext}}$ is the specific external entropy input rate. Equation (2.157) is called the *Clausius–Duhem inequality*. The specific internal entropy production rate, $\dot{s}^{\text{int}}$, follows as

$$\dot{s}^{\text{int}} \equiv \dot{s} - \dot{s}^{\text{ext}} = \dot{s} - \frac{r}{T} + \frac{1}{\rho} \operatorname{div} \frac{\boldsymbol{q}}{T}. \tag{2.158}$$

The Clausius–Duhem inequality is then simply

$$\dot{s}^{\text{int}} \geq 0. \tag{2.159}$$

This is the local analog to Eqn. (2.134).

This concludes our overview of thermodynamics. We have introduced the important concepts of energy, temperature and entropy that will remain with us for the rest of the book. In the next section we turn to the remaining piece of the continuum puzzle, the establishment of constitutive relations that govern the behavior of materials.

[69] However, some authors have argued that a different temperature should be used; see, for example, [GW66].

2.5 Constitutive relations

In the previous two sections, we laid out the physical laws that govern the behavior of continuum systems. The result was the following set of partial differential equations expressed in the deformed configuration taken from Eqns. (2.76), (2.90) and (2.151):

conservation of mass:	$\dot{\rho} + \rho(\operatorname{div} \boldsymbol{v}) = 0$	(1 equation),
balance of linear momentum:	$\operatorname{div} \boldsymbol{\sigma} + \rho \boldsymbol{b} = \rho \boldsymbol{a}$	(3 equations),
conservation of energy (first law):	$\boldsymbol{\sigma} : \boldsymbol{d} + \rho r - \operatorname{div} \boldsymbol{q} = \rho \dot{u}$	(1 equation),

along with the algebraic Eqns. (2.94) and the differential inequality (2.157):

balance of angular momentum:	$\boldsymbol{\sigma}^T = \boldsymbol{\sigma}$	(3 equations),
Clausius–Duhem inequality (second law):	$\dot{s} \geq (r/T) - (1/\rho)\operatorname{div}(\boldsymbol{q}/T)$	(1 equation).

Excluding the balance of angular momentum and the Clausius–Duhem inequality, which provide constraints on material behavior but are not governing equations, a continuum thermomechanical system is therefore governed by five differential equations. These are called the *field equations* or *governing equations* of continuum mechanics.

The independent fields entering into these equations are:

$$\rho \;\; \text{(1 unknown)}, \quad \boldsymbol{\sigma} \;\; \text{(6 unknown)}, \quad u \;\; \text{(1 unknown)}, \quad T \;\; \text{(1 unknown)},$$
$$\boldsymbol{x} \;\; \text{(3 unknowns)}, \quad \boldsymbol{q} \;\; \text{(3 unknowns)}, \quad s \;\; \text{(1 unknown)},$$

where we have imposed the symmetry of the stress due to the the balance of angular momentum. The result is a total of sixteen unknowns. The heat source r and body force $\boldsymbol{b}$ are assumed to be known external interactions of the body with its environment. The velocity, acceleration and the rate of deformation tensor are not independent fields, but are given by

$$\boldsymbol{v} = \dot{\boldsymbol{x}}, \qquad \boldsymbol{a} = \ddot{\boldsymbol{x}}, \qquad \boldsymbol{d} = \frac{1}{2}(\nabla \dot{\boldsymbol{x}} + (\nabla \dot{\boldsymbol{x}})^T).$$

Consequently, a continuum thermomechanical system is characterized by five equations with sixteen unknowns. The missing equations are the *constitutive relations* (or *response functions*) that describe the response of the material to the mechanical and thermal loading imposed on it. Constitutive relations are required for u, T, $\boldsymbol{\sigma}$ and $\boldsymbol{q}$ [CN63]. These provide the additional eleven equations required to close the system.

Constitutive relations cannot be selected arbitrarily. They must conform to certain constraints imposed on them by physical laws and they must be consistent with the structure of the material. In addition, certain simplifications may be adopted to further restrict the allowable forms of constitutive relations. We describe these restrictions in the next section. We note, however, that these restrictions do not lead to actual constitutive relations for specific materials. A specific constitutive relation is obtained either by performing experiments or by direct computation using an atomistic model. The latter approach is discussed in Chapter 11. In such a case, one starts with an "atomistic constitutive relation," describing how individual atoms interact based on their kinematic description, and then uses certain averaging techniques to obtain the continuum-level relations.

2.5.1 Constraints on constitutive relations

The continuum constitutive relations considered in this book satisfy the following seven constraints:[70]

(I) **Principle of determinism** *The current value of any physical variable can be determined from knowledge of the present and past values of other variables.* For example, we assume that the stress at a material particle $\boldsymbol{X}$ in a body at time t can be determined from the history of the motion of the body, its temperature history and so on [Jau67]:

$$\boldsymbol{\sigma}(\boldsymbol{X},t) = \boldsymbol{f}(\boldsymbol{\varphi}^t(\cdot), T^t(\cdot), \dots, \boldsymbol{X}, t). \tag{2.160}$$

Here, $\boldsymbol{\varphi}^t(\cdot)$ and $T^t(\cdot)$ represent the time histories of the deformation mapping and temperature at all points in the body. A material that depends on the past as well as the present is called a material with memory.

(II) **Principle of local action** *The material response at a point depends only on the conditions within an arbitrarily small region about that point.* For a material without memory, the stress function is

$$\boldsymbol{\sigma}(\boldsymbol{X},t) = \boldsymbol{h}(\boldsymbol{\varphi}(\boldsymbol{X},t), \boldsymbol{F}(\boldsymbol{X},t), \dots, T(\boldsymbol{X},t), \nabla_0 T(\boldsymbol{X},t), \dots, \boldsymbol{X}, t). \tag{2.161}$$

An example of such a model is the generalized Hooke's law for a hyperelastic material[71] under conditions of infinitesimal deformations, where the stress is a linear function of the small strain tensor at a point

$$\sigma_{ij}(\boldsymbol{X}) = c_{ijkl}(\boldsymbol{X})\epsilon_{kl}(\boldsymbol{X}).$$

Here c is the small strain elasticity tensor. It is important to point out that the principle of local action is not universally accepted. There are *nonlocal* continuum theories that reject this hypothesis. In such theories, the constitutive response at a point is obtained by integrating over the volume of the body. For example in Eringen's nonlocal continuum theory the Cauchy stress $\boldsymbol{\sigma}$ at a point is [Eri02]

$$\sigma_{ij}(\boldsymbol{X}) = \int_{B_0} K(\|\boldsymbol{X} - \boldsymbol{X}'\|) t_{ij}(\boldsymbol{X}')\, dV(\boldsymbol{X}'),$$

where the kernel $K(r)$ is an *influence function* (often taken to be a Gaussian of finite support, i.e. it is identically zero for all $r > r_{\text{cut}}$ for some cutoff distance $r_{\text{cut}} > 0$) and $t_{ij} = c_{ijkl}\epsilon_{kl}$ are the usual local stresses. Silling has developed a nonlocal continuum theory called *peridynamics* formulated entirely in terms of forces [Sil02].

Nonlocal theories can be very useful in certain situations, such as in the presence of discontinuities; however, local constitutive relations tend to be the dominant choice due to their simplicity and their ability to adequately describe most phenomena of interest. In particular, in the context of the multiscale methods discussed in this book

[70] See Chapter 6 of [TME12] for a more detailed discussion.

[71] We define what we mean by "elastic" and "hyperelastic" materials below.

continuum theories are applied only in regions where gradients are sufficiently smooth to warrant the local action approximation. Regions where such approximations break down are described using atomistic methods that are naturally nonlocal. For more on this see Chapter 12.

(III) **Second law restrictions** *A constitutive relation cannot violate the second law of thermodynamics which states that the entropy of an isolated system remains constant for a reversible process and increases for an irreversible one.* For example a constitutive model for heat flux must ensure that heat flows from hot to cold regions and not vice versa.

(IV) **Principle of material frame-indifference (objectivity)** *All physical variables for which constitutive relations are required must be objective tensors.* An objective tensor is a tensor which is physically the same in all frames of reference. For example the relative position between two physical points is an objective vector, whereas the velocity of a physical point is not objective since it will change depending on the frame of reference in which it is measured.

(V) **Material symmetry** *A constitutive relation must respect any symmetries that the material possesses.* For example the stress in a uniformly strained homogeneous isotropic material (i.e. a material that has the same mechanical properties in all directions at all points) is the same regardless of how the material is rotated before the strain is applied.

In addition to the five general principles described above, in this book we will restrict the discussion further to the most commonly encountered types of constitutive relations with two additional constraints:

(VI) *Only materials without memory and without aging are considered.* This, along with the principle of local action, means that the constitutive relations for the variables u, T, $\boldsymbol{\sigma}$ and $\boldsymbol{q}$ only depend on the local values of other state variables (including possibly a finite number of higher-order gradients) and their time rates of change. Thus in Eqn. (2.161) the explicit dependence on time is dropped.

(VII) *Only materials whose internal energy depends solely on the entropy and deformation gradient are considered.* That is we explicitly exclude the possibility of dependence on any rates of deformation as well as the higher-order gradients of the deformation. This is consistent with the thermodynamic definition in Eqn. (2.113).

In the next three sections we see the implications of the restrictions described above on allowable forms of the constitutive relations.

2.5.2 Local action and the second law of thermodynamics

Referring back to Eqn. (2.113), changing from an extensive to an intensive representation, and applying principles (I) and (II) and constraints (VI) and (VII), we obtain the functional form for the specific internal energy constitutive relation:[72]

$$u = \bar{u}(s, \boldsymbol{F}), \tag{2.162}$$

which is referred to as the *caloric equation of state*. A material whose constitutive relation depends on the deformation only through the history of the local value of $\boldsymbol{F}$ is called a *simple material*. A simple material without memory (depending only on the instantaneous value of $\boldsymbol{F}$) is called an *elastic simple material*.

Before continuing, we note that a set of possible constitutive relations, which we have excluded from discussion via constraint (VII), are those that include a dependence on higher-order gradients of the deformation,[73]

$$u = \widehat{u}(s, \boldsymbol{F}, \nabla_0 \boldsymbol{F}, \dots).$$

The result is a *strain gradient theory*. This approach has been successfully used to study length scale[74] dependence in plasticity [FMAH94] and localization of deformation in the form of shear bands [TA86]. See the discussion in Section 6.6 of [TME12]. An alternative approach is the polar *Cosserat theory* in which nonuniform local deformation is characterized by associating a triad of orthonormal director vectors with each material point [Rub00]. These approaches are beyond the scope of the present discussion.

Coleman–Noll procedure The functional forms for the temperature, heat flux vector and stress tensor are obtained by careful consideration of the implications of the second law of thermodynamics. As a first step, the Clausius–Duhem inequality (repeated at the start of this section) can be combined with the energy equation to give the following inequality:[75]

$$\rho \left[T - \frac{\partial \bar{u}}{\partial s} \right] \dot{s} + \left[\boldsymbol{\sigma} \boldsymbol{F}^{-T} - \rho \frac{\partial \bar{u}}{\partial \boldsymbol{F}} \right] : \dot{\boldsymbol{F}} - \frac{1}{T} \boldsymbol{q} \cdot \nabla T \geq 0. \tag{2.163}$$

In an important paper, Coleman and Noll [CN63] made the argument that this equation must be satisfied for *every* admissible process. By selecting special cases, insight is gained into the relation between the different continuum fields. This line of thinking is referred to as the *Coleman–Noll procedure*. By following this procedure,[75] it ispossible to infer the

[72] As in Section 2.4, a bar or (other accent) over a variable, as in $\bar{u}$, is used to denote the response function in terms of a particular set of arguments (as opposed to the actual quantity).

[73] Interestingly, it is not possible to simply add on a dependence on higher-order gradients without introducing additional variables that are conjugate with the higher-order gradient fields and modifying the energy equation and the Clausius–Duhem inequality [Gur65]. For example a second-gradient theory requires the introduction of couple stresses. Therefore, classical continuum thermodynamics is by necessity limited to simple materials.

[74] Each higher-order gradient introduced into the formulation is associated with a length scale. For example a second-order gradient has units of 1/length. It must therefore be multiplied by a parameter with units of length to cancel this out in the energy expression. In contrast, the classical continuum mechanics of simple materials has no length scale. This qualitative difference has sometimes led to authors calling these strain gradient theories "nonlocal." However, this terminology does not appear to be consistent with the original definition of the term "local."

[75] See Section 6.2 of [TME12] for details.

specific functional dependence of the temperature T and heat flux $\boldsymbol{q}$ response functions:

$$T = \bar{T}(s, \boldsymbol{F}) \equiv \frac{\partial \bar{u}}{\partial s}, \qquad \boldsymbol{q} = \bar{\boldsymbol{q}}(s, \boldsymbol{F}, \nabla T). \tag{2.164}$$

In addition, it can be shown that the stress tensor $\boldsymbol{\sigma}$ can be divided into a conservative elastic part $\boldsymbol{\sigma}^{(e)}$ and an irreversible viscous part $\boldsymbol{\sigma}^{(v)}$ (see Eqn. (2.147)) with the following functional forms:

$$\boldsymbol{\sigma}^{(e)} = \bar{\boldsymbol{\sigma}}^{(e)}(s, \boldsymbol{F}) \equiv \rho \frac{\partial \bar{u}}{\partial \boldsymbol{F}} \boldsymbol{F}^T, \qquad \boldsymbol{\sigma}^{(v)} = \bar{\boldsymbol{\sigma}}^{(v)}(s, \boldsymbol{F}, \boldsymbol{d}). \tag{2.165}$$

A material for which $\boldsymbol{\sigma}^{(v)} = \boldsymbol{0}$, and for which an energy function exists, such that the stress is entirely determined by Eqn. $(2.165)_1$, is called a *hyperelastic* material.

Constitutive relations for alternative stress variables Continuum formulations for solids are often expressed in a Lagrangian description, where the appropriate stress variables are the first or second Piola–Kirchhoff stress tensors. The constitutive relation for the elastic part of the first Piola–Kirchhoff stress is obtained by substituting Eqn. $(2.165)_1$ into Eqn. $(2.96)_1$. The result after using Eqn. (2.75) is

$$P_{iJ}^{(e)} = \rho_0 \frac{\partial \bar{u}}{\partial F_{iJ}} \quad \Leftrightarrow \quad \boldsymbol{P}^{(e)} = \rho_0 \frac{\partial \bar{u}}{\partial \boldsymbol{F}}. \tag{2.166}$$

The second Piola–Kirchhoff stress is obtained in similar fashion from Eqn. (2.99) as $S_{IJ}^{(e)} = \rho_0 F_{Ii}^{-1}(\partial \bar{u}/\partial F_{iJ})$, which can be rewritten in terms of the Lagrangian strain tensor $\boldsymbol{E}$ as

$$S_{IJ}^{(e)} = \rho_0 \frac{\partial \tilde{u}}{\partial E_{IJ}} \quad \Leftrightarrow \quad \boldsymbol{S}^{(e)} = \rho_0 \frac{\partial \tilde{u}}{\partial \boldsymbol{E}}. \tag{2.167}$$

Equations $(2.165)_1$, (2.166) and (2.167) provide the constitutive relations for the elastic parts of the Cauchy and Piola–Kirchhoff stress tensors. These expressions provide insight into the work conjugate pairs obtained earlier in the derivation of the deformation power in Section 2.4.6. That analysis identified three pairs of conjugate variables: $(\boldsymbol{\sigma}^{(e)}, \boldsymbol{\epsilon})$, $(\boldsymbol{P}^{(e)}, \boldsymbol{F})$ and $(\boldsymbol{S}^{(e)}, \boldsymbol{E})$. From Eqns. (2.166) and (2.167) we see that the elastic parts of the first and second Piola–Kirchhoff stress tensors are conservative thermodynamic tensions conjugate with their respective kinematic variables. In contrast, the elastic part of the Cauchy stress tensor *cannot* be written as the derivative of the energy with respect to the small strain tensor $\boldsymbol{\epsilon}$. The reason is that unlike $\boldsymbol{F}$ and $\boldsymbol{E}$, the small strain tensor $\boldsymbol{\epsilon}$ is not a state variable. Rather it is an incremental deformation measure. The conclusion is that $\boldsymbol{\sigma}^{(e)}$ is not a conservative thermodynamic tension. Consequently, a calculation of the change in internal energy using the conjugate pair $(\boldsymbol{\sigma}, \boldsymbol{\epsilon})$ requires an integration over the time history.[76]

Thermodynamic potentials and connection with experiments The mathematical description of a process can be significantly simplified by an appropriate choice of independent state variables. A process occurring at constant entropy ($\dot{s} = 0$) is called *isentropic*. A process where $\boldsymbol{F}$ is controlled is subject to *displacement control*. Thus, $u = \overline{u}(s, \boldsymbol{F})$ is the appropriate energy variable for isentropic processes under displacement control. If in addition to being isentropic, the process is also reversible, it can be shown that[77]

$$\rho r - \operatorname{div} \boldsymbol{q} = 0. \tag{2.168}$$

A process satisfying this condition is called *adiabatic*. It is important to note that for continuum systems, adiabatic conditions are not ensured by thermally isolating the system from its environment, which given Eqn. (2.155), only ensures that

$$\mathcal{R}(B) = \int_B \rho r \, dV - \int_{\partial B} \boldsymbol{q} \cdot \boldsymbol{n} \, dA = \int_B [\rho r - \operatorname{div} \boldsymbol{q}] \, dV = 0. \tag{2.169}$$

This does not translate to the local requirement in Eqn. (2.168), unless Eqn. (2.169) is assumed to hold for every subbody of the body. This implies that there is no transfer of heat between different parts of the body. The assumption is that such conditions can be approximately satisfied if the loading is performed "rapidly" on time scales associated with heat transfer [Mal69]. For example if a tension test in the elastic regime is performed in a laboratory where the sample is thermally isolated from its environment and is loaded (sufficiently fast) by applying a fixed displacement to its end, the engineering stress (i.e. the first Piola–Kirchhoff stress) measured in the experiment will be $\rho_0 \partial \overline{u}(s, \boldsymbol{F}) / \partial \boldsymbol{F}$.

In many cases, the loading conditions will be different. For example if the tension test mentioned above is performed at constant temperature (i.e. the sample is not insulated and the laboratory has a thermostat) or (instead of displacement control) a *load control* device that maintains a specified force is used, the results will be different. The suitable energy variable in these cases is not the specific internal energy, but other thermodynamic potentials that are obtained by the application of Legendre transformations.[78] We write the expressions below in generic form for arbitrary kinematic variables $\boldsymbol{\Gamma}$ and thermodynamic tensions $\boldsymbol{\gamma}$ and then give the results for two particular choices of $\boldsymbol{\Gamma}$: $\boldsymbol{F}$ and $\boldsymbol{E}$.

The *specific Helmholtz free energy* ψ is the appropriate energy variable for processes where T and $\boldsymbol{\Gamma}$ are the independent variables:

$$\psi = u - Ts, \qquad \widehat{\psi}(T, \boldsymbol{\Gamma}) = \overline{u}(\widehat{s}(T, \boldsymbol{\Gamma}), \boldsymbol{\Gamma}) - T\widehat{s}(T, \boldsymbol{\Gamma}). \tag{2.170}$$

[76] This has important implications for the application of constant stress boundary conditions in atomistic simulations as explained in Section 9.5.4.

[77] See Section 6.2.5 of [TME12] for details.

[78] For more details on the derivation of the thermodynamics potentials, see Section 6.2.5 of [TME12].

The first expression is the generic definition and the second shows the explicit dependence on variables of the response functions. The entropy and stress variables at constant temperature for the two choices of $\boldsymbol{\Gamma}$ are

$$s = -\frac{\partial \widehat{\psi}(T, \boldsymbol{\Gamma})}{\partial T}, \qquad \boldsymbol{P}^{(e)} = \rho_0 \frac{\partial \widehat{\psi}(T, \boldsymbol{F})}{\partial \boldsymbol{F}}, \qquad \boldsymbol{S}^{(e)} = \rho_0 \frac{\partial \widetilde{\psi}(T, \boldsymbol{E})}{\partial \boldsymbol{E}}. \tag{2.171}$$

The *strain energy density function* W is closely related to the specific Helmholtz free energy. This is simply the free energy per unit reference volume instead of per unit mass:

$$W = \rho_0 \psi. \tag{2.172}$$

In some atomistic simulations, where calculations are performed at "zero temperature," the strain energy density is directly related to the internal energy,[79] $W = \rho_0 u$. In this way strain energy density can be used as a catch-all for both zero-temperature and finite-temperature conditions. The stress variables follow as

$$\boldsymbol{P}^{(e)} = \frac{\partial \widehat{W}(T, \boldsymbol{F})}{\partial \boldsymbol{F}}, \qquad \boldsymbol{S}^{(e)} = \frac{\partial \widetilde{W}(T, \boldsymbol{E})}{\partial \boldsymbol{E}}. \tag{2.173}$$

The *specific enthalpy* h is the appropriate energy variable for processes where s and $\boldsymbol{\gamma}$ are the independent variables:

$$h = u - \boldsymbol{\gamma} \cdot \boldsymbol{\Gamma}, \qquad \widehat{h}(s, \boldsymbol{\gamma}) = \bar{u}(s, \widehat{\boldsymbol{\Gamma}}(s, \boldsymbol{\gamma})) - \boldsymbol{\gamma} \cdot \widehat{\boldsymbol{\Gamma}}(s, \boldsymbol{\gamma}). \tag{2.174}$$

The temperature and continuum deformation measures at constant entropy are

$$T = \frac{\partial \widehat{h}(s, \boldsymbol{\gamma})}{\partial s}, \qquad \boldsymbol{F} = -\rho_0 \frac{\partial \widehat{h}(s, \boldsymbol{P}^{(e)})}{\partial \boldsymbol{P}^{(e)}}, \qquad \boldsymbol{E} = -\rho_0 \frac{\partial \widehat{h}(s, \boldsymbol{S}^{(e)})}{\partial \boldsymbol{S}^{(e)}}. \tag{2.175}$$

The *specific Gibbs free energy* (or *specific Gibbs function*) g is the appropriate energy variable for processes where T and $\boldsymbol{\gamma}$ are the independent variables:

$$g = u - Ts - \boldsymbol{\gamma} \cdot \boldsymbol{\Gamma}, \qquad \widehat{g}(T, \boldsymbol{\gamma}) = \bar{u}(\widehat{s}(T, \boldsymbol{\gamma}), \widehat{\boldsymbol{\Gamma}}(T, \boldsymbol{\gamma})) - T\widehat{s}(T, \boldsymbol{\gamma}) - \boldsymbol{\gamma} \cdot \widehat{\boldsymbol{\Gamma}}(T, \boldsymbol{\gamma}). \tag{2.176}$$

[79] At $T = 0$ K, u is just the interatomic potential energy per unit mass. See Section 11.5 for more details.

The entropy and continuum deformation measures at constant temperature are

$$s = -\frac{\partial \widetilde{g}(T,\gamma)}{\partial T}, \qquad F = -\rho_0 \frac{\partial \widetilde{g}(T, P^{(e)})}{\partial P^{(e)}}, \qquad E = -\rho_0 \frac{\partial \widetilde{g}(T, S^{(e)})}{\partial S^{(e)}}. \tag{2.177}$$

2.5.3 Material frame-indifference

Constitutive relations provide a connection between a material's deformation and its entropy, stress and temperature. A fundamental assumption in continuum mechanics is that this response is intrinsic to the material and should therefore be independent of the frame of reference used to describe the motion of the material. This hypothesis is referred to as the *principle of material frame-indifference*. Explicitly, it states that (intrinsic) constitutive relations must be invariant with respect to changes of frame.

The application of the principle of material frame-indifference to constitutive relations is a two-step process. First, it must be established how different variables transform under a change of frame of reference. Variables that are unaffected, in a certain sense, by such transformations are called *objective*. Second, variables for which constitutive relations are necessary are required to be objective. The second step imposes constraints on the allowable form of the constitutive relations.[80]

Material frame-indifference is a complicated subject that involves subtle arguments regarding the nature of frames of reference and the transformations between them. The interested reader is referred to Section 6.3 of [TME12] for an in-depth discussion. Here we only give the final results. A constitutive relation for a scalar s, vector u and second-order tensor T is frame-indifferent provided that the following relations are satisfied for all proper orthogonal tensors Q and for all values of the argument γ:

$$\widehat{s}(\mathcal{L}_0^{-1}\mathcal{L}_t\gamma) = \widehat{s}(\gamma), \qquad \widehat{u}(\mathcal{L}_0^{-1}\mathcal{L}_t\gamma) = Q\widehat{u}(\gamma), \qquad \widehat{T}(\mathcal{L}_0^{-1}\mathcal{L}_t\gamma) = Q\widehat{T}(\gamma)Q^T. \tag{2.178}$$

When expressed in this form, material frame-indifference is sometimes referred to as *invariance with respect to superposed rigid-body motion*. In Eqn. (2.178), γ represents a generic argument of the constitutive relation that can itself be a scalar, vector or tensor quantity. The operator $\mathcal{L}_0^{-1}\mathcal{L}_t$ is related to the way in which the argument γ transforms between frames of reference.[81] For objective scalars (like the entropy s and temperature T),

80 Material frame-indifference is generally accepted as a fundamental principle of continuum mechanics. However, this point of view is not without controversy. Some authors claim that this principle is not a principle at all, but an approximation which is valid as long as macroscopic time and length scales are large relative to microscopic phenomena. Our view is that this is essentially a debate over semantics. Material frame-indifference is a principle for *intrinsic* constitutive relations as they are defined in continuum mechanics. However, these relations are an idealization of a more complex physical reality that is not necessarily frame-indifferent. See Section 6.3.7 of [TME12] for more details on this controversy.

81 The variables γ and γ^+ are given relative to two frames of reference $\mathcal{F}$ and $\mathcal{F}^+$. The operator $\mathcal{L}_t$ is the mapping taking γ to γ^+ at time t, i.e. $\gamma^+ = \mathcal{L}_t\gamma$. The inverse mapping is $\gamma = \mathcal{L}_t^{-1}\gamma^+$. See Section 6.3.4 of [TME12] for a detailed discussion.

objective vectors (like the temperature gradient ∇T and relative position vectors $\boldsymbol{r}_{12} = \boldsymbol{x}_1 - \boldsymbol{x}_2$) and objective tensors (like the rate of deformation tensor $\boldsymbol{d}$ and stress tensor $\boldsymbol{\sigma}$) the transformation relations are

$$\mathcal{L}_0^{-1}\mathcal{L}_t s = s, \qquad \mathcal{L}_0^{-1}\mathcal{L}_t \boldsymbol{u} = \boldsymbol{Q}\boldsymbol{u}, \qquad \mathcal{L}_0^{-1}\mathcal{L}_t \boldsymbol{T} = \boldsymbol{Q}\boldsymbol{T}\boldsymbol{Q}^T, \tag{2.179}$$

where as before $\boldsymbol{Q}$ is a proper orthogonal tensor. We will also require the transformations for the position of a point (and later an atom) $\boldsymbol{x}$, the deformation gradient $\boldsymbol{F}$ and the right Cauchy–Green deformation tensor $\boldsymbol{C}$. These are

$$\mathcal{L}_0^{-1}\mathcal{L}_t \boldsymbol{x} = \boldsymbol{Q}\boldsymbol{x} + \boldsymbol{c}, \qquad \mathcal{L}_0^{-1}\mathcal{L}_t \boldsymbol{F} = \boldsymbol{Q}\boldsymbol{F}, \qquad \mathcal{L}_0^{-1}\mathcal{L}_t \boldsymbol{C} = \boldsymbol{C}, \tag{2.180}$$

where in the leftmost relation, $\boldsymbol{c}$ is an arbitrary vector related to the relative translation between the frames. So, for example, the heat flux response function $\bar{\boldsymbol{q}}(s, \boldsymbol{F}, \nabla T)$ defined in Eqn. (2.164) is frame-indifferent only if

$$\bar{\boldsymbol{q}}(s, \boldsymbol{Q}\boldsymbol{F}, \boldsymbol{Q}\nabla T) = \boldsymbol{Q}\bar{\boldsymbol{q}}(s, \boldsymbol{F}, \nabla T)$$

for all proper orthogonal tensors $\boldsymbol{Q}$ and all values of s, $\boldsymbol{F}$ and ∇T. This clearly places constraints on the allowable functional forms for $\bar{\boldsymbol{q}}$. It can be shown (see Section 6.3.5 of [TME12]) that the material frame-indifference constraints on the response functions for u, T and $\boldsymbol{q}$ lead to the following *reduced constitutive relations*:[82]

$$u = \breve{u}(s, \boldsymbol{C}), \qquad T = \frac{\partial \breve{u}(s, \boldsymbol{C})}{\partial s}, \qquad \boldsymbol{q} = \boldsymbol{R}\breve{\boldsymbol{q}}(s, \boldsymbol{C}, \boldsymbol{R}^T \nabla T), \tag{2.181}$$

where $\boldsymbol{R}$ in the rightmost equation is the finite rotation part of the polar decomposition of the deformation gradient $\boldsymbol{F}$ (see Eqn. (2.57)). The simplest heat flux response function that satisfies Eqn. $(2.181)_3$ is Fourier's law, $\boldsymbol{q} = -k\nabla T$, where k is the thermal conductivity of the material. In this case the $\boldsymbol{R}$ terms cancel out. In more complex models the explicit dependence on $\boldsymbol{R}$ remains.

The elastic and viscous parts of the Cauchy stress tensor $\boldsymbol{\sigma}$ can also be expressed in reduced form:

$$\boldsymbol{\sigma}^{(e)} = 2\rho \boldsymbol{F} \frac{\partial \breve{u}(s, \boldsymbol{C})}{\partial \boldsymbol{C}} \boldsymbol{F}^T, \qquad \boldsymbol{\sigma}^{(v)} = \boldsymbol{R}\breve{\boldsymbol{\sigma}}^{(v)}(s, \boldsymbol{C}, \boldsymbol{R}^T \boldsymbol{d}\boldsymbol{R})\boldsymbol{R}^T. \tag{2.182}$$

It is straightforward to show from Eqn. $(2.182)_1$ that the elastic part of the stress is symmetric. This implies that $\boldsymbol{\sigma}^{(v)}$ must also be symmetric in order not to violate the balance of

[82] Alternative expressions in terms of the right stretch tensor $\boldsymbol{U}$ or Lagrangian strain tensor $\boldsymbol{E}$ (instead of the right Cauchy–Green deformation tensor $\boldsymbol{C}$) can also be written.

angular momentum. The simplest constitutive relation for the viscous stress that satisfies Eqn. (2.182)$_2$ is a linear response model where the components of $\boldsymbol{\sigma}^{(v)}$ are proportional to those of $\boldsymbol{d}$. (A fluid exhibiting this behavior is called a *Newtonian fluid*.)

The stress expressions given above are for an isentropic process where the internal energy density is the appropriate energy variable. More commonly, experiments are performed under isothermal conditions for which the specific Helmholtz free energy ψ must be used, or more conveniently the strain energy density W defined in Eqn. (2.172). The reduced relations in this case for the Cauchy stress and the first and second Piola–Kirchhoff stresses are

$$\boldsymbol{\sigma}^{(e)} = \frac{2}{J}\boldsymbol{F}\frac{\partial \widetilde{W}(T,\boldsymbol{C})}{\partial \boldsymbol{C}}\boldsymbol{F}^T, \qquad \boldsymbol{P}^{(e)} = 2\boldsymbol{F}\frac{\partial \widetilde{W}(T,\boldsymbol{C})}{\partial \boldsymbol{C}}, \qquad \boldsymbol{S}^{(e)} = 2\frac{\partial \widetilde{W}(T,\boldsymbol{C})}{\partial \boldsymbol{C}}, \tag{2.183}$$

where $J = \det \boldsymbol{F}$ is the Jacobian of the deformation. As for the other constitutive variables, expressions in terms of $\boldsymbol{U}$ or $\boldsymbol{E}$ can also be written.

2.5.4 Material symmetry

Most materials possess symmetries which are reflected by their constitutive relations. Consider, for example, the deformation of a material with the two-dimensional square lattice structure shown in Fig. 2.10.[83] The unit cell and lattice vectors of the crystal are shown. In Fig. 2.10(a), the material is uniformly deformed with a deformation gradient $\boldsymbol{F}$, so that a particle $\boldsymbol{X}$ in the reference configuration is mapped to $\boldsymbol{x} = \boldsymbol{F}\boldsymbol{X}$ in the deformed configuration. The response of the material to the deformation is given by a constitutive relation, $\bar{g}(\boldsymbol{F})$, where $\bar{g}$ can be the internal energy density function $\bar{u}$, the temperature function $\bar{T}$, etc. Now consider a second scenario, shown in Fig. 2.10(b), where the material is first rotated by 90° counter-clockwise, represented by the proper orthogonal tensor (rotation) $\boldsymbol{H}$,

$$[\boldsymbol{H}] = \begin{bmatrix} 0 & -1 \\ 1 & 0 \end{bmatrix},$$

and then deformed by $\boldsymbol{F}$. One can think of this as a two-stage process. First, particles in the reference configuration are rotated to an intermediate stage with coordinates $\boldsymbol{y} = \boldsymbol{H}\boldsymbol{X}$. Second, the final positions in the deformed configuration are obtained by applying $\boldsymbol{F}$, so that $\boldsymbol{x} = \boldsymbol{F}\boldsymbol{y} = \boldsymbol{F}\boldsymbol{H}\boldsymbol{X}$. The constitutive relation is therefore evaluated at the deformation $\boldsymbol{F}\boldsymbol{H}$, the composition of the rotation followed by deformation. However, due to the symmetry of the crystal, the 90° rotation does not affect its response to the subsequent deformation. In fact, unless arrows are drawn on the material (as in the figure) it would be impossible to know whether the material was rotated or not prior to its deformation. Therefore, we must have that $\bar{g}(\boldsymbol{F}) = \bar{g}(\boldsymbol{F}\boldsymbol{H})$ for all $\boldsymbol{F}$. This is a constraint on the form of the constitutive relation due to the symmetry of the material.

[83] The concepts of lattice vectors and crystal structures are discussed extensively in Chapter 3. Here we will assume the reader has some basic familiarity with these concepts.

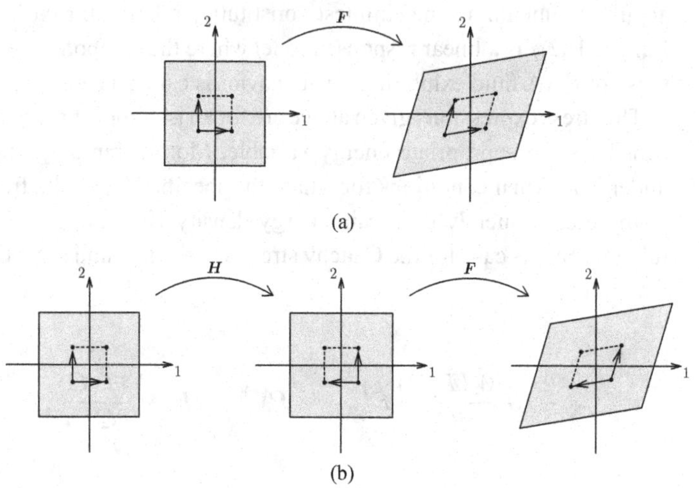

Fig. 2.10 A two-dimensional example of material symmetry. A material with a square lattice structure is (a) subjected to a homogeneous deformation F, or (b) first subjected to a rotation H by 90 degrees in the counter-clockwise direction and then deformed by F.

In general, depending on the symmetry of the material, there will be multiple transformations $\boldsymbol{H}$ that leave the constitutive relations invariant. We define the *material symmetry group* G of a material as the set of uniform density-preserving changes of its reference configuration that leave all of its constitutive relations unchanged [CN63]. Thus, G is the set of all second-order tensors $\boldsymbol{H}$ for which $\det \boldsymbol{H} = 1$ (density preserving) and for which

$$\begin{aligned} &\breve{u}(s, \boldsymbol{F}) = \breve{u}(s, \boldsymbol{F}\boldsymbol{H}), \qquad \breve{T}(s, \boldsymbol{F}) = \breve{T}(s, \boldsymbol{F}\boldsymbol{H}), \qquad \breve{\boldsymbol{q}}(s, \boldsymbol{F}, \nabla T) = \breve{\boldsymbol{q}}(s, \boldsymbol{F}\boldsymbol{H}, \nabla T), \\ &\breve{\boldsymbol{\sigma}}^{(e)}(s, \boldsymbol{F}) = \breve{\boldsymbol{\sigma}}^{(e)}(s, \boldsymbol{F}\boldsymbol{H}), \qquad \breve{\boldsymbol{\sigma}}^{(v)}(s, \boldsymbol{F}, \boldsymbol{d}) = \breve{\boldsymbol{\sigma}}^{(v)}(s, \boldsymbol{F}\boldsymbol{H}, \boldsymbol{d}), \end{aligned} \tag{2.184}$$

for all s, ∇T, $\boldsymbol{d}$ and $\boldsymbol{F}$ (i.e. all second-order tensors with positive determinants). Note that the symmetry relations for mixed and material tensors take slightly different forms than those shown in Eqn. (2.184). For example the relations for the elastic part of the first and second Piola–Kirchhoff stress tensors are

$$\breve{\boldsymbol{P}}^{(e)}(s, \boldsymbol{F}) = \breve{\boldsymbol{P}}^{(e)}(s, \boldsymbol{F}\boldsymbol{H})\boldsymbol{H}^T \quad \text{and} \quad \breve{\boldsymbol{S}}^{(e)}(s, \boldsymbol{F}) = \boldsymbol{H}\breve{\boldsymbol{S}}^{(e)}(s, \boldsymbol{F}\boldsymbol{H})\boldsymbol{H}^T.$$

These may be obtained directly from Eqn. (2.184) by substituting Eqns. $(2.96)_2$ and (2.100).[84]

An important material symmetry group for solids is the *proper orthogonal group* $SO(3)$ already encountered in Section 2.1.4. A member of this group represents a rigid-body rotation of the material. Materials possessing this symmetry are *isotropic*. They have the same constitutive response regardless of how they are rotated before being deformed.

[84] When substituting on the right-hand side of Eqn. (2.185) do not forget that Eqns. $(2.96)_2$ and (2.100) relate $\breve{\boldsymbol{\sigma}}(s, \boldsymbol{F})$ to $\breve{\boldsymbol{P}}(s, \boldsymbol{F})$ and $\breve{\boldsymbol{S}}(s, \boldsymbol{F})$, respectively.

The symmetry constraints described above together with material frame-indifference can be used to derive simplified stress constitutive relations. See Section 6.4 of [TME12] for examples of how this is done.

2.5.5 Linearized constitutive relations for anisotropic hyperelastic solids

An *anisotropic* material has different properties along different directions and therefore has less symmetry than the isotropic materials discussed above. The term *hyperelastic*, defined on page 94, means that the material has no dissipation and that an energy function exists for it. The stress then follows as the gradient of the energy function with respect to a conjugate strain variable. For example the Piola–Kirchhoff stress tensors for a hyperelastic material are given in Eqn. (2.173) and reproduced here for convenience (dropping the functional dependence on T for notational simplicity):

$$\boldsymbol{S}^{(e)} = \frac{\partial \widetilde{W}(\boldsymbol{E})}{\partial \boldsymbol{E}}, \qquad \boldsymbol{P}^{(e)} = \frac{\partial \widehat{W}(\boldsymbol{F})}{\partial \boldsymbol{F}}. \tag{2.185}$$

Additional constraints on these functional forms can be obtained by considering material symmetry (as discussed in the previous section). This together with carefully planned experiments can then be used to construct *phenomenological* (i.e. fitted) models for the nonlinear material response (see examples in Section 6.4 of [TME12]). Alternatively, $\widetilde{\boldsymbol{S}}(\boldsymbol{E})$ can be computed directly from an atomistic model as explained in Chapter 11. A third possibility that is often used in numerical solutions to continuum boundary-value problems (see Section 2.6) is an incremental approach where the equations are linearized. This requires the calculation of *linearized constitutive relations* for the material which involve the definition of elasticity tensors. When the linearization is about the reference configuration of the material this approach leads to the well-known generalized Hooke's law.

The linearized form of Eqn. $(2.185)_1$ relates the increment of the second Piola–Kirchhoff stress $d\boldsymbol{S}$ to the increment of the Lagrangian strain $d\boldsymbol{E}$ and is given by

$$dS_{IJ} = C_{IJKL} dE_{KL}, \quad \Leftrightarrow \quad d\boldsymbol{S} = \boldsymbol{C} : d\boldsymbol{E}, \tag{2.186}$$

where $\boldsymbol{C}$ is a fourth-order tensor called the *material elasticity tensor* defined as

$$C_{IJKL} = \frac{\partial \widetilde{S}_{IJ}(\boldsymbol{E})}{\partial E_{KL}} = \frac{\partial^2 \widetilde{W}(\boldsymbol{E})}{\partial E_{IJ} \partial E_{KL}} \quad \Leftrightarrow \quad \boldsymbol{C} = \frac{\partial \widetilde{\boldsymbol{S}}(\boldsymbol{E})}{\partial \boldsymbol{E}} = \frac{\partial^2 \widetilde{W}(\boldsymbol{E})}{\partial \boldsymbol{E}^2}. \tag{2.187}$$

$\boldsymbol{C}$ has the following symmetries:

$$C_{IJKL} = C_{JIKL} = C_{IJLK}, \qquad C_{IJKL} = C_{KLIJ}. \tag{2.188}$$

The equalities on the left are called the *minor* symmetries of $\boldsymbol{C}$ and are due to the symmetry of $\boldsymbol{S}$ and $\boldsymbol{E}$. The equality on the right is called the *major* symmetry of $\boldsymbol{C}$ and follows from the definition of $\boldsymbol{C}$ as the second derivative of an energy with respect to strain where the order of differentiation is unimportant.

Similarly, we may obtain the relationship between increments of the first Piola–Kirchhoff stress $d\boldsymbol{P}$ and the deformation gradient $d\boldsymbol{F}$ by linearizing Eqn. $(2.185)_2$:

$$dP_{iJ} = D_{iJkL}\, dF_{kL}, \quad \Leftrightarrow \quad d\boldsymbol{P} = \boldsymbol{D} : d\boldsymbol{F}, \tag{2.189}$$

where $\boldsymbol{D}$ is the *mixed elasticity tensor* given by

$$D_{iJkL} = \frac{\partial \widehat{P}_{iJ}(\boldsymbol{F})}{\partial F_{kL}} = \frac{\partial^2 \widehat{W}(\boldsymbol{F})}{\partial F_{iJ} \partial F_{kL}} \quad \Leftrightarrow \quad \boldsymbol{D} = \frac{\partial \widehat{\boldsymbol{P}}(\boldsymbol{F})}{\partial \boldsymbol{F}} = \frac{\partial^2 \widehat{W}(\boldsymbol{F})}{\partial \boldsymbol{F}^2}. \tag{2.190}$$

$\boldsymbol{D}$ does not have the minor symmetries that $\boldsymbol{C}$ does since $\boldsymbol{P}$ and $\boldsymbol{F}$ are not symmetric. However, for a hyperelastic material it does possess the major symmetry, $D_{iJkL} = D_{kLiJ}$, due to invariance with respect to the order of differentiation. The mixed and material elasticity tensors are related by[85]

$$D_{iJkL} = C_{IJKL} F_{iI} F_{kK} + \delta_{ik} S_{JL}. \tag{2.191}$$

For practical reasons, it is often useful to treat the deformed configuration as a new reference configuration and then consider increments of deformation and stress measured from this configuration. This leads to the following relationship:[85]

$$\mathring{\sigma}_{ij} = c_{ijkl} \dot{\epsilon}_{kl}, \tag{2.192}$$

where

$$\mathring{\boldsymbol{\sigma}} \equiv \dot{\boldsymbol{\sigma}} - \boldsymbol{l}\boldsymbol{\sigma} - \boldsymbol{\sigma}\boldsymbol{l}^T + \boldsymbol{\sigma}\, \mathrm{tr}\, \boldsymbol{l} \tag{2.193}$$

is the objective *Truesdell stress rate* of the Cauchy stress tensor [Hol00], $\dot{\boldsymbol{\epsilon}}$ is the time rate of change of the small strain tensor and $\mathbf{c}$ is the *spatial elasticity tensor*. Note that $\mathbf{c}$ has the same minor and major symmetries as its material counterpart:

$$c_{ijkl} = c_{jikl} = c_{ijlk} = c_{klij}. \tag{2.194}$$

[85] See Section 6.5 of [TME12] for the derivation.

It can be shown that $\mathbf{c}$ is related to the material and mixed elasticity tensors, $\boldsymbol{C}$ and $\boldsymbol{D}$, by[86]

$$c_{ijkl} = J^{-1} F_{iI} F_{jJ} F_{kK} F_{lL} C_{IJKL}, \qquad c_{ijkl} = J^{-1} \left(F_{jJ} F_{lL} D_{iJkL} - \delta_{ik} F_{lL} P_{jL} \right). \tag{2.195}$$

In these relations it is understood that $\boldsymbol{C}$ and $\boldsymbol{D}$ are evaluated at the deformed configuration corresponding to the new reference configuration and $\boldsymbol{F}$ is the deformation gradient from the original reference configuration to the new one.[86]

Generalized Hooke's law and the elasticity matrix When the new reference configuration considered above is taken to be the same as the original reference configuration (which is assumed to be stress free), then $\boldsymbol{F} = \boldsymbol{I}$, $J = 1$ and $\boldsymbol{\sigma} = \boldsymbol{P} = \boldsymbol{S} = \mathbf{0}$. In this case, the relations in Eqn. (2.195) show that all of the elasticity tensors are the same. For the corresponding linearized stress–strain relations the distinctions between the various stress measures and the various conjugate deformation measures vanish. Accordingly, the single relation can be written as

$$\sigma_{ij} = c_{ijkl} \epsilon_{kl} \quad \Leftrightarrow \quad \boldsymbol{\sigma} = \mathbf{c} : \boldsymbol{\epsilon}, \tag{2.196}$$

which is valid for small stresses and small strains. This is called the *generalized Hooke's law*. The fourth-order tensor $\mathbf{c}$ is the *elasticity tensor*. (The epithet "spatial" is dropped since all elasticity tensors are the same. The term "small strain elasticity tensor" is also used.) Hooke's law can also be inverted to relate strain to stress:

$$\epsilon_{ij} = s_{ijkl} \sigma_{kl} \quad \Leftrightarrow \quad \boldsymbol{\epsilon} = \boldsymbol{s} : \boldsymbol{\sigma}, \tag{2.197}$$

where $\boldsymbol{s}$ is the *compliance tensor*. The corresponding strain energy density function, W, is

$$W = \frac{1}{2} \sigma_{ij} \epsilon_{ij} = \frac{1}{2} c_{ijkl} \epsilon_{ij} \epsilon_{kl} = \frac{1}{2} s_{ijkl} \sigma_{ij} \sigma_{kl}. \tag{2.198}$$

In the above relations, we assumed a stress-free reference configuration. If this is not the case, then an additional constant stress term $\boldsymbol{\sigma}^0$ is added to Eqn. (2.196), $\boldsymbol{\sigma}$ is replaced by $\boldsymbol{\sigma} - \boldsymbol{\sigma}^0$ in Eqn. (2.197) and the energy expression has an additional term linear in strain,[87] $(\boldsymbol{\sigma}^0 : \boldsymbol{\epsilon})/2$. In addition, a constant reference strain energy density W_0 can always be added to W.

[86] See Section 6.5 of [TME12].

[87] Note, however, that the resulting stress–strain relations are no longer linear.

Due to the symmetry of the stress and strain tensors, it is convenient to write Eqn. (2.196) in a contracted matrix notation referred to as *Voigt notation*, where pairs of indices in the tensor notation are replaced with a single index in the matrix notation (see also Tab. 2.3):

tensor indices ij:	11	22	33	23, 32	13, 31	12, 21
matrix index m:	1	2	3	4	5	6

Using this notation, the generalized Hooke's law (Eqn. (2.196)) is

$$\begin{bmatrix}\sigma_{11}\\ \sigma_{22}\\ \sigma_{33}\\ \sigma_{23}\\ \sigma_{13}\\ \sigma_{12}\end{bmatrix} = \begin{bmatrix} \mathsf{c}_{11} & \mathsf{c}_{12} & \mathsf{c}_{13} & \mathsf{c}_{14} & \mathsf{c}_{15} & \mathsf{c}_{16}\\ \mathsf{c}_{21} & \mathsf{c}_{22} & \mathsf{c}_{23} & \mathsf{c}_{24} & \mathsf{c}_{25} & \mathsf{c}_{26}\\ \mathsf{c}_{31} & \mathsf{c}_{32} & \mathsf{c}_{33} & \mathsf{c}_{34} & \mathsf{c}_{35} & \mathsf{c}_{36}\\ \mathsf{c}_{41} & \mathsf{c}_{42} & \mathsf{c}_{43} & \mathsf{c}_{44} & \mathsf{c}_{45} & \mathsf{c}_{46}\\ \mathsf{c}_{51} & \mathsf{c}_{52} & \mathsf{c}_{53} & \mathsf{c}_{54} & \mathsf{c}_{55} & \mathsf{c}_{56}\\ \mathsf{c}_{61} & \mathsf{c}_{62} & \mathsf{c}_{63} & \mathsf{c}_{64} & \mathsf{c}_{65} & \mathsf{c}_{66} \end{bmatrix} \begin{bmatrix}\epsilon_{11}\\ \epsilon_{22}\\ \epsilon_{33}\\ 2\epsilon_{23}\\ 2\epsilon_{13}\\ 2\epsilon_{12}\end{bmatrix}, \qquad (2.199)$$

where c is the *elasticity matrix*.[88] The entries c_{mn} of the elasticity matrix are referred to as the *elastic constants*. c is therefore also called the "elastic constants matrix". The stress and strain tensors can also be expressed in compact notation by defining the column matrices,

$$\boldsymbol{\sigma} = [\sigma_{11}, \sigma_{22}, \sigma_{33}, \sigma_{23}, \sigma_{13}, \sigma_{12}]^T, \qquad \boldsymbol{\epsilon} = [\epsilon_{11}, \epsilon_{22}, \epsilon_{33}, 2\epsilon_{23}, 2\epsilon_{13}, 2\epsilon_{12}]^T.$$

Hooke's law is then

$$\sigma_m = \mathsf{c}_{mn}\epsilon_n \quad \text{or} \quad \epsilon_m = \mathsf{s}_{mn}\sigma_n, \qquad (2.200)$$

where $\mathsf{s} = \mathsf{c}^{-1}$ is the *compliance matrix*.[89]

The minor symmetries of c_{ijkl} (and s_{ijkl}) are automatically accounted for in c_{mn} (and s_{mn}) by the Voigt notation. The major symmetry of c_{ijkl} (and s_{ijkl}) implies that c_{mn} (and s_{mn}) are symmetric, i.e. $\mathsf{c}_{mn} = \mathsf{c}_{nm}$ (and $\mathsf{s}_{mn} = \mathsf{s}_{nm}$). Therefore in the most general case a material can have 21 independent elastic constants. The number of independent constants is reduced as the symmetry of the material increases (see Section 6.5.1 of [TME12]). In particular for materials with cubic symmetry (which will often be considered in this book)

[88] Note that we use a sans serif font for the elasticity matrix. This stresses the fact that the numbers that constitute this 6×6 matrix are not the components of a second-order tensor in a six-dimensional space and therefore do not transform according to standard transformation rules.

[89] Note, however, that the fourth-order tensor $\boldsymbol{s} \neq \boldsymbol{c}^{-1}$. This is because, strictly speaking, $\boldsymbol{c}$ is not invertible, since $\boldsymbol{c} : \boldsymbol{w} = \boldsymbol{0}$, where $\boldsymbol{w} = -\boldsymbol{w}^T$ is any anti-symmetric second-order tensor. This indicates that $\boldsymbol{w}$ is an "eigen-tensor" of $\boldsymbol{c}$ associated with the eigenvalue 0, and further, implies that $\boldsymbol{c}$ is not invertible. However, if $\boldsymbol{c}$ and $\boldsymbol{s}$ are viewed as linear mappings from the space of all *symmetric* second-order tensors to itself (as opposed to the space of *all* second-order tensors), then $\boldsymbol{c} : \boldsymbol{w}$ is not a valid operation. In this sense, $\boldsymbol{c}$ is invertible and only then do we have that $\boldsymbol{s} = \boldsymbol{c}^{-1}$.

there are only three independent elastic constants, c_{11}, c_{12} and c_{44}, with

$$\mathbf{c} = \begin{bmatrix} c_{11} & c_{12} & c_{12} & 0 & 0 & 0 \\ & c_{11} & c_{12} & 0 & 0 & 0 \\ & & c_{11} & 0 & 0 & 0 \\ & & & c_{44} & 0 & 0 \\ & \text{sym} & & & c_{44} & 0 \\ & & & & & c_{44} \end{bmatrix}. \tag{2.201}$$

For isotropic symmetry, $c_{44} = (c_{11} - c_{22})/2$, and so an isotropic material has only two independent elastic constants. The elasticity tensor for this special case can be written as

$$c_{ijkl} = \lambda \delta_{ij}\delta_{kl} + \mu(\delta_{ik}\delta_{jl} + \delta_{il}\delta_{jk}), \tag{2.202}$$

where $\lambda = c_{12}$ and $\mu = c_{44} = (c_{11} - c_{12})/2$ are called the Lamé constants (μ is also called the *shear modulus*). Substituting Eqn. (2.202) into Eqn. (2.196), we obtain *Hooke's law* for an isotropic linear elastic material:

$$\sigma_{ij} = \lambda(\epsilon_{kk})\delta_{ij} + 2\mu\epsilon_{ij} \quad \Leftrightarrow \quad \boldsymbol{\sigma} = \lambda(\operatorname{tr}\boldsymbol{\epsilon})\boldsymbol{I} + 2\mu\boldsymbol{\epsilon}. \tag{2.203}$$

Equation (2.203) can be inverted, in which case it is more conveniently expressed in terms of two other material parameters, *Young's modulus*, E, and *Poisson's ratio*, ν,

$$\boldsymbol{\epsilon} = -\frac{\nu}{E}(\operatorname{tr}\boldsymbol{\sigma})\boldsymbol{I} + \frac{1+\nu}{E}\boldsymbol{\sigma}. \tag{2.204}$$

The two sets of material parameters are related through

$$\mu = \frac{E}{2(1+\nu)}, \quad \lambda = \frac{\nu E}{(1+\nu)(1-2\nu)} \quad \text{or} \quad E = \frac{\mu(3\lambda + 2\mu)}{\lambda + \mu}, \quad \nu = \frac{\lambda}{2(\lambda + \mu)}. \tag{2.205}$$

Equation (2.204) can be reduced to the simple and familiar one-dimensional case by setting all stresses to zero, except $\sigma_{11} = \sigma$, and solving for the strains. The result is the one-dimensional Hooke's law:

$$\sigma = E\epsilon,$$

where $\epsilon = \epsilon_{11}$ is the strain in the 1-direction.

2.6 Boundary-value problems and the principle of minimum potential energy

At this stage, we have a complete framework for continuum mechanics including the mechanical balance laws of Section 2.3, the laws of thermodynamics of Section 2.4 and the constitutive relations of Section 2.5. We now pull all of these together into a formal *problem statement* which consists of three distinct parts: (1) the partial differential *field equations* to be satisfied; (2) the *unknown fields* that constitute the solution of the problem and the relations between them; and (3) the *prescribed data* which include everything else that is required to make the problem one which *can* be solved. If we are interested in the dynamic response of a system, then the problem is referred to as an *initial boundary-value problem* and its three parts will all have a temporal component. If we are only interested in the static equilibrium state of our system, then the term *boundary-value problem* is used.

Continuum mechanics problems can be formulated within the spatial description (Eulerian description) or the material description (Lagrangian description). The former category is most useful for fluid mechanics problems and the latter category is most applicable to solid mechanics problems. In this section we focus on purely mechanical static problems (i.e. in the limit of 0 K) in the material description. We also limit the discussion to hyperelastic materials where the stress in the material is given by the derivative of a strain energy density function with respect to strain (Eqn. (2.173)). A more general description for all cases is given in Section 7.1 of [TME12].

In the material description the balance of linear momentum is given by Eqn. (2.95) in terms of the first Piola–Kirchhoff stress $\boldsymbol{P}$. Under static conditions this equation reduces to

$$\operatorname{Div} \boldsymbol{P} + \rho_0 \boldsymbol{b} = \boldsymbol{0}, \quad \boldsymbol{X} \in \mathcal{B}_0, \tag{2.206}$$

where ρ_0 is the reference mass density, $\boldsymbol{b}$ is the body force and $\mathcal{B}_0$ is the domain occupied by the body in the reference configuration. Since the material is hyperelastic, $\boldsymbol{P} = \partial W/\partial \boldsymbol{F}$, where W is a known strain energy density function (see Eqn. (2.172)) and $\boldsymbol{F}_\varphi = \nabla_0 \varphi$. (Note that in this section, a subscript φ (as in $\boldsymbol{F}_\varphi$) is used to indicate the explicit dependence of a variable on the deformation mapping function.)

Equation (2.206) has to be solved for the deformation mapping field $\varphi(\boldsymbol{X})$. To do so, boundary conditions must be specified at *each* point on the boundary of $\mathcal{B}_0$ *one* quantity for *each unknown field component*. Since the deformation mapping is a vector quantity we must specify three values associated with the deformation mapping at each boundary point; one value for each spatial direction. These values can correspond to either the components of the traction or the position:

$$\boldsymbol{P}(\boldsymbol{F}_\varphi)\boldsymbol{N}(\boldsymbol{X}) = \bar{\boldsymbol{T}}(\boldsymbol{X}) \quad \text{or} \quad \varphi(\boldsymbol{X}, t) = \bar{\boldsymbol{x}}(\boldsymbol{X})$$

for all $\boldsymbol{X} \in \partial\mathcal{B}_0$. Here $\boldsymbol{N}(\boldsymbol{X})$ is the outward unit normal of $\partial\mathcal{B}_0$ and $\bar{\boldsymbol{T}}(\boldsymbol{X})$ and $\bar{\boldsymbol{x}}(\boldsymbol{X})$ are specified fields of external reference tractions and positions applied to the surfaces of the body, respectively. It is worth pointing out that "free surfaces," i.e. parts of the body where no forces and no positions are applied, are described as traction boundary conditions with $\bar{\boldsymbol{T}} = \boldsymbol{0}$. Often position boundary conditions are provided in terms of displacements from the reference configuration. In this case, the position boundary condition reads

$$\varphi(\boldsymbol{X}) = \boldsymbol{X} + \bar{\boldsymbol{u}}(\boldsymbol{X}), \qquad \boldsymbol{X} \in \partial\mathcal{B}_0,$$

where $\bar{\boldsymbol{u}}(\boldsymbol{X})$ is the specified boundary displacement field. Clearly the two forms are related by $\bar{\boldsymbol{x}}(\boldsymbol{X}) \equiv \boldsymbol{X} + \bar{\boldsymbol{u}}(\boldsymbol{X})$. It is also possible to combine traction and displacement boundary conditions. In this case the boundary is divided into a part $\partial\mathcal{B}_{0t}$ where traction boundary conditions are applied and a part $\partial\mathcal{B}_{0u}$ where displacement boundary conditions are applied, such that $\partial\mathcal{B}_{0t} \cup \partial\mathcal{B}_{0u} = \partial\mathcal{B}_0$ and $\partial\mathcal{B}_{0t} \cap \partial\mathcal{B}_{0u} = \emptyset$. The resulting *mixed boundary conditions* are

$$\boldsymbol{P}(\boldsymbol{F}_\varphi)\boldsymbol{N}(\boldsymbol{X}) = \bar{\boldsymbol{T}}(\boldsymbol{X}), \qquad \boldsymbol{X} \in \partial\mathcal{B}_{0t},$$
$$\boldsymbol{\varphi}(\boldsymbol{X}) = \bar{\boldsymbol{x}}(\boldsymbol{X}), \qquad \boldsymbol{X} \in \partial\mathcal{B}_{0u}.$$

Finally, *mixed-mixed boundary conditions* are also possible where a point on the surface may have a position boundary condition along some directions and traction boundary conditions along the others. See Section 7.1 of [TME12] for details.

Principles of stationary and minimum potential energy The boundary-value problem described above can be reformulated as a *variational* problem. This means that we seek to write the problem in such a way that its solution is the stationary point (maximum, minimum or saddle point) of some energy functional. We will see that *stable* equilibrium solutions correspond to minima of this functional.

The appropriate energy functional for a static continuum mechanics boundary-value problem is the *total potential energy* Π. The total potential energy is defined as the strain energy stored in the body together with the potential of the applied loads:

$$\Pi_\varphi = \int_{\mathcal{B}_0} W(\boldsymbol{F}_\varphi)\, dV_0 - \int_{\mathcal{B}_0} \rho_0 \boldsymbol{b} \cdot \boldsymbol{\varphi}\, dV_0 - \int_{\partial\mathcal{B}_{0t}} \bar{\boldsymbol{T}} \cdot \boldsymbol{\varphi}\, dA_0. \tag{2.207}$$

We postulate the following variational principle.

Principle of stationary potential energy Given the set of admissible deformation mapping fields for a conservative continuum system, an equilibrium state will correspond to one that stationarizes the total potential energy.

Here, an *admissible* deformation mapping field is one that satisfies the position boundary conditions. The proof of this principle (given in Section 7.2 of [TME12]) shows that stationary points of Π identically satisfy the static equilibrium equation in Eqn. (2.206).

The principle of stationary potential energy does not distinguish between stable and unstable forms of equilibrium. Normally, however, we are interested in states of stable equilibrium. To identify these we supplement the principle of stationary potential energy with the following principle.

Principle of minimum potential energy If a stationary point of the potential energy corresponds to a (local) isolated minimum, then the equilibrium is stable.

A proof for *finite-dimensional systems*[90] (although not the first) was given by Warner Koiter, the "father" of modern stability theory [Koi65c]. Mathematically, the principle of minimum potential energy for a finite-dimensional system with total potential energy $\Pi(\boldsymbol{z})$, where $\boldsymbol{z} \in \mathbb{R}^n$ and n is the number of degrees of freedom, corresponds to the requirement that the Hessian (second gradient) of the potential energy is positive definite. We write

$$\left(\frac{\partial^2 \Pi}{\partial \boldsymbol{z} \partial \boldsymbol{z}} \delta \boldsymbol{z}\right) \cdot \delta \boldsymbol{z} > 0 \quad \forall \delta \boldsymbol{z} \neq \mathbf{0}.$$

This is guaranteed to be satisfied if all of the eigenvalues of the Hessian, $\partial^2 \Pi / \partial \boldsymbol{z}^2$, are positive. The principle of minimum potential energy can be extended to static continuum systems at a constant (nonzero) temperature (although the proof is nontrivial). In this case, the strain energy density is understood to be the Helmholtz free energy per unit volume. For a more detailed discussion of stability, see Sections 5.5.3 and 7.3 of [TME12].

The principle of minimum potential energy is used extensively in computational and theoretical mechanics, as well as atomistic simulations and multiscale methods. We will revisit it many times later in this book.

Further reading

- This chapter is a summary of the book *Continuum Mechanics and Thermodynamics: from Fundamental Concepts to Governing Equations*, written by the authors together with Ryan Elliott and also published by Cambridge University Press [TME12]. That book provides a stand-alone in-depth introduction to the subject consistent in spirit and notation with the rest of this book.
- Although published in 1969, Malvern's book [Mal69] continues to be considered the classic text in the field. It is not the best organized of books, but it is thorough and correct. It will be found on most continuum mechanicians' book shelves.
- A mathematically rigorous presentation is provided by Truesdell and Toupin's volume in the *Handbuch der Physik* [TT60]. This authoritative and comprehensive book presents the foundations of continuum mechanics in a deep and readable way. The companion book [TN65] (currently available as [TN04]) continues where [TT60] left off and discusses everything known (up to the original date of publication) regarding all manner of constitutive laws. Surprisingly approachable and in-depth, both of these books are a must read for those interested in the foundations of continuum mechanics and constitutive theory, respectively.
- Ogden's book [Ogd84] has long been considered to be an important classic text on the subject of nonlinear elastic materials. Mathematical in nature, it provides a high-level authoritative discussion of many topics not covered in other books.

[90] The situation for infinite-dimensional continuum systems is much more complex and, in the most general setting, the rigorous status of the principle of minimum potential energy is not known. However, despite the lack of a rigorous proof, the principle has long been applied to continuum problems with much empirical success [Koi65a, Koi65b].

- An excellent *very* concise and yet complete introduction to continuum mechanics is given by Chadwick [Cha99]. This book takes a self-work approach, where many details and derivations are left to the reader as exercises along the way.
- Another excellent, concise presentation of the subject aimed at the more advanced reader is that of Gurtin [Gur95]. More recently, Gurtin together with Fried and Anand have published a much larger book covering many advanced topics [GFA10], which can serve as an excellent reference for the advanced practitioner.
- Holzapfel's book [Hol00] presents a clear derivation of equations and provides a good review of tensor algebra. It also has a good presentation of constitutive relations used in different applications.
- Salençon's book [Sal01] provides a complete introduction from the viewpoint of the French school. The interested reader will find a number of differences in the philosophical approach to developing the basic theory. In this sense, the book nicely complements the above treatments.
- Truesdell's *A First Course in Rational Continuum Mechanics* [Tru77] is a highly mathematical treatment of the most basic foundational ideas and concepts on which the theory is based. This title is for the more mathematically inclined and/or advanced reader.
- Marsden and Hughes' book [MH94] is a modern, authoritative and highly mathematical presentation of the subject.
- Finally, we would like to mention a book by Jaunzemis [Jau67] that is not well known in the continuum mechanics community.[91] Published at about the same time as Malvern's book, Jaunzemis takes a completely different tack. Written with humor (a rare quality in a continuum text) it is a pleasure to read. Since the terminology and some of the principles are inconsistent with modern theory, it is not recommended for the beginner, but a more advanced reader will find it a refreshing read.

Exercises

2.1 [SECTION 2.1] Expand the following indicial expressions (all indices range from 1 to 3). Indicate the rank and the number of resulting expressions.

1. $a_i b_i$.
2. $a_i b_j$.
3. $\sigma_{ik} n_k$.
4. $A_{ij} x_i x_j$ ($\boldsymbol{A}$ is symmetric, *i.e.* $A_{ij} = A_{ji}$).

2.2 [SECTION 2.1] Simplify the following indicial expressions as much as possible (all indices range from 1 to 3).

1. $\delta_{mm} \delta_{nn}$.
2. $X_I \delta_{IK} \delta_{JK}$.
3. $B_{ij} \delta_{ij}$ ($\boldsymbol{B}$ is antisymmetric, i.e. $B_{ij} = -B_{ji}$).
4. $(A_{ij} B_{jk} - 2A_{im} B_{mk})\delta_{ik}$.
5. Substitute $A_{ij} = B_{ik} C_{kj}$ into $\phi = A_{mk} C_{mk}$.
6. $\epsilon_{ijk} a_i a_j a_k$.

[91] We'd like to thank Roger Fosdick for pointing out this book to us. Professor Fosdick studied with Walter Jaunzemis as an undergraduate and still has the original draft of the book that also included a discussion of the electrodynamics of continuous media which was dropped from the final book due to length constraints.

2.3 [SECTION 2.1] Write out the following expressions in indicial notation.
1. $A_{11} + A_{22} + A_{33}$.
2. $\mathbf{A}^T \mathbf{A}$ where $\mathbf{A}$ is a 3×3 matrix.
3. $A_{11}^2 + A_{22}^2 + A_{33}^2$.
4. $(u_1^2 + u_2^2 + u_3^2)(v_1^2 + v_2^2 + v_3^2)$.
5. $A_{11} = B_{11}C_{11} + B_{12}C_{21}, \quad A_{12} = B_{11}C_{12} + B_{12}C_{22},$
$A_{21} = B_{21}C_{11} + B_{22}C_{21}, \quad A_{22} = B_{21}C_{12} + B_{22}C_{22}.$

2.4 [SECTION 2.1] Obtain an expression for $\partial \boldsymbol{A}^{-1}/\partial \boldsymbol{A}$, where $\boldsymbol{A}$ is a second-order tensor. This expression turns up in Section 8.1.3 when computing stress in statistical mechanics systems. **Hint:** Start with the identity $A_{ik}^{-1}A_{kj} = \delta_{ij}$. Use indicial notation in your derivation.

2.5 [SECTION 2.1] Solve the following problems related to indicial notation for tensor field derivatives. In all cases indices range from 1 to 3. All variables are tensors and functions of the variables that they are differentiated by unless explicitly noted. The comma notation refers to differentiation with respect to $\boldsymbol{x}$.
1. Write out explicit expressions (i.e. ones that only have numbers as indices) for the following indicial expressions. In each case, indicate the rank and the number of the resulting expressions.
   a $\dfrac{\partial u_i}{\partial z_k}\dfrac{\partial z_k}{\partial x_j}$.
   b $\sigma_{ij,j} + \rho b_i = \rho a_i$.
   c $u_{k,j}\delta_{jk} - u_{i,i}$.
2. Expand out and then simplify the following indicial expressions as much as possible. Leave the expression in indicial form.
   a $(T_{ij}x_j)_{,i} - T_{ii}$.
   b $(x_m x_m x_i A_{ij})_{,k}$ ($\boldsymbol{A}$ is constant).
   c $(S_{ij}T_{jk})_{,ik}$.
3. Write out the following expressions in indicial notation.
   a $B_{i1}\dfrac{\partial c_1}{\partial x_j} + B_{i2}\dfrac{\partial c_2}{\partial x_j} + B_{i3}\dfrac{\partial c_3}{\partial x_j}$.
   b $\operatorname{div} \boldsymbol{v}$, where $\boldsymbol{v}$ is a vector.
   c $\dfrac{\partial^2 T_{11}}{\partial x_1^2} + \dfrac{\partial^2 T_{12}}{\partial x_1 \partial x_2} + \dfrac{\partial^2 T_{13}}{\partial x_1 \partial x_3} + \dfrac{\partial^2 T_{21}}{\partial x_2 \partial x_1} + \dfrac{\partial^2 T_{22}}{\partial x_2^2} + \dfrac{\partial^2 T_{23}}{\partial x_2 \partial x_3} + \dfrac{\partial^2 T_{31}}{\partial x_3 \partial x_1} + \dfrac{\partial^2 T_{32}}{\partial x_3 \partial x_2} + \dfrac{\partial^2 T_{33}}{\partial x_3^2}$.

2.6 [SECTION 2.2] The most general two-dimensional homogeneous finite strain deformation is defined by giving the spatial coordinates as linear homogeneous functions,

$$x_1 = X_1 + aX_1 + bX_2, \quad x_2 = X_2 + cX_1 + dX_2.$$

1. Express the components of the right Cauchy–Green deformation tensor $\boldsymbol{C}$ and Lagrangian strain $\boldsymbol{E}$ in terms of the given constants a, b, c, d. Display your answers in two matrices.
2. Calculate ds^2 and $ds^2 - dS^2$ for $d\boldsymbol{X}$ with components (dL, dL).

2.7 [SECTION 2.2] Consider a pure two-dimensional rotation by angle θ about the 3-axis. The deformation gradient for this case is:

$$[\boldsymbol{F}] = \begin{bmatrix} \cos\theta & -\sin\theta & 0 \\ \sin\theta & \cos\theta & 0 \\ 0 & 0 & 1 \end{bmatrix}.$$

1. Show that the Lagrangian strain tensor $\boldsymbol{E}$ is zero for this case.
2. Compute the small strain tensor and show that it is *not* zero for $\theta > 0$.
3. As an example, consider the case where $\theta = 30°$. Compute the small strain tensor for this case. Discuss the applicability of the small strain approximation.

2.8 [SECTION 2.2] Consider the following motion:

$$x_1 = (1 + p(t))X_1 + q(t)X_2, \quad x_2 = q(t)X_1 + (1 + p(t))X_2, \quad x_3 = X_3,$$

where $p(t) > 0$ and $q(t) > 0$ are parameters.

1. Compute the time-dependent deformation gradient $\boldsymbol{F}(t)$.
2. Compute the components of the rate of change of the deformation gradient $\dot{\boldsymbol{F}}$.
3. Compute the inverse deformation mapping, $\boldsymbol{X} = \varphi^{-1}(\boldsymbol{x}, t)$.
4. Verify that $\dot{F}_{iJ} = l_{ij}F_{jJ}$. **Hint:** You will need to compute $\boldsymbol{l}$ for this deformation and show that the result obtained from $l_{ij}F_{jJ}$ is equal to the result obtained above.

2.9 [SECTION 2.3] Show that the continuity equation in Eqn. (2.76) is identically satisfied for any deformation of the form

$$x_1 = \alpha_1(t)X_1, \qquad x_2 = \alpha_2(t)X_2, \qquad x_3 = \alpha_3(t)X_3,$$

where $\alpha_i(t)$ are differentiable scalar functions of time. The mass density field in the reference configuration is $\rho_0(\boldsymbol{X})$.

2.10 [SECTION 2.3] A bar made of a homogeneous *incompressible* material is stretched in the 1-direction by a pair of equal and opposite forces with magnitude R applied to its ends. Assume the deformation is uniform, the stretch in the 1-direction is α and that the bar contracts in the other two directions by an equal amount. Due to the (isotropic) symmetry of the material, no shearing takes place relative to the Cartesian coordinate system. The initial cross-section area of the bar is A_0.
1. Express the deformation gradient in the bar in terms of α.
2. Determine the components of the Cauchy stress and the first and second Piola–Kirchhoff stress tensors. Express your results in terms of α, R and A_0. What is the physical significance of these three stress measures?
3. Determine the plane of maximum shear stress in the deformed configuration and the value of the Cauchy shear stress on this plane.
4. Determine the material plane in the reference configuration corresponding to the plane of maximum shear stress found above. Plot the angle Θ between the normal to this plane and the horizontal axis as a function of α. Which plane does this tend to as $\alpha \to \infty$?

2.11 [SECTION 2.4] A closed isolated cylinder of volume V contains n moles of an ideal gas. The cylinder has a removable, frictionless piston that can be inserted at the end, quasistatically and adiabatically moved to a position where the available volume is $V/2$ and then quickly removed to allow the gas to freely expand back to the full volume of the cylinder. This procedure is repeated k times. The gas has a molar heat capacity at constant volume of C_v and a reference internal energy $\mathcal{U}_0$. The gas initially has temperature, T_{init}, internal energy, $\mathcal{U}_{\text{init}}$, pressure, p_{init}, and entropy, $\mathcal{S}_{\text{init}}$.
1. Obtain expressions for the temperature $T(k)$, pressure $p(k)$, internal energy $\mathcal{U}(k)$, and entropy $\mathcal{S}(k)$, after k repetitions of the procedure.
2. Plot $T(k)/T_{\text{init}}$ and $p(k)/p_{\text{init}}$ as a function of k. Use material constants for air.

2.12 [SECTION 2.4] Consider a one-dimensional system with temperature $T(x)$, heat flux $q(x)$, heat source density $r(x)$, mass density $\rho(x)$ and entropy density $s(x)$. Construct a one-dimensional differential element and show that for a reversible process the balance of entropy is

$$\rho\dot{s} = \frac{\rho r}{T} - \frac{\partial}{\partial x}\left(\frac{q}{T}\right),$$

in agreement with the Clausius–Duhem inequality. **Hint:** You will need to use the following expansion: $1/(1+\delta) \approx 1 - \delta + \delta^2 - \cdots$, where $\delta = dT/T \ll 1$, and retain only first-order terms.

2.13 [SECTION 2.5] A tensile test is a one-dimensional experiment where a material sample is stretched in a controlled manner to measure its response. The loading machine can control either the displacement, u, applied to the end of the sample (displacement control) or the force, f, applied to its end (load control). If displacement is controlled, the output is f/A_0, where A_0 is the reference cross-section area. If load is controlled, the output is $L/L_0 = (L_0 + u)/L_0$, where L_0 and L are the reference and deformed lengths of the sample. The mass of the sample is m.

Describe different experiments where the relevant thermodynamic potentials are: (i) internal energy density, u; (ii) Helmholtz free energy density, ψ; (iii) enthalpy density, h; (iv) Gibbs free energy density, g. In each case indicate what quantity is measured in the experiment (i.e. force or length) and provide an explicit expression for it in terms of m and the appropriate potential. **Hint:** You will need to consider thermal boundary conditions when setting up your experiments.

2.14 [SECTION 2.5] A material undergoes a homogeneous, time-dependent, simple shear motion with deformation gradient

$$[\boldsymbol{F}] = \begin{bmatrix} 1 & \gamma(t) & 0 \\ 0 & 1 & 0 \\ 0 & 0 & 1 \end{bmatrix},$$

where $\gamma(t) = \dot{\gamma}t$ is the shear parameter and the shear rate $\dot{\gamma}$ is constant. The material is elastic, incompressible and rubberlike with a Helmholtz free energy density given by $\psi = A(\operatorname{tr} \boldsymbol{B} - 3)$, where $\boldsymbol{B} = \boldsymbol{F}\boldsymbol{F}^T$ is the left Cauchy–Green deformation tensor, and A is a material constant. A material of this type is called a *neo-Hookean*.

1. For constant temperature conditions, show that the Cauchy stress for a neo-Hookean material is given by, $\boldsymbol{\sigma} = -p\boldsymbol{I} + \mu\boldsymbol{B}$, where p is the pressure, $\boldsymbol{I}$ is the identity tensor, $\mu = 2\rho_0 A$ is the shear modulus, and ρ_0 is the reference mass density.
2. Compute the Cauchy stress due to the imposed simple shear. Present your results as a 3×3 matrix of the components of $\boldsymbol{\sigma}$. Explicitly show the time dependence.

2.15 [SECTION 2.5] Under conditions of hydrostatic loading, $\boldsymbol{\sigma} = -p\boldsymbol{I}$, where p is the pressure, the *bulk modulus* B is defined as the negative ratio of the pressure and dilatation, $e = \operatorname{tr} \boldsymbol{\epsilon}$, so that $p = -Be$.

1. Starting with the generalized Hooke's law in Voigt notation in Eqn. (2.200), show that the bulk modulus for materials with cubic symmetry (see Eqn. (2.201)) is

$$B = \frac{c_{11} + 2c_{12}}{3}.$$

2. Show that for isotropic symmetry, the bulk modulus can also be expressed in terms of the Lamé constants as $B = \lambda + 2\mu/3$.

PART II

ATOMISTICS

3 Lattices and crystal structures

Crystalline materials were known from ancient times for their beautiful regular shapes and useful properties. Many materials of important technological value are crystalline, and we now understand that their characteristics are a result of the regular, repeating arrangement of atoms making up the crystal structure. The details of this crystal structure determine, for example, the elastic anisotropy of the material. It helps determine whether the crystal is ductile or brittle (or both depending on the direction of the applied loads). The crystallinity manifests itself in structural phase transformations, where materials change from one crystal structure to another under applied temperature or stress. Defects in crystals (discussed in Chapters 1, 6 and 12) determine the electrical and mechanical response of the material. Indeed, we saw in Chapter 1 that the starting point for understanding any of the properties of crystalline materials is the understanding of the underlying crystal structure itself.

3.1 Crystal history: continuum or corpuscular?

The evolution of the modern science of crystallography was a long time in coming. Here, we present a brief overview, partly based on the fascinating detailed history of this science in the article by J. N. Lalena [Lal06].

Prehistoric man used flint, a microcrystalline form of quartz peppered with impurities, to make tools and weapons. Most likely he never concerned himself with the inner structure of the material he was using. If pressed he would probably have adopted a continuum view of his material since clearly as he formed his tools the chips flying off were always just smaller pieces of flint. The first suggestion on record that materials have a discrete internal structure came in the fifth century BC with the work of Leucippus that was later extended by his student Democritus. Democritus viewed the world as being composed of atoms (from the Greek adjective *atomos* meaning "indivisible"), moving about ceaselessly in a void of nothingness. In Democritus' own words [Bak39]:

> by convention sweet is sweet, by convention bitter is bitter, by convention hot is hot, by convention cold is cold, by convention color is color. But in reality there are atoms and the void. That is, the objects of sense are supposed to be real and it is customary to regard them as such, but in truth they are not. Only the atom and the void are real.

These are words that ring true even to modern ears. Democritus' atoms came in a multitude of shapes and sizes that were related to the macroscopic properties we observe. So, for example, sweet materials were made of "round atoms which are not too small" and sour materials from "large, many angled atoms with the minimum of roundness" [Tay99]. Atoms repelled each other when coming too close together, otherwise forming clusters when tiny hooks on their surfaces become entangled [Ber04]. This was a completely materialistic view of nature devoid of a need for divine creation or purpose.

Democritus' view was strongly contested by Plato who in his book *Timaeus* lays out a world wholly in terms of a beneficent divine *craftsman*. Plato also differs from Democritus in his model for material structure. He agrees with the atomist view that materials are made of more basic particles, but suggests that these are particles made of four basic elements: earth, fire, water and air. The particles themselves have very definite shapes: cube for earth, tetrahedron for fire, icosahedron for water and octahedron for air. A fifth shape, the dodecahedron is associated with the universe as a whole. These are the five *platonic solids*. The faces of these solids can all be constructed from two kinds of right triangles: a scalene with angles 30°, 60°, 90° and an isosceles with angles 45°, 45°, 90°. These fundamental triangles were taken to be the indivisible atoms of the material. As was typical of Greek science, these conclusions were not drawn from observation of actual materials, but rather from a philosophical view of how things ought to be. Plato's dominance was such that this worldview eclipsed that of Democritus and the other atomists well into the seventeenth century. This was particularly due to the secular nature of the atomists' philosophy. In fact, all of Democritus' books were burned in the third and fourth centuries so that current knowledge of his work comes entirely from references to his work by others.

In 1611 the German astronomer Johannes Kepler wrote a short booklet *On the Six-cornered Snowflake* [Kep66] as a New Year's gift to his friend and patron Johann Matthäus Wacker von Wackenfels at the court of Emperor Rudolph II of Prague. In the booklet Kepler ponders the persistent six-fold symmetry of snowflakes (see Fig. 1.7(b)): "There must be some definite cause why, whenever snow begins to fall, its initial formations invariably display the shape of a six-cornered starlet." Kepler knew nothing about the atomic structure of crystalline materials. He discarded the possibility that this symmetry has something to do with the internal structure of snow since "the stuff of snow is vapor" that has no definite structure. He therefore attempted to answer the question by comparing with other structures that have similar symmetries in nature such as honeycombs and the packing of seeds in a pomegranate. In a pomegranate, for example, he postulates that the seeds are arranged in the way that they are in order to obtain a maximum packing density.[1] He postulated that the highest possible packing density is obtained from cubic or hexagonal close packing.[2] This became known as *Kepler's conjecture*, a conjecture that was proven by the American mathematician Thomas Hales using an exhaustive computer search [Hal05]. Kepler suggested that perhaps snowflakes are similarly composed of "balls of vapor" packed together in a hexagonal pattern, but admitted that in this case efficiency cannot be the reason. Instead this could just be an innate property of the material, a "formative faculty," to be beautiful instilled in it by God. It is interesting that although Kepler did not have sufficient

[1] This line of thinking arose from the optimal manner to stack cannon balls raised by his English colleague Thomas Hariot. In fact Thomas Hariot pondered many of the issues raised by Kepler in his monograph including some early thoughts on the relation between close packing and the corpuscular theory of matter.

[2] The corresponding crystal structures, face-centered cubic and hexagonal close-packed are discussed in Section 3.6.2.

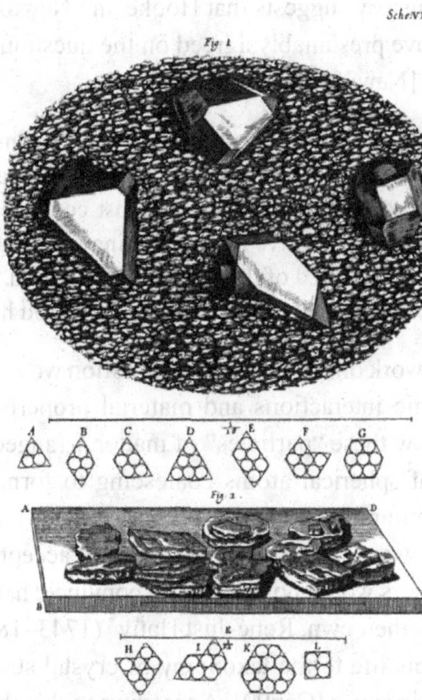

Fig. 3.1 A figure from Hooke's *Micrographia* [Hoo87], showing sketches of some crystals he observed and his proposal for the stacking of spheres to explain the crystal shapes.

information to answer his question, and actually made an incorrect assumption about the structure of snow, he did hit on the idea of sphere packing that is related to the correct answer although in a way completely different than he expected.

Fifty years later, the English scientist Robert Hooke came to a similar conclusion using the microscope newly invented by the Dutch Anton van Leeuwenhoek. Hooke's researches with the microscope are documented in his book *Micrographia* that appeared in 1665 [Hoo87]. In this rambling book, Hooke used the microscope to look at everything from needle points to "eels in vinegar."[3] Among this multitude he also considers the "crystalline bodies" found embedded in the fracture surfaces of flint stones (Fig. 3.1). He commented on the regularity of their shapes and the fact that similar shapes are observed in metals, minerals, precious stones and salts. Like Kepler, he then made a remarkable leap by suggesting that all such shapes can be constructed by packing spheres ("globular particles") together (Fig. 3.1). This insight appears to have been motivated from the observations, made earlier in the book, that immiscible liquids mixed together form spherical drops. Whatever his inspiration, Hooke admitted that he did not know what these particles are and proposed an eight-step research program to investigate this matter.[4]

[3] Presumably some sort of parasite.

[4] When making this proposal he laments the fact that he does not have the necessary time or the necessary assistance to carry out his proposed program. Apparently not much has changed in the scientific community in the 350 years since Hooke was active.

Although history suggests that Hooke and Newton were generally not on friendly terms, they would have presumably agreed on the question of the existence of atoms. In Newton's book *Opticks* [New30] he wrote

> it seems probable to me, that God in the beginning form'd matter in solid, massy, hard, impenetrable, moveable particles, of such sizes and figures, and with such properties, and in such proportions to space, as most conduced to the end for which he form'd them; and that these primitive particles being solids, are incomparably harder than any porous bodies compounded of them; even so very hard, as never to wear or to break in pieces; no ordinary power being able to divide what God himself made one in the first Creation.

Although his work on the three laws of motion would one day help to explain the connection between atomic interactions and material properties, Newton himself was silent on the question of how these "particles" of matter arranged themselves to form crystals.

The idea of spherical atoms coalescing to form solids continued to drift through the scientific community for nearly another century, until John Dalton (1766–1844) put forth in 1808 what would ultimately become the accepted atomic theory of matter. However, crystallographers would not be easy to convince, having been so strongly influenced by the work of one of their own. René-Just Haüy[5] (1743–1822) was a French crystallographer who had devoted his life to the taxonomy of crystal structures and to advancing his theory of *molécules intégrantes* [Cah99]. According to this theory, crystals were composed of small polyhedral particles that neatly slotted together to create the different observed crystal shapes, and Haüy strongly believed that crystallography could not be explained using the Kepler–Hooke idea of sphere packing. He was so convinced that he simply could not accept the findings of the German chemist Eilhardt Mitscherlich in 1818 who discovered that in some cases the same material can crystallize in different structures (a phenomenon that came to be called *polymorphism*), something that is impossible with *molécules intégrantes*. Haüy's influence was so strong that despite Mitscherlich's work, another 100 years would pass until in 1908 Jean-Baptiste Perrin's experimental studies of Brownian motion validated Albert Einstein's theoretical work on the subject, conclusively proving the atomic structure of matter and silencing all doubts.

It is often the case in science that when a theory proves difficult to overturn, it is because its successor, although clearly superior at explaining most of the experimental evidence, is still not telling the whole story elsewhere. There was no denying the atomic nature of matter, but spherical atoms could still not fully explain the variety of crystal shapes. It was ultimately the mathematicians, in the mid 1800s, who reconciled Haüy's *molécules intégrantes* with Dalton's atoms. Gabriel Delafosse (1796–1878) proposed that Haüy's molecules could be replaced by polyhedra with spherical atoms at the vertices. Thus, crystals were not composed of indivisible polyhedra, but spherical atoms forced to respect rigid polyhedral configurations with their neighbors. These polyhedra became the *unit cells* of the various crystal structures that would ultimately be classified by Bravais, and that we will discuss in detail in this chapter.

[5] The pronunciation of Haüy leaves the English tongue feeling like there is surely something missing from the word. The IPA notation is [aɥi], which is something like "a-ew-ee" with the opening syllable sounding *vaguely* like the "a" in "atom."

3.2 The structure of ideal crystals

A great many materials that are of interest to engineering applications adopt a crystalline structure. These include metals, ceramics, semiconductors and minerals. Their description must therefore be based on a clear mathematical framework for defining crystal structures. Let us start by clearly defining what we mean by a crystal:

> **Ideal crystal** An ideal crystal is an infinite structure formed by regularly repeating an atom or group of atoms, called a *basis*, on a space filling lattice.

This definition involves two new terms: *basis* and *lattice*. In the next section, we discuss at length the concept of the lattice before returning to general crystals and the introduction of the basis in Section 3.6.

As the term *ideal* suggests, this definition is the standard of perfection to which real crystals are compared. It describes a perfect crystal of infinite extent possessing no defects of any kind. It is important to realize that the requirement for infinity here is not arbitrary, since a finite crystal will not possess the ideal structure near its boundaries. It is possible today to manufacture crystals that are essentially free of internal defects[6] but these are, of course, finite in size. As such, the ideal crystal remains an unattainable concept.

3.3 Lattices

A lattice is an infinite space filling arrangement of points in a regular pattern. To be a lattice, the arrangement and orientation of all points viewed relative to any one point must be the same, no matter which vantage point is chosen. In other words, the arrangement must have *translational symmetry.* An example of a two-dimensional lattice and an arrangement of points that is not a lattice are shown in Fig. 3.2. It is important to note that the lattice points are *not* atoms. Rather, they are locations in space around which groups of atoms will be placed to form a physical crystal.

The properties of lattices were established by researchers in the nineteenth century. A particularly influential figure was the French physicist Auguste Bravais, whose name has become permanently linked with the lattice concept. In fact, the term *Bravais lattice* is commonly used interchangeably with *lattice*, especially in the context of three-dimensional lattices and crystals. We will often refer to the lattice points as *lattice sites* or *Bravais sites*.

[6] An example of this is the single crystal silicon that is used in the microelectronics industry. These crystals take the form of cylinders with diameters as large as 30 cm (astronomical by atomic standards), which are then sliced into the wafers that serve as the foundation for integrated circuits.

Fig. 3.2 The two-dimensional arrangement of points in (a) satisfies the definition of a lattice given in the text. However, the honeycomb pattern in (b) is not a lattice; the arrangement around each point is the same, but the orientation changes.

3.3.1 Primitive lattice vectors and primitive unit cells

How are lattices described mathematically? In other words, what equations generate a set of points that satisfy the definition of a lattice given above? The only possibility is to define the lattice points $\boldsymbol{R}$ using the following equation:[7]

$$\boldsymbol{R}^{[\boldsymbol{\ell}]} = \ell_i \hat{\boldsymbol{A}}_i, \quad \ell_i \in \mathbb{Z}, \tag{3.1}$$

where $\hat{\boldsymbol{A}}_i$ are three linearly independent vectors and $\mathbb{Z}$ is the set of all integers. As usual we use the summation convention (see Section 2.1.1). It is sometimes convenient to write this equation in the form

$$\boldsymbol{R}^{[\boldsymbol{\ell}]} = \hat{\mathsf{H}}\boldsymbol{\ell}, \tag{3.2}$$

where each column of the matrix $\hat{\mathsf{H}}$ is one of the vectors $\hat{\boldsymbol{A}}_i$:

$$\hat{\mathsf{H}} = \begin{bmatrix} [\hat{\boldsymbol{A}}_1]_1 & [\hat{\boldsymbol{A}}_2]_1 & [\hat{\boldsymbol{A}}_3]_1 \\ [\hat{\boldsymbol{A}}_1]_2 & [\hat{\boldsymbol{A}}_2]_2 & [\hat{\boldsymbol{A}}_3]_2 \\ [\hat{\boldsymbol{A}}_1]_3 & [\hat{\boldsymbol{A}}_2]_3 & [\hat{\boldsymbol{A}}_3]_3 \end{bmatrix}. \tag{3.3}$$

At the heart of this definition is a set of three vectors $\hat{\boldsymbol{A}}_1$, $\hat{\boldsymbol{A}}_2$, $\hat{\boldsymbol{A}}_3$ called the *primitive lattice vectors*. (The "hat" indicate that these are primitive lattice vectors as opposed to the "nonprimitive" lattice vectors described later.) These vectors are generally not orthogonal to each other and they must not all lie in the same plane.[8] The lattice is generated by taking all possible integer combinations of the primitive lattice vectors. As is the convention throughout this book, the use of capital letters $\boldsymbol{R}$ and $\hat{\boldsymbol{A}}_i$ indicates that they refer to a given reference configuration of the lattice.

To prove that Eqn. (3.1) defines a lattice we must show that it generates an identical arrangement of points about each point. To demonstrate this it is enough to show that the lattice is unchanged if we translate all points by $\boldsymbol{t} = n_i \hat{\boldsymbol{A}}_i$ $(n_i \in \mathbb{Z})$:

[7] The brackets indicate that the superscript is a lattice site. They are used to differentiate this case from the notation used other places in the book where $\boldsymbol{R}^{\alpha}$ is the position of atom number α in a finite set of atoms.

[8] Mathematically the condition is that $\hat{\boldsymbol{A}}_1$ and $\hat{\boldsymbol{A}}_2$ are not collinear and $\hat{\boldsymbol{A}}_3$ is not coplanar with the plane defined by $\hat{\boldsymbol{A}}_1$ and $\hat{\boldsymbol{A}}_2$.

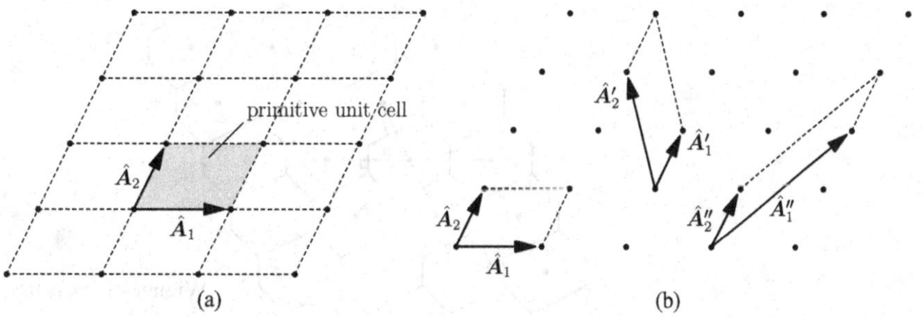

Fig. 3.3 A two-dimensional Bravais lattice. (a) The lattice vectors $\hat{\boldsymbol{A}}_i$ define a primitive cell that fills space when repeated. (b) Alternative choices for primitive lattice vectors.

Proof

$$\boldsymbol{R}^{[\boldsymbol{\ell}]} + \boldsymbol{t} = \ell_i \hat{\boldsymbol{A}}_i + n_i \hat{\boldsymbol{A}}_i = (\ell_i + n_i)\hat{\boldsymbol{A}}_i = \ell'_i \hat{\boldsymbol{A}}_i = \boldsymbol{R}^{[\boldsymbol{\ell}']}, \quad \ell'_i \in \mathbb{Z}. \tag{3.4}$$

□

This property of *translational invariance* is the most basic property of lattices, and indeed of all ideal crystals. Any vector, such as $\boldsymbol{t}$, taken as an integer combination of primitive lattice vectors is referred to as a *translation vector* of the lattice.

The primitive lattice vectors define a *unit cell* that, when repeated through space, generates the lattice. A two-dimensional example of lattice vectors and the unit cell they define is given in Fig. 3.3(a). The concept of a unit cell is a convenient idea that helps visualize the structure of the lattice. We can think of the lattice as being composed of an infinite number of primitive unit cells packed together in a space-filling pattern.

Algorithm 3.1 Producing a set of primitive lattice vectors

1: Select a lattice point P.
2: Pass a line through P that passes through other points in the lattice.
3: Define $\hat{\boldsymbol{A}}_1$ as the vector connecting P with one of its nearest neighbors on the line (there are two options).
4: Choose a point P' that is as close to the line defined in step 2 as any other (there is an infinite number of possibilities) and define $\hat{\boldsymbol{A}}_2$ as the vector connecting P with P'.
5: Choose a point P'' that is as close as any other point to the plane defined by $\hat{\boldsymbol{A}}_1$ and $\hat{\boldsymbol{A}}_2$ and passing through P (there is an infinite number of possibilities). P'' cannot be *on* the $\hat{\boldsymbol{A}}_1 - \hat{\boldsymbol{A}}_2$ plane. Define $\hat{\boldsymbol{A}}_3$ as the vector connecting P with P''.
6: The vectors $\hat{\boldsymbol{A}}_1$, $\hat{\boldsymbol{A}}_2$, $\hat{\boldsymbol{A}}_3$ form a set of primitive lattice vectors for the lattice.

What complicates the mathematics of crystal lattices is the fact that the choice of primitive lattice vectors for a given lattice is not unique. Figure 3.3(b) shows a number of examples in two dimensions. There are in fact an infinite number of possibilities, but the choice is not arbitrary. Primitive lattice vectors must connect lattice points and the primitive unit cell they define must contain only one lattice point. When calculating the number of points contained in a unit cell, the lattice points at the corners of the cell are shared equally amongst all cells

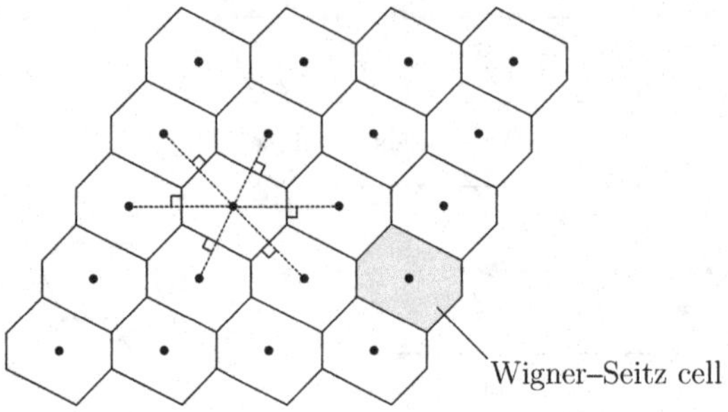

Fig. 3.4 Voronoi tessellation for a lattice. The Voronoi cell is the Wigner–Seitz cell of the lattice.

in contact with that point. In two dimension the corner points will contribute $4 \times 1/4 = 1$ lattice points and similarly in three dimensions they contribute $8 \times 1/8 = 1$. This means that a primitive unit cell cannot contain internal lattice points. Algorithm 3.1 (given as an exercise in [AM76]) provides a simple recipe for obtaining all of the possible primitive lattice vectors for a given lattice. The convention is to work with the set of primitive lattice vectors that are most orthogonal to each other. Algorithms for identifying this set, referred to as *lattice reduction*, are described in [AST09].

An important property of the primitive unit cell is its volume, given by

$$\widehat{\Omega}_0 = |\hat{\boldsymbol{A}}_1 \cdot \hat{\boldsymbol{A}}_2 \times \hat{\boldsymbol{A}}_3|. \tag{3.5}$$

It is easy to show that this volume remains the same for any choice of primitive lattice vectors. In two dimensions:

$$\widehat{\Omega}_0 = |\hat{\boldsymbol{A}}_1 \times \hat{\boldsymbol{A}}_2| = |\hat{\boldsymbol{A}}_1||\hat{\boldsymbol{A}}_2| \sin\theta = d|\hat{\boldsymbol{A}}_1|.$$

Here θ is the angle between $\hat{\boldsymbol{A}}_1$ and $\hat{\boldsymbol{A}}_2$, and $d = |\hat{\boldsymbol{A}}_2| \sin\theta$ is the distance between the line defined in step 2 of Algorithm 3.1 and the line parallel to it passing through P'. Clearly $\widehat{\Omega}_0$ will be the same for any choice of P'. The generalization to three dimensions is straightforward. Since each primitive unit cell contains only one lattice point, $\widehat{\Omega}_0$ is the volume associated with a single lattice point. This quantity plays an important role in mapping discrete lattice properties onto continuum field measures.

3.3.2 Voronoi tessellation and the Wigner–Seitz cell

A concept that is used time and again in various guises in solid-state physics is the so-called Wigner–Seitz cell. This cell is related to the Voronoi tessellation, and in the field of computational geometry it is in fact often referred to as a Voronoi cell.[9] Illustrated in Fig. 3.4, the Wigner–Seitz cell associated with a particular lattice site is the volume of space closer to that particular lattice site than to any other. It can be generated by drawing

[9] The principal difference is that a Voronoi tessellation is applied to a random arrangement of points, such that the size and shape of each Voronoi cell can be different. In a perfect lattice, all Wigner–Seitz cells are identical.

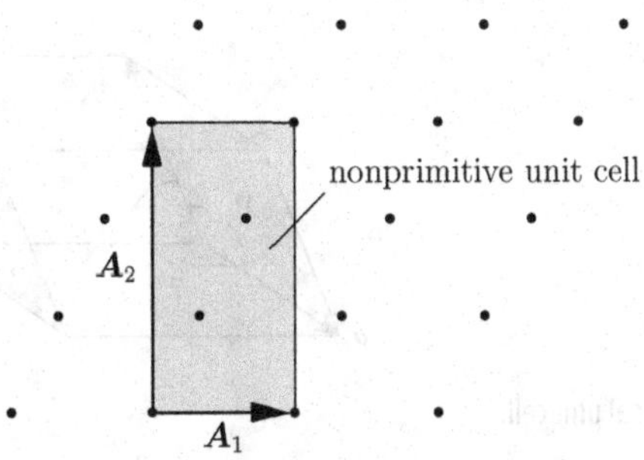

Fig. 3.5 A nonprimitive set of lattice vectors and unit cell. In this case an orthogonal set of vectors has been selected that more clearly highlights the two-fold rotational symmetry of the lattice.

a line segment joining the site of interest to each of its neighbors, starting from the nearest ones, and forming the perpendicular bisecting plane for each segment. This is repeated for all neighbors, eventually forming a closed polyhedron around the site from the intersection of the bisector planes. It is easy to see that, once formed, distant neighbors will have no impact on the shape of this polyhedron, which becomes the Wigner–Seitz cell.

The Wigner–Seitz cell has the same volume as a primitive cell of the lattice, but unlike the primitive cell it also has the same symmetries as the underlying lattice. We revisit it again in the discussion of reciprocal lattices and the first Brillouin zone in Section 3.7.2.

3.3.3 Conventional unit cells

The primitive lattice vectors and unit cell described above provide the most basic definition for a lattice. However, it is often more convenient to work with larger unit cells that more obviously reveal the symmetries of the lattice they generate. Consider, for example, the lattice in Fig. 3.3. This lattice is symmetric with respect to a two-fold rotation about an axis perpendicular to the page. In other words if we rotate the lattice by 180° about an axis passing through one of the lattice points, we obtain the same lattice again. This symmetry is not immediately apparent if we consider only the primitive unit cell in Fig. 3.3(a). We must first use the cell to generate the lattice and then apply the symmetry operation to the lattice as a whole. A better choice for this lattice is the *nonprimitive* unit cell given in Fig. 3.5. By "nonprimitive" we mean a unit cell that is larger than the minimal primitive cell, but still generates the lattice when repeated through space. This particular nonprimitive unit cell clearly reveals the two-fold rotational symmetry of the lattice, and indeed if we rotate the unit cell by 180° it remains unchanged. We denote nonprimitive lattice vectors by $\boldsymbol{A}_i$ ($i = 1, 2, 3$), in contrast to the primitive lattice vectors, $\hat{\boldsymbol{A}}_i$, that have hats.

It is clear from its definition that a nonprimitive unit cell has a larger volume than the primitive unit cell and that it contains more than one lattice point. For example, the nonprimitive cell in Fig. 3.5 contains three lattice sites ($4 \times \frac{1}{4} = 1$ from the corners and two internally). Similarly to primitive unit cells, there is an infinite number of possible non-primitive unit cells for a lattice. In principle, nonprimitive cells can have complex shapes,

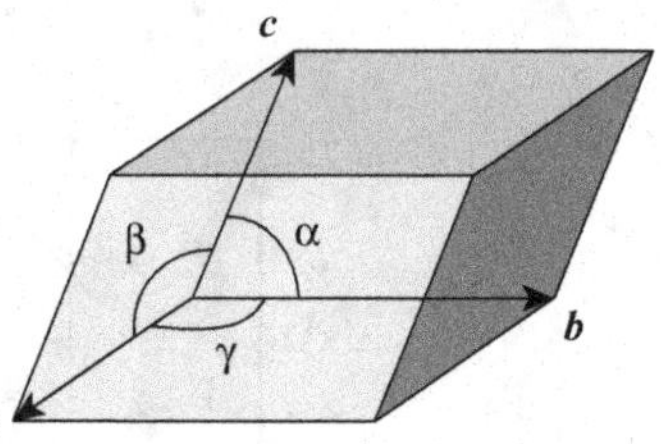

Fig. 3.6 The conventional unit cell.

however, the convention is to select the minimal parallelepiped that shares the symmetry properties of the lattice.[10] This is called the *conventional unit cell* of the lattice.

Crystallographers refer to the nonprimitive lattice vectors of the conventional unit cell as the *crystal axes*, and denote them $\boldsymbol{a}$, $\boldsymbol{b}$, $\boldsymbol{c}$. The magnitudes of the crystal axes, $a = |\boldsymbol{a}|$, $b = |\boldsymbol{b}|$, $c = |\boldsymbol{c}|$, and the angles between them α, β, γ (see Fig. 3.6), are called *lattice constants* (or *lattice parameters*). For the remainder of this section and the next two sections we will adopt the crystallography notation, which clarifies the explanations, and return to our standard continuum notation ($\boldsymbol{A}_i$) in Section 3.6 and thereafter.

Conventionally, lattice vectors form a right-handed set, such that the triple products

$$\boldsymbol{a} \cdot (\boldsymbol{b} \times \boldsymbol{c}) = \boldsymbol{b} \cdot (\boldsymbol{c} \times \boldsymbol{a}) = \boldsymbol{c} \cdot (\boldsymbol{a} \times \boldsymbol{b}) = \Omega_0$$

are always positive. (Note that $\Omega_0 \geq \widehat{\Omega}_0$, since the conventional cell is often a nonprimitive unit cell.) In the most general case of a lattice possessing no symmetry the vectors will be nonorthogonal and of different lengths. In more symmetric cases the crystal axes satisfy certain relations as we shall discuss shortly.

An alternative to the conventional unit cell is the Wigner–Seitz cell described in Section 3.3.2. This choice seems ideal in that the Wigner–Seitz cell is a primitive unit cell that has the same symmetries as the lattice. The problem is that Wigner–Seitz cells have complicated shapes that make them difficult to use. For this reason, conventional unit cells are used even though the primitive property is lost.

3.3.4 Crystal directions

Directions in the lattice are given relative to the crystal axes. So a direction $\boldsymbol{d}$ is

$$\boldsymbol{d} = u\boldsymbol{a} + v\boldsymbol{b} + w\boldsymbol{c},$$

where u, v, w are dimensionless numbers. The convention is to scale $\boldsymbol{d}$ in such a way that these numbers are integers (this can always be done for directions connecting lattice sites)

[10] The exceptions are trigonal and hexagonal lattices that possess three-fold and six-fold symmetries, respectively, which cannot be reproduced by their parallelepiped unit cells.

and then to write the direction in shorthand notation as $[uvw]$, using square brackets as shown. For example, $[112] = \boldsymbol{a} + \boldsymbol{b} + 2\boldsymbol{c}$. Negative indices are denoted with an overhead bar, $[\bar{1}1\bar{2}] = -\boldsymbol{a} + \boldsymbol{b} - 2\boldsymbol{c}$. Another useful notation is $\langle uvw \rangle$, which denotes all possible order and sign permutations of uvw. For example, $\langle 110 \rangle = [110], [101], [011], [\bar{1}10], [1\bar{1}0], [\bar{1}\bar{1}0], \ldots$ In highly symmetric lattices all of these directions are equivalent, which is what makes this notation convenient.

3.4 Crystal systems

It turns out that there are only 14 unique types of lattices in three dimensions. This was first established by Bravais in 1845, which is why we refer to these 14 crystal classes as *Bravais lattices* to this day.[11] Most materials science books tend to avoid the subject of *why* there are 14 unique lattices, simply stating the fact and showing a table of the 14 lattice classes. To understand this better is the task of crystallographers and mathematicians who study symmetry carefully. Here we would like to go a little deeper than most materials texts, without quite the rigor or depth of the crystallographers. We will try to show why the 14 lattices are each unique, and why there can in fact be only 14.

Classification of lattices proceeds in a top-down fashion. Rather than starting from a lattice and identifying its symmetries, the procedure begins with symmetry operations and determines the resulting constraints on the lattice structure. The objective is then to systematically consider possible symmetries and to group lattices based on those they share.

3.4.1 Point symmetry operations

If we imagine that the two-dimensional lattice in Fig. 3.4 is infinite, it is not hard to see that rotating it by 180° around any of the points will leave the lattice indistinguishable from the initial picture. We say that this lattice has a two-fold axis of symmetry, an example of a *point symmetry operation*. A point symmetry operation is a transformation of the lattice specified with respect to a single point that remains unchanged during the process. Translations of the lattice like Eqn. (3.4) are not point symmetry operations since they do not leave any points in the lattice unchanged.

The point symmetry operations consist of three basic types (and combinations thereof): rotation, reflection and inversion. To describe these operations succinctly, it is convenient to define a shorthand notation, although there are two competing notations that are widely used: the *international notation* (also called Hermann–Maugin notation) and the *Schoenflies notation*. We will follow the approach of [BG90] and indicate both notations: first the international, followed by the Schoenflies inside parentheses.

[11] It is only due to a small error that we do not call them Frankenheim lattices. Three years earlier than Bravais, the German physicist Moritz Frankenheim concluded erroneously that there were 15 distinct lattices. In 1850, Bravais explained the discrepancy by showing that two of Frankenheim's proposed lattice types were actually equivalent.

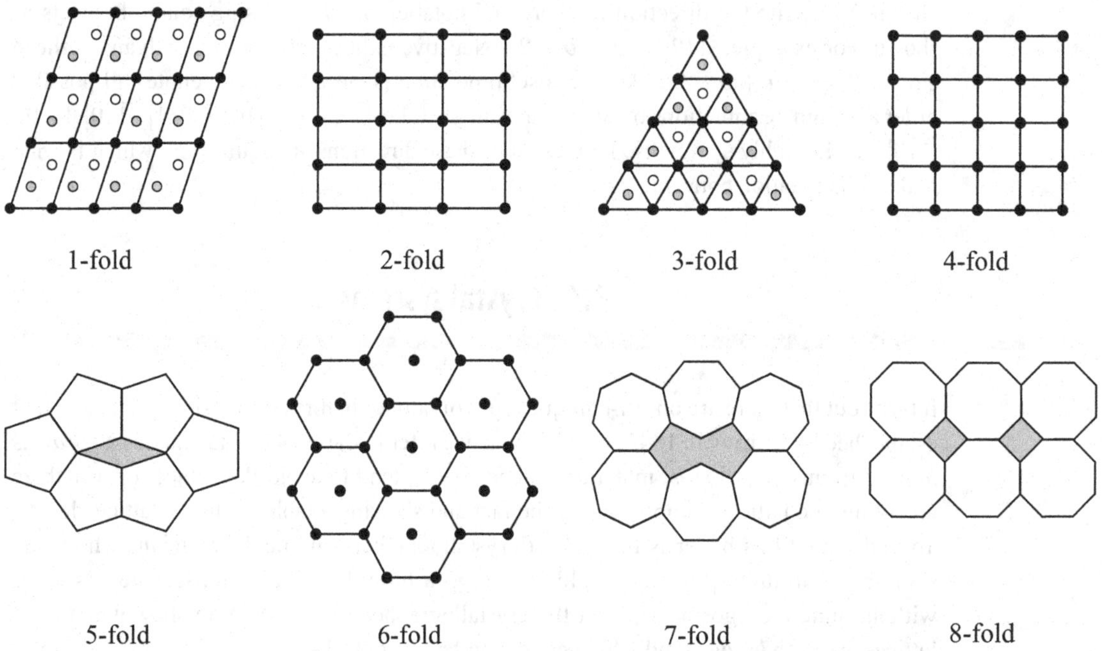

Fig. 3.7 Projections of three-dimensional crystals onto the plane of the paper showing different possible rotational symmetries. The circles are lattice points and the shading indicates different heights above the plane of the paper. The lines connecting the points are a guide to the eye to help identify the symmetry in each case. It is clear from the figure that five-fold, seven-fold, and eight-fold symmetries (and above) are not possible for a crystal since it is not possible to fill space with a basic building block possessing the required symmetry. The gray areas drawn in these cases show the resulting gaps in the packing. This figure is based on Fig. 1.13 in [VS82].

Rotations A rotation operator rotates the lattice by some angle about an axis passing through a lattice point. A lattice is said to possess an n-fold rotational symmetry about a given axis if the lattice remains unchanged after a rotation of $2\pi/n$ about it. In the international notation, rotations are denoted simply as n, while in the Schoenflies notation it is C_n. We will write them both as $n(C_n)$. For the trivial case of $n = 1$, we recover an identity operator, since we simply rotate the entire lattice 360°. This has a special symbol, E, in the Schoenflies notation, so we denote it as $1(E)$.

For an isolated molecule or other finite structure, n can be any integer value. However, for infinite lattices with translational symmetry, n can only take on the values 1, 2, 3, 4 and 6. One can get a sense of the reasons for this from Fig. 3.7. In this figure, we start with basic building blocks (unit cells) that, on their own, possess n-fold symmetry, and try to assemble them in a way that fills all space with no gaps or overlaps. This is not possible for $n = 5, 7, 8$. A rigorous proof of this is straightforward and left to the exercises, but already we are seeing a narrowing of the possible crystal types on the grounds of symmetry: there can be no crystal systems with $n = 5, 7$ or 8 as symmetry operations.

The rotation axis generally coincides with some convenient crystal direction. Take, for example, the 90° rotation $4(C_4)$ about the $\mathbf{c}$ lattice vector (which we can assume to lie along the 3-axis in some global coordinate system). As discussed on page 30 in Section 2.1.2, this rotation transforms the lattice point (R_1, R_2, R_3) to (R'_1, R'_2, R'_3) as

$$\begin{bmatrix} R'_1 \\ R'_2 \\ R'_3 \end{bmatrix} = \begin{bmatrix} \cos\frac{2\pi}{4} & -\sin\frac{2\pi}{4} & 0 \\ \sin\frac{2\pi}{4} & \cos\frac{2\pi}{4} & 0 \\ 0 & 0 & 1 \end{bmatrix} \begin{bmatrix} R_1 \\ R_2 \\ R_3 \end{bmatrix} = \begin{bmatrix} 0 & -1 & 0 \\ 1 & 0 & 0 \\ 0 & 0 & 1 \end{bmatrix} \begin{bmatrix} R_1 \\ R_2 \\ R_3 \end{bmatrix}. \tag{3.6}$$

We denote the rotation matrix above as $\mathbf{Q}_{4[001]}$, where the subscript indicates the four-fold rotation and the rotation axis:

$$\mathbf{Q}_{4[001]} = \begin{bmatrix} 0 & -1 & 0 \\ 1 & 0 & 0 \\ 0 & 0 & 1 \end{bmatrix}. \tag{3.7}$$

If we take one point on the lattice as the origin and apply $\mathbf{Q}_{4[001]}$ to all lattice points, we have applied the $4(C_4)$ rotation operation. If, further, the lattice points before and after this operation are indistinguishable, then $4(C_4)$ is a point symmetry operator of the lattice.

Similarly, one can see that a two-fold rotation is given by

$$\mathbf{Q}_{2[001]} = \begin{bmatrix} -1 & 0 & 0 \\ 0 & -1 & 0 \\ 0 & 0 & 1 \end{bmatrix}. \tag{3.8}$$

Reflection across a plane A reflection or "mirror" operation, denoted $m(\sigma)$, corresponds to what the name intuitively suggests. We define a plane passing through (at least) one lattice point. Then for every point in the lattice we draw the perpendicular line from the point to the plane, and determine the distance, *d*, from point to plane along this line. We then move the lattice point perpendicularly to the other side of the mirror plane at distance *d*. This is easiest to see mathematically if the mirror plane coincides with a coordinate direction. For example, reflection in the 1–2-plane (with normal along the 3-axis) is

$$\begin{bmatrix} R'_1 \\ R'_2 \\ R'_3 \end{bmatrix} = \begin{bmatrix} 1 & 0 & 0 \\ 0 & 1 & 0 \\ 0 & 0 & -1 \end{bmatrix} \begin{bmatrix} R_1 \\ R_2 \\ R_3 \end{bmatrix} = \begin{bmatrix} R_1 \\ R_2 \\ -R_3 \end{bmatrix}. \tag{3.9}$$

Inversion The inversion operator, denoted $\bar{1}(i)$, has the straightforward effect of transforming any lattice point (R_1, R_2, R_3) to $(-R_1, -R_2, -R_3)$. The origin is left unchanged, and is referred to as the *center of inversion* or sometimes the *center of symmetry*. A crystal structure which possesses a center of inversion is said to possess *centrosymmetry*.[12] Sometimes,

[12] Note that it is common to refer to crystals with inversion symmetry in their space group as "centrosymmetric." However, we will reserve that term to mean crystals in which *all* atoms are located at a center of symmetry. This distinction becomes important for multilattice crystals (see Fig. 11.4).

this operator is referred to as the *parity operator*. We can write

$$\begin{bmatrix} R_1' \\ R_2' \\ R_3' \end{bmatrix} = \begin{bmatrix} -1 & 0 & 0 \\ 0 & -1 & 0 \\ 0 & 0 & -1 \end{bmatrix} \begin{bmatrix} R_1 \\ R_2 \\ R_3 \end{bmatrix} = \begin{bmatrix} -R_1 \\ -R_2 \\ -R_3 \end{bmatrix} \tag{3.10}$$

to show the effect of the inversion operator. This means that if a lattice includes a point at $(-R_1, -R_2, -R_3)$ for every point (R_1, R_2, R_3), it must have inversion (or parity) symmetry. A quick look at the definition of a lattice in Eqn. (3.1) shows that all lattices must possess at least this symmetry.

Combined operations Operators applied to the lattice one after the other can be indicated using the typical operator notation of mathematics. That is to say, for example, that a two-fold rotation followed by two four-fold rotations can be indicated as $442(C_4C_4C_2)$, or using powers for shorthand as $4^2 2(C_4^2 C_2)$. Of course, the effect of these particular operators is a $360°$ rotation, so that $4^2 2 = 1$.

Improper rotations An improper rotation is the combination of two basic operations. It can be viewed as either a rotation followed by an inversion, or a rotation followed by a reflection. Many structures will have the compound improper rotation as a symmetry operator even though neither of the two basic operators (rotation and reflection) is a symmetry operator of the structure.

Because the international and Schoenflies approaches treat improper rotations in slightly different ways, the correspondence between the two notations is more subtle than for the basic operations. The international approach treats improper rotation as a rotation followed by an inversion, thus $\bar{1}n$ or simply $\bar{n}$ for short. For example, we can combine Eqns. (3.6) and (3.10) to produce the $\bar{4}$ improper rotation about the 3-axis:

$$\begin{bmatrix} R_1' \\ R_2' \\ R_3' \end{bmatrix} = \begin{bmatrix} -1 & 0 & 0 \\ 0 & -1 & 0 \\ 0 & 0 & -1 \end{bmatrix} \begin{bmatrix} 0 & -1 & 0 \\ 1 & 0 & 0 \\ 0 & 0 & 1 \end{bmatrix} \begin{bmatrix} R_1 \\ R_2 \\ R_3 \end{bmatrix} = \begin{bmatrix} 0 & 1 & 0 \\ -1 & 0 & 0 \\ 0 & 0 & -1 \end{bmatrix} \begin{bmatrix} R_1 \\ R_2 \\ R_3 \end{bmatrix}. \tag{3.11}$$

In the Schoenflies approach, an improper rotation is treated as a rotation followed by a mirror reflection, $\sigma_h C_n$, where the subscript h on the mirror operation emphasizes that the reflection is in the "horizontal" plane relative to the vertical rotation axis. The further shorthand S_n is introduced to represent the combination $\sigma_h C_n$.

An analogous improper rotation to the example just given for the international approach is S_4 (or $\sigma_h C_4$), which we can obtain by combining Eqns. (3.6) and (3.9):

$$\begin{bmatrix} R_1' \\ R_2' \\ R_3' \end{bmatrix} = \begin{bmatrix} 1 & 0 & 0 \\ 0 & 1 & 0 \\ 0 & 0 & -1 \end{bmatrix} \begin{bmatrix} 0 & -1 & 0 \\ 1 & 0 & 0 \\ 0 & 0 & 1 \end{bmatrix} \begin{bmatrix} R_1 \\ R_2 \\ R_3 \end{bmatrix} = \begin{bmatrix} 0 & -1 & 0 \\ 1 & 0 & 0 \\ 0 & 0 & -1 \end{bmatrix} \begin{bmatrix} R_1 \\ R_2 \\ R_3 \end{bmatrix}. \tag{3.12}$$

Comparing Eqn. (3.11) with Eqn. (3.12), we see that the two transformations are not the same, i.e. $\bar{4}(\neq S_4)$. However, it is left as an exercise to show that three $\bar{4}$ operators are equivalent to S_4, or conversely that three S_4 operators are the same as $\bar{4}$. In other words, we can denote improper rotations as $\bar{4}(S_4^3)$ or $\bar{4}^3(S_4)$. While the operations are slightly different, a lattice which has $\bar{4}$ as a symmetry operation will also have S_4, and vice versa.

3.4.2 The seven crystal systems

We now have at our disposal four types of operations that we can perform on a lattice: rotations, reflections, inversions and improper rotations. The possible crystal systems are classified by asking what conditions are imposed on a lattice if it is to have any of these operations as point symmetry operators. These conditions will come in the form of restrictions on the relative lengths of $\boldsymbol{a}$, $\boldsymbol{b}$ and $\boldsymbol{c}$ and on the angles between them. Next, we catalog the seven crystal systems that result from these symmetry considerations, but later show that some of these can be satisfied by more than one unique arrangement of lattice points. The end result is the 14 unique Bravais lattices we have mentioned previously.

Triclinic The triclinic crystal system is the least symmetric lattice that is possible. Other than the trivial identity operator, $1(E)$, the triclinic system has only one symmetry, the inversion $\bar{1}(i)$. As we already mentioned when we introduced the inversion operator, inversion symmetry imposes only the condition that for every lattice point at (R_1, R_2, R_3) there exists a point at $(-R_1, -R_2, -R_3)$. A quick look at the definition of a lattice in Eqn. (3.1) shows that all lattices must possess at least this symmetry, regardless of the lengths of the lattice vectors or the angles between them. Fig. 3.6 illustrates the triclinic unit cell.

> The triclinic crystal system:
>
> $\bar{1}(i)$ symmetry.
> No conditions on lengths a, b or c.
> No conditions on angles α, β, or γ.

Monoclinic The next crystal system in order of increasing symmetry is one with a single two-fold rotation axis $2(C_2)$ or (equivalently) a reflection plane of symmetry, $m(\sigma)$. The axis of symmetry lies along one of the lattice vectors, conventionally chosen to be $\boldsymbol{c}$. The mirror plane, in this case, has its normal along the $\boldsymbol{c}$ direction. $2(C_2)$ symmetry imposes no restrictions on the lengths of the lattice vectors, but it requires that $\alpha = \beta = 90°$. This can be seen by considering Fig. 3.8(a). If $\boldsymbol{a}$ and $\boldsymbol{b}$ are at right angles to $\boldsymbol{c}$, then rotating vectors $\boldsymbol{a}$ and $\boldsymbol{b}$ by 180° about the c-axis will make $\boldsymbol{a} \to -\boldsymbol{a}$ and $\boldsymbol{b} \to -\boldsymbol{b}$ and the same lattice is produced. However, any vector like $\boldsymbol{b}'$ for which $\alpha' \neq 90°$ will rotate to $\boldsymbol{b}''$ (such that its tip remains above the $\boldsymbol{a}$–$\boldsymbol{b}$ plane), rather than rotating to $-\boldsymbol{b}'$ as required to restore the lattice. As such, the lattice is restored only if $\alpha = 90°$. Similar arguments apply to the vector $\boldsymbol{a}$, and therefore $\beta = 90°$. It is clear that neither the lengths a, b and c, nor the angle γ between $\boldsymbol{a}$ and $\boldsymbol{b}$ entered into these considerations, and thus they are not constrained by the $2(C_2)$ symmetry.

More formally, we take Eqn. (3.1) and apply $\mathbf{Q}_{2[001]}$ from Eqn. (3.8) to get

$$\begin{aligned} \boldsymbol{R}'^{[\ell]} &= \mathbf{Q}_{2[001]}\boldsymbol{R}^{[\ell]} \\ &= \ell_1 \mathbf{Q}_{2[001]}\boldsymbol{a} + \ell_2 \mathbf{Q}_{2[001]}\boldsymbol{b} + \ell_3 \mathbf{Q}_{2[001]}\boldsymbol{c} \\ &= \ell_1 \boldsymbol{a}' + \ell_2 \boldsymbol{b}' + \ell_3 \boldsymbol{c}'. \end{aligned}$$

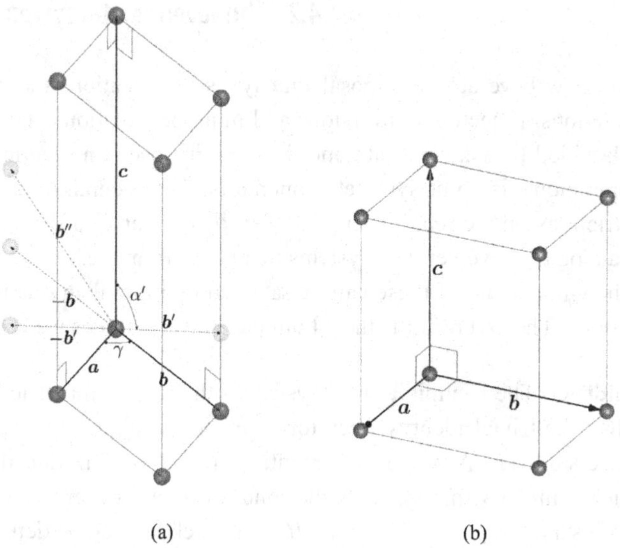

Fig. 3.8 (a) The monoclinic unit cell, illustrating the conditions imposed by $2(C_2)$ symmetry on the $\boldsymbol{b}$ vector and (b) the orthorhombic unit cell.

It is easy to establish that the effect of $\mathbf{Q}_{2[001]}$ on the lattice vectors is such that

$$[\boldsymbol{a}'] \equiv \begin{bmatrix} a'_1 \\ a'_2 \\ a'_3 \end{bmatrix} = \begin{bmatrix} -a_1 \\ -a_2 \\ a_3 \end{bmatrix}, \qquad [\boldsymbol{b}'] \equiv \begin{bmatrix} b'_1 \\ b'_2 \\ b'_3 \end{bmatrix} = \begin{bmatrix} -b_1 \\ -b_2 \\ b_3 \end{bmatrix}, \qquad \boldsymbol{c}' = \boldsymbol{c}.$$

By inspecting the components of $\boldsymbol{a}'$ and $\boldsymbol{b}'$, we see that $\boldsymbol{a}' = -\boldsymbol{a}$ and $\boldsymbol{b}' = -\boldsymbol{b}$ only if $a_3 = -a_3$ and $b_3 = -b_3$. This is only possible if these components are zero. This implies that $\boldsymbol{a}$ and $\boldsymbol{b}$ must be orthogonal to the rotation axis, $\boldsymbol{c}$.

The monoclinic crystal system:

$2(C_2)$ and $m(\sigma)$ symmetry.
No conditions on lengths a, b or c.
$\alpha = \beta = 90°$ is required.
No conditions on angle γ.

Orthorhombic The orthorhombic system has two two-fold axes (or equivalently, two mirror planes). The arguments just applied to the monoclinic system can be used again to determine the conditions on the lattice vectors. If we take the first two-fold axis to be along $\boldsymbol{c}$, then $\boldsymbol{b}$ and $\boldsymbol{a}$ must be perpendicular to $\boldsymbol{c}$ ($\alpha = \beta = 90°$) as in the monoclinic system. Next, we can take the second two-fold axis to be along $\boldsymbol{b}$, in which case the conclusion will be that $\alpha = \gamma = 90°$. Thus, all three lattice vectors must be orthogonal for the orthorhombic crystal system. However, as with the monoclinic case, no conditions were imposed on the lengths of the lattice vectors. Therefore, the orthorhombic unit cell is a rectangular box with three, unequal, perpendicular edge lengths, as shown in Fig. 3.8(b).

Finally, we note that having two two-fold axes automatically implies a third: the fact that $\beta = \gamma = 90°$ implies that $\boldsymbol{a}$ is also a two-fold symmetry axis.

The orthorhombic crystal system:

Two $2(C_2)$ symmetry axes (which implies a third).
No conditions on lengths a, b or c.
$\alpha = \beta = \gamma = 90°$ is required.

Tetragonal The tetragonal crystal system requires a single four-fold axes of symmetry, $4(C_4)$, or $\bar{4}(S_4^3)$. Although this may sound less restrictive than the orthorhombic case, it is actually a higher symmetry and imposes further restrictions on the lattice vectors. Let us choose the four-fold axis to lie along $\boldsymbol{c}$, and consider rotating the orthorhombic unit cell in Fig. 3.8(b) about $\boldsymbol{c}$ by $90°$ as a four-fold axis implies. It soon becomes clear that the only hope for $4(C_4)$ symmetry rests in the vector $\boldsymbol{a}$ rotating to $\boldsymbol{b}$ and $\boldsymbol{b}$ rotating to $-\boldsymbol{a}$. By the same arguments used to discuss the conditions of the monoclinic system, this imposes orthogonality between $\boldsymbol{c}$ and $\boldsymbol{a}$, and between $\boldsymbol{c}$ and $\boldsymbol{b}$. It is also clear that $\boldsymbol{a}$ and $\boldsymbol{b}$ must be at right angles and of the same length for them to line up with one another after a $90°$ degree rotation.

Compared with the orthorhombic system, the only additional condition imposed on the tetragonal system is that $a = b$. Therefore, the tetragonal crystal unit cell is also a rectangular box, with two of the three perpendicular edge lengths being equal.

The tetragonal crystal system:

One $4(C_4)$ or $\bar{4}(S_4^3)$ symmetry axis.
$a = b$ is required.
There is no restriction on the length, c.
$\alpha = \beta = \gamma = 90°$ is required.

Cubic The cubic crystal system is the crystal system most often encountered in this book and is likely the one most familiar to readers. It is also the most symmetric system, requiring four three-fold symmetry axes. This is not the symmetry condition one may intuitively guess for a cubic system, as our eyes are naturally drawn to the three four-fold axes along the edges of the unit cell. In fact, the key symmetry axes are the cube *diagonals* $[111]$, $[\bar{1}11]$, $[1\bar{1}1]$ and $[11\bar{1}]$. Incidentally, it is possible to prove that having two three-fold axes implies having exactly four, at angles of $109°$ from each other.

A cubic unit cell is shown in Fig. 3.9(a), and is shown again looking directly down the $[111]$ and $[1\bar{1}1]$ directions in (b) and (c). It is clear from (b) that a rotation of $120°$ about the $[111]$ direction will bring $\boldsymbol{a} \rightarrow \boldsymbol{b}$, $\boldsymbol{b} \rightarrow \boldsymbol{c}$ and $\boldsymbol{c} \rightarrow \boldsymbol{a}$. This will only be true provided the lengths of the three lattice vectors are the same, and that the angles between them are the same as well, but note that a *single* three-fold axis can be satisfied for any value of this common angle. That the angles α, β and γ can be *only* $90°$ is imposed by considering a second three-fold axis, as shown in the projection along the $[1\bar{1}1]$ direction. In this case,

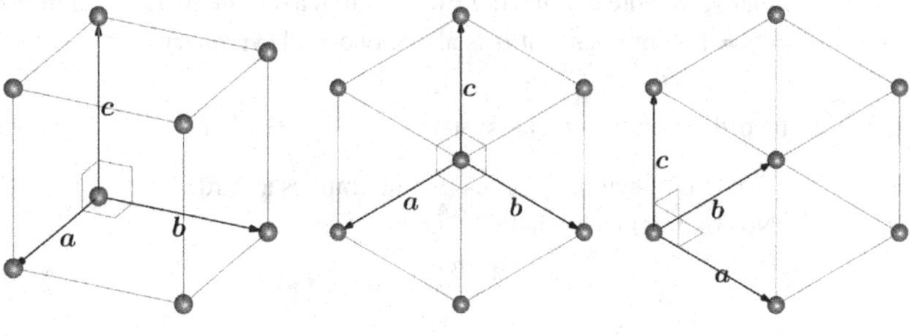

Fig. 3.9 The cubic unit cell in (a) is shown looking down the $[111]$ direction in (b) and the $[1\bar{1}1]$ direction in (c).

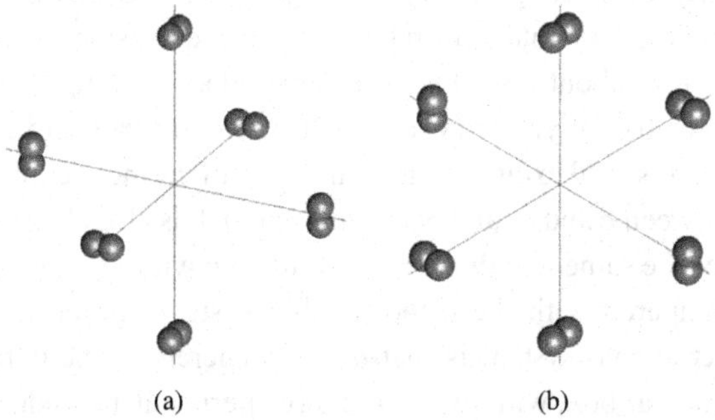

Fig. 3.10 (a) An example of an arrangement of dumbbells that possesses cubic symmetry but no four-fold symmetry axes. (b) The same arrangement of dumbbells is viewed along $[111]$, which is one of the three-fold axes.

a rotation of 120° brings $\boldsymbol{a} \to \boldsymbol{c}$, $\boldsymbol{c} \to -\boldsymbol{b}$ and $\boldsymbol{b} \to -\boldsymbol{a}$. Rotations about two independent $\langle 111 \rangle$ axes can only *both* be symmetry operations if the angles $\alpha = \beta = \gamma = 90°$.

The cubic crystal system:

Two $3(C_3)$ symmetry axes (which imply two more).
$a = b = c$ is required.
$\alpha = \beta = \gamma = 90°$ is required.

The notion that an object can have cubic symmetry without the four-fold symmetry axes is confusing, at least in part because it is easy to lose sight of what objects are the focus of our symmetry discussion. It is the symmetry of *crystals* that we are classifying here, although we are focusing on only the underlying *lattice* in the present discussion. While the cubic *lattice* will indeed exhibit four-fold axes of symmetry, cubic *crystals* may not, once we consider the basis within each lattice cell (to be discussed in Section 3.6). For now, we provide a simple example in Fig. 3.10 of an object that has cubic symmetry, but does not have any four-fold symmetry axes. Each identical dumbbell is located at the same distance along three orthogonal directions, and is oriented parallel to one of the other directions. This leads to three-fold, but not four-fold, symmetry as one can see by imagining 120°

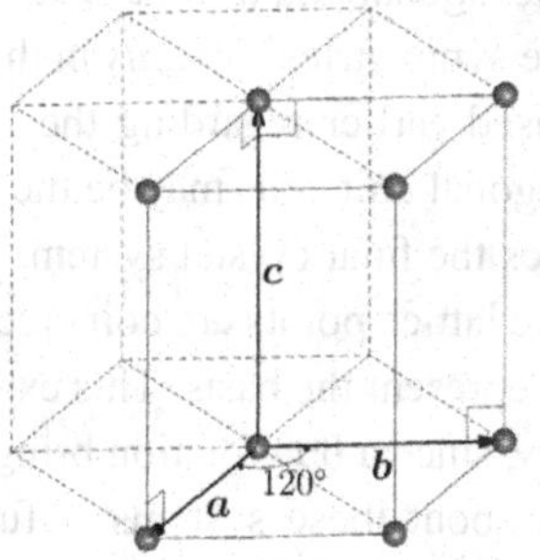

Fig. 3.11 The hexagonal or trigonal unit cell.

and 90° rotations of the figure. If such a cluster of six dumbbells were centered on every point in a cubic lattice, they would define a crystal with three-fold cubic symmetry but no four-fold symmetry axes.

Trigonal and hexagonal It may have been more systematic to discuss trigonal and hexagonal systems before cubic systems, to keep with the order of progressively more symmetric systems. However, there is some confusion and lack of consensus about these systems in the literature, and so we present them last, and simultaneously. The hexagonal system possesses a single $6(C_6)$ symmetry axis, while the trigonal system possesses a single $3(C_3)$ symmetry operation. The confusion comes mainly from the fact that these two symmetry conditions lead to the identical set of restrictions on the lattice vectors, *and thus we cannot distinguish between the trigonal and hexagonal lattices*.

The hexagonal (or trigonal) unit cell is shown in Fig. 3.11. As with monoclinic symmetry, we consider the $\mathbf{c}$ lattice vector to be the axis of $6(C_6)$ or $3(C_3)$ symmetry for the hexagonal and trigonal systems, respectively. In order for a rotation of 60° or 120° about the $\mathbf{c}$ axis to bring the $\mathbf{a}$ and $\mathbf{b}$ vectors back onto themselves, they must be orthogonal to $\mathbf{c}$. Further, if lengths $a = b$ and $\gamma = 120°$, then a rotation of 60° (six-fold) brings $\mathbf{a} \to \mathbf{a} + \mathbf{b}$ and $\mathbf{b} \to -\mathbf{a}$. Similarly, a rotation of 120° (three-fold) brings $\mathbf{a} \to \mathbf{b}$ and $\mathbf{b} \to -\mathbf{a} - \mathbf{b}$.

The hexagonal crystal system:

One $6(C_6)$ symmetry axis.
No restriction on the length of c.
$a = b$ is required.
$\alpha = \beta = 90°$ is required.
$\gamma = 120°$ is required.

The trigonal crystal system:

One $3(C_3)$ symmetry axis.
Same conditions on the lattice vectors as for the hexagonal system.

Although the hexagonal and trigonal lattices have the same conditions imposed on them by their respective symmetries, *crystals* in the two systems are not the same. This is related to the example used earlier regarding the four-fold axes in the cubic system; while the trigonal and hexagonal unit cells may be the same, the symmetry of the basis atoms within the cell determines the final crystal system. This can be seen in the three-fold example in Fig. 3.7, where the lattice points are coincident with the black atoms, while the black, gray and white atoms represent the basis. This example with three-fold symmetry does not have six-fold symmetry, since a 60° rotation brings gray atoms on top of white ones.

The confusion about these systems is further confounded by the introduction of the *rhombohedral system* in the literature, which has $3(C_3)$ symmetry and can be viewed as a special case of the trigonal system. As such, some authors identify only six crystal systems (classifying these three cases as hexagonal), while others replace trigonal with rhombohedral as the seventh crystal class. The rhombohedral unit cell is characterized by $a = b = c$ and $\alpha = \beta = \gamma$; it is the same as the cubic unit cell except that the angles need not be 90°. We will discuss the rhombohedral system further in Section 3.5.5.

3.5 Bravais lattices

The seven crystal systems described in the previous section were established based on their symmetry, and for each we identified a primitive lattice consistent with the symmetry requirements. These formed six unique lattices,[13] but these are not the only possibilities. Next, we can consider adding lattice points within these unit cells, referred to as "centering" points inside the cell. Centering creates nonprimitive unit cells, but this is not a problem so long as three conditions are satisfied:

(i) the combination of corner and centered points together still form a lattice,
(ii) the additional points do not lower the symmetry of the system and
(iii) this new lattice is truly unique from all other possible lattices.

We shall see that there are three possible centered lattices for each crystal system, in addition to the primitive (P) lattice. There is also one special type of centering for trigonal symmetry. However, not all of these combinations satisfy the three conditions above, and in the end only 14 remain.

3.5.1 Centering in the cubic system

We start with the most familiar and easy to visualize system; the cubic system. There are at least four ways that we can imagine adding points to the cubic cell (there are probably an infinite number of ways, but the discussion of these four will make clear why no others need be pursued). These are illustrated in Fig. 3.12. We can add a point at the center, as

[13] The seventh lattice, trigonal, was indistinguishable from hexagonal.

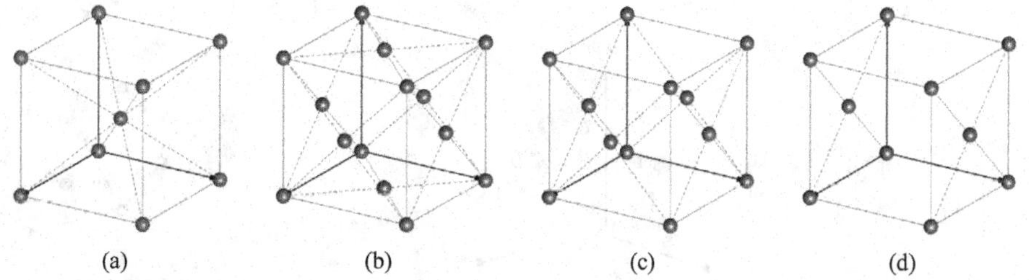

Fig. 3.12 Four alternative centerings for the cubic unit cell: (a) body, (b) 3-face, (c), 2-face and (d) 1-face centering.

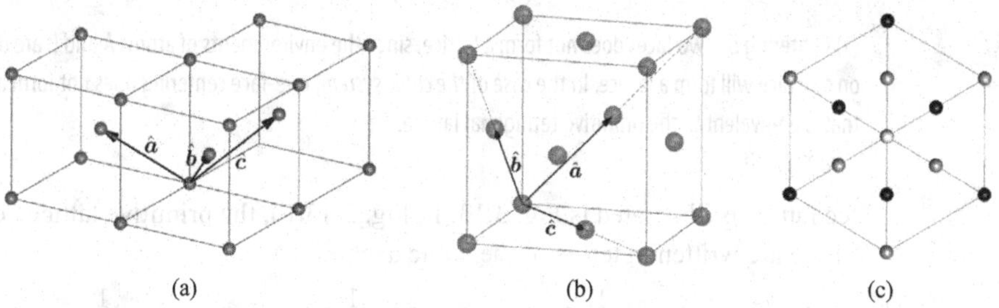

Fig. 3.13 (a) The bcc (cubic-I) Bravais lattice. (b) The fcc (cubic-F) Bravais lattice. (c) The fcc lattice viewed along one of the $3(C_3)$ axes, with lattice sites shaded depending on their depth into the page.

depicted in Fig. 3.12(a), or add points at the center of all three pairs of opposing faces as in (b), to two pairs as in (c) or to one pair as in (d). Let us determine which of these centerings form unique lattices with cubic symmetry.

The body-centered cubic lattice The first centering option (Fig. 3.12(a)), called body centering, satisfies our three criteria, and so is identified as either the "cubic-I" lattice or the *body-centered cubic* (bcc) lattice. The bcc lattice is illustrated in Fig. 3.13(a).

The fact that the bcc arrangement is a lattice (criterion (i) on page 134) is illustrated in Fig. 3.13(a), where we show an alternative set of lattice vectors that are primitive. These are given in terms of the original cubic axes as

$$\hat{a} = \frac{1}{2}(a - b + c), \qquad \hat{b} = \frac{1}{2}(a + b + c), \qquad \hat{c} = \frac{1}{2}(-a + b + c).$$

A little reflection will satisfy the reader that translations in integral multiples of these new vectors will in fact restore the same lattice. That the bcc lattice retains cubic symmetry (criterion (ii)) is easily seen since the central lattice site sits at the intersection point of the three $3(C_3)$ axes of cubic symmetry and therefore is not moved under rotations about these axes. Finally, we note that the bcc lattice is actually a distinct lattice from the six primitive lattices identified so far (criterion (iii)). Since it has cubic symmetry, it must be unique from the others except perhaps the primitive cubic lattice. Clearly, though, the angles between $\hat{a}$, $\hat{b}$ and $\hat{c}$ are not right angles and a new lattice type is therefore formed.

The face-centered cubic lattice The centering illustrated in Fig. 3.12(b) also satisfies our criteria for a new lattice, identified as the "cubic-F" or *face-centered cubic* (fcc) lattice. The

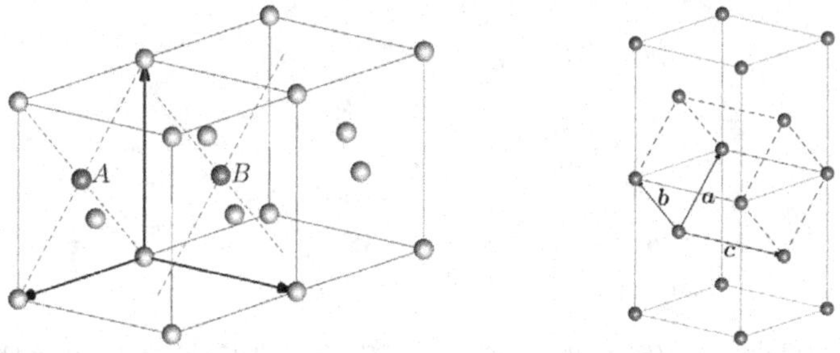

(a) (b)

Fig. 3.14 (a) Centering on two faces does not form a lattice, since the environments of atoms A and B are different. (b) Centering on one face will form a lattice. In the case of *the cubic system*, one-face centering does not form a new lattice, but one that is equivalent to the primitive tetragonal lattice.

fcc lattice is illustrated is Fig. 3.13. In Fig. 3.13(b), the primitive lattice vectors are shown. These are written in terms of the cubic axes as

$$\hat{a} = \frac{1}{2}(\boldsymbol{b}+\boldsymbol{c}), \qquad \hat{b} = \frac{1}{2}(\boldsymbol{a}+\boldsymbol{c}), \qquad \hat{c} = \frac{1}{2}(\boldsymbol{a}+\boldsymbol{b})$$

and it is a simple matter to verify that translations by these vectors restore the same lattice. That the symmetry remains cubic is best seen by Fig. 3.13(c), where we view the unit cell along one of the $3(C_3)$ symmetry axes. The lattice sites are shaded according to their depth into the paper. It is clear that a rotation of 120° brings the gray sites onto other gray sites, and the same can be said for the black sites. Finally, it is also clear that this is a unique lattice. It is different from all but the primitive cubic lattice (denoted cubic-P) and cubic-I (bcc) lattices by virtue of its cubic symmetry, and distinct from the cubic-P and cubic-I lattices due to the different angles formed between its primitive lattice vectors.

Centering on two faces Centering on two faces does not form a lattice. This is true for two-face centering on any primitive unit cell, although we only illustrate this on the cubic system in Fig. 3.14(a), where it is clear that the environments of atoms A and B (indicated by the dashed lines) are not equivalent. Since centering on two faces does not produce a lattice, we will no longer explore it in the context of other crystal systems.

Centering on one face: base-centered lattices As illustrated by the primitive unit vectors in Fig. 3.14(b), centering on one face of the cubic unit cell (or, for that matter of any primitive cell) will indeed form a lattice, satisfying our criterion (i). For some crystal systems, criteria (ii) and (iii) are also satisfied, and a new "base-centered" Bravais lattice is formed. In these cases, we denote the base-centered variant by either "-A", "-B" or "-C" to indicate which of the three faces is centered: the A-face is formed by the plane of the vectors $\boldsymbol{b}$ and $\boldsymbol{c}$, the B-face by $\boldsymbol{a}$ and $\boldsymbol{c}$ and the C-face by $\boldsymbol{a}$ and $\boldsymbol{b}$. However, for the cubic system this type of centering violates both criteria (ii) and (iii). The new lattice formed is, in fact, a tetragonal primitive cell as shown in Fig. 3.14(b). This lattice has already been shown to have lower symmetry than the cubic system.

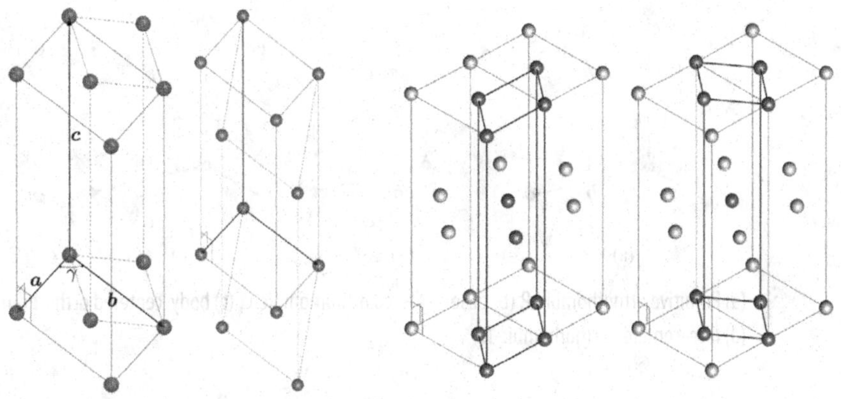

Fig. 3.15 (a) Base-centering on the face whose normal is along the axis of symmetry simply produces another monoclinic-P. (b) Base-centering on one of the other faces in the monoclinic system forms a new lattice. (c) The monoclinic-F face-centered lattice is equivalent to the monoclinic-B lattice. (d) The monoclinic-F lattice is also equivalent to the monoclinic-I body-centered lattice.

3.5.2 Centering in the triclinic system

Body-, face- and base-centering in the triclinic system all lead to lattices, but ones that are not unique; all can be recast as new triclinic systems. The new primitive lattice vectors always satisfy triclinic symmetry conditions since these are trivially satisfied by all lattices. Thus, there is only the primitive (triclinic-P) Bravais lattice in the triclinic crystal system.

3.5.3 Centering in the monoclinic system

Body-, face- and base-centering in the monoclinic system all lead to lattices, but only one of these is unique. Starting with base-centering, we see that unlike the more symmetric cubic system, we need to distinguish between the pairs of faces to which the centered site is added. In principle, there are monoclinic-A, monoclinic-B and monoclinic-C lattices for this reason. Adopting the same convention as in Fig. 3.8(a) (i.e. assuming that the two-fold axis is the c-axis), there is no difference between A and B centering in the monoclinic system because the vectors $\boldsymbol{a}$ and $\boldsymbol{b}$ are essentially interchangeable. As such, there are only two base-centering options to consider in this case: monoclinic-C and monoclinic-B.

In Figs. 3.15(a) and 3.15(b) we show the two base-centerings. Monoclinic-C is shown in Fig. 3.15(a), but we see that centering on the face whose normal is along the symmetry axis is equivalent to a primitive monoclinic lattice with a different choice of unit vectors. On the other hand, centering on one or the other of the remaining faces produces a new primitive cell as shown in Fig. 3.15(b). The reader can verify that two-fold rotations of this new lattice around the original c-axis still satisfy $2(C_2)$ symmetry, but it is also clear that the relation between the lengths and angles of the lattice vectors are distinct from monoclinic-P. Since this centering forms a lattice, the lattice is unique and it retains the symmetry of its crystal

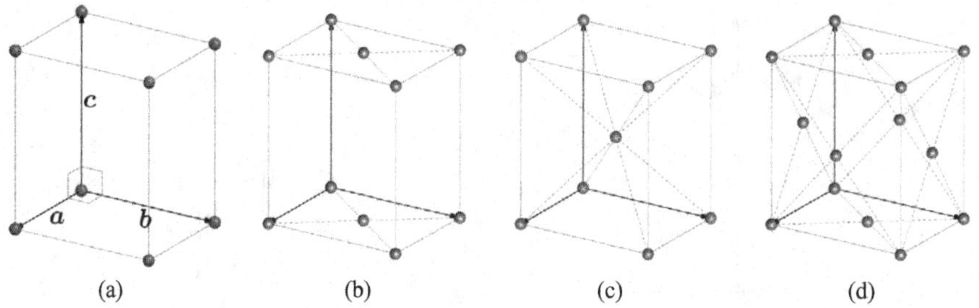

Fig. 3.16 (a) Primitive orthorhombic-P, (b) base-centered orthorhombic-C, (c) body-centered orthorhombic-I and (d) face-centered orthorhombic-F.

system, all our criteria (i)–(iii) are satisfied. Conventionally, this new lattice is denoted monoclinic-B, although monoclinic-A is equivalent.

Body-centering and face-centering are alternative ways of producing the same lattice as monoclinic-B, as shown in Figs. 3.15(c) and 3.15(d). There are therefore only two unique Bravais lattices in the monoclinic system, the primitive monoclinic-P and one of monoclinic-B, monoclinic-I or monoclinic-F. There appears to be no universal convention as to which of the three to choose, which can lead to some confusion in the literature.

3.5.4 Centering in the orthorhombic and tetragonal systems

The discussion of centering in the orthorhombic system is similar to the discussion we presented about centering in the cubic system; body-, face- or base-centering of the primitive orthorhombic unit cell all produce new lattices. However, whereas the base-centered cubic lattice violates the requirement of three-fold symmetry in the cubic system, it does not effect the two-fold symmetry of the orthorhombic cell (for the same reasons that we could base-center the monoclinic unit cell). As a result, there are four unique orthorhombic Bravais lattices: the primitive (orthorhombic-P), base-centered (orthorhombic-C), body-centered (orthorhombic-I) and face-centered (orthorhombic-F).[14] To see this, the reader is referred to Fig. 3.16, where the four unique orthorhombic lattices are shown.

In the tetragonal system, there are only two unique lattices, tetragonal-P and the body-centered tetragonal-I. This is straightforward to see and therefore left to the exercises.

3.5.5 Centering in the hexagonal and trigonal systems

It is left as an exercise to verify that the body-, face- and base-centerings discussed so far violate the symmetry requirements of the trigonal and hexagonal systems. However, there is one special centering of the hexagonal unit cell that leads to a unique Bravais lattice with three-fold symmetry. This centering was already illustrated in the three-fold example of Fig. 3.7 and is shown in more detail in Fig. 3.17. In Fig. 3.17(a), the view down the c-axis

[14] It is conventional to consider the base-centered variant of orthorhombic to have the centered atom on the C face. Of course, the choice of A, B and C is arbitrary in this case due to the interchangeability of $\boldsymbol{a}$, $\boldsymbol{b}$ and $\boldsymbol{c}$ in defining the orthorhombic lattice.

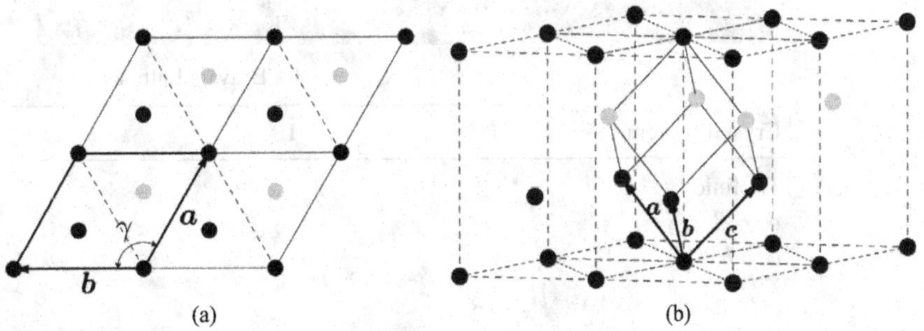

Fig. 3.17 (a) A projection along the c-axis of the trigonal unit cell, with special centering. Black sites are in the $a-b$ plane, dark gray sites are above this plane by $c/3$, light gray sites are at $2c/3$ (b) The rhombohedral unit cell.

shows the primitive sites in black and sites centered at $\boldsymbol{r} = (\boldsymbol{a} - \boldsymbol{b} + \boldsymbol{c})/3$ in dark gray and $\boldsymbol{r} = (2\boldsymbol{a} + \boldsymbol{b} + 2\boldsymbol{c})/3$ in light gray. This centering does not preserve the hexagonal six-fold symmetry, since 60° rotations about the $\boldsymbol{c}$-axis move dark gray atoms onto light gray atoms. However, the three-fold symmetry of the trigonal system remains. The result of this special centering is therefore a nonprimitive trigonal unit cell. This cell can be recast as a primitive *rhombohedral* cell as shown in Fig. 3.17(b), where the angles $(\alpha = \beta = \gamma) \neq 90^\circ$ and the lengths $a = b = c$. We classify this as the trigonal-R Bravais lattice (recall that the trigonal-P is indistinguishable from hexagonal-P as discussed in Section 3.4.2).

3.5.6 Summary of the fourteen Bravais lattices

In the discussion above, we identified each of the possible Bravais lattices. We have opted for geometrical arguments and diagrams over mathematical rigor, with the goal of motivating how there come to be exactly 14 Bravais lattices. In short, each lattice is a unique combination of a crystal system determined by the *symmetry* of the structure and symmetry-preserving *centerings* within the unit cell. In the end, exactly 14 unique Bravais lattices remain, as summarized in Tab. 3.1.

3.6 Crystal structure

In Section 3.2, we stated that a crystal structure can be described as a combination of a lattice and a basis:

$$\text{crystal} = \text{lattice} + \text{basis}.$$

The discussion of the first ingredient, the lattice, took us on a long departure into symmetry considerations in order to identify all possible lattices. It is easy to lose sight of the fact that a lattice is not a crystal; there are no atoms in a lattice. It is only once we attach a *basis*

Table 3.1. Summary of the 14 Bravais Lattices

	Bravais Lattices			
Crystal System	P	I	F	C
Triclinic $\alpha \neq \beta \neq \gamma$ $a \neq b \neq c$	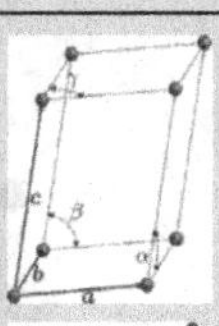			
Monoclinic $\alpha = \beta = 90°$ $\gamma \neq \alpha, \beta$ $a \neq b \neq c$	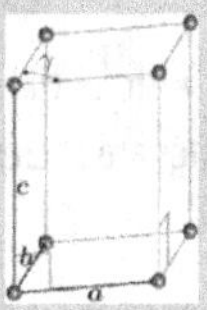	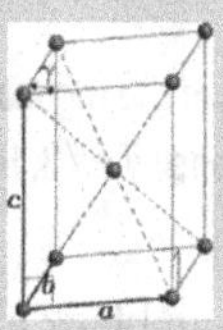		
Orthorhombic $\alpha = \beta = \gamma = 90°$ $a \neq b \neq c$	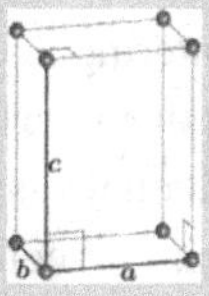	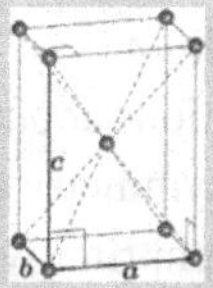	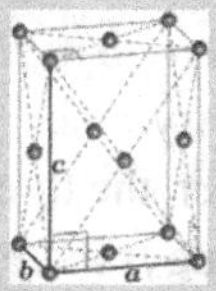	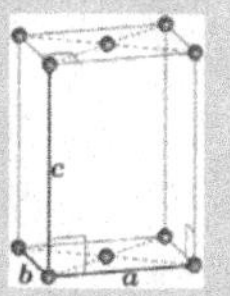
Tetragonal $\alpha = \beta = \gamma = 90°$ $a = b \neq c$	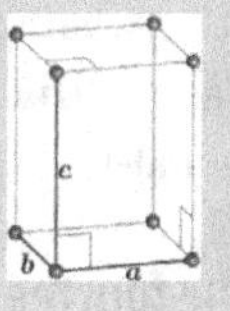	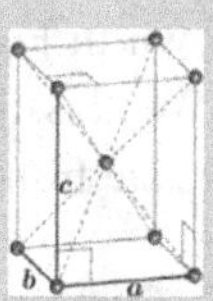		
Rhombohedral (trigonal-R) $(\alpha{=}\beta{=}\gamma) \neq 90°$ $a = b = c$	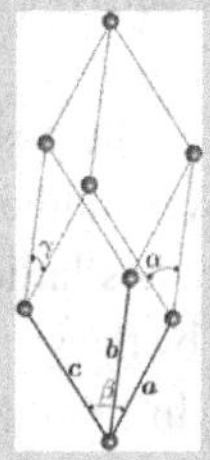			
Hexagonal $\alpha = \beta = 90°$ $\gamma = 60°$ $a = b \neq c$	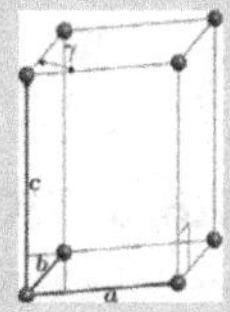			
Cubic $\alpha = \beta = \gamma = 90°$ $a = b = c$	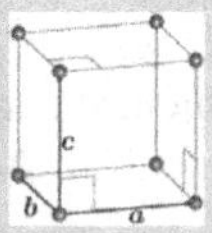	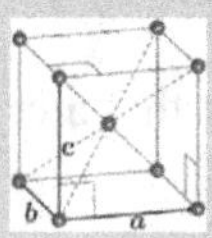	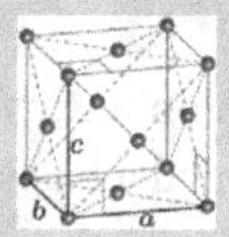	

of atoms to each lattice site that we have a physical crystal structure of a real material. We denote the number of basis atoms by $\hat{N}_B$. If $\hat{N}_B = 1$, the result is referred to as a *simple crystal*. Otherwise, the term *complex crystal* or *multilattice crystal* is used.[15]

[15] There are many other terminologies in the literature for crystals with $\hat{N}_B > 1$, such as "lattice with a basis," "multiatomic lattice" (versus a "monoatomic lattice"), "general crystal" (versus a "parameter-free crystal") "composite crystal" and so on. We will avoid these terms in the interest of clarity.

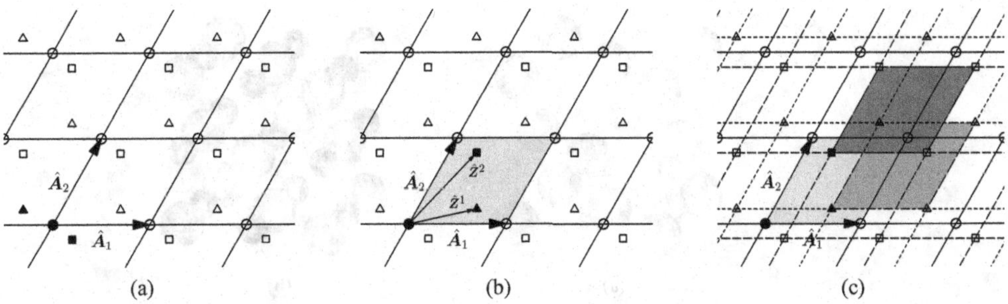

Fig. 3.18 A two-dimensional multilattice crystal. (a) A basis of three atoms (circle, square and triangle) is placed at each lattice site. (b) The basis atoms at the origin are mapped into the unit cell (gray) defined by the lattice vectors. (c) The multilattice can also be represented as three interpenetrating simple lattices. The unit cells of the different lattices are shaded with different intensity.

The basis is a motif of atoms that is translationally invariant from one lattice site to the next. It is sometimes described as a "molecule" attached to each lattice site, but this misleadingly suggests a special bonding or affinity between the atoms within the motif compared to neighboring motifs. The lattice + basis scheme is only a convenient tool to systematically describe the crystal structure.

An example of a two-dimensional crystal with a basis of three atoms is shown in Fig. 3.18. One basis is highlighted (in black). The positions of the basis atoms relative to their lattice site can be arbitrary. However, it can be convenient to identify the basis atoms of a given lattice site with the atoms falling inside the unit cell associated with it. This is shown in Fig. 3.18(b). Another way of thinking of a crystal with a basis of more than one atom is as a set of $\hat{N}_B$ interpenetrating simple lattices, referred to as *sublattices*, as shown in Fig. 3.18(c). This leads to the "multilattice" terminology mentioned above. More specifically, the term $\hat{N}_B$*-lattice* can be used to indicate the number of atoms in the basis. (The crystal in Fig. 3.18 is thus a "3-lattice.")

It is convenient to introduce a special notation for multilattices, so that the crystal is the set of all atoms with positions defined by[16]

$$\boldsymbol{R}^{[\boldsymbol{\ell}\lambda]} = \ell_i \hat{\boldsymbol{A}}_i + \hat{\boldsymbol{Z}}^\lambda, \tag{3.13}$$

where $\hat{\boldsymbol{A}}_i$ are the primitive lattice vectors, $\ell_i \in \mathbb{Z}$ and $\lambda = 0 \ldots \hat{N}_B - 1$. The vectors $\hat{\boldsymbol{Z}}^\lambda$ denote the positions of the basis atoms in the unit cell as shown in Fig. 3.18(b). Without loss of generality, we can always take $\hat{\boldsymbol{Z}}^0 = \boldsymbol{0}$, and the position vectors $\hat{\boldsymbol{Z}}^\lambda$ can be expressed

[16] The notation introduced here is closely related to the notation used by Wallace in [Wal72]. Wallace denotes the position of basis atom λ associated with lattice site $\boldsymbol{L} = (L_1, L_2, L_3)$ by $\boldsymbol{R}(\boldsymbol{L}\lambda)$. Our notation is essentially identical, except that we raise the lattice/atom pair to a superscript to be consistent with the rest of the book, where $\boldsymbol{R}^\alpha$ denotes the position of atom number α in a finite set of atoms.

Another important notation commonly used in the literature is the so-called "Born notation" introduced in [BH54]. In this notation, the position of a basis atom is given by $\boldsymbol{R}\binom{\ell}{\lambda}$. Born notation is very useful for advanced applications of lattice dynamics (see for example [ETS06b]), however, for our purposes the modified Wallace notation suffices and provides a consistent notation across the topics of this book.

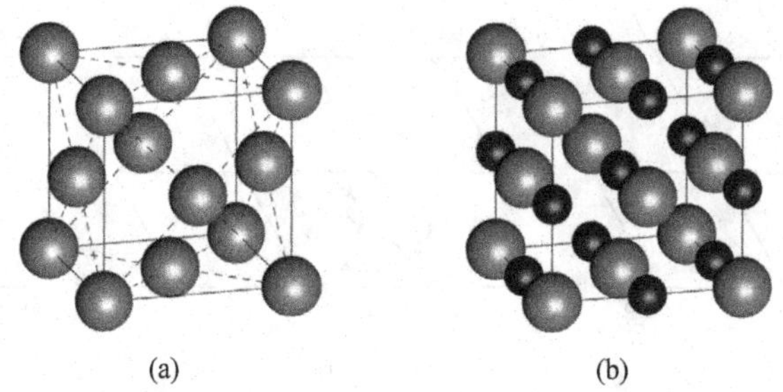

Fig. 3.19 (a) The fcc unit cell. (b) The rock salt (NaCl) crystal structure, also called B1, showing the smaller Na ions and larger Cl ions.

relative to the lattice vectors so that

$$\hat{\boldsymbol{Z}}^\lambda = \zeta_i^\lambda \hat{\boldsymbol{A}}_i, \tag{3.14}$$

where ζ_i^λ are called *fractional coordinates* [San93] and satisfy $0 \le \zeta_i^\lambda < 1$. Some crystal structures contain multiple elements, in which case additional information must be recorded, such as the masses of the basis atoms and the atomic species. For the simple crystal with $\hat{N}_B = 1$, we can drop the basis atom index (λ) and our notation reverts back to the simpler notation in Eqn. (3.1).

3.6.1 Essential and nonessential descriptions of crystals

The discussion of crystal structure given in the previous section did not address the issue of uniqueness. We know from Sections 3.3.1 that the primitive lattice vectors are not unique; however, more than that we have the option of selecting nonprimitive lattice vectors as explained in Section 3.3.3. Thus, Eqn. (3.13) can be rewritten as

$$\boldsymbol{R}^{[\ell\lambda]} = \ell_i \boldsymbol{A}_i + \boldsymbol{Z}^\lambda, \qquad \lambda = 0, \ldots, N_B - 1, \tag{3.15}$$

where $\boldsymbol{A}_i$ are nonprimitive lattice vectors, and the number of basis atoms is now $N_B = n\hat{N}_B$ (where n is the integer number of primitive unit cells contained in the non-primitive cell and $\hat{N}_B$ is the number of basis atoms in the primitive unit cell).

The definition of a crystal with primitive lattice vectors is called the *essential description* as opposed to a *nonessential description* obtained using nonprimitive lattice vectors [Pit98]. The problem of determining the essential description for a given multilattice is an interesting one that we do not discuss here (see, for example, [Par04]).

3.6.2 Crystal structures of some common crystals

Face-centered cubic (fcc) structure On page 135, we saw that the fcc arrangement of points (see Fig. 3.19(a)) forms one of the Bravais lattices. Many metals adopt the *fcc crystal structure* which corresponds to an fcc Bravais lattice with a one-atom basis, i.e. with one atom at each lattice site. Some technologically-important metals with an fcc crystal structure are Al, Ni, Cu, Au, Ag, Pd and Pt. The high-temperature phase of Fe (austenite) is also fcc.

If we take the cube edges in Fig. 3.13(b) to lie along the coordinate axes with unit vectors e_i, then

$$\hat{A}_1 = \frac{a}{2}(e_2 + e_3), \qquad \hat{A}_2 = \frac{a}{2}(e_1 + e_3), \qquad \hat{A}_3 = \frac{a}{2}(e_1 + e_2), \tag{3.16}$$

where a is the length of the side of the cube and is called the *lattice constant* (or *lattice parameter*). It is often more convenient to treat the fcc structure using using a nonessential description, as a simple cubic multilattice with a four-atom basis, in order to make the cubic symmetry more apparent. In this case we have

$$A_1 = ae_1, \qquad A_2 = ae_2, \qquad A_3 = ae_3, \tag{3.17}$$

and four basis atoms at

$$Z^0 = \mathbf{0}, \qquad Z^1 = \frac{a}{2}(e_1 + e_2), \qquad Z^2 = \frac{a}{2}(e_1 + e_3), \qquad Z^3 = \frac{a}{2}(e_2 + e_3). \tag{3.18}$$

Rock salt (B1) structure Rock salt (NaCl) adopts an fcc crystal structure with a two-atom basis that is also called the "B1" structure . It can be thought of as two interpenetrating fcc lattices, with one centered on the edge of the unit cube of the other. In the nonessential description of Eqn. (3.17), it requires eight basis atoms:

$$\begin{aligned}
Z^0 &= \mathbf{0} \ \text{Cl}, & Z^1 &= \frac{a}{2}(e_1 + e_2) \ \text{Cl}, \\
Z^2 &= \frac{a}{2}(e_1 + e_3) \ \text{Cl}, & Z^3 &= \frac{a}{2}(e_2 + e_3) \ \text{Cl}, \\
Z^4 &= \frac{a}{2}e_1 \ \text{Na}, & Z^5 &= \frac{a}{2}e_2 \ \text{Na}, \\
Z^6 &= \frac{a}{2}e_3 \ \text{Na}, & Z^7 &= \frac{a}{2}(e_1 + e_2 + e_3) \ \text{Na}.
\end{aligned} \tag{3.19}$$

Body-centered cubic (bcc) structure Like the fcc structure, we also saw that the bcc arrangement of points forms one of the Bravais lattices. Many metals, including such technologically-important examples as Fe, Cr, Ta, W and Mo, adopt the bcc crystal structure.

Using the primitive bcc lattice, we can describe the bcc crystal as a simple crystal with a one-atom basis. If we imagine the cube edges in Fig. 3.13(a) to lie along the coordinate axes with unit vector e_i, then

$$\hat{A}_1 = \frac{a}{2}(-e_1+e_2+e_3), \qquad \hat{A}_2 = \frac{a}{2}(e_1-e_2+e_3), \qquad \hat{A}_3 = \frac{a}{2}(e_1+e_2-e_3). \tag{3.20}$$

Once again, we may prefer to make plain the cubic symmetry by using a nonessential description. If we take the simple cubic lattice we must use a two-atom basis, as follows:

$$A_1 = ae_1, \qquad A_2 = ae_2, \qquad A_3 = ae_3, \tag{3.21}$$

and two basis atoms at

$$Z^0 = \mathbf{0}, \qquad Z^1 = \frac{a}{2}(e_1 + e_2 + e_3). \tag{3.22}$$

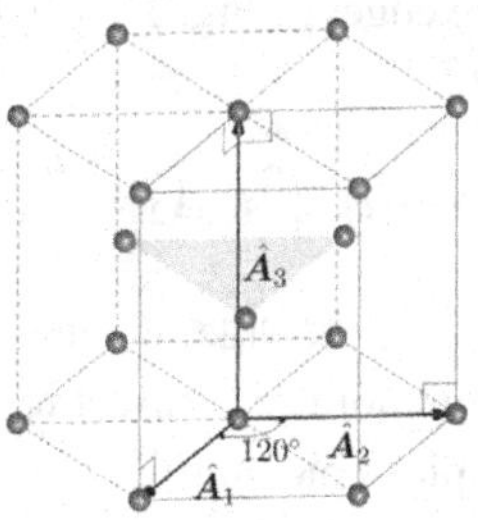

Fig. 3.20 The hcp crystal structure.

Hexagonal close-packed (hcp) structure Many materials exhibit a hexagonal crystal structure: most of them are ceramic or natural minerals containing multiple elements. These are typically multilattices with large numbers of atoms in their basis. However, one relatively simple hexagonal crystal which is important to engineering applications is the *hcp* structure. Pure metals that adopt an hcp structure include Mg, Zn, Ti and Zr.

The hcp structure is a multilattice with the hexagonal unit cell and a two-atom basis. The hexagonal lattice vectors (illustrated in Fig. 3.11) are given by[17]

$$\hat{\boldsymbol{A}}_1 = a\boldsymbol{e}_1, \qquad \hat{\boldsymbol{A}}_2 = -\frac{a}{2}\boldsymbol{e}_1 + \frac{\sqrt{3}a}{2}\boldsymbol{e}_2, \qquad \hat{\boldsymbol{A}}_3 = c\boldsymbol{e}_3, \tag{3.23}$$

where a and c are the hcp lattice constants. It is easy to see from Eqn. (3.23) that $\|\hat{\boldsymbol{A}}_1\| = \|\hat{\boldsymbol{A}}_2\| = a$. The two basis atoms are located at

$$\hat{\boldsymbol{Z}}^0 = \boldsymbol{0}, \qquad \hat{\boldsymbol{Z}}^1 = \frac{2\hat{\boldsymbol{A}}_1}{3} + \frac{\hat{\boldsymbol{A}}_2}{3} + \frac{\hat{\boldsymbol{A}}_3}{2}, \tag{3.24}$$

where we have written the basis atoms in terms of the lattice vectors instead of an orthogonal coordinate system for clarity. The hcp crystal structure is shown in Fig. 3.20. It is important to note that the hcp structure gives close-packing only for the *ideal* hcp structure for which $c = a\sqrt{8/3}$ (see Exercise 3.5).

Diamond cubic structure Silicon and germanium are two extremely important materials for the microelectronics industry. Carbon has multiple solid phases, but it is perhaps most valuable and useful as diamond due to both its aesthetic beauty and high hardness. All of these elements adopt the diamond cubic crystal structure.

The diamond structure is an fcc lattice with a two-atom basis. As such, we can describe it using the primitive lattice vectors of Eqns. (3.16) with two basis atoms at

$$\hat{\boldsymbol{Z}}^0 = \boldsymbol{0}, \qquad \hat{\boldsymbol{Z}}^1 = \frac{a}{4}(\boldsymbol{e}_1 + \boldsymbol{e}_2 + \boldsymbol{e}_3) = \frac{1}{4}\left(\hat{\boldsymbol{A}}_1 + \hat{\boldsymbol{A}}_2 + \hat{\boldsymbol{A}}_3\right). \tag{3.25}$$

[17] Note that in Fig. 3.11 $\boldsymbol{a} = \hat{\boldsymbol{A}}_1$, $\boldsymbol{b} = \hat{\boldsymbol{A}}_2$ and $\boldsymbol{c} = \hat{\boldsymbol{A}}_3$.

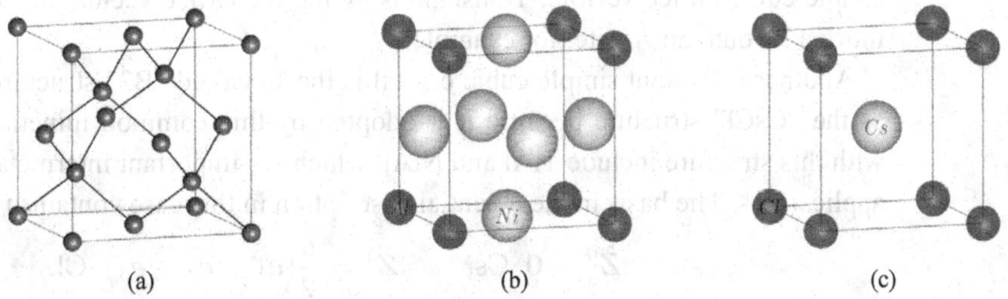

Fig. 3.21 (a) The diamond cubic crystal structure. Nearest-neighbor bonds are drawn to show the tetrahedral arrangement of atoms. (b) The Ni_3Al crystal structure. (c) The CsCl (B2) crystal structure.

Alternatively, we can use a nonessential description, with the cubic lattice vectors of Eqn. (3.17) and an eight-atom basis:

$$\begin{aligned}
\boldsymbol{Z}^0 &= \mathbf{0}, & \boldsymbol{Z}^1 &= \frac{a}{2}(\mathbf{e}_1 + \mathbf{e}_2),\\
\boldsymbol{Z}^2 &= \frac{a}{2}(\mathbf{e}_1 + \mathbf{e}_3), & \boldsymbol{Z}^3 &= \frac{a}{2}(\mathbf{e}_2 + \mathbf{e}_3),\\
\boldsymbol{Z}^4 &= \frac{a}{4}(3\mathbf{e}_1 + 3\mathbf{e}_2 + \mathbf{e}_3), & \boldsymbol{Z}^5 &= \frac{a}{4}(3\mathbf{e}_1 + \mathbf{e}_2 + 3\mathbf{e}_3),\\
\boldsymbol{Z}^6 &= \frac{a}{4}(\mathbf{e}_1 + 3\mathbf{e}_2 + 3\mathbf{e}_3), & \boldsymbol{Z}^7 &= \frac{a}{4}(\mathbf{e}_1 + \mathbf{e}_2 + \mathbf{e}_3).
\end{aligned} \tag{3.26}$$

Each atom in the diamond structure has four nearest neighbors in a tetrahedral arrangement (the angle between any pair of near-neighbor bonds is 109.5°). The diamond cubic structure is shown in Fig. 3.21(a).

Zincblende structure An important variation on the diamond cubic crystal structure is the so-called "zincblende" or B3 structure adopted by important semi-conductor alloys like GaAs. It can be described by the same nonessential fcc lattice and eight-atom basis as above for diamond cubic, except that $\boldsymbol{Z}^0$ through $\boldsymbol{Z}^3$ are one atomic species and $\boldsymbol{Z}^4$ through $\boldsymbol{Z}^7$ are another.

Simple cubic structure Crystals for which the primitive cell is the cubic-P Bravais lattice are often referred to as *simple cubic* crystals. Many intermetallic materials have the simple cubic structure. For example, the intermetallic compound Ni_3Al is an important component in gas turbine blades. In the essential description, its structure is given by the simple cubic lattice of Eqn. (3.17) and a four-atom basis:

$$\begin{aligned}
\hat{\boldsymbol{Z}}^0 &= \mathbf{0}\ \text{Al}, & \hat{\boldsymbol{Z}}^1 &= \frac{a}{2}(\mathbf{e}_1 + \mathbf{e}_2)\ \text{Ni},\\
\hat{\boldsymbol{Z}}^2 &= \frac{a}{2}(\mathbf{e}_1 + \mathbf{e}_3)\ \text{Ni}, & \hat{\boldsymbol{Z}}^3 &= \frac{a}{2}(\mathbf{e}_2 + \mathbf{e}_3)\ \text{Ni}.
\end{aligned} \tag{3.27}$$

The Ni_3Al structure is shown in Fig. 3.21(b). Note that this is *not* an fcc structure, even though the atomic sites coincide with the fcc lattice positions. This is because there are two different atomic species in the crystal, and translational symmetry only holds for the simple cubic lattice vectors. Translations by the fcc lattice vectors of Eqns. (3.16) would move a Ni onto an Al site, for example.

Another important simple cubic crystal is the so-called "B2" structure, also referred to as the "CsCl" structure because it is adopted by this common mineral. Other materials with this structure include TiAl and NiAl, which are important intermetallics in aerospace applications. The basis in the essential description in this case contains two atoms:

$$\hat{\boldsymbol{Z}}^0 = \mathbf{0} \ \text{Cs}, \qquad \hat{\boldsymbol{Z}}^1 = \frac{a}{2}(\boldsymbol{e}_1 + \boldsymbol{e}_2 + \boldsymbol{e}_3) \ \text{Cl}. \tag{3.28}$$

As with the Ni_3Al structure, it is important to recognize that CsCl is not bcc due to the presence of two different atomic species, as shown in Fig. 3.21(c).

3.7 Some additional lattice concepts

3.7.1 Fourier series and the reciprocal lattice

The periodic structure of lattices and crystals means that Fourier analysis is often an ideal tool for studying them. Indeed, we will make use of Fourier techniques numerous times throughout this text, including in discussions of quantum mechanics (Section 4.4.2), density functional theory (Section 4.5.5) and tight-binding methods (Section 4.6.6). The Fourier approach is inextricably linked to the important physical concept of the *reciprocal lattice*, a key concept in understanding density functional theory implementations, diffraction, crystalline vibrations and phase transformations.

Imagine that we have a function, $g(\boldsymbol{R})$, that is related to an underlying crystal structure and that therefore shares the same periodicity as the crystal. This periodicity will therefore be commensurate with the unit cell of the lattice, with cell volume Ω_0. The Fourier series of this function can be written as

$$g(\boldsymbol{R}) = \frac{1}{\sqrt{\Omega_0}} \sum_{\boldsymbol{k}} \widehat{g}_{\boldsymbol{k}} e^{i\boldsymbol{k}\cdot\boldsymbol{R}}, \tag{3.29}$$

where $i = \sqrt{-1}$, $\boldsymbol{k}$ is a *wave vector* and $\widehat{g}_{\boldsymbol{k}}$ is a complex coefficient. In the physics terminology that we will use later (most notably in our discussion of quantum mechanics in Chapter 4), this is sometimes referred to as a *plane-wave expansion* of a function. As we shall see in later chapters, the exponential in this series is the mathematical form that describes a standing planar wave in three dimensions.

Reciprocal lattice We return shortly to a method for determining the coefficients $\widehat{g}_{\boldsymbol{k}}$, but for now we note that the periodicity of the function $g(\boldsymbol{R})$ imposes constraints on which wave vectors can be admitted in the series. Specifically, we must have that

$$g(\boldsymbol{R}) = g(\boldsymbol{R} + \boldsymbol{t}),$$

where $\boldsymbol{t}$ is a translation vector of the lattice

$$\boldsymbol{t} = n_i \boldsymbol{A}_i \quad (n_i \in \mathbb{Z}). \tag{3.30}$$

Using the expanded form of $g(\boldsymbol{R})$ we therefore have

$$\sum_{\boldsymbol{k}} \widehat{g}_{\boldsymbol{k}} e^{i\boldsymbol{k}\cdot\boldsymbol{R}} = \sum_{\boldsymbol{k}} \widehat{g}_{\boldsymbol{k}} e^{i\boldsymbol{k}\cdot(\boldsymbol{R}+\boldsymbol{t})}. \tag{3.31}$$

In order for this to be true for any arbitrary function $g(\boldsymbol{R})$, it must be true term-by-term throughout the series,[18] meaning that the condition

$$e^{i\boldsymbol{k}\cdot\boldsymbol{t}} = 1$$

determines the allowable wave vectors $\boldsymbol{k}$. This is satisfied for

$$2m\pi = \boldsymbol{k}\cdot\boldsymbol{t} = n_1(\boldsymbol{k}\cdot\boldsymbol{A}_1) + n_2(\boldsymbol{k}\cdot\boldsymbol{A}_2) + n_3(\boldsymbol{k}\cdot\boldsymbol{A}_3),$$

where m is any integer and we have substituted the definition of $\boldsymbol{t}$ from Eqn. (3.30). Since this must hold for *all possible* translation vectors, it follows that each term in parentheses must by itself be an integer multiple of 2π, so that

$$\boldsymbol{k}\cdot\boldsymbol{A}_i = 2m_i\pi. \tag{3.32}$$

By expanding this expression, it is easy to see that it can be equivalently written in matrix notation (see Eqn. (3.2)) as

$$\mathbf{H}^T\boldsymbol{k} = 2\pi\mathbf{m},$$

where $\mathbf{m}$ is a vector of three integers, unique for each $\boldsymbol{k}$. Inverting this equation makes it clear that the vectors $\boldsymbol{k}$, like their counterparts $\boldsymbol{t}$, also form a lattice

$$\boldsymbol{k} = 2\pi\mathbf{H}^{-T}\mathbf{m} = \mathbf{B}\mathbf{m},$$

where we have defined the matrix of the new lattice vectors to be

$$\mathbf{B} = 2\pi\mathbf{H}^{-T}. \tag{3.33}$$

The columns of this matrix are the lattice vectors of a new *reciprocal lattice*:

$$\mathbf{B} = \begin{bmatrix} [\boldsymbol{B}_1]_1 & [\boldsymbol{B}_2]_1 & [\boldsymbol{B}_3]_1 \\ [\boldsymbol{B}_1]_2 & [\boldsymbol{B}_2]_2 & [\boldsymbol{B}_3]_2 \\ [\boldsymbol{B}_1]_3 & [\boldsymbol{B}_2]_3 & [\boldsymbol{B}_3]_3 \end{bmatrix}. \tag{3.34}$$

Equation (3.33) can also be written directly in terms of the individual lattice vectors as

$$\boldsymbol{B}_i\cdot\boldsymbol{A}_j = 2\pi\delta_{ij}. \tag{3.35}$$

[18] Each term can, by itself, be considered a possible choice for the function $g(\boldsymbol{R})$, in which case Eqn. (3.31) can be satisfied only if the term-by-term correspondence holds.

One can then readily verify that the reciprocal lattice vectors take the form,[19]

$$\boldsymbol{B}_1 = 2\pi\frac{\boldsymbol{A}_2 \times \boldsymbol{A}_3}{\Omega_0}, \qquad \boldsymbol{B}_2 = 2\pi\frac{\boldsymbol{A}_3 \times \boldsymbol{A}_1}{\Omega_0}, \qquad \boldsymbol{B}_3 = 2\pi\frac{\boldsymbol{A}_1 \times \boldsymbol{A}_2}{\Omega_0}, \tag{3.36}$$

where $\Omega_0 = \boldsymbol{A}_1 \cdot (\boldsymbol{A}_2 \times \boldsymbol{A}_3)$ is the unit cell volume.

We will sometimes refer to the original lattice as the "direct lattice" when it is necessary to make the clear distinction from the reciprocal lattice.

Example 3.1 (The reciprocal lattice of an fcc Bravais lattice) Starting from the direct lattice vectors for the fcc structure given in Eqns. (3.16), it is straightforward to apply Eqn. (3.36) to show that the reciprocal lattice vectors are

$$\boldsymbol{B}_1 = \frac{2\pi}{a}(-\boldsymbol{e}_1 + \boldsymbol{e}_2 + \boldsymbol{e}_3), \qquad \boldsymbol{B}_2 = \frac{2\pi}{a}(\boldsymbol{e}_1 - \boldsymbol{e}_2 + \boldsymbol{e}_3), \qquad \boldsymbol{B}_3 = \frac{2\pi}{a}(\boldsymbol{e}_1 + \boldsymbol{e}_2 - \boldsymbol{e}_3).$$

Comparing with Eqn. (3.20), we see that this forms a bcc reciprocal lattice, with cube side length $4\pi/a$.

The Fourier series coefficients We have identified the allowable wave vectors in the Fourier series of Eqn. (3.29) as translation vectors on the *reciprocal lattice*. This allows us to derive an expression for the Fourier coefficients, $\widehat{g}_{\boldsymbol{k}}$. First, we note that once we have confined ourselves to wave vectors of this form, the following orthogonality holds:

$$\int_{\Omega_0} e^{i\boldsymbol{k}\cdot\boldsymbol{R}} e^{-i\boldsymbol{k}'\cdot\boldsymbol{R}}\, d\boldsymbol{R} = \begin{cases} \Omega_0 & \text{for } \boldsymbol{k} = \boldsymbol{k}', \\ 0 & \text{for } \boldsymbol{k} \neq \boldsymbol{k}', \end{cases} \tag{3.37}$$

where the integral is over the domain of a single unit cell of the Bravais lattice. This orthogonality is tedious but straightforward to prove if one recognizes the restrictions we have imposed on the periodicity of the integrand by our choices of wave vectors and notes that $\boldsymbol{k} + \boldsymbol{k}' = \boldsymbol{K}$, where $\boldsymbol{K}$ must also be a translation vector on the reciprocal lattice.

Multiplying both sides of Eqn. (3.29) by $\exp(-i\boldsymbol{k}' \cdot \boldsymbol{R})$, integrating over the unit cell and making use of Eqn. (3.37) allows us to determine that the Fourier coefficients are

$$\widehat{g}_{\boldsymbol{k}} = \frac{1}{\sqrt{\Omega_0}} \int g(\boldsymbol{R}) e^{-i\boldsymbol{k}\cdot\boldsymbol{R}}\, d\boldsymbol{R}. \tag{3.38}$$

We will make use of this result at several junctures in the following chapters.

3.7.2 The first Brillouin zone

Just as for the direct lattice, the reciprocal lattice is defined by its primitive cell. The volume of this cell is of course $\boldsymbol{B}_1 \cdot (\boldsymbol{B}_2 \times \boldsymbol{B}_3)$, but its shape is somewhat arbitrary. The most

[19] See page 30 for the connection to the reciprocal basis vectors of continuum mechanics.

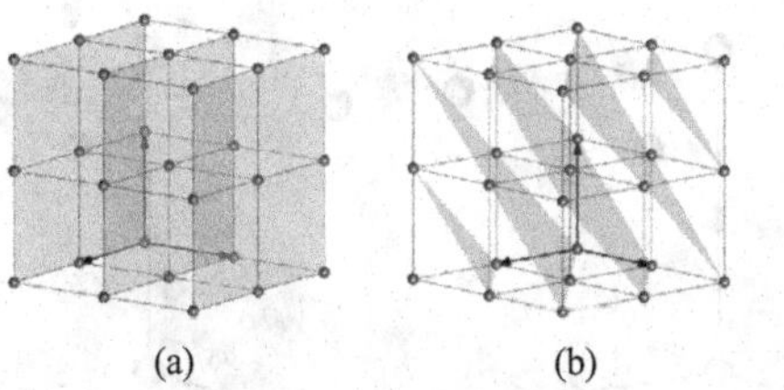

Fig. 3.22 The (a) (100) and (b) $(1\bar{1}1)$ families of planes in the simple cubic lattice.

natural choice, as shown in Fig. 3.3, is the parallelepiped defined by the lattice vectors. But it is also possible to choose the Voronoi cell, as shown in Fig. 3.4. Whereas this was referred to as the Wigner–Seitz cell in the direct lattice, it is called the *first Brillouin zone* in the reciprocal space.[20] We will make use of the Brillouin zone in our discussion of density functional theory in Section 4.5 and in the discussion of wave propagation in Section 13.3.2.

3.7.3 Miller indices

The reciprocal lattice has a connection with the definition of planes of atoms within a lattice. A *lattice plane* is a geometric plane defined by any three noncollinear Bravais sites. Due to translational symmetry, such a plane contains an infinite number of other Bravais lattice sites and is one of a family of equally-spaced lattice planes that together contain all the points of the lattice. In Fig. 3.22, examples of lattice planes are shown.

A lattice plane can be defined by choosing any two vectors, $\boldsymbol{X}$ and $\boldsymbol{Y}$, that lie in the plane. If the heads and tails of these vectors are lattice sites, the vectors will be translation vectors, so that

$$\boldsymbol{X} = m_i \boldsymbol{A}_i, \qquad \boldsymbol{Y} = n_j \boldsymbol{A}_j,$$

where m_i and n_j are integers. The cross-product of these vectors lies along the normal to the lattice plane, which we denote $\boldsymbol{N}$

$$\boldsymbol{N} = m_i n_j \boldsymbol{A}_i \times \boldsymbol{A}_j.$$

This is a sum of nine terms, but of course the cross-product of a vector with itself ($i = j$) is zero. Further, since $\boldsymbol{A}_i \times \boldsymbol{A}_j = -\boldsymbol{A}_j \times \boldsymbol{A}_i$ we can collect the remaining terms into just three terms as:

$$\boldsymbol{N} = l_1 \boldsymbol{A}_2 \times \boldsymbol{A}_3 + l_2 \boldsymbol{A}_1 \times \boldsymbol{A}_3 + l_3 \boldsymbol{A}_1 \times \boldsymbol{A}_2,$$

[20] As the name suggests there are higher-order Brillouin zones. They need not concern us here, but they are important for some applications. The interested reader may like to look at [AM76]. The "Brillouin zone" is named after the French physicist Léon Nicolas Brillouin, who introduced the concept in the 1930s. The name "Brillouin" is a bit of a tongue twister that leaves many non-French speakers at a loss as to how to pronounce it. The correct pronunciation in IPA notation is [ˈbʁilwɛ̃], where the "ʁ" is the French "R", the "i" is pronounced like the vowel in "field", and the "ɛ̃" is a nasalized version of the vowel in "when", which sounds like it falls between the vowels in the words "then" and "than".

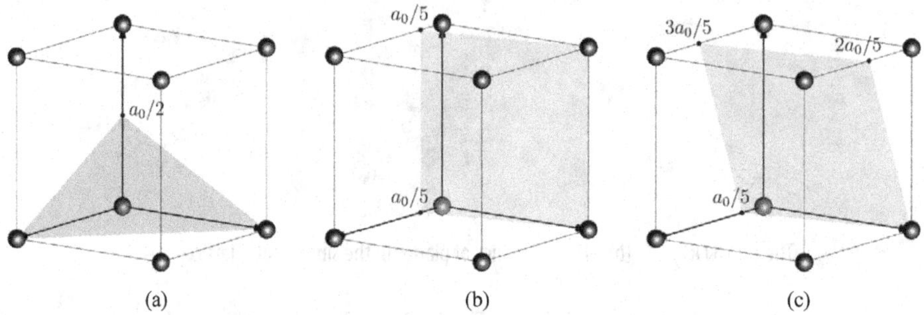

Fig. 3.23 The (a) (112), (b) (510) and (c) $(51\bar{2})$ planes in the simple cubic lattice. Labels indicate axis intercepts.

where the l_i are integers comprising combinations of the original integers m_i and n_j. It becomes clear, when comparing with Eqn. (3.36), that this can be written in terms of the reciprocal lattice vectors:

$$\boldsymbol{N} = \frac{\Omega_0}{2\pi}\left(l_1\boldsymbol{B}_1 + l_2\boldsymbol{B}_2 + l_3\boldsymbol{B}_3\right).$$

In other words, the normal to any direct lattice plane is always parallel to a translation vector in the corresponding reciprocal lattice. Of course, there are an infinite number of these normal vectors, but they differ only in their length. This presents a convenient way to classify lattice planes in terms of the reciprocal lattice. Namely, if we take the *shortest* reciprocal lattice vector, $\boldsymbol{B}$, that is normal to a given set of planes:

$$\boldsymbol{B} = h_1\boldsymbol{B}_1 + h_2\boldsymbol{B}_2 + h_3\boldsymbol{B}_3,$$

then the integers h_i uniquely define these planes. The conventional notation is to call these integers h, k and l, and identify them as the *Miller indices*,[21] denoting the plane within parentheses as (hkl).

Miller indices in the simple cubic system Miller indices can be used to describe lattice planes within *any* Bravais lattice; they are not tied to cubic crystals in particular, although they are most commonly presented, used and understood within the context of the cubic lattice.

For a simple cubic lattice, both the direct and reciprocal lattice vectors are orthogonal. In addition, each of the three direct lattice vectors is parallel to one of the reciprocal lattice vectors. This means that the intercepts of a plane with the crystal axes will be inversely proportional to the Miller indices, and a simple procedure for determining the Miller indices can be written as in Algorithm 3.2. Some examples in the simple cubic lattice are shown in Fig. 3.23.

Note the important difference between the notation for planes and that for directions (see Section 3.3.4) is the shape of the brackets. Just as with directions, families of planes that are equivalent after symmetry operations of the crystal can be denoted collectively using

[21] Miller indices were introduced by the British mineralogist William Hallowes Miller in his book *Crystallography* published in 1839. His new notation revolutionized the field and continues to be the basis for modern crystallographic theory.

Algorithm 3.2 Miller indices of a given plane in the simple cubic system

1: If the given plane contains the origin, choose any (equivalent) parallel plane that does not.
2: Identify the intercept, d_i of the plane with each of the crystal axes. If an axis does not intersect the plane, call the corresponding intercept $d_i = \infty$.
3: Take the reciprocals $h_i = 1/d_i$.
4: Use a common factor to convert h_1, h_2, h_3 to the smallest possible set of integers.
5: Write the resulting integers in parentheses as (hkl), with negatives as overbars.

$\{\cdot\}$ as the bounding brackets. For example, the $\{111\}$ family of planes in the simple cubic system includes all of the planes (111), $(\bar{1}11)$, $(1\bar{1}1)$, $(11\bar{1})$, ...

Finally, it is left as an exercise to show that in the simple cubic lattice the distance between adjacent planes with Miller indices (hkl) is given by

$$d = \frac{a}{\sqrt{h^2 + k^2 + l^2}}. \tag{3.39}$$

Further reading

- The book by Verma and Srivastava [VS82] is out of print, but useful if you can find it. It provides a very clear introduction with many examples to crystal structure and crystal symmetries. Excellent for the total novice.
- The book by Burns and Glazer [BG90] is similar to that of Verma and Srivastava, but adopts a somewhat more mathematical approach.
- Compared with other books on the subject of crystal symmetry, the one by Kelly *et al.* [KGK00] uses an approach that is less mathematical, and more geometrical and physical, to explain the crystal systems and Bravais lattices.
- *Solid State Physics* by Ashcroft and Mermin [AM76] is a classical reference on solid-state physics. It contains two chapters on Bravais lattices. Chapter 4 provides a very basic introduction to the concept of lattices. Chapter 7 discusses the classification of lattices according to their symmetries.
- A very useful website has been produced by the US Naval Research Laboratory [NRL09]. The site catalogs the details of hundreds of crystal structures and provides interactive visualization tools.

Exercises

3.1 [Section 3.4] Prove that n-fold rotations are symmetry operations of space filling lattices only for $n = 1, 2, 3, 4, 6$ (after [BG90]). To do this, consider two lattice points, A and A' separated by a lattice translation t as shown in Fig. 3.24. Now, imagine two rotations and their effects: the effect on A' of a rotation of the lattice through angle $\alpha = 2\pi/n$ about A, and the effect on

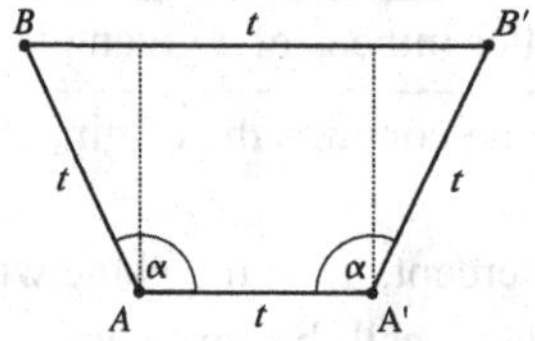

Fig. 3.24 Geometry to demonstrate that only certain n-fold rotations can be symmetry operators on an infinite lattice.

A of a rotation of the lattice through angle $-\alpha$ about A'. These move the points as $A \to B'$ and $A' \to B$. If a lattice has $n(C_n)$ symmetry, both rotations α and $-\alpha$ must restore the lattice since one operation is merely the inverse of the other. This means that the points B and B' must both be lattice sites and must therefore be an integral multiple of t apart. Use this condition to determine the allowable angles, α.

3.2 [SECTION 3.4] Show that $\bar{4}$ in the international notation is the same operation as S_4^3 in the Schoenflies notation, and similarly that $\bar{4}^3$ is equivalent to S_4. Do this both mathematically and "intuitively," considering the effect on an arbitrary lattice point with a "handedness." For example think of the lattice point as represented by a three-dimensional object that can be made backwards or inverted by the operations of reflection or inversion, respectively (for example one might consider the letter "F," with different colors distinguishing the symbol's "front" from its "back").

3.3 [SECTION 3.5] For the tetragonal crystal system, do the following:
1. Show with a sketch that base-centering on either the A or B faces will violate the required four-fold symmetry of this system.
2. Show with a sketch that the base-centered tetragonal-C is equivalent to tetragonal-P with an appropriately redefined set of lattice vectors.
3. Show with a sketch that the face-centered tetragonal-F lattice is equivalent to the body-centered tetragonal-I lattice after redefining the lattice vectors.

3.4 [SECTION 3.5] Convince yourself, using a sketch of the hexagonal system projected along the c-axis, that none of the face-, body- or base-centerings will produce a new lattice that has hexagonal or trigonal symmetry.

3.5 [SECTION 3.6] The (001) planes in the hcp structure (the plane containing vectors $\hat{\boldsymbol{A}}_1$ and $\hat{\boldsymbol{A}}_2$ in Fig. 3.20) are referred to as the "basal planes" because they form the base of the hexagonal prism defining the structure. Treat the atoms in each plane as perfect spheres touching along the $\hat{\boldsymbol{A}}_1$ and $\hat{\boldsymbol{A}}_2$ directions, and show that the ratio $c/a = \sqrt{8/3}$ corresponds to the most densely packed hcp arrangement of spheres that is possible.

3.6 [SECTION 3.7] Sketch an arbitrary (nonorthogonal) two-dimensional direct lattice and identify its primitive cell and lattice vectors. Determine its reciprocal lattice. Draw a sketch of the reciprocal lattice to scale and overlapping the direct lattice.

3.7 [SECTION 3.7] Show that a simple cubic direct lattice has a simple cubic reciprocal lattice.

3.8 [SECTION 3.7] Show that a bcc direct lattice has an fcc reciprocal lattice.

3.9 [SECTION 3.7] Derive Eqn. (3.39).

4 Quantum mechanics of materials

4.1 Introduction

In the preface to his book on the subject, Peierls says of the quantum theory of solids [Pei64]

> [It] has sometimes the reputation of being rather less respectable than other branches of modern theoretical physics. The reason for this view is that the dynamics of many-body systems, with which it is concerned, cannot be handled without the use of simplifications, or without approximations which often omit essential features of the problem.

This book, in essence, is all about ways to *further* approximate the quantum theory of solids; it is about ways to build approximate models that describe the collective behavior of quantum mechanical atoms in materials. Whether it is the development of density functional theory (DFT), empirical classical interatomic potentials, continuum constitutive laws or multiscale methods, one can only conclude that, by Peierls' measure, modeling materials is a science that is a great deal less respectable than even the quantum theory of solids itself. However, Peierls goes on to say that

> Nevertheless, the [quantum] theory [of solids] contains a good deal of work of intrinsic interest in which one can either develop a solution convincingly from first principles, or at least give a clear picture of the features which have been omitted and therefore discuss qualitatively the changes which they are likely to cause.

It is our view that these redeeming qualities are equally applicable to the modern science of modeling materials.

To understand the underpinnings of materials modeling, we need to understand a little of the quantum theory of solids and, by extension, the basics of quantum mechanics itself. These topics are the subject of this chapter. Our intention is to provide a very simple introduction for a reader who knows little or nothing about quantum mechanics. The hope is that by the end of the chapter, such a reader will have an appreciation of the complexity of the quantum theory, in anticipation of the models to follow. For instance, an empirical atomistic model seems little more than a curve-fitting exercise, unless its basis in the quantum theory can be appreciated. With a little of this background, the reader can see that materials modeling, despite its approximate nature, indeed passes Peierls' standard for a respectable branch of physics. A more practical objective of the chapter is to enable readers with no quantum mechanics background to be able to read and understand the results of DFT calculations (see Section 4.5) even if they are not able to appreciate all of the subtleties. Papers on DFT results are rife with jargon that can be impenetrable without the sort of introduction we are providing here.

Of the expert in quantum theory, we ask a little leeway regarding our brief discussion of the subject that will follow. Indeed, we have skipped or glossed over great swathes of the field, adopting a "strictly on a need to know basis" approach with the goal of rudely constructing a bridge to materials modeling. The quantum expert may skip the chapter or, we hope, take from it a different perspective on what comprise the important aspects of quantum mechanics for materials science.

4.2 A brief and selective history of quantum mechanics

The essence of quantum mechanics rests on the notion that energy, light and matter are not continuous, but rather made up of indivisible "quanta" that can only exist in discrete states. While the *atomic* nature of matter had been well understood since the work of John Dalton at the dawn of the nineteenth century (see Section 3.1), it was more than a century later before the quantization of light was fully accepted and understood.

Like the history of many branches of science, the origins of quantum mechanics are somewhat murky, and many textbooks present an apocryphal and inaccurate version of the origins of the subject [Kra08]. The common story is that Planck proposed the quantization of light in order to solve the so-called "ultraviolet catastrophe" (a failure of classical mechanics to predict the emissivity of a black body at short wavelengths via the "Rayleigh–Jeans" law). While Planck was indeed interested in black body radiation, his goal was an accurate model of black body radiation as a means to shed light on the fundamental question of entropy and the second law of thermodynamics. Having failed in previous attempts to accurately describe the low-frequency spectrum, Planck was forced to revisit a theory which he had strongly resisted: the probabilistic interpretation of entropy advocated by Boltzmann. Along the way, Planck found it necessary to introduce his famous constant ($h = 6.6261 \times 10^{-34}$ J · s) and the notion of "countable" energy states mainly as a mathematical convenience. He was not thinking of material or light as quanta, and in fact was at the time likely in the camp of scientists who preferred the view of all matter as continuous. It was only afterwards that Planck and others appreciated the significance of the quantum, and he ultimately came around to the idea of quantized matter.

Soon after Planck introduced his theory for black body radiation, Albert Einstein burst on the scene publishing three landmark papers on the subject: in 1905 on the theory of light quanta (photons), in 1907 on the quantum theory of specific heats of solids, and in 1909 on the dual wave–particle nature of electromagnetic radiation. Einstein's 1905 paper took the idea of quantized radiation and used it to explain the photoelectric effect. Briefly, it was known that when light struck a metallic plate, electrons were ejected from the plate in a measurable current, but the way in which this effect comes about could not be explained by classical mechanics. One manifestation of the problem is that experiments show a critical value of the radiation frequency, ν_0, below which no current is detected no matter how intense the radiation source. The classical theory contradicted this, predicting that there should be no lower bound on the frequency to activate the effect as long as the radiation intensity was high enough. Einstein's contribution was to show that quantized radiation

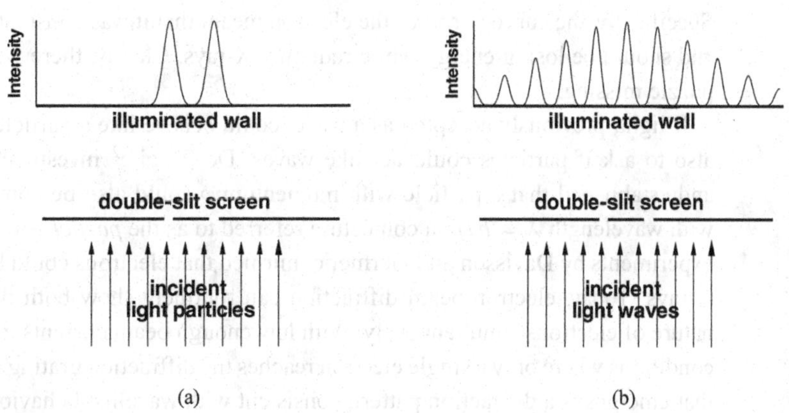

Fig. 4.1 Light passing through two narrow slits. If light were to behave as particles, the pattern of illumination intensity would be as in (a), whereas the actual observed pattern is consistent with light behaving as waves, as shown in (b).

explained this, ultimately earning him a Nobel prize.[1] To make the theory work, Einstein had to take a much bolder step than Planck, who introduced quanta of light only as a mathematical convenience. Einstein's work required him to change the very *nature* of light itself. To explain the photoelectric effect Einstein required that radiation existed not as continuous waves, but as discrete "grains" or "particles" of light (which came to be known as "photons").

On the other hand, it is clear that light diffracts in the way that waves do. In the canonical experiment (originally performed by Thomas Young in 1803), light is directed at a grating of two narrow slits and a film or screen records the pattern of light emerging from the other side. As shown schematically in Fig. 4.1, the diffraction pattern that results can only be explained if light comprises waves; particles would show a different pattern based on their simple ballistic trajectories. As such, Einstein's grains of light were very difficult to accept. It was not until 1923 that the particle nature of light was fully accepted, thanks to experiments by Arthur Compton. These results showed that X-rays "recoiled" like a particle when diffracted by electrons. It was then undeniable that light apparently had a dual nature, behaving as either particles or waves depending on the circumstances.

Shortly after Einstein's contribution, a clearer picture of the structure of the atom began to emerge. Experiments by Geiger and Marsden in 1909 showed that alpha particles were deflected as they passed through a thin metal sheet in a way that was inconsistent with the picture of atomic structure at the time. In 1911, Rutherford determined that these results could be explained by a new picture of the atom in which there was a small, heavy, positively-charged nucleus surrounded by much lighter electrons. Bohr built a more precise picture of the atom by arguing that, like the quantum levels of radiation energy that Planck introduced, there were certain discrete energy levels for the electrons in the atom, corresponding to a set of permissible circular orbital trajectories around the central nucleus. This picture had problems that would prove difficult to resolve if electrons were indeed charged particles.

[1] Indeed, there are some historians who argue that Einstein, not Planck, should be considered the "father" of quantum mechanics [Kuh78].

Specifically, the curved orbit of the electron meant that it was constantly being accelerated, and should be losing energy while radiating X-rays. Clearly, there were still pieces of the puzzle missing.

If light, previously accepted as a wave, could behave like a particle, it was only natural also to ask if particles could act like waves. De Broglie[2] investigated this theoretically, and established that a particle with momentum p could also be considered to be a *wave* with wavelength $\lambda = h/p$; a conjecture referred to as the *particle–wave duality*.[3] In 1927, experiments by Davisson and Germer confirmed that electrons could be diffracted just like X-rays. Today, electron-beam diffraction can routinely show both the particle and wave nature of electrons simultaneously. With low enough beam currents, it is possible to create conditions where only a single electron reaches the diffraction grating at a time. The pattern that emerges is a diffraction pattern consistent with wavelike behavior, although it is built up statistically: a point here, a point there until eventually the accumulation shows the familiar pattern of light and dark bands. One must conclude that the individual electron "particles" are somehow passing through multiple slits simultaneously.

The particle–wave duality of the electron paved the way for a complete understanding of the structure of the atom, including a resolution of the problems with the Bohr atom. As it turns out, the positively-charged nuclei of atoms are usually large enough that we can treat them as classical objects. However, the much smaller electrons that surround the nucleus need to be treated using a wavelike description. This wavelike description permits an understanding of electronic "orbitals" as a manifestation of eigensolutions to a wave equation. In 1926, Schrödinger[4] laid out the framework for finding the electronic *wave function*, $\chi(\boldsymbol{x}, t)$, as a function of position $\boldsymbol{x}$ in space and time t, the eigenvalues of which determine the quantized states of the electron around the atomic nucleus.

Quantum mechanics shook the foundations of physics, introducing a host of apparent paradoxes and new challenges of how to correctly interpret the results of the theory. Even today, understanding and resolving these paradoxes is an active area of research. Quantum mechanics is both a practical science and a symbolic, abstract and philosophical one. At one extreme, numerical implementations of quantum mechanics are used to explain and predict

[2] The name "de Broglie" is not pronounced as written. The correct French pronunciation in IPA notation is [dəˈbʁœj], where "də" is pronounced like the first syllable in "demise", the "ʁ" is the French "R" (rolled on the back of the tongue), and "œj" is similar to the "oy" in the Yiddish "oy vey".

[3] The particle–wave duality of nature is a concept that is outside the bounds of our normal understanding. *Everything* is simultaneously a particle and a wave – surely this is an absurd statement. How can something be two things, and two very different things at that, simultaneously? Yet it is true, verified by experimental observations and theoretical predictions. All materials are composed of indivisible elementary objects, of which the electron is one example. Electrons are indivisible and discrete, with a finite mass and charge. In this sense, electrons behave as particles. But at the same time, electrons exhibit diffraction and interference phenomena, and thus electrons also behave as waves. Composite objects, such as protons, molecules or crystals also exhibit particle–wave duality, but the wavelike behaviors become more difficult to observe as the objects become larger. This is why the wave behavior of objects is often so difficult for us to understand and to accept; the wavelike characteristics of matter only become apparent when things are so small as to be well below the range of normal human experience.

[4] Erwin Schrödinger was an Austrian theoretical physicist who made seminal contributions to the field of quantum mechanics. The name "Schrödinger" can be hard to pronounce for English speakers. The correct German pronunciation in IPA notation is [ˈʃrøːdɪŋɐ], where the ʃ is equivalent to the English "sh", the "øː" is similar to a lengthened version of the vowel in "boat", "dɪŋ" is pronounced like the English word "ding", and "ɐ" is like the final vowel in "quota" (the final "r" is silent).

the behavior of materials. At the other, philosophers study the consequences of quantum mechanics for questions on the existence of free will (see the further reading section at end of this chapter). Our focus is certainly on the former of these two extremes, and we have paused only briefly here to acknowledge the profound beauty of the latter.

At the end of the chapter, we provide a selective list of a few of our favorite accounts of quantum mechanics for further reading.

4.3 The Hamiltonian formulation

The starting point for quantum mechanics of solids is the Hamiltonian formulation of classical mechanics. Schrödinger himself found the inspiration for his famous equation in the work of Hamilton and in the fact that Hamiltonian mechanics are exactly analogous to the mechanics of light waves [Sch26]. For this reason, the notion of the "Hamiltonian" of an electron is central to Schrödinger's framework, which is built around the premise that classical quantities like the kinetic and potential energy become operators on the electronic wave function at the quantum level. The relationship between the classical quantities and the quantum operators is dictated by the requirement that quantum and classical mechanics must agree in the macroscopic limit.

In this section, we briefly review the essentials of the Hamiltonian formulation. For more insight, the reader is directed to the classical literature on the subject, in books such as [Lan70, Gol80]. For a more mathematical treatment see, for example, [Arn89, Eva98]. We consider a system of N particles as described in Section 2.3.2. The positions $\boldsymbol{r}^\alpha$, velocities $\boldsymbol{v}^\alpha$ and momenta $\boldsymbol{p}^\alpha$ of the particles (with $\alpha = 1, \dots, N$) are defined in Eqn. (2.81). For brevity, we sometimes use $\boldsymbol{r}(t)$ and $\boldsymbol{p}(t)$ to denote the vectors $(\boldsymbol{r}^1(t), \boldsymbol{r}^2(t), \dots, \boldsymbol{r}^N(t))$ and $(\boldsymbol{p}^1(t), \boldsymbol{p}^2(t), \dots, \boldsymbol{p}^N(t))$, respectively.

The time evolution of the system of particles can be studied using three common approaches referred to as the *Newtonian formulation*, the *Lagrangian formulation* and the *Hamiltonian formulation*. The Newtonian formulation was introduced in Section 2.3.2. It involves the application of Newton's second law in Eqn. (2.82) to the motion of the particles. In the special case of a *conservative system*, the forces acting on the particles are derived from a scalar potential energy function,[5] $\mathcal{V}(\boldsymbol{r}^1, \dots \boldsymbol{r}^N)$. In this case,

$$\boldsymbol{f}^\alpha = -\frac{\partial \mathcal{V}(\boldsymbol{r}^1, \dots, \boldsymbol{r}^N)}{\partial \boldsymbol{r}^\alpha}. \tag{4.1}$$

The system evolution is then obtained by solving the coupled set of differential equations in Eqn. (2.82). This is exactly the approach taken in molecular dynamics simulations as explained in Chapter 9. The Lagrangian and Hamiltonian formulations are more elegant and can sometimes be used to obtain useful information from systems in the absence of closed-form solutions. The Hamiltonian formulation, in particular, plays an important role in the quantum mechanics framework discussed in this chapter and serves as the basis of classical statistical mechanics in Chapter 7.

[5] The classical potential energy function is discussed in depth in Section 5.3.

In the Lagrangian formulation, a system is characterized by the vector $\boldsymbol{r}(t)$ and a *Lagrangian function*, $\mathcal{L}$, defined as

$$\mathcal{L}(\boldsymbol{r}, \dot{\boldsymbol{r}}; t) = \mathcal{T}(\dot{\boldsymbol{r}}) - \mathcal{V}(\boldsymbol{r}), \tag{4.2}$$

where $\mathcal{T}$ is the kinetic energy of the system and $\mathcal{V}$ (already introduced above) is the potential energy. For a system of particles, we have

$$\mathcal{L}(\boldsymbol{r}, \dot{\boldsymbol{r}}; t) = \sum_{\alpha=1}^{N} \frac{1}{2} m^{\alpha} \left\| \dot{\boldsymbol{r}}^{\alpha} \right\|^2 - \mathcal{V}(\boldsymbol{r}^1, \dots \boldsymbol{r}^N). \tag{4.3}$$

The time evolution of $\boldsymbol{r}(t)$ is described by a variational principle called *Hamilton's principle*. Hamilton's principle states that the time evolution of $\boldsymbol{r}(t)$ is the extremum of the *action integral* defined as a functional[6] of $\boldsymbol{r}$ by

$$\mathcal{A}[\boldsymbol{r}] = \int_{t_1}^{t_2} \mathcal{L}(\boldsymbol{r}, \dot{\boldsymbol{r}}; t)\, dt, \tag{4.4}$$

where t_1, t_2, $\boldsymbol{r}(t_1)$ and $\boldsymbol{r}(t_2)$ are held fixed with respect to the class of variations being considered [Lan70, Section V.1]. In mathematical terms, we require

$$\delta \mathcal{A} = 0, \tag{4.5}$$

while keeping the ends fixed as described above. The Euler–Lagrange equation associated with Eqn. (4.5) is[7]

$$\frac{d}{dt}\left(\frac{\partial \mathcal{L}}{\partial \dot{\boldsymbol{r}}^{\alpha}}\right) - \frac{\partial \mathcal{L}}{\partial \boldsymbol{r}^{\alpha}} = \boldsymbol{0}, \qquad \alpha = 1, \dots, N. \tag{4.6}$$

This is a system of $3N$ coupled second-order ordinary differential equations for the scalar components of $\boldsymbol{r}$. The Lagrangian formulation is commonly used as a calculation tool in solving simple problems.

The Lagrangian is a function of a single set of variables, $\boldsymbol{r}$, and their time derivatives, $\dot{\boldsymbol{r}}$. It turns out to be very useful to recast the problem in terms of two sets of *independent*

[6] A functional is a map that assigns a real number to every function that belongs to a given class.

[7] The Euler–Lagrange equation is the differential equation associated with a variational principle. The solution to this equation gives the function that satisfies the variational requirement. By analogy, consider a function $f(x)$, where x is a scalar variable, and the "variational" requirement that $f(x)$ be at an extremum with respect to x. Clearly, the condition on x for satisfying this requirement is $f'(x) = 0$. This is the analogy to the Euler–Lagrange equation. With a variational principle, the difference is that we do not seek a variable for which a function is extremal, but a *function* for which a *functional* is extremal. An example would be to find the equation of a closed curve of a given length that encloses the maximum area. (Here the function is the equation of the curve and the functional is the area enclosed by it.) The theory of the calculus of variations provides a standardized approach for constructing the Euler–Lagrange equation for a given variational principle. It is beyond the scope of this book to develop this theory. The interested reader is referred to the many fine books on this topic; see, for example, [Lan70, GF00]. For a brief popular introduction to the subject, see [Men00].

variables, $\boldsymbol{r}$ and $\boldsymbol{p}$, and impose a constraint to enforce the connection between them stated in Eqn. (2.81). This can be done by noting that the Lagrangian can be expressed as the Legendre transformation of a new function called the *Hamiltonian*, $\mathcal{H}(\boldsymbol{r},\boldsymbol{p};t)$ [Eva98, Section 3.2.2]:

$$\mathcal{L}(\boldsymbol{r},\dot{\boldsymbol{r}};t) = \max_{\boldsymbol{p}} \left[\boldsymbol{p}\cdot\dot{\boldsymbol{r}} - \mathcal{H}(\boldsymbol{r},\boldsymbol{p};t)\right]. \tag{4.7}$$

For conservative systems, the Hamiltonian is the total energy. For a system of particles it is given by

$$\mathcal{H}(\boldsymbol{r},\boldsymbol{p};t) = \sum_{\alpha=1}^{N} \frac{\|\boldsymbol{p}^\alpha\|^2}{2m^\alpha} + \mathcal{V}(\boldsymbol{r}^1,\dots,\boldsymbol{r}^N) = \mathcal{T}(\boldsymbol{p}^1,\dots,\boldsymbol{p}^N) + \mathcal{V}(\boldsymbol{r}^1,\dots,\boldsymbol{r}^N). \tag{4.8}$$

To see the connection between the definition of the Hamiltonian in Eqn. (4.8) and the Legendre transformation in Eqn. (4.7), consider the following one-dimensional example.

Example 4.1 (Legendre transformation of a one-dimensional particle) For a system containing one particle moving in one dimension, the Hamiltonian is

$$\mathcal{H}(r,p) = p^2/2m + \mathcal{V}(r).$$

The Lagrangian is then

$$\mathcal{L}(r) = \max_p \left[p\dot{r} - \mathcal{H}(r,p)\right] = \max_p \left[p\dot{r} - p^2/2m - \mathcal{V}(r)\right].$$

The maximum occurs at $\dot{r} = p/m$. We see that the maximization condition[8] imposes the constraint relating p and $\dot{r}$. Substituting this into the above equation, we have that

$$\mathcal{L}(r) = m\dot{r}^2/2 - \mathcal{V}(r).$$

This is exactly the Lagrangian defined in Eqn. (4.3).

Using the new definition of the Lagrangian in terms of the Hamiltonian in Eqn. (4.7), Eqn. (4.5) can be rewritten as

$$\delta \int_{t_1}^{t_2} \left[\boldsymbol{p}\cdot\dot{\boldsymbol{r}} - \mathcal{H}(\boldsymbol{r},\boldsymbol{p};t)\right] dt = 0. \tag{4.9}$$

Note that in Eqn. (4.5), the variation is only with respect to $\boldsymbol{r}$, whereas in Eqn. (4.9) the variations are taken with respect to both $\boldsymbol{r}$ and $\boldsymbol{p}$ independently. In both cases, t_1, t_2, $\boldsymbol{r}(t_1)$ and $\boldsymbol{r}(t_2)$ are held fixed. The variational principle given in Eqn. (4.9) is referred to as the *modified Hamilton's principle* [Gol80] or simply as the "Hamiltonian formulation." The advantage of the Hamiltonian formulation lies not in its use as a calculation tool, but rather

[8] More generally, this corresponds to an extremal condition. However, since $\mathcal{H}$ (and $\mathcal{L}$) is a convex function of $\boldsymbol{p}$, the extremum corresponds to a maximum. This is the reason for the max operation in Eqn. (4.7) [Arn89, Section 14].

in the deeper insight it affords into the formal structure of mechanics. The Euler–Lagrange equations associated with Eqn. (4.9) are

$$\dot{\boldsymbol{p}}^\alpha + \frac{\partial \mathcal{H}}{\partial \boldsymbol{r}^\alpha} = \boldsymbol{0}, \qquad -\dot{\boldsymbol{r}}^\alpha + \frac{\partial \mathcal{H}}{\partial \boldsymbol{p}^\alpha} = \boldsymbol{0}, \tag{4.10}$$

which are called the *canonical equations of motion* or *Hamilton's equations*.[9] It is important to note that the Hamiltonian formulation is more general than the Lagrangian formulation, since it accords the coordinates and momenta independent status, thus providing the analyst with far greater freedom in selecting generalized coordinates (see Section 8.1.1). The Hamiltonian description is used heavily in the derivation of statistical mechanics. For more on this and a discussion of the properties of conservative (Hamiltonian) systems, see Section 7.1. Here we continue with the derivation of quantum mechanics.

4.4 The quantum theory of bonding

In this section, we give a brief overview of the essentials of quantum mechanics necessary to understand bonding. Then, we derive the time-independent Schrödinger equation and solve it for two simple systems: the hydrogen atom and the hydrogen molecule. The solutions for the hydrogen *atom* illustrate the fundamental role of quantum mechanics in establishing the electronic shells and orbitals familiar from introductory chemistry. The method used to approximately solve for the bonding in the hydrogen *molecule* serves as the archetype for all the subsequent methods we will discuss in the remainder of the chapter.

4.4.1 Dirac notation

Physicists have adopted a special notation that is quite convenient for expressing the ideas of quantum mechanics. It sometimes goes by the name "bra-ket" notation to acknowledge its form, and is otherwise called Dirac notation to acknowledge its originator. Because our need for quantum mechanics in this book is focused on issues related to building materials models, we avoid much of the formal quantum theory and therefore need Dirac notation only occasionally. In many ways, the notation is a generalization of the tensor notation which is summarized in Section 2.1 and described in detail in Chapter 2 of [TME12]. We will try to point the reader to those analogous concepts as we briefly introduce Dirac notation here.

[9] These equations are named after the Irish mathematician William Hamilton, who used this approach as the basis of his theoretical investigations of mechanics published in 1835. Hamilton did not discover these equations (they were first written down by the Italian-born mathematician Joseph-Louis Lagrange in 1809 and their connection with the equations of motion was first understood by the French mathematician Augustin-Louis Cauchy in 1831), however, Hamilton's work demonstrated the great utility of this framework for the study of mechanics [Lan70, page 167].

In Section 2.1, one thing we considered was the rank 1 tensor or the vector, $\boldsymbol{a}$. We thought of a vector as an entity that existed in a finite-dimensional (usually three-dimensional) real space. We could think of $\boldsymbol{a}$ in terms of its components, a_i, but this only had meaning in the context of a specific choice of basis vectors, $\boldsymbol{e}_i$, such that $a_i = \boldsymbol{a} \cdot \boldsymbol{e}_i$. In quantum mechanics, the state of a system is described by a generalization of the concept of a vector called a "ket", $|a\rangle$. It usually exists in an infinite-dimensional, complex vector space called the *Hilbert space*.[10] Note the key differences compared with Section 2.1: infinite versus finite dimensionality and complex- versus real-valued functions. These differences are the impetus for the differences in the notation.

Every ket $|a\rangle$ has a dual state called a "bra" and denoted $\langle a|$. We use the combination of a bra and a ket to define an inner product, $\langle a|b\rangle$, which must have exactly the same properties as the inner product defined on page 28 except that the complex Hilbert space generalizes condition 2 to be

$$\langle \boldsymbol{a}|\boldsymbol{b}\rangle = (\langle \boldsymbol{b}|\boldsymbol{a}\rangle)^*$$

where $*$ denotes the complex conjugate.[11]

An operator $\mathcal{A}$ can be thought of as a tensor of rank 2. Its effect is to transform the ket to its right into a different ket:

$$|b\rangle = \mathcal{A}\,|a\rangle\,.$$

Because the result of an operation is simply a new ket, operators are associative but not necessarily commutative:

$$\mathcal{A}_1(\mathcal{A}_2\,|a\rangle) = (\mathcal{A}_1\mathcal{A}_2)\,|a\rangle \neq (\mathcal{A}_2\mathcal{A}_1)\,|a\rangle\,.$$

Kets and bras must commute with scalars, meaning that for the scalar λ

$$\lambda\,|a\rangle = |a\rangle\,\lambda, \qquad \lambda\,\langle a| = \langle a|\,\lambda,$$

and since an inner product is a scalar by definition this means that

$$\langle a|b\rangle\,|c\rangle = |c\rangle\,\langle a|b\rangle\,. \tag{4.11}$$

The vector space can sometimes be finite-dimensional; in our case this happens when we introduce approximate methods to solve for the state of the electronic system. Once we refer to a specific basis in a finite-dimensional space, $|e_i\rangle$, $(i = 1, \ldots, n)$, a general ket $|a\rangle$ becomes a column matrix of components $|a\rangle \to [a_1, a_2, \ldots, a_n]^T$ and the bra $\langle a|$ becomes a row matrix of complex conjugates $\langle a| \to [a_1^*, a_2^*, \ldots, a_n^*]$. Now, the interpretation of the relationship between a ket and a bra is exactly like that between a vector and a dual vector (or covector) as discussed in more detail in Section 2.3.6 of [TME12].

The components of $|a\rangle$ are found in exact analogy to Eqn. (2.11),

$$a_i = \langle e_i|a\rangle\,,$$

[10] A Hilbert space can also be real instead of complex.

[11] Recall that a *complex number* z is defined as $z = a + ib$, where $a, b \in \mathbb{R}$ and $i = \sqrt{-1}$. The complex conjugate is defined as $z^* = a - ib$.

such that the ket $|a\rangle$ is expressed in terms of the basis vectors as

$$|a\rangle = \sum_{i=1}^{n} \langle e_i|a\rangle |e_i\rangle . \tag{4.12}$$

Here, we will start to see the advantages of the bra-ket notation. Given that commutative property of Eqn. (4.11), Eqn. (4.12) becomes

$$|a\rangle = \sum_{i=1}^{n} |e_i\rangle \langle e_i|a\rangle = \left(\sum_{i=1}^{n} |e_i\rangle \langle e_i|\right) |a\rangle = \mathcal{I}\, |a\rangle , \tag{4.13}$$

where we have defined the identity as the operator that transforms ket $|a\rangle$ into itself

$$\mathcal{I} = \sum_{i=1}^{n} |e_i\rangle \langle e_i| . \tag{4.14}$$

This operator is analogous to a sum of dyads as defined in Eqn. (2.23). The combination $|a\rangle \langle b|$ forms the tensor product in the vector space, and the specific tensor product of a basis ket $|e_i\rangle$ with itself is called *the projection operator*, Λ_i:

$$\Lambda_i \equiv |e_i\rangle \langle e_i| , \qquad \sum_{i=1}^{n} \Lambda_i = \mathcal{I}. \tag{4.15}$$

In the infinite-dimensional Hilbert space, the countable set of basis kets $|e_i\rangle$ becomes a continuous variable such as $|\boldsymbol{x}\rangle$ and the "components" of an arbitrary ket $|a\rangle$ become a continuous complex function of $\boldsymbol{x}$,

$$\langle \boldsymbol{x}|a\rangle = a(\boldsymbol{x}).$$

At the same time, the sum of Eqn. (4.14) rolls over to an integral over all space as

$$\mathcal{I} = \int |\boldsymbol{x}\rangle \langle \boldsymbol{x}| \, d\boldsymbol{x}.$$

Here $\boldsymbol{x}$ can belong to the usual Euclidean three-dimensional vector space associated with physical space or it can be more general. For our purposes, it will occasionally be replaced by a $3N^{\text{el}}$-dimensional vector of the coordinates of N^{el} electrons in our system i.e. $\boldsymbol{x} \rightarrow \boldsymbol{x}^{\text{el}} \equiv (\boldsymbol{x}_1, \boldsymbol{x}_2, \ldots, \boldsymbol{x}_{N^{\text{el}}})$. In this case, the identity operator becomes

$$\mathcal{I} = \int \left|\boldsymbol{x}^{\text{el}}\right\rangle \left\langle \boldsymbol{x}^{\text{el}}\right| \, d\boldsymbol{x}_1 d\boldsymbol{x}_2 \cdots d\boldsymbol{x}_{N^{\text{el}}}.$$

Just as we can carry out a quantitative analysis on a three-dimensional vector only after it has been referred to a basis and reduced to components, a state $|a\rangle$ in an infinite-dimensional space is rendered tractable through projection onto an infinite basis such as $\boldsymbol{x}$. For example, we can evaluate the inner product $\langle a|b\rangle$ by inserting the identity operator:

$$\langle a|b\rangle = \int \langle a|\boldsymbol{x}\rangle \langle \boldsymbol{x}|b\rangle \, d\boldsymbol{x} = \int a^*(\boldsymbol{x}) b(\boldsymbol{x}) \, d\boldsymbol{x},$$

and similarly for a general operator, $\mathcal{A}$:

$$\langle a|\, \mathcal{A}\, |b\rangle = \int \langle a|\boldsymbol{x}\rangle \langle \boldsymbol{x}|\, \mathcal{A}\, |b\rangle \, d\boldsymbol{x} = \int a^*(\boldsymbol{x}) \mathcal{A} b(\boldsymbol{x}) \, d\boldsymbol{x}. \tag{4.16}$$

When considering a system of N^{el} electrons, this extends naturally as

$$\begin{aligned}\langle a|\, \mathcal{A}\, |b\rangle &= \int \langle a|\boldsymbol{x}^{\text{el}}\rangle \langle \boldsymbol{x}^{\text{el}}|\, \mathcal{A}\, |b\rangle \, d\boldsymbol{x}^{\text{el}} \\ &= \int a^*(\boldsymbol{x}_1, \boldsymbol{x}_2, \ldots, \boldsymbol{x}_{N^{\text{el}}}) \mathcal{A} b(\boldsymbol{x}_1, \boldsymbol{x}_2, \ldots, \boldsymbol{x}_{N^{\text{el}}}) \, d\boldsymbol{x}_1 d\boldsymbol{x}_2 \cdots d\boldsymbol{x}_{N^{\text{el}}}.\end{aligned}$$

Working in a different basis presents no special difficulty. For example, it is often convenient to work in the space of the Fourier transform, denoted by the variable $\boldsymbol{k}$. We can equivalently evaluate the scalar $\langle a|b\rangle$ in this basis as

$$\langle a|b\rangle = \int \langle a|\boldsymbol{k}\rangle \langle \boldsymbol{k}|b\rangle \, d\boldsymbol{k} = \int \widehat{a}^*(\boldsymbol{k}) \widehat{b}(\boldsymbol{k}) \, d\boldsymbol{k},$$

where $\widehat{a}(\boldsymbol{k})$ is the Fourier transform of $a(\boldsymbol{x})$.

4.4.2 Electron wave functions

Treating an electron as a wave rather than as a particle requires the introduction of a *wave function*, $\chi(\boldsymbol{x}, t)$, to characterize the electron at position $\boldsymbol{x}$ in space at time t. In general, χ is a complex number, for reasons that will be discussed shortly. For systems of N^{el} electrons (or a combination of electrons and other quantum particles, such as protons), the wave function depends on all the particle positions, i.e. $\chi(\boldsymbol{x}_1, \boldsymbol{x}_2, \ldots, \boldsymbol{x}_{N^{\text{el}}}, t)$. In this section we focus on a single electron wave function.

One interpretation of the wave function is that it serves as a measure of the probability of finding the electron in the vicinity of $\boldsymbol{x}$ at a certain time t. The disadvantage of this interpretation is that it still suggests that the electron is a classical particle, traveling in some complex trajectory through space such that it occasionally visits location $\boldsymbol{x}$, with a certain probability described by the wave function. This picture is not really correct. In fact, the electron exists in an undetermined state until such time as we attempt to measure[12] its location and thereby interact with it. It is this interaction that causes the electron to "materialize" at some position in space, determined randomly but weighted by the probability distribution of the wave function. Practically speaking, we can reconcile this probabilistic behavior at the atomic scale with what we perceive as a deterministic macroscopic world, at least in part, because our observations typically involve extremely large numbers of particles. In this circumstance, the statistics of large numbers make probable events into near certainties, and a classical, deterministic model will accurately describe our observations (the statistics of large numbers of particles is discussed in more detail in Section 7.2.3).

[12] The nature of this measurement refers to any process which renders the quantum mechanical state observable in the classical sense. It is not meant to imply that a sentient being ("we") must be involved. More discussion of the philosophical nature of quantum mechanics may be found in, for example, [Omn99].

Specifically, the wave function can be interpreted such that the probability, $P(\boldsymbol{x}, t)$, of finding the electron at $\boldsymbol{x}$ and t is proportional to

$$P(\boldsymbol{x}, t) \propto \|\chi(\boldsymbol{x}, t)\|^2 = \chi^*(\boldsymbol{x}, t)\chi(\boldsymbol{x}, t). \tag{4.17}$$

Note that this is strictly a probability *density*, and we should really speak of the probability of finding a particle in the infinitesimal volume $d\boldsymbol{x}$ around $\boldsymbol{x}$, which is proportional to $P(\boldsymbol{x}, t)d\boldsymbol{x}$. In order to make sense as a probability, the integral of P over all space must be unity, so we normalize it as

$$P(\boldsymbol{x}, t) = \frac{\chi^*(\boldsymbol{x}, t)\chi(\boldsymbol{x}, t)}{\int \chi^*(\boldsymbol{x}', t)\chi(\boldsymbol{x}', t)\, d\boldsymbol{x}'}. \tag{4.18}$$

It is often convenient, although not required, to include this normalization directly in the wave function itself by introducing a normalization constant inside χ such that

$$\int \chi^*(\boldsymbol{x}, t)\chi(\boldsymbol{x}, t)\, d\boldsymbol{x} = \langle \chi | \chi \rangle = 1, \tag{4.19}$$

where we have noted the equivalence with the more concise bra-ket notation that we will often use in the following. We will sometimes use normalized wave functions and other times use wave functions that are not normalized, as will be convenient and clear from the context. Wave functions are complex numbers as a matter of mathematical convenience; this allows us to effectively manage two properties of the electron (its position and its momentum) within a single function. At the end of any manipulations, however, we must remember that only the *real* quantity in Eqn. (4.17) has physical significance.

As we shall see, it is often convenient to describe a general wave function as a weighted sum of simpler wave function expressions that constitute a basis. One particular wave function which is commonly used as the basis is the *plane wave*, given in three dimensions by

$$\chi(\boldsymbol{x}, t) = \chi_0 \exp\left(i\boldsymbol{k} \cdot \boldsymbol{x} - i\omega t\right), \tag{4.20}$$

where $i = \sqrt{-1}$ and the complex number χ_0 serves as an appropriate normalization constant, or less precisely as the amplitude of the wave. The wave vector, $\boldsymbol{k}$ points in the direction of propagation of the wave, as illustrated by the two-dimensional section of a plane wave shown schematically in Fig. 4.2. In the figure, the dark and light variations imply "peaks" and "valleys" in the wave amplitude. Because it is a plane wave, any two-dimensional section like this one will show these peaks and valleys as infinite parallel bands. The magnitude of $\boldsymbol{k}$ is related to the wavelength, λ, by $\|\boldsymbol{k}\| = 2\pi/\lambda$, while the frequency of oscillation, ω, tells us how quickly the value of χ fluctuates at a given position in space. In other circumstances, plane waves may not be a convenient description. For example, in a spherically-symmetric problem, spherical waves will be more useful.

Earlier, we briefly touched on de Broglie's result relating the *particlelike* properties of the electron (its energy and momentum) to its *wavelike* properties (its frequency and wave vector). This connection was shown by de Broglie to be

$$\epsilon = \hbar\omega, \qquad \boldsymbol{p} = \hbar\boldsymbol{k}, \tag{4.21}$$

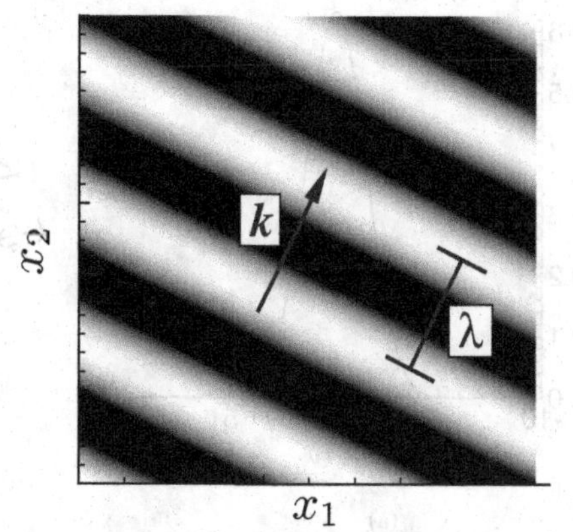

Fig. 4.2 A plane wave traveling along the $\boldsymbol{k} = [1, 2, 0]$ direction, with wavelength λ. Light and dark represent "hills" and "valleys" in the wave.

where $\boldsymbol{p}$ is the momentum and ϵ the energy of the electron. The constant of proportionality is the reduced Planck's constant, $\hbar = h/2\pi = 1.055 \times 10^{-34}$ J · s in SI units, where h is Planck's constant. For our purposes, we will not stray into the origin of these relations. Rather, we take them as experimentally determined facts and leave the details for more in-depth treatises (for example, [FLS06]).[13]

A single plane wave is a convenient mathematical idea, but it is not a physically sensible one or one that can be easily reconciled with the probabilistic interpretation of the wave function. For example, one can quickly see that it is not possible to satisfy Eqn. (4.19) for a plane wave, since the integral is unbounded. Instead, we must use the totality of all plane waves (i.e. all values of $\boldsymbol{k}$) to serve as a complete basis to represent a more general wave function. In effect, plane waves of different wavelength and amplitude can combine such that they cancel out over all but a finite region of space. For simplicity, we shall temporarily put aside the time dependence by assuming[14] that it is possible to separate the time and space dependence as $\chi(\boldsymbol{x}, t) = \tau(t)\psi(\boldsymbol{x})$ and work only with the spatial component. Also, let us focus on a one-dimensional space just to simplify the notation temporarily. Then, a general one-dimensional wave function can be represented as the sum over all possible plane waves, each weighted appropriately:

$$\psi(x) = \frac{1}{\sqrt{2\pi}} \int_{-\infty}^{\infty} \widehat{\psi}(k) \exp(ikx)\, dk. \tag{4.22}$$

The quantity $\widehat{\psi}\, dk$ weights the relative importance of the plane waves with wave vectors between k and $k + dk$. The factor of $\sqrt{2\pi}$ is introduced as a mathematical convenience

[13] Deriving the particle–wave duality relations of Eqns. (4.21) is a bit of a chicken and egg problem unless one is willing to go deeper into the path integral formulation of quantum mechanics. Many authors take the approach that we follow here, where Eqns. (4.21) are taken as axioms in order to establish the momentum operator and then "derive" the Schrödinger equation, while one can alternatively take the Schrödinger equation as a "given" and then "prove" Eqns. (4.21). To do the latter, one must first derive the Schrödinger equation. This requires a deeper exploration of quantum mechanics than we want to pursue here. See, for example, [FH65, ZJ05].

[14] We will see later that this assumption is often valid.

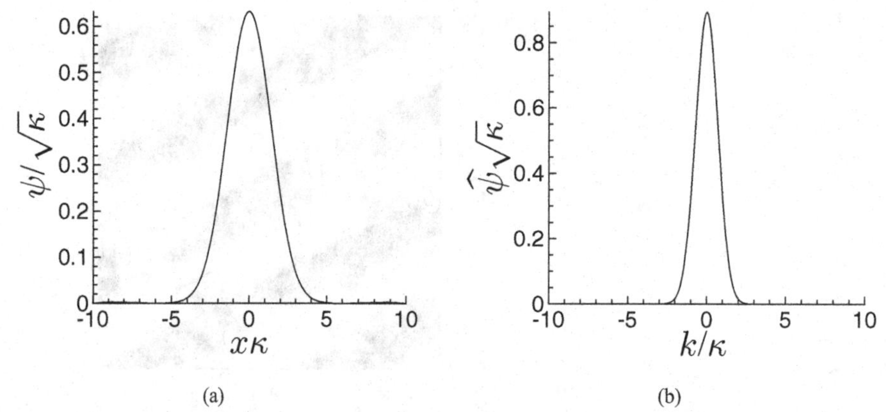

Fig. 4.3 (a) A Gaussian wave function and (b) its transform corresponding to Eqn. (4.25).

(taking advantage of the liberty granted by the ever-present normalization constant within ψ). This is simply the concept of Fourier transforms applied to our electron description, $\widehat{\psi}(k)$ being the Fourier transform of $\psi(x)$:

$$\widehat{\psi}(k) = \frac{1}{\sqrt{2\pi}} \int_{-\infty}^{\infty} \psi(x) \exp(-ikx)\, dk. \tag{4.23}$$

The transform of the wave function, $\widehat{\psi}(k)$, has an interpretation similar to that of $\psi(x)$. Above, we alluded to the fact that the wave function is complex because it is really representing two properties of the electron: its position and its momentum. Although we will state this without proof (for a proof see, for example, [Har00]) it turns out that the transform of the wave function, properly normalized, provides the probability of finding the electron with a certain momentum $\hbar k$. Similarly to Eqn. (4.18), we write the probability density of finding the electron with wave vector k as

$$\widehat{P} = \frac{\widehat{\psi^*}(k)\widehat{\psi}(k)}{\left\langle \widehat{\psi} \middle| \widehat{\psi} \right\rangle}. \tag{4.24}$$

Let us now consider a fictitious electron wave function to illustrate how we may see the connection between the wave and particle description of electrons. Imagine that ψ is described by a simple bounded function of the form[15]

$$\psi(x) = \left(\frac{\kappa^2}{2\pi}\right)^{1/4} \exp\left(-\kappa^2 x^2/4\right). \tag{4.25}$$

This is a normalized wave function with a Gaussian distribution centered on the origin, as shown in Fig. 4.3(a). Its width is determined by the value of the constant κ. For values of $\|x\|$ greater than about $2/\kappa$, the wave function goes rapidly to zero, meaning that the

[15] We mention in passing that we will often consider wave functions that have no imaginary part (like the example in Eqn. (4.25)) as a matter of convenience. This corresponds to a special case of an electron with an expectation value of its momentum equal to zero. Based on the definition of an expectation value found in the following pages, it is left as an exercise for you to verify that if the wave function is either purely real or purely imaginary, then $\langle p \rangle = 0$. Obviously, in circumstances where the momentum or the time evolution of the electron is important, the complex character of the wave function becomes apparent.

probability of finding the electron is highest at the origin and becomes vanishingly small as we move in the positive or negative x directions. Such a wave function is consistent with our notion of an electron as a particle, in the sense that the electron is confined to a finite region of space. Viewed from a sufficient distance, the electron will appear as a well-defined particle, but on close inspection its "edges" are "fuzzy," as there is no clear line demarcating the end of the electron. This is a common feature of electronic wave functions. The transform of $\psi(x)$ is a representation of the particle in k-space (or equivalently, from Eqn. $(4.21)_2$, momentum space)

$$\widehat{\psi}(k) = \frac{1}{\sqrt{2\pi}} \int_{-\infty}^{\infty} \psi(x) \exp(-ikx)\, dx = \left(\frac{2}{\pi\kappa^2}\right)^{1/4} \exp\left(-k^2/\kappa^2\right),$$

and this function is shown in Fig. 4.3(b). From Eqn. (4.24), the probability of finding the electron with its momentum greater than about κ is quite small, and indeed one can verify that the expectation value (defined below) of the momentum is zero. Viewed from a classical perspective, the particle is "stationary" even though its quantum mechanical momentum spans a range of nonzero values.[16]

A more precise correspondence between the wave function and macroscopic observables like the position and momentum of the electron is made through the definition of expectation values. These are analogous to ensemble averages in statistical mechanics (see Chapter 7). For example, the expectation value of the particle's position, denoted $\langle x \rangle$, is found by averaging over all possible positions x, weighted by the probability of finding the electron there. Specifically,

$$\langle x \rangle = \int P(x) x\, dx = \frac{\int \psi^*(x) x \psi(x)\, dx}{\langle \psi | \psi \rangle} = \frac{\langle \psi | x | \psi \rangle}{\langle \psi | \psi \rangle}. \tag{4.26}$$

For the simple example wave function discussed here, it is easy to show that $\langle x \rangle$ is zero, consistent with the classical notion of a stationary particle at the origin. This definition of the expectation value extends to any function of position, i.e.

$$\langle f(x) \rangle = \frac{\langle \psi | f(x) | \psi \rangle}{\langle \psi | \psi \rangle}. \tag{4.27}$$

For the expectation value of the momentum, it is most directly determined from the Fourier transform of the wave function and Eqn. (4.24):

$$\langle p \rangle = \langle \hbar k \rangle = \hbar \langle k \rangle = \hbar \frac{\left\langle \widehat{\psi} \middle| k \middle| \widehat{\psi} \right\rangle}{\left\langle \widehat{\psi} \middle| \widehat{\psi} \right\rangle}. \tag{4.28}$$

It is easy to verify that $\langle p \rangle = 0$ for our example wave function.

[16] The relationship between position and real space on the one hand, and momentum and Fourier space on the other also relates to an important quantum mechanical concept that we will not cover here: the Heisenberg uncertainly principle. In short, the more localized a particle is in space, the more uncertain its momentum. Conversely, the more narrowly defined a particle's momentum, the more diffuse its position.

4.4.3 Schrödinger's equation

We have formally defined the wave function of an electron or collection of electrons, and now we need to determine the governing equation that predicts the form this wave function must take in the presence of some potential energy field.

We can make use of the properties of Fourier transforms to rewrite the expectation value of an electron's momentum in terms of $\psi(x)$ rather than $\widehat{\psi}(k)$ as in Eqn. (4.28). Differentiation of Eqn. (4.22) with respect to x leads to the observation that the Fourier transform of $d\psi/dx$ is simply ik times the Fourier transform of the original function:

$$\widehat{\left(\frac{d\psi(x)}{dx}\right)} = ik\widehat{\psi}(k).$$

Also, Parseval's well-known theorem for Fourier transforms states that for any two wave functions, $\psi_1(x)$ and $\psi_2(x)$,

$$\int_{-\infty}^{\infty} \psi_1^*(x)\psi_2(x)\, dx = \int_{-\infty}^{\infty} \widehat{\psi}_1^*(k)\widehat{\psi}_2(k)\, dk.$$

If we choose $\widehat{\psi}_1 = \widehat{\psi}$ and $\widehat{\psi}_2 = k\widehat{\psi}$, we can use Parseval's theorem to transform Eqn. (4.28) from momentum space to coordinate space, obtaining

$$\langle p\rangle = \hbar\frac{\left\langle\widehat{\psi}\middle| k \middle|\widehat{\psi}\right\rangle}{\left\langle\widehat{\psi}\middle|\widehat{\psi}\right\rangle} = \frac{\hbar}{i}\frac{\left\langle\psi\middle|\frac{d}{dx}\middle|\psi\right\rangle}{\langle\psi|\psi\rangle}.$$

Thus, we see the correspondence between the electron momentum, $\boldsymbol{p}$, and the differentiation operator in real space:

$$p \rightarrow \frac{\hbar}{i}\frac{\partial}{\partial x}. \tag{4.29}$$

In three dimensions, this simply becomes the gradient operator[17]

$$\boldsymbol{p} \rightarrow \frac{\hbar}{i}\boldsymbol{\nabla}. \tag{4.30}$$

For any polynomial functions of p, it is straightforward to extend this correspondence to appropriate higher-order differentiation. For example, an important quantity is p^2, which appears in the kinetic energy of a classical particle. We see that from the probabilistic interpretation of the wave function

$$\left\langle p^2\right\rangle = \hbar^2\left\langle k^2\right\rangle = \hbar^2\frac{\left\langle\widehat{\psi}\middle| k^2 \middle|\widehat{\psi}\right\rangle}{\left\langle\widehat{\psi}\middle|\widehat{\psi}\right\rangle}. \tag{4.31}$$

[17] In footnote 14 on page 40, we discussed the confusion that can be created when using the del differential operator $\boldsymbol{\nabla} \equiv \boldsymbol{e}_i\partial/\partial x_i$. However, this confusion only arises when the operator is applied to tensors of rank 1 or higher. Since we only apply the gradient to scalars in the quantum mechanics derivation, we find it convenient to use this notation.

A second differentiation of Eqn. (4.22) with respect to x leads to the observation that

$$\widehat{\left(\frac{d^2\psi}{dx^2}\right)} = -k^2\widehat{\psi}(k),$$

and so an application of Parseval's theorem to Eqn. (4.31) leads to

$$\langle p^2 \rangle = \frac{\left\langle \psi \middle| -\hbar^2 \frac{d^2}{dx^2} \middle| \psi \right\rangle}{\left\langle \widehat{\psi} \middle| \widehat{\psi} \right\rangle}. \tag{4.32}$$

It is straightforward to extend this to three dimensions and find $\langle \boldsymbol{p} \cdot \boldsymbol{p} \rangle$, for which we adopt the shorthand $\langle p^2 \rangle$. In this case, rather than two applications of the differentiation operator we have two applications of the gradient operator, $\boldsymbol{\nabla} \cdot \boldsymbol{\nabla}$, i.e. the Laplacian operator ∇^2:

$$p^2 \rightarrow -\hbar^2 \nabla^2. \tag{4.33}$$

Expectation values of a function of position, for example a potential energy field $\mathcal{U}(x)$, are similarly found by weighting the value of $\mathcal{U}(x)$ at every position x by the probability of finding an electron there. Thus

$$\langle \mathcal{U} \rangle = \frac{\langle \psi | \mathcal{U}(x) | \psi \rangle}{\langle \psi | \psi \rangle}.$$

An especially important expectation value is that of the total energy of an electron. In a classical framework, an electron with mass $m^{\text{el}} = 9.1095 \times 10^{-31}$ kg has an energy that is simply the sum of the potential and kinetic energies given by the Hamiltonian,[18]

$$\mathcal{H} = \mathcal{U}(\boldsymbol{x}) + \frac{p^2}{2m^{\text{el}}}. \tag{4.34}$$

In order to have correspondence between the classical and quantum descriptions, the expectation value of this Hamiltonian, $\langle \mathcal{H} \rangle$, should be equal to the energy of the particle.[19]

[18] Note that in discussing electrons we have reserved the symbol $\mathcal{U}$ for the potential energy operator, whereas we use $\mathcal{V}$ for the classical potential energy in Eqn. (4.8) and in the remainder of the book. This is because, as we will see in the beginning of Chapter 5, we can formally think of the classical potential energy as a quantity that encompasses all the electronic effects within its scope, reducing them to a function of only the positions of the atomic nuclei via the Born–Oppenheimer approximation (see Sections 4.4.6 and 5.1). As such, the two quantities are physically different, but the role of $\mathcal{U}$ with respect to electrons in an environment of fixed atomic nuclei is effectively the same as that of $\mathcal{V}$ with respect to moving atoms interacting with each other and experiencing an external potential.

[19] It is not obvious, from what we have presented here, why the quantum and classical Hamiltonians should be the same. However, it can be shown that this form is required to ensure that Newton's laws are recovered in the classical limit. To show this, we would have either to delve into the *matrix mechanics* formulation of quantum mechanics or to introduce *Ehrenfest's theorem* (which would also provide a more rigorous way of establishing the momentum operator). However, these details are beyond the scope of this book. More advanced treatments are referenced in the *Further reading* section at the end of this chapter.

Making use of the operator identified in Eqn. (4.33), we have

$$\langle \mathcal{H} \rangle = \left\langle \psi \middle| \mathcal{U}(\boldsymbol{x}) - \frac{\hbar^2}{2m^{\text{el}}} \nabla^2 \middle| \psi \right\rangle. \tag{4.35}$$

Next, we recall the idea that a general electron wave function can be represented by Eqn. (4.22) and that we had temporarily put aside the time dependence, $\tau(t)$, when we introduced the Fourier sum. Reintroducing the time dependence, we have for each plane wave in the basis:

$$\chi(\boldsymbol{x}, t) = \frac{1}{\sqrt{(2\pi)^3}} \int_{-\infty}^{\infty} \widehat{\psi}(\boldsymbol{k}) \exp\left(i\boldsymbol{k} \cdot \boldsymbol{x} - i\omega(\boldsymbol{k})t\right) d\boldsymbol{k}, \tag{4.36}$$

where the factor $1/\sqrt{2\pi}$ has been replaced by $1/\sqrt{(2\pi)^3}$ because we are now working in three dimensions. It would be unreasonable to expect that all plane waves necessarily have the same frequency, so we include in the frequency a dependence on the wave vector, $\omega(\boldsymbol{k})$. The relation between ω and $\boldsymbol{k}$ is the *dispersion relation* for the wave. We also know, from Eqn. $(4.21)_1$ that the energy of a wave is $\epsilon = \hbar\omega$, so the expectation value of $\langle \epsilon \rangle$ is

$$\langle \epsilon \rangle = \hbar \langle \omega(\boldsymbol{k}) \rangle = \hbar \langle \widehat{\chi} | \, \omega(\boldsymbol{k}) \, | \widehat{\chi} \rangle. \tag{4.37}$$

To put this in terms of the real-space wave function $\chi(\boldsymbol{x}, t)$, we must again determine the appropriate operator representing $\omega(\boldsymbol{k})$. We rearrange Eqn. (4.36) as

$$\chi(\boldsymbol{x}, t) = \frac{1}{\sqrt{(2\pi)^3}} \int_{-\infty}^{\infty} \left[\widehat{\psi}(\boldsymbol{k}) \exp\left(-i\omega(\boldsymbol{k})t\right)\right] \exp\left(i\boldsymbol{k} \cdot \boldsymbol{x}\right) d\boldsymbol{k}.$$

This shows that the quantity inside the square brackets is the Fourier transform of $\chi(\boldsymbol{x}, t)$ when we include the time dependence. The time derivative of this equation is

$$\frac{\partial \chi(\boldsymbol{x}, t)}{\partial t} = \frac{1}{\sqrt{(2\pi)^3}} \int_{-\infty}^{\infty} \left[-i\omega\widehat{\psi}(\boldsymbol{k}) \exp\left(-i\omega(\boldsymbol{k})t\right)\right] \exp\left(i\boldsymbol{k} \cdot \boldsymbol{x}\right) d\boldsymbol{k},$$

and thus the Fourier transform of the time derivative of χ is simply $-i\omega$ times the transform of χ itself. Similarly to Eqn. (4.29), then, we identify the operator associated with ω to be

$$\omega \rightarrow i\frac{\partial}{\partial t}. \tag{4.38}$$

Thus we can revert to a real-space representation of the wave vector, by transforming Eqn. (4.37) to

$$\langle \epsilon \rangle = \left\langle \chi \middle| i\hbar\frac{\partial}{\partial t} \middle| \chi \right\rangle. \tag{4.39}$$

At this point, we have developed two expressions for the expectation value of the energy of the electron: Eqn. (4.35) and Eqn. (4.39). Since these two quantities are the same, the effect of their related operators on the wave function should also be the same. Equating these two operations we obtain *Schrödinger's equation*:[20]

$$i\hbar\frac{\partial}{\partial t}\chi(\boldsymbol{x},t) = \left(\mathcal{U}(\boldsymbol{x},t) - \frac{\hbar^2}{2m^{\mathrm{el}}}\nabla^2\right)\chi(\boldsymbol{x},t). \tag{4.40}$$

The mastery of this equation over the behavior of electrons is a central tenet of quantum mechanics. Its predictions are supported by all the experimental evidence we can muster. For a system of N^{el} electrons, it is straightforward to extend this derivation to

$$\begin{aligned} &i\hbar\frac{\partial}{\partial t}\chi(\boldsymbol{x}_1,\boldsymbol{x}_2,\ldots,\boldsymbol{x}_{N^{\mathrm{el}}},t) \\ &= \left[\mathcal{U}(\boldsymbol{x}_1,\boldsymbol{x}_2,\ldots,\boldsymbol{x}_{N^{\mathrm{el}}},t) - \frac{\hbar^2}{2m^{\mathrm{el}}}\left(\nabla_1^2+\nabla_2^2+\cdots+\nabla_{N^{\mathrm{el}}}^2\right)\right]\chi(\boldsymbol{x}_1,\boldsymbol{x}_2,\ldots,\boldsymbol{x}_{N^{\mathrm{el}}},t), \end{aligned}$$

where ∇_i^2 refers to the Laplacian with respect to the coordinates of the ith electron.

Schrödinger's equation determines the wave function of an electron subject to a potential field $\mathcal{U}(\boldsymbol{x}_1,\boldsymbol{x}_2,\ldots,\boldsymbol{x}_{N^{\mathrm{el}}},t)$. The form of this potential is determined by the circumstances in which the electron finds itself, and includes such effects as the Coulomb interactions between the electron and the nuclei of nearby atoms. Closed-form, exact solutions to this equation are available for only the simplest of potential fields, and so the science of materials modeling has invested heavily in the development of approximate methods of solution and even more approximate effective models that eliminate the need for a full consideration of the electronic degrees of freedom. These models are the subject of Sections 4.5 and 4.6, and all of Chapter 5.

4.4.4 The time-independent Schrödinger equation

Often, the external potential experienced by the electron is not a function of time. In that case, the wave function can be written as a product of two functions, one depending only on time and the other depending only on space, $\chi(\boldsymbol{x},t) = \tau(t)\psi(\boldsymbol{x})$. Inserting this into Eqn. (4.40) permits a rearrangement[21] so that all terms on one side of the equation depend only on t, while all terms on the other depend only on $\boldsymbol{x}$. If an expression depending only on one variable is equal to an expression depending only on another variable, we must conclude that both expressions are equal to a constant. Specifically, we have

$$i\hbar\frac{\partial\tau}{\partial t}\frac{1}{\tau} = \mathcal{U}(\boldsymbol{x}) - \frac{\hbar^2}{2m^{\mathrm{el}}}\frac{1}{\psi}\nabla^2\psi = \epsilon.$$

[20] Tacit in this informal derivation of the Schrödinger equation is the assumption that the Hamiltonian and frequency operators are linear, so that we can write a general wave function as a superposition of plane waves (through a Fourier transform) and then expect the effect of the operator to be equal to the sum of the effects on the plane wave basis. For the interested reader, a rigorous derivation of the Schrödinger equation from the path integral formulation of quantum mechanics can be found in [Fey85].

[21] This separation is not possible if $\mathcal{U}$ depends on both space and time.

This is easily solved for the time-dependent part, which is

$$\tau(t) = \exp(-i\omega t).$$

The fact that the constant value of the equation is equal to the energy, ϵ, follows from $\epsilon = \hbar\omega$. The remaining equation for the positional part of the wave function is known as the *time-independent Schrödinger equation*:

$$\left(\mathcal{U}(\boldsymbol{x}) - \frac{\hbar^2}{2m^{\mathrm{el}}}\nabla^2\right)\psi(\boldsymbol{x}) = \epsilon\psi(\boldsymbol{x}). \tag{4.41}$$

For a given potential, $\mathcal{U}$, the eigenvector solutions to this equation form a complete, orthogonal basis (see the discussion in Section 2.1). In Section 4.4.5, we will see how this equation can be solved exactly for the case of the hydrogen atom, while the idealized case of a Dirac delta-function potential is left to the exercises. When modeling materials, the primary origin of the external potential is the interaction between the electrons and the atomic nuclei in the solid. For applications of interest in this book, it is always a good assumption that this external potential is time-independent (we elaborate on this in Sections 4.4.6 and 5.1).

"Atoms" versus "ions" versus "nuclei" A brief word about terminology is in order here. Our primary interest in modeling materials is the behavior of the atomic nuclei; their motion and arrangement lead to the mechanical properties that we want to study. In much of the book, our focus will be on atomistic models where the electrons have been explicitly eliminated from the picture, replaced by effective interactions between the nuclei. It is the convention in atomistic models to refer to the "atoms" as the only particles in the system. This is an unfortunate term, since in more fundamental physics and chemistry the atom refers specifically to the combination of a single nucleus and the appropriate number of charge-neutralizing electrons. As we make the transition from quantum to classical models of materials, it will also be necessary to segue between terminologies. As a result the reader will notice a certain interchangeability between "atom," "ion," and "nucleus," and an evolving definition of the term "atom." We hope that it will be clear from the context what is meant at any point in the discussion.

4.4.5 The hydrogen atom

One example of a simple form of the potential $\mathcal{U}(\boldsymbol{x})$ that permits an analytic solution to the time-independent Schrödinger equation is the problem of an isolated hydrogen atom. Such an atom consists of a single proton in the nucleus, surrounded by a single electron. We can treat the proton as a stationary potential experienced by the electron:[22]

$$\mathcal{U}(\boldsymbol{x}) = \frac{-\tilde{e}^2}{4\pi\epsilon_0}\frac{1}{\|\boldsymbol{x}\|} = \frac{-e^2}{\|\boldsymbol{x}\|}, \tag{4.42}$$

[22] Later, when we consider the bonding between two hydrogen atoms to form a molecule in Section 4.4.6, we will say more about the assumptions which we must make to treat the proton as a fixed potential field.

where $\tilde{e} = 1.6022 \times 10^{-19}$ C is the fundamental unit charge of an electron, $\epsilon_0 = 8.854 \times 10^{-12}$ $\text{C}^2/(\text{J}\cdot\text{m})$ is the permittivity of free space and the proton is assumed to sit at the origin. We have also defined a new unit "charge," e, such that

$$e^2 \equiv \frac{\tilde{e}^2}{4\pi\epsilon_0} = 14.4\,\text{eV}\cdot\text{Å}. \tag{4.43}$$

This definition will simplify the notation throughout the coming chapters. Now, let us identify the possible states in which the electron can exist around the nucleus at the origin by solving Eqn. (4.41) given $\mathcal{U}$ from Eqn. (4.42).

Due to the spherical symmetry of the potential in this case, it will be convenient to adopt the usual spherical coordinates where r is the radial distance from the origin, θ is the angle between the x_3-axis and the radial vector, and ϕ is the angle between the x_1-axis and the projection of the radial vector into the x_1–x_2 plane. In addition, as we did in simplifying the Schrödinger equation to the time-independent Schrödinger equation in Eqn. (4.41), we will assume that we can write the wave function as a product of three separate functions of the three spatial coordinates:

$$\varphi(r,\theta,\phi) = \varrho(r)\Theta(\theta)\Phi(\phi). \tag{4.44}$$

Note that we have introduced a new symbol for the total wave function, φ. We use this to distinguish waves that are a solution to a specific form of the Hamiltonian (in this case, that of the hydrogen atom) from a general wave function, ψ, because the functions φ will later serve as a set of basis functions when building approximate solutions for ψ.

We will need the Laplacian operator in spherical coordinates:

$$\nabla^2 = \frac{1}{r^2}\frac{\partial}{\partial r}\left(r^2\frac{\partial}{\partial r}\right) + \frac{1}{r^2\sin\theta}\frac{\partial}{\partial\theta}\left(\sin\theta\frac{\partial}{\partial\theta}\right) + \frac{1}{r^2\sin^2\theta}\frac{\partial^2}{\partial\phi^2}. \tag{4.45}$$

Using Eqn. (4.44) and Eqn. (4.45) in Eqn. (4.41), we can rearrange the terms so that everything on one side of the equality is a function of only ϕ, while everything on the other side is a function of either θ or r. Once again, we must conclude that both sides of this equation are equal to a constant, which we will for convenience call m^2. Thus, we have

$$-\frac{1}{\Phi}\frac{\partial^2\Phi}{\partial\phi^2} = m^2 = \frac{\sin^2\theta}{\varrho}\frac{\partial}{\partial r}\left(r^2\frac{\partial\varrho}{\partial r}\right) + \frac{\sin\theta}{\Theta}\frac{\partial}{\partial\theta}\left(\sin\theta\frac{\partial\Theta}{\partial\theta}\right) + \frac{2m^{\text{el}}r^2\sin^2\theta}{\hbar^2}\left(\frac{e^2}{r}+\epsilon\right).$$

Taking the last equality, we can again rearrange the equation so that terms which depend on r are separated from those depending on θ, and equate each collection of terms to another constant, λ. After some algebraic rearrangements, we can thereby write three ordinary differential equations for the three unknown functions, $\Phi(\phi)$, $\Theta(\theta)$ and $\varrho(r)$:

$$\frac{\partial^2\Phi}{\partial\phi^2} + m^2\Phi = 0,$$

$$\frac{1}{\sin\theta}\frac{\partial}{\partial\theta}\left(\sin\theta\frac{\partial\Theta}{\partial\theta}\right) + \left(\lambda - \frac{m^2}{\sin^2\theta}\right)\Theta = 0,$$

$$\frac{\hbar^2}{2m^{\text{el}}r^2}\frac{\partial}{\partial r}\left(r^2\frac{\partial\varrho}{\partial r}\right) + \left(\frac{e^2}{r} + \epsilon - \frac{\lambda\hbar^2}{2r^2m^{\text{el}}}\right)\varrho = 0.$$

If we can solve these three equations, we can combine the three solutions to form the electron wave function surrounding an isolated hydrogen atom. It is at this stage that we can begin to clearly see the "quantum" in quantum mechanics, since these equations are in fact eigenvalue problems. Only certain values of the constants m, λ and the energy ϵ will lead to solutions that satisfy the following two requirements: (1) we require that the wave function be everywhere continuous, finite, and single-valued in order to be able to make sense of the probabilities associated with it; (2) physically, we are only interested in solutions that correspond to a localized electron that is bound to the nucleus, meaning that the wave function must decay to zero as $r \to \infty$. Considering these requirements, we will find that three integer values uniquely identify the discrete states that the electron can occupy. These are the *quantum numbers* of the electron, m, l and n. One of these is the integer m that already appears in the equations. Another will be obtained from $\lambda = l(l+1)$ with l an integer, a constraint which must be imposed on λ in order that $\Theta(0)$ and $\Theta(\pi)$ remain finite. Finally, the integer n identifies the permissible[23] discrete energy levels:

$$\epsilon_n = -\frac{m^{\text{el}} e^4}{2\hbar^2 n^2}. \tag{4.46}$$

It is readily verified that solutions to the equation for Φ take the form

$$\Phi(\phi) = \exp(im\phi). \tag{4.47}$$

For $\Phi(\phi)$ to be single-valued it must have 2π periodicity. As such, m must be restricted to integer values. Solutions to the equations for ϱ and Θ are less obvious, and the details are beyond what we require. We therefore discuss them only briefly (full details can be found in [AW95]). The solutions $\Theta(\theta)$ are the *associated Legendre polynomials*,

$$\Theta(\theta) = P_{lm}(\cos\theta), \tag{4.48}$$

where the integer quantum numbers l and m determine the polynomial through

$$P_{lm}(x) = \frac{(1-x^2)^{m/2}}{2^l l!} \frac{d^{l+m}}{dx^{l+m}} (x^2-1)^l \qquad 0 \le |m| \le l, \tag{4.49}$$

and d/dx is the derivative operation, applied $(l+m)$ times as indicated by the superscripts. Note that since both m and l appear in the equation for Θ, it happens that they are interrelated through $0 \le |m| \le l$. The solutions $\varrho(r)$ take the form of *associated Laguerre polynomials*, $L_i^j(x)$, where the quantum numbers n and l determine the indices i and j, which further determine the polynomial. Specifically,

$$\varrho_{nl}(r) = \left(\frac{r}{nr_0}\right)^l \exp(-r/nr_0)\, L_{n-l-1}^{2l+1}\left(\frac{2r}{nr_0}\right), \tag{4.50}$$

where the associated Laguerre polynomials are generated from

$$L_j^i(x) = (-1)^i \frac{d^i}{dx^i} L_{i+j}(x), \qquad L_k(x) = \frac{\exp(x)}{k!} \frac{d^k}{dx^k}\left(x^k \exp(-x)\right).$$

[23] It can be shown that for other values of ϵ, the wave functions will not decay to zero for large r.

The length scale in Eqn. (4.50) is set by the *Bohr radius*

$$r_0 = \frac{\hbar^2}{m^{\text{el}} e^2}, \tag{4.51}$$

which evaluates[24] to $r_0 = 0.529$ Å. Finally, there are restrictions imposed on m and l based on the value of n in order to have a bounded wave function. For a given n, $0 \leq l < n$, which imposes further restrictions on m as discussed previously.

Combining Eqns. (4.47), (4.48) and (4.50), the total (unnormalized) wave function is

$$\varphi_{nlm}(r, \theta, \phi) = \left(\frac{r}{nr_0}\right)^l \exp(im\phi) \exp(-r/nr_0) L^{2l+1}_{n-l-1}\left(\frac{2r}{nr_0}\right) P_{lm}(\cos\theta). \tag{4.52}$$

Summarizing the quantum numbers, n, l and m, we have:

1. The integer n is normally referred to as the *first quantum number*. It must be an integer greater than zero, from which the energy of the state is determined through Eqn. (4.46).
2. Possible values of the second quantum number, l, for a given n are $l = 0, 1, 2, \ldots, n-1$.
3. For each value of l the possible values of m are $m = 0, \pm 1, \pm 2, \ldots, \pm l$.

What do these solutions show? First, they demonstrate the fact that the electron can only adopt certain *quantized* states, φ_{nlm}, identified by the three quantum numbers n, l and m. This quantization comes about from the physical conditions required of the wave function, i.e. that it be single-valued, finite, and bounded to the hydrogen atom's nucleus. The wavelike nature of the electron is responsible for this quantization, and there are direct parallels to the discrete mode shapes that give rise to overtones in a vibrating guitar string. We note that one of these solutions is the ground state electronic structure of the hydrogen atom, which is the lowest energy level the electron can reach. This corresponds to the wave function φ_{100} given by $n = 1$ (which, by the restrictions on l and m, means that $l = m = 0$), with an energy value of

$$\epsilon_1 = -\frac{m^{\text{el}} e^4}{2\hbar^2} = -\frac{1}{2}\frac{e^2}{r_0},$$

relative to the datum of an isolated electron in vacuum having energy equal to zero.

A second observation regarding these solutions is that they include all possible *excited* states of the electron. Many of these states are energetically equivalent, since the energy in Eqn. (4.46) is independent of l or m. Consider an electron in its ground state. To make it occupy a higher energy state, and thus to "transform" its wave function from φ_{100} to some other φ_{nlm}, we must impart to it the energy difference between the two states. Conversely, if an excited electron is allowed to drop down from state $\varphi_{n_2 l_2 m_2}$ to $\varphi_{n_1 l_1 m_1}$, it will emit an X-ray with precisely the energy difference between the two states:

$$\Delta\epsilon = \epsilon_{n_2} - \epsilon_{n_1} = \frac{m^{\text{el}} e^4}{2\hbar^2}\left(\frac{1}{n_1^2} - \frac{1}{n_2^2}\right). \tag{4.53}$$

[24] Recall $e^2 = 14.4$ eV · Å, $\hbar = 1.055 \times 10^{-34}$ J · s and $m^{\text{el}} = 9.1095 \times 10^{-31}$ kg.

A third important observation regarding the wave function solutions is that they are orthogonal to each other, i.e.

$$\langle \varphi_j | \varphi_{j'} \rangle = 0, \qquad j \neq j', \tag{4.54}$$

where (for the purpose of a more concise notation) we introduce the integer indices j and j' such that each has a unique one-to-one mapping to a triplet of quantum numbers n, m and l. This orthogonality is a consequence of the fact that the Hamiltonian is Hermitian; one can show that eigenfunctions of a Hermitian operator can always be chosen to be orthogonal [Sak94].

The orthogonality of the solutions for the hydrogen atom wave functions will be useful in our discussion of bonding between atoms. Specifically, we will make use of the orthogonality to represent a general, unknown wave function as a series approximation, summing orthogonal wave functions weighted by coefficients,

$$\psi(\boldsymbol{x}) \approx \sum_{j=1}^{N_{\text{basis}}} c_j \varphi_j(\boldsymbol{x}). \tag{4.55}$$

This is analogous to the plane wave representation using Fourier transforms presented in Section 4.4.2. In principle, this approximation becomes exact if N_{basis}, the number of terms in the expansion, is taken to infinity.

More protons or electrons So far we have focused on a single proton (a hydrogen ion) as the potential. A single electron in the presence of the nucleus of a heavier element (e.g., two protons for He, three of Li, and so on) has very similar wave functions to those described above, with the larger positive charge having the effect of reducing the energy levels (due to greater Coulomb interaction) and contracting the extent of the wave functions along the r directions. However, the essential features of the wave functions remain the same. The addition of other electrons, on the other hand, introduces additional complications, some of which will be discussed in Sections 4.4.6 and 4.5. Here, we note that the most important extra twist introduced by there being more than one electron comes about because electrons are *fermions* that obey the Pauli exclusion principle. This behavior is another of the fundamental postulates of quantum mechanics, which we accept based on overwhelming supporting experimental evidence. In essence, the Pauli exclusion principle says that two electrons cannot occupy the same "state," where a state is defined here as one of the solutions, φ_{nlm}, we have obtained. For simplicity, we have neglected the fact that electrons have an additional property known as "spin" that can take a value of either $\pm 1/2$, effectively making each φ_{nlm} into two possible electronic states. At this stage, let us just say that the Pauli exclusion principle means that *up to two* electrons can occupy each of the orbitals of Eqn. (4.52), each with opposite spin.

Now let us consider an element from the periodic table with a much larger nucleus, say Cu with atomic number $Z = 29$ (the number of protons in the nucleus of a Cu atom). Although there are many important subtleties that we would miss, a reasonably accurate picture of the electronic structure of Cu can be obtained by imagining the wave functions of Eqn. (4.52) (adjusted slightly to reflect the stronger positive nuclear charge) as "bins" into which we can drop electrons. Starting from a naked nucleus with no electrons, we add one electron and find that it occupies the lowest energy state φ_{100}. The second electron also occupies this state with opposite spin. By the Pauli exclusion principle, the third electron would occupy one of the eight available φ_{2lm} states because there is no room left in the φ_{100} states. The order of filling these states cannot be predicted by the simple single-electron picture we used to obtain Eqn. (4.46). Careful consideration of electron–electron interactions would reveal a splitting of the degenerate φ_{nlm} energies, but this is beyond the scope of our current discussion.

The quantum numbers we have identified as n, l and m correspond to what the chemists historically refer to as the shells and orbitals around an atom. Thus, the series $n = 1, 2, 3, 4, \ldots$ corresponds to the electronic shells, and n is referred to as the *principle quantum number*. Next, the series $l = 0, 1, 2, 3, 4, \ldots$ corresponds to the orbitals within each shell and l is referred to as the *orbital momentum number*. For example, for the $n = 4$ shell, we are restricted to the values $l = 0, 1, 2, 3$, which correspond to the orbitals usually labeled s, p, d and f. We recall that there is a limit to the number of electrons that can reside within a given orbital. For an s orbital this maximum is two, for a p orbital it is six and for a d orbital it is ten. These maxima are dictated by the permissible values of the final quantum number, m and the electron spin. For example, for a p orbital, we have $l = 1$, and thus there are three permissible values, $m = -1, 0, 1$. Each of these quantum states can be occupied by two electrons with opposite spins, for a total of six.

The probabilistic interpretation of the wave function can be used to visualize the shape of the electronic orbitals around the nucleus. Let us consider, for example, $n = 3$, the third electron shell around the nucleus, and examine the s, p and d orbitals from that shell in turn. It is straightforward to show that the $3s$ wave function has no angular dependence, taking the simple form

$$\varphi_{300} = \exp\left(-r/3r_0\right)\left(\frac{2}{9}\left(\frac{r}{r_0}\right)^2 - 2\frac{r}{r_0} + 3\right).$$

The quantity $P_{300}(r) = r^2 \langle \varphi_{300} | \varphi_{300} \rangle = r^2 \varphi_{300}^2$, gives the probability of finding the electron at a distance r from the center of the atom,[25] and is plotted in Fig. 4.4.

The $3p$ and $3d$ orbitals are somewhat more interesting to visualize, since they are no longer radially symmetric and each have multiple possible states corresponding to the same energy level. Figure 4.5 illustrates the shape of the $3p$ orbitals. Figure 4.5(a) shows the volume of space in which the probability $P(\boldsymbol{x})$ of finding a φ_{310} ($3p$) electron is relatively high. In Fig. 4.5(b) the $x = 0$ section through the same $3p$ orbital is shown, with contours indicating how the probability varies. This particular orbital is symmetric about the x_3-axis, and we can envision the other two p orbitals to be the same as this one, but rotated to lie along either the x_1- or x_2-axis. The $3p$ orbitals show two distinct lobes of high probability near the nucleus, and then a region of moderate probability a short distance further away. Finally, Fig. 4.6 shows the shape of the φ_{320} ($3d$) electronic orbital.

[25] The factor of r^2 comes from the fact that we are considering the probability of finding the electron at any ϕ and θ but fixed r, so we are effectively looking for the electron on the surface of a sphere of radius r.

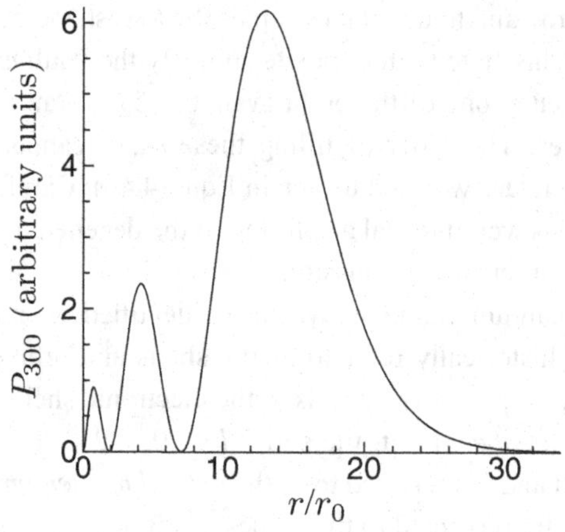

Fig. 4.4 Probability $P_{300}(r)$ of finding a $3s$ electron at r for the hydrogen atom.

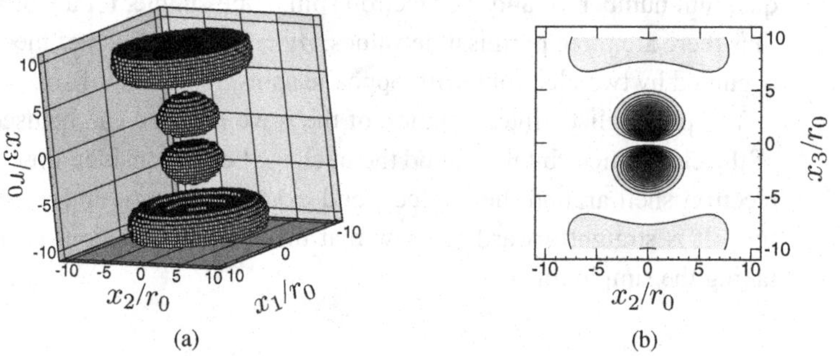

Fig. 4.5 Probability $P(x)$ of finding a $3p$ electron at x around the hydrogen atom in (a) three dimensions and (b) on the $x_1 = 0$ plane.

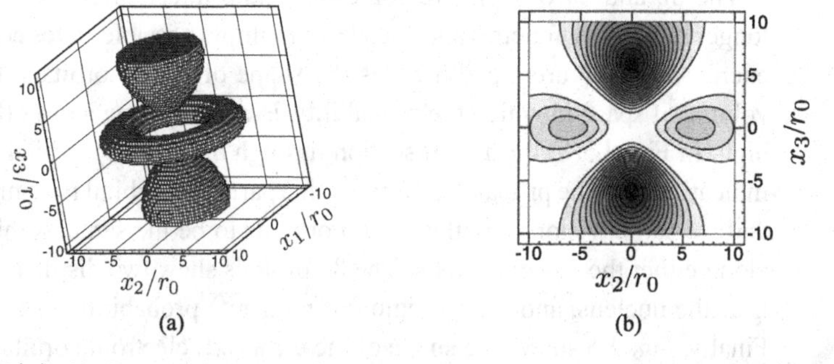

Fig. 4.6 Probability $P(x)$ of finding a $3d$ electron at x for the hydrogen atom (a) in three dimensions or (b) on the $x_1 = 0$ plane.

4.4.6 The hydrogen molecule

A main goal of this chapter on quantum mechanics is to provide a brief glimpse into the physics underlying bonding in solids. We started from a discussion of the wavelike behavior of electrons and the governing equation – the Schrödinger equation – that dictates their behavior in the presence of a potential field, $\mathcal{U}$. This allowed us to find the ground state and excited states of the electron in a hydrogen atom, and now brings us to the simplest bonding problem that we can discuss: what happens when two hydrogen atoms are brought together? We know from basic chemistry that the H_2 molecule is the stable configuration for hydrogen. In this section we will see how quantum mechanics predicts this stability. Through a series of approximations, we will solve the Schrödinger equation for the interaction between two hydrogen atoms, and show that the total energy of the system is lowered by bringing them together to form a bond. We will also see how the energy as a function of the bond length can be used to develop a simple *interatomic potential* for hydrogen. The method that we will use in this section will serve as the basis for solutions in DFT in Section 4.5, which permits practical, accurate calculations of the electronic structure for systems of hundreds or even thousands of atoms.

The Born–Oppenheimer approximation The problem of two electrons and two protons is already too complex to solve exactly, and requires us to make a number of assumptions. The first assumption concerns the motion of the protons relative to the electrons. The rest mass of a proton is roughly 1836 times that of an electron, while the force imposed on one by the other is the same. From a simple classical perspective, it seems reasonable that the electron would respond quickly to any motion of the proton, while the proton would react sluggishly to electron motion. Based on this, it is common in calculations of bonding and deformation in materials to assume that electron motion is always "fast" compared with proton motion. This allows us to treat the protons as fixed in space and solve for the resulting electronic structure. In problems where the motion of the protons is of interest (for example, in the deformation of solids), this approximation remains. In this case, we assume that as the protons move, the electrons find the appropriate ground state configuration by responding instantaneously to the gradual evolution of the proton positions. This assumption is known as the *Born–Oppenheimer approximation* (BOA), or sometimes simply as the adiabatic approximation, and it serves as a fundamental assumption in most materials modeling. We will explore the consequences of the BOA in more detail in Section 5.1.

Of course, the BOA does not always apply, and we need to examine the level of error that the approximation introduces in order to know when to avoid it. In materials science, we are often interested in the deformation of materials, which involves the relative motion of the nuclei in the material. Such motion means a continual variation in the electronic structure. In the discussion of the hydrogen molecule which follows, the notion of "deformation" is equivalent to a change in the interproton distance, R. As we shall see, this variation of the electronic structure also implies variation in the energy of the electronic ground state. Thus, if we assume that the electron is always in the ground state, it must constantly be emitting or absorbing photons (i.e. electromagnetic radiation, usually X-rays) to maintain the correct energy level during deformation. It is assumed with the BOA that this X-ray exchange with the ambient electromagnetic field can always occur. For most problems related to the mechanics of materials, the energy lost or absorbed in this way is quite small when compared, for example, with energy dissipated as heat, and so the BOA is reasonable.

There are, however, some circumstances when electrons *do not* move quickly from an excited state to their ground state, thus invalidating the BOA. Most relevant to our focus on materials is the case of rapid bond-breaking due to fracture, where the response time of the electrons cannot be viewed as instantaneous compared with the speed at which atomic nuclei are moving. This applies to metals, but it is made worse in semi-conductors and insulators where strongly covalent bonding further impedes the electronic rearrangement. For molecules containing hydrogen atoms, the relatively light hydrogen nucleus is small enough that the BOA will lead to appreciable error in the geometry and vibrational frequency of the molecule. Certain electrical and optical properties of molecules and solids are also dependent on so-called "non-Born–Oppenheimer" effects. For example, the experimental technique known as "ultrafast spectroscopy" involve analyzing the spectrum of light emitted by molecules when they are excited by lasers over very short time scales. The correct interpretation of these results can sometimes require the consideration of the transitional states that are inherently beyond the BOA. A review of problems in which non-Born–Oppenheimer effects can be important, as well as methods to take these effects into account, are discussed in [CBA03].

The hydrogen molecule Hamiltonian Given the BOA, the problem of solving for the electronic structure amounts to solving the Schrödinger equation with an appropriate potential function for two electrons. The full Hamiltonian is

$$\mathcal{H}(\boldsymbol{x}_1, \boldsymbol{x}_2, \boldsymbol{p}_1, \boldsymbol{p}_2) = \frac{1}{2m^{\text{el}}}\left(\boldsymbol{p}_1^2 + \boldsymbol{p}_2^2\right) + \mathcal{U}(\boldsymbol{x}_1, \boldsymbol{x}_2)$$

with

$$\mathcal{U}(\boldsymbol{x}_1, \boldsymbol{x}_2) = \frac{e^2}{\|\boldsymbol{x}_1 - \boldsymbol{x}_2\|} - \frac{e^2}{\|\boldsymbol{x}_1 - \boldsymbol{r}^\alpha\|} - \frac{e^2}{\|\boldsymbol{x}_1 - \boldsymbol{r}^\beta\|} - \frac{e^2}{\|\boldsymbol{x}_2 - \boldsymbol{r}^\alpha\|} - \frac{e^2}{\|\boldsymbol{x}_2 - \boldsymbol{r}^\beta\|}, \tag{4.56}$$

where $\boldsymbol{r}^\alpha$ and $\boldsymbol{r}^\beta$ are the coordinates of the two protons and $\boldsymbol{x}_1$ and $\boldsymbol{x}_2$ are the coordinates of the two electrons. The five terms in this potential energy represent the Coulomb interactions between the two electrons, and between each of the four possible electron–proton pairs. An assumption we have made at this point is to neglect magnetic interactions between the electrons, which we can assume to be small for our purposes. Since the potential $\mathcal{U}$ is assumed to be independent of time (on the time scale of the electrons, at least, thanks to the BOA) we can use the time-independent Schrödinger equation; our goal is to find a wave function $\chi(\boldsymbol{x}_1, \boldsymbol{x}_2)$ which satisfies the equation

$$\left(-\frac{\hbar^2}{2m^{\text{el}}}\left(\nabla_1^2 + \nabla_2^2\right) + \mathcal{U}\right)\chi(\boldsymbol{x}_1, \boldsymbol{x}_2) = \epsilon\chi(\boldsymbol{x}_1, \boldsymbol{x}_2), \tag{4.57}$$

where ∇_i^2 denotes the Laplacian differential operator with respect to $\boldsymbol{x}_i$.

To further simplify the problem, we assume that the two electrons do not interact, so that the first term in the potential energy of Eqn. (4.56) can be neglected. This means that the Hamiltonian now involves only terms in *either* $\boldsymbol{x}_1$ or $\boldsymbol{x}_2$, but not both. This permits a complete separation of the wave function $\chi(\boldsymbol{x}_1, \boldsymbol{x}_2) = \psi(\boldsymbol{x}_1)\psi(\boldsymbol{x}_2)$, where $\psi(\boldsymbol{x})$ is the solution to the single-particle Schrödinger equation in the presence of two positively charged protons. Now we can solve the much simpler problem for ψ; our goal is to find $\psi(\boldsymbol{x})$ which satisfies the equation

$$\mathcal{H}^{\mathrm{sp}}\psi(\boldsymbol{x}) = \epsilon\psi(\boldsymbol{x}), \tag{4.58}$$

where

$$\mathcal{H}^{\mathrm{sp}} = -\frac{\hbar^2}{2m^{\mathrm{el}}}\nabla^2 + \left(-\frac{e^2}{\|\boldsymbol{x} - \boldsymbol{r}^{\alpha}\|} - \frac{e^2}{\|\boldsymbol{x} - \boldsymbol{r}^{\beta}\|}\right) \tag{4.59}$$

is the Hamiltonian for each electron separately. Clearly, neglecting the electron–electron interactions is a bad assumption, but when the particles are considered noninteracting, it is always possible to treat the problem by solving a single-particle Schrödinger equation.[26] In Section 4.5 we shall see that it is possible, in principle, to reformulate multiple electron problems as single-particle problems without any approximation, as long as a suitable effective potential can be determined. As such, the method described here is the same as the one that we use to solve more complex multielectron problems later.

The variational method Despite the simplifications we have made, it is *still* not possible to solve this problem exactly. In order to make progress, we adopt a *variational approach*, whereby solving Schrödinger's equation is replaced with a minimization problem. Consider the energy of an electron with wave function ψ:

$$\epsilon = \frac{\langle\psi|\,\mathcal{H}^{\mathrm{sp}}\,|\psi\rangle}{\langle\psi|\psi\rangle} = \frac{\int \psi^*(\boldsymbol{x})\mathcal{H}^{\mathrm{sp}}(\boldsymbol{x})\psi(\boldsymbol{x})\,d\boldsymbol{x}}{\int \psi^*(\boldsymbol{x})\psi(\boldsymbol{x})\,d\boldsymbol{x}}, \tag{4.60}$$

where $\mathcal{H}^{\mathrm{sp}}$ was defined in Eqn. (4.59). Minima of this energy are found by taking a functional derivative with respect to the complex conjugate of the wave function, ψ^*, and setting it equal to zero[27]

$$\frac{\delta\epsilon}{\delta\psi^*(\boldsymbol{x})} = \frac{\left(\int \psi^*(\boldsymbol{x}')\psi(\boldsymbol{x}')\,d\boldsymbol{x}'\right)\mathcal{H}^{\mathrm{sp}}(\boldsymbol{x})\psi(\boldsymbol{x}) - \left(\int \psi^*(\boldsymbol{x}')\mathcal{H}^{\mathrm{sp}}(\boldsymbol{x}')\psi(\boldsymbol{x}')\,d\boldsymbol{x}'\right)\psi(\boldsymbol{x})}{\left(\int \psi^*(\boldsymbol{x}')\psi(\boldsymbol{x}')\,d\boldsymbol{x}'\right)^2} = 0, \tag{4.61}$$

which we can simplify using Eqn. (4.60) to get

$$\mathcal{H}^{\mathrm{sp}}(\boldsymbol{x})\psi(\boldsymbol{x}) = \epsilon\psi(\boldsymbol{x}). \tag{4.62}$$

Thus, we see that minimizing the energy is equivalent to the solution of Schrödinger's equation.[28] This is analogous to the *principle of minimum potential energy* that is routinely

[26] We are neglecting important subtleties relating to electron spin and the Pauli exclusion principle, but the single-particle approximation remains an essential tool for making progress with the solutions to bonding problems.

[27] We can treat ψ^* and ψ as independent variables since this is effectively equivalent to considering the independent real and imaginary components of ψ. Recall that for functional differentiation, $\delta f(x)/\delta f(y) = \delta(x-y)$ and that it is possible to exchange the order of integration and differentiation. Therefore $(\delta/\delta f(x))\int g(y)f(y)\,dy = g(x)$.

[28] Of course, the procedure we have just outlined will also find saddle points or maxima, which correspond to unstable solutions to the Schrödinger equation.

invoked in continuum mechanics.[29] As in classical mechanics, we see that in quantum mechanics, valid physical solutions are those that minimize the energy of the system.

The variational approach provides a method of solving the Schrödinger equation that is more amenable to computational methods. The idea is to build an approximation to the wave function $\psi(\boldsymbol{x})$ as a linear combination of some basis set that we will denote $\varphi_j(\boldsymbol{x})$, as in Eqn. (4.55). With the basis set chosen and fixed, our approximate wave function depends only on the weighting coefficients c_j, and hence we expect that the set of these coefficients that minimizes ϵ will provide the best approximate solution to the Schrödinger equation. We can infer that the energy obtained in this way using a finite (and therefore incomplete) basis set is an *upper bound* to the exact energy.[30] Of course, the fidelity of the approximation is dependent on the nature of the basis functions we choose and the number of terms in the series. A rigorous choice would be to use a complete, orthonormal basis set, but this could require an exceedingly large number of terms in the sum for sufficient accuracy if the basis set is sufficiently different in form from the expected solution.[31]

The variational approach also leads to fundamental questions about the uniqueness of the solutions so obtained. For complicated systems, there will inevitably be multiple local minima and only one global minimum that will generally be difficult to find. Some methods of solution might lead to unstable stationary points (maxima or saddle points) that do not correspond to physically realistic results. Like any nonlinear, multidimensional minimization, the solution will depend on the initial "guess" and the details of the iterative method used. It will also depend on the "ruggedness" of the energy surface in multidimensional space (see the discussion of potential energy surfaces in Chapter 6). Unfortunately, the computational expense associated with most fully quantum mechanical calculations makes it difficult to carefully explore the robustness of an obtained solution.

A common basis set used in the variational approach, and the one we use next, is the so-called linear combination of atomic orbitals (LCAO). In LCAO, the basis set is the wave function solutions we found for the isolated hydrogen atom in Eqn. (4.52) centered on each nucleus in the system (in our present example, the two nuclei of the hydrogen molecule). These wave functions are orthogonal for orbitals centered on the same nucleus, but not for orbitals centered on different nuclei (see Exercise 4.6). Let us take the simplest LCAO approximation where the electron wave function for the molecule is a linear combination of only the $1s$ orbitals for the isolated hydrogen atom. Thus

$$\psi(\boldsymbol{x}) = c^{\mathsf{A}}\varphi^{\mathsf{A}}(\boldsymbol{x}) + c^{\mathsf{B}}\varphi^{\mathsf{B}}(\boldsymbol{x}), \tag{4.63}$$

[29] The principle of minimum potential energy is discussed in Section 2.6. In continuum mechanics, this principle shows that minimization of the energy is equivalent to solving the equation for pointwise (stable) equilibrium of the stresses in the body. The equilibrium equation is the governing partial differential equation in that case, analogous to the Schrödinger equation in quantum mechanics. The variational approach paves the way for approximate, numerical solutions in both continuum and quantum mechanics.

[30] This argument is effectively the Ritz theorem. For instance, the same argument can be used to show that in finite element analysis, the energy of the approximate solution is an upper bound to the exact energy.

[31] For example a basis set of plane waves may require a large number of terms to accurately describe a highly localized wave function.

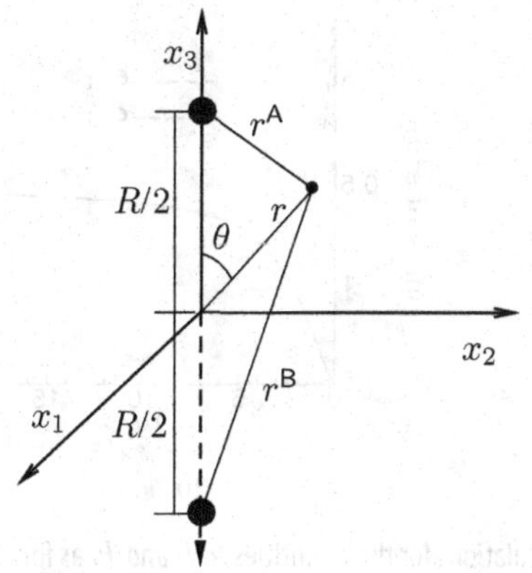

Fig. 4.7 Coordinate system for the hydrogen molecule.

where $\varphi^A(\boldsymbol{x})$ and $\varphi^B(\boldsymbol{x})$ refer to the $1s$ orbital centered on protons A and B respectively. Referring to Eqn. (4.52), we have

$$\varphi^A(\boldsymbol{x}) = \varphi_{100}(\boldsymbol{x} - \boldsymbol{r}^A) = \frac{1}{\sqrt{r_0^3 \pi}} \exp\left(-r^A/r_0\right), \tag{4.64}$$

and similarly for $\varphi^B(\boldsymbol{x})$, where the factor $\sqrt{r_0^3\pi}$ ensures that the wave functions are normalized and $r^A = \|\boldsymbol{x} - \boldsymbol{r}^A\|$ is the distance from the electron to proton A. For definiteness, let us assume that the two protons lie along the x_3-axis, are symmetric about the origin and are separated by a distance R (they are thus located at $[0, 0, \pm R/2]$ as shown in Fig. 4.7.) In spherical coordinates the distances from the electron to the protons are

$$r^A = \sqrt{r^2 + \frac{R^2}{4} - Rr\cos\theta}, \qquad r^B = \sqrt{r^2 + \frac{R^2}{4} + Rr\cos\theta}.$$

One can readily verify by using Eqn. (4.64) in Eqn. (4.54) that this is *not* an orthogonal basis; the hope is that the deviation from orthogonality is small enough for this to be only a minor transgression.

For a fixed value of R, we need to minimize the expectation value of the Hamiltonian for our choice of basis. Inserting Eqn. (4.63) into Eqn. (4.60) we have

$$\epsilon = \frac{c^{A^2}\left\langle \varphi^A \middle| \mathcal{H}^{sp} \middle| \varphi^A \right\rangle + c^A c^B \left\langle \varphi^A \middle| \mathcal{H}^{sp} \middle| \varphi^B \right\rangle + c^B c^A \left\langle \varphi^B \middle| \mathcal{H}^{sp} \middle| \varphi^A \right\rangle + c^{B^2}\left\langle \varphi^B \middle| \mathcal{H}^{sp} \middle| \varphi^B \right\rangle}{c^{A^2}\left\langle \varphi^A \middle| \varphi^A \right\rangle + c^A c^B \left\langle \varphi^A \middle| \varphi^B \right\rangle + c^B c^A \left\langle \varphi^B \middle| \varphi^A \right\rangle + c^{B^2}\left\langle \varphi^B \middle| \varphi^B \right\rangle} \tag{4.65}$$

where we have used the fact that $\langle \varphi^A + \varphi^B | \mathcal{H} | \psi \rangle = \langle \varphi^A | \mathcal{H} | \psi \rangle + \langle \varphi^B | \mathcal{H} | \psi \rangle$, and have taken c^A and c^B to be real numbers. The following definitions will be convenient:

$$\begin{aligned} E(R) &= \left\langle \varphi^A \middle| \mathcal{H}^{sp} \middle| \varphi^A \right\rangle = \left\langle \varphi^B \middle| \mathcal{H}^{sp} \middle| \varphi^B \right\rangle, \\ h(R) &= \left\langle \varphi^A \middle| \mathcal{H}^{sp} \middle| \varphi^B \right\rangle = \left\langle \varphi^B \middle| \mathcal{H}^{sp} \middle| \varphi^A \right\rangle, \\ s(R) &= \left\langle \varphi^A \middle| \varphi^B \right\rangle = \left\langle \varphi^B \middle| \varphi^A \right\rangle. \end{aligned} \tag{4.66}$$

It is straightforward to show that the second equality in each of these expressions follows from the definitions of $\mathcal{H}^{sp}$ and the wave functions φ^A and φ^B. Also, since φ^A and φ^B are

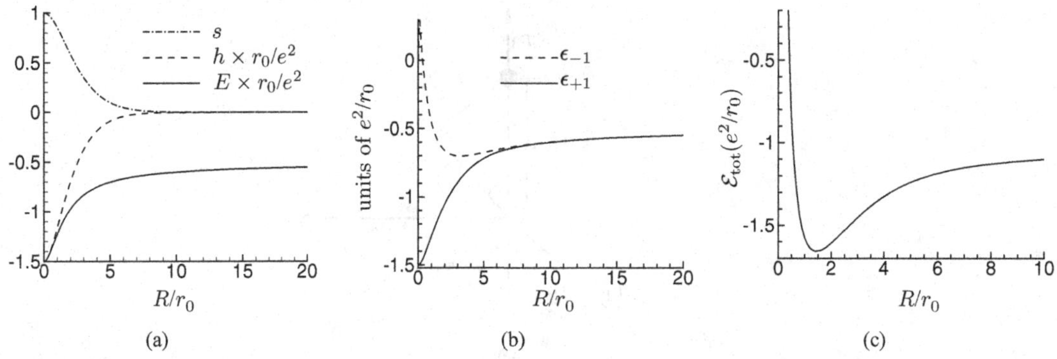

Fig. 4.8 (a) Numerical calculations for the quantities s, h and E as functions of the proton separation. (b) Energy of the two possible electron wave functions as a function of the proton separation. (c) The energy of the full hydrogen molecule as a function of proton separation.

properly normalized, we have that $\langle \varphi^A | \varphi^A \rangle = \langle \varphi^B | \varphi^B \rangle = 1$. The quantity h is sometimes referred to as the "hopping integral," as it is related to the probability of the electron hopping between states φ^A and φ^B. The quantity s is called the "overlap integral," which for an orthogonal basis set is identically zero. For improved accuracy, s should be as close to zero as possible. We shall see that this condition is not well satisfied in this case, giving rise to some of the error in the final solution. Using these definitions simplifies Eqn. (4.65) to

$$\epsilon = \frac{((c^A)^2 + (c^B)^2)E + 2c^A c^B h}{((c^A)^2 + (c^B)^2) + 2c^A c^B s}. \tag{4.67}$$

It is possible to evaluate the integrals of Eqns. (4.66) directly by making a few clever changes of variables. Here we simply state the results and look at them graphically to see their relative values and dependencies on R. The three quantities E, h and s evaluate to

$$E(R) = -\frac{e^2}{r_0}\left[\frac{1}{2} + \frac{r_0}{R} - \left(1 + \frac{r_0}{R}\right)\exp(-2R/r_0)\right], \tag{4.68}$$

$$h(R) = -\frac{e^2}{r_0}\left[\frac{s}{2} + \left(1 + \frac{R}{r_0}\right)\exp(-R/r_0)\right], \tag{4.69}$$

$$s(R) = \exp(-R/r_0)\left[1 + \left(\frac{R}{r_0}\right) + \frac{1}{3}\left(\frac{R}{r_0}\right)^2\right], \tag{4.70}$$

and are plotted in Fig. 4.8(a). From the figure, we note that in the limit of large spacing between the protons, the overlap integral and the hopping integral both go to zero, while E approaches the ground state energy of an isolated hydrogen atom, $-0.5e^2/r_0$. We also see that both h and E are always negative, with $E < h$, while s is always positive and lies between unity at $R = 0$ (complete overlap) and zero at $R = \infty$ (no overlap).

Taking the derivative of Eqn. (4.67) with respect to the coefficients c^A and c^B gives

$$\begin{bmatrix} E & h \\ h & E \end{bmatrix}\begin{bmatrix} c^A \\ c^B \end{bmatrix} = \epsilon \begin{bmatrix} 1 & s \\ s & 1 \end{bmatrix}\begin{bmatrix} c^A \\ c^B \end{bmatrix}. \tag{4.71}$$

This is a generalized eigenvalue problem for which it is easily shown that there are two eigenvector solutions corresponding to $c^{\mathsf{A}}/c^{\mathsf{B}} = \pm 1$, with unnormalized wave functions

$$\psi_{+1}(\boldsymbol{x}) = \varphi^{\mathsf{A}}(\boldsymbol{x}) + \varphi^{\mathsf{B}}(\boldsymbol{x}), \qquad \psi_{-1}(\boldsymbol{x}) = \varphi^{\mathsf{A}}(\boldsymbol{x}) - \varphi^{\mathsf{B}}(\boldsymbol{x}).$$

The energy for each of these wave functions is found from the corresponding eigenvalues of Eqn. (4.71) to be

$$\epsilon_{+1} = \frac{E+h}{1+s}, \qquad \epsilon_{-1} = \frac{E-h}{1-s}.$$

These are plotted in Fig. 4.8(b). It is clear from the figure that the state corresponding to $c^{\mathsf{A}}/c^{\mathsf{B}} = +1$ has the lower energy, and thus it will be adopted by the electron. It is straightforward to show (see the exercises) that in fact ϵ_{+1} is a minimum and ϵ_{-1} is a maximum with respect to the ratio $c^{\mathsf{A}}/c^{\mathsf{B}}$.

We now see how a bond between two atoms can form, and how it is possible to develop a simple pair potential description of the hydrogen molecule, which we will describe in more detail in Section 5.4.2. We imagine two isolated hydrogen atoms ($R = \infty$) in their ground state, with their electrons in $1s$ ($n = 1$) states. We recall from Eqn. (4.46) that each of these has an energy of $-m^{\text{el}}e^4/2\hbar^2 = -e^2/2r_0$, for a total energy of $-e^2/r_0$. Bringing the atoms together, each electron follows the energy curve of ϵ_{+1} in Fig. 4.8(b). Note that since there are two possible spins, $\pm 1/2$, for the electrons, it is permissible for them both to occupy the same state, ψ_{+1}. At the same time, there is a Coulomb interaction between the two protons of the form e^2/R. Summing these energy contributions, we have

$$\mathcal{E}_{\text{tot}}(R) = 2\epsilon_{+1}(R) + \frac{e^2}{R}, \qquad (4.72)$$

where the factor of 2 represents the two electrons in the model. This energy as a function of the bond length in the hydrogen molecule is plotted in Fig. 4.8(c).[32]

We can see that the energy of the system is lowered as we bring the two atoms from far apart ($R = \infty$). This lowering of the energy is the essence of bond formation: the two atoms are "happier" together than apart because their energy has been reduced. Our model predicts a minimum energy of about $-1.657e^2/r_0$ corresponding to a reduction of $0.657e^2/r_0$ relative to the isolated atoms. Since $e^2/r_0 = 27.2114$ eV, the prediction for the bonding energy is 17.9 eV, which is rather high compared with the experimentally known value of about 4.52 eV. On the other hand, the minimum energy occurs at $R = 1.45r_0 = 0.767$ Å, which agrees well with the correct value of about 0.74 Å. Effects that we have neglected, such as the Coulomb interactions between the two electrons and the electron exchange interactions (which are discussed in more detail in Section 4.5.1) contribute to correcting

[32] In Section 5.4.2 we will revisit the energy function of Fig. 4.8(c) as an example of developing a simple pair potential. As we shall see, such models are extremely powerful for their simplicity, but at the same time must be used only within the domain to which they are fit, lest they be incorrectly interpreted.

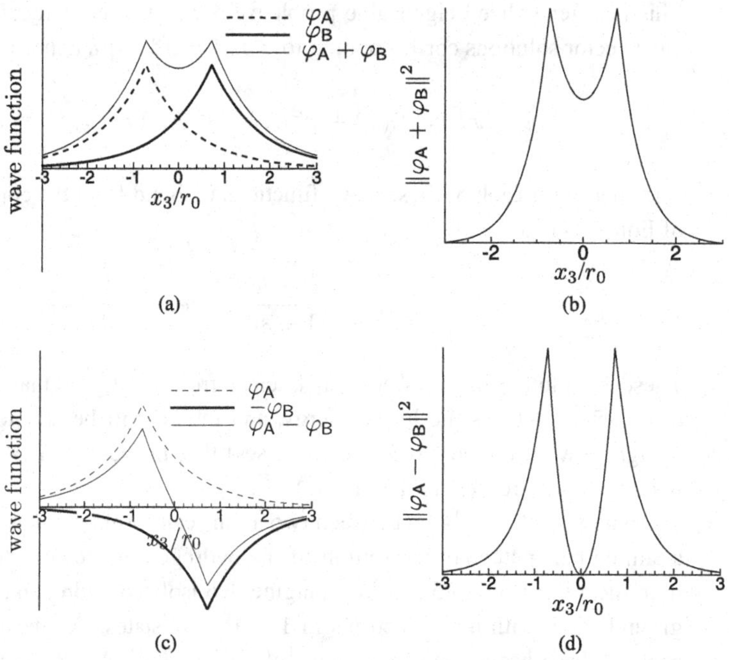

Fig. 4.9 (a) Wave function for the bonding state, ψ_{+1}, (b) probability distribution for the bonding state, $\|\psi_{+1}\|^2$, (c) wave function for the antibonding state, ψ_{-1}, (d) probability distribution for the antibonding state, $\|\psi_{-1}\|^2$.

these discrepancies.[33] We could also improve the approximation by adding more terms to our guess in Eqn. (4.63) or by choosing a different basis set altogether.

We can see the nature of the resulting bond between the atoms by considering the probability distribution of the electrons. In Fig. 4.9 we plot the wave function, ψ and the probability $\|\psi\|^2$ along the x_3-axis. Figure 4.9(a) shows the lower-energy solution, ψ_{+1}, and Fig. 4.9(b) shows the probability distribution of the electrons for that wave function. The peaks correspond to the locations of the nuclei at $\approx\pm 0.72 r_0$. Between the two peaks, there is a high probability of finding the electron in the bond that has formed between the two nuclei. This is the signature of a covalent bond. On the other hand, the second solution, ψ_{-1} shown in Fig. 4.9(c) and Fig. 4.9(d), leads to a clear reduction in the probability of finding the electron between the two nuclei. In fact, the probability goes to zero at $x_3 = 0$. In this case no bond is formed. A general terminology referring to electronic states is to define "bonding" states as those which lower the energy relative to the isolated atoms, while states with energy higher than the isolated state are called "antibonding." In this case, then,

[33] The Coulomb interactions between the electrons, in particular, contributes in large part to lowering this artificially high bond energy. On average, the two electrons lie slightly closer to each other than the two protons (see probability distributions discussed below). If we simply imagine the two electrons as fixed point charges at some equilibrium spacing slightly less than $R = 1.45 r_0 = 0.767$ Å, the energy of their Coulomb interaction almost exactly cancels the excess bond energy predicted here. The exchange interactions, on the other hand, tend to lower the depth of the energy minimum and thus raise the bond energy.

we have found one bonding and one antibonding state, as shown by the charge distributions in Fig. 4.9.

The antibonding solution and the higher energy associated with it give us some insight into why the H_3 molecule is not stable. Since only two electrons can occupy the same state, a third electron would have to be in a state resembling the antibonding configuration.[34] The energy reduction from the bonding state is not great enough to overcome the higher energy of the antibonding state and the repulsion between the three nuclei.

4.4.7 Summary of the quantum mechanics of bonding

The quantum mechanics of bonding, for our purposes, boils down to solving Schrödinger's equation for the electronic wave functions in the presence of the potential field introduced by the stationary atoms. Knowing these wave functions, in principle, tells us everything about the bonding, including the energy of the filled ground states, and the discrete energy levels of possible excited states. These energetics and how they change as atoms are brought together are the essence of chemical bonding.

The wave nature of electrons and the probabilistic interpretation of these wave functions are difficult to reconcile with our typical understanding of the world. However, notions of "particles" emerge when wave functions localize in space and we consider the expectation values of such things as electron position and momentum.

In Section 4.4.5, we saw how the Schrödinger equation can be solved exactly for the simple case of a single isolated hydrogen atom. While this is a long way from a "material," it serves as an essential building block for understanding more complex bonding, as the features of the discrete electronic orbitals (their energy, shape and number) emerge. The resulting solutions in Eqn. (4.52) also serve as a convenient basis set for approximate methods. In Section 4.4.6, we moved on to the hydrogen molecule, and solved for its ground state. To do so, we had to make a number of simplifying assumptions, and even so we could not obtain an exact solution. We could improve this solution, for example by choosing more or "better" basis functions, but clearly the computational effort will increase, and obtaining solutions for more atoms and more electrons becomes correspondingly more expensive.

The direct solution of the Schrödinger equation (employed for the single hydrogen atom) is clearly not a practical approach for the large, complex systems of atoms that we call materials. Rather, the solution process we used for the hydrogen molecule is the basis for the DFT and tight-binding methods that follow; first, a basis set is chosen from which to build an approximation to the wave function, then a matrix eigenvalue equation is set up to solve for the coefficients weighting these basis functions. This gives multiple solutions that represent all the possible electronic states. The lowest-energy states are filled first until all the electrons are accounted for, at which time the total electronic energy can be computed.

34 We say "resembling" because the introduction of a third proton would modify the Hamiltonian and the resulting solutions, but the bonding and antibonding features will remain.

The inversion (or diagonalization) of the matrix equation, an $O\left(N_{\text{basis}}^3\right)$ operation,[35] is the key computational expense in these methods.

4.5 Density functional theory (DFT)

4.5.1 Exact formulation

In the last section, we made a rather bold simplification in order to solve for the bonding in the hydrogen molecule. Specifically, we assumed that the two electrons in the problem did not interact, and we therefore neglected a term in the Hamiltonian of the form

$$\mathcal{E}^{\text{ee}} = \left\langle \psi \middle| \frac{e^2}{2} \sum_{\substack{i,j \\ i \neq j}} \frac{1}{\|\boldsymbol{x}_i - \boldsymbol{x}_j\|} \middle| \psi \right\rangle,$$

where in our simple example i and j ran only from 1 to 2. In a complete formulation of quantum mechanics, this simple looking term hides enormous complexity, including not only the classical Coulomb interaction between the electrons, but additional effects that are strictly quantum mechanical in nature. While the full details are beyond where we wish to go here, we must briefly discuss these *exchange* and *correlation* effects. Exchange effects are a manifestation of the Pauli exclusion principle, which imposes restrictions on the mathematical form of the wave function and thus affects the energy states of the electrons. Specifically, it requires that a many-electron wave function be antisymmetric with respect to the exchange of position between any two electrons. For example

$$\psi(\boldsymbol{x}_1, \boldsymbol{x}_2, \ldots, \boldsymbol{x}_{N^{\text{el}}}) = -\psi(\boldsymbol{x}_2, \boldsymbol{x}_1, \ldots, \boldsymbol{x}_{N^{\text{el}}}), \tag{4.73}$$

where we have interchanged the positions of electrons 1 and 2 and N^{el} is the number of electrons. At the same time, electrons carry charge and interact electrostatically. They will thus have to move in such a way as to avoid one another, leading to correlation in their motion and correlation energy effects. We completely ignored all of these interactions in the previous section in the interest of simplicity. Incorporating them into the model would require careful choice of form for the wave functions,[36] and would make it more difficult to solve the resulting equations.

Fortunately, it is possible to treat these effects with reasonable accuracy in a formulation that does not explicitly involve wave functions. We will see below that the problem can be reformulated in terms of the *electron density*, ρ, at each point in space. The method will allow us to find the ground state energy for complex many-electron problems, including

[35] Here $O(\cdot)$ is the "big O" notation that gives the order of growth of the algorithm. So under worst case conditions, the diagonalization operation will scale like N_{basis}^3.

[36] For example the total wave function can be written as a determinant of a matrix whose elements are single-particle wave functions in such a way as to ensure the required antisymmetry. See, for example, [Fin03, Kax03] and the exercises at the end of this chapter.

electron–electron interactions, without ever having to find the many-electron wave function itself. We start by defining the electron density for a general system of N^{el} electrons as

$$\rho(\boldsymbol{x}) \equiv N^{\text{el}} \int \psi^*(\boldsymbol{x}, \boldsymbol{x}_2, \ldots, \boldsymbol{x}_{N^{\text{el}}})\psi(\boldsymbol{x}, \boldsymbol{x}_2, \ldots, \boldsymbol{x}_{N^{\text{el}}})\, d\boldsymbol{x}_2 d\boldsymbol{x}_3 \ldots d\boldsymbol{x}_{N^{\text{el}}}. \quad (4.74)$$

Note that the definition of ρ is independent of which electrons we choose to integrate out in Eqn. (4.74). This follows from the required antisymmetry of the wave function in Eqn. (4.73). The normalization of the wave function ensures that

$$\int \rho(\boldsymbol{x})\, d\boldsymbol{x} = N^{\text{el}}. \quad (4.75)$$

If, in fact, we can work in terms of density instead of the full wave function, it is an enormous simplification. In its original form, the wave function depends on $3N^{\text{el}}$ electronic degrees of freedom, whereas the density is a scalar field over merely a three-dimensional space. The key development that allows us to do this is DFT, which is due to Kohn, Sham and Hohenberg [HK64, KS65], and it is the goal of this section to describe the essential details of how DFT works. Kohn, Sham and Hohenberg not only showed that it is possible to replace wave functions with electron density, but also that it is possible to reduce the solution of any Hamiltonian to the solution of a Hamiltonian of *a single, noninteracting electron, without any approximations being made*. This is achieved by the introduction of an effective potential that replaces the external potential. Since we know how to solve the problem of a single electron in an arbitrary potential quite accurately, this is an enormous breakthrough. Indeed, it has made possible the quantum mechanical calculation of the structure and dynamics of complex systems involving hundreds or thousands of atoms, and ultimately won Kohn the 1998 Nobel prize, shared with John Pople.

There are essentially three "steps" in going from the full solution of the Schrödinger equation to the DFT solution. Step 1 shows that solving for the electron density is, practically speaking, equivalent to finding the full many-electron wave function. Step 2 involves a rearrangement of terms which demonstrates that the energy of N^{el} electrons in a *fixed* external potential is equivalent to the energy of *one* electron in an effective external potential that depends on the electron density itself. Finally, in step 3, we return to solving for wave functions, but this time for the much simpler case of a single electron (where we know from step 1 that as long as the single electron solution leads to the same electron density, it will produce the same physics as the N^{el}-electron result). The main trade-off in the procedure is that mathematical complexity is exchanged for an iterative solution procedure. Since the effective potential itself is a function of the electron density, we must start with an initial guess and iterate to a self-consistent final result. The computational paradigm is especially well suited to this trade-off, since computers are much better at repetitive iterations than at seeking closed-form solutions.

Step 1 Replacing wave functions by electron densities Recall that the potential, $\mathcal{U}$, completely determines the Hamiltonian and thus the Schrödinger equation to be solved. More specifically, since the electron–electron interactions are always the same, it is the *external potential* due to the ions, $\mathcal{U}^{\text{ext}}$, that uniquely determines the Hamiltonian. The first step to building DFT is to show that $\mathcal{U}$ also uniquely determines the resulting electron density. This is the *essential DFT theorem*, which was established by Hohenberg and Kohn in 1964 [HK64]. It states:

> *The DFT theorem* The external potential $\mathcal{U}^{\text{ext}}$ is determined to within an unimportant arbitrary constant by the electron density, $\rho(\boldsymbol{x})$.

This is important because it means that, conversely, if we solve for the electron density we can be sure that it is the unique solution for the potential from which we started. We omit the proof here, but note that it is relatively simple (in hindsight), following a *reductio ad absurdum* approach. A succinct proof is given by, for example, Kaxiras [Kax03].

The Hamiltonian as a function of density Confident that we can leave behind the full wave function, we proceed to rewrite the Hamiltonian in terms of the electron density, and now focus on developing a method whereby we can find the density $\rho(\boldsymbol{x})$ that minimizes the energy of the system. We write the energy as

$$\langle\psi|\,\mathcal{H}\,|\psi\rangle = \mathcal{T}^{\text{el}} + \mathcal{E}^{\text{ZZ}} + \mathcal{E}^{\text{ext}} + \mathcal{E}^{\text{ee}}, \tag{4.76}$$

where

$$\mathcal{T}^{\text{el}} = \left\langle \psi \middle| -\frac{\hbar^2}{2m^{\text{el}}}\sum_i \nabla_i^2 \middle| \psi \right\rangle, \tag{4.77}$$

$$\mathcal{E}^{\text{ZZ}} = \left\langle \psi \middle| \frac{1}{2}\sum_{\substack{\alpha,\beta \\ \alpha\neq\beta}} \frac{e^2 Z^\alpha Z^\beta}{\|\boldsymbol{r}^\alpha - \boldsymbol{r}^\beta\|} \middle| \psi \right\rangle = \frac{1}{2}\sum_{\substack{\alpha,\beta \\ \alpha\neq\beta}} \frac{e^2 Z^\alpha Z^\beta}{\|\boldsymbol{r}^\alpha - \boldsymbol{r}^\beta\|}, \tag{4.78}$$

$$\mathcal{E}^{\text{ext}} = \left\langle \psi \middle| \sum_i \sum_\beta \frac{-e^2 Z^\beta}{\|\boldsymbol{x}_i - \boldsymbol{r}^\beta\|} \middle| \psi \right\rangle, \tag{4.79}$$

$$\mathcal{E}^{\text{ee}} = \left\langle \psi \middle| \frac{1}{2}\sum_{\substack{i,j \\ i\neq j}} \frac{e^2}{\|\boldsymbol{x}_i - \boldsymbol{x}_j\|} \middle| \psi \right\rangle. \tag{4.80}$$

The first term represents the kinetic energy of the electrons. At this point, we cannot write a simple expression for $\mathcal{T}^{\text{el}}$ in terms of the electron density, but we assume that some functional, $\mathcal{T}^{\text{el}}[\rho(\boldsymbol{x})]$, exists. We will deal with this term in Section 4.5.1 using the approach of Kohn and Sham. The second term, $\mathcal{E}^{\text{ZZ}}$, is simply the interactions between the atomic nuclei. Other than the wave functions themselves, terms in the integrand of $\mathcal{E}^{\text{ZZ}}$ are independent of the electrons and as such we can rearrange terms and use $\langle\psi|\psi\rangle = 1$ to get the last equality in Eqn. (4.78). The third term, $\mathcal{E}^{\text{ext}}$, is the interaction between the electrons and the external potential, here assumed to be provided by an arbitrary collection of nuclei with positions $\boldsymbol{r}^\beta$ and charges[37] eZ^β. Because each term in the sum only involves

[37] It is common practice to treat only the valence electrons in DFT, and replace the core electrons and the nuclei with *pseudopotentials*, about which more will follow in Section 4.5.2.

one electron at a time, the external potential becomes

$$\mathcal{E}^{\text{ext}} = \sum_{i=1}^{N^{\text{el}}} \left(\int \psi^*(\boldsymbol{x}_1, \boldsymbol{x}_2, \ldots, \boldsymbol{x}_{N^{\text{el}}}) \mathcal{U}^{\text{ext}}(\boldsymbol{x}_i) \psi(\boldsymbol{x}_1, \boldsymbol{x}_2, \ldots, \boldsymbol{x}_{N^{\text{el}}}) \, d\boldsymbol{x}_1 \ldots d\boldsymbol{x}_{N^{\text{el}}} \right), \tag{4.81}$$

where

$$\mathcal{U}^{\text{ext}}(\boldsymbol{x}_i) = \sum_{\beta} \frac{-e^2 Z^{\beta}}{\|\boldsymbol{x}_i - \boldsymbol{r}^{\beta}\|}.$$

If we now visit each term in the sum on i, and separate out the integration with respect to $\boldsymbol{x}_i$ from all the other integrations, what remains is the definition of the electron density in Eqn. (4.74). For example, for $i = 1$ we have

$$\int \mathcal{U}^{\text{ext}}(\boldsymbol{x}_1) \left[\int \psi^*(\boldsymbol{x}_1, \boldsymbol{x}_2, \ldots, \boldsymbol{x}_{N^{\text{el}}}) \psi(\boldsymbol{x}_1, \boldsymbol{x}_2, \ldots, \boldsymbol{x}_{N^{\text{el}}}) \, d\boldsymbol{x}_2, \ldots, d\boldsymbol{x}_{N^{\text{el}}} \right] d\boldsymbol{x}_1. \tag{4.82}$$

Using Eqn. (4.74) to replace the term in the square brackets and repeating this for each term in the sum on i gives N^{el} identical terms, so

$$\mathcal{E}^{\text{ext}}[\rho] = \int \rho(\boldsymbol{x}) \mathcal{U}^{\text{ext}}(\boldsymbol{x}) \, d\boldsymbol{x}. \tag{4.83}$$

Turning to the final term in the energy, $\mathcal{E}^{\text{ee}}$, it is convenient to divide it into two parts. The first, denoted $\mathcal{E}^{\text{H}}$, is the Coulomb interactions between the electrons (called the *Hartree energy*). It is straightforward to express this energy in terms of the electron density as follows. Consider two infinitesimal electronic charges $dC_{\boldsymbol{x}} = e\rho(\boldsymbol{x})d\boldsymbol{x}$ and $dC_{\boldsymbol{x}'} = e\rho(\boldsymbol{x}')d\boldsymbol{x}'$ located at positions $\boldsymbol{x}$ and $\boldsymbol{x}'$ respectively. The energy of their Coulomb interactions is

$$d\mathcal{E}^{\text{H}} = \frac{dC_{\boldsymbol{x}} dC_{\boldsymbol{x}'}}{\|\boldsymbol{x} - \boldsymbol{x}'\|} = \frac{e^2 \rho(\boldsymbol{x}) \rho(\boldsymbol{x}')}{\|\boldsymbol{x} - \boldsymbol{x}'\|} d\boldsymbol{x} d\boldsymbol{x}'.$$

The total Coulomb energy associated with the charge $dC_{\boldsymbol{x}}$ requires us to integrate over all charges $dC_{\boldsymbol{x}'}$. A second integral over all charges $dC_{\boldsymbol{x}}$ will give us the total Coulomb energy, except that we will have "double counted" each contribution. Thus the total energy $\mathcal{E}^{\text{H}}$ is

$$\mathcal{E}^{\text{H}}(\rho) = \frac{1}{2} \int \frac{e^2 \rho(\boldsymbol{x}) \rho(\boldsymbol{x}')}{\|\boldsymbol{x} - \boldsymbol{x}'\|} d\boldsymbol{x} d\boldsymbol{x}', \tag{4.84}$$

where the factor of $1/2$ takes care of the double counting. It is also convenient to define a *Hartree potential*

$$\mathcal{U}^{\text{H}}(\boldsymbol{x}) \equiv \frac{\delta \mathcal{E}^{\text{H}}}{\delta \rho} = \int \frac{e^2 \rho(\boldsymbol{x}')}{\|\boldsymbol{x} - \boldsymbol{x}'\|} d\boldsymbol{x}', \tag{4.85}$$

such that

$$\mathcal{E}^{\text{H}} = \frac{1}{2} \int \rho(\boldsymbol{x}) \mathcal{U}^{\text{H}}(\boldsymbol{x}) \, d\boldsymbol{x}. \tag{4.86}$$

The second part of $\mathcal{E}^{ee}$, denoted $\tilde{\mathcal{E}}^{xc}$, is not as straightforward: it comes from the exchange and correlation effects mentioned earlier.[38] We will see how to evaluate this term later. For now we assume that it exists and define it as

$$\tilde{\mathcal{E}}^{\mathrm{xc}}[\rho] \equiv \mathcal{E}^{\mathrm{ee}}[\rho] - \mathcal{E}^{\mathrm{H}}[\rho]. \tag{4.87}$$

The total energy is then

$$\mathcal{E}[\rho] = \mathcal{T}^{\mathrm{el}} + \mathcal{E}^{\mathrm{ZZ}} + \mathcal{E}^{\mathrm{ext}} + \mathcal{E}^{\mathrm{H}} + \tilde{\mathcal{E}}^{\mathrm{xc}}. \tag{4.88}$$

We need to find the electron density that minimizes this total energy, subject to the constraint that the total number of electrons is conserved (i.e. that Eqn. (4.75) is enforced). The necessary condition, expressed as functional derivative with respect to ρ, is

$$\frac{\delta}{\delta\rho(\boldsymbol{x})}\left[\mathcal{E}[\rho] - \mu\left(\int \rho(\boldsymbol{x})\,d\boldsymbol{x} - N^{\mathrm{el}}\right)\right] = 0,$$

where μ is a Lagrange multiplier used to impose constraint Eqn. (4.74). By Eqn. (4.83) and the definition of the functional derivative (see footnote 27 on page 181) this becomes

$$\frac{\delta\mathcal{T}^{\mathrm{el}}}{\delta\rho} + \frac{\delta\mathcal{E}^{\mathrm{ZZ}}}{\delta\rho} + \frac{\delta\mathcal{E}^{\mathrm{ext}}}{\delta\rho} + \frac{\delta\mathcal{E}^{\mathrm{H}}}{\delta\rho} + \frac{\delta\tilde{\mathcal{E}}^{\mathrm{xc}}}{\delta\rho} = \mu. \tag{4.89}$$

Equation (4.89) determines ρ for the ground state, but there are still two outstanding issues: we do not know how to find either $\mathcal{T}^{\mathrm{el}}[\rho]$ or $\tilde{\mathcal{E}}^{\mathrm{xc}}[\rho]$. This is addressed next.

The effective potential Kohn and Sham [KS65] argued that the kinetic energy can be divided into a part arising from treating the electrons as though they were *noninteracting* and a correction to reflect their interactions. The first part (which we will call $\mathcal{T}^{\mathrm{s}}$) turns out to be the main contribution to the kinetic energy. Unfortunately, it also turns out to be difficult to write down as a functional of ρ. Luckily, we will see that this does not matter, since we will be able to find our solution without ever explicitly computing $\mathcal{T}^{\mathrm{s}}$. The remaining kinetic energy, arising from the fact that the electrons do, in fact, interact, is then lumped into the exchange-correlation term, which we must therefore redefine, as follows. Adding and subtracting $\mathcal{T}^{\mathrm{s}}$ from Eqn. (4.88) and rearranging the terms we have

$$\mathcal{E} = \mathcal{T}^{\mathrm{s}} + \mathcal{E}^{\mathrm{ZZ}} + \mathcal{E}^{\mathrm{ext}} + \mathcal{E}^{\mathrm{H}} + \tilde{\mathcal{E}}^{\mathrm{xc}} + \left(\mathcal{T}^{\mathrm{el}} - \mathcal{T}^{\mathrm{s}}\right),$$

and we now define a new "exchange-correlation" energy as[39]

$$\mathcal{E}^{\mathrm{xc}} = \tilde{\mathcal{E}}^{\mathrm{xc}} + \left(\mathcal{T}^{\mathrm{el}} - \mathcal{T}^{\mathrm{s}}\right), \tag{4.90}$$

such that

$$\mathcal{E} = \mathcal{T}^{\mathrm{s}} + \mathcal{E}^{\mathrm{ZZ}} + \mathcal{E}^{\mathrm{ext}} + \mathcal{E}^{\mathrm{H}} + \mathcal{E}^{\mathrm{xc}}, \tag{4.91}$$

[38] One can think of $\mathcal{E}^{\mathrm{H}}$ as the contribution to the electron energy from the purely classical Coulomb interactions, and $\tilde{\mathcal{E}}^{\mathrm{xc}}$ as a quantum mechanical correction.

[39] Although $\mathcal{E}^{\mathrm{xc}}$ is traditionally referred to as the exchange-correlation energy, this is no longer strictly correct since it includes the kinetic energy due to electron interactions. Instead, it is more of a convenient "catch-all" for all the parts we do not know how to deal with yet. Later, we will see how approximations are made for its calculation. Efforts to improve the accuracy of the exchange-correlation term remain an active area of DFT research.

and Eqn. (4.89) becomes

$$\frac{\delta \mathcal{T}^{\mathrm{s}}}{\delta \rho}+\frac{\delta \mathcal{E}^{\mathrm{ZZ}}}{\delta \rho}+\frac{\delta \mathcal{E}^{\mathrm{ext}}}{\delta \rho}+\frac{\delta \mathcal{E}^{\mathrm{H}}}{\delta \rho}+\frac{\delta \mathcal{E}^{\mathrm{xc}}}{\delta \rho}=\mu. \tag{4.92}$$

Carrying out the differentiation and noting that $\mathcal{E}^{\mathrm{ZZ}}$ does not depend on ρ, we get

$$\frac{\delta \mathcal{T}^{\mathrm{s}}[\rho]}{\delta \rho}+\mathcal{U}^{\mathrm{ext}}(\boldsymbol{x})+\mathcal{U}^{\mathrm{H}}(\boldsymbol{x};\rho)+\mathcal{U}^{\mathrm{xc}}(\boldsymbol{x};\rho)=\mu. \tag{4.93}$$

We have indicated the dependence on ρ in Eqn. (4.93) to emphasize this fact; later we will see that this leads to the need for an iterative solution procedure. Note that we have assumed the existence of an exchange-correlation potential, $\mathcal{U}^{\mathrm{xc}}$, defined as

$$\mathcal{U}^{\mathrm{xc}}=\frac{\delta \mathcal{E}^{\mathrm{xc}}}{\delta \rho}. \tag{4.94}$$

At this stage we do not know how to evaluate this function.

So far, all this may seem to be a mere rearrangement of terms and a barrage of definitions. All we have done, remember, is to divide the total energy into several terms, each as either a known or an assumed function of the electron density. Then, we differentiated these terms on our way to finding the energy-minimizing density. The result of these differentiations is Eqn. (4.93), which contains two terms that we know how to compute for a given electron density ($\mathcal{U}^{\mathrm{H}}$ and $\mathcal{U}^{\mathrm{ext}}$) and two which we do not ($\mathcal{U}^{\mathrm{xc}}$ and $\delta\mathcal{T}^{\mathrm{s}}/\delta\rho$). The last of these four terms was defined in such a way as to enable the following important step. Note that Eqn. (4.93) can be rewritten as

$$\frac{\delta \mathcal{T}^{\mathrm{s}}[\rho]}{\delta \rho}+\mathcal{U}^{\mathrm{eff}}(\boldsymbol{x};\rho)=\mu, \tag{4.95}$$

where we have collected all the potentials into one *effective potential* as

$$\mathcal{U}^{\mathrm{eff}}(\boldsymbol{x};\rho)=\mathcal{U}^{\mathrm{H}}(\boldsymbol{x};\rho)+\mathcal{U}^{\mathrm{xc}}(\boldsymbol{x};\rho)+\mathcal{U}^{\mathrm{ext}}(\boldsymbol{x}). \tag{4.96}$$

Now recall that $\mathcal{T}^{\mathrm{s}}$ is, by definition, the kinetic energy of a noninteracting system of electrons: we shall see next that this is exactly the same equation as one would solve for a *single* electron in the new effective potential!

Step 2 Replacing the multiparticle problem with an equivalent single-particle system We will now show that a single-electron system, subject to the same potential $\mathcal{U}^{\mathrm{eff}}$ as we have just defined, leads to an equation for the electron density that is identical to Eqn. (4.95). Since the *DFT theorem* introduced on page 190 tells us that there is only one unique density solution for a given potential, this allows us to solve the much simpler single-electron problem instead.

Consider a *single electron* with the Hamiltonian

$$\mathcal{H}^{\text{sp}} = \mathcal{U}^{\text{eff}}(\boldsymbol{x};\rho) + \frac{p^2}{2m^{\text{el}}}. \tag{4.97}$$

We denote the single-particle wave functions as ψ^{sp} in order to distinguish them from the wave functions for the real system (which we called ψ). Analogous to Eqn. (4.76) (but considerably simpler), the energy of the single-particle system is

$$\langle\psi^{\text{sp}}|\,\mathcal{H}^{\text{sp}}\,|\psi^{\text{sp}}\rangle = \mathcal{T}^{\text{s}} + \mathcal{E}^{\text{eff}}, \tag{4.98}$$

where

$$\mathcal{T}^{\text{s}} = \left\langle \psi^{\text{sp}} \middle| -\frac{\hbar^2}{2m^{\text{el}}}\nabla^2 \middle| \psi^{\text{sp}} \right\rangle \tag{4.99}$$

is the same $\mathcal{T}^{\text{s}}$ introduced earlier by construction, and the effective potential plays the role of the external potential so that

$$\mathcal{E}^{\text{eff}}[\rho] = \langle\psi^{\text{sp}}|\,\mathcal{U}^{\text{eff}}\,|\psi^{\text{sp}}\rangle = \int \rho(\boldsymbol{x})\mathcal{U}^{\text{eff}}(\boldsymbol{x};\rho)\,d\boldsymbol{x}, \tag{4.100}$$

in analogy to Eqn. (4.83). Even though Eqn. (4.98) is not the same energy as that of Eqn. (4.76) for the real system, we can still choose to minimize $\langle\psi^{\text{sp}}|\,\mathcal{H}^{\text{sp}}\,|\psi^{\text{sp}}\rangle$, subject to the constraint that $N^{\text{el}} = 1$ in Eqn. (4.75), by taking a functional derivative and solving

$$\frac{\delta\mathcal{T}^{\text{s}}[\rho]}{\delta\rho} + \mathcal{U}^{\text{eff}}(\boldsymbol{x};\rho) = \mu. \tag{4.101}$$

This is exactly the same as Eqn. (4.95) for our real system, which brings us to the key point. *The electron density obtained from the solution for our new system of noninteracting particles governed by the Hamiltonian of Eqn. (4.97) is the same as that for the original system of interacting particles because it satisfies the same governing equation in Eqn. (4.101). Furthermore, we know from the DFT theorem that this density is a unique solution. We can therefore replace the original multiparticle problem with the new single-particle one which is, relatively speaking, much easier to deal with.*

Step 3 Solving the single-particle problem We now proceed to find single-particle solutions exactly as we did in Section 4.4.6, by minimizing the function

$$\Pi = \langle\psi^{\text{sp}}|\,\mathcal{H}^{\text{sp}}\,|\psi^{\text{sp}}\rangle - \epsilon\left(\langle\psi^{\text{sp}}|\psi^{\text{sp}}\rangle - 1\right), \tag{4.102}$$

for an approximate wave function obtained from a linear combination of basis vectors

$$\psi_I^{\text{sp}} = \sum_{n=1}^{N_{\text{basis}}} c_{In}\varphi_n, \qquad I = 1,\ldots,N_{\text{basis}}. \tag{4.103}$$

For N_{basis} members of the basis set, there will be N_{basis} eigensolutions,[40] and using the definition in Eqn. (4.74), the density field resulting from each can be found from

$$\rho_I(\boldsymbol{x}) = \sum_{n=1}^{N_{\text{basis}}} \|c_{In}\|^2 \|\varphi_n(\boldsymbol{x})\|^2. \tag{4.104}$$

Note that since each solution is obtained for a *single-particle* system, the density terms must satisfy Eqn. (4.75) with $N^{\text{el}} = 1$. Since there are generally $N^{\text{el}} > 1$ electrons in our original system we need to superimpose N^{el} single-particle solutions, but this is straightforward since they are noninteracting by construction. Given the fact that only two electrons can occupy each state, we systematically fill the ψ_I^{sp} states, from the lowest energy up, until the number of filled states equals the number of electrons in the system. The total density is then

$$\rho(\boldsymbol{x}) = \sum_{I \in \text{filled}} \rho_I(\boldsymbol{x}) = \sum_{I \in \text{filled}} \sum_{n=1}^{N_{\text{basis}}} \|c_{In}\|^2 \|\varphi_n(\boldsymbol{x})\|^2, \tag{4.105}$$

for which $\int \rho(\boldsymbol{x})\, d\boldsymbol{x} = N^{\text{el}}$. This density can now be used to evaluate an approximation for the energy of the original *multiparticle* system, since we know that the density is a unique solution independent of the basis functions.

Should we need it, the kinetic energy term, T^{s}, can be evaluated from

$$T^{\text{s}} = \sum_{I \in \text{filled}} \left\langle \psi_I^{\text{sp}} \middle| -\frac{\hbar^2 \nabla^2}{2m^{\text{el}}} \middle| \psi_I^{\text{sp}} \right\rangle, \tag{4.106}$$

so long as the basis wave functions used to approximate ψ_I^{sp} in Eqn. (4.103) are readily amenable to the integration required in Eqn. (4.106).

The total energy of the multiparticle system The single-particle system was merely a tool: a fictitious set of wave functions that nevertheless gives us the correct electron density for our more complex system through the DFT theorem and Eqn. (4.105). Having the correct density as a function of space is a useful quantity in its own right, but an even more useful quantity is the total energy in Eqn. (4.91) that we are now nearly in a position to evaluate. Starting from Eqn. (4.91) we insert expressions for the quantities we know:

$$\begin{aligned}\mathcal{E} &= T^{\text{s}} + \mathcal{E}^{\text{ZZ}} + \mathcal{E}^{\text{ext}} + \mathcal{E}^{\text{H}} + \mathcal{E}^{\text{xc}},\\ &= \sum_{I \in \text{filled}} \left\langle \psi_I^{\text{sp}} \middle| -\frac{\hbar^2 \nabla^2}{2m^{\text{el}}} \middle| \psi_I^{\text{sp}} \right\rangle + \mathcal{E}^{\text{ZZ}} + \int \rho(\boldsymbol{x}) \mathcal{U}^{\text{ext}}(\boldsymbol{x};\rho)\, d\boldsymbol{x}\\ &\quad + \frac{1}{2}\int \rho(\boldsymbol{x}) \mathcal{U}^{\text{H}}(\boldsymbol{x};\rho)\, d\boldsymbol{x} + \mathcal{E}^{\text{xc}},\end{aligned} \tag{4.107}$$

40 Recall that the system of equations resulting from the minimization of Eqn. (4.102) will be an eigenvalue problem for an $N_{\text{basis}} \times N_{\text{basis}}$ matrix. There will be N_{basis} eigenvector solutions for this equation, each containing values for the c_{In}, $n = 1, \ldots, N_{\text{basis}}$ to form the Ith eigenfunction solution, ψ_I^{sp}, from Eqn. (4.103). Corresponding to each of these is the eigenvalue (or energy), ϵ_I.

where we have used Eqn. (4.106) for $\mathcal{T}^{\mathrm{s}}$, Eqn. (4.83) for $\mathcal{E}^{\mathrm{ext}}$ and Eqn. (4.86) for $\mathcal{E}^{\mathrm{H}}$. Next, we add and subtract the quantity $\mathcal{E}^{\mathrm{eff}}$ in its two different forms from Eqn. (4.100):

$$\mathcal{E} = \sum_{I \in \text{filled}} \left\langle \psi_I^{\mathrm{sp}} \left| -\frac{\hbar^2 \nabla^2}{2m^{\mathrm{el}}} + \mathcal{U}^{\mathrm{eff}} \right| \psi_I^{\mathrm{sp}} \right\rangle - \int \rho(\boldsymbol{x}) \mathcal{U}^{\mathrm{eff}}(\boldsymbol{x}; \rho)\, d\boldsymbol{x}$$
$$+ \mathcal{E}^{\mathrm{ZZ}} + \int \rho(\boldsymbol{x}) \mathcal{U}^{\mathrm{ext}}(\boldsymbol{x})\, d\boldsymbol{x} + \frac{1}{2} \int \rho(\boldsymbol{x}) \mathcal{U}^{\mathrm{H}}(\boldsymbol{x}; \rho)\, d\boldsymbol{x} + \mathcal{E}^{\mathrm{xc}}.$$

Here, we recognize the first term as simply the sum of the energies of the filled *single-particle* states, something we know directly from solving the single-particle eigenvalue problem. Inserting Eqn. (4.96) for $\mathcal{U}^{\mathrm{eff}}$ and using Eqn. (4.86) allows us to simplify this equation to a final form that will prove useful as we move forward with implementation:

$$\mathcal{E} = \sum_{I \in \text{filled}} \epsilon_I - \mathcal{E}^{\mathrm{H}} - \int \rho \mathcal{U}^{\mathrm{xc}}(\boldsymbol{x}; \rho)\, d\boldsymbol{x} + \mathcal{E}^{\mathrm{xc}}(\rho) + \mathcal{E}^{\mathrm{ZZ}}. \tag{4.108}$$

Summary of the formulation A brief, reorienting summary is in order. We started out by replacing the task of solving for wave functions with the task of solving for density. Then, we effectively *went back* to solving for a fictitious, but much simpler, set of single-particle wave functions that gave us the correct density.[41] The cost of this exchange has been a much more complex potential, $\mathcal{U}^{\mathrm{eff}}(\boldsymbol{x}; \rho)$, but one that we know how to deal with entirely except for the exchange-correlation part, $\mathcal{U}^{\mathrm{xc}}$, that will be discussed next. Another important cost associated with the introduction of $\mathcal{U}^{\mathrm{eff}}(\boldsymbol{x}; \rho)$ is the need to iterate until a self-consistent density field is reached. Thus, the procedure will be to start from an initial guess for ρ, compute $\mathcal{U}^{\mathrm{eff}}$, minimize Eqn. (4.102) for a new ρ and repeat until convergence. Although writing efficient and robust code to do these iterations is no trivial matter, the problem, at this stage, has been reduced to one of optimizing the computer implementation and waiting patiently for convergence to be achieved.

It will now be necessary to make some approximations in order to develop a viable computational scheme. These approximations involve the exchange-correlation energy and the form of the external potential (the so-called pseudopotentials) due to the atomic nuclei.

4.5.2 Approximations necessary for computational progress

Most active areas of research in DFT involve efforts to improve the speed and accuracy of the approximations made to compute $\mathcal{U}^{\mathrm{xc}}$ and $\mathcal{U}^{\mathrm{ext}}$. Here, we present the simplest methods, with the goal of completing the picture of a DFT implementation. Once these simple approaches are understood, we hope that you will be comfortably positioned to understand the additional layers of complexity found in the modern DFT literature and techniques.

[41] In practice, we will do this by finding the eigenvalues of a matrix form of Eqn. (4.102), just as we did in Section 4.4.6. The eigenvalues give us a direct way to calculate the energy of our full system, in Eqn. (4.108), in principle without any approximations.

The exchange-correlation energy and the local density approximation (LDA) Until now, we have remained deliberately vague about the details of computing the exchange and correlation energy, which we recall was discussed at the start of Section 4.5.1 and then formally defined in Eqn. (4.90). This is a difficult energy contribution to compute, and developing improved approximations to it is an active area of research. The starting point to computing $\mathcal{E}^{\text{xc}}$ is the so-called *local density approximation* or LDA, a simple approach that gives remarkably good results. The idea is to assume that *gradients* in the electron density do not matter, and therefore if we know the exchange-correlation energy for electrons in a uniform density field, we also know $\mathcal{E}^{\text{xc}}$ pointwise in a varying density field.[42] Thus, we postulate that

$$\mathcal{E}^{\text{xc}}[\rho] = \int \rho(\boldsymbol{x})\epsilon^{\text{xc}}[\rho(\boldsymbol{x})]\, d\boldsymbol{x}, \tag{4.109}$$

where $\epsilon^{\text{xc}}(\rho)$ is the exchange-correlation energy per electron in a uniform density field. From Eqn. (4.94) we therefore have

$$\mathcal{U}^{\text{xc}}(\boldsymbol{x}) = \rho(\boldsymbol{x})\frac{\delta\epsilon^{\text{xc}}(\rho[\boldsymbol{x}])}{\delta\rho} + \epsilon^{\text{xc}}(\rho[\boldsymbol{x}]). \tag{4.110}$$

Even to compute ϵ^{xc} is no simple matter. The typical approach used in DFT is to curve-fit data obtained via high-quality quantum Monte Carlo calculations [CA80, PZ81, PW92]. That the LDA works so well is a surprise, given the fact that gradients in the electron density are often quite severe. This is mitigated (at least in part) by the use of pseudopotentials to eliminate the details of the core electron wave functions, as we will discuss next. Further discussion of why the LDA works as well as it does can be found in [JG89].

More advanced formulations of DFT incorporate gradients of density into the formulation. These methods fall under the umbrella of "generalized gradient approximations" (GGAs), and there are several of them (see, for example, chapter 8 of [FGU96]). Many of the GGA formulations show clear improvements over LDA for specific systems, but there is currently no universally applicable GGA approach.

Pseudopotentials The composition of engineering materials is rarely limited to light elements with few electrons. Consider, for example, Al with 13 electrons per atom or Cu with 29. Even though we can, in principle, treat all of these electrons using DFT, it is difficult to get good accuracy in this way. Fortunately, the main reason for this difficulty also offers an easy way out. Most electrons in materials are tightly bound *core electrons* that remain close to the nucleus and do not participate in any significant way in the bonding process. The electron density in the core experiences rapid fluctuations between regions of very high and very low density, and the LDA thus becomes quite unreliable there. These rapid fluctuations also mean that we must choose a very large basis set of wave functions in Eqn. (4.103) to realistically capture the density field. Luckily, the fact that core electrons do not contribute to the bonding means that, for our purposes, we can neglect them and only solve for the electronic structure of the *valence electrons* as they interact to form bonds. Away from the

[42] This is analogous to the *principle of local action* for constitutive relations in continuum mechanics. See Section 2.5.1.

core, where valence electrons exist, the variations in the electron density become much less severe and a reduced basis set of wave functions can be used.

We cannot completely neglect the core electrons, of course. As a bare minimum, we must acknowledge that they interact electrostatically with the valence electrons, effectively screening some of the Coulomb potential of the atomic nucleus. To be more accurate, we must also incorporate some model of other interactions (like exchange and correlation effects) between the valence and core electrons. Incorporating these effects means replacing the naked Coulomb potential in $\mathcal{U}^{\text{ext}}$ with what is known as a *pseudopotential*.[43] In effect, this means that we are solving a modified Schrödinger equation for *only* the valence electrons, where the external potential has changed due to the presence of the core electrons. This modification is not known exactly, and is instead approximated by a pseudopotential.

There are many different approaches to pseudopotential development, and the literature is filled with many good pseudopotentials for virtually every element in the periodic table. This is evident from an examination of the features of various commercially (or otherwise) available codes (see footnote 48 on page 210). Perhaps the simplest pseudopotential to understand is the *empty-core pseudopotential* of Ashcroft [Ash66]. The approach is to define a core radius, r_{core}, within which the potential is taken to be identically zero. Outside of the core, the potential is simply the Coulomb attraction between an electron and a point charge of $+ez$, where z is the valence of the atom (i.e. the difference between the number of protons in the nucleus and the number of electrons in the core). Thus, we have

$$\mathcal{U}^{\text{empty}}(r) = \begin{cases} 0 & \text{for } r < r_{\text{core}}, \\ -ze^2/r & \text{for } r \geq r_{\text{core}}, \end{cases} \tag{4.111}$$

where r is the distance between the electron and the center of the nucleus. The parameter r_{core} is chosen so that the lowest-energy s-state in the presence of the isolated pseudopotential has the experimentally known energy of the valence s-state for the element. Now Al, for example, can be modeled as having only three electrons in the presence of the Al pseudopotential, and Cu only one. Not surprisingly, pseudopotentials are usually significantly weaker than the Coulomb potential,[44] which helps to explain, for instance, why valence electrons in metals behave essentially as free electrons. It also sheds light on why the LDA works reasonably well for the $\mathcal{E}^{\text{xc}}$ contribution to the Hamiltonian in DFT calculations; a weak potential means relatively slow variations in $\rho(\boldsymbol{x})$.

[43] At the same time, the Coulomb interaction between the nuclei is usually replaced with a screened Coulomb interaction between nuclei that are screened by the core electrons. As a result, $\mathcal{E}^{\text{ZZ}}$ of Eqn. (4.78) is replaced by a simple, empirically determined pair potential of the form

$$\mathcal{E}^{\text{ZZ}} = \frac{1}{2} \sum_{\substack{\alpha,\beta \\ \alpha \neq \beta}}^{N} \phi^{\alpha\beta}(r^{\alpha\beta}).$$

For large $r^{\alpha\beta}$ this recovers the $1/r$ Coulomb interaction, while at short distances it mimics the effect of core electrons overlapping. See Section 5.4.2 for a discussion of pair potentials.

[44] Consider, for instance, the empty-core pseudopotential. Outside the core, it is equivalent to the Coulomb potential. However, for any solution there will always be electron density inside the core region, which will contribute less to the external energy of Eqn. (4.83) than if the Coulomb potential were used there.

A more realistic pseudopotential is the so-called "evanescent core" pseudopotential [FPA+95]. It is suitable for the description of simple metals and takes the analytical form

$$\mathcal{U}^{\mathrm{ec}}(r, Z) = -\frac{e^2 z_Z}{B_Z}\left\{\frac{1}{y}[1-(1+b_Z y)\exp(-a_Z y)] - A_Z \exp(-y)\right\}, \tag{4.112}$$

where $y = r/B_Z$ and the constants A_Z and b_Z are given in terms of parameter a_Z. The valence of an ion with atomic number Z is denoted by z_Z. Two parameters, B_Z and a_Z, completely determine the potential for atomic number Z, and can be found by fitting to known properties of simple metals. The main advantage of this potential over the empty-core potential is the smooth functional form, which permits a simple closed-form expression for the Fourier transform. The utility of Fourier transforms in DFT will become apparent in Section 4.5.5.

In more accurate implementations of DFT, the pseudopotentials are typically of nonlocal form. That is to say, the potential at $\boldsymbol{x}$, $\mathcal{U}^{\mathrm{ext}}(\boldsymbol{x})$, depends on the electron density everywhere. This requires an integral over all space, which we can write as

$$\mathcal{U}^{\mathrm{ext}}(\boldsymbol{x}) = \int v(\boldsymbol{x}, \boldsymbol{x}')\rho(\boldsymbol{x}')\, d\boldsymbol{x}'.$$

We will briefly revisit the consequence of this in Section 4.5.5. For more details, the interested reader is encouraged to explore references such as [FPA+95, MH00].

4.5.3 The choice of basis functions

The starting point of a DFT calculation is a convenient approximation to a single-electron wave function, exactly as in Section 4.4.6. But whereas in Section 4.4.6 we were mostly interested in something simple and analytically tractable, now we are interested in something that is more accurate, easily implemented on a computer and easily expandable (at only the expense of more CPU time and memory).

The DFT community mostly uses one of two basis sets:[45] "atomic orbitals" or plane waves. Atomic orbitals are the hydrogen atom electronic orbitals we found in Eqn. (4.52), or some sensible approximation of them. The wave function is then approximated with a generalization of Eqn. (4.103):

$$\psi^{\mathrm{sp}}(\boldsymbol{x}) \approx \sum_{\alpha=1}^{N}\sum_{i=1}^{N_{\mathrm{orb}}} c_i^{\alpha}\varphi_i^{\alpha}(\boldsymbol{x} - \boldsymbol{r}^{\alpha}),$$

where the first sum is over the N atomic nuclei and i represents a unique combination of the quantum numbers n, l and m defined in Section 4.4.5. The wave functions φ_i^{α} making up the basis are centered on the atoms, and may differ from atom to atom to reflect different atomic numbers. The coefficients c_i^{α} appropriately weight the contribution of each basis function. This is the so-called "LCAO" that we introduced to solve the hydrogen molecule in Section 4.4.6.

[45] But not always. There are many other ways to develop a wave function basis, as summarized by Marx and Hutter in [MH00], including the so-called real-space grid and "orbital-free" methods. A good introduction to the real-space approach is its "translation" into the language of finite elements as done by Phillips [Phi01].

LCAO is especially suited to the study of molecules. Its main advantage is that the basis functions are relatively short-range (bound) states that are likely to be a good approximation to the real wave function of the molecule with a relatively small number of basis functions. The main disadvantage is that these basis functions are nonorthogonal. As we saw in Section 4.4.6 this nonorthogonality complicates the calculations, but also introduces a practical inconvenience in that there is no longer a systematic way to increase the basis set that is *guaranteed* to improve accuracy. Without this, it is more difficult to verify that the solution is adequately converged.

The other common basis choice, and the one we focus on here, is a basis set of plane waves. This is a natural choice for the study of periodic systems of atoms (crystals), and can be adapted for isolated molecules by making the periodicity of the simulation box large enough that the interaction between periodic copies is minimal. Either way, however, we will be modeling an infinite system by studying a single periodic cell within that system, and this has some subtle consequences for the electronic wave functions. To better understand these, we take a brief detour into the subjects of free electrons and Bloch's theorem.

4.5.4 Electrons in periodic systems

Free electrons and Bloch's theorem Bloch's theorem is a statement about the form that an electronic wave function must take when the electron is subject to a periodic potential. In one-dimensional space, suppose we have some potential that satisfies

$$\mathcal{U}^{\text{ext}}(x) = \mathcal{U}^{\text{ext}}(x + na)$$

for some fixed length a and for n an integer. Bloch's theorem states that regardless of the specifics of the potential, the electronic wave function must take the form

$$\psi^{\text{sp}}(x) = \exp(ikx)\, u(x), \tag{4.113}$$

where $u(x)$ is an arbitrary periodic function with the same periodicity as $\mathcal{U}^{\text{ext}}(x)$ and k is the wave vector (i.e. the momentum) of a plane wave, as discussed in previous sections.

The proof of Bloch's theorem is relatively straightforward, as follows (note that it is also easily generalizable to three dimensions).

Proof First, we note that Eqn. (4.113) also implies that

$$\psi^{\text{sp}}(x + a) = \exp(ikx + ika)\, u(x + a) = \exp(ikx + ika)\, u(a) = \exp(ika)\, \psi^{\text{sp}}(x), \tag{4.114}$$

where the second to last equality follows from the periodicity of $u(x)$. Next, we write Schrödinger's equation twice, where the second is merely translated by a:

$$\frac{-\hbar^2}{2m^{\text{el}}} \frac{d^2\psi^{\text{sp}}}{dx^2}(x) = \left(\epsilon - \mathcal{U}^{\text{ext}}(x)\right) \psi^{\text{sp}}(x),$$
$$\frac{-\hbar^2}{2m^{\text{el}}} \frac{d^2\psi^{\text{sp}}}{dx^2}(x + a) = \left(\epsilon - \mathcal{U}^{\text{ext}}(x)\right) \psi^{\text{sp}}(x + a).$$

In the second equation, we have invoked the periodicity of $\mathcal{U}^{\text{ext}}$. We assume that the wave function ψ^{sp} satisfies the first equation. The second equation is only satisfied if

$$\psi^{\text{sp}}(x+a) = C\psi^{\text{sp}}(x),$$

where C is an imaginary constant (independent of x). This is easy to see. Since C is a constant, we have that

$$\frac{d^2\psi^{\text{sp}}}{dx^2}(x+a) = C\frac{d^2\psi^{\text{sp}}}{dx^2}(x).$$

Substituting this into the second Schrödinger equation recovers the first (omitting the trivial solution $C = 0$). If C were dependent on x, the second equation could not be satisfied.

Next, we require that the *electron density* be periodic in a. This is necessary on physical grounds, lest a perfect infinite crystal have different electron densities around two otherwise identical lattice sites. The electron density is given by $\psi^{\text{sp}*}\psi^{\text{sp}}$, so we require $\psi^{\text{sp}*}(x+a)\psi^{\text{sp}}(x+a) = \psi^{\text{sp}*}(x)\psi^{\text{sp}}(x)$, which implies

$$C^*C\psi^{\text{sp}*}(x)\psi^{\text{sp}}(x) = \psi^{\text{sp}*}(x)\psi^{\text{sp}}(x),$$

so that $C^*C = 1$. This can only be satisfied if $C = \exp(iA)$, where A is a real constant, or without loss of generality, $C = \exp(ika)$, where k is another real constant. Thus,

$$\psi^{\text{sp}}(x+a) = \exp(ika)\,\psi^{\text{sp}}(x),$$

and Eqn. (4.114) holds. □

Next, we explore why we need Bloch's theorem in practical calculations done on a periodic simulation box. To do this, we will look at the solution for free electrons. That is, we consider a number of electrons in a box that neither interact with each other nor experience an external potential, the so-called "free-electron gas." Bloch's theorem does not normally come up in the context of the free-electron gas problem, because the potential $\mathcal{U}^{\text{ext}}(x) = 0$. But in fact the potential $\mathcal{U}^{\text{ext}}(x) = 0$ *is* periodic, if only in a trivial sense. In our case, Bloch's theorem will come up not so much because the *potential* is periodic, but because our *solution method* will be periodic.

First, let us consider the free-electron gas in its traditional form, because its solution is exact and straightforward. Imagine a one-dimensional "box" of length L. We call it a "box," but it is of course really a line, and we call it a line even though we will think of it as a *periodic* line. That it to say we think of the end of the box at $x = 0$ being joined to the other end at $x = L$ so that the wave function must satisfy

$$\psi^{\text{sp}}(x) = \psi^{\text{sp}}(x+L). \tag{4.115}$$

Since the potential is zero, the Schrödinger equation is simply

$$-\frac{\hbar^2}{2m^{\text{el}}}\frac{\partial^2\psi^{\text{sp}}}{\partial x^2} = \epsilon\psi^{\text{sp}},$$

for which the solutions are of the form

$$\psi^{\text{sp}}(x) = A\exp(ikx), \tag{4.116}$$

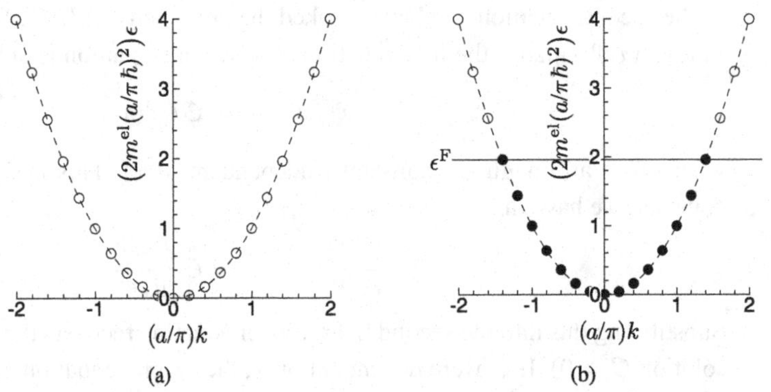

Fig. 4.10 (a) Eigenvalues (ϵ) and eigenvectors (k) of the free-electron gas in a periodic box of length $L = 10a$ shown by the open circles, while the dashed line shows Eqn. (4.118). (b) The filled circles are the filled eigenstates for $\bar{\rho} = 3/a$.

with k and A determined from the boundary conditions and the normalization of the wave function. In order for the wave function to obey the periodicity of Eqn. (4.115), the wave vector k must take the values

$$k = 0, \pm\frac{2\pi}{L}, \pm\frac{4\pi}{L}, \pm\frac{6\pi}{L}, \dots, \tag{4.117}$$

giving us the discrete eigenvectors for the solution (the normalization constant A is not of any concern at the moment, so let us leave it alone). Inserting the solution back into the Schrödinger equation allows us to find the eigenvalues of the energy

$$\epsilon = \frac{\hbar^2}{2m^{\text{el}}}k^2. \tag{4.118}$$

To make things definite, we can choose $L = 10a$, where a is some unit of length that we will make more use of shortly. For this choice of L, the eigenvalues of the energy are plotted versus their wave vectors, k, in Fig. 4.10(a). The dashed line is Eqn. (4.118), and the allowed eigenstates are shown as the open circles.

Now imagine that we populate the box with electrons, and just for the purpose of fixing ideas we will choose to fill the box in such a way that the average electron density throughout the box is $\bar{\rho} = 3/a$. Since $L = 10a$, this requires a total of $N^{\text{el}} = 30$ electrons, which each fall into the lowest eigenstates that are not already occupied. Remembering that electron spin effectively allows two electrons at each k-value, and also that there is a corresponding state at $k = -j\pi/L$ for every $k = j\pi/L$, the electrons fill the states as indicated in Fig. 4.10(b). The highest energy state that is filled is called the *Fermi energy*, ϵ^{F}, which is directly related to the largest wave vector, k_{F}, of any filled eigenstate ($k_{\text{F}} = 14\pi/10a$ for this example).

One thing of interest may be the average energy of an electron in the system, which we could compute by simply summing all energy states up to the Fermi level and then dividing

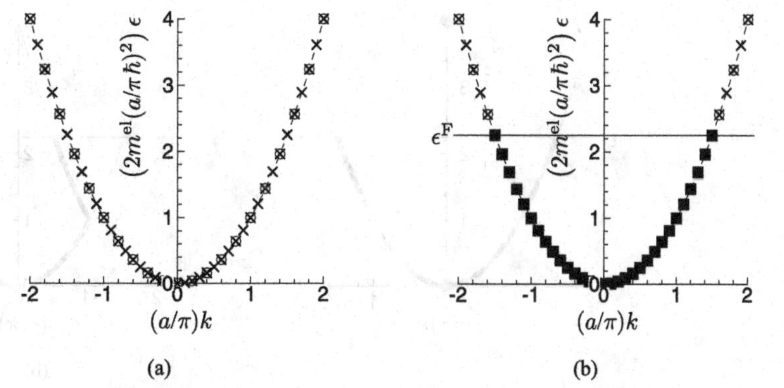

Fig. 4.11 (a) Eigenvalues (ϵ) and eigenvectors (k) of the free-electron gas in a periodic box of length $L = 20a$ shown by the $\times$ symbols, superimposed on the results for $L = 10a$ shown by the open squares. The curve shows Eqn. (4.118). (b) The filled squares are the filled eigenstates for $\bar{\rho} = 3/a$.

by the number of electrons:

$$\bar{\epsilon} = \frac{1}{N^{\rm el}} \frac{\hbar^2}{2m^{\rm el}} \sum_{i=1}^{N^{\rm el}} k_i^2 = \frac{1}{N^{\rm el}} \frac{\hbar^2}{2m^{\rm el}} 2 \left(\frac{2\pi}{L}\right)^2 \left(0^2 + (1)^2 + (-1)^2 + \cdots + (7)^2 + (-7)^2\right)$$

$$= 0.7467 \frac{\hbar^2 \pi^2}{2m^{\rm el} a^2}, \tag{4.119}$$

where the last result is specific to this example with 30 electrons and $L = 10a$.

What if our periodic box was longer, say $L = 20a$? To make the system similar to the previous one, we now need 60 electrons to keep $\bar{\rho} = 3/a$. Looking at the eigenvectors in Eqn. (4.117), we see that they will now be spaced half as far apart on the k-axis as they were previously, as shown by the symbols in Fig. 4.11(a). Filling the 60 lowest-energy states leads to roughly the same Fermi energy as before as shown in Fig. 4.11(b). It is "roughly" the same only because there are two electrons left over that go into the next available state with $k_{\rm F} = 3\pi/2a$. Computing the average energy of the electrons also yields "roughly" the same value (0.7467 becomes 0.7517 in Eqn. (4.119)), since we are sampling twice as many points along the parabola, but these points are half as far apart and ultimately averaged over twice as many electrons.

For larger and larger numbers of electrons, $N^{\rm el}$, and correspondingly larger $L = N^{\rm el}a/3$ to maintain our chosen average electron density, the Fermi wavevector is

$$k_{\rm F} = \frac{3\pi}{2a} \frac{N^{\rm el} - 2}{N^{\rm el}}.$$

Now let us make this an infinite system by letting $L \to \infty$. Clearly, $k_{\rm F}$ goes to

$$k_{\rm F} = \frac{3\pi}{2a} = \frac{\pi \bar{\rho}}{2},$$

and from Eqn. (4.118) this gives us the limiting Fermi energy of

$$\epsilon^{\rm F} = \frac{\hbar^2 \pi^2}{2m^{\rm el}} \left(\frac{\bar{\rho}}{2}\right)^2.$$

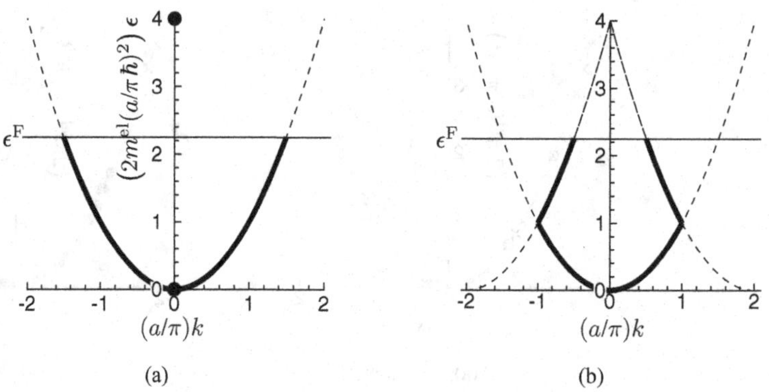

Fig. 4.12 (a) Eigenvalues (ϵ) and eigenvectors (k) of the free-electron gas in a periodic box of length $L = \infty$ form a continuous curve. The heavy line shows the filled states and the two filled circles show the predicted eigenstates for the naïve solution attempt. (b) Band structure and filled states for the correct Bloch solution.

Since the spacing between the k points, $\Delta k = 2\pi/L$, becomes smaller and smaller, the discrete k points eventually merge into a continuous curve as shown in Fig. 4.12(a). At the same time, the average energy per electron goes from a sum to an integral. We can multiply and divide the sum of Eqn. (4.119) by Δk, to allow us to take the infinitesimal limit as

$$\bar{\epsilon} = \lim_{\Delta k \to 0} \frac{L}{2\pi}\frac{1}{N^{\rm el}}\sum_{i=1}^{N^{\rm el}} \frac{\hbar^2}{2m^{\rm el}} k_i^2 \Delta k = \frac{L}{2\pi}\frac{2}{N^{\rm el}}\int_{-3\pi/2a}^{3\pi/2a} \frac{\hbar^2}{2m^{\rm el}} k^2\, dk$$
$$= \frac{1}{\pi\bar{\rho}}\int_{-3\pi/2a}^{3\pi/2a} \frac{\hbar^2}{2m^{\rm el}} k^2\, dk = \frac{3}{4}\frac{\hbar^2\pi^2}{2m^{\rm el}a^2}. \tag{4.120}$$

Note that we have explicitly included the negative and positive values of k in the integration limits and the factor of 2 to account for spin. Our ability to evaluate this integral is dependent on our knowledge of the Fermi level, which determines the integration bounds.

Given this exact solution for the eigenstates and energy of the free-electron gas, we will now try to solve this problem using a finite simulation box and periodic boundary conditions. We will see that we can only get the correct result if we correctly incorporate Bloch's theorem.

A periodic DFT calculation of the free electron gas (done incorrectly) In DFT calculations with plane waves, we are going to want to make use of a periodic simulation box of finite size to simulate an infinite system. Let us examine how this will work on the infinite system we just described: an electron gas of a certain electron density.

We approximate the solution by studying a periodic box of length a and filling this box with $N^{\rm el}$ electrons such that $N^{\rm el}/a = \bar{\rho}$, which in our example means that $N^{\rm el} = 3$. One might think (incorrectly) that since our potential is periodic in a (in the trivial sense, remember, because $\mathcal{U}^{\rm ext} = 0$), the wave function of the electrons should be periodic too.

So let us approximate our wave function as

$$\psi^{\text{sp}} = \frac{1}{\sqrt{a}} c_j \varphi_j \qquad \text{(wrong)}, \tag{4.121}$$

where Einstein's summation convention is applied on repeating indices. The coefficients c_j form a vector of constants to be determined ($j = 1, \ldots, N_{\text{basis}}$), $\sqrt{a}$ is just a convenient normalization, and φ_j is the jth component of a vector of the plane waves in the basis:

$$\boldsymbol{\varphi} = (e^0, e^{i(2\pi/a)x}, e^{-i(2\pi/a)x}, e^{i(4\pi/a)x}, e^{-i(4\pi/a)x}, \ldots). \tag{4.122}$$

Note that each plane wave in the basis is periodic in multiples of a in order to fit the periodic cell. By increasing the number of terms in the sum (and thus the number of elements in the vectors $\mathbf{c}$ and $\boldsymbol{\varphi}$), we can hopefully get a better and better approximation to the exact solution. However, this chosen form for ψ^{sp} *will not work*, because it is not complete. It gives us one solution, for $k = 0$, but the form chosen in Eqn. (4.121) is missing an infinite number of other possible solutions as implied by Bloch's theorem. Nevertheless, it is instructive to continue a little further on this naïve path.

We introduce a set of wave numbers, Γ_j ($j = 1, \ldots, N_{\text{basis}}$) that define the plane waves used in the expansion. We can then write the components of $\boldsymbol{\varphi}$ as

$$\varphi_j = \exp(i\Gamma_j x),$$

where allowable values of Γ_j are those that produce waves commensurate with the periodicity of the simulation cell

$$\Gamma = 0, 2\pi/a, -2\pi/a, 4\pi/a, -4\pi/a, \ldots. \tag{4.123}$$

Following the methodology of Section 4.4.6, we aim to minimize

$$\Pi = \langle \psi^{\text{sp}} | \mathcal{H}^{\text{sp}} | \psi^{\text{sp}} \rangle - \epsilon \left(\langle \psi^{\text{sp}} | \psi^{\text{sp}} \rangle - 1 \right) \tag{4.124}$$

with respect to the constants c_j. For our simple system Π becomes

$$\Pi = \frac{1}{a} \frac{\hbar^2}{2m^{\text{el}}} \int_0^a c_j^* \varphi_j^* \gamma_{kl} c_k \varphi_l \, dx - \epsilon \left(\frac{1}{a} \int_0^a c_j^* \varphi_j^* c_k \varphi_k \, dx - 1 \right), \tag{4.125}$$

where we can integrate only over $[0, a]$ since we expect the contribution to Π to be the same from each periodic copy of the box. The matrix $[\gamma_{jl}]$ comes from the differentiation of the basis functions with respect to x. It is a diagonal matrix, defined as

$$\gamma_{jl} \equiv \begin{cases} 0 & \text{for } j \neq l, \\ (\Gamma_j)^2 & \text{for } j = l. \end{cases}$$

Differentiating Eqn. (4.125) with respect to the complex conjugate $\mathbf{c}^*$ and setting the result equal to zero to find the minimizer yields the eigenvalue equation for $\mathbf{c}$:

$$\frac{\hbar^2}{2m^{\text{el}}} \gamma_{ij} c_{\underline{I}j} = \epsilon_{\underline{I}} c_{\underline{I}i}, \tag{4.126}$$

where we have introduced an additional subscript I (and where the underbar indicates that the summation convention is not enforced) to take care of the fact that there will be N_{basis}

eigensolutions to this equation, each representing a possible electronic state, ψ_I^{sp}. We have also taken advantage of the fact that the plane waves are orthonormal, i.e.

$$\frac{1}{a}\int_0^a \varphi_i^* \varphi_j \, dx = \delta_{ij},$$

as one can readily verify. Since the matrix $[\gamma_{ij}]$ is already diagonal, the eigenvalues and eigenvectors are obtainable by inspection, and the three lowest-energy solutions are

$$\begin{aligned}
\psi_1^{\text{sp}} &= \frac{1}{\sqrt{a}}, & \epsilon_1 &= 0, \\
\psi_2^{\text{sp}} &= \frac{1}{\sqrt{a}} \exp\left(i(2\pi/a)x\right), & \epsilon_2 &= \frac{\hbar^2}{2m^{\text{el}}}\left(\frac{2\pi}{a}\right)^2, \\
\psi_3^{\text{sp}} &= \frac{1}{\sqrt{a}} \exp\left(i(4\pi/a)x\right), & \epsilon_3 &= \frac{\hbar^2}{2m^{\text{el}}}\left(\frac{4\pi}{a}\right)^2.
\end{aligned}$$

Note that there are also the wave functions containing $\exp(-i(2\pi/a)x)$ and $\exp(-i(4\pi/a)x)$ that are degenerate in energy with ψ_2^{sp} and ψ_3^{sp}, respectively.

Since these functions do not depend on k, we plot the first two as filled circles on the line $k = 0$ in Fig. 4.12(a). Clearly, these energy levels are inconsistent with the correct values shown by the heavy line. Our assumption that the electron wave function can be approximated as a function with periodicity a does not work for the free electron gas, and the problem will persist when the electron also experiences a nonzero periodic potential.

A periodic DFT calculation of the free electron gas (done correctly) The problem with the above approach is that in an infinite crystal (even a *periodic* infinite crystal), electrons can assume wave functions with wavelengths longer than the periodicity of the lattice, and these longer wavelengths are of course not represented by a simple periodic function. The correction is the form dictated by Bloch's theorem, which means we must modify Eqn. (4.121) and assume that the electron wave functions are of the form

$$\psi_I^{\text{sp}} = \frac{1}{\sqrt{a}} \exp(ikx)\, c_{Ij}\varphi_j. \tag{4.127}$$

The effect of this change on the equations is straightforward and readily seen by rederiving Eqn. (4.125). The change affects only γ_{jl}, which becomes

$$\gamma_{jl} = \begin{cases} 0 & \text{for } j \neq l, \\ (k + \Gamma_j)^2 & \text{for } j = l, \end{cases} \tag{4.128}$$

and thus the desired k-dependence of the eigenvalues is introduced:

$$\begin{aligned}
\psi_1^{\text{sp}} &= \frac{1}{\sqrt{a}} \exp(ikx), & \epsilon_1 &= \frac{\hbar^2}{2m^{\text{el}}}(k)^2, \\
\psi_2^{\text{sp}} &= \frac{1}{\sqrt{a}} \exp\left(i(k - 2\pi/a)x\right), & \epsilon_2 &= \frac{\hbar^2}{2m^{\text{el}}}\left(k - \frac{2\pi}{a}\right)^2, \\
\psi_3^{\text{sp}} &= \frac{1}{\sqrt{a}} \exp\left(i(k + 2\pi/a)x\right), & \epsilon_3 &= \frac{\hbar^2}{2m^{\text{el}}}\left(k + \frac{2\pi}{a}\right)^2,
\end{aligned}$$

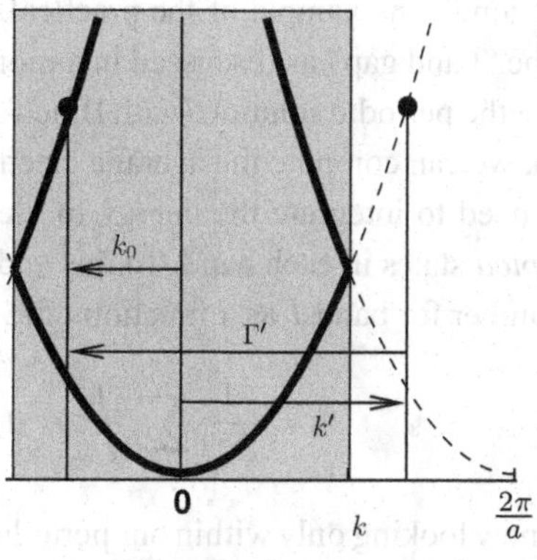

Fig. 4.13 A wave vector k' has its first eigenvalue at the value shown by the filled circle on the right. This value is the same as the second eigenvalue at k_0 which is within the first Brillouin zone (shaded). The values k' and k_0 are related by a reciprocal lattice vector, $k' = k_0 + \Gamma'$, where in this case $\Gamma' = -2\pi/a$.

where we see the splitting of the energy levels associated with $\Gamma_j = \pm 2\pi/a$. Each eigensolution of our model defines a different energy "band" as a function of k. In Fig. 4.12(b), we plot the three lowest energy bands, and fill them only within the *first Brillouin zone* of the reciprocal lattice.[46] In this simple one-dimensional case, the reciprocal lattice is just the series of points $0, \pm 2\pi/a, \pm 4\pi/a, \ldots$, so that the first Brillouin zone extends from $k = -\pi/a$ to $k = \pi/a$. This is useful because, as we shall see, it is possible to work entirely within this range of k in our computations, instead of over all of k-space.

To see why we can confine our attention to the first Brillouin zone, imagine that we have a wave vector k' that is outside of it (Fig. 4.13)). This k' can always be written as a sum $k' = k_0 + \Gamma'$ of a wave vector within the first Brillouin zone, k_0, and one of the reciprocal lattice vectors, $\Gamma' = 0, \pm 2\pi/a, \pm 4\pi/a, \ldots$. Solving for the free electron gas as we did above for this particular k', there will be a set of wave functions and energies

$$\psi_j^{\mathrm{sp}} = \frac{1}{\sqrt{a}} \exp\left(i(k_0 + \Gamma' + \Gamma_j)x\right), \qquad \epsilon_j = \frac{\hbar^2}{2m^{\mathrm{el}}}(k_0 + \Gamma' + \Gamma_j)^2,$$

where Γ_j are the wave vectors in the plane wave expansion used in the solution. However, so long as the wave number $\Gamma'' \equiv \Gamma' + \Gamma_j$ appears in the list of basis functions (Eqn. (4.123)), then the identical solution will also be found by considering the point k_0 inside the first Brillouin zone instead of k' outside of it. The solution will correspond to one of the higher-energy eigensolutions, as shown in Fig. 4.13.

The fact that all the possible solutions can be represented in the first Brillouin zone has practical computational utility as well, since integrals over an infinite k-space can be confined to a well-defined region, and replaced by a sum of integrals over multiple bands.

Returning to our specific example of an electron density of $3/a$, we show the filled states in Fig. 4.12(b) by the heavy black line. Our periodic model is now exactly reconciled with the exact solution (Fig. 4.12(a)), but with the curve folded into two bands in the

[46] The Brillouin zone and reciprocal lattice were defined in Section 3.7.2 and Section 3.7.1, respectively.

first Brillouin zone. An example of the practical implications of this band structure is the existence of the "band gap" as discussed in a moment.

To show that the periodic solution (with Bloch's correction) yields the same results as the exact solution, we can compute the average electron energy to compare with Eqn. (4.120). To do so, we need to integrate the energy of electrons within a single periodic box over only the *occupied* states in each band. To this end we introduce a function $f_I(k)$ that is the occupation number for band I as a function of k. The average electron energy is then

$$\bar{\epsilon} = \frac{1}{N^{\text{el}}} \sum_I \int_{-\pi/a}^{\pi/a} f_I(k) \epsilon_I(k)\, dk, \tag{4.129}$$

where we are now looking only within our periodic cell from 0 to a, and averaging over only $N^{\text{el}} = 3$ electrons. At zero temperature, states are either filled or not, so the occupation number is a simple step function for each band,[47] appropriately scaled:

$$f_I = \begin{cases} f^0 & \text{for } k \text{ filled,} \\ 0 & \text{for } k \text{ unfilled.} \end{cases} \tag{4.130}$$

The value of f^0 is set so that a filled band, integrated over the entire Brillouin zone, yields the correct two electrons

$$\int_{-\pi/a}^{\pi/a} f dk = f^0 \int_{-\pi/a}^{\pi/a} dk = 2,$$

and so for the one-dimensional example, $f^0 = a/\pi$. More generally, $f^0 = 2/\widehat{\Omega}$, where $\widehat{\Omega}$ is the volume of the first Brillouin zone in k-space. We can now see that in a more general problem, the determination of the filled and unfilled states is made by the requirement that the lowest-energy states fill first, and that the total number of filled states equals the number of electrons in the periodic computational cell:

$$\sum_I \int_{\mathcal{B}} f_I(k)\, dk = N^{\text{el}},$$

where the notation $\mathcal{B}$ indicates that the integration is over the first Brillouin zone.

Returning now to Eqn. (4.129), we can evaluate the average energy for our specific example. Looking at Fig. 4.12(b), we see that there are three filled or partly filled bands

$$\bar{\epsilon} = \frac{1}{N^{\text{el}}} \frac{a}{\pi} \frac{\hbar^2}{2m^{\text{el}}} \left(\int_{-\pi/a}^{\pi/a} k^2\, dk + \int_{\pi/2a}^{\pi/a} \left(k - \frac{2\pi}{a}\right)^2 dk + \int_{-\pi/a}^{-\pi/2a} \left(k - \frac{2\pi}{a}\right)^2 dk \right).$$

It is straightforward to evaluate this expression and verify that it is the same as Eqn. (4.120). In this case, the exact solution can be obtained because the free electron gas is completely uniform, and plane waves are exactly the same as the correct electronic states found in Eqn. (4.116). For any nontrivial external potential, this approach is approximate.

[47] At finite temperature, some states above the zero-temperature Fermi level are filled as electrons fluctuate thermally (at the expense of emptying some lower-energy states to conserve the number of electrons), but the available electronic states are the same. See [AM76, Kit96] for more details.

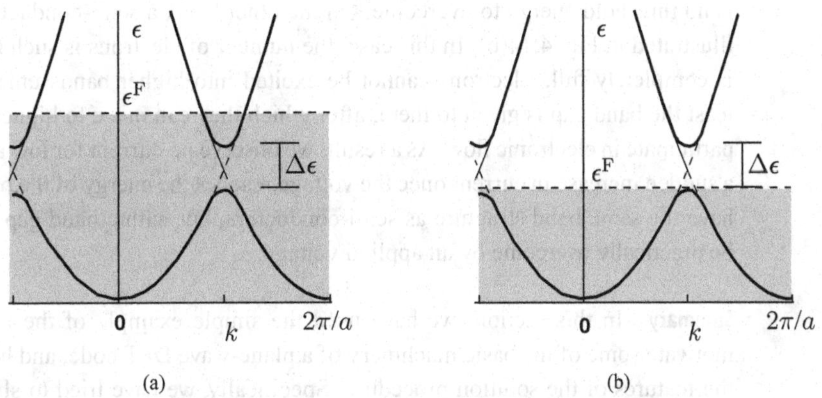

Fig. 4.14 The effect of a weak external potential on a free electron gas is to introduce gaps, $\Delta\epsilon$, defining energy levels that cannot be obtained by the electrons. In (a), the gap is of little consequence in a metal with the Fermi level above the gap. In (b), a semi-conductor has the Fermi level at the bottom of the gap.

Note that the Bloch form, although not periodic itself, still leads to a periodic electron density, since

$$\begin{aligned}\rho(x) &\equiv \sum_I \int_{\mathcal{B}} f_I(k)\psi_I^{\text{sp}*}\,\psi_I^{\text{sp}}\,dk \\ &= \sum_I \int_{\mathcal{B}} f_I(k)\exp(-ikx)\,c_{Ij}^*(k)\varphi_j(x)\exp(ikx)\,c_{Il}(k)\varphi_l(x)\,dk,\end{aligned}$$

and the part that is not periodic in a (meaning $\exp(ikx - ikx)$) clearly cancels out. This is physically sensible, since external observables are properties of the electron wave functions through the electron density. It would be inconsistent with observation if, in a periodic system, the electron density were not also periodic.

All of what we discussed here is of course readily generalizable to higher dimensions. In three dimensions, the reciprocal lattice and the first Brillouin zone may take the shape of a simple cubic or rectangular box if the real-space periodic cell is rectangular, or a more complex shape for nonorthogonal periodic arrangements (see Section 3.7.1).

Band structure: metals versus semi-conductors Band structure arises naturally from the energetic ranking of the eigenvector solutions at each k point, but the bands are not simply a computational artifact: in the limit of an infinite number of basis vectors the solution is exact and reproduces an infinite number of possible energy bands. In the ground state, only the lowest bands will be filled, but the higher-energy bands tell us about the behavior of the electrons in their excited states. Figure 4.14 schematically illustrates the effect of a weak periodic potential on the shape of the free electron band structure. There will be energy ranges in which no electrons can exist, the so-called band gaps of the crystal. A metal is illustrated in Fig. 4.14(a), where the number of electrons is such that the Fermi level is somewhere within a continuous band. Applying a voltage to such a crystal excites electrons into a higher energy level to produce a current regardless of the level of the voltage, i.e. there is no threshold energy to overcome. On the other hand, a semi-conductor is schematically illustrated in Fig. 4.14(b). In this case, the number of electrons is such that the lower band is completely full. Electrons cannot be excited into higher bands unless an energy of at least the band gap is given to them, after which they can move to higher energy levels and participate in electronic flow. As a result, we observe no current for low applied voltage and a sudden increase in current once the voltage reaches the energy of the band gap. Insulators have the same band structure as semi-conductors, but with a band gap that is too large to be practically overcome by an applied voltage.

Summary In this section, we have used the simple example of the free electron gas to motivate some of the basic machinery of a plane-wave DFT code, and highlighted some of the features of the solution procedure. Specifically, we have tried to show why the Bloch form for the approximate wave function is essential, even though we have not proven the Bloch theorem rigorously. In addition, we have discussed how an infinite system can be modeled as a periodic system, and how this results in a band structure of allowable electron states. This has practical implications for plane-wave DFT calculations: since the system we are modeling is infinite we need to integrate over an infinite number of states in the first Brillouin zone to get the correct energy (or other property) of the system. Also, the N_{basis} eigensolutions at a given point in $\boldsymbol{k}$-space correspond to the N_{basis} lowest-energy bands. Now, we proceed with our goal of outlining the implementation of a simple DFT code.

4.5.5 The essential machinery of a plane-wave DFT code

There are many powerful DFT codes available[48] and it is unlikely that anyone reading this book is interested in sitting down and writing their own DFT implementation. Nonetheless, the implementational details are illuminating in their own right. First, they permit a more complete understanding of the features contained in a commercial DFT code and their related strengths and weaknesses. Second, understanding the details of a basic computer implementation is often a good way to improve one's understanding of the theory. In a less practical but no less important way, the implementational details illuminate the cleverness and beauty of an elegant meshing between theory and computations. Here we take a step-by-step walk through the typical elements of a plane-wave DFT code.

Defining the plane-wave basis The starting point of a plane-wave DFT code is the plane-wave expansion of the wave function. The three-dimensional extension of Eqn. (4.127) is

$$\psi_I^{\text{sp}} = \frac{1}{\sqrt{\Omega}} \exp\left(i\boldsymbol{k}\cdot\boldsymbol{x}\right) c_{Ij}(\boldsymbol{k})\varphi_j, \tag{4.131}$$

where Ω is the volume of the periodic simulation cell (a region of space we will denote as $\mathcal{C}$) and the c_{Ij} components can generally have real and imaginary parts. Components of the vector $\boldsymbol{\varphi}$ are of the form

$$\varphi_j = \exp\left(i\boldsymbol{\Gamma}_j\cdot\boldsymbol{x}\right) \tag{4.132}$$

[48] Available, research-grade DFT codes include Abinit [ABI09], CASTEP [CAS10], Gaussian [Gau10], NWChem [PNN10], SeqQuest [Seq09], SIESTA [SIE10], and VASP [VAS09].

and $\boldsymbol{\Gamma}_j$ is a reciprocal lattice vector. We emphasize the $\boldsymbol{k}$-dependence of the coefficients c_{Ij}, since we will be computing these eigenvectors at several points in $\boldsymbol{k}$-space. The vector $\boldsymbol{c}$ is of finite dimension, with N_{basis} components corresponding to the N_{basis} basis vectors in φ. We further emphasize the fact that there are different values of c_{Ij} for each band I, corresponding to the N_{basis} eigensolutions that are obtained at each $\boldsymbol{k}$ point.

To keep things as simple as possible, let us imagine that our periodic simulation cell C is a simple cubic box with sides of length a (and thus a volume of a^3). In this case the jth *reciprocal lattice vector* is defined by taking any three integers[49] (l_j, m_j, n_j), such that

$$\boldsymbol{\Gamma}_j = \frac{2\pi}{a}\left[l_j, m_j, n_j\right]. \tag{4.133}$$

Note that this forms a simple cubic reciprocal lattice. For a concise notation, we denote each vector as $\boldsymbol{\Gamma}_j$ and assume a unique mapping between each integer j and a triplet of integers (l_j, m_j, n_j). In what follows, we will select N_{basis} basis functions, numbered from $j = 0, \ldots, (N_{\text{basis}} - 1)$ with $j = 0$ specifically reserved for the triplet $(l_0, m_0, n_0) = (0, 0, 0)$. For the remainder of the discussion of DFT, any sums on lower-case Roman indices are assumed to run over $j = 0 \ldots (N_{\text{basis}} - 1)$ unless otherwise indicated.

It will sometimes be convenient to write the summation in Eqn. (4.131) more explicitly, rather than with the assumed sum on j. Note that Eqn. (4.131) is equivalent to

$$\psi_I^{\text{sp}} = \frac{1}{\sqrt{\Omega}} \exp\left(i\boldsymbol{k}\cdot\boldsymbol{x}\right) \sum_{j=0}^{N_{\text{basis}}-1} c_{Ij}(\boldsymbol{k}) \exp\left(i\boldsymbol{\Gamma}_j\cdot\boldsymbol{x}\right). \tag{4.134}$$

It is worth remembering at this point that Eqns. (4.131) and (4.134) are Fourier expansions of an unknown periodic function multiplied by $\exp(i\boldsymbol{k}\cdot\boldsymbol{x})$. This is the key to the entire plane-wave approach to DFT; it allows us to rely heavily on Fourier transforms and thereby speed up the computations. At the same time, the use of Fourier space can make it difficult for the beginner to understand what is going on in a DFT implementation. It is helpful to recall (see Section 3.7.1) that we can approximate any function that is periodic in C as

$$g(\boldsymbol{x}) = \frac{1}{\sqrt{\Omega}} \sum_{j=0}^{N_{\text{basis}}-1} \widehat{g}_j \exp\left(i\boldsymbol{\Gamma}_j\cdot\boldsymbol{x}\right), \tag{4.135}$$

where

$$\widehat{g}_j = \frac{1}{\sqrt{\Omega}} \int_C g(\boldsymbol{x}) \exp(-i\boldsymbol{\Gamma}_j\cdot\boldsymbol{x})\, d\boldsymbol{x}$$

are the Fourier coefficients for the function $g(\boldsymbol{x})$. Knowing $\widehat{g}_j$ is equivalent to knowing $g(\boldsymbol{x})$ once we evaluate Eqn. (4.135).

A question that naturally arises is how big to make the basis. A sensible way to make this decision is to define a cutoff wavelength for $\boldsymbol{\Gamma}$. That is to say, we can include in the basis every vector $\boldsymbol{\Gamma}$ for which

$$\|\boldsymbol{\Gamma}\| = \frac{2\pi}{a}\sqrt{l^2 + m^2 + n^2} \leq \Gamma_{\max}, \tag{4.136}$$

[49] Note that these integers have nothing to do with the quantum numbers introduced earlier, despite the notational similarity.

thus taking all $\mathbf{\Gamma}$ within a sphere of radius $\Gamma_{\max}$ in $\boldsymbol{k}$-space. Increasing $\Gamma_{\max}$ will increase the number of terms in the Fourier approximation for the electron wave functions. This improves the accuracy of the calculation, but simultaneously increases the workload. Note that since $\Gamma_{\max}$ is the modulus of a wave vector, it is related to a maximum kinetic energy

$$\mathcal{T}_{\max} = \frac{\hbar^2 \Gamma^2_{\max}}{2m^{\text{el}}}.$$

Therefore it is common in the literature to talk about the cutoff energy that is used to determine the basis set in a simulation.

Defining the $\boldsymbol{k}$-points The band structure is a continuous function in $\boldsymbol{k}$-space. Quantities of interest (like the simple example of energy in Eqn. (4.120)) require integration over all of $\boldsymbol{k}$-space, or more practically, over all the filled bands in the first Brillouin zone. From the outset, we will replace this by a numerical approximation, where we plan to evaluate our equations at only a pre-defined set of points in $\boldsymbol{k}$-space and weight each one to approximate the true integral. There are a number of different approaches to choosing the $\boldsymbol{k}$-points in an optimal way. A common choice is the so-called Monkhorst–Pack grid [MP76], which has strong analogies with Gauss quadrature of the finite element method (discussed in Chapter 9 of the companion volume to this one [TME12]). In essence, one can cleverly choose the number, location and weight of integration points to improve accuracy by taking advantage of one's knowledge of the expected characteristics of the integrand. A simple (but extremely inaccurate) approach would be to divide the first Brillouin zone (which is a cube in the simple example we are outlining here) into $N_k \times N_k \times N_k$ smaller cubes and choose a $\boldsymbol{k}$-point at the center of each. The volume of each such smaller cube is $\Delta\widehat{\Omega} = \widehat{\Omega}/N_k^3$ and the volume of the Brillouin zone for our example is $\widehat{\Omega} = (2\pi/a)^3$. We can now approximate any integral over $\boldsymbol{k}$-space, say a quantity Y, as

$$Y = \sum_{I \in \text{filled}} \int_{\mathcal{B}} y(\boldsymbol{k})\, d\boldsymbol{k} \approx \sum_{I \in \text{filled}} \sum_{\boldsymbol{k}} f_I(\boldsymbol{k}) y(\boldsymbol{k}),$$

where the domain of integration is the first Brillouin zone $\mathcal{B}$. For simple zero-temperature calculations $f_I(\boldsymbol{k})$ is zero if the Ith band at $\boldsymbol{k}$-point $\boldsymbol{k}$ is unfilled. If it is filled, then

$$f_I(\boldsymbol{k}) = 2\Delta\widehat{\Omega}/\widehat{\Omega}.$$

This definition of the occupation function f includes the weighting due to the discretization of the $\boldsymbol{k}$-space integral.[50] For a total of N^{el} electrons per periodic cell, we have

$$N^{\text{el}} = \sum_I \sum_{\boldsymbol{k}} f_I(\boldsymbol{k}). \tag{4.137}$$

Equation (4.137) allows us to determine for which I and $\boldsymbol{k}$ the value of f is nonzero: once we know all the bands (energy levels) I at each $\boldsymbol{k}$-point, we will systematically fill the lowest-energy $\boldsymbol{k}$-states until Eqn. (4.137) is satisfied.

In a careful calculation, it would be appropriate to choose a number of $\boldsymbol{k}$-points, obtain a solution for the electron density, and then repeat the calculation with a larger number

[50] The factor of 2 is introduced since each state can have both a spin-up and a spin-down electron.

of $\boldsymbol{k}$-points until the additional $\boldsymbol{k}$-points have no great effect on the solution. As such, the solution is deemed to have "converged" with respect to the discretization of $\boldsymbol{k}$-space.

The electron density Recall that the power of DFT is that we can solve an equivalent single-electron system for our multiple-electron problem. For each $\boldsymbol{k}$-point, we need to find the N_{basis} possible single-particle wave functions that satisfy

$$-\frac{\hbar^2}{2m^{\text{el}}}\nabla^2\psi^{\text{sp}} + \mathcal{U}^{\text{eff}}\psi^{\text{sp}} = \epsilon\psi^{\text{sp}}, \tag{4.138}$$

and then determine which of these are filled by ranking them in terms of their corresponding energy values. Once we know the filled states, the electron density will be the superposition of the electron density of each filled state. The electron density of each state was defined in Eqn. (4.74), although since we are dealing with a single-electron wave function it reduces to simply $\rho = \psi^{\text{sp}*}\psi^{\text{sp}}$. Summing over the filled states yields

$$\rho(\boldsymbol{x}) = \sum_{I,\boldsymbol{k}} f_I(\boldsymbol{k})\psi^{\text{sp}*}(\boldsymbol{x};\boldsymbol{k})\psi^{\text{sp}}(\boldsymbol{x};\boldsymbol{k}) = \frac{1}{\Omega}\sum_{I,\boldsymbol{k}}\sum_{i,j} f_I(\boldsymbol{k})c^*_{Ii}(\boldsymbol{k})\varphi^*_i c_{Ij}(\boldsymbol{k})\varphi_j.$$

It is useful to rewrite this more explicitly using Eqn. (4.134)

$$\rho(\boldsymbol{x}) = \frac{1}{\Omega}\sum_{I,\boldsymbol{k}}\sum_{i,j} f_I(\boldsymbol{k})c^*_{Ii}(\boldsymbol{k})c_{Ij}(\boldsymbol{k})\exp\left(i(\boldsymbol{\Gamma}_j - \boldsymbol{\Gamma}_i)\cdot\boldsymbol{x}\right) \tag{4.139}$$

and note that any difference between reciprocal lattice vectors $(\boldsymbol{\Gamma}_j - \boldsymbol{\Gamma}_i)$ is in fact equal to another reciprocal lattice vector that we can call $\boldsymbol{\Gamma}_l$ (corresponding to a component in the eigenvector[51] c_{Il}). We can now make a change of variables $\boldsymbol{\Gamma}_l = \boldsymbol{\Gamma}_j - \boldsymbol{\Gamma}_i$ so that

$$\rho(\boldsymbol{x}) = \frac{1}{\Omega}\sum_{I,\boldsymbol{k}}\sum_{j,l} f_I(\boldsymbol{k})c^*_{I,j-l}(\boldsymbol{k})c_{I,j}(\boldsymbol{k})\exp\left(i\boldsymbol{\Gamma}_l\cdot\boldsymbol{x}\right). \tag{4.140}$$

The notation $c^*_{I,j-l}$ is meant to imply the component of the vector $\boldsymbol{c}^*$ that corresponds to the component of vector $\boldsymbol{\varphi}$ containing the term $\exp(i[\boldsymbol{\Gamma}_j - \boldsymbol{\Gamma}_l]\cdot\boldsymbol{x})$. This component is not generally going to have an index numerically equal to $j - l$. Writing the density in this form allows us to see directly the same form as Eqn. (4.135), and therefore we can identify the Fourier coefficients of the density as

$$\widehat{\rho}_l = \frac{1}{\sqrt{\Omega}}\sum_{I,\boldsymbol{k}}\sum_{j} f_I(\boldsymbol{k})c^*_{I,j-l}(\boldsymbol{k})c_{I,j}(\boldsymbol{k}). \tag{4.141}$$

As such, knowing the eigenvectors $c_{I,j}(\boldsymbol{k})$ allows us to find the electron density (actually its Fourier coefficients) through a direct and rapid summation. Should we need the real-space density $\rho(\boldsymbol{x})$, we can take the inverse Fourier transform of Eqn. (4.135).

While we are trying to *solve* for $\rho(\boldsymbol{x})$, we also *need* $\rho(\boldsymbol{x})$ to evaluate $\mathcal{U}^{\text{eff}}$. The process will be iterative, whereby we will start with an initial guess for ρ, evaluate $\mathcal{U}^{\text{eff}}$, solve for the new ρ and repeat until the change in ρ during an iteration is deemed sufficiently small that we can call the solution "converged." In practice, this is a subtle business. First, the

[51] Of course we are using a finite number of eigenvectors. If $\boldsymbol{\Gamma}_l$ is not explicitly in our chosen basis, then by definition $c_{Il} = 0$ and we must take care of such cases when implementing these expressions on a computer.

initial guess to $\rho(\boldsymbol{x})$ is very important and a bad choice may make convergence very slow or even impossible. Second, the DFT problem is essentially a multidimensional, nonlinear minimization problem, and so all the challenges of multiple local minima, discussed on page 182 and in more detail in Chapter 6, are also potential problems in DFT.

The DFT solution process involves an exceedingly complex and nonlinear minimization of the total energy with respect to the set of coefficients c_{Ij}. Discussion of general strategies for minimization will appear in Chapter 6, and in principle any of the methods discussed there can be applied to DFT.[52] The simplest and most common minimization approach used in DFT is equivalent to the steepest descent algorithm discussed in Section 6.2.3, and proceeds as follows. The initial guess for iteration n is taken as a mixture of the solution and the initial guess from iteration $n-1$:

$$\rho_{(n)}^{\text{guess}} = \beta \rho_{(n-1)}^{\text{soln}} + (1-\beta)\rho_{(n-1)}^{\text{guess}}. \tag{4.142}$$

Unfortunately, finding the optimal value of β is a bit of a black art, and typically it must be quite small to permit convergence. To see that this is indeed equivalent to a steepest descent search, we rearrange Eqn. (4.142) to the form

$$\rho_{(n)}^{\text{guess}} = \rho_{(n-1)}^{\text{guess}} + \beta\left(\rho_{(n-1)}^{\text{soln}} - \rho_{(n-1)}^{\text{guess}}\right),$$

identifying the difference $(\rho_{(n-1)}^{\text{soln}} - \rho_{(n-1)}^{\text{guess}})$ as a search direction and β as a step-size. In problems where the total energy is relatively insensitive to changes in the electron density, the step-size β must be small to prevent wild oscillations in the solution from one iteration to the next. Unfortunately, the computation expense and complexity of DFT make more sophisticated minimization algorithms prohibitive and difficult to implement.

The eigenvalue problem Given a specific $\boldsymbol{k}$-point and an initial guess for the electron density (or the blended electron density from a previous iteration from Eqn. (4.142)), we proceed as we did in Section 4.5.4. Now, however, we start from the full three-dimensional Schrödinger equation and use the correct Bloch form of the wave functions in Eqn. (4.131). Using Eqn. (4.131) in Eqn. (4.124) yields[53]

$$\begin{aligned}\Pi = &\frac{1}{\Omega}\frac{\hbar^2}{2m^{\text{el}}}\int_C c_{\underline{I}j}^* \varphi_j^* \gamma_{kl} c_{\underline{I}k}\varphi_l \, d\boldsymbol{x} \\ &+ \frac{1}{\Omega}\int_C c_{\underline{I}j}^*\varphi_j^*(\mathcal{U}^{\text{ext}} + \mathcal{U}^{\text{H}} + \mathcal{U}^{\text{xc}}) c_{\underline{I}k}\varphi_k \, d\boldsymbol{x} - \epsilon_{\underline{I}}\left(\frac{1}{\Omega}\int_C c_{\underline{I}j}^*\varphi_j^* c_{\underline{I}k}\varphi_k \, d\boldsymbol{x} - 1\right).\end{aligned}$$

The matrix $[\gamma_{jl}]$ is the three-dimensional analog of Eqn. (4.128),

$$\gamma_{jl} = \begin{cases} 0 & \text{for } j \neq l, \\ \|\boldsymbol{k} + \boldsymbol{\Gamma}_j\|^2 & \text{for } j = l, \end{cases}$$

and the three contributions to the potential are respectively the external potential, $\mathcal{U}^{\text{ext}}$, due to the atomic nuclei (modeled by pseudopotentials), the Hartree potential, $\mathcal{U}^{\text{H}}$, due

[52] Indeed, many of them have been applied to DFT. See, for example, the review article in [PTA+92] and also Section 9.3 of the excellent book by Martin [Mar04].

[53] In anticipation of the N_{basis} eigensolutions that we now come to expect after studying the free electron gas example, we will introduce the subscript I from the outset.

to Coulomb interactions between electron density at different points, and the exchange-correlation potential, $\mathcal{U}^{xc}$. Because the coefficients $c_{Ij}(\boldsymbol{k})$ are independent of the spatial coordinate, we can take them outside the integrals. Then, since we are once again minimizing Π with respect to c_{Ij}, we take derivatives with respect to c^*_{Ij} and set them equal to zero. This leads to the eigenvector equation

$$\left(-\frac{\hbar^2}{2m}\gamma_{ij}(\boldsymbol{k}) + U^{\mathrm{eff}}_{ij}\right) c_{Ij}(\boldsymbol{k}) = \epsilon_I c_{Ii}(\boldsymbol{k}), \tag{4.143}$$

where we have used the orthonormal property of the basis

$$\frac{1}{\Omega}\int_C \varphi^*_i \varphi_j \, d\boldsymbol{x} = \delta_{ij}, \tag{4.144}$$

and defined

$$U^{\mathrm{eff}}_{ij} \equiv \frac{1}{\Omega}\int_C \varphi^*_i \varphi_j \mathcal{U}^{\mathrm{ext}} \, d\boldsymbol{x} + \frac{1}{\Omega}\int_C \varphi^*_i \varphi_j \mathcal{U}^{\mathrm{H}} \, d\boldsymbol{x} + \frac{1}{\Omega}\int_C \varphi^*_i \varphi_j \mathcal{U}^{\mathrm{xc}} \, d\boldsymbol{x}. \tag{4.145}$$

Note that in Eqn. (4.143), we have emphasized the dependence of γ_{ij} on the wave vector $\boldsymbol{k}$, whereas there is no such dependence in U^{eff}_{ij}. This is important since the solution of Eqn. (4.143) at multiple $\boldsymbol{k}$-points requires only one evaluation of U^{eff}_{ij}. In Eqn. (4.145), we see the power of using a plane-wave basis. Each integral is just one of the Fourier coefficients of each real-space potential. To see this, we use the same notation as in Eqn. (4.140), and denote φ_{i-j} as the component of φ containing $\exp(i[\boldsymbol{\Gamma}_i - \boldsymbol{\Gamma}_j]\cdot\boldsymbol{x})$. Thus, a Fourier coefficient for one of the potentials $\mathcal{U}(\boldsymbol{x})$ can be identified as

$$\frac{1}{\Omega}\int \varphi^*_i \varphi_j \mathcal{U} d\boldsymbol{x} = \frac{1}{\sqrt{\Omega}}\frac{1}{\sqrt{\Omega}}\int \exp\left(-i(\boldsymbol{\Gamma}_i - \boldsymbol{\Gamma}_j)\cdot\boldsymbol{x}\right)\mathcal{U}\, d\boldsymbol{x} = \frac{1}{\sqrt{\Omega}}\widehat{\mathcal{U}}_{i-j}.$$

In effect, all the entries in the matrix U^{eff}_{ij} are just appropriately indexed Fourier coefficients of the potentials. If we can work in Fourier space as much as possible, the assembly of this matrix will be very rapid. Next, we look at how the Fourier coefficients of each of the three components of the potential can be determined.

The external potential As discussed in Section 4.5.2, the effect of the electron–ion interactions is normally treated using a pseudopotential. The pseudopotential represents the net effect on the *interesting* valence electrons due to the nuclei and the *chemically inert* core electrons. Considering the local evanescent core pseudopotential ($\mathcal{U}^{\mathrm{ec}}$ of Eqn. (4.112)) as an example for the simplified implementation discussed here, the Fourier coefficients $\widehat{\mathcal{U}}^{\mathrm{ext}}_j$ are obtained by integrating

$$\widehat{\mathcal{U}}^{\mathrm{ext}}_j = \frac{1}{\sqrt{\Omega}}\int_C \exp\left(-i\boldsymbol{\Gamma}_j\cdot\boldsymbol{x}\right)\sum_\alpha \mathcal{U}^{\mathrm{ec}}(\|\boldsymbol{x} - \boldsymbol{r}^\alpha\|, Z^\alpha)\, d\boldsymbol{x}.$$

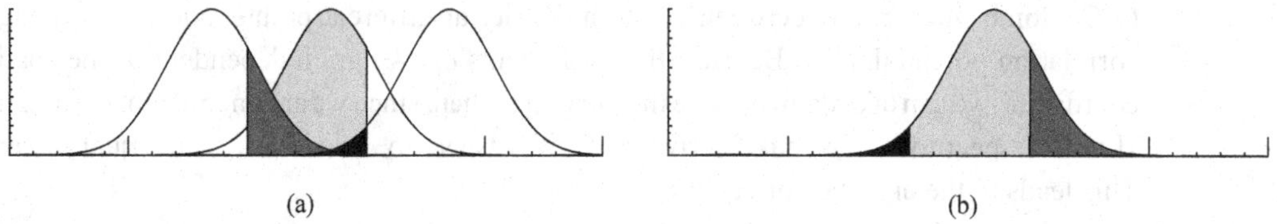

Fig. 4.15 (a) An integral over the periodic cell of a potential (inside the cell) and two periodic copies of the potential. (b) The equivalent integral of just one copy of the potential over all space.

The sum over α is over the ions at positions $\boldsymbol{r}^\alpha$ with atomic number Z^α. Although the integral is over a finite volume, note that the sum is infinite, recognizing the contributions from every periodic copy of the ions in the fundamental simulation box. However, because the contributions are linearly superimposed, we can make a clever exchange: we sum *only* over the finite number of ions in the fundamental box, but now perform the integral over infinite space. This equivalence is shown schematically in Fig. 4.15. From each ion α in the simulation box, there is a contribution $\widehat{\mathcal{U}}_j^\alpha$ that can be evaluated in closed form [FPA+95]:

$$\widehat{\mathcal{U}}_j^\alpha = \frac{1}{\sqrt{\Omega}} \int_\infty \exp\left(-i\boldsymbol{\Gamma}_j \cdot \boldsymbol{x}\right) \mathcal{U}^{\text{ec}}(\|\boldsymbol{x} - \boldsymbol{r}^\alpha\|, Z^\alpha)\, d\boldsymbol{x},$$

so that $\widehat{\mathcal{U}}_j^{\text{ext}} = \sum_\alpha \widehat{\mathcal{U}}_j^\alpha$. Usually, the external potential is singular at $Q_0 = \|\boldsymbol{\Gamma}_0\| = 0$. This singularity is real in the sense that it reflects the interaction between the positive ions and a uniform background electron density; physically it is compensated for by a similar term in the Hartree potential (discussed next). Taken together, these two divergent terms sum to a constant that depends only on the total valence of the ions in the periodic cell, the result for which will be given below. As a practical matter of implementation, we simply disregard this term at this point by setting $\widehat{\mathcal{U}}_0^\alpha = 0$ (recall that $j = 0$ corresponds to $\boldsymbol{\Gamma}_0 = \boldsymbol{0}$) and then add the appropriate correction to the total energy later (see Eqn. (4.157)).

In Section 4.5.2 we touched briefly on the more accurate nonlocal pseudopotentials typically used in modern DFT codes. The main effect of this approach, once the form of the nonlocal pseudopotential is set, is a change to the form of the term in Eqn. (4.145) due to the external potential. Formally, this becomes

$$\frac{1}{\Omega} \int_C \varphi_i^* \varphi_j \mathcal{U}^{\text{ext}}\, d\boldsymbol{x} \to \frac{1}{\Omega} \int_C \int_{C'} \varphi_i^*(\boldsymbol{x}) \varphi_j(\boldsymbol{x}') \mathcal{U}^{\text{nl}}(\boldsymbol{x}, \boldsymbol{x}')\, d\boldsymbol{x} d\boldsymbol{x}'.$$

Clearly, this will increase the computational effort, but will not change the overall structure of the solution process. The form also permits a relatively modular code structure where local or nonlocal pseudopotentials can be interchanged easily.

The Hartree potential The second term in Eqn. (4.145) is made up of the Fourier coefficients of the Hartree potential. These are known from the Fourier coefficients of the electron density, as we shall now see. Recall that the Hartree potential is given by Eqn. (4.85) for a known electron density, ρ. This integral is the solution to Poisson's equation [AW95]. That is to say, if $\mathcal{U}^{\text{H}}$ is given by Eqn. (4.85), it must also satisfy

$$\nabla^2 \mathcal{U}^{\text{H}}(\boldsymbol{x}) = -4\pi e^2 \rho(\boldsymbol{x}).$$

Taking the Fourier transform of both sides of this equation gives a straightforward relation between the coefficients $\widehat{\mathcal{U}}_j^{\mathrm{H}}$ and $\widehat{\rho}_j$:

$$\widehat{\mathcal{U}}_j^{\mathrm{H}} = 4\pi e^2 \frac{\widehat{\rho}_j}{\|\boldsymbol{\Gamma}_j\|^2}. \tag{4.146}$$

Recall that for a given iteration, we have at hand a guess for the electron density and its Fourier coefficients, and so the $\widehat{\mathcal{U}}_j^{\mathrm{H}}$ terms follow directly.[54] As previously mentioned, there is a singular term at $\|\boldsymbol{\Gamma}_j\| = 0$ corresponding to the interaction between an electron and a uniform background electronic density, but it will combine with an analogous term in the external potential to produce a finite constant in the energy. We treat this by ignoring the singular term here (setting $\widehat{\mathcal{U}}_0^{\mathrm{H}} = 0$) and adding the energy constant in Eqn. (4.157).

The exchange-correlation potential The last part of the DFT equation, as discussed in Section 4.5.1, is the "catch-all" for effects not explicitly present in the other terms. To keep our example simple, we use the LDA and therefore Eqn. (4.110). There exist several accurate parameterizations of the exchange-correlation energy density ϵ^{xc}, for example [PZ81, PW92], and for our purposes we treat ϵ^{xc} as a known scalar function of the scalar value ρ. However, we see that Eqn. (4.110) requires the electron density in real space, whereas we have so far been able to limit our calculations to $\widehat{\rho}$ in $\boldsymbol{k}$-space via Eqn. (4.141). At this point in the evaluation of U_{ij}^{eff} it is therefore necessary to make a rather painful computational step; we have to numerically evaluate the three-dimensional integral over the real-space simulation box for each basis vector $\boldsymbol{\Gamma}_j$:

$$\widehat{\mathcal{U}}_j^{\mathrm{xc}} = \frac{1}{\sqrt{\Omega}} \int_C \varphi_j^* \mathcal{U}^{\mathrm{xc}}(\rho[\boldsymbol{x}])\, d\boldsymbol{x}. \tag{4.147}$$

This requires a discretization of the simulation box into a number of points $\boldsymbol{x}_i$ and the subsequent evaluation of the electron density at each $\boldsymbol{x}_i$ via the inverse Fourier sum

$$\rho(\boldsymbol{x}_i) = \frac{1}{\sqrt{\Omega}} \sum_j \widehat{\rho}_j \varphi_j(\boldsymbol{x}_i).$$

It is worth noting that even though the size of U_{ij}^{eff} is $N_{\mathrm{basis}} \times N_{\mathrm{basis}}$, all the exchange-correlation contributions are linear combinations of the N_{basis} Fourier coefficients of $\mathcal{U}^{\mathrm{xc}}$. This mitigates the pain of Eqn. (4.147) somewhat, and of course a clever choice of integration points and weights can improve efficiency and accuracy a great deal.[55]

The solution at each $\boldsymbol{k}$-point At this stage, we have built the matrix on the left-hand side of Eqn. (4.143), comprising the diagonal matrix $[\gamma_{ij}]$ and appropriately indexed Fourier coefficients of the three potential terms. The eigenvectors $c_{Ij}(\boldsymbol{k})$ and corresponding eigenvalues $\epsilon_I(\boldsymbol{k})$ can now be found using standard linear algebra routines. At each $\boldsymbol{k}$-point, there will

[54] As an interesting side note, the approach outlined here only works in three dimensions. In one or two dimensions, the Hartree integral is singular and there is no analogous approach to using the three-dimensional Poisson equation.

[55] Choosing an efficient set of integration points is discussed in the context of finite elements and Gauss quadrature in Section 9.3.2 of the companion volume to this one [TME12].

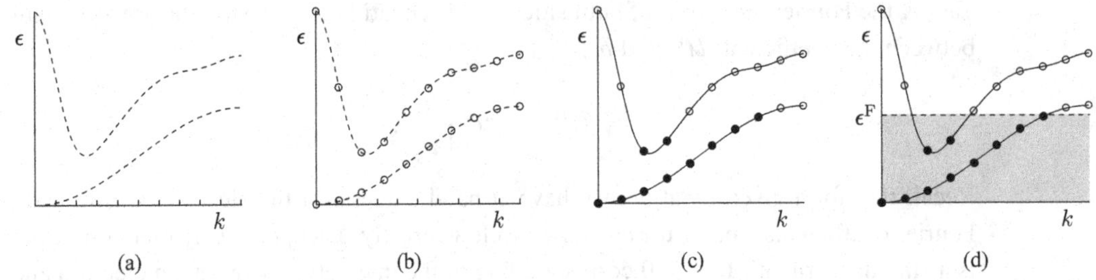

Fig. 4.16 Schematic illustration of the filling of eigenstates. (a) The unknown band structure (dashed lines). (b) Eigenvalues at discrete k-points approximate the band structure. (c) We fill the lowest energy states until we have enough electrons. (In this example, there are 2 electrons and 10 k-points, so there must be 20 filled states.) (d) The Fermi energy is the energy of the highest filled state. The number of filled states at each k-point varies. From left to right, the numbers of filled states at each k-point are respectively $2, 2, 4, 4, 2, 2, 2, 2, 0, 0$ (recall that spin allows two states per energy level).

be N_{basis} solutions, arranged in order of increasing energy $\epsilon_I(\boldsymbol{k})$ and indexed by the band number $I = 1 \ldots N_{\text{basis}}$. Depending on the number of electrons in the simulation box, some number of the lowest energy bands will be filled at each $\boldsymbol{k}$-point. One can roughly think of the N^{el} electrons in the simulation box filling the lowest $N^{\text{el}}/2$ bands at each $\boldsymbol{k}$-point, although this will not always be true for a system with a more complicated band structure. It is possible that some points in $\boldsymbol{k}$-space will have more than $N^{\text{el}}/2$ filled bands (and others proportionally less) if the global ranking of their energies so dictates. This is illustrated schematically in Fig. 4.16.

Essentially then, we need to build Eqn. (4.143) at each $\boldsymbol{k}$-point, find the resulting eigenvectors and energies, and store them.[56] After all the $\boldsymbol{k}$-points have been evaluated, we determine the filled bands by comparing all the energy levels at all $\boldsymbol{k}$-points and systematically filling the lowest energy states.

Once we have filled the states, we can recompute the Fourier coefficients of the electron density from Eqn. (4.141). To decide if the solution has converged, these final $\widehat{\rho}_j$ must not differ significantly from the guess at the start of the iteration. Thus we can take

$$\|\widehat{\boldsymbol{\rho}} - \widehat{\boldsymbol{\rho}}^{\text{guess}}\|^2 = \sum_j (\widehat{\rho}_j - \widehat{\rho}_j^{\text{guess}})^* (\widehat{\rho}_j - \widehat{\rho}_j^{\text{guess}}) < tol$$

as the convergence criterion for some suitable value of tol. If the solution has not converged, we return to Eqn. (4.142) and iterate again.

Computing the total energy (and correctly cancelling divergent terms) Once the electron density has converged to self-consistency, we can compute the total energy of the system, which is the fundamental quantity we seek in modeling materials. We recall Eqn. (4.108), which provides the energy expression we want to evaluate. However, we cannot compute the terms

56 In practice, it may not be necessary to store all the solutions, but just the few with the lowest energies. The number of solutions will be N_{basis} at each $\boldsymbol{k}$-point, but N_{basis} is typically much larger than the number of electrons in the simulation, and it is the number of electrons that dictates how many bands will be filled.

in this energy directly for two reasons. The first is that we have neglected the singular terms in both the external potential and the Hartree potential. The second is that $\mathcal{E}^{ZZ}$ must be evaluated carefully since it involves long-range Coulomb interactions between like charges in an infinite crystal and is therefore also divergent.

In developing our equations for the Hamiltonian matrix, we essentially made use of a modified $\mathcal{U}^{\rm eff}$, since we threw away the singular terms in the external and Hartree potentials. As such, the eigenstates we have found are really solutions for the equation

$$\mathcal{H}^{\rm sp} = \mathcal{U}_0^{\rm eff}(\boldsymbol{x}) + \frac{p^2}{2m^{\rm el}},$$

where

$$\mathcal{U}_0^{\rm eff} = \mathcal{U}_0^{\rm H} + \mathcal{U}^{\rm xc} + \mathcal{U}_0^{\rm ext}.$$

Here, the new Hartree and external potentials ($\mathcal{U}_0^{\rm H}$ and $\mathcal{U}_0^{\rm ext}$) are related to their correct counterparts ($\mathcal{U}^{\rm H}$ and $\mathcal{U}^{\rm ext}$) as

$$\mathcal{U}^{\rm H} = \mathcal{U}_0^{\rm H} + \mathcal{U}_\infty^{\rm H}, \qquad \mathcal{U}^{\rm ext} = \mathcal{U}_0^{\rm ext} + \mathcal{U}_\infty^{\rm ext},$$

where the subscripts ∞ indicate the singular terms that we have ignored. The electron density we have found in this way is still correct up to a uniform constant, because the terms we have neglected are associated with the reciprocal lattice vector $\mathbf{\Gamma} = \mathbf{0}$ (i.e. with a plane wave of "infinite wavelength"), but we need to make corrections to the total energy to account for the uniform background density.

We must now repeat the process leading Eqn. (4.107) to Eqn. (4.108) using the modified effective potential, $\mathcal{U}_0^{\rm eff}$, in place of $\mathcal{U}^{\rm eff}$. We see that the total energy in this case is

$$\mathcal{E} = \sum_{I \in filled} \epsilon_I - \mathcal{E}_0^{\rm H} - \int \rho\, \mathcal{U}^{\rm xc}\, d\boldsymbol{x} + \mathcal{E}^{\rm xc}(\rho) + \mathcal{E}^{ZZ} + \mathcal{E}_\infty^{\rm H} + \mathcal{E}_\infty^{\rm ext}, \tag{4.148}$$

where the newly introduced subscripts on the energy terms follow from the singular and nonsingular parts of the potentials, for example

$$\mathcal{E}_0^{\rm ext} = \int \rho\, \mathcal{U}_0^{\rm ext}\, d\boldsymbol{x}, \qquad \mathcal{E}_\infty^{\rm ext} = \int \rho\, \mathcal{U}_\infty^{\rm ext}\, d\boldsymbol{x}.$$

Essentially, the single-particle energies ϵ_I that we have found are missing the two (singular) energy terms, $\mathcal{E}_\infty^{\rm H} + \mathcal{E}_\infty^{\rm ext}$. However, Yin and Cohen [YC82] showed that it is possible to also decompose the ion–ion energy into nonsingular and singular parts, $\mathcal{E}^{ZZ} = \mathcal{E}_0^{ZZ} + \mathcal{E}_\infty^{ZZ}$, in such a way that the resulting three singular terms in the total energy combine to a finite constant. As such, the final energy becomes

$$\mathcal{E} = \sum_{I \in filled} \epsilon_I - \mathcal{E}_0^{\rm H} - \int \rho\, \mathcal{U}^{\rm xc}\, d\boldsymbol{x} + \mathcal{E}^{\rm xc}(\rho) + \mathcal{E}_0^{ZZ} + \mathcal{E}^{\rm rep}, \tag{4.149}$$

where

$$\mathcal{E}^{\rm rep} = \mathcal{E}_\infty^{ZZ} + \mathcal{E}_\infty^{\rm H} + \mathcal{E}_\infty^{\rm ext}.$$

We can now evaluate each of the terms in the energy. The first term in Eqn. (4.149) is a simple sum of the eigenvalues associated with the filled states,

$$\sum_{I \in filled} \epsilon_I = \sum_{I,\boldsymbol{k}} f_I(\boldsymbol{k})\epsilon(\boldsymbol{k}), \tag{4.150}$$

and is straightforward to evaluate. The second and third terms are of the same general form, since they are both derived from potentials: $\mathcal{U}_0^{\mathrm{H}}$ and $\mathcal{U}^{\mathrm{xc}}$. Thus they take the form

$$\mathcal{E} = \int \rho \mathcal{U}\, d\boldsymbol{x}. \tag{4.151}$$

It is left as an exercise to show that this can be written in Fourier space as

$$\int \rho \mathcal{U}\, d\boldsymbol{x} = \sum_{j=0}^{N_{\text{basis}}-1} \widehat{\rho}_j^* \widehat{\mathcal{U}}_j, \tag{4.152}$$

where the $\widehat{\rho}_j$ are known from Eqn. (4.141). As such the Hartree term becomes

$$\mathcal{E}_0^{\mathrm{H}} = \frac{1}{2} \sum_{j=1}^{N_{\text{basis}}-1} \widehat{\rho}_j^* \widehat{\mathcal{U}}_j^{\mathrm{H}} = 2\pi e^2 \sum_{j=1}^{N_{\text{basis}}-1} \frac{\widehat{\rho}_j^* \widehat{\rho}_j}{\|\boldsymbol{\Gamma}_j\|^2}, \tag{4.153}$$

where we used Eqn. (4.146), and moved the singular $\boldsymbol{\Gamma} = \boldsymbol{0}$ term (corresponding to the missing $j = 0$ term) into $\mathcal{E}^{\mathrm{rep}}$ as noted above. Similarly, the exchange-correlation term is

$$\int \rho \mathcal{U}^{\mathrm{xc}}\, d\boldsymbol{x} = \sum_{j=0}^{N_{\text{basis}}-1} \widehat{\rho}_j^* \widehat{\mathcal{U}}_j^{\mathrm{xc}}. \tag{4.154}$$

Usually the next term in Eqn. (4.149), $\mathcal{E}^{\mathrm{xc}}$, has to be evaluated in real space, using numerical quadrature of Eqn. (4.109):

$$\mathcal{E}^{\mathrm{xc}} = \int \rho(\boldsymbol{x}) \epsilon^{\mathrm{xc}}[\rho(\boldsymbol{x})]\, d\boldsymbol{x}. \tag{4.155}$$

The details of computing $\mathcal{E}_0^{\mathrm{ZZ}}$ and $\mathcal{E}^{\mathrm{rep}}$ are rather involved and beyond where we want to go here. Instead, we point the reader to the literature [YC82] and simply state the final results. The ion–ion energy is easily evaluated in Fourier space using the expression

$$\mathcal{E}_0^{\mathrm{ZZ}} = \frac{e^2}{2} \sum_{\substack{\alpha,\beta \\ \alpha \neq \beta}} z^\alpha z^\beta \left[\frac{4\pi}{\Omega} \sum_{j=1}^{N_{\text{basis}}-1} \frac{\cos[\boldsymbol{\Gamma}_j \cdot (\boldsymbol{r}^\alpha - \boldsymbol{r}^\beta)]}{\|\boldsymbol{\Gamma}_j\|^2} \exp\left(-\frac{\|\boldsymbol{\Gamma}_j\|^2}{4\eta^2} \right) \right.$$
$$\left. + \sum_C \frac{\operatorname{erfc}(\eta \|\boldsymbol{x}_C^0 + \boldsymbol{r}^\alpha - \boldsymbol{r}^\beta\|)}{\|\boldsymbol{x}_C^0 + \boldsymbol{r}^\alpha - \boldsymbol{r}^\beta\|} - \frac{\pi}{\eta^2 \Omega} - \frac{2\eta}{\sqrt{\pi}} \delta_{\alpha\beta} \right]. \tag{4.156}$$

Note once again that we have explicitly left out the $\boldsymbol{\Gamma} = \boldsymbol{0}$ ($j = 0$) term from the first sum inside the square brackets. In this expression z^α is the valence of an ion with atomic number Z^α, $\boldsymbol{r}^\alpha$ is the coordinate of ion α within the simulation cell and $\boldsymbol{x}_C^0$ is the origin of the Cth simulation cell. Also, the sum on C implies a sum over every periodic copy of the simulation cell, and so it is an infinite sum, but the singular term for which $\boldsymbol{x}_C^0 + \boldsymbol{r}^\alpha - \boldsymbol{r}^\beta = \boldsymbol{0}$ is omitted. The parameter η is arbitrary; its value does not affect the final result but can be used to optimize the rate of convergence of the infinite sum.[57] The complementary error function is represented by erfc. Note that because Eqn. (4.156) is independent of the electron density, it can be computed once at the start of a calculation (for a fixed set of ion positions).

Finally, we need to include the sum of the divergent terms we have omitted ($\mathcal{E}^{\text{rep}} = \mathcal{E}^{\text{H}}_{\infty} + \mathcal{E}^{\text{ext}}_{\infty} + \mathcal{E}^{\text{ZZ}}_{\infty}$). These add up to a finite constant as shown by Yin and Cohen [YC82]:

$$\mathcal{E}^{\text{rep}} = \frac{z^{\text{tot}}}{\Omega} \sum_{\alpha} \Lambda_{\alpha}, \tag{4.157}$$

where

$$\Lambda_{\alpha} = \int \left[\mathcal{U}^{\text{ec}}(\|\boldsymbol{x}\|, Z^{\alpha}) + \frac{e^2 z^{\alpha}}{\|\boldsymbol{x}\|} \right] d\boldsymbol{x} = 4\pi e^2 z^{\alpha} B_{\alpha}^2 \left(\frac{1}{a_{\alpha}^2} + \frac{2b_{\alpha}}{a_{\alpha}^3} + 2A_{\alpha} \right).$$

In Eqn. (4.157), z^{tot} is the total valence of one periodic simulation cell and so the prefactor is simply the average electron density. The second equality for Λ_{α} follows from the evanescent core pseudopotential that we are using in our example (Eqn. (4.112)).

The overall algorithm The flowchart in Fig. 4.17(a) shows the DFT solution process. Accuracy depends to some extent on the number and location of the $\boldsymbol{k}$-points, with the computational effort scaling linearly with the number of $\boldsymbol{k}$-points. However, the main improvement in accuracy comes from the number of plane waves in the basis, N_{basis}, for which the effort scales as $O(N_{\text{basis}}^3)$. This is due to the diagonalization of the eigenvalue equation which happens on every pass through the loop to achieve a self-consistent electron density. This bottleneck is the main motivation for efforts in DFT research to find approximations and improved algorithms that lead to linear scaling [KF96a, KF96b, SAG$^+$02, GStVB98, WGC98, ZLC05].

4.5.6 Energy minimization and dynamics: forces in DFT

It is one thing to compute the energy of a configuration of atomic nuclei and the accompanying electrons, and quite another to use that tool to determine a minimum energy configuration of the nuclei or to evolve the positions of the nuclei dynamically. Both goals are commonly sought using the methods of DFT. Energy minimization uses the machinery of molecular statics (MS) discussed in more detail in Chapter 6, while molecular dynamics (MD) is discussed in Chapter 9. Both methods require the forces on the nuclei.

Taking derivatives of Eqn. (4.149) with respect to the nuclear positions to find forces is indeed possible, but it is not a trivial task. We consider these forces beyond the scope of our current discussion and refer the reader to [MH00] as a good source for the details. Given the force expressions, however, an energy minimization procedure would require us to compute the energy of the initial structure by iteratively finding the self-consistent electron density as described above. Then, the forces would be computed and the nuclear positions moved in response to the forces.[58] The new electron density would need to be

[57] While η is arbitrary in principle, in an implementation this convergence can be sensitive to numerical precision issues and so η must be chosen carefully.

[58] Based on a suitable minimization algorithm as discussed in Chapter 6.

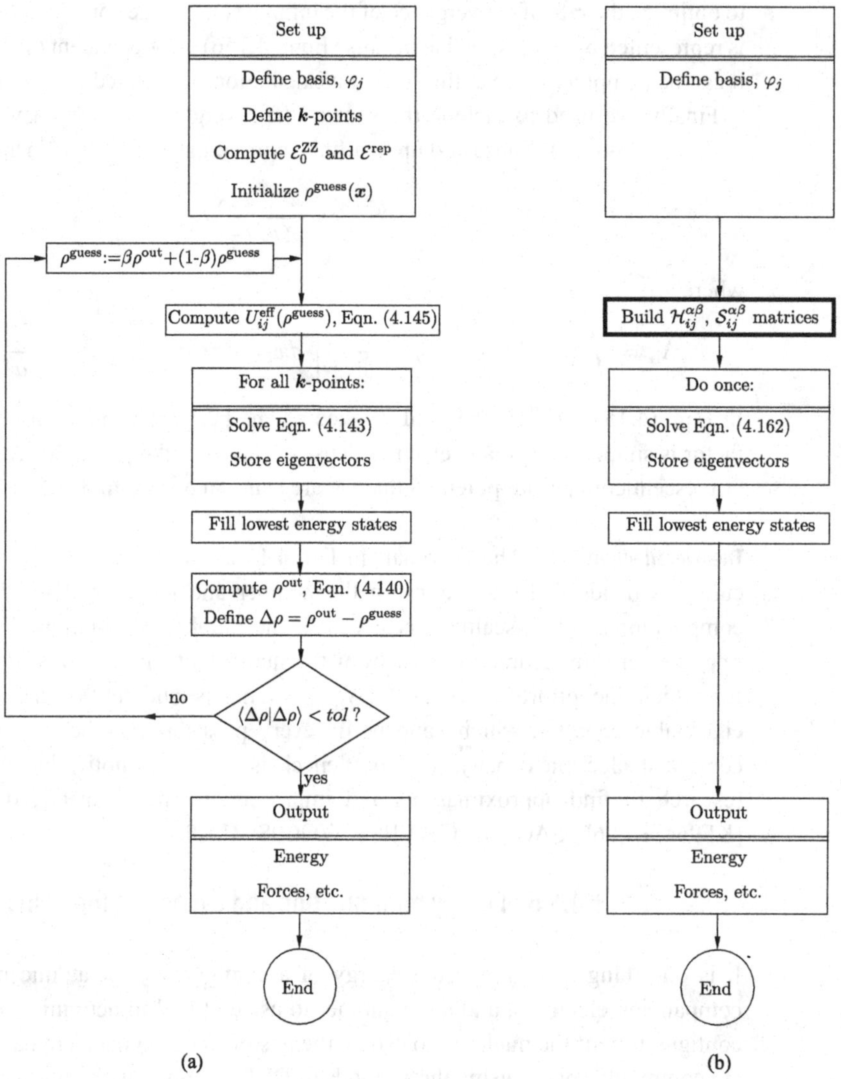

Fig. 4.17 (a) Flow chart of the DFT solution process. (b) Flow chart of the TB solution process that will be discussed in Section 4.6. The two charts are compared in the discussion of Section 4.6.4.

computed, again by iteration, although most times this would converge quickly since the change to nuclear positions between steps is small and the previously computed electron density provides a good initial guess. This would be repeated until the forces on the nuclei fell below some convergence tolerance.

MD simulations with DFT, on the other hand, have two major variants: so-called Born–Oppenheimer MD (BOMD) [MH00] and Car–Parrinello MD (CPMD) [CP85]. The former is essentially the dynamic equivalent of the minimization procedure just described: the electron density is found, the forces computed and the nuclei moved according to those forces, and the process is repeated. As the name implies, the BOA is assumed, allowing for the relaxation of the electronic degrees of freedom between each update of the nuclear positions. By contrast, CPMD is a method for evolving the electronic wave functions simultaneously with the nuclei by assigning a fictitious mass to each basis wave and computing configurational forces that act to "move" the basis waves. BOMD and CPMD have their strengths and weaknesses, depending on the application.

As one might imagine, the number of atoms and number of timesteps that can be studied with DFT using either energy minimization or MD are extremely limited. For typical computational resources available at the time of writing, these simulations are limited to several hundreds of atoms for less than about 10 picoseconds. These limitations motivate the need for more approximate methods. In the last section of this chapter, we look at the next level of approximation: the tight-binding (TB) method.

4.6 Semi-empirical quantum mechanics: tight-binding (TB) methods

We introduce the TB method mainly as a way to make a bridge between DFT and the seemingly ad hoc curve-fitting procedure that is the world of atomistic models presented in Chapter 5.[59] We will see in that chapter that through a series of approximations to TB, one can justify the forms of some of the atomistic models, and give plausibility to some of the others.

It is worth noting that while the TB model does remarkably well at describing structures close to those to which it was fit, it tends to suffer from a lack of *transferability*. By this, we mean that the model is parametrically fit to certain structural arrangements of atoms (for example bulk diamond cubic crystals), but does not necessarily make accurate predictions about the same species of atoms in very different arrangements (such as near a free surface).[60] Traditionally, TB has been applied mainly to semi-conductor systems (Si and Ge) and to carbon. Strongly covalent systems like these are consistent with the picture implied by the TB framework: that the valence electrons are "tightly bound" to the atoms and the electronic orbitals are not strongly altered during the bonding process. However, the parametric fitting that is used in TB makes it applicable to other systems as well. TB is arguably the simplest method that we can devise while still claiming that we are doing *quantum mechanics*; the empirical models of Chapter 5 can only rightly be called classical models of atomic interactions, as all of the electronic details have, at that point, been replaced by effective force laws.

4.6.1 LCAO

In our presentation of DFT, we chose to use plane waves as our basis set, but we mentioned in passing that one could equally well have used a basis set of orbitals centered on each

[59] The presentation in this section follows rather closely that of [Erc08a] who had a similar objective.
[60] Transferability is discussed in more detail in Section 5.7.2.

atomic nucleus: the so-called LCAO. These are the orbitals we found analytically in the discussion of the hydrogen atom; we will denote them here as $\varphi_i^\alpha(\boldsymbol{x})$. The notation serves to remind us that each nucleus α can have several orbitals centered on it, where i represents one of the s, p or d orbitals.[61] For a fixed atomic nucleus α, these orbitals are orthonormal, i.e.

$$\langle \varphi_i^\alpha | \varphi_j^\alpha \rangle = \delta_{ij} \qquad \text{(no sum on } \alpha\text{)}.$$

The main disadvantage of this basis is that orthonormality does not hold for two orbitals centered on different nuclei (i.e. $\langle \varphi_i^\alpha | \varphi_j^\beta \rangle$ does not equal δ_{ij} unless $\alpha = \beta$). In Section 4.4.6, we saw that this introduces both a practical inconvenience (in that we must compute the overlap integrals in Eqn. (4.71)) and a source of error. However, for systems where we expect the electrons to remain relatively tightly bound to the nuclei, the overlap between orbitals on different nuclei will be small and the LCAO serves as a reasonable approximation. Furthermore, the localization of the orbitals means that atomic interactions will be short-range, offering an advantage in efficiency.

In essence, the method we used to solve for the hydrogen molecule in Section 4.4.6 was really a form of TB calculation, and the steps we take here will closely parallel the steps of that discussion. We start from an LCAO approximation to the unknown electronic wave function ψ of our system:

$$\psi(\boldsymbol{x}) = \sum_{j,\alpha} c_j^\alpha \varphi_j^\alpha(\boldsymbol{x}), \tag{4.158}$$

which is just a slight modification of Eqn. (4.55) to emphasize that each basis function in our set depends on an atom α and an orbital of that atom, j. Note also that we assume each orbital for atom α is centered on atom α, and so the wave functions used here relate to the exact wave functions presented in Section 4.4.6 through

$$\varphi_j^\alpha(\boldsymbol{x}) = \varphi_j(\boldsymbol{x} - \boldsymbol{r}^\alpha).$$

Except where it is really necessary for clarity, we will drop the explicit dependence on $\boldsymbol{x}$ in the wave functions. Now, we can proceed to minimize the energy of the system as a function of the coefficients, c_j^α, that we have introduced.[62]

4.6.2 The Hamiltonian and overlap matrices

The goal of TB is to solve the single-particle Schrödinger equation approximately. We start from the DFT single-particle Hamiltonian (Eqn. (4.97)), and generally proceed as we did in the DFT solution with two important differences: (1) we will try to evaluate all the components of the Hamiltonian matrix without any integrals, and (2) we will avoid iterating

[61] Most TB treatments stop at these nine orbitals, but in principle one could include more.

[62] Equation (4.158) implies an aperiodic system. Periodic systems within TB must still satisfy Bloch's theorem, but this presents no major conceptual difficulty. In this case, we must modify the wave function as

$$\psi(\boldsymbol{x}) = \exp(i\boldsymbol{k}\cdot\boldsymbol{x}) \sum_{j,\alpha} c_j^\alpha \varphi_j^\alpha(\boldsymbol{x}),$$

to satisfy Eqn. (4.113) and perform the calculation at a discrete number of $\boldsymbol{k}$-points on a reciprocal space grid.

to a self-consistent electron density. This will be achieved through the use of empirically-fit approximations for the elements of the Hamiltonian matrix, as we will describe soon.

Returning to bra-ket notation for the time being and recalling our starting point at Eqn. (4.76), we have the energy of the system as

$$\langle\psi|\mathcal{H}|\psi\rangle = \mathcal{E}^{\rm e} + \mathcal{E}^{\rm ZZ}, \tag{4.159}$$

where we have combined all the *electronic energy terms* into $\mathcal{E}^{\rm e}$ and left the energy due to interactions between the nuclei as a separate term, $\mathcal{E}^{\rm ZZ}$. The electronic energy is

$$\mathcal{E}^{\rm e} = \left\langle \psi \middle| \mathcal{U} + \frac{p^2}{2m^{\rm el}} \middle| \psi \right\rangle, \tag{4.160}$$

where $\mathcal{U}$ encompasses all electron–electron and electron–nucleus interactions. Because we have introduced pseudopotentials, we take $\mathcal{E}^{\rm ZZ}$ to be a pair potential (as discussed in footnote 43 on page 198):

$$\mathcal{E}^{\rm ZZ} = \frac{1}{2}\sum_{\substack{\alpha,\beta \\ \alpha\neq\beta}}^{N} \phi^{\alpha\beta}(r^{\alpha\beta}), \tag{4.161}$$

where $r^{\alpha\beta}$ is the distance between nuclei α and β.

Assuming that we have a sensible form for the pair potential $\phi^{\alpha\beta}$, the repulsive term $\mathcal{E}^{\rm ZZ}$ presents no real difficulty and we focus our attention on the electronic contribution. Our goal is to compute this energy as a function of the atomic positions, with a still greater goal of using this energy to find equilibrium structures or perform MD. To find the energy, we must proceed as we did in Section 4.4.6: guided by the variational principle, we find the coefficients c_j^α for Eqn. (4.158) that minimize the energy in Eqn. (4.159). As before, this will lead to a generalized eigenvalue problem like Eqn. (4.71), the lowest-energy solutions to which will be filled states consistent with the number of valence electrons in the system. *But here is the big difference*; we want to avoid doing any integrations while building the matrices of the eigenvalue problem. We will therefore replace them with parameterized analytic functions dependent only on the positions of the nuclei.

Recalling the steps in the hydrogen molecule problem that took us from the energy expression to the eigenvalue problem of Eqn. (4.71), we can similarly write

$$\Pi = \left\langle \psi \middle| \mathcal{U} + \frac{p^2}{2m^{\rm el}} \middle| \psi \right\rangle - \epsilon\left(\langle\psi|\psi\rangle - 1\right),$$

and insert Eqn. (4.158) to obtain the matrix equation

$$\sum_\beta \mathcal{H}_{ij}^{\alpha\beta} c_{jI}^\beta = \sum_\beta \mathcal{S}_{ij}^{\alpha\beta} \epsilon_I c_{jI}^\beta \quad \text{(no sum on } I\text{)}, \tag{4.162}$$

where the summation convention is applied to j, while I reminds us that there will be as many solutions to this problem, ψ_I, as there are members in our LCAO basis and

$$\mathcal{H}_{ij}^{\alpha\beta} = \left\langle \varphi_i^\alpha \middle| \mathcal{U} + \frac{p^2}{2m^{\rm el}} \middle| \varphi_j^\beta \right\rangle, \qquad \mathcal{S}_{ij}^{\alpha\beta} = \left\langle \varphi_i^\alpha \middle| \varphi_j^\beta \right\rangle, \tag{4.163}$$

are the *Hamiltonian matrix* and *overlap matrix*, respectively. Later, we will want to compute the energy of the system by filling the lowest-energy eigenstates

$$\mathcal{E}^{\mathrm{e}} = \sum_{I=1}^{N_{\mathrm{basis}}} f_I \epsilon_I, \tag{4.164}$$

where f_I is 2 for energy states up to the Fermi energy, and zero otherwise. The 2, as usual, accounts for electron spin.

Let us look more closely at the terms that appear in $\mathcal{H}_{ij}^{\alpha\beta}$. We can break the Hamiltonian into parts and see the two essential types of terms that appear in $\mathcal{H}_{ij}^{\alpha\beta}$:

$$\mathcal{T}_{ij}^{\alpha\beta} \equiv \left\langle \varphi_i^\alpha \middle| \frac{p^2}{2m^{\mathrm{el}}} \middle| \varphi_j^\beta \right\rangle = \frac{-\hbar^2}{2m^{\mathrm{el}}} \left\langle \varphi_i^\alpha \middle| \nabla^2 \middle| \varphi_j^\beta \right\rangle,$$
$$\mathcal{U}_{ij}^{\alpha\beta} \equiv \left\langle \varphi_i^\alpha \middle| \mathcal{U} \middle| \varphi_j^\beta \right\rangle,$$

so that, $\mathcal{H}_{ij}^{\alpha\beta} = \mathcal{T}_{ij}^{\alpha\beta} + \mathcal{U}_{ij}^{\alpha\beta}$. Integrals like the components of $\mathcal{T}_{ij}^{\alpha\beta}$ are called "two-center" integrals when $\alpha \neq \beta$ and "one-center" or "on-site" integrals if $\alpha = \beta$. For this particular two-center integral, it is easy to see that its value depends only on the positions of atoms α and β, and not on any other nearby atoms. Since it is an integral over all of space, we can define the origin to be located at atom α, in which case the position of atom β and the vector from α to β are the same thing. Then

$$\mathcal{T}_{ij}^{\alpha\beta} = \frac{-\hbar^2}{2m^{\mathrm{el}}} \int \varphi_i^\alpha(\boldsymbol{x}) \nabla^2 \varphi_j^\beta(\boldsymbol{x} - \boldsymbol{r}^{\alpha\beta})\, d\boldsymbol{x}$$

is clearly a function only of the relative positions of the two atoms, $\boldsymbol{r}^{\alpha\beta} = \boldsymbol{r}^\beta - \boldsymbol{r}^\alpha$. This term, at least, seems readily amenable to treatment by some sort of simple parameterization of atomic positions.

The integrals defining the components of $\mathcal{U}_{ij}^{\alpha\beta}$ are not as straightforward, and require us to make the first *big assumption* of TB. If we think back to the DFT section, we recall that $\mathcal{U}$ is a complicated function of the electron density, and in fact depends on the positions of all the atoms in the system. Without some serious simplification, it seems doubtful that we could parameterize it in a simple yet meaningful way. The assumption that is made in TB is to write $\mathcal{U}(\boldsymbol{x})$ as a sum of atom-centered potentials $\mathcal{U}^\alpha(\boldsymbol{x} - \boldsymbol{r}^\alpha)$, each of which is independent of all the other atoms in the system:

$$\mathcal{U}(\boldsymbol{x}) = \sum_{\alpha=1}^{N} \mathcal{U}^\alpha(\boldsymbol{x} - \boldsymbol{r}^\alpha), \tag{4.165}$$

where N is the number of atoms in the system. Again, we emphasize that this is a bold simplification, and it is not clear that it should be especially accurate if we revisit the form of $\mathcal{U}$ from our discussion of DFT. But by making this assumption, we simplify the terms in the Hamiltonian matrix so that they become

$$\mathcal{U}_{ij}^{\alpha\beta} = \sum_{\gamma=1}^{N} \left\langle \varphi_i^\alpha \middle| \mathcal{U}^\gamma(\boldsymbol{x} - \boldsymbol{r}^\gamma) \middle| \varphi_j^\beta \right\rangle. \tag{4.166}$$

These terms are either "one-center" terms, when $\alpha = \beta = \gamma$, "two-center" if $\gamma = \alpha$ or $\gamma = \beta$, or finally "three-center" integrals when all three of α, β and γ are different. As we saw with the kinetic energy term, the two-center terms are straightforward because they only depend on the relative positions of the two atoms, independent of their environment. The three-center terms, however, present another problem, since it will be difficult to devise a scheme to parameterize them in terms of atomic positions. Instead, we are led to make the second *big assumption* that is common in TB methods: that is, we assume that all three-center integrals are zero.

As with our first big assumption, the second is difficult to quantify or justify. For justification, we can certainly imagine that if atom γ is far from both atoms α and β, then the integral of Eqn. (4.166) should be negligible. But on the other hand it seems that there will be many nearby atoms γ for which this justification will not work, and in fact there are TB models in the literature that include the three-center terms in various ways (see, for example, [TPT93, HWF$^+$98] and references therein). Here, we retain the common assumption that the three-center terms are zero. This means that the terms in $\mathcal{H}_{ij}^{\alpha\beta}$ are each only dependent on vectors joining atoms α and β, and it seems reasonable to hope for a robust parameterization in terms of this single vector.

4.6.3 Slater–Koster parameters for two-center integrals

A typical TB formulation may include the s, p and d orbitals on each atom, for a total of nine orbitals. In the simple case of all atoms being the same type, this is a total of 9×9 possible two-center integrals, each of which needs to be parameterized is some sensible way as a function of both bond direction and bond length between two atoms. The sheer number of combinations may have been the "deal-breaker" for TB calculations, if it were not for the fact that symmetries between them reduce the number of nonzero integrals. In fact, only 19 are nonzero, and some of these are related. The 19 comprise a mere 11 independent integrals if atoms α and β are the same species, 15 if α and β are different.

The breakthrough that made parameterization of these integrals possible came in the form of a paper by Slater and Koster [SK54], who made the assumption that the integrals could be decomposed into a radial and an angular dependence, after which the integrals of the latter could be evaluated in closed form. The radial (bond length) dependence can then be empirically approximated by a pair potential to complete the parameterization. Details of this parameterization are beyond the scope of this book, and the interested reader is directed to, for example, the book by Finnis [Fin03]. The important message from our perspective is that the two-center integrals can be replaced by a handful of function evaluations depending only on the vector joining the two atoms being considered and the orientation of the basis orbitals relative to the direction of the bond.[63]

Finally, it is typical to cut-off the radial dependence to first- or second-neighbor distances in the crystal structure of interest to speed up the computations. This has to be done carefully, since abrupt truncation leads to numerical problems (this is discussed more generally in

[63] The Slater–Koster derivation tacitly incorporates rotational invariance into TB. In other words, the results of a TB simulation are independent of a rigid rotation of the coordinate system defining the atomic positions, as required on physical grounds. See further discussion of rotational invariance in Chapter 5.

Section 5.3). Keeping in mind that we are already resigned to choosing an approximate form for the radial dependence, and further that we have already neglected much physics in ignoring the three-center integrals, it seems unlikely that the longer-range interactions will be worth the computational effort.[64]

4.6.4 Summary of the TB formulation

In essence, this completes the task of formulating a TB model, and we summarize the main steps in the flowchart in Fig. 4.17(b). All the integrals in the Hamiltonian matrix are parameterized, and given the positions of the atoms and the LCAO basis we can find the total energy as follows. We first loop over all elements in the Hamiltonian matrix and compute their values by summing over suitable linear combinations of the Slater–Koster parameters. The eigenvalue problem is then solved as usual, to obtain the constants c^{β}_{jI} and the energies ϵ_I defining the n eigenstates. These are then filled in ascending order of energy, until the total number of filled states is equal to the number of electrons in the system.

Clearly, this is still not a trivial calculation, and in fact has many of the same elements as the DFT computations previously discussed. We still need to build the Hamiltonian matrix (indicated by the bold-ruled box in Fig. 4.17(b)), but it is now a smaller matrix (assuming we are using a smaller number of orbitals than we would plane waves in DFT). The elements of the Hamiltonian matrix are more rapid to compute than in DFT (since integrals are replaced by empirical fits), but they still involve a large number of function evaluations for each element. By far the most dominant component of the computation is the need to diagonalize the Hamiltonian, although we note that the TB assumptions make this a rather sparse, banded matrix amenable to efficient $O(N^2)$ algorithms. Finally, there is no need to iterate to find a self-consistent electron density in this case, and so a single energy calculation requires only one, rather than several, passes through the above steps.

4.6.5 TB molecular dynamics

Although it is still significantly more expensive than an empirical atomistic model (to be discussed in Chapter 5), TB is sufficiently fast to permit the real possibility of MS (Chapter 6) and MD (Chapter 9) simulations within the framework. Currently, calculations of on the order of hundreds of atoms over a few picoseconds in TB-MD simulations are routine, and considerably larger simulations are possible with large-scale parallelization (see Tab. 5.2). Of course, for either MS or MD, one requires the forces on the atoms, or more mathematically the derivatives of the total energy with respect to atomic positions. Fortunately, the TB formalism permits a rather expedient force calculation in an analytical form derived directly from the energy expression. As such, TB models are not confronted with the potential pitfalls of using numerically (or otherwise) approximated forces. Here, we sketch out the main ideas, while details of the force calculation can be found, for example, in [Col05].

[64] This assumes, of course, that we are applying TB to materials for which it is appropriate, i.e. covalently bonded systems, where the bonding is, in fact, relatively short-range.

The force on an ion is the derivative of the total energy with respect to its position:

$$\boldsymbol{f}^{\alpha} = -\frac{\partial}{\partial \boldsymbol{r}^{\alpha}}\left(\mathcal{E}^{\mathrm{e}} + \mathcal{E}^{\mathrm{ZZ}}\right).$$

The derivative of the pair potential $\mathcal{E}^{\mathrm{ZZ}}$ presents no difficulty (it is discussed in detail in Section 5.8), while the energy of the electrons is given in Eqn. (4.164) as a sum of the eigenvalues of Eqn. (4.162). Recall that these are

$$\epsilon_I = \langle \psi_I | \mathcal{H} | \psi_I \rangle = \sum_{\alpha,i,\beta,j} (c_{iI}^{\alpha})^* c_{jI}^{\beta} \mathcal{H}_{ij}^{\alpha\beta} \qquad \text{(no sum on } I\text{)},$$

where the last step made use of Eqn. (4.158). The coefficients c_{iI}^{α} are the eigenvector solutions, and so it is not obvious how to take their derivatives with respect to the ion positions since the parameterization in terms of $\boldsymbol{r}$ was applied to the matrices $\boldsymbol{\mathcal{H}}$ and $\boldsymbol{\mathcal{S}}$. However, we can rearrange things so that the derivatives are only applied to the Hamiltonian and overlap matrices. We start by formally taking the derivative of the last equation

$$\frac{\partial \epsilon_I}{\partial r} = \sum_{\alpha,i,\beta,j} \left[\frac{\partial (c_{iI}^{\alpha})^*}{\partial r} c_{jI}^{\beta} \mathcal{H}_{ij}^{\alpha\beta} + (c_{iI}^{\alpha})^* \frac{\partial c_{jI}^{\beta}}{\partial r} \mathcal{H}_{ij}^{\alpha\beta} + (c_{iI}^{\alpha})^* c_{jI}^{\beta} \frac{\partial \mathcal{H}_{ij}^{\alpha\beta}}{\partial r} \right], \tag{4.167}$$

where r represents any one component of one of the ion positions. We then use the normalization of the eigensolutions, ψ_I, which gives us

$$\langle \psi_I | \psi_I \rangle = \sum_{\alpha,i,\beta,j} (c_{iI}^{\alpha})^* c_{jI}^{\beta} \mathcal{S}_{ij}^{\alpha\beta} = 1.$$

Differentiating this equation with respect to r and then multiplying all terms by ϵ_I, we get

$$\sum_{\alpha,i,\beta,j} \left[\epsilon_I \frac{\partial (c_{iI}^{\alpha})^*}{\partial r} c_{jI}^{\beta} \mathcal{S}_{ij}^{\alpha\beta} + \epsilon_I (c_{iI}^{\alpha})^* \frac{\partial c_{jI}^{\beta}}{\partial r} \mathcal{S}_{ij}^{\alpha\beta} + \epsilon_I (c_{iI}^{\alpha})^* c_{jI}^{\beta} \frac{\partial \mathcal{S}_{ij}^{\alpha\beta}}{\partial r} \right] = 0. \tag{4.168}$$

The first term of this equation is equal to the first term on the right-hand side of Eqn. (4.167), which is apparent after making use of Eqn. (4.162). The same is true for the second terms in the two equations. Thus, we can subtract Eqn. (4.168) from Eqn. (4.167) to get

$$\frac{\partial \epsilon_I}{\partial r} = \sum_{\alpha,i,\beta,j} \left[(c_{iI}^{\alpha})^* \left(\frac{\partial \mathcal{H}_{ij}^{\alpha\beta}}{\partial r} - \epsilon_I \frac{\partial \mathcal{S}_{ij}^{\alpha\beta}}{\partial r} \right) c_{jI}^{\beta} \right].$$

Because the Hamiltonian and overlap matrices are parameterized in terms of the ionic positions, it is relatively straightforward to compute their derivatives, and the forces on the ions can subsequently be found.

4.6.6 From TB to empirical atomistic models

Above, we introduced the TB approximation to quantum mechanics. Although the TB formalism represents a big reduction in the workload compared to a DFT calculation, it is still a long way from being able to model the hundreds of thousands or more atoms required to look at deformation problems of "engineering" interest. It is for this reason that we have the multitude of classical interatomic models described in Chapter 5. However, the TB formalism does offer a theoretical bridge between DFT and many of the simpler empirical forms. We sketch this bridging process here, but refer the reader to the books in the further reading section at the end of the chapter for more detail.

An alternative way of discussing electronic energy contributions is through the *density of states* (DOS), $D(\epsilon)$, which is defined as the number of available electronic states with energy lying between ϵ and $\epsilon + d\epsilon$ for a given system. In this way, the total electronic energy can be expressed as the integral

$$\mathcal{E}^{\mathrm{e}} = \int_{-\infty}^{\infty} \epsilon f(\epsilon) D(\epsilon)\, d\epsilon, \tag{4.169}$$

where $f(\epsilon)$ plays the role of the "filling function," f_I, as a function of energy:

$$f(\epsilon) = \begin{cases} 2 & \text{when } \epsilon = \epsilon_I \text{ is a filled state,} \\ 0 & \text{otherwise.} \end{cases}$$

We can also define a DOS on an orbital-by-orbital basis. This provides information about the filled electronic states associated with a given orbital and the contribution of those states to the total energy. Thus we can define a DOS, $D_j^\alpha(\epsilon)$, associated with orbital j attached to atom α, from which we find an orbital-based energy:

$$\mathcal{E}_j^\alpha = \int_{-\infty}^{\infty} \epsilon f(\epsilon) D_j^\alpha(\epsilon)\, d\epsilon. \tag{4.170}$$

For consistency, we require that

$$D(\epsilon) = \sum_\alpha \sum_j D_j^\alpha(\epsilon),$$

so that

$$\mathcal{E}^{\mathrm{e}} = \sum_\alpha \sum_j \mathcal{E}_j^\alpha. \tag{4.171}$$

In principle, all the information we need about the bonding of a material is encoded in this DOS, and much work has been done to study the relation between the DOS and alternative energy expressions. Cyrot-Lackmann and her coworkers [CL68, DCL70, DCL71, GCL73] were the first to propose that the DOS can be approximated by a function of its *moments*. We define the mth moment, $\mu(m)$, of the DOS, $D(\epsilon)$, as

$$\mu(m) \equiv \int_{-\infty}^{\infty} \epsilon^m D(\epsilon)\, d\epsilon, \tag{4.172}$$

where ϵ^m is ϵ raised to the mth power. The zeroth moment is just the area under $D(\epsilon)$, which should be equal to the number of filled states in the system (the number of valence electrons), while the first moment is the total electronic energy of the system, $\mathcal{E}^{\mathrm{e}}$ (see Eqn. (4.169)). Higher-order moments provide information about the shape and symmetries of the DOS function. It is also natural to define moments of the orbital DOS as

$$\mu_j^\alpha(m) \equiv \int_{-\infty}^{\infty} \epsilon^m D_j^\alpha(\epsilon)\, d\epsilon, \tag{4.173}$$

where $D_j^\alpha(\epsilon)$ is the orbital-based DOS.

The utility of defining these moments comes from the fact that one can rigorously write the DOS (and therefore the total electronic energy) as an infinite series of terms dependent only on the moments $\mu(0), \ldots, \mu(\infty)$. This has inspired much work to find approximations to the total energy based on functions of a finite number of low-order moments, much like a Taylor series polynomial expansion approximates more complex mathematical functions. The TB formulation is especially well suited to this approach, since the moments of the DOS turn out to be very easily computed from the existing TB Hamiltonian. Omitting the derivation for brevity,[65] the orbital DOS moments can be computed as

$$\mu_j^\alpha(m) = \langle \varphi_j^\alpha | \mathcal{H}^m | \varphi_j^\alpha \rangle . \tag{4.174}$$

Here $\mathcal{H}^m$ refers to the Hamiltonian raised to the mth power – in other words the Hamiltonian operator applied m times. As with the total DOS, the orbital DOS has a zeroth moment (equal to 1 for normalized orbitals) and first moment (equal to the average electron energy, relative to some arbitrary scale) that are not especially useful for characterizing the DOS. However, second and higher moments provide additional information that allows us to build approximate models. Evaluating the first moment is straightforward since

$$\mu_i^\alpha(1) = \langle \varphi_i^\alpha | \mathcal{H} | \varphi_i^\alpha \rangle , \tag{4.175}$$

which we recognize as the diagonal elements of the Hamiltonian matrix from Eqn. $(4.163)_1$. The second moment can be readily evaluated by inserting the identity operator (Eqn. (4.14)) between the two Hamiltonian operators

$$\mu_i^\alpha(2) = \sum_{j,\beta} \left\langle \varphi_i^\alpha \middle| \mathcal{H} \middle| \varphi_j^\beta \right\rangle \left\langle \varphi_j^\beta \middle| \mathcal{H} \middle| \varphi_i^\alpha \right\rangle , \tag{4.176}$$

and likewise third and higher moments are systematically evaluated by

$$\mu_i^\alpha(3) = \sum_{k,\gamma} \sum_{j,\beta} \left\langle \varphi_i^\alpha \middle| \mathcal{H} \middle| \varphi_j^\beta \right\rangle \left\langle \varphi_j^\beta \middle| \mathcal{H} \middle| \varphi_k^\gamma \right\rangle \left\langle \varphi_k^\gamma \middle| \mathcal{H} \middle| \varphi_i^\alpha \right\rangle . \tag{4.177}$$

It is hard to immediately see the power of these forms, but a little reflection will help. First, we see that they present a potentially rapid way to evaluate the energy, since all of these moments can be evaluated directly from the Hamiltonian matrix without the need for matrix diagonalization. This is clearly an advantage for large systems where matrix diagonalization is going to be expensive. The computational effort to compute the moments is even easier than the equations suggest, and can be described as a "linked path" method. In essence, $\mu_j^\alpha(m)$ is the sum over all possible paths comprising exactly m links that start and end on orbital φ_j^α. A "link" exists between φ_j^α and φ_k^β only if $\mathcal{H}_{jk}^{\alpha\beta} \neq 0$, representing an energetic bond between the two orbitals for the given configuration of the atomic nuclei.

Armed with this method for rapid calculation of the moments, the next step in computing the energy of the system is to solve the "inverse problem" of estimating the DOS from

[65] See Finnis' book [Fin03] or the original references [CL68, DCL70, DCL71, GCL73] for the derivation of Eqn. (4.174).

knowledge of only the first few moments. Conceptually, we might try to replace the unknown function D_j^α with an approximation parameterized by its moments:

$$D_j^\alpha(\epsilon) \approx \hat{D}(\epsilon; \mu_j^\alpha(0), \mu_j^\alpha(1), \mu_j^\alpha(2), \ldots). \tag{4.178}$$

For a small, fixed number of moments, this might be achieved by assuming a simple functional form for the DOS and fitting it to the calculated moments from the Hamiltonian matrix (see Exercise 4.12). In a more general framework, the DOS can be optimally fitted with a linear combination of basis functions [KV95, VKS96]. Finally, the electronic energy can be estimated by using the fitted DOS in Eqn. (4.169). In this way, "linear scaling" ($O(N_{\text{basis}})$) TB is possible, although its accuracy is not nearly as good as the full $O(N_{\text{basis}}^3)$ TB with matrix inversion unless a very large number of moments are calculated. These ideas are the driving force behind the bond-order potentials discussed in Section 5.6, whereby a method is developed that effectively selects optimal combinations of the moments to use in approximating the TB energy.

From the perspective of this book, the moment expansion approach serves as a bridge to the empirical forms that will be presented in Chapter 5. Here, we elaborate on two examples: the generic cluster potential and the Finnis–Sinclair form of the pair functional.

The "cluster potential" form Just to simplify things, let us consider a case where there is only one orbital attached to every atom. Results for larger numbers of orbitals are exactly analogous, but we have to carry around summations over the orbitals that clutter the notation. Recall the definition of the Hamiltonian matrix in Eqn. $(4.163)_1$ and consider Eqn. (4.176). We see that this is simply a contraction on $\mathcal{H}^{\alpha\beta}$, i.e.

$$\mu^\alpha(2) = \sum_\beta \mathcal{H}^{\alpha\beta}\mathcal{H}^{\beta\alpha},$$

where the references to orbitals have been removed for clarity. For a given atom, α, one adds a contribution for each nonzero entry in row α of the Hamiltonian matrix. These entries are nonzero because they represent an energetic interaction between atom α and another atom; these contributions are then *pairwise interactions between atoms*. Likewise, the third moment (Eqn. (4.177)) represents energetic contributions due to triplets of atoms:

$$\mu^\alpha(3) = \sum_\gamma \sum_\beta \mathcal{H}^{\alpha\beta}\mathcal{H}^{\beta\gamma}\mathcal{H}^{\gamma\alpha}.$$

Visiting an atom α and finding a neighbor β with which there is an interaction (i.e. a nonzero $\mathcal{H}^{\alpha\beta}$), one multiplies this contribution by the contribution of interactions between a third atom γ and both atoms α and β (nonzero entries $\mathcal{H}^{\beta\gamma}$ and $\mathcal{H}^{\gamma\alpha}$). Thus, this is a sum of all contributions for which one can make a path from α to β to γ and back to α along bonds for which the Hamiltonian matrix entries are nonzero; these are the three-body interactions. Similarly, the fourth moment reflects four-body interactions, etc. In other words, we can write the moments as

$$\mu^\alpha(1) = \mathcal{H}^{\alpha\alpha}, \qquad \mu^\alpha(2) = [\mu^\alpha(1)]^2 + \sum_{\substack{\beta \\ \beta\neq\alpha}} \widetilde{\phi}_{\alpha\beta}(\boldsymbol{r}^\alpha, \boldsymbol{r}^\beta),$$

$$\mu^\alpha(3) = \mu^\alpha(1)\mu^\alpha(2) + \sum_{\substack{\beta,\gamma \\ \beta\neq\gamma\neq\alpha}} \widetilde{\phi}_{\alpha\beta\gamma}(\boldsymbol{r}^\alpha, \boldsymbol{r}^\beta, \boldsymbol{r}^\gamma), \qquad \dots$$

where

$$\widetilde{\phi}_{\alpha\beta} = \mathcal{H}^{\alpha\beta}\mathcal{H}^{\beta\alpha}, \qquad \widetilde{\phi}_{\alpha\beta\gamma} = \mathcal{H}^{\alpha\beta}\mathcal{H}^{\beta\gamma}\mathcal{H}^{\gamma\alpha}, \qquad \dots$$

and these are functions of atomic positions through the Slater–Koster parameters and the bond length that went into evaluating the elements of the Hamiltonian. Now, recall that we have an approximate DOS parameterized by these moments (Eqn. (4.178)) that we can put back into Eqn. (4.171). Again, dropping the reference to orbitals for simplicity we have

$$\mathcal{E}^{\mathrm{e}} \approx \sum_{\alpha=1}^{N} \int_{-\infty}^{\infty} \epsilon f(\epsilon) \hat{D}^\alpha(\epsilon; \mu^\alpha(0), \mu^\alpha(1), \mu^\alpha(2), \dots)\, d\epsilon. \tag{4.179}$$

The final form of this expression after the evaluation of the integral will depend on how $\hat{D}^\alpha$ is parameterized by the moments. If, for instance, the parameterization allows $\hat{D}^\alpha$ to be written as a linear combination of the form

$$\hat{D}^\alpha = \sum_{m=0}^{m_{\max}} G_m(\epsilon)\mu^\alpha(m),$$

where G_m is a general function of ϵ, then the electronic energy can be written as

$$\mathcal{E}^{\mathrm{e}}(\boldsymbol{r}^1, \dots, \boldsymbol{r}^N) = \sum_{\alpha} \widehat{\phi}_1^\alpha + \frac{1}{2!} \sum_{\substack{\alpha,\beta \\ \alpha\neq\beta}}^{N} \widehat{\phi}_2(\boldsymbol{r}^\alpha, \boldsymbol{r}^\beta) + \frac{1}{3!} \sum_{\substack{\alpha,\beta,\gamma \\ \alpha\neq\beta\neq\gamma}}^{N} \widehat{\phi}_3(\boldsymbol{r}^\alpha, \boldsymbol{r}^\beta, \boldsymbol{r}^\gamma) + \cdots, \tag{4.180}$$

where the $\widehat{\phi}$ functions depend only on the $\widetilde{\phi}$ expressions and the results of the integrals

$$\int_{-\infty}^{\infty} \epsilon f(\epsilon) G_m(\epsilon)\, d\epsilon.$$

While these expressions are not likely to be evaluated easily, the discussion is sufficient to motivate the empirical form in Eqn. (4.180), which is precisely the form of the *cluster potential* that will be pursued as one of the important empirical models presented in Chapter 5. Recall that the electronic energy must be combined with another pair potential as in Eqn. (4.159) to get the total energy.

The "Finnis–Sinclair" form Unlike cluster potentials, a *cluster functional* approach (see Section 5.6) seeks to write the energy in terms of *functionals* of groups of two, three or more atoms. It is clear from Eqn. (4.179) that we can write the electronic energy $\mathcal{E}^{\mathrm{e}}$ as

$$\mathcal{E}^{\mathrm{e}} = \sum_{\alpha} F(\mu^\alpha(0), \mu^\alpha(1), \mu^\alpha(2), \mu^\beta(3), \dots),$$

where we have defined the moment per atom as a sum over the moments of orbitals centered on the atom:

$$\mu^\alpha(m) = \sum_j \mu_j^\alpha(m). \tag{4.181}$$

This form of $\mathcal{E}^e$ would be convenient computationally, but it is difficult to derive given the form of Eqn. (4.179). The simplest such approach is to neglect higher moments so that only the second moments of the DOS play a role,[66] so that we can write the electronic energy as simply

$$\mathcal{E}^e = \sum_\alpha U(\mu^\alpha(2)). \tag{4.182}$$

Now the second moment will comprise a sum over each orbital for each neighbor β to atom α, as we can see by summing Eqn. (4.176) over i and rearranging the sums:

$$\mu^\alpha(2) = \sum_\beta \left[\sum_{i,j} \left\langle \varphi_i^\alpha \middle| \mathcal{H} \middle| \varphi_j^\beta \right\rangle \left\langle \varphi_j^\beta \middle| \mathcal{H} \middle| \varphi_i^\alpha \right\rangle \right].$$

The quantity inside the square brackets depends only the pairwise positions of two ions, $\boldsymbol{r}^\alpha$ and $\boldsymbol{r}^\beta$. In fact, it only depends on the scalar inter-ion *distance*, $r^{\alpha\beta} = \left\| \boldsymbol{r}^\beta - \boldsymbol{r}^\alpha \right\|$, although we did not go into sufficient detail in the discussion of the Slater–Koster parameterization to make this obvious. This restriction is a necessary and sufficient condition for the moment $\mu^\alpha(2)$ and the energy which depends on it to be rotationally and translationally invariant (we prove this statement in Chapter 5), so it is important that this restriction holds. It also means that the second moment can be computed as

$$\mu^\alpha(2) = \sum_\beta g(r^{\alpha\beta}), \tag{4.183}$$

where g also depends on some combination of Slater–Koster parameters.

Finally, it is evident from Eqn. (4.173) that the units of $\mu^\alpha(2)$ will be energy-squared. From a purely dimensional argument then, the simplest choice for the functional form of U in Eqn. (4.182) is a $\sqrt{\mu^\alpha(2)}$ dependence:

$$\mathcal{E}^e = -A \sum_\alpha \sqrt{\mu^\alpha(2)}, \tag{4.184}$$

where A is some positive fitting parameter.[67] This is precisely the form of the Finnis–Sinclair empirical model described in Section 5.5.

In this chapter, we have essentially talked about Schrödinger's equation and how it pertains to materials science. This single equation holds the key to understanding how materials behave, but its relatively simple form hides so much complexity that exact solutions do not exist for systems of more than a single electron. As such, we looked at the principal methods of approximate solution, all of which are variational techniques that make use

[66] The zeroth and first moments only add a constant term to the energy.

[67] The square-root form can be more rigorously motivated from a more thorough treatment of TB, but it is beyond the scope of our discussion. The interested reader may try Finnis' book [Fin03].

of a finite set of basis functions to approximate the real electronic structure. DFT and its more approximate cousin TB are the main tools we have to investigate materials quantum mechanically, but they are limited to relatively small systems because of their computational expense. In the next chapter, we look at more approximate, empirical methods that permit simulations involving billions, even trillions, of atoms, thereby making the study of complex materials systems possible.

Further reading

- For a thought-provoking (delightfully controversial) discussion of the role of quantum mechanics in the concept of "free will" and materials science, amongst other things, see *The Emperor's New Mind* by Roger Penrose [Pen99].
- For readers who are completely unfamiliar with quantum mechanics, and who seek an even more elementary starting point, the undergraduate-level physics textbook *Introduction to Quantum Mechanics* by David J. Griffiths [Gri05] is a lucid explanation, written with an engaging style.
- *The Feynmann Lectures on Physics* [FLS06], originally published in 1963, are legendary for their clarity and humor. The last 21 of the 115 lectures focus on quantum mechanics specifically.
- More details about quantum mechanical approaches to atomic bonding can be found in Finnis' *Interatomic Forces in Condensed Matter* [Fin03].
- For an historical account of the interpretation of quantum mechanics, there is *Understanding Quantum Mechanics* by Roland Omnès [Omn99].
- Schrödinger's more philosophical books, *What is Life* [Sch44] and *My View of the World* [Sch83b] are compelling reads, revealing how Schrödinger's depth of thought went beyond pure mathematics and physics.
- For a more "practical" view of quantum mechanics aimed at engineers and materials scientists, Harrison's *Applied Quantum Mechanics* [Har00] and Kroemer's *Quantum Mechanics: For Engineering, Materials Science, and Applied Physics* [Kro94] are good places to look, although they require some knowledge of the basics of quantum theory.

Exercises

4.1 [SECTION 4.4] Verify that the expectation value of the momentum, Eqn. (4.29), is identically zero for a bounded wave function that is purely real or purely imaginary, i.e. for $\psi(x) = a(x)$ or $\psi(x) = ib(x)$, where a and b are real functions of x that vanish at $x = \pm\infty$.

4.2 [SECTION 4.4] Solve the time-independent Schrödinger's equation (Eqn. (4.41)) in a one-dimensional setting for the case of a Dirac delta-function potential, $\mathcal{U} = -A\delta(x)$, with A a constant with units of eV × Å. To make progress, note that this potential is zero everywhere except at the origin, where it is infinite. As such, you can solve the equation independently for $x < 0$ and $x > 0$, and then enforce continuity of the wave function at $x = 0$. It is also necessary to integrate Schrödinger's equation from $-\Delta$ to Δ and take the limit as $\Delta \to 0$, to determine the jump in the derivative $d\psi/dx$ at the origin. Show that there is only one bound state (defined as a state with $\epsilon < 0$), for which $\epsilon = -m^{\text{el}} A^2/2\hbar^2$, and plot the wave function.

4.3 [SECTION 4.4] Write down the $3s$ and $3p$ wave functions of the hydrogen atom (φ_{300} and φ_{310}) and show that:
1. $\langle r \rangle = 27r_0/2$ for φ_{300};
2. $\langle r \rangle = 25r_0/2$ for φ_{310};
3. $\langle \theta \rangle = \pi/2$ for both wave functions;
4. the location where the electron is most likely to be found for φ_{310} is at $r = 1.7489r_0$ and $\theta = (0, \pi)$.

4.4 [SECTION 4.4] Verify that the wave function of Eqn. (4.64) is normalized.

4.5 [SECTION 4.4] Verify the second equality in each of the relations shown in Eqns. (4.66).

4.6 [SECTION 4.4] Evaluate the expression for the overlap integral, s, in Eqn. (4.66) to verify that Eqn. (4.70) is correct. To do this, it is helpful to first evaluate the (trivial) ϕ integral, then evaluate the θ integral by making the change of variables $y^2 = r^2 + R^2 - 2rR\cos\theta$.

4.7 [SECTION 4.4] Consider the single-electron hydrogen molecule with Eqn. (4.63) as the approximate wave function. Write the expression for the expectation value of the Hamiltonian in terms of the ratio $A = c^{\mathsf{A}}/c^{\mathsf{B}}$ and verify that ϵ_{+1} and ϵ_{-1} are in fact the minimum and maximum of this function with respect to A.

4.8 [SECTION 4.5] Consider a system of three electrons with coordinates $\boldsymbol{x}_i$ $(i = 1, 2, 3)$ and imagine that we have found a set of three orthonormal single-particle wave functions $\psi_i(\boldsymbol{x})$. Verify that the determinant of the matrix $\mathbf{S}$ with components $\mathsf{S}_{ij} \equiv \psi_i(\boldsymbol{x}_j)$ satisfies the anti-symmetry condition of Eqn. (4.73). This is the so-called "Slater determinant" and it is generalizable to any number of electrons.

4.9 [SECTION 4.5] Using Eqn. (4.73), show that the definition of the electron density in Eqn. (4.74) is independent of which electrons we choose to integrate out.

4.10 [SECTION 4.5] Show that Eqn. (4.152) follows from Eqn. (4.151) and the discrete Fourier transform of Eqn. (4.135). It helps to recall that the electron density is real, so that $\rho = \rho^*$.

4.11 [SECTION 4.5] Following the steps from Eqn. (4.107) to Eqn. (4.108), derive Eqn. (4.148).

4.12 [SECTION 4.6] Consider a simple DOS of the following double-peaked form:

$$D(\epsilon) = \begin{cases} a & \text{for } c < \epsilon < (c + \Delta\epsilon), \\ b & \text{for } d < \epsilon < (d + \Delta\epsilon), \\ 0, & \text{otherwise.} \end{cases}$$

Assume that all parameters, a, b, c, d and $\Delta\epsilon$ are positive real numbers.
1. Write a general expression for the mth moment of this DOS.
2. Write a computer routine that performs a least-squares fit of the moments $\mu(0)$ through $\mu(m)$ to an arbitrary set of n real numbers.
3. Write a computer program that takes an arbitrary DOS, numerically evaluates its moments and fits the simple DOS proposed above to these moment values. Compare the actual DOS to the fitted results for different functions and numbers of moments and consider the limitations of the simple DOS suggested here.

5 Empirical atomistic models of materials

In the previous chapter, we saw the enormous complexity that governs the bonding of solids. Electronic interactions and structure in the presence of the positively charged ionic cores can only be fully understood using the machinery of Schrödinger's equation and quantum mechanics. But the simplest of bonding problems, that of two hydrogen atoms, is already too complex to solve exactly. Density functional theory (DFT), whereby Schrödinger's equation is recast in a form amenable to numerical solution, provides us with a powerful tool for accurately solving complex bonding problems, but at the expense of significant computational effort. Tight-binding (TB) reduces the burden by parameterizing many of the integrals in DFT into simple analytic forms, but still requires expensive matrix inversion. To have any hope of modeling the complex deformation problems of interest to us (see, for example, Chapter 6), we must rely on much more approximate approaches for describing the atomic interactions using fitted functional forms.

As we saw in Section 4.6, the TB formulation provides a bridge between DFT and empirical potentials. However, given the boldness of some of the approximations that take us from full quantum mechanics to TB, one might question why empirical models should work at all. It is true that we must view all empirical results suspiciously. In some cases they can be quite accurate, while in others we may only be able to use them as idealized models that capture the general trends or basic mechanisms of the real systems. In their latter role such models can still be quite useful since they allow us to systematically change a material parameter to isolate the effect of that parameter on material behavior. For example, we could use this approach to study the effect of the stacking fault energy (see Section 6.5.5) on the plasticity of a metal by varying the value of this parameter through appropriate changes to the parameterization of the empirical model.

It is interesting that the current trend towards the development of accurate, fitted, empirical atomistic models essentially brings the field full circle back to its origins. In an early discussion of interatomic forces written by Lennard-Jones [Fow36, Chapter X] the following optimistic assessment was made:

> It will some day be possible, thanks to the work initiated by Heitler and London, to derive the interatomic energy and so the forces for any atoms or molecules from their electronic structure. But though no difficulties of principle remain, the day is far distant when such calculations will be a practical possibility. At present we must still rely on indirect methods for such knowledge as we have of intermolecular fields.

Clearly a great deal of progress has been made since those words were originally written in 1929 and much of Lennard-Jones' prediction has come through. It is indeed possible to use quantum mechanical methods to compute the energy and forces for any arrangement of atoms. However, the computational burdens associated with such calculations are so formidable that the "indirect methods" (empirical models) that Lennard-Jones mentions still play a dominant role today.

5.1 Consequences of the Born–Oppenheimer approximation (BOA)

In Section 4.4.6, we introduced the BOA, which states that the electrons can always respond instantaneously to changes in the atomic positions. This assumption is reasonable for many problems of interest in materials science. In this section we show that a key consequence of the BOA is that it leads to a Hamiltonian for the nuclei that does not depend directly on the electrons.[1] Instead, the effect of the electrons can be embodied in a single potential energy function dependent only on the nuclear positions, $\mathcal{U}(\boldsymbol{r})$, where $\boldsymbol{r} = (\boldsymbol{r}^1, \dots, \boldsymbol{r}^N)$ and N is the number of atomic nuclei. This result is important because it paves the way for the entire field of atomistic modeling, whereby we seek approximate analytical forms for the unknown function $\mathcal{U}(\boldsymbol{r})$.

In principle, when we are solving a problem involving both electrons and atomic nuclei, we are searching for a wave function in terms of all the electronic and nuclear coordinates. Thus we want to solve the full, time-dependent Schrödinger equation

$$\mathcal{H}\chi(\boldsymbol{x}^{\mathrm{el}}, \boldsymbol{r}, t) = i\hbar\frac{\partial\chi}{\partial t}, \tag{5.1}$$

where χ is the time-dependent wave function for all particles in the system (electrons and nuclei), $\boldsymbol{x}^{\mathrm{el}} = (\boldsymbol{x}_1, \dots, \boldsymbol{x}_{N^{\mathrm{el}}})$ are the coordinates of the N^{el} electrons and $\mathcal{H}$ is the Hamiltonian for *all particles*:[2]

$$\mathcal{H} = \mathcal{U}^{\mathrm{ee}}(\boldsymbol{x}^{\mathrm{el}}) + \mathcal{U}^{\mathrm{ZZ}}(\boldsymbol{r}) + \mathcal{U}^{\mathrm{eZ}}(\boldsymbol{x}^{\mathrm{el}}, \boldsymbol{r}) - \sum_{i=1}^{N^{\mathrm{el}}} \frac{\hbar^2\nabla_i^2}{2m^{\mathrm{el}}} - \sum_{\alpha=1}^{N} \frac{\hbar^2\nabla_\alpha^2}{2m^\alpha}.$$

Here m^α is the mass of nucleus α, and the Laplacian operators are over the coordinates of a single electron i or nucleus α as indicated by the subscripts. The potential energy terms represent the electron–electron ($\mathcal{U}^{\mathrm{ee}}$), nucleus–nucleus ($\mathcal{U}^{\mathrm{ZZ}}$) and electron–nucleus ($\mathcal{U}^{\mathrm{eZ}}$) interactions:

$$\mathcal{U}^{\mathrm{ZZ}} = \frac{1}{2}\sum_{\substack{\alpha,\beta \\ \alpha\neq\beta}}^{N} \frac{e^2 Z^\alpha Z^\beta}{\|\boldsymbol{r}^\alpha - \boldsymbol{r}^\beta\|}, \qquad \mathcal{U}^{\mathrm{eZ}} = \sum_{i=1}^{N^{\mathrm{el}}}\sum_{\beta=1}^{N} \frac{-e^2 Z^\beta}{\|\boldsymbol{x}_i - \boldsymbol{r}^\beta\|}, \qquad \mathcal{U}^{\mathrm{ee}} = \frac{1}{2}\sum_{\substack{i,j \\ i\neq j}}^{N^{\mathrm{el}}} \frac{e^2}{\|\boldsymbol{x}_i - \boldsymbol{x}_j\|},$$

where Z^β is the atomic number of nucleus β. Throughout this discussion, sums over lower-case Roman letters are over all electrons, while sums over Greek letters are over all nuclei.

[1] In our discussion, we follow closely the arguments of [Wei83].

[2] Note that when we introduced Schrödinger's equation in the context of only electrons in Chapter 4, the nuclei were treated as a fixed potential. In effect, we had already tacitly invoked the BOA at that time. However, all particles, including the protons and neutrons in the atomic nuclei, are governed by the same equation.

The BOA suggests that the solution can be represented in separable form:

$$\chi(\boldsymbol{x}^{\text{el}}, \boldsymbol{r}, t) = \psi^{\text{el}}(\boldsymbol{x}^{\text{el}}, \boldsymbol{r})\psi^{\text{nuc}}(\boldsymbol{r}, t),$$

where the wave function of the electrons, ψ^{el}, depends on the positions of the nuclei (since they provide the external potential), but the assumption of instantaneous electronic response means it is not directly dependent on time. The wave function of the nuclei, ψ^{nuc}, is time-dependent. Applying the Hamiltonian operator to this wave function and putting it in Eqn. (5.1) leads to

$$\frac{-\hbar^2}{2}\left[\sum_{i=1}^{N^{\text{el}}} \frac{\psi^{\text{nuc}}\nabla_i^2\psi^{\text{el}}}{m^{\text{el}}} + \sum_{\alpha=1}^{N} \frac{\psi^{\text{nuc}}\nabla_\alpha^2\psi^{\text{el}}}{m^\alpha} + \sum_{\alpha=1}^{N} \frac{2\nabla_\alpha\psi^{\text{el}} \cdot \nabla_\alpha\psi^{\text{nuc}}}{m^\alpha} + \sum_{\alpha=1}^{N} \frac{\psi^{\text{el}}\nabla_\alpha^2\psi^{\text{nuc}}}{m^\alpha}\right] + (\mathcal{U}^{\text{ee}} + \mathcal{U}^{\text{ZZ}} + \mathcal{U}^{\text{eZ}})\psi^{\text{el}}\psi^{\text{nuc}} = i\hbar\psi^{\text{el}}\frac{\partial\psi^{\text{nuc}}}{\partial t}, \quad (5.2)$$

where we have rearranged the terms to group all the derivatives of the electronic part of the wave function together. Considering the four sums inside the square brackets, we ask: what are the expected magnitudes of the key quantities? Since m^α is about 10 000 times larger than m^{el}, we can neglect the second sum relative to the first (if we assume that all second derivatives of ψ^{el} are of comparable magnitude). Also, the gradients of ψ^{el} which appear in the third sum will be zero if the BOA holds; the wave functions are at a minimum with respect to nuclear positions. This leaves only the first and fourth sums in the square brackets, and the Schrödinger equation now simplifies to

$$\frac{-\hbar^2}{2}\left[\psi^{\text{nuc}}\sum_{i=1}^{N^{\text{el}}} \frac{\nabla_i^2\psi^{\text{el}}}{m^{\text{el}}} + \psi^{\text{el}}\sum_{\alpha=1}^{N} \frac{\nabla_\alpha^2\psi^{\text{nuc}}}{m^\alpha}\right] + (\mathcal{U}^{\text{ee}} + \mathcal{U}^{\text{ZZ}} + \mathcal{U}^{\text{eZ}})\psi^{\text{el}}\psi^{\text{nuc}} = i\hbar\psi^{\text{el}}\frac{\partial\psi^{\text{nuc}}}{\partial t}. \quad (5.3)$$

Since the BOA implies that the electrons respond instantaneously to the positions of the nuclei, it further implies two results of relevance here. First, we expect that the electronic wave function itself satisfies a time-independent Schrödinger equation of the form

$$\frac{-\hbar^2}{2m^{\text{el}}}\sum_{i=1}^{N^{\text{el}}} \nabla_i^2\psi^{\text{el}} + (\mathcal{U}^{\text{eZ}} + \mathcal{U}^{\text{ee}})\psi^{\text{el}} = \epsilon\psi^{\text{el}}, \quad (5.4)$$

where the positions of the nuclei now play the role of parameters in $\mathcal{U}^{\text{eZ}}$, instead of being independent variables. Second, the BOA implies that the electrons will always find their ground state, and as such the energy ϵ in Eqn. (5.4) becomes the ground state energy, $\epsilon_0(\boldsymbol{r})$, which depends only on the nuclear positions. Examining Eqn. (5.3), we can replace the terms common to the left-hand side of Eqn. (5.4) with $\epsilon_0(\boldsymbol{r})\psi^{\text{el}}$ from the right-hand side, leading to the Schrödinger equation for the nuclear wave functions, ψ^{nuc}:

$$\frac{-\hbar^2}{2}\sum_{\alpha=1}^{N} \frac{\nabla_\alpha^2\psi^{\text{nuc}}}{m^\alpha} + (\mathcal{U}^{\text{ZZ}}(\boldsymbol{r}) + \epsilon_0(\boldsymbol{r}))\psi^{\text{nuc}} = i\hbar\frac{\partial\psi^{\text{nuc}}}{\partial t}. \quad (5.5)$$

We recognize this as the time-dependent Schrödinger equation for the nuclei in the presence of a potential field defined by

$$\mathcal{U}(\boldsymbol{r}) = \mathcal{U}^{\mathrm{ZZ}}(\boldsymbol{r}) + \epsilon_0(\boldsymbol{r}). \tag{5.6}$$

The explicit dependence on the electrons is gone and is represented by an unknown function of the nuclear positions, $\epsilon_0(\boldsymbol{r})$. If, in fact, we can treat the nuclei as classical particles, then the function $\mathcal{U}(\boldsymbol{r})$ becomes a classical interatomic model $\mathcal{U}(\boldsymbol{r}) \to \mathcal{V}(\boldsymbol{r})$, and the BOA serves as a fundamental justification for the field of classical atomistic modeling. Now we can see that the thing we call the "potential energy" in classical systems is really a combination of the Coulomb interactions between the atoms and both the *potential* and the *kinetic* energy of electrons bound to their ground state by the BOA. The BOA allows us to say confidently that this new "potential energy of the system," $\mathcal{V}(\boldsymbol{r})$, is dependent only on the positions of the atomic nuclei, which we will now start calling simply "atoms" for short.

The potential energy of a system of atoms, $\mathcal{V}$, defined above, can be divided into an internal part, $\mathcal{V}^{\mathrm{int}}$, and an external part, $\mathcal{V}^{\mathrm{ext}}$. The external potential energy can be further divided into contributions due to external fields and contact with atoms outside the system:

$$\mathcal{V} = \mathcal{V}^{\mathrm{int}} + \mathcal{V}^{\mathrm{ext}}; \qquad \mathcal{V}^{\mathrm{ext}} = \mathcal{V}^{\mathrm{ext}}_{\mathrm{fld}} + \mathcal{V}^{\mathrm{ext}}_{\mathrm{con}}. \tag{5.7}$$

Specifically, the *internal* potential energy, $\mathcal{V}^{\mathrm{int}}$, is due to the short-range interactions between the atoms making up the system. This term can be defined as the potential energy of the system when it is isolated from external fields and there are no other atoms outside it. We will sometimes refer to $\mathcal{V}^{\mathrm{int}}$ as the *interatomic potential energy*. The *field* part of the external potential energy, $\mathcal{V}^{\mathrm{ext}}_{\mathrm{fld}}$, is the contribution to the potential energy due to the interaction of the system's atoms with external fields, such as gravity or electromagnetic radiation. External field interactions are marked by being long-range and affecting all atoms in the system at a distance. Finally, the *contact* part of the external potential energy, $\mathcal{V}^{\mathrm{ext}}_{\mathrm{con}}$, is due to the interaction of the system's atoms with other atoms outside of it. Formally, it can be defined by substituting Eqn. $(5.7)_2$ into Eqn. $(5.7)_1$ and inverting the resulting relation:

$$\mathcal{V}^{\mathrm{ext}}_{\mathrm{con}} \equiv \mathcal{V} - \mathcal{V}^{\mathrm{int}} - \mathcal{V}^{\mathrm{ext}}_{\mathrm{fld}}. \tag{5.8}$$

The contact energy represents short-range interactions near the boundaries of the system. It therefore scales with surface area, whereas $\mathcal{V}^{\mathrm{int}}$ and $\mathcal{V}^{\mathrm{ext}}_{\mathrm{fld}}$ scale with the volume of the system. As a result, $\mathcal{V}^{\mathrm{ext}}_{\mathrm{con}}$ becomes negligible with increasing system size and is, in fact, neglected under the assumption of *weak interaction* in statistical mechanics (see Chapter 7).

5.2 Treating atoms as classical particles

Our goal in this chapter is a simple classical model of the energy based only on the positions of the atoms, since this will permit us to compute the forces on the atoms for use in Newton's equations of motion. We have just shown how the BOA allows us to replace the electrons with an effective potential, even if we do not yet know what form the potential should take. Next, we justify the treatment of the atoms as *classical particles*, rather than quantum mechanical ones, by consideration of the well-known *de Broglie*[3] *thermal wavelength*:[4]

$$\Lambda = \sqrt{\frac{h^2}{2\pi m k_B T}} = \sqrt{\frac{2\pi\hbar^2}{m k_B T}}. \tag{5.9}$$

If this wavelength is much smaller than the average interatomic spacing, then the waves are spatially localized and the atoms can be treated as classical particles. On the other hand, if Λ is on the order of (or greater than) the interatomic distances, then the wave-like behavior of the atoms is relevant. All atoms are classical at high enough temperatures, but Λ gives us a way to estimate the minimum temperature for classical mechanics to be valid for a certain atom. Consider, for example, crystalline Al (mass 26.98 g/mol), with a near-neighbor distance of roughly $b = 2.9$ Å, then the temperature at which $\Lambda = b$ is

$$T = \frac{2\pi\hbar^2}{m k_B b^2} \approx 1.34 \text{ K},$$

illustrating that even at temperatures of only a few degrees Kelvin, Al atoms in a solid are expected to behave classically.

This is not to say that quantum mechanics can be disregarded, since it still governs the behavior of the electrons and therefore determines the form of the interatomic potential, $\mathcal{V}(\boldsymbol{r})$. What we have shown here is that other than for a few rare exceptions, such as solid He at a few degrees Kelvin, the atoms in solid materials can be treated as classical particles subjected to forces that, although arising from quantum effects, may be approximated by a potential energy function, $\mathcal{V}(\boldsymbol{r})$. Of course, finding such a potential function that applies to a broad range of configurations can be very difficult. Such potentials are called "empirical," since the approach involves the selection of a functional form (usually based on a mix of theory and intuition) with parameters that are obtained by fitting the predictions of the model to experiments and first-principles[5] calculations. This is the "business" of empirical potentials development which is discussed in the remainder of this chapter.

5.3 Sensible functional forms

Developing an interatomic model, $\mathcal{V}^{\text{int}}$, can be generically summed up as trying to design a function that depends on the positions of the atoms, such that we can accurately approximate the energy of the electrons. Other contributions to the energy of the atomic system, such as

[3] For a pronunciation of the name "de Broglie" see footnote 2 on page 156.

[4] Recalling the particle–wave duality of atoms, there is a direct relation between particle momentum and wave number (which is inverse to wavelength). The de Broglie thermal wavelength is roughly the average wavelength associated with the ensemble of particle momenta in an ideal gas at a certain temperature [HM86, Erc08b]. This follows directly from Eqn. (4.21) and the equipartition theorem (see Section 7.4.4).

[5] "First principles" generally refers to DFT results or results obtained with other highly accurate solutions to Schrödinger's equation.

the Coulomb interactions between the atoms or the kinetic energy due to their motion are well understood and described by classical mechanics notions; it is the electronic part that is the hardest to address.

In principle, then, we are searching for a general function of N atomic coordinates:

$$\mathcal{V}^{\text{int}} = \widehat{\mathcal{V}}^{\text{int}}(\boldsymbol{r}^1, \boldsymbol{r}^2, \dots, \boldsymbol{r}^N). \tag{5.10}$$

Stated this way, the task seems quite intractable. What are some suitable forms to start with, and what are some obvious forms to avoid? In this section, we take a look at a few of the features that, on physical grounds, we can require in an atomistic model.

5.3.1 Interatomic distances

In what follows, we will make extensive use of distances between atoms and derivatives of these distances. Specifically, we define the vector pointing from atom α to atom β as

$$\boldsymbol{r}^{\alpha\beta} = \boldsymbol{r}^\beta - \boldsymbol{r}^\alpha, \qquad \boldsymbol{R}^{\alpha\beta} = \boldsymbol{R}^\beta - \boldsymbol{R}^\alpha, \tag{5.11}$$

reserving the lower-case $\boldsymbol{r}$ for the deformed position of atoms and (when such a distinction is necessary) upper-case $\boldsymbol{R}$ for a reference, undeformed set of atomic sites. We refer to $\boldsymbol{r}^{\alpha\beta}$ and $\boldsymbol{R}^{\alpha\beta}$ as the deformed and reference *relative position vectors*. From these vectors, the deformed and reference interatomic distances in indicial form are simply

$$r^{\alpha\beta} = \sqrt{r_i^{\alpha\beta} r_i^{\alpha\beta}}, \qquad R^{\alpha\beta} = \sqrt{R_I^{\alpha\beta} R_I^{\alpha\beta}}. \tag{5.12}$$

Note that throughout this section, the Einstein summation convention is assumed for component indices (Roman subscripts), but not atom numbers (Greek superscripts). For example in Eqn. $(5.12)_1$, the double i subscript implies summation, but the double α and double β superscripts do not. We use the same symbol for the interatomic spacing and the vector joining the two atoms, but the coordinate index or bold face for the vector will generally prevent confusion.

When computing forces, we will encounter the derivative of the interatomic distance with respect to the deformed coordinate of an atom. This is straightforward to derive from Eqns. $(5.11)_1$ and $(5.12)_1$, and can be written as

$$\frac{\partial r^{\alpha\beta}}{\partial r_i^\gamma} = \frac{r_i^{\alpha\beta}}{r^{\alpha\beta}} \left(\delta^{\beta\gamma} - \delta^{\alpha\gamma}\right), \tag{5.13}$$

where δ is the Kronecker delta and once again there is no summation on Greek indices. This equation emphasizes the rather obvious fact that the distance between two atoms depends only on the positions of those two atoms and that this is only nonzero when either $\gamma = \alpha$ or $\gamma = \beta$. The result is simply a unit vector, along the line between the two atoms, pointing toward the atom with respect to which the derivative is taken. We will also need the derivative of one distance with respect to another:

$$\frac{\partial r^{\alpha\beta}}{\partial r^{\gamma\epsilon}} = \delta^{\alpha\gamma}\delta^{\beta\epsilon} + \delta^{\alpha\epsilon}\delta^{\beta\gamma}. \tag{5.14}$$

This derivative equals 1 if either $\alpha = \gamma$ and $\beta = \epsilon$, or $\alpha = \epsilon$ and $\beta = \gamma$. It reflects the fact that distances are symmetric with respect to swapping of their indices, i.e. $r^{\alpha\beta} = r^{\beta\alpha}$.

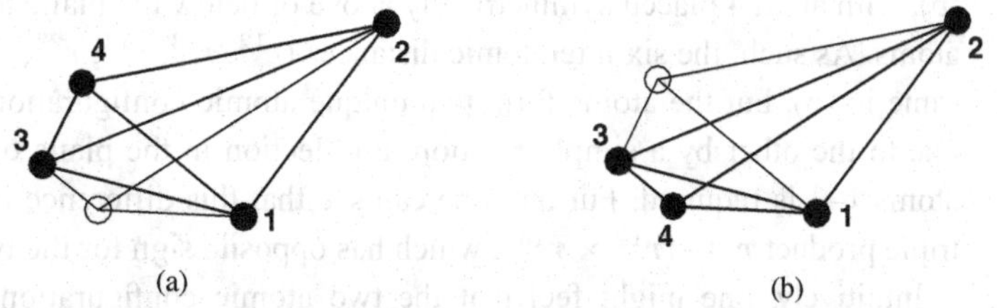

Fig. 5.1 Two atomic configurations that are the same but for a parity operation. In (a) atom 4 is above the plane defined by atoms 1, 2 and 3, and in (b) it has been reflected to a position below this plane.

5.3.2 Requirement of translational, rotational and parity invariance

The internal potential energy (or the *interatomic potential energy*) defined in Eqn. (5.10) must satisfy certain invariances based on the nature of the laws of physics. In this section, we introduce these invariances and explain (without proof) their implications for the form of the potential energy function. A rigorous mathematical derivation discussing important subtleties related to the definition of interatomic potentials is given in Appendix A.

First, we expect $\widehat{\mathcal{V}}^{\text{int}}(\boldsymbol{r}^1, \dots, \boldsymbol{r}^N)$ to be invariant with respect to superposed rigid-body translation and rotation. This invariance is a consequence of the *principle of material frame-indifference*, which loosely states that the response of a material to deformation should be invariant with respect to changes of reference frame, and which was discussed in the context of continuum constitutive laws in Section 2.5.3. The same principle also applies to atomistic models. Mathematically, we require[6] (see Eqns. (2.178)$_1$ and (2.180)$_1$):

$$\widehat{\mathcal{V}}^{\text{int}}(\boldsymbol{Q}\boldsymbol{r}^1 + \boldsymbol{c}, \boldsymbol{Q}\boldsymbol{r}^2 + \boldsymbol{c}, \dots, \boldsymbol{Q}\boldsymbol{r}^N + \boldsymbol{c}) = \widehat{\mathcal{V}}^{\text{int}}(\boldsymbol{r}^1, \boldsymbol{r}^2, \dots, \boldsymbol{r}^N), \tag{5.15}$$

for all rotations (proper orthogonal tensors) $\boldsymbol{Q} \in SO(3)$ and vectors $\boldsymbol{c} \in \mathbb{R}^3$. It can be shown using the *basic representation theorem*, proved by Cauchy in 1850 [Cau50] and discussed in [TN65, Section 11], that Eqn. (5.15) implies that the internal potential energy can only depend on the distances between the particles and the triple products of the relative position vectors. The dependence on triple products can be understood by considering the next invariance.

Second, we expect $\widehat{\mathcal{V}}^{\text{int}}(\boldsymbol{r}^1, \dots, \boldsymbol{r}^N)$ to be invariant with respect to the parity operator, i.e. the inversion of space, as defined in Section 3.4.1. Therefore, we require

$$\widehat{\mathcal{V}}^{\text{int}}(-\boldsymbol{r}^1, -\boldsymbol{r}^2, \dots, -\boldsymbol{r}^N) = \widehat{\mathcal{V}}^{\text{int}}(\boldsymbol{r}^1, \boldsymbol{r}^2, \dots, \boldsymbol{r}^N). \tag{5.16}$$

It is often convenient to combine the parity operator with rotation which gives improper rotations, or reflections,[7] of a cluster of atoms. For the simplest example, consider the cluster of four atoms shown in Fig. 5.1.[8] Atoms 1–3 form the same triangle in (a) and

[6] Note that only the internal part of the potential energy is required to be frame-indifferent. The external potential energy, $\mathcal{V}^{\text{ext}}$, will not be frame-indifferent in general since it depends on absolute positions in space.

[7] A reflection is a parity operation combined with a 180° rotation about the normal to the plane of the reflection.

[8] Pairs or triplets of atoms do not exhibit parity ambiguity. A triangle of three atoms is indeed uniquely described by the three interatomic distances forming its sides. Problems arise only for clusters of four or more.

(b), with atom 4 placed symmetrically above or below the plane formed by the other three atoms. As such, the six interatomic distances (r^{12}, r^{13}, r^{14}, r^{23}, r^{24}, r^{34}) in (a) are all the same in (b), but the atoms form two unique atomic configurations. One cannot get from one to the other by a simple rotation; a reflection in the plane of the triangle formed by atoms 1–3 is required. Further, one can see that this difference is in fact resolved by the triple product $\boldsymbol{r}^{14} \cdot (\boldsymbol{r}^{12} \times \boldsymbol{r}^{13})$, which has opposite sign for the two configurations.

Intuitively, one might feel that the two atomic configurations in Fig. 5.1 should be energetically equivalent. It turns out that if we return to Schrödinger's equation and quantum mechanics, one can show [BMW03] that our intuition is correct, and all Hermitian Hamiltonians have parity symmetry. That is to say that any Hamiltonian of the form

$$\frac{-\hbar^2}{2m^{\mathrm{el}}}\nabla^2 + \mathcal{U}(\boldsymbol{x}^{\mathrm{el}}, \boldsymbol{r}),$$

with a real potential energy function $\mathcal{U}(\boldsymbol{x}^{\mathrm{el}}, \boldsymbol{r})$ will have the same energy eigenvalues regardless of parity operations. Since the Hamiltonians that describe interatomic bonding are Hermitian, we can rest assured that parity symmetry is a genuine property of the energy that empirical models must possess.

The invariance requirements in Eqns. (5.15) and (5.16) can be combined into a single invariance requirement that can be stated as follows.

Principle of interatomic potential invariance The internal potential energy of a system of particles is invariant with respect to the Euclidean group[9] $G \equiv \{\boldsymbol{x} \mapsto \boldsymbol{Q}\boldsymbol{x} + \boldsymbol{c} \mid \boldsymbol{x} \in \mathbb{R}^3, \boldsymbol{Q} \in O(3), \boldsymbol{c} \in \mathbb{R}^3\}$, where $O(3)$ denotes the full orthogonal group. (5.17)

Cauchy's basic representation theorem, mentioned above, can be used to show that this invariance principle implies that the internal potential energy can *only be a function of the distances between the particles.*[10] This is proved in Section A.1 based on a derivation originally given in [AT10].[11] Thus, we have

$$\mathcal{V}^{\mathrm{int}} = \mathcal{V}^{\mathrm{int}}(r^{12}, r^{13}, \ldots, r^{1N}, r^{23}, \ldots, r^{N-1,N}) = \mathcal{V}^{\mathrm{int}}(\{r^{\alpha\beta}\}). \tag{5.18}$$

[9] The notation defining the Euclidean group, "$G \equiv \{\boldsymbol{x} \mapsto \boldsymbol{Q}\boldsymbol{x} + \boldsymbol{c} \mid \boldsymbol{x} \in \mathbb{R}^3, \boldsymbol{Q} \in O(3), \boldsymbol{c} \in \mathbb{R}^3\}$", states that G consists of all mappings taking $\boldsymbol{x}$ to $\boldsymbol{Q}\boldsymbol{x} + \boldsymbol{c}$, where $\boldsymbol{x}$, $\boldsymbol{Q}$ and $\boldsymbol{c}$ are defined after the vertical line.

[10] The simplest example is a system containing only two particles in which case the potential energy can only depend on the distance between them: $\mathcal{V}^{\mathrm{int}} = \widehat{\phi}(\boldsymbol{r}^1, \boldsymbol{r}^2) = \phi(r^{12})$, where $\phi(r)$ is a pair potential function. This means that "noncentral" pair potentials, which depend on more than just the distance between the particles, violate the invariance principle. An example of such a potential is given by Johnson in [Joh72]. Johnson defined the following potential, $\phi(\boldsymbol{r}) = p(r) + q(r)w(\boldsymbol{r})$, where $\boldsymbol{r}$ is a vector connecting two atoms, $p(r)$ and $q(r)$ are functions of $r = \|\boldsymbol{r}\|$, and $w(\boldsymbol{r})$ is the cubic harmonic function, $w(\boldsymbol{r}) = (r_1^4 + r_2^4 + r_3^4)/r^4 - 3/5$. While the function $\phi(\boldsymbol{r})$ is translationally invariant it is clearly not invariant with respect to orthogonal transformations and therefore not a viable physical model.

[11] A different proof for the special case of just two particles is given in Example 6.2 of [TME12].

The list of arguments includes the $N(N-1)/2$ terms for which $\alpha < \beta$, since $r^{\alpha\beta} = r^{\beta\alpha}$. In short-hand notation, we denote this set of distances as[12]

$$\{r^{\alpha\beta}\} = \{r^{\alpha\beta} \mid \alpha, \beta \in (1, \ldots, N), \alpha < \beta\}. \tag{5.19}$$

It is easy to see that the representation in Eqn. (5.18) satisfies the invariance principle in Eqn. (5.17), which requires that

$$\mathcal{V}^{\text{int}}(\{\widetilde{r}^{\alpha\beta}\}) = \mathcal{V}^{\text{int}}(\{r^{\alpha\beta}\}), \tag{5.20}$$

where $\widetilde{r}^{\alpha\beta}$ are the distances between the atoms following an arbitrary Euclidean transformation. Let us compute $\widetilde{r}^{\alpha\beta}$:

$$\begin{aligned}
\widetilde{r}^{\alpha\beta} &= \left\| \widetilde{\boldsymbol{r}}^{\beta} - \widetilde{\boldsymbol{r}}^{\alpha} \right\| = \left\| \boldsymbol{Q}\boldsymbol{r}^{\beta} + \boldsymbol{c} - \boldsymbol{Q}\boldsymbol{r}^{\alpha} - \boldsymbol{c} \right\| = \left\| \boldsymbol{Q}(\boldsymbol{r}^{\beta} - \boldsymbol{r}^{\alpha}) \right\| \\
&= \left[(\boldsymbol{Q}(\boldsymbol{r}^{\beta} - \boldsymbol{r}^{\alpha}))^T \boldsymbol{Q}(\boldsymbol{r}^{\beta} - \boldsymbol{r}^{\alpha}) \right]^{1/2} = \left[(\boldsymbol{r}^{\beta} - \boldsymbol{r}^{\alpha})^T \boldsymbol{Q}^T \boldsymbol{Q}(\boldsymbol{r}^{\beta} - \boldsymbol{r}^{\alpha}) \right]^{1/2} \\
&= \left\| \boldsymbol{r}^{\beta} - \boldsymbol{r}^{\alpha} \right\| = r^{\alpha\beta},
\end{aligned} \tag{5.21}$$

where we have made use of the fact that $\boldsymbol{Q}$ is an orthogonal tensor ($\boldsymbol{Q}^{-1} = \boldsymbol{Q}^T$). The conclusion is that any interatomic potential that depends only on the interatomic distances, as given in Eqn. (5.18), automatically satisfies the invariance principle in Eqn. (5.17). As noted above, the proof that a representation in terms of distances is not just sufficient but *necessary* is given in Appendix A.

5.3.3 The cutoff radius

A basic assumption in the development of many interatomic potentials is that atomic interactions are inherently local.[13] The idea is that beyond some bond length, atoms interact so weakly that they make essentially no contribution to the total energy. To be consistent with this view, the interatomic potential energy function in Eqn. (5.18) must be constructed in such a way that the energy contribution due to a given atom is only affected by atoms within a specified distance called the *cutoff radius* and denoted r_{cut}. For example, for the pair potential introduced in Eqn. (5.35), we would have

$$\phi_{\alpha\beta}(r) = \begin{cases} f_{\alpha\beta}(r), & r < r_{\text{cut}}, \\ 0, & r \geq r_{\text{cut}}. \end{cases} \tag{5.22}$$

The details of how to apply this cutoff can be important. Specifically, forces on atoms ultimately depend on derivatives of the functions defining the model; in this case derivatives

[12] The notation "$\{r^{\alpha\beta} \mid \alpha, \beta \in (1, \ldots, N), \alpha < \beta\}$" denotes the set of all distances $r^{\alpha\beta}$ such that α and β are in the range 1–N and $\alpha < \beta$.

[13] The main place where the assumption of local bonding breaks down is in ionic systems dominated by long-range Coulomb interactions. We discuss ionically-bonded materials in Section 5.4.3. However, for metallic, covalent and van der Waals bonding, we can usually assume that bonding is weak beyond a relatively short cutoff distance. Note that there are *always* Coulomb interactions between the protons and electrons in a solid; however, in many materials the cancellations between like-charge and unlike-charge interactions mean that they can be neglected.

of $\phi_{\alpha\beta}$ with respect to r. In a molecular dynamics (MD) simulation, it is likely that atoms will regularly pass near each other with interatomic distances that are equal or very close to r_{cut}. This presents a problem if the function $f_{\alpha\beta}(r)$ does not itself go to zero at $r = r_{\text{cut}}$, in which case $\phi'_{\alpha\beta}(r_{\text{cut}})$ becomes undefined. (See Section 5.8.5 for a discussion of the relationship between the cutoff radius and interatomic forces.)

For some applications, it may be necessary that higher derivatives of the potential be continuous. For example, elastic constants involve second derivatives of the potential. A general modification to the potential can be applied such that

$$\phi_{\alpha\beta}(r) = \begin{cases} f_{\alpha\beta}(r) - t_{\alpha\beta}(r - r_{\text{cut}}), & r < r_{\text{cut}}, \\ 0, & r \geq r_{\text{cut}}, \end{cases} \tag{5.23}$$

where $f_{\alpha\beta}(r)$ is a smooth, continuous pair potential defined for all $r > 0$ and $t_{\alpha\beta}(x)$ is polynomial that provides smooth truncation,

$$t_{\alpha\beta}(x) = \sum_{i=0}^{m} \frac{1}{i!} x^i f_{\alpha\beta}^{(i)}(r_{\text{cut}}).$$

Here $f_{\alpha\beta}^{(i)}(r)$ is the ith derivative of $f_{\alpha\beta}(r)$ and x^i is x raised to the ith power. This form ensures that all derivatives up to order m are identically zero at $r = r_{\text{cut}}$. A similar procedure can be applied to n-body terms of any order by setting the potential to zero whenever any of the distances exceeds the cutoff radius.[14]

It is important to include any smoothing term during the fitting procedure of the potential. Simply tacking on a smoothing function to an existing potential can be dangerous since it can strongly affect the predictions. For example, for crystalline materials it can change the predicted ground state structure. This is common in metals that have a face-centered cubic (fcc) ground state structure. Small changes to the potentials can sometimes lead to the nearby hexagonal close packed (hcp) structure becoming the ground state. See, for example, the discussion in [JBA02] (and Fig. 1 in that paper) showing the effect of different cutoff strategies for the Lennard-Jones potential (Section 5.4.2) on the predicted ground state energy versus density curve for fcc and hcp structures.

5.4 Cluster potentials

The discussion in the previous section clarified some of the properties that interatomic potentials must have, but it did not provide an explicit form for these potentials. In this section, we show that a formally exact representation of the potential energy of a system of N atoms can be constructed as a series of n-body terms ($n = 1, \ldots, N$), each of which depends on the positions of a *cluster* of n atoms. An approximate model can then be defined by terminating this series at a desired order. The result is the so-called *cluster potential*.[15]

[14] Other smoothing functions are of course possible. See, for example, the discussion of "mollifying functions" in Section 8.2.5 and in [Mur07].

[15] Instead of "cluster potential," the term "cluster expansion" is sometimes used to describe the decomposition of the energy into a series of cluster terms [Mar75a]. However, this can be confusing since "cluster expansion"

5.4.1 Formally exact cluster potentials

Consider a system of N atoms. For simplicity, we initially focus on systems containing only a single species of atoms. The total potential energy of such a system can always be expressed in the following form [Fis64, Mar75a],

$$\mathcal{V}_{(N)} = \phi_0 + \sum_{\alpha=1}^{N} \widehat{\phi}_1(\boldsymbol{r}^\alpha) + \sum_{\substack{\alpha,\beta \\ \alpha<\beta}}^{N} \widehat{\phi}_2(\boldsymbol{r}^\alpha, \boldsymbol{r}^\beta) + \sum_{\substack{\alpha,\beta,\gamma \\ \alpha<\beta<\gamma}}^{N} \widehat{\phi}_3(\boldsymbol{r}^\alpha, \boldsymbol{r}^\beta, \boldsymbol{r}^\gamma) + \cdots, \tag{5.24}$$

where ϕ_0 is a reference term, $\widehat{\phi}_n$ are n-body potential functions, and for a system of N atoms the series terminates with the $\widehat{\phi}_N$ term. The summations for the two-body terms and higher denote multiple sums subject to the indicated constraints (e.g. for the three-body term, the summation is $\sum_\alpha \sum_\beta \sum_\gamma$ subject to the constraint $\alpha < \beta < \gamma \leq N$). In general, for the n-body potential $\widehat{\phi}_n$, there are $\binom{N}{n} = N!/(N-n)!n!$ terms in the corresponding sums.

The potential functions satisfy permutation symmetry with respect to their arguments. Thus, the pair term satisfies, $\widehat{\phi}_2(\boldsymbol{r}^\alpha, \boldsymbol{r}^\beta) = \widehat{\phi}_2(\boldsymbol{r}^\beta, \boldsymbol{r}^\alpha)$, the three-body term satisfies, $\widehat{\phi}_3(\boldsymbol{r}^\alpha, \boldsymbol{r}^\beta, \boldsymbol{r}^\gamma) = \widehat{\phi}_3(\boldsymbol{r}^\beta, \boldsymbol{r}^\alpha, \boldsymbol{r}^\gamma) = \widehat{\phi}_3(\boldsymbol{r}^\alpha, \boldsymbol{r}^\gamma, \boldsymbol{r}^\beta) = \ldots$, and so on.

We separate the terms in Eqn. (5.24) into two parts. The "one-body" terms, $\widehat{\phi}_1$, represent the interaction of the atoms with an external potential and other atoms lying outside of the system, which together are included in the external potential energy (see Eqn. $(5.7)_2$):

$$\mathcal{V}^{\text{ext}}_{(N)} = \sum_{\alpha=1}^{N} \widehat{\phi}_1(\boldsymbol{r}^\alpha). \tag{5.25}$$

The remaining terms correspond to the internal potential energy of the system, which includes a reference energy and the energy associated with interactions between the atoms:

$$\mathcal{V}^{\text{int}}_{(N)} = \phi_0 + \sum_{\substack{\alpha,\beta \\ \alpha<\beta}}^{N} \widehat{\phi}_2(\boldsymbol{r}^\alpha, \boldsymbol{r}^\beta) + \sum_{\substack{\alpha,\beta,\gamma \\ \alpha<\beta<\gamma}}^{N} \widehat{\phi}_3(\boldsymbol{r}^\alpha, \boldsymbol{r}^\beta, \boldsymbol{r}^\gamma) + \cdots. \tag{5.26}$$

The reference energy term corresponds to the energy of the atoms in isolation from each other. We can therefore write

$$\phi_0 = N E_{\text{free}}(Z), \tag{5.27}$$

where Z is the atomic number of the atoms in the system and $E_{\text{free}}(Z)$ is a function giving the energy of a free (isolated) atom with atomic number Z.

is also used to describe a closely-related method of decomposing the energy of a lattice of atoms to study ordering in alloys or magnetic systems [SDG84]. We therefore stick to the term "cluster potentials."

It is important to point out that the series in Eqn. (5.26) is not unique. In fact, for a system of N atoms, the functional forms for $\widehat{\phi}_n$ $(n = 2, \dots, N-1)$ could be set arbitrarily. The satisfaction of Eqn. (5.26) would then be enforced by *defining* the N-body term as

$$\widehat{\phi}_N(\boldsymbol{r}^1, \dots, \boldsymbol{r}^N) \equiv \widehat{\mathcal{V}}^{\text{int}}_{(N)}(\boldsymbol{r}^1, \dots, \boldsymbol{r}^N) - \phi_0 - \sum_{\substack{\alpha,\beta \\ \alpha<\beta}}^{N} \widehat{\phi}_2(\boldsymbol{r}^\alpha, \boldsymbol{r}^\beta) - \sum_{\substack{\alpha,\beta,\gamma \\ \alpha<\beta<\gamma}}^{N} \widehat{\phi}_3(\boldsymbol{r}^\alpha, \boldsymbol{r}^\beta, \boldsymbol{r}^\gamma) - \cdots.$$

Obviously, a cluster potential constructed in this fashion would not be very illuminating. There is, however, a simple recursive procedure for defining the potential functions, which we describe now, that provides them with clear physical significance.

Consider sets of increasing numbers of atoms

$$(\boldsymbol{r}^1, \boldsymbol{r}^2), \quad (\boldsymbol{r}^1, \boldsymbol{r}^2, \boldsymbol{r}^3), \quad (\boldsymbol{r}^1, \boldsymbol{r}^2, \boldsymbol{r}^3, \boldsymbol{r}^4), \quad \dots,$$

and the internal potential energy functions that go with them

$$\widehat{\mathcal{V}}^{\text{int}}_{(2)}(\boldsymbol{r}^1, \boldsymbol{r}^2), \quad \widehat{\mathcal{V}}^{\text{int}}_{(3)}(\boldsymbol{r}^1, \boldsymbol{r}^2, \boldsymbol{r}^3), \quad \widehat{\mathcal{V}}^{\text{int}}_{(4)}(\boldsymbol{r}^1, \boldsymbol{r}^2, \boldsymbol{r}^3, \boldsymbol{r}^4), \quad \dots,$$

which are assumed to be known. For two atoms, Eqns. (5.26) and (5.27) give

$$\widehat{\phi}_2(\boldsymbol{r}^1, \boldsymbol{r}^2; Z) \equiv \widehat{\mathcal{V}}^{\text{int}}_{(2)}(\boldsymbol{r}^1, \boldsymbol{r}^2) - 2E_{\text{free}}(Z),$$

which defines the pair potential as the change in internal energy in a two-atom system relative to the energy of the atoms when isolated. For three atoms, Eqns. (5.26) and (5.27) give

$$\widehat{\mathcal{V}}^{\text{int}}_{(3)}(\boldsymbol{r}^1, \boldsymbol{r}^2, \boldsymbol{r}^3) = 3E_{\text{free}}(Z) + \sum_{\substack{\alpha,\beta \\ \alpha<\beta}}^{3} \widehat{\phi}_2(\boldsymbol{r}^\alpha, \boldsymbol{r}^\beta) + \widehat{\phi}_3(\boldsymbol{r}^1, \boldsymbol{r}^2, \boldsymbol{r}^3).$$

Inverting this relation and writing the sum out explicitly, we have

$$\widehat{\phi}_3(\boldsymbol{r}^1, \boldsymbol{r}^2, \boldsymbol{r}^3; Z) \equiv \widehat{\mathcal{V}}^{\text{int}}_{(3)}(\boldsymbol{r}^1, \boldsymbol{r}^2, \boldsymbol{r}^3) - \widehat{\phi}_2(\boldsymbol{r}^1, \boldsymbol{r}^2) - \widehat{\phi}_2(\boldsymbol{r}^1, \boldsymbol{r}^3) - \widehat{\phi}_2(\boldsymbol{r}^2, \boldsymbol{r}^3) - 3E_{\text{free}}(Z).$$

Defined in this way, $\widehat{\phi}_3$ characterizes the *additional* energy in a three-atom system, not captured by the pair interactions. In a similar fashion, the four-body term is

$$\begin{aligned}
&\widehat{\phi}_4(\boldsymbol{r}^1, \boldsymbol{r}^2, \boldsymbol{r}^3, \boldsymbol{r}^4) \\
&\quad \equiv \widehat{\mathcal{V}}^{\text{int}}_{(4)}(\boldsymbol{r}^1, \boldsymbol{r}^2, \boldsymbol{r}^3, \boldsymbol{r}^4) - \sum_{\substack{\alpha,\beta \\ \alpha<\beta}}^{4} \widehat{\phi}_2(\boldsymbol{r}^\alpha, \boldsymbol{r}^\beta) - \sum_{\substack{\alpha,\beta \\ \alpha<\beta<\gamma}}^{4} \widehat{\phi}_3(\boldsymbol{r}^\alpha, \boldsymbol{r}^\beta, \boldsymbol{r}^\gamma) - 4E_{\text{free}}(Z),
\end{aligned}$$

and so on to $\widehat{\phi}_N$. The recursive procedure described above is known as a *Moebius inversion* in number theory [And94, DFS04].[16] Building the terms in the cluster potential recursively like this removes the ambiguity in their definition on good physical grounds. In particular,

[16] The Moebius inversion is well defined as long as $\widehat{\mathcal{V}}^{\text{int}}_{(n)}$ $(n < N)$ is not infinite for some combination of its arguments [Fis64]. If it is, then the higher-order functions $\widehat{\phi}_{n+1}, \widehat{\phi}_{n+2}, \dots$, are ill defined for the corresponding combination of arguments. However, such configurations are normally not important physically and therefore not of concern.

the potential functions, $\widehat{\phi}_n(\boldsymbol{r}^\alpha, \boldsymbol{r}^\beta, \dots)$, so defined, have the property that they tend to zero whenever any one of the atoms is removed to infinity. This is consistent with the idea that the n-body term represents the excess energy for an isolated system of exactly n atoms after all interactions between smaller clusters of $m < n$ within the system are considered.

Equation (5.26) can also be written in the following symmetric form, which is often convenient for theoretical derivations:

$$\mathcal{V}^{\text{int}}_{(N)} = \phi_0 + \frac{1}{2!}\sum_{\substack{\alpha,\beta \\ \alpha\neq\beta}}^{N} \widehat{\phi}_2(\boldsymbol{r}^\alpha, \boldsymbol{r}^\beta) + \frac{1}{3!}\sum_{\substack{\alpha,\beta,\gamma \\ \alpha\neq\beta\neq\gamma}}^{N} \widehat{\phi}_3(\boldsymbol{r}^\alpha, \boldsymbol{r}^\beta, \boldsymbol{r}^\gamma) + \cdots, \tag{5.28}$$

where we have used the symmetry of the potential functions with respect to permutations of their arguments. In cases where the atoms are not all of the same species, the potential functions will be different depending on the atoms to which they are applied. We denote this explicitly by adding the atom subscripts to the potential functions:

$$\mathcal{V}^{\text{int}}_{(N)} = \phi_0 + \frac{1}{2!}\sum_{\substack{\alpha,\beta \\ \alpha\neq\beta}}^{N} \widehat{\phi}_{\alpha\beta}(\boldsymbol{r}^\alpha, \boldsymbol{r}^\beta) + \frac{1}{3!}\sum_{\substack{\alpha,\beta,\gamma \\ \alpha\neq\beta\neq\gamma}}^{N} \widehat{\phi}_{\alpha\beta\gamma}(\boldsymbol{r}^\alpha, \boldsymbol{r}^\beta, \boldsymbol{r}^\gamma) + \cdots. \tag{5.29}$$

The reference energy also depends on the atomic numbers of the atoms in the system:

$$\phi_0 = \sum_{\alpha=1}^{N} E_{\text{free}}(Z^\alpha), \tag{5.30}$$

where Z^α is the atomic number of atom α. We can make this notation explicit by writing,

$$\widehat{\phi}_{\alpha\beta}(\boldsymbol{r}^\alpha, \boldsymbol{r}^\beta) = \widehat{\phi}_2(\boldsymbol{r}^\alpha, \boldsymbol{r}^\beta; Z^\alpha, Z^\beta),$$
$$\widehat{\phi}_{\alpha\beta\gamma}(\boldsymbol{r}^\alpha, \boldsymbol{r}^\beta, \boldsymbol{r}^\gamma) = \widehat{\phi}_3(\boldsymbol{r}^\alpha, \boldsymbol{r}^\beta, \boldsymbol{r}^\gamma; Z^\alpha, Z^\beta, Z^\gamma),$$
$$\vdots$$

where $\{Z^\alpha\}$ serve as parameters that set the functional form of the functions $\widehat{\phi}_m$.

The potential functions that depend on species also possess permutation symmetry, like the single-species functions. However, when permuting arguments in this case it is also necessary to permute the species subscripts. Thus, the pair term satisfies

$$\widehat{\phi}_{\alpha\beta}(\boldsymbol{r}^\alpha, \boldsymbol{r}^\beta) = \widehat{\phi}_{\beta\alpha}(\boldsymbol{r}^\beta, \boldsymbol{r}^\alpha),$$

the three-body term satisfies

$$\widehat{\phi}_{\alpha\beta\gamma}(\boldsymbol{r}^\alpha, \boldsymbol{r}^\beta, \boldsymbol{r}^\gamma) = \widehat{\phi}_{\beta\alpha\gamma}(\boldsymbol{r}^\beta, \boldsymbol{r}^\alpha, \boldsymbol{r}^\gamma) = \widehat{\phi}_{\alpha\gamma\beta}(\boldsymbol{r}^\alpha, \boldsymbol{r}^\gamma, \boldsymbol{r}^\beta) = \dots,$$

and so on. The symmetric forms in Eqns. (5.28) and (5.29) provide a definition for the contribution of an individual atom α to the potential energy of the system. For example, for the general case of multiple species, we could define

$$E^{\alpha} \equiv E_{\text{free}}(Z^{\alpha}) + \frac{1}{2!}\sum_{\substack{\beta \\ \beta\neq\alpha}}^{N} \widehat{\phi}_{\alpha\beta}(\boldsymbol{r}^{\alpha}, \boldsymbol{r}^{\beta}) + \frac{1}{3!}\sum_{\substack{\beta,\gamma \\ \beta\neq\gamma\neq\alpha}}^{N} \widehat{\phi}_{\alpha\beta\gamma}(\boldsymbol{r}^{\alpha}, \boldsymbol{r}^{\beta}, \boldsymbol{r}^{\gamma}) + \cdots, \tag{5.31}$$

with the total potential energy following as $\mathcal{V}^{\text{int}}_{(N)} = \sum_{\alpha=1}^{N} E^{\alpha}$, However, since the cluster potential itself is not unique, this definition for the energy of an individual atom is also not unique and we must take care when trying to attach to it a physical significance. (See the discussion regarding this in [AT11].)

So far, we have expressed the n-body potentials in terms of the absolute atomic positions. However, we know from Section 5.3.2 that the potential energy must be expressible in terms of interatomic distances in order to satisfy the invariance principle in Eqn. (5.17). The same condition applies to each of the terms in the series. We therefore have that

$$\begin{aligned}
\mathcal{V}^{\text{int}}_{(N)} &= \phi_0 + \frac{1}{2!}\sum_{\substack{\alpha,\beta \\ \alpha\neq\beta}}^{N} \phi_{\alpha\beta}(r^{\alpha\beta}) + \frac{1}{3!}\sum_{\substack{\alpha,\beta,\gamma \\ \alpha\neq\beta\neq\gamma}}^{N} \phi_{\alpha\beta\gamma}(r^{\alpha\beta}, r^{\alpha\gamma}, r^{\beta\gamma}) \\
&\quad + \frac{1}{4!}\sum_{\substack{\alpha,\beta,\gamma,\delta \\ \alpha\neq\beta\neq\gamma\neq\delta}}^{N} \phi_{\alpha\beta\gamma\delta}(r^{\alpha\beta}, r^{\alpha\gamma}, r^{\alpha\delta}, r^{\beta\gamma}, r^{\beta\delta}, r^{\gamma\delta}) + \cdots \\
&= \phi_0 + \Phi_{(2)} + \Phi_{(3)} + \Phi_{(4)} + \cdots,
\end{aligned} \tag{5.32}$$

where $\Phi_{(n)}$ denotes the complete sum over the n-body terms. Many atomistic models seek to include the effect of bond angles, which play an important role in materials with covalent character. Effectively, a bond angle is determined by a cluster of three atoms, and as such it may be more convenient to write the atomistic model in terms of an angle formed between each group of three atoms. For example, $\theta^{\beta\alpha\gamma}$ is the angle formed by bonds α–β and α–γ with the vertex at atom α. This is equivalent to writing an expression in terms of the three distances, $r^{\alpha\beta}$, $r^{\alpha\gamma}$ and $r^{\beta\gamma}$, since from the law of cosines in trigonometry we have

$$\cos\theta^{\beta\alpha\gamma} = \frac{(r^{\alpha\beta})^2 + (r^{\alpha\gamma})^2 - (r^{\beta\gamma})^2}{2r^{\alpha\beta}r^{\alpha\gamma}}. \tag{5.33}$$

Clearly, a representation of the interatomic potential energy in terms of bond angles also satisfies the invariance principle. Later we will talk about models which include dependence on the relative arrangement of clusters of three, four, or more atoms. These models can always be written in terms of the interatomic distances, even if it is sometimes more convenient to use bond angles directly.

The cluster potential given above is formally exact. However, since the construction of this series requires knowledge of the exact potential energies, one may wonder what, if anything, has been achieved by this procedure. Specifically, if $\widehat{\mathcal{V}}^{\text{int}}_{(N)}(\boldsymbol{r}^1, \ldots, \boldsymbol{r}^N)$ were known, then this function could be used directly without computing the n-body potentials. There are two reasons why the cluster potential form is useful. First, it can be convenient for analytical derivations. An example is the statistical mechanics derivation in Section 7.4.5, which identifies specific constraints on allowable forms for $\widehat{\phi}_n$ that ensure that the atomistic system behaves correctly in the thermodynamic limit.

Second, the cluster potential can be used as a jumping off point for the creation of approximate interatomic models. In this case, the series is artificially terminated at some level retaining only terms up to $\Phi_{(n)}$ where n is normally a small number. Typically, ϕ_0 is set to zero as an arbitrary reference energy. If $n = 2$ the result is a pair potential model, $n = 3$ is a three-body potential model, and so on. The forms of the corresponding potential energy functions, which for a single-species system are ϕ_k $(k = 2, \ldots, n)$, are then "guessed" based on physical intuition or simplified theoretical models and fitted to properties measured in experiments or computed using first-principles methods. The justification for this approach is based on the observation that bonding in materials is usually quite local in nature. It therefore seems reasonable that models based on clusters of atoms surrounding the bonding sites will be able to capture the behavior of the system. This idea goes hand in hand with the introduction of a cutoff radius for each k-body term as explained in Section 5.3.3. The example for a pair potential in Eqn. (5.22) is generalized to any order potential by requiring that the k-body term (for $k \leq n$) goes to zero if *any* of its distance arguments exceeds r_{cut}. For example, for a three-body term,

$$\phi_{\alpha\beta\gamma}(r^{\alpha\beta}, r^{\alpha\gamma}, r^{\beta\gamma}) = \begin{cases} f_{\alpha\beta\gamma}(r^{\alpha\beta}, r^{\alpha\gamma}, r^{\beta\gamma}), & \text{if } r^{\alpha\beta}, r^{\alpha\gamma}, r^{\beta\gamma} < r_{\text{cut}}, \\ 0, & \text{otherwise}, \end{cases} \tag{5.34}$$

and similarly for higher-order terms.

5.4.2 Pair potentials

Retaining only the $\phi_{\alpha\beta}$ (or ϕ_2 for a single species) sum in Eqn. (5.32) and setting $\phi_0 = 0$ gives us the pair potential description:

$$\mathcal{V}^{\text{int}} = \frac{1}{2!} \sum_{\substack{\alpha,\beta \\ \alpha \neq \beta}} \phi_{\alpha\beta}(r^{\alpha\beta}). \tag{5.35}$$

The most common and widely proliferated pair potential is the *Lennard-Jones potential*[17] (see Fig. 5.2) which takes the form

$$\phi(r) = 4\varepsilon \left[\left(\frac{\sigma}{r}\right)^{12} - \left(\frac{\sigma}{r}\right)^{6} \right], \tag{5.36}$$

[17] From its name, many people assume that the Lennard-Jones potential was the collaborative venture of two gentlemen, Dr. Lennard and Mr. Jones. Actually, the Lennard-Jones potential was developed in the 1920s by Professor Sir John Edward Lennard-Jones (see Fig. 5.2(b)), a professor of theoretical chemistry at the University of Cambridge, to study the interactions of the noble (closed-shell) elements Ne, Ar, Kr and Xe [Jon24a, Jon24b, LJ25]. It was certainly not the first interatomic potential, but due to its simple form it has persisted and is probably the most well-known (and overused) interatomic potential today.

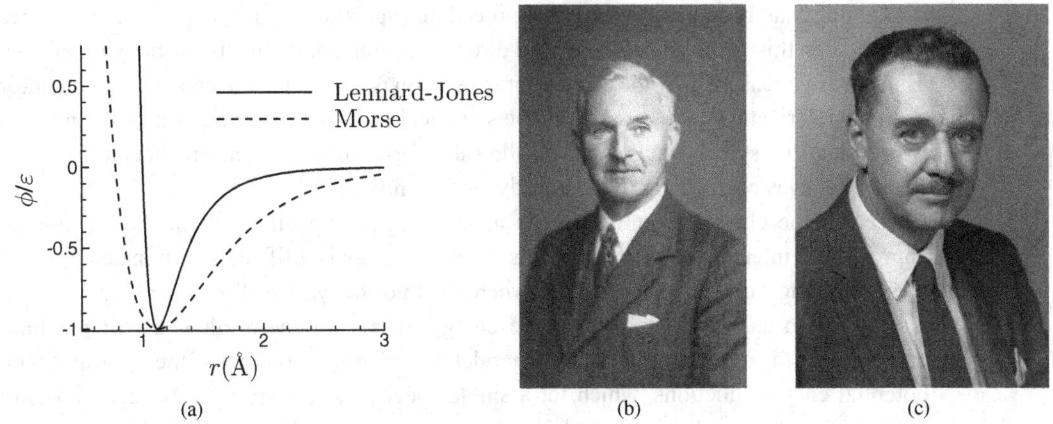

Fig. 5.2 (a) Comparison of the Lennard-Jones potential (with $\sigma = 1.0$ Å) and the Morse potential (with $\sigma = 0.5$ Å, $r_e = 1.2246$ Å). (b) Professor Sir John Edward Lennard-Jones, a professor of theoretical chemistry at the University of Cambridge. Reprinted from [UoC09], with permission from the Department of Chemistry, University of Cambridge. See footnote 17 on page 251. (c) Professor Philip Morse, a professor of physics at MIT. Reprinted with permission, AIP Emilio Segré Visual Archives.

where the constants σ and ε set the length and energy scales, respectively.[18] It is easy to verify that the minimum of the potential is $\phi_{\min} = -\varepsilon$, occurring at $r = 2^{1/6}\sigma$.

This form is only physically realistic for studying solidified ideal gases, a system of mainly scientific (as opposed to engineering) interest. Since it is possible to show that van der Waals forces between noble gas atoms lead to an attractive force that goes as $1/r^6$, the attractive part of this potential can be exactly fitted to the attraction between noble gas atoms. Although the repulsive force is not as easily determined in a simple form, it is clear that it should have a somewhat stronger dependence on the inverse distance than the attractive part, and so the exponent 12 is chosen to give the function a pleasing mathematical simplicity. The Lennard-Jones parameters for the noble gases are given in Tab. 5.1.

Even though the simple pair potential form cannot accurately describe metallic or covalent bonding, the 6–12 Lennard-Jones model has become somewhat of a standard potential for MD simulations.[19] There is a vast literature of applications for the Lennard-Jones potential, on problems that include liquids, gases, solid phase transitions, amorphous metals and fracture of crystals. However, it must always be treated as a *model system*, used to study the fundamental behavior of a system of discrete particles as opposed to the actual behavior of a real material. Its role as a model system must always be kept in mind, and great care must be taken in extrapolating its behavior to real materials. We will see this when we

[18] Since a pair potential has only two-body terms, we drop the "2" subscript on ϕ_2 and just use ϕ for the generic pair potential function. Also, the form in Eqn. (5.36) is sometimes described as the "Lennard-Jones 6–12" potential. In its most general form, the Lennard-Jones potential allows for the exponents 6 and 12 to be any whole number n and $2n$. The values of 6 and 12 were specifically chosen to model the van der Waals interactions of noble gas atoms.

[19] The 6–12 Lennard-Jones potential is often used with complete disregard for the fact that the 6 exponent was chosen specifically to model van der Waals interactions. Our fitting exercise in Example 5.1 suggests that this is a very bad model of covalent bonding, and even the 1–2 Lennard-Jones model is not a great fit in that case.

Table 5.1. Noble gas parameters for the Lennard-Jones potential[a]

Element	Nearest-neighbor distance [Å]	Cohesive energy [eV/atom]	Lennard-Jones parameters ε [eV]	σ [Å]
He	N/A	N/A	0.00088	2.56
Ne	3.13	0.02	0.0031	2.74
Ar	3.76	0.080	0.0104	3.40
Kr	4.01	0.116	0.0140	3.65
Xe	4.35	0.17	0.0200	3.98

[a] The experimental nearest-neighbor distance and cohesive energy correspond to the fcc structure and are extrapolated to 0 K and zero pressure. Note that He is a liquid at zero pressure, so experimental data are not provided for it. The Lennard-Jones parameters are obtained by fitting to the thermodynamic properties of the low-density gaseous phase. Data taken from [Kit96].

discuss, for example, elastic constants and the Cauchy relations (Section 11.5.3), surfaces (Section 6.5.4) and the energy to form a vacancy (Sections 5.5 and 6.5.3).

Example 5.1 (An atomistic model of the hydrogen molecule) The hydrogen molecule studied in Section 4.4.6 provides a simple case study for constructing an empirical model. Recall Fig. 4.8(c) and Eqn. (4.72), and imagine that this is the "exact" quantum mechanical result (remember that it is not exact, for the reasons already discussed in Section 4.4.6, but it will suffice for our purposes). Let us try to approximate the results of Fig. 4.8(c) with a Lennard-Jones potential.

For two identical hydrogen atoms at a distance r apart, there is only one bond and the general pair potential sum reduces to

$$\mathcal{V}^{\text{int}} = \phi(r) = 4\varepsilon\left[\left(\frac{\sigma}{r}\right)^{2n} - \left(\frac{\sigma}{r}\right)^{n}\right], \tag{5.37}$$

which is the general form of the Lennard-Jones potential (see footnote 18 on page 252). There are three free parameters: the exponent n, the length scale σ and the energy scale ε. The exponent need not be a whole number, but there would be certain computational advantages if it were, so let us limit ourselves to this class of functions.

One rational approach to the fitting process would be to insist that the minimum energy and corresponding bond length exactly agree with the results of Fig. 4.8(c). Then, we will choose the integer n that produces the best overall fit to the energy versus atomic distance curve, which will be easy to determine by eye in this case. We can readily verify that $\phi_{\min} = -\varepsilon$ and $r_{\min} = 2^{1/n}\sigma$ are the minimum of $\phi(r)$ and corresponding bond length, so we choose[20] $\varepsilon = 0.657e^2/r_0$ and $\sigma = 2^{-1/n} \times 1.45r_0$, where e and r_0 are as defined in Eqns. (4.43) and (4.51). In Fig. 5.3, we see how well the empirical model fits for $n = 1$. (It is straightforward to verify that larger values of n fare considerably worse.) Thus, within the confines of the functional form we have chosen, this is the best we can do.[21]

[20] Notice that the energy is written with respect to the energy of the pair of atoms when they are separated to $r \to \infty$. Since the pair potential form we have chosen goes to 0 at $r \to \infty$, whereas the original quantum result goes to $-e^2/r_0$, we have added a constant shift to the pair potential energy.

[21] Note that in Chapter 4, we referred to the energy of the pair of hydrogen atoms as $\mathcal{E}_{\text{tot}}$ to remind us that this is the sum of the electrostatic potential energy and the kinetic energy of the electrons. Here we have introduced the notation $\mathcal{V}^{\text{int}}$ as discussed on page 240.

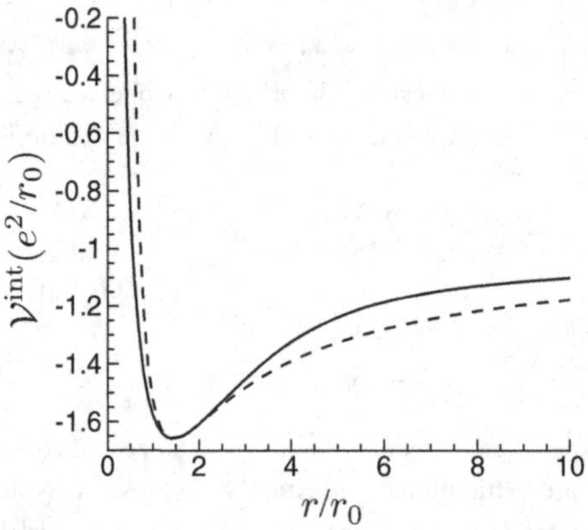

Fig. 5.3 Fitting the exact energy of the hydrogen molecule (solid line) with a simple pair potential (dashed line).

Next we might like to test the model and ask how *transferable* it is to hydrogen atoms in different environments. To study the behavior of a single isolated H_2 molecule, we might decide we are quite happy with the model we have just developed. But, how well does it predict properties that we did not fit?

One quantity of interest would be the "stiffness" of the bond, i.e. the second derivative of the energy with respect to bond length. The stiffness will determine such things as the vibrational frequency of the molecule. Note that this is something we can no longer fit, since all our free parameters are used up. Instead, we are stuck with a stiffness value that may or may not agree well with experimental or more fundamental theoretical results.

A tedious but straightforward analysis of Eqn. (4.72) tells us that the "exact" stiffness we want to reproduce is $d^2\mathcal{V}^{\rm int}/dr^2 = 0.5117e^2/r_0^3$, while it is also not difficult to find that for the model $d^2\phi_2/dr^2 = \varepsilon/2\sigma^2 = 0.625e^2/r_0^3$. This model stiffness is only about 20% too high, a reasonably good fit considering that we made no effort to match this quantity when choosing the functional form.

On the other hand, it is clear that in some respects the transferability of this model will be quite poor. For example, it is easy to see that it would predict a stable H_3 molecule, because the model does not account for the antibonding state associated with the Pauli exclusion principle. See Section 5.7.2 for more on transferability.

Another simple pair potential that is widely used as a model system is the Morse potential [Mor29] (see Fig. 5.2), which takes the form

$$\phi(r) = \varepsilon\left[\left(1 - e^{-(r-r_{\rm e})/\sigma}\right)^2 - 1\right]. \tag{5.38}$$

As with the Lennard-Jones potential, the minimum of the well is again $-\varepsilon$, at $r = r_{\rm e}$, while σ controls the width of the well and therefore the curvature or bond stiffness. Originally proposed by Philip Morse (see Fig. 5.2(c)) as a simple way to introduce anharmonicity effects into the study of the quantum mechanics of a diatomic molecule, it has since been used to study solids and liquids. However, there is no reason why it should be predictive of any real material behavior; like the Lennard-Jones potential it must be used only as an idealized model system.

While our discussion of pair potentials up to this point suggests that they are only useful as a model system (or to study the solidification of the noble gases), there is still something to be learned from pair potentials if their lack of transferability is acknowledged and they are used only as a first-order expansion of the energy around some reference state. This works well for simple metals, whose bonding is loosely described as "metallic" and more precisely is formed by valence electrons in the s and p shells. The transition metals do not fit this category, as their bonding includes d-band electrons, although the noble metals (Ni, Pd, Pt, Cu, Ag, Au) can sometimes be studied to a limited extent with pair potentials because the d-bands are filled. For these systems, the bonding electrons can be viewed as a perturbation of the free electron gas, and as such it is possible to derive a pair potential as a second-order perturbation of the energy around a reference state defining a baseline electron gas density [Car90, Fin03]. Thus, the energy can be written as a sum of a *volume-dependent but structurally independent* energy plus an effective pair potential:

$$\mathcal{V}^{\text{int}} = \mathcal{V}^0(\Omega) + \frac{1}{2} \sum_{\alpha,\beta \neq \alpha} \phi_{\text{eff}}(r^{\alpha\beta}, \Omega), \tag{5.39}$$

where Ω represents the average atomic volume about which the perturbation was imposed. The main disadvantage of this form is the volume dependence (which, we emphasize, appears in the effective pair potential ϕ_{eff} as well as in $\mathcal{V}^0$). Clearly, transferability is only possible if we can develop a pair potential at every conceivable average volume. The form is even more restrictive than this suggests, since the constant volume restriction must be satisfied in a local sense. For example, a box containing N atoms in the gaseous state or the same N atoms condensed into a solid will have the same global average volume, but the effective pair potential will not be transferable between the two. As such, the effective pair potential approach is not especially useful for studying configurations like free surfaces.

In Fig. 5.4, we reproduce a figure from [Car90] that illustrates dramatically how the form of an effective pair potential depends on the reference structure to which the potential is fitted. The curve labeled "molecule" is the pair potential for an isolated Cu_2 molecule [ABRT65]; it is in effect the analog of the curve we produced for the H_2 molecule in Fig. 4.8(c). The other extreme, perhaps, is a pair potential generated based on a perfect infinite crystal of fcc Cu. This is shown by the curve labeled "equation of state," which is generated by fitting the pair potential to the universal binding energy relation (UBER) form of the cohesive energy for bulk fcc Cu [CGE80] (see Section 6.5.2). The curve labeled "phonons" is also fitted to bulk fcc Cu, but is designed to match the phonon spectrum of the crystal [ES79], while the curve labeled "defects" is fitted to the vacancy formation energy and other defects in the bulk lattice [Joh64]. While it is perhaps not surprising that the "molecule" curve is dramatically different from the "equation of state" curve, it is somewhat shocking that the other three curves, all fitted to environments that are close to perfect fcc Cu, are so different. There is clearly little or no transferability to be expected from an effective pair potential. For example, using the "phonon" potential is unlikely to accurately reproduce the UBER from of the "equation of state" curve, or vice versa. This points to the need for more sophisticated models if we hope to achieve some level of transferability, even for a simple metal like Cu. The effective pair potential form is only

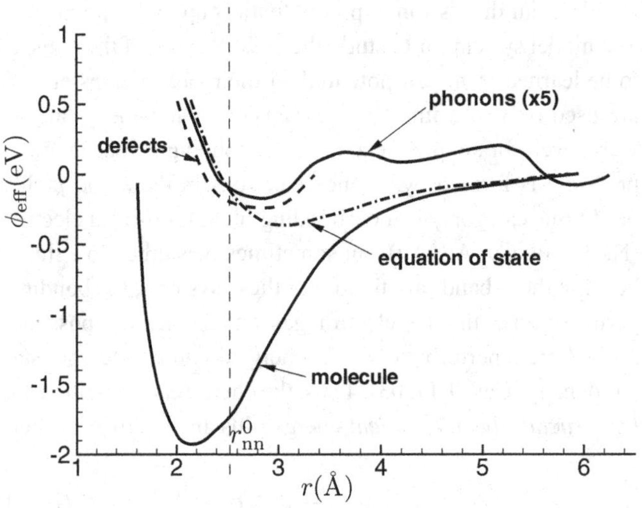

Fig. 5.4 Effective pair potentials for Cu fit to different properties, as explained in the text. r^0_{nn} indicates the nearest-neighbor spacing in fcc Cu at equilibrium. Adapted from [Car90], with permission from Elsevier.

useful for understanding specific properties and trends in properties that do not involve large atomic density changes or structural rearrangement.

The Cauchy relations A revealing test of pair potentials is to use them to compute the elastic constants of a perfect crystal. This exercise sheds light on the limitations of the pair potential, while simultaneously showing that they can indeed work well when confined to systems where pairwise interactions are dominant (i.e. where van der Waals or ionic bonding dominates). This is discussed in detail in Section 11.5.3.

5.4.3 Modeling ionic crystals: the Born–Mayer potential

Ionic crystals are a class of materials for which a simple pair potential description is actually quite accurate [ARK$^+$94], at least insofar as the assumption of pairwise, environmentally independent interactions is a good approximation. Combinations of alkali or alkali-earth elements with group VI or VII elements typically result in such a strong electron transfer that the principal cohesion comes from Coulomb interactions between the resulting charged ions. In this case, a pair potential is a sufficient model of the energetics.

The classical example of such an ionic model is the Born–Mayer potential. It is really a combination of two pair potentials: a long-range Coulomb interaction and a short-range repulsive potential that describes the overlap of filled orbitals in the ion cores. The entire model takes the form

$$\mathcal{V}^{\mathrm{int}} = \frac{1}{2}\sum_{\substack{\alpha,\beta\\ \alpha\neq\beta}}^{N} \frac{Z^\alpha Z^\beta e^2}{r^{\alpha\beta}} + \frac{1}{2}\sum_{\substack{\alpha,\beta\\ \alpha\neq\beta}}^{\mathrm{near}} A_{\alpha\beta}\left(1+\frac{Z^\alpha}{z^\alpha}+\frac{Z^\beta}{z^\beta}\right)\exp\left[(\sigma^\alpha+\sigma^\beta-r^{\alpha\beta})/\rho^{\alpha\beta}\right], \tag{5.40}$$

where Z is the atomic number of an ion, e is the electron charge (defined in Eqn. (4.43)), z is the number of electrons in the outer shell of an ion and σ, A and ρ are fitting parameters that depend only on the ionic species. The second sum is obviously the more empirical of the two, as it includes the fitting parameters. As the notation "near" over the second sum is meant to suggest, it is taken only over nearest neighbors in the equilibrium crystal structure (or some other short cutoff distance). This sum represents repulsion due to core overlap, something that is difficult to quantify exactly. The first sum, however, is the well-understood Coulomb energy. It contains both repulsive and attractive terms (depending on the charge of each ion) and is long-range. While it looks simple, it is extremely subtle to use in a practical simulation. This is because the potential decays slowly, as $1/r$, while the number of ions found at a distance r from a given point tends to increase as r^2 (as the surface area of an expanding sphere). As such, this sum only converges because oppositely charged particles arrange themselves in a crystal to ensure charge neutrality on the local scale. This means that the gaps between neighboring shells decrease with distance[22] and the ions in these shells tend to be of alternate charge. This ensures a cancellation of terms and convergence in the limit of $r \to \infty$. The practical implication of this is that we cannot simply truncate the summation at some cutoff radius and obtain a correct result; we must resort to special summation techniques when modeling an infinite crystal using periodic boundary conditions (PBCs). A similar situation in DFT led to the expressions introduced in Eqn. (4.156) to allow for exact evaluation of these sums, but these methods are not practical when there are large numbers of atoms in the periodic simulation box. In the late 1980s, fast multipole methods [GR87] were developed which make it possible to accurately evaluate these sums with $O(N \log N)$ efficiency, but the fact that the interactions are long-range will always impede the speed of calculations on ionic systems compared with short-range covalent or metallic bonding.

The Born–Mayer potential is also called the "rigid-ion" model, as it neglects the possibility of ion polarization induced by neighboring ions. Higher-order effects of "deformable ions" can be included in more advanced ionic models as discussed by Finnis [Fin03].

5.4.4 Three- and four-body potentials

Materials with a strong covalent-bonding character need a model with bond-angle dependence. This means that it is necessary to include at least the $\phi_{\alpha\beta\gamma}$ (or ϕ_3 for a single atomic species) contributions in Eqn. (5.32).[23] In this section, we discuss cluster potentials that include three- or four-body interactions. These potentials are used primarily for covalent materials like Si or mixed metallic-covalent systems like Mo.

Probably the most famous three-body potential is the Stillinger–Weber model for silicon [SW85]. The functions included in the Stillinger–Weber potential were chosen based mainly

[22] To elaborate a little on this point, consider an fcc arrangement of atoms. Atoms in the nth shell are all at some distance r_n from an atom at the origin, where for fcc $r_n = a\sqrt{n/2}$ and a is the lattice constant. Clearly, the distance between two neighboring shells, $(r_{n+1} - r_n) \to 0$ as $n \to \infty$.

[23] Even simple metals require more than a pair potential description if the model is to be reasonably transferable. In the case of the simple metals, however, the cluster potentials discussed in this section are not the most effective way to extend the pair potential, and instead the *pair functional* approach of Section 5.5 is more useful.

on ease of implementation and a few key physical observations, rather than any rigorous justification from first principles, however, the model is a reasonable first cut at a potential with bond-angle dependence. The Stillinger–Weber potential includes only the pair and triplet contributions of Eqn. (5.32), which take the form

$$\phi_2(r^{\alpha\beta}) = \varepsilon f_2(r^{\alpha\beta}/\sigma), \tag{5.41}$$

$$\phi_3(r^{\alpha\beta}, r^{\alpha\gamma}, r^{\beta\gamma}) = \varepsilon f_3(r^{\alpha\beta}/\sigma, r^{\alpha\gamma}/\sigma, r^{\beta\gamma}/\sigma), \tag{5.42}$$

where

$$f_2(\hat{r}) = \begin{cases} A\,(B\hat{r}^{-p} - \hat{r}^{-q})\, e^{(\hat{r}-a)^{-1}}, & \hat{r} < a, \\ 0, & \hat{r} \geq a, \end{cases} \tag{5.43}$$

$$f_3(\hat{r}^{\alpha\beta}, \hat{r}^{\alpha\gamma}, \hat{r}^{\beta\gamma}) = h(\hat{r}^{\alpha\beta}, \hat{r}^{\alpha\gamma}, \theta^{\beta\alpha\gamma}) + h(\hat{r}^{\beta\gamma}, \hat{r}^{\alpha\beta}, \theta^{\gamma\beta\alpha}) + h(\hat{r}^{\alpha\gamma}, \hat{r}^{\beta\gamma}, \theta^{\alpha\gamma\beta}), \tag{5.44}$$

and the hats indicate that distances are normalized by σ. In Eqn. (5.44), $\theta^{\beta\alpha\gamma}$ is the angle formed by the bonds α–β and α–γ with the vertex at atom α. It can be written in terms of the bond lengths as shown in Eqn. (5.33), but using the angle directly provides a more convenient functional form. The other angles appearing in Eqn. (5.44) are similarly constructed. The function h is designed to penalize bond angles that differ from the ideal tetrahedral angle, $\theta_t = 109.47°$, observed in diamond-cubic materials like Si. Specifically,

$$h(\hat{r}_1, \hat{r}_2, \theta) = \begin{cases} \lambda \exp\left[\mu(\hat{r}_1 - a)^{-1} + \mu(\hat{r}_2 - a)^{-1}\right](\cos\theta + \frac{1}{3})^2, & \hat{r}_1 < a \text{ and } \hat{r}_2 < a, \\ 0, & \text{otherwise.} \end{cases} \tag{5.45}$$

Note that $\cos\theta_t = -1/3$, so that $h = 0$ is the minimum of this function when $\theta = \theta_t$.

This set of functions gives a cluster potential with seven free parameters, which in principle could be modified to study other materials like Ge or C in the diamond cubic phase. In most cases, however, the Stillinger–Weber model is used to study Si with the original parameters determined by Stillinger and Weber, which are

$$A = 7.049556277, \qquad B = 0.6022245584, \qquad p = 4, \qquad q = 0,$$
$$a = 1.80, \qquad \lambda = 21.0, \qquad \mu = 1.2.$$

The energy and length scales are determined entirely by the constants ε and σ. Stillinger and Weber [SW 85] established these values by fitting the Si lattice constant as $a_0 = 5.429$ Å and an (incorrect) equilibrium cohesive energy of the diamond structure as $E^0_{\mathrm{coh}} = 4.37$ eV. This gives[24]

$$\sigma = 2.0951 \text{ Å}, \qquad \varepsilon = 2.1682 \text{ eV}, \tag{5.46}$$

to complete the definition of the model.

A more fundamental approach to the development of cluster potentials was taken by Moriarty in a series of papers [Mor88, Mor90, Mor94]. Originally, the so-called "generalized pseudopotential theory (GPT)" was developed to provide a rigorous expansion of the

[24] The cohesive energy is defined in Section 6.5.1. Note that the value used in [SW 85] differs from the experimentally accepted value of 4.63 eV for Si.

Fig. 5.5 A simple ball-and-stick model of the ethane molecule.

energy of a system of atoms into interatomic potentials up to the four-body cluster terms in Eqn. (5.32). Later, "model GPT" (MGPT) was developed in which the interatomic potentials were approximated by simplified analytic forms. To retain their accuracy, the forms remain considerably more complex than models like the Stillinger–Webber model, but are nonetheless simple enough to allow MD simulations involving millions of atoms over hundreds of picoseconds. The focus of the MGPT efforts has been on the transition metals, like Mo and Ta [XM96, XM98, CBM99, YSM01a, YSM01b, MBR+02, MBG+06].

Efforts to develop cluster potentials beyond the ϕ_4 terms in Eqn. (5.32) have not met with much success, mainly because it becomes very difficult to conceive and parameterize appropriate functions of the 20 interatomic distances that characterize a five-atom cluster as discussed earlier.

5.4.5 Modeling organic molecules: CHARMM and AMBER

In materials science, we tend to focus on "materials" consisting of a large collection of atoms, usually in the solid state, often in a crystalline form. But atomistic modeling is also essential for the study of individual macromolecules, ranging from the organic polymers of man-made plastics to DNA. Understanding the structure and configuration of these molecules is critical for understanding such things as the strength and reactivity of a plastic, or the interactions between a drug and proteins in the body. A collaborative effort by many researchers has led to the development of interatomic cluster potentials for these purposes, referred to in the biology and chemistry fields as "force field" models.

The two most commonly used force field models are the classes of potentials known as CHARMM (Chemistry at HARvard Macromolecular Mechanics) [BBO+83, SDKF08] and AMBER (Assisted Model Building with Energy Refinement) [PC03, CDC+08]. The CHARMM and AMBER potentials are, at their essence, generic cluster potentials. However, because they are developed with macromolecules in mind, they are constructed differently to optimize them for this application. We focus our discussion on CHARMM, but the AMBER model is essentially the same.

Macromolecules tend to have strong covalent bonds between atoms along the chain and out to side groups, together with weaker interactions such as van der Waals and Coulomb effects. Thus, macromolecules lend themselves naturally to a separation into two kinds of atomic interaction: "bonded" and "nonbonded". The former refers to covalent bonds; within CHARMM these are assumed to be unbreakable. The latter are the weaker, breakable interactions. This is illustrated in Fig. 5.5, which shows a simple model of an ethane molecule. Bonded interactions are shown by thick lines (cylinders). Atoms like those labeled α and δ (for example) also experience a nonbonded interaction through Coulomb and van der Waals forces, which play a role in determining the conformation of the molecule. The total energy of the CHARMM model includes these effects as

$$\mathcal{V}^{\text{int}}_{\text{CHARMM}} = \mathcal{V}^{\text{bonded}} + \mathcal{V}^{\text{nonbonded}}. \tag{5.47}$$

The bonded part is simply treated as a sum of independent harmonic terms due to three physical effects on the bonds (stretching, bending and rotation) as

$$\mathcal{V}^{\text{bonded}} = \mathcal{V}^{\text{stretch}} + \mathcal{V}^{\text{bend}} + \mathcal{V}^{\text{rotate}}. \tag{5.48}$$

Bond stretching is described by a simple pair potential of the form

$$\mathcal{V}^{\text{stretch}} = \frac{1}{2} \sum_{\substack{\alpha,\beta \\ \alpha \neq \beta}}^{\text{bonded}} k_{\text{b}}^{\alpha\beta} (r^{\alpha\beta} - r_0^{\alpha\beta})^2. \tag{5.49}$$

The notation reminds us that the sum is only over atoms identified as "bonded" during the initial set-up of the problem. The coefficients $k_{\text{b}}^{\alpha\beta}$ and $r_0^{\alpha\beta}$ are fitting parameters representing the stiffness of the bond and the equilibrium length of the bond. They depend on atoms α and β only through their atomic species.

Bond bending is essentially a three-body cluster potential of harmonic form, again summed only over atoms that are identified as "bonded." It takes the form

$$\mathcal{V}^{\text{bend}} = \frac{1}{6} \sum_{\substack{\alpha,\beta,\gamma \\ \alpha \neq \beta \neq \gamma}}^{\text{bonded}} k_{\theta}^{\alpha\beta\gamma} (\theta^{\alpha\beta\gamma} - \theta_0^{\alpha\beta\gamma})^2, \tag{5.50}$$

where $\theta^{\alpha\beta\gamma}$ is the angle formed between bonds α-β and β-γ. As with the bond-stretching energy, the fitting parameters $k_{\theta}^{\alpha\beta\gamma}$ and $\theta_0^{\alpha\beta\gamma}$ depend on the species of atoms.

Determining the bond *rotation* requires knowing the position of at least four bonded atoms. For instance, consider the four atoms labeled α, β, γ and δ in Fig. 5.5. Atoms α, β and γ all lie in a plane, as do atoms β, γ and δ. The angle between these planes characterizes the rotation of the bond β–γ. Although the actual machinery of a CHARMM code would be set up to compute the bond-rotation energy more efficiently, it can be written as a four-body cluster potential:

$$\mathcal{V}^{\text{rotate}} = \frac{1}{24} \sum_{\alpha,\beta,\gamma,\delta}^{\text{bonded}} k_{\phi}^{\alpha\beta\gamma\delta} (\phi^{\alpha\beta\gamma\delta} - \phi_0^{\alpha\beta\gamma\delta})^2, \tag{5.51}$$

and this completes the description of the bonded part of the energy.

The nonbonded part is treated using two simple pair potentials that we have already introduced, namely the Lennard-Jones 6–12 potential of Eqn. (5.36) to mimic the van der Waals interactions and the Coulomb part of the interactions from the Born–Mayer ionic model of Eqn. (5.40). These contributions are summed over all atom pairs, bonded or not, although the Lennard-Jones part typically includes a cutoff radius.

From this description, we can see that the basic CHARMM model is quite simple, making use of easy-to-implement harmonic, Lennard-Jones and Coulomb potentials. No effort is made to include quantum mechanical effects since the bonded atoms remain bonded by construction. Instead, harmonic approximations are used to describe what are known from quantum mechanics to be much more complex functions. This means that the bonded interactions are only valid near the equilibrium bond configuration (length, angle or rotation). It also means that it is necessary to know a great deal about the type of bonds that are expected when setting up a CHARMM/AMBER model. For example, it is not enough to simply state that there is a "bond" between two carbon atoms. It is necessary to know whether that C–C bond is a so-called "σ" or "π" bond,[25] in order to determine the harmonic part of the bond energy. For many important molecules, these details are known. However, for the deformation problems of interest in materials science, the absence of bond breaking and formation make these models extremely limited.

Due to the above, a great deal of effort is spent on accurately determining the large number of fitting parameters necessary to characterize a CHARMM (or AMBER) force field. This includes determining the different k_{b}, k_θ and k_ϕ parameters which are required for every possible combination of atomic species, in addition to accurately fitting the van der Waals and Coulomb parts. Indeed, many clever uses of experimental and more fundamental (DFT) results have made it possible to build a vast set of bonding libraries for the CHARMM and AMBER programs. For example, the force constants k can be related to frequencies of vibration in simple molecules. What makes CHARMM and AMBER feasible is the relatively small number of atomic species that participate in the bonding of interest in macromolecular science,[26] which keeps the number of fitting parameters manageable.

It is important to recognize that CHARMM and AMBER models cannot be used to study structural rearrangements and chemical reactions that involve bond breaking. However, that is not the goal of these methods. Instead, they aim to shed light on the structure, configuration and motion of stable, nonreacting molecules. An extension to the CHARMM/AMBER approach of fixed bonding is the so-called "ReaxFF" model discussed in Section 5.6.3. ReaxFF permits the breaking and healing of bonds. This essentially removes the distinction between bonded and nonbonded pairs, as all pairs are treated on an equal footing. The ReaxFF model *decides* whether two atoms are bonded or not based on their proximity and an estimate of the *bond order*. The general class of bond-order potentials is discussed in Section 5.6.

5.4.6 Limitations of cluster potentials and the need for interatomic functionals

The cluster potentials discussed in this section are based on the assumption that the energy of a system of atoms can be approximated by terminating the series in Eqn. (5.32) at a low order. This has led us to the definition of pair, three-body and four-body potentials. Unfortunately, there are many materials in which the bonding is such that the series of

[25] Geometrical constraints in atomic arrangements mean that the same two orbitals can form bonds of different strength and nature depending on their relation in space. The difference can be visualized using Fig. 4.5(b). Placing two such p-orbitals "side-by-side" in that figure, with the upper and lower lobes overlapping between the orbitals, will form a π-bond. By contrast, a σ-bond between two p-orbitals occurs when the two images in Fig. 4.5(b) are stacked end-to-end along the bond's z-axis.

[26] Consider, for example, organic molecules that are dominated by C, O, H and N atoms with a few other more "exotic" elements only occasionally playing a role.

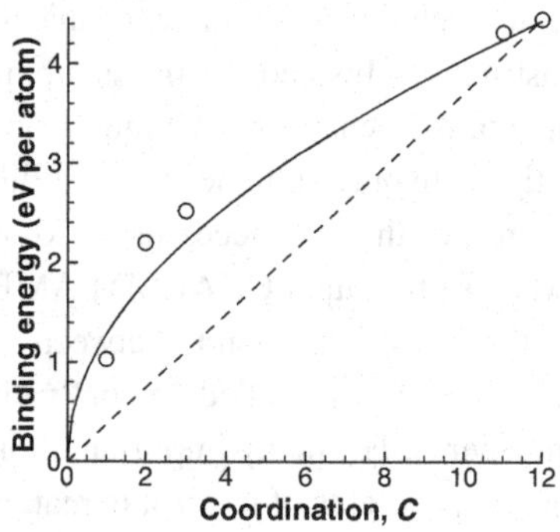

Fig. 5.6 The binding energy (in absolute value per atom) of Ni as a function of the coordination. $C = 1, 11$ and 12 are experimental results from [Car90], while $C = 2$ and 3 are DFT results from [MPDJ98]. The square-root dependence on coordination (solid line) is a much better fit than the linear dependence predicted by a pair potential.

n-body terms, where n ranges from 2 to the number of atoms, converges far more slowly. As a result cluster potentials can be very inefficient. This has led to the development of alternative functional forms which we discuss in the next section.

For the purpose of organizing ideas, we follow the nomenclature of Carlsson [Car90], who used the term "functionals" to denote any interatomic model that did not fit Eqn. (5.32). Also following Carlsson, we can further divide interatomic functionals into "pair functionals" and "cluster functionals." The distinction, as the names suggest, is in how the functionals depend on atomic positions. Pair functionals require information about atom positions pairwise, while cluster functionals involve atoms in groups of three or more.

These functionals take many forms, and it is difficult to write a general expression that is especially enlightening.[27] It is of course always possible to represent any interatomic functional in the form of Eqn. (5.32) by following the recursive procedure outlined in Section 5.4.1. (This is shown for a pair functional in Section 5.5.4.) The distinction between a "cluster potential" and an "interatomic functional" is that an interatomic functional in its simplest "standard" form is different from the representation in Eqn. (5.32). This will become more apparent as we explore specific functional forms for pair and cluster functionals in the following two sections.

5.5 Pair functionals

The *coordination*, C, of an atom is defined as the number of its nearest neighbors. Now consider a relatively simple question. How does the energy per atom of a material change as C increases? This depends, clearly, on the type of bonding between the atoms and is especially pronounced in metals. Following Carlsson [Car90], we demonstrate this for the example of Ni in Fig. 5.6. Three experimental values and two DFT results allow us to

[27] The form proposed by Carlsson [Car90] is not general enough for some of the more complicated functionals like the Tersoff or ReaxFF models discussed in Section 5.6.

estimate the dependence of the energy on C. From experiments, we know the binding energy of a Ni dimer at one extreme ($C = 1$) and the cohesive energy of solid fcc Ni ($C = 12$) at the other. A third experimental value, the energy of vacancy formation, can be used to estimate the energy of an atom in the vicinity of a vacancy, which would have approximately $C = 11$. Finally, DFT results for isolated Ni_3 and Ni_4 molecules round out the data.

A pair potential, which we have already seen to be inadequate for simple metals for other reasons, is only further condemned by this exercise. It is easy to verify that the predicted C dependence will be linear for a near-neighbor pair potential, as follows. In a near-neighbor model, the equilibrium lattice constant for the perfect fcc crystal will be that which sets the near-neighbor distance to the minimum point on the pair potential curve. Lowering the coordination, say by removing a single atom to form a vacancy, simply removes the (negative) energy contribution for that bond, but does not affect any of the other bonds since any further distortions would only raise the energy of the system. There is therefore a one-to-one correspondence between the number of near-neighbor bonds (the coordination) and the energy. For a model with second (or further) neighbor interactions, the broken bonds due to the formation of a vacancy can be somewhat compensated for by rearranging the remaining atoms, but only rather weakly. So while the dependence is no longer linear for second-neighbor pair potentials, it tends to be only weakly nonlinear and nowhere near the dependence shown in Fig. 5.6. The coordination dependence is in fact much better fitted by a square-root function (a form that emerges from the discussion of moment approximations to TB in Section 4.6.6). Clearly, bonds formed in the presence of other bonds are weaker than those formed in relative isolation: there is a strong environmental dependence of the bond energy that a simple pair potential cannot describe.

5.5.1 The generic pair functional form: the glue–EAM–EMT–FS model

Despite the shortcomings of pair potentials, it is hard to let them go because of one clear advantage: they are markedly faster to use than any cluster potential will be. Luckily, pair *functionals* can be designed to be almost as fast to use as pair potentials, while providing a vast improvement in their accuracy for the case of simple metals. Pair functionals appear under numerous titles, including the "glue potential" [ETP86], "effective medium theory (EMT)" [NL80, JNP87, Jac88, RD91], the "embedded atom method (EAM)" [DB84, DFB93, Daw89], and the Finnis–Sinclair model [FS84]. While they are all derived from slightly different underlying perspectives, they all lead to the form

$$\mathcal{V}^{\text{int}} = \frac{1}{2} \sum_{\substack{\alpha,\beta \\ \alpha \neq \beta}} \phi_{\alpha\beta}(r^{\alpha\beta}) + \sum_{\alpha} U_{\alpha}(\rho^{\alpha}), \tag{5.52}$$

where $\phi_{\alpha\beta}$ is a pair potential, and U_α is the energy associated with placing an atom at a point whose environment is defined by an intermediary function, ρ^α. This function is found by linearly superposing a third function, g_β, which is again a function of interatomic distances $r^{\alpha\beta}$ (the physical interpretations of U, g and ρ will be discussed in Section 5.5.2).

Thus

$$\rho^\alpha = \sum_{\substack{\beta \\ \beta \neq \alpha}} g_\beta(r^{\alpha\beta}). \tag{5.53}$$

It is often useful to write the internal potential energy as

$$\mathcal{V}^{\text{int}} = \sum_{\alpha} E^\alpha,$$

where E^α is the contribution of atom α to the total energy:

$$E^\alpha = U_\alpha(\rho^\alpha) + \frac{1}{2} \sum_{\substack{\beta \\ \beta \neq \alpha}} \phi_{\alpha\beta}(r^{\alpha\beta}), \tag{5.54}$$

although, as noted after Eqn. (5.31), the physical significance of E^α is suspect.

Efficiency of pair functionals The advantage of the pair functional scheme is the significant improvement over pair potentials for relatively little cost. A pair functional ameliorates the problems of a simple pair potential: for fcc metals it can restore the prediction of the elastic constants and the relation between bond energy and coordination. At the same time, it removes the ill-defined volume dependence of an effective pair potential and, as a result, permits reasonable models of surface energetics. Computing ρ requires exactly the same data structure and computer operations as the pair potential ϕ. Once ρ is computed, there is only one additional function evaluation for each atom, $U(\rho^\alpha)$, and so the overall cost of a pair functional is essentially only a factor of 2 more expensive than a pair potential.

5.5.2 Physical interpretations of the pair functional

Depending on the details of the pair functional model being considered, the functions U, ρ and g have different physical interpretations [Vot94]. The most intuitive is probably the interpretation used to motivate the EAM and EMT models. In these approaches, the pair potential part of the expression is viewed as the Coulomb interactions between the atom cores (each comprising the nucleus and the nonvalence electrons), while U_α is the electronic energy associated with "embedding" atom α in a homogeneous electron gas of density ρ^α. This idea has its roots in DFT, where (as we have shown) the electron density determines the energy. If, as is the case with the simple metals, the environment of each atom is a nearly uniform electron gas, this notion of the embedding energy is a reasonable approximation. Naturally, the electron density at the point where atom α is to be embedded is determined in some way by contributions from neighboring atoms, and to a first approximation we may assume a simple linear superposition of electron density from each neighbor through the function g_β. Although we know from our discussion of atomic orbitals in Chapter 4 that the electron density can be a function of both distance from the nucleus and direction, some orbitals (the s-orbitals) have no angular dependence and others can at least be approximated in this way. As such, g_β is a function only of distance and represents a spherically-averaged electron density field around an isolated atom, β.

A different interpretation of the pair functional form is taken in the FS model, where

$$U = -A\sqrt{\rho}, \tag{5.55}$$

and A is a positive fitting parameter. This simple form was chosen because it mimics the results of the second-moment approximation to TB (Section 4.6.6), and therefore permits the interpretation of $g(r^{\alpha\beta})$ as a sum of squares of overlap integrals between electronic orbitals centered on atoms α and β. This form also directly gives us the $\sqrt{C}$ dependence of the energy on atom coordination, as observed in Fig. 5.6.

For configurations far from the equilibrium electron density of the perfect crystal, the assumption that the atom is embedded in a uniform electron gas will start to become less accurate. Many developers of pair functional models follow the proposal of [Foi85], where the functional form of U is taken to exactly fit the so-called UBER described in Section 6.5.2.[28] The UBER gives the energy of a crystal as a function of the lattice constant or, within the simple linear superposition used in the pair functional, as a direct function of the electron density at a lattice site. Based on this, one can *define* the embedding energy so that the UBER energy is exactly followed for the equilibrium crystal structure of the material once a form for the pair potential, $\phi_{\alpha\beta}$, is chosen.

5.5.3 Fitting the pair functional model

Pair functionals have come to be treated much more generally than the original formulations, and they are often developed with the functional forms themselves included as part of the fitting process. For example, the force-matching method of Ercolessi and coworkers [EA94, LSAL03] uses splines as the functional forms with the spline knots taken as degrees of freedom to be fit. Then, a wide range of atomic configurations is considered and the forces on the atoms are computed using DFT. The parameters of the functional forms are then optimized to best fit the DFT forces. With this as one example, there are many such interatomic models that are referred to as "EAM" or having the "EAM-form", while they are in fact based on a much more generalized pair functional scheme.

Efforts to fit the pair functional form have met with a great deal of success in describing fcc metals with filled or nearly filled d-orbitals. They have had more limited success with body-centered cubic (bcc) and hcp transition metals, because the spherically symmetric nature of the pair functional form makes it difficult to correctly describe these systems when the d-band is only partially filled. A partially filled d-band implies a partially covalent, directional nature to the bonding that cannot be captured by the pair functional form.

The approach of completely generalizing the functional forms is a double-edged sword, and can only be used in conjunction with sound judgment and experience. While completely arbitrary forms of the functions might seem like a way to extend the applicability of pair functionals beyond the pure transition metals for which they excel, they do not guarantee a good model. A simple example where fitting can fail is the elastic properties of crystals. In Section 11.5.3, we will revisit this in more detail, but for now we can simply note that

[28] This approach can break down in certain cases [MMP+ 01]. See the discussion in Section 6.5.2 regarding different points of view on using the UBER curve as part of the fitting process for empirical potentials.

there is an elastic parameter known as the "Cauchy pressure," which for many crystals is a positive quantity. It turns out that it is straightforward to design a pair functional that fits this value. However, there are many crystals for which the Cauchy pressure is negative, and there is nothing stopping us from fitting a pair functional to a negative Cauchy pressure. The problem, however, is that doing so requires a completely unphysical form for the embedding function U. Such a model would successfully match the Cauchy pressure, but would be a very poor description of the material in almost any other respect. This is reflected by the fact that the crystals for which the pair functional form works well, notably fcc metals like Al, Ni or Cu, all have a positive Cauchy pressure. On the other hand, bcc Cr, for example, has a negative Cauchy pressure, and all attempts to describe it with a pair functional have been unsuccessful.

In short, the message here is that while generalizing the functional forms of a certain type of model may seem to be a good way to improve its fitting capacity it may not lead to a better model outside the fit. The functional forms must retain the essential features that serve as physical reality checks; we are not free to simply choose *any* functions that seem to fit the dataset we have chosen.

Another difficulty in fitting pair functionals is the presence of more than one atomic species. In principle, it should be straightforward to develop pair functional models for alloy systems involving many different elements. One needs a unique embedding function, U_α, and density, g_α for each element, as well as pair potentials $\phi_{\alpha\beta}$ for each possible combination of elements. In practice, however, it can be difficult to model such alloy systems accurately. Models can be built that are relatively successful with simpler intermetallic compounds like Ni_3Al and NiAl (for example, [VC87, Mis04]). However, complex intermetallics typically have a significant amount of covalent bonding character that cannot be described by spherically symmetric pair descriptions.

For *dilute* solutions of one simple metal in another, it is common to model the alloy by combining two pair functionals that were originally designed for the pure metals. In this case, all the functional forms for species A (g_A, U_A, ϕ_{AA}) and B (g_B, U_B, ϕ_{BB}) remain the same, and only ϕ_{AB} needs to be determined. This is taken as a geometric mean of the two pure element pair potentials:

$$\phi_{AB}(r) = \sqrt{\phi_{AA}(r)\phi_{BB}(r)}.$$

However, this approach is limited to dilute solid solutions and cannot be considered transferable to general crystal structures between the two elements.

5.5.4 Comparing pair functionals to cluster potentials

In Section 5.4.1, we derived the formally exact expression of the energy of a cluster of atoms in Eqn. (5.28). Here, we follow Drautz *et al.* [DFS04] and use this exact expression to evaluate the nature of the approximation inherent in a pair functional. At the same time, this allows for a comparison of the efficiency of pair functionals versus cluster potentials.

Let us assume for a moment that the pair functional of the form Eqn. (5.52) is exact for a particular material. What does this tell us about the nature of the terms in the exact cluster potential of Eqn. (5.28)? Setting the two expressions equal to each other we have

$$\frac{1}{2}\sum_{\substack{\alpha,\beta\\ \alpha\neq\beta}}\phi_{\rm PF}(r^{\alpha\beta})+\sum_{\alpha}U(\rho^{\alpha})=\frac{1}{2}\sum_{\substack{\alpha,\beta\\ \alpha\neq\beta}}\widehat{\phi}_2(\boldsymbol{r}^{\alpha},\boldsymbol{r}^{\beta})+\frac{1}{3!}\sum_{\substack{\alpha,\beta,\gamma\\ \alpha\neq\beta\neq\gamma}}\widehat{\phi}_3(\boldsymbol{r}^{\alpha},\boldsymbol{r}^{\beta},\boldsymbol{r}^{\gamma})+\cdots. \tag{5.56}$$

We have limited our attention to a system containing only one atomic species for clarity, and we label the two-body function from Eqn. (5.52) as $\phi_{\rm PF}$ (for "pair functional") to distinguish it from the exact two-body function. We have also explicitly written the cluster potential functions in a way that does not show their invariance properties for notational convenience, although the invariance principle in Eqn. (5.17) must still hold for these terms. For definiteness, we assume that the embedding energy is

$$U(\rho)=-D\rho\ln\rho,$$

which is a commonly used form within the EAM formalism. This form has the key physical features required of U but retains analytical simplicity. We set $D=1$ in this discussion.

Now, we follow Drautz *et al.* [DFS04] and use the recursive process described in Section 5.4.1 to determine the potential functions. The results are

$$\widehat{\phi}_2(\boldsymbol{r}^1,\boldsymbol{r}^2)=\phi_{\rm PF}(r^{12})-g_{(1)2}\ln g_{(1)2}-g_{(2)1}\ln g_{(2)1}, \tag{5.57}$$

$$\widehat{\phi}_3(\boldsymbol{r}^1,\boldsymbol{r}^2,\boldsymbol{r}^3)=\left[g_{(1)2}\ln\left(\frac{g_{(1)2}}{g_{(1)23}}\right)+g_{(1)3}\ln\left(\frac{g_{(1)3}}{g_{(1)32}}\right)+\cdots\right], \tag{5.58}$$

$$\widehat{\phi}_4(\boldsymbol{r}^1,\boldsymbol{r}^2,\boldsymbol{r}^3,\boldsymbol{r}^4)=-\left[g_{(1)2}\ln\left(\frac{g_{(1)2}g_{(1)234}}{g_{(1)23}g_{(1)24}}\right)+\cdots\right], \tag{5.59}$$

$$\widehat{\phi}_5(\boldsymbol{r}^1,\boldsymbol{r}^2,\boldsymbol{r}^3,\boldsymbol{r}^4,\boldsymbol{r}^5)=\left[g_{(1)2}\ln\left(\frac{g_{(1)2}g_{(1)234}g_{(1)235}g_{(1)245}}{g_{(1)23}g_{(1)24}g_{(1)25}g_{(1)2345}}\right)+\cdots\right], \tag{5.60}$$

$$\vdots$$

where we have adopted the short-hand, $g_{(\alpha)\beta_1\beta_2\ldots\beta_M}=\sum_{j=1}^{M}g(r^{\alpha\beta_j})$, and have used explicit integer indices to clarify the notation.

There is a large but finite number of terms in each of Eqns. (5.58)–(5.60), but they are all of the form shown (the details are left to Exercise 5.8). We expect $g(r)$ to go to zero over some radial distance, so the cluster potential functions $\widehat{\phi}_n$ go to zero as n increases. This is because all the terms are either pre-multiplied by $g_{(\alpha)\beta}$ (which is zero for large $r^{\alpha\beta}$) or include the logarithm of a ratio which goes to unity for large enough clusters.

Writing the pair functional in this form allows us to clearly see the efficiency of this approach compared with cluster potentials. If, for example, we consider an fcc lattice and a cutoff, such that $g(r)=0$ between the first- and second-neighbor distances, we would require clusters up to $\widehat{\phi}_{13}$ to write a cluster potential that is fully equivalent to the pair functional [DFS04]. This comes from the nature of the embedding term, whereby the energy of an atom is a function of its own position and the positions of its 12 nearest neighbors through the electron density they contribute. This is far more efficiently evaluated through the pair functional form than through the cluster potential.

5.6 Cluster functionals

Pair functionals represent the simplest method of incorporating an environmental dependence into the strength of each bond; bonds formed in isolation are stronger than those formed in the presence of other bonds. This varying bond strength is related to the *bond order*, a quantum mechanical notion that we discuss in this section by building on our previous discussion of TB in Section 4.6. This is a necessary precursor to the discussion of *cluster functionals* in general, since most cluster functionals are built around some notion of estimating the bond order to compute the bond energy.

Cluster functionals are the most general of the empirical atomistic models, as they can depend on groups of three or more atoms through any general functional dependence (not simply n-body potentials like Eqn. (5.32)). In this section, we discuss the spectrum of cluster functional, ranging from so-called "bond-order potentials" (BOPs), which are built as a systematic approximation to TB, to a myriad of empirical forms that postulate ad hoc functional forms for estimating the bond order.

The most rigorous of what we classify as cluster functionals is the BOP approach developed by Pettifor and many coworkers [HBF$^+$96]. This procedure is a rigorous and systematic set of approximations to TB that aim to make TB much faster while maintaining good accuracy. The biggest disadvantage of TB methods (and of DFT methods, for that matter) is the cost of diagonalizing the Hamiltonian matrix. In general, this is a $O(N_{\text{basis}}^3)$ process, where N_{basis} is the number of orbitals used to define the electronic structure. In TB, the number of orbitals scales directly with the number of atoms, so the algorithm's cost will grow as the cube of the number of atoms.

However, as we discussed in detail in Section 4.6.6, all of the information necessary for computing the energy is contained in the moments of the electronic density of states (DOS). Further, these moments can be determined from the Hamiltonian directly, without the need to diagonalize. Working in terms of the moments is also advantageous because the moment calculation can be decomposed into an atom-by-atom process. For a single atom, calculating M moments will scale as $O(M)$, and so for N atoms we can achieve $O(MN)$ efficiency. The problem with this approach, as Kress and Voter [KV95] showed, is that simply taking more and more moments leads to extremely slow convergence; while such an approach does indeed scale linearly with the number of atoms, the pre-factor associated with the number of moments required for accuracy is too large to be practical.

As a physical justification for many cluster functional forms, we need to understand the quantum mechanical notion of the "bond order" and how it can be related to the ingredients of the TB formalism.

5.6.1 Introduction to the bond order: the Tersoff potential

Basic chemistry courses might describe how, for example, carbon atoms can form so-called *single, double or triple* bonds, where single bonds are generally the weakest and triple the strongest. Exactly which bond forms depends on what *other* bonds the carbon atom is participating in – the bonds are *environmentally dependent*. The nomenclature "single, double or triple" is a rough description of the *bond order* that we discuss in some detail in the next few pages; it is defined such that higher bond order generally correlates with stronger bonds. If we think of building a crystal atom by atom, the first two atoms form a strong bond with high bond order. Add a third atom and the overall energy of the three atoms is reduced, but the strength of each of the bonds (including the original two-atom bond) is now less than the two-atom bond in isolation. Adding more atoms continues this trend of lowering the strength of the individual bonds while still lowering the energy overall.

One can imagine that depending on *how much* each individual bond strength is reduced by neighboring bonds, this effect could favor the more open crystal structures like bcc or diamond-cubic over a close-packed fcc. Clearly, this is a behavior that cannot be captured by a pair potential where each bond is modeled as independent of all others. A pair functional could in principle be designed to mimic this effect; one could use the embedding energy to penalize higher coordination. But in practice, this makes the functional unreliable for other reasons,[29] and pair functionals that actually work have the opposite effect: the embedding energy is always of bonding character and favors higher coordination. This relates to the physical motivation of the pair functional, which is suited to metallic bonding but not to covalent bonding where bond-order effects are most prevalent. Simple cluster potentials fail to "see" the effect of bond order for the same reasons as pair potentials, since each cluster energy is computed independently of other clusters.

A popular model of the covalent, diamond-cubic structures of C, Ge and Si is the Tersoff potential [Ter86, Ter88a, Ter88b]. Tersoff set out to develop a potential that explicitly incorporates the effect of bond order into the strength of the bond. The form resembles a pair potential, but one where the attractive part is modified by a three-body cluster functional:[30]

$$\mathcal{V}^{\text{int}} = \frac{1}{2} \sum_{\substack{\alpha,\beta \\ \alpha \neq \beta}} \left[\phi_R(r^{\alpha\beta}) + b(z^{\alpha\beta}) \phi_A(r^{\alpha\beta}) \right]. \tag{5.61}$$

Here ϕ_R and ϕ_A are repulsive and attractive pair potentials, respectively. The key improvement to these pair potentials is from the bond-order function b, which changes the *strength* of an attractive interaction depending on the environment in which that bond exists. The bond order depends on a *coordination function*, z, which quantifies the bond coordination of the environment. This environment is determined from neighboring atoms not only through their distance away (as in pair functionals) but also through their bond angles. Since defining a bond angle requires at least three atoms, the Tersoff potential involves clusters

[29] For example, a pair functional cannot accurately fit both the cohesive energy and the vacancy formation energy unless the embedding energy *favors* higher coordination (see Exercise 6.7 and Section 6.5.3).

[30] In fact, there are two different functional forms that Tersoff introduced. The first appeared in his earlier paper [Ter86], but Tersoff later dismissed this form because it does not reproduce diamond cubic silicon as the ground state structure [Ter88b]. The form we present here is the same as the *alternative* form Tersoff proposes to rectify this problem [Ter88a, Ter88b]. Finally, Tersoff offered two alternative parameter sets for this particular form (see [Ter88a, Tab. I]), which one could interpret as two distinct potentials. These three forms are sometimes referred to as T1, T2 and T3.

of three atoms in its determination of the attractive bond strengths. The detailed functional forms are

$$\phi_R(r) = Ae^{-\lambda_1 r}, \qquad \phi_A(r) = -Be^{-\lambda_2 r},$$

$$b(z) = (1+\delta^n z^n)^{-1/2n}, \qquad z^{\alpha\beta} = \sum_{\substack{\gamma \\ \gamma\neq(\alpha,\beta)}} f(r^{\alpha\gamma})g(\theta^{\alpha\beta\gamma})e^{\lambda_3^3(r^{\alpha\beta}-r^{\alpha\gamma})^3},$$

$$f(r) = \begin{cases} 1 & \text{if } r < R-D, \\ \frac{1}{2} - \frac{1}{2}\sin\left(\frac{\pi}{2}\frac{r-R}{D}\right) & \text{if } R-D < r < R+D, \\ 0 & \text{if } r > R+D, \end{cases}$$

$$g(\theta) = 1 + \frac{c^2}{d^2} - \frac{c^2}{d^2+(h-\cos\theta)^2},$$

where $\theta^{\alpha\beta\gamma}$ is the angle formed between bonds α–β and α–γ, and the functions contain 12 fitting parameters:[31] $A, B, \lambda_1, \lambda_2, \lambda_3, \delta, n, R, D, c, d$, and h.

Brenner [Bre90] extended the Tersoff form to better treat hydrocarbon systems. Because of the disparate characteristics of H and C, Brenner found that the Tersoff form could not be satisfactorily fitted to this system. He then showed that by somewhat generalizing the bond-order function, b, a new and improved potential (the "Tersoff–Brenner" potential) could be obtained.[32] However, Pettifor and Oleynik [PO04] pointed out two major limitations of the Tersoff–Brenner form: it cannot correctly choose between certain structural arrangements and it does not accurately model the so-called "π-bond" in hydrocarbons (see footnote 25 on page 261). These are limitations in the transferability of the model because it was, in essence, designed to model certain structural arrangements at the expense of others.

Tersoff's incorporation of bond order into his potential was an important first step towards realistic modeling of covalent bonding, and it has paved the way for a wide range of models that include an empirical bond order function of the Tersoff–Brenner variety. The term "BOP" has come to loosely imply any model of this type. However, we prefer to reserve the term "BOP" for the most rigorous, quantum-theory-based BOP developed by Pettifor and coworkers, and refer to the rest as *empirical* BOPs, a class that would include the Tersoff and Tersoff–Brenner forms, as well as potentials like the "environment-dependent interatomic potential" (EDIP) for Si [BKJ97] and the reactive bond order (REBO) potential for hydrocarbons [BSH$^+$02]. The ReaxFF and MEAM models, which we will discuss in Sections 5.6.3 and 5.6.4, are also examples of models that try to capture the bond-order effect empirically.

In the interest of better appreciating the challenge of developing and using such models, let us look at the bond order from the more fundamental perspective of the TB approximation to quantum mechanics.

[31] Probably the most commonly implemented parameter values are the ones Tersoff developed for Si and called "Si(B)" in [Ter88a]. These are $A = 3264.7$ eV, $B = 95.373$ eV, $\lambda_1 = 3.2394$ Å^{-1}, $\lambda_2 = \lambda_3 = 1.3258$ Å^{-1} $\delta = 0.33675$, $n = 22.956$, $R = 3.0$ Å, $D = 0.2$ Å, $c = 4.8381$, $d = 2.0417$ and $h = 0.0$. For alternative numerical values of these parameters, see [Ter88a, Tab. I].

[32] Horsfield *et al.* [HBF$^+$96] showed that the Tersoff–Brenner form is essentially a second-moment expansion of the TB formalism.

5.6.2 Bond energy and bond order in TB

The cohesive energy of a collection of atoms is the sum of the energy of the electrons in the filled bands of the bonded state (which we will call the "electron" energy) and the energy due to core–core repulsion, less the original energy of the unbonded atoms in isolation:

$$\mathcal{E}^{\text{coh}} = \mathcal{E}^{\text{el}} + \mathcal{E}^{\text{rep}} - \mathcal{E}^{\text{atoms}}.$$

The electron energy $\mathcal{E}^{\text{el}}$ is given by Eqn. (4.164), the repulsive part $\mathcal{E}^{\text{rep}}$ can be handled accurately using a pair potential and the energy of the unbonded atoms $\mathcal{E}^{\text{atoms}}$ is simply the number of electrons in each basis orbital, N_i^α, times the energy of that orbital. Thus,

$$\mathcal{E}^{\text{coh}} = \sum_I f_I \epsilon_I + \frac{1}{2} \sum_{\substack{\alpha,\beta \\ \alpha \neq \beta}} \phi(r^{\alpha\beta}) - \sum_{i\alpha} N_i^\alpha \epsilon_i^\alpha.$$

Here the filling function, f_I, is 2 for all states up to the Fermi energy, and zero otherwise, and ϵ_I is the energy of electronic state I. We use the notation $i\alpha$ to indicate a double sum over all atoms and orbitals, where the Greek index refers to an atom and the lower-case Roman index refers to an orbital centered on that atom. The eigenvalues ϵ_I are found during the solution of Eqn. (4.162), but they can be determined from the eigenvectors, c_{iI}^α, as well. Specifically, we can contract both sides of Eqn. (4.162) with c_{iI}^α and rearrange it to find

$$\epsilon_I = \sum_{i\alpha,j\beta,k\gamma,l\delta} \frac{c_{iI}^\alpha \mathcal{H}_{ij}^{\alpha\beta} c_{jI}^\beta}{c_{kI}^\gamma S_{kl}^{\gamma\delta} c_{lI}^\delta} = \sum_{i\alpha,j\beta} c_{iI}^\alpha \mathcal{H}_{ij}^{\alpha\beta} c_{jI}^\beta, \qquad \text{(no sum on } I), \tag{5.62}$$

where the last step only applies if we assume an orthonormal set of basis orbitals.[33] Thus,

$$\mathcal{E}^{\text{el}} = \sum_{i\alpha,j\beta} \sum_I f_I c_{iI}^\alpha \mathcal{H}_{ij}^{\alpha\beta} c_{jI}^\beta, \tag{5.63}$$

so that

$$\mathcal{E}^{\text{coh}} = \sum_{i\alpha,j\beta} \sum_I f_I c_{iI}^\alpha \mathcal{H}_{ij}^{\alpha\beta} c_{jI}^\beta + \frac{1}{2} \sum_{\substack{\alpha,\beta \\ \alpha \neq \beta}} \phi(r^{\alpha\beta}) - \sum_{i\alpha} N_i^\alpha \epsilon_i^\alpha.$$

Next, we can separate the first sum into two parts: one containing the terms for which $i\alpha = j\beta$ and the other containing the rest. We then rearrange the terms so that we can partition the energy differently:

$$\mathcal{E}^{\text{coh}} = \mathcal{E}^{\text{prom}} + \mathcal{E}^{\text{bond}} + \mathcal{E}^{\text{rep}}, \tag{5.64}$$

where

$$\mathcal{E}^{\text{prom}} = \sum_{i\alpha} \sum_I f_I c_{iI}^\alpha c_{iI}^\alpha \mathcal{H}_{ii}^{\alpha\alpha} - \sum_{i\alpha} N_0^{i\alpha} \epsilon_i^\alpha, \qquad \mathcal{E}^{\text{bond}} = \sum_{\substack{i\alpha,j\beta \\ i\alpha \neq j\beta}} \sum_I f_I c_{iI}^\alpha c_{jI}^\beta \mathcal{H}_{ij}^{\alpha\beta}. \tag{5.65}$$

[33] Note that we can always choose atomic orbitals to be real functions, and therefore the components of the eigenvectors c_{iI}^α can always be made real. Thus, to simplify notation we will drop any concern about complex conjugates in this section.

The first of these is the "promotion" energy, which is the part involving the change of energy within an orbital from the unbonded to the bonded state, while the second is the "bond" energy representing the interaction between orbitals. Note that since ϵ_i^α is defined as the energy of an electron in an isolated orbital $i\alpha$, this can be viewed as an eigenvalue corresponding to an eigenvector

$$c_{jI}^\beta = \begin{cases} 1, & \text{when } j\beta = i\alpha, \\ 0, & \text{otherwise,} \end{cases}$$

so we see from Eqn. (5.62) that $\epsilon_i^\alpha = \mathcal{H}_{ii}^{\alpha\alpha}$ (no sum on i, α), and the promotion energy becomes

$$\mathcal{E}^{\text{prom}} = \sum_{i\alpha} \left[\left(\sum_I f_I c_{iI}^\alpha c_{iI}^\alpha \right) - N_0^{i\alpha} \right] \epsilon_i^\alpha. \tag{5.66}$$

We can now *define* the bond-order matrix as

$$BO_{ij}^{\alpha\beta} \equiv \sum_I f_I c_{iI}^\alpha c_{jI}^\beta, \tag{5.67}$$

and write

$$\mathcal{E}^{\text{prom}} = \sum_{i\alpha} \left[BO_{ii}^{\alpha\alpha} - N_0^{i\alpha} \right] \epsilon_i^\alpha, \qquad \mathcal{E}^{\text{bond}} = \sum_{\substack{i\alpha, j\beta \\ i\alpha \neq j\beta}} BO_{ij}^{\alpha\beta} \mathcal{H}_{ij}^{\alpha\beta}. \tag{5.68}$$

The physical interpretation of $\boldsymbol{BO}$ starts to emerge from these expressions. The diagonal terms represent the number of electrons occupying each orbital in the bonded state, and the promotion energy accounts for the change in this number compared to the unbonded state ($N_0^{i\alpha}$). The off-diagonal terms in $\boldsymbol{BO}$ are in fact the bond order between different orbitals; the total bond energy is a sum of energy-per-orbital terms (Hamiltonian components $\mathcal{H}_{ij}^{\alpha\beta}$), each weighted by $BO_{ij}^{\alpha\beta}$. To further understand the interpretation of $BO_{ij}^{\alpha\beta}$ as the bond order, let us consider a simple example.

Example 5.2 (The bond order for two s-valent atoms) Consider the TB equivalent of the hydrogen molecule we studied in Section 4.4.6. As we did there, we use a single s-orbital centered on each atom (which we will denote by simply φ_{A} and φ_{B}), build the Hamiltonian from Eqn. (4.163)$_1$ (we will treat the orbitals as orthonormal for simplicity) and diagonalize it to find the eigenvectors and eigenvalues. Within TB, the computation of the Hamiltonian is reduced to the evaluation of a Slater–Koster parameter and an empirical function of distance, the product of which we simply denote as β_{ss}. The set of equations we are solving is then

$$\beta_{\text{ss}} \begin{bmatrix} 0 & -1 \\ -1 & 0 \end{bmatrix} \begin{bmatrix} c_A \\ c_B \end{bmatrix} = \epsilon \begin{bmatrix} c_A \\ c_B \end{bmatrix}, \tag{5.69}$$

where we have assumed an arbitrary energy datum of 0 for the isolated atoms, leading to the 0 in the diagonal of the Hamiltonian matrix. It is straightforward to see that the first of the two eigenvectors is the bonding state:

$$\begin{bmatrix} c_A \\ c_B \end{bmatrix} = \frac{1}{\sqrt{2}} \begin{bmatrix} 1 \\ 1 \end{bmatrix}, \qquad \epsilon_1 = -\beta_{ss}, \tag{5.70}$$

and the second is the antibonding state:

$$\begin{bmatrix} c_A \\ c_B \end{bmatrix} = \frac{1}{\sqrt{2}} \begin{bmatrix} 1 \\ -1 \end{bmatrix}, \qquad \epsilon_2 = +\beta_{ss}. \tag{5.71}$$

With a single electron, this is a model of an H_2^+ ion. If the atom pair share a single electron, it will fall into the lower-energy bonding state. It is straightforward, then, to evaluate the bond-order matrix from Eqns. (5.67) and (5.70) as

$$[BO]^{(N^{el}=1)} = \begin{bmatrix} 1/2 & 1/2 \\ 1/2 & 1/2 \end{bmatrix}. \tag{5.72}$$

It is the off-diagonal term that tells us the bond order between the two atoms, suggesting that an H_2^+ ion has a bond order of $1/2$.

The diagonal terms of BO are also meaningful. Recall the general definition of the electron density in Eqn. (4.74). For a single-particle system in the linear combination of atomic orbitals (LCAO) approach this is

$$\rho(\boldsymbol{x}) = \psi^*(\boldsymbol{x})\psi(\boldsymbol{x}) = \sum_{i\alpha,j\beta} c_i^\alpha c_j^\beta \varphi_i^\alpha(\boldsymbol{x})\varphi_j^\beta(\boldsymbol{x}). \tag{5.73}$$

Now, if we superimpose all the occupied single-electron states and integrate over all space, we must recover the total number of electrons in the system, so that

$$\begin{aligned} N^{el} = \int \rho(\boldsymbol{x})d\boldsymbol{x} &= \int \sum_I f_I \sum_{i\alpha,j\beta} c_{iI}^\alpha c_{jI}^\beta \varphi_i^\alpha(\boldsymbol{x})\varphi_j^\beta(\boldsymbol{x})\, d\boldsymbol{x} \\ &= \sum_I f_I \sum_{i\alpha,j\beta} c_{iI}^\alpha c_{jI}^\beta \int \varphi_i^\alpha(\boldsymbol{x})\varphi_j^\beta(\boldsymbol{x})\, d\boldsymbol{x}. \end{aligned} \tag{5.74}$$

Assuming orthonormal orbitals, this reduces to simply

$$N^{el} = \sum_{i\alpha} BO_{ii}^{\alpha\alpha}, \tag{5.75}$$

and we see that the diagonal elements, $BO_{ii}^{\alpha\alpha}$, can be interpreted as the number of electrons residing in orbital $i\alpha$. Summing over all the orbitals naturally gives the total number of electrons in the system. Returning to Eqn. (5.72), this interpretation implies that the single electron is equally shared between the orbitals centered on each atom, consistent with the expected covalent nature of the H_2^+ ion.

If we add a second electron, we have the neutral H_2 molecule, which we would expect to bond more strongly than the H_2^+ ion. Indeed, since each eigenstate can hold two electrons due to spin, the second electron will also fall into the lower-energy bonding state. As such the bond order is simply twice the bond order for the one-electron case:

$$[BO]^{(N^{el}=2)} = \begin{bmatrix} 1 & 1 \\ 1 & 1 \end{bmatrix}. \tag{5.76}$$

The bond strength between the two atoms, embodied in the off-diagonal elements of BO, is about twice as strong as if there is only one electron, and the diagonal elements show that the two electrons are equally shared between the two atoms. (Note that the exact bond strength cannot be known without also considering the other terms in the cohesive energy and determining the equilibrium bond length that results, since β_{ss} depends on the bond length.) Here, we are simply looking at the change in the $\mathcal{E}^{\rm bond}$ term.

If we try to add a third and then fourth electron, they must occupy antibonding states since the bonding state is filled by the first two. This leads to a reduction in the bond order:

$$[BO]^{(N^{\rm el}=3)} = \begin{bmatrix} 3/2 & 1/2 \\ 1/2 & 3/2 \end{bmatrix}, \qquad [BO]^{(N^{\rm el}=4)} = \begin{bmatrix} 2 & 0 \\ 0 & 2 \end{bmatrix}$$

The case of $N^{\rm el} = 4$ is something like a crude model of two He atoms; there is no bonding between the atoms (the off-diagonal terms are 0) and each takes two of the four available electrons (the diagonal terms are 2).

And so, the challenge in BOP and empirical BOP methods is to capture the subtle and complex behavior of the bond-order matrix in an atomistic model that is sufficiently fast to deal with practical system sizes. Recall that up to this point, we have illustrated examples where we have computed the bond order exactly to within the approximations of TB: we have found the eigenvectors of the Hamiltonian and performed the sum in Eqn. (5.67). This, in principle, requires the $O(N^3)$ scaling associated with matrix diagonalization.

The BOP methods developed by Pettifor and coworkers [HBF$^+$96] achieve this in a rigorous way that blurs the distinction between TB calculations and empirical atomistic models. The idea of BOP revolves around developing a fast and accurate way to estimate the bond order without diagonalizing the Hamiltonian. It is based on the idea of moments of the DOS and their direct evaluation from elements of the Hamiltonian matrix, already introduced in Section 4.6.6. A good discussion of the connections between BOP and moments can be found in Finnis' book [Fin03].

In the interest of brevity, we do not discuss the details of BOP, noting only that it requires considerable computation compared with empirical BOP methods because it retains as much of the quantum mechanical detail as possible. Recognizing this limitation, much work has been invested in making the BOP approach as efficient as possible while remaining sufficiently accurate. The development of closed-form analytic expressions to approximate the bond-order expressions, for example, has been an area of much research. The BOP formalism has been used to model metals [ANMPV07], semiconductors [DZM$^+$07], intermetallics [NMVH07] and hydrocarbons [MMEG07].

We close this section by discussing two important cluster functional methods that are additional examples of empirical BOP models.

5.6.3 ReaxFF

The ReaxFF model [vDDLG01], as the name is meant to suggest, is a "reactive force-field" model. It is meant to replace models like CHARMM and AMBER (Section 5.4.5) by automatically determining whether a pair of atoms is bonded or nonbonded, essentially treating all atomic interactions on an equal footing.

To the materials science community, this "reactivity" is nothing new. Indeed, all of the models we discuss in this chapter with the *exception* of CHARMM and AMBER are reactive. However, whereas reactions in metals, intermetallics and (to some extent) ceramics and semi-conductors are relatively simple, reactions in biological systems involve extremely subtle and complex changes in the energy. In ReaxFF this is addressed by building a general energy functional with a large number of fitting parameters. We focus our discussion on the specific case of ReaxFF for hydrocarbons [vDDLG01] (i.e. for molecules comprising only the elements C and H). Since 2001, ReaxFF parameters have been determined for a wide variety of systems, including such things as metal hydrides [ZCvD$^+$04], silicon, and silicon dioxide [vDSS$^+$03].

First, the ReaxFF method postulates an energy similar to that used in CHARMM or AMBER, but with additional terms:

$$\mathcal{V}^{\text{int}} = \mathcal{V}^{\text{bond}} + \mathcal{V}^{\text{over}} + \mathcal{V}^{\text{under}} + \mathcal{V}^{\text{val}} + \mathcal{V}^{\text{pen}} + \mathcal{V}^{\text{tors}} + \mathcal{V}^{\text{vdWaals}} + \mathcal{V}^{\text{Coul}} + \mathcal{V}^{\text{conj}}.$$

Here each term seeks to represent a physical attribute of the bonding. For example, the first term accounts for the energy of all two-atom bonds, while the second and third terms aim to correct this energy based on the over- or under-coordination of each bond (determined by estimates of the bond order). The other terms are

1. $\mathcal{V}^{\text{val}}$: an energy associated with body angles (effectively a three-body potential),
2. $\mathcal{V}^{\text{pen}}$: a penalty to correct the bond-angle energy for certain special cases,
3. $\mathcal{V}^{\text{tors}}$: the energy of bond torsion angles (effectively a four-body potential for bond rotation),
4. $\mathcal{V}^{\text{vdWaals}}$: the van der Waals interaction energy,
5. $\mathcal{V}^{\text{Coul}}$: the Coulomb interaction energy, and finally
6. $\mathcal{V}^{\text{conj}}$: a term to deal with conjugated bonds.

Each of these terms is an empirical function with fitting parameters. Some (such as the van der Waals and Coulomb terms) have functional forms that are rigorously motivated from theory, but most are convenient mathematical forms that roughly mimic other bonding features. In total, there are 93 fitting parameters in the ReaxFF model for the C–H system.[34]

The ReaxFF strategy is centered on the premise that the bond order is the key parameter in determining bond energy. At the same time, it is recognized that the estimates of the bond order from a quantum mechanical formulation will be too expensive to compute, and so ReaxFF starts from an elaborate curve fit to develop an empirical formula for the bond order based only on the atomic species and interatomic distances. For hydrocarbons, the order of the bond between atoms α and β is first estimated from the empirical expression

$$\widehat{BO}^{\alpha\beta} = \sum_{i=1}^{3} \exp\left[\lambda_i \left(\frac{r^{\alpha\beta}}{r_{0,i}}\right)^{\mu_i}\right], \tag{5.77}$$

[34] This calls to mind the quote, "Give me four parameters and I can fit an elephant, give me five and I can wiggle its trunk." The quote has been attributed to many, including Gauss, Bohr, Kelvin, Fermi and von Neumann, but a verifiable attribution seems to have been lost in science folklore. In any event, it would appear from our discussion that hydrocarbons are considerably more complex than elephants, even though elephants predominantly comprise carbon, hydrogen and oxygen. Perhaps the four–five parameter fit is an example of an extremely clever multiscale model (see Part IV).

where $i = 1$ corresponds to the σ-bond, $i = 2$ to the first π-bond, and $i = 3$ to the second π-bond (see footnote 25 on page 261). The λ_i, μ_i and $r_{0,i}$ are fitting parameters, and they take different values depending on whether the two atoms involved in the bond are C–C, C–H or H–H. Finally, the π-bonds ($i = 2, 3$) are only included for C–C bond interactions. As a second step, the bond orders are corrected by another empirical expression of the form

$$BO^{\alpha\beta} = BO^{\alpha\beta}(\widehat{BO}^{\alpha\beta}, \Delta^\alpha, \Delta^\beta), \tag{5.78}$$

where Δ^α is the deviation of atom α from its ideal valence, z^α. This is determined by summing all bond orders associated with atom α:

$$\Delta^\alpha = \sum_{\substack{\beta \\ \beta \neq \alpha}} \widehat{BO}^{\alpha\beta} - z^\alpha. \tag{5.79}$$

The bond orders, $BO^{\alpha\beta}$, and deviations, Δ^α, so determined are then used as the primary parameters to estimate each contribution to the energy. (The exceptions to this are the van der Waals and Coulomb energies, which are simple pair potentials as discussed in Section 5.4.5.) As one example, the bond energy, $\mathcal{V}^{\text{bond}}$, is written as

$$\mathcal{V}^{\text{bond}} = \sum_{\text{bonds}} -D^{\alpha\beta} BO^{\alpha\beta} \exp\{p^{\alpha\beta}[1 - (BO^{\alpha\beta})^{p^{\alpha\beta}}]\},$$

where $D^{\alpha\beta}$ and $p^{\alpha\beta}$ are parameters that depend on the species of the two atoms involved in a given bond.

Note that Eqn. (5.77) *appears* to be at odds with our previous discussions of the bond order, since it suggests that the bond order is only a function of bond length. We showed earlier that the essence of the bond order is to characterize the *environment* of a given bond. However, the environmental dependence of the bond order comes in through Eqn. (5.78) and the valence deviations Δ^α and Δ^β of the two atoms in each bond. In addition, the energy terms $\mathcal{V}^{\text{under}}$ and $\mathcal{V}^{\text{over}}$ aim to further incorporate many-body environmental dependence of the bonds through the valence deviations. In this way, there is a great deal of fitting flexibility built into the ReaxFF approach. Compared with a pair functional like EAM, ReaxFF models are on the order of 100 times slower to compute. Full details of the expressions for each term in the ReaxFF energy are left to the literature [vDDLG01].

5.6.4 The modified embedded atom method

Another cluster functional form is taken by the "modified embedded atom method" (MEAM) originally developed by Baskes and coworkers [BNW89, Bas92]. Its starting point is the EAM discussed earlier, as embodied in Eqn. (5.52), and the same form remains: the energy is the sum of a pair potential term and a embedding energy term, the latter being dependent on the background electron density, ρ, at each atomic site. MEAM extends the original EAM exclusively through the use of a more sophisticated functional form for ρ. In EAM, the assumption is made that ρ at a point is a linear superposition of spherically-averaged electron densities from each nearby atom. In essence, one can think of this assumption as a zeroth order approximation to the real electron density. MEAM uses a more elaborate scheme for determining ρ, one that can be physically interpreted as the corrective terms arising from a perturbative expansion of the density starting from the spherically-averaged approximation. The goal is to incorporate directional dependence in the electron density.

The focus in MEAM is on the embedding energy. The pair potential term is essentially defined by the embedding energy, since the pair potential is used to ensure that the MEAM model exactly fits the UBER form for the cohesive energy as a function of lattice constant.[35] The exact functional forms for the MEAM embedding functions have evolved over the years as researchers fine tuned the fitting of the model. Here, we reproduce the forms of [JHK$^+$07], where full details can be found for anyone interested in implementing MEAM.

The energy of an atom is as in Eqn. (5.54), with the embedding energy, U_α given by

$$U_\alpha(\rho^\alpha) = A^\alpha E_0^\alpha \rho^\alpha \ln \rho^\alpha,$$

where A^α and E_0^α are fitting parameters. However, the calculation of ρ^α from Eqn. (5.53) is replaced by

$$\rho^\alpha = \frac{\rho_{(0)}^\alpha}{g_0^\alpha} G(\Gamma^\alpha), \tag{5.80}$$

where g_0^α is a fitting parameter determined by the reference crystal structure, and G and Γ are

$$G(\Gamma^\alpha) = \sqrt{1+\Gamma^\alpha}, \qquad \Gamma^\alpha = \sum_{l=1}^{3} t_{(l)}^\alpha \left(g_{(l)}^\alpha / g_{(0)}^\alpha \right)^2 .$$

Here the $t_{(l)}^\alpha$ are weighting functions with the following form:

$$t_{(l)}^\alpha = \frac{1}{\rho_{(0)}^\alpha} \sum_{\substack{\beta \\ \beta \neq \alpha}} t_{0,(l)}^\beta g_{(0)}^\alpha S^{\alpha\beta},$$

and $t_{0,(l)}^\beta$ is an element-dependent fitting parameter. The term $S^{\alpha\beta}$ is a screening function (described below) that depends on all triplets of atoms $\alpha\beta\gamma$ for γ within the cutoff distance from atoms α and β. The electron density at site α, ρ^α, is determined from four intermediate

[35] UBER is discussed in more detail in Section 6.5.2. Note that the MEAM approach is the opposite of the common approach used in EAM as discussed in Section 5.5. There, the *embedding energy* was fitted to the UBER curve minus the pair potential contribution. Here, it is the pair potential function that is fitted, by the opposite process.

density functions that take the forms

$$\rho^{\alpha}_{(0)} = \sum_{\substack{\beta \\ \beta\neq\alpha}} S^{\alpha\beta} g^{\beta}_{(0)}, \qquad \left(\rho^{\alpha}_{(1)}\right)^2 = \sum_{i=1}^{3}\left(\sum_{\substack{\beta \\ \beta\neq\alpha}} S^{\alpha\beta}\frac{r_i^{\alpha\beta}}{r^{\alpha\beta}} g^{\beta}_{(1)}\right)^2,$$

$$\left(\rho^{\alpha}_{(2)}\right)^2 = \sum_{i,j=1}^{3}\left(\sum_{\substack{\beta \\ \beta\neq\alpha}} S^{\alpha\beta}\frac{r_i^{\alpha\beta} r_j^{\alpha\beta}}{(r^{\alpha\beta})^2} g^{\beta}_{(2)}\right)^2 - \frac{1}{3}\left(\sum_{\substack{\beta \\ \beta\neq\alpha}} S^{\alpha\beta} g^{\beta}_{(2)}\right)^2,$$

$$\left(\rho^{\alpha}_{(3)}\right)^2 = \sum_{i,j,k=1}^{3}\left(\sum_{\substack{\beta \\ \beta\neq\alpha}} S^{\alpha\beta}\frac{r_i^{\alpha\beta} r_j^{\alpha\beta} r_k^{\alpha\beta}}{(r^{\alpha\beta})^3} g^{\beta}_{(3)}\right)^2 - \frac{3}{5}\sum_{i=1}^{3}\left(\sum_{\substack{\beta \\ \beta\neq\alpha}} S^{\alpha\beta}\frac{r_i^{\alpha\beta}}{r^{\alpha\beta}} g^{\beta}_{(3)}\right)^2,$$

and the sums on i, j, k are over the vector components of $\boldsymbol{r}^{\alpha\beta}$. The atomic electron densities, $g^{\beta}_{(k)}$, are all functions of distance from the atom only and take the form

$$g^{\beta}_{(k)}(r^{\alpha\beta}) = g^{\beta}_0 \exp\left[-\beta^{\beta}_{(k)}\left(\frac{r^{\alpha\beta}}{r^{\beta}_0} - 1\right)\right],$$

where $\beta^{\beta}_{(k)}$ and r^{β}_0 are fitting parameters that depend on the element (the latter is simply the near-neighbor distance in the equilibrium reference structure of the pure element).

The screening function, as the name suggests, creates a screening effect. If two atoms, α and β, are bonding with other atoms that lie, in some sense, "between" atoms α and β, the strength of the α–β bonding is reduced. In particular, $S^{\alpha\beta} = 0$, if the bond is completely screened by other atoms, and it varies from 0 up to 1 as the screening becomes less. These functions take the form

$$S^{\alpha\beta} = \bar{S}^{\alpha\beta} f_{\text{cut}}\left(\frac{r_{\text{cut}} - r^{\alpha\beta}}{\Delta r}\right), \qquad \bar{S}^{\alpha\beta} = \prod_{\substack{\gamma \\ \gamma\neq\alpha,\beta}} S^{\alpha\gamma\beta},$$

$$S^{\alpha\gamma\beta} = f_{\text{cut}}\left(\frac{C^{\alpha\gamma\beta} - C^{\alpha\gamma\beta}_{\text{min}}}{C^{\alpha\gamma\beta}_{\text{max}} - C^{\alpha\gamma\beta}_{\text{min}}}\right), \qquad C^{\alpha\gamma\beta} = 1 + 2\frac{(r^{\alpha\beta}r^{\alpha\gamma})^2 + (r^{\alpha\beta}r^{\beta\gamma})^2 - (r^{\alpha\beta})^4}{(r^{\alpha\beta})^4 - [(r^{\alpha\gamma})^2 - (r^{\beta\gamma})^2]^2},$$

$$f_{\text{cut}}(x) = \begin{cases} 1, & x \geq 1, \\ [1-(1-x)^4]^2, & 0 < x < 1, \\ 0, & x \leq 0. \end{cases}$$

In these expressions $C^{\alpha\gamma\beta}_{\text{min}}$ and $C^{\alpha\gamma\beta}_{\text{max}}$ are fitting parameters, and Δr controls the radial length over which the cutoff function goes from 0 to 1.

We see that $\rho^{\alpha}_{(0)}$ is similar to what was used for ρ^{α} in the original EAM, and thus choosing $S^{\alpha\beta} = 0$ and all $t^{\alpha}_{(l)} = 0$ will recover the simpler model. Although the forms of $\rho^{\alpha}_{(l)}$ for $l = 1, 2, 3$ appear complicated, each is in fact equivalent to a dependence on only interatomic distances. As such, the MEAM forms satisfy the invariance principle in Eqn. (5.17).

For a single element, these functional forms have a total of 14 fitting parameters that must be determined in order to define the model. For a binary alloy system, this number grows to

41 (14 for each element and 13 to fit the interspecies interactions). The MEAM literature (e.g. [LSB03]) describes relatively straightforward fitting procedures for determining these parameters by fitting to well-defined physical quantities that can be determined experimentally or from first principles.[36] The emphasis in the MEAM development has been to avoid completely general functional forms and fitting procedures. Instead, the developers have tried to keep the fitting parameters grounded in a clear connection to important physical quantities. The idea is to make the fitting procedure transparent and easy, with the advantage that models for new elements or alloys can be produced by anyone who is interested in doing a little work. The disadvantage of this philosophy is that it restricts the MEAM formalism somewhat. When the fitting procedure leads to an unsatisfactory model for some application, there is no clear recourse for improving the fit.

5.7 Atomistic models: what can they do?

Starting with the detailed discussion of the theory and implementation of DFT in Section 4.5, and ending with the overview of a few of the more common empirical and semi-empirical methods in this chapter, we have presented a flavor of the atomistic approaches available to the materials modeler. We have not tried to be exhaustive in this presentation, but rather aimed at two goals. First, a demonstration of the logical systematic progression from the most "exact" DFT model to the most approximate empirical pair potentials. Second, we have tried to include at least one model for each major class of materials.

The real point of all of these methods, for the most part, is to do molecular statics (MS) or MD simulations as described in detail in Chapters 6 and 9, respectively. MS consists of computational methods to explore the energy landscape created by a given atomistic model. This may include finding equilibrium structures through energy minimization, or finding saddle points and transition paths between two local energy minima. MD is the time integration of Newton's equations of motion for a system of particles, the forces on which are determined by the empirical atomistic model in order to compute time-averaged quantities or to discover dynamic deformation mechanisms.

Not all of the models can be treated on an equal footing in MS or MD, however. The most important distinctions between the methods are the computational effort they require, their transferability and the class of materials to which they are appropriate to apply. We discuss each of these in turn, next.

5.7.1 Speed and scaling: how many atoms over how much time?

An important practical concern is the computational cost of an atomistic model. Whether in a static energy minimization or a MD simulation, the basic requirements will always be the same: we will need to compute the energy and forces on atoms in a series of configurations.

[36] These quantities are specifically the cohesive energy, the lattice constant, the bulk modulus, two shear constants, two energy differences between alternative crystal structures, the vacancy formation energy, the activation energy for vacancy migration, the coefficient of thermal expansion and surface reconstructions [LSB03].

However, the underlying atomistic model we use determines the computational cost of these operations, and thus, we would like to have some idea of the limitations of each model in two respects: the largest practical system size (number of atoms, N_{max}), and the longest simulated period of time in a dynamic setting, t_{max}, that can be computed practically on the computers of the day.

These questions are not straightforward to answer for a number of reasons:

1. The values of N_{max} and t_{max} are not independent (one decreases as the other increases).
2. Computing power is always increasing, and any attempt to pin down the speed of a model will be obsolete almost as soon as it is made.
3. There are many researchers working to improve the efficiency of the codes used in these methods, leading to steady increments in their capabilities.
4. The methods themselves are always subject to additional approximations and improvements that are aimed at improving efficiency. Sometimes, promising new ideas arise that suggest a way to greatly enhance the speed of a particular method. It is not always clear if a current research direction will live up to its apparent promise or if some unexpected limitation will arise.

There is some interest in pursuing "world record" system sizes. As of the day we are writing this, it would seem that the current MD simulation record is *one trillion atoms*[37] (albeit for only 40 timesteps as a test of system performance) [GK08]. This calculation was performed on the BlueGene/L at Lawrence Livermore National Laboratories. At the time the simulation was performed, this was the world's largest computer, with 212, 922 processors and a peak performance rate of 596 Tflop/s. Such calculations are often of interest as much for the technological challenge as for the expected scientific insight into material behavior. The mere tasks of managing and analyzing the huge resulting dataset will push the limits of computational technology. Usually, the real "practical" limit of system size lags significantly behind the world record, as there must be a balance between wanting to run large systems but not wanting to process too much data. In what follows we will try to quote more "practical" limits, commensurate with these competing demands.

An important question is one of scaling. If two different models are currently both able to handle about the same number of atoms, N_{max}, they may appear to be about the same speed. However, imagine that the computational effort for one of the models scales as $O(N)$ and the other as $O(N^3)$. A future, ten-fold increase in computer power would lead to a ten-fold increase in N_{max} for the first method, but just barely a doubling of N_{max} for the second. As such, the best methods (in terms of speed) are always those of $O(N)$.

In general, the cost[38] of a single force evaluation can be written as

$$\Xi_F = \kappa_{\mathrm{f}} N^p, \tag{5.81}$$

where p is the scaling exponent and κ_{f} is some constant. We must evaluate the forces once every timestep, and so there is obviously a trade-off between the longest possible simulation time, t_{max}, and the maximum number of atoms, N_{max}. Fewer atoms can be run for longer

[37] !

[38] By "cost" we mean the amount of CPU time expended on a calculation.

Table 5.2. Comparison of maximum practical system size (N_{max}), simulation time (t_{max}) on modest parallel clusters (32–64 CPUs), and the scaling of various atomistic models.

Method	N_{max}	t_{max} (ps)	Scaling
Quantum mechanics			
DFT	10^3	10	$O(N)$–$O(N^3)$
Tight-binding			
TB	10^4	100	$O(N)$–$O(N^3)$
TB-BOP	10^5	100	$O(N)$
Empirical and semi-empirical			
Pair potentials (Lennard-Jones)	10^8	10000	$O(N)$
Pair potentials (ionic models)	10^7	1000	$O(N)$–$O(N^2)$
Pair functionals (EAM)	10^8	10000	$O(N)$
Cluster potentials (CHARMM/AMBER)	10^6	1000	$O(N)$–$O(N^2)$
Cluster potentials (MGPT)	10^7	1000	$O(N)$
Cluster functionals (ReaxFF)	10^6	100	$O(N)$–$O(N^2)$
Cluster functionals (MEAM)	10^7	1000	$O(N)$
Cluster functionals (Tersoff)	10^7	10000	$O(N)$

times, and we can imagine defining a "total computation cost," Ξ_{tot}, that would be related to both the number of atoms and the simulation time. While the scaling with the number of atoms varies between models, the scaling with simulation time is always linear if we reasonably assume that the timestep size is not affected by the choice of atomistic model. As such, we might write

$$\Xi_{\mathrm{tot}} = \Xi_F \kappa_t t = \kappa_f \kappa_t N^p t, \tag{5.82}$$

where Ξ_{tot} is the total computational cost and κ_t is the cost per unit simulation time t additional to any cost of the force calculation already included in Ξ_F. All this is meant to emphasize that one cannot really quote a N_{max} and t_{max} independently; one can extend the simulation time by reducing the number of atoms, to a degree determined by the scaling of a particular atomistic model.

Table 5.2 is adapted from [Wim96] and [NKN$^+$07], and updated to reflect computer power as of the publication date of this book. We have listed a "typical" N_{max} that one might reasonably expect to achieve with a well-parallelized code on a modest cluster on the order of 32–64 CPUs, rather than the current world record or theoretical maximum.

Some comments about the contents of Tab. 5.2 are in order. In Section 4.5, we have presented the most straightforward version of DFT, which requires the diagonalization of a general symmetric Hamiltonian matrix and is therefore $O(N^3)$. However, there are many advances beyond the scope of this book that provide linear scaling to DFT under certain simplifying assumptions and reductions in accuracy (for example [HSMP06]). Similarly, the straightforward TB model we have presented in Section 4.6 is $O(N^3)$, but certain specialized treatments and approximations can make this scaling nearly $O(N)$. For example, Tjima *et al.* [TIP$^+$05] have reported a variation of the TB method that is specially designed to treat carbon atoms and which has linear scaling. With this code, they have been

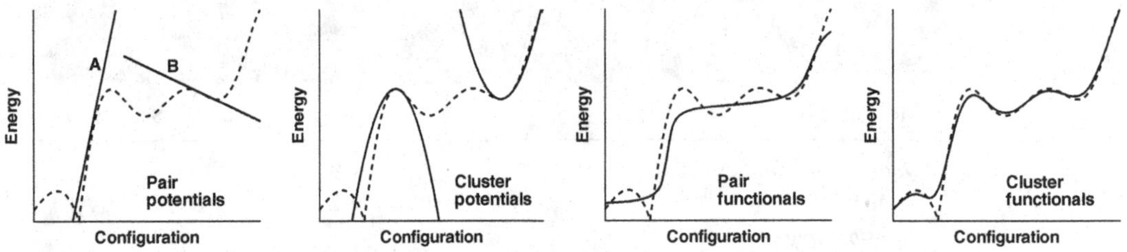

Fig. 5.7 Schematic effect of the type of model on its transferability. The dashed line is the exact energy, each solid line represents a different model. (Adapted from [Car90].)

able to study nearly 50 000 carbon atoms in a carbon nanotube. Ionic models, CHARMM, AMBER and ReaxFF have slightly worse performances than apparently similar methods because of their treatment of long-range Coulomb interactions.

Since the value of κ_f can vary by several orders of magnitude, Tab. 5.2 and Eqn. (5.81) also emphasize that not all $O(N)$ methods are created equal. This is evident in comparing, for example, the maximum number of atoms one can model using $O(N)$ pair potentials versus $O(N)$ TB-BOP.

5.7.2 Transferability: predicting behavior outside the fit

We have already brought up the issue of transferability a few times in passing. This refers to the ability of a model, once its functional form is chosen and its fitting parameters are decided, to make predictions about structures and behavior outside of the fitted region of configuration space. The idea of transferability was summed up schematically by Carlsson [Car90] in the figure we reproduce in Fig. 5.7. An energy landscape is illustrated as a function of a single generalized "configuration" coordinate. The exact DFT energy surface would have some complex form: nonlinear, nonmonotonic, and replete with discontinuities and cusps at places where dramatic changes in bonding character take place. A simple pair potential generally starts from one point on this energy landscape and is fitted to local data, as illustrated schematically in Fig. 5.7 by the straight line labeled A.[39] The resulting model is quite accurate around this local point, but not very transferable, and only a short distance away in configuration space it is completely unphysical. If one were to fit a pair potential starting from a different point in configurational space, one would get a different "straight line" (labeled B in the figure). Both A and B are pair potential descriptions of the same system, but the two models disagree with each other over most of the configurational space.

A pair potential needs to be thought of as a first-order expansion of behavior about some reference structure. In other words, we can use a pair potential to study any material system, so long as the configuration does not change enough to significantly affect the bond order. Finnis [Fin03] discussed perturbation treatments of the full quantum mechanical interaction that allow the derivation of pair potentials that, while admittedly not transferable, are still able to shed much light on the behavior of the simple metals. Similarly, Bush *et al.*

[39] The straight line is meant to be schematic, of course, since pair potentials are not literally linear functions of their arguments.

[BGCB94] showed how simple pair potentials can be fitted to properties of binary and ternary oxides to use in predicting other properties of the equilibrium structure.

There are some systems for which the pair potential form turns out to be quite good. For example, the Lennard-Jones 6–12 pair potential can be used for the solidified noble gases and the Born–Mayer potential is a good model of ionic systems. In the context of Fig. 5.7, the exact energy curve turns out to be quite close to a straight line (schematically speaking) for these systems. But the pair potential has no way to address bond-order effects: it is constructed using the explicit assumption that each bond is independent of all others. As such, a pair potential cannot describe covalent or even metallic bonding where bond order affects bond strength. The assumption that each bond's contribution is independent means that a pair potential is also not able to sense the effects of *bond angles,* which are defined by at least two bonds simultaneously.

Of course, pair potentials often form an important *component* of more complicated models. They describe Coulomb interactions exactly, van der Waals interactions quite well and even the repulsive part of a TB calculation with good fidelity. In this context they are indispensable and accurate, but they are incomplete without the other (usually much more complex) parts of the model. This is seen in the DFT, TB, ionic, pair functional and ReaxFF models we have discussed earlier.

Cluster potentials provide more flexibility in the fitting parameters, but are still similar to pair potentials in that they are essentially fitted to local regions of configuration space. They will be more transferable than pair potentials, but any two cluster potentials of the same functional form fitted to different local environments will likely disagree with each other most of the time. In Fig. 5.7, this is schematically represented by the parabolic fits; these are better than linear but still of limited range of application. Such potentials work well for modeling situations where even though the exact energy function is known to be quite complex, the region of configuration space that is accessed during modeling is also known to be quite small. This is seen in the study of macromolecules using CHARMM or AMBER, for instance. By making the assumption that certain bonds do not break, the modelers effectively confine their systems to the fitted configurational domain.

Pair and cluster functionals allow more freedom of the form of the model and can therefore capture a broader range of configurations, although still in an approximate "smeared out" way. For example, the cusps in the energy function associated with abrupt changes in quantum states are unlikely to be fitted exactly. Functionals provide a better *global fit,* although a specifically parameterized model will often still be quite inaccurate outside a rather narrow region of configurational space.

It is unlikely that any model can be made fully transferable without including the full complexity of the electronic structure calculation. On the other hand, certain models will still be more transferable than others. In addition, each model is based on a certain set of assumptions that are essentially specialized for certain types of bonding and certain types of materials. As such, they may fit a fairly broad region of configurational space for one particular type of material (for example pair functionals fit the transition metals), but the same model will be quite poor for other material types (pair functionals do not fit covalent systems). Work to quantify, predict and control the extent to which a given model is transferable is an active area of research.

Quantifying transferability The notion of transferability is difficult to quantify and even more difficult to predict and control as new interatomic models are developed. For example interatomic models fitted to bulk metal properties can give dislocation core structures that are completely wrong [IBA00, LTK06]. Even for a "simple" liquid like water, no interatomic model exists that is able to reproduce all of the properties of real water, despite over 30 years of attempts [Gui02]. Any success that interatomic models currently enjoy relies heavily on the experience of researchers in determining when a predicted behavior is reliable.

One way to quantify transferability may be to evaluate whether a model fitted to one set of data has validity for a select set of "regions" in configuration space that are of importance to a given application. Challenges to such an approach are mainly: (a) the availability of a wide range of reliable data against which to test the models; and (b) a suitable measure of "closeness" in configuration space that will accurately identify important regions. The necessary data are likely to become more and more available as researchers join forces to share interatomic models, experimental data and DFT results using the internet. The launch of the KIM project (knowledgebase of interatomic models) [TESM09] is one promising example where more insight may be gained as users add new models or atomic configurations to a central repository. The question of how to measure proximity in configuration space remains an open challenge, and we provide some speculative ideas in that direction next.

Transferability and configurational similarity Users of interatomic models normally have a specific application in mind, but often there may be several interatomic models in the literature that purport to describe the same material system.[40] The best model is likely the one that is specifically fitted to the properties that are most relevant to a particular application. Experienced users often have an intuitive sense for this – dislocation depinning and twin boundary energy for plasticity applications, for example. But developing a similarity measure which unambiguously identifies the most relevant test configurations would be invaluable. This is especially true for complex multicomponent systems and the exploration of new regimes of materials behavior. This similarity measure should have two key properties:

1. It should be able to compare systems with different numbers of atoms and boundary conditions, and be translationally and rotationally invariant. Hence, direct configurational (nuclei position vector) comparisons are not helpful. Instead, we are led to consider correlation functions, coordination topologies or shape coordinates [LR95] for describing configurations. Ultimately, an appropriate vector space description of atomic configurations should be identified.
2. Within this vector space, the similarity measure should be able to identify and give special weights to key rare atomic arrangements buried inside a weakly disordered crystalline background. Surface simulations, dislocation simulations and the calculation of diffusion barriers are all examples that have a few interesting atoms buried in a large crystalline environment. This type of property has been very successfully exploited in

[40] For example, over 40 different interatomic models for crystalline silicon have been published between 1970 and 2010, an average of one per year. Even a simple fcc metal like Cu has a half-dozen or so widely used models that describe it, all of which are different fits with the EAM (pair functional) form.

the BLAST algorithm [AGM+90] for comparing biological sequences such as amino acids, proteins, and DNA sequences. One might also be able to adapt visualization methods to define a similarity measure. For example, by discarding the crystalline atoms and illuminating the defects of interest using strain [SOL07, GHBF08, ZBG09], centrosymmetric measures [KPH98] or coordination shells [HA87].

A robust similarity measure would help quantify transferability. For example, it would allow one to identify clusters in the space of atomic configurations that are closely related and therefore relevant to one another. This would provide insight as to which simple properties may be most important to a particular application. Conversely, it would suggest which properties one must use in the fitting process when developing a model so as to achieve the desired transferability to specific applications.

5.7.3 Classes of materials and our ability to model them

Although we have already mentioned many of the applications of each type of model, it is useful to summarize these applications briefly in one place.

The noble gases The simplest of solid systems is made from noble gas atoms, since they interact primarily by van der Waals forces. Solid crystals of the nobles gases form fcc structures and can be reasonably well described by pair potentials in general and in particular by the 6–12 Lennard-Jones model.

It is not surprising, perhaps, that a pair potential would meet the demands of modeling the noble gases, whose electronic structures make them essential inert. In the context of Fig. 5.7, the dashed line representing the energy landscape of an inert gas would essentially be a straight line (schematically speaking, of course). There are no abrupt changes in bonding characteristics as different electronic shells compete for electrons. There is instead only the predictable, radially symmetric interactions of induced dipoles in neighboring atoms.

Despite the reasonable agreement between the Lennard-Jones potential and the behavior of the noble gases, it is still an approximation. More sophisticated models that incorporate many-body effects for argon continue to be proposed [LS97, RPFS00, SGK+06, AK07]. In this book, however, we find that the Lennard-Jones model suffices for many of our objectives and provides good agreement with experiment. For example, we consider the ability of the Lennard-Jones model to reproduce the elastic constants of argon in Section 8.1.4.

Ionic crystals Compounds where the cohesive forces are dominated by ionic bonding can be accurately modeled using the Born–Mayer formalism described earlier. Pair potentials work in this case because ionic bonding is dominated by the well-understood pairwise Coulomb interaction between point charges. A major difference between the pair potentials describing ionic crystals and other pair potentials is the long-range nature of Coulomb interactions, requiring a large cutoff radius and special treatment of periodic systems (see, for example, [GR87]). This makes the computation more cumbersome, requiring large neighbor lists and additional calculations, but does not (in principle) limit the physical accuracy and transferability.

Metallic crystals: pure elements and solid solutions One of the great successes of empirical interatomic models has been in the description of metallic bonding. For simple metals and transition metals with filled or nearly filled d-orbitals, simple pair functionals have proven remarkably accurate. Of particular technological interest are elements such as Al, Ni, Cu, Pd, Ag, Pt and Au. These are all fcc crystals under standard room temperature and pressure conditions, a feature that arises from their nondirectional, metallic bonding character. While a simple pair potential is not enough, the addition of a pairwise, spherically symmetric environmental term in the pair functional is sufficient to capture coordination effects like those illustrated in Fig. 5.6.

Pair functionals do less well for metals that have a bcc or hcp structure as their ground state at low temperature. Examples include Fe, Mo, W, Mg, Zr, and Ti. Pair functionals have been developed for these systems, but their accuracy is disappointing relative to the pair functional models of the fcc systems listed earlier. This is not surprising, as the tendency of the systems to stabilize to a bcc or hcp phase already indicates some degree of covalent bonding. Covalent bonds are characterized by directionality – bond angles tend to matter – and this cannot be captured by the pair functional form. This is less of a problem in the hcp metals, especially those with a c/a ratio that is close to the ideal hard-sphere packing value. While it is possible to fit a pair functional so that, for example, the energy of a bcc phase is lower than that of the fcc phase, the magnitude of the energy difference tends to be smaller than predicted by DFT calculations. As a result, these models tend to be unrealistically inclined to transformation from the ground state to fcc under deformation.

Work by Ackland and coworkers (for example [AR03]) is a promising extension of the pair functional approach in an effort to improve its fidelity for bcc and hcp crystals. In essence, this so-called "two-band" model aims to describe the bonding of the system as the sum of two pair functional forms representing contributions from two different electron bands. Each pair functional is weighted according to a simple model to determine the electronic occupancy in the s- and d-bands respectively.

The MGPT model (see references in Section 5.4.4) is a cluster potential that does a good job of treating hcp and bcc metals, including Mo, Ta, V and Cr.

Metallic alloy systems Pair functionals can do a reasonable job of modeling dilute solid solutions and, to a *limited* extent, intermetallic alloys of simple metals. However, the development of reliable, efficient models to describe alloy systems remains one of the current challenges of interatomic modeling. Initially, one of the main difficulties in this regard was a lack of data against which to fit such models. While basic properties like the equilibrium structure and lattice constant were known, there were very few experimental data points that could be used to robustly fit a model to alloy behavior. Today, the role of DFT in model design is becoming much more central, as high-quality DFT calculations are used to provide the fitting and validation data for the empirical models. However, while this has improved the fitting data, it has demonstrated that the simple pair functional form is not general enough to be transferable across the entire phase diagram of most alloy systems. Successful applications of the pair functional form to alloys [Far94, Vot94, HPM99, Mis04, MMP02, ZM03, RFA02] usually work within the relatively small part of the phase diagram for which they have been optimized. For example, the Voter–Chen

EAM functional describing Al–Ni interactions [VC87] is not to be treated as a general model for the Al–Ni binary alloy system. Rather, it is optimized to the Ni-rich end of the phase diagram. It does a reasonably good job modeling alloys ranging from dilute solutions of Al in Ni to the Ni_3Al structure. Limitations of this type are usually imposed on any intermetallic model.

Intermetallic alloys tend to exhibit covalent bonding character, and so require cluster functionals or cluster potentials. Indeed, they are modeled with some success using the MEAM [AB96, CD97, HXS+00, BH01, CBS03] and MGPT [MW97] formalisms, but the number of fitting parameters and the complexity of the functional forms required for alloys makes model design and testing extremely challenging.

Hydrogen and carbon are two important elements in dilute concentrations in many metals. Hydrogen is usually undesirable, leading to the formation of brittle hydrides, whereas carbon is essential for the formation of the cementite phase that gives steel its strength. Although efforts have been made to model the effect of both of these elements on metals using pair functionals (see, for example, [DBBW86, AMB95, RFG09]), the reliability of these models is limited. Especially in the case of carbon, with the strongly covalent nature of its bonding, but also for hydrogen, cluster functionals such as MEAM [AB96, LJ07] or ReaxFF [CDvDG05] are necessary to study these systems.

Covalent crystals: pure elements Due to the huge technological importance of Si, it is not surprising that dozens of models have been developed in efforts to describe its behavior [BHT92, CC93]. Unfortunately, the highly covalent nature of Si bonding and its tendency toward different orbital hybridizations in different environments makes it extremely difficult to model correctly.

Most models of Si are based on the cluster potential framework, meaning that they tend to be well fitted to certain regions of configuration space but are not highly transferable. The canonical Stillinger–Weber potential, for example, was fitted to the melting of diamond cubic Si, and as such is best suited for modeling the solid–liquid interface at the melting temperature. But this potential has been used in a host of applications, all the way down to 0 K, often without a great deal of success.

The Tersoff potential for Si [Ter88a] offers a reasonably accurate model, thanks to its ability to approximately capture some of the effects of orbital hybridization through its empirical description of the bond order. On the other hand, the Tersoff form still tends to be less reliable for Si than the pair functionals are for simple metals. The Tersoff form can similarly treat Ge or C in various crystalline states to some degree, and it has been extended to model alloys and hydrocarbons [Bre90].

For the transition metals with partially-filled d-bands, the bonding tends to be a complex mix of metallic and covalent bonding. Cluster potentials are well suited to these materials since the effects of bond angle are clearly interpretable for the three- and four-body interactions. This is demonstrated by the MGPT potentials of Moriarty and coworkers (for example [Mor94]) and the MEAM form [JHK+07]. However, the issue of limited transferability with cluster potentials remains for these systems.

Diamond and other ordered states of carbon (such as nanotubes and bucky-balls) can be modeled using the cluster potentials of the CHARMM or AMBER form, but only in the limit where no bond breaking is expected. As this limitation is quite severe, it is more common to use the Tersoff–Brenner [Bre90] model for carbon. Although this model is highly empirical (see the discussion in [PO04]), it provides a reasonable fit to the behavior of ordered carbon. More transferable behavior can be obtained with the more costly TB or TB-BOP approaches, or with the more recently developed ReaxFF model [vDDLG01].

Covalent crystals: ceramics, geological materials and other complex solids Ceramics, minerals and other complex structures of covalent or mixed covalent–metallic character are really only treatable by the TB methods described in Section 4.6 or by a full DFT analysis of Section 4.5, although some of the simpler ones are now being studied using the ReaxFF framework [vDMH$^+$08]. The challenge with such systems is the massive complexity of the fitting procedure as more elements are added; such materials often contain several different elements in significant concentrations.

Organic molecules Modeling organic molecules has been the primary focus of models of the CHARMM or AMBER variety that were discussed in Section 5.4.5. Since the interest in these fields lies often in molecular conformations as opposed to reactions, the lack of reactivity in these models is not always a major limitation, and the models can easily be fitted around a few known bonding states. More recently, the ReaxFF model discussed in Section 5.6.3 has introduced the possibility of reactivity in such systems. ReaxFF has been applied to a variety of complex organic molecules [EMM$^+$08, ZJZ$^+$08, JvDGD09, SvDL$^+$09, CvDDG09].

5.8 Interatomic forces in empirical atomistic models

Having completed our zoography of empirical interatomic models, we now turn to the practical matter of calculating the forces resulting from these models. These forces are key to the simulation methods of Chapters 6 and 9. We begin by discussing the generic properties of forces in atomistic systems and then derive specific expressions for the forces for a number of important empirical atomistic models.

5.8.1 Weak and strong laws of action and reaction[41]

In this section, we prove that the force on an atom due to its interaction with other atoms in a system can be decomposed as a sum of pairwise forces, $\boldsymbol{f}^{\alpha\beta}$, which satisfy $\boldsymbol{f}^{\alpha\beta} = -\boldsymbol{f}^{\beta\alpha}$ (this is called the "weak law of action and reaction") and are aligned with the direction $\boldsymbol{r}^{\alpha\beta} = \boldsymbol{r}^{\beta} - \boldsymbol{r}^{\alpha}$ between the atoms (this is called the "strong law of action and reaction"). The derivation hinges on the fact that in a material system, the balance laws of linear and angular momentum must be satisfied for any part of the body.

[41] This derivation is due to Roger Fosdick [Fos09].

Consider a system of N particles with masses m^α ($\alpha = 1, \dots, N$). The total force on particle α is

$$\boldsymbol{f}^\alpha = \boldsymbol{f}^{\text{ext},\alpha} + \boldsymbol{f}^{\text{int},\alpha}, \tag{5.83}$$

where $\boldsymbol{f}^{\text{ext},\alpha}$ is the external force acting on the atom due to external fields and atoms outside of the system, and $\boldsymbol{f}^{\text{int},\alpha}$ is the internal force resulting from interactions with other atoms in the system. Without loss of generality, the internal force can be expressed as

$$\boldsymbol{f}^{\text{int},\alpha} = \sum_{\substack{\beta \\ \beta \neq \alpha}} \boldsymbol{f}^{\alpha\beta}, \tag{5.84}$$

where[42] $\boldsymbol{f}^{\alpha\beta}$ can generally be said to be the contribution to the force on atom α due to the presence of atom β. At this stage, no assumptions are made regarding the forces $\boldsymbol{f}^{\alpha\beta}$ or the interatomic potential from which they are derived.

We define a "part" $\wp_t$ of the system as a subsystem consisting of $K \leq N$ particles. We suppose $\boldsymbol{x}_0$ is a fixed point in space. Let $\boldsymbol{F}^{\text{ext}}(\wp_t)$ denote the total force on the part $\wp_t$ external to the part. Let $\boldsymbol{M}^{\text{ext}}(\wp_t; \boldsymbol{x}_0)$ denote the total external moment on $\wp_t$ about $\boldsymbol{x}_0$. Let $\boldsymbol{L}(\wp_t)$ be the linear momentum of the part $\wp_t$ and $\boldsymbol{H}(\wp_t; \boldsymbol{x}_0)$ be the angular momentum of $\wp_t$ about $\boldsymbol{x}_0$.

We adopt the following balance laws, which are valid for all parts of the system:[43]

$$\boldsymbol{F}^{\text{ext}}(\wp_t) = \frac{d\boldsymbol{L}}{dt}(\wp_t), \qquad \boldsymbol{M}^{\text{ext}}(\wp_t; \boldsymbol{x}_0) = \frac{d\boldsymbol{H}}{dt}(\wp_t; \boldsymbol{x}_0). \tag{5.85}$$

We now show that by applying these balance laws to particular parts of the system, the strong law of action and reaction can be established. As a first observation, let $\wp_t$ consist of the single particle α. The external force and linear momentum for $\wp_t = \{\alpha\}$ are

$$\boldsymbol{F}^{\text{ext}}(\{\alpha\}) = \boldsymbol{f}^{\text{ext},\alpha}(t) + \sum_{\substack{\gamma \\ \gamma \neq \alpha}} \boldsymbol{f}^{\alpha\gamma}(t), \tag{5.86}$$

$$\boldsymbol{L}(\{\alpha\}) = m^\alpha \dot{\boldsymbol{r}}^\alpha(t) \qquad \text{(no sum)}. \tag{5.87}$$

The balance of linear momentum in Eqn. $(5.85)_1$ requires

$$\boldsymbol{f}^{\text{ext},\alpha} + \sum_{\substack{\gamma \\ \gamma \neq \alpha}} \boldsymbol{f}^{\alpha\gamma} = m^\alpha \ddot{\boldsymbol{r}}^\alpha. \tag{5.88}$$

The external moment of $\wp_t$ is

$$\boldsymbol{M}^{\text{ext}}(\{\alpha\}; \boldsymbol{x}_0) = (\boldsymbol{r}^\alpha(t) - \boldsymbol{x}_0) \times (\boldsymbol{f}^{\text{ext},\alpha} + \sum_{\substack{\gamma \\ \gamma \neq \alpha}} \boldsymbol{f}^{\alpha\gamma}) = m^\alpha (\boldsymbol{r}^\alpha(t) - \boldsymbol{x}_0) \times \ddot{\boldsymbol{r}}^\alpha(t), \tag{5.89}$$

[42] For a pair potential model, $\boldsymbol{f}^{\alpha\beta}$ can be more narrowly interpreted as the force that atom β exerts on atom α. For more general interatomic models, the physical interpretation of $\boldsymbol{f}^{\alpha\beta}$ may not be so obvious, but it is always possible to decompose the force according to Eqn. (5.84).

[43] The view that the balance of linear momentum and the balance of angular momentum are fundamental laws of mechanics lies at the basis of continuum mechanics. See Section 2.3 and also Truesdell's article "Whence the law of moment and momentum?" in [Tru68].

where we have used Eqn. (5.88), and the angular momentum of $\wp_t$ is

$$\boldsymbol{H}(\{\alpha\};\boldsymbol{x}_0) = (\boldsymbol{r}^\alpha - \boldsymbol{x}_0) \times m^\alpha \dot{\boldsymbol{r}}^\alpha(t). \tag{5.90}$$

The balance of angular momentum in Eqn. $(5.85)_2$ is satisfied identically, since

$$\begin{aligned} m^\alpha(\boldsymbol{r}^\alpha(t) - \boldsymbol{x}_0) \times \ddot{\boldsymbol{r}}^\alpha(t) &= \frac{d}{dt}\left[(\boldsymbol{r}^\alpha(t) - \boldsymbol{x}_0) \times m^\alpha \dot{\boldsymbol{r}}^\alpha(t)\right] \\ &= \dot{\boldsymbol{r}}^\alpha(t) \times m^\alpha \dot{\boldsymbol{r}}^\alpha(t) + (\boldsymbol{r}^\alpha(t) - \boldsymbol{x}_0) \times m^\alpha \ddot{\boldsymbol{r}}^\alpha(t) \\ &= m^\alpha(\boldsymbol{r}^\alpha(t) - \boldsymbol{x}_0) \times \ddot{\boldsymbol{r}}^\alpha(t). \end{aligned}$$

As a second observation, let $\wp_t$ consist of the union of the two particles α and β. The external force and linear momentum are

$$\boldsymbol{F}^{\text{ext}}(\{\alpha,\beta\}) = \boldsymbol{f}^{\text{ext},\alpha} + \boldsymbol{f}^{\text{ext},\beta} + \sum_{\substack{\gamma \\ \gamma\neq\alpha\neq\beta}} (\boldsymbol{f}^{\alpha\gamma} + \boldsymbol{f}^{\beta\gamma}), \tag{5.91}$$

$$\boldsymbol{L}(\{\alpha,\beta\}) = m^\alpha \dot{\boldsymbol{r}}^\alpha + m^\beta \dot{\boldsymbol{r}}^\beta. \tag{5.92}$$

The balance of linear momentum in Eqn. $(5.85)_1$ requires

$$\boldsymbol{f}^{\text{ext},\alpha} + \boldsymbol{f}^{\text{ext},\beta} + \sum_{\substack{\gamma \\ \gamma\neq\alpha\neq\beta}} (\boldsymbol{f}^{\alpha\gamma} + \boldsymbol{f}^{\beta\gamma}) = m^\alpha \ddot{\boldsymbol{r}}^\alpha + m^\beta \ddot{\boldsymbol{r}}^\beta. \tag{5.93}$$

Subtracting Eqn. (5.88) for particles α and β gives

$$\sum_{\substack{\gamma \\ \gamma\neq\alpha\neq\beta}} (\boldsymbol{f}^{\alpha\gamma} + \boldsymbol{f}^{\beta\gamma}) - \sum_{\substack{\gamma \\ \gamma\neq\alpha}} \boldsymbol{f}^{\alpha\gamma} - \sum_{\substack{\gamma \\ \gamma\neq\beta}} \boldsymbol{f}^{\beta\gamma} = \boldsymbol{0}, \tag{5.94}$$

from which

$$\boldsymbol{f}^{\alpha\beta} + \boldsymbol{f}^{\beta\alpha} = \boldsymbol{0}. \tag{5.95}$$

This relation is called the *weak law of action and reaction* [Gol80]. It shows that $\boldsymbol{f}^{\alpha\beta} = -\boldsymbol{f}^{\beta\alpha}$, but does not guarantee that $\boldsymbol{f}^{\alpha\beta}$ lies along the line connecting particles α and β.

Next, the external moment of $\wp_t$ is

$$\begin{aligned} &\boldsymbol{M}^{\text{ext}}(\{\alpha,\beta\};\boldsymbol{x}_0) \\ &= (\boldsymbol{r}^\alpha - \boldsymbol{x}_0) \times (\boldsymbol{f}^{\text{ext},\alpha} + \sum_{\substack{\gamma \\ \gamma\neq\alpha\neq\beta}} \boldsymbol{f}^{\alpha\gamma}) + (\boldsymbol{r}^\beta - \boldsymbol{x}_0) \times (\boldsymbol{f}^{\text{ext},\beta} + \sum_{\substack{\gamma \\ \gamma\neq\beta\neq\alpha}} \boldsymbol{f}^{\beta\gamma}) \\ &= (\boldsymbol{r}^\alpha - \boldsymbol{x}_0) \times (m^\alpha \ddot{\boldsymbol{r}}^\alpha - \boldsymbol{f}^{\alpha\beta}) + (\boldsymbol{r}^\beta - \boldsymbol{x}_0) \times (m^\beta \ddot{\boldsymbol{r}}^\beta - \boldsymbol{f}^{\beta\alpha}), \end{aligned} \tag{5.96}$$

where we have used Eqn. (5.88), and its angular momentum is

$$\boldsymbol{H}(\{\alpha,\beta\};\boldsymbol{x}_0) = (\boldsymbol{r}^\alpha - \boldsymbol{x}_0) \times m^\alpha \dot{\boldsymbol{r}}^\alpha + (\boldsymbol{r}^\beta - \boldsymbol{x}_0) \times m^\beta \dot{\boldsymbol{r}}^\beta. \tag{5.97}$$

The balance of angular momentum in Eqn. (5.85)$_2$ requires

$$\begin{aligned}(\boldsymbol{r}^\alpha - \boldsymbol{x}_0) \times (m^\alpha \ddot{\boldsymbol{r}}^\alpha - \boldsymbol{f}^{\alpha\beta}) + (\boldsymbol{r}^\beta - \boldsymbol{x}_0) \times (m^\beta \ddot{\boldsymbol{r}}^\beta - \boldsymbol{f}^{\beta\alpha}) \\ = \frac{d}{dt}\left[(\boldsymbol{r}^\alpha - \boldsymbol{x}_0) \times m^\alpha \dot{\boldsymbol{r}}^\alpha + (\boldsymbol{r}^\beta - \boldsymbol{x}_0) \times m^\beta \dot{\boldsymbol{r}}^\beta\right] \\ = \dot{\boldsymbol{r}}^\alpha \times m^\alpha \dot{\boldsymbol{r}}^\alpha + (\boldsymbol{r}^\alpha - \boldsymbol{x}_0) \times m^\alpha \ddot{\boldsymbol{r}}^\alpha + \dot{\boldsymbol{r}}^\beta \times m^\beta \dot{\boldsymbol{r}}^\beta + (\boldsymbol{r}^\beta - \boldsymbol{x}_0) \times m^\beta \ddot{\boldsymbol{r}}^\beta \\ = (\boldsymbol{r}^\alpha - \boldsymbol{x}_0) \times m^\alpha \ddot{\boldsymbol{r}}^\alpha + (\boldsymbol{r}^\beta - \boldsymbol{x}_0) \times m^\beta \ddot{\boldsymbol{r}}^\beta,\end{aligned} \tag{5.98}$$

which simplifies to

$$(\boldsymbol{r}^\alpha - \boldsymbol{x}_0) \times \boldsymbol{f}^{\alpha\beta} + (\boldsymbol{r}^\beta - \boldsymbol{x}_0) \times \boldsymbol{f}^{\beta\alpha} = \boldsymbol{0},$$

and after using Eqn. (5.95), we obtain

$$(\boldsymbol{r}^\alpha - \boldsymbol{r}^\beta) \times \boldsymbol{f}^{\alpha\beta} = \boldsymbol{0}. \tag{5.99}$$

This shows that the force $\boldsymbol{f}^{\alpha\beta}$ must be *parallel* to the line joining particles α and β, so that

$$\boldsymbol{f}^{\alpha\beta} = \varphi^{\alpha\beta} \frac{\boldsymbol{r}^{\alpha\beta}}{r^{\alpha\beta}}, \tag{5.100}$$

where $r^{\alpha\beta} = \left\|\boldsymbol{r}^{\alpha\beta}\right\|$ and $\varphi^{\alpha\beta}$ is a scalar. Forces satisfying this condition are called *central*. Equation (5.99) is called the *strong law of action and reaction* [Gol80]. We have shown that this law must hold for any force decomposition in order for the balance of linear and angular momentum to hold for any subset of a system of particles.

5.8.2 Forces in conservative systems

We now consider the case where the system is conservative and internal forces are derived from a general multibody interaction potential energy, $\mathcal{V}^{\text{int}} = \widehat{\mathcal{V}}^{\text{int}}(\boldsymbol{r}^1, \ldots, \boldsymbol{r}^N)$. The internal force is

$$\boldsymbol{f}^{\text{int},\alpha} = -\frac{\partial \widehat{\mathcal{V}}^{\text{int}}(\boldsymbol{r}^1, \ldots, \boldsymbol{r}^N)}{\partial \boldsymbol{r}^\alpha}. \tag{5.101}$$

Let us first consider the special case where the potential energy is limited to pairwise interactions:

$$\mathcal{V}^{\text{int}} = \frac{1}{2} \sum_{\substack{\alpha,\beta \\ \alpha \neq \beta}} \widehat{\phi}_{\alpha\beta}(\boldsymbol{r}^\alpha, \boldsymbol{r}^\beta) = \frac{1}{2} \sum_{\substack{\alpha,\beta \\ \alpha \neq \beta}} \phi_{\alpha\beta}(r^{\alpha\beta}) = \sum_{\substack{\alpha,\beta \\ \alpha < \beta}} \phi_{\alpha\beta}(r^{\alpha\beta}). \tag{5.102}$$

The second equality follows since, due to the invariance principle in Eqn. (5.17), the potential energy can only depend on the distances between the atoms (see Section 5.3.2). The third equality is a result of the permutation symmetry of the potential. It is easy to show that for a pair potential the decomposition in Eqn. (5.84) can be made and that the resulting interatomic forces, $\boldsymbol{f}^{\alpha\beta}$, satisfy the strong law of action and reaction. The internal force on atom α is obtained by applying Eqn. (5.101) to Eqn. (5.102) and using the chain rule:

$$\begin{aligned}
\boldsymbol{f}^{\mathrm{int},\alpha} &= -\frac{\partial}{\partial \boldsymbol{r}^{\alpha}}\left[\sum_{\substack{\beta,\gamma\\ \beta<\gamma}}^{N} \phi_{\beta\gamma}(r^{\beta\gamma})\right] \\
&= -\sum_{\substack{\beta,\gamma\\ \beta<\gamma}}^{N} \phi'_{\beta\gamma}(r^{\beta\gamma})\frac{\partial r^{\beta\gamma}}{\partial \boldsymbol{r}^{\alpha}} = -\sum_{\substack{\beta,\gamma\\ \beta<\gamma}}^{N} \phi'_{\beta\gamma}(r^{\beta\gamma})\frac{\boldsymbol{r}^{\beta\gamma}}{r^{\beta\gamma}}(\delta^{\gamma\alpha}-\delta^{\beta\alpha}) \\
&= -\left[\sum_{\beta=1}^{\alpha-1} \phi'_{\beta\alpha}(r^{\beta\alpha})\frac{\boldsymbol{r}^{\beta\alpha}}{r^{\beta\alpha}} - \sum_{\gamma=\alpha+1}^{N} \phi'_{\alpha\gamma}(r^{\alpha\gamma})\frac{\boldsymbol{r}^{\alpha\gamma}}{r^{\alpha\gamma}}\right] = \sum_{\substack{\beta\\ \beta\neq\alpha}} \phi'_{\alpha\beta}(r^{\alpha\beta})\frac{\boldsymbol{r}^{\alpha\beta}}{r^{\alpha\beta}},
\end{aligned} \tag{5.103}$$

where we have used Eqn. (5.13), the property, $\boldsymbol{r}^{\alpha\beta} = -\boldsymbol{r}^{\beta\alpha}$ and the permutation symmetry of the potential function. Comparing Eqn. (5.103) with Eqn. (5.84), we see that for a pair potential the interatomic force term is

$$\boldsymbol{f}^{\alpha\beta} = \phi'_{\alpha\beta}(r^{\alpha\beta})\frac{\boldsymbol{r}^{\alpha\beta}}{r^{\alpha\beta}}. \tag{5.104}$$

This expression clearly satisfies the strong law of action and reaction ($\boldsymbol{f}^{\alpha\beta} = -\boldsymbol{f}^{\beta\alpha}$ and $\boldsymbol{f}^{\alpha\beta}$ is central).

The results for the pair potential are instructive, but we saw above that modern interatomic models normally involve more than simple pairwise interactions and therefore cannot be written in the form of Eqn. (5.102). Despite this, it is still possible to express the internal forces as a sum over pairwise terms (as in Eqn. (5.84)) that satisfy the strong law of action and reaction.

In the derivation given above for the pair potential force, we used the fact that due to the principle of interatomic potential invariance in Eqn. (5.17), the pair potential function can only depend on the distances between the atoms. However, as stated in Section 5.3.2 and proved in Appendix A, the same conclusion applies to atomistic models of arbitrary form. Thus, as given in Eqn. (5.18), and repeated here for convenience,

$$\mathcal{V}^{\mathrm{int}} = \mathcal{V}^{\mathrm{int}}(r^{12}, r^{13}, \ldots, r^{1N}, r^{23}, \ldots, r^{N-1,N}) = \mathcal{V}^{\mathrm{int}}(\{r^{\alpha\beta}\}). \tag{5.105}$$

The force can then be computed using a similar procedure to the one that led to Eqn. (5.103):

$$\boldsymbol{f}^{\mathrm{int},\alpha} = -\sum_{\substack{\beta,\gamma\\ \beta<\gamma}} \frac{\partial \mathcal{V}^{\mathrm{int}}}{\partial r^{\beta\gamma}}\frac{\partial r^{\beta\gamma}}{\partial \boldsymbol{r}^{\alpha}} = \sum_{\substack{\beta\\ \beta\neq\alpha}} \frac{\partial \mathcal{V}^{\mathrm{int}}}{\partial r^{\alpha\beta}}\frac{\boldsymbol{r}^{\alpha\beta}}{r^{\alpha\beta}}, \tag{5.106}$$

where we have made use of Eqn. (5.13). It is clear from the above result that the force can be written as a sum of central forces that satisfy the strong law of action and reaction:

$$\boldsymbol{f}^{\text{int},\alpha} = \sum_{\substack{\beta \\ \beta \neq \alpha}} \boldsymbol{f}^{\alpha\beta}, \qquad \boldsymbol{f}^{\alpha\beta} \equiv \varphi^{\alpha\beta} \frac{\boldsymbol{r}^{\alpha\beta}}{r^{\alpha\beta}}, \qquad \varphi^{\alpha\beta} \equiv \frac{\partial \mathcal{V}^{\text{int}}(r^{12}, r^{13}, \dots)}{\partial r^{\alpha\beta}}. \tag{5.107}$$

We derived this relation for a representation given in terms of the $N(N-1)/2$ distances, $r^{\alpha\beta}$ $(\alpha < \beta)$. However, it also applies to representations in terms of all $N(N-1)$ distances as long as the symmetric partial derivative definition in Eqn. (5.14) is used. We see this in Section 5.8.3 where we derive an explicit expression for the forces using the full representation for different potentials. (See also Exercise 5.15.)

In deriving Eqn. (5.107), we have not specified anything about the form of the interatomic model except that it satisfies the invariance principle in Eqn. (5.17). This is a very strong result that requires some further clarification.

1. The result that the force on an atom, modeled using any interatomic model, can be decomposed as sum of central forces may seem strange to some readers. This may be due to the common confusion in the literature of using the term "central-force model" to refer to simple pair potentials. In fact, we see that due the principle of interatomic potential invariance specified in Eqn. (5.17), *all* interatomic potentials (including those with explicit bond-angle dependence) are central-force models. By this we mean that the force on any atom (say α) can always be decomposed as a sum of terms $\boldsymbol{f}^{\alpha\beta}$ aligned with the vectors joining atom α with its neighbors and satisfying action and reaction.
   The difference between the general case and that of a pair potential is that for a pair potential the magnitude of the interatomic force, $\varphi^{\alpha\beta}$, depends *only* on the distance $r^{\alpha\beta}$ between the atoms (as shown in Eqn. (5.104)), whereas for a general potential the magnitude depends on a larger set of distances, $\varphi^{\alpha\beta} = \varphi^{\alpha\beta}(r^{12}, r^{13}, \dots, r^{N-1,N})$, i.e. it depends on the *environment* of the "bond" between α and β. For this reason, $\boldsymbol{f}^{\alpha\beta}$ for a pair potential is a property of atoms α and β alone and can be physically interpreted as the "force exerted on atom α by atom β." Whereas, in the more general case of arbitrary interatomic potentials, the physical significance of the interatomic force is less clear and at best we can say that $\boldsymbol{f}^{\alpha\beta}$ is the "contribution to the force on atom α due to the presence of atom β."
2. As pointed out by Admal and Tadmor [AT10], the partial differentiation of $\mathcal{V}^{\text{int}}$ with respect to the distance arguments in Eqns. (5.106) and (5.107) formally involves the definition of a *continuously differentiable extension* to the potential energy function. This issue is discussed at length in Appendix A, where a more rigorous derivation of Eqn. (5.107) is presented in Section A.4. The conclusion from this analysis is that the decomposition in Eqn. (5.107) is *not* unique. By that we mean that given an interatomic potential energy function, $\mathcal{V}^{\text{int}}$, it is possible to construct a completely equivalent function, $\mathcal{V}^{\text{int}}_*$, which gives exactly the same energy and forces for all atomic configurations,

but gives *different* values for the $\boldsymbol{f}^{\alpha\beta}$ terms appearing in Eqn. (5.107). In other words, we can have two decompositions for the same force satisfying

$$\boldsymbol{f}^{\text{int},\alpha} = \sum_{\substack{\beta \\ \beta\neq\alpha}} \boldsymbol{f}^{\alpha\beta} = \sum_{\substack{\beta \\ \beta\neq\alpha}} \boldsymbol{f}_*^{\alpha\beta},$$

$$\boldsymbol{f}^{\alpha\beta} \equiv \left(\frac{\partial \mathcal{V}^{\text{int}}}{\partial r^{\alpha\beta}}\right)\frac{\boldsymbol{r}^{\alpha\beta}}{r^{\alpha\beta}}, \qquad \boldsymbol{f}_*^{\alpha\beta} \equiv \left(\frac{\partial \mathcal{V}_*^{\text{int}}}{\partial r^{\alpha\beta}}\right)\frac{\boldsymbol{r}^{\alpha\beta}}{r^{\alpha\beta}},$$

where $\boldsymbol{f}^{\alpha\beta} \neq \boldsymbol{f}_*^{\alpha\beta}$. This conclusion is true even for pair potentials. A simple one-dimensional example demonstrating this is presented in Section A.4. This is an important result because we will see in Section 8.2 that force decomposition is used in the derivation of the pointwise stress and therefore since force decomposition is nonunique this implies that the pointwise stress tensor is also not unique.

3. Another question is whether it is also possible to construct *noncentral* force decompositions for $\boldsymbol{f}^{\alpha}$. Mathematically, the answer to this is clearly yes, since it is always possible to decompose a $3N$-dimensional vector as a sum of terms along the "basis" of directions defined by the interatomic vectors $\{\boldsymbol{r}^{\alpha\beta}\}$ as long as $\{\boldsymbol{r}^{\alpha\beta}\}$ span the $\mathbb{R}^{3N}$ space in which the vector is defined. (One example of a noncentral force decomposition is given in Exercise 5.13.) We discount these noncentral decompositions on physical grounds since they violate the strong law of action and reaction proved in Section 5.8.1.
4. It is important to emphasize that our discussion is limited to classical potential energy functions that only depend on the positions of the particles. In situations where the potential energy depends on additional variables, Eqn. (5.107) will not hold in general and the forces may violate the strong law of action and reaction. Goldstein in his book *Classical Mechanics* [Gol80], specifically states: "If [the potential energy] were also a function of the differences of some other pair of vectors associated with the particles, such as their velocities or (to step into the domain of modern physics) their intrinsic 'spin' angular momenta, then the forces would still be equal and opposite, but would not necessarily lie along the direction between the particles." An example of this are the forces derived from the Biot–Savart law for moving charges [Gol80, Page 8]. Such effects, although clearly present in the original quantum mechanical system, are masked in classical potential energy models because they neglect the effects of electron spin.

In the next section, we evaluate Eqn. (5.107) for some of the main potential energy functional forms.

5.8.3 Atomic forces for some specific interatomic models

The central force decomposition in Eqn. (5.107) provides a convenient recipe for computing the force on an atom for any interatomic potential. Specific models differ from each other only by the force magnitude, $\varphi^{\alpha\beta} = \partial\mathcal{V}^{\text{int}}/\partial r^{\alpha\beta}$. We can therefore simply insert the appropriate expressions for the derivative of the interaction energy and simplify for each specific model.

Cluster potentials The cluster potential form is given in Eqn. (5.32). The potential is exact for a system of N atoms if all N terms are retained. However, in practice, approximate potentials are constructed by terminating the series at $n = 2$, 3 or 4, which correspond to pair, three-body and four-body potentials. We derive each term in the sum separately, denoting by $\Phi_{(k)}$ $(k = 1, \dots, n)$ the contribution to the potential energy coming from the k-body terms. From these, we obtain thc the contributions to the force magnitude, $\varphi^{\alpha\beta}_{(k)}$, and internal force, $\boldsymbol{f}^{\text{int},\alpha}_{(k)}$, resulting from the k-body terms. The total force magnitude and force for an n-body potential follow from

$$\varphi^{\alpha\beta} = \sum_{k=1}^{n} \varphi^{\alpha\beta}_{(k)}, \qquad \boldsymbol{f}^{\text{int},\alpha} = \sum_{k=1}^{n} \boldsymbol{f}^{\text{int},\alpha}_{(k)}.$$

Similar expressions for forces in cluster potentials are given in [Mor94, Appendix].

For pair interactions,

$$\Phi_{(2)} = \frac{1}{2} \sum_{\substack{\gamma,\delta \\ \gamma\neq\delta}} \phi_{\gamma\delta}(r^{\gamma\delta}),$$

and we have

$$\varphi^{\alpha\beta}_{(2)} = \frac{\partial \Phi_{(2)}}{\partial r^{\alpha\beta}} = \frac{1}{2} \sum_{\substack{\gamma,\delta \\ \gamma\neq\delta}} \phi'_{\gamma\delta}(r^{\gamma\delta}) \frac{\partial r^{\gamma\delta}}{\partial r^{\alpha\beta}} = \frac{1}{2} [\phi'_{\alpha\beta}(r^{\alpha\beta}) + \phi'_{\beta\alpha}(r^{\beta\alpha})] = \phi'_{\alpha\beta}(r^{\alpha\beta}), \tag{5.108}$$

where Eqn. (5.14) and the permutation symmetry of the potential function were used. The force on atom α is therefore

$$\boldsymbol{f}^{\text{int},\alpha}_{(2)} = \sum_{\substack{\beta \\ \beta\neq\alpha}} \phi'_{\alpha\beta}(r^{\alpha\beta}) \frac{\boldsymbol{r}^{\alpha\beta}}{r^{\alpha\beta}}. \tag{5.109}$$

This is the same result that we found in Eqn. (5.104).

For three-body interactions,

$$\Phi_{(3)} = \frac{1}{6} \sum_{\substack{\gamma,\delta,\epsilon \\ \gamma\neq\delta\neq\epsilon}} \phi_{\gamma\delta\epsilon}(r^{\gamma\delta}, r^{\gamma\epsilon}, r^{\delta\epsilon}),$$

and we have

$$\varphi^{\alpha\beta}_{(3)} = \frac{\partial \Phi_{(3)}}{\partial r^{\alpha\beta}} = \frac{1}{6} \left[\sum_{\substack{\epsilon \\ \epsilon\neq\alpha\neq\beta}} \left(\frac{\partial \phi_{\alpha\beta\epsilon}}{\partial r^{\alpha\beta}} + \frac{\partial \phi_{\beta\alpha\epsilon}}{\partial r^{\beta\alpha}} \right) + \sum_{\substack{\delta \\ \delta\neq\alpha\neq\beta}} \left(\frac{\partial \phi_{\alpha\delta\beta}}{\partial r^{\alpha\beta}} + \frac{\partial \phi_{\beta\delta\alpha}}{\partial r^{\beta\alpha}} \right) \right.$$

$$\left. + \sum_{\substack{\gamma \\ \gamma\neq\alpha\neq\beta}} \left(\frac{\partial \phi_{\gamma\alpha\beta}}{\partial r^{\alpha\beta}} + \frac{\partial \phi_{\gamma\beta\alpha}}{\partial r^{\beta\alpha}} \right) \right] = \sum_{\substack{\gamma \\ \gamma\neq\alpha\neq\beta}} \frac{\partial \phi_{\alpha\beta\gamma}}{\partial r^{\alpha\beta}}, \tag{5.110}$$

where Eqn. (5.14) was used. The six terms inside the square brackets are equal due to permutation symmetry and therefore combine to give the final result. The force follows as

$$\boldsymbol{f}_{(3)}^{\mathrm{int},\alpha} = \sum_{\substack{\beta,\gamma \\ \beta\neq\gamma\neq\alpha}} \frac{\partial \phi_{\alpha\beta\gamma}}{\partial r^{\alpha\beta}} \frac{\boldsymbol{r}^{\alpha\beta}}{r^{\alpha\beta}}. \tag{5.111}$$

Since three-body potentials are often expressed in terms of bond angles, obtaining an explicit relation for the forces may require an additional chain rule between the angles and distances using Eqn. (5.33). This can be tedious, but does not constitute a limitation on the approach.

For four-body interactions,

$$\Phi_{(4)} = \frac{1}{24} \sum_{\substack{\gamma,\delta,\epsilon,\zeta \\ \gamma\neq\delta\neq\epsilon\neq\zeta}} \phi_{\gamma\delta\epsilon\zeta}(r^{\gamma\delta}, r^{\gamma\epsilon}, r^{\gamma\zeta}, r^{\delta\epsilon}, r^{\delta\zeta}, r^{\epsilon\zeta}),$$

and we have

$$\varphi_{(4)}^{\alpha\beta} = \frac{\partial \Phi_{(4)}}{\partial r^{\alpha\beta}} = \frac{1}{24}\left[\sum_{\substack{\epsilon,\zeta \\ \epsilon\neq\zeta\neq\alpha\neq\beta}} \left(\frac{\partial \phi_{\alpha\beta\epsilon\zeta}}{\partial r^{\alpha\beta}} + \frac{\partial \phi_{\beta\alpha\epsilon\zeta}}{\partial r^{\beta\alpha}} \right) + \cdots \right] = \frac{1}{2} \sum_{\substack{\gamma,\delta \\ \gamma\neq\delta\neq\alpha\neq\beta}} \frac{\partial \phi_{\alpha\beta\gamma\delta}}{\partial r^{\alpha\beta}}. \tag{5.112}$$

The square brackets in the third expression contain six sums each with two terms. All of these are equal due to permutation symmetry and can be combined into a single sum with a factor of 12 brought out in front. The result is the single sum on the right with a $12/24 = 1/2$ prefactor. The force follows as

$$\boldsymbol{f}_{(4)}^{\mathrm{int},\alpha} = \frac{1}{2} \sum_{\substack{\beta,\gamma,\delta \\ \beta\neq\gamma\neq\delta\neq\alpha}} \frac{\partial \phi_{\alpha\beta\gamma\delta}}{\partial r^{\alpha\beta}} \frac{\boldsymbol{r}^{\alpha\beta}}{r^{\alpha\beta}}. \tag{5.113}$$

The $\frac{1}{2}$ factor suggests that an additional symmetry exists in the expression. Indeed, the computational burden of computing the forces can be reduced by a factor of 2 by respecting $\gamma < \delta$ during the summation process (see [Mor94, Appendix]).

For five-body interactions and above, it can be shown that the generic expression for the force due the n-body term is [Mor94, Appendix]

$$\boldsymbol{f}_{(n)}^{\mathrm{int},\alpha} = \frac{1}{(n-2)!} \sum_{\substack{\beta,\gamma,\dots \\ \beta\neq\gamma\neq\cdots\neq\alpha}} \frac{\partial \phi_{\alpha\beta\dots}}{\partial r^{\alpha\beta}} \frac{\boldsymbol{r}^{\alpha\beta}}{r^{\alpha\beta}}. \tag{5.114}$$

However, due to the complexity of constructing such potentials and their computational cost, few (if any) potentials of this form are being used in practice at the time of writing.

Pair functionals For pair functionals, we have a combination of a pair potential and an additional term of the form

$$\mathcal{V}_{\text{embed}} = \sum_{\gamma} U_{\gamma}(\rho^{\gamma}), \qquad \rho^{\gamma} = \sum_{\substack{\delta \\ \delta \neq \gamma}} g_{\delta}(r^{\gamma\delta}).$$

For this additional term, we have

$$\begin{aligned}
\varphi^{\alpha\beta}_{\text{embed}} = \frac{\partial \mathcal{V}_{\text{embed}}}{\partial r^{\alpha\beta}} &= \sum_{\gamma} U'_{\gamma}(\rho^{\gamma}) \frac{\partial \rho^{\gamma}}{\partial r^{\alpha\beta}} = \sum_{\gamma} U'_{\gamma}(\rho^{\gamma}) \sum_{\substack{\delta \\ \delta \neq \gamma}} g'_{\delta}(r^{\gamma\delta}) \frac{\partial r^{\gamma\delta}}{\partial r^{\alpha\beta}} \\
&= U'_{\alpha}(\rho^{\alpha}) g'_{\beta}(r^{\alpha\beta}) + U'_{\beta}(\rho^{\beta}) g'_{\alpha}(r^{\beta\alpha}),
\end{aligned} \tag{5.115}$$

where we have used Eqn. (5.14). Combining this with Eqn. (5.108), the force is

$$\boldsymbol{f}^{\text{int},\alpha}_{\text{PF}} = \sum_{\substack{\beta \\ \beta \neq \alpha}} \left[\phi'_{\alpha\beta}(r^{\alpha\beta}) + U'_{\alpha}(\rho^{\alpha}) g'_{\beta}(r^{\alpha\beta}) + U'_{\beta}(\rho^{\beta}) g'_{\alpha}(r^{\alpha\beta}) \right] \frac{\boldsymbol{r}^{\alpha\beta}}{r^{\alpha\beta}}, \tag{5.116}$$

where the "PF" stands for "pair functional."

Similar derivations could be carried out for the other potentials discussed in this chapter, but these expressions can be found in the literature and we stop here.

5.8.4 Bond stiffnesses for some specific interatomic models

In Sections 8.1.4, 11.3.5 and 11.5, we will discuss the microscopic underpinnings of stress and the elasticity tensor that relates stress to strain. A quantity that will appear often in that discussion is the *bond stiffness*, $\kappa^{\alpha\beta\gamma\delta}$, which we define as

$$\kappa^{\alpha\beta\gamma\delta} \equiv \frac{\partial \varphi^{\alpha\beta}}{\partial r^{\gamma\delta}} = \frac{\partial^2 \mathcal{V}^{\text{int}}}{\partial r^{\alpha\beta} \partial r^{\gamma\delta}}. \tag{5.117}$$

This quantity is clearly interpreted as the bond stiffness in a simple pairwise potential, where the force on atom α due to atom β, $\boldsymbol{f}^{\alpha\beta}$, depends only on the distance $r^{\alpha\beta}$. In more complicated systems where $\varphi^{\alpha\beta}$ can depend on other interatomic distances, the interpretation is not as clear, but we use the name out of convenience. Here, we catalog $\kappa^{\alpha\beta\gamma\delta}$ for a few common interatomic models.

Pair potentials From Eqn. (5.108), we obtain

$$\kappa^{\alpha\beta\gamma\delta}_{(2)} = \phi''_{\alpha\beta}(r^{\alpha\beta})(\delta^{\alpha\gamma}\delta^{\beta\delta} + \delta^{\alpha\delta}\delta^{\beta\gamma}). \tag{5.118}$$

Three-body cluster potentials From Eqn. (5.110) we find

$$\kappa_{(3)}^{\alpha\beta\gamma\delta} = \frac{\partial^2 \phi_{\alpha\beta\delta}}{\partial r^{\alpha\beta}\partial r^{\alpha\delta}}\delta^{\alpha\gamma} + \frac{\partial^2 \phi_{\alpha\beta\gamma}}{\partial r^{\alpha\beta}\partial r^{\alpha\gamma}}\delta^{\alpha\delta} + \frac{\partial^2 \phi_{\alpha\beta\delta}}{\partial r^{\alpha\beta}\partial r^{\beta\delta}}\delta^{\beta\gamma} + \frac{\partial^2 \phi_{\alpha\beta\gamma}}{\partial r^{\alpha\beta}\partial r^{\alpha\gamma}}\delta^{\beta\delta} + \sum_{\substack{\epsilon \\ \epsilon\neq\alpha,\beta}} \frac{\partial^2 \phi_{\alpha\beta\epsilon}}{\partial (r^{\alpha\beta})^2}(\delta^{\alpha\gamma}\delta^{\beta\delta} + \delta^{\alpha\delta}\delta^{\beta\gamma}). \tag{5.119}$$

Note the two different kinds of terms. The first four are derivatives with respect to two different arguments of $\phi_{\alpha\beta\gamma}$, while the sum involves terms that are a second derivative with respect to a single argument.

Pair functionals Starting from Eqn. (5.115), we have

$$\begin{aligned}\kappa_{\text{embed}}^{\alpha\beta\gamma\delta} &= \left[U'_\alpha(\rho^\alpha)g''_\beta(r^{\alpha\beta}) + U'_\beta(\rho^\beta)g''_\alpha(r^{\alpha\beta})\right]\left[\delta^{\alpha\gamma}\delta^{\beta\delta} + \delta^{\alpha\delta}\delta^{\beta\gamma}\right] \\ &\quad + g'_\beta(r^{\alpha\beta})U''_\alpha(\rho^\alpha)\left[g'_\delta(r^{\alpha\delta})\delta^{\alpha\gamma} + g'_\gamma(r^{\alpha\gamma})\delta^{\alpha\delta}\right] \\ &\quad + g'_\alpha(r^{\alpha\beta})U''_\beta(\rho^\beta)\left[g'_\delta(r^{\beta\delta})\delta^{\beta\gamma} + g'_\gamma(r^{a\gamma})\delta^{\beta\delta}\right].\end{aligned} \tag{5.120}$$

5.8.5 The cutoff radius and interatomic forces

We close with a comment regarding the relationship between the cutoff radius, r_{cut}, of an interatomic potential and the internal forces derived from it. The generic expression in Eqn. (5.107) shows that the force on a given atom α depends on all atoms β for which $\varphi^{\alpha\beta}$ is nonzero. Since $\varphi^{\alpha\beta}$ is derived from $\mathcal{V}^{\text{int}}$ it may seem obvious that the force on an atom will be affected by all atoms up to distance r_{cut} from it. This is indeed the case for cluster potentials but is not correct in general.

First, consider the example of a pair potential. In this case,

$$\varphi^{\alpha\beta} = \phi'_{\alpha\beta}(r^{\alpha\beta}).$$

If $\phi_{\alpha\beta}(r^{\alpha\beta}) = 0$ for $r^{\alpha\beta} \geq r_{\text{cut}}$, then clearly $\varphi^{\alpha\beta}$ satisfies the same condition. However, now consider the case of a pair functional. The expression for the embedding contribution to $\varphi^{\alpha\beta}$ is given in Eqn. (5.115). Let us assume that the electron density functions, g_α, and the embedding functions, U_α, have the same cutoff radius r_{cut}. Despite this, the force on atom α will be affected by all atoms within $2r_{\text{cut}}$ (*twice* the cutoff radius). This dependence comes from the second term in Eqn. (5.115):

$$U'_\beta(\rho^\beta)g'_\alpha(r^{\beta\alpha}).$$

In other words, the force on atom α depends directly on the electron density, ρ^β, at each neighbor β. But this quantity is determined by all the neighbors of atom β, as shown in Fig. 5.8, and the effective range of influence on the atomic forces is twice the cutoff radius.

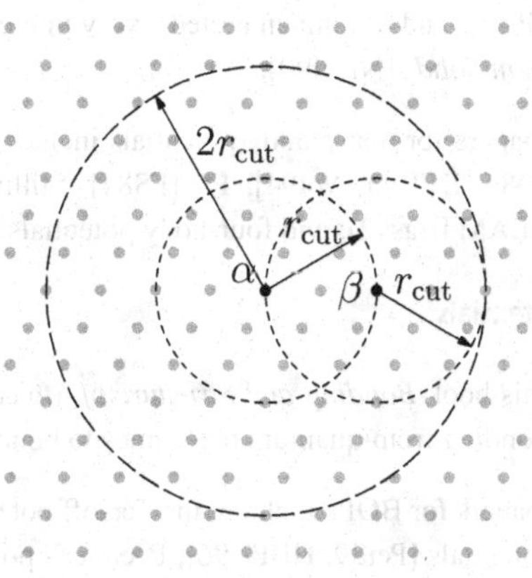

Fig. 5.8 The force on atom α, $f^{int,\alpha}$, for a pair functional model depends on the electron density at atom β, which in turn depends on all atoms within r_{cut} of β. This means that $f^{int,\alpha}$ effectively depends on all atoms within $2r_{cut}$ of atom α.

This result shows that care must be taken when evaluating the range of interactions that contribute to the force for a given interatomic potential energy function.

Further reading

Pair potentials

- An interesting early discussion of interatomic forces, written by Lennard-Jones, appears in Chapter X of Fowler's classic book on statistical mechanics [Fow36].
- Ian Torrens' book *Interatomic Potentials* [Tor72] is a bit dated at this point, but still provides a good introduction to pair potentials.
- More discussion of the development of effective pair potentials can be found in Finnis' book *Interatomic Forces in Condensed Matter* [Fin03].
- Some key papers on pair potentials include Lennard-Jones [Jon24a, Jon24b, LJ25], Morse [Mor29], Born–Mayer [BM32] potentials and potentials for ionic systems [LC85, SH86, BGCB94].

Empirical models for organic materials

- MacKerell reviews the field of interatomic models for biological systems in [Mac04].
- Jensen's *Introduction to Computational Chemistry* [Jen06] includes a chapter on the interatomic potentials used to model organic molecules.

Many-body interactions

- The review article "Beyond pair potentials in elemental transition metals and semi-conductors" by Carlsson [Car90] provides a more detailed discussion of the quantum mechanical and TB justifications for empirical models.

- Nieminen, Puska and Manninen edited a very interesting collection of articles called *Many-Atom Interactions in Solids* [NPM90].

- Some key papers for many-body potentials include: EMT [NL80, JNP87, Jac88]; EAM [DB83, DB84, FBD86, DFB93, Vot94]; FS [FS84], Stillinger–Weber three-body potential for silicon [SW85]; MEAM [Bas92]; and four-body potentials [Mor88, Mor90, Mor94].

Bond-order potentials

- Pettifor in his book *Bonding and Structure of Molecules and Solids* [Pet95] presents an in depth review of bonding from quantum mechanics to bond-order potentials.

- Some key papers for BOPs include; the Tersoff potentials for silicon and carbon [Ter88a, Ter89]; Pettifor's potentials [Pet89, HBF+96]; Brenner's potentials for carbon and hydrocarbons [Bre90, Bre92, BSH+02]; EDIP for silicon [BKJ97] and the ReaxFF potential for hydrocarbons, silicon and silicon oxides [vDDLG01, vDSS+03].

General reviews

- The February 1996 issue of *MRS Bulletin* edited by A. F. Voter is devoted to the subject of "Interatomic potentials for atomistic simulations" and contains a number of interesting articles on the subject.

- Chapter 2 in the *Handbook of Materials Modeling (*Part A, *Methods)* edited by Sidney Yip [Yip06] has a number of articles on interatomic atomistic models.

- An early reference on the development of pseudopotentials is that of Harrison [Har66]. Another good reference is [Haf87].

- Finally, general articles on the art of developing interatomic models are [SS92, Bre00].

Exercises

5.1 [SECTION 5.3] Demonstrate that the noncentral potential defined in footnote 10 on page 244 violates the frame-indifference condition in Eqn. (5.15).

5.2 [SECTION 5.3] Invent a new pair potential $\phi(r)$ which has the following characteristics: (1) $\lim_{r\to 0}\phi(r) = \infty$; (2) $\phi(r)$ has only one minimum in the range $r \in [0, \infty]$; (3) $\phi(r_{\rm cut}) = \phi'(r_{\rm cut}) = \phi''(r_{\rm cut}) = 0$, where $r_{\rm cut}$ is a user-defined cutoff radius. Clearly indicate the functional form of the potential and the parameters on which it depends. Briefly explain the motivation for the functional form you are proposing. Notes: (1) Your function should *not* be similar to the Lennard-Jones, Morse or ionic potentials discussed in Sections 5.4.2 and 5.4.3. Be creative! (2) Your potential should have at least three free parameters *after* satisfying the smooth cutoff requirements at $r_{\rm cut}$. These free parameters provide flexibility for fitting the physical properties of the material being modeled by $\phi(r)$. In future exercises (Exercises 6.14, 6.15 and 11.11), we will explore the behavior and predictions of your potential.

5.3 [SECTION 5.4] Repeat the process of Example 5.1 using the Morse potential in Eqn. (5.38) instead of the Lennard-Jones potential. Fit the potential to the equilibrium bond energy, bond length and bond stiffness. Compare the overall shape of the Morse potential to the exact result.

5.4 [SECTION 5.4] Verify that the minimum of the 6–12 Lennard-Jones potential (Eqn. (5.36)) is $\phi_{2,\min} = -\varepsilon$, occurring at $r = 2^{1/6}\sigma$.

5.5 [SECTION 5.4] A modified Lennard-Jones potential with the following form is proposed:

$$\phi(r) = 4\varepsilon\left[\left(\frac{\sigma}{r}\right)^{12} - \left(\frac{\sigma}{r}\right)^{6}\right] + ar^2 + b.$$

The polynomial $ar^2 + b$ is added in order to obtain smooth behavior at the cutoff radius. Take $r_{\text{cut}} = 2.5\sigma$.

1. Compute the discontinuity in the energy and force on an atom when a new neighbor moves into its cutoff sphere for the standard Lennard-Jones potential ($a = b = 0$).
2. What numerical problems can result from a potential which is not smooth at the cutoff?
3. Compute the parameters a and b (in terms of ε and σ) that enforce the smooth cutoff conditions: $\phi(r_{\text{cut}}) = 0$ and $\phi'(r_{\text{cut}}) = 0$.
4. On a single graph, plot the standard Lennard-Jones potential and the modified Lennard-Jones potential with the a and b parameters computed above. Use the Lennard-Jones σ and ε parameters for Ar given in Tab. 5.1.

5.6 [SECTION 5.5] Verify that the pair functional for single-element systems is such that the energy is invariant to the following transformations of the functional form:

$$U(\rho) \to U\left(\frac{\rho}{A}\right), \qquad g \to Ag$$

and

$$\phi(r) \to \phi(r) + 2Bg(r), \qquad U(\rho) \to U(\rho) - B\rho$$

for arbitrary constants A and B. (This is often a source of confusion in practical circumstances such as when one is trying to implement a potential as described in the literature, or when comparing the relative form of different potentials. Two identical potentials can appear very different due to these transformations.)

5.7 [SECTION 5.5] Johnson [Joh88] developed a simple nearest-neighbor EAM potential, as defined in Eqn. (5.52), for single-species fcc metals, where

$$\phi(r) = \phi_0 e^{-\hat{\gamma}(r/r_0 - 1)}, \qquad g(r) = g_0 e^{-\hat{\beta}(r/r_0 - 1)},$$

$$U(\rho) = -E^0_{\text{coh}}\left(1 - \frac{\hat{\alpha}}{\hat{\beta}}\ln\frac{\rho}{\rho_0}\right)\left(\frac{\rho}{\rho_0}\right)^{\hat{\alpha}/\hat{\beta}} - 6\phi_0\left(\frac{\rho}{\rho_0}\right)^{\hat{\gamma}/\hat{\beta}}.$$

Here r_0 is the equilibrium near-neighbor distance, $\rho_0 = 12g_0$ is the electron density in an fcc crystal at the equilibrium lattice constant, E^0_{coh} is the equilibrium cohesive energy and the rest are fitted parameters. For Cu, $\phi_0 = 0.59$ eV, $g_0 = 0.30$ eV, $\hat{\alpha} = 5.09$, $\hat{\beta} = 5.85$, $\hat{\gamma} = 8.00$, $r_0 = 2.556$ Å, $E^0_{\text{coh}} = 3.54$ eV.

1. Plot the functions $\phi(r)$ and $g(r)$ for $r \in [r_0/2, 3r_0/2]$, and $U(\rho)$ for $\rho \in [\rho_0/3, 3\rho_0]$.
2. Demonstrate that Johnson's EAM potential exhibits the correct relation between bond strength and coordination number. Assume that the equilibrium distance remains r_0 at all coordinations. **Hint:** Compute the energy of a single atom as a function of its coordination number (i.e. the number of near neighbors), then plot the bond strength (the energy per atom divided by the number of neighbors) as a function of the coordination number. Explain how the graph you obtained shows that Johnson's model gives the correct behavior. Note that the maximum possible coordination is 12, which is obtained for the close-packed fcc structure. There is no point looking at coordinations larger than this.

5.8 [SECTION 5.5] Derive the full expression for Eqn. (5.58).

5.9 [SECTION 5.7] A four-body cluster potential for a single species has the form:

$$\mathcal{V}^{\text{int}} = \frac{1}{2!}\sum_{\substack{\alpha,\beta \\ \alpha\neq\beta}} \phi_2(r^{\alpha\beta}) + \frac{1}{3!}\sum_{\substack{\alpha,\beta,\gamma \\ \alpha\neq\beta\neq\gamma}} \phi_3(r^{\alpha\beta}, r^{\alpha\gamma}, r^{\beta\gamma})$$

$$+ \frac{1}{4!}\sum_{\substack{\alpha,\beta,\gamma,\delta \\ \alpha\neq\beta\neq\gamma\neq\delta}} \phi_4(r^{\alpha\beta}, r^{\alpha\gamma}, r^{\alpha\delta}, r^{\beta\gamma}, r^{\beta\delta}, r^{\gamma\delta}).$$

All of the n-body potentials have the same cutoff radius.

1. For a system containing N atoms with each atom having m neighbors inside the cutoff, compute the number of ϕ_2, ϕ_3 and ϕ_4 function calls necessary to compute $\mathcal{V}^{\text{int}}$. Make your computations without accounting for potential symmetries that could reduce the number of function calls.
2. Repeat part 1, but this time account for symmetries. Write down the reduced summation rule for this case. When writing down the summation rules make use of the data structure $n(\alpha, j)$ which contains the list of neighbors of each atom (i.e. $n(\alpha, j)$ is the jth neighbor of atom α, where $j = 1, 2, \ldots, m$). Assume all m neighbors that an atom has are inside the cutoff radius. For example, due to the $\alpha\beta$ symmetry of ϕ_2, we have
$$\frac{1}{2!}\sum_{\substack{\alpha,\beta\\ \alpha\neq\beta}} \phi_2(r^{\alpha\beta}) = \sum_{\alpha=1}^{N-1} \sum_{\substack{j=1\\ n(\alpha,j)>\alpha}}^{m} \phi_2(r^{\alpha,n(\alpha,j)}),$$
with the result that the number of function calls is cut by a factor of 2.
3. Assuming that it takes the same amount of computer time to compute ϕ_2, ϕ_3 and ϕ_4, use the results in part 2 to estimate the computation time per atom (= total computation time/N) for: (a) a pair potential; (b) a three-body potential (includes both ϕ_2 and ϕ_3); (c) a four-body potential (includes ϕ_2, ϕ_3 and ϕ_4). Give your results in terms of numbers of function calls. How do your results scale with m in each case?
4. As a specific example, use the expressions from part 3 to compute the number of function calls per atom for the three potentials for the case $m = 42$ (these are the number of atoms in a third-neighbor fcc model).

5.10 [Section 5.8] The N atoms in a one-dimensional chain interact via a pair potential $\phi(r)$. The positions of the atoms are r^α ($\alpha = 1, 2, \ldots, N$).
1. Accounting for interactions up to second neighbor, find an analytical expression for the force on a generic atom α away from the ends of the chain.
2. Can the forces on the atom be viewed as a sum of forces exerted on it by its neighbors? If yes, identify these forces.

5.11 [Section 5.8] Show that the weak law of action and reaction implies that for a system of N atoms, the following identity is satisfied:
$$\sum_{\alpha=1}^{N} \boldsymbol{f}^{\text{int},\alpha} = \mathbf{0}.$$

5.12 [Section 5.8] Prove that for any n-body cluster potential ($2 \le n \le N$) in which the first n terms in Eqn. (5.29) are retained, the following identity is satisfied:
$$\sum_{\alpha=1}^{n} \frac{\partial \mathcal{V}^{\text{int}}}{\partial \boldsymbol{r}^\alpha} = \mathbf{0}.$$
Hint: You will need to use the permutation symmetry of the potentials, the result in Exercise 5.11 and an inductive approach beginning with two particles and progressing from there.

5.13 [Section 5.8] In this problem we explore an example of a noncentral force decomposition.
1. Show that the following interatomic force:
$$\widetilde{\boldsymbol{f}}^{\alpha\beta} = -\frac{1}{N}\left(\frac{\partial \mathcal{V}^{\text{int}}}{\partial \boldsymbol{r}^\alpha} - \frac{\partial \mathcal{V}^{\text{int}}}{\partial \boldsymbol{r}^\beta}\right)$$
satisfies Eqn. (5.84). **Hint:** You will need to use the result in Exercise 5.11.
2. Derive the explicit form for $\widetilde{\boldsymbol{f}}^{\alpha\beta}$ for the pair potential form in Eqn. (5.35) and show that this definition does not satisfy the strong law of action and reaction. What does this result mean?

5.14 [Section 5.8] Derive the final relation in Eqn. (5.106).

5.15 [SECTION 5.8] In this problem, you will derive the force decomposition relation for the terms in a cluster potential and show that the same result is obtained regardless of which representation for the potential energy is used.

1. Fill in the details in the derivation of Eqns. (5.109), (5.111) and (5.113). In each case start with the appropriate potential energy, $\Phi_{(k)}$, given in the text and follow the outlined procedure to obtain the force.
2. Repeat the above derivations using a more efficient representation in terms of just $N(N-1)/2$ distances, and show that the same results are obtained. For example, for a pair potential, show the same result is obtained for

$$\Phi_{(2)} = \frac{1}{2}\sum_{\substack{\alpha,\beta \\ \alpha\neq\beta}} \phi_{\alpha\beta}(r^{\alpha\beta}) \qquad \text{and} \qquad \Phi_{(2)} = \sum_{\substack{\alpha,\beta \\ \alpha<\beta}} \phi_{\alpha\beta}(r^{\alpha\beta}).$$

Do the same for the three-body and four-body terms. **Hint:** The key is to use the symmetric partial derivative expression in Eqn. (5.14).

5.16 [SECTION 5.8] Show that the general expression for the force of an n-body potential is given by Eqn. (5.114).

5.17 [SECTION 5.8] Derive the magnitude of the interatomic force, $\varphi^{\alpha\beta}$, defined in Eqn. (5.107), for the following potentials:

1. the Lennard-Jones potential defined in Eqn. (5.36);
2. the Morse potential in Eqn. (5.38);
3. Johnson's nearest-neighbor EAM potential given in Exercise 5.7;
4. the Stillinger–Weber potential for silicon given in Section 5.4.4.

6 Molecular statics

Chapters 4 and 5 were essentially about ways to estimate the potential energy of configurations of atoms. In this chapter, we discuss how we can use these potential energy landscapes to understand phenomena in materials science. Generally, this is done by studying key features in the potential energy landscape (local minima and the transition paths between them) using the methods of molecular statics (MS). After discussing some of the details of implementing MS algorithms, we will turn to several example applications in crystalline materials in Section 6.5.

6.1 The potential energy landscape

Using quantum mechanics, density functional theory (DFT) or tight-binding (TB) we are able to compute the energy of the electrons given the fixed positions of the atomic nuclei, and add this to the Coulomb interactions between the nuclei to get the total potential energy. In developing empirical methods, we approximated this electronic energy as a potential energy function dependent on only the interatomic distances. In either case, we are able to compute the potential energy, $\mathcal{V} = \mathcal{V}(\boldsymbol{r})$, of any arbitrary configuration of N atoms, with positions $\boldsymbol{r} = (\boldsymbol{r}^1, \dots, \boldsymbol{r}^N)$. We refer to the set of all possible coordinates $\{\boldsymbol{r}^\alpha\}$ as the *configuration space* of our system. Much of this configuration space is likely to be unphysical or at least impractical; we can create virtual atomic configurations on a computer that are very unlikely ever to occur in nature. On the other hand, there are significant portions of the phase space that tell us a great deal about how materials will behave.

It is not possible to plot the potential energy function for a system of more than one atom. However, it is still useful to think of it as a $3N$-dimensional *potential energy landscape*, with topography analogous to the mountains, craters, valleys and mountain passes familiar to us from our terrestrial landscape. In particular, much can be learned about the predictions of a potential energy function from the extremal points (maxima, minima and saddle points) of this landscape.

Given any point $\boldsymbol{r}_0$ in configuration space, one can (almost always[1]) associate with it a unique minimum on the potential energy surface by performing a steepest descent (SD) minimization (as described in Section 6.2.3). Mathematically, this corresponds to solving

$$\dot{\boldsymbol{r}} = -\nabla_{\boldsymbol{r}} \mathcal{V}, \tag{6.1}$$

[1] The exceptions form a set of zero measure [SW82].

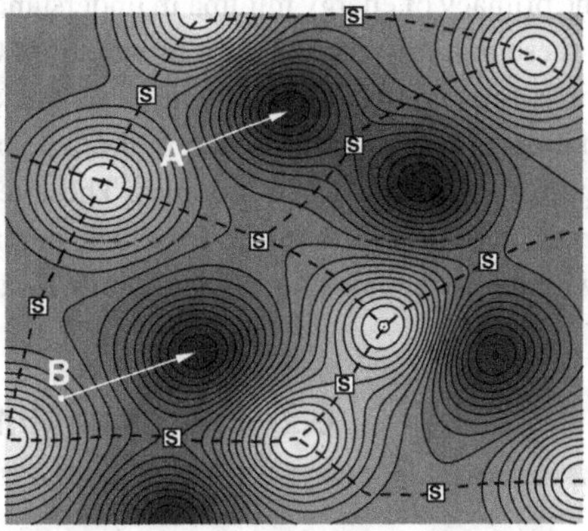

Fig. 6.1 A schematic representation of a $3N$-dimensional energy landscape. Minima are the darkest regions, maxima the lightest regions and saddle points are indicated by "s". The solid lines correspond to contours of constant energy. The dashed lines are the "dividing surfaces" separating the different minima. The region enclosed by dashed lines around each minimum is its "basin of attraction." (Adapted from [SW84].)

starting from the initial guess $\boldsymbol{r} = \boldsymbol{r}_0$. Physically, this corresponds to an infinitely rapid quench at zero temperature [SW82]. Configuration space is divided in this way into disjoint regions, each of which corresponds to the set of all points that quench to the same minimum. These regions are referred to as *basins of attraction* (or "basins" for short). They are separated by *dividing surfaces* and connected by saddle points. This is depicted schematically in Fig. 6.1.

The deepest minima on the energy landscape are associated with defect-free crystalline structures.[2] Minima with somewhat higher energies are associated with crystals containing internal defects, such as vacancies, interstitials, dislocations, grain boundaries and so on. Next, come a host of amorphous structures and finite clusters of every conceivable size and configuration. Just how many minima are there on this rugged landscape? A rough estimate for a system containing only one atomic species gives the number of minima as $N! \exp(\nu N)$, where ν is some positive number [SW83]. The first term counts the number of ways a given minimum energy structure can be made with N indistinguishable particles. It accounts for the permutation symmetry of the atoms. The second term counts the number of distinct structural minima. It is based on the assumption that nearby minima are related through local rearrangements and therefore the total number of such minima is exponential in N [SW82]. Using this relation, Stillinger and Weber [SW84] estimated that 1 gram of argon would have of the order of $10^{10^{22}}$ distinct structural minima!

The minima and the basins around them are perhaps the most physically relevant of the features in $\mathcal{V}(\boldsymbol{r})$. These show us the energetically stable configurations that the system is likely to adopt and how these atomic arrangements will respond to small perturbations. The curvature of the energy function around a minimum determines such things as the elastic moduli of a solid.

[2] Surprisingly, perhaps, it is still not completely clear why the most stable states are associated with crystalline structures. The existence of crystals is clear from experiments, however, a theoretical explanation for this fact is still lacking [Rad87, BL05].

Given the primacy of energy minima in understanding materials, it is no surprise that the search for these minima receives a lot of attention. One key type of MS simulation refers to this search for minima in the atomic potential energy landscape. Another important implementation of MS uses methods to find saddle points and *transition paths* (the lowest-energy trajectory connecting two minima). These tell us something about the likely ways that materials deform and the rates at which deformation takes place. In the next few sections, we explore methods for finding minima, saddle points, and transition paths.

6.2 Energy minimization

The search for the minima of a nonconvex, multi-dimensional function is one of the great challenges of computational mathematics. It is interesting that the human eye and brain can look at a hilly landscape and almost immediately find the point of lowest elevation, in addition to establishing where most of the other local minima lie. To do the same with a high-dimensional mathematical function on a computer is much more difficult. Although finding *any minimum* is not too difficult, finding it *quickly* is a bit more of a challenge, and finding the *global* minimum confidently and quickly is still an open field of research. The entire branch of numerical mathematics known as "optimization" is essentially dedicated to these goals. It is not our intention to discuss nonlinear optimization exhaustively. Rather, we discuss the theory and implementation of the most common workhorses of the field: the steepest descent (SD), conjugate gradient (CG) and Newton–Raphson (NR) methods. The ideas behind these methods serve as the basis for more advanced optimization approaches. Entire textbooks on optimization exist (see, for example [Pol71, Rus06]), and the interested reader may find more rigorous detail and advanced topics there.

The CG and NR methods are also classified as "local" minimization methods in the sense that they will only find *some "nearby"* energy minimum; there is no guarantee that this is the global minimum or even the closest energy minimum to the starting configuration. For atomistic systems, there are also *global optimization* techniques such as *simulated annealing*. Simulated annealing uses so-called *Monte Carlo sampling* to randomly change the configuration of atoms in order to probe new regions of phase space that might be closer to the global minimum. Acceptance of such a stochastic change is probabilistic, with higher-energy configurations being more likely to be rejected. We will not pursue global optimization methods in this book.

In MS, it is usual to refer to the "unrelaxed" and the "relaxed" structure. The unrelaxed structure is the starting configuration or the initial guess for the equilibrium structure. Running the algorithm to search for a minimum is the "relaxation" process that takes us to the local minimum, where we find the relaxed structure.

6.2.1 Solving nonlinear problems: initial guesses

All the methods we describe seek to minimize some energy function that we will generically call $\mathcal{V}(\mathbf{u})$, where $\mathbf{u}$ represents an n_{dof}-dimensional column matrix of variables (constituting the degrees of freedom of the problem) upon which the energy depends. In the context of atomistic methods, $\mathbf{u}$ typically represents the set of atomic coordinates, but it can also be more generally extended to include variables that impose constraints on the solution. We will refer to $\mathbf{u}$ as the "coordinates" or the "configuration" for short. Our methods proceed by evolving the system from some initial configuration $\mathbf{u}^{(0)}$ (with energy $\mathcal{V}(\mathbf{u}^{(0)})$) to the configuration $\mathbf{u}_{\min}$ that minimizes $\mathcal{V}$.

Figure 6.1 shows an energy landscape for a system with $n_{\text{dof}} = 2$, $[\mathbf{u}] = [X, Y]$ and several local minima. A minimization method invariably converges to different minima depending on the initial guess $\mathbf{u}^{(0)}$. In Fig. 6.1, for instance, starting from point "A" is likely to take the system to a different minimum than starting from point "B."

In physical terms, the need for an initial guess, and the dependence of the solution on that guess may or may not be problematic. Sometimes, we may not even be interested in the true global minimum, but rather a local minimum that is nearby some physically motivated starting point for the system. For example, in solids, the global minimum is almost always a perfect crystal. But we are typically more interested in defect structures: dislocations, vacancies and grain boundaries. An initial guess close to the defective structure will hopefully take us to a nearby local minimum which is physically relevant, but we can rarely be absolutely sure. Rerunning a simulation from multiple starting configurations is an inelegant but often unavoidable technique for testing the robustness of a particular solution. For many problems it is not even clear what the initial guess should be, and the dependence of the solution on this arbitrary choice is disconcerting.

6.2.2 The generic nonlinear minimization algorithm

Given an energy function $\mathcal{V}(\mathbf{u})$, we seek a configuration that is a local energy minimum:

$$\mathbf{u}_{\min} = \underset{\mathbf{u} \in \mathcal{C}^{(0)}}{\arg\min}\, \mathcal{V}(\mathbf{u}), \tag{6.2}$$

where $\mathcal{C}^{(0)}$ is a set of configurations in the vicinity of the initial guess $\mathbf{u}^{(0)}$. For most atomistic systems, we can compute the gradient (forces) $\mathbf{f}(\mathbf{u}) = -\nabla_{\mathbf{u}}\mathcal{V}(\mathbf{u})$, as well.[3] Since at the minimum the forces on all the degrees of freedom are zero,[4] we can test for convergence using a prescribed tolerance on the force norm

$$\text{if } \|\mathbf{f}(\mathbf{u})\| < \text{tol}, \qquad \mathbf{u}_{\min} := \mathbf{u}. \tag{6.3}$$

The notation ":=" denotes "is assigned the value" to distinguish it from an equality.

If the forces are nonzero, we can lower the energy by iteratively moving the system along some *search direction*, $\mathbf{d}$, that is determined from the energy and forces at the current and possibly past configurations visited during the minimization. As such, all the minimization methods we will discuss are based on the simple algorithm presented in Algorithm 6.1. The methods differ principally in how they determine the search direction, $\mathbf{d}$ at line 4 and

[3] For some of the more complex models, like ReaxFF (Section 5.6.3) or MEAM (Section 5.6.4), the force expressions require a lot of perseverance to derive and code, but it can be done.

[4] Of course, forces are zero at maxima and saddle points as well, so care must be taken that the algorithm avoids these points.

the step size α at line 5. In this book, we discuss two types of methods. The first type, embodied in the SD and CG methods, works using only information about the energy $\mathcal{V}$ and the forces. The second type, referred to as the NR method and quasi-Newton methods, utilizes the *stiffness* matrix or Hessian of the system (i.e. the second derivative of the energy with respect to the independent variables) or an approximation to it. For the atomistic systems of interest here, the extra complexity of the second derivatives often overshadows their benefits, and methods based on first derivatives are more common. However, for the finite element method,[5] the Newton methods are the most popular, because they converge more rapidly and the structure of a finite element model is such that the stiffness matrix is relatively easy to compute and store. Later, we will discuss multiscale methods that combine atomistic and finite element techniques. In these models, both first and second gradient methods are common.

Algorithm 6.1 A generic minimization algorithm

1: $n := 0$
2: $\mathbf{f}^{(0)} := -\nabla_{\mathbf{u}}\mathcal{V}(\mathbf{u}^{(0)})$
3: **while** $\left\|\mathbf{f}^{(n)}\right\| > \text{tol}$ **do**
4:     find the search direction $\mathbf{d}^{(n)}$
5:     find step size $\alpha^{(n)} > 0$
6:     $\mathbf{u}^{(n+1)} := \mathbf{u}^{(n)} + \alpha^{(n)}\mathbf{d}^{(n)}$
7:     $\mathbf{f}^{(n+1)} := -\nabla_{\mathbf{u}}\mathcal{V}(\mathbf{u}^{(n+1)})$
8:     $n := n + 1$
9: **end while**
10: $\mathbf{u}_{\min} := \mathbf{u}^{(n)}$

6.2.3 The steepest descent (SD) method

The SD method is generally an inefficient approach to finding a local minimum, but it has several advantages. First and foremost, it is a very simple algorithm to code without error. If one is more concerned with reliability than speed (or if one wants to do as little coding and debugging as possible) it is a good choice. Second, the SD trajectory followed in going from the initial configuration to the minimized state has a clear physical interpretation as an overdamped dynamic system. This can be important when one is actually interested in entire pathways in configurational space, as opposed to just the minimum (see Section 6.3). Finally, the SD method is pedagogically useful as an introduction to energy minimization. Once you understand the SD method, you are equipped to understand the more complicated methods discussed later.

As the name "steepest descent" suggests, the idea is simply to choose the search direction at each iteration to be along the direction of the forces. This corresponds to the steepest

[5] The finite element method is reviewed briefly in Section 12.1 and discussed in more detail in the companion volume to this book [TME12].

"downhill" direction at that particular point in the energy landscape. Referring to the generic minimization method in Algorithm 6.1, line 4 becomes

$$\mathbf{d} \equiv \mathbf{f} \quad \text{for steepest descent.}$$

In the absolutely simplest implementation of the SD method, the step-size α may be prescribed to be some fixed, small value, although a check needs to be made to ensure that taking the full step does not lead to an increase in energy (something that could happen if $\mathbf{u}$ is already near the minimum and the full step $\alpha\mathbf{f}$ overshoots it). In more sophisticated implementations, the system is moved some variable amount α along the direction of the forces until the one-dimensional minimum along that direction is found. In other words, the multi-dimensional minimization problem is replaced by a series of constrained one-dimensional minimizations. Details of this *line minimization* process are discussed below, for now we note the essential idea: for a fixed $\mathbf{u}$ and $\mathbf{d}$ we seek a positive real number α such that $\mathcal{V}(\mathbf{u} + \alpha\mathbf{d})$ is minimized with respect to α, and then update the system as

$$\mathbf{u} := \mathbf{u} + \alpha\mathbf{d}.$$

The new $\mathbf{u}$ is used to compute a new force, and the process repeats until the force norm is below the set tolerance. The SD method is summarized in Algorithm 6.2.

Algorithm 6.2 The SD Algorithm

$n := 0$
$\mathbf{f}^{(0)} := -\nabla_{\mathbf{u}}\mathcal{V}(\mathbf{u}^{(0)})$
while $\left\|\mathbf{f}^{(n)}\right\| > \text{tol}$ **do**
    $\mathbf{d}^{(n)} = \mathbf{f}^{(n)}$
    find $\alpha^{(n)} > 0$ such that $\phi(\alpha^{(n)}) \equiv \mathcal{V}(\mathbf{u}^{(n)} + \alpha^{(n)}\mathbf{d}^{(n)})$ is minimized
    $\mathbf{u}^{(n+1)} := \mathbf{u}^{(n)} + \alpha^{(n)}\mathbf{d}^{(n)}$
    $\mathbf{f}^{(n+1)} := -\nabla_{\mathbf{u}}\mathcal{V}(\mathbf{u}^{(n+1)})$
    $n := n + 1$
end while
$\mathbf{u}_{\text{min}} := \mathbf{u}^{(n)}$

The SD algorithm is an "intuitive" one: from where you are, move in the local direction of steepest descent, determine the new direction of steepest descent and repeat. It is not especially fast, however, because for many landscapes the most direct route to the minimum is not in the direction of steepest descent.[6] Fortunately, there is a reasonably straightforward modification to the SD approach, in the form of the CG method, that significantly improves the rate of convergence. We discuss this approach in Section 6.2.5, after fleshing out the details of the line minimization that is germane to all of these algorithms.

[6] Consider a long narrow trench dug straight down the side of a mountain. At a short distance up either side of the trench, the steepest descent direction is back towards the trench bottom, which is almost at right angles to the "global" downhill direction taking us down the mountain. Using the SD algorithm results in many short hops back and forth across the trench floor, gradually moving us down the mountain.

6.2.4 Line minimization

Most multi-dimensional minimization algorithms are carried out by a series of one-dimensional constrained minimizations (for example, see line 5 of Algorithm 6.2). Since it is used many times, the efficiency of the *line minimization* (or *line search*) is important.

Line minimization is an interesting area of computational mathematics because we can actually gain overall efficiency in this part of the algorithm through sloppiness; it is not necessary to find the line minimum exactly, so long as each line minimization does a reasonable job of lowering the energy of the system. In other words, we would like to replace line 5 of Algorithm 6.2 with

$$\text{find } \alpha^{(n)} > 0 \text{ such that } \phi(\alpha^{(n)}) \equiv \mathcal{V}(\mathbf{u}^{(n)} + \alpha^{(n)}\mathbf{d}^{(n)}) \text{ is } \textit{sufficiently reduced.}$$

If we can quantify the term "sufficiently reduced" we can avoid wasting time on polishing our estimate of $\alpha^{(n)}$ when starting along a new search direction.

There are a number of approaches to this, but a successful one (described in more detail in [NW99]) is a combination of backtracking and the so-called "sufficient decrease" condition,[7] as follows. Given

$$\phi(\alpha) \equiv \mathcal{V}(\mathbf{u} + \alpha\mathbf{d}),$$

we must first choose some sensible initial guess for minimizing α. This is tricky, since $\mathbf{d}$ often has units that are different from $\mathbf{u}$. In the SD algorithm, for example, this means that α will have units of length/force and its scale will be problem-dependent. One approach is to choose some arbitrary step in α and march along $\mathbf{d}$ until we find two points such that $0 < \alpha_1 < \alpha_2$ and

$$\phi(0) > \phi(\alpha_1) < \phi(\alpha_2), \tag{6.4}$$

so that there must be a minimum in the interval $(0, \alpha_2)$. We can then approximate the function ϕ as a parabola passing through $\phi(0)$, $\phi(\alpha_1)$ and $\phi(\alpha_2)$, and through simple algebra arrive at the minimum of the parabola at α_p:

$$\alpha_p = \frac{\phi(0)(\alpha_2^2 - \alpha_1^2) - \phi(\alpha_1)\alpha_2^2 + \phi(\alpha_2)\alpha_1^2}{2\left(\phi(0)(\alpha_2 - \alpha_1) - \phi(\alpha_1)\alpha_2 + \phi(\alpha_2)\alpha_1\right)}. \tag{6.5}$$

Now, we can make $\alpha := \alpha_p$ our initial guess and determine whether $\phi(\alpha)$ is sufficiently decreased compared with $\phi(0)$. This condition requires α to satisfy

$$\phi(\alpha) \le \phi(0) - c_1\alpha\mathbf{f}(\mathbf{u}) \cdot \mathbf{d}(\mathbf{u}),$$

for some value of $c_1 \in (0, 1)$. Note that $-\mathbf{f}(\mathbf{u}) \cdot \mathbf{d}(\mathbf{u}) = \phi'(0)$, i.e. this is equivalent to

$$\phi(\alpha) \le \phi(0) + c_1\alpha\phi'(0),$$

and as such it is just a way to estimate the expected decrease in ϕ based on the slope at $\alpha = 0$. When $c_1 = 1$, the last term is exactly the expected decrease in the energy based on a linear interpolation from the point $\mathbf{u}$, and this limits α to very small values as shown in Fig. 6.2. Typically, a value of c_1 on the order of 10^{-4} is chosen. Figure 6.2 shows how

[7] The sufficient decrease condition, incidentally, makes up part of the so-called "Wolfe conditions" described in more detail in [NW99].

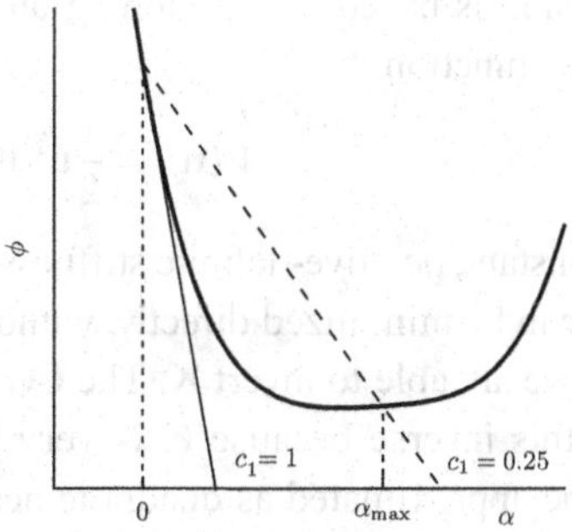

Fig. 6.2 The sufficient reduction condition determines a maximum value for α. In this case, it is pictured for a value of $c_1 = 0.25$. The region where the function ϕ is less than the dashed line is the range of acceptable values for α.

this condition imposes a *maximum* value on α, and provides a way to decide when to quit searching along a particular direction $\mathbf{d}$, as outlined in Algorithm 6.3. This algorithm would replace, for example, line 5 in Algorithm 6.2.

Algorithm 6.3 Line minimization using quadratic interpolation

---

choose ρ, such that $0 < \rho < 1$, and a tolerance tol
find $0 < \alpha_1 < \alpha_2$, such that $\phi(0) > \phi(\alpha_1) < \phi(\alpha_2)$
find α_p using Eqn. (6.5).
$\alpha := \alpha_p$
while $\phi(\alpha) > \phi(0) - c_1 \alpha \mathbf{f}(\mathbf{u}) \cdot \mathbf{d}(\mathbf{u})$ **do**
    $\alpha := \rho\alpha$
    if $\alpha \leq$ tol **then**
        exit with error code {Line minimization has failed.}
    end if
end while
$\alpha_{\text{min}} := \alpha$

In Algorithm 6.3, ρ is a scaling factor $\rho \in (0, 1)$, typically chosen on the order of $\rho = 0.5$, whose role is to systematically backtrack along the minimization line to find α. More sophisticated algorithms are discussed in [NW99], but are beyond the scope of our present discussion.

6.2.5 The conjugate gradient (CG) method

While the SD method is simple to understand, it is generally quite slow to converge. The underpinnings of the CG method are more difficult to understand, but its implementation is relatively straightforward and it is considerably more efficient than the SD method. The CG method is one of the most widely used energy minimization methods in MS. It is particularly well suited to molecular systems, probably because such systems lead naturally to a well-conditioned stiffness matrix (we will return to this shortly), atomic forces are usually readily available and further because the CG method only requires $O(n_{\text{dof}})$ storage as opposed to $O(n_{\text{dof}}^2)$ for the Newton and quasi-Newton methods discussed later.

The CG method is based on developing an optimal method for minimizing a general quadratic energy function

$$\mathcal{V}(\mathbf{u}) = \frac{1}{2}\mathbf{u}^T\mathbf{K}\mathbf{u} + \mathbf{c}^T\mathbf{u},$$

where $\mathbf{K}$ is a constant, positive-definite stiffness matrix and $\mathbf{c}$ is a constant vector. Of course such a system can be minimized directly, without the need for an iterative method like the CG method, if we are able to invert $\mathbf{K}$. The CG method is useful in cases where we cannot afford to find this inverse because $\mathbf{K}$ is very large. It is also useful since most general functions can be approximated as quadratic near their minima, and so the CG framework is a good scheme for tackling the iterative minimization of more general functions.

Although we will skip the details here, one can show [Pol71] that for any $n_{\text{dof}} \times n_{\text{dof}}$ matrix $\mathbf{K}$, one can choose an optimal set of search directions such that the minimum will be reached in at most n_{dof} line minimizations.[8] The savings over the SD method are significant; referring back to the trench analogy in the footnote on page 309, the trip down the mountain would be completed in only two steps! The set of optimal directions are the so-called *conjugate* directions, a set of n_{dof} nonzero, linearly independent vectors satisfying

$$\mathbf{d}_k^T\mathbf{K}\mathbf{d}_j = 0 \quad \text{for all } k \neq j. \tag{6.6}$$

Intuitively, perhaps, it seems that each new search direction should somehow "avoid" previous search directions, to avoid traversing the same regions of configuration space more than once. The orthogonality condition of Eqn. (6.6) ensures this optimal series of search directions. Further, it is possible to construct these directions iteration by iteration, with each new search direction built from the last one. This means that it is not necessary to perform any direct manipulations of $\mathbf{K}$ or even to know $\mathbf{K}$ at all (which is especially important in the case of general nonquadratic functions). Algorithm 6.4 summarizes the CG method.

The key difference between the CG and SD methods is the modification of the search directions using β in lines 7–8 (setting $\beta = 0$ recovers the SD method). The fact that each search direction is orthogonal to all previous directions can be shown for a quadratic energy function, but the details are beyond the scope of this text (see, for example, [Pol71, Rus06]).

6.2.6 The condition number

Of course, an energy function that is sufficiently smooth around a certain configuration $\mathbf{u}_0$ can always be approximated by a Taylor series to second order:

$$\mathcal{V}(\mathbf{u}) \approx \frac{1}{2}(\Delta\mathbf{u})^T\mathbf{K}\Delta\mathbf{u} - \mathbf{f}^T\Delta\mathbf{u} + \mathcal{V}(\mathbf{u}_0), \tag{6.7}$$

where $\Delta\mathbf{u} = \mathbf{u} - \mathbf{u}_0$ and

$$\mathbf{f} = -\left.\frac{\partial\mathcal{V}(\mathbf{u})}{\partial\mathbf{u}}\right|_{\mathbf{u}=\mathbf{u}_0}, \qquad \mathbf{K} = \left.\frac{\partial^2\mathcal{V}(\mathbf{u})}{\partial\mathbf{u}\partial\mathbf{u}}\right|_{\mathbf{u}=\mathbf{u}_0}.$$

[8] Notice also that each line minimization can be solved exactly in one step since we know that the function is quadratic.

Algorithm 6.4 The CG algorithm

$n := 0$
$\mathbf{f}^{(0)} := -\nabla_{\mathbf{u}}\mathcal{V}(\mathbf{u}^{(0)})$
while $\left\|\mathbf{f}^{(n)}\right\| >$ tol **do**
  if $n = 0$ **then**
    $\mathbf{d}^{(n)} = \mathbf{f}^{(n)}$
  else
    $\beta^{(n)} = (\mathbf{f}^{(n)} \cdot \mathbf{f}^{(n)} - \mathbf{f}^{(n)} \cdot \mathbf{f}^{(n-1)}) / \left\|\mathbf{f}^{(n-1)}\right\|^2$
    $\mathbf{d}^{(n)} = \mathbf{f}^{(n)} + \beta^{(n)}\mathbf{d}^{(n-1)}$
  end if
  find $\alpha^{(n)} > 0$ using line minimization (Algorithm 6.3)
  $\mathbf{u}^{(n+1)} := \mathbf{u}^{(n)} + \alpha^{(n)}\mathbf{d}^{(n)}$
  $\mathbf{f}^{(n+1)} := -\nabla_{\mathbf{u}}\mathcal{V}(\mathbf{u}^{(n+1)})$
  $n := n + 1$
end while
$\mathbf{u}_{\min} := \mathbf{u}^{(n)}$

The Hessian or stiffness matrix K can be used to determine the convergence properties of the CG method. The ratio of the largest to the smallest eigenvalues of K is called the *condition number*[9] of the matrix:

$$\aleph = \frac{\lambda_{\max}}{\lambda_{\min}}. \tag{6.8}$$

For large values of $\aleph$, the CG method converges more slowly than for $\aleph \approx 1$. There is a large body of optimization literature focused on so-called "pre-conditioning" techniques, whereby the function to be minimized is transformed in such a way as to improve the conditioning of the system (see the on-line article by Shewchuk [She94] for a good introduction to pre-conditioning, as well as to the CG method in general).

It is well known that, *in general*, the CG algorithm is not very efficient without pre-conditioning. Despite this, many MS codes do not employ pre-conditioners, and yet remain stubbornly efficient. This is likely related to the physical nature of the systems being studied using MS. The eigenvalues of the stiffness matrix, K, are related to the stiffness of the interatomic bonds in the system, and, as long as the interatomic force models and the initial atomic coordinates are reasonable, it is perhaps no surprise that the ratio of the "stiffest" bond to the "softest" bond in the system is only, at worst, a couple of orders of magnitude.

6.2.7 The Newton–Raphson (NR) method

Suppose again that the energy is a simple quadratic function of the coordinates

$$\mathcal{V}(\mathbf{u}) = \frac{1}{2}\mathbf{u}^T\mathsf{K}\mathbf{u} - \mathbf{f}^T\mathbf{u}, \tag{6.9}$$

[9] This definition assumes a positive-definite K so that $\aleph \geq 1$.

where $\mathbf{K}$ is a constant, positive-definite matrix and $\mathbf{f}$ is a constant vector. Stationarization of this function amounts to finding $\mathbf{u}$ such that

$$\nabla_{\mathbf{u}}\mathcal{V} = \mathbf{K}\mathbf{u} - \mathbf{f} = \mathbf{0}.$$

If we are willing to invert the stiffness matrix, we can solve this directly:

$$\mathbf{u} = \mathbf{K}^{-1}\mathbf{f}.$$

The NR method applies this same approach iteratively to more general functions. We start from the Taylor expansion of the energy about the current guess to the configuration, $\mathbf{u}^{(n)}$:

$$\mathcal{V}(\mathbf{u}) \approx \frac{1}{2}(\mathbf{u} - \mathbf{u}^{(n)})^T\mathbf{K}^{(n)}(\mathbf{u} - \mathbf{u}^{(n)}) - (\mathbf{f}^{(n)})^T(\mathbf{u} - \mathbf{u}^{(n)}) + \mathcal{V}(\mathbf{u}^{(n)}),$$

where as usual

$$\mathbf{f}^{(n)} = -\left.\frac{\partial\mathcal{V}(\mathbf{u})}{\partial\mathbf{u}}\right|_{\mathbf{u}=\mathbf{u}^{(n)}}, \qquad \mathbf{K}^{(n)} = \left.\frac{\partial^2\mathcal{V}(\mathbf{u})}{\partial\mathbf{u}\partial\mathbf{u}}\right|_{\mathbf{u}=\mathbf{u}^{(n)}},$$

so that

$$\nabla_{\mathbf{u}}\mathcal{V}(\mathbf{u}) \approx \mathbf{K}^{(n)}(\mathbf{u} - \mathbf{u}^{(n)}) - (\mathbf{f}^{(n)}). \tag{6.10}$$

Now instead of solving for the next approximation to $\mathbf{u}$ by setting the above expression to zero (which could result in convergence to a maximum or saddle point rather than a minimum), we search for a solution by heading in the direction of the minimum of this quadratic approximation to the real energy. Thus we set Eqn. (6.10) equal to zero to obtain the search direction as

$$\mathbf{d}^{(n)} \equiv \mathbf{u} - \mathbf{u}^{(n)} = (\mathbf{K}^{(n)})^{-1}\mathbf{f}^{(n)}, \tag{6.11}$$

and then move this system by a line minimization in the usual way

$$\mathbf{u}^{(n+1)} = \mathbf{u}^{(n)} + \alpha^{(n)}\mathbf{d}^{(n)}, \tag{6.12}$$

where $\alpha^{(n)} > 0$ is obtained from line minimization (Algorithm 6.3). This approach can fail when $\mathbf{K}^{(n)}$ is not positive definite, in which case $\mathbf{d}^{(n)}$ may not be a descent direction. In this case, one option is to abandon the NR method for the current step and set $\mathbf{d}^{(n)}$ to the SD direction. Alternatively, $\mathbf{K}^{(n)}$ can be modified in some way to force it to be positive definite (see for example [FF77]). The NR method is summarized in Algorithm 6.5.[10]

Clearly, the main extra cost in the NR method compared with the CG method is the need to compute, store and invert $\mathbf{K}$. The finite element method (FEM) (see Section 12.1) is particularly well suited to NR minimization, since the stiffness matrix takes on a relatively simple form that permits efficient storage and inversion. Essentially, the foundations of finite elements revolve around developing an efficient way to compute the stiffness matrix. For the atomic systems discussed in this chapter, building the stiffness matrix is usually a significant effort that tips the balance back in favor of the CG method.

[10] Strictly speaking this is a *modified Newton–Raphson* approach since the standard Newton–Raphson method does not involve a line minimization and is therefore a method for finding roots (i.e. stationary points of a function) and not a minimization approach. However, for simplicity we refer to the method as "Newton–Raphson".

Algorithm 6.5 The NR algorithm

$n := 0$
$\mathbf{f}^{(0)} := -\nabla_{\mathbf{u}}\mathcal{V}(\mathbf{u}^{(0)})$
$\mathbf{K}^{(0)} := \partial^2\mathcal{V}(\mathbf{u}^{(0)})/\partial\mathbf{u}\partial\mathbf{u}$
while $\left\|\mathbf{f}^{(n)}\right\| >$ tol **do**
  $\mathbf{d}^{(n)} := (\mathbf{K}^{(n)})^{-1}\mathbf{f}^{(n)}$
  find $\alpha^{(n)} > 0$ using line minimization (Algorithm 6.3). if the minimization fails, set $\mathbf{d}^{(n)} := \mathbf{f}^{(n)}$ and retry
  $\mathbf{u}^{(n+1)} := \mathbf{u}^{(n)} + \alpha^{(n)}\mathbf{d}^{(n)}$
  $\mathbf{f}^{(n+1)} := -\nabla_{\mathbf{u}}\mathcal{V}(\mathbf{u}^{(n+1)})$
  $\mathbf{K}^{(n+1)} := \partial^2\mathcal{V}(\mathbf{u}^{(n+1)})/\partial\mathbf{u}\partial\mathbf{u}$
  $n := n + 1$
end while
$\mathbf{u}_{\min} := \mathbf{u}^{(n)}$

The NR method is also generally better for the minimization of systems that are poorly conditioned, as it makes direct use of the stiffness matrix to determine the search directions. FEM is an important case where the condition number can be quite large due to wide variations in element sizes. In this case, the NR method is much more efficient than the CG algorithm without pre-conditioning. Similarly, multiscale methods (the subject of Chapters 12 and 13) can benefit from choosing the NR method rather than the CG method. This is because multiscale methods, by their nature, involve both large and small finite elements in addition to atomic bonds.

Quasi-Newton methods Often, one may want to use Eqn. (6.12), but it is too expensive or difficult to obtain and invert the Hessian matrix. There are several methods to produce approximations to $\mathbf{K}^{-1}$, or more generally to provide an algorithm for generating conjugate search directions of the form of a matrix multiplying the force vector. These methods are broadly classed as "quasi-Newton methods," and they can be advantageous for problems where the second derivatives required for the Hessian are sufficiently complex to make the code either tedious to implement or slow to execute.[11] For more details, the interested reader may try [Pol71, PTVF08, Rus06].

6.3 Methods for finding saddle points and transition paths

Chemical reactions, diffusion, dislocation motion and fracture are just a few important materials processes that can be viewed as pathways through the potential energy landscape.

[11] The more sophisticated of the quasi-Newton methods are amongst the fastest algorithms for finding minima. Wales [Wal03] argued that one such method in particular, Nocedal's limited memory Broyden–Fletcher–Goldfarb–Shanno (L-BFGS) method, is in fact currently the fastest method that can be applied to relatively large systems.

Specifically, each of these processes take us between two configurations of atoms that are local minima. The transition path between them can be important for our understanding of how these processes take place. It also tells us the activation energy of the process (which sets its rate as explained in Section 1.1.7 and is important for temporal multiscale methods as discussed in Section 10.4) and what intermediate equilibrium configurations may exist along the transition. Here, we briefly introduce the numerical methods used to determine minimum energy transition paths and saddle points along their length.

Mathematically speaking, local minima are much simpler entities than saddle points. Both have zero first derivatives, but whereas minima have positive curvature in all directions, saddle points have at least one direction along which the curvature of the energy landscape is negative. Most methods for finding saddle points depend on analyzing the Hessian matrix, which can be expensive for many systems.

Saddle point searches are usually based on methods of eigenvector-following (see [Wal03, Sect. 6.2.1] for a review of the literature), whereby search directions along eigenvectors of the Hessian with the lowest eigenvalues are chosen to systematically locate saddle points. In systems where the Hessian is too expensive to compute, there are methods whereby only the lowest eigenvalues need to be found, or where numerical approximations are used to estimate the search directions. Techniques such as that of Wales and coworkers [Wal94, WW96a, WW96b, MW99], the activation–relaxation technique (ART nouveau) of Barkema and Mousseau [BM96, MM00], and the dimer method of Henkelman and Jónsson [HJ99] are all methods based on eigenvector following.

Once a saddle point is determined, it is relatively simple to find transition paths through the saddle point by systematically perturbing the configuration and following SD pathways to the neighboring minima (see Section 6.2.3). It is really the search for the saddle points themselves that can be costly and difficult for complex energy landscapes involving many atoms. For some physical phenomena, we may already know (or be willing to guess) the two local minima at the start and end of a transition path. In this case, the search for the path and any intervening saddle point is simplified and can be carried out without any recourse to the Hessian matrix or its eigenvectors. In the next section, we elaborate on how this is done by describing the "nudged elastic band (NEB)" method.

6.3.1 The nudged elastic band (NEB) method

The NEB method was first proposed in the mid-1990s by Jónsson and coworkers [MJ94, MJS95]. Since then there have been a number of improvements and optimizations of the method [HJJ98, HJJ00]. This method can be traced back to the original "elastic band" approach of Elber and Karplus [EK87]. More recently, E *et al.* [ERVE02] have devised the "string method" in which NEB is reformulated as a continuous curve evolving according to a differential equation. Below, we describe the essential ideas of NEB, and encourage the reader to examine the literature for the more subtle details.

The discretized transition path First, we define a *replica*, ϱ, of a system of N atoms as a $3N$-dimensional point in its configuration space, $\varrho = (\boldsymbol{r}^1, \boldsymbol{r}^2, \ldots, \boldsymbol{r}^N)$. We can build a set of R such replicas, the first and last of which (ϱ^1 and ϱ^R) are at local energy minima

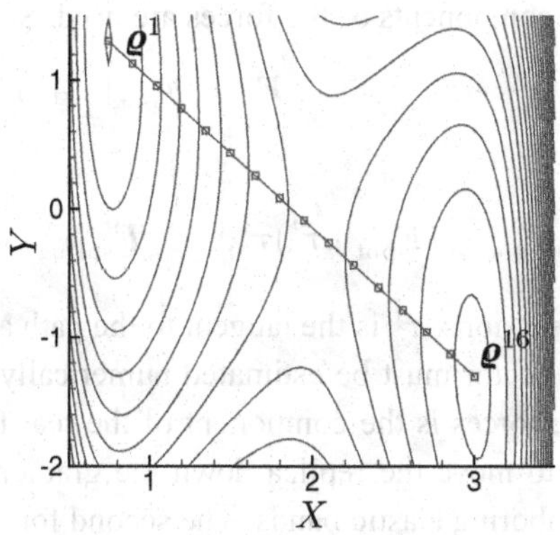

Fig. 6.3 Initial configuration of a string of replicas for the two-dimensional NEB example.

in the landscape. We can think of ϱ^1 as the "reactants" and ϱ^R as the "products" of a particular reaction or process in our system, and treat the intervening $R-2$ replicas as discrete degrees of freedom in our search; we will move these replicas until they form a discrete approximation to the true transition path.

Our initial guess for the series of R replicas might be as simple as a linear interpolation between the reactants and the products:

$$\varrho^i = \varrho^1 + \frac{i-1}{R-1}\left(\varrho^R - \varrho^1\right), \tag{6.13}$$

as illustrated in Fig. 6.3 with $R = 16$. Other than replicas ϱ^1 and ϱ^{16}, none of these are in equilibrium, and there are therefore forces on the atoms of each replica coming from the gradients of the potential energy

$$\boldsymbol{F}^i_{\text{pot}} = -\nabla_{\varrho}\mathcal{V}(\varrho^i) = \{-\nabla_{\boldsymbol{r}^1}\mathcal{V}|_{\varrho^i}, \ldots, -\nabla_{\boldsymbol{r}^N}\mathcal{V}|_{\varrho^i}\}. \tag{6.14}$$

Minimizing these forces would of course move each replica into one of the local minima, and would not help us to find the transition path. Instead, we fix the reactant and product replicas and imagine that the string of all replicas is joined by "elastic bands" with zero unstretched length and spring constants k. The springs constrain the replicas to remain spread out along the path. The force acting on replica i due to the spring is

$$\boldsymbol{F}^i_{\text{spring}} = k(\varrho^{i+1} - \varrho^i) - k(\varrho^i - \varrho^{i-1}), \qquad i = 2, \ldots, R-1, \tag{6.15}$$

where $\boldsymbol{F}^i_{\text{spring}}$ is a $3N$-dimensional vector of forces on all the atoms in replica ϱ^i.

One could now try to zero the combined forces $\boldsymbol{F}^i_{\text{pot}} + \boldsymbol{F}^i_{\text{spring}}$ on the intermediate replicas, $i = 2, \ldots, R-1$, to obtain a path between reactants and products. However, such an approach would in general not lead to the correct transition path and would be very sensitive to the choice of the spring constant, k. If k is very small (relative to some appropriate scale for $\mathcal{V}$), the spring forces will have little effect and the replicas will fall into the minima. If k is very large, the stiff elastic bands will force the path to cut corners and travel along a higher-energy part of the landscape. To avoid these problems, in NEB only certain components of the forces are used. Specifically, the force on replica ϱ^i is

$$\boldsymbol{F}^i = \boldsymbol{F}^i_{\mathrm{pot}}\big|_{\perp} + \boldsymbol{F}^i_{\mathrm{spring}}\big|_{\parallel}, \tag{6.16}$$

where

$$\boldsymbol{F}^i_{\mathrm{pot}}\big|_{\perp} = \boldsymbol{F}^i_{\mathrm{pot}} - (\boldsymbol{F}^i_{\mathrm{pot}} \cdot \widehat{\boldsymbol{\tau}}^i)\widehat{\boldsymbol{\tau}}^i, \qquad \boldsymbol{F}^i_{\mathrm{spring}}\big|_{\parallel} = k(\|\varrho^{i+1} - \varrho^i\| - \|\varrho^i - \varrho^{i-1}\|)\widehat{\boldsymbol{\tau}}^i.$$

In these expressions, $\widehat{\boldsymbol{\tau}}^i$ is the tangent to the path at replica i. Since the path is discretized, this tangent vector must be estimated numerically (as we will describe shortly). The first of the above forces is the component of the real forces that is perpendicular to the path; it will tend to move the replica down the gradient of the energy landscape but will not stretch neighboring elastic bands. The second force is an approximation to the component of the spring force acting parallel to the path tangent (it is exact only if the angle between $\varrho^{i+1} - \varrho^i$ and $\varrho^i - \varrho^{i-1}$ is zero). Even though the exact expression for this parallel force, $(\boldsymbol{F}^i_{\mathrm{spring}} \cdot \widehat{\boldsymbol{\tau}}^i)\widehat{\boldsymbol{\tau}}^i$, could easily be evaluated, the approximate form turns out to improve performance by keeping the distance between the replicas about equal.

Estimating the tangent vector The tangent vector to the path at each replica, $\widehat{\boldsymbol{\tau}}^i$, can be estimated by bisecting the two unit vectors along adjacent segments of the path

$$\boldsymbol{\tau}^i = \frac{\varrho^i - \varrho^{i-1}}{\|\varrho^i - \varrho^{i-1}\|} + \frac{\varrho^{i+1} - \varrho^i}{\|\varrho^{i+1} - \varrho^i\|}, \tag{6.17}$$

and then converting the result to a unit vector as $\widehat{\boldsymbol{\tau}}^i = \boldsymbol{\tau}^i / \|\boldsymbol{\tau}^i\|$. However, this estimate can lead to poor convergence if the transition path traverses an especially rugged region of the energy landscape. Henkelman and Jónsson [HJ00] described an improved tangent estimate that changes for each replica depending on the surrounding energy landscape. If the energy of replica ϱ^i lies between the energy of its two neighboring replicas, the tangent is taken as the vector from replica ϱ^i to the higher-energy neighbor:

$$\boldsymbol{\tau}^i = \begin{cases} \boldsymbol{\tau}^i_+, & \text{if } \mathcal{V}(\varrho^{i-1}) < \mathcal{V}(\varrho^i) < \mathcal{V}(\varrho^{i+1}), \\ \boldsymbol{\tau}^i_-, & \text{if } \mathcal{V}(\varrho^{i-1}) > \mathcal{V}(\varrho^i) > \mathcal{V}(\varrho^{i+1}), \end{cases} \tag{6.18}$$

where

$$\boldsymbol{\tau}^i_+ = \varrho^{i+1} - \varrho^i \qquad \text{and} \qquad \boldsymbol{\tau}^i_- = \varrho^i - \varrho^{i-1}. \tag{6.19}$$

On the other hand, if replica ϱ^i is a minimum or maximum relative to replicas ϱ^{i-1} and ϱ^{i+1}, then $\boldsymbol{\tau}^i$ is a weighted sum of the two vectors $\boldsymbol{\tau}^i_-$ and $\boldsymbol{\tau}^i_+$,

$$\boldsymbol{\tau}^i = \begin{cases} \boldsymbol{\tau}^i_+ \Delta\mathcal{V}^i_{\mathrm{max}} + \boldsymbol{\tau}^i_- \Delta\mathcal{V}^i_{\mathrm{min}}, & \text{if } \mathcal{V}(\varrho^{i-1}) < \mathcal{V}(\varrho^{i+1}), \\ \boldsymbol{\tau}^i_+ \Delta\mathcal{V}^i_{\mathrm{min}} + \boldsymbol{\tau}^i_- \Delta\mathcal{V}^i_{\mathrm{max}}, & \text{if } \mathcal{V}(\varrho^{i-1}) > \mathcal{V}(\varrho^{i+1}), \end{cases} \tag{6.20}$$

where

$$\Delta\mathcal{V}^i_{\mathrm{max}} = \max(|\mathcal{V}(\varrho^{i+1}) - \mathcal{V}(\varrho^i)|, |\mathcal{V}(\varrho^{i-1}) - \mathcal{V}(\varrho^i)|) \tag{6.21}$$

and

$$\Delta\mathcal{V}^i_{\mathrm{min}} = \min(|\mathcal{V}(\varrho^{i+1}) - \mathcal{V}(\varrho^i)|, |\mathcal{V}(\varrho^{i-1}) - \mathcal{V}(\varrho^i)|). \tag{6.22}$$

Finally, the unit tangent vector is determined as $\hat{\boldsymbol{\tau}}^i = \boldsymbol{\tau}^i / \|\boldsymbol{\tau}^i\|$. This procedure ensures a smooth transition in the tangent estimate and has been shown to improve convergence in rugged landscapes [HJ00].

Equilibrating the replica forces The transition path can now be estimated by running an algorithm that moves the replicas in configuration space until all the forces of Eqn. (6.16) are reduced to zero. This is not the same as a minimization algorithm, since there is no well-defined objective function (i.e. there is no total energy) to minimize. Instead, a routine such as so-called "quenched dynamics (QD)" can be used. Briefly, QD treats the forces computed here as physical forces in Newton's second law, such that they determine the instantaneous acceleration of each degree of freedom (in this case, the position of each atom in all of the replicas):

$$\boldsymbol{F}^i = \boldsymbol{M}\ddot{\boldsymbol{\varrho}}^i, \tag{6.23}$$

where $\boldsymbol{M}$ is an arbitrarily assigned mass matrix. The system is allowed to evolve according to this equation of motion, but with modifications made to the "velocity," $\dot{\boldsymbol{\varrho}}^i$, in order to systematically bleed energy from the system so that it settles into a minimum energy path. These modifications to remove energy are essential, since Eqn. (6.23) is a dynamic equation with no inherent damping; its predicted evolution would continually oscillate around the correct solution. Full details of QD are presented in Section 9.3.2.

An alternative method for finding equilibrium (zero-force) configurations in the absence of a well-defined total energy is the so-called "force-based conjugate gradient (CG-FB)" method. This uses the steps described in Section 6.2.5 to build a series of search directions, **d**, but because there is no energy function to minimize in this case, each line search is terminated by the condition that

$$|\boldsymbol{F} \cdot \mathbf{d}| < \text{tol}, \tag{6.24}$$

for some tolerance, tol. Specifically, on line 10 of Algorithm 6.4, we change the criterion for convergence of the line minimization so that a search direction is used until the force vector is perpendicular to the search direction. The procedure is exactly as in Algorithm 6.4, except, of course, that the forces are computed through some means other than the gradient of an energy at lines 2 and 12 and from Eqn. (6.24) line 10 becomes[12]

$$\text{find } \alpha^n > 0 \text{ such that } \mathbf{f}(\mathbf{u}^n + \alpha^n \mathbf{d}^n) \cdot \mathbf{d}^n = 0.$$

In the NEB example below,[13] we show that the CG-FB method suffers from instabilities if the system of equations for the forces has a "Hessian" which is not positive definite.[14] This can occur for the NEB method, as well as for force-based multiscale methods [DLO10a, DLO10b] (see also Section 12.5).

[12] The notation in Algorithm 6.4 is consistent with the previous discussion. Of course, we must use the correspondences between $\mathbf{f} \leftrightarrow \boldsymbol{F}$ and $\mathbf{u} \leftrightarrow \boldsymbol{\varrho}$ to put it into the context of NEB.

[13] The authors thank Tsvetanka Sendova for running the calculations for this example.

[14] Normally, the term "Hessian" refers to the matrix of second-order partial derivatives of a scalar function like the energy. Here, by "Hessian" we mean the matrix of first-order partial derivatives computed from the NEB forces, which themselves are *not* the derivatives of an energy. Such a Hessian may not be symmetric like the Hessian derived from an energy function. See Example 6.1 for more details.

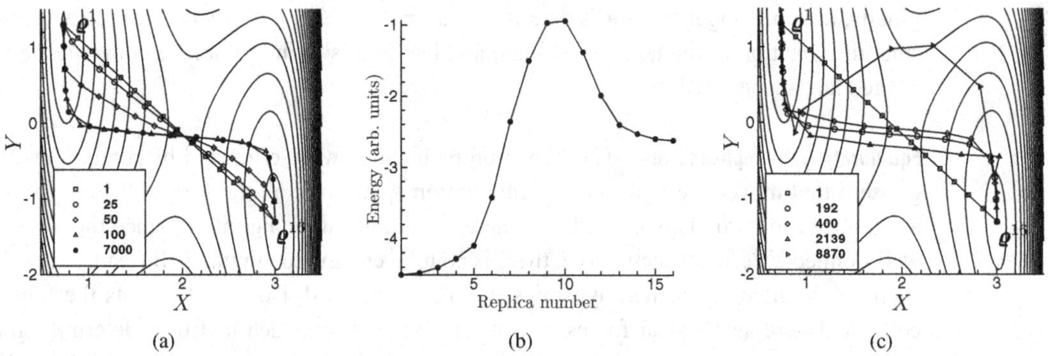

Fig. 6.4 (a) Evolution of the minimum energy path using the QD algorithm. The key indicates the iteration number for each path shown, with the filled circles indicating the final converged result. (b) The energy along the final converged path from (a). In (c), we see the unsatisfactory results using the CG-FB algorithm, where the solution diverges after initially approaching the correct path.

Example 6.1 (An NEB calculation in two dimensions) Consider the two-dimensional potential energy surface shown in Fig. 6.3 and originally given in [HJJ98]. This energy surface represents the bonding between a set of three atoms, confined to move along a line, and interacting harmonically with a fourth atom. See [HJJ98, Appendix A] for details of the potential energy function. The straight line connecting the two minima on the potential energy surface is the initial guess at the transition path containing 16 replicas (the 2 endpoints and 14 that are free to move).

Figure 6.4(a) shows the evolution of the path to the final, minimum energy path as obtained using the QD algorithm. In Fig. 6.4(b), the energy as a function of replica number shows the landscape along the minimum energy path. The saddle point (maximum energy along the transition path) is crossed at replica 10. The difference between this energy and that of one of the endpoints gives the energy barrier that dictates the transition rate according to transition state theory. Figure 6.4(c) shows the failure of the CG-FB routine to find the minimum in this case. The path approaches the correct solution, but then becomes unstable and diverges away from the solution. This behavior can be understood by studying the eigenmodes of the Hessian

$$\mathsf{K} = -\frac{\partial \boldsymbol{F}}{\partial \boldsymbol{\varrho}},$$

where we define the vectors $\boldsymbol{F} = (\boldsymbol{F}^2, \ldots, \boldsymbol{F}^{R-1})$ and $\boldsymbol{\varrho} = (\boldsymbol{\varrho}^2, \ldots, \boldsymbol{\varrho}^{R-1})$, and the negative sign is introduced to be consistent with previous definitions of the Hessian as the second derivative of the energy (the force is the *negative* of the gradient). Notice that we have omitted the fixed replicas 1 and R to eliminate eigenmodes associated with rigid motion of the chain of replicas.

Suppose we are at a solution $\boldsymbol{\varrho}_{\min}$ for which $\boldsymbol{F}_{\min} = \mathbf{0}$. Then the forces on the system due to any perturbation can be found as a Taylor series expansion to first order:

$$\boldsymbol{F}(\boldsymbol{\varrho}_{\min} + \Delta\boldsymbol{\varrho}) \approx -\mathsf{K}_{\min}\Delta\boldsymbol{\varrho}, \tag{6.25}$$

where $\mathsf{K}_{\min}$ is the Hessian evaluated at $\boldsymbol{\varrho} = \boldsymbol{\varrho}_{\min}$. If the Hessian has any eigenvalues for which the real part is negative, this means there is an eigenvector $\Delta\boldsymbol{\varrho}$ for which

$$\Delta\boldsymbol{\varrho} \cdot (\mathsf{K}_{\min}\Delta\boldsymbol{\varrho}) < 0,$$

and this further implies

$$\Delta \varrho \cdot F > 0$$

from Eqn. (6.25). This means that a perturbation from the solution in the direction of $\Delta \varrho$ will generate a force component in the same direction. This force will tend to push the system further from ϱ_{min}, leading to a divergence of the solution. Indeed, a numerical evaluation of the Hessian in the vicinity of the solution for this example reveals such unstable eigenmodes.

As in the case of energy minimization (discussed in Section 6.2.1), the NEB solution will depend on the initial guess. In practice, it is often necessary to rerun the NEB simulation with multiple initial configurations to test for alternative, lower-energy transition paths.

6.4 Implementing molecular statics

In this section, we discuss some of the "practitioner's points" related to implementing an MS solution algorithm.

6.4.1 Neighbor lists

For every atomistic model described in Chapter 5 (and, for that matter, for practical TB implementations), the potential energy can be written as a sum of terms that depend on the interactions of each atom in the system with its surrounding neighbor atoms. Practically speaking, there is a cutoff radius built into most of these interactions,[15] and so what are really needed are sums over only the atoms within r_{cut} of each other. Consider the simple pair potential for definiteness:

$$\mathcal{V}^{\text{int}} = \frac{1}{2} \sum_{\substack{\alpha,\beta \\ \alpha \neq \beta}} \phi_{\alpha\beta}(r^{\alpha\beta}).$$

A naïve approach would be to implement this double sum directly, as in Algorithm 6.6. The result of this calculation would be the total energy, as well as the force on every atom. But there are clearly a lot of inefficiencies in this routine. An obvious one is that we should compare squared distances $(r^{\alpha\beta})^2$ and $(r_{\text{cut}})^2$ and only take the square-root when it is needed at line 8. This avoids taking $O(N^2)$ unnecessary and expensive square roots. Another improvement is related to the force calculation at line 10, where the term being added on each pass through the loop is $\boldsymbol{f}^{\alpha\beta}$. We showed in Section 5.8.1 (see Eqn. (5.95)) that $\boldsymbol{f}^{\alpha\beta} = -\boldsymbol{f}^{\beta\alpha}$, and so we would be smart to also add appropriate contributions to the force on atom β at the same time. In this way, the inner loop on β could be from $\beta = \alpha + 1$ to N, the factor of $1/2$ on line 8 can be removed, and we simply add another calculation

[15] The exception is long-range Coulomb interactions in ionic solids, where the summation must employ special tricks like multipole methods to gain efficiency (see Section 5.4.3).

Algorithm 6.6 The N^2 neighbor search routine

1: Initialize $n := 0$
2: Initialize $\boldsymbol{f}^\alpha := \boldsymbol{0}, \ \forall\, \alpha$
3: **for** $\alpha = 1$ to N **do**
4:   **for** $\beta = 1$ to N **do**
5:     **if** $\beta \neq \alpha$ **then**
6:       Compute $r^{\alpha\beta}$
7:       **if** $r^{\alpha\beta} \leq r_{\text{cut}}$ **then**
8:         $\mathcal{V}^{\text{int}} := \mathcal{V}^{\text{int}} + \phi_{\alpha\beta}(r^{\alpha\beta})/2$
9:         **if** Forces are needed **then**
10:           $\boldsymbol{f}^\alpha := \boldsymbol{f}^\alpha + \phi'_{\alpha\beta}(r^{\alpha\beta})\boldsymbol{r}^{\alpha\beta}/r^{\alpha\beta}$
11:         **end if**
12:       **end if**
13:     **end if**
14:   **end for**
15: **end for**

at line 10 to take care of $\boldsymbol{f}^\beta$. This will increase the speed on the routine by a factor of 2. But there are much greater inefficiencies, mainly because as this routine is repeatedly called during a minimization procedure, or transition path calculation (or even dynamical evolution in MD), the positions of the atoms do not change very much. The chances are good that two atoms that were well out of range of each other in one iteration will remain out of range in the next, and we will spend a lot of time checking distances between atoms that are never going to interact. There are two methods that are generally used to address this: Verlet lists and binning.

Verlet neighbor lists The idea of the Verlet list method [Ver67] is to use Algorithm 6.6 occasionally (hopefully rarely) and store information about which atoms are neighbors during its execution. Subsequent energy or force calculations use the resultant neighbor lists, updating only the quantities $\boldsymbol{r}^{\alpha\beta}$ as necessary due to the motion of atoms at each iteration. Thus, for each atom α, there is stored a list of integers identifying its neighbors. We will need to update the neighbor list of atom α whenever any atom that is not currently in its neighbor list moves within r_{cut} of α. But how can we know when this will be?

The trick of the Verlet method is to store neighbors, not within r_{cut} of each atom but within some larger radius, r_{neigh}:

$$r_{\text{neigh}} = (1 + \epsilon_{\text{neigh}})r_{\text{cut}}, \tag{6.26}$$

where ϵ_{neigh} is typically on the order of 0.2. Then, when the energy and forces are computed we will waste a little time on the neighbors in the "padding" between r_{cut} and r_{neigh}, but considerably less time than if we visited every atom in the system. Now, a conservative approach to determining when we need to update the neighbor lists is to keep track of the two largest distances ($\delta_{\max,1}$ and $\delta_{\max,2}$) moved by any two atoms since the last neighbor

Algorithm 6.7 The Verlet neighbor search routine

1: First, the neighbor search, if necessary:
2: **if** $\delta_{\max,1} + \delta_{\max,2} > \epsilon_{\text{neigh}} r_{\text{cut}}$ **then**
3:   Store current atomic positions for future comparison.
4:   **for** $\alpha = 1$ to $(N-1)$ **do**
5:     $N^{\alpha}_{\text{neigh}} := 0$, $\mathcal{N}^{\alpha} := \emptyset$
6:     **for** $\beta = \alpha + 1$ to N **do**
7:       Compute $(r^{\alpha\beta})^2$
8:       **if** $(r^{\alpha\beta})^2 \leq r^2_{\text{neigh}}$ **then**
9:         $N^{\alpha}_{\text{neigh}} := N^{\alpha}_{\text{neigh}} + 1$
10:         Store β in the neighbor list of α, $\mathcal{N}^{\alpha} := \mathcal{N}^{\alpha} \cup \{\beta\}$
11:       **end if**
12:     **end for**
13:   **end for**
14: **end if**
15: Then, compute the energy and forces
16: **for** $\alpha = 1$ to $(N-1)$ **do**
17:   **for all** $\beta \in \mathcal{N}^{\alpha}$ **do**
18:     Compute $(r^{\alpha\beta})^2$
19:     **if** $(r^{\alpha\beta})^2 \leq r^2_{\text{cut}}$ **then**
20:       $\mathcal{V}^{\text{int}} := \mathcal{V}^{\text{int}} + \phi_{\alpha\beta}(r^{\alpha\beta})$
21:       **if** forces are needed **then**
22:         $\boldsymbol{f}^{\alpha} := \boldsymbol{f}^{\alpha} + \phi'_{\alpha\beta}(r^{\alpha\beta})\boldsymbol{r}^{\alpha\beta}/r^{\alpha\beta}$
23:         $\boldsymbol{f}^{\beta} := \boldsymbol{f}^{\beta} - \phi'_{\alpha\beta}(r^{\alpha\beta})\boldsymbol{r}^{\alpha\beta}/r^{\alpha\beta}$
24:       **end if**
25:     **end if**
26:   **end for**
27: **end for**

list computations.[16] If

$$\delta_{\max,1} + \delta_{\max,2} \leq \epsilon_{\text{neigh}} r_{\text{cut}}, \tag{6.27}$$

it is not possible that any neighbor list has changed in any consequential way; neighbors may have moved in or out of the padding, but not into the critical range of less than r_{cut}. If

$$\delta_{\max,1} + \delta_{\max,2} > \epsilon_{\text{neigh}} r_{\text{cut}}, \tag{6.28}$$

it is *possible* that a neighbor has moved into range, we update all the lists. Algorithm 6.7 summarizes the Verlet list algorithm, where we have also incorporated some of the other efficiency measures discussed in passing above. By line 14, the set of neighbors for each

[16] This requires us to store the positions of all atoms at the time of the last neighbor update, $\{\boldsymbol{r}^{\alpha}_{\text{save}}\}$, so that we can compare the current positions during each energy or force calculation to them. Thus, $\delta_{\max,1} = \max(\|\boldsymbol{r}^1 - \boldsymbol{r}^1_{\text{save}}\|, \dots, \|\boldsymbol{r}^N - \boldsymbol{r}^N_{\text{save}}\|)$, and $\delta_{\max,2}$ is computed in the same way after removing the term associated with $\delta_{\max,1}$ from the max function list.

atom α, numbering $N^{\alpha}_{\text{neigh}}$, has been stored in $\mathcal{N}^{\alpha}$. The second half of the algorithm uses those neighbor lists to compute the energy and forces.

Clearly, this is still quite a conservative approach. For one, we have essentially assumed that the two most mobile atoms have moved directly toward each other in Eqn. (6.28). Further, we have not checked whether the two most mobile atoms are in fact anywhere near each other. However, the practical reality of most MS runs on solids and liquids is that the simple criterion of Eqn. (6.28) reduces the number of neighbor list updates substantially because atoms do not move much within each iteration. More sophisticated criteria are often only marginally better, and may have requirements for additional calculation and storage that outweigh their benefit.

The improvement due to the Verlet approach is difficult to know precisely, as it depends on both how much the atoms move during minimization and the value of ϵ_{neigh}. Often, only a handful of Verlet lists need to be generated over thousands of force evaluations. In this case, the savings are clearly substantial. Consider a system of N atoms for which the minimization requires M force evaluations. The simple N^2 routine requires MN^2 evaluations, whereas the Verlet method with a single neighbor list update requires about $N^2 + (M-1)PN$ operations, where P is the average number of neighbors per atom (within r_{neigh}). Note that P is more or less fixed, and is typically much smaller than N for large systems of atoms. As such the Verlet approach is $O(N)$ with the occasional $O(N^2)$ calculation. For intermediate values of N on the order of a few thousand, Verlet lists are a good approach. However, for larger N, even occasionally having to compute $O(N^2)$ lists dominates the calculation. In this case, a binning approach is needed.

Binning Binning involves dividing the physical space containing the atoms into cubic bins of side-length r_{cut} along each coordinate direction. It is then clear that a given atom can only interact with atoms within its own bin or one of its 26 immediate neighbor bins. Much like the Verlet method, only nearby atoms are checked when searching for the neighbors of a certain atom. The process of assigning each atom to a bin is $O(N)$, and the number of atoms per bin is typically about constant, so the binning approach as a whole is also ultimately $O(N)$, although the bin assignment must, in principle, be done before every force evaluation to avoid missing any interactions. The binning approach is particularly well suited to parallel implementations of atomistic simulations, where each processor is responsible for a region in physical space. (For a discussion of parallel algorithms, see, for example, [Pli95, BvdSvD95, KSB$^+$99, Ref00].)

Further improvements can be obtained by combining binning and Verlet lists. This is necessary because although the binning approach is $O(N)$, the number of neighbors sampled for each atom is significantly larger than in Verlet (see Exercise 6.1.) When combining the two approaches, the machinery of the Verlet lists and the Verlet criterion for updating these lists remains the same, but the binning concept is used at the time of the Verlet neighbor list update. For each neighbor list update, atoms are assigned to bins, although in this case of width r_{neigh} rather than r_{cut}. Now, neighbor lists are generated by searching over only the atoms in adjacent bins. This has the benefit of making the entire neighbor finding effort $O(N)$, like simple binning, while also avoiding the unnecessary regeneration of neighbor lists.

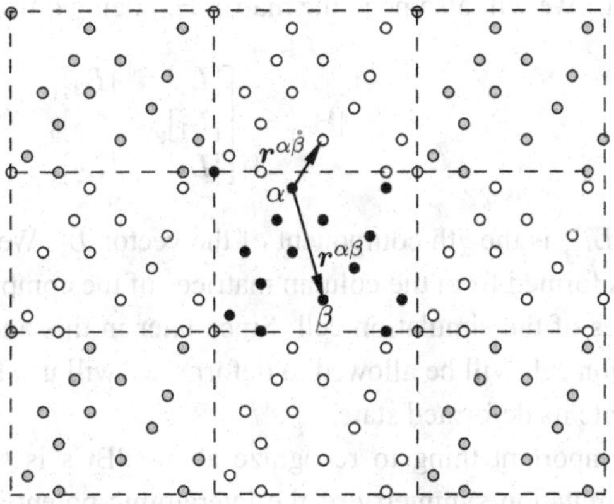

Fig. 6.5 PBCs: the black atoms that are simulated in the simulation cell are periodically copied in all directions (gray and white atoms). The vectors $r^{\alpha\beta}$ and $r^{\alpha\mathring{\beta}}$ illustrate how atom α is nearer to the periodic copy of atom β than to atom β itself.

6.4.2 Periodic boundary conditions (PBCs)

In MS and MD, perhaps the most common boundary conditions in use are PBCs.[17] We have already encountered PBCs in the context of DFT in Section 4.5. PBCs are conceptually simple. The simulation is performed on a set of atoms contained within a finite simulation cell, where the positions of atoms outside the simulation cell are obtained by generating periodic images of the simulated atoms in accordance with the periodicity of the cell. An example for a two-dimensional rectangular cell is illustrated in Fig. 6.5. The main advantage of PBCs, and the reason that they are usually applied, is that they can be used to eliminate surface effects for the atoms in the simulation cell and in this manner mimic the behavior of a "bulk" material.

The simulation cell can be of any shape that can be used to fill space when regularly repeated [AT87, Section 1.5.2]. Most commonly, a rectangular box is used, although non-orthogonal parallelepiped cells can be important, especially when the cell is allowed to change its shape in response to applied stress (see Section 6.4.3). In the general non-orthogonal case, the periodic cell is defined by three nonorthogonal vectors,[18] $\boldsymbol{L}_1$, $\boldsymbol{L}_2$ and $\boldsymbol{L}_3$. By repeating this cell through space, an infinite set of atoms is generated from each atom α in the simulation cell, where the positions of the periodic images are given by

$$\boldsymbol{r}^{\alpha}_{\text{per}}(n_1, n_2, n_3) = \boldsymbol{r}^{\alpha} + n_i \boldsymbol{L}_i. \tag{6.29}$$

Here, n_i are integers ranging from $-\infty$ to ∞ and Einstein's summation convention is observed. (For $n_1 = n_2 = n_3 = 0$ the atom in the simulation cell is obtained from this

[17] These are sometimes referred to as the Born–von Karman boundary conditions, as they were first introduced in a paper by these two authors [BvK12].

[18] Note the similarity between the periodic simulation cell and the definition of a multilattice crystal in Section 3.6.

function.) We will also need the matrix $\mathbf{H}_0$, defined from the periodic cell vectors as

$$[\mathbf{H}_0] = \begin{bmatrix} [\boldsymbol{L}_1]_1 & [\boldsymbol{L}_2]_1 & [\boldsymbol{L}_3]_1 \\ [\boldsymbol{L}_1]_2 & [\boldsymbol{L}_2]_2 & [\boldsymbol{L}_3]_2 \\ [\boldsymbol{L}_1]_3 & [\boldsymbol{L}_2]_3 & [\boldsymbol{L}_3]_3 \end{bmatrix}, \tag{6.30}$$

where $[\boldsymbol{L}_i]_j$ is the jth component of the vector $\boldsymbol{L}_i$. We see that the matrix of components of $\mathbf{H}_0$ is formed from the column matrices of the components of the three vectors defining the edges of the simulation cell. Since later in this and subsequent chapters the periodic simulation cell will be allowed to deform, we will use $\mathbf{H}_0$ to indicate the initial cell and $\mathbf{H}$ to indicate its deformed state.

The important thing to recognize about PBCs is that due to translational invariance and permutation symmetry of the interatomic potential function, the force computed on an atom in the periodic cell, $\boldsymbol{f}^{\mathrm{int},\alpha}$, is identical to the force that would be computed on any of its images located at $\boldsymbol{r}^{\alpha}_{\mathrm{per}}(n_1, n_2, n_3)$ (see [DJ07] for a proof of this for the more general boundary conditions imposed to build *objective structures* [Jam06], of which translationally-invariant periodicity is one special case). This means that by performing a static relaxation or dynamical evolution of *only* the atoms in the periodic cell, with the periodic images of these atoms moving in accordance with the applied PBCs, a self-consistent solution is obtained. In a static simulation, where the energy of the atoms in the periodic cell is minimized, the force on *all* atoms, both in and out of the periodic cell, will be zero in the final configuration. For a dynamic simulation, *all* atoms will be moving according to a solution of Newton's equations of motion (see [DJ07] for a proof of this property and further discussion). Of course, nothing comes without a price. In this case, an eye must be kept out for artifacts associated with the periodicity of the PBCs themselves. We discuss this a bit later for the case where the periodic cell contains a defect. Sometimes these effects can be accounted for explicitly. At the very least, the effect of varying the system size should be investigated before conclusions are drawn.

From an implementational standpoint, the use of PBCs requires a modification of the search for neighbors, since the search is now no longer just over the actual simulated atoms but also over their periodic copies. The necessary changes to Algorithm 6.7 are minor if we enforce the rule[19] that the perpendicular distance between any two parallel sides of the simulation box is at least $2r_{\mathrm{neigh}}$; lines 6 and 17 need to be adjusted so that the nearest copy of β is used and the periodicity is factored into the calculation of $\boldsymbol{r}^{\alpha\beta}$. Let us define a new type of superscript, such that $\boldsymbol{r}^{\alpha\mathring{\beta}}$ is the vector connecting atom α to the closest among atom β and its periodic copies. (Figure 6.5 shows an example, where a periodic image of β is closer to α than the actual atom β in the simulation cell.) We define $\boldsymbol{r}^{\alpha\mathring{\beta}}$ as

$$\boldsymbol{r}^{\alpha\mathring{\beta}} = \min_{n_1, n_2, n_3} \left\| \boldsymbol{r}^{\beta}_{\mathrm{per}}(n_1, n_2, n_3) - \boldsymbol{r}^{\alpha} \right\|, \tag{6.31}$$

where $\boldsymbol{r}^{\beta}_{\mathrm{per}}$ is defined as in Eqn. (6.29). In a moment, we will discuss the calculation of $\boldsymbol{r}^{\alpha\mathring{\beta}}$ for orthogonal and nonorthogonal periodic cells.

[19] The restriction on minimum box size makes sure that an atom interacts with, at most, one periodic copy of any other atom. This restriction can be relaxed, but the coding becomes more complicated.

Within a periodic framework, the statement of the MS minimization problem changes slightly. The internal potential energy of the system of atoms is now $\mathcal{V}^{\text{int}} = \mathcal{V}^{\text{int}}(\{r^{\alpha\mathring{\beta}}\})$, where $\{r^{\alpha\mathring{\beta}}\}$ represents the set of all interatomic distances between an atom α and the nearest periodic image of atom β. Now, we seek

$$\boldsymbol{r}_{\text{min}} = \arg\min_{\boldsymbol{r}} \widehat{\mathcal{V}}^{\text{int}}(\boldsymbol{r}), \tag{6.32}$$

where $\widehat{\mathcal{V}}^{\text{int}}(\boldsymbol{r}) = \mathcal{V}^{\text{int}}(\{r^{\alpha\mathring{\beta}}(\boldsymbol{r})\})$ and $\boldsymbol{r}$ represents all atomic positions in the simulation box $\boldsymbol{r}^{\alpha}$, $\alpha = 1, \ldots, N$. Calculation of forces in the periodic system is a straightforward extension of the force expressions in Section 5.8.3, where all Greek superscripts except α are replaced with their periodic counterpart (e.g. $\mathring{\beta}$ replaces β).

Orthogonal periodicity For the special case of a rectangular box, the periodic cell vectors defined above are

$$\boldsymbol{L}_1 = L_1 \boldsymbol{e}_1, \qquad \boldsymbol{L}_2 = L_2 \boldsymbol{e}_2, \qquad \boldsymbol{L}_3 = L_3 \boldsymbol{e}_3, \tag{6.33}$$

where L_i are the lengths of the box along the axis directions. Substituting Eqn. (6.33) in Eqn. (6.30), the matrix of components of $\mathbf{H}_0$ is

$$[\mathbf{H}_0] = \begin{bmatrix} L_1 & 0 & 0 \\ 0 & L_2 & 0 \\ 0 & 0 & L_3 \end{bmatrix}. \tag{6.34}$$

The computation of $\boldsymbol{r}^{\alpha\mathring{\beta}}$ for this case is straightforward to implement due to the orthogonality of the box. The vector $\boldsymbol{r}^{\alpha\mathring{\beta}}$ is found directly as

$$\boldsymbol{r}^{\alpha\mathring{\beta}} = \boldsymbol{r}^{\alpha\beta} - \mathbf{H}_0 \boldsymbol{n}, \qquad \boldsymbol{n} = \text{nint}\left(\mathbf{H}_0^{-1} \boldsymbol{r}^{\alpha\beta}\right). \tag{6.35}$$

Making use of Eqn. (6.34), the expression for $\boldsymbol{n}$ in component form is

$$n_i = \text{nint}\left(\frac{r_i^{\alpha\beta}}{L_i}\right). \tag{6.36}$$

The function nint(x) returns the nearest integer to the argument (or the column matrix of integers nearest each component of a vector argument). This approach is referred to as the *minimum image convention.*

Nonorthogonal periodicity It is conceptually straightforward to extend the neighbor-finding framework discussed above to nonorthogonal cells, but it requires a more carefully written code to ensure that all periodic images of an atom's neighbors are correctly located. In particular, Eqn. (6.35) cannot simply be used with the general expression for $\mathbf{H}_0$ in Eqn. (6.30). Such an approach would not guarantee that the closest periodic image is found. It can fail for rather extreme simulation boxes, in which there is a very small angle between two of the vectors $\boldsymbol{L}_i$ or if the ratio of vector lengths is very different from 1.

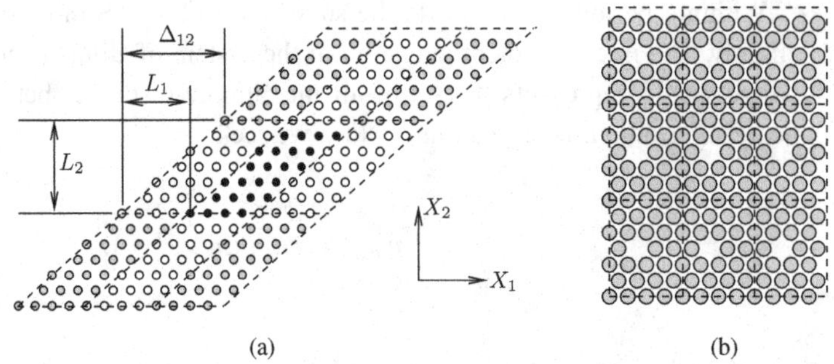

Fig. 6.6 (a) Nonorthogonal PBCs applied to a crystal. (b) Modeling a vacancy with PBCs really means modeling a periodic array of vacancies.

The following somewhat slower procedure for determining the nearest periodic image [Pli09] will work regardless of the shape of the simulation box. (Note that this is related to the problem of *lattice reduction* discussed in [AST09].) First, we choose a basis in which the components of the cell vectors are

$$[\boldsymbol{L}_1] = [L_1, 0, 0]^T, \qquad [\boldsymbol{L}_2] = [\Delta_{12}, L_2, 0]^T, \qquad [\boldsymbol{L}_3] = [\Delta_{13}, \Delta_{23}, L_3]^T. \tag{6.37}$$

This still allows for a box of general size and shape, but fixes its orientation in space relative to a global coordinate system. This, of course, does not affect the atomic interactions. Defined in this way, the three *positive* scalars L_i define the width of the box along the axis direction, i, while Δ_{ij} has the effect of shearing the box as illustrated in two dimensions in Fig. 6.6(a). Second, Algorithm 6.8 can then be used to find the vector $\boldsymbol{r}^{\alpha\mathring{\beta}}$.

Modeling free surfaces and defects PBCs are designed to avoid free surface effects, but sometimes we might actually *want* to model free surfaces. Doing so within a code designed to use PBCs presents no real difficulty, as we can merely set one periodic length large enough to create a gap of at least r_{cut}, as shown later in Fig. 6.12(b). As such, we can model an infinite slab of material with a free surface on either side. (See also Exercise 6.3.) Using this idea in all three periodic directions permits modeling of a finite collection of atoms rather than a window on an infinite system.

PBCs can have disadvantages if we truly want to examine the behavior of a single isolated defect or microstructural feature in an infinite crystal. With PBCs, we are always modeling a periodic array of defects, as illustrated for a vacancy in Fig. 6.6(b). This may or may not present a significant problem depending on the type of defect we are modeling; the severity of the effect is determined by the rate of decay of the stress field generated by the defect. In Sections 6.5.3 and 6.5.4, we talk more about the modeling of point defects and vacancies in crystalline systems.

6.4.3 Applying stress and pressure boundary conditions

In our discussion of PBCs in the previous section, we assumed that the shape of the periodic cell, defined by the three vectors $\boldsymbol{L}_i$ ($i = 1, 2, 3$) in Eqn. (6.29), was fixed. Minimizing the energy of the system subject to this constraint leads to an equilibrium state

Algorithm 6.8 Finding the nearest image of atom β to atom α.

$\boldsymbol{r}^{\alpha\mathring{\beta}} := \boldsymbol{r}^{\beta} - \boldsymbol{r}^{\alpha}$
if $|r_3^{\alpha\mathring{\beta}}| > L_3/2$ **then**
  if $r_3^{\alpha\mathring{\beta}} < 0.0$ **then**
    $\boldsymbol{r}^{\alpha\mathring{\beta}} := \boldsymbol{r}^{\alpha\mathring{\beta}} + \boldsymbol{L}_3$
  else
    $\boldsymbol{r}^{\alpha\mathring{\beta}} := \boldsymbol{r}^{\alpha\mathring{\beta}} - \boldsymbol{L}_3$
  end if
end if
if $|r_2^{\alpha\mathring{\beta}}| > L_2/2$ **then**
  if $r_2^{\alpha\mathring{\beta}} < 0.0$ **then**
    $\boldsymbol{r}^{\alpha\mathring{\beta}} := \boldsymbol{r}^{\alpha\mathring{\beta}} + \boldsymbol{L}_2$
  else
    $\boldsymbol{r}^{\alpha\mathring{\beta}} := \boldsymbol{r}^{\alpha\mathring{\beta}} - \boldsymbol{L}_2$
  end if
end if
if $|r_1^{\alpha\mathring{\beta}}| > L_1/2$ **then**
  if $r_1^{\alpha\mathring{\beta}} < 0.0$ **then**
    $\boldsymbol{r}^{\alpha\mathring{\beta}} := \boldsymbol{r}^{\alpha\mathring{\beta}} + \boldsymbol{L}_1$
  else
    $\boldsymbol{r}^{\alpha\mathring{\beta}} := \boldsymbol{r}^{\alpha\mathring{\beta}} - \boldsymbol{L}_1$
  end if
end if

in which the forces on all atoms are zero. However, in doing so we have not considered the "thermodynamic tension" (see Eqn. (2.125)) which is conjugate with the shape of the periodic cell itself. This thermodynamic tension is, in fact, the external stress that must be applied in order to impose the desired periodicity. One reason why this is of interest is that in some cases we may wish to control the applied stress and not the shape of the periodic cell. The difference between the forces on the atoms in a periodic cell and the force conjugate with the cell itself can be readily demonstrated by the following one-dimensional example.

Example 6.2 (The thermodynamic tension in a chain of atoms) Consider a periodic one-dimensional chain of N identical atoms, interacting via a pair potential, $\phi(r)$. For simplicity, assume that atoms only interact with their nearest neighbors. Let the periodic length of the chain be L and assume that the atoms are uniformly distributed along its length. The chain and boundary conditions are illustrated in Fig. 6.7. The total potential energy of the chain is

$$\mathcal{V}^{\text{int}} = \phi(r^2 - r^1) + \phi(r^3 - r^2) + \cdots + \phi(r^N - r^{N-1}) + \phi(r^1 + L - r^N), \tag{6.38}$$

where r^α is the (scalar) position of atom α along the chain. The force on any atom α is

$$\begin{aligned} f^\alpha &= -\frac{\partial \mathcal{V}^{\text{int}}}{\partial r^\alpha} = -\frac{\partial}{\partial r^\alpha}\left[\phi(r^{\alpha+1} - r^\alpha) + \phi(r^\alpha - r^{\alpha-1})\right] \\ &= \phi'(r^{\alpha+1} - r^\alpha) - \phi'(r^\alpha - r^{\alpha-1}), \end{aligned} \tag{6.39}$$

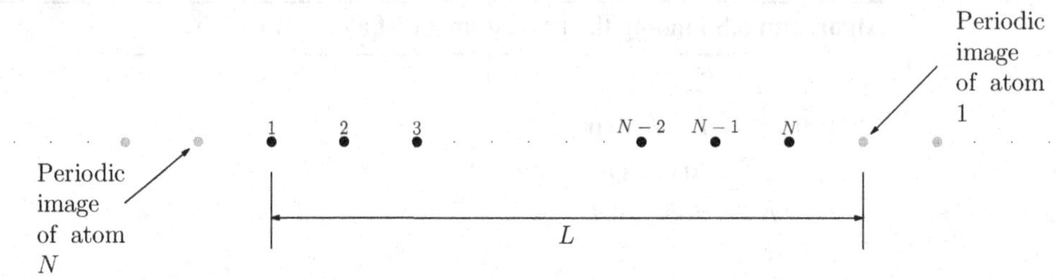

Fig. 6.7 Periodic one-dimensional chain.

where due to the periodicity, we set $r^0 = r^N - L$ and $r^{N+1} = r^1 + L$. Since the atoms are uniformly distributed along the chain, we have that

$$r^{\alpha+1} - r^{\alpha} = L/N \tag{6.40}$$

for all α. The force on atom α is then

$$f^{\alpha} = \phi'(L/N) - \phi'(L/N) = 0.$$

We see that the force on the atoms is identically zero for any length of chain! The effect of changing L is captured by the thermodynamic tension, which in this case is created by the force, P, that has to be applied to maintain the imposed periodicity. Substituting Eqn. (6.40) into Eqn. (6.38), the potential energy for a uniformly-spaced chain is $\mathcal{V}^{\text{int}} = N\phi(L/N)$, and the tension conjugate with L is

$$P = \frac{\partial \mathcal{V}^{\text{int}}}{\partial L} = \frac{\partial}{\partial L}(N\phi(L/N)) = \phi'(L/N).$$

Referring back to Fig. 5.2(a), which shows typical interatomic pair potentials, we note that $\phi(r)$ has one minimum. Let us denote the value of r at this minimum as r_{min}. If L is selected so that $L/N = r_{\text{min}}$, then $P = 0$. For $L/N > r_{\text{min}}$, the slope $\phi'(L/N)$ is positive and $P > 0$, which means that the chain will be in tension. For $L/N < r_{\text{min}}$, $P < 0$ and the chain will be in compression.

The above example is instructive. The question now is how can we reformulate MS so that instead of controlling the shape of the periodic cell, we control the thermodynamic tension conjugate with it? For example, in the one-dimensional chain example above, we may require, $P = \widetilde{P}$, where $\widetilde{P}$ is a specified constant. In the general three-dimensional case, the thermodynamic tension is the second Piola–Kirchhoff stress tensor $\boldsymbol{S}$. In order to carry out this reformulation, the variables controlling the shape of the periodic cell must be included as degrees of freedom in addition to the positions of the atoms in the cell.

In Section 9.5, we present the methodology for running MD simulations at a constant applied stress. Because there is much overlap between the derivation of that approach and the analogous approach for MS, we reserve this discussion for later. In Section 9.5.4, we will revisit and fully explain how an arbitrary stress state can be imposed in MS.

6.4.4 Boundary conditions on atoms

The simplest way to add external forces to individual atoms in an MS system is to modify the potential energy to include these applied forces. Let $\mathcal{C}_{\text{f}}$ be the set of atoms on which external forces are to be applied. The total potential energy becomes

$$\mathcal{V} = \mathcal{V}^{\text{int}} - \sum_{\alpha \in \mathcal{C}_f} \tilde{\boldsymbol{f}}^{\alpha} \cdot (\boldsymbol{r}^{\alpha} - \boldsymbol{R}^{\alpha}), \tag{6.41}$$

where $\tilde{\boldsymbol{f}}^{\alpha}$ is the constant force applied to atom α and $\boldsymbol{R}^{\alpha}$ is its reference position. Differentiating this energy, we obtain the force on atom $\alpha \in \mathcal{C}_f$:

$$\boldsymbol{f}^{\alpha} = -\frac{\partial \mathcal{V}^{\text{int}}}{\partial \boldsymbol{r}^{\alpha}} + \tilde{\boldsymbol{f}}^{\alpha}. \tag{6.42}$$

At equilibrium we will have a balance between the internal forces coming from the gradient of $\mathcal{V}^{\text{int}}$ and the external applied forces.

Next, consider the case where we want to constrain the position of an atom α to a fixed value $\boldsymbol{R}^{\alpha}$. Usually, this is handled within MS simulations by building an initial configuration in which the constrained atom is already in the desired position, and then setting the force on that atom identically to zero. The SD and CG minimization methods discussed previously work such that a degree of freedom α will never move during the minimization procedure if $\boldsymbol{f}^{\alpha} = \boldsymbol{0}$. This introduces some inefficiencies in that the forces on the constrained atoms are computed and then effectively thrown away, but the losses are not usually worth the extra coding required to avoid these force calculations.

Now, if we want to isolate a small subregion of a much larger solid body, we are no longer limited to regions that can be described using PBCs. Instead, we can model a finite collection of atoms, holding fixed any atom within some distance of the free surfaces created by truncating the body to finite size. The number of atoms that we need to hold fixed is determined by the range of the interatomic model being used. This will be discussed further in the context of an example application in Section 6.5.5.

Mixed boundary conditions are possible. For example, a single component of force can be applied to an atom while its displacement in orthogonal directions can be constrained. Alternatively, an atom can be constrained to move only along a certain line by resolving the total force on the atom into components along and perpendicular to the line, after which the perpendicular component can be set to zero. In this case the force on atom α is modified to

$$\boldsymbol{f}^{\alpha} \equiv (\boldsymbol{f}^{\alpha} \cdot \boldsymbol{e})\boldsymbol{e}, \tag{6.43}$$

where $\boldsymbol{e}$ is a unit vector along the line to which atom α will be confined.

6.5 Application to crystals and crystalline defects

Here, we look at the use of MS to study crystalline materials. We start with perfect crystals, but focus mainly on defects due to their ubiquity in all but the most idealized and carefully prepared specimens. Defects such as vacancies, free surfaces, grain boundaries and dislocations play a central role in the response of real materials, most notably in affecting their strength, ductility and toughness. MS simulations have helped materials scientists to understand some of the most fundamental questions about these properties, and these models will continue to help us predict and explain material behavior in the future.

6.5.1 Cohesive energy of an infinite crystal

While the fundamental quantity for a continuum constitutive law is the strain energy density $W(\boldsymbol{F})$ (see Section 2.5.2), the parallel quantity at the atomic scale is the cohesive energy of a crystal. The two concepts are closely related.

The cohesive energy of a crystal, E_{coh}, is the difference between the energy of a collection of atoms bonded in a crystalline structure and the energy of those same atoms infinitely separated and isolated from each other, divided by the number of atoms in the crystal:

$$E_{\mathrm{coh}} = -\lim_{N\to\infty} \frac{\mathcal{V}^{\mathrm{int}}_{(N)} - \sum_{\alpha=1}^{N} E_{\mathrm{free}}(Z^\alpha)}{N}. \tag{6.44}$$

In this equation we use the notation introduced in Section 5.4.1; $\mathcal{V}^{\mathrm{int}}_{(N)}$ is the internal energy of the bonded crystal composed of N atoms and $E_{\mathrm{free}}(Z^\alpha)$ is the energy of a free (isolated) atom with atomic number Z^α. This definition applies equally to simple lattices and multilattices, where $E_{\mathrm{free}}(Z^\alpha)$ allows each interpenetrating lattice to comprise a different species. The negative sign is introduced by convention, so that the final E_{coh} will be positive for stable crystals.

Within an empirical atomistic model it is convenient and typical to set the reference energy such that any isolated atom has $E_{\mathrm{free}} = 0$. This is possible since we are only considering the energy *changes* due to bonding in an empirical model. Also it is impossible, of course, to model an infinite crystal, but we can use PBCs and just consider the energy of the atoms in the simulation cell.[20] Under these circumstances, the cohesive energy becomes simply

$$E_{\mathrm{coh}} = \frac{-E_{\mathrm{cell}}}{n_{\mathrm{cell}}}, \tag{6.45}$$

where n_{cell} is the number of atoms in the simulation cell.

Now consider a system of atoms that are constrained to remain in a particular crystal structure, but for which the volume can change. For example, imagine atoms that are arranged in a face-centered cubic (fcc) or body-centered cubic (bcc) crystal structure, with only the length of the cube side, a, allowed to vary. For more complex crystal structures, we hold all atomic positions fixed such that the interatomic distances scale with a single length parameter a; for instance the hexagonal close-packed (hcp) structure is defined for a fixed c/a ratio. Subject to these constraints and at zero temperature, we can relate the cohesive energy to the definition of the strain energy density in Eqn. (2.172) as

$$E_{\mathrm{coh}}(a) = W(\boldsymbol{F}(a))\frac{\Omega(a)}{n_{\mathrm{cell}}}, \tag{6.46}$$

where Ω is the volume of the periodic cell. In Eqn. (6.46), we have restricted the deformation gradient $\boldsymbol{F}(a)$ to deformations of the form

$$\boldsymbol{F}(a) = \frac{a}{a_0}\boldsymbol{I}, \tag{6.47}$$

[20] For a perfect crystal, the periodic cell can be taken as small as a single unit cell of the crystal.

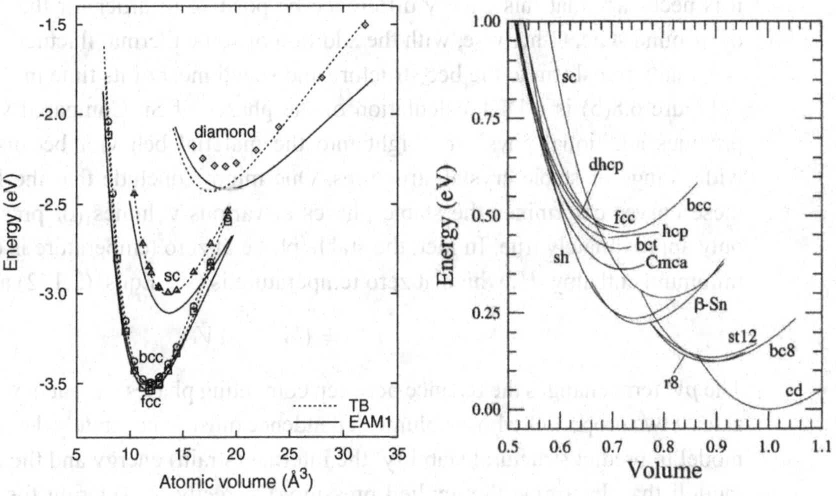

Fig. 6.8 (a) Comparison of TB, DFT and pair functional (EAM1, described in [MMP+01]) results for the cohesive energy of Cu. The datum points are from DFT. (Image kindly provided by Yuri Mishin.) (b) DFT-LDA calculations of the energy per atom for different structural phases of Si. The volume is normalized by the volume at the equilibrium cubic diamond (cd) phase. See footnote 21 on page 334 for details of the crystal structures. Reprinted from [NM02], by permission of the publisher (Taylor & Francis Group, http://www.informaworld.com).

where $\boldsymbol{I}$ is the identity tensor and a_0 is the reference lattice constant normally taken as the value which minimizes W for a particular crystal structure.

Plots of the cohesive energy versus volume are examples of the simplest potential energy surface, where the configurational space is one-dimensional and characterized by the lattice constant (or alternatively the volume). Such plots provide a number of important insights into materials and our ability to model them. For example, they prove invaluable as a test against first-principles data for the development of empirical interatomic models. This is because it is relatively inexpensive to compute the DFT result for a variety of crystal structures. Even if such structures are not observed in nature (for example, a bcc phase of pure Cu), they provide a good test by probing atomic environments different from the ground state crystal structure. Figure 6.8(a) compares DFT, TB and pair functional results for four different crystal structures of Cu. We see that for relatively close-packed phases like fcc and bcc, the pair functional agrees well with DFT. However, the graph points to some of the limitations of the pair functional formalism that we have already mentioned. Specifically, open structures like simple cubic or diamond, which indicate a more covalent bond character, are not as well described by the pair functional form.

The so-called "structural energy differences" are the differences between the minima of the various curves on these graphs. For example for Cu, we can define $\Delta E_{\text{fcc}\to\text{bcc}}$ as the energy difference between the minimum of the bcc curve and the minimum of the fcc curve in Fig. 6.8(a). This is the *additional* energy that would have to be supplied to a system in the fcc structure to take it to the bcc structure. In designing a pair functional to model Cu, it is necessary that this energy difference be positive in order for the fcc structure to be the ground state. Otherwise, with the addition of some thermal fluctuation, the system will eventually transform to the bcc structure and spend most of its time in that phase.

Figure 6.8(b) is a DFT calculation of the phases of Si. Compared with Fig. 6.8(a), it provides additional physical insight into the material behavior because Si can adopt a wider range of stable crystal structures. One might conclude that the lower envelope of these curves determines the stable phases at various volumes (or pressures), but this is only approximately true. In fact, the stable phase at zero temperature is determined by the minimum enthalpy, H, which at zero temperature is (see Eqns. (2.172) and (2.174)):

$$H = (W + p)\,V. \tag{6.48}$$

The pV term changes the balance between competing phases. As such, we see that there are at least two properties whose volume dependence must be accurately described by an atomic model to predict structural stability: the internal (strain) energy and the (nonlinear) elastic moduli that determine the applied pressure. Correctly considering the enthalpy leads to the correct experimental order of phase transformations with decreasing pressure: diamond cubic $\rightarrow$ β-tin $\rightarrow$ Imma $\rightarrow$ simple hexagonal $\rightarrow$ Cmca $\rightarrow$ hcp $\rightarrow$ fcc.[21]

6.5.2 The universal binding energy relation (UBER)

In the early 1980s, a series of papers by Rose and coworkers [RFS81, FSR83, RSGF84] revealed a remarkable universality in the cohesive energy curves for metallic and covalently bonded crystals. They found that only two parameters were needed to describe, reasonably well, the entire cohesive energy versus lattice constant curve for any such crystal.

Returning to the plots of Figs. 6.8 and Figs. 6.8(b), we note that these graphs show a limited range of volumes for each structure, focused around the basin of the curve near the equilibrium value. In Fig. 6.9(a), the curve for fcc is extended far from equilibrium and plotted against lattice constant instead of volume.[22] At large tensile strains the energy asymptotically goes to zero as the atoms eventually become fully isolated from each other at large distances. Large compressive strains (not shown) will continually increase

[21] Some of these crystal structures were described in Chapter 3 including simple cubic (sc), fcc, bcc, hcp, body-centered tetragonal (bct), simple hexagonal (sh) and cubic diamond (cd). The less common ones can all be viewed at [NRL09] and include:

1. dhcp (double hexagonal close-packed): this is like hcp, but requires two hexagonal unit cells with alternating locations of the atom in the center of each cell. In the language of the discussion on page 358, the stacking of the planes is $ABAC$;
2. Imma: a body-centered orthorhombic structure (not shown in Figure 6.8(b));
3. Cmca: an orthorhombic structure with centering on the C-faces and 16 atoms per unit cell;
4. β-Sn: a tetragonal crystal with four atoms per unit cell, a prototypical example of which is found in tin at room temperature and normal pressures;
5. st12: a tetragonal structure with 12 atoms per unit cell and an atomic arrangement similar to that of diamond cubic, but more efficiently packed;
6. bc8: a bcc primitive unit cell containing eight atoms [PMC+ 95];
7. r8: closely related to bc8, this is a rhombohedral unit cell containing eight atoms [PMC+ 95].

[22] Note that for fcc $\Omega = a^3/4$.

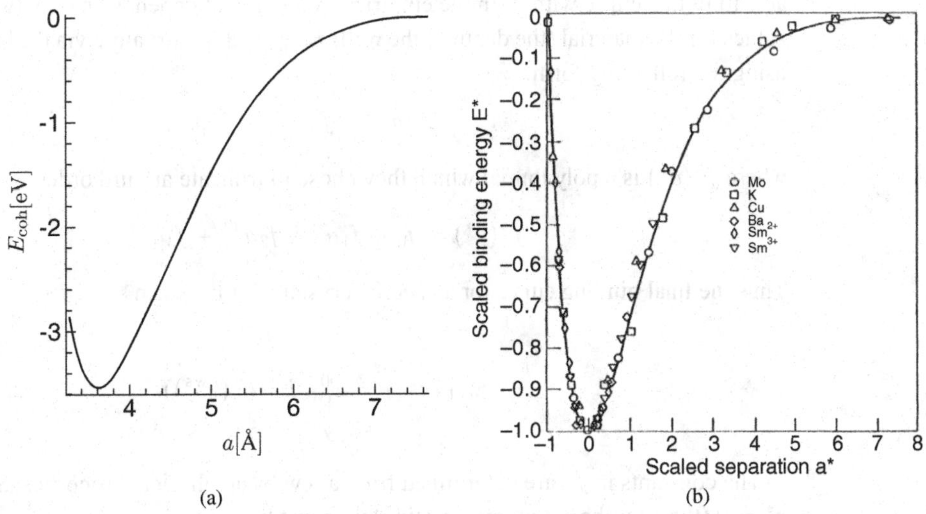

Fig. 6.9 (a) Cohesive energy versus lattice constant for fcc Cu modeled using the pair functional called EAM1 in [MMP+01]. (b) The fit of Eqn. (6.51) to several metal crystals. (Reprinted figure with permission from [RSGF84], copyright 1984 by the American Physical Society.)

the energy as atomic cores start to overlap. Note that to produce this figure, we simple apply a deformation of the form of Eqn. (6.47) without allowing any additional atomic rearrangement; no changes from the perfect fcc crystal structure are permitted.

The UBER functional form Rose and coworkers showed that curves like the one in Fig. 6.9(a) can be collapsed onto a master curve by rescaling the axes based on only two material parameters. Specifically, we obtain a single master curve if we plot E^* versus a^*, where[23]

$$E^* = -\frac{E_{\rm coh}}{E^0_{\rm coh}}, \qquad a^* = \frac{r^{\rm ws} - r^{\rm ws}_0}{l}. \tag{6.49}$$

Here $E^0_{\rm coh}$ is the equilibrium cohesive energy, $r^{\rm ws}$ is the radius of a sphere containing the same volume[24] as the crystal's average volume per atom, $\Omega^{\rm ws}$, ($r^{\rm ws}_0$ is this same radius at the equilibrium volume), and l is a material length scale defined by Rose and coworkers:

$$r^{\rm ws} = \left(\frac{3\Omega^{\rm ws}}{4\pi}\right)^{1/3}, \qquad l = \sqrt{\frac{E^0_{\rm coh}}{12\pi B r^{\rm ws}_0}}, \tag{6.50}$$

where B is the bulk modulus. In Fig. 6.9(b), we reproduce the results from one of the original universal binding energy relation (UBER) papers [RSGF84], which shows the effect of this rescaling on several crystal structures. Although the points (obtained from first-principles quantum calculations) do not *exactly* collapse onto a single smooth line, the results are certainly compelling and suggest that universal features exist. Rose and coworkers were

[23] Note that we have added a minus sign to the definition of E^*. This is because we define the cohesive energy in Eqn. (6.44) as a positive quantity.
[24] This is the so-called "Wigner–Seitz" volume, which explains the superscripts chosen.

able to fit this curve with a simple empirical relation that depends on only two *equilibrium* values for the material (the depth of the well, E_{coh}, and its curvature via the bulk modulus), using the following form:

$$E^*(a^*) = f^*(a^*)e^{-a^*}, \tag{6.51}$$

where $f^*(a^*)$ is a polynomial which they chose to truncate at third order:

$$f^*(a^*) = f_0 + f_1 a^* + f_2 a^{*2} + f_3 a^{*3}. \tag{6.52}$$

Thus the final binding curve for a specific crystal is of the form

$$E_{\text{coh}}(r^{\text{ws}}) = E^0_{\text{coh}} E^*(a^*(r^{\text{ws}})). \tag{6.53}$$

The constants in f^* are determined from a few basic physical properties. Specifically, at $a^* = 0$ (the equilibrium atomic spacing) we must have

$$E^*(0) = -1, \qquad (E^*)'(0) = 0, \qquad (E^*)''(0) = 1, \tag{6.54}$$

where the primes indicate derivatives. The first condition ensures that $E_{\text{coh}} = E^0_{\text{coh}}$ at the equilibrium atomic spacing. The second ensures that $a^* = 0$ is, in fact, an equilibrium configuration. The final condition ensures that the linear elastic response of the crystal is correctly fitted by the curve. To see this, we must recall the definition of the bulk modulus (see Exercise 2.15) in terms of the strain energy density, which we have already related to the cohesive energy in Eqn. (6.46). The bulk modulus is defined as

$$B = \Omega_0^{\text{ws}} \left. \frac{\partial^2 E_{\text{coh}}}{\partial(\Omega^{\text{ws}})^2} \right|_{\Omega^{\text{ws}} = \Omega_0^{\text{ws}}}, \tag{6.55}$$

where Ω_0^{ws} is the equilibrium volume per atom. Through a few chain rules applied to Eqn. (6.53), a bit of algebra, and the definition of the length scale l in Eqn. $(6.50)_2$, it is relatively straightforward to show that condition Eqn. $(6.54)_3$ must hold in order to fit B exactly. One must also invoke Eqn. $(6.54)_2$ during this derivation.

Applying the conditions in Eqn. (6.54) to the form of f^* above leads to $f_0 = -1$, $f_1 = -1$ and $f_2 = 0$, leaving only f_3 undetermined. Rose and coworkers fitted this final constant somewhat arbitrarily to the thermal expansion of Cu, which can be shown to be related to the cubic term in an expansion of the binding curve about the equilibrium lattice constant. This leads to a value of $f_3 = 0.05$ and the final form of the UBER becomes

$$E^*(a^*) = -\left(1 + a^* + 0.05(a^*)^3\right) e^{-a^*}. \tag{6.56}$$

There is some discrepancy in the literature, as most authors discard the $(a^*)^3$ term as negligible. Even the curve shown in the original paper [RSGF84] (reproduced in our

Fig. 6.9(b)) is of the form

$$E^*(a^*) = -(1 + a^*)\, e^{-a^*}. \tag{6.57}$$

We will refer to Eqns. (6.56) and (6.57) as the "third-order UBER" and "first-order UBER" respectively. The difference between the two curves is only significant far from equilibrium where the higher-order term has an appreciable effect. As we shall see, it seems that the first-order UBER is a better fit for large expansion, while the third-order UBER is better in high compression. (An alternative form for the first-order UBER is given in Exercise 6.5.)

The UBER concept has been shown to be applicable to a wide variety of material equations of state, including the energy of covalent diatomic molecules (the energy of the hydrogen molecule shown in Fig. 6.8(c) is one such result), the cleaving of a crystal to create new surfaces and the energetics of crystalline slip [BFS91]. A complete literature review can be found in [BS88].

Banerjea and Smith [BS88] shed some light on the underpinnings of the UBER, showing that it can be motivated from the relationship between host electron density and energy. First, the host electron density at each atomic site in a wide range of solid or molecular configurations is well described by a simple exponential dependence on atomic spacing (although this functional form is also empirical). Then, it seems that the energy of an atom embedded in a certain host electron density also has a more or less universal form. As a result, the UBER form works well for materials that are well described by the picture of atom-based orbitals overlapping to build the electron density field as atoms come together. This helps explain why the UBER does not work very well for ionic or van der Waals bonding, but is a reasonably good fit for metallic and covalent systems.

Examining the UBER's agreement with the data Closer inspection of the UBER curve reveals that it is really only an approximate fit at best. In Fig. 6.10(a), we show the first- and third-order UBER curves versus data for Cu computed using DFT with the generalized gradient approximation (GGA) exchange-correlation energy. Clearly, neither UBER curve is a very good fit far from $a^* = 0$. The GGA-DFT data suggest that perhaps the curve should be considerably less smooth than the UBER form far from equilibrium.

Mishin *et al.* [MMP$^+$01] also computed the cohesive energy curve for high compression of Cu using GGA, and we reproduce their results in Fig. 6.10(b). It is clear that in the highly compressive regime, the fit of the UBER curves is less good than in the tensile regime. The third-order UBER is a better fit than the first-order UBER, but it seems that higher-order terms may be necessary under high compression.

UBER: concluding remarks In 1957, Varshni [Var57] tried to find a universal function to describe the potential energy of the diatomic molecule by examining 25 different functions and testing their ability to reproduce the experimental results that were available for diatomic molecules at the time. One of the functions he examined was, in fact, essentially the UBER form (Varshni and Rose and coworkers also refer to this as the Rydberg function).

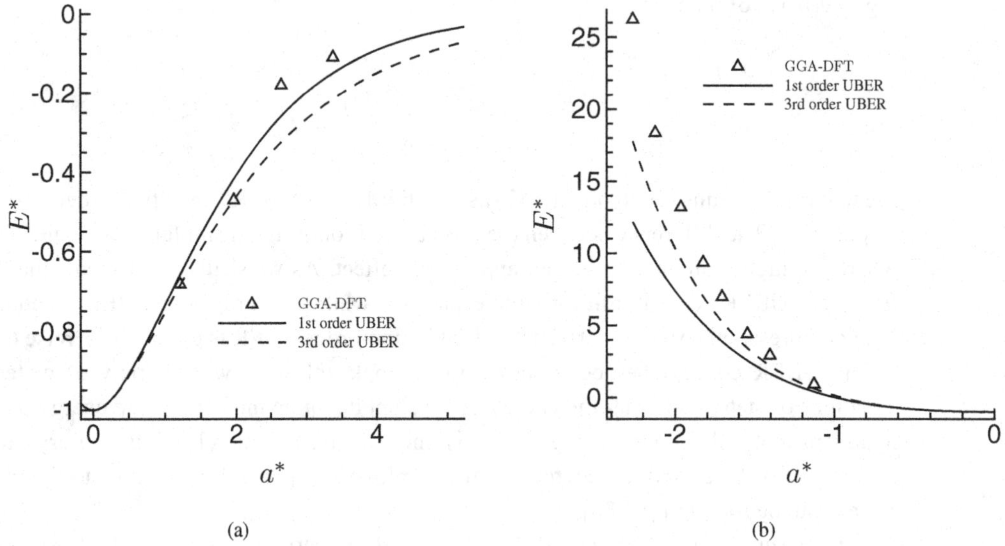

Fig. 6.10 UBER curves versus first-principles data for fcc Cu in (a) the tensile state and (b) the compressive state. First-principles GGA-DFT results from [MMP+01].

Varshni's conclusion was that there really was no such universal function, yet the current understanding is that UBER is a good universal fit. What has changed?

Perhaps the answer is that nothing has changed, and it is only that Varshni's standard of a "good universal function" was higher than the standard we apply to the UBER today. Also, Varshi was trying to fit experimental data, each with its own uncertainty and accuracy. Today, there is some debate as to how good a fit UBER really is and as to how it should be used. Some developers of interatomic models make an exact fit to UBER an uncompromising constraint of their models. For example some pair functionals (see Section 5.5, [Foi85] and Exercise 6.6) and the MEAM model (see Section 5.6.4 and references such as [BNW89, Bas92]) *identically define* part of their functional form to exactly fit the first order UBER curve. On the other hand, Mishin *et al.* [MMP+01] have argued that fitting the UBER curve means a rather poor fit to DFT results, and as such may not be advisable.

6.5.3 Crystal defects: vacancies

For someone starting to learn about atomistic simulations of crystalline systems, the first calculation he or she might perform is to find the cohesive energy of Eqn. (6.45). The second would likely be determining the *vacancy formation energy*. This is the energy "cost" for removing one atom from a crystal. It is defined as

$$E_{\text{vac}} \equiv \lim_{n_{\text{cell}} \to \infty} \{E_{\text{cell}}(n_{\text{cell}}) - (-n_{\text{cell}} E_{\text{coh}})\}, \tag{6.58}$$

where E_{cell} is the relaxed energy of the simulation cell containing a single vacancy in an otherwise perfect crystal. The awkward combination of negative signs is deliberate to remind us that E_{coh} is defined as a positive quantity while E_{cell} is negative. The limiting

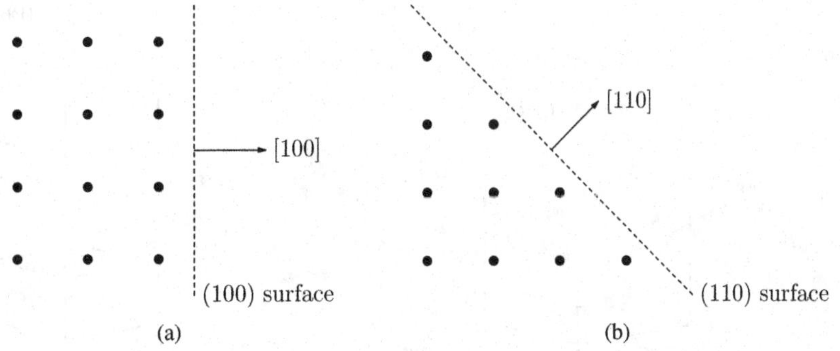

Fig. 6.11 (a) The (100) and (b) (110) surfaces in a simple cubic crystal.

process is meant to imply a systematic increase in the size of the simulation cell such that all periodic images of the vacancy become infinitely removed from each other (see Fig. 6.6(b)). The actual maximum n_{cell} required to get a good estimate of E_{vac} depends on the nature of the bonding in the crystal and the accuracy required.[25]

6.5.4 Crystal defects: surfaces and interfaces

The crystal structures discussed in Chapter 3 are ideal structures that are by definition infinite and uniform. Real crystals differ from the ideal in a number of ways. First, by necessity, real crystals are finite and therefore terminated by *free surfaces* which have structures and properties that differ from those found deep in the crystal. Second, most crystalline materials are not *single crystals*, which have a single uniform orientation at all points. Instead, they consist of a large number of *grains*, each of which is a single crystal but with a different orientation from its neighbors. These grains are separated from each other with internal interfaces called *grain boundaries*. Third, the grains themselves contain defects, such as the vacancies described in the previous section and dislocations discussed in the next. The structure and energies of free surfaces and grain boundaries play an important role in determining the physical properties of the material (recall Chapter 1).

Surface structure and energy The structure of *ideal* crystalline surfaces is entirely defined by the crystal structure and plane at which the crystal is terminated. For example, Fig. 6.11 shows the two surfaces obtained by terminating a simple cubic crystal normal to the [100] and [110] directions. These surfaces have different structures and will therefore have different properties. The term "ideal" used above refers to the fact that the surface obtained from the ideal crystal structure is not the one observed in practice. In real crystals, the atoms near the surface move away from their ideal positions to reduce the energy of the

[25] Taking the example of metals, vacancy formation energies are typically on the order of 1 eV, whereas practical calculation with Eqn. (6.58) may require hundreds or thousands of atoms, with total energies on the order of thousands of electron volts, to reasonably approximate the "infinite" limit. This means sufficient numerical precision must be retained in evaluating the vacancy formation energy, which comes from a small difference between two large numbers.

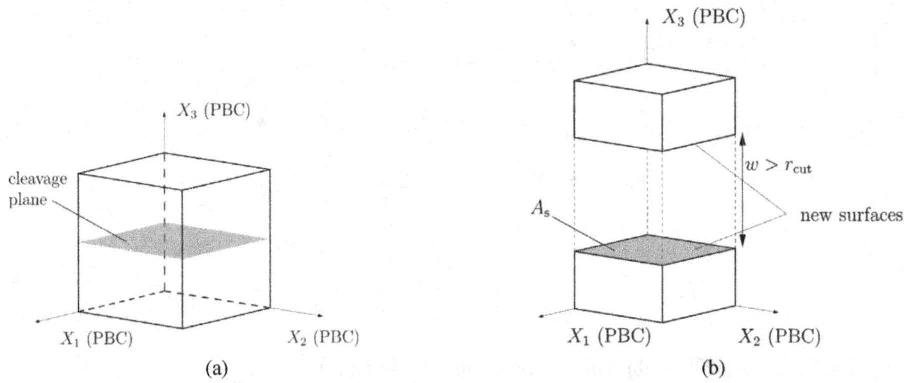

Fig. 6.12 MS simulations for computing surface energy: (a) the bulk energy calculation; (b) the cleaved system calculation. See discussion in text.

crystal. This is called *surface relaxation* or *surface reconstruction* (depending on the extent of the structural changes), and was discussed in Chapter 1.

A key property of a surface is the *surface energy* denoted by γ_s. This is defined as the change in the energy of a system when it is cleaved into two parts along a specified cleavage plane per unit area of the newly created surfaces. Strictly speaking this is a "surface energy density" with typical units of millijoules per square meter (or electron volts per square ångström), however the term "surface energy" is normally used. Surface energetics plays a key role in fracture, as lower-energy crystal surfaces tend to form preferential cleavage planes. The relative energies of various crystal facets also determine the shape of small atomic clusters as they coalesce from a liquid or gas, which in turn will determine the pattern of crystal growth and microstructure.

Computing the surface energy involves two straightforward MS calculations. First, a bulk calculation is performed for a block with PBCs applied in all directions as shown in Fig. 6.12(a). The energy obtained in this calculation is E_{bulk}. Next, the block is separated into two parts as defined by the cleavage plane which is taken to be normal to one of the axes directions as shown in Fig. 6.12(b). PBCs are applied in all directions here as well. However, as long as the separation between the two parts is sufficiently large, the resulting surfaces will not interact and we can think of the model as a single isolated slab with two free surfaces.[26] The result of this calculation is E_{surf}. The surface energy is then defined as

$$\gamma_s \equiv \frac{E_{surf} - E_{bulk}}{2A_s}, \tag{6.59}$$

where A_s is the area of the plane normal to the cleavage direction as shown in Fig. 6.12(b). The factor of 2 in the denominator accounts for the fact that two new surfaces have been formed. If the calculations described above are performed without minimizing the energy of the periodic cells, the result obtained from Eqn. (6.59) is the *unrelaxed* surface energy. If minimization is carried out, then the *relaxed* surface energy is obtained.

[26] The figure suggests two slabs, but remember that the PBC in the X_3 direction means that the two halves are in direct contact at the top and bottom edges of the simulation cell.

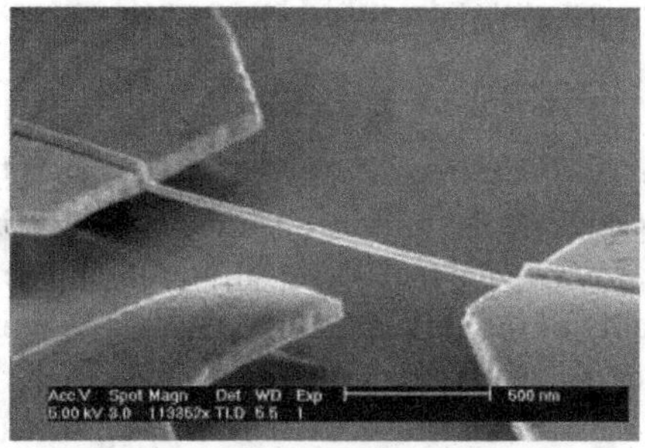

Fig. 6.13 A nano-scale resonator, approximately 1.5 μm long. (Reproduced from [vdZ07] with the kind permission of Dr. Herre van der Zant. Copyright 2007 by TU Delft.)

Computing the correct relaxed surface energy and associated relaxed structure suffers from the usual problems of a complex potential energy landscape. Simply cleaving the bulk crystal structure along some plane and allowing the system to relax will certainly provide a local minimum, but there is always the possibility that another lower energy minimum exists. Further, many crystal surfaces can have multiple possible reconstructions that are close enough in energy for more than one to play a role in material behavior. This is illustrated dramatically by the scanning tunneling micrograph of a (111) surface in diamond cubic Si shown in Fig. 1.6(b). This arrangement of atoms is referred to as the 7×7 reconstruction to indicate the number of atoms from the unrelaxed (111) plane that ultimately form the periodic cell after relaxation.[27] Obtaining this structure in a simulation is sensitive to the periodicity of the simulation cell, since it must be commensurate with the equilibrium 7×7 cell in order to even permit such a symmetry to form.

Surface effects in nanostructures Advances in nanotechnology have enabled us to fabricate mechanical structures and devices on the submicron scale. For example Fig. 6.13 shows an example of a nanoresonator; a simple beam structure that vibrates at a frequency characteristic of its stiffness and mass. However, these beams are extremely small, with lengths on the order of a micrometer and thicknesses on the order of 100 nm or less. In such a beam, a significant fraction of the atoms are not in a bulk environment, but rather in a *surface-dominated* environment, and as such the response of the beam is like that of a composite beam whose core exhibits one elastic modulus and whose surface layers exhibit a different elastic response. Miller and Shenoy [MS00] showed how this atomic-scale effect could be modeled using continuum mechanics if it was extended to include surface elasticity as presented, for example, by Gurtin and Murdoch [GM75] and Cammarata [Cam94]. Whereas the linear elasticity of the bulk material is described by the usual generalized form of Hooke's law (see Eqn. (2.196)):

$$\sigma_{ij} = c_{ijkl}\epsilon_{kl}, \tag{6.60}$$

[27] Note that careful study of Fig. 1.6(b) will suggest that the unit cell contains only 12 atoms. This is considerably fewer than 7×7 atoms because the reconstruction involves the motion of atoms into and out of the plane of the surface, while the scanning tunneling microscope only images the outermost layer.

the surface elasticity is described by an analogous expression of reduced dimensionality:

$$\tau_{\alpha\beta} = \tau^0_{\alpha\beta} + \Sigma_{\alpha\beta\gamma\delta}\epsilon_{\gamma\delta}, \tag{6.61}$$

where $\Sigma_{\alpha\beta\gamma\delta}$ is the *surface elasticity tensor*, $\tau_{\alpha\beta}$ is the *surface stress tensor* and $\tau^0_{\alpha\beta}$ is the surface stress when the bulk is unstrained. Here, Greek indices refer to two-dimensional curvilinear coordinates that are locally in the plane of a surface. Because of the reduced dimensionality of the surface, surface stress has units of *force per unit length*, rather than the force per unit area of the usual bulk stress.

Like the small strain elasticity tensor, c_{ijkl}, the surface elasticity tensor, $\Sigma_{\alpha\beta\gamma\delta}$, can be obtained for a specific surface from lattice statics simulations. For example, an infinite surface can be created as in Fig. 6.12(b), to which a uniform strain can be applied in the X_2 direction. The total force which must be applied to maintain the strain, ϵ, is the sum of the bulk stress times the cross-sectional area of the slab, plus the surface stress times the depth of the slab, d, in the X_1 direction. The force per unit depth is then

$$\frac{f(\epsilon)}{d} = Ew\epsilon + 2\tau^0_{22} + 2\Sigma_{2222}\epsilon = 2\tau^0_{22} + \underbrace{(Ew + 2\Sigma_{2222})}_{D}\epsilon, \tag{6.62}$$

where the factor of 2 comes from the fact that there are two free surfaces. E is Young's modulus for the plate, which is a combination of the components of $\mathbf{c}$ depending on the anisotropy of the bulk crystal.[28] Since $\mathbf{c}$ can be determined from a separate calculation on an infinite crystal (see Chapter 11), the slope of the force versus strain (the *plate modulus, D*) can be used to determine Σ_{2222}. Other strained states can be used to find the remaining components of the surface elasticity tensor.

Miller and Shenoy showed that the characteristic elastic property of a nano-sized beam or plate, D, can be written as

$$\frac{D - D_c}{D_c} = \alpha\frac{w_o}{w}, \tag{6.63}$$

where D_c is the equivalent property in the large-scale continuum limit, α is a dimensionless geometric factor, w is a characteristic length of the beam or plate and w_0 is a *material length scale*. For example in the case of the plate stretching just described, the property of interest was identified as the *plate modulus*, and we see that

$$\frac{D - D_c}{D_c} = 2\frac{\Sigma_{2222}}{E}\frac{1}{w}. \tag{6.64}$$

In this case, then, the material length scale is the ratio $w_0 = \Sigma_{2222}/E$. The results of Miller and Shenoy for single crystal plates of Al and Si in tension are reproduced in Fig. 6.14. In both cases, the $\langle 100\rangle$ crystal directions are oriented along X_1, X_2 and X_3. The solid curves are Eqn. (6.64) with appropriate values of Σ_{2222}/E for these crystals, while the circles are direct molecular statics simulations of the response of plates of varying width. We can see that Eqn. (6.64) works extremely well, down to plates as narrow as 10 Å in width. In addition, there is a 10% deviation from the bulk response for plates thinner than about 20 Å. This effect is more significant for plates or beams in bending, where curvature makes the

[28] The deformation is plane strain, but there can be a strain in the X_3 direction due to Poisson's effect.

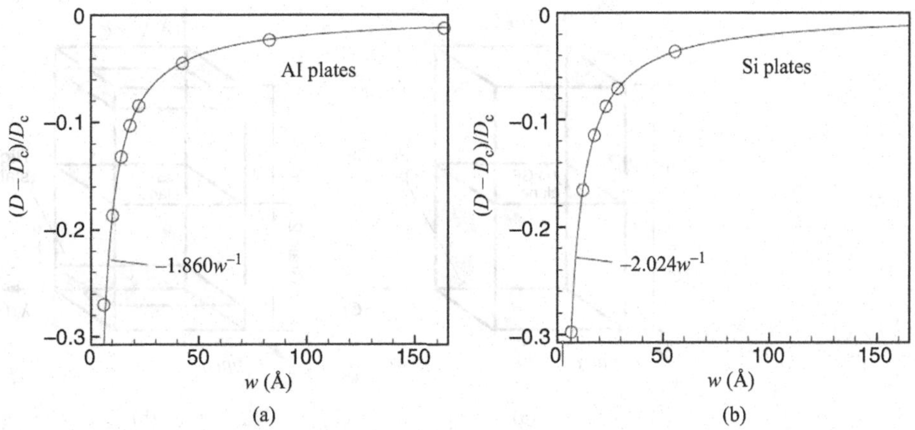

Fig. 6.14 Size effects on the elastic response of single-crystal plates in tension for: (a) Al and (b) Si. The circles represent direct atomistic simulation and the solid line comes from the continuum theory with surface elasticity. (Adapted from [MS00].)

surface the place where the strains are largest. In that case, Miller and Shenoy report a 10% deviation for thicknesses below about 60–80 Å. Of course, what constitutes a "significant" deviation in the properties depends on the application in mind, while the characteristic length scale controlling this deviation varies from material to material [MS00, MS07]. Even the sign of the deviation (i.e. whether nano-scale systems are stiffer or softer than the bulk system) depends on the material and the surface being studied. It is clear, at least, that atomic-scale surface effects play a role in the behavior of nano-scale devices.

The results of Miller and Shenoy show that a continuum formulation that accounts for surface effects, through the surface stress and surface elasticity tensor as in Eqn. (6.61), does exceptionally well in capturing surface effects in nano-scale structures. Capitalizing on this, Bar On *et al.* [BAT10] have formulated a composite beam theory model for nanobeams, where the surfaces are treated as separate layers with their own properties. An interesting aspect of the model is that it leads to the definition of an *effective surface thickness* which is found to be on the order of the lattice spacing. The approach works quite well for uniform nanobeams under arbitrary loading. When surface heterogeneity exists, for example due to the presence of different coatings on different parts of the beam surfaces, significant deviations from the continuum predictions are observed. These can be addressed in a phenomenological manner by introducing an appropriate correction factor. Others have extended the ideas of surface elasticity at the nano-scale beyond the linear domain, for example by using an extension of the Cauchy–Born idea (see Section 11.2.2) to free surfaces [PKW06, PK07, PK08].

Grain boundary (GB) structure The structure and energy of a GB strongly depend on the relative orientation of the two grains. This can serve as a driving force for microstructural transformation, as reorientation of the grains (through diffusion or deformation) may lower the overall energy of the system by removing or changing GB structures. At the same time, the details of the structure itself can play an important role in material behavior.

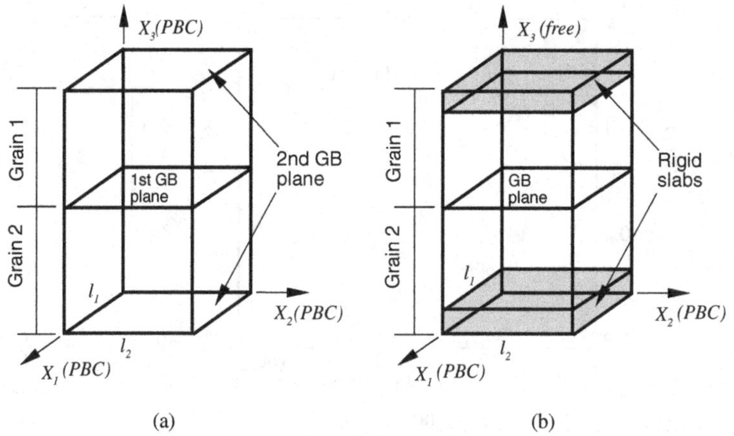

Fig. 6.15 MS simulations of GB structures can be performed using fully periodic boundary conditions as in (a), or with rigid slabs and free surfaces in one direction as in (b).

Some boundary structures are more amenable to *GB sliding*, where two grains move relative to one another, parallel to the plane of the GB. Under an appropriate driving force, other boundaries may *migrate*, moving normal to their plane as one grain is essentially transformed into the other. The details of these mechanisms and their relative energetic costs are determined by the GB structure [RPW$^+$91, WM92, SWPF97].

The details of MS simulations of GBs are discussed at some length in the book by Sutton and Balluffi [SB06], while more specific studies include, for example, [RS96, PMP01, SM05]. Reviewing some of the details of these studies serves as a good example of the use of MS in materials science. The goal is to determine the equilibrium arrangement of atoms at the GB, given the orientation of the two grains defining it and a model for the atomic interactions. This is done by building a bi-crystal and running an MS simulation to find the minimum energy structure.

We can build a GB by putting two crystals into a simulation box like the one illustrated in Fig. 6.15(a). The two crystal orientations can be specified, say, through a rotation transformation from some reference orientation. This suggests a total of six degrees of freedom, as each crystal requires three parameters to specify a rotation: two for a unit vector specifying the axis of rotation and one for the rotation angle. In fact, there are only five degrees of freedom insofar as the structure is concerned, since we can always rotate the two crystals by the same amount around the X_3-axis in the figure without changing the relative orientation of the two grains.

How to treat the boundary conditions can be tricky. Leaving the sides of the box free will lead to unwanted surface effects unless we make the system very large. On the other hand, holding the sides fixed will prevent the GB from correctly relaxing during minimization. To make the two crystals into infinite "slabs," it is necessary to choose a periodic length in the X_1 and X_2 directions that is commensurate with the periodicity of both crystals in those directions. This means, in effect, that only the so-called "rational" GBs[29] can be modeled in

[29] See [SB06] for more details about the differences between rational and irrational interfaces.

this way. Periodicity in the X_3 direction will lead to another unwanted effect: a second GB will be created at the edge of the periodic cell as shown in Fig. 6.15(a). This only matters for so-called "enantiomorphic" crystals, or crystals that have no mirror-symmetry planes. Otherwise, the two boundaries created in the periodic cell will be crystallographically indistinguishable. In this case, a calculation using the periodic cell in Fig. 6.15(a) will give twice the GB energy as long as the periodic length in the X_3 direction is sufficiently large to minimize the interaction between the GBs.

However, the methodology described above does not account for an additional *microscopic* degree of freedom which is associated with a GB. The two crystals defining the GB can rigidly translate one with respect to the other by a vector $\boldsymbol{s}$ called the "rigid body translation" (RBT). This does not change the relative orientation, but will locally change the neighbor environment of atoms near the interface. The optimal RBT during relaxation is not necessarily the same for the two GBs in the periodic simulation box, except for cases of rather special, high-symmetry boundaries. To circumvent this problem, a more complex model can be built as illustrated in Fig. 6.15(b). Here, periodicity is maintained only in X_1 and X_2, with free surfaces on the top and bottom, X_3, directions. The finite slabs of atoms on the top and bottom are not held fixed, but rather constrained to "float" as rigid slabs during relaxation. As long as these slabs are sufficiently thick, the free atoms see only a perfect crystal above and below the GB, and that crystal is free to move to the optimal RBT vector dictated by relaxation. Additionally, GBs often exhibit a higher free volume per atom than the perfect crystal. The technique of Fig. 6.15(b) permits free expansion of the system along X_3 (although this can also be accomplished with the simulation box of Fig. 6.15(a) as well if the approach of Section 9.5.4 is implemented).

In order to impose the rigid slab constraint, imagine there is a subset $\mathcal{S}$ of the atoms that define a slab. The position of these atoms is constrained to be

$$\boldsymbol{r}^\alpha = \boldsymbol{R}^\alpha + \boldsymbol{s}, \qquad \forall \alpha \in \mathcal{S}, \tag{6.65}$$

where $\boldsymbol{R}^\alpha$ is the reference position of atom α and $\boldsymbol{s}$ is a single displacement vector common to all the atoms in the slab.[30] The configurational force on the slab displacement can be found by a simple chain rule to be

$$\boldsymbol{f}^{\text{slab}} = \frac{\partial \mathcal{V}^{\text{int}}}{\partial \boldsymbol{s}} = \sum_\alpha \frac{\partial \mathcal{V}^{\text{int}}}{\partial \boldsymbol{r}^\alpha} \frac{\partial \boldsymbol{r}^\alpha}{\partial \boldsymbol{s}} = \sum_{\alpha \in \mathcal{S}} \boldsymbol{f}^\alpha. \tag{6.66}$$

To implement this within an existing MS code, it may be more expedient to modify the force routine to compute the sum in Eqn. (6.66) and then replace the force on each atom $\alpha \in \mathcal{S}$ by $\boldsymbol{f}^{\text{slab}}/n_{\mathcal{S}}$, where $n_{\mathcal{S}}$ is the number of atoms comprising the slab.[31] This can be handled by a few lines of code confined to the end of the force routine, and will give the

[30] Since one of the two slabs can always be constrained to have zero displacement, $\boldsymbol{s}$ of the other slab will be equivalent to the RBT vector defined earlier, so we use the same notation.

[31] This trick of dividing the slab force equally amongst the atoms in the slab will not work for all minimization algorithms. It *will* work for common gradient solvers like the CG method, because the search direction is a linear combination of vectors that are proportional to the forces. Other methods may allow the slab atoms to move independently of each other unless they are specifically constrained. For example, in NR methods two atoms in the slab may move a different amount under the same force because of varying atomic stiffness throughout the slab.

same result without the need to modify the minimization routine to explicitly account for the slab constraint.

An important result of GB simulations, in addition to the GB structure itself, is the GB energy per unit area. We can extract this value from either of the simulation cells described in Fig. 6.15. In (a), the GB energy is simply the excess energy per unit area of GB:

$$\gamma_{\mathrm{GB}} \equiv \frac{E_{\mathrm{cell}} - N(-E_{\mathrm{coh}})}{2l_1 l_2}, \tag{6.67}$$

where E_{cell} is the total relaxed energy of the simulation cell and the factor of 2 takes care of the fact that there are two such boundaries. However, care must be taken in these simulations that the two boundaries are in fact the same; as we have already mentioned a certain RBT may lead to two different structures at the GBs, or even introduce an elastic strain energy into the crystals that will erroneously affect the result. For the simulation cell of Fig. 6.15(b), the simplest way of determining the GB energy is to consider the excess energy of *only* the free atoms:

$$\gamma_{\mathrm{GB}} \equiv \frac{\sum_{\alpha \notin \mathcal{S}} (E^{\alpha} - (-E_{\mathrm{coh}}))}{l_1 l_2}. \tag{6.68}$$

As usual with MS simulation, the challenge is the existence of multiple minima in an extremely rugged energy landscape. Depending on the initial guess for the RBT, there will likely be several different relaxed GB energies and structures. What we really want is the *lowest-energy* GB structure, since this is what we are likely to see in real crystals. As we have already touched upon at the start of this chapter, there is no guarantee that we will find the global minimum, and all that can really be done is the brute force approach of trying many different values of the RBT for the initial guess.

Rittner and Seidman [RS96] illustrated this dramatically with a relatively simple family of boundaries, the $[110]$ symmetric tilt boundaries in fcc metals. We can understand the construction of a "tilt" boundary by imagining the two crystals in Fig. 6.15 to be initially in exactly the same orientation, with a particular direction aligned with the X_1-axis (in this case, $X_1 = [110]$). The two crystals are then rotated by different angles about this axis before they are used to fill the two halves of the simulation box. Since the X_2-axis lies in the GB plane, all of the misorientation between the boundaries is due to "tilting" the grains with respect to each other. This is as opposed to a "twist" boundary, which is produced by rotating the two grains by different amounts around the X_3-axis. Finally, the special case of a *symmetric* tilt boundary refers to the situation where the tilt angle for the two grains is equal but opposite; grain 1 is rotated α degrees clockwise, and grain 2 is rotated by α degrees counter-clockwise.

In Fig. 6.16(a), we reproduce the results from [RS96] for the energy of symmetric tilt boundaries as a function of the angle of tilt in fcc Ni. The shape of this curve is characteristic of such plots found throughout the literature. Specifically, there are cusps at a number of special, low-energy boundaries separating ranges of angles over which the energy changes more smoothly. Pawaskar *et al.* [PMP01] showed that these curves can be somewhat misleading because they tend to show the energetics of only relatively short-period rational boundaries. There are indeed an infinite number of long-period (rational and irrational) boundaries with misorientation angles lying between the points shown in Fig. 6.16(a), and their energy levels do not necessarily follow the smooth curve suggested

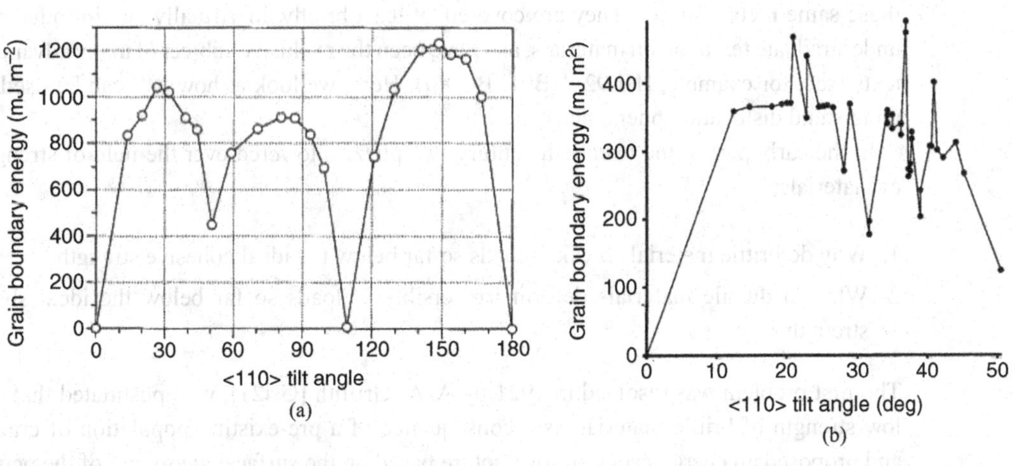

Fig. 6.16 (a) The energy of a family of grain boundaries as a function of tilt angle. (Reprinted with permission from [RS96], copyright 1996 by the American Physical Society.) (b) A closer look at the grain boundary energy dependence on tilt angle for several long-period boundaries. (Reprinted with permission from [PMP01], copyright 2001 by the American Physical Society.)

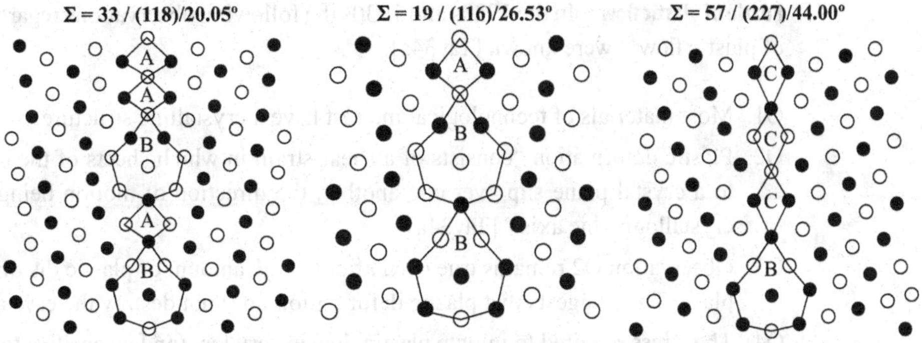

Fig. 6.17 The structures of some tilt boundaries in fcc Al. Black and white atoms indicate different planes of atoms, and the lines and letters denote characteristic structural units that comprise the grain boundaries. (Reprinted with permission from [RS96], copyright 1996 by the American Physical Society.)

by their intervening short-period cousins. This is shown in Fig. 6.16(b) for the same family of boundaries as in Fig. 6.16(a), but for fcc Al instead of Ni. Some representative boundary structures from this family are shown in Fig. 6.17. They can be viewed in two dimensions because they are tilt boundaries, and so there is no variation in the structure in the out-of-plane direction other than the AB stacking illustrated by the black and white atoms.

6.5.5 Crystal defects: dislocations

We now turn to the most prevalent defect in crystalline materials – the *dislocation* – as introduced in Chapter 1. Dislocations are recognized as the primary carriers of plastic deformation in crystalline solids, giving rise to the ductility that makes metals workable and, through their multiplication and entanglement, to the work hardening that makes those same metals strong. They are covered, at least briefly, in virtually any introductory undergraduate textbook on materials, and have been the exclusive subject of many advanced texts (see, for example, [HL92, HB01, BC06]). Here, we look at how MS can be used to understand dislocation phenomena.

In the early part of the twentieth century two puzzles towered over the field of strength of materials:

1. Why do brittle materials break at loads so far below the ideal cohesive strength?
2. Why do ductile materials deform irreversibly at loads so far below the ideal shear strength?

The first problem was resolved in 1921 by A. A. Griffith [Gri21], who postulated that the low strength of brittle materials is a consequence of a pre-existing population of cracks and proposed an energy criterion for fracture based on the surface energy, γ_s, of the newly formed surfaces following fracture. We do not discuss fracture mechanics in this book.[32] The second puzzle, that of plastic flow, remained unsolved until 1934 when G. I. Taylor, M. Polanyi and E. Orowan gave a simultaneous explanation.

Puzzle of plastic flow In the 1920s and 1930s the following observations regarding the nature of plastic flow[33] were known [Tay34a]:

O1. Most materials of technological interest have a crystalline structure.

O2. Plastic deformation "consists of a shear strain in which sheets of the crystal parallel to a crystal plane slip over one another, the direction of motion being some simple crystallographic axis." [Tay34a].

O3. Observation O2 remains true even after a large amount of plastic distortion has taken place. This suggests that plastic deformation does not destroy the crystal structure.

O4. The stress required to initiate plastic flow is very low (and insensitive to stress normal to the stress plane). However, as the deformation increases so does the stress required to continue the flow – a process referred to as *hardening*.

O5. Plastic deformation is athermal, i.e. the stress required for plastic deformation does not change significantly at very low temperature.

What is the microscopic mechanism responsible for this behavior? A hypothesis which is consistent with observations O1–O3 above is that plastic flow occurs when one part of a crystal is rigidly displaced over another as schematically illustrated in Fig. 6.18. The mathematical plane lying between two adjacent planes of atoms across which the slip occurs is called the *slip plane*. This form of deformation is called *rigid slip*. Is this the explanation for plasticity?

[32] The interested reader is referred to the many books on this subject. Brian Lawn's excellent book [Law93] is a good place to start.

[33] The term "plasticity" is used to describe irreversible deformation in materials. Unlike "elastic" deformation, which is recovered when the load is removed, plastic deformation is permanent.

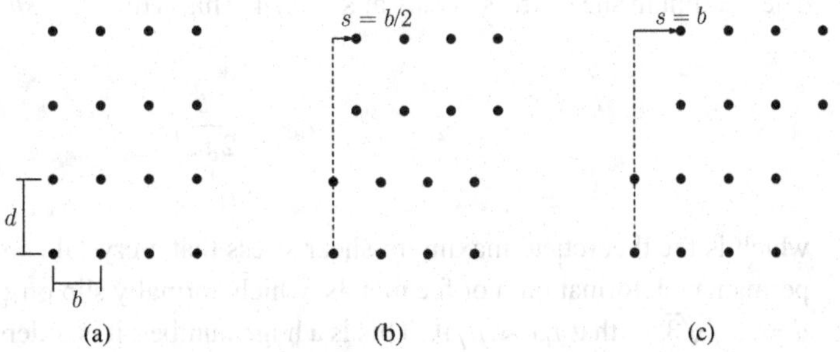

Fig. 6.18 A schematic of rigid slip. The top half of the crystal is rigidly displaced relative to the bottom by (a) $s = 0$, (b) $s = b/2$, (c) $s = b$. In the final configuration the perfect crystal structure is recovered since the atoms in the slipped layers are situated on lattice sites.

Frenkel model for rigid slip and the ideal shear strength In 1926, Jacov Frenkel (also known as Yakov Il'ich Frenkel) constructed a simple yet profoundly important model to estimate the strength of a perfect crystal in shear [Fre26]. Frenkel's model can be derived from the following reasoning. The shear stress, $\tau(s)$, associated with a rigid slip, s, of the rectangular lattice, depicted in Fig. 6.18, must satisfy the following conditions:

1. $\tau(s) = 0$ at $s = nb/2$ ($n = 0, 1, 2, \dots$). This condition reflects the fact that a rigid slip of $b/2 + nb$ corresponds to a state of unstable equilibrium where the slipped atoms are exactly midway between the unslipped ones as shown in Fig. 6.18(b). Similarly, a rigid slip of b (or its integer multiple) corresponds to a state of stable equilibrium since the perfect crystal structure is restored as shown in Fig. 6.18(c).
2. $\tau(s) > 0$ for $nb < s < (n + \frac{1}{2})b$ ($n = 0, 1, 2, \dots$). In this range a positive shear stress must be applied or the upper half of the crystal will return to the left to its original position at $s = nb$.
3. $\tau(s) < 0$ for $(n + \frac{1}{2})b < s < (n + 1)b$ ($n = 0, 1, 2, \dots$). In this range a negative shear stress must be applied to keep the upper half from sliding to the new equilibrium configuration at $s = (n + 1)b$.

A simple function with these properties is

$$\tau = k \sin \frac{2\pi s}{b}. \tag{6.69}$$

The constant k can be obtained approximately from Hooke's law as follows. The shear strain across the layer where the rigid slip is applied is s/d, where d is the interplanar spacing (see Fig. 6.18). For small s/d, we expect Hooke's law to hold, so that

$$\tau = \mu s/d. \tag{6.70}$$

Comparing Eqns. (6.69) and (6.70) in the limit of small strain, where $\sin(2\pi s/b) \approx 2\pi s/b$, we find that $k = b\mu/2\pi d$. Therefore, Frenkel's model for rigid slip becomes

$$\tau = \frac{\mu b}{2\pi d} \sin \frac{2\pi s}{b}. \tag{6.71}$$

The maximum shear stress occurs at $s = b/4$. This is the *ideal shear strength* of the crystal:

$$\tau_{\text{id}} = \frac{\mu b}{2\pi d}, \tag{6.72}$$

which is the theoretical maximum shear stress that a crystal can sustain before exhibiting permanent deformation. For fcc metals, which normally slip on $\{111\}$ planes, $b = a_0/\sqrt{6}$, $d = a_0/\sqrt{3}$, so that $\tau_{\text{id}} \approx \mu/9$. This is a huge number; it is orders of magnitude larger than the yield stresses of typical crystals.[34] This result is clearly at odds with observation O4 regarding plastic flow outlined at the start of this section.[35]

The invention of the dislocation Frenkel's model appeared to suggest that rigid slip was not the explanation for plastic flow. An alternative explanation emerged at this point, which is reminiscent of the explanation due to Griffith for brittle fracture. The suggestion was that real crystals possess defects which served as stress concentrations that locally facilitate plastic flow. The nucleation, interaction and multiplication of these defects should explain hardening. Remarkably, in 1934 three papers were published simultaneously by G. I. Taylor [Tay34a, Tay34b], E. Orowan [Oro34a, Oro34b, Oro34c] and M. Polanyi [Pol34], which postulated the existence of "dislocations."[36] The basic idea was that slip did not occur simultaneously across the entire slip plane but instead occurred "over a limited region, which is propagated from side to side of the crystal in a finite time" [Tay34a]. The defect separating the slipped region from the unslipped region was called a "dislocation" by Taylor (and a "*versetzung*," which means "transfer," by Orowan and Polanyi). Plastic deformation could then be described by the motion of the dislocations.

The structure and motion of a dislocation is illustrated in Fig. 6.19. The slip plane is denoted ∂B in Fig. 6.19(a) (this notation will be made clear when dislocations are discussed from a continuum perspective below). Slip is then introduced on the left and

[34] The "yield stress" is the stress at which plastic deformation is initiated. Although Frenkel's model fails to predict this material parameter for macroscopic ductile materials, it turns out that in highly-pure nanostructures with very low defect densities, strengths approaching this value are indeed observed. See, for example, [SL08].

[35] One possibility for the lack of agreement between Frenkel's model and experimental observations is that this is due to the highly-approximate nature of the Frenkel model. However, it turns out that detailed atomistic calculations lead to results that differ in detail from the Frenkel predictions but not in its conclusions. See Section 6.5.6 for more details.

[36] As might be expected, when multiple papers are simultaneously published on the same topic – there is a story [NA95]. Orowan and Polanyi worked independently but were aware of each other's work. Orowan who was less senior, just out of graduate school, wanted to publish a joint paper with Polanyi. However, Polanyi thought that Orowan's work was independent and in the end they agreed to publish simultaneously. Both acknowledged Prandtl as having anticipated the dislocation in some of his work in the later 1920s. Taylor worked independently of Orowan and Polanyi and was not aware of their work until it came out in *Z. Phys.* He had submitted his paper to *Proc. Roy. Soc. London* before them but it came out later. When Taylor saw Orowan's and Polanyi's publications, he sent his draft to Orowan for comment. Orowan replied that "unfortunately his theory was all wrong." (He was referring to Taylor's assumptions that: (1) dislocations were produced abundantly through thermal activation; (2) crystals possessed built-in obstacles to dislocation motion which were responsible for hardening; (3) only one slip system was active at a time.) Taylor replied that Orowan "was unable to follow a mathematical argument." Eventually the two met, and according to Orowan, Taylor conceded that he was wrong.

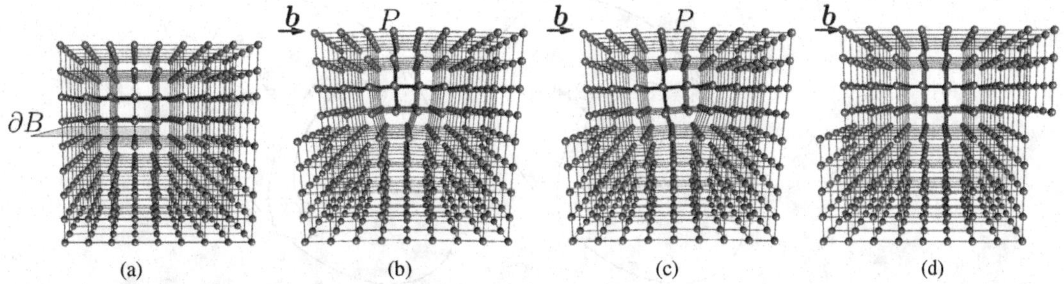

Fig. 6.19 A dislocation introduced on the slip plane shown in (a) will move through the crystal as in (b)–(c) to ultimately displace the two half crystals by one Burgers vector (d).

gradually propagates to the right. In Fig. 6.19(b) and (c) the slip has only propagated part way across the crystal. The end of the slipped region is marked by the presence of an apparent *extra half plane* P. The dislocation is the line located at the end of the extra half plane and extending into the paper; it separates the slipped and unslipped regions. As Taylor explains: "the stresses produced in the material by slipping over a portion of a plane are necessarily such as to give rise to increased stresses in the part of the plane near the edge of the region where slipping has already occurred, so that the propagation of slip is readily understandable and is analogous to the propagation of a crack." [Tay34a].

Referring again to Fig. 6.19, we see that a dislocation carries with it a quantum of plastic deformation (much like an electron carries a set electric charge). The amount and direction of slip carried by a dislocation is called its *Burgers vector*,[37] and is denoted by $\boldsymbol{b}$. Note that passage of the dislocation from one side of the crystal to the other has resulted in the slip of the top half relative to the bottom by an amount equal to the magnitude of $\boldsymbol{b}$. Thus the end effect is identical to that of rigid slip, but the process is gradual. In fact, the motion of the extra half plane (and thus the dislocation) by one atomic spacing involves a minor rearrangement of a single row of bonds. This explains why it is so easy to move a dislocation relative to the stress that would be required for rigid slip.

The characterization of dislocations and their geometry can be made more precise by adopting a continuum perspective as illustrated in Fig. 6.20.[38] We pass through the body with an arbitrary plane with normal $\boldsymbol{n}$. On part of this plane we make an internal cut, ∂B, to create two surfaces, one on each side of the cut and each enclosed by the contour line C. We now slip the two sides of the cut plane relative to one another by the Burgers vector $\boldsymbol{b}$. Note that $\boldsymbol{b}$ *usually* lies in the plane of ∂B so that $\boldsymbol{b} \cdot \boldsymbol{n} = 0$. After one side of the cut plane is translated with respect to the other by $\boldsymbol{b}$, the two free surfaces are welded back together and what remains is a *dislocation loop* along the line C. In this way, we see that the dislocation can be thought of as the boundary separating the slipped region of the plane within ∂B from the unslipped region without. The line C (with local tangent vector $\boldsymbol{l}$) is referred to as the *dislocation line*.

37 The Burgers vector is named after Dutch physicist Jan Burgers. Note that it is not "Burger's" vector, a common error.

38 In fact, we will see below that the possibility of dislocations had already been studied by Volterra from a purely mathematical perspective 30 years before its role was recognized in materials science.

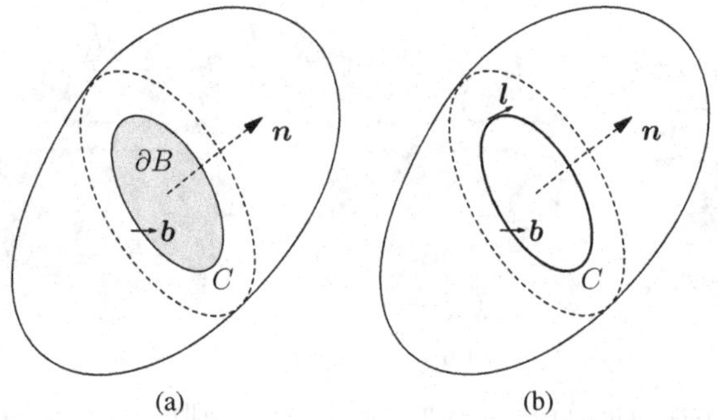

Fig. 6.20 Creation of a dislocation. Cutting along a surface ∂B and slipping the two faces by $\boldsymbol{b}$ in (a) leaves a dislocation loop after the faces are rewelded in (b). The dislocation line is along C with local tangent $\boldsymbol{l}$.

Some features of this simple dislocation loop turn out to be true for dislocations in general. First, we see that at every point along the dislocation, the Burgers vector is constant; it is *defined* by the slip that we introduced. However, the line direction $\boldsymbol{l}$ changes as we go around the loop. At some places it is at right angles to $\boldsymbol{b}$, where we say that the dislocation is *an edge dislocation*. At others, $\boldsymbol{b}$ and $\boldsymbol{l}$ are parallel and we say that the dislocation is *a screw dislocation*. At most places along the line, the situation is somewhere in between, and we say that the dislocation is *mixed*. (From this definition it is easy to see that the dislocation illustrated in Fig. 6.19 is an edge dislocation.) Constructed in this way, we also see that the dislocation line cannot abruptly end within the crystal, it must either form a perfect loop or meet one of the free surfaces of the solid. Indeed, this is true of all dislocations.

Elastic fields around dislocations – Volterra solutions The "cutting and welding" process can be applied as an internal boundary condition on an elastic continuum, and the resulting stress and displacement fields can be solved exactly for certain geometries (see, for example, [HL92, HB01]). This problem was initially studied by the Italian mathematician and professor of mechanics Vito Volterra in 1905.[39] Volterra's solutions were obtained using small strain isotropic elasticity theory. For instance, consider a straight dislocation with its line direction along X_3. If we assume that the slipped region is the X_1–X_3 plane for $X_1 < 0$ and the dislocation is of mixed character, with an edge component of its Burgers vector b_e and screw component b_s, then the total Burgers vector is $\boldsymbol{b} = b_e \boldsymbol{e}_1 + b_s \boldsymbol{e}_3$. In this case the boundary conditions are:

1. a continuous displacement field everywhere except on ∂B, where the jump in the displacements is $\boldsymbol{u}^+ - \boldsymbol{u}^- = \boldsymbol{b}$;
2. stresses must all go to zero at "infinity" (far from the dislocation core at $(X_1, X_2) = (0, 0)$); and
3. continuous tractions across ∂B.

[39] Volterra studied this problem as a purely theoretical exercise in the elasticity theory of solids with multi-valued displacement fields. This is described in Love's book [Lov27]. Volterra named deformations of this kind "distorsioni," which Love translated as "dislocations." This is perhaps the origin of the English term.

The displacement field from Volterra's solution for this problem is

$$
\begin{aligned}
u_1 &= \frac{b_e}{2\pi}\left[\tan^{-1}\frac{X_2}{X_1} + \frac{X_1X_2}{2(1-\nu)(X_1^2+X_2^2)}\right],\\
u_2 &= -\frac{b_e}{2\pi}\left[\frac{1-2\nu}{4(1-\nu)}\ln(X_1^2+X_2^2) + \frac{(X_1^2-X_2^2)}{4(1-\nu)(X_1^2+X_2^2)}\right],\\
u_3 &= \frac{b_s}{2\pi}\tan^{-1}\frac{X_2}{X_1},
\end{aligned}
\tag{6.73}
$$

where ν is Poisson's ratio. Note the complete decoupling between the edge and screw components: u_1 and u_2 depend only on b_e, while u_3 depends only on b_s. The corresponding stress field is

$$
\begin{aligned}
&\sigma_{11} = -\frac{\mu b_e}{2\pi(1-\nu)}\frac{X_2(3X_1^2+X_2^2)}{(X_1^2+X_2^2)^2}, \qquad \sigma_{22} = \frac{\mu b_e}{2\pi(1-\nu)}\frac{X_2(X_1^2-X_2^2)}{(X_1^2+X_2^2)^2},\\
&\sigma_{33} = \nu(\sigma_{11}+\sigma_{22}) = -\frac{\mu b_e\nu}{\pi(1-\nu)}\frac{X_2}{X_1^2+X_2^2},\\
&\sigma_{12} = \frac{\mu b_e}{2\pi(1-\nu)}\frac{X_1(X_1^2-X_2^2)}{(X_1^2+X_2^2)^2}, \qquad \sigma_{13} = -\frac{\mu b_s}{2\pi}\frac{X_2}{X_1^2+X_2^2}, \qquad \sigma_{23} = \frac{\mu b_s}{2\pi}\frac{X_1}{X_1^2+X_2^2},
\end{aligned}
\tag{6.74}
$$

where μ is the shear modulus of the elastic body. As for the displacement field, there is a decoupling between edge and screw effects since each stress component depends only on b_e or b_s but not both (at least for this isotropic case). The fields for a pure edge or pure screw case are easily found by setting the other Burgers vector component to zero.

The strain energy due to a straight, infinite dislocation is not well defined, even on a per-unit-length basis. If we consider an annular region surrounding the core with inner radius r_0 and outer radius R, it is straightforward to compute the total energy (per unit length, L) in the region to be

$$
\frac{\mathcal{E}_{\text{elast}}}{L} = \int_0^{2\pi}\int_{r_0}^{R}\frac{1}{2}(\boldsymbol{\sigma}:\boldsymbol{\epsilon})r\,dr\,d\theta = \left(\frac{\mu b_e^2}{4\pi(1-\nu)} + \frac{\mu b_s^2}{4\pi}\right)(\ln R - \ln r_0). \tag{6.75}
$$

We see that the energy of a dislocation is proportional to b^2. This indicates that dislocations with shorter Burgers vectors are energetically more favorable, which is indeed observed in materials. It also leads to dislocations splitting into smaller "partial dislocations" as is discussed below for fcc crystals.

The singularity in the stresses at the core mean that the energy will diverge if we try to take the limit $r_0 \to 0$. We must instead think of r_0 as some estimate of the "core radius" within which elasticity ceases to be valid. In contrast, the divergence in the limit of $R \to \infty$ is a real effect which indicates that a dislocation in an infinite crystal has infinite energy. In real crystals, the finite dimensions of the crystal, and more importantly the mean spacing between dislocations of opposite sign whose fields cancel, provide a length scale for R.

The energy of a typical dislocation in a metal is of the order of 10^{-9} J/m [Gor76]. This may not seem very large, but a sugar-cube-sized piece of a typical engineering alloy contains about 10^5 km of dislocation line [AJ05]!! This gives an energy of about 0.1 J.

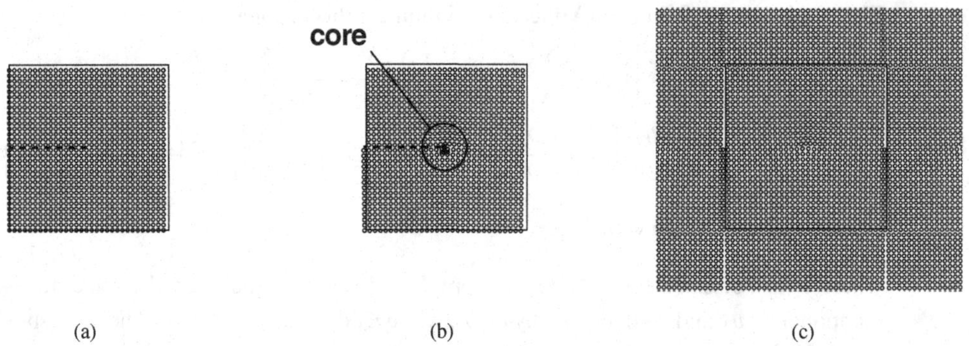

Fig. 6.21 The periodic simulation box for a perfect crystal in (a) is "cut" along the dotted line to create the dislocation in (b). Periodicity cannot be satisfied with the dislocation, as shown in (c). Viewing the page edge-on from below is helpful in visualizing the dislocation cores.

Compare this with the surface energy of the new surfaces created when cleaving the same cube of fcc metal, which is only about 5×10^{-4} J. This explains the discrepancy between brittle cleavage and ductile fracture.

Within the elastic continuum framework, the slipped region that we have called ∂B is merely a mathematical branch cut with no physical meaning. The displacement fields u_1 and u_3 jump by b_e and b_s respectively as we cross the branch cut, while all other fields are continuous across it.[40] As far as the continuous elastic fields are concerned, the location of the branch cut is arbitrary, and for that reason there are alternative versions of u_1 and u_3 in the literature that represent the same dislocation and produce the same stresses.[41] In crystals composed of discrete atoms, however, the branch cut has physical meaning; it is the slip plane over which atoms have slipped as the dislocation passes through the crystal.

Dislocation core structure Modeling dislocations using MS and MD is an important tool in our understanding of plasticity. Indeed, the subject constitutes a large part of the book by Bulatov and Cai [BC06], where the reader may find more details. Such simulations are critical for understanding the core structure of these defects and for revealing how multiple dislocations interact and how they move under various applied stresses. Here, we touch on the subject briefly.

The process of building an infinite, straight, edge dislocation in a rectangular simulation cell is illustrated in Fig. 6.21. While the simple cutting and welding process previously described can be used to initialize the unrelaxed dislocation core, care must be taken to avoid bringing atoms unphysically close to one another. Instead, we commonly make use of the solution for the displacements around a dislocation in an elastic continuum given in Eqn. (6.73). The effect of applying the displacement field is similar to the "cut" and slip process previously described, but with a better distribution of the resulting strains in the

[40] However, none of the displacements or stresses are defined at the core $(X_1, X_2) = (0, 0)$.

[41] This can be the source of some confusion. For example the displacement fields given for an edge dislocation in [HL92] and [Nab87] appear different because they assume different branch cuts.

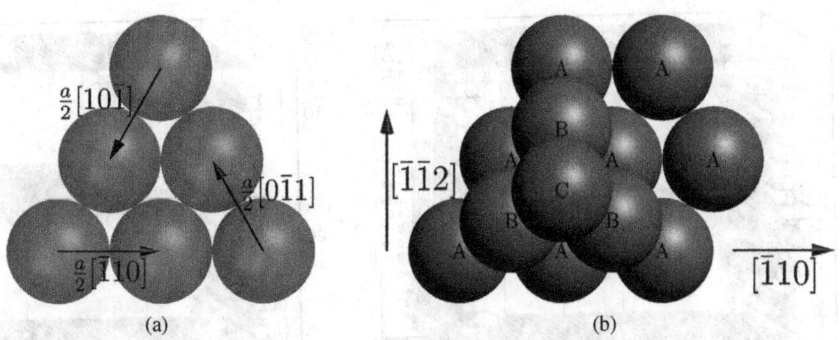

Fig. 6.22 (a) A single (111) plane in the fcc structure, showing the three dense-packed directions, and (b) the stacking arrangement of the (111) planes in the fcc structure.

crystal. Far from the core, we expect that the elastic solution will be exact provided that the elastic constants and correct anisotropy of the crystal are used.[42]

Once the initial structure is created, boundary conditions must be applied. As explained in the caption, PBCs cannot be used directly in this case, since the resulting atomic structure in the simulation cell no longer "fits" with its neighbors in a periodically repeating infinite model: there are gaps and overlaps at the edges.[43] The simplest approach for simulating a dislocation core is to fix a layer of the outermost atoms to the positions dictated by linear elasticity. In this way, free atoms in the interior do not "see" beyond the fixed region and do not experience surface effects. The total energy of this structure does not have physical meaning, as it includes an artificially created surface energy. But the final structure of the dislocation core is correct, so long as we have created a large enough model that the assumptions of linear elasticity are valid on the fixed outer region. We can always check this, of course, by systematically increasing the size of the simulation box until the results of interest do not change.

The $(111)[\bar{1}10]$ edge dislocation in Al As a simple example, let us consider the most common dislocation in fcc metals (for definiteness, we consider fcc Al modeled using the pair functional of Ercolessi and Adams [EA94] with $a_0 = 4.032$ Å). The planes in the fcc crystal with the highest atomic density per unit area are the family of $\{111\}$ planes. In general, dense-packed planes are the most favorable for dislocation glide in a crystal structure, and this is indeed the case for $\{111\}$ in the fcc structure. It is also the case that slip occurs most favorably along dense directions within a slip plane, i.e. with the Burgers vector oriented along directions of a high number of atoms per unit length. In the (111) plane shown in Fig. 6.22(a) there are three equivalent dense-packed directions, with the shortest repeat distances $a_0[1\bar{1}0]/2$, $a_0[10\bar{1}]/2$ and $a_0[01\bar{1}]/2$ forming the Burgers vectors. Since there are four sets of planes in the $\{111\}$ family, there are 24 possible *slip systems* in fcc: 4 planes times 3 Burgers vectors times 2 directions for each ($\boldsymbol{b}$ or $-\boldsymbol{b}$).

[42] In many instances, only a few atomic spacings is already "far enough" from the core for linear elasticity to be valid [MP96].

[43] Periodic approaches for modeling dislocations do exist but require care as explained in [CBC$^+$ 03].

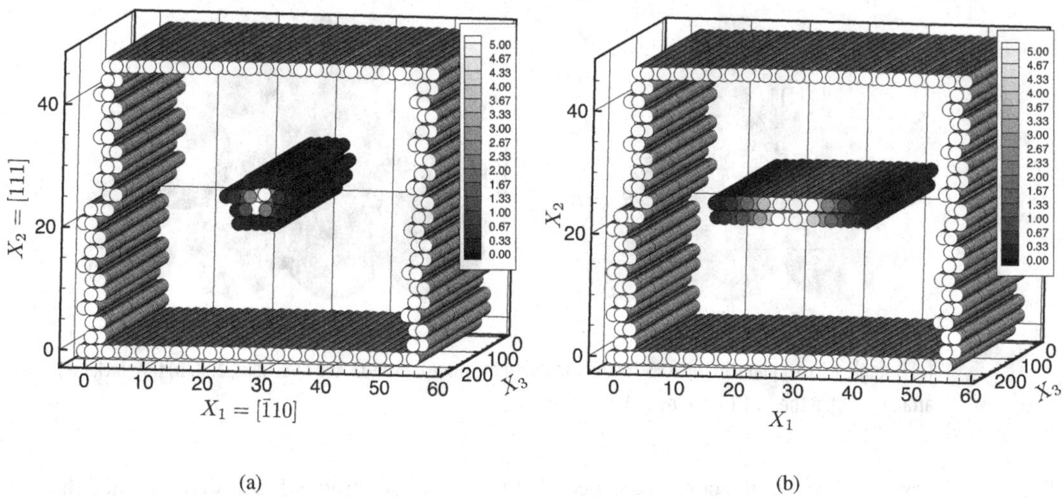

Fig. 6.23 The initially imposed perfect edge dislocation in (a) dissociates into two partials and a stacking fault in (b) upon relaxation. See text for details.

To perform an MS simulation of the core structure of an infinite straight edge dislocation of this type, we orient an fcc crystal as shown in Fig. 6.23(a), with the X_1 direction along the Burgers vector $[\bar{1}10]a/2$, the X_2 direction along the slip plane normal $[111]$ and the X_3 direction along the dislocation line $[11\bar{2}]$. The model shown comprises 21 (111) planes, each of which is 20 atoms wide along X_1 and 100 atoms along X_3. A dislocation is introduced by applying the displacements of Eqn. (6.73) with $b_e = a_0/\sqrt{2}$ and $b_s = 0$. We use PBCs only along X_3, with the other two directions free. To avoid the influence of the free surfaces, atoms within 7.0 Å of the free surfaces are fixed to the linear elastic positions (the cutoff radius of the atomistic model is 5.55 Å).

Figure 6.23(a) shows only the "interesting atoms" that are either near the free surfaces defining the model or near the core of the dislocation. This is done by erasing atoms with a value of the *centrosymmetry parameter* ($P_{\rm CS}$) less than 0.1 Å^2. The centrosymmetry parameter is a measure of the local deviation from perfect centrosymmetry, and is a common tool used for visualization of defects in simple lattices (which are centrosymmetric when defect-free). Recalling the definition of centrosymmetry on page 127, $P_{\rm CS}$ was originally defined by Kelchner *et al.* [KPH98] as

$$P^{\alpha}_{\rm CS} = \sum_{\beta=1}^{N_{\rm pair}} \left\| \boldsymbol{d}^{\beta} + \boldsymbol{d}^{\beta+N_{\rm pair}} \right\|^2, \tag{6.76}$$

where $\boldsymbol{d}^{\beta}$ and $\boldsymbol{d}^{\beta+N_{\rm pair}}$ are vectors pointing from atom α to a pair of its neighbors. The pairs are chosen such that in the perfect lattice, they are at equal and opposite vectors from atom α and $P^{\alpha}_{\rm CS}$ goes to zero. The number of pairs, $N_{\rm pair}$, depends on the crystal structure and the number of neighbor shells to be considered in the analysis. For example, in fcc crystals there are 12 nearest neighbors that can be arranged into six offsetting pairs. In bcc crystals, there are four such pairs. The second-neighbor shell would add three pairs in either the fcc or bcc structure. Since any homogeneous deformation preserves centrosymmetry in these crystals, only deformation that is heterogeneous on the atomic scale make a contribution to P_{CS}. Large values of P_{CS} are registered in dislocation cores and at free surfaces (where some neighbors are in fact missing for which we define $\boldsymbol{d}^{\beta} = \mathbf{0}$). We can see in Fig. 6.23(a) that the unrelaxed core of the dislocation is confined to a small region at the center of the simulation box.

Relaxation using the CG method (Section 6.2.5) leads to the final core structure of Fig. 6.23(b). We see that the core has spread out over the slip plane. This is the commonly observed *dissociation* of the full dislocation into two *Shockley partial dislocations*[44] which bound a stacking fault. The dissociation reaction is

$$\frac{a_0}{2}[\bar{1}10] \rightarrow \frac{a_0}{6}[\bar{2}11] + \frac{a_0}{6}[\bar{1}2\bar{1}], \tag{6.77}$$

meaning that the two partial dislocations have the Burgers vectors of the right-hand side. Each of these is a mixed dislocation, since $\boldsymbol{b} \cdot \boldsymbol{l} \neq 0$. These are termed "partial" dislocations because either one by itself does not create enough slip to restore atomic registry across the slip plane. The remaining misregistry is the stacking fault, which carries with it an associated energy per unit area, γ_{SF}. The width of the separation between the two dislocations is determined by the stacking fault energy and the elastic moduli of the crystal. This is because having two dislocations in close proximity leads to an interaction energy due to the overlap of their elastic strain fields. Dislocations of like sign[45] can lower this interaction energy by moving apart. However, in doing so there is an energetic penalty because the width of the stacking fault grows, and the equilibrium spacing of the partials reflects this energetic trade-off.

It turns out that much simpler MS simulations than those described above can be used to explain dislocation behavior such as core splitting. These alternative approaches are based on the concept of the γ-surface which is discussed in the next section.

6.5.6 The γ-surface

The γ-surface is a concept that was introduced by Vitek [Vit68] as a generalization of the rigid slip model of Frenkel discussed on page 349. Imagine an infinite perfect crystal, for which we define the potential energy to be zero. Now, choose a single plane (with normal $\boldsymbol{n}$) within that crystal and rigidly displace all the atoms on one side of this plane by a vector $\boldsymbol{s}$, such that $\boldsymbol{s} \cdot \boldsymbol{n} = 0$. The effect will be a single plane of disregistry within an otherwise perfect crystal, for which there will be a resulting energy per unit area, γ. We can now imagine repeating this calculation for all possible $\boldsymbol{s}$ lying in the slip plane to get an energy landscape $\gamma(\boldsymbol{s})$. This is the so-called γ-surface.

[44] The Shockley partial is named for William Shockley who, among other things, shared the Nobel prize in physics in 1956 for inventing the transistor with John Bardeen and Walter Houser Brattain. His work to commercialize the transistor is largely credited for the rise of Silicon Valley in California. Later in life, he became a controversial advocate of eugenics and was accused of racism. The historical record of the link between his namesake and partial dislocations is limited to a one-paragraph contribution called "Half-dislocations" in the minutes of the 1947 APS meeting [Sho48].

[45] The Shockley partials are predominantly of like sign, since the like-sign edge components are larger than the opposite-sign screw components.

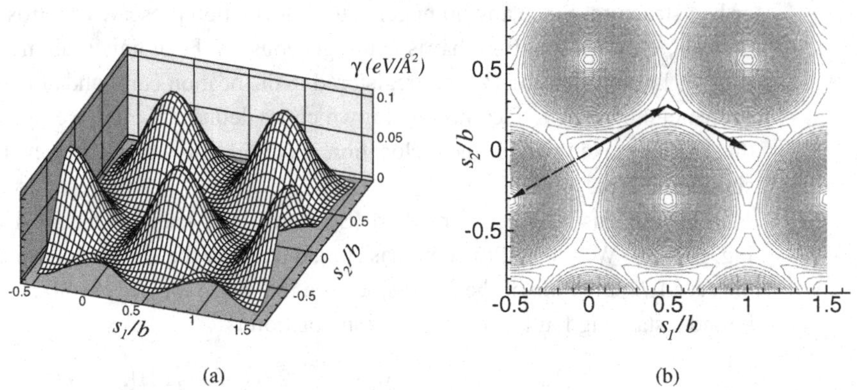

(a) (b)

Fig. 6.24 (a) The (111) γ-surface for fcc Al. X_1 is along $[\bar{1}10]$, X_2 is along $[\bar{1}\bar{1}2]$. (b) A two-dimensional projection of the surface in (a). The Burgers vector length is $b = a_0/\sqrt{2}$.

An example of a γ-surface is shown in Fig. 6.24 for the case of the (111) plane in Al.[46] There is of course a periodicity to the γ-surface, since any displacements of the form

$$s = m\frac{a_0}{2}[\bar{1}10] + n\frac{a_0}{2}[\bar{1}01], \tag{6.78}$$

will restore the perfect lattice when m and n are integers.

We can understand this energy surface, and with it the formation of partial dislocations, when considering the stacking of (111) planes in the fcc structure as shown in Fig. 6.22(b). The bottom (111) plane is the triangular arrangement of atoms labeled "A" in the figure, which we can think of as hard spheres for simplicity. To build an fcc crystal, we add the "B" plane of atoms on the "divots" between the A atoms, like the stacking of neatly arranged oranges in a crate. There are two options for the next plane (the "C" atoms). Depending on which set of divots in the B plane we use, the C atoms can lie directly above the A atoms or (as shown in the figure) between them. If we choose to stack them as shown we create the fcc unit cell, and we can build an fcc crystal by repeating this stacking sequence (ABCABC...).[47]

This can be related to the γ-surface if we suppose that the slip plane is between the A and B atoms in Fig. 6.22(b), such that any slip will rigidly carry the B and C atoms over the A plane. We can see that slip directly along $[\bar{1}10]$ will not be the lowest-energy path, taking us partially up the side of one of the peaks in the landscape. A much lower-energy path is to follow the two solid arrows shown in Fig. 6.24(b). The first arrow, $a_0[\bar{2}11]/6$, takes the B planes into the alternative set of divots in the A planes, and puts the C planes directly above the A planes, to form a local region of hcp within the otherwise fcc lattice. This is

[46] This γ-surface was computed using the Ercolessi–Adams pair functional [EA94], and is "unrelaxed" (see the discussion of extensions to the Peierls–Nabarro model on page 370).

[47] On the other hand, stacking the C atoms directly over the A atoms leads to the stacking sequence ABAB.... This forms the hcp crystal structure of Fig. 3.20, with the A planes forming the top and bottom of the hexagonal cell, and the B plane in between.

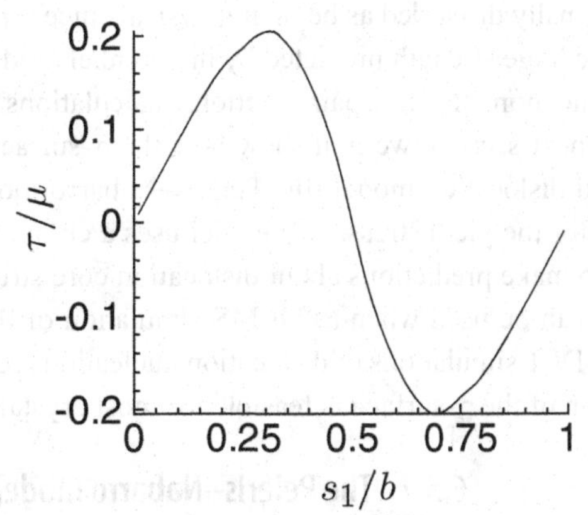

Fig. 6.25 The shear stress versus slip along $[\bar{1}10]$ in the (111) plane of fcc Al. The maximum, 0.20μ, is an estimate of the ideal shear strength of the crystal. The shear modulus is $\mu = 0.2291\,\mathrm{eV/\AA}^3 = 36.7\,\mathrm{GPa}$ and $b = a_0/\sqrt{2}$.

the *stacking fault* configuration . . . ABC AC ABC. . . .[48] Note that the stacking fault energy is small compared with the maxima, but it is not zero. For this particular pair functional it is about $7.6\,\mathrm{meV/\AA}^2$. The second arrow, $a_0[1\bar{2}\bar{1}]/6$, restores the perfect lattice. These two arrows are the Burgers vectors of the two partial dislocations formed during the dissociation described earlier.

The γ-surface reveals the asymmetry of slip in the fcc structure. For instance, we have already seen that sliding the B planes over the A planes along $[\bar{2}11]$ is a low-energy path. However, sliding along the opposite direction, $[2\bar{1}\bar{1}]$, as shown by the dashed arrow in Fig. 6.24(b) takes us up the high peak as B atoms are moved directly on top of the A atoms below them.

Revisiting the ideal shear strength Earlier we obtained the estimate in Eqn. (6.72) for the ideal shear strength of a crystal from Frenkel's model of rigid slip. It is of interest to compare this simple prediction with the ideal shear strength computed from the γ-surface. The stress required to impart a slip $\boldsymbol{s}$ can be determined directly from γ as

$$\tau = \frac{\partial \gamma}{\partial \boldsymbol{s}}, \tag{6.79}$$

which can be used as an estimate of the ideal shear strength of the crystal. In Fig. 6.25, we take this derivative along the direction of s_1 from Fig. 6.24, to see the stress required to slip the crystal along the direction of the full $a_0[\bar{1}10]/2$ Burgers vector.

The maximum stress along this curve is about 0.20 times the shear modulus, μ. In comparison the Frenkel model gives $\tau_{\mathrm{id}} = 0.195\mu$ (where we have used $b = a_0/\sqrt{2}$ and $d = a_0/\sqrt{3}$ for the spacing between (111) planes). The predictions of the Frenkel model

[48] Specifically, this is referred to as an *intrinsic* stacking fault. It can alternatively be thought of as forming due to the removal of a plane from the sequence. Conversely, the *extrinsic* stacking fault can be thought of as forming due to the insertion of an extra plane. The extrinsic fault can also be formed by the same slip occurring on two adjacent planes.

were originally discarded as being unphysical since it predicted such a large ideal strength. In fact the large strength predicted by the Frenkel model is realistic and in this case is very close to the more accurate pair functional calculations.

In the next section, we will show how the γ-surface can be used as an ingredient in a simplified dislocation model (the Peierls–Nabarro model) that significantly improves the estimate for the yield stress. The model uses a combination of linear elasticity and the γ-surface to make predictions about dislocation core structure, nucleation and motion under stress. It can be used when a full MS simulation of the dislocation is too expensive. For instance, DFT simulations of dislocation nucleation are prohibitive, but a high-quality DFT calculation of the γ-surface is feasible for most crystal structures.

6.5.7 The Peierls–Nabarro model of a dislocation

Volterra's solution for the elastic field around a dislocation shows that Taylor's premise – expressed in the quote on page 351 – was correct: a dislocation creates a strong stress concentration in the material. However, Volterra's solution cannot be used to verify the second premise that the presence of a stress concentration implies that the dislocation can be moved easily. In other words, it cannot be used to compute the stress required to move a dislocation – the *Peierls stress*. This is because the motion of a dislocation involves atomic rearrangements at the core of the dislocation which is exactly where the elastic solution breaks down with a singularity. In order to make progress, the Peierls–Nabarro method was developed to take into account the effects of the discrete lattice.

The model starts from the assumption that dislocation slip is confined to a single plane, and that away from the slip plane, the distortions of the lattice are relatively minor. These assumptions are borne out reasonably well in many important dislocations; we have already seen one such example in the glissile $\{111\}\langle 110\rangle$ dislocation of fcc metals. On the other hand, screw dislocations in bcc structures and sessile dislocations in fcc structures (such as the Lomer dislocation) exhibit nonplanar cores. Nevertheless, dislocations with planar cores are an important subset of all possible dislocations, and so the model, first proposed by Peierls [Pei40] (who attributes the idea to Orowan), remains widely useful. In 1947, Nabarro revisited the problem, corrected a minor error in Peierls original paper and fleshed out the details considerably [Nab47].

The energy of a continuum containing a single slip plane The assumption that all slip is confined to a single slip plane allows the model depicted in Fig. 6.26 to be used. Two semi-infinite continuous half spaces, B^+ and B^- are joined at the slip plane, where the traction exerted between them is determined by a law relating the energy of slip to the relative motion of the two half spaces across the slip plane. As such, the total energy is

$$\mathcal{E}_{\text{tot}} = \mathcal{E}_{\text{elast}} + \mathcal{E}_{\text{slip}}, \tag{6.80}$$

where $\mathcal{E}_{\text{elast}}$ accounts for the energy of the elastic strain fields in the two half spaces, while $\mathcal{E}_{\text{slip}}$ is the energy of slip. Since the behavior of the half spaces is confined to linear elasticity, the model effectively assumes that all nonlinear behavior is confined to a single plane, and that the associated nonlinear energy is embodied by $\mathcal{E}_{\text{slip}}$.

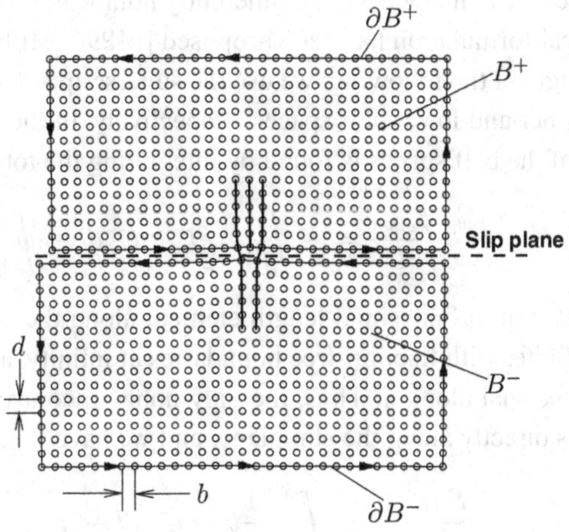

Fig. 6.26 The Peierls–Nabarro model.

Although it is relatively straightforward to extend the model to curved dislocations, loops, and dislocations with mixed character, we follow the original formulation and confine our attention to a straight, infinite edge dislocation in a simple rectangular lattice with lattice spacing b along the slip plane and lattice spacing d between planes. As a result the problem is two-dimensional and the surfaces ∂B^+ and ∂B^- illustrated in Fig. 6.26 can now be considered to be closed circuits in the plane of the figure. Energies calculated from this two-dimensional model will be per-unit-length, L, along the dislocation line.

To obtain $\mathcal{E}_{\text{slip}}$, we write

$$\frac{\mathcal{E}_{\text{slip}}}{L} = \int_{-\infty}^{\infty} \gamma(s(x))\, dx, \tag{6.81}$$

where $\gamma(s)$ is a one-dimensional cut through the generalized stacking fault energy surface discussed in the last section and x measures the position on the slip plane. Because it is more convenient for the Peierls–Nabarro model, we change our notational convention slightly here, using x, y and z instead of X_i. This serves to free the subscript for other uses later in the discussion and should not present any confusion.

This last equation illustrates how the Peierls–Nabarro model can be considered an early multiscale modeling approach: γ is an atomic-scale quantity injected into an otherwise continuum framework. Since γ is the energy per unit area of slip plane due to a *uniform* slip, s, over the entire plane, there is an inherent assumption that the slip distribution varies slowly on the scale of the atomic spacing. This implies rather wide dislocation cores, although we shall see that the Peierls–Nabarro model solution is not self-consistent in this respect. There are additional assumptions inherent in Eqn. (6.81). One is that *opening* displacements are neglected, in that we ignore any possible dependence of γ on the relative motion of the two half spaces normal to the slip plane. It is relatively straightforward to lift this restriction from the model, and the effects of this so-called "shear-tension coupling" have been studied [SBR93, BK97]. A more subtle approximation is that of a *local* model; the energy at a point x in the integrand is determined entirely by the slip at that point, $s(x)$.

Since atomistic interactions are inherently nonlocal, this assumption can be questioned and a nonlocal formulation has been proposed [MP96, MPBO96].

Throughout the following, we use the superscripts $+$ and $-$ to indicate quantities related to the upper and lower half spaces, respectively. In the absence of body forces, the elastic energy of the half spaces is found by integrating the total work done over ∂B^+ and ∂B^-,

$$\frac{\mathcal{E}_{\text{elast}}}{L} = -\oint_{\partial B^+} \frac{1}{2}\boldsymbol{t}^+ \cdot \boldsymbol{u}^+ dl - \oint_{\partial B^-} \frac{1}{2}\boldsymbol{t}^- \cdot \boldsymbol{u}^- dl, \tag{6.82}$$

where dl is an infinitesimal length element along the curve.[49] We assume the half spaces to be infinite, with traction-free boundaries at infinity, and further that the slip plane is flat with its normal along y. Thus, the only nonzero contribution to the integrals comes from the paths directly along the slip plane, and we have

$$\begin{aligned}\frac{\mathcal{E}_{\text{elast}}}{L} &= -\int_{-\infty}^{\infty} \frac{1}{2}\boldsymbol{t}^+ \cdot \boldsymbol{u}^+ \, dl - \int_{\infty}^{-\infty} \frac{1}{2}\boldsymbol{t}^- \cdot \boldsymbol{u}^- \, dl,\\ &= -\int_{-\infty}^{\infty} \frac{1}{2}\boldsymbol{t}^+ \cdot \boldsymbol{u}^+ \, dx - \int_{\infty}^{-\infty} \frac{1}{2}\boldsymbol{t}^- \cdot \boldsymbol{u}^- \, (-dx).\end{aligned}$$

Note that the bounds on the second integral on the first line are from ∞ to $-\infty$. That is because of the direction of the ∂B^- circuit in Fig. 6.26. Now we use the Cauchy relation (Eqn. (2.87)) so that

$$\frac{\mathcal{E}_{\text{elast}}}{L} = -\int_{-\infty}^{\infty} \frac{1}{2}(\boldsymbol{\sigma}^+\boldsymbol{n}^+) \cdot \boldsymbol{u}^+ \, dx - \int_{-\infty}^{\infty} \frac{1}{2}(\boldsymbol{\sigma}^-\boldsymbol{n}^-) \cdot \boldsymbol{u}^- \, dx,$$

and substitute $[\boldsymbol{n}^+] = [0, -1, 0]^T$ and $[\boldsymbol{n}^-] = [0, 1, 0]^T$ to get

$$\frac{\mathcal{E}_{\text{elast}}}{L} = \int_{-\infty}^{\infty} \frac{1}{2}\begin{bmatrix}\sigma_{xy}^+ & \sigma_{yy}^+ & \sigma_{zy}^+\end{bmatrix}\begin{bmatrix}u_x^+\\ u_y^+\\ u_z^+\end{bmatrix} dx - \int_{-\infty}^{\infty} \frac{1}{2}\begin{bmatrix}\sigma_{xy}^- & \sigma_{yy}^- & \sigma_{zy}^-\end{bmatrix}\begin{bmatrix}u_x^-\\ u_y^-\\ u_z^-\end{bmatrix} dx.$$

As discussed in the last section, tractions at the interface are determined from the derivatives of the interplanar potential with respect to the components of slip in the three coordinate directions. If we confine ourselves to a purely edge dislocation so that $u_z^+ = u_z^- = 0$, the stress components σ_{zy} are not a factor. Further, since we have simplified $\gamma(s)$ to be a function of only the x-component of slip, an additional assumption is typically made that only the shear components of the tractions are nonzero at the interface. This implies that $\sigma_{yy} = 0$. Lastly, in order for the system to be in equilibrium the jump in tractions across

[49] Each elastic half space can be considered a body subject to a traction loading, $\boldsymbol{t}$, on its boundary. The total potential energy of the system (body + loads) is then (see Eqn. (2.207)):

$$\mathcal{E}_{\text{elast}} = \int_B W dV - \int_{\partial B} t_i u_i dS = \int_B \frac{1}{2}\sigma_{ij}\epsilon_{ij} dV - \int_{\partial B} t_i u_i \, dS.$$

To put this in terms of only traction and displacement, we note that the integrand in the volume integral is $\sigma_{ij}\epsilon_{ij} = \sigma_{ij}u_{i,j} = (\sigma_{ij}u_i)_{,j}$, where the last step makes use of the equilibrium condition $\sigma_{ij,j} = 0$. Inserting this and applying the divergence theorem in Eqn. $(2.51)_2$ and the Cauchy relation ($t_i = \sigma_{ij}n_j$) recovers the form from which we have started in Eqn. (6.82), including the negative sign and the factor of $1/2$.

the interface must be zero, implying that the remaining traction components must be equal, $\sigma_{xy}^{+} = \sigma_{xy}^{-} = \tau$. These steps let us combine the integrals such that

$$\frac{\mathcal{E}_{\text{elast}}}{L} = \int_{-\infty}^{\infty} \frac{1}{2}\tau(u_x^+ - u_x^-)\,dx = \int_{-\infty}^{\infty} \frac{1}{2}\tau(x)s(x)\,dx, \tag{6.83}$$

where the slip s is the displacement jump across the slip plane:

$$s \equiv u_x^+ - u_x^-. \tag{6.84}$$

Turning to the contribution to the energy due to slip, we can, in principle, obtain $\gamma(s)$ from any atomistic model ranging from DFT to pair potentials, and use it as the atomistic input to the otherwise entirely continuum-based Peierls–Nabarro model. However, analytical progress was made in the original formulation by using the Frenkel form for $\gamma(s)$. Referring back to Eqn. (6.79) and using the Frenkel expression for the shear stress in Eqn. (6.71), we have

$$\gamma(s) = \int \tau(s)\,ds = \frac{\gamma_{\max}}{2}\left(1 - \cos\frac{2\pi s}{b}\right), \tag{6.85}$$

where we have imposed the condition, $\gamma(0) = 0$, and where

$$\gamma_{\max} = \frac{\mu b^2}{2\pi^2 d} \equiv \gamma_{\max}^{\text{LE}}. \tag{6.86}$$

The superscript "LE" reminds us that the amplitude of the Frenkel form was obtained by fitting the small strain limit to linear elasticity. Other ways to fit the Frenkel form to real crystals have been proposed. For example, Joós and Duesbery [JD97] argue that the small-strain regime is not relevant to dislocations, and a more appropriate fit might be to ensure that $\gamma(s)$ produces a realistic value for the maximum shear stress that the slip plane can sustain. This amounts to fitting the maximum slope of $\gamma(s)$ to some value, τ_0, and leads to

$$\gamma_{\max}^{\text{JD}} = \frac{b\tau_0}{\pi}. \tag{6.87}$$

Note that τ_0 is related to the maximum ideal shear strength of the crystal, which is not the same thing as the Peierls stress.

Given an appropriate $\gamma(s)$ and the elastic moduli for the half spaces, the total energy of the system is then

$$\frac{\mathcal{E}_{\text{tot}}}{L} = \int_{-\infty}^{\infty} \frac{1}{2}\tau(x)s(x)\,dx + \int_{-\infty}^{\infty} \gamma(s(x))\,dx. \tag{6.88}$$

The next step is to determine the relationship between the stress, τ, and the slip distribution, s, so that we can determine the function $s(x)$ that minimizes the total energy of Eqn. (6.88).

Continuously distributed dislocations We have already discussed how a dislocation core can be viewed as the boundary between the part of a crystal plane that has slipped and the part that has not. This idealized picture of a dislocation suggests the slip distribution shown in Fig. 6.27(a); the so-called Volterra dislocation. In this case, as we move from $x = -\infty$ to $x = 0$, we pass through the slipped region, after which there is a sudden jump to $s = 0$ in the

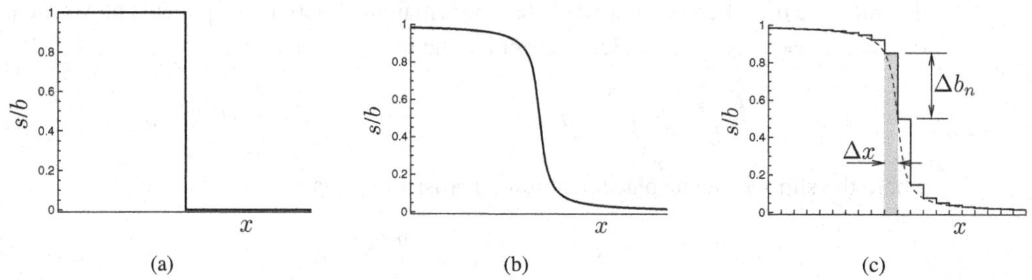

Fig. 6.27 Slip distributions: (a) the Volterra solution; (b) the Peierls–Nabarro model; and (c) a continuous slip distribution approximated by an array of discrete dislocations.

unslipped region. It is the discontinuous nature of this distribution that leads to singularities in the stress field at the core. By introducing the interplanar potential at the slip plane, the Peierls–Nabarro model permits distributed slip profiles like the one shown in Fig. 6.27(b).

To determine the stresses in the continuum half spaces due to a distributed slip distribution like Fig. 6.27(b), we can make use of the concept shown in Fig. 6.27(c), where the slip distribution is approximated by an array of Volterra dislocations with Burgers vector Δb_n located at $x_n = n\Delta x$, where n represents all integers. From the figure, we can see the relation between these Volterra dislocations and the slope of the slip distribution:

$$\frac{\Delta b_n(x_n)}{\Delta x} \approx \left.\frac{ds}{dx}\right|_{x=x_n}, \tag{6.89}$$

which becomes exact in the infinitesimal limit $\Delta x \to 0$, such that

$$db(x) = \frac{ds(x)}{dx}dx. \tag{6.90}$$

We can now determine the total stress at any point (x, y) in either half space by superimposing the stresses due to each infinitesimal dislocation at x' with Burgers vector $db(x')$. As we are interested only in the stress at the slip plane, we set $y = 0$ and use the result for the shear stress around a dislocation from Eqns. (6.74) to write

$$\tau(x) = \int_{-\infty}^{\infty} \frac{K}{2\pi}\frac{db(x')}{x - x'} = \frac{K}{2\pi}\int_{-\infty}^{\infty}\frac{ds(x')}{dx'}\frac{dx'}{x - x'}, \tag{6.91}$$

where for an isotropic continuum $K = \mu/(1-\nu)$, and elastic anisotropy can be introduced by using a different value of K. Using this in Eqn. (6.88) gives us

$$\frac{\mathcal{E}_{\text{tot}}}{L} = \frac{K}{4\pi}\int_{-\infty}^{\infty}\frac{ds(x')}{dx'}\left[\int_{-\infty}^{\infty}\frac{s(x)}{x - x'}dx\right]dx' + \int_{-\infty}^{\infty}\gamma(s(x))\,dx. \tag{6.92}$$

It is helpful to do an integration by parts on the inner integral of the first term,[50] so that we have

$$\frac{\mathcal{E}_{\text{tot}}}{L} = \frac{\mathcal{E}_{\text{Volt}}}{L} - \frac{K}{4\pi}\int_{-\infty}^{\infty}\int_{-\infty}^{\infty}\frac{ds(x')}{dx'}\frac{ds(x)}{dx}\ln|x - x'|\,dx dx' + \int_{-\infty}^{\infty}\gamma(s(x))\,dx, \tag{6.93}$$

[50] Recall the formula $\int_a^b fg'\,dx = fg|_a^b - \int_a^b f'g\,dx$. In this case, $f = s(x)$ and $g' = 1/(x - x')$.

where we have introduced a term $\mathcal{E}_{\text{Volt}}$ (for reasons that will soon be clear) that is defined as

$$\frac{\mathcal{E}_{\text{Volt}}}{L} = \frac{K}{4\pi} \int_{-\infty}^{\infty} \frac{ds(x')}{dx'} \left[s(x) \ln |x - x'| \right]_{x=-\infty}^{x=\infty} dx'. \tag{6.94}$$

A little thought can simplify this term considerably. Let us replace the evaluation limits for x with $\pm R$ (we will later take the limit as $R \to \infty$):

$$\frac{\mathcal{E}_{\text{Volt}}}{L} = \lim_{R\to\infty} \frac{K}{4\pi} \int_{-\infty}^{\infty} \frac{ds(x')}{dx'} \left[s(R) \ln |R - x'| - s(-R) \ln |-R - x'| \right] dx'. \tag{6.95}$$

We expect the dislocation core to be bounded near the origin, such that $ds(x')/dx'$ is zero far from $x' = 0$. Further, as $R \to \infty$, the logarithmic terms are nearly constant over the range of x' for which $ds(x')/dx'$ is appreciably different from zero, and can be taken as equal to $\ln |R|$. Finally, since

$$\int_{-\infty}^{\infty} \frac{ds}{dx'} dx' = -b \tag{6.96}$$

must hold if the slip distribution corresponds to a full dislocation, we can evaluate

$$\frac{\mathcal{E}_{\text{Volt}}}{L} = \lim_{R\to\infty} -\frac{Kb}{4\pi} \left[s(R) \ln |R| - s(-R) \ln |R| \right]. \tag{6.97}$$

At $R = -\infty$, we expect $s = b$, whereas at $R = \infty$ we expect $s = 0$, so that

$$\frac{\mathcal{E}_{\text{Volt}}}{L} = \lim_{R\to\infty} \frac{Kb^2}{4\pi} \ln R. \tag{6.98}$$

It now becomes clear that this term represents the energy of the Volterra dislocation solution (see Eqn. (6.75)), and it is independent of the form of $s(x)$ beyond the reasonable assumptions about how $s(x)$ behaves at $x = \pm\infty$. Putting this all together one more time gives us the final expression for the total energy:

$$\frac{\mathcal{E}_{\text{tot}}}{L} = -\frac{K}{4\pi} \int_{-\infty}^{\infty} \int_{-\infty}^{\infty} \frac{ds(x')}{dx'} \frac{ds(x)}{dx} \ln |x - x'| \, dx dx' + \int_{-\infty}^{\infty} \gamma(s(x)) \, dx + \lim_{R\to\infty} \frac{Kb^2}{4\pi} \ln R. \tag{6.99}$$

Solving for the slip distribution The energy of Eqn. (6.99) is a function of only the slip distribution, and equilibrium will correspond to an energy minimum. We can now seek the function $s(x'')$ such that the functional derivative of $\mathcal{E}_{\text{tot}}$ satisfies

$$\frac{\delta(\mathcal{E}_{\text{tot}}/L)}{\delta s(x'')} = 0. \tag{6.100}$$

Evaluating this functional derivative using Eqn. (6.99) and simplifying leads to[51]

$$-\frac{d\gamma(s(x))}{ds} = \frac{K}{2\pi}\int_{-\infty}^{\infty} \frac{ds(x')}{dx'}\frac{1}{x-x'}\,dx'. \tag{6.101}$$

We see that this is simply a statement that the tractions due to the interplanar potential along the interface must be equal to the traction on the boundaries of the half spaces due to the superposition of the infinitesimal dislocations.

At this point, one could proceed to solve this equation numerically for any general $\gamma(s)$, and this has been carried out by a number of researchers (see, for example, [SBR93, BF94, RB94, JRD94, GB95, XAO95, BK97, LKBK00a]). However, an analytic solution is possible if we assume the Frenkel form of Eqn. (6.85) for the interplanar potential. Using Eqns. (6.71) and Eqn. (6.79) in Eqn. (6.101) we have

$$-\frac{\pi\gamma_{\max}}{b}\sin\frac{2\pi s}{b} = \frac{K}{2\pi}\int_{-\infty}^{\infty} \frac{ds(x')}{dx'}\frac{1}{x-x'}\,dx', \tag{6.102}$$

where we have used the definition of $\gamma_{\max}$ from Eqn. (6.86) to emphasize the fitting procedure discussed previously. This can be solved for $s(x)$ subject to the conditions that $s = b$ at $x = -\infty$ and $s = 0$ at $x = \infty$. One can verify[52] that the slip distribution that satisfies this equation is

$$s(x) = \frac{b}{2} - \frac{b}{\pi}\tan^{-1}\frac{x}{\zeta}, \tag{6.103}$$

where

$$\zeta = \frac{Kb^2}{4\pi^2\gamma_{\max}} \tag{6.104}$$

is a measure of the width of the dislocation. Taking isotropic elasticity and Eqn. (6.86) for $\gamma_{\max}$ yields

$$\zeta_{\mathrm{LE}} = \frac{d}{2(1-\nu)}. \tag{6.105}$$

Figure 6.28(a) shows this solution in order to give a sense of the effect of the ζ, which is often referred to as the half width of the core. Since ν is usually between about 0.2 and 0.4 for metals, we can take an average value of $\nu = 0.3$ and conclude that the core width is $2\zeta_{\mathrm{LE}} \approx 1.4d$. This is clearly not very wide, calling into question one of the key assumptions used to define the slip energy. This lack of consistency in the model was noted from the start by Peierls himself, but the model nonetheless allows us to see the effects of lattice periodicity, specifically with respect to the stress field of the dislocation and the Peierls stress required to move it.

[51] Some details regarding functional differentiation need to be recalled in this derivation. First, recall that $\delta f(x)/\delta f(y) = \delta(x-y)$ and that it is possible to exchange the order of integration and differentiation. Therefore $(\delta/\delta f(x))\int g(y)f(y)\,dy = g(x)$. Also, derivatives of Dirac delta functions that appear can be turned into the Dirac delta function itself using integration by parts.

[52] Although Eqn. (6.103) is, in fact, an exact solution to Eqn. (6.101), it is not trivial to see that this is so. The proof is outlined briefly in [HL92] and makes use of the method of residues [AW95].

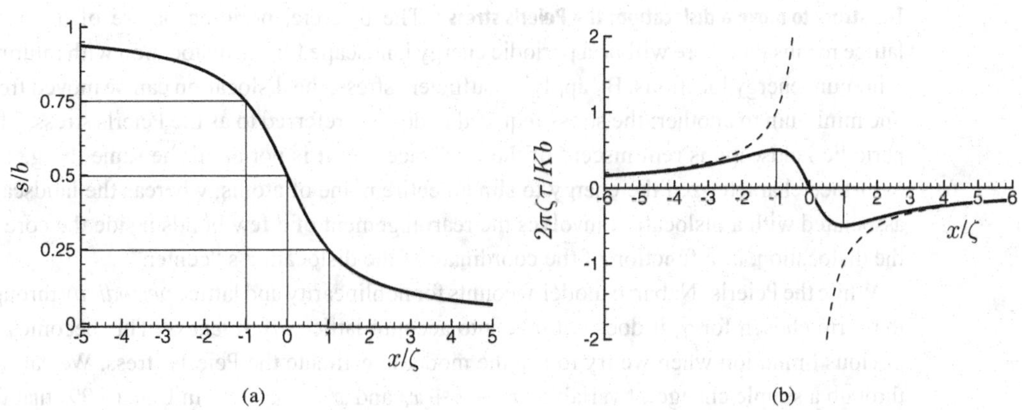

Fig. 6.28 (a) The solution for the slip distribution in the Peierls–Nabarro model. (b) The shear stress along the slip plane predicted by the Peierls–Nabarro model (solid) line compared to that of the Volterra solution (dashed lines).

The stress field of a Peierls–Nabarro dislocation Once the slip distribution has been found, its derivative provides complete information about the continuously distributed dislocations that reside on the slip plane. In this case

$$\frac{ds}{dx} = -\frac{b}{\pi}\frac{\zeta}{x^2+\zeta^2}, \tag{6.106}$$

and we can determine the stress anywhere in the body by integrating the effects of the distributed dislocations. For example, Eqn. (6.91) gives the shear stress along the slip plane

$$\tau(x) = \frac{-Kb\zeta}{2\pi^2}\int_{-\infty}^{\infty}\frac{1}{(x')^2+\zeta^2}\frac{1}{x-x'}\,dx', \tag{6.107}$$

which can be solved in closed form:

$$\tau(x) = \frac{-Kb}{2\pi}\frac{x}{x^2+\zeta^2}. \tag{6.108}$$

This stress distribution is plotted in Fig. 6.28(b), and compared with the result for the Volterra solution (which can be found by setting $\zeta = 0$ in Eqn. (6.108)). The most important effect, of course, is that this stress distribution is no longer singular, reaching a well-defined maximum at $\pm\zeta$. However, the value of this maximum (which corresponds to the Peierls–Nabarro prediction for the ideal shear strength) is

$$\tau_{\text{id}} = \frac{Kb}{4\pi\zeta} = \frac{\pi\gamma_{\max}}{b}, \tag{6.109}$$

where in the last step we substituted in Eqn. (6.104). If we assume isotropic elasticity and substitute Eqn. (6.86) for $\gamma_{\max}$ into Eqn. (6.109), we obtain $\tau_{\text{id}}^{\text{LE}} = \mu b/2\pi d$. Note that despite the more elaborate Peierls–Nabarro analysis, this prediction for the ideal shear strength is identical to that of the Frenkel model in Eqn. (6.72)! However, we shall see in what follows that the stress required to move a dislocation is considerably lower.

The stress to move a dislocation: the Peierls stress The discrete, periodic nature of a crystal lattice means that there will be a periodic energy landscape for the dislocation with multiple minimum energy locations. By applying sufficient stress, the dislocation can be moved from one minimum to another; the stress required to do so is referred to as the Peierls stress. This periodic landscape is reminiscent of the γ-surface, but it is not at all the same thing. The γ-surface characterized the energy to slip an entire plane of atoms, whereas the landscape associated with a dislocation involves the rearrangement of a few bonds inside the core of the dislocation, as a function of the coordinate of the dislocation's "center."

While the Peierls–Nabarro model accounts for nonlinearity and lattice *periodicity* through the form chosen for γ, it does not take into account lattice *discreteness*. This becomes an obvious limitation when we try to use the model to estimate the Peierls stress. We can see through a simple change of variables $x \to x + x_c$ and $x' \to x' + x_c$ in Eqn. (6.99) that the energy of the dislocation is indifferent to x_c, the location of the center of the core.[53] This indifference stems from the treatment of $\mathcal{E}_{slip}$ as an integral, which is only correct in the limit where the lattice constant is infinitesimal. In a sense, we would really like to sum over energy contributions due to slip between atomic pairs across the slip plane as

$$\frac{\mathcal{E}_{slip}}{L} = \int_{-\infty}^{\infty} \gamma(s(x))\, dx \to \sum_{n=-\infty}^{\infty} \gamma(nb)\Delta x, \tag{6.110}$$

which samples the slip distribution only at discrete positions that are $\Delta x (= b)$ apart. It is now possible to introduce a dependence on the core location by shifting the slip distribution relative to the sampling points:

$$\frac{\mathcal{E}_{slip}(x_c)}{L} = \sum_{n=-\infty}^{\infty} \gamma(s(nb + x_c))\Delta x. \tag{6.111}$$

The consistent thing to do at this point would be to reevaluate the solution to the model with this version of the slip energy, leading to a slip distribution (and energy) that depends on the core location. Considering variations with respect to the core location would then allow us to obtain the maximum stress required to move the dislocation between neighboring wells. This has been solved [BK97]. The approximate approach taken in the original Peierls–Nabarro work is simpler, but still enlightening for our purposes.

The Peierls–Nabarro approximation to the Peierls stress calculation was to assume that the slip distribution of Eqn. (6.103) remains unchanged if the discrete form for $\mathcal{E}_{slip}$ is used. This is consistent with the assumption of wide dislocation cores (although we have already seen that this is not what the model predicts), and it has the convenient feature that the elastic energy $\mathcal{E}_{elast}$ also remains constant. As such, all one needs to do is to substitute Eqn. (6.103) into Eqn. (6.111) to determine the dependence of the total energy on x_c. Using the Frenkel form (Eqn. (6.85)) for γ we therefore have

$$\frac{\mathcal{E}_{slip}(x_c)}{L} = \frac{b\gamma_{max}}{2} \sum_{n=-\infty}^{\infty} \left\{1 + \cos\left[2\tan^{-1}\left(\frac{nb - x_c}{\zeta}\right)\right]\right\}. \tag{6.112}$$

[53] We define x_c as the location at which $s(x) = 0.5b$.

Joós and Duesbery [JD97] have shown how to obtain this energy in closed form. Defining the nondimensional variables $\Gamma = \zeta/b$ and $z = x_c/b$, and using the identity

$$1 + \cos\left[2\tan^{-1}\frac{x}{\zeta}\right] = \frac{2\zeta^2}{x^2+\zeta^2}, \tag{6.113}$$

we can write $\mathcal{E}_{\text{slip}}(z)/L$ explicitly as

$$\frac{\mathcal{E}_{\text{slip}}(z)}{L} = b\Gamma\gamma_{\max}\sum_{n=-\infty}^{\infty}\frac{\Gamma}{\Gamma^2+(n-z)^2}. \tag{6.114}$$

Since $\mathcal{E}_{\text{slip}}(z)$ is an even periodic function with period 1, it can be represented as a Fourier series

$$\frac{\mathcal{E}_{\text{slip}}(z)}{L} = \frac{a_0}{2} + \sum_{m=1}^{\infty} a_m\cos(2\pi m z),$$

where

$$a_m = \int_0^1 \frac{\mathcal{E}_{\text{slip}}(z)}{L}\cos(2\pi m z)\,dz.$$

These integrals can be evaluated in closed form such that

$$\frac{\mathcal{E}_{\text{slip}}(z)}{L} = \pi b\Gamma\gamma_{\max} + \sum_{m=1}^{\infty} 2\pi b\Gamma\gamma_{\max}e^{-2\pi m\Gamma}\cos(2\pi m z), \tag{6.115}$$

and the resulting series can be readily summed. Replacing the nondimensional variables z and Γ with their definitions leads to the final expression for the energy:

$$\frac{\mathcal{E}_{\text{slip}}(x_c)}{L} = \zeta\gamma_{\max}\pi\frac{\sinh(2\pi\zeta/b)}{\left(\cosh(2\pi\zeta/b)-\cos(2\pi x_c/b)\right)^2}. \tag{6.116}$$

The force on the dislocation is the derivative of this energy per unit length with respect to the core location, while the stress, τ, must be equal to the applied force divided by the norm of the Burgers vector [HL92]:

$$\tau = \frac{1}{bL}\frac{d\mathcal{E}_{\text{slip}}}{dx_c}. \tag{6.117}$$

The Peierls stress is the maximum value of the applied stress:

$$\sigma_P = \max\{\tau\} = \max\left\{\frac{1}{Lb}\frac{d\mathcal{E}_{\text{slip}}}{dx_c}\right\}. \tag{6.118}$$

This maximization is straightforward and leads to

$$\sigma_P = \tau(x_{c,\max}), \tag{6.119}$$

where $x_{c,\max}$ is obtained from

$$2\cos\left(\frac{2\pi x_{c,\max}}{b}\right) = -\cosh\left(\frac{2\pi\zeta}{b}\right) + \sqrt{9+\sinh^2\left(\frac{2\pi\zeta}{b}\right)}. \tag{6.120}$$

It is simplest to examine this expression in the limit of wide dislocation cores. To obtain this result, we return to Eqn. (6.115), and note that the core width appears in the exponent (through $\Gamma = \zeta/b$), multiplied by m. If the core is wide, these exponential terms decay rapidly with increasing m, and we therefore need only to retain the first term in the series to approximate the slip energy:

$$\frac{\mathcal{E}_{\text{slip}}(x_c)}{L} \approx \pi\zeta\gamma_{\max}\left[1 + 2e^{-2\pi\zeta/b}\cos\left(\frac{2\pi x_c}{b}\right)\right], \tag{6.121}$$

from which a straightforward maximization (and substitution of Eqn. (6.104)) leads to

$$\sigma_P = Ke^{-2\pi\zeta/b}. \tag{6.122}$$

If we assume isotropic elasticity this becomes

$$\sigma_P = \frac{\mu}{1-\nu}e^{-2\pi\zeta/b}. \tag{6.123}$$

This result differs (by a factor of 2 in the exponent) from the original derivation and more recent presentations. However, those results are in error due to a subtlety in how they treat the misfit at the interface. Specifically, the original derivation treated slip of the upper and lower half spaces, u^+ and u^-, independently; linearly combining the two results during the later stages of the derivation. It turns out that this linear combination is not correct for the nonlinear slip energy, and the correct exponent is as in Eqn. (6.123) [JD97].

We saw above that the ideal strength was on the order of the shear modulus, which is several orders of magnitude higher than experimental values. The Peierls–Nabarro dislocation provides a mechanism for plastic flow with a much lower Peierls stress, which we can now estimate. Assuming an fcc metal, isotropic elasticity and Eqn. (6.86) for the maximum surface energy, and taking $\nu = 0.3$, $b = a_0/\sqrt{2}$ and $d = a_0/\sqrt{3}$ leads to

$$\sigma_P \approx 3.7 \times 10^{-2}\mu.$$

While this is much higher than experimental results (for example in fcc Cu $\sigma_P \approx \times 10^{-6}\mu$), it is better by two orders of magnitude than the ideal shear stress calculation. Due to the exponential dependence of σ_P on ζ, the model is quite sensitive to small errors in the estimate of the core width. Given the numerous approximations that contribute to the estimate of ζ, it is perhaps unrealistic to expect much better agreement in the prediction of σ_P.

Extensions of the Peierls–Nabarro model A number of enhancements can, and have, been made to the Peierls–Nabarro model that improve the estimate for the Peierls stress. Most of these enhancements add sufficient complexity to the model to make an analytical solution intractable, so that numerical techniques must be used. For this reason we have chosen to discuss only the simplest form of the Peierls–Nabarro model. For example, one can recognize that the γ-surface is at least two-dimensional, to allow for the possibility of mixed and partial dislocations. The way we estimated the Peierls stress was to assume no change in the form of the slip distribution as the dislocation moves, yet it has been shown [BK97, LKBK00a] that relaxations during the motion are important. The exact details of computing the γ-surface can also be improved. For example, a constrained minimization of the atomic positions (such that the atoms can relax only perpendicular to the slip plane) is sometimes used to find $\gamma(\boldsymbol{s})$ [LBK02]. "Shear-tension coupling" effects [SBR93] make the γ-surface a three-dimensional energy landscape by also including the effect of rigidly separating the two halves of the crystals, such that the slip, $\boldsymbol{s}$, becomes a vector with three components and $\gamma = \gamma(\boldsymbol{s})$. Finally, the computation of the misfit energy can be improved by recognizing the discreteness of the lattice instead of treating it as a continuous integral [BK97, LKBK00b], as well as by treating the misfit energy as a nonlocal integral [MPBO96]. Rice [Ric92] has also used the Peierls–Nabarro model to obtain an analytical nucleation criterion for dislocations at a crack tip. This work was later extended to dislocation nucleation beneath a nanoindenter [SPT00] and deformation twinning at a crack tip [TH03].

6.6 Dealing with temperature and dynamics

This chapter has presented examples of what we can do with atomistic models at zero temperature (i.e. 0 K). The search for local equilibrium structures and transition paths between them can tell us a great deal about the structure of materials. Simple energy landscapes like the γ-surface can be used to understand mechanisms of deformation, and can also be incorporated into models like the Peierls–Nabarro model to explain observed behavior.

However, through all of this discussion the elephant in the room has been temperature. Our interest in real materials is always at finite temperature, and the fact that the atoms in crystals are always vibrating cannot be overlooked. In continuum mechanics, two of the field variables we introduced were the velocity and the temperature at every point in a body. We treated these as if they were independent variables. However, at the atomic scale, the velocity of the atoms is directly related to temperature, and the two phenomena are inextricably linked. Reconciling these two different worldviews is no easy task.

On the scale of atoms, these thermal fluctuations appear random and chaotic, occurring on a femtosecond time scale. This means that we often must resort to statistical methods to understand the effect of this motion on the macroscopic length and time scales of materials science. This is our approach in the discussion that follows. We begin in Chapter 7 with the critical foundations of statistical mechanics and then in Chapter 9 discuss equilibrium molecular dynamics, which is effectively a numerical realization of this method. In Chapter 8, we see how the statistical mechanics of discrete atoms gives rise to the continuum notions of stress and leads to the thermoelastic constitutive laws, as described in Chapter 11, which are so central to our modeling of material behavior.

Further reading

- The entire book by Wales [Wal03] is dedicated to understanding and computing features of energy landscapes, and relating these features to physical phenomena. It is an excellent, comprehensive treatment of the subject.

- For more details about methods for finding transition paths, the reader is directed to the chapter reviewing the subject in [HJJ00].
- Optimization (i.e. minimization) methods are discussed in many books. A classic, practical reference is the *Numerical Recipes* series [PTVF92]. Other, more mathematical treatments include [Pol71, Rus06]. More modern techniques, beyond what we touched upon in this chapter are discussed in the book by Deuflhard [Deu04].
- The book by Allen and Tildesley [AT87] is a good reference for the implementation details of MS (as well as MD). For a more modern discussion of issues surrounding implementation in a parallel-processing environment, see, for example, the paper by Plimpton [Pli95] which discusses the LAMMPS MD code [Pli09].
- For applications to crystals, a broadly accessible discussion is given in Phillips' book [Phi01]. A friendly introduction to dislocations is given in the book by Hull and Bacon [HB01], while a more comprehensive reference on the subject is that of [HL92]. For a thorough discussion of grain boundaries and other interfaces in materials (including atomistic simulation), Sutton and Balluffi's book [SB06] is a good source.

Exercises

6.1 [Section 6.4] Let V_{Verlet} be the volume around an atom contributing to a Verlet neighbor list with padding defined by ϵ_{neigh}. Similarly, let V_{binning} be the volume around an atom sampled in the binning method.
1. Obtain an expression for the ratio $V_{\text{binning}}/V_{\text{Verlet}}$ and plot it in the range $\epsilon_{\text{neigh}} \in [0, 1]$. Comment on the relative efficiency of the two methods.
2. At what value of ϵ_{neigh} does binning become more efficient than Verlet lists?

6.2 [Section 6.4] Consider a periodic one-dimensional chain interacting through a pair potential, $\phi(r)$, with n-neighbor interactions (i.e. each atom interacts with its n-closest neighbors on either side).
1. Show that the equilibrium chain spacing, a_0, is obtained by solving the following equation:

$$\sum_{k=1}^{n} k\phi'(ka_0) = 0.$$

This is the one-dimensional *zero-pressure condition.* **Hint** Refer to Example 6.2 and consider the thermodynamic tension conjugate with the periodic length of the chain $L = Na$, where N is the number of atoms in the periodic chain and a is the spacing between atoms.
2. Solve the above equation for the Lennard-Jones potential given in Eqn. (5.36) and obtain a closed-form expression for a_0 as a function of n. **Hint** In your solution, you will need to use $\sum_{k=1}^{n} k^{-m} = H_{n,m}$, where $H_{n,m}$ is the generalized harmonic number.

6.3 [Section 6.4] In the discussion of PBCs, it was stated that free surfaces can be modeled within a periodic system by making sure that the periodic direction normal to the surface is large enough to created a gap of width r_{cut}. However, in Section 5.8.5, it was demonstrated that for pair functionals, the force on an atom is influenced by atoms up to $2r_{\text{cut}}$ away. Does this mean that, when modeling surfaces in a pair functional system with PBCs, a $2r_{\text{cut}}$ gap is necessary? Explain.

6.4 [Section 6.5] Verify that the UBER form of Eqn. (6.53) exactly returns the bulk modulus of Eqn. (6.55).

6.5 [Section 6.5] Show that an alternative form for the UBER expression in Eqn. (6.57) is

$$E^*(r_{\text{nn}}) = -\left[1 + \eta\left(\frac{r_{\text{nn}}}{r_{\text{nn}}^0} - 1\right)\right] e^{-\eta(r_{\text{nn}}/r_{\text{nn}}^0 - 1)},$$

where $r_{\rm nn}$ is the nearest-neighbor distance, $r^0_{\rm nn}$ is the equilibrium nearest-neighbor distance, and $\eta = 3[\Omega_0^{\rm ws} B/E^0_{\rm coh}]^{1/2}$ is called the *anharmonicity parameter*.

6.6 [SECTION 6.5] Show that Johnson's nearest-neighbor pair functional given in Exercise 5.7 identically satisfies the UBER form given above in Exercise 6.5 for the fcc crystal structure.

6.7 [SECTION 6.5] Consider an fcc crystal and a nearest-neighbor pair functional model (see Section 5.5)

$$\mathcal{V}^{\rm int} = \frac{1}{2}\sum_{\substack{\alpha,\beta\\ \alpha\neq\beta}} \phi(r^{\alpha\beta}) + \sum_{\alpha} U(\rho^\alpha), \qquad \rho^\alpha = \sum_{\substack{\beta\\ \beta\neq\alpha}} g(r^{\alpha\beta}).$$

1. Show that if the model is reduced to a pair potential (by setting $U = 0$), both the cohesive energy and the vacancy formation energy are given by $-6\phi(r_{\rm min})$, where $r_{\rm min}$ is the distance corresponding to the minimum of $\phi(r)$.
2. Compute expressions for the cohesive energy and vacancy formation energy when $U \neq 0$.
3. Real metals typically have a vacancy formation energy less than half the cohesive energy (which cannot be fitted by the pair potential as shown in part 1). Show that if $U(\rho) = -A\sqrt{\rho}$, the vacancy formation energy is less than the cohesive energy only if A is positive (i.e. only if the embedding energy favors higher coordination). Neglect the effects of lattice relaxation, which will be small in this problem.

6.8 [SECTION 6.5] Consider a finite one-dimensional chain of N atoms interacting through a pair potential, $\phi(r)$, with second-neighbor interactions. The equilibrium atomic spacing for an *infinite* chain interacting with the same potential is a_0. Assume that $N > 4$.
1. Compute the ideal surface energy associated with each end of the chain.
2. Obtain an expression for the forces acting on all the atoms in the chain; use the zero-pressure condition in Exercise 6.2 to simplify your result as much as possible. Do you expect surface relaxation in this case? How many atoms from the end of the chain do you think would move if the energy of the chain were relaxed?
3. Repeat the above parts for the case where the potential is limited to near-neighbor interactions? Will there be surface relaxation in this case?

6.9 [SECTION 6.5] The atoms in an fcc crystal interact via a nearest-neighbor pair potential, $\phi(r)$. The equilibrium lattice parameter is a_0.
1. Show that the ideal surface energies for the (100), (110), and (111) surfaces are:

$$\gamma_{\rm s}^{(100)} = -4\frac{\phi(r^0_{\rm nn})}{a_0^2}, \qquad \gamma_{\rm s}^{(110)} = -5\frac{\sqrt{2}}{2}\frac{\phi(r^0_{\rm nn})}{a_0^2}, \qquad \gamma_{\rm s}^{(111)} = -2\sqrt{3}\frac{\phi(r^0_{\rm nn})}{a_0^2},$$

where $r^0_{\rm nn} = a_0/\sqrt{2}$ is the equilibrium nearest-neighbor distance.
2. Compute the above surface energies for Cu by estimating $\phi(r^0_{\rm nn})$ from Fig. 5.4. (Use an average value inferred from the different potentials plotted there.) Note that the equilibrium lattice parameter for Cu is $a_0 = 3.61$ Å.
3. Compare the pair potential surface energies computed above with the first-principles values from [VRSK98]: $\gamma_{\rm s}^{\rm DFT(100)} = 2166\ {\rm mJ/m^2}$, $\gamma_{\rm s}^{\rm DFT(110)} = 2237\ {\rm mJ/m^2}$, $\gamma_{\rm s}^{\rm DFT(111)} = 1952\ {\rm mJ/m^2}$, and the experimental value from [TM77]: $\gamma_{\rm s}^{\rm exp} = 1770\ {\rm mJ/m^2}$. (Note that the experimental value is an average over all orientations.)
4. Two main factors that affect the ordering of the surface energy are: (i) the surface coordination number, z_s, i.e. the number of nearest neighbors of atoms on the surface; (ii) the area-per-atom, $A_{\rm atom}$, on the surface. How do you expect to surface energy to change with each of these factors when considered separately (increase or decrease)? Use these results to discuss the ordering you obtained above from the pair potential and DFT results.

6.10 [SECTION 6.5] Derive the dislocation energy expression given in Eqn. (6.75).

6.11 [SECTION 6.5] Covalently bonded crystals like Si tend to have much narrower dislocation cores than metallic crystals. In the Peierls–Nabarro model, if the core is very narrow then only one term contributes to the sum in Eqn. (6.112). (To see this, consider that the slip energy is periodic in $x_{\rm c}$ with period b. We are seeking the maximum slope which will occur somewhere between $x_{\rm c} = 0$ and $x_{\rm c} = b/2$. In this range, only the $n = 0$ term contributes if ζ is very small.) Show that in this limit, the Peierls–Nabarro model prediction for the Peierls stress is inversely proportional to core width:

$$\sigma_{\mathrm{P}} = \frac{3\sqrt{3}}{8} \frac{\gamma_{\max}}{\zeta}.$$

6.12 [SECTION 6.5] Fill in the steps to get from Eqn. (6.100) to Eqn. (6.101).

6.13 [SECTION 6.5] For this exercise, use the MiniMol program provided on the companion website to this book, using the Ercolessi–Adams pair functional model for aluminum (see full directions on the website). All calculations should be performed with PBCs in all directions. Does the size of the periodic cell affect any of the results? If yes, increase the cell size until the result has converged.

1. Verify that the fcc structure is the most stable of the three structures fcc, bcc and hcp by generating a graph like Fig. 6.8(a). You should find that the cohesive energy of this model is 3.36 eV. Do not use any relaxation – simply build ideal periodic crystals and compute their energy as a function of varying the lattice constant. For the hcp structure, assume an ideal $c/a = \sqrt{8/3}$.
2. Test the stability of the bcc structure by adding small random displacements to the perfect lattice positions and then relaxing. Ensure that the final relaxed structure is still bcc.
3. Draw an E^* versus a^* UBER curve, similar to Fig. 6.10, for this fcc aluminum model. Plot the first- and third-order UBER models in the same figure. Discuss your results.
4. Calculate the fcc vacancy formation energy for this model, with and without relaxation. Without relaxation means that all atoms remain in their perfect lattice positions except the atom that is removed to create the vacancy, whereas relaxation requires energy minimization after the vacancy is removed before computing E_{cell} in Eqn. (6.58).

6.14 [SECTION 6.5] Using the argon pair potential implementation in MiniMol as a template (see companion website), implement the pair potential that you invented in Exercise 5.2. Make it so that you can easily modify your fitting parameters. Choose fitting parameters so that your potential gives the correct lattice constant and cohesive energy of fcc argon (see Tab. 5.1). If you require additional fitting parameters, set them arbitrarily for now.

6.15 [SECTION 6.5] Using your implemented potential from Exercise 6.14, repeat Exercise 6.13. If you have sufficient free parameters, adjust them so that your model predicts a bulk modulus of 2.54 GPa to match the experiments of [PBS66] (see Eqn. (6.55)). (When repeating Exercise 6.13, note that for argon the cohesive energy is 80 meV/atom and the experimental vacancy formation energy is in the range of 60–90 meV [Sch76].)

PART III

ATOMISTIC FOUNDATIONS OF CONTINUUM CONCEPTS

7 Classical equilibrium statistical mechanics

Statistical mechanics provides a bridge between the atomistic world and continuum models. It capitalizes on the fact that continuum variables represent averages over huge numbers of atoms. But why is such a connection necessary? The theory of continuum mechanics[1] is an incredibly successful theory; its application is responsible for most of the engineered world that surrounds us in our daily lives. This fact, combined with the internal consistency of continuum mechanics, has led some of its proponents to adopt the view that there is no need to attempt to connect this theory with more "fundamental" models of nature. (See, for example, the discussion of Truesdell and Toupin's view on this in Section 2.2.1.) However, there are a number of reasons why making such a connection is important.

First, continuum mechanics is not a complete theory since in the end there are more unknowns than the number of equations provided by the basic physical principles. To close the theory it is necessary to import external "constitutive relations" that in engineering applications are obtained by fitting functional forms to experimental measurements of materials. Continuum mechanics places constraints on these functional forms (see Section 2.5) but it cannot be used to derive them. A similar state of affairs exists for failure criteria, such as fracture and plasticity, which are add-ons to the theory. There is a strong emphasis in modern engineering to go beyond the phenomenology of classical continuum mechanics to a theory that can also predict the material constitutive response and failure. This clearly requires connections to be forged between the continuum and atomistic descriptions.

Second, the powerful computers that are available today make it possible to directly simulate materials at the atomic scale using molecular dynamics (MD, see Chapter 9). In this approach the positions and velocities of the atoms making up the system are obtained by numerically integrating Newton's equations of motion. Simulations of systems with in excess of one billion atoms are routinely performed on today's parallel supercomputers. The result is a vast amount of data, which must be processed in some way to become useful. Clearly, an engineer designing an airplane has no interest in knowing the positions and velocities of all its atoms. But if that information can be translated into the stress and temperature fields in the vicinity of a fatigue crack making its way through the wing material, it becomes a different story. Again, what is needed is a way to connect atomistic information and continuum macroscopic concepts.

Third, and of particular interest in this book, any attempts at designing "multiscale methods" that concurrently couple continuum and atomistic descriptions naturally require a methodology for connecting the concepts in these two regimes. The methods described in

[1] We include thermodynamics under this heading.

Chapter 13 connect atomistic regions modeled using MD with surrounding continuum regions. Difficulties emerge at the interface, where the motion of the atoms beating up against an artificial barrier must be converted to appropriate mechanical and thermal boundary conditions for the continuum. Similarly, the flux of heat and momentum in the continuum at the interface needs to be converted to forces on the atoms nearby.

As noted above, the key to connecting continuum and atomistic descriptions of materials is the fact that continuum variables constitute averages over the dynamical behavior of huge numbers of atoms. How can the fact that there are so many atoms become an asset rather than the liability that it is in MD? The answer is that systems containing large numbers of members often exhibit highly regular *statistical* behavior. Looking at the sidewalk of a busy street in a city it is impossible to predict the individual actions of this or that person. However, it is possible to predict to a high degree of accuracy the average flow of pedestrians at a particular time of day. In similar fashion, the motions of an individual atom in a container of gas may seem erratic, but the pressure exerted by a large number of atoms on the container walls is remarkably steady. It is exactly these special steady properties of the atomistic system that constitute the variables in a continuum theory. It is the business of *statistical mechanics* to predict these variables, the so-called macroscopic observables, from the dynamics of the underlying atomistic jungle.

In this chapter we discuss the basic principles of equilibrium[2] statistical mechanics, which serve as the foundation for much of the rest of this book. The treatment of stress in nonequilibrium systems is discussed in Sections 8.2 and 13.3.3.[3] Since statistical mechanics deals with averages over the motion of atoms, we begin with a detailed discussion of the dynamical behavior of a system of atoms.

7.1 Phase space: dynamics of a system of atoms

The concept of a *phase space* was loosely introduced in Section 2.4.1 where the relation between macroscopic variables and the underlying atomistics was discussed. Here this concept is revisited and made more explicit.

7.1.1 Hamilton's equations

In classical mechanics, a system of N atoms is completely defined by the positions and momenta of the atoms, $\boldsymbol{r}^\alpha$ and $\boldsymbol{p}^\alpha = m^\alpha \dot{\boldsymbol{r}}^\alpha$ ($\alpha = 1, \dots, N$), where m^α is the mass of atom α. The dynamical behavior of the system is governed by Newton's equations of motion (see page 54), which for a conservative system are conveniently represented by Hamilton's equations as shown in Section 4.3. Hamilton's equations (see Eqn. (4.10)) are

$$\dot{\boldsymbol{r}}^\alpha = \frac{\partial \mathcal{H}}{\partial \boldsymbol{p}^\alpha}, \quad \dot{\boldsymbol{p}}^\alpha = -\frac{\partial \mathcal{H}}{\partial \boldsymbol{r}^\alpha}, \quad (\alpha = 1, \dots, N). \tag{7.1}$$

[2] By "equilibrium," we mean "thermodynamic equilibrium" as defined in Section 2.4.1.

[3] See the "Further reading" section at the end of this chapter for other sources on nonequilibrium statistical mechanics.

We denote by $\{\mathbf{r}^\alpha\}$ the set of positions of all atoms in the system, so that $\{\mathbf{r}^\alpha\} = (\mathbf{r}^1, \dots, \mathbf{r}^N)$. A similar notation applies to the momentum variables. The Hamiltonian

$$\mathcal{H}(\{\mathbf{r}^\alpha\}, \{\mathbf{p}^\alpha\}; \mathbf{\Gamma}) = \mathcal{T}(\{\mathbf{p}^\alpha\}) + \mathcal{V}(\{\mathbf{r}^\alpha\}; \mathbf{\Gamma}) \tag{7.2}$$

is the total energy of the system, which is the sum of the kinetic energy

$$\mathcal{T}(\{\mathbf{p}^\alpha\}) = \sum_{\alpha=1}^{N} \frac{\mathbf{p}^\alpha \cdot \mathbf{p}^\alpha}{2m^\alpha}, \tag{7.3}$$

and the potential energy (see Eqn. (5.7)),

$$\mathcal{V}(\{\mathbf{r}^\alpha\}; \mathbf{\Gamma}) = \mathcal{V}^{\text{int}}(\{\mathbf{r}^\alpha\}) + \mathcal{V}^{\text{ext}}_{\text{fld}}(\{\mathbf{r}^\alpha\}) + \mathcal{V}^{\text{ext}}_{\text{con}}(\{\mathbf{r}^\alpha\}; \mathbf{\Gamma}), \tag{7.4}$$

resulting from internal short-range interactions between the atoms in the system, external interactions due to long-range fields and short-range external contact between atoms near the boundaries of the system and atoms outside it. The last of these reflects the constraints imposed on the system by the extensive kinematic state variables $\mathbf{\Gamma} = (\Gamma_1, \dots, \Gamma_{n_\Gamma})$, defined on page 63. For example, for gas in a container of volume V, $\Gamma_1 = V$ ($n_\Gamma = 1$), the contact term can be approximated by a potential confining the gas atoms to the container:

$$\mathcal{V}^{\text{ext}}_{\text{con}}(\{\mathbf{r}^\alpha\}; V) = \begin{cases} 0 & \text{if all } \mathbf{r}^\alpha \text{ are inside } V, \\ \infty & \text{otherwise.} \end{cases} \tag{7.5}$$

More generally, $\mathbf{\Gamma}$ can represent a state of strain as we shall see in the next chapter, where the derivative with respect to strain leads to a definition for stress (the thermodynamic tension conjugate to strain). For notational brevity, we will normally not write the explicit dependence of the Hamiltonian on $\mathbf{\Gamma}$ unless needed for the discussion. Throughout this chapter we limit consideration to potential energy functions (and hence Hamiltonians) that do not depend explicitly on time.

7.1.2 Macroscopic translation and rotation

The above equations fully characterize the motion of the atoms, but they also include rigid-body translation and rotation that must be separated out when considering vibrational properties of the system. It can be shown that a system in thermodynamic equilibrium can only have uniform macroscopic translational and rotational motion [LL80, §10].[4] Uniform translations can be removed from the system by adopting center of mass coordinates $\mathbf{r}^\alpha_{\text{rel}}$ and $\mathbf{p}^\alpha_{\text{rel}}$ as explained below. The kinetic energy associated with rotation of the body as a whole can also be removed from the system (see [JL89]). The result is that the kinetic energy can be uniquely decomposed into translation, rotation and vibrational parts.

$$\mathcal{T} = \mathcal{T}^{\text{trans}} + \mathcal{T}^{\text{rot}} + \mathcal{T}^{\text{vib}}. \tag{7.6}$$

[4] The proof is obtained by dividing the system into many small parts and maximizing the total entropy subject to the constraints of constant linear and angular momentum of the full system. The result is that the velocity of the center of mass of each part i has the form, $\mathbf{v}^{(i)} = \mathbf{u} + \mathbf{\Omega} \times \mathbf{r}^{(i)}$, where $\mathbf{u}$ and $\mathbf{\Omega}$ are constant vectors.

The thermodynamic properties of a system are related to its vibrational kinetic energy. For example, we will see in Section 7.3.6 that temperature is related to the mean vibrational kinetic energy (since it is intuitively clear that simply moving a body rigidly at a constant velocity does not increase its temperature). For this reason, statistical mechanics texts usually assume from the start that the system has zero linear and angular momentum and that $\mathcal{T}$ represents only the vibrational part of the kinetic energy. (This is equivalent to assuming that the angular momentum is zero and that $\boldsymbol{r}^\alpha$ and $\boldsymbol{p}^\alpha$ are center of mass coordinates.) We make the same assumption here. However, we will generally be more careful adding the explicit "vib" subscript to the kinetic energy, since elsewhere in the book we also refer to the rigid-body portions of the kinetic energy.

7.1.3 Center of mass coordinates

Hamilton's equations in Eqn. (7.1) fully characterize the dynamics of a system of particles subjected to conservative forces. However, more insight into the dynamics can be gained by considering the motion of the particles relative to the system's center of mass $\boldsymbol{R}$:

$$\boldsymbol{R} = \frac{1}{M}\sum_{\alpha=1}^{N} m^\alpha \boldsymbol{r}^\alpha, \tag{7.7}$$

where $M = \sum_{\alpha=1}^{N} m^\alpha$ is the total mass of the system. The linear momentum of the system (Eqn. (2.83)) is then

$$\boldsymbol{L} = \sum_{\alpha=1}^{N} m^\alpha \dot{\boldsymbol{r}}^\alpha = \frac{d}{dt}(M\boldsymbol{R}) = M\dot{\boldsymbol{R}}, \tag{7.8}$$

and the balance of linear momentum (Eqn. (2.80)) follows as

$$M\ddot{\boldsymbol{R}} = \boldsymbol{F}^{\text{ext}}, \tag{7.9}$$

where $\boldsymbol{F}^{\text{ext}} = \sum_{\alpha=1}^{N} \boldsymbol{f}^\alpha$ is the force resultant (the total external force acting on the system). This result shows that the center of mass moves as if the entire mass of the system were concentrated at the center of mass with the resultant force acting there. To obtain the dynamics of the particles relative to this global motion, we introduce the *center of mass coordinates* $\boldsymbol{r}^\alpha_{\text{rel}}$, defined by

$$\boldsymbol{r}^\alpha = \boldsymbol{R} + \boldsymbol{r}^\alpha_{\text{rel}}. \tag{7.10}$$

The momentum of particle α follows as

$$\boldsymbol{p}^\alpha = m^\alpha \dot{\boldsymbol{R}} + \boldsymbol{p}^\alpha_{\text{rel}}. \tag{7.11}$$

Here $\dot{\boldsymbol{R}}$ is the velocity of the center of mass and $\boldsymbol{p}^\alpha_{\text{rel}} = m^\alpha \boldsymbol{v}^\alpha_{\text{rel}}$, where $\boldsymbol{v}^\alpha_{\text{rel}} = \dot{\boldsymbol{r}}^\alpha_{\text{rel}}$. Substituting Eqn. (7.11) into the Hamiltonian in Eqn. (7.2), we have

$$\begin{aligned}\mathcal{H} &= \sum_{\alpha=1}^{N} \frac{\|m^\alpha \dot{\boldsymbol{R}} + \boldsymbol{p}^\alpha_{\text{rel}}\|^2}{2m^\alpha} + \mathcal{V}(\boldsymbol{R} + \boldsymbol{r}^1_{\text{rel}}, \dots, \boldsymbol{R} + \boldsymbol{r}^N_{\text{rel}}) \\ &= \frac{1}{2}M\|\dot{\boldsymbol{R}}\|^2 + \dot{\boldsymbol{R}} \cdot \sum_{\alpha=1}^{N} \boldsymbol{p}^\alpha_{\text{rel}} + \sum_{\alpha=1}^{N} \frac{\|\boldsymbol{p}^\alpha_{\text{rel}}\|^2}{2m^\alpha} + \mathcal{V}(\boldsymbol{R} + \boldsymbol{r}^1_{\text{rel}}, \dots, \boldsymbol{R} + \boldsymbol{r}^N_{\text{rel}}).\end{aligned} \tag{7.12}$$

The second term drops out since

$$\sum_{\alpha=1}^{N} \boldsymbol{p}_{\text{rel}}^{\alpha} = \sum_{\alpha=1}^{N} m^{\alpha} \dot{\boldsymbol{r}}_{\text{rel}}^{\alpha} = \frac{d}{dt}\left[\sum_{\alpha=1}^{N} m^{\alpha} \boldsymbol{r}_{\text{rel}}^{\alpha}\right]$$

$$= \frac{d}{dt}\left[\sum_{\alpha=1}^{N} m^{\alpha}(\boldsymbol{r}^{\alpha} - \boldsymbol{R})\right] = \frac{d}{dt}[M\boldsymbol{R} - M\boldsymbol{R}] = \boldsymbol{0}. \tag{7.13}$$

Substituting Eqns. (7.8) and (7.13) into Eqn. (7.12) gives the Hamiltonian in center of mass coordinates,

$$\mathcal{H} = \frac{\|\boldsymbol{L}\|^2}{2M} + \sum_{\alpha=1}^{N} \frac{\|\boldsymbol{p}_{\text{rel}}^{\alpha}\|^2}{2m^{\alpha}} + \mathcal{V}(\boldsymbol{R} + \boldsymbol{r}_{\text{rel}}^{1}, \ldots, \boldsymbol{R} + \boldsymbol{r}_{\text{rel}}^{N}). \tag{7.14}$$

The Hamiltonian form for the equation of motion of the center of mass in Eqn. (7.9) is

$$\dot{\boldsymbol{R}} = \frac{\partial \mathcal{H}}{\partial \boldsymbol{L}}, \qquad \dot{\boldsymbol{L}} = -\frac{\partial \mathcal{H}}{\partial \boldsymbol{R}} = -\sum_{\alpha=1}^{N}(-\boldsymbol{f}^{\alpha}) = \boldsymbol{F}^{\text{ext}}, \tag{7.15}$$

where Eqns. (2.83) and (4.1) were used. The equations of motion for the particles are obtained from Eqn. (7.1) after substituting in Eqn. (7.14) and using Eqn. (7.8) and the fact that $\partial \mathcal{H}/\partial \boldsymbol{r}^{\alpha} = \partial \mathcal{H}/\partial \boldsymbol{r}_{\text{rel}}^{\alpha}$:

$$\dot{\boldsymbol{r}}_{\text{rel}}^{\alpha} = \frac{\partial \mathcal{H}}{\partial \boldsymbol{p}_{\text{rel}}^{\alpha}}, \qquad \dot{\boldsymbol{p}}_{\text{rel}}^{\alpha} = -\frac{\partial \mathcal{H}}{\partial \boldsymbol{r}_{\text{rel}}^{\alpha}} - \frac{m^{\alpha}}{M}\dot{\boldsymbol{L}} = \boldsymbol{f}^{\alpha} - \frac{m^{\alpha}}{M}\boldsymbol{F}^{\text{ext}}. \tag{7.16}$$

An important special case is $\boldsymbol{F}^{\text{ext}} = \boldsymbol{0}$, for which the linear momentum of the system (and hence the velocity of the center of mass) is constant. This occurs, for example, when the forces on the atoms result only from internal interactions. In this case, Eqns. (7.15) and (7.16) are decoupled and the dynamics of the particles can be solved from Eqn. (7.16) without regard to the constant linear momentum of the system. It is common practice in this case to remove the linear momentum of the system from the Hamiltonian and to write

$$\mathcal{H} = \sum_{\alpha=1}^{N} \frac{\|\boldsymbol{p}_{\text{rel}}^{\alpha}\|^2}{2m^{\alpha}} + \mathcal{V}(\boldsymbol{r}_{\text{rel}}^{1}, \ldots, \boldsymbol{r}_{\text{rel}}^{N}), \tag{7.17}$$

where $\boldsymbol{R}$ has been removed from $\mathcal{V}$ due the translational invariance of the interatomic potential (see Section 5.3.2), and the equations of motion reduce to

$$\dot{\boldsymbol{r}}_{\text{rel}}^{\alpha} = \frac{\partial \mathcal{H}}{\partial \boldsymbol{p}_{\text{rel}}^{\alpha}}, \qquad \dot{\boldsymbol{p}}_{\text{rel}}^{\alpha} = -\frac{\partial \mathcal{H}}{\partial \boldsymbol{r}_{\text{rel}}^{\alpha}}. \tag{7.18}$$

7.1.4 Phase space coordinates

In statistical mechanics, it is often convenient to replace the positions and momenta of the atoms with a set of *generalized* positions and momenta, $\boldsymbol{q} = (q_1, \ldots, q_n)$ and $\boldsymbol{p} = (p_1, \ldots, p_n)$, where $n = 3N$ is the number of degrees of freedom. We will see in

Section 8.1.1 that this gives great freedom in the selection of the canonical variables. In this chapter we take $\boldsymbol{q}$ and $\boldsymbol{p}$ to correspond to the concatenated list of the center of mass positions and momenta of the atoms:

$$\boldsymbol{q} = (r^1_{\text{rel},1}, r^1_{\text{rel},2}, r^1_{\text{rel},3}, r^2_{\text{rel},1}, \dots, r^N_{\text{rel},3}), \boldsymbol{p} = (p^1_{\text{rel},1}, p^1_{\text{rel},2}, p^1_{\text{rel},3}, p^2_{\text{rel},1}, \dots, p^N_{\text{rel},3}).$$

A particular realization of the system with coordinates $(\boldsymbol{q}, \boldsymbol{p})$ can be identified as a point in a $2n$-dimensional space called the *phase space* of the system. A point in phase space is therefore a complete snapshot of the system, which includes the positions and momenta of all the particles. We denote the phase space by the symbol[5] Γ. Using the concatenated notation, Hamilton's equations and the Hamiltonian are

$$\dot{q}_i = \frac{\partial \mathcal{H}}{\partial p_i}, \quad \dot{p}_i = -\frac{\partial \mathcal{H}}{\partial q_i}, \quad \mathcal{H} = \sum_{i=1}^{n} \frac{p_i^2}{2m_i} + \mathcal{V}(q_1, \dots, q_n), \tag{7.19}$$

where the concatenated mass vector is $\boldsymbol{m} = (m^1, m^1, m^1, \dots, m^N, m^N, m^N)$.

7.1.5 Trajectories through phase space

A motion of the atoms making up the system is fully characterized by their positions and momenta as a function of time, $(\boldsymbol{q}(t), \boldsymbol{p}(t))$. Such a motion can be represented as a continuous line through phase space called a *trajectory*. A trajectory cannot be arbitrarily specified since it constitutes a solution to the equations of motion in Eqn. (7.19), subject to the initial conditions $(\boldsymbol{q}(0), \boldsymbol{p}(0))$. Trajectories have several important properties:

1. The solution to Hamilton's equations of motion is unique due to the determinism of classical mechanics. This means that only a single trajectory passes through each point in phase space and therefore trajectories cannot intersect.
2. The Hamiltonian is constant along a trajectory. This is another way of saying that the total energy of a Hamiltonian system is conserved. The proof for this is straightforward.

   Proof The time rate of change of $\mathcal{H}$ (which does not depend explicitly on time) is

   $$\frac{d\mathcal{H}}{dt} = \sum_{i=1}^{n} \left[\frac{\partial \mathcal{H}}{\partial q_i}\dot{q}_i + \frac{\partial \mathcal{H}}{\partial p_i}\dot{p}_i \right] = \sum_{i=1}^{n} [-\dot{p}_i \dot{q}_i + \dot{q}_i \dot{p}_i] = 0,$$

   where we have used Eqn. (7.19). Since $d\mathcal{H}/dt = 0$, we have $\mathcal{H} =$ constant. □

   This means that each trajectory in phase space is confined to a $2n - 1$ *hypersurface* S_E on which $\mathcal{H}(\boldsymbol{q}, \boldsymbol{p}) = E =$ constant.
3. Given sufficient time, a Hamiltonian system with a bounded phase space will return to a state arbitrarily close to its initial state. This is referred to as the *Poincaré recurrence theorem*, named after the French mathematician Henri Poincaré who published the proof in 1890. Roughly, the proof is based on the idea that an infinite trajectory passing through

[5] Note that there is no connection between the phase space Γ and the set of extensive kinematic state variables $\boldsymbol{\Gamma}$ discussed earlier.

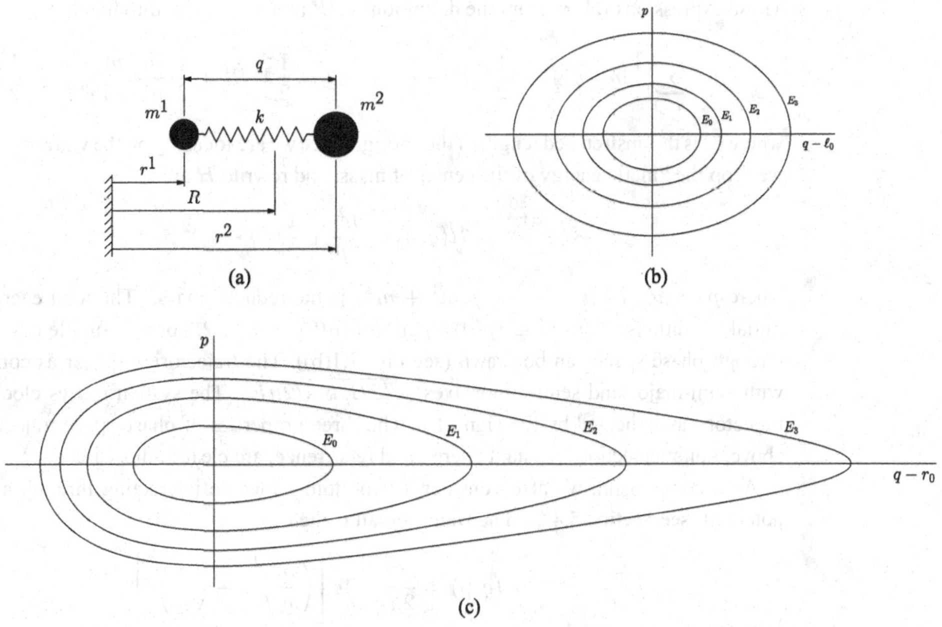

Fig. 7.1 Harmonic and Lennard-Jones oscillators in Example 7.1. (a) Two atoms interact through a linear spring with constant k. The distance between the atoms is q. The positions of the atoms and their center of mass are indicated. (b) The phase space of the harmonic oscillator. Its trajectories, plotted for $E_0 < E_1 < E_2 < E_3$, appear as concentric ellipses. (c) For comparison, the trajectories of a Lennard-Jones oscillator.

a finite incompressible phase space (see Liouville's theorem in the next section) will fill phase space after some period of time and must then recur to continue. See, for example, [Arn89, page 71] for a proof. This is a surprising result, but the "catch" is that the recurrence time is larger than the age of the universe for realistic systems. For this reason Poincaré recurrence does not have practical significance. It does, however, raise interesting philosophical questions regarding the nature of reversibility and the validity of the second law of thermodynamics (see, for example, [Bri95] and [Cal01]).

Example 7.1 (Phase space of a two-atom molecule) A simple example of a Hamiltonian system is a freely vibrating two-atom molecule as shown in Fig. 7.1(a). The masses of the atoms are m^1 and m^2. We denote the total mass by $M = m^1 + m^2$. Let us first assume that the interaction between the atoms can be modeled as a linear spring with constant k. The effect of gravity is neglected. A linear mass-spring system like this is referred to as a *harmonic oscillator*.

Our first step is to determine appropriate coordinates to describe the system. For simplicity, we will assume that the atoms are constrained to move along a line. The positions of the atoms relative to an arbitrary origin are r^1 and r^2 (see Fig. 7.1(a)). To separate out rigid-body translation, we rewrite these positions in terms of the center of mass $R = (m^1 r^1 + m^2 r^2)/M$ and the distance between the atoms $q = r^2 - r^1$:

$$r^1 = R - \frac{m^2}{m^1 + m^2} q, \qquad r^2 = R + \frac{m^1}{m^1 + m^2} q.$$

These expressions follow from the definitions of R and q. The Hamiltonian is

$$\mathcal{H} = \sum_{\alpha=1}^{2} \frac{1}{2} m^\alpha (\dot{r}^\alpha)^2 + \frac{1}{2} k(r^2 - r^1 - \ell_0)^2 = \frac{1}{2} M \dot{R}^2 + \frac{1}{2} \frac{m^1 m^2}{m^1 + m^2} \dot{q}^2 + \frac{1}{2} k(q - \ell_0)^2$$

where ℓ_0 is the unstretched length of the spring. Since we are focusing on the vibration of the molecule, we drop the kinetic energy of the center of mass, and rewrite $\mathcal{H}$ as

$$\mathcal{H}(q, p) = \frac{p^2}{2\mu} + \frac{1}{2} k(q - \ell_0)^2,$$

where $p = \mu \dot{q}$ and $\mu = m^1 m^2/(m^1 + m^2)$ is the reduced mass.[6] The total energy is set by the initial conditions, $\mathcal{H} = E \equiv (p(0))^2/2\mu + k(q(0) - \ell_0)^2/2$. For this simple case the trajectories through phase space can be drawn (see Fig. 7.1(b)). The trajectories appear as concentric ellipses with semi-major and semi-minor axes $(\sqrt{2E/k}, \sqrt{2\mu E})$. The system moves clockwise around a trajectory as indicated by the sign of p. The three properties of phase space trajectories discussed above, nonintersection, constant energy and recurrence, are clear in this case.

As a comparison, we also consider a two-atom molecule interacting through a Lennard-Jones potential (see Section 5.4.2). The Hamiltonian is then

$$\mathcal{H}(q, p) = \frac{p^2}{2\mu} + 4\epsilon \left[\left(\frac{\sigma}{q}\right)^{12} - \left(\frac{\sigma}{q}\right)^6 \right],$$

where σ and ϵ are the Lennard-Jones parameters. The total energy of the system must lie between $E = -\epsilon$, at which the molecule is stationary at its equilibrium spacing of $r_0 = 2^{1/6}\sigma$, and $E = 0$, at which the molecule separates as $q \to \infty$. The trajectories of the Lennard-Jones molecule are plotted in Fig. 7.1(c). For small values of E the trajectories resemble the ellipses of the harmonic oscillator, but as E increases anharmonic effects dominate creating the distorted shapes seen in the figure. Without solving the equations of motion we do not know how quickly the system traverses different portions of the trajectory, but the shape suggests that more time is spent on the $q > r_0$ side than on the other. This is indeed the case; an effect we refer to as thermal expansion.

7.1.6 Liouville's theorem

We now turn to a more abstract property of phase space that is of central importance in statistical mechanics. So far we have focused on a single Hamiltonian system represented by a point in phase space moving along a trajectory. Now imagine expanding this view to include all of the systems lying at a given time, say $t = t_1$, within a specified region $R(t_1)$ of phase space. Each point in this region corresponds to a copy of the Hamiltonian system with different initial conditions. The systems move along their trajectories until at time $t = t_2$ they occupy region $R(t_2)$, as shown schematically in Fig. 7.2. We define the "volume" of region $R(t)$ as $V_R(t)$. The question is how is the volume $V_R(t_2)$ related to the volume $V_R(t_1)$, or more generally, how does the volume associated with the systems occupying a region of phase space at some given time change with time?

[6] The variable $q\sqrt{\mu}$ is the "Jacobi vector" for this two-atom cluster (see [LR97]).

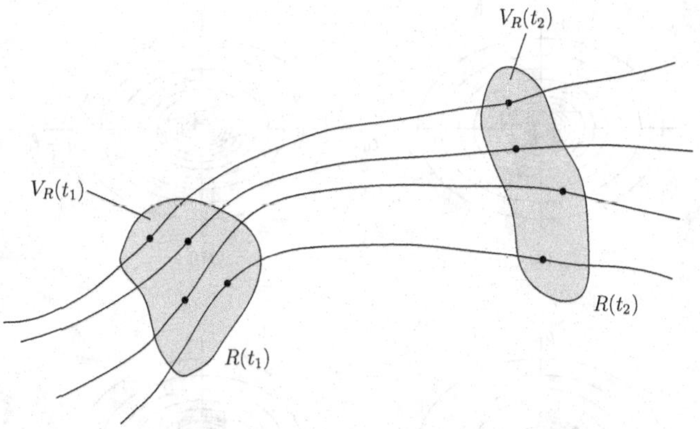

Fig. 7.2 Schematic of the trajectories traversing phase space. The systems occupying region $R(t_1)$ with volume $V_R(t_1)$ at time $t = t_1$, at a later time $t = t_2$, occupy region $R(t_2)$ with volume $V_R(t_2)$.

The simplest way to answer this question is to make an analogy with fluid flow in continuum mechanics [Wei83]. Each trajectory can be thought of as a pathline[7] in a $2n$-dimensional space. A point in this space is defined as

$$\boldsymbol{y} = (y_1, \dots, y_{2n}) = (q_1, \dots, q_n, p_1, \dots, p_n).$$

The volume associated with region $R(t)$ of phase space at time t is

$$V_R(t) = \int_{R(t)} dy_1 \cdots dy_{2n} = \int_{R(t)} d\boldsymbol{y},$$

where $d\boldsymbol{y}$ is shorthand for $dy_1 \cdots dy_{2n}$. We seek to compute DV_R/Dt, the *material* time derivative of $V_R(t)$, which is the time rate of change of the volume while following the same set of systems. (Material time derivatives are discussed in Section 2.2.5.) This can be computed immediately by applying the Reynolds transport theorem in Eqn. (2.74),

$$\frac{DV_R}{Dt} = \frac{D}{Dt}\int_{R(t)} d\boldsymbol{y} = \int_{R(t)} (\operatorname{div}_y \boldsymbol{v})\, d\boldsymbol{y}, \tag{7.20}$$

where the "velocity" vector $\boldsymbol{v}$ is given by

$$\boldsymbol{v} = \dot{\boldsymbol{y}} = (\dot{q}_1, \dots, \dot{q}_n, \dot{p}_1, \dots, \dot{p}_n) = \left(\frac{\partial \mathcal{H}}{\partial p_1}, \dots, \frac{\partial \mathcal{H}}{\partial p_n}, -\frac{\partial \mathcal{H}}{\partial q_1}, \dots, -\frac{\partial \mathcal{H}}{\partial q_n}\right), \tag{7.21}$$

and we have used Eqn. (7.19). Substituting Eqn. (7.21) into the integrand on the right-hand side of Eqn. (7.20) we have

$$\operatorname{div}_y \boldsymbol{v} = \sum_{i=1}^{2n} \frac{\partial v_i}{\partial y_i} = \sum_{i=1}^{n} \left[\frac{\partial \dot{q}_i}{\partial q_i} + \frac{\partial \dot{p}_i}{\partial p_i}\right] = \sum_{i=1}^{n} \left[\frac{\partial^2 \mathcal{H}}{\partial p_i \partial q_i} - \frac{\partial^2 \mathcal{H}}{\partial q_i \partial p_i}\right] = 0. \tag{7.22}$$

[7] A *pathline* in fluid mechanics is the trajectory traced by a moving fluid particle. The analogy with fluid flow is not complete, though, since systems flowing through phase space do not interact with each other as do physical fluid particles.

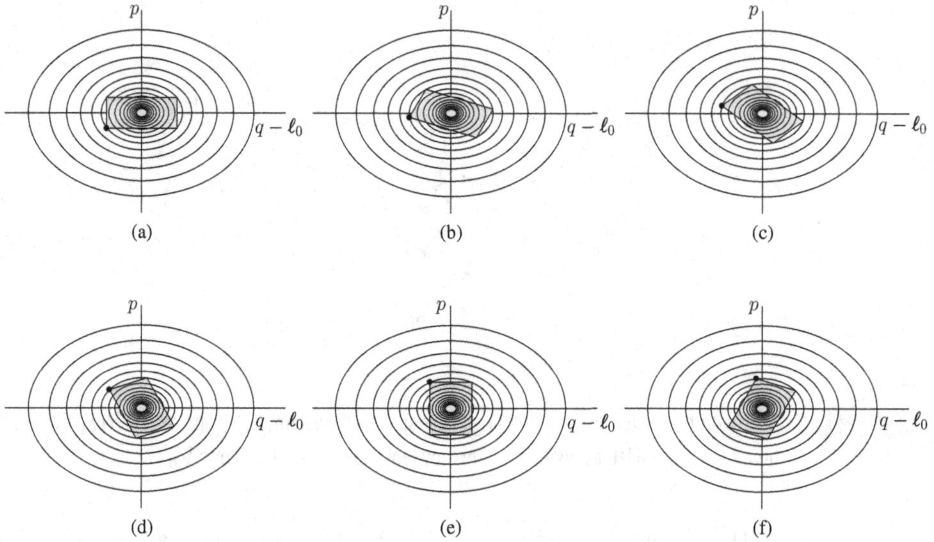

Fig. 7.3 The evolution of a region of phase space as a function of time for a harmonic oscillator (see Example 7.2). The different frames correspond to different times: (a) $\omega t = 0$; (b) $\omega t = \pi/8$; (c) $\omega t = \pi/4$; (d) $\omega t = 3\pi/8$; (e) $\omega t = \pi/2$; (f) $\omega t = 5\pi/8$. One particular system is highlighted as a visual aid.

We have shown that

$$\frac{DV_R}{Dt} = 0, \tag{7.23}$$

which is *Liouville's theorem*, proven by the French mathematician Joseph Liouville in 1838 [Lio38]. In words, the theorem states that although the shape of a region R of phase space associated with a collection of systems can change, the volume remains constant as the systems evolve. Continuing the fluid flow analogy, we say that the flow of systems through phase space is *incompressible*. This is a rather abstract theorem. Let us try to make it more clear by revisiting the harmonic oscillator discussed in Example 7.1.

Example 7.2 (Liouville's theorem for a harmonic oscillator) We consider the family of harmonic oscillators (introduced in Example 7.1) that at time $t = 0$ have initial conditions in the range $-a \le q(0) - \ell_0 \le a$ and $-b \le p(0) \le b$. This appears as a rectangle in phase space as shown in Fig. 7.3(a). We denote this rectangular region R. The phase space volume of region R is $V_R(0) = 4ab$. Assuming that the center of mass of the system is not accelerating, the equation of motion for the harmonic oscillator is

$$\mu\ddot{q} + k(q - \ell_0) = 0.$$

The solution, subject to initial conditions $q(0)$ and $p(0) = \mu\dot{q}(0)$, is

$$q(t) - \ell_0 = (q(0) - \ell_0)\cos\omega t + \frac{p(0)}{\mu\omega}\sin\omega t,$$
$$p(t) = -(q(0) - \ell_0)\mu\omega\sin\omega t + p(0)\cos\omega t,$$

where $\omega = \sqrt{k/\mu}$. Using this solution it is straightforward to see that the initial rectangle at $t = 0$ is mapped into a parallelogram. For example, the lower-left corner $(-a, -b)$ (which appears as a bold dot in Fig. 7.3(a)) is mapped to the new time-dependent position $(-a \cos \omega t - b/(\mu\omega) \sin \omega t, a\mu\omega \sin \omega t - b \cos \omega t)$. The other corners can similarly be found and it can readily be shown that the lines connecting the corners remain straight.

The area of a parallelogram is equal to the absolute value of the determinant of an array whose columns are equal to the vectors defining two (nonparallel) sides of the parallelogram. Taking the sides emanating from the lower-left corner, this is

$$V_R(t) = \det \begin{bmatrix} 2a \cos \omega t & (2b/\mu\omega) \sin \omega t \\ -2a\mu\omega \sin \omega t & 2b \cos \omega t \end{bmatrix} = 4ab.$$

Thus Liouville's theorem is satisfied. Figures 7.3(b)–(f) show the region of phase space associated with the initial rectangle at increasing times. The region changes its shape as a result of the motion of the systems contained in it. As a visual aid, the system originally located at the lower-left corner is highlighted, and its motion along its trajectory is clear. The parallelogram reverts back to the original rectangular shape at times $\omega t = 2n\pi$ $(n \in \mathbb{Z})$ as all systems return to their initial conditions, but this is not shown in the figure.

7.2 Predicting macroscopic observables

Phenomenological theories, such as thermodynamics and continuum mechanics, are phrased in terms of variables that can be measured and manipulated at the macroscopic scale, like volume, energy, temperature, stress and entropy. Such variables are referred to as *macroscopic observables*.[8] The macroscopic observables used in continuum mechanics and thermodynamics are called *state variables*. (See the discussion in Section 2.4.1.) We have already discussed the fact that at the microscopic scale, a system is fully characterized by the set of positions and momenta $(\boldsymbol{q}, \boldsymbol{p})$ of its atoms. It is therefore reasonable to assume that a macroscopic observable $\mathcal{A}$ is related to a function $A(\boldsymbol{q}, \boldsymbol{p})$ that depends on the phase space coordinates; such a function is called a *phase function*. For example $\mathcal{A}$ could be the temperature, and $A(\boldsymbol{q}, \boldsymbol{p})$, the instantaneous kinetic energy of the system. The objective of statistical mechanics is to make this connection explicit.

7.2.1 Time averages

On the face of it, computing macroscopic observables from the underlying microscopic dynamics appears to be straightforward. The laws of mechanics are deterministic, therefore

[8] The term *macroscopic observable* is a bit dated since today it is possible to measure with a precision unimaginable in the early days of thermodynamics or statistical mechanics. A dynamical transmission electron microscope (DTEM) can take snapshots with nanosecond exposure times and with spatial resolution in the nanometer range. In the future, it is likely that it will be possible to image individual atoms at femtosecond resolution [KCF+ 05]. This is certainly not the traditional view of a *macroscopic measurement*, suggesting a gross lumbering instrument providing a spatially and temporally smeared view of a huge microscopic system. Nevertheless, despite the advances in experimental science, this is exactly how we must continue to think of macroscopic observables. For example, the macroscopic concept of temperature is related in some way to the vibrations of a huge number of atoms and not to the vibration of a single atom that DTEM may be able to image in the future.

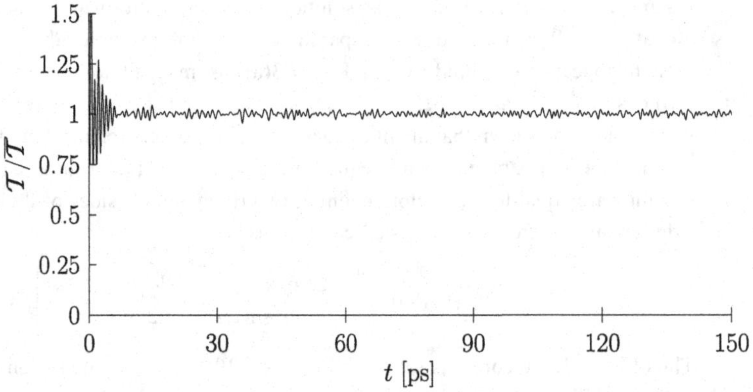

Fig. 7.4 The instantaneous kinetic energy $\mathcal{T}$ normalized by its time average obtained from a constant energy MD simulation (see Chapter 9) of a periodic box of 4000 aluminum atoms with an average temperature of 308 K. The mean kinetic energy per atom for this system was $\bar{\mathcal{T}} = 0.0398\,\text{eV}$. The system takes about 10 ps to equilibrate at the start of the simulation before settling into small fluctuations about a well-defined average kinetic energy.

given a set of initial conditions $(\boldsymbol{q}(t_0), \boldsymbol{p}(t_0))$ at time t_0 when a measurement is initiated, the subsequent trajectory of the system is known. The macroscopic observation is presumably an average over some characteristic time Δt representative of the measurement. Thus a reasonable assumption is that

$$\mathcal{A}(t_0) = \frac{1}{\Delta t} \int_{t_0}^{t_0+\Delta t} A(\boldsymbol{q}(\tau), \boldsymbol{p}(\tau))\, d\tau, \tag{7.24}$$

where Δt is large compared with the times associated with microscopic fluctuations, so that $\mathcal{A}$ is well behaved. A typical example of such fluctuations is illustrated in Fig. 7.4.

For a system in thermodynamic equilibrium (see Section 2.4.1), the measurement does not depend on the time that it is taken, hence we set $t_0 = 0$ for simplicity. It is also insensitive to Δt provided that this interval is large enough. This is the case in Fig. 7.4 once the system is in equilibrium and with Δt taken to be much larger than the fluctuation wavelength, which is about 1–2 ps in this case. We therefore define the time average $\overline{A}$ of the phase function $A(\boldsymbol{q}, \boldsymbol{p})$ as

$$\mathcal{A} = \overline{A} \equiv \lim_{\Delta t \to \infty} \frac{1}{\Delta t} \int_0^{\Delta t} A(\boldsymbol{q}(\tau), \boldsymbol{p}(\tau))\, d\tau. \tag{7.25}$$

This limit exists for almost all initial conditions (except for a "certain set of measure zero").[9] This is *Birkhoff's theorem*, proven in 1931 by George David Birkhoff using Liouville's theorem and ideas from the metric theory of sets. See details of the proof in [Khi49]. The

[9] The fact that the limit exists for almost all cases does not necessarily mean that the effect of the cases where it does not exist is negligible. This issue in the foundations of statistical mechanics is called the "measure-zero problem". See [Skl93] for an in-depth discussion.

following example demonstrates how Eqn. (7.25) is computed for the very simple case of a harmonic oscillator.

Example 7.3 (Average kinetic energy of a harmonic oscillator) Referring back to Examples 7.1 and 7.2, the kinetic energy of a harmonic oscillator is

$$\mathcal{T}(p) = \frac{p^2}{2\mu} = \frac{1}{2\mu}\left[-(q(0) - \ell_0)\mu\omega \sin \omega t + p(0)\cos \omega t\right]^2.$$

The motion of the oscillator repeats with period $T_{\text{p}} = 2\pi/\omega$, so we set $\Delta t = T_{\text{p}}$ in Eqn. (7.25). The time average of the kinetic energy is therefore

$$\begin{aligned}\overline{\mathcal{T}} &= \frac{1}{T_{\text{p}}}\int_0^{T_{\text{p}}} \mathcal{T}(p(\tau))\, d\tau \\ &= \frac{\omega}{4\mu\pi}\int_0^{2\pi/\omega} \left[-(q(0) - \ell_0)\mu\omega \sin \omega\tau + p(0)\cos \omega\tau\right]^2 d\tau \\ &= \frac{1}{2}\left[\frac{(p(0))^2}{2\mu} + \frac{1}{2}k(q(0) - \ell_0)^2\right] = \frac{1}{2}\mathcal{H},\end{aligned}$$

where we have used $\omega = \sqrt{k/\mu}$. Thus, a long-time measurement of the kinetic energy of a harmonic oscillator is equal to half of its total energy.

The time average definition in Eqn. (7.24) is conceptually attractive; however, except for trivial cases, such as the harmonic oscillator in Example 7.3, it is not conducive to further analytical progress since it requires a solution for the trajectory through phase space of the system. This is impossible to do analytically for realistic systems. It is not even possible to do numerically, given the overwhelmingly huge number of atoms in a real system, the exponential sensitivity of trajectories to initial conditions that implies the need for infinite numerical precision[10] and the inability to exactly integrate the equations of motion.

Faced with a situation where a prediction needs to be made with incomplete information, it is natural to turn to a probabilistic approach. This is embodied in the ensemble techniques pioneered by Josiah Willard Gibbs and codified in his landmark book [Gib02] that serves as the foundation for modern statistical mechanics.

7.2.2 The ensemble viewpoint and distribution functions

It is impossible to control microscopic conditions at the macroscopic level. Imagine trying to design a machine that sets the positions and velocities of 10^{23} atoms. What we can do is far more modest. We can fix the volume of a container that a gas is placed in. We can (attempt to) mechanically and thermally isolate a system and in this manner set its

[10] This is referred to as the *Lyapunov instability*. Consider two trajectories that begin from initial conditions that are different by an infinitesimal amount. It may seem logical that the trajectories will remain close together over time. However, it turns out that Hamiltonian systems can exhibit chaotic behavior where the trajectories rapidly diverge. The rate of this separation is measured by the (nonvanishing) Lyapunov exponents. See, for example, [Mey85] for an instructive presentation of this subject.

energy. We can place a system in an oven and control its temperature. In each of these cases, we have imposed macroscopic constraints on the system. These constraints guide the behavior of the system but do not control it because there are many (an infinite number in a classical system) microstates that are compatible with a given macroscopic constraint. By *microstate*, we mean one particular snapshot of the system with a given set of positions and momenta. For example, if the energy of the system is constrained to E, the set of compatible microstates satisfy

$$\sum_{i=1}^{n} \frac{p_i^2}{2m_i} + \mathcal{V}(q_1, \ldots, q_n) = E. \tag{7.26}$$

The set of all microstates that are consistent with the applied macroscopic constraints is called an *ensemble*. This definition makes it clear that ensembles are not unique. Different macroscopic constraints lead to different ensembles. The two ensembles that we will deal with are the *microcanonical ensemble*, associated with the constant energy constraint in Eqn. (7.26), and the *canonical ensemble* associated with constant temperature.[11]

Phase averages It is reasonable to assume that an experimental measurement made on a system subject to macroscopic constraints will be some function over the members of the corresponding ensemble. Since it is not known which microstates are visited during the measurement, it is natural to adopt a probabilistic approach. Each time a measurement is made, the system follows a trajectory through a sequence of microstates that are members of the imposed ensemble. The entire trajectory is uniquely defined by the initial microstate, and so the result of the measurement is a function of this initial condition. To predict this result, we imagine that we perform the measurement many times one after another, or equivalently, in parallel on a very large number of exact copies of the system subjected to the same macroscopic constraints. The probable outcome of the measurement is then

$$\langle A \rangle = \frac{1}{\nu} \sum_{i=1}^{\nu} A(\boldsymbol{q}^i, \boldsymbol{p}^i), \tag{7.27}$$

where ν measurements are performed and $(\boldsymbol{q}^i, \boldsymbol{p}^i)$ are the initial conditions for the ith measurement. If we take ν to infinity, we can replace the sum with an integral over all possible initial conditions. In this case each initial condition will be visited an infinite number of times, and so it is necessary to weight each measurement by the probability that

[11] This rather uninspiring terminology is due to Gibbs. In his book [Gib02], Gibbs first defines the canonical ensemble and states: "this distribution, on account of its unique importance in the theory of statistical equilibrium, I have ventured to call *canonical*". Gibbs then obtains the microcanonical distribution from the canonical distribution by limiting it to a range of energies that is collapsed onto a single energy. The suffix "micro" is added to reflect this limiting operation. Gibbs states: "we shall call the limiting distribution at which we arrive by this process *microcanonical*". When we derive the microcanonical and canonical distributions, we will actually take the opposite approach.

a system subject to the imposed constraints will be in this microstate. Thus the observable $\mathcal{A}$ is given by the following *phase average* (or *ensemble average*):[12]

$$\begin{aligned}\mathcal{A}(t) = \langle A; f\rangle &\equiv \int_{q_1}\cdots\int_{q_n}\int_{p_1}\cdots\int_{p_n} A(\boldsymbol{q},\boldsymbol{p})f(\boldsymbol{q},\boldsymbol{p};t)\,dq_1\cdots dq_n\,dp_1\cdots dp_n\\ &= \int_{\Gamma} A(\boldsymbol{q},\boldsymbol{p})f(\boldsymbol{q},\boldsymbol{p};t)\,d\boldsymbol{q}d\boldsymbol{p},\end{aligned} \tag{7.28}$$

where Γ represents the entire phase space and $f(\boldsymbol{q},\boldsymbol{p};t) \geq 0$ is an appropriate *distribution function*, which satisfies the normalization condition

$$\int_{\Gamma} f(\boldsymbol{q},\boldsymbol{p};t)\,d\boldsymbol{q}d\boldsymbol{p} = 1. \tag{7.29}$$

The distribution function $f(\boldsymbol{q},\boldsymbol{p};t)$ is a probability density, i.e. the probability per unit phase space volume. Thus, the product $f(\boldsymbol{q},\boldsymbol{p};t)d\boldsymbol{q}d\boldsymbol{p}$ is the probability of finding the system in the region $([\boldsymbol{q},\boldsymbol{q}+d\boldsymbol{q}],[\boldsymbol{p},\boldsymbol{p}+d\boldsymbol{p}])$ at time t. If the system is in equilibrium, f does not depend on time and the probability distribution is called *stationary*. The microcanonical and canonical distribution functions are examples of stationary distributions.

Liouville's equation We saw in Section 7.1.6 that according to Liouville's theorem, the phase space volume of a collection of systems moving through phase space is preserved. This has important implications for the time evolution of distribution functions.

Consider a region of phase space $R(t)$ at time t. The probability that the system is in this region is

$$p_R(t) = \int_{R(t)} f(\boldsymbol{y};t)\,d\boldsymbol{y},$$

where, as in Section 7.1.6, $\boldsymbol{y} = (y_1,\ldots,y_{2n}) = (q_1,\ldots,q_n,p_1,\ldots,p_n)$. The material time derivative of p_R follows from the Reynolds transport theorem in Eqn. (2.74) as

$$\dot{p}_R = \int_{R(t)} \left[\dot{f} + f(\operatorname{div}_{\boldsymbol{y}}\boldsymbol{v})\right]\,d\boldsymbol{y} = \int_{R(t)} \dot{f}\,d\boldsymbol{y},$$

where $\boldsymbol{v} = \partial\boldsymbol{y}/\partial t$ and in the second equality we have used the fact that $\operatorname{div}_{\boldsymbol{y}}\boldsymbol{v} = 0$ for a Hamiltonian system[13] (see Eqn. (7.22)). According to Liouville's theorem, phase space volume is preserved and therefore the probability associated with $R(t)$ remains constant with time, i.e. $\dot{p}_R = 0$. Since $R(t)$ is arbitrary this implies that $\dot{f} = 0$, which is referred to as *Liouville's equation* (see [AT10] for a more detailed proof). Given the form of the material time derivative in Eqn. (2.67), we have

$$\dot{f} = \frac{\partial f}{\partial t} + \boldsymbol{v}\cdot(\nabla_{\boldsymbol{y}} f) = 0.$$

[12] We discuss the concepts of distribution functions and phase averages further in Section 7.2.3. For a very clear and detailed explanation of these concepts, see Oliver Penrose's book [Pen05].

[13] We revisit Liouville's equation for non-Hamiltonian (i.e. nonconservative) systems in Section 9.4.5.

Now, substituting in Eqn. (7.21), Liouville's equation takes on the explicit form

$$\frac{\partial f}{\partial t} + \sum_{i=1}^{n} \left[\frac{\partial f}{\partial q_i} \frac{\partial \mathcal{H}}{\partial p_i} - \frac{\partial f}{\partial p_i} \frac{\partial \mathcal{H}}{\partial q_i} \right] = 0. \tag{7.30}$$

An alternative form for Liouville's equation is obtained by using Eqn. (7.19),

$$\frac{\partial f}{\partial t} + \sum_{i=1}^{n} \left[\frac{\partial f}{\partial q_i} \dot{q}_i + \frac{\partial f}{\partial p_i} \dot{p}_i \right] = 0. \tag{7.31}$$

Liouville's equation is an evolution equation for the distribution function. If the distribution is stationary then $\partial f / \partial t = 0$, and the distribution function satisfies

$$\sum_{i=1}^{n} \left[\frac{\partial f}{\partial q_i} \frac{\partial \mathcal{H}}{\partial p_i} - \frac{\partial f}{\partial p_i} \frac{\partial \mathcal{H}}{\partial q_i} \right] = 0. \tag{7.32}$$

Equation (7.32) is satisfied identically if f is assumed to be a function of the Hamiltonian, i.e. $f(\boldsymbol{q}, \boldsymbol{p}) = \widehat{f}(\mathcal{H}(\boldsymbol{q}, \boldsymbol{p}))$, since in this case

$$\frac{\partial f}{\partial q_i} = \frac{d\widehat{f}}{d\mathcal{H}} \frac{\partial \mathcal{H}}{\partial q_i}, \qquad \frac{\partial f}{\partial p_i} = \frac{d\widehat{f}}{d\mathcal{H}} \frac{\partial \mathcal{H}}{\partial p_i}.$$

For an isolated system, the Hamiltonian is constant and therefore the distribution function is also constant. The implication is that, for an isolated system, if all microstates of the same energy are equally probable then the distribution is time-independent. The fundamental hypothesis of statistical mechanics is the converse of this statement – for an isolated system in equilibrium, all states of the same energy have the same probability density. This is often referred to as the *postulate of equal a priori probabilities* and forms the basis of the microcanonical ensemble discussed later in Section 7.3.

The phase average approach described in this section may seem odd to you. The averaging procedure tells us what to expect if we were to perform a given experiment an infinite number of times, but in practice we do not do this, so why is this approach relevant? This question goes to the heart of the foundations of modern statistical mechanics. Answers based on the concept of ergodicity and probability theory are discussed in the next section.

7.2.3 Why does the ensemble approach work?

The ensemble approach to statistical mechanics, in which macroscopic observables are identified with phase averages, is empirically known to be highly successful. However, the explanation for why it works so well is still an area of active research. We present a brief discussion below. The interested reader is referred to the many excellent books and review articles on this subject [Pen79, Gra87, Skl93, Alb00, vL01, Pen05, Uff07].

The ergodic hypothesis The original motivation for phase averages, going back to Ludwig Boltzmann, came from the idea that physical systems were *ergodic*.[14] This was understood to mean that given sufficient time a system will visit all points in phase space consistent with the imposed macroscopic constraints. As Boltzmann put it [Skl93, p. 44]:

> The great irregularity of thermal motion, and the multiplicity of forces that act on the body from the outside, make it probable that the atoms themselves, by virtue of the motion that we call heat, pass through all possible positions and velocities consistent with the equation of kinetic energy.

If this is true then a macroscopic observable, identified with a long-time average over a phase function, would in the limit of the time going to infinity be equivalent to a suitably weighted average over all of phase space consistent with the imposed constraints, i.e. the *phase average*. The reasoning goes as follows [Gra87, Section 1.D],

$$\langle A \rangle = \overline{\langle A \rangle} = \langle \overline{A} \rangle = \overline{A},$$

where it helps to recall the definitions of Eqns. (7.25) and (7.28). The first equality is satisfied for a system in equilibrium since the distribution function is stationary. The second equality reflects the fact that the order of time and phase averaging can be interchanged. The third equality is the tricky one. It is satisfied if the system is ergodic, since in that case the time average is the same for all systems in the ensemble.

This initial interpretation of statistical mechanics had to be discarded in 1913 when Arthur Rosenthal [Ros13] and Michel Plancherel [Pla13] independently showed that it is topologically impossible for a one-dimensional trajectory to fill a higher-dimensional hypersurface in phase space [BH03]. In fact, this result had already been anticipated by the husband and wife team of Paul and Tatiana Ehrenfest in their influential encyclopedia article summarizing the state of statistical mechanics in 1911 [EE11]. As an alternative to ergodicity, which they doubted, the Ehrenfests proposed the weaker concept of *quasi-ergodicity* according to which a system will come arbitrarily close to any point in phase space given sufficient time. This explanation was in turn discarded when it was shown that a system can be quasiergodic and not satisfy the condition $\langle A \rangle = \overline{A}$ [Skl93].

The "foundations issue" as it has come to be known, appeared to be resolved at last with the *ergodic theorem* of John von Neumann [vN32] and George Birkhoff [Bir31].[15] This theorem showed that phase averages equal infinite time averages for all systems that are *metrically indecomposable*. A system is metrically indecomposable if all parts of its phase space, consistent with the imposed macroscopic constraints, are accessible from an

[14] The term *ergodic* refers to the concept of an *ergode* introduced by Boltzmann in 1884. The term appears to be derived from the Greek words "ergo" (work) and "hodos" (path). Boltzmann's ergode was equivalent to Gibbs' concept of a microcanonical ensemble with the additional requirement that all members of the ensemble lie on a single trajectory so that any initial condition chosen from the ensemble eventually leads to all others. This secondary requirement is what the Ehrenfests used when they coined the term *ergodic* in their encyclopedia article [EE11]. See [Uff07] for more details.

[15] Although Birkhoff's paper appeared first, his work appears to have been based on prior work due to von Neumann that was awaiting publication. Apparently, Birkhoff refused to wait for von Neumann's paper to come out before publishing his proof even though von Neumann asked him to. If you are interested in this ancient feud, a detailed discussion appears in [Zun02].

initial condition consistent with these constraints. This is obviously satisfied if the system is ergodic in the original definition of the word. However, the ergodic theorem shows that it is satisfied for any system that is metrically indecomposable. The assumption at the time was that realistic systems would quickly be proven to be metrically indecomposable, and thus the foundations issue would be settled. This optimism, it seems, was misplaced.

The ergodic theorem encountered several difficulties. First, the proofs that real systems were metrically indecomposable were not forthcoming. There was some progress when in 1963 Yakov Sinai announced that he had proven that a box of N hard spheres (a model for an ideal gas) was metrically indecomposable [Sin63].[16] However, the work of Andrey Kolmogorov [Kol54], Vladimir Arnold [Arn63] and Jürgen Moser [Mos62], which goes by the name of the KAM theorem [Bro04], showed that some realistic systems are not metrically indecomposable at sufficiently low energies.

A second criticism of the ergodic theorem approach to justifying statistical mechanics was that equating *infinite* time averages with a macroscopic observation was not reasonable. Macroscopic measurements can be long on microscopic time scales, but certainly not infinite. Often they are not even all that long on the microscale. Another problem is that the idea of infinite time averages does not allow for nonequilibrium processes, where the value of the observable may change with time.

The application of ergodic theory as an explanation for statistical mechanics appears to have stalled at that point. Instead, mathematicians interested in the problem continued to study ergodic theory as a subject of interest in its own right. The ideas of a "dynamical system" and "phase space" were generalized to include any evolving system and the set of structures that it can assume. The theorems originally proven for Hamiltonian systems were adapted to this far broader stage and today ergodic theory is applied to fields as far removed from statistical mechanics as number theory. At the same time, physicists and other statistical mechanics practitioners adopted the equality of ensemble averages with macroscopic observables as a basic postulate of statistical mechanics. (A postulate sometimes referred to as the *explanandum* in philosophy circles.) The *ergodic hypothesis* was limited to the statement that phase averages equal time averages and was no longer considered in a foundational sense. In fact, since all statistical mechanics is based on phase averages, the equality of phase averages and time averages is not necessary outside of its use in motivating phase averages. This is the state of affairs as it is represented in many modern books on statistical mechanics.

The probabilistic approach and the law of large numbers An alternative motivation for using phase averages to predict macroscopic observables is based on probability theory without explicitly considering the dynamical behavior of the system. This was the approach originally taken by Gibbs, who did not invoke the concept of ergodicity as a basis for his theory or even mention it in his writing [Gra87, p. 26].

Probability theory deals with trials, probabilities of possible outcomes, and the implications for the behavior of the system being studied. A *trial* is any well-defined procedure that

[16] This announcement has since been retracted. A proof that a system of three hard spheres in a box is metrically indecomposable was published. But the full proof was repeatedly delayed until Sinai announced that the original announcement was "premature." Work on this front continues. See [Sza93] for a review.

results in a measurable quantity; for example, throwing a fair six-sided die on a horizontal surface. The result of a trial is normally not a number. A function called a *random variable* is used to translate the result of a trial to a numerical value. For example, a random variable associated with the die throwing experiment can be the number of dots on the uppermost face of the die when it comes to a complete stop. In this case, there are six possible *outcomes* to the trial, for which the random variable takes on numerical values from 1 to 6. If we denote this random variable by T, then the possible outcomes for throwing a die are

$$T(1) = 1, \quad T(2) = 2, \quad T(3) = 3, \quad T(4) = 4, \quad T(5) = 5, \quad T(6) = 6.$$

This is an example of a *discrete* random variable. When the possible outcomes of a trial form a continuous spectrum, the corresponding random variable is called *continuous*. An example would be the length of time it takes the die to come to a halt after being thrown.

Assuming that the outcomes of repeated trials are independent, the probability of a given outcome can be defined as [Pen05]

$$\Pr(T(k)) = \lim_{N\to\infty} \frac{n(T(k), N)}{N}, \tag{7.33}$$

where $n(T(k), N)$ is the number of times outcome $T(k)$ is observed in N trials.[17] The *expectation value* of T is defined as the average outcome in an infinite series of trials:

$$\langle T\rangle \equiv \mathrm{E}(T) \equiv \lim_{N\to\infty} \frac{1}{N}\sum_{i=1}^{N} T^i, \tag{7.34}$$

where T^i is the outcome of the ith trial. The sum on the right-hand side of Eqn. (7.34) can be grouped in terms of occurrences of the possible outcomes. For a discrete random variable with κ possible outcomes this gives,

$$\langle T\rangle = \sum_{k=1}^{\kappa} T(k) \lim_{N\to\infty} \frac{n(T(k), N)}{N} = \sum_{k=1}^{\kappa} T(k) \Pr(T(k)). \tag{7.35}$$

This expression corresponds to the definition of the phase average in statistical mechanics.[18] For example, for a fair six-sided die, we have $\Pr(T(k)) = 1/6$, $k = 1, \ldots, 6$, and the expectation value is

$$\langle T\rangle = 1 \times \frac{1}{6} + 2 \times \frac{1}{6} + 3 \times \frac{1}{6} + 4 \times \frac{1}{6} + 5 \times \frac{1}{6} + 6 \times \frac{1}{6} = 3.5.$$

[17] The probability defined in Eqn. (7.33) is called a *physical probability*, suggesting that it is a physical reproducible property of the system. It is not always possible to define probabilities in this manner. For example a weather forecast stating that there is a 30% probability for rain is not the result of the meteorologist reliving the day an infinite number of times and measuring the relative frequency of rain. Rather it is the result of a "reasonable assessment by a knowledgeable person" based on the available information. This form of probability is called *subjective probability* [Pen05]. Obviously the definition of probability is of great importance for a foundational view of statistical mechanics based on probability theory and this continues to be an area of active debate. See, for example, [Skl93] for a discussion of this issue.

[18] For a continuous random variable the sum is replaced with an integral and the probability with a probability density as in Eqn. (7.28).

The spread of the trial outcomes about the expectation value is quantified by the *variance*:

$$\begin{aligned}\mathrm{Var}(T) &\equiv \lim_{N\to\infty} \frac{1}{N}\sum_{i=1}^{N}(T^i - \langle T\rangle)^2 \\ &= \lim_{N\to\infty} \frac{1}{N}\sum_{i=1}^{N}(T^i)^2 - 2\langle T\rangle \lim_{N\to\infty}\frac{1}{N}\sum_{i=1}^{N} T^i + \langle T\rangle^2 \\ &= \langle T^2\rangle - \langle T\rangle^2, \end{aligned} \tag{7.36}$$

where Eqn. (7.34) was used and the expectation value of T^2 is defined as

$$\langle T^2\rangle = \lim_{N\to\infty}\frac{1}{N}\sum_{i=1}^{N}(T^i)^2 = \sum_{k=1}^{\kappa}(T(k))^2\,\mathrm{Pr}(T(k)). \tag{7.37}$$

The squaring operation is necessary in the definition of the variance since without it any symmetric distribution would have a variance of zero. The *standard deviation* σ_T of the variable T is defined as

$$\sigma_T \equiv \sqrt{\mathrm{Var}(T)}. \tag{7.38}$$

This definition is useful since it has the same units as the original random variable and can therefore be directly compared with it.

Continuing with the six-sided die example, the expectation value of T^2 is

$$\langle T^2\rangle = 1\times\frac{1}{6} + 4\times\frac{1}{6} + 9\times\frac{1}{6} + 16\times\frac{1}{6} + 25\times\frac{1}{6} + 36\times\frac{1}{6} = 15.167,$$

and so the variance follows as

$$\mathrm{Var}(T) = 15.167 - (3.5)^2 = 2.917.$$

The standard deviation is $\sigma_T = \sqrt{2.917} = 1.708$. Note that this result does not mean that a particular throw will be within 1.708 of the expectation value of 3.5. Rather, σ_T is the deviation from the expectation value averaged over an infinite series of throws. It is therefore a property of the system just like the expectation value.

These results from probability theory may be interesting, but it is still not clear how they justify the use of phase averaging in statistical mechanics. The six-sided die example appears particularly damning. The expectation value, $\langle T\rangle = 3.5$, computed for this case is of little use in predicting the result of a particular throw of the die, which is how phase averages are used in statistical mechanics. There is, however, an additional fact that we have not used yet. Statistical mechanics is meant to be applied to systems containing a huge number of particles. In the die example this translates to throwing multiple dice at each trial rather than just one. Let us examine how this changes our conclusions.

Consider a new trial in which n dice are thrown at the same time. Define the random variable $A_T^{(n)}$ as the total score thrown divided by n:

$$A_T^{(n)} = \frac{1}{n}\sum_{i=1}^{n} T^i. \tag{7.39}$$

Here T is the random variable associated with the throw of a single die. We divide by n so that we can directly compare the expectation value and standard deviation of $A_T^{(n)}$ with the single-die experiment. For example, if two dice are thrown, there are 36 die combinations leading to 11 different possible outcomes:

		Die #1					
		1	2	3	4	5	6
	1	1	1.5	2	2.5	3	3.5
	2	1.5	2	2.5	3	3.5	4
Die #2	3	2	2.5	3	3.5	4	4.5
	4	2.5	3	3.5	4	4.5	5
	5	3	3.5	4	4.5	5	5.5
	6	3.5	4	4.5	5	5.5	6

The probabilities of the 11 outcomes are obtained from the relative frequency of their appearance in the above table:

k	1	2	3	4	5	6	7	8	9	10	11
$A_T^{(2)}(k)$	1	1.5	2	2.5	3	3.5	4	4.5	5	5.5	6
$\Pr\left[A_T^{(2)}(k)\right]$	1/36	2/36	3/36	4/36	5/36	6/36	5/36	4/36	3/36	2/36	1/36

The expectation values for $A_T^{(2)}$ and its square for the two-dice experiment can now be computed. The results are

$$\left\langle A_T^{(2)} \right\rangle = \sum_{k=1}^{11} A_T^{(2)}(k) \Pr\left[A_T^{(2)}(k)\right] = 3.5,$$

$$\left\langle (A_T^{(2)})^2 \right\rangle = \sum_{k=1}^{11} \left(A_T^{(2)}(k)\right)^2 \Pr\left[A_T^{(2)}(k)\right] = 13.708$$

and the standard deviation is

$$\sigma_{A_T}^{(2)} = \left[\left\langle (A_T^{(2)})^2 \right\rangle - \left\langle A_T^{(2)} \right\rangle^2\right]^{1/2} = [13.708 - (3.5)^2]^{1/2} = 1.208.$$

Note that the standard deviation for two dice, $\sigma_{A_T}^{(2)}$, is smaller than it is for one die ($\sigma_T = 1.708$). This trend continues. Let us quantify this effect and explore its implications.

The expectation value of $A_T^{(n)}$ for any value of n is

$$\left\langle A_T^{(n)} \right\rangle = \mathrm{E}\left(\frac{1}{n}\sum_{i=1}^{n} T^i\right) = \frac{1}{n}\sum_{i=1}^{n} \mathrm{E}(T^i) = \frac{1}{n} n \langle T \rangle = \langle T \rangle . \tag{7.40}$$

For the example of the dice experiment, the expectation value is $\langle A_T^{(n)}\rangle = \langle T\rangle = 3.5$ regardless of the number of dice. The variance of $A_T^{(n)}$ is given by

$$\mathrm{Var}\left(A_T^{(n)}\right) = \mathrm{Var}\left(\frac{1}{n}\sum_{i=1}^{n} T^i\right) = \frac{1}{n^2}\mathrm{Var}\left(\sum_{i=1}^{n} T^i\right) = \frac{1}{n^2}\sum_{i=1}^{n}\mathrm{Var}(T^i), \tag{7.41}$$

where the last equality is a consequence of the independence of the T^is. However, $\mathrm{Var}(T^i) = \sigma_T^2$ for all i, therefore

$$\mathrm{Var}\left(A_T^{(n)}\right) = \frac{\sigma_T^2}{n} \quad \Rightarrow \quad \sigma_{A_T}^{(n)} = \frac{\sigma_T}{\sqrt{n}}. \tag{7.42}$$

For example, for the two-dice experiment, $\sigma_{A_T}^{(2)} = \sigma_T/\sqrt{2} = 1.708/\sqrt{2} = 1.208$, which is exactly the result found above. Equation (7.42) shows that $\sigma_{A_T}^{(n)}$ decreases with increasing n. In particular, as $n \to \infty$, the standard deviation $\sigma_{A_T}^{(n)}$ goes to zero. This suggests that the larger the value of n, the less likely it is to obtain a measurement far from the expectation value. In fact, the law of large numbers states that as $n \to \infty$ the probability of detecting any value other than the expectation value is zero. Let us prove that this is indeed the case.

As a first step, we derive an important inequality (known as Chebyshev's inequality) that is correct for any random variable X. We will then apply it to the special case where $X = A_T^{(n)}$. Our objective is to compute the probability that X deviates from its expectation value $\langle X\rangle$ by more than some prescribed value $\epsilon > 0$. To measure the deviation, define a new random variable $Y = (X - \langle X\rangle)^2$, which by definition is always positive. The expectation value of Y can be divided into two parts coming from the terms where Y is smaller and larger than ϵ^2:

$$\langle Y\rangle = \sum_{k=1}^{\kappa} Y(k)\,\mathrm{Pr}(Y(k)) = \sum_{k:Y(k)\le\epsilon^2} Y(k)\,\mathrm{Pr}(Y(k)) + \sum_{k:Y(k)>\epsilon^2} Y(k)\,\mathrm{Pr}(Y(k)). \tag{7.43}$$

As before, κ is the number of possible outcomes. The two sums on the right-hand side of Eqn. (7.43) are both nonnegative since $Y(k) \ge 0$ and $\mathrm{Pr}(Y(k)) \ge 0$ for all k. If we drop the first nonnegative sum, we can replace Eqn. (7.43) with an inequality,

$$\langle Y\rangle \ge \sum_{k:Y(k)>\epsilon^2} Y(k)\,\mathrm{Pr}(Y(k)). \tag{7.44}$$

Next, consider the left- and right-hand sides of Eqn. (7.44) separately. On the left,

$$\langle Y\rangle = \langle (X - \langle X\rangle)^2\rangle = \langle X^2\rangle - 2\langle X\rangle^2 + \langle X\rangle^2 = \langle X^2\rangle - \langle X\rangle^2 = \mathrm{Var}(X) = \sigma_X^2. \tag{7.45}$$

On the right,

$$\sum_{k:Y(k)>\epsilon^2} Y(k)\,\mathrm{Pr}(Y(k)) \geq \sum_{k:Y(k)>\epsilon^2} \epsilon^2\,\mathrm{Pr}(Y(k)) = \epsilon^2 \sum_{k:Y(k)>\epsilon^2} \mathrm{Pr}(Y(k))$$
$$= \epsilon^2\,\mathrm{Pr}(Y \geq \epsilon^2)$$
$$= \epsilon^2\,\mathrm{Pr}((X - \langle X\rangle)^2 \geq \epsilon^2)$$
$$= \epsilon^2\,\mathrm{Pr}(|X - \langle X\rangle\,| \geq \epsilon). \quad (7.46)$$

The inequality is obtained by replacing $Y(k)$ with ϵ^2 and noting that $Y(k) > \epsilon^2$ for all terms in the sum. The equality follows from the definition of cumulative probability. Substituting Eqns. (7.45) and (7.46) into Eqn. (7.44) and rearranging gives

$$\mathrm{Pr}(|X - \langle X\rangle\,| \geq \epsilon) \leq \frac{\sigma_X^2}{\epsilon^2}, \quad (7.47)$$

which is called *Chebyshev's inequality*. This inequality, which applies to any random variable, provides a bound on the probability that a particular observation of the variable will deviate by more than a stated amount from its expectation value.

Applying Chebyshev's inequality to the case where $X = A_T^{(n)} = (1/n)\sum_{i=1}^{n} T^i$, we have

$$\mathrm{Pr}\left(\left|A_T^{(n)} - \langle T\rangle\right| \geq \epsilon\right) \leq \frac{\sigma_T^2}{n\epsilon^2}, \quad (7.48)$$

where Eqns. (7.40) and (7.42) were used. In the limit $n \to \infty$ (with ϵ fixed), the right-hand side goes to zero and we obtain the important result

$$\lim_{n\to\infty} \mathrm{Pr}\left(\left|A_T^{(n)} - \langle T\rangle\right| \geq \epsilon\right) = 0 \quad \forall\epsilon > 0, \quad (7.49)$$

which is called the *weak law of large numbers*.[19]

We began this discussion of the law of large numbers by noting that the phase average idea does not appear to make sense when an individual trial is associated with a small number of random variables. Thus the expectation value of a single throw of the die, $\langle T\rangle = 3.5$, is of no use in predicting the result of a particular throw. However, the law of large numbers shows that if the result of a trial is the average of a series of independent random variables, then as the number of variables increases the deviation of the trial result from its expected value decreases. For example, if each trial were the average over $n = 10\,000$ dice, the probability that the result would deviate by more than $\epsilon = 0.1$ from the expectation value $\langle T\rangle = 3.5$ is less than 2.9% according to Chebyshev's inequality. Of course, this number can be made as small as desired by increasing n. Thus as n increases the expectation value becomes more and more relevant as a predictor for individual trials.

A good way to visualize the implications of the law of large numbers is to plot the probability distribution function $f(X)$ for a given random variable X. For a discrete random variable, the probability distribution function is a discrete function that for every outcome $X(k)$ is equal to the relative number of occurrences of that outcome. Thus, $f(X)$

19 It is also possible to prove the *strong* law of large numbers, which states directly that $A_T^{(n)}$ converges to $\langle T\rangle$ as $n \to \infty$, and not just convergence in probability as in the weak law.

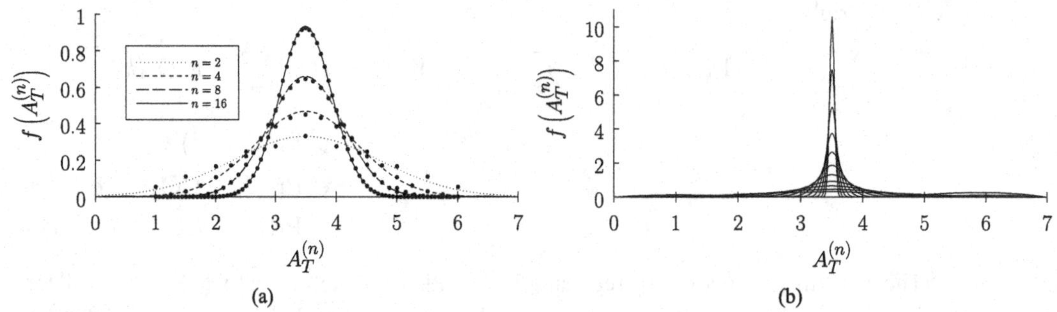

Fig. 7.5 Probability distribution plots for an experiment in which n dice are thrown: (a) details for low values of n; (b) distributions for large values of n. See text for an explanation.

is simply equal to $\Pr(X(k))$ as defined in Eqn. (7.33). For a continuous random variable, it is necessary to change to a probability *density*, which is defined as

$$f(X) = \lim_{\Delta X \to 0} \frac{\Pr(X - \Delta X/2 < X < X + \Delta X/2)}{\Delta X}. \tag{7.50}$$

A remarkable result in probability theory is that the distribution of the sum of a series of independent, identically-distributed, random variables will approach the *normal distribution* (also called the *Gaussian distribution*),

$$f(X) = \frac{1}{\sigma_X \sqrt{2\pi}} \exp\left(-\frac{(X - \langle X \rangle)^2}{2\sigma_X^2} \right), \tag{7.51}$$

as the number of summed variables becomes large. This is called the *central limit theorem*. A basic proof of this theorem is not too hard and can be found in any elementary book on statistics. Figure 7.5 presents the results for the dice throwing experiment. Figure 7.5(a) shows a comparison between the theoretically predicted normal distribution for $X = A_T^{(n)}$ (lines) and the exact distribution density computed in the manner shown above for $n = 2, 4, 8, 16$ (dots). We see that as n increases the agreement between the theoretical curve and the exact results improves as expected. Calculation of the exact distribution function becomes very expensive for large n, therefore for $n > 16$ we only plot the normal distribution curves. Figure 7.5(b) presents the normal distributions for $n = 2, 4, 8, 16, 32, 64, \ldots, 2048$. This plot clearly demonstrates that as n increases the distribution becomes more concentrated about the expectation value as discussed above. In the limit $n \to \infty$ the distribution collapses onto the expectation value becoming a Dirac delta function.

The idea that the law of large numbers and the limit theorems of probability theory can be applied as a rationale for the foundations of statistical mechanics was first proposed by Aleksandr Khinchin and summarized in his 1949 book [Khi49]. In the book, Khinchin begins by reviewing the ergodic theorem, which he attributes to Birkhoff, and then states:

> All the results obtained by Birkhoff and his followers [...] pertain to the most general type of dynamic systems, and consider different problems connected with them. The authors of these studies have been working, as a rule, on the development of so-called "general dynamics" – an important and interesting branch of modern mechanics. They

> have not been interested in the problem of the foundation of statistical mechanics which is our primary interest in the present book. Their aim was to obtain the results in the most general form; in particular all these results pertain equally to the systems with a very large number of degrees of freedom.
>
> From our point of view we must deviate from this tendency. We would unnecessarily restrict ourselves by neglecting the special properties of the systems considered in statistical mechanics (first of all their fundamental property of having a very large number of degrees of freedom), and demanding the applicability of the obtained results to any dynamical system. Furthermore, we do not have any basis for demanding the possibility of substituting phase averages for the time averages of *all* functions; in fact the functions for which such substitution is desirable have many specific properties which make such a substitution apparent in these cases.

Following this reasoning, Khinchin limits himself to a special class of observables that can be described as *sum-functions*, i.e. "the sums of functions each depending on the dynamical coordinates of only one [atom]":

$$A = \sum_{\alpha=1}^{N} A^{\alpha}(\boldsymbol{r}^{\alpha}, \boldsymbol{p}^{\alpha}).$$

He is then able to prove that the phase average of sum-functions converges to their infinite time average when the number of atoms goes to infinity. Specifically, he proves that the relative volume of phase space for which

$$\left| \frac{\overline{A} - \langle A \rangle}{\langle A \rangle} \right| > K_1 N^{-1/4}$$

is less than $K_2 N^{-1/4}$, where K_1 and K_2 are positive constants. Thus, as $N \to \infty$ the fraction of phase space associated with trajectories that violate the equality of phase averages and time averages goes to zero. This result is similar to the von Neumann–Birkhoff ergodic theorem, but has the advantage that it does not require the system be metrically indecomposable. The only requirement is that the number of atoms be large.

Theory of the thermodynamic limit Although Khinchin's work was conceptually groundbreaking, the limitation to sum-functions is overly restrictive since it rules out properties depending on correlations between particles and is limited to systems where the atoms do not interact. In addition, Khinchin's approach retains the problem of infinite time averages that the ergodic theorem has and is therefore not suitable as a foundation for non-equilibrium statistical mechanics. These limitations were addressed by subsequent development along the same lines by Ruelle [Rue69], Fisher [Fis64], Lanford [Lan73] and others starting in the 1960s. The results have led to an approach called the *theory of the thermodynamic limit*. The following discussion is based on [Lan73] and [Uff07].

Unlike Khinchin's approach, the theory of the thermodynamic limit does not attempt to prove that the phase average of a phase function is equal to its infinite time average. Rather the objective is to show that the value of an *extensive* observable of the system (i.e. an observable that scales with the number of atoms) converges to a single value with zero variance as the size of the system goes to infinity.[20] This property is proved for a special class of observables called *finite-range observables*. A finite-range observable A is an extensive property that is associated with a sequence of functions $A^{(N)}(\boldsymbol{r}^1, \ldots, \boldsymbol{r}^N)$, where N ranges from 1 to ∞, with the normalization condition[21] $A^{(1)}(\boldsymbol{r}^1) = 0$. An example for $A^{(N)}$ is the potential energy $\mathcal{V}(\boldsymbol{r}^1, \ldots, \boldsymbol{r}^N)$ of N atoms. The function $A^{(N)}$ is assumed to have the following physically-motivated properties:

1. Continuity with respect to its arguments.
2. Invariance with respect to any permutation of its arguments.
3. Invariance with respect to translation,

$$A^{(N)}(\boldsymbol{r}^1 + \boldsymbol{a}, \ldots, \boldsymbol{r}^N + \boldsymbol{a}) = A^{(N)}(\boldsymbol{r}^1, \ldots, \boldsymbol{r}^N) \qquad \forall \boldsymbol{a} \in \mathbb{R}^3.$$

4. A finite range of interaction: atoms that are separated by more than the interaction cutoff radius r_{cut} do not interact.[22] The consequence of this is that if we have two clusters of atoms, cluster I with m atoms and cluster II with n atoms, that are separated by more than r_{cut}, then

$$A^{(N)}(\boldsymbol{r}_{\text{I}}^1, \ldots, \boldsymbol{r}_{\text{I}}^m, \boldsymbol{r}_{\text{II}}^1, \ldots, \boldsymbol{r}_{\text{II}}^n) = A^{(m)}(\boldsymbol{r}_{\text{I}}^1, \ldots, \boldsymbol{r}_{\text{I}}^m) + A^{(n)}(\boldsymbol{r}_{\text{II}}^1, \ldots, \boldsymbol{r}_{\text{II}}^n),$$

where $N = m + n$.

The *thermodynamic limit* of the sequence $A^{(N)}$ is

$$\lim_{N \to \infty} \frac{A^{(N)}}{N}, \tag{7.52}$$

where this limit must be taken in a manner that preserves the density ρ, while keeping the energy per unit volume[23] fixed to a prescribed value w:

$$\lim_{N \to \infty} \frac{Nm}{V^{(N)}} = \rho, \qquad \frac{\mathcal{V}(\boldsymbol{r}^1, \ldots, \boldsymbol{r}^N)}{V^{(N)}} = w,$$

where m is the mass of an atom and $V^{(N)}$ is the volume in physical space associated with the system of N atoms. If these conditions are not satisfied, the limit in Eqn. (7.52) is ill defined and may not exist. For example, if atoms are added without proportionally increasing the volume, the energy of the system and other observables will diverge with N. An example of the calculation of the thermodynamic limit for an ideal gas is shown in Section 7.3.5. The thermodynamic limit of the free energy in the canonical ensemble is discussed in Section 7.4.5.

[20] In this theory, intensive variables, such as temperature and pressure, are obtained as derivatives of extensive variables in the thermodynamic limit [Uff07].

[21] Note that observables are limited to functions of position only. This is done to simplify the presentation but is not a limitation of the theory [Lan73].

[22] This condition can be relaxed to the weaker requirement that the interaction between atoms drops off to zero sufficiently quickly with separation to infinity [Lan73].

[23] Note that w is an energy density per unit volume. It is related to the internal energy per unit mass, u, which is used in continuum mechanics, through $w = \rho u$.

The main results of the theory of the thermodynamic limit are that the limit in Eqn. (7.52) exists and that there are two possibilities for its limiting value. The first possibility is that there exists a unique value A_0 for which

$$\Pr\left(\left|\lim_{N\to\infty}\frac{A^{(N)}(\boldsymbol{r}^1,\ldots,\boldsymbol{r}^N)}{N} - A_0\right| > \epsilon\right) = 0,$$

where ϵ is any positive number. This means that the distribution of observable A collapses onto a delta function centered on an expectation value of A_0. This of course makes the entire question of the equality of phase averages with time averages trivial, since any average of a constant is equal to the constant. This result also leads to another important conclusion, which is the equivalence of ensembles in the thermodynamic limit since the same expectation value is obtained regardless of the distribution function.

The second possibility is that instead of converging onto a single value, the thermodynamic limit converges to a range of values. This behavior is interpreted as an indication of the existence of phase transformations. This is a very important result since it provides a means for treating phase transformations within statistical mechanics. See [Leb99] for a review of this topic and additional references.

The theory of the thermodynamic limit appears to offer a promising avenue for motivating statistical mechanics. The fact that phase averages work is simply a reflection of the fact that the phase function being averaged is very close to the observable in almost all of phase space except for a tiny fraction of unlikely states. This is in fact the essence of equilibrium. A system tends to equilibrium for the simple reason that almost the entire phase space is associated with this state. If this is the case, then the "ergodic hypothesis" follows in a trivial fashion. Of course, the theory of the thermodynamic limit has its own issues and work continues (see [Uff07] for a discussion).

Having discussed the theoretical justification for phase averaging in statistical mechanics, we are now ready to derive the distribution functions for two important cases: (1) the microcanonical ensemble of an isolated system; and (2) the canonical ensemble for a system in weak interaction with a heat bath.

7.3 The microcanonical (NVE) ensemble

An isolated system has a constant number N of atoms occupying a volume V and a constant energy E. The set of all microstates consistent with these constraints is called the *microcanonical*[24] or NVE ensemble. Isolated systems play a fundamental role in statistical mechanics, since (as we will see below) the distribution function for this case can be obtained without approximation.

7.3.1 The hypersurface and volume of an isolated Hamiltonian system

Before proceeding to derive the microcanonical distribution function, we begin by introducing some geometrical variables in phase space associated with an isolated system that

[24] See footnote 11 on page 390 for a discussion of the origin of the term "microcanonical."

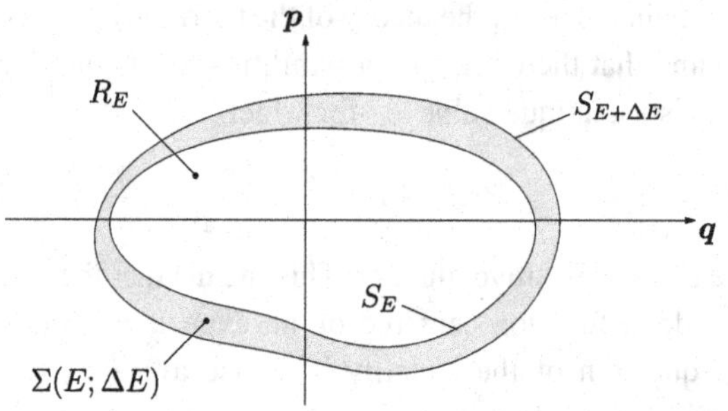

Fig. 7.6 A schematic representation of phase space and various phase space regions.

we will need. Figure 7.6 presents a schematic representation of phase space. We recall that an isolated Hamiltonian system has a constant energy E. In phase space, all microstates satisfying this constraint form a hypersurface S_E presumed to be closed. The region of phase space enclosed by this hypersurface is denoted R_E. The phase space volume of this region, denoted $V_R(E)$, is given by

$$V_R(E) = \int_\Gamma U(E - \mathcal{H}(\boldsymbol{q},\boldsymbol{p}))d\boldsymbol{q}d\boldsymbol{p} = \int_{\mathcal{H}<E} d\boldsymbol{q}d\boldsymbol{p}, \tag{7.53}$$

where Γ is the entire phase space and U is the unit step function (Heaviside function)

$$U(z) = \begin{cases} 0 & z < 0, \\ 1 & z \geq 0. \end{cases} \tag{7.54}$$

The Heaviside function confines the integration in Eqn. (7.53) to the region inside the surface S_E, where $\mathcal{H}(\boldsymbol{q},\boldsymbol{p}) \leq E$. The second expression in Eqn. (7.53) is a shorthand notation that stresses this point.

In Fig. 7.6, we also see a second hypersurface $S_{E+\Delta E}$ associated with energy $E + \Delta E$. We may regard ΔE as the precision with which energy can be measured in an experimental system (see, for example, [Pen05, page 6]). Alternatively, one can take the view that a real system can never be truly isolated and therefore its energy can only be specified to within some tolerance ΔE [Hil56, page 12]. These two points of view are related, since an act of measurement necessarily brings a system into contact with its environment. In either view, ΔE is assumed to be very small relative to E, and so the hypersurfaces S_E and $S_{E+\Delta E}$ are close together. The region of phase space between these two hypersurfaces is the *hypershell* $\Sigma(E;\Delta E)$ defined by

$$\Sigma(E;\Delta E) = \{(\boldsymbol{q},\boldsymbol{p}) \mid E \leq \mathcal{H}(\boldsymbol{q},\boldsymbol{p}) \leq E + \Delta E\}. \tag{7.55}$$

The volume of $\Sigma(E;\Delta E)$ is denoted $\Omega(E;\Delta E)$. Since by definition, the phase space volume $V_R(E)$ is a monotonically increasing function of E and since hypershells cannot intersect, the following relation between Ω and V_R holds:

$$\Omega(E;\Delta E) = V_R(E+\Delta E) - V_R(E). \tag{7.56}$$

Assuming $\Delta E \ll E$, we have

$$V_R(E+\Delta E) = V_R(E) + D(E)\Delta E + O(\Delta E^2), \tag{7.57}$$

where

$$D(E) = \frac{dV_R(E)}{dE}, \tag{7.58}$$

is the *density of states* (DOS) of the system.[25] Substituting Eqn. (7.57) into Eqn. (7.56), we have

$$\Omega(E;\Delta E) = D(E)\Delta E. \tag{7.59}$$

We will see below that the equilibrium properties of Hamiltonian systems are closely related to the geometric properties of the phase space volumes described above. It will therefore be helpful to briefly consider the nature of $V_R(E)$, $\Omega(E;\Delta E)$ and $D(E)$, in particular as the number of atoms N becomes very large. (Recall that the thermodynamic limit discussed at the end of the previous section corresponds to the case of $N \to \infty$.)

The most important property of these three geometric measures is that for large values of N these are extremely rapidly increasing functions of E. This observation is easy to demonstrate in a quantum mechanical system where energy is discretized and therefore the number of quantum states corresponding to a given energy level is countable and can be estimated (see, for example, [Rei85, Section 2.5]). For a classical system, let us demonstrate this by example. Consider the simplest case of a system of N identical noninteracting atoms. The total energy of the system is equal to the kinetic energy of the atoms:

$$\mathcal{H}(\boldsymbol{p}) = \frac{1}{2m}\sum_{i=1}^{n} p_i^2,$$

where m is the mass of an atom and $n = 3N$. This is the case of an ideal gas discussed in more detail in Section 7.3.5. The region of (momentum) phase space for which $\mathcal{H}(\boldsymbol{p}) < E$ corresponds to an n-dimensional hypersphere (n-sphere) of radius $r = \sqrt{2mE}$, defined by

$$p_1^2 + p_2^2 + \cdots + p_n^2 = r^2.$$

[25] This DOS is conceptually equivalent to the DOS introduced in the context of TB in Section 4.6.6

The volume of an n-sphere can be shown to be $C_n r^n$, where C_n is a bounded function of n. We therefore see that $V_R(E)$ is proportional to $E^{n/2}$:

$$V_R(E) = C_n (2m)^{n/2} E^{n/2} = D_n E^{n/2}. \tag{7.60}$$

Clearly, for a macroscopic system where n is of order 10^{23}, $V_R(E)$ is a wildly increasing function of E as noted above. The same holds for $D(E)$:

$$D(E) = V_R'(E) = \frac{n}{2} D_n E^{n/2-1} = \frac{n}{2E} V_R(E), \tag{7.61}$$

and for $\Omega(E;\Delta E) = D(E)\Delta E$. The rapid growth of phase space volume with energy is key to the success of statistical mechanics; it is in fact the origin of the notion of equilibrium upon which thermodynamics is based (as we shall see later). We now turn to the derivation of the microcanonical distribution function.

7.3.2 The microcanonical distribution function

To derive the microcanonical distribution function, we combine two properties of Hamiltonian systems that were obtained earlier:

1. In Section 7.1.5, we showed that the total energy E of an isolated Hamiltonian system is conserved. This means that the trajectories of this system are confined to a $2n-1$ hypersurface $S_E = \{\boldsymbol{y} \mid \mathcal{H}(\boldsymbol{y}) = E\}$, where $\boldsymbol{y} = (q_1, \dots, q_n, p_1, \dots, p_n)$.
2. At the end of Section 7.2.2, we postulated that a stationary distribution function is a function of the Hamiltonian, $f(\boldsymbol{y}) = \widehat{f}(\mathcal{H}(\boldsymbol{y}))$.

Combining 1 and 2, we conclude that for an isolated system the distribution function is constant on S_E. This means that all microstates consistent with the constant energy constraint are equally likely, an idea referred to as the *postulate of equal a priori probabilities*. However, since $f(\boldsymbol{y})$ is a volume density, it would be incorrect simply to write

$$f(\boldsymbol{y};E) = \begin{cases} \widehat{f}(E) = \text{const} & \text{if } \boldsymbol{y} \in S_E, \\ 0 & \text{otherwise,} \end{cases} \qquad \text{(wrong)}$$

since the integral of a volume density on a lower-dimensional surface is identically zero. Instead we stretch the hypersurface S_E into a hypershell $\Sigma(E;\Delta E)$ of thickness ΔE as defined in Section 7.3.1 and shown in Fig. 7.6. As noted above, ΔE represents uncertainty in the value of E due to unavoidable external interactions. Later in the derivation, we will take the limit as $\Delta E \to 0$, so the results will not depend on a particular choice of ΔE.

Next, we define a distribution function $f_\Sigma(\boldsymbol{y};E,\Delta E)$, which is constant within $\Sigma(E;\Delta E)$ and zero outside it. Since f_Σ also satisfies the normalization condition,

$$\int_\Gamma f_\Sigma(\boldsymbol{y};E,\Delta E)\, d\boldsymbol{y} = 1,$$

where as before $d\boldsymbol{y} = dy_1 \cdots dy_{2n}$, we have

$$f_\Sigma(\boldsymbol{y};E,\Delta E) = \begin{cases} 1/\Omega(E;\Delta E) & \text{if } \boldsymbol{y} \in \Sigma(E;\Delta E), \\ 0 & \text{otherwise,} \end{cases}$$

where $\Omega(E;\Delta E)$ is the phase space volume of the hypershell $\Sigma(E;\Delta E)$. In quantum mechanics, energy is quantized and so $\Omega(E;\Delta E)$ represents a finite countable number of microstates. For this reason $\Omega(E;\Delta E)$ is often referred to as the *number of microstates* consistent with the constraint that the energy of the system lies between E and $E+\Delta E$.

The phase average of a function $A(\boldsymbol{y})$ of the isolated system, can now be written as

$$\langle A\rangle = \lim_{\Delta E\to 0}\int_{\Sigma(E;\Delta E)} A(\boldsymbol{y}) f_\Sigma(\boldsymbol{y};E,\Delta E)\,d\boldsymbol{y} = \lim_{\Delta E\to 0}\frac{\int_{\Sigma(E;\Delta E)} A(\boldsymbol{y})\,d\boldsymbol{y}}{V_R(E+\Delta E)-V_R(E)},$$

where we have used Eqn. (7.56). Dividing the numerator and denominator by ΔE and using the definition of $\Sigma(E;\Delta E)$, we have

$$\langle A\rangle = \frac{\lim_{\Delta E\to 0}\dfrac{1}{\Delta E}\left[\displaystyle\int_{R_{E+\Delta E}} A(\boldsymbol{y})\,d\boldsymbol{y} - \int_{R_E} A(\boldsymbol{y})\,d\boldsymbol{y}\right]}{\lim_{\Delta E\to 0}\dfrac{V_R(E+\Delta E)-V_R(E)}{\Delta E}}.$$

The expressions in the numerator and denominator translate to derivatives, so that

$$\langle A\rangle = \frac{1}{D(E)}\frac{\partial}{\partial E}\int_{R_E} A(\boldsymbol{q},\boldsymbol{p})\,d\boldsymbol{q}d\boldsymbol{p}, \tag{7.62}$$

where we have reverted from the concatenated $\boldsymbol{y}$ notation back to $\boldsymbol{q}$ and $\boldsymbol{p}$, and where $D(E)$ is the DOS defined in Eqn. (7.58). The integral in Eqn. (7.62) can be expanded to the entire phase space Γ by making use of the unit step function U in Eqn. (7.54),

$$\langle A\rangle = \frac{1}{D(E)}\frac{\partial}{\partial E}\int_\Gamma A(\boldsymbol{q},\boldsymbol{p})U(E-\mathcal{H}(\boldsymbol{q},\boldsymbol{p}))\,d\boldsymbol{q}d\boldsymbol{p}.$$

As in Section 7.3.1, U confines the integration to the region inside the surface S_E, where $\mathcal{H}(\boldsymbol{q},\boldsymbol{p})\le E$. The differentiation with respect to E can now be carried out with the result

$$\langle A\rangle = \frac{1}{D(E)}\int_\Gamma A(\boldsymbol{q},\boldsymbol{p})\delta(E-\mathcal{H}(\boldsymbol{q},\boldsymbol{p}))\,d\boldsymbol{q}d\boldsymbol{p} = \int_\Gamma A(\boldsymbol{q},\boldsymbol{p}) f_{\mathrm{mc}}(\boldsymbol{q},\boldsymbol{p};E)\,d\boldsymbol{q}d\boldsymbol{p}, \tag{7.63}$$

where $\delta(z) = dU(z)/dz$ is the Dirac delta function,[26] and where we have defined the *microcanonical distribution function*:

$$f_{\mathrm{mc}}(\boldsymbol{q},\boldsymbol{p};E) \equiv \frac{\delta(E-\mathcal{H}(\boldsymbol{q},\boldsymbol{p}))}{D(E)}. \tag{7.64}$$

The microcanonical distribution function was obtained by a limiting operation which collapses the hypershell $\Sigma(E;\Delta E)$ onto the hypersurface S_E. It is important to realize that a hypershell, like $\Sigma(E;\Delta E)$, generally has a nonuniform thickness as shown schematically

[26] The Dirac delta function satisfies the relation $\int_{-\infty}^{\infty}\delta(z-a)h(z)\,dz = h(a)$ for any function $h(z)$.

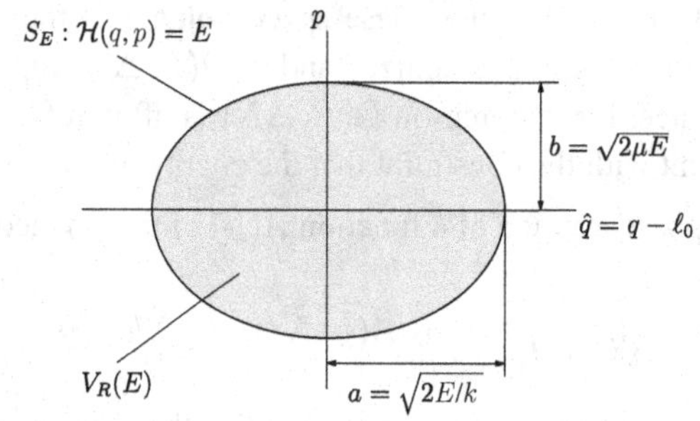

Fig. 7.7 The hypersurface S_E associated with the harmonic oscillator of Example 7.1 is a closed ellipse with semi-major and semi-minor axes, a and b, as shown. The region of phase space enclosed by S_E is R_E and its volume, indicated in the figure, is $V_R(E)$. The calculation of microcanonical averages for this system is discussed in Example 7.4.

in Fig. 7.6. As a result the probability density on the hypersurface S_E is not uniform and varies to account for the changing thickness. This information is encoded in the Dirac delta distribution function, which has units of 1/energy, and weights the configurations associated with "thick" parts of the hypersurface more heavily. Due to Liouville's theorem, this means that the system will spend more time on portions of S_E where S_E and $S_{E+\Delta E}$ are far apart, relative to parts of S_E where they are close together. See, for example, [Wei83, page 68] and [Set06, Section 3.1] for a discussion of this point.

Equation (7.63) also provides a useful expression for the DOS. We see from the special case $A(\boldsymbol{q}, \boldsymbol{p}) = 1$ that $D(E)$ can be written as

$$D(E) = \int_\Gamma \delta(E - \mathcal{H}(\boldsymbol{q}, \boldsymbol{p}))\, d\boldsymbol{q}d\boldsymbol{p}, \qquad (7.65)$$

which highlights the fact that $f_{\mathrm{mc}}(\boldsymbol{q}, \boldsymbol{p}; E)$ defined in Eqn. (7.64) satisfies the normalization condition in Eqn. (7.29).

Although Eqn. (7.63) is the formal expression for the microcanonical phase average, Eqn. (7.62) can be more convenient in practice as shown in the following example.

Example 7.4 (Microcanonical average of a harmonic oscillator) Our objective is to compute the microcanonical ensemble average for the harmonic oscillator in Examples 7.1 and 7.2 using Eqn. (7.62). For simplicity, we limit ourselves to phase functions that depend only on the position of the oscillator.

The Hamiltonian of the harmonic oscillator is $\mathcal{H} = p^2/2\mu + k\hat{q}^2/2$, where $\hat{q} = q - \ell_0$. The hypersurface S_E is a closed ellipse as shown in Fig. 7.7. On this surface,

$$p(\hat{q}; E) = \left[2\mu(E - k\hat{q}^2/2)\right]^{1/2} \quad \text{and} \quad \partial p(\hat{q}; E)/\partial E = \left[2\mu^{-1}(E - k\hat{q}^2/2)\right]^{-1/2}, \qquad (7.66)$$

which we will need later. The phase space volume of the region R_E enclosed by S_E is

$$V_R(E) = \pi ab = \pi\sqrt{2E/k}\sqrt{2\mu E} = 2\pi E\sqrt{\mu/k},$$

where a and b are the ellipse semi-major and semi-minor axes given in Fig. 7.7. The DOS follows as

$$D(E) = V_R'(E) = 2\pi\sqrt{\mu/k} = 2\pi/\omega = T_{\mathrm{p}},$$

where $\omega = \sqrt{k/\mu}$ is the angular velocity of the oscillator and T_p is its period of oscillation (see Examples 7.1 and 7.2). The microcanonical phase average follows from Eqn. (7.62) as

$$\langle A \rangle = \frac{1}{T_\mathrm{p}} \frac{\partial}{\partial E} \int_{R_E} A(\hat{q})\, d\hat{q} dp = \frac{2}{T_\mathrm{p}} \frac{\partial}{\partial E} \int_{-\sqrt{2E/k}}^{\sqrt{2E/k}} A(\hat{q}) p(\hat{q}; E)\, d\hat{q}, \tag{7.67}$$

where $p(\hat{q}; E)$ is defined in Eqn. $(7.66)_1$. To compute the derivative in Eqn. (7.67), we use Leibniz's rule:

$$\frac{d}{d\alpha} \int_{\phi_1(\alpha)}^{\phi_2(\alpha)} G(x, \alpha)\, dx = \int_{\phi_1(\alpha)}^{\phi_2(\alpha)} \frac{\partial G}{\partial \alpha}\, dx + G(\phi_2, \alpha) \frac{d\phi_2}{d\alpha} - G(\phi_1, \alpha) \frac{d\phi_1}{d\alpha}.$$

The last two terms are zero in our case, since $p(\pm\sqrt{2E/k}; E) = 0$, so

$$\langle A \rangle = \frac{2}{T_\mathrm{p}} \int_{-\sqrt{2E/k}}^{\sqrt{2E/k}} \frac{A(\hat{q})}{[2\mu^{-1}(E - k\hat{q}^2/2)]^{1/2}}\, d\hat{q},$$

where we have used $\partial p/\partial E$ from Eqn. $(7.66)_2$. Changing the integration variable to $\xi = \hat{q}/\sqrt{2E/k}$, this simplifies to

$$\langle A \rangle = \frac{1}{\pi} \int_{-1}^{1} \frac{A(\xi)}{\sqrt{1 - \xi^2}}\, d\xi. \tag{7.68}$$

For example the mean position and position squared of the oscillator are

$$\langle \hat{q} \rangle = \frac{a}{\pi} \int_{-1}^{1} \frac{\xi}{\sqrt{1-\xi^2}}\, d\xi = 0, \qquad \langle \hat{q}^2 \rangle = \frac{a^2}{\pi} \int_{-1}^{1} \frac{\xi^2}{\sqrt{1-\xi^2}}\, d\xi = \frac{a^2}{\pi} \frac{\pi}{2} = \frac{a^2}{2}.$$

In general, we have

$$\langle \hat{q}^n \rangle = \begin{cases} 0 & n \text{ odd}, \\ \dfrac{a^n}{2^{n/2}} \dfrac{(n-1)!!}{(n/2)!} & n \text{ even}, \end{cases} \tag{7.69}$$

where $(n-1)!! \equiv 1 \times 3 \times 5 \times \cdots \times (n-1)$. The standard deviation of the motion of the oscillator is $\sigma = \sqrt{\langle \hat{q}^2 \rangle - \langle \hat{q} \rangle^2} = a/\sqrt{2}$. We learn from this that the two masses making up the oscillator vibrate about the mean position $\hat{q} = 0$ $(q = \ell_0)$ with a rather large standard deviation of $a/\sqrt{2} = \sqrt{E/k}$.

The microcanonical ensemble is of fundamental importance in classical statistical mechanics. It describes the equilibrium behavior of an isolated Hamiltonian system without any approximations. It is also of fundamental importance in equilibrium MD (where it is more commonly referred to as the NVE ensemble) as discussed in Section 9.3. However, to use this theory, we must make a connection with macroscopic thermodynamic concepts, such as internal energy, entropy and temperature, introduced in Section 2.4. It turns out that in order to do so it is first necessary to introduce the concept of "weak interaction" discussed next.

7.3.3 Systems in weak interaction

Imagine that an isolated Hamiltonian system is composed of two subsystems, A and B, as illustrated in Fig. 7.8. Systems A and B can be physically distinct, such as a material and surrounding gas, or simply a conceptual division of the isolated system into parts. In

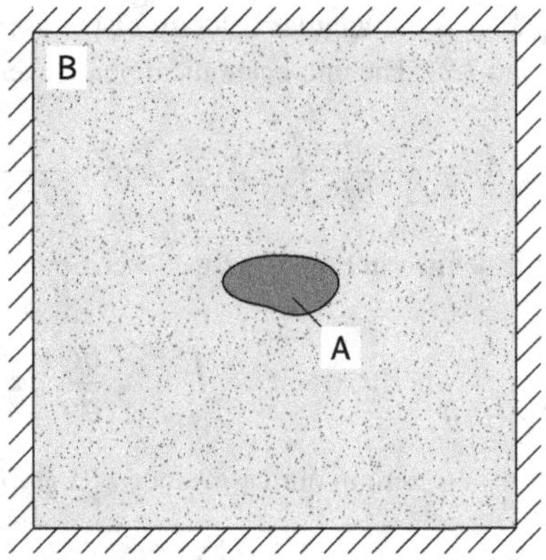

Fig. 7.8 An isolated system is divided into two subsystems A and B. The hatching around the perimeter indicates that the combined system A + B is thermally and mechanically isolated from the rest of the world.

either case, both A and B are assumed to be "macroscopic" so that the basic principles of statistical mechanics apply. The Hamiltonian $\mathcal{H}$ of the total system is

$$\mathcal{H}(\boldsymbol{q}, \boldsymbol{p}) = \sum_{i=1}^{n} \frac{(p_i)^2}{2m_i} + \mathcal{V}(\boldsymbol{q}).$$

Let us rewrite this expression explicitly accounting for the membership of the atoms in the two subsystems:

$$\mathcal{H} = \sum_{i=1}^{n^{\mathsf{A}}} \frac{(p_i^{\mathsf{A}})^2}{2m_i^{\mathsf{A}}} + \sum_{j=1}^{n^{\mathsf{B}}} \frac{(p_j^{\mathsf{B}})^2}{2m_j^{\mathsf{B}}} + \mathcal{V}(\boldsymbol{q}^{\mathsf{A}}, \boldsymbol{q}^{\mathsf{B}}), \tag{7.70}$$

where the superscripts A and B on $\boldsymbol{q}$ and $\boldsymbol{p}$ denote the subsystem to which the atom belongs and n^{A} and n^{B} are the number of degrees of freedom in systems A and B (three times the number of atoms, N^{A} and N^{B}). In general, it is not possible to divide the Hamiltonian expression in Eqn. (7.70) into two separate contributions from the two subsystems due to the coupling introduced by the potential energy function. Nevertheless, we write

$$\mathcal{H} = \mathcal{H}^{\mathsf{A}} + \mathcal{H}^{\mathsf{B}} + \mathcal{H}^{\mathsf{A} \leftrightarrow \mathsf{B}}. \tag{7.71}$$

Here $\mathcal{H}^{\mathsf{A}}$ and $\mathcal{H}^{\mathsf{B}}$ are the Hamiltonians of systems A and B *as if* they were isolated:

$$\mathcal{H}^{\mathsf{A}} = \sum_{i=1}^{n^{\mathsf{A}}} \frac{(p_i^{\mathsf{A}})^2}{2m_i^{\mathsf{A}}} + \mathcal{V}^{\mathsf{A}}(\boldsymbol{q}^{\mathsf{A}}), \qquad \mathcal{H}^{\mathsf{B}} = \sum_{j=1}^{n^{\mathsf{B}}} \frac{(p_j^{\mathsf{B}})^2}{2m_j^{\mathsf{B}}} + \mathcal{V}^{\mathsf{B}}(\boldsymbol{q}^{\mathsf{B}}), \tag{7.72}$$

where $\mathcal{V}^{\mathsf{A}}$ is the potential energy function for the atoms in system A in the absence of system B. Similarly, $\mathcal{V}^{\mathsf{B}}$ is the potential energy function for system B on its own. Clearly the sum of $\mathcal{H}^{\mathsf{A}}$ and $\mathcal{H}^{\mathsf{B}}$ does not give $\mathcal{H}$, since the interactions between A and B atoms are not taken into accounts. This is accounted for by the interaction term $\mathcal{H}^{\mathsf{A} \leftrightarrow \mathsf{B}}$. Formally, it is defined as

$$\mathcal{H}^{\mathsf{A} \leftrightarrow \mathsf{B}} \equiv \mathcal{H} - \mathcal{H}^{\mathsf{A}} - \mathcal{H}^{\mathsf{B}}. \tag{7.73}$$

Substituting in the definitions of $\mathcal{H}$, $\mathcal{H}^{\mathsf{A}}$ and $\mathcal{H}^{\mathsf{B}}$ given above, we see that the kinetic energy terms cancel and we are left with a difference of potential energy functions:

$$\mathcal{H}^{\mathsf{A}\leftrightarrow\mathsf{B}} = \mathcal{V}(\boldsymbol{q}^{\mathsf{A}}, \boldsymbol{q}^{\mathsf{B}}) - \mathcal{V}^{\mathsf{A}}(\boldsymbol{q}^{\mathsf{A}}) - \mathcal{V}^{\mathsf{B}}(\boldsymbol{q}^{\mathsf{B}}) \equiv \mathcal{V}^{\mathsf{A}\leftrightarrow\mathsf{B}}(\boldsymbol{q}^{\mathsf{A}}, \boldsymbol{q}^{\mathsf{B}}), \tag{7.74}$$

where we have defined the difference, $\mathcal{V}^{\mathsf{A}\leftrightarrow\mathsf{B}}(\boldsymbol{q}^{\mathsf{A}}, \boldsymbol{q}^{\mathsf{B}})$, as the potential energy interaction function.[27] The form of this function depends on the model used to describe the atomic interactions. For example, for the simplest case of an isolated system with pair potential interactions (see Section 5.4.2), we have

$$\mathcal{V} = \frac{1}{2}\sum_{\substack{\alpha,\beta \\ \alpha\neq\beta}}^{N^{\mathsf{A}}} \phi_{\mathsf{AA}}(r^{\alpha\beta}) + \frac{1}{2}\sum_{\substack{\alpha,\beta \\ \alpha\neq\beta}}^{N^{\mathsf{B}}} \phi_{\mathsf{BB}}(r^{\alpha\beta}) + \sum_{\alpha=1}^{N^{\mathsf{A}}}\sum_{\beta=1}^{N^{\mathsf{B}}} \phi_{\mathsf{AB}}(r^{\alpha\beta}),$$

$$\mathcal{V}^{\mathsf{A}} = \frac{1}{2}\sum_{\substack{\alpha,\beta \\ \alpha\neq\beta}}^{N^{\mathsf{A}}} \phi_{\mathsf{AA}}(r^{\alpha\beta}), \qquad \mathcal{V}^{\mathsf{B}} = \frac{1}{2}\sum_{\substack{\alpha,\beta \\ \alpha\neq\beta}}^{N^{\mathsf{B}}} \phi_{\mathsf{BB}}(r^{\alpha\beta}),$$

where $\phi_{\mathsf{AA}}(r)$, $\phi_{\mathsf{BB}}(r)$ and $\phi_{\mathsf{AB}}(r)$ are the pair potentials for A–A, B–B and A–B interactions, respectively. The interaction term is obtained by subtracting the last two terms from the first,

$$\mathcal{V}^{\mathsf{A}\leftrightarrow\mathsf{B}} = \sum_{\alpha=1}^{N^{\mathsf{A}}}\sum_{\beta=1}^{N^{\mathsf{B}}} \phi_{\mathsf{AB}}(r^{\alpha\beta}).$$

Assuming a finite range of interaction, this term scales with the area of the surface separating systems A and B in physical space. This term is small relative to the total energy as long as systems A and B are macroscopic (which was assumed from the start) since in this case the surface-to-volume ratio is small. This does *not* mean that $\mathcal{V}^{\mathsf{A}\leftrightarrow\mathsf{B}}$ can be neglected when studying the dynamics of the combined system A + B. The dynamical processes at the separation surface are vital for maintaining equilibrium between systems A and B. However, under conditions of *weak interaction* it is possible to neglect the interaction term in the computation of ensemble averages. Thus, weak interaction is defined as follows.

Weak interaction Two systems A and B are said to be in *weak interaction* if the interaction term $\mathcal{H}^{\mathsf{A}\leftrightarrow\mathsf{B}}$ can be neglected in the computation of any integral over phase space.

Let us compute the ensemble average for a phase function in systems undergoing weak interaction. The combined system A + B is isolated, therefore according to Eqn. (7.63) the ensemble average of a phase function $A(\boldsymbol{y})$, where $\boldsymbol{y}$ is combined set of positions and momenta of atoms in A and B, is given by

$$\langle A \rangle = \frac{1}{D(E)} \int_{\Gamma} A(\boldsymbol{y})\delta\left(\mathcal{H}(\boldsymbol{y}) - E\right) d\boldsymbol{y},$$

[27] Referring back to Eqn. (7.4), we see that $\mathcal{V}^{\mathsf{A}}$ and $\mathcal{V}^{\mathsf{B}}$ correspond to the internal and external field parts of the potential energies of each of the subsystems, i.e. $\mathcal{V}^{\mathsf{A}} = \mathcal{V}^{\mathrm{int},\mathsf{A}} + \mathcal{V}^{\mathrm{ext},\mathsf{A}}_{\mathrm{fld}}$ and $\mathcal{V}^{\mathsf{B}} = \mathcal{V}^{\mathrm{int},\mathsf{B}} + \mathcal{V}^{\mathrm{ext},\mathsf{B}}_{\mathrm{fld}}$. The interaction term corresponds to the external contact terms of the two subsystems, i.e. $\mathcal{V}^{\mathsf{A}\leftrightarrow\mathsf{B}} = \mathcal{V}^{\mathrm{ext},\mathsf{A}}_{\mathrm{con}} = \mathcal{V}^{\mathrm{ext},\mathsf{B}}_{\mathrm{con}}$ (where we have neglected any coupling between system B and its surroundings).

where $D(E)$ is the DOS for the combined system and E is the total energy. Assuming weak interaction, we replace $\mathcal{H}(\boldsymbol{y})$ in the above integral with $\mathcal{H}^{\mathsf{A}}(\boldsymbol{y}^{\mathsf{A}}) + \mathcal{H}^{\mathsf{B}}(\boldsymbol{y}^{\mathsf{B}})$, so that

$$\langle A \rangle = \frac{1}{D(E)} \int_{\Gamma} A(\boldsymbol{y}) \delta \left(\mathcal{H}^{\mathsf{A}}(\boldsymbol{y}^{\mathsf{A}}) + \mathcal{H}^{\mathsf{B}}(\boldsymbol{y}^{\mathsf{B}}) - E \right) d\boldsymbol{y}^{\mathsf{A}} d\boldsymbol{y}^{\mathsf{B}}. \tag{7.75}$$

The constraint introduced by the Dirac δ confines the integration to a hypersurface S_E^*, where $\mathcal{H}^{\mathsf{A}}(\boldsymbol{y}^{\mathsf{A}}) + \mathcal{H}^{\mathsf{B}}(\boldsymbol{y}^{\mathsf{B}}) = E$. This is not the desired hypersurface, S_E, on which $\mathcal{H} = E$, since the interaction term is missing. The assumption is that under conditions of weak interaction, integrating on S_E^* provides a good approximation for the exact integral on S_E.

7.3.4 Internal energy, temperature and entropy

It is of interest to see how the quantities associated with the microcanonical ensemble are related to the thermodynamic variables we encountered in Section 2.4. In particular, we are interested in the temperature T (Section 2.4.2), internal energy $\mathcal{U}$ (Section 2.4.3) and entropy S (Section 2.4.5). We are concerned here with systems in *uniform* thermodynamic equilibrium. This means that thermodynamic variables are single-valued and well defined, and that fields have no spatial dependence and are constant in time.

Under conditions of uniform thermodynamic equilibrium the internal energy is identified with the phase average of the Hamiltonian:[28]

$$\mathcal{U} = \langle \mathcal{H} \rangle . \tag{7.76}$$

For an isolated system, the Hamiltonian is constant, $\mathcal{H} = E$, and therefore

$$\mathcal{U} = E. \tag{7.77}$$

We will see in Section 7.4.2 that for a system in contact with a heat bath (which is described by the canonical ensemble), Eqn. (7.76) still holds. However, the energy of the system now fluctuates about $\mathcal{U}$ and therefore Eqn. (7.77) is no longer correct.

Next, we turn to temperature and entropy. The zeroth law of thermodynamics, discussed in Section 2.4.2, introduces the concept of thermal equilibrium. Two systems, initially in thermodynamic equilibrium, are said to be in *thermal equilibrium* if they remain in thermodynamic equilibrium when brought into contact. Using this concept it is possible to construct an empirical temperature scale by bringing together two systems, one of which only has a single state variable that is used to define temperature (for example the height of the mercury in a glass thermometer). In this way, the thermal equilibrium of two systems A and B is established if they both have the same temperature (see Eqn. (2.103)):

$$T^{\mathsf{A}} = T^{\mathsf{B}}. \tag{7.78}$$

[28] See Appendix A of [TME12] for a heuristic derivation of the total energy based on time averages that motivates the connection between the Hamiltonian and the internal energy.

Exactly the same procedure of bringing two systems into contact can be explored from the perspective of the second law of thermodynamics (as explained starting on page 75). In this case, equilibrium is identified with the state that maximizes the entropy of the combined system. The resulting condition for thermal equilibrium is

$$\frac{\partial \mathcal{S}^{\mathsf{A}}}{\partial \mathcal{U}^{\mathsf{A}}} = \frac{\partial \mathcal{S}^{\mathsf{B}}}{\partial \mathcal{U}^{\mathsf{B}}}. \tag{7.79}$$

Comparing this relation with Eqn. (7.78) and accounting for the fact that heat flows from hot to cold it can be inferred that[29]

$$\frac{\partial \mathcal{S}}{\partial \mathcal{U}} = \frac{1}{T}. \tag{7.80}$$

We now seek to do the same thing for our statistical mechanics systems A and B. The basic idea is the following.

Two initially isolated systems A and B are brought into contact through a diathermal partition (i.e. a partition that transmits thermal energy but does not allow mass transfer). The combined system remains isolated from the rest of the world. The energies of the systems before they are brought in contact are E_0^{A} and E_0^{B}. Once the systems are brought into contact, the total energy, $E = E_0^{\mathsf{A}} + E_0^{\mathsf{B}}$, remains constant, but the energies E^{A} and E^{B} of the individual systems will change until the systems arrive at a state of thermal equilibrium. This seems straightforward, however, it is important to note that we have implicitly made an assumption of weak interaction. Without this it is not possible to refer separately to the energy of system A or system B since they are coupled. The fact that an assumption of weak interaction is necessary in the discussion of thermal equilibrium of statistical mechanics systems was already recognized by Gibbs in 1902 [Gib02, page 37]:

> The most simple test of the equality of temperature of two bodies is that they remain in equilibrium when brought into thermal contact. Direct thermal contact implies molecular forces acting between the bodies. Now the test will fail unless the energy of these forces can be neglected in comparison with the other energies of the bodies. Thus, in the case of energetic chemical action between the bodies, or when the number of particles affected by the forces acting between the bodies is not negligible in comparison with the whole number of particles (as when the bodies have the form of exceedingly thin sheets), the contact of bodies of the same temperature may produce considerable thermal disturbance, and thus fail to afford a reliable criterion of the equality of temperature.

Assuming weak interaction, we seek to find the energies $E_{\mathrm{eq}}^{\mathsf{A}}$ and $E_{\mathrm{eq}}^{\mathsf{B}}$ that systems A and B adopt when thermal equilibrium is established, i.e. when Eqn. (7.78) is satisfied. However, at this point we still do not have a microscopic definition for temperature. Instead, we need to obtain some other indicator for equilibrium. Recall, that we were in a similar predicament when we introduced the concept of entropy in Section 2.4.5. Entropy was introduced in a completely formal manner as a variable designed to predict the direction of thermodynamic processes and equilibrium states. It was postulated that the entropy of any isolated system can never decrease and is a maximum at equilibrium. This is the second

[29] Any monotonic function of $1/T$ would do here. See footnote on page 77.

law of thermodynamics. Now consider our microcanonical ensemble. Is there something that plays a similar role to entropy in this case?

From a statistical perspective we argue that the equilibrium state will be the one that is most likely to be observed in an experimental measurement. We therefore seek to compute the probability $\Pr(E^{\mathsf{A}})$ of observing the combined system A+B in a state where the energy of system A is E^{A}. This is defined as

$$\Pr(E^{\mathsf{A}}) = \frac{\text{phase space volume when system A has energy } E^{\mathsf{A}}}{\text{phase space volume when combined system A + B has energy } E}, \tag{7.81}$$

where by "phase space" we are referring to the phase space of the combined system. When system A has energy E^{A} it is confined to a hypershell $\Sigma^{\mathsf{A}}(E^{\mathsf{A}}; \Delta E)$ (see Section 7.3.1 and Fig. 7.6) with volume $\Omega^{\mathsf{A}}(E^{\mathsf{A}}; \Delta E)$. At the same time, system B has energy $E^{\mathsf{B}} = E - E^{\mathsf{A}}$ and is confined to its own hypershell $\Sigma^{\mathsf{B}}(E - E^{\mathsf{A}}; \Delta E)$ with volume $\Omega^{\mathsf{B}}(E - E^{\mathsf{A}}; \Delta E)$. Since the systems are energetically decoupled by the assumption of weak interaction, system B can occupy *any* state in $\Sigma^{\mathsf{B}}(E - E^{\mathsf{A}}; \Delta E)$ for a given state of system A in $\Sigma^{\mathsf{A}}(E^{\mathsf{A}}; \Delta E)$. This means that the volume in phase space of the combined system (when system A has energy E^{A}) is just the product

$$\Omega^{\mathsf{A}}(E^{\mathsf{A}}; \Delta E)\Omega^{\mathsf{B}}(E - E^{\mathsf{A}}; \Delta E). \tag{7.82}$$

This is the numerator of Eqn. (7.81). The denominator is simply the volume $\Omega(E; 2\Delta E)$ of the hypershell defined by the condition $E \leq \mathcal{H} < E + 2\Delta E$. Thus, Eqn. (7.81) is

$$\Pr(E^{\mathsf{A}}) = \frac{\Omega^{\mathsf{A}}(E^{\mathsf{A}}; \Delta E)\Omega^{\mathsf{B}}(E - E^{\mathsf{A}}; \Delta E)}{\Omega(E; 2\Delta E)}. \tag{7.83}$$

Calculation of $\Omega(E; 2\Delta E)$ would require us to sum Eqn. (7.82) for all values of E^{A} between 0 and E for a given energy increment ΔE (or equivalently to take the limit as $\Delta E \to 0$ and replace the sum with an integral). However, since this term is a constant that does not depend on E^{A}, we will see that this is not necessary.

The most probable state for system A is the one that maximizes $\Pr(E^{\mathsf{A}})$. At the maximum, $\partial \Pr(E^{\mathsf{A}})/\partial E^{\mathsf{A}} = 0$, thus from Eqn. (7.83),

$$\frac{\partial \Omega^{\mathsf{A}}}{\partial E^{\mathsf{A}}}\Omega^{\mathsf{B}} - \frac{\partial \Omega^{\mathsf{B}}}{\partial E^{\mathsf{B}}}\Omega^{\mathsf{A}} = 0. \tag{7.84}$$

The solution to this equation is the energy $E^{\mathsf{A}}_{\mathrm{eq}}$ of system A that is most likely to be observed in an experimental measurement. This property on its own is not sufficient to identify $E^{\mathsf{A}}_{\mathrm{eq}}$ with the equilibrium energy of system A. An additional requirement is that the fluctuations of system A about this value are extremely small. This is exactly what happens when the numbers of atoms, N^{A} and N^{B}, are large. As the numbers of atoms tend to infinity the distribution about $E^{\mathsf{A}}_{\mathrm{eq}}$ collapses to a delta function in the same manner shown in Fig. 7.5. This means that there is an overwhelming probability that system A will have energy $E^{\mathsf{A}}_{\mathrm{eq}}$ in any observation. This defines the equilibrium state. The reason that this occurs is that for large N the function $\Omega(E)$ is an extremely rapidly increasing function of E as we argued in Section 7.3.1. (We see this explicitly in the next section for the special case of an ideal gas.) Consequently, the product $\Omega(E^{\mathsf{A}})\Omega(E - E^{\mathsf{A}})$, which is composed of one

Fig. 7.9 Ludwig Boltzmann's grave in the Central Cemetery (Zentralfriedhof) in Vienna, Austria, with his famous entropy principle serving as his epitaph at the top. Reprinted from [Wik05] under the GNU Free Documentation License.

rapidly increasing function of E^{A} and one rapidly decreasing function of E^{A}, will have a very sharp maximum. For a more formal derivation see for example [Hua63, Section 7.2].

Dividing Eqn. (7.84) through by $\Omega^{\mathrm{A}}\Omega^{\mathrm{B}}$, we have

$$\frac{1}{\Omega^{\mathrm{A}}}\frac{\partial \Omega^{\mathrm{A}}}{\partial E^{\mathrm{A}}} = \frac{1}{\Omega^{\mathrm{B}}}\frac{\partial \Omega^{\mathrm{B}}}{\partial E^{\mathrm{B}}}.$$

This relation can be rewritten as

$$\frac{\partial}{\partial E^{\mathrm{A}}}\ln \Omega^{\mathrm{A}} = \frac{\partial}{\partial E^{\mathrm{B}}}\ln \Omega^{\mathrm{B}}.$$

Comparing this relation with Eqn. (7.79) and recalling Eqn. (7.77), we obtain the following key relation linking the macroscopic concept of entropy with its microscopic analog:

$$S = k_{\mathrm{B}} \ln \Omega, \tag{7.85}$$

where $k_{\mathrm{B}} = 1.3807 \times 10^{-23}$ J/K $= 8.617 \times 10^{-5}$ eV/K is a constant of nature called the *Boltzmann constant* whose value relates the units of energy and temperature. Equation (7.85) was referred to as *Boltzmann's principle* by Einstein and is of such central importance that it was inscribed on Boltzmann's grave in Vienna as shown in Fig. 7.9.[30]

[30] It is interesting to note that Boltzmann never actually wrote down this famous equation. He described the connection between probability and entropy in 1877, but the actual equation was first written down by Max Planck in 1901 [Pla01]. It was also Planck who introduced "Boltzmann's constant." According to Planck, Boltzmann actually doubted whether this constant could ever be measured [Pla20]: "This constant is often referred to as Boltzmann's constant, although, to my knowledge, Boltzmann himself never introduced it – a peculiar state of affairs, which can be explained by the fact that Boltzmann, as appears from his occasional utterances, never gave thought to the possibility of carrying out an exact measurement of the constant." Planck obtained a first estimate for k_{B} (off by about 5% from current values) by fitting his theoretical model for radiation of a black body (called Planck's law) to experimental data. A more accurate measurement was made by the French physicist Jean Perrin in 1908 as part of his experiments on Brownian motion that verified Einstein's theory and settled the dispute on the existence of atoms.

It may seem problematic that the entropy expression in Eqn. (7.85) is stated in terms of $\Omega(E; \Delta E)$, which depends on the arbitrary value of ΔE. This would appear to suggest an indeterminacy in the value of the entropy. In fact, it is easy to show that this indeterminacy is negligible when the number of atoms is large. The entropy can be rewritten in terms of the DOS using Eqn. (7.59):

$$\mathcal{S} = k_{\mathrm{B}} \ln \Omega(E; \Delta) = k_{\mathrm{B}} \ln D(E)\Delta E = k_{\mathrm{B}} (\ln D(E) + \ln \Delta E).$$

Clearly, $\ln \Delta E$ is utterly negligible with respect to $\ln D(E)$ for macroscopically large values of N, since ΔE is a constant whereas $\ln D(E)$ is $O(N)$. (We know this since we showed in Section 7.3.1 that $D(E)$ grows exponentially with N.) Therefore,

$$\mathcal{S}(E) = k_{\mathrm{B}} \ln D(E). \tag{7.86}$$

In situations where $D(E)$ grows exponentially with N, so does $V_R(E)$. In fact, from Eqn. (7.61), $D(E) = (3N/2E)V_R(E)$, therefore

$$\mathcal{S}(E) = k_{\mathrm{B}} \ln \frac{3N}{2E} V_R(E) = k_{\mathrm{B}} \left(\ln \frac{3N}{2E} + \ln V_R(E) \right).$$

The term $\ln(3N/2E)$ is $O(1)$ since energy is an extensive quantity that scales with N (see the discussion of the thermodynamic limit at the end of Section 7.2.3). In contrast $\ln(V_R(E))$ is $O(N)$ as explained above and therefore for a macroscopic system

$$\mathcal{S}(E) = k_{\mathrm{B}} \ln V_R(E). \tag{7.87}$$

Thus, the entropy of a macroscopic system can be computed using any of the Eqns. (7.85), (7.86) or (7.87), with negligible numerical change in the result.

The derivation of the entropy given above is based on classical statistical mechanics. We have adopted this approach since the classical theory provides most of the necessary physical insight into statistical mechanics without the mental gymnastics introduced by quantum mechanics. The world, however, is quantum in nature and the classical model is simply a useful approximation. It is therefore desirable that the classical expressions correspond to those of quantum statistical mechanics in the appropriate limit. It turns out that for this to be the case, Eqn. (7.85) must be rewritten as (see, for example, [Pen05]) as

$$\mathcal{S} = k_{\mathrm{B}} \ln \frac{\Omega}{N!\, h^{3N}}, \tag{7.88}$$

where h is Planck's constant and N is the number of atoms. Planck's constant has units of momentum $\times$ distance, therefore the argument of the log function is now dimensionless. The $N!$ term corrects for the fact that in a quantum mechanical system atoms are inherently indistinguishable, whereas Ω contains $N!$ configurations that are identical except for the swapping of atoms. The $1/N!$ correction (referred to as "correct Boltzmann counting") turns out to be important even in a classical model to ensure that the entropy of mixing of two gases of the same species has the correct physical behavior (see Exercise 7.7). Equation

(7.88) can also be written as

$$\mathcal{S} = k_B \ln \Omega - k_B \ln N!\, h^{3N}.$$

The second term is a constant and therefore since only the derivative of the entropy appears in thermodynamic expressions, the addition of the denominator in Eqn. (7.88) has no direct effect at the thermodynamic level. For example, temperature is defined by Eqn. (7.80). Thus, using Eqn. (7.77) we have

$$T = \left[\frac{\partial \mathcal{S}}{\partial E}\right]^{-1} = \left[k_B \frac{1}{\Omega(E;\Delta E)} \frac{\partial \Omega(E;\Delta E)}{\partial E}\right]^{-1}.$$

Substituting in the first-order approximation $\Omega(E;\Delta E) = D(E)\Delta E$ (Eqn. (7.59)), we obtain the following microscopic definition for temperature:

$$T = \frac{1}{k_B} \frac{D(E)}{D'(E)}. \tag{7.89}$$

Given the equivalence of Eqns. (7.85), (7.86) and (7.87) shown above, the corrected entropy expression can be written in terms of the phase space volume or density of states:

$$\mathcal{S}(E) = k_B \ln \frac{V_R(E)}{N!\, h^{3N}} = k_B \ln \frac{D(E)}{N!\, h^{3N}}. \tag{7.90}$$

The last expression can be inverted to provide a useful expression for the DOS,

$$D(E) = N! h^{3N} e^{\mathcal{S}(E)/k_B}. \tag{7.91}$$

We will use this relation later in the derivation of the canonical ensemble.

Boltzmann's principle provides a clear explanation for the role that entropy plays in thermodynamics. At the macroscopic level, entropy was built into the theory in order to provide a means for predicting the direction of thermodynamic processes as explained in Section 2.4.5. Equation (7.80) provides a connection between entropy, temperature and internal energy, but it does not give a clear physical picture as to what entropy actually is. For most people who study thermodynamics, entropy is a rather mysterious concept that they eventually just accept. Boltzmann's principle provides a far more satisfying explanation for entropy. Consider, for example, the isolated cylinder in Fig. 2.9 with an internal partition separating two subsystems A and B containing ideal gases. Now for simplicity assume that subsystem B contains no gas. In this case we expect the gas in subsystem A to expand to fill the cylinder when the internal piston is released. However, as explained in Section 2.4.5, it is not possible to predict this by only considering the first law of thermodynamics (conservation of energy). The second law has to be introduced for this purpose. Here the explanation is obvious. The number of states Ω available to the gas increases monotonically with the volume V that it occupies in physical space. (We show in the next section that for an ideal gas, Ω is proportional to V^N.) The maximum possible value for the expanding gas is the volume of the entire container which, as explained above, is overwhelmingly probable compared to any other volume. Consequently, this is the one that is observed and the one we refer to as the equilibrium state.[31]

7.3.5 Derivation of the ideal gas law

The definition and properties of an ideal gas from a thermodynamic perspective were given in Section 2.4. We noted there that an ideal gas is a gas composed of noninteracting atoms. No such material exists in nature, but as the density of a real gas is reduced, the accuracy of this idealized model improves to the point where it provides an excellent approximation for most gases under moderate conditions. In this section, we compute the internal energy $\mathcal{U}$, entropy S and temperature T (derived in the previous section) for an ideal gas and derive the ideal gas law.

The Hamiltonian of an ideal gas composed of N atoms is

$$\mathcal{H} = \frac{1}{2m}\sum_{\alpha=1}^{N} \|\boldsymbol{p}^\alpha\|^2 + \mathcal{V}^{\mathrm{ext}}_{\mathrm{con}}(\{\boldsymbol{r}^\alpha\}; V) = \mathcal{H}_p + \mathcal{H}_q,$$

where $\boldsymbol{r}^\alpha$ and $\boldsymbol{p}^\alpha$ are the position and momentum of atom α, and $\mathcal{V}^{\mathrm{ext}}_{\mathrm{con}}$, defined in Eqn. (7.5), is a potential confining the atoms to a volume V. Assuming that the gas is isolated from the rest of the world, it has a constant energy E and is therefore restricted to the closed hypersurface S_E in phase space (see Fig. 7.6 and accompanying discussion). According to Eqn. (7.53), the phase space volume enclosed by S_E is

$$V_R(E) = \int_{\mathcal{H}<E} d\boldsymbol{q} d\boldsymbol{p} = \left[\prod_{\alpha=1}^{N} \int_{\boldsymbol{r}^\alpha \in V} d\boldsymbol{r}^\alpha\right] \int_{\mathcal{H}_p(\boldsymbol{p})<E} d\boldsymbol{p} = V^N \int_{\mathcal{H}_p(\boldsymbol{p})<E} d\boldsymbol{p},$$

where we have evaluated the integral over the position degrees of freedom. The domain of integration for the remaining integral is a $3N$-dimensional hypersphere defined by

$$\sum_{i=1}^{3N} p_i^2 = 2mE,$$

where we have reverted to the concatenated notation $\boldsymbol{p} = (p_1, \ldots, p_{3N})$. The volume of an n-dimensional hypersphere of radius R is $\pi^{n/2}R^n/\hat{\Gamma}(n/2+1)$, where $\hat{\Gamma}(x)$ is the Gamma function. Substituting this into the phase space volume expression we have

$$V_R(E) = V^N \frac{\pi^{3N/2}}{\hat{\Gamma}(3N/2+1)}(2mE)^{3N/2}.$$

The number of states associated with the hypershell $\Sigma(E;\Delta E)$ is $\Omega(E;\Delta E) = D(E)\Delta E$, where $D(E) = V_R'(E)$ is the DOS. Differentiating $V_R(E)$ with respect to E, we obtain

[31] See the very interesting explanation of the relationship between Boltzmann entropy and thermodynamic entropy (also called Gibbs entropy) in Joel Lebowitz's article [Leb93].

after simplification

$$\Omega(E;\Delta E) = V^N \frac{(2\pi m E)^{3N/2}}{\hat{\Gamma}(3N/2)} \frac{\Delta E}{E}, \tag{7.92}$$

where we have used the identity $\hat{\Gamma}(x+1) = x\hat{\Gamma}(x)$. We see that $\Omega \propto E^{3N/2-1}$, which for large N is a very rapidly increasing function of E as expected in the previous section.

The entropy of an ideal gas is obtained by substituting Eqn. (7.92) into Eqn. (7.88). Using the identity $\hat{\Gamma}(n) = (n-1)!$, we have after some manipulation,

$$S(N,V,E) = k_B \left[\frac{3N}{2} \ln \frac{2\pi m E V^{2/3}}{h^2} - \ln N! - \ln \left(\frac{3N}{2} - 1 \right)! + \ln \frac{\Delta E}{E} \right].$$

Agreement with the macroscopic concept of entropy is expected in the thermodynamic limit defined in Eqn. (7.52),

$$s = \lim_{N\to\infty} \frac{S(N,V,E)}{mN}.$$

Here s is the specific entropy (entropy per unit mass) at equilibrium.[32] As explained after Eqn. (7.52), this limit must be taken while maintaining the specific energy, $u = E/(mN)$, and the mass density, $\rho = mN/V$, fixed. Therefore

$$\begin{aligned} s &= \lim_{N\to\infty} \frac{S(N, V = mN/\rho, E = mNu)}{mN} \\ &= \lim_{N\to\infty} \frac{k_B}{mN} \left[\frac{3N}{2} \ln \frac{2\pi m^2 N u (mN/\rho)^{2/3}}{h^2} - \ln N! - \ln \left(\frac{3N}{2} - 1 \right)! + \ln \frac{\Delta E}{mNu} \right]. \end{aligned} \tag{7.93}$$

For large N, we neglect the -1 in the third term, apply Stirling's approximation, $\ln n! \approx n \ln n - n$, and neglect the last term that goes to zero with N. After simplification, we have

$$s = \frac{k_B}{m} \ln \left[\frac{e^{5/2} m^4}{\rho h^3} \left(\frac{4\pi u}{3} \right)^{3/2} \right]. \tag{7.94}$$

This expression for the specific entropy of an ideal gas is called the *Sackur–Tetrode* equation. The total entropy of the gas is then $S = mNs$. Since s is not a function of N, we see that S is proportional to N, which indicates that entropy is an extensive property as postulated at the macroscopic level.

[32] It is more common in statistical mechanics to use specific variables defined per unit volume or per particle instead of per unit mass as we do here. We do so in order to be consistent with the variables used in continuum mechanics.

Once the entropy is known, all other thermodynamic variables can be computed. First, we invert Eqn. (7.94) to obtain an expression for the internal energy:

$$\mathcal{U} = mNu = \frac{3h^2 N}{4\pi m}\left(\frac{N}{V}\right)^{2/3} \exp\left[\frac{2\mathcal{S}}{3Nk_B} - \frac{5}{3}\right]. \tag{7.95}$$

The temperature follows as the derivative with respect to entropy (the inverse of Eqn. (7.80)),

$$T = \frac{\partial \mathcal{U}}{\partial \mathcal{S}} = \frac{2\mathcal{U}}{3Nk_B} = \frac{2mu}{3k_B}. \tag{7.96}$$

Thus, the temperature of an ideal gas is a measure of its internal energy. This is a special case of the more general result, $T = (2m/3k_B) \times$ (kinetic energy per particle), which holds for any classical system of particles in equilibrium, as we shall see in the next section.

The pressure in the gas follows from Eqn. (2.126) as

$$p = -\frac{\partial \mathcal{U}}{\partial V} = \frac{2\mathcal{U}}{3V}. \tag{7.97}$$

Combining Eqns. (7.96) and Eqn. (7.97), we obtain a relationship between pressure, volume and the number of atoms,

$$p = \frac{2}{3V}\left(\frac{3Nk_B T}{2}\right) = \frac{Nk_B T}{V}, \tag{7.98}$$

which is called the *ideal gas law* (given earlier in Eqn. (2.129)). At the macroscopic level, it is more convenient to express the ideal gas law in terms of moles of gas instead of the number of atoms,

$$pV = nR_g T, \tag{7.99}$$

where $n = N/N_A$ is the number of moles of gas, $N_A = 6.022 \times 10^{23}$ is Avogadro's number, and $R_g = 8.314472\ \mathrm{J \cdot K^{-1} \cdot mol^{-1}}$.

7.3.6 Equipartition and virial theorems: microcanonical derivation

In the previous section we obtained specific expressions for the temperature and pressure of an ideal gas. In this section, we obtain general expressions for these quantities. The derivation follows that of [Hua63, Section 7.4].

We begin by computing the microcanonical ensemble average of $x_i(\partial\mathcal{H}/\partial x_j)$ where i and j are in the range $1, \dots n$, and where $\boldsymbol{x}$ is either the set of positions $\boldsymbol{q}$ or the set of momenta $\boldsymbol{p}$. Using Eqn. (7.62), we have

$$\left\langle x_i \frac{\partial \mathcal{H}}{\partial x_j}\right\rangle = \frac{1}{D(E)}\frac{\partial}{\partial E}\int_{R_E} x_i \frac{\partial \mathcal{H}}{\partial x_j}\, d\boldsymbol{q}d\boldsymbol{p},$$

where R_E is the portion of phase space satisfying $\mathcal{H}(\boldsymbol{q}, \boldsymbol{p}) < E$. The energy E is a constant and can therefore be subtracted from the Hamiltonian in the derivative expression without changing the expression. We then have

$$\begin{aligned}\left\langle x_i \frac{\partial \mathcal{H}}{\partial x_j}\right\rangle &= \frac{1}{D(E)}\frac{\partial}{\partial E}\int_{R_E} x_i \frac{\partial(\mathcal{H}-E)}{\partial x_j}\,d\boldsymbol{q}d\boldsymbol{p},\\ &= \frac{1}{D(E)}\frac{\partial}{\partial E}\left[\int_{R_E}\frac{\partial}{\partial x_j}(x_i(\mathcal{H}-E))\,d\boldsymbol{q}d\boldsymbol{p} - \int_{R_E}\delta_{ij}(\mathcal{H}-E)\,d\boldsymbol{q}d\boldsymbol{p}\right]\\ &= \frac{1}{D(E)}\frac{\partial}{\partial E}\left[\int_{S_E} x_i(\mathcal{H}-E)n_j\,dA - \int_{R_E}\delta_{ij}(\mathcal{H}-E)\,d\boldsymbol{q}d\boldsymbol{p}\right],\end{aligned}$$

where in the second line δ_{ij} is the Kronecker delta (Eqn. (2.1)) and we have used the identity $\partial x_i/\partial x_j = \delta_{ij}$. To pass from the second to the third line we used the divergence theorem in Eqn. $(2.51)_1$ to change the first volume integral to a surface integral on the hypersurface S_E with normal $\boldsymbol{n}$ and hypersurface area element dA. We now note that the surface integral is identically zero since $\mathcal{H} = E$ on S_E. Therefore only the second integral remains. We expand the domain of integration to the entire phase space by introducing the unit step function (Eqn. (7.54)) and then bring the derivative with respect to E inside the integral,

$$\begin{aligned}\left\langle x_i \frac{\partial \mathcal{H}}{\partial x_j}\right\rangle &= \frac{\delta_{ij}}{D(E)}\frac{\partial}{\partial E}\int_\Gamma (E-\mathcal{H})U(E-\mathcal{H})\,d\boldsymbol{q}d\boldsymbol{p}\\ &= \frac{\delta_{ij}}{D(E)}\left[\int_\Gamma U(E-\mathcal{H})\,d\boldsymbol{q}d\boldsymbol{p} + \int_\Gamma (E-\mathcal{H})\delta(E-\mathcal{H})\,d\boldsymbol{q}d\boldsymbol{p}\right].\end{aligned}$$

The first integral is by definition $V_R(E)$ (see Eqn. (7.53)). The second integral is zero since the Dirac delta limits the integration to the region in phase space where the integrand is zero. We therefore have,

$$\left\langle x_i \frac{\partial \mathcal{H}}{\partial x_j}\right\rangle = \delta_{ij}\frac{V_R(E)}{D(E)} = \delta_{ij}\frac{V_R(E)}{V_R'(E)} = \delta_{ij}\left[\frac{\partial}{\partial E}\ln V_R(E)\right]^{-1} = \delta_{ij}\frac{k_B}{\partial S/\partial E} = \delta_{ij}k_B T, \tag{7.100}$$

where in the second to last step we used Eqn. (7.87),[33] and in the final step we used Eqns. (7.80) and (7.77). Now, recall that $\boldsymbol{x}$ represents either $\boldsymbol{p}$ or $\boldsymbol{q}$. With $\boldsymbol{x} = \boldsymbol{p}$, we obtain the *equipartition theorem* for the momentum degrees of freedom, which states that

$$\left\langle p_i \frac{\partial \mathcal{H}}{\partial p_j}\right\rangle = \delta_{ij}k_B T. \tag{7.101}$$

Substituting in $\partial\mathcal{H}/\partial p_j = \dot{q}_j$ (Eqn. $(7.19)_2$), we have for $i = j$, $\langle p_i \dot{q}_i\rangle = k_B T$ (no sum), and since $\dot{q}_i = p_i/m_i$, we have

$$\left\langle \frac{p_i^2}{2m_i}\right\rangle = \frac{1}{2}k_B T. \tag{7.102}$$

This important result shows that the average kinetic energy of each degree of freedom at equilibrium is fixed and equal to $k_B T/2$. Summing over all $n = 3N$ momentum degrees

[33] Note that the normalization terms in Eqn. (7.90) do not make a difference since they drop out upon differentiation.

of freedom, we obtain the fundamental relationship between the average vibrational kinetic energy of a system,[34] $\langle \mathcal{T}^{\text{vib}} \rangle$, and its temperature:

$$\langle \mathcal{T}^{\text{vib}} \rangle = \sum_{i=1}^{3N} \left\langle \frac{p_i^2}{2m_i} \right\rangle = \sum_{i=1}^{3N} \left\langle \frac{1}{2} m_i \dot{q}_i^2 \right\rangle = \frac{3N}{2} k_B T. \tag{7.103}$$

Inverting this relation, we obtain an expression for the temperature of the system,

$$T = \frac{2 \langle \mathcal{T}^{\text{vib}} \rangle}{3N k_B}. \tag{7.104}$$

Compare this relation with the temperature of an ideal gas in Eqn. (7.96). In an ideal gas, the average kinetic energy and internal energy are equal (since there is no potential energy) and the two relations are identical. Equation (7.104), however, is a general expression that holds for all systems.

A second important relation is obtained from Eqn. (7.100) for $\boldsymbol{x} = \boldsymbol{q}$:

$$\left\langle q_i \frac{\partial \mathcal{H}}{\partial q_j} \right\rangle = \delta_{ij} k_B T. \tag{7.105}$$

This is the equipartition theorem for the position degrees of freedom. (This relation is sometimes called the "virial theorem," but we will avoid this terminology to prevent confusion with another relation given below which goes by the same name.) Summing Eqn. (7.105) over the three degrees of freedom associated with atom α, we have

$$-\langle \boldsymbol{r}^\alpha \cdot \boldsymbol{f}^\alpha \rangle = 3 k_B T, \tag{7.106}$$

where $\boldsymbol{f}^\alpha$ is the total force acting on atom α:

$$\boldsymbol{f}^\alpha = -\frac{\partial \mathcal{V}}{\partial \boldsymbol{r}^\alpha} = -\frac{\partial \mathcal{H}}{\partial \boldsymbol{r}^\alpha}. \tag{7.107}$$

Summing over all atoms, we have

$$-\langle \mathcal{W} \rangle = -\sum_{\alpha=1}^{N} \langle \boldsymbol{r}^\alpha \cdot \boldsymbol{f}^\alpha \rangle = 3N k_B T, \tag{7.108}$$

where $\mathcal{W} = \sum_{\alpha=1}^{N} \boldsymbol{r}^\alpha \cdot \boldsymbol{f}^\alpha$ is called the *virial* of the system. Comparing Eqns. (7.108) with Eqn. (7.103), we see that

$$\langle \mathcal{W} \rangle = -2 \langle \mathcal{T}^{\text{vib}} \rangle. \tag{7.109}$$

[34] Recall that $\boldsymbol{q}$ and $\boldsymbol{p}$ are assumed to be center-of-mass coordinates and therefore represent vibrational behavior as explained in Section 7.1.

This relation is referred to as the "virial theorem" (leading to some minor confusion in the literature with Eqn. (7.105) as explained above). Equation (7.109) was originally derived by Rudolf Clausius in 1870 using time averages instead of phase averages. See [AT10, Appendix A] for a discussion of the Clausius derivation and an extension to tensor quantities leading to the *tensor virial theorem*.

The virial theorem can be used to obtain an expression for the pressure in a thermodynamic system (which was the original objective of Clausius). Basically, the idea is that the virial can be divided into an internal part, which only involves forces due to atoms in the system, and an external part related to forces from the surrounding world. After some manipulation (see [AT10, Appendix A] for details) and using Eqn. (7.103), one obtains

$$pV = Nk_{\rm B}T + \frac{1}{3}\langle \mathcal{W}^{\rm int}\rangle ,$$

where p is the pressure, V is the volume of the system, $\mathcal{W}^{\rm int} = \sum_{\alpha=1}^{N} \boldsymbol{r}^{\alpha} \cdot \boldsymbol{f}^{{\rm int},\alpha}$ is the internal virial and $\boldsymbol{f}^{{\rm int},\alpha}$ is the internal force on atom α due to other atoms in the system. For an ideal gas, $\mathcal{W}^{\rm int} = 0$, and this relation reduces to the ideal gas law in Eqn. (7.97). We give this derivation in an abbreviated manner here, since we will be discussing the derivation of the stress tensor (for which pressure is a special case) for both equilibrium and non-equilibrium systems in a far more general manner in Chapter 8.

It is important to point out that the equipartition theorem only applies to classical systems. For quantum systems (which for our purposes here can be narrowly interpreted as atoms at very low temperatures), the quantization of the possible energy levels available to the system leads to a breakdown in the equipartition theorem. For example, equipartition predicts that the specific heat of crystalline solids is independent of temperature (Dulong–Petit law), whereas experiments clearly show that the specific heat reduces at low temperatures, tending towards zero as the temperature approaches 0 K. Another important example is the so-called "ultraviolet catastrophe" predicted by equipartition (and introduced on page 154), where the intensity of the radiation emitted by a black body grows without bounds with increasing frequency. This is in stark contrast to the experimental result, which shows the curve turning over and decreasing to zero. The latter failure was sufficiently disturbing that in a famous lecture in 1900, Lord Kelvin referred to it as one of the "nineteenth-century clouds" overshadowing classical mechanics. (The other "cloud" was the Michelson–Morley experiment which eventually led to relativity theory.) Failures of equipartition were undoubtedly instrumental in driving the development of quantum mechanics, although it was usually easier to doubt equipartition rather than classical mechanics as a whole. (See Section 4.2 for a discussion of the origins of quantum mechanics.)

7.4 The canonical (NVT) ensemble

Real physical systems are normally not isolated. Depending on the imposed constraints, they may be able to exchange mass and energy with their surroundings. A particularly important case is that of a system in contact with a heat bath. The heat bath maintains the system at a specified temperature by providing and removing energy as necessary. The result is a system with a constant number of particles N occupying a volume V at a temperature T set by the heat bath. The set of all microstates consistent with these constraints is called the *canonical*[35] or NVT ensemble.

We discussed the interaction of two systems in Section 7.3.3, where the concept of weak interaction was introduced. We make the same assumption here regarding the interaction of the system of interest (system A) and the heat bath (system B). A schematic depiction was shown in Fig. 7.8. As explained in Section 7.3.3, the combined system A + B is taken to be an isolated system with constant energy E. The assumption of weak interaction implies that the total energy can be written as the sum of the energies of each of the systems as if they were isolated (the part of the energy due to the interaction between the two systems is neglected), i.e. $E^{\mathsf{A}} + E^{\mathsf{B}} = E$, where $E^{\mathsf{A}} = \mathcal{H}^{\mathsf{A}}(\boldsymbol{y}^A)$ and $E^{\mathsf{B}} = \mathcal{H}^{\mathsf{B}}(\boldsymbol{y}^B)$ are the energies of systems A and B which are given by their Hamiltonians.[36]

The key step which is different from the derivation given earlier for the microcanonical ensemble is that we now assume that system B is much larger than system A and therefore $E^{\mathsf{B}} \gg E^{\mathsf{A}}$. This allows us to make certain approximations that lead to a distribution function for system A on its own with system B only featuring in terms of its temperature. This is exactly the role we expect a heat bath to play. We derive the canonical distribution function in the next section beginning from where the microcanonical derivation for weak interaction concluded.

7.4.1 The canonical distribution function

Equation (7.75) provides an approximate expression for the expectation value of a phase function, $A(\boldsymbol{y}) = A(\boldsymbol{y}^{\mathsf{A}}, \boldsymbol{y}^{\mathsf{B}})$, under conditions of weak interaction between systems A and B. We now focus on system A and limit consideration to phase functions, $A(\boldsymbol{y}^{\mathsf{A}})$, that only depend on the positions and momenta of atoms in A. In this case, we can recast Eqn. (7.75) as an integral over the phase space of system A alone:

$$
\begin{aligned}
\langle A \rangle &= \frac{1}{D^{\mathsf{A+B}}(E)} \int_{\Gamma^{\mathsf{A}}} \int_{\Gamma^{\mathsf{B}}} A(\boldsymbol{y}^{\mathsf{A}}) \delta \left(\mathcal{H}^{\mathsf{A}}(\boldsymbol{y}^{\mathsf{A}}) + \mathcal{H}^{\mathsf{B}}(\boldsymbol{y}^{\mathsf{B}}) - E \right) d\boldsymbol{y}^{\mathsf{A}} d\boldsymbol{y}^{\mathsf{B}} \\
&= \frac{1}{D^{\mathsf{A+B}}(E)} \int_{\Gamma^{\mathsf{A}}} A(\boldsymbol{y}^{\mathsf{A}}) \left[\int_{\Gamma^{\mathsf{B}}} \delta \left(\mathcal{H}^{\mathsf{A}}(\boldsymbol{y}^{\mathsf{A}}) + \mathcal{H}^{\mathsf{B}}(\boldsymbol{y}^{\mathsf{B}}) - E \right) d\boldsymbol{y}^{\mathsf{B}} \right] d\boldsymbol{y}^{\mathsf{A}} \\
&\equiv \int_{\Gamma^{\mathsf{A}}} A(\boldsymbol{y}^{\mathsf{A}}) f_{\mathrm{c}}^{\mathsf{A}}(\boldsymbol{y}^{\mathsf{A}})\, d\boldsymbol{y}^{\mathsf{A}},
\end{aligned}
$$

where $D^{\mathsf{A+B}}$ is the DOS for the combined system A + B, and where we have defined the *canonical distribution function*, $f_{\mathrm{c}}^{\mathsf{A}}$, for system A as

$$
f_{\mathrm{c}}^{\mathsf{A}}(\boldsymbol{y}^{\mathsf{A}}) \equiv \frac{1}{D^{\mathsf{A+B}}(E)} \int_{\Gamma^{\mathsf{B}}} \delta \left(\mathcal{H}^{\mathsf{A}}(\boldsymbol{y}^{\mathsf{A}}) + \mathcal{H}^{\mathsf{B}}(\boldsymbol{y}^{\mathsf{B}}) - E \right) d\boldsymbol{y}^{\mathsf{B}}.
$$

[35] See footnote 11 on page 390 for a discussion of the origin of the term "canonical".

[36] As explained in footnote 27 on page 411, the potential energy of a system in weak interaction includes only contributions due to internal interactions and external fields, but not "contact" interactions with atoms outside the system. Here we also assume that there are no long-range fields. Thus, $\mathcal{V}^{\mathsf{A}} = \mathcal{V}^{\mathrm{int},\mathsf{A}}$, and similarly for system B.

Multiplying and dividing by $D^{\mathrm{B}}(E^{\mathrm{B}})$, where $E^{\mathrm{B}} = E - \mathcal{H}^{\mathrm{A}}(\boldsymbol{y}^{\mathrm{A}})$, we have

$$f_{\mathrm{c}}^{\mathrm{A}}(\boldsymbol{y}^{\mathrm{A}}) = \frac{D^{\mathrm{B}}(E^{\mathrm{B}})}{D^{\mathrm{A+B}}(E)} \int_{\Gamma^{\mathrm{B}}} \frac{\delta\left(\mathcal{H}^{\mathrm{B}}(\boldsymbol{y}^{\mathrm{B}}) - E^{\mathrm{B}}\right)}{D^{\mathrm{B}}(E^{\mathrm{B}})} d\boldsymbol{y}^{\mathrm{B}} = \frac{D^{\mathrm{B}}(E^{\mathrm{B}})}{D^{\mathrm{A+B}}(E)} \int_{\Gamma^{\mathrm{B}}} f_{\mathrm{mc}}^{\mathrm{B}}(\boldsymbol{y}; E^{\mathrm{B}})\, d\boldsymbol{y}^{\mathrm{B}},$$

where we have used the definition of the microcanonical distribution function in Eqn. (7.64). The last integral is equal to 1 due to the normalization of the distribution function, so

$$f_{\mathrm{c}}^{\mathrm{A}}(\boldsymbol{y}^{\mathrm{A}}) = \begin{cases} \frac{D^{\mathrm{B}}(E - \mathcal{H}^{\mathrm{A}}(\boldsymbol{y}^{\mathrm{A}}))}{D^{\mathrm{A+B}}(E)} & \text{if } \mathcal{H}^{\mathrm{A}}(\boldsymbol{y}^{\mathrm{A}}) \le E, \\ 0 & \text{otherwise.} \end{cases} \tag{7.110}$$

The conditional phrasing ensures that the energy of system A is not greater than the energy of the combined system. (We assume that the energy of a system has a lower bound that can be arbitrarily taken to be zero.) We can rewrite Eqn. (7.110) in a more convenient form by multiplying and dividing by $D^{\mathrm{B}}(E)$:

$$f_{\mathrm{c}}^{\mathrm{A}}(\boldsymbol{y}^{\mathrm{A}}) = \frac{D^{\mathrm{B}}(E)}{D^{\mathrm{A+B}}(E)} \frac{D^{\mathrm{B}}(E - E^{\mathrm{A}})}{D^{\mathrm{B}}(E)} = C \frac{D^{\mathrm{B}}(E - E^{\mathrm{A}})}{D^{\mathrm{B}}(E)}, \tag{7.111}$$

where $E^{\mathrm{A}} = \mathcal{H}^{\mathrm{A}}(\boldsymbol{y}^{\mathrm{A}}) \le E$ is the energy of system A. The fraction $D^{\mathrm{B}}(E)/D^{\mathrm{A+B}}(E)$ does not depend on $\boldsymbol{y}^{\mathrm{A}}$, we can therefore replace it with a constant C that will be determined later through the normalization condition for $f_{\mathrm{c}}^{\mathrm{A}}$. Substituting Eqn. (7.91) into Eqn. (7.111), we have

$$f_{\mathrm{c}}^{\mathrm{A}}(\boldsymbol{y}^{\mathrm{A}}) = C \frac{e^{\mathcal{S}^{\mathrm{B}}(E - E^{\mathrm{A}})/k_{\mathrm{B}}}}{e^{\mathcal{S}^{\mathrm{B}}(E)/k_{\mathrm{B}}}} = C \exp\left[\frac{1}{k_{\mathrm{B}}}\left(\mathcal{S}^{\mathrm{B}}(E - E^{\mathrm{A}}) - \mathcal{S}^{\mathrm{B}}(E)\right)\right]. \tag{7.112}$$

Up until this point we have treated both systems A and B on an equal footing. Both are assumed to be large enough to be considered macroscopic and the assumption of weak interaction applies equally to both. We now consider the case where system B is much larger than A. So large that $E \approx E^{\mathrm{B}}$ and the energy of system A is negligible relative to both ($E^{\mathrm{A}} \ll E$). In this case, the energy difference in the exponent of Eqn. (7.112) can be approximated as

$$\mathcal{S}^{\mathrm{B}}(E - E^{\mathrm{A}}) - \mathcal{S}^{\mathrm{B}}(E) \approx -\left.\frac{\partial \mathcal{S}^{\mathrm{B}}}{\partial E^{\mathrm{B}}}\right|_{E^{\mathrm{B}}=E} E^{\mathrm{A}} = -\frac{E^{\mathrm{A}}}{T^{\mathrm{B}}},$$

where we have used[37] Eqn. (7.80) and T^{B} is the temperature of system B. Substituting this into Eqn. (7.112), we have

$$f_{\mathrm{c}}^{\mathrm{A}}(\boldsymbol{y}^{\mathrm{A}}) = C e^{-E^{\mathrm{A}}/k_{\mathrm{B}} T^{\mathrm{B}}}.$$

C is computed from the normalization condition, $\int_{\Gamma^{\mathrm{A}}} f_{\mathrm{c}}^{\mathrm{A}}(\boldsymbol{y}^{\mathrm{A}})\, d\boldsymbol{y}^{\mathrm{A}} = 1$, so that

$$C^{-1} = \int_{\mathcal{H}^{\mathrm{A}} \le E} e^{-\mathcal{H}^{\mathrm{A}}(\boldsymbol{y}^{\mathrm{A}})/k_{\mathrm{B}} T^{\mathrm{B}}}\, d\boldsymbol{y}^{\mathrm{A}}. \tag{7.113}$$

[37] Since $E \approx E^{\mathrm{B}}$, system B is essentially an isolated system and therefore $E^{\mathrm{B}} = \mathcal{U}^{\mathrm{B}}$, where $\mathcal{U}^{\mathrm{B}}$ is the internal energy of system B, and hence Eqn. (7.80) can be used.

Note that the integration is confined to the portion of phase space of system A for which f_c^A is nonzero. (Recall the conditional definition of the canonical distribution function in Eqn. (7.110).) The explicit dependence of the distribution function on the total energy E is inconvenient since this value is not normally known and should not be important when considering the behavior of a small system whose energy is negligible compared with that of the surrounding system. We can show that this is indeed the case by changing the integration variable in Eqn. (7.113) to E^A and making use of the DOS,[38]

$$C^{-1} = \int_0^E e^{-E^A/k_B T^B} D^A(E^A)\, dE^A. \tag{7.114}$$

The integrand of Eqn. (7.114) has a very sharp maximum, since it is a product of two functions of E^A, one rapidly *decreasing*, $e^{-E^A/k_B T^B}$, and one rapidly *increasing*, $D^A(E^A)$, as explained in Section 7.3.1. The maximum is located at the equilibrium value E^A_{eq} defined by the equation

$$\frac{d}{dE^A}\left[e^{-E^A/k_B T^B} D^A(E^A)\right] = e^{-E^A/k_B T^B}\left[\frac{dD^A(E^A)}{dE^A} - \frac{D^A(E^A)}{k_B T^B}\right] = 0. \tag{7.115}$$

Systems A and B are in thermal equilibrium, therefore E^A_{eq} obtained here is the same as the equilibrium value obtained by solving Eqn. (7.84) in Section 7.3.4. We can see this, since from the above equation we obtain, $k_B T = D(E)/D'(E)$, which is exactly the relation obtained in Eqn. (7.89) for thermal equilibrium. The fact that one particular value of the integrand of Eqn. (7.114), at $E^A = E^A_{eq}$, dominates all others means that the upper bound on the integration can be extended to infinity with negligible effect on the result. Equivalently, the integration in Eqn. (7.113) can be expanded to the entire phase space of system A.

The above results mean that system B enters into the behavior of system A only through its temperature, fulfilling the role of a heat bath in thermodynamics. To stress this fact, we rewrite the canonical distribution function for a generic system, dropping the system superscripts. By convention, the normalization constant C is replaced by a new constant, $Z = C^{-1}/N!h^{3N}$, called the *partition function*:

$$Z = \frac{1}{N!h^{3N}} \int_\Gamma e^{-\mathcal{H}(\boldsymbol{q},\boldsymbol{p})/k_B T}\, d\boldsymbol{q}d\boldsymbol{p}, \tag{7.116}$$

where we have dropped the explicit references to system A, which is now tacitly implied (for example $(\boldsymbol{q},\boldsymbol{p})$ implies $\boldsymbol{y}^A$ and $\mathcal{H}$ implies $\mathcal{H}^A$). Thus,[39]

$$\mathcal{H}(\boldsymbol{q},\boldsymbol{p}) = \sum_{\alpha=1}^{N} \frac{\boldsymbol{p}^\alpha \cdot \boldsymbol{p}^\alpha}{2m^\alpha} + \mathcal{V}^{int}(\boldsymbol{q}), \tag{7.117}$$

[38] In effect, what we are doing in Eqn. (7.114) is dividing the phase space of system A into hypershells of constant energy E^A and weighting each shell by its volume $D^A(E^A)dE^A$.

[39] As explained in footnote 36 on page 424, only the *internal* potential energy contributes to the Hamiltonian in the canonical ensemble.

where $N = N^{\mathsf{A}}$ is the number of atoms in the system of interest and $\mathcal{V}^{\text{int}} = \mathcal{V}^{\mathsf{A}}$. Note that as in the definition of the entropy in Eqn. (7.88), $N!$ and h^{3N} are added to Eqn. (7.116) to obtain agreement with quantum mechanics in the appropriate limit. The canonical distribution function is then

$$f_c(\boldsymbol{q}, \boldsymbol{p}; T) = \frac{1}{N! h^{3N} Z} e^{-\mathcal{H}(\boldsymbol{q},\boldsymbol{p})/k_B T}. \tag{7.118}$$

Given the derivation that led to these expressions, we understand that T is the temperature of the surrounding heat bath. The expectation value of a phase function of the system is

$$\langle A \rangle = \int_{\Gamma} A(\boldsymbol{q}, \boldsymbol{p}) f_c(\boldsymbol{q}, \boldsymbol{p}; T)\, d\boldsymbol{q} d\boldsymbol{p}, \tag{7.119}$$

where for the reasons explained above the integration is carried out over the entire phase space of the system regardless of the energy of the heat bath.

We will see later that the partition function plays a key role in describing the thermodynamic properties of the system. For now, we just point out that due to the decoupling of the position and momentum terms in the Hamiltonian, the partition function can be decomposed into kinetic and potential parts:

$$Z = Z^{\text{K}} Z^{\text{V}}, \tag{7.120}$$

with

$$Z^{\text{K}} = \frac{1}{N! h^{3N}} \int_{\Gamma_p} e^{-\sum_i p_i^2/(2m_i k_B T)}\, d\boldsymbol{p}, \qquad Z^{\text{V}} = \int_{\Gamma_q} e^{-\mathcal{V}^{\text{int}}(q_1,\dots,q_n)/kT}\, d\boldsymbol{q}. \tag{7.121}$$

The potential part of the partition function is often referred to as the *configurational integral*. The kinetic part of the partition function can be explicitly integrated to obtain a closed-form analytical expression:

$$Z^{\text{K}} = \frac{1}{N! h^{3N}} \prod_{i=1}^{3N} \int_{-\infty}^{\infty} e^{-p_i^2/(2m_i k_B T)}\, dp_i = \frac{1}{N!} \left(\frac{2\pi k_B T}{h^2} \right)^{3N/2} \left(\prod_{i=1}^{3N} m_i \right)^{1/2}. \tag{7.122}$$

For a monoatomic system, $m_i = m$, and $Z^{\text{K}} = 1/(N!\Lambda^{3N})$, where $\Lambda = \sqrt{h^2/2\pi m k_B T}$ is the de Broglie thermal wavelength defined in Eqn. (5.9). The resulting partition function in this case is

$$Z = \frac{1}{N! \Lambda^{3N}} Z^{\text{V}}. \tag{7.123}$$

7.4.2 Internal energy and fluctuations

As explained in Section 7.3.4, the internal energy $\mathcal{U}$ corresponds to the expectation value of the Hamiltonian:

$$\mathcal{U} = \langle \mathcal{H} \rangle = \frac{1}{N! h^{3N} Z} \int_\Gamma \mathcal{H} e^{-\mathcal{H}/k_B T}\, dq dp. \tag{7.124}$$

Substituting in Eqn. (7.116), we have

$$\mathcal{U} = \langle \mathcal{H} \rangle = \frac{\int_\Gamma \mathcal{H} e^{-\mathcal{H}/k_B T}\, dq dp}{\int_\Gamma e^{-\mathcal{H}/k_B T}\, dq dp} = \frac{\int_0^\infty E e^{-E/k_B T} D(E)\, dE}{\int_0^\infty e^{-E/k_B T} D(E)\, dE}, \tag{7.125}$$

where in the second term we have changed the integration variable to E (as was done in Eqn. (7.114) for C). The integrands of the numerator and denominator have sharp maxima for the reason explained in the previous section. The denominator has a maximum at E_{eq}, obtained from solving Eqn. (7.115), and it is easy to show that the numerator has a maximum at the same value (as $N \to \infty$). The condition for the maximum of the numerator integrand is

$$\frac{d}{dE}\left[E e^{-E/k_B T} D(E)\right] = e^{-E/k_B T} D(E) \left[E \frac{D'(E)}{D(E)} - \frac{E}{k_B T} + 1\right] = 0.$$

Since E is proportional to N, the 1 term is negligible for large N and we obtain the same condition as in Eqn. (7.115). We now replace the integrals in the numerator and denominator with their dominant terms:

$$\mathcal{U} = \langle \mathcal{H} \rangle = \frac{E_{eq} e^{-E_{eq}/k_B T} D(E_{eq})}{e^{-E_{eq}/k_B T} D(E_{eq})} = E_{eq}. \tag{7.126}$$

We have established that the equilibrium value of the energy of the system corresponds to its internal energy. However, unlike for an isolated system, the energy of a system in contact with a heat bath fluctuates about this value. Let us compute the magnitude of the expected fluctuations. We start by rewriting Eqn. (7.125) in the following way:

$$0 = \mathcal{U} - \langle \mathcal{H} \rangle = \mathcal{U} - \frac{\int_\Gamma \mathcal{H} e^{-\mathcal{H}/k_B T}\, dq dp}{\int_\Gamma e^{-\mathcal{H}/k_B T}\, dq dp} = \frac{\int_\Gamma (\mathcal{U} - \mathcal{H}) e^{-\mathcal{H}/k_B T}\, dq dp}{\int_\Gamma e^{-\mathcal{H}/k_B T}\, dq dp}, \tag{7.127}$$

where the last step is possible since $\mathcal{U}$ is a constant. Taking the derivative with respect to T and using Eqn. (7.127) to cancel one of the terms we have

$$0 = k_B T^2 \frac{\partial \mathcal{U}}{\partial T} + \mathcal{U} \frac{\int_\Gamma \mathcal{H} e^{-\mathcal{H}/k_B T}\, dq dp}{\int_\Gamma e^{-\mathcal{H}/k_B T}\, dq dp} - \frac{\int_\Gamma \mathcal{H}^2 e^{-\mathcal{H}/k_B T}\, dq dp}{\int_\Gamma e^{-\mathcal{H}/k_B T}\, dq dp}.$$

The second term on the right is simply $\mathcal{U} \langle \mathcal{H} \rangle = \langle \mathcal{H} \rangle^2$, where we have used Eqn. (7.124). The third term is $\langle \mathcal{H}^2 \rangle$. We therefore have after using Eqn. (2.107)

$$\langle \mathcal{H}^2 \rangle - \langle \mathcal{H} \rangle^2 = k_B T^2 n C_v, \tag{7.128}$$

where C_v is the molar heat capacity and n is the number of moles. The expression on the left is the variance of $\mathcal{H}$ (see Eqn. (7.36)), which is a measure of the magnitude of the fluctuations. What happens to the fluctuations in the thermodynamic limit? To answer this, we compute the normalized standard deviation of $\mathcal{H}$:

$$\frac{\sigma_{\mathcal{H}}}{\mathcal{U}} = \frac{\sqrt{\langle \mathcal{H}^2 \rangle - \langle \mathcal{H} \rangle^2}}{\mathcal{U}} = \frac{1}{\mathcal{U}} \sqrt{k_B T^2 n C_v}.$$

We must now take the limit of the right-hand side as $N \to \infty$ while keeping the average specific energy, $u = \mathcal{U}/(mN)$, constant:

$$\lim_{N\to\infty} \frac{1}{mNu} \sqrt{k_B T^2 (N/N_A) C_v} = \left[\frac{k_B T^2 C_v}{N_A}\right]^{1/2} \frac{1}{\sqrt{N}} = 0.$$

We see that the normalized fluctuations go to zero as $1/\sqrt{N}$ as N becomes large. The conclusion is that the canonical and microcanonical ensembles coincide in the thermodynamic limit.

Next we turn to the derivation of the Helmholtz free energy for the canonical ensemble, which plays a central role in the thermodynamics of the system.

7.4.3 Helmholtz free energy

The partition function Z introduced in Section 7.4.1 as a normalization constant turns out to be of fundamental importance to the thermodynamic properties of the canonical ensemble. We introduce a new variable Ψ defined by the relation

$$\Psi = -k_B T \ln Z. \tag{7.129}$$

We show below that Ψ, defined in this manner, is the extensive counterpart of the specific Helmholtz free energy introduced earlier in Eqn. (2.170). Inverting Eqn. (7.129), we have

$$Z = e^{-\Psi/k_B T},$$

which when substituted into the canonical distribution function in Eqn. (7.118) gives

$$f_c = \frac{1}{N!h^{3N}} e^{-(\mathcal{H}-\Psi)/k_B T}.$$

We now use the normalization condition for f_c to obtain

$$\frac{1}{N!h^{3N}} \int_\Gamma e^{-(\mathcal{H}-\Psi)/k_B T} \, dqdp = 1. \tag{7.130}$$

Taking the derivative with respect to T, we have

$$\frac{1}{N!h^{3N}} \frac{\partial}{\partial T} \int_\Gamma e^{-(\mathcal{H}-\Psi)/k_B T} \, dqdp$$
$$= \frac{1}{N!h^{3N}} \int_\Gamma \frac{\frac{\partial \Psi}{\partial T} k_B T + (\mathcal{H} - \Psi) k_B}{(k_B T)^2} e^{-(\mathcal{H}-\Psi)/k_B T} \, dqdp = 0. \tag{7.131}$$

Multiplying by $k_B T^2$ and rearranging gives

$$\left(T\frac{\partial \Psi}{\partial T} - \Psi\right)\left[\frac{1}{N!h^{3N}}\int_\Gamma e^{-(\mathcal{H}-\Psi)/k_B T}\, dqdp\right] + \left[\frac{1}{N!h^{3N}}\int_\Gamma \mathcal{H} e^{-(\mathcal{H}-\Psi)/k_B T}\, dqdp\right] = 0. \tag{7.132}$$

The first integral is equal to 1 due to the normalization condition (Eqn. (7.130)). The second integral is the expectation value of the Hamiltonian, which is equal to the internal energy $\mathcal{U}$ (Eqn. (7.124)). We therefore have,

$$\Psi = \mathcal{U} + T\frac{\partial \Psi}{\partial T}.$$

This relation is consistent with the identification of Ψ with the Helmholtz free energy, since we know from thermodynamics that $\mathcal{S} = -\partial\Psi/\partial T$ (see Eqn. (2.171)$_1$), and therefore

$$\Psi = \mathcal{U} - T\mathcal{S}, \tag{7.133}$$

which is the definition of the Helmholtz free energy.

The reader may be wondering whether the definition given in Eqn. (7.129) for the free energy is consistent with the earlier definition for entropy in Boltzmann's principle (Eqn. (7.85)). It is instructive to show that this is indeed the case. We start by rewriting the partition function as an integral over energy as explained above (see Eqn. (7.114)),

$$Z = \frac{1}{N!h^{3N}}\int_0^\infty e^{-E/k_B T} D(E)\, dE. \tag{7.134}$$

Substituting in Eqn. (7.91), we have

$$Z = \int_0^\infty e^{-(E-T\mathcal{S})/k_B T}\, dE. \tag{7.135}$$

As explained in Section 7.4.1, this integrand has a very sharp peak at the equilibrium value $E_{\rm eq} = \mathcal{U}$ (see Eqn. (7.126)). Expanding $E - T\mathcal{S}$ about this value, we have

$$E - T\mathcal{S} = [\mathcal{U} - T\mathcal{S}(\mathcal{U})] + \left[1 - T\left.\frac{\partial \mathcal{S}}{\partial E}\right|_{E=\mathcal{U}}\right]\epsilon + \frac{1}{2}\left[-T\left.\frac{\partial^2 \mathcal{S}}{\partial E^2}\right|_{E=\mathcal{U}}\right]\epsilon^2 + O(\epsilon^3),$$

where $\epsilon = E - \mathcal{U}$. The first term in square brackets is the thermodynamic definition for the Helmholtz free energy Ψ (Eqn. (7.133)). The term multiplying ϵ is zero, since $\partial\mathcal{S}/\partial E|_{E=\mathcal{U}} = 1/T$ (Eqn. (7.80)). The term multiplying ϵ^2 can be rewritten in terms of the molar heat capacity at constant volume, C_v, using Eqn. (2.122). We therefore have

$$E - T\mathcal{S} = \Psi + \frac{1}{2TnC_v}\epsilon^2 + O(\epsilon^3), \tag{7.136}$$

where n is the number of moles. Substituting Eqn. (7.136) into Eqn. (7.135) and integrating gives

$$Z = e^{-\Psi/k_B T}\int_0^\infty e^{-\epsilon^2/2k_B T^2 nC_v}\, d\epsilon = e^{-\Psi/k_B T}\sqrt{2\pi k_B T^2 nC_v}.$$

Thus,

$$-k_B T \ln Z = \Psi - \frac{1}{2}k_B T \ln\frac{2\pi k_B T^2 C_v}{N_A}N,$$

where we have used $n = N/N_A$ and N_A is Avogadro's number. The second term is proportional to $\ln N$ and is therefore negligible relative to Ψ (which is proportional to N) as $N \to \infty$, and so $\Psi = -k_B T \ln Z$ as we showed above. Thus, the definitions of entropy in the microcanonical and canonical ensembles coincide in the thermodynamic limit.

7.4.4 Equipartition theorem: canonical derivation

In Section 7.3.6 we derived the equipartition theorem for the microcanonical ensemble. Here we rederive it for the canonical ensemble and show that the same results are obtained. This derivation is based on [Tuc11].

We wish to compute the canonical average, $\langle x_i(\partial\mathcal{H}/\partial x_j)\rangle$, where $\boldsymbol{x}$ is either the set of positions $\boldsymbol{q}$ or the set of momenta $\boldsymbol{p}$. From Eqns. (7.118)–(7.119), we have

$$\left\langle x_i \frac{\partial \mathcal{H}}{\partial x_j} \right\rangle = \frac{1}{N!h^{3N}Z} \int_\Gamma x_i \frac{\partial \mathcal{H}}{\partial x_j} e^{-\mathcal{H}/k_B T}\, d\boldsymbol{q}d\boldsymbol{p}.$$

Next, making use of the identity

$$\frac{\partial}{\partial x_j}\left(x_i e^{-\mathcal{H}/k_B T}\right) = \delta_{ij} e^{-\mathcal{H}/k_B T} - \frac{1}{k_B T} x_i \frac{\partial \mathcal{H}}{\partial x_j} e^{-\mathcal{H}/k_B T},$$

we have

$$\begin{aligned}\left\langle x_i \frac{\partial \mathcal{H}}{\partial x_j} \right\rangle &= \frac{k_B T}{N!h^{3N}Z} \int_\Gamma \left[\delta_{ij} e^{-\mathcal{H}/k_B T} - \frac{\partial}{\partial x_j}\left(x_i e^{-\mathcal{H}/k_B T}\right)\right] d\boldsymbol{q}d\boldsymbol{p} \\ &= k_B T \left[\delta_{ij} - \frac{1}{N!h^{3N}Z} \int_\Gamma \frac{\partial}{\partial x_j}\left(x_i e^{-\mathcal{H}/k_B T}\right) d\boldsymbol{q}d\boldsymbol{p}\right], \end{aligned} \tag{7.137}$$

where we have used the normalization condition of the canonical distribution function. We would now like to show that the remaining integral is zero and in this manner obtain the equipartition theorem. Let us denote that integral as I:

$$I = \int_\Gamma \frac{\partial}{\partial x_j}\left(x_i e^{-\mathcal{H}/k_B T}\right) d\boldsymbol{q}d\boldsymbol{p}.$$

First, let us consider the case where $\boldsymbol{x} = \boldsymbol{p}$. Then, separating the Hamiltonian into its kinetic energy and potential parts, $\mathcal{H}(\boldsymbol{q},\boldsymbol{p}) = \mathcal{T}^{\text{vib}}(\boldsymbol{p}) + \mathcal{V}^{\text{int}}(\boldsymbol{q})$, we have

$$\begin{aligned} I &= \int_{\Gamma_q} e^{-\mathcal{V}^{\text{int}}(\boldsymbol{q})/k_B T} d\boldsymbol{q} \int_{\Gamma_p} \frac{\partial}{\partial p_j}\left(p_i e^{-\mathcal{T}^{\text{vib}}(\boldsymbol{p})/k_B T}\right) d\boldsymbol{p} \\ &= \int_{\Gamma_q} e^{-\mathcal{V}^{\text{int}}(\boldsymbol{q})/k_B T} d\boldsymbol{q} \int_{-\infty}^{\infty} \cdots \int_{-\infty}^{\infty} \left[\int_{-\infty}^{\infty} \frac{\partial}{\partial p_j}\left(p_i e^{-\mathcal{T}^{\text{vib}}(\boldsymbol{p})/k_B T}\right) dp_j\right] \ldots dp_{j-1} dp_{j+1} \ldots \\ &= \int_{\Gamma_q} e^{-\mathcal{V}^{\text{int}}(\boldsymbol{q})/k_B T} d\boldsymbol{q} \int_{-\infty}^{\infty} \cdots \int_{-\infty}^{\infty} \left[p_i e^{-\mathcal{T}^{\text{vib}}(\boldsymbol{p})/k_B T}\right]\Bigg|_{p_j=-\infty}^{\infty} dp_1 \ldots dp_{j-1} dp_{j+1} \ldots dp_n, \end{aligned}$$

where in the last step we explicitly integrated the p_j term. We now note that $p_i e^{-\mathcal{T}^{\text{vib}}(\boldsymbol{p})/k_B T}$ is zero for any $p_j = \pm\infty$, therefore $I = 0$, and we obtain from Eqn. (7.137) that

$$\left\langle p_i \frac{\partial \mathcal{H}}{\partial p_j} \right\rangle = \delta_{ij} k_B T. \tag{7.138}$$

Thus, the equipartition theorem is satisfied for canonical phase averages, just as shown earlier for microcanonical averages in Eqn. (7.101).

Next, consider the case $\boldsymbol{x} = \boldsymbol{q}$. A similar derivation leads to an integral I of the form

$$I = \int_{\Gamma_p} e^{-\mathcal{T}^{\text{vib}}(\boldsymbol{p})/k_B T} \, d\boldsymbol{p} \int dq_1 \cdots \int dq_{j-1} \int dq_{j+1} \cdots \int dq_n \left[q_i e^{-\mathcal{V}^{\text{int}}(\boldsymbol{q})/k_B T} \right] \Big|_{q_j},$$

where $[\cdot]|_{q_j}$ indicates that the quantity in the square brackets is evaluated at the integration bounds of q_j. This term is zero under certain conditions. For example:

1. For an unbounded system where the atomic interactions decrease to zero with distance. In this case the integration bounds of q_j are $\pm\infty$. Moving a particle to $-\infty$ or $+\infty$ has the same effect of removing the atom from the system. In both cases the energy of the remaining atoms is unaffected by the missing atom, so that $\mathcal{V}^{\text{int}}(\ldots, q_j \to -\infty, \ldots) = \mathcal{V}^{\text{int}}(\ldots, q_j \to +\infty, \ldots)$, and therefore $I = 0$.
2. For a periodic system described by a supercell with periodic boundary conditions (PBCs) (for example an infinite crystalline solid). In this case, the lower and upper limits of q_j correspond to the same point on the periodic boundary and therefore, as for the unbounded system, $\mathcal{V}^{\text{int}}(\boldsymbol{q})$ is the same at both ends, and $I = 0$.

In these cases,

$$\left\langle q_i \frac{\partial \mathcal{H}}{\partial q_j} \right\rangle = \delta_{ij} k_B T, \tag{7.139}$$

and the equipartition theorem for the position degrees of freedom, shown for the microcanonical ensemble in Eqn. (7.105), is established. The case of a system confined to a finite physical domain is more complicated since in this case the surface contributions in I lead to a stress in the system [Tuc11]. This case is treated in detail in Chapter 8.

The reader is referred to Section 7.3.6 for further discussion of the equipartition theorem relations and subsequent derivations resulting from them.

7.4.5 Helmholtz free energy in the thermodynamic limit

The thermodynamic free energy, $\Psi = \mathcal{U} - T\mathcal{S}$, is an extensive variable, i.e. it scales with the size of the system, but does not depend on its shape. This follows from the basic properties of the entropy, $\mathcal{S}$, introduced in Section 2.4.5. In addition, we know from the same section that the stability of thermodynamic systems requires that the entropy has to be a *concave* function of the internal energy and volume. For the Helmholtz free energy, this translates to the requirements that Ψ must be a *concave* function of the temperature and a *convex* function of the volume [Cal85, Section 8-2], i.e.

$$\left.\frac{\partial^2 \Psi}{\partial T^2}\right|_{V,N} \leq 0, \qquad \left.\frac{\partial^2 \Psi}{\partial V^2}\right|_{T,N} \geq 0.$$

On the face of it, there are no guarantees that the statistical mechanics free energy defined in Eqn. (7.129) satisfies these requirements. The statistical free energy appears to explicitly depend on shape through the integration bounds of the partition function. It is *assumed* that this dependence becomes negligible with increasing system size and that, in the thermodynamic limit, the statistical free energy converges to a constant that depends only on the temperature and number density, N/V. However, it is not clear a priori whether the functional dependence of this limiting value on its arguments satisfies the convexity/concavity requirements. Clearly, the assertion (let us call it the "thermodynamic assertion") that the statistical free energy acquires the properties of the thermodynamic free energy in the thermodynamic limit is central to the validity of statistical mechanics. Proving this assertion is therefore of central importance.

When pursuing such a proof, it must first be recognized that the thermodynamic assertion is not always true. It is easy to find counter-examples, where either the statistical free energy is unbounded in the thermodynamic limit, or it exhibits nonphysical behavior. An example of the former case occurs for a potential energy function that is always attractive as in a self-gravitating system. In this case, the potential energy divided by the number of particles diverges with increasing N [HH03, page 17]. The key is then to identify the most general conditions on the potential energy and sequence of volumes to be taken to infinity for which the thermodynamic assertion is valid. This question was first addressed by van Hove in 1949 [vH49] and later by Yang and Lee [YL52] and Ruelle [Rue63]. A particularly clear and comprehensive proof is given by Fisher [Fis64]. An extension of the proof to periodic systems is given in [FL70]. Without going into all the details, we list the main assumptions in Fisher's analysis in [Fis64] and report on his key results below.

Main assumptions The potential energy of a system of N atoms is $\mathcal{V} = \widehat{\mathcal{V}}_{(N)}(\boldsymbol{r}^1, \ldots, \boldsymbol{r}^N)$. As shown in Section 5.4.1 (and pointed out by Fisher), such functions can always be uniquely decomposed into a series of many-body potentials (Eqn. (5.26)):

$$\widehat{\mathcal{V}}_{(N)}(\boldsymbol{r}^1, \ldots, \boldsymbol{r}^N) = \sum_{\alpha<\beta}^{N} \widehat{\phi}_2(\boldsymbol{r}^\alpha, \boldsymbol{r}^\beta) + \sum_{\alpha<\beta<\gamma}^{N} \widehat{\phi}_3(\boldsymbol{r}^\alpha, \boldsymbol{r}^\beta, \boldsymbol{r}^\gamma) + \cdots,$$

where for a system of N atoms the series terminates with the $\widehat{\phi}_N$ term. In his proof, Fisher makes the following assumptions regarding the potential energy functions:

1. *Symmetry* $\widehat{\mathcal{V}}_{(N)}(\boldsymbol{r}^1, \ldots, \boldsymbol{r}^N)$ is symmetric with respect to permutations of its arguments.
2. *Translational invariance* $\widehat{\mathcal{V}}_{(N)}(\boldsymbol{r}^1, \ldots, \boldsymbol{r}^N) = \widehat{\mathcal{V}}_{(N)}(\boldsymbol{r}^1 + \boldsymbol{t}, \ldots, \boldsymbol{r}^N + \boldsymbol{t})$, $\forall \boldsymbol{t} \in \mathbb{R}^3$.

3. *Boundedness* $\exp(-\widehat{\mathcal{V}}_{(N)}(\boldsymbol{r}^1,\ldots,\boldsymbol{r}^N)/k_{\mathrm{B}}T)$ is bounded and piecewise continuous.
4. *Stability*

$$\widehat{\mathcal{V}}_{(N)}(\boldsymbol{r}^1,\ldots,\boldsymbol{r}^N) \geq -Nw_1, \tag{7.140}$$

where w_1 is a finite constant. We will see that this is the key requirement for the existence of the thermodynamic limit of the free energy.

5. *Weak tempering* This condition ensures that the interaction between two sets of atoms that are pulled apart decays sufficiently rapidly with the distance. It is defined in the following way. Divide a system of N atoms into two subsystems A and B with N^{A} and N^{B} atoms (so that $N^{\mathsf{A}} + N^{\mathsf{B}} = N$). The interaction energy between two subsystems (defined as in Eqn. (7.74) but with slightly different notation) is

$$\begin{aligned}
&\widehat{\mathcal{V}}_{(N^{\mathsf{A}},N^{\mathsf{B}})}(\boldsymbol{r}^{\mathsf{A},1},\ldots,\boldsymbol{r}^{\mathsf{A},N^{\mathsf{A}}};\boldsymbol{r}^{\mathsf{B},1},\ldots,\boldsymbol{r}^{\mathsf{B},N^{\mathsf{B}}}) \\
&\quad = \widehat{\mathcal{V}}_{(N)}(\boldsymbol{r}^1,\ldots,\boldsymbol{r}^N) - \widehat{\mathcal{V}}_{(N^{\mathsf{A}})}(\boldsymbol{r}^{\mathsf{A},1},\ldots,\boldsymbol{r}^{\mathsf{A},N^{\mathsf{A}}}) - \widehat{\mathcal{V}}_{(N^{\mathsf{B}})}(\boldsymbol{r}^{\mathsf{B},1},\ldots,\boldsymbol{r}^{\mathsf{B},N^{\mathsf{B}}}).
\end{aligned} \tag{7.141}$$

Define the distance between the two sets, R, as the minimum distance between an atom in system A and an atom in system B:

$$R = \min_{\alpha,\beta} \left\| \boldsymbol{r}^{\mathsf{A},\alpha} - \boldsymbol{r}^{\mathsf{B},\beta} \right\|.$$

Weak tempering requires that for all N^{A} and N^{B} and $R \geq R_0$,

$$\widehat{\mathcal{V}}_{(N^{\mathsf{A}},N^{\mathsf{B}})}(\boldsymbol{r}^{\mathsf{A},1},\ldots,\boldsymbol{r}^{\mathsf{A},N^{\mathsf{A}}};\boldsymbol{r}^{\mathsf{B},1},\ldots,\boldsymbol{r}^{\mathsf{B},N^{\mathsf{B}}}) \leq N^{\mathsf{A}}N^{\mathsf{B}}w_2/R^{3+\epsilon}, \tag{7.142}$$

where R_0, w_2 and ϵ are fixed positive constants. In terms of the n-body potentials this condition is

$$\widehat{\phi}_n(\boldsymbol{r}^1,\ldots,\boldsymbol{r}^\gamma,\boldsymbol{r}^{\gamma+1},\ldots,\boldsymbol{r}^n) \leq D_1\mu^{n-2}/R^{(n-1)(3+\epsilon)}, \tag{7.143}$$

where D_1, μ and ϵ are positive constants, whenever $\left\|\boldsymbol{r}^\alpha - \boldsymbol{r}^\beta\right\| \geq R \geq R_0$ for all $\alpha \leq \gamma$, $\beta \geq \gamma + 1$ and $\gamma = 1,\ldots,n$. To show that the conditions on $\widehat{\phi}_n$ in Eqn. (7.143) are equivalent to Eqn. (7.142) is not trivial. See [Fis64, Appendix C] for a proof.

Main results Fisher's main results are based on two properties of a system, satisfying the assumptions listed above, that can be readily proved:

1. *Lower bound for the free energy*[40] The Helmholtz free energy is defined in Eqn. (7.129). Substituting in Eqns. (7.123) and $(7.121)_2$, we have

$$\Psi(N) = -k_{\mathrm{B}}T \ln \frac{1}{N!\Lambda^{3N}} \int \cdots \int e^{-\widehat{\mathcal{V}}_{(N)}(\boldsymbol{r}^1,\ldots,\boldsymbol{r}^N)/k_{\mathrm{B}}T}\, d\boldsymbol{r}^1 \ldots \boldsymbol{r}^N.$$

[40] Note that Fisher computed bounds for variables related to the negative of the free energy and hence refers to upper bounds (rather than lower bounds) for these quantities.

A lower bound on the free energy is obtained by inserting the stability requirement of Eqn. (7.140) into this expression and carrying out the integration:

$$\begin{aligned}\Psi(N) &\geq -k_B T \ln\left(V^N e^{Nw_1/kT}/N!\Lambda^{3N}\right)\\ &= -k_B T\left(\ln V^N/\Lambda^{3N} + Nw_1/k_B T - \ln N!\right)\\ &\geq -k_B T\left(N \ln V/\Lambda^3 + Nw_1/k_B T - N\ln N + N\right)\\ &= -k_B T N\left(\ln V/N\Lambda^3 + w_1/k_B T + 1\right),\end{aligned}$$

where in passing from the second to the third lines we have used Stirling's formula, $\ln N! \geq N \ln N - N$. The specific free energy (i.e. the free energy per unit mass) is

$$\psi = \frac{\Psi(N)}{Nm} \geq -\frac{k_B T}{m}\left(\ln m/\rho\Lambda^3 + w_1/k_B T + 1\right), \quad (7.144)$$

where $\rho = mN/V$ is the mass density. Thus, the specific free energy is bounded from below by a constant which does not depend on system size.

2. *Basic partition function inequality* Consider a system of N atoms confined to a region Γ_q in configuration space. The partition function for the system, defined in Eqn. (7.123), is

$$Z(N;\Gamma_q) = Z^{V}(N;\Gamma_q)/N!\Lambda^{3N},$$

where Γ_q sets the integration bounds for the configurational integral. Next, consider the case where the atoms are confined to two nonoverlapping[41] regions, Γ_q^{A} and Γ_q^{B}, separated by a distance $R \geq R_0$ and contained within Γ_q. Denote the partition function for this case as $Z(N^{A}, N^{B}; \Gamma_q^{A}, \Gamma_q^{B})$, where N^{A} and N^{B} are the number of atoms in Γ_q^{A} and Γ_q^{B}, respectively, and $N^{A} + N^{B} = N$. Since the partition function scales with volume, we expect

$$Z(N;\Gamma_q) \geq Z(N^{A}, N^{B}; \Gamma_q^{A}, \Gamma_q^{B}). \quad (7.145)$$

To compute $Z(N^{A}, N^{B}; \Gamma_q^{A}, \Gamma_q^{B})$, we must account for all the ways in which N atoms can be divided into the two domains. The result is

$$Z(N^{A}, N^{B}; \Gamma_q^{A}, \Gamma_q^{B}) = \frac{1}{N!\Lambda^{3N}} \sum_{N^{A}+N^{B}=N}$$

$$\int_{\Gamma_q^{A}} d\boldsymbol{r}^{A,1} \cdots \int_{\Gamma_q^{A}} d\boldsymbol{r}^{A,N^{A}} \int_{\Gamma_q^{B}} d\boldsymbol{r}^{B,1} \cdots \int_{\Gamma_q^{B}} d\boldsymbol{r}^{B,N^{B}} e^{-(\mathcal{V}_{(N^{A})} + \mathcal{V}_{(N^{B})} + \mathcal{V}_{(N^{A},N^{B})})/k_B T},$$

where $\mathcal{V}_{(N^{A})} = \widehat{\mathcal{V}}_{(N^{A})}(\boldsymbol{r}^{A,1}, \ldots, \boldsymbol{r}^{A,N^{A}})$, $\mathcal{V}_{(N^{B})} = \widehat{\mathcal{V}}_{(N^{B})}(\boldsymbol{r}^{B,1}, \ldots, \boldsymbol{r}^{B,N^{A}})$, and $\mathcal{V}_{(N^{A},N^{B})}$ is the interaction potential defined in Eqn. (7.141). Since we have stipulated that $R \geq R_0$, the weak tempering assumption in Eqn. (7.142) applies and the

[41] Fisher in his derivation is more careful, allowing for domains to have "walls" of finite thickness. In this case "nonoverlapping" means that the internal portions of the domains do not overlap.

interaction energy can be replaced with $N^{\mathsf{A}}N^{\mathsf{B}}w_2/R^{3+\epsilon}$ and removed from the integral. We therefore have the following inequality:

$$Z(N^{\mathsf{A}}, N^{\mathsf{B}}; \Gamma_q^{\mathsf{A}}, \Gamma_q^{\mathsf{B}}) \geq \frac{1}{N!\Lambda^{3N}} \sum_{N^{\mathsf{A}}+N^{\mathsf{B}}=N} Z^{\mathrm{V}}(N^{\mathsf{A}}; \Gamma_q^{\mathsf{A}}) Z^{\mathrm{V}}(N^{\mathsf{B}}; \Gamma_q^{\mathsf{B}}) e^{-N^{\mathsf{A}}N^{\mathsf{B}}w_2/k_{\mathrm{B}}TR^{3+\epsilon}}. \tag{7.146}$$

Substituting Eqn. (7.146) into Eqn. (7.145) and noting that since the terms in the sum in Eqn. (7.146) are positive, the inequality holds also for a single term in the sum, we therefore have

$$Z(N^{\mathsf{A}} + N^{\mathsf{B}}; \Gamma_q) \geq Z(N^{\mathsf{A}}; \Gamma_q^{\mathsf{A}}) Z(N^{\mathsf{B}}; \Gamma_q^{\mathsf{B}}) e^{-N^{\mathsf{A}}N^{\mathsf{B}}w_2/k_{\mathrm{B}}TR^{3+\epsilon}}. \tag{7.147}$$

This is referred to as the *basic partition function inequality*.

Using the inequalities in Eqns. (7.144) and (7.147), the following properties of the Helmholtz free energy in the thermodynamic limit are established for arbitrarily-shaped domains. These are Fisher's main results:

1. *Existence* The specific free energy ψ (free energy per unit mass) of an infinite system exists and is bounded by Eqn. (7.144).
2. *Continuity and differentiability* The specific free energy ψ is a continuous and differentiable function of the density. Likewise, it is a continuous and differentiable function of the volume.
3. *Convexity* The free energy per volume $\psi_V = \Psi/V$ satisfies the following convexity relation in the thermodynamic limit where it is a function of the mass density $\rho = mN/V$:

$$\psi_V(\rho) \leq \sum_i \omega_i \psi_V(\rho_i), \tag{7.148}$$

where $\rho = \sum_i \omega_i \rho_i$ and $\omega_i = V_i/V$ are positive numbers satisfying $\sum_i \omega_i = 1$. Equation (7.148) leads to the important result that the free energy per volume is a convex function of the density (see Exercise 7.12):

$$\psi_V''(\rho) \geq 0. \tag{7.149}$$

4. *Monotonicity* The free energy per particle $\psi_N = \Psi/N$ in the thermodynamic limit is a monotonically decreasing (non-increasing to be precise) function of the specific

volume $v = V/N$:

$$\psi_N(v) \geq \psi_N(\lambda v), \qquad \lambda \geq 1. \tag{7.150}$$

The monotonicity of the free energy implies that the pressure (defined as the negative of the derivative with respect to volume) is always positive and monotonically decreasing with volume.

Summarizing, we see that Fisher's results show that the "thermodynamic assertion" is satisfied for a broad class of potential energy functions (as detailed under "main assumptions"). Two important conclusions can be drawn from Fisher's analysis of the thermodynamic limit. First, the free energy computed from statistical mechanics has the same properties as those of the macroscopic free energy defined in the theory of thermodynamics. Thus, the two theories are consistent. Second, the limiting value of the free energy per particle is independent of the shape of the system. This implies that statistical mechanics is really a theory of fluids that cannot sustain shear stresses. This issue is noted in the following chapter where expressions for the stress and elasticity tensors are derived from statistical mechanics and discussed further in Chapter 11, where a *restricted ensemble* approach for metastable systems (like solids) is adopted.

Further reading

- There are many good references on mechanics. Herbert Goldstein's book on *Classical Mechanics* [Gol80] is considered a classic, providing a trustworthy and complete introduction to the subject. A less formal text that provides many insights is Cornelius Lanczos' book on *The Variational Principles of Mechanics* [Lan70]. For the more mathematically minded, V. I. Arnold's book on *Mathematical Methods of Classical Mechanics* [Arn89] provides a rigorous presentation full of many interesting examples.

- A good introduction to the subject of the foundations of statistical mechanics is given in Lawrence Sklar's book *Physics and Chance: Philosophical Issues in the Foundations of Statistical Mechanics* [Skl93]. This is mostly a descriptive book that lays out the key issues. Along the same lines is the review article by Jos Uffink called "Compendium of the foundations of classical statistical physics" published in the *Handbook for Philosophy of Physics* [Uff07]. Another excellent source is Oliver Penrose's book, *Foundations of Statistical Mechanics: A Deductive Treatment* [Pen05], which attempts to construct a statistical mechanics theory in axiomatic fashion from a small number of well-defined physical assumptions. Finally, David Ruelle's short book *Statistical Mechanics: Rigorous Results* [Rue69] is a standard text on thermodynamic limit theory.

- There are many books on the principles and applications of equilibrium statistical mechanics. We list a few that we found ourselves returning to often. Early books on statistical mechanics are often worth studying since they cover topics that more recent books consider "settled" and no longer discuss. Fowler's book [Fow36] is an authoritative text that summarizes the state of the field in 1936. An excellent book by Hill [Hil56] written 20 years later covers further developments. Kerson Huang's book, *Statistical Mechanics* [Hua63] is a classic that provides a clear and concise introduction to the subject. Another important book is *Statistical Physics* in the Landau and Lifshitz series on theoretical physics [LL80]. The book is a bit dated in some respects, however, it often covers topics missing in other books. A lesser known book in the statistical mechanics community, which was particularly helpful given our goal of connecting with continuum mechanics, is Jerome H. Weiner's book called *Statistical Mechanics of Elasticity* [Wei83]. This book provides a very clear and readable introduction to statistical mechanics from the perspective of time averages. Another excellent and concise introduction to equilibrium statistical mechanics is Toda, Kubo and Saitô's book [TKS92]. James Sethna's book on *Statistical Mechanics: Entropy, Order Parameters, and Complexity* [Set06], provides a very lively, entertaining introduction to the subject replete with physical intuition and insights. Finally, Mark Tuckerman's book *Statistical Mechanics: Theory and Molecular Simulation* [Tuc10] takes the interesting approach of weaving a theoretical discussion of statistical mechanics with the computational aspect of molecular simulation.

- We have not discussed nonequilibrium statistical mechanics in this chapter. We will return to this topic in Section 8.2, where the Irving–Kirkwood formalism for transport processes is discussed, and in Section 13.3.3, where the generalized Langevin equation is discussed. For a more detailed introduction to the subject, the reader is referred to the following excellent references: [Cha43, dGM62, EM90, TKH91, CH09].

Exercises

7.1 [SECTION 7.3] Rewrite Eqn. (7.68) in Example 7.4 for the case where the phase function A depends on the momentum of the oscillator. Use this relation to compute the microcanonical ensemble average for the kinetic energy of the system. Compare your result with the time average for the same quantity in Example 7.3.

7.2 [SECTION 7.3] Derive Eqn. (7.69) in Example 7.4.

7.3 [SECTION 7.3] Two containers of ideal gases are brought into contact across a diathermal partition. Both containers contain the same number of atoms N of the same species with mass m. The volume and initial energy of the two containers are $(V^{\mathsf{A}}, E_0^{\mathsf{A}})$ and $(V^{\mathsf{B}}, E_0^{\mathsf{B}})$. Assuming weak interaction, do the following:
   1. Using Eqns. (7.84) and (7.92), obtain expressions for the energies $E_{\mathrm{eq}}^{\mathsf{A}}$ and $E_{\mathrm{eq}}^{\mathsf{B}}$ of the systems at equilibrium.
   2. Compute the equilibrium temperature T_{eq}.

7.4 [SECTION 7.3] Obtain the Sackur–Tetrode equation in Eqn. (7.94) from Eqn. (7.93).

7.5 [SECTION 7.3] Compute the temperature of an ideal gas using Eqn. (7.89) and verify that the result is identical to the one given in Eqn. (7.96).

7.6 [SECTION 7.3] Consider two containers A and B of ideal gas at the same temperature T and pressure p. The two containers are connected and the atoms are allowed to mix.
   1. Show that if the two containers contain *different* gases, the change in entropy ΔS due to the mixing is

$$\Delta S = k_{\mathrm{B}} \left[N^{\mathsf{A}} \ln \frac{V}{V^{\mathsf{A}}} + N^{\mathsf{B}} \ln \frac{V}{V^{\mathsf{B}}} \right], \qquad (7.151)$$

   where N^{A} and N^{B} are the numbers of atoms initially in containers A and B respectively, V^{A} and V^{B} are the volumes of the two containers, and the combined volume is $V = V^{\mathsf{A}} + V^{\mathsf{B}}$.

Assume that the number of atoms is large so that the Sackur–Tetrode equation (Eqn. (7.94)) can be used. (Equation (7.151) is called the *entropy of mixing*.)

2. Repeat the derivation in the previous part for the case that the two containers contain the *same* gas. Show that the change in entropy in this case is identically zero. Explain why this result is sensible.

7.7 [SECTION 7.3] In Section 7.3.4 we argued that an $N!$ term must be added to the denominator of the entropy expression in order to ensure agreement with quantum mechanics (see Eqn. (7.88)). We see in this exercise, that the presence of this term is also necessary to prevent unphysical results.

1. Rederive the Sackur–Tetrode equation (Eqn. (7.94)) using an expression for entropy that does not include the $N!$ term.
2. Use your new expression to repeat Exercise 7.6. Show that the entropy of mixing for different gases remains unchanged, but the change in entropy when two containers of the same gas are mixed is no longer zero. This nonphysical result is called *Gibbs paradox* and it is what prompted Gibbs to introduce the additional $N!$ term into the definition of the entropy.

7.8 [SECTION 7.3] Compute the specific heat capacity at constant volume, c_v, and the molar heat capacity at constant volume, C_v, for an ideal gas described by the Sackur–Tetrode equation (Eqn. (7.94)). **Hint:** See the definition in Eqn. (2.107) and the relationship between the specific heat capacity and molar heat capacity in Eqn. (2.108).

7.9 [SECTION 7.3] Show that Joule's observation that $\partial\mathcal{U}/\partial V|_T = 0$ (Eqn. (2.109)) is satisfied identically for an ideal gas. **Hint:** Note that the derivative is taken at constant *temperature*. Make sure you use the appropriate internal energy expression in Section 7.3.5.

7.10 [SECTION 7.3] Prove that an ideal gas which is initially confined to part of a container will expand to fill the entire container when the partition is removed. The container isolates the gas from the rest of the world so that no energy or mass is exchanged.

7.11 [SECTION 7.4] Restate the convexity relation in Eqn. (7.148) in terms of specific free energy, $\psi(v) = \Psi/mN$, where $v = V/N$ is the specific volume.

7.12 [SECTION 7.4] Derive Eqn. (7.149). **Hint:** In Eqn. (7.148), take the special case, $\omega_1 = \omega_2 = 1/2$ and set $N_1 = N_2 = N/2$. Introduce a small perturbation where $V_1 = V/2 - \Delta V$ and $V_2 = V/2 + \Delta V$. The result follows by expanding Eqn. (7.148) to second-order in $\Delta\rho$.

7.13 [SECTION 7.4] It is sometimes convenient to define an "instantaneous temperature," T_{inst}, in terms of the vibrational kinetic energy as

$$T_{\text{inst}} = \frac{2}{3Nk_B}\left(\sum_{\alpha=1}^{N}\frac{\boldsymbol{p}^\alpha\cdot\boldsymbol{p}^\alpha}{2m^\alpha}\right).$$

Show that the variance of T_{inst} in the canonical ensemble is (see Eqn. (7.36)):

$$\text{Var}(T_{\text{inst}}) = \frac{2T^2}{3N}.$$

(For simplicity, you can assume all the atoms have the same mass, m.)

8 Microscopic expressions for continuum fields

The governing equations of continuum mechanics and thermodynamics were derived in Chapter 2 based on the fundamental laws of physics and the assumption of local thermodynamic equilibrium. These equations, summarized at the start of Section 2.5, provide relationships between a number of different continuum fields: density $\rho(\boldsymbol{x})$, velocity $\boldsymbol{v}(\boldsymbol{x})$, Cauchy stress $\boldsymbol{\sigma}(\boldsymbol{x})$, heat flux $\boldsymbol{q}(\boldsymbol{x})$, temperature $T(\boldsymbol{x})$, entropy $s(\boldsymbol{x})$, and the internal energy density $u(\boldsymbol{x})$. In the continuum worldview these entities are primitive quantities that emerge as part of the framework of the theory. When solving a continuum problem it is not necessary to know "what" they are as long as experiments can be devised to measure the constitutive relations that connect them. This view of continuum mechanics was strongly held by its early developers as evidenced by the quote in footnote 17 on page 43.

The objective of this chapter is to go beyond the phenomenological approach of classical continuum mechanics and thermodynamics by establishing a direct connection with the underlying atomistic system. Our motivation for doing so is not to prove that the continuum theories are correct (there is ample proof for this by their success), but rather to provide a mechanism for computing continuum measures in molecular simulations. This is important in order to be able to extract constitutive relations from "computer experiments" and to help rationalize the results of such simulations in the language of continuum mechanics. Of course, it is also of philosophical interest to understand what the continuum fields correspond to on the microscopic scale. We saw in Chapter 7 that by deriving a microscopic expression for entropy, this rather elusive concept on the macroscopic level obtained a clear physical significance; this is extremely helpful to have at the macroscopic scale even if it is not used in practice. In similar fashion, in this chapter we obtain microscopic definitions for other continuum measures, and in particular for the stress tensor. We then derive practical expressions from these definitions that can be used in molecular simulations and compare their predictions for a prototypical numerical example. The chapter is partly based on [AT10] which describes a unified framework for stress in molecular systems. That article extends the discussion to possible nonsymmetric definitions for the stress tensor for systems consisting of particles with internal structure.

Our derivation starts with the special case of systems in thermodynamic equilibrium. We use the equilibrium statistical mechanics framework developed in the previous chapter to derive expressions for the stress tensor (often called the *virial stress*) and elasticity tensor (elastic constants). It is important to note that strictly speaking these expressions apply only to fluid systems. This conclusion can be drawn from the analysis of Section 7.4.5, which shows that the free energy of an equilibrium system in the thermodynamic limit depends on only its density and not its shape. Since stress and elasticity correspond to the derivatives of the free energy, this implies that a system in thermodynamic equilibrium cannot sustain

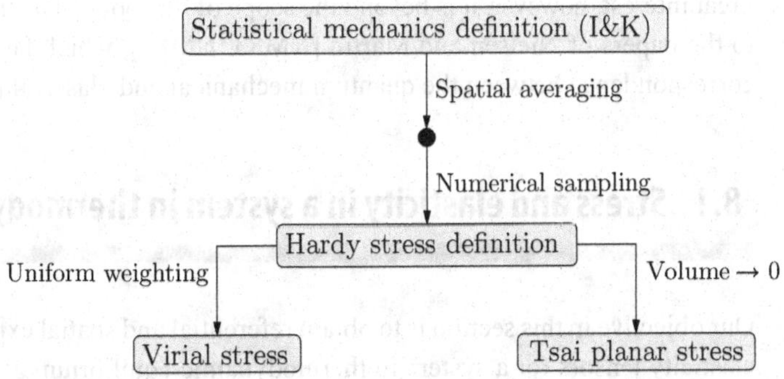

Fig. 8.1 A flow chart showing how the different microscopic stress definitions in common use are closely related and can be obtained in a consistent fashion from the Irving–Kirkwood–Noll statistical mechanics definition.

shear deformations. We address this point in greater detail in Chapter 11, where solids are treated as *metastable* systems. Despite the formal limitation of the equilibrium expressions, they are commonly used in molecular dynamics (MD) simulations to compute stress and elasticity for solids as well. The reason that this works is that in MD, phase averages are replaced by time averages, which de facto sample the portion of phase space that is more formally described by the *restricted ensemble* statistical mechanics of metastable systems described in Chapter 11.

Following the equilibrium derivation, we turn to the more general case of systems under *nonequilibrium* conditions. We apply the nonequilibrium statistical mechanics framework pioneered in the landmark paper of Irving and Kirkwood [IK50]. In this approach, continuum field variables (density, momentum and energy) are defined as expectation values of certain microscopic expressions, which automatically satisfy the continuum balance and conservation laws. Combining this analysis with a pair of lemmas proposed by Noll in [Nol55], closed-form microscopic expressions for the pointwise stress tensor and heat flux vector can be obtained. We refer to this as the *Irving–Kirkwood–Noll procedure*. In this chapter we only discuss the derivation of the stress tensor. The application of the Irving–Kirkwood–Noll procedure to the energy equation and consequent derivation of an expression for the heat flux vector are discussed in [AT11]. Following the derivation of the pointwise stress tensor, we show that by a process of spatial averaging and numerical approximation the standard microscopic measures for stress in use today are seamlessly derived. These include Hardy's stress definition [Har82], Tsai's traction definition [Tsa79] and the virial stress obtained at the start of the chapter. These definitions are therefore shown to all flow from the same source and to be closely related to each other as shown schematically in Fig. 8.1. We note that an alternative approach, which involves only spatial averaging and is not directly based on statistical mechanics, has been proposed independently by Hardy [Har82] and Murdoch in a series of articles [Mur83, Mur85, MB93, MB94, Mur03, Mur07]. This approach and its connections with the Irving–Kirkwood–Noll procedure is discussed in [AT10].

Finally, we note that the Irving–Kirkwood–Noll procedure discussed above is based on *classical* statistical mechanics. The issue of stress in quantum mechanical systems is also of great interest, however, it is beyond the scope of this book. The interested reader is referred to the papers of Nielsen and Martin [NM83, NM85], which lay out the foundations for a correspondence between the quantum mechanical and classical notions of stress.

8.1 Stress and elasticity in a system in thermodynamic equilibrium

Our objective in this section is to obtain referential and spatial expressions for the stress and elasticity tensors for a system in thermodynamic equilibrium as canonical phase averages over suitable phase space functions. Recall that thermodynamic equilibrium implies that all state variables are single-valued and well defined and that any fields must be independent of position and time (see Section 2.4.1). This means that a system in thermodynamic equilibrium must be in a state of uniform deformation (i.e. any strain measures must be constant in space), and therefore two such systems must be related through an affine transformation. At the continuum level, we identify such transformations with the deformation gradient $\boldsymbol{F}$ (see Section 2.2.4). In the language of thermodynamics, we identify $V_0\boldsymbol{F}$ (where V_0 is the reference volume) with the extensive kinematic state variables $\boldsymbol{\Gamma}$ characterizing the thermodynamic system (see discussion on page 63). As a first step, to help define some important concepts in a simpler setting, we consider the case of stress in a finite system at zero temperature, i.e. a system in static equilibrium. We will then turn to the more general dynamical case of thermodynamic equilibrium at finite temperature. In order to describe the dependence of the energy on $\boldsymbol{F}$ and to compute derivatives with respect to $\boldsymbol{F}$ (which lead to the stress and elasticity of the system), it is necessary to change the phase space variables to new variables where the dependence on $\boldsymbol{F}$ is made explicit. The appropriate change of phase space variables is the subject of canonical transformations discussed next.

8.1.1 Canonical transformations

Hamilton's equations of motion are often referred to as the canonical equations of motion. The phase space variables $\boldsymbol{q}$ and $\boldsymbol{p}$ appearing in these equations are called *canonical variables*.[1] The choice of variables is not unique but neither is it arbitrary. Any transformation of variables must be such that the canonical nature of the equations is maintained. A transformation satisfying this condition is called a *canonical transformation*.[2]

The main motivation for the introduction of canonical transformations in mechanics textbooks is to identify transformations that simplify the solution of the equations of motion. As simple as Hamilton's equations appear, in general they cannot be directly integrated.

[1] The term "canonical" in this context has nothing to do with the canonical ensemble. The terminology was introduced by Jacobi to indicate that Hamilton's equations constitute the simplest form of the equations of motion.

[2] This definition suffices for our purpose, but a more correct definition can be found in [Arn89] using *differential forms*.

There are, however, cases where through a clever change of variables, the Hamiltonian of the system is rendered sufficiently simple to be integrable. Our motivation is a bit different. We are interested in computing the stress and elasticity tensors, which correspond to the derivative of the free energy with respect to a measure of the deformation of the system. We will see that these derivatives can be readily computed following a canonical transformation that changes both the position and momentum degrees of freedom. The former leads to the "potential" part of the stress tensor, while the latter is responsible for the "kinetic" part of the stress. We will also use a canonical transformation in Section 9.5 to derive a formulation for applying constant stress in MD simulations.

Objective of a canonical transformation We noted above that the variables $(\boldsymbol{q}, \boldsymbol{p})$ are called canonical if they satisfy Hamilton's equations of motion,

$$\dot{q}_i = \frac{\partial \mathcal{H}}{\partial p_i}, \quad \dot{p}_i = -\frac{\partial \mathcal{H}}{\partial q_i}, \quad \mathcal{H} = \sum_{i=1}^{n} \frac{p_i^2}{2m_i} + \mathcal{V}(q_1, \ldots, q_n),$$

where $i = 1, \ldots, n$ and $n = 3N$, with N the number of atoms. Another set of variables $(\widehat{\boldsymbol{q}}, \widehat{\boldsymbol{p}})$ for the same system are also canonical if they satisfy equations of the same form:

$$\dot{\widehat{q}}_i = \frac{\partial \widehat{\mathcal{H}}}{\partial \widehat{p}_i}, \quad \dot{\widehat{p}}_i = -\frac{\partial \widehat{\mathcal{H}}}{\partial \widehat{q}_i}, \quad \widehat{\mathcal{H}} = \sum_{i=1}^{n} \frac{\widehat{p}_i^2}{2m_i} + \widehat{\mathcal{V}}(\widehat{q}_1, \ldots, \widehat{q}_n),$$

although possibly with a different Hamiltonian $\widehat{\mathcal{H}}$. Our objective is to obtain a transformation of the form

$$\boldsymbol{q} = \boldsymbol{q}(\widehat{\boldsymbol{q}}, \widehat{\boldsymbol{p}}, t), \qquad \boldsymbol{p} = \boldsymbol{p}(\widehat{\boldsymbol{q}}, \widehat{\boldsymbol{p}}, t),$$

where both $(\widehat{\boldsymbol{q}}, \widehat{\boldsymbol{p}})$ and $(\boldsymbol{q}, \boldsymbol{p})$ are canonical.

An alternative definition for canonical variables that provides great insight into canonical transformations and a practical method for constructing them is based on the modified Hamilton's principle. This principle and the Hamiltonian formulation in general were introduced in Section 4.3 for the special case where the variables $\boldsymbol{q}$ and $\boldsymbol{p}$ are the Cartesian positions and momenta of a system of particles. We do not repeat the discussion here, but only rewrite Hamilton's principle for the more general case where the canonical variables are generalized coordinates.

The modified Hamilton's principle states that the time evolution of the generalized coordinates, $\boldsymbol{q}(t)$ and $\boldsymbol{p}(t)$, in phase space is the extremum of the *action integral*, so that

$$\delta \mathcal{A} = \delta \int_{t_1}^{t_2} [\boldsymbol{p} \cdot \dot{\boldsymbol{q}} - \mathcal{H}(\boldsymbol{q}, \boldsymbol{p}, t)] \, dt = 0, \tag{8.1}$$

where $\mathcal{L}$ is the Lagrangian of the system, and the variation is taken in both $\boldsymbol{q}$ and $\boldsymbol{p}$ independently, and t_1, t_2, $\boldsymbol{q}(t_1)$ and $\boldsymbol{q}(t_2)$ are held fixed. The Euler–Lagrange equations associated with Eqn. (8.1) are

$$\dot{p}_i + \frac{\partial \mathcal{H}}{\partial q_i} = 0, \qquad -\dot{q}_i + \frac{\partial \mathcal{H}}{\partial p_i} = 0, \tag{8.2}$$

which are exactly Hamilton's equations. This demonstrates that the modified Hamilton's principle in Eqn. (8.1) leads to the canonical equations of motion.

Generating functions The modified Hamilton's principle in Eqn. (8.1) is much easier to work with than Hamilton's equations of motion when attempting to construct canonical transformations, since it involves a single scalar equation instead of a set of $2n$ coupled differential equations.

Consider two sets of canonical variables $(\boldsymbol{q}, \boldsymbol{p})$ and $(\widehat{\boldsymbol{q}}, \widehat{\boldsymbol{p}})$. Since the variables are canonical, they satisfy the modified Hamilton's principle in Eqn. (8.1):

$$\delta \int_{t_1}^{t_2} [p_i \dot{q}_i - \mathcal{H}(\boldsymbol{q}, \boldsymbol{p}, t)] \, dt = 0, \qquad \delta \int_{t_1}^{t_2} \left[\widehat{p}_i \dot{\widehat{q}}_i - \widehat{\mathcal{H}}(\widehat{\boldsymbol{q}}, \widehat{\boldsymbol{p}}, t)\right] dt = 0. \tag{8.3}$$

The integrands of Eqns. $(8.3)_1$ and $(8.3)_2$ can therefore only differ by a quantity whose variation after integration is identically zero. A possible solution is

$$\delta \int_{t_1}^{t_2} \left[p_i \dot{q}_i - \widehat{p}_i \dot{\widehat{q}}_i - (\mathcal{H} - \widehat{\mathcal{H}})\right] dt = \delta \int_{t_1}^{t_2} \frac{dG}{dt} \, dt, \tag{8.4}$$

where G is an arbitrary scalar function of the canonical variables and time with continuous second derivatives. The integral on the right is only evaluated at the fixed integration bounds and hence its variation is identically zero. This is not obvious since there is no restriction on the variation of the momenta at the ends, but we assume this to be true to avoid the introduction of differential forms.[3] The difference between the integrands of Eqns. $(8.3)_1$ and $(8.3)_2$ therefore satisfies

$$dG - p_i dq_i + \widehat{p}_i d\widehat{q}_i + (\mathcal{H} - \widehat{\mathcal{H}}) dt = 0, \tag{8.5}$$

where we have multiplied through by dt. Now, consider the case where $G = G_1(\boldsymbol{q}, \widehat{\boldsymbol{q}}, t)$. (We consider other possible functional dependences later.) The total differential of G is

$$dG = \frac{\partial G_1}{\partial q_i} dq_i + \frac{\partial G_1}{\partial \widehat{q}_i} d\widehat{q}_i + \frac{\partial G_1}{\partial t} dt. \tag{8.6}$$

Substituting Eqn. (8.6) into Eqn. (8.5) gives

$$\left(\frac{\partial G_1}{\partial q_i} - p_i\right) dq_i + \left(\frac{\partial G_1}{\partial \widehat{q}_i} + \widehat{p}_i\right) d\widehat{q}_i + \left(\frac{\partial G_1}{\partial t} + \mathcal{H} - \widehat{\mathcal{H}}\right) dt = 0. \tag{8.7}$$

Since the variables q_i, $\widehat{q}_i$ and t are independent, Eqn. (8.7) is satisfied provided that

$$p_i = \frac{\partial G_1}{\partial q_i}, \qquad \widehat{p}_i = -\frac{\partial G_1}{\partial \widehat{q}_i}, \qquad \widehat{\mathcal{H}} = \mathcal{H} + \frac{\partial G_1}{\partial t}. \tag{8.8}$$

These relations define the canonical transformation. Recall that $G_1(\boldsymbol{q}, \widehat{\boldsymbol{q}}, t)$ is an arbitrary function. A given function G_1 will generate a transformation between $\boldsymbol{q}$ and $\widehat{\boldsymbol{q}}$ with the definitions of $\boldsymbol{p}$, $\widehat{\boldsymbol{p}}$ and $\widehat{\mathcal{H}}$ given in Eqn. (8.8). For this reason G_1 is called the *generating*

[3] For a mathematically rigorous argument, refer to [Arn89, Section 45]. Briefly, the proof is based on the symmetry present in the geometry of any Hamiltonian system commonly called *symplectic geometry*.

function of the canonical transformation. Note that if G_1 does not depend on t, then $\widehat{\mathcal{H}} = \mathcal{H}$. Explicitly, in terms of the variable dependence of the Hamiltonian functions, this last equality states that $\widehat{\mathcal{H}}(\widehat{\boldsymbol{q}},\widehat{\boldsymbol{p}}) = \mathcal{H}(\boldsymbol{q}(\widehat{\boldsymbol{q}},\widehat{\boldsymbol{p}}),\boldsymbol{p}(\widehat{\boldsymbol{q}},\widehat{\boldsymbol{p}}))$.

As suggested above, generating functions of the form $G = G_1(\boldsymbol{q},\widehat{\boldsymbol{q}},t)$ do not generate all possible canonical transformations. In general, there are four primary classes of generating functions, where the functional dependence is $(\boldsymbol{q},\widehat{\boldsymbol{q}})$, $(\boldsymbol{q},\widehat{\boldsymbol{p}})$, $(\boldsymbol{p},\widehat{\boldsymbol{q}})$ or $(\boldsymbol{p},\widehat{\boldsymbol{p}})$. We have already encountered the first class, where $G = G_1(\boldsymbol{q},\widehat{\boldsymbol{q}},t)$. The remaining classes can be obtained from the first through Legendre transformations. Consider, for example, the following definition:

$$G = G_2(\boldsymbol{q},\widehat{\boldsymbol{p}},t) - \widehat{q}_i\widehat{p}_i. \tag{8.9}$$

The total differential of this expression is

$$dG = \frac{\partial G_2}{\partial q_i}dq_i + \frac{\partial G_2}{\partial \widehat{p}_i}d\widehat{p}_i + \frac{\partial G_2}{\partial t}dt - \widehat{q}_i d\widehat{p}_i - \widehat{p}_i d\widehat{q}_i. \tag{8.10}$$

Substituting this into Eqn. (8.5) gives

$$\left(\frac{\partial G_2}{\partial q_i} - p_i\right)dq_i + \left(\frac{\partial G_2}{\partial \widehat{p}_i} - \widehat{q}_i\right)d\widehat{p}_i + \left(\frac{\partial G_2}{\partial t} + \mathcal{H} - \widehat{\mathcal{H}}\right)dt = 0, \tag{8.11}$$

which leads to the following canonical transformation:

$$p_i = \frac{\partial G_2}{\partial q_i}, \qquad \widehat{q}_i = \frac{\partial G_2}{\partial \widehat{p}_i}, \qquad \widehat{\mathcal{H}} = \mathcal{H} + \frac{\partial G_2}{\partial t}. \tag{8.12}$$

In similar fashion, two additional Legendre transformations are possible. Defining

$$G = G_3(\boldsymbol{p},\widehat{\boldsymbol{q}},t) + q_i p_i \tag{8.13}$$

leads to

$$q_i = -\frac{\partial G_3}{\partial p_i}, \qquad \widehat{p}_i = -\frac{\partial G_3}{\partial \widehat{q}_i}, \qquad \widehat{\mathcal{H}} = \mathcal{H} + \frac{\partial G_3}{\partial t}. \tag{8.14}$$

Finally, the definition

$$G = G_4(\boldsymbol{p},\widehat{\boldsymbol{p}},t) + q_i p_i - \widehat{q}_i\widehat{p}_i \tag{8.15}$$

leads to

$$q_i = -\frac{\partial G_4}{\partial p_i}, \qquad \widehat{q}_i = \frac{\partial G_4}{\partial \widehat{p}_i}, \qquad \widehat{\mathcal{H}} = \mathcal{H} + \frac{\partial G_4}{\partial t}. \tag{8.16}$$

In addition to the four classes of transformation discussed above, it is possible to have a mixed dependence, where each degree of freedom can belong to a different class. These cases are discussed in detail in [Gol80].

Example 8.1 (Exchanging coordinates and momenta) As an example, consider the simple case where $G = G_1(\boldsymbol{q}, \widehat{\boldsymbol{q}}) = q_i \widehat{q}_i$. From Eqn. (8.8), we have

$$p_i = \widehat{q}_i, \qquad \widehat{p}_i = -q_i, \qquad \widehat{\mathcal{H}} = \mathcal{H}.$$

This canonical transformation exchanges the coordinates with the momenta. The last relation requires some care. The Hamiltonian $\widehat{\mathcal{H}}$ is a function of the variables $(\widehat{\boldsymbol{q}}, \widehat{\boldsymbol{p}})$, therefore

$$\begin{aligned}\widehat{\mathcal{H}}(\widehat{\boldsymbol{q}}, \widehat{\boldsymbol{p}}) = \mathcal{H}(\boldsymbol{q}(\widehat{\boldsymbol{q}}, \widehat{\boldsymbol{p}}), \boldsymbol{p}(\widehat{\boldsymbol{q}}, \widehat{\boldsymbol{p}})) &= \sum_i \frac{(p_i(\widehat{\boldsymbol{q}}, \widehat{\boldsymbol{p}}))^2}{2m_i} + \mathcal{V}(q_1(\widehat{\boldsymbol{q}}, \widehat{\boldsymbol{p}}), \dots, q_n(\widehat{\boldsymbol{q}}, \widehat{\boldsymbol{p}})) \\ &= \sum_i \frac{\widehat{q}_i^2}{2m_i} + \mathcal{V}(-\widehat{p}_1, \dots, -\widehat{p}_n).\end{aligned}$$

This example demonstrates the great generality of the Hamiltonian formulation. The generalized coordinates $(\boldsymbol{q}, \boldsymbol{p})$ can really be anything; they are not necessarily positions and momenta. In this example, solving the original problem or the new problem, where the roles of "position" and "momenta" are reversed, leads to the same results (see Exercise 8.1). Many other examples of canonical transformations exist (see, for example, [Gol80]).

Invariance of phase space volume We conclude the discussion of canonical transformations by proving the important property that the element of phase space volume is preserved by a canonical transformation. Thus, if $d\boldsymbol{q}d\boldsymbol{p} = dq_1 \cdots dq_n dp_1 \cdots dp_n$ and $d\widehat{\boldsymbol{q}}d\widehat{\boldsymbol{p}} = d\widehat{q}_1 \cdots d\widehat{q}_n d\widehat{p}_1 \cdots d\widehat{p}_n$, we would like to show that

$$d\boldsymbol{q}d\boldsymbol{p} = d\widehat{\boldsymbol{q}}d\widehat{\boldsymbol{p}}. \tag{8.17}$$

The following proof is based on the derivation in [Gol80, Section 9-3].

Proof We limit ourselves to canonical transformations which do not explicitly depend on time, i.e.

$$\boldsymbol{q} = \boldsymbol{q}(\widehat{\boldsymbol{q}}, \widehat{\boldsymbol{p}}), \qquad \boldsymbol{p} = \boldsymbol{p}(\widehat{\boldsymbol{q}}, \widehat{\boldsymbol{p}}).$$

We introduce the compact notation,

$$\boldsymbol{y} = (q_1, \dots, q_n, p_1, \dots, p_n), \qquad \boldsymbol{Y} = (\widehat{q}_1, \dots, \widehat{q}_n, \widehat{p}_1, \dots, \widehat{p}_n),$$

so that $d\boldsymbol{y} = d\boldsymbol{q}d\boldsymbol{p}$ and $d\boldsymbol{Y} = d\widehat{\boldsymbol{q}}d\widehat{\boldsymbol{p}}$. From the change of variables theorem, we know that [Apo69]:

$$d\boldsymbol{Y} = |\det \mathbf{J}| d\boldsymbol{y}. \tag{8.18}$$

We need to show that $|\det \mathbf{J}| = 1$, where $\mathbf{J}$ is the Jacobian matrix of the transformation:

$$\mathbf{J} = \begin{bmatrix} \partial Y_1/\partial y_1 & \cdots & \partial Y_1/\partial y_n \\ \vdots & & \vdots \\ \partial Y_n/\partial y_1 & \cdots & \partial Y_n/\partial y_n \end{bmatrix}. \tag{8.19}$$

The vectors $\boldsymbol{y}$ and $\boldsymbol{Y}$ satisfy Hamilton's equations, which compactly are

$$\dot{\boldsymbol{y}} = \mathbf{A}\frac{\partial \mathcal{H}}{\partial \boldsymbol{y}}, \qquad \dot{\boldsymbol{Y}} = \mathbf{A}\frac{\partial \widehat{\mathcal{H}}}{\partial \boldsymbol{Y}}, \tag{8.20}$$

where $\mathbf{A}$ is a constant antisymmetric matrix with the following form:

$$\mathbf{A} = \begin{bmatrix} \mathbf{0} & \mathbf{I} \\ -\mathbf{I} & \mathbf{0} \end{bmatrix}. \tag{8.21}$$

Here $\mathbf{0}$ is an $n \times n$ matrix of zeroes and $\mathbf{I}$ is the $n \times n$ identity matrix. Now we start with $\dot{\boldsymbol{Y}}$, apply the chain rule and use the definition of the Jacobian matrix in Eqn. (8.19):

$$\dot{Y}_i = \frac{\partial Y_i}{\partial y_j}\dot{y}_j = \mathsf{J}_{ij}\dot{y}_j = \mathsf{J}_{ij}\left(\mathsf{A}_{jk}\frac{\partial \mathcal{H}}{\partial y_k}\right) = \mathsf{J}_{ij}\mathsf{A}_{jk}\frac{\partial \widehat{\mathcal{H}}}{\partial Y_l}\frac{\partial Y_l}{\partial y_k} = \mathsf{J}_{ij}\mathsf{A}_{jk}\mathsf{J}_{lk}\frac{\partial \widehat{\mathcal{H}}}{\partial Y_l}.$$

In the third equality, we used Eqn. $(8.20)_1$, then the chain rule, and finally the definition of the Jacobian matrix. In direct notation, we have shown that

$$\dot{\boldsymbol{Y}} = \mathbf{J}\mathbf{A}\mathbf{J}^T\frac{\partial \widehat{\mathcal{H}}}{\partial \boldsymbol{Y}}. \tag{8.22}$$

Comparing this equation with Eqn. $(8.20)_2$, we see that[4]

$$\mathbf{J}\mathbf{A}\mathbf{J}^T = \mathbf{A}. \tag{8.23}$$

Taking the determinant of both sides and using the fact that $\det \mathbf{A} = 1$ gives $(\det \mathbf{J})^2 = 1$, which means that $|\det \mathbf{J}| = 1$ as required. □

8.1.2 Microscopic stress tensor in a finite system at zero temperature

In the next section we will be deriving an expression for the stress tensor from equilibrium statistical mechanics. The derivation starts with a finite system and is then extended to periodic systems. To help understand this derivation and, in particular, to understand the role of internal and external forces, it is helpful to begin with a derivation of the stress tensor in finite system under static equilibrium conditions at zero temperature ($T = 0$ K).

We consider a system of particles, $\mathsf{S} = \mathsf{A} + \mathsf{B}$, consisting of an inner region A embedded in an outer region B as shown in Fig. 8.2(a). System S can be either periodic or isolated (we consider both cases below). We identify a reference configuration of S with a set of positions,

$$\boldsymbol{R}^{\mathsf{A}} = \{\boldsymbol{R}^\alpha \mid \alpha \in \mathsf{A}\}, \qquad \boldsymbol{R}^{\mathsf{B}} = \{\boldsymbol{R}^\beta \mid \beta \in \mathsf{B}\}, \tag{8.24}$$

for which the force on all atoms is zero. Next, let the positions of the atoms in S be deformed relative to the reference configuration according to

$$\boldsymbol{r}^\gamma = \boldsymbol{F}\boldsymbol{R}^\gamma + \boldsymbol{u}^\gamma \qquad \gamma \in \mathsf{S}, \tag{8.25}$$

where $\boldsymbol{F}$ is a specified deformation gradient (see Section 2.2.4) and $\boldsymbol{u}^\gamma$ is the displacement of atom γ. If S is a periodic system, the uniform part of the deformation defined by $\boldsymbol{F}$ is

[4] Equation (8.23) is called the *symplectic condition* and matrices $\mathbf{J}$ satisfying it are called *symplectic matrices*.

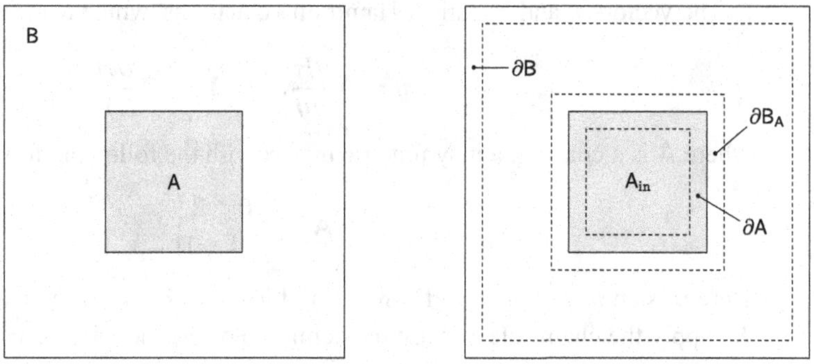

Fig. 8.2 Schematic diagram of the systems being used in the static stress derivation: (a) system A is embedded in system B; (b) various boundaries and subregions. Note system A (shown in frame (a)) consists of A_{in} and $\partial\mathsf{A}$, i.e. $\mathsf{A} = \mathsf{A}_{\text{in}} + \partial\mathsf{A}$.

imposed by changing the shape of the periodic supercell as explained in Section 6.4.2. If S is finite, the atoms in the outer boundary of B, denoted $\partial\mathsf{B}$ (see Fig. 8.2(b)), must be fixed to the deformed positions in Eqn. (8.25) with the displacements $\boldsymbol{u}^\beta$ ($\beta \in \partial\mathsf{B}$) initialized to suitable values (or simply set to zero).

As in Section 7.3.3, the potential energy of the combined system can be written as

$$\mathcal{V}^{\mathsf{A+B}}(\boldsymbol{r}^{\mathsf{A}}, \boldsymbol{r}^{\mathsf{B}}) = \mathcal{V}^{\mathsf{A}}(\boldsymbol{r}^{\mathsf{A}}) + \mathcal{V}^{\mathsf{B}}(\boldsymbol{r}^{\mathsf{B}}) + \mathcal{V}^{\mathsf{A}\leftrightarrow\mathsf{B}}(\boldsymbol{r}^{\mathsf{A}}, \boldsymbol{r}^{\mathsf{B}}), \tag{8.26}$$

where $\boldsymbol{r}^{\mathsf{A}}$ and $\boldsymbol{r}^{\mathsf{B}}$ denote concatenated positions in analogy to Eqn. (8.24), $\mathcal{V}^{\mathsf{A}}(\boldsymbol{r}^{\mathsf{A}})$ and $\mathcal{V}^{\mathsf{B}}(\boldsymbol{r}^{\mathsf{B}})$ are the energies of systems A and B each in the absence of the other and $\mathcal{V}^{\mathsf{A}\leftrightarrow\mathsf{B}}$ is the interaction energy between the systems defined by the relation:

$$\mathcal{V}^{\mathsf{A}\leftrightarrow\mathsf{B}}(\boldsymbol{r}^{\mathsf{A}}, \boldsymbol{r}^{\mathsf{B}}) \equiv \mathcal{V}^{\mathsf{A+B}}(\boldsymbol{r}^{\mathsf{A}}, \boldsymbol{r}^{\mathsf{B}}) - \mathcal{V}^{\mathsf{A}}(\boldsymbol{r}^{\mathsf{A}}) - \mathcal{V}^{\mathsf{B}}(\boldsymbol{r}^{\mathsf{B}}). \tag{8.27}$$

This definition holds for any interatomic potential.

The equilibrium positions of the atoms in S are then obtained by minimizing $\mathcal{V}^{\mathsf{A+B}}$ with respect to the positions of the atoms, subject to either periodic boundary conditions (PBCs) or fixed atom positions in $\partial\mathsf{B}$ depending on the nature of system S. At equilibrium, the forces on all unconstrained atoms are zero, i.e.

$$\boldsymbol{f}^\gamma = -\frac{\partial \mathcal{V}^{\mathsf{A+B}}(\boldsymbol{r}^{\mathsf{A}}, \boldsymbol{r}^{\mathsf{B}})}{\partial \boldsymbol{r}^\gamma} = \boldsymbol{0} \qquad \forall \text{ unconstrained } \gamma \in \mathsf{S}. \tag{8.28}$$

This system of equations is solved for the unconstrained displacements of the atoms $\boldsymbol{u}^\gamma$.

Let us now focus on system A. In terms of the definitions in Eqn. (7.4), $\mathcal{V}^{\mathsf{A}}$ is the internal energy $\mathcal{V}^{\text{int}}$ of system A, and $\mathcal{V}^{\mathsf{A}\leftrightarrow\mathsf{B}}$ is the external energy $\mathcal{V}^{\text{ext}}_{\text{con}}$ due to contact with atoms outside A.[5] The force in Eqn. (8.28) for atoms in system A (i.e. $\forall\alpha \in \mathsf{A}$) is

$$\begin{aligned} \boldsymbol{f}^\alpha = \boldsymbol{0} &= -\frac{\partial \mathcal{V}^{\mathsf{A}}(\boldsymbol{r}^{\mathsf{A}})}{\partial \boldsymbol{r}^\alpha} - \frac{\partial \mathcal{V}^{\mathsf{A}\leftrightarrow\mathsf{B}}(\boldsymbol{r}^{\mathsf{A}}, \boldsymbol{r}^{\mathsf{B}})}{\partial \boldsymbol{r}^\alpha} \\ &= -\frac{\partial \mathcal{V}^{\text{int}}(\boldsymbol{r}^{\mathsf{A}})}{\partial \boldsymbol{r}^\alpha} - \frac{\partial \mathcal{V}^{\text{ext}}_{\text{con}}(\boldsymbol{r}^{\mathsf{A}}, \boldsymbol{r}^{\mathsf{B}})}{\partial \boldsymbol{r}^\alpha} \equiv \boldsymbol{f}^{\text{int},\alpha} + \boldsymbol{f}^{\text{ext},\alpha}, \end{aligned} \tag{8.29}$$

[5] See footnote 27 on page 411.

where we have defined the internal and external parts of the force acting on atom α (from the perspective of system A). Due to the short-range nature of interatomic potentials, only atoms within a certain distance of the interface between A and B, represented by the boundaries ∂A and ∂B^A in Fig. 8.2(b), will contribute to the external part of the force.[6] Therefore,

$$\boldsymbol{f}^{\text{ext},\alpha} = \boldsymbol{f}^{\text{ext},\alpha}(\boldsymbol{r}^{\partial A}, \boldsymbol{r}^{\partial B^A}) \qquad \forall \alpha \in A, \tag{8.30}$$

where $\boldsymbol{r}^{\partial A}$ and $\boldsymbol{r}^{\partial B^A}$ are the concatenated positions of the atoms within the boundaries ∂A and ∂B^A. Note that $\boldsymbol{f}^{\text{ext},\alpha} = \boldsymbol{0}$ for atoms in $A_{\text{in}} = A \setminus \partial A$, i.e. atoms in A but not in ∂A.

We now define a potential energy function which only includes the internal energy of system A and represents system B through the (known) forces that it applies to A:

$$\mathcal{V}(\boldsymbol{r}^A) = \mathcal{V}^{\text{int}}(\boldsymbol{r}^A) - \sum_{\gamma \in \partial A} \boldsymbol{f}^{\text{ext},\gamma} \cdot \boldsymbol{r}^\gamma. \tag{8.31}$$

Referring to Eqn. (8.29), the equilibrium equations are

$$\boldsymbol{f}^\alpha = -\frac{\partial \mathcal{V}}{\partial \boldsymbol{r}^\alpha} = \boldsymbol{0} = \begin{cases} \boldsymbol{f}^{\text{int},\alpha} & \alpha \in A_{\text{in}}, \\ \boldsymbol{f}^{\text{int},\alpha} + \boldsymbol{f}^{\text{ext},\alpha} & \alpha \in \partial A. \end{cases} \tag{8.32}$$

Equation (8.32) provides a microscopic perspective on the equilibrium of the atoms in system A. Let us now adopt a *macroscopic* viewpoint. The equilibrium positions obtained from Eqn. (8.28) are determined by the applied deformation gradient $\boldsymbol{F}$, thus

$$\boldsymbol{r}^\gamma(\boldsymbol{F}) = \boldsymbol{F}\boldsymbol{R}^\gamma + \boldsymbol{u}^\gamma(\boldsymbol{F}) \qquad \forall \gamma \in S. \tag{8.33}$$

We can therefore define $W(\boldsymbol{F})$, a *homogenized continuum energy density* [ETS06a] (which plays the role of the strain energy density), for system A:

$$W(\boldsymbol{F}) \equiv \frac{1}{V_0}\mathcal{V}^{\text{int}}(\boldsymbol{r}^A(\boldsymbol{F})), \tag{8.34}$$

where V_0 is the volume of system A in the reference configuration. The corresponding continuum potential energy is

$$\Pi(\boldsymbol{F}) = V_0\left(W(\boldsymbol{F}) - \boldsymbol{P} : \boldsymbol{F}\right), \tag{8.35}$$

where $\boldsymbol{P}$ is first Piola–Kirchhoff stress which is work conjugate with $\boldsymbol{F}$ and ":" denotes a tensor contraction ($\boldsymbol{P} : \boldsymbol{F} = P_{iJ}F_{iJ}$, where as usual the summation convention is applied to repeating spatial indices). At equilibrium $\Pi(\boldsymbol{F})$ is at a minimum with respect to $\boldsymbol{F}$ and therefore satisfies, $\partial\Pi/\partial\boldsymbol{F} = \boldsymbol{0}$, which gives (see Eqn. (2.173))

$$\boldsymbol{P}(\boldsymbol{F}) = \frac{\partial W(\boldsymbol{F})}{\partial \boldsymbol{F}}. \tag{8.36}$$

[6] Note that depending on the nature of the atomic interactions, the distance defining the boundaries ∂A and ∂B^A may exceed the cutoff radius of the potential (r_{cut}). For example, for pair functionals, $\boldsymbol{f}^{\text{ext},\alpha}$ will depend on all atoms within a distance of $2r_{\text{cut}}$ of atom α (see Section 5.8.5 for an explanation). Strictly speaking, for ionic crystals ∂A and ∂B^A must extend to all atoms in A and B due to the presence of long-range electrostatic fields. However, since the contribution to the stress of the electrostatic fields decays with distance, the approach can be used by selecting a size for ∂A and ∂B^A that provides an acceptable level of accuracy.

Substituting Eqn. (8.34) into Eqn. (8.36) and applying the chain rule gives

$$P_{iJ} = \frac{1}{V_0} \sum_{\alpha \in \mathsf{A}} \frac{\partial \mathcal{V}^{\text{int}}}{\partial r_k^\alpha} \frac{\partial r_k^\alpha}{\partial F_{iJ}} = -\frac{1}{V_0} \left[\sum_{\alpha \in \mathsf{A}} f_i^{\text{int},\alpha} R_J^\alpha + \sum_{\alpha \in \mathsf{A}} f_k^{\text{int},\alpha} \frac{\partial u_k^\alpha}{\partial F_{iJ}} \right], \tag{8.37}$$

where Eqn. (8.33) was used in the second step. The second term in the square brackets in Eqn. (8.37) can be neglected in certain situations. Weiner discusses this in his book [Wei83, Section 4.3] and shows that if S is a (periodic) multilattice crystal (see Section 3.6) and A is large so that edge and corner effects can be neglected, then this term cancels due to the overall equilibrium of system A. (See Exercise 8.4 for an insight into this from a one-dimensional chain problem.) We discuss the case of multilattice crystals in much more detail in Chapter 11. Assuming that the second term in Eqn. (8.37) is indeed negligible, then the final expression for the first Piola–Kirchhoff stress tensor is

$$\boldsymbol{P} = -\frac{1}{V_0} \sum_{\alpha \in \mathsf{A}} \boldsymbol{f}^{\text{int},\alpha} \otimes \boldsymbol{R}^\alpha = \frac{1}{V_0} \sum_{\alpha \in \partial \mathsf{A}} \boldsymbol{f}^{\text{ext},\alpha} \otimes \boldsymbol{R}^\alpha . \tag{8.38}$$

The second equality follows by using Eqn. (8.32) from which we see that only boundary terms in the sum remain, where $\boldsymbol{f}^{\text{int},\alpha} = -\boldsymbol{f}^{\text{ext},\alpha}$.

It is important to note that although Eqn. (8.38) for $\boldsymbol{P}$ was derived without the requirement that A be "large" (in the sense that the width of the boundary ∂A is small relative to the size of A), this expression only corresponds to a *macroscopic* stress if this condition is satisfied. Otherwise $\boldsymbol{P}$ computed from Eqn. (8.38) will be affected by finite-size effects of the domain of A. An example demonstrating this effect is given in Exercise 8.6.

8.1.3 Microscopic stress tensor at finite temperature: the virial stress

We now turn to the derivation of the stress tensor for a system in thermodynamic equilibrium at a fixed temperature T. Just as in the static case considered in the previous section, the system is subjected to a uniform deformation characterized by the deformation gradient $\boldsymbol{F}$. In analogy to the potential energy in Eqn. (8.35), the Gibbs free energy G is the relevant thermodynamic potential for a system at constant temperature (see Eqn. (2.176)):

$$G(T, \boldsymbol{F}) = \Psi(T, \boldsymbol{F}) - V_0\, \boldsymbol{P} : \boldsymbol{F}, \tag{8.39}$$

where Ψ is the Helmholtz free energy of the system, $\boldsymbol{P}$ is the first Piola–Kirchhoff stress tensor, and V_0 is the volume of the system in the reference configuration. The equilibrium state is associated with the minimum of the Gibbs free energy with respect to the deformation gradient. At the minimum, $\partial G / \partial \boldsymbol{F} = \boldsymbol{0}$, from which we obtain the following definitions for the stress tensor:

$$\boldsymbol{P}(T, \boldsymbol{F}) = \frac{1}{V_0} \frac{\partial \Psi}{\partial \boldsymbol{F}}, \qquad \boldsymbol{S}(T, \boldsymbol{F}) = \boldsymbol{F}^{-1} \boldsymbol{P}, \qquad \boldsymbol{\sigma}(T, \boldsymbol{F}) = \frac{1}{\det \boldsymbol{F}} \boldsymbol{P} \boldsymbol{F}^T. \tag{8.40}$$

The first definition in Eqn. (8.40) is exactly the one given in Eqn. $(2.171)_2$. The other two expressions, for the second Piola–Kirchhoff stress, $\boldsymbol{S}$, and the Cauchy stress, $\boldsymbol{\sigma}$, follow from standard continuum relations in Eqns. (2.99) and $(2.96)_2$. (Note that we limit the discussion to conservative systems where there is no viscous (dissipative) contribution to the total stress as discussed in Section 2.5.2.)

The Helmholtz free energy in the canonical ensemble is given by (Eqn. (7.129)):

$$\Psi(T, \boldsymbol{F}) = -k_B T \ln Z = -k_B T \ln \left[\frac{1}{N! h^{3N}} \int_{\Gamma(\boldsymbol{F})} e^{-\mathcal{H}(\boldsymbol{q},\boldsymbol{p})/k_B T} \, d\boldsymbol{q} d\boldsymbol{p} \right], \tag{8.41}$$

where we have used the expression for the partition function Z from Eqn. (7.116). Note that exactly as in the definition of the strain energy density in Eqn. (8.34), the Hamiltonian only depends on the *internal* potential energy of the system, $\mathcal{V}^{\text{int}}$, and does not include contributions due to the surrounding container:

$$\mathcal{H}(\boldsymbol{q}, \boldsymbol{p}) = \sum_{\alpha} \frac{\boldsymbol{p}^\alpha \cdot \boldsymbol{p}^\alpha}{2m^\alpha} + \mathcal{V}^{\text{int}}(\boldsymbol{q}). \tag{8.42}$$

This is consistent with the assumption of weak interaction inherent in the canonical ensemble description.[7]

The free energy depends on $\boldsymbol{F}$ through the integration bounds of the position degrees of freedom. If we assume that the system is finite and confined to a physical container (as was done in the previous section), then the shape of that container, set by $\boldsymbol{F}$, dictates the integration bounds. (We will consider the case of PBCs a bit later.) The fact that Ψ depends on $\boldsymbol{F}$ through the integration bounds makes the calculation of the derivatives in Eqn. (8.40) difficult. We would therefore like to reformulate the problem relative to a fixed reference configuration (i.e. adopt a Lagrangian perspective in the language of continuum mechanics as explained in Section 2.2.3) in which the deformation gradient appears explicitly in the integrand. The necessary change of variables requires us to apply a canonical transformation (see Section 8.1.1) to the variables $\boldsymbol{q}$ and $\boldsymbol{p}$ as we now explain.[8]

For specificity, let us take the container to be a parallelepiped defined by the three vectors $\boldsymbol{a}_1$, $\boldsymbol{a}_2$ and $\boldsymbol{a}_3$, which need not be orthogonal. This selection is done for convenience and does not limit the generality of the derivation as we shall see below. The position of an atom, $\boldsymbol{r}^\alpha$ ($\alpha = 1, \dots, N$), can be expressed in terms of scaled coordinates $s_i^\alpha \in [0, 1]$ as

$$\boldsymbol{r}^\alpha = s_i^\alpha \boldsymbol{a}_i. \tag{8.43}$$

The deformation of the container is defined relative to a reference configuration where the cell vectors are $\boldsymbol{A}_1$, $\boldsymbol{A}_2$ and $\boldsymbol{A}_3$. (We follow the continuum mechanics convention of using capital letters to refer to the reference configuration.) The deformed and reference cell vectors are related through an affine mapping defined by $\boldsymbol{F}$:

$$\boldsymbol{a}_i = \boldsymbol{F} \boldsymbol{A}_i. \tag{8.44}$$

[7] Recall that in the canonical ensemble, the Hamiltonian $\mathcal{H}$ neglects the interaction term of the system with the surrounding "heat bath." In Section 7.4.1, we referred to $\mathcal{H}$ specifically as $\mathcal{H}^{\text{A}}$ to stress this fact. This means that the potential energy $\mathcal{V}$ in $\mathcal{H}$ only includes the *internal* energy of the system.

[8] The approach described here is equivalent to the one used by Ray and Rahman [RR84] and later by Lutsko [Lut89]. However, we express the necessary canonical transformation directly in terms of continuum mechanics deformation measures, which makes it easier to understand what stress (and later elasticity) measures are being computed.

Equations (8.43) and (8.44) can be combined to relate the position $\boldsymbol{r}^\alpha$ of atom α in the deformed configuration its position in the reference configuration $\boldsymbol{R}^\alpha$ to:

$$\boldsymbol{r}^\alpha = s_i^\alpha(\boldsymbol{F}\boldsymbol{A}_i) = \boldsymbol{F}(s_i^\alpha \boldsymbol{A}_i) = \boldsymbol{F}\boldsymbol{R}^\alpha. \tag{8.45}$$

Note that the assumed (parallelepiped) shape of the container does not enter into Eqn. (8.45), $\boldsymbol{r}^\alpha = \boldsymbol{F}\boldsymbol{R}^\alpha$, which means that this relation holds for a container of any shape.

Note, also, that Eqn. (8.45) does *not* impose a kinematic constraint on the position of atom α in the deformed configuration based on its position in the reference configuration (as does the Cauchy–Born rule discussed in Section 11.2.2). As we explained above, Eqn. (8.45) is merely used as a change of variables, where instead of integrating over the deformed configuration with the variables $\boldsymbol{r}^\alpha$, the integration is carried out over the reference configuration using the variables $\boldsymbol{R}^\alpha$. In both cases the same result is obtained. However, by using the referential variables the dependence on the deformation gradient is made explicit.

We now identify the positions of the atoms in the reference and deformed configurations as the position coordinates which we seek to relate through a canonical transformation,

$$\boldsymbol{q} = (r_1^1, r_2^1.r_3^1, r_1^2, \dots, r_3^N), \qquad \widehat{\boldsymbol{q}} = (R_1^1, R_2^1.R_3^1, R_1^2, \dots, R_3^N).$$

It is clear that if we enforce the mapping in Eqn. (8.45) on our system, with no change in the momentum coordinates, then the newly obtained variables will not satisfy Hamilton's equations. Therefore any change of variables should be governed by a canonical transformation. The following generator function provides the desired transformation:

$$G_3(\boldsymbol{p}, \widehat{\boldsymbol{q}}) = -\sum_{\alpha=1}^{N} \boldsymbol{p}^\alpha \cdot (\boldsymbol{F}\boldsymbol{R}^\alpha). \tag{8.46}$$

Substituting G_3 into Eqn. (8.14) (evaluated for each atom at a time) gives

$$\boldsymbol{r}^\alpha = -\frac{\partial G_3}{\partial \boldsymbol{p}^\alpha} = \boldsymbol{F}\boldsymbol{R}^\alpha, \qquad \mathfrak{P}^\alpha = -\frac{\partial G_3}{\partial \boldsymbol{R}^\alpha} = \boldsymbol{F}^T \boldsymbol{p}^\alpha, \qquad \widehat{\mathcal{H}} = \mathcal{H}, \tag{8.47}$$

where $\mathfrak{P}^\alpha$ is the momentum of atom α in the reference configuration. The first relation in the above equation is exactly the desired transformation in Eqn. (8.45). The second relation is the corresponding transformation that the momentum degrees of freedom must satisfy, so that the new set of reference variables $(\widehat{\boldsymbol{q}}, \widehat{\boldsymbol{p}})$ are canonical. We also define the following "pull-back momenta":[9]

$$\boldsymbol{\Pi}^\alpha = \boldsymbol{F}^{-1}\boldsymbol{p}^\alpha, \tag{8.48}$$

where the momentum vector is mapped to the reference configuration as if it were a spatial quantity. We stress that $\boldsymbol{\Pi}^\alpha$ is *not* conjugate with $\boldsymbol{R}^\alpha$; however, it turns out to be useful

[9] The terms "pull-back" and "push-forward" are standard terminology in continuum mechanics for transferring tensor quantities between the reference and deformed configurations. See, for example, 3.4.4 in [TME12].

later in the derivation. The Hamiltonian of the system in terms of the reference variables is

$$\widehat{\mathcal{H}}(\widehat{\boldsymbol{q}},\widehat{\boldsymbol{p}};\boldsymbol{F}) = \mathcal{H}(\boldsymbol{q}(\widehat{\boldsymbol{q}};\boldsymbol{F}),\boldsymbol{p}(\widehat{\boldsymbol{p}};\boldsymbol{F})) = \sum_\alpha \frac{(\boldsymbol{F}^{-T}\boldsymbol{\mathfrak{P}}^\alpha)\cdot(\boldsymbol{F}^{-T}\boldsymbol{\mathfrak{P}}^\alpha)}{2m^\alpha} + \mathcal{V}^{\text{int}}(\{\boldsymbol{F}\boldsymbol{R}^\alpha\}). \tag{8.49}$$

The dependence of the positions and momenta on $\boldsymbol{F}$ has now been made explicit and the integration bounds for the position degrees of freedom are now fixed to the region of space confined by the container in the reference configuration. We denote the accessible phase space in the reference configuration as Γ_0. The partition function appearing in Eqn. (8.41) can now be written relative to the reference configuration:

$$Z(T,\boldsymbol{F}) = \frac{1}{N!h^{3N}}\int_{\Gamma(\boldsymbol{F})} e^{-\mathcal{H}(\boldsymbol{q},\boldsymbol{p})/k_BT}\,d\boldsymbol{q}d\boldsymbol{p} = \frac{1}{N!h^{3N}}\int_{\Gamma_0} e^{-\widehat{\mathcal{H}}(\widehat{\boldsymbol{q}},\widehat{\boldsymbol{p}};\boldsymbol{F})/k_BT}\,d\widehat{\boldsymbol{q}}d\widehat{\boldsymbol{p}}. \tag{8.50}$$

Note that in changing the integration domain from Γ to Γ_0 in Eqn. (8.50), the Jacobian of the change of variables is unity, i.e. $d\boldsymbol{q}d\boldsymbol{p} = d\widehat{\boldsymbol{q}}d\widehat{\boldsymbol{p}}$. This is a basic property of canonical transformations (see Eqn. (8.17)). Similarly to Eqn. (8.50), canonical phase averages (see Section 7.4.1) can be expressed in both the deformed and reference configurations:

$$\langle A\rangle = \int_{\Gamma(\boldsymbol{F})} A(\boldsymbol{q},\boldsymbol{p})f_c(\boldsymbol{q},\boldsymbol{p};T,\boldsymbol{F})\,d\boldsymbol{q}d\boldsymbol{p} = \int_{\Gamma_0}\widehat{A}(\widehat{\boldsymbol{q}},\widehat{\boldsymbol{p}};\boldsymbol{F})\widehat{f}_c(\widehat{\boldsymbol{q}},\widehat{\boldsymbol{p}};T,\boldsymbol{F})\,d\widehat{\boldsymbol{q}}d\widehat{\boldsymbol{p}}, \tag{8.51}$$

where f_c and $\widehat{f}_c$ are the canonical distribution functions expressed in terms of the deformed and reference coordinates:

$$f_c(\boldsymbol{q},\boldsymbol{p};T,\boldsymbol{F}) = \frac{e^{-\mathcal{H}(\boldsymbol{q},\boldsymbol{p})/k_BT}}{N!h^{3N}Z(T,\boldsymbol{F})}, \qquad \widehat{f}_c(\widehat{\boldsymbol{q}},\widehat{\boldsymbol{p}};T,\boldsymbol{F}) = \frac{e^{-\widehat{\mathcal{H}}(\widehat{\boldsymbol{q}},\widehat{\boldsymbol{p}};\boldsymbol{F})/k_BT}}{N!h^{3N}Z(T,\boldsymbol{F})}. \tag{8.52}$$

We now derive the expression for first Piola–Kirchhoff stress using Eqn. $(8.40)_1$:

$$\boldsymbol{P}(T,\boldsymbol{F}) = \frac{1}{V_0}\frac{\partial}{\partial\boldsymbol{F}}(-k_BT\ln Z) = -\frac{k_BT}{V_0Z}\left(\frac{\partial Z}{\partial\boldsymbol{F}}\right) = \frac{1}{V_0}\left\langle\frac{\partial\widehat{\mathcal{H}}}{\partial\boldsymbol{F}}\right\rangle, \tag{8.53}$$

where in the last step we have used

$$\frac{\partial Z}{\partial\boldsymbol{F}} = \frac{\partial}{\partial\boldsymbol{F}}\left[\frac{1}{N!h^{3N}}\int_{\Gamma_0} e^{-\widehat{\mathcal{H}}/k_BT}\,d\widehat{\boldsymbol{q}}d\widehat{\boldsymbol{p}}\right] = -\frac{1}{k_BTN!h^{3N}}\int_{\Gamma_0}\frac{\partial\widehat{\mathcal{H}}}{\partial\boldsymbol{F}}e^{-\widehat{\mathcal{H}}/k_BT}\,d\widehat{\boldsymbol{q}}d\widehat{\boldsymbol{p}}. \tag{8.54}$$

Next, we compute $\partial\widehat{\mathcal{H}}/\partial\boldsymbol{F}$. In indicial notation, we have

$$\frac{\partial\widehat{\mathcal{H}}}{\partial F_{iJ}} = \frac{\partial}{\partial F_{iJ}}\left[\sum_\alpha\frac{p_k^\alpha p_k^\alpha}{2m^\alpha} + \mathcal{V}^{\text{int}}(\boldsymbol{r}^1,\dots,\boldsymbol{r}^N)\right] = \sum_\alpha\left[\frac{1}{m^\alpha}\frac{\partial p_k^\alpha}{\partial F_{iJ}}p_k^\alpha + \frac{\partial\mathcal{V}^{\text{int}}}{\partial r_k^\alpha}\frac{\partial r_k^\alpha}{\partial F_{iJ}}\right], \tag{8.55}$$

where as usual the summation convention is applied. From Eqn. (8.47), we have

$$\frac{\partial r_k^\alpha}{\partial F_{iJ}} = \frac{\partial}{\partial F_{iJ}}(F_{kL}R_L^\alpha) = \delta_{ik}R_J^\alpha, \tag{8.56}$$

$$\frac{\partial p_k^\alpha}{\partial F_{iJ}} = \frac{\partial}{\partial F_{iJ}}(F_{Lk}^{-1}\mathfrak{P}_L^\alpha) = -F_{Jk}^{-1}F_{Li}^{-1}\mathfrak{P}_L^\alpha = -F_{Jk}^{-1}p_i^\alpha, \tag{8.57}$$

where in Eqn. (8.57) we have used the following identity:[10]

$$\frac{\partial F^{-1}_{Lk}}{\partial F_{iJ}} = -F^{-1}_{Li} F^{-1}_{Jk}. \tag{8.58}$$

Substituting Eqns. (8.56) and (8.57) into Eqn. (8.55), we have

$$\frac{\partial \widehat{\mathcal{H}}}{\partial F_{iJ}} = -\sum_{\alpha} \left[\frac{p^\alpha_i \Pi^\alpha_J}{m^\alpha} + f^{\mathrm{int},\alpha}_i R^\alpha_J \right], \tag{8.59}$$

where $\boldsymbol{\Pi}^\alpha$ is the "pull-back" momentum defined in Eqn. (8.48) and $\boldsymbol{f}^{\mathrm{int},\alpha} = -\partial \mathcal{V}^{\mathrm{int}}/\partial \boldsymbol{r}^\alpha$ is the internal force on atom α in the deformed configuration. Finally, substituting Eqn. (8.59) into Eqn. (8.53), we obtain the following expression for the first Piola–Kirchhoff stress:

$$P_{iJ} = -\frac{1}{V_0} \sum_{\alpha} \left\langle \frac{p^\alpha_i \Pi^\alpha_J}{m^\alpha} + f^{\mathrm{int},\alpha}_i R^\alpha_J \right\rangle. \tag{8.60}$$

Note the similarity of this equation to Eqn. (8.38) obtained in the static case. The difference is that the finite temperature expression includes a kinetic term and involves a phase average that accounts for the vibrations of the atoms.

The results of the statistical mechanics calculation can be understood in terms of system A and its boundary ∂A used in the static case (see Section 8.1.2 and Fig. 8.2) as follows. In the dynamic case, we interpret the assumption of "weak interaction" inherent in the canonical ensemble to mean that the atoms in ∂A are fixed and therefore do not contribute to the phase average. The atoms in $\mathsf{A}_{\mathrm{in}} = \mathsf{A} \setminus \partial\mathsf{A}$ (i.e. the atoms in A which are not in ∂A) are confined by ∂A to a region of configuration space. They explore this space and contribute to the potential part of the stress as they interact with the fixed atoms in ∂A. Furthermore, it is implicitly assumed that the size of ∂A is negligible relative to the size of A since the atoms in ∂A are not explicitly considered. The fact that ∂A is neglected explains the absence of the $\partial \boldsymbol{u}/\partial \boldsymbol{F}$ term that appears in Eqn. (8.37). It also implies that the microscopic stress expression in Eqn. (8.60) only corresponds to a macroscopic stress in the thermodynamic limit where the size of A goes to infinity (while ∂A is held fixed). The same point was made earlier for the static case after Eqn. (8.38) and explored in Exercise 8.6.

Given the expression for the first Piola–Kirchhoff stress in Eqn. (8.60), the Cauchy stress and second Piola–Kirchhoff stresses follow immediately from Eqns. $(8.40)_2$ and $(8.40)_3$ as

$$S_{IJ} = -\frac{1}{V_0} \sum_{\alpha} \left\langle \frac{\Pi^\alpha_I \Pi^\alpha_J}{m^\alpha} + (F^{-1}_{Ii} f^{\mathrm{int},\alpha}_i) R^\alpha_J \right\rangle, \tag{8.61}$$

$$\sigma_{ij} = -\frac{1}{V} \sum_{\alpha} \left\langle \frac{p^\alpha_i p^\alpha_j}{m^\alpha} + f^{\mathrm{int},\alpha}_i r^\alpha_j \right\rangle, \tag{8.62}$$

where $V = V_0 \det \boldsymbol{F}$ is the deformed volume and Eqn. (8.47) was used.

[10] The proof for this is straightforward. Start with the identity $\boldsymbol{F}^{-1}\boldsymbol{F} = \boldsymbol{I}$. Take the derivative of both sides with respect to $\boldsymbol{F}$ and then right multiply by $\boldsymbol{F}^{-1}$ to obtain $\partial \boldsymbol{F}^{-1}/\partial \boldsymbol{F}$. This is most easily done in indicial notation. See Exercise 2.4.

The expression for the Cauchy stress tensor in Eqn. (8.62) is sometimes called the *virial stress*.[11] Although, the derivation here made use of the canonical ensemble, it is expected to apply to any ensemble in the thermodynamic limit where all ensembles are equivalent.

Continuum mechanics also tells us that the Cauchy stress tensor is symmetric – something which is not apparent in Eqn. (8.62). We also see that the stress is composed of two parts: a "kinetic" part related the momenta of the atoms and a "potential" part related to the forces in the system. (This second part is related to the virial of the system from which the stress gets its name.) For students of continuum mechanics, the presence of a kinetic term in the stress tensor may be surprising. We first discuss the symmetry of the virial stress tensor, before going on to explore the roles of its kinetic and potential parts.

Symmetry of the stress tensor and PBCs The Cauchy stress in a local continuum theory has to be symmetric due to the balance of angular momentum (see Section 2.3.3). Does the virial stress in Eqn. (8.62) have this property? The kinetic part is clearly symmetric, but the potential part does not appear to be so. We show below that due to the *principle of interatomic potential invariance* (see Section 5.3.2) the potential part of the stress is also symmetric and therefore the virial stress tensor as a whole is symmetric.

The implication of invariance principle in Eqn. (5.17) for the forces in an atomistic system was discussed in Section 5.8.1. We showed there that the force on an atom, regardless of the nature of the interatomic potential from which it is derived, can always be written as a sum of central terms (see Eqn. (5.107)):

$$\boldsymbol{f}^{\text{int},\alpha} = \sum_{\substack{\beta \\ \beta\neq\alpha}} \varphi^{\alpha\beta}\frac{\boldsymbol{r}^\beta - \boldsymbol{r}^\alpha}{r^{\alpha\beta}} = \sum_{\substack{\beta \\ \beta\neq\alpha}} \varphi^{\alpha\beta}\frac{\boldsymbol{r}^{\alpha\beta}}{r^{\alpha\beta}} \equiv \sum_{\substack{\beta \\ \beta\neq\alpha}} \boldsymbol{f}^{\alpha\beta}, \qquad \varphi^{\alpha\beta} \equiv \frac{\partial \mathcal{V}^{\text{int}}}{\partial r^{\alpha\beta}}. \tag{8.63}$$

Note that $\boldsymbol{f}^{\alpha\beta} = -\boldsymbol{f}^{\beta\alpha}$. Referring back to the Hamiltonian derivative in Eqn. (8.59) and focusing on the force term, we have using Eqn. $(8.63)_1$

$$-\sum_\alpha \boldsymbol{f}^{\text{int},\alpha}\otimes\boldsymbol{R}^\alpha = -\sum_{\substack{\alpha,\beta \\ \alpha\neq\beta}} \boldsymbol{f}^{\alpha\beta}\otimes\boldsymbol{R}^\alpha.$$

In the expression on the right, $\boldsymbol{f}^{\alpha\beta}$ is a relative force between two atoms, whereas $\boldsymbol{R}^\alpha$ is the absolute position of a single atom. A more symmetric expression is obtained by noting:

$$\begin{aligned} -\sum_{\substack{\alpha,\beta \\ \alpha\neq\beta}} \boldsymbol{f}^{\alpha\beta}\otimes\boldsymbol{R}^\alpha &= -\frac{1}{2}\left[\sum_{\substack{\alpha,\beta \\ \alpha\neq\beta}} \boldsymbol{f}^{\alpha\beta}\otimes\boldsymbol{R}^\alpha + \sum_{\substack{\beta,\alpha \\ \beta\neq\alpha}} \boldsymbol{f}^{\beta\alpha}\otimes\boldsymbol{R}^\beta\right] \\ &= \frac{1}{2}\sum_{\substack{\alpha,\beta \\ \alpha\neq\beta}} \boldsymbol{f}^{\alpha\beta}\otimes(\boldsymbol{R}^\beta - \boldsymbol{R}^\alpha) = \frac{1}{2}\sum_{\substack{\alpha,\beta \\ \alpha\neq\beta}} \boldsymbol{f}^{\alpha\beta}\otimes\boldsymbol{R}^{\alpha\beta}, \end{aligned}$$

[11] For a derivation based on time averages (closer to the original derivation of Clausius [Cla70]), see, for example, [AT10, Appendix A].

where $\boldsymbol{R}^{\alpha\beta} = \boldsymbol{R}^{\beta} - \boldsymbol{R}^{\alpha}$. We therefore have the following identity:

$$-\sum_{\alpha} \boldsymbol{f}^{\text{int},\alpha} \otimes \boldsymbol{R}^{\alpha} = \frac{1}{2} \sum_{\substack{\alpha,\beta \\ \alpha \neq \beta}} \boldsymbol{f}^{\alpha\beta} \otimes \boldsymbol{R}^{\alpha\beta}. \tag{8.64}$$

Substituting this into Eqn. (8.59) and using the definition in Eqn. (8.63), we have

$$\frac{\partial \widehat{\mathcal{H}}}{\partial F_{iJ}} = \sum_{\alpha} \left[-\frac{p_i^{\alpha} \Pi_J^{\alpha}}{m^{\alpha}} + \frac{1}{2} \sum_{\substack{\beta \\ \beta \neq \alpha}} \varphi'^{\alpha\beta} \frac{r_i^{\alpha\beta} R_J^{\alpha\beta}}{r^{\alpha\beta}} \right]. \tag{8.65}$$

The first and second Piola–Kirchhoff stresses, $\boldsymbol{P}$ and $\boldsymbol{S}$, and the Cauchy stress $\boldsymbol{\sigma}$ are then

$$P_{iJ} = \frac{1}{V_0} \left\langle -\sum_{\alpha} \frac{p_i^{\alpha} \Pi_J^{\alpha}}{m^{\alpha}} + \frac{1}{2} \sum_{\substack{\alpha,\beta \\ \alpha \neq \beta}} \varphi'^{\alpha\beta} \frac{r_i^{\alpha\beta} R_J^{\alpha\beta}}{r^{\alpha\beta}} \right\rangle, \tag{8.66}$$

$$S_{IJ} = \frac{1}{V_0} \left\langle -\sum_{\alpha} \frac{\Pi_I^{\alpha} \Pi_J^{\alpha}}{m^{\alpha}} + \frac{1}{2} \sum_{\substack{\alpha,\beta \\ \alpha \neq \beta}} \varphi'^{\alpha\beta} \frac{R_I^{\alpha\beta} R_J^{\alpha\beta}}{r^{\alpha\beta}} \right\rangle, \tag{8.67}$$

$$\sigma_{ij} = \frac{1}{V} \left\langle -\sum_{\alpha} \frac{p_i^{\alpha} p_j^{\alpha}}{m^{\alpha}} + \frac{1}{2} \sum_{\substack{\alpha,\beta \\ \alpha \neq \beta}} \varphi'^{\alpha\beta} \frac{r_i^{\alpha\beta} r_j^{\alpha\beta}}{r^{\alpha\beta}} \right\rangle. \tag{8.68}$$

The symmetry of $\boldsymbol{S}$ and $\boldsymbol{\sigma}$ is clear in these expressions. $\boldsymbol{P}$ need not be symmetric; however, it satisfies Eqn. (2.98), which is a consequence of the balance of angular momentum.

The stress tensors given above were computed for a system confined to a physical container. Another important case is an infinite system subjected to PBCs (see Section 6.4.2). Our derivation does not apply to this case; however; it turns out that the final expression is correct provided that the minimum image convention is obeyed [EW77, Tsa79, LB06].[12] Explicitly, for a periodic system the virial stress is

$$\sigma_{ij} = \frac{1}{V} \left\langle -\sum_{\alpha} \frac{p_i^{\alpha} p_j^{\alpha}}{m^{\alpha}} + \frac{1}{2} \sum_{\substack{\alpha,\beta \\ \alpha \neq \beta}} \varphi'^{\alpha\mathring{\beta}} \frac{r_i^{\alpha\mathring{\beta}} r_j^{\alpha\mathring{\beta}}}{r^{\alpha\mathring{\beta}}}. \right\rangle. \tag{8.69}$$

where a circle over an atom number indicates that the minimum image convention is applied to the position of that atom (see Section 6.4.2).

[12] A direct derivation for stress in a restricted statistical mechanics ensemble under PBCs is given in Section 11.3.4. In particular, see the discussion on page 572. See also Exercise 8.7.

Atomic-level stress The phase averaging in the stress expressions is an integral part of their definition and cannot be removed. Nevertheless, some authors use the virial definition to define an instantaneous *atomic-level stress tensor* for each atom in the system [CY91]:

$$\hat{\sigma}^{\alpha}_{ij} = \frac{1}{V^{\alpha}}\left[-\frac{p^{\alpha}_i p^{\alpha}_j}{m^{\alpha}} + \frac{1}{2}\sum_{\substack{\beta \\ \beta\neq\alpha}} \varphi^{\alpha\beta}\frac{r^{\alpha\beta}_i r^{\alpha\beta}_j}{r^{\alpha\beta}}\right]. \tag{8.70}$$

Here V^{α} is the volume of atom α in the deformed configuration,[13] defined so that $\sum_{\alpha} V^{\alpha} = V$. In similar fashion, atomic-level definitions can be given for the Piola–Kirchhoff stresses,

$$\hat{P}^{\alpha}_{iJ} = \frac{1}{V^{\alpha}_0}\left[-\frac{p^{\alpha}_i \Pi^{\alpha}_J}{m^{\alpha}} + \frac{1}{2}\sum_{\substack{\beta \\ \beta\neq\alpha}} \varphi^{\alpha\beta}\frac{r^{\alpha\beta}_i R^{\alpha\beta}_J}{r^{\alpha\beta}}\right], \tag{8.71}$$

$$\hat{S}^{\alpha}_{IJ} = \frac{1}{V^{\alpha}_0}\left[-\frac{\Pi^{\alpha}_I \Pi^{\alpha}_J}{m^{\alpha}} + \frac{1}{2}\sum_{\substack{\beta \\ \beta\neq\alpha}} \varphi^{\alpha\beta}\frac{R^{\alpha\beta}_I R^{\alpha\beta}_J}{r^{\alpha\beta}}\right], \tag{8.72}$$

where V^{α}_0 is the volume of atom α in the reference configuration ($\sum_{\alpha} V^{\alpha}_0 = V_0$). From the atomic-level stresses, instantaneous expressions for the total stress can be defined:

$$\boldsymbol{\sigma}^{\text{inst}} = \frac{1}{V}\sum_{\alpha} V^{\alpha}\hat{\boldsymbol{\sigma}}^{\alpha}, \qquad \boldsymbol{P}^{\text{inst}} = \frac{1}{V_0}\sum_{\alpha} V^{\alpha}_0\hat{\boldsymbol{P}}^{\alpha}, \qquad \boldsymbol{S}^{\text{inst}} = \frac{1}{V_0}\sum_{\alpha} V^{\alpha}_0\hat{\boldsymbol{S}}^{\alpha}. \tag{8.73}$$

The atomic-level and instantaneous stress tensors can provide useful qualitative information and give zero tractions near a free surface in line with continuum expectations [CY91], but it is important to realize that they are not theoretically motivated as macroscopic stress measures. The instantaneous stress tensors are related to stress in the same way that the instantaneous kinetic energy is related to temperature. Thus,

$$\boldsymbol{\sigma} = \langle\boldsymbol{\sigma}^{\text{inst}}\rangle, \qquad \boldsymbol{P} = \langle\boldsymbol{P}^{\text{inst}}\rangle, \qquad \boldsymbol{S} = \langle\boldsymbol{S}^{\text{inst}}\rangle. \tag{8.74}$$

The instantaneous expressions are used in the elasticity tensor derivation in Section 8.1.4.

Physical significance of the kinetic stress The presence of a kinetic contribution to the stress tensor appears at odds with the continuum definition of stress that is stated solely in terms of the forces acting between different parts of the body (see Section 2.3.2). This discrepancy has led to controversy in the past as to whether the kinetic term belongs in the stress definition (see, for example, [Zho03]). The confusion is related to the difference between absolute and relative velocity. (Recall that the phase space variables in statistical mechanics are assumed to be center-of-mass coordinates as explained in Section 7.1.) The kinetic stress reflects the momentum flux associated with the vibrational kinetic energy portion of the internal energy (see [AT11]). This is clearly demonstrated in Section 8.3.3, where a microscopic definition for the traction due to Tsai (and originating in the work of Cauchy) is discussed. See also the discussion of the recovery of kinetic terms in the Cauchy stress tensor derived for solid systems on page 574 and footnote 18 on the same page.

[13] We can define the volume of an atom as the Voronoi cell associated with it. See Section 3.3.2.

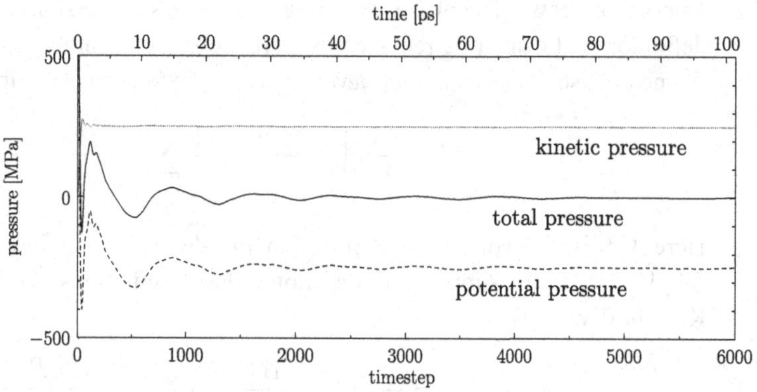

Fig. 8.3 Pressure as a function of time in an isolated cube of aluminum. The total pressure is the sum of the potential and kinetic pressures. The values plotted are the time averages from the start of a constant energy MD simulation versus the number of MD timesteps (or actual time in picoseconds as shown on top).

The need for the kinetic term becomes apparent when considering an ideal gas, where the potential interaction term is zero by definition and therefore it is the kinetic term that is wholly responsible for the transmission of pressure. To demonstrate that the kinetic term in the stress tensor does indeed exist and plays an important role in solids as well, consider the following example of an equilibrium MD simulation of a cube consisting of 4000 aluminum atoms in an fcc arrangement floating freely in a vacuum. The atoms interact according to a pair functional [EA94]. The initial positions of the atoms are randomly perturbed by a small amount relative to their zero-temperature equilibrium positions and the system is evolved by integrating the equations of motion. The initial perturbations are selected so that the temperature of the cube is 300 K. Since the block is unconstrained, we expect the stress in the box and consequently the pressure to be zero, $p = -\frac{1}{3}\,\mathrm{tr}\,\boldsymbol{\sigma} = 0$. The virial expression for the pressure follows from Eqn. (8.68) as[14]

$$p = \frac{1}{3N_\tau V}\sum_{i=1}^{N_\tau}\left[\sum_{\alpha}\frac{\left\|\boldsymbol{p}^{\alpha(i)}\right\|^2}{m^\alpha} - \frac{1}{2}\sum_{\substack{\alpha,\beta\\ \alpha\neq\beta}}\varphi^{\alpha\beta(i)}r^{\alpha\beta(i)}\right],$$

where the phase average has been replaced with a time average over the molecular dynamics trajectory. Here N_τ is the number of timesteps and a superscript (i) refers to the value of a given variable at timestep i. Figure 8.3 shows the results from this simulation. The three curves are the potential and kinetic parts of the pressure and the total pressure as a function of time. As expected the total pressure tends to zero as the system equilibrates. However, the potential and kinetic parts of the pressure are *not* zero, rather they converge to values that are equal and opposite, such that their sum is zero. This clearly shows that the kinetic term belongs in the stress tensor. This example also demonstrates the interesting and perhaps surprising fact that the kinetic part of the stress is not insignificant even when considering solid systems. This can be quantified by using the equipartition theorem to obtain an explicit

[14] Note that the sign convention for pressure is opposite to that of stress, i.e. a positive pressure corresponds to a negative compressive stress.

expression for the kinetic part of the stress. Equation (7.103) can be written in the following way

$$\sum_{\alpha} \frac{\|\boldsymbol{p}^{\alpha}\|^2}{m^{\alpha}} = 3Nk_{\mathrm{B}}T.$$

Therefore, the pressure expression is

$$p = \frac{1}{V}\left[Nk_{\mathrm{B}}T - \frac{1}{6N_{\tau}} \sum_{i=1}^{N_{\tau}} \sum_{\substack{\alpha,\beta \\ \alpha \neq \beta}} \varphi^{\alpha\beta(i)} r^{\alpha\beta(i)} \right].$$

For example at 300 K, $k_{\mathrm{B}}T = 0.02585$ eV, and a typical volume per atom in a solid is $V/N = 10 - 20$ $\mathring{\mathrm{A}}^3$. So the kinetic pressure term is 1.3–2.6 meV/$\mathring{\mathrm{A}}^3$, which translates to 200–400 MPa. The result from the simulation falls in this range with a kinetic pressure of 249 MPa, a sizable stress.

Another interesting example is an infinite system with PBCs applied to a supercell of fixed volume as the temperature is increased by ΔT. The rising temperature would normally cause the material to thermally expand, but due to the volume constraint that prevents this, a compressive stress is set up instead. It is instructive to consider how this problem is solved in elementary mechanics. The total small strain tensor, $\boldsymbol{\epsilon}$, is written as a sum of the strain due to the elastic stress and strain due to thermal expansion:

$$\boldsymbol{\epsilon} = \boldsymbol{s} : \boldsymbol{\sigma} + \boldsymbol{I}\alpha\Delta T, \tag{8.75}$$

where $\boldsymbol{s}$ is the elastic compliance tensor, α is the coefficient of thermal expansion and $\boldsymbol{I}$ is the identity tensor. Since the periodic supercell cannot change its size, $\boldsymbol{\epsilon} = \boldsymbol{0}$. Inverting this relation for an fcc crystal (which has cubic symmetry) oriented along the crystallographic axes, we have

$$\sigma_{11} = \sigma_{22} = \sigma_{33} = -(\mathrm{c}_{11} + 2\mathrm{c}_{12})\alpha\Delta T, \tag{8.76}$$

where c_{ij} are the entries of the elasticity matrix in Voigt notation (see Section 2.5.5). The rest of the stress components are zero. The pressure is then $p = -\sigma_{kk}/3 = (\mathrm{c}_{11} + 2\mathrm{c}_{12})\alpha\Delta T$. It is exactly the equality between the potential and kinetic parts of the stress that determines the amount of thermal expansion the system will experience and in this case the amount of compressive stress. For example, consider a periodic aluminum system constrained to constant size as its temperature is increased. The aluminum is modeled using the Ercolessi–Adams pair functional [EA94] with[15] $\mathrm{c}_{11} = 118.1$ GPa and $\mathrm{c}_{12} = 62.3$ GPa. Figure 8.4 shows the total pressure as well as the kinetic and potential parts as a function of the temperature. As expected from Eqn. (8.76), the stress is linearly proportional to the change in temperature. The value for α measured from the σ versus T curve is 1.57×10^{-5} K^{-1}, which is very close to the reported value of 1.6×10^{-5} K^{-1} for this model [EA94]. (Note that the experimental value is about 2.4×10^{-5} K^{-1}, which reflects a limitation of the atomistic model used in the simulation.) The kinetic part of the pressure is about 25% of the total pressure over the entire temperature range. This is a significant fraction, demonstrating that

[15] Strictly speaking, we should be accounting for the temperature dependence of the elastic constants. However, as shown in Section 8.1.4, the dependence is weak. We therefore use the 0 K constants.

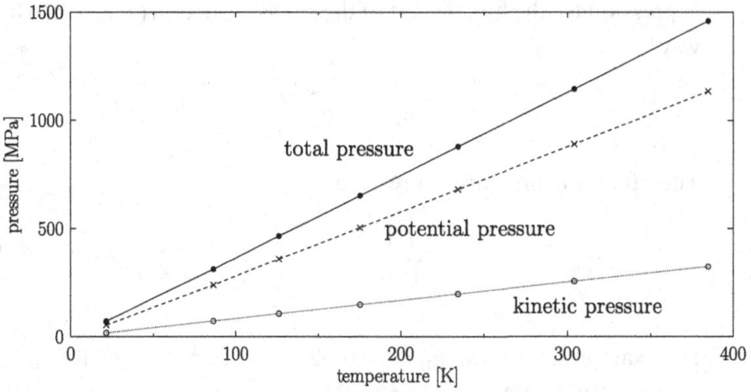

Fig. 8.4 The pressure as a function of temperature in an aluminum system with PBCs consisting of 4000 atoms modeled using a pair functional [EA94]. The size of the periodic box is set according to the zero temperature lattice parameter. As the temperature of the system is increased a compressive stress (positive pressure) builds up within it. The dots are constant energy MD time averages over 20 000 timesteps. (The standard deviation is comparable to the size of the symbols.)

the kinetic stress plays an important role in the total stress of solid systems. The issue of stresses due to constrained thermal expansion is discussed further in Section 8.3.3.

8.1.4 Microscopic elasticity tensor

In the previous section, we derived expressions for the stress tensor by taking derivatives of the free energy with respect to the deformation gradient $\boldsymbol{F}$. This procedure can be applied again to obtain the elasticity tensor from the stress tensor.[16] It is important to point out that depending on the loading conditions there are two different elasticity tensors. The *adiabatic* elasticity tensor is derived from the internal energy and is associated with rapid loading of the material (where the material does not have sufficient time to exchange heat internally or with its surroundings). The *isothermal* elasticity tensor is derived from the Helmholtz free energy and is associated with slow loading where the material has time to reach thermal equilibrium with the surrounding heat bath.[17] At $T = 0$ K the two tensors are equal, but the discrepancy increases with temperature [SHH69]. In our case, the elasticity tensor is computed from the Helmholtz free energy and is therefore isothermal.

The mixed elasticity tensor $\boldsymbol{D}$, defined in Eqn. (2.190), is given by

$$D_{iJkL} = \frac{\partial P_{iJ}}{\partial F_{kL}} = \frac{1}{V_0}\frac{\partial^2 \Psi}{\partial F_{iJ} F_{kL}} = \frac{1}{V_0}\frac{\partial}{\partial F_{kL}}\left\langle \frac{\partial \widehat{\mathcal{H}}}{\partial F_{iJ}} \right\rangle, \tag{8.77}$$

[16] An early derivation along the same lines for the bulk modulus is given by Siegert and Meeron [SM66]. Several years later, Squire *et al.* [SHH69] obtained the elastic constants for a cubic crystal including the "fluctuation terms, omitted by Born, which make important contributions at nonzero temperatures." More recently, Lutsko [Lut89] derived general expressions for the elastic constants equivalent to the ones given here. However, since we adopt a standard continuum mechanics framework for the derivation, the identity of the various referential and spatial elasticity tensors is made clear.

[17] See page 95 for a discussion of adiabatic and isothermal conditions.

where we have used Eqn. (8.53). Explicitly writing out the phase average, taking the derivative with respect to $\boldsymbol{F}$ and using $\partial Z/\partial \boldsymbol{F}$ in Eqn. (8.54), we obtain

$$D_{iJkL} = \frac{1}{V_0}\left\{\left\langle \frac{\partial^2 \widehat{\mathcal{H}}}{\partial F_{iJ}\partial F_{kL}}\right\rangle - \frac{1}{k_B T}\left[\left\langle \frac{\partial \widehat{\mathcal{H}}}{\partial F_{ij}}\frac{\partial \widehat{\mathcal{H}}}{\partial F_{kL}}\right\rangle - \left\langle \frac{\partial \widehat{\mathcal{H}}}{\partial F_{ij}}\right\rangle\left\langle \frac{\partial \widehat{\mathcal{H}}}{\partial F_{kL}}\right\rangle\right]\right\}. \tag{8.78}$$

We already have an expression for $\partial\widehat{\mathcal{H}}/\partial \boldsymbol{F}$ in Eqn. (8.65). To compute the second derivative, we will use the following identities:

$$\frac{\partial r^{\alpha\beta}}{\partial F_{kL}} = \frac{r_k^{\alpha\beta} R_L^{\alpha\beta}}{r^{\alpha\beta}}, \qquad \frac{\partial \Pi_J^\alpha}{\partial F_{kL}} = -\left(F_{Jk}^{-1} p_i^\alpha + F_{Ji}^{-1} p_k^\alpha\right) F_{Li}^{-1}, \tag{8.79}$$

where Eqn. (8.58) was used to obtain Eqn. $(8.79)_2$. The second derivative follows after some algebra as

$$\frac{\partial^2 \widehat{\mathcal{H}}}{\partial F_{iJ}\partial F_{kL}} = \sum_\alpha \left\{ \frac{1}{m^\alpha}\left(F_{Li}^{-1} p_k^\alpha \Pi_J^\alpha + F_{Jk}^{-1} p_i^\alpha \Pi_L^\alpha + F_{Jj}^{-1} F_{Lj}^{-1} p_i^\alpha p_k^\alpha\right) \right.$$
$$\left. + \sum_{\substack{\beta \\ \beta\neq\alpha}} \left(\frac{1}{2}\varphi^{\alpha\beta}\left[\delta_{ik} - \frac{r_i^{\alpha\beta} r_k^{\alpha\beta}}{(r^{\alpha\beta})^2}\right]\frac{R_J^{\alpha\beta} R_L^{\alpha\beta}}{r^{\alpha\beta}} + \frac{1}{4}\sum_{\substack{\gamma,\delta \\ \gamma\neq\delta}}\left[\kappa^{\alpha\beta\gamma\delta}\frac{r_i^{\alpha\beta} R_J^{\alpha\beta} r_k^{\gamma\delta} R_L^{\gamma\delta}}{r^{\alpha\beta} r^{\gamma\delta}}\right]\right)\right\}, \tag{8.80}$$

where $\kappa^{\alpha\beta\gamma\delta}$ is the *bond stiffness* defined in Eqn. (5.117). The mixed elasticity tensor is related to the spatial stiffness tensor $\mathbf{c}$ through Eqn. $(2.195)_2$. Inverting this relation gives

$$c_{ijkl} = \frac{1}{\det \boldsymbol{F}} F_{jJ} F_{lL} D_{iJkL} - \delta_{ik}\sigma_{jl}. \tag{8.81}$$

Substituting into this the expression for $\boldsymbol{D}$ in Eqn. (8.78), together with the derivatives in Eqns. (8.65) and (8.80), as well as the Cauchy stress in Eqn. (8.68), gives after simplification the following expression for the spatial elasticity tensor:

$$c_{ijkl} = \frac{1}{V}\left[\sum_\alpha \left\langle \frac{1}{m^\alpha}\left(\delta_{il} p_j^\alpha p_k^\alpha + \delta_{jk} p_i^\alpha p_l^\alpha + \delta_{jl} p_i^\alpha p_k^\alpha + \delta_{ik} p_j^\alpha p_l^\alpha\right)\right\rangle \right.$$
$$\left. + \left\langle c_{ijkl}^0\right\rangle - \frac{V^2}{k_B T}\operatorname{Cov}(\sigma_{ij}^{\text{inst}}, \sigma_{kl}^{\text{inst}})\right], \tag{8.82}$$

where $\boldsymbol{\sigma}^{\text{inst}}$ is defined in Eqn. (8.73), $\operatorname{Cov}(\cdot,\cdot)$ is the covariance operator defined by

$$\operatorname{Cov}(A, B) \equiv \langle AB\rangle - \langle A\rangle\langle B\rangle \tag{8.83}$$

and

$$c_{ijkl}^0 = \frac{1}{V}\left[\frac{1}{4}\sum_{\substack{\alpha,\beta \\ \alpha\neq\beta}}\sum_{\substack{\gamma,\delta \\ \gamma\neq\delta}} \kappa^{\alpha\beta\gamma\delta}\frac{r_i^{\alpha\beta} r_j^{\alpha\beta} r_k^{\gamma\delta} r_l^{\gamma\delta}}{r^{\alpha\beta} r^{\gamma\delta}} - \frac{1}{2}\sum_{\substack{\alpha,\beta \\ \alpha\neq\beta}} \varphi^{\alpha\beta}\frac{r_i^{\alpha\beta} r_j^{\alpha\beta} r_k^{\alpha\beta} r_l^{\alpha\beta}}{(r^{\alpha\beta})^3}\right] \tag{8.84}$$

is the zero-temperature expression for the elasticity originally given by Born [BH54]. (We will see below that as $T \to 0\ K$, the elasticity tensor reduces to $\mathbf{c}^0$.) At finite temperature, additional kinetic terms are present in the elasticity tensor in Eqn. (8.82). Note in particular, the presence of the fluctuation (covariance) term of the instantaneous stress components. This term reflects changes in probabilities of states available to the system due to the deformation [Hoo86, Section II.C].

Direct calculation of the momentum terms The kinetic terms in Eqn. (8.82) can be directly evaluated. The first term, with the four Kronecker deltas, can be computed using the equipartition theorem (Eqn. (7.138)),

$$\sum_{\alpha}\left\langle \frac{1}{m^\alpha}\left(\delta_{il}p_j^\alpha p_k^\alpha + \delta_{jk}p_i^\alpha p_l^\alpha + \delta_{jl}p_i^\alpha p_k^\alpha + \delta_{ik}p_j^\alpha p_l^\alpha\right)\right\rangle$$
$$= Nk_{\mathrm{B}}T(\delta_{il}\delta_{jk} + \delta_{jk}\delta_{il} + \delta_{jl}\delta_{ik} + \delta_{ik}\delta_{jl}) = 2Nk_{\mathrm{B}}T(\delta_{il}\delta_{jk} + \delta_{jl}\delta_{ik}). \tag{8.85}$$

This gives the conventional expression for the elasticity tensor,

$$c_{ijkl} = \frac{1}{V}\left[2Nk_{\mathrm{B}}T(\delta_{il}\delta_{jk} + \delta_{jl}\delta_{ik}) + \langle c^0_{ijkl}\rangle - \frac{V^2}{k_{\mathrm{B}}T}\,\mathrm{Cov}(\sigma_{ij}^{\mathrm{inst}}, \sigma_{kl}^{\mathrm{inst}})\right]. \tag{8.86}$$

Equation (8.86) can be further simplified by splitting the instantaneous stress terms into kinetic and potential parts,

$$\sigma_{ij}^{\mathrm{K,inst}} = -\frac{1}{V}\sum_{\alpha}\frac{p_i^\alpha p_j^\alpha}{m^\alpha}, \qquad \sigma_{ij}^{\mathrm{V,inst}} = \frac{1}{2V}\sum_{\substack{\alpha,\beta\\ \alpha\neq\beta}}\varphi^{\alpha\beta}\frac{r_i^{\alpha\beta}r_j^{\alpha\beta}}{r^{\alpha\beta}}. \tag{8.87}$$

Substituting $\boldsymbol{\sigma}^{\mathrm{inst}} = \boldsymbol{\sigma}^{\mathrm{K,inst}} + \boldsymbol{\sigma}^{\mathrm{V,inst}}$ into the third term (see the definition of covariance in Eqn. (8.83)), and noting that the cross-terms cancel, since

$$\left\langle \sigma_{ij}^{\mathrm{K,inst}}\sigma_{ij}^{\mathrm{V,inst}}\right\rangle = \left\langle \sigma_{ij}^{\mathrm{K,inst}}\right\rangle\left\langle \sigma_{ij}^{\mathrm{V,inst}}\right\rangle,$$

we have

$$\mathrm{Cov}(\sigma_{ij}^{\mathrm{inst}}, \sigma_{kl}^{\mathrm{inst}}) = \mathrm{Cov}(\sigma_{ij}^{\mathrm{K,inst}}, \sigma_{kl}^{\mathrm{K,inst}}) + \mathrm{Cov}(\sigma_{ij}^{\mathrm{V,inst}}, \sigma_{kl}^{\mathrm{V,inst}}). \tag{8.88}$$

Let us focus on the kinetic terms:

$$\mathrm{Cov}(\sigma_{ij}^{\mathrm{K,inst}}, \sigma_{kl}^{\mathrm{K,inst}}) = \left\langle \sum_{\alpha}\sum_{\beta}\frac{p_i^\alpha p_j^\alpha p_k^\beta p_l^\beta}{m^\alpha m^\beta}\right\rangle - \left\langle \sum_{\alpha}\frac{p_i^\alpha p_j^\alpha}{m^\alpha}\right\rangle\left\langle \sum_{\beta}\frac{p_k^\beta p_l^\beta}{m^\beta}\right\rangle$$
$$= \left\langle \sum_{\alpha}\frac{p_i^\alpha p_j^\alpha p_k^\alpha p_l^\alpha}{(m^\alpha)^2}\right\rangle + \left\langle \sum_{\substack{\alpha,\beta\\ \alpha\neq\beta}}\left(\frac{p_i^\alpha p_j^\alpha}{m^\alpha}\right)\left(\frac{p_k^\beta p_l^\beta}{m^\beta}\right)\right\rangle - \left\langle \sum_{\alpha}\frac{p_i^\alpha p_j^\alpha}{m^\alpha}\right\rangle\left\langle \sum_{\beta}\frac{p_k^\beta p_l^\beta}{m^\beta}\right\rangle$$
$$= (\delta_{ik}\delta_{jl} + \delta_{il}\delta_{jk} + \delta_{ij}\delta_{kl})N(k_{\mathrm{B}}T)^2 + \delta_{ij}\delta_{kl}N(N-1)(k_{\mathrm{B}}T)^2 - \delta_{ij}\delta_{kl}(Nk_{\mathrm{B}}T)^2$$
$$= (\delta_{ik}\delta_{jl} + \delta_{il}\delta_{jk})N(k_{\mathrm{B}}T)^2. \tag{8.89}$$

In the third line, we divide the first expression into two parts, one where $\alpha = \beta$ and the other where $\alpha \neq \beta$. In the fourth line, the equipartition theorem is used to evaluate the last two terms and the first term is evaluated by integration and consideration of the different possibilities (this appears to have been first shown by Lutsko [Lut89]).

Substituting Eqns. (8.85), (8.88) and (8.89) into Eqn. (8.82), gives the elasticity tensor in a simpler form:

$$c_{ijkl} = \frac{1}{V}\left[\langle c^0_{ijkl}\rangle - \frac{V^2}{k_B T}\mathrm{Cov}(\sigma^{\mathrm{V,inst}}_{ij}, \sigma^{\mathrm{V,inst}}_{kl}) + Nk_B T(\delta_{ik}\delta_{jl} + \delta_{il}\delta_{jk})\right]. \quad (8.90)$$

As $T \to 0\ K$, the last term clearly goes to zero and it is also possible to show that the fluctuation term vanishes by expanding the stress and potential terms [Lut89]. In this case, $\mathbf{c} = \mathbf{c}^0$, where $\mathbf{c}^0$ is given in Eqn. (8.84).

As pointed out in the introduction to this chapter, it is important to keep in mind that in the thermodynamic limit the elastic constants associated with shear vanish. This is shown, for example, by Bavaud *et al.* [BCF86]. Nevertheless, the above expressions can be used to compute the elastic constants of solids (including shear moduli) by replacing the phase averages with time averages. This is demonstrated in the example below where we compute the elastic constants of argon and explore the magnitudes of the various terms in Eqn. (8.86). A more formally correct treatment for solids is discussed in Chapter 11.

Example 8.2 (Elastic constants of argon: a numerical example and comparison with experiment) We saw above that the expression for the elasticity tensor consists of three parts: a kinetic term, the zero-temperature (Born) expression and a stress fluctuation term. What is the relative contribution of each of these terms to the total elasticity? To explore this and to see the dependence of the elastic constants on temperature, we performed zero-pressure, constant temperature MD simulations[18] and computed the elasticity tensor of solid argon at various temperatures using Eqn. (8.86). At sufficiently low temperatures argon crystallizes into a cubic fcc structure. The elasticity tensor therefore has three independent elastic constants: c_{11}, c_{12} and c_{44} in Voigt notation (see Section 2.5.5). We chose argon because the simple nature of bonding in this material (short-range van der Waals forces) means it is well described by the Lennard-Jones pair potential (see Eqn. (5.36)).[19] We use Lennard-Jones parameters ($\epsilon = 0.01025$ eV and $\sigma = 3.4$ Å) proposed by Horton and Leech [HL63], who performed an extensive study of the optimal parameter choice for solid argon. These parameters were also used by Squire *et al.* [SHH69] to compute the elastic constants of argon at several temperatures using Monte Carlo simulations. A cutoff radius of $r_{\mathrm{cut}} = 15$ Å was used, with the smoothing function of Eqn. (5.23) applied to second order ($m = 2$). In the fcc ground state, this gives a lattice parameter of $a = 5.247$ Å that includes atoms up to sixteenth neighbor distance.

The results are presented in Fig. 8.5. The dependence of the elastic constants on temperature is plotted in Fig. 8.5(a). The elastic constants decrease linearly with temperature in agreement with the

[18] The constant temperature and pressure were enforced using the NσT ensemble of MD described in Section 9.6. In particular, a Nosé–Hoover thermostat was used to maintain the temperature (see Section 9.4.4).

[19] The noble gas solids are the ideal case for the Lennard-Jones potential (see Section 5.7.3).

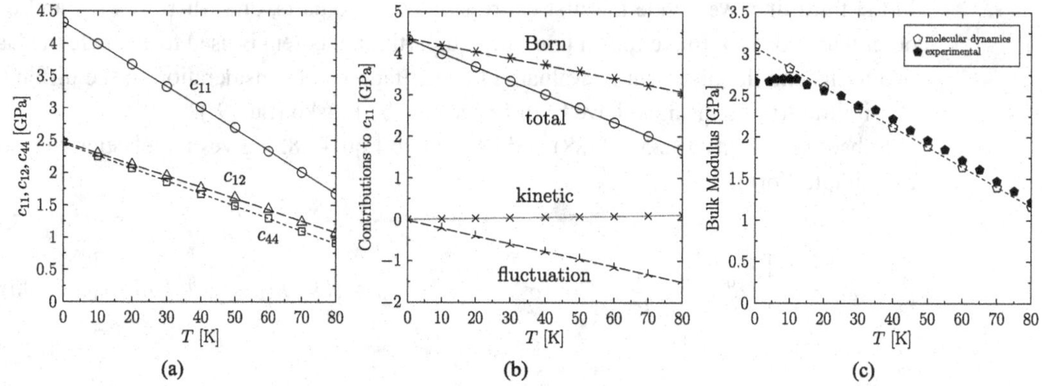

Fig. 8.5 Isothermal elastic constants of argon at zero pressure as a function of temperature. (a) Molecular dynamics results obtained using the NσT ensemble for a Lennard-Jones potential (see text for details). The simulation cell contained 864 atoms. The results are time averages over 3000 ps after a 10 ps equilibration time. (b) The different contributions to c_{11} as a function of temperature. The results were obtained from the MD calculations described above. (c) The bulk modulus computed from the MD results versus the experimental results of Peterson *et al.* [PBS66].

predictions of the quasiharmonic approximation for the elasticity tensor discussed further in Section 11.4.3. The constants c_{12} and c_{44} start off equal at $T = 0$ K (as a consequence of the fact that pair potentials satisfy the Cauchy relations as explained in Section 11.5.3) and then separate somewhat due to thermal fluctuations.

Figure 8.5(b) shows the contributions of the three terms (kinetic, Born and fluctuation), defined in Eqn. (8.86), to the total value of c_{11}. (The relative contributions of the separate terms in the other two elastic constants are similar and therefore not displayed.) It is clear that the contribution of the kinetic term is quite small; however, the contribution from the fluctuation term, $-(V/k_B T)\,\mathrm{Cov}(\sigma_{ij}^{\mathrm{inst}}, \sigma_{kl}^{\mathrm{inst}})$, becomes substantial with increasing temperature. At 70 K (which is approaching the melting temperature of argon of 83.8 K), the magnitude of the fluctuation term is 40% of the Born term.

Figure 8.5(c) shows a comparison of the MD and experimental results for the isothermal bulk modulus, $B = (c_{11} + 2c_{12})/3$. (See Exercise 2.15 for a derivation of the bulk modulus.) Although argon is relatively simple from a theoretical perspective, obtaining experimental values for its elastic constants is quite difficult. First, the experiments must be performed at cryogenic temperatures given the melting temperature of argon. Second, argon crystals are extremely brittle at low temperatures and have a mushy consistency close to the melting temperature [SHH69]. Therefore rather than direct mechanical testing, indirect techniques, such as ultrasonic measurements and scattering experiments, are used to infer the elastic constants. There is a great deal of scatter in the available results, and it appears that the elastic constants of argon are still not known with any certainty. It is also important to point out that ultrasonic and scattering techniques yield the *adiabatic* elastic constants, since they represent rapid loading of the material, whereas the expression for the elasticity tensor derived above is isothermal (as explained at the start of this section). The only set of experiments that we could find for the isothermal constants is that of Peterson *et al.* from 1966 [PBS66]. These authors report the isothermal bulk modulus of argon as a function of temperature. The compressibility of the material was measured by applying hydrostatic pressure via surrounding liquid helium and measuring the change in lattice parameter by means of X-ray diffraction. The MD results are compared with the experimental measurements of Peterson *et al.* in Fig. 8.5(c), showing good agreement except near 0 K, where the experimental results level off due to quantum effects (see [Wei83, Section 10.3] for a discussion of quantum effects in elastic constants).[20]

In the next section, we turn to the treatment of systems under *nonequilibrium* conditions. We obtain an expression for the stress tensor and show that under equilibrium conditions, in the thermodynamic limit, it converges to the virial stress obtained above.

8.2 Continuum fields as expectation values: nonequilibrium systems

From the mid 1940s to early 1950s, John Gamble Kirkwood and coworkers published a series of important papers on the fundamentals of nonequilibrium statistical mechanics and the application of this theory to transport processes.[21] One paper in particular, published in 1950 with Jack Irving[22] [IK50], laid the foundation for the microscopic definition of continuum measures. This work was extended several years later by Walter Noll [Nol55], who showed how closed-form analytical solutions can be obtained for some expressions, such as the stress tensor, that were solved only approximately by Irving and Kirkwood. The derivation we provide below combines these approaches into a unified framework [AT10].

The derivation begins with the assumption that continuum fields $\mathcal{A}(\boldsymbol{x};t)$ can be defined as expectation values of certain phase functions $A(\boldsymbol{r},\boldsymbol{v};\boldsymbol{x})$:[23]

$$\mathcal{A}(\boldsymbol{x};t) = \iint A(\boldsymbol{r},\boldsymbol{v};\boldsymbol{x}) f(\boldsymbol{r},\boldsymbol{v};t)\, d\boldsymbol{r} d\boldsymbol{v} \equiv \langle A; f\rangle\,, \tag{8.91}$$

where $f(\boldsymbol{r},\boldsymbol{v};t)$ is an appropriate distribution function and $\boldsymbol{r}$ and $\boldsymbol{v}$ are the concatenated positions, $\boldsymbol{r}^\alpha$, and velocities, $\boldsymbol{v}^\alpha$, of the N atoms in the system ($\alpha = 1,\dots,N$):

$$\boldsymbol{r} = (\boldsymbol{r}^1,\dots,\boldsymbol{r}^N), \qquad \boldsymbol{v} = (\boldsymbol{v}^1,\dots,\boldsymbol{v}^N).$$

[20] The large deviation at zero temperature creates a problem when fitting interatomic potentials. If the potential is to be used for zero-temperature simulations it should be fitted to the zero-temperature experimental elastic constants. However, if the potential is to be used at finite temperature, it would be better to match the linear portions of elastic constant versus temperature curves at higher temperature, which would mean fitting the empirical potential to the values of the elastic constants at 0 K obtained by extrapolating the experimental curve back to zero along the linear portion of the curve. See Chapter 5 for further discussion of empirical interatomic models and the way in which they are fit.

[21] We limit ourselves to nonequilibrium systems that remain close to equilibrium as explained in Section 2.4.6. This means that we accept the postulate of local thermodynamic equilibrium. For more on this, see [Kir46, dGM62, EM90, LJCV08].

[22] Jack Howard Irving was a physics graduate student at Princeton University who interrupted his studies to go Caltech to work with Kirkwood on the development of innovative statistical mechanics techniques. Instead of returning to Princeton, he accepted a position at Hughes Aircraft Company and remained in the aerospace industry. He returned to Princeton in 1963 on sabbatical to complete his PhD. studies [Irv08].

[23] In the standard statistical mechanics formulation, the integration is performed with respect to generalized positions and momenta, $\boldsymbol{q}$ and $\boldsymbol{p}$. Here, we choose to work with absolute positions and velocities. The phase average expression retains the same form due to the normalization of the distribution function.

Note that $(\boldsymbol{r}, \boldsymbol{v})$ represent the *absolute* positions and velocities of the atoms, whereas in the previous section and in Chapter 7, which dealt with equilibrium statistical mechanics, $(\boldsymbol{q}, \boldsymbol{p})$ represent center-of-mass coordinates.

The idea is to choose the functions $A(\boldsymbol{r}, \boldsymbol{v}; \boldsymbol{x})$ in Eqn. (8.91) in such a way that the resulting continuum fields identically satisfy the balance and conservation equations of continuum mechanics and thermodynamics. In doing so, the microscopic definitions for density, stress, heat flux and energy density are revealed. Since the balance and conservation laws involve time derivatives it will be necessary to compute expressions of the form:

$$\frac{\partial \mathcal{A}}{\partial t} = \frac{\partial}{\partial t} \langle A; f \rangle .$$

The key to the Irving–Kirkwood–Noll derivation is an identity derived next from Liouville's theorem that shows that the time derivative of the expectation value of a phase function is equal to the expectation value of a different phase function related to it.

8.2.1 Rate of change of expectation values

Consider a phase function $A(\boldsymbol{r}, \boldsymbol{v})$. Assuming that A does not have a direct dependence on time, it is apparent from Eqn. (7.28) that the time derivative of its expectation value is

$$\frac{\partial}{\partial t} \langle A; f \rangle = \left\langle A; \frac{\partial f}{\partial t} \right\rangle . \tag{8.92}$$

The rate of change of the distribution function, $\partial f / \partial t$, is given by Liouville's equation (Eqn. (7.31)). We rewrite this expression in terms of atomic positions and velocities:

$$\frac{\partial f}{\partial t} = - \sum_{\alpha} \left[\dot{\boldsymbol{r}}^{\alpha} \cdot \frac{\partial f}{\partial \boldsymbol{r}^{\alpha}} + \dot{\boldsymbol{v}}^{\alpha} \cdot \frac{\partial f}{\partial \boldsymbol{v}^{\alpha}} \right] , \tag{8.93}$$

where here, and in all sums, α runs from 1 to N. Substituting in $\dot{\boldsymbol{r}}^{\alpha} = \boldsymbol{v}^{\alpha}$ and $m^{\alpha} \dot{\boldsymbol{v}}^{\alpha} = -\partial \mathcal{H} / \partial \boldsymbol{r}^{\alpha} = -\partial \mathcal{V} / \partial \boldsymbol{r}^{\alpha}$ (Eqn. (7.19)), we have

$$\frac{\partial f}{\partial t} = \sum_{\alpha} \left[-\boldsymbol{v}^{\alpha} \cdot \frac{\partial f}{\partial \boldsymbol{r}^{\alpha}} + \frac{1}{m^{\alpha}} \frac{\partial \mathcal{V}}{\partial \boldsymbol{r}^{\alpha}} \cdot \frac{\partial f}{\partial \boldsymbol{v}^{\alpha}} \right] . \tag{8.94}$$

Next, substituting Eqn. (8.94) into Eqn. (8.92) yields

$$\frac{\partial}{\partial t} \langle A; f \rangle = \sum_{\alpha} \left[\left\langle A; -\boldsymbol{v}^{\alpha} \cdot \frac{\partial f}{\partial \boldsymbol{r}^{\alpha}} \right\rangle + \left\langle A; \frac{1}{m^{\alpha}} \frac{\partial \mathcal{V}}{\partial \boldsymbol{r}^{\alpha}} \cdot \frac{\partial f}{\partial \boldsymbol{v}^{\alpha}} \right\rangle \right] . \tag{8.95}$$

The phase averages of A on the right-hand side are taken with respect to unknown distribution functions that involve the gradient of f and other functions. We would like to obtain equivalent expressions where the phase averages are taken with respect to f with A replaced with some appropriate function. Consider the first term in the square brackets:

$$\left\langle A; -\boldsymbol{v}^{\alpha} \cdot \frac{\partial f}{\partial \boldsymbol{r}^{\alpha}} \right\rangle = - \iint A(\boldsymbol{r}, \boldsymbol{v}) \left(\boldsymbol{v}^{\alpha} \cdot \frac{\partial f}{\partial \boldsymbol{r}^{\alpha}} \right) d\boldsymbol{r} d\boldsymbol{v} . \tag{8.96}$$

To simplify this expression, we use the following identity:

$$\boldsymbol{v}^{\alpha} \cdot \frac{\partial (Af)}{\partial \boldsymbol{r}^{\alpha}} = \left(\boldsymbol{v}^{\alpha} \cdot \frac{\partial A}{\partial \boldsymbol{r}^{\alpha}} \right) f + \left(\boldsymbol{v}^{\alpha} \cdot \frac{\partial f}{\partial \boldsymbol{r}^{\alpha}} \right) A . \tag{8.97}$$

Isolating the last term on the right, and substituting this back into Eqn. (8.96) gives

$$\left\langle A; -\boldsymbol{v}^\alpha \cdot \frac{\partial f}{\partial \boldsymbol{r}^\alpha} \right\rangle = -\int \boldsymbol{v}^\alpha \cdot \left[\int \frac{\partial}{\partial \boldsymbol{r}^\alpha}(Af)\, d\boldsymbol{r} \right] d\boldsymbol{v} + \left\langle \boldsymbol{v}^\alpha \cdot \frac{\partial A}{\partial \boldsymbol{r}^\alpha}; f \right\rangle. \tag{8.98}$$

The integral inside the square brackets on the right-hand side of Eqn. (8.98) can be carried out for each of the components of $\boldsymbol{r}^\alpha$,

$$\int \frac{\partial}{\partial r_j^\alpha}(Af)\, d\boldsymbol{r} = \int \cdots \int (Af) \bigg|_{r_j^\alpha=-\infty}^{r_j^\alpha=\infty} \cdots dr_1^{\alpha-1} dr_2^{\alpha-1} dr_3^{\alpha-1} dr_i^\alpha dr_k^\alpha dr_1^{\alpha+1} dr_2^{\alpha+1} dr_3^{\alpha+1} \cdots, \tag{8.99}$$

where on the right $i \neq j$ and $k \neq j$. If we assume that $Af \to 0$ as $r_j^\alpha \to \pm\infty$, then this integral is zero and Eqn. (8.98) becomes

$$\left\langle A; -\boldsymbol{v}^\alpha \cdot \frac{\partial f}{\partial \boldsymbol{r}^\alpha} \right\rangle = \left\langle \boldsymbol{v}^\alpha \cdot \frac{\partial A}{\partial \boldsymbol{r}^\alpha}; f \right\rangle. \tag{8.100}$$

Thus, the phase average of A on the left taken with respect to a complicated distribution function is replaced with the standard phase average of a more complex function. A similar procedure on the second term in the square brackets in Eqn. (8.95) leads to

$$\left\langle A; \frac{1}{m^\alpha} \frac{\partial \mathcal{V}}{\partial \boldsymbol{r}^\alpha} \cdot \frac{\partial f}{\partial \boldsymbol{v}^\alpha} \right\rangle = -\left\langle \frac{1}{m^\alpha} \frac{\partial \mathcal{V}}{\partial \boldsymbol{r}^\alpha} \cdot \frac{\partial A}{\partial \boldsymbol{v}^\alpha}; f \right\rangle. \tag{8.101}$$

Substituting Eqns. (8.100) and (8.101) into Eqn. (8.95) gives

$$\frac{\partial}{\partial t}\langle A; f \rangle = \sum_\alpha \left\langle \boldsymbol{v}^\alpha \cdot \frac{\partial A}{\partial \boldsymbol{r}^\alpha} - \frac{1}{m^\alpha} \frac{\partial \mathcal{V}}{\partial \boldsymbol{r}^\alpha} \cdot \frac{\partial A}{\partial \boldsymbol{v}^\alpha}; f \right\rangle. \tag{8.102}$$

This is quite an interesting result since it shows that the rate of change of the expectation value of phase function A is equal to the expectation value of a different phase function. Equation (8.102) can be readily extended to vector phase functions $\boldsymbol{a}(\boldsymbol{r}, \boldsymbol{v})$ by replacing A with the components a_j. The resulting expression is

$$\frac{\partial}{\partial t}\langle \boldsymbol{a}; f \rangle = \sum_\alpha \left\langle \left(\boldsymbol{v}^\alpha \cdot \frac{\partial}{\partial \boldsymbol{r}^\alpha} \right) \boldsymbol{a} - \left(\frac{1}{m^\alpha} \frac{\partial \mathcal{V}}{\partial \boldsymbol{r}^\alpha} \cdot \frac{\partial}{\partial \boldsymbol{v}^\alpha} \right) \boldsymbol{a}; f \right\rangle. \tag{8.103}$$

8.2.2 Definition of pointwise continuum fields

The classical model of matter is discrete with atoms occupying positions $\boldsymbol{r}^\alpha$ and possessing velocities $\boldsymbol{v}^\alpha$. Nevertheless, a continuous picture can emerge as an average over atomic motion. For example, the probability per unit volume of finding atom α at point $\boldsymbol{x}$ is

$$\Pr(\boldsymbol{r}^\alpha = \boldsymbol{x}) = \langle \delta(\boldsymbol{r}^\alpha - \boldsymbol{x}); f \rangle = \int_\Gamma \delta(\boldsymbol{r}^\alpha - \boldsymbol{x}) f(\boldsymbol{r}, \boldsymbol{v}; t)\, d\boldsymbol{r} d\boldsymbol{v},$$

where $f(\boldsymbol{r}, \boldsymbol{v}; t)$ is a time-dependent distribution function (see Section 7.2.2) and δ is the Dirac delta.[24] The expectation of the total mass per unit volume at point $\boldsymbol{x}$ is then

$$\sum_\alpha m^\alpha \left\langle \delta(\boldsymbol{r}^\alpha - \boldsymbol{x}); f \right\rangle.$$

Given this, we define the *pointwise* continuum mass density field at point $\boldsymbol{x}$ as

$$\rho^{\text{pt}}(\boldsymbol{x}; t) \equiv \sum_\alpha m^\alpha \left\langle \delta(\boldsymbol{r}^\alpha - \boldsymbol{x}); f \right\rangle. \tag{8.104}$$

The term "pointwise" stresses the fact that although $\rho^{\text{pt}}(\boldsymbol{x}; t)$ is a continuous field over $\boldsymbol{x}$, it does not directly correspond to the "continuum" density field, $\rho(\boldsymbol{x}; t)$, defined in Section 2.3.1. The reason for this is that the continuum formulation includes the concept of a "continuum particle" which represents some finite length scale ℓ on the microscopic scale as explained in Section 2.2.1. To connect $\rho^{\text{pt}}(\boldsymbol{x}; t)$ and $\rho(\boldsymbol{x}; t)$, it is necessary to spatially average $\rho^{\text{pt}}(\boldsymbol{x}; t)$ over a domain of size ℓ. This is done in Section 8.2.5.

Similarly to $\rho^{\text{pt}}(\boldsymbol{x}; t)$, other pointwise continuum fields can be defined. Consider the velocity field. It may seem that the most natural way to define a continuous velocity field from the discrete atom velocities is

$$\boldsymbol{v}^{\text{pt}}(\boldsymbol{x}; t) = \sum_\alpha \left\langle \boldsymbol{v}^\alpha \delta(\boldsymbol{r}^\alpha - \boldsymbol{x}); f \right\rangle \qquad \text{(wrong)}.$$

However, this is incorrect since the expression on the right has units of velocity per unit volume (due to the presence of the Dirac delta) not velocity. Instead, we define velocity from the requirement that the expectation value of the atomic momenta agrees with the continuum momentum density field which includes the definition of density given earlier:

$$\rho^{\text{pt}}(\boldsymbol{x}; t)\boldsymbol{v}^{\text{pt}}(\boldsymbol{x}; t) = \sum_\alpha \left\langle m^\alpha \boldsymbol{v}^\alpha \delta(\boldsymbol{r}^\alpha - \boldsymbol{x}); f \right\rangle, \tag{8.105}$$

so that

$$\boldsymbol{v}^{\text{pt}}(\boldsymbol{x}; t) \equiv \frac{1}{\rho^{\text{pt}}(\boldsymbol{x}; t)} \sum_\alpha m^\alpha \left\langle \boldsymbol{v}^\alpha \delta(\boldsymbol{r}^\alpha - \boldsymbol{x}); f \right\rangle. \tag{8.106}$$

This definition identifies the pointwise continuum velocity with the velocity of the center of mass of the sampled atoms.

[24] The introduction of the Dirac delta here is just a convenience. It is conveying the idea that the integration is done over all other degrees of freedom, while the position of atom α is fixed to $\boldsymbol{x}$:

$$\langle A(\boldsymbol{r}, \boldsymbol{v})\delta(\boldsymbol{r}^\alpha - \boldsymbol{x}); f \rangle = \int_\Gamma A(\boldsymbol{r}^1, \dots, \boldsymbol{r}^{\alpha-1}, \boldsymbol{x}, \boldsymbol{r}^{\alpha+1}, \dots, \boldsymbol{r}^N, \boldsymbol{v}^1, \dots, \boldsymbol{v}^N)$$
$$\times f(\boldsymbol{r}^1, \dots, \boldsymbol{r}^{\alpha-1}, \boldsymbol{x}, \boldsymbol{r}^{\alpha+1}, \dots, \boldsymbol{r}^N, \boldsymbol{v}^1, \dots, \boldsymbol{v}^N)\, d\boldsymbol{r}^1 \cdots d\boldsymbol{r}^{\alpha-1} d\boldsymbol{r}^{\alpha+1} \cdots d\boldsymbol{r}^N d\boldsymbol{v}^1 \cdots d\boldsymbol{v}^N.$$

8.2.3 Continuity equation

The pointwise continuum fields defined above appear plausible, but do they satisfy the continuum balance equations? Let us begin by inspecting the mass balance or continuity equation. The rate of change of density follows from Eqn. (8.104) as

$$\frac{\partial}{\partial t}\rho^{\mathrm{pt}}(\boldsymbol{x};t) = \sum_{\alpha} m^{\alpha} \frac{\partial}{\partial t} \langle \delta(\boldsymbol{r}^{\alpha} - \boldsymbol{x}); f \rangle .$$

Using Eqn. (8.102) this becomes

$$\frac{\partial}{\partial t}\rho^{\mathrm{pt}}(\boldsymbol{x};t) = \sum_{\alpha} m^{\alpha} \sum_{\beta} \left\langle \boldsymbol{v}^{\beta} \cdot \frac{\partial \delta(\boldsymbol{r}^{\alpha} - \boldsymbol{x})}{\partial \boldsymbol{r}^{\beta}} - \frac{1}{m^{\beta}} \frac{\partial \mathcal{V}}{\partial \boldsymbol{r}^{\beta}} \cdot \frac{\partial \delta(\boldsymbol{r}^{\alpha} - \boldsymbol{x})}{\partial \boldsymbol{v}^{\beta}}; f \right\rangle .$$

The second term in the phase average cancels since $\delta(\boldsymbol{r}^{\alpha} - \boldsymbol{x})$ is not a function of $\boldsymbol{v}^{\beta}$, and in the first term $\partial \delta(\boldsymbol{r}^{\alpha} - \boldsymbol{x})/\partial \boldsymbol{r}^{\beta}$ is only nonzero when $\beta = \alpha$, therefore[25]

$$\frac{\partial}{\partial t}\rho^{\mathrm{pt}}(\boldsymbol{x};t) = \sum_{\alpha} m^{\alpha} \left\langle \boldsymbol{v}^{\alpha} \cdot \frac{\partial \delta(\boldsymbol{r}^{\alpha} - \boldsymbol{x})}{\partial \boldsymbol{r}^{\alpha}}; f \right\rangle = -\mathrm{div} \sum_{\alpha} m^{\alpha} \langle \boldsymbol{v}^{\alpha} \delta(\boldsymbol{r}^{\alpha} - \boldsymbol{x}); f \rangle , \tag{8.107}$$

where we used the identity

$$\partial \delta(\boldsymbol{r}^{\alpha} - \boldsymbol{x})/\partial \boldsymbol{r}^{\alpha} = -\partial \delta(\boldsymbol{r}^{\alpha} - \boldsymbol{x})/\partial \boldsymbol{x}. \tag{8.108}$$

Substituting Eqn. (8.105) into Eqn. (8.107), we have

$$\frac{\partial}{\partial t}\rho^{\mathrm{pt}} + \mathrm{div}\,(\rho^{\mathrm{pt}} \boldsymbol{v}^{\mathrm{pt}}) = 0. \tag{8.109}$$

This is exactly the continuity equation (see Eqn. (2.77)). Thus the pointwise definitions for density and velocity in Eqns. (8.104) and (8.106) identically satisfy conservation of mass.

8.2.4 Momentum balance and the pointwise stress tensor

The rate of change of momentum follows from Eqn. (8.105) as

$$\frac{\partial}{\partial t}(\rho^{\mathrm{pt}} \boldsymbol{v}^{\mathrm{pt}}) = \frac{\partial}{\partial t} \left\langle \sum_{\beta} m^{\beta} \boldsymbol{v}^{\beta} \delta(\boldsymbol{r}^{\beta} - \boldsymbol{x}); f \right\rangle .$$

[25] Note that the divergence operation here and elsewhere in the chapter, denoted by div, is taken with respect to the spatial variable $\boldsymbol{x}$.

Applying Eqn. (8.103), we have

$$\begin{aligned}\frac{\partial}{\partial t}(\rho^{\rm pt}\boldsymbol{v}^{\rm pt}) &= \sum_\alpha \left\langle \left(\boldsymbol{v}^\alpha \cdot \frac{\partial}{\partial \boldsymbol{r}^\alpha}\right)\sum_\beta m^\beta \boldsymbol{v}^\beta \delta(\boldsymbol{r}^\beta - \boldsymbol{x}) \right.\\ &\qquad \left. - \left(\frac{1}{m^\alpha}\frac{\partial \mathcal{V}}{\partial \boldsymbol{r}^\alpha}\cdot\frac{\partial}{\partial \boldsymbol{v}^\alpha}\right)\sum_\beta m^\beta \boldsymbol{v}^\beta \delta(\boldsymbol{r}^\beta - \boldsymbol{x}); f\right\rangle \\ &= \sum_\alpha \left\langle m^\alpha (\boldsymbol{v}^\alpha \otimes \boldsymbol{v}^\alpha)\frac{\partial \delta(\boldsymbol{r}^\alpha - \boldsymbol{x})}{\partial \boldsymbol{r}^\alpha} - \frac{\partial \mathcal{V}}{\partial \boldsymbol{r}^\alpha}\delta(\boldsymbol{r}^\alpha - \boldsymbol{x}); f\right\rangle \\ &= -\operatorname{div}\sum_\alpha m^\alpha \left\langle \boldsymbol{v}^\alpha \otimes \boldsymbol{v}^\alpha \delta(\boldsymbol{r}^\alpha - \boldsymbol{x}); f\right\rangle - \sum_\alpha \left\langle \frac{\partial \mathcal{V}}{\partial \boldsymbol{r}^\alpha}\delta(\boldsymbol{r}^\alpha - \boldsymbol{x}); f\right\rangle,\end{aligned} \tag{8.110}$$

where we have used Eqn. (8.108) to establish the last line. The first term in Eqn. (8.110) can be modified by noting that

$$\begin{aligned}&\sum_\alpha m^\alpha \left\langle (\boldsymbol{v}^\alpha - \boldsymbol{v}^{\rm pt}) \otimes (\boldsymbol{v}^\alpha - \boldsymbol{v}^{\rm pt})\, \delta(\boldsymbol{r}^\alpha - \boldsymbol{x}); f\right\rangle \\ &\quad = \sum_\alpha m^\alpha \left\langle \boldsymbol{v}^\alpha \otimes \boldsymbol{v}^\alpha \delta(\boldsymbol{r}^\alpha - \boldsymbol{x}); f\right\rangle - \boldsymbol{v}^{\rm pt} \otimes \sum_\alpha m^\alpha \left\langle \boldsymbol{v}^\alpha \delta(\boldsymbol{r}^\alpha - \boldsymbol{x}); f\right\rangle \\ &\qquad + \boldsymbol{v}^{\rm pt} \otimes \boldsymbol{v}^{\rm pt} \sum_\alpha m^\alpha \left\langle \delta(\boldsymbol{r}^\alpha - \boldsymbol{x}); f\right\rangle - \sum_\alpha m^\alpha \left\langle \boldsymbol{v}^\alpha \delta(\boldsymbol{r}^\alpha - \boldsymbol{x}); f\right\rangle \otimes \boldsymbol{v}^{\rm pt} \\ &\quad = \sum_\alpha m^\alpha \left\langle \boldsymbol{v}^\alpha \otimes \boldsymbol{v}^\alpha \delta(\boldsymbol{r}^\alpha - \boldsymbol{x}); f\right\rangle - \rho^{\rm pt}\boldsymbol{v}^{\rm pt} \otimes \boldsymbol{v}^{\rm pt},\end{aligned} \tag{8.111}$$

where we have used Eqns. (8.104) and (8.105) to pass from the first equality to the second. Substituting Eqn. (8.111) into Eqn. (8.110) gives

$$\begin{aligned}\frac{\partial}{\partial t}(\rho^{\rm pt}\boldsymbol{v}^{\rm pt}) + \operatorname{div}(\rho^{\rm pt}\boldsymbol{v}^{\rm pt} \otimes \boldsymbol{v}^{\rm pt}) &= -\sum_\alpha \left\langle \frac{\partial \mathcal{V}}{\partial \boldsymbol{r}^\alpha}\delta(\boldsymbol{r}^\alpha - \boldsymbol{x}); f\right\rangle \\ &\quad - \operatorname{div}\left[\sum_\alpha m^\alpha \left\langle (\boldsymbol{v}^\alpha - \boldsymbol{v}^{\rm pt}) \otimes (\boldsymbol{v}^\alpha - \boldsymbol{v}^{\rm pt})\, \delta(\boldsymbol{r}^\alpha - \boldsymbol{x}); f\right\rangle\right].\end{aligned} \tag{8.112}$$

The left-hand side of the above equation is exactly the right-hand side of the conservation of mass and linear momentum relation given in Eqn. (2.91) and reproduced here:

$$\operatorname{div}\boldsymbol{\sigma}^{\rm pt} + \rho^{\rm pt}\boldsymbol{b}^{\rm pt} = \frac{\partial(\rho^{\rm pt}\boldsymbol{v}^{\rm pt})}{\partial t} + \operatorname{div}(\rho^{\rm pt}\boldsymbol{v}^{\rm pt} \otimes \boldsymbol{v}^{\rm pt}). \tag{8.113}$$

In Eqn. (8.113), $\boldsymbol{\sigma}^{\rm pt}$ is the pointwise Cauchy stress tensor and $\boldsymbol{b}^{\rm pt}$ is the pointwise body force per unit mass. We therefore have that

$$\operatorname{div}\boldsymbol{\sigma}^{\rm pt} + \rho^{\rm pt}\boldsymbol{b}^{\rm pt} = -\operatorname{div}\left[\sum_\alpha m^\alpha \left\langle \boldsymbol{v}^\alpha_{\rm rel} \otimes \boldsymbol{v}^\alpha_{\rm rel}\delta(\boldsymbol{r}^\alpha - \boldsymbol{x}); f\right\rangle\right] - \sum_\alpha \left\langle \frac{\partial \mathcal{V}}{\partial \boldsymbol{r}^\alpha}\delta(\boldsymbol{r}^\alpha - \boldsymbol{x}); f\right\rangle, \tag{8.114}$$

where

$$\boldsymbol{v}^{\alpha}_{\text{rel}} \equiv \boldsymbol{v}^{\alpha} - \boldsymbol{v}^{\text{pt}} \tag{8.115}$$

is the velocity of atom α relative to the pointwise velocity field.

To proceed, we divide the potential energy $\mathcal{V}(\boldsymbol{r})$ into internal and external parts as described in Section 5.4.1:[26]

$$\mathcal{V}(\boldsymbol{r}) = \mathcal{V}^{\text{int}}(\boldsymbol{r}) + \mathcal{V}^{\text{ext}}_{\text{fld}}(\boldsymbol{r}).$$

The internal potential energy, $\mathcal{V}^{\text{int}}$, is associated with short-range atomic interactions, whereas the external potential energy, $\mathcal{V}^{\text{ext}}_{\text{fld}}$, is associated with long-range interactions due to external fields, such as gravity or electromagnetic radiation.[27] It is natural to identify $\mathcal{V}^{\text{ext}}_{\text{fld}}$ with the pointwise body force field $\boldsymbol{b}^{\text{pt}}$ in Eqn. (8.114). We thus define $\boldsymbol{b}^{\text{pt}}(\boldsymbol{x};t)$ as

$$\boldsymbol{b}^{\text{pt}}(\boldsymbol{x};t) \equiv -\sum_{\alpha} \left\langle \frac{\partial \mathcal{V}^{\text{ext}}_{\text{fld}}}{\partial \boldsymbol{r}^{\alpha}} \delta(\boldsymbol{r}^{\alpha} - \boldsymbol{x}); f \right\rangle. \tag{8.116}$$

Substituting Eqn. (8.116) into Eqn. (8.114), we have

$$\operatorname{div} \boldsymbol{\sigma}^{\text{pt}} = -\operatorname{div}\left[\sum_{\alpha} m^{\alpha} \left\langle \boldsymbol{v}^{\alpha}_{\text{rel}} \otimes \boldsymbol{v}^{\alpha}_{\text{rel}} \delta(\boldsymbol{r}^{\alpha} - \boldsymbol{x}); f \right\rangle\right] - \sum_{\alpha} \left\langle \frac{\partial \mathcal{V}^{\text{int}}}{\partial \boldsymbol{r}^{\alpha}} \delta(\boldsymbol{r}^{\alpha} - \boldsymbol{x}); f \right\rangle. \tag{8.117}$$

From Eqn. (8.117) we see that the pointwise stress tensor has two contributions:

$$\boldsymbol{\sigma}^{\text{pt}}(\boldsymbol{x};t) = \boldsymbol{\sigma}^{\text{pt,K}}(\boldsymbol{x};t) + \boldsymbol{\sigma}^{\text{pt,V}}(\boldsymbol{x};t). \tag{8.118}$$

The kinetic part of the stress is[28]

$$\boldsymbol{\sigma}^{\text{pt,K}}(\boldsymbol{x};t) = -\sum_{\alpha} m^{\alpha} \left\langle \boldsymbol{v}^{\alpha}_{\text{rel}} \otimes \boldsymbol{v}^{\alpha}_{\text{rel}} \delta(\boldsymbol{r}^{\alpha} - \boldsymbol{x}); f \right\rangle, \tag{8.119}$$

which in indicial form is

$$\sigma^{\text{pt,K}}_{ij} = -\sum_{\alpha} m^{\alpha} \left\langle v^{\alpha}_{\text{rel},i} v^{\alpha}_{\text{rel},j} \delta(\boldsymbol{r}^{\alpha} - \boldsymbol{x}); f \right\rangle.$$

[26] Recall from Eqn. (5.7) that the external potential energy consists of two parts: $\mathcal{V}^{\text{ext}} = \mathcal{V}^{\text{ext}}_{\text{fld}} + \mathcal{V}^{\text{ext}}_{\text{con}}$. As with the weak interaction assumption of the canonical ensemble discussed in Sections 7.3.3 and 7.4, interactions with atoms outside of the system are neglected. We therefore have that $\mathcal{V}^{\text{ext}} = \mathcal{V}^{\text{ext}}_{\text{fld}}$.

[27] There can be some ambiguity in the definition of "short-range" and "long-range" by which the potential energy is divided into internal and external contributions. See [Pit86] for a discussion and proposed criterion for defining short- and long-range interactions.

[28] See Section 8.1.3 for a discussion of the physical significance of the kinetic part of the stress tensor. Also, see Sections 8.3.3 for the role of kinetic stress in the Tsai traction definition for stress.

We see that the kinetic part of the stress tensor is symmetric. The potential part of the stress must satisfy the following differential equation:

$$\operatorname{div} \boldsymbol{\sigma}^{\mathrm{pt,V}} = \sum_{\alpha} \left\langle -\frac{\partial \mathcal{V}^{\mathrm{int}}}{\partial \boldsymbol{r}^{\alpha}} \delta(\boldsymbol{r}^{\alpha} - \boldsymbol{x}); f \right\rangle. \tag{8.120}$$

The term $-\partial \mathcal{V}^{\mathrm{int}} / \partial \boldsymbol{r}^{\alpha}$ in Eqn. (8.120) is the internal force $\boldsymbol{f}^{\mathrm{int},\alpha}$ on atom α due to interaction with its neighbors. We have shown that due to translational, rotational and parity symmetry (the so-called "principle of interatomic potential invariance" in Eqn. (5.17)), the internal force on an atom (in conservative systems) can be decomposed as a sum over central terms associated with its neighbors (see Section 5.8.1):

$$-\frac{\partial \mathcal{V}^{\mathrm{int}}}{\partial \boldsymbol{r}^{\alpha}} = \boldsymbol{f}^{\mathrm{int},\alpha} = \sum_{\substack{\beta \\ \beta \neq \alpha}} \boldsymbol{f}^{\alpha\beta}, \tag{8.121}$$

where as given in Eqn. (5.107) and reproduced here

$$\boldsymbol{f}^{\alpha\beta} \equiv \varphi^{\alpha\beta} \frac{\boldsymbol{r}^{\beta} - \boldsymbol{r}^{\alpha}}{\|\boldsymbol{r}^{\beta} - \boldsymbol{r}^{\alpha}\|}, \qquad \varphi^{\alpha\beta} \equiv \frac{\partial \mathcal{V}^{\mathrm{int}}(r^{12}, r^{13}, \ldots,)}{\partial r^{\alpha\beta}}. \tag{8.122}$$

An important issue pointed out in Section 5.8.2 (and discussed at length in Appendix A and more specifically in Section A.4) is that the central force decomposition in Eqn. (8.121) is *not* unique. This is related to the fact that the partial differentiation in Eqn. $(8.122)_2$ requires the definition of a *continuously differentiable extension* to the potential energy function and this extension is not unique. The implication of this result for the uniqueness of the stress tensor will be discussed later.

Substituting Eqn. (8.121) into Eqn. (8.120) gives

$$\operatorname{div} \boldsymbol{\sigma}^{\mathrm{pt,V}} = \sum_{\substack{\alpha,\beta \\ \alpha \neq \beta}} \left\langle \boldsymbol{f}^{\alpha\beta} \delta(\boldsymbol{r}^{\alpha} - \boldsymbol{x}); f \right\rangle. \tag{8.123}$$

In their derivation, Irving and Kirkwood continue from this point to obtain an approximate solution for $\boldsymbol{\sigma}^{\mathrm{pt,V}}$ for the special case of pair potential interactions, where $\boldsymbol{f}^{\alpha\beta}$ depends only on the distance $r^{\alpha\beta}$ between atoms α and β, using a Taylor expansion for the difference between two Dirac delta functions that emerges. However, in 1955 Noll [Nol55] showed that it is possible to obtain a closed-form solution to Eqn. (8.123). The first step in Noll's derivation is based on the following identity which holds for any phase function A:

$$\langle A\delta(\boldsymbol{r}^{\alpha} - \boldsymbol{x}); f \rangle = \int_{\boldsymbol{y} \in \mathbb{R}^3} \langle A\delta(\boldsymbol{r}^{\alpha} - \boldsymbol{x})\delta(\boldsymbol{r}^{\beta} - \boldsymbol{y}); f \rangle \, d\boldsymbol{y}. \tag{8.124}$$

This merely transfers the integration with respect to $\boldsymbol{r}^{\beta}$ that occurs in $\langle A\delta(\boldsymbol{r}^{\alpha} - \boldsymbol{x}); f \rangle$ as part of the definition of the phase average to an explicit integral in terms of $\boldsymbol{y}$. Applying Eqn. (8.124) to Eqn. (8.123) gives

$$\operatorname{div} \boldsymbol{\sigma}^{\mathrm{pt,V}} = \int_{\boldsymbol{y} \in \mathbb{R}^3} \sum_{\substack{\alpha,\beta \\ \alpha \neq \beta}} \left\langle \boldsymbol{f}^{\alpha\beta} \delta(\boldsymbol{r}^{\alpha} - \boldsymbol{x})\delta(\boldsymbol{r}^{\beta} - \boldsymbol{y}); f \right\rangle d\boldsymbol{y}. \tag{8.125}$$

Substituting in Eqn. $(8.122)_1$, we have

$$\operatorname{div}\boldsymbol{\sigma}^{\mathrm{pt,V}} = -\int_{\boldsymbol{y}\in\mathbb{R}^3} \frac{\boldsymbol{x}-\boldsymbol{y}}{\|\boldsymbol{x}-\boldsymbol{y}\|} \sum_{\substack{\alpha,\beta \\ \alpha\neq\beta}} \left\langle \varphi^{\alpha\beta}\delta(\boldsymbol{r}^\alpha-\boldsymbol{x})\delta(\boldsymbol{r}^\beta-\boldsymbol{y});f\right\rangle d\boldsymbol{y}. \tag{8.126}$$

We now note that the integrand in Eqn. (8.126) is antisymmetric with respect to $\boldsymbol{x}$ and $\boldsymbol{y}$, i.e. swapping $\boldsymbol{x}$ and $\boldsymbol{y}$ changes the sign of the integrand. This is always satisfied regardless of the interatomic potential or boundary conditions. The proof is very simple.

Proof The prefactor $(\boldsymbol{x}-\boldsymbol{y})/\|\boldsymbol{x}-\boldsymbol{y}\|$ in Eqn. (8.126) is clearly antisymmetric with respect to $\boldsymbol{x}$ and $\boldsymbol{y}$. That means we just need to show that the sum of phase averages is symmetric. For every pair of atoms α and β, the sum contains two terms:

$$\left\langle \varphi^{\alpha\beta}\delta(\boldsymbol{r}^\alpha-\boldsymbol{x})\delta(\boldsymbol{r}^\beta-\boldsymbol{y});f\right\rangle + \left\langle \varphi^{\beta\alpha}\delta(\boldsymbol{r}^\beta-\boldsymbol{x})\delta(\boldsymbol{r}^\alpha-\boldsymbol{y});f\right\rangle.$$

Swapping $\boldsymbol{x}$ and $\boldsymbol{y}$ leaves this expression unchanged provided that $\varphi^{\alpha\beta}=\varphi^{\beta\alpha}$. But this is true by definition since $\varphi^{\alpha\beta}$ depends only on the distances between pairs of atoms not their absolute positions.[29] □

The antisymmetry of the integrand can be exploited by using the first of two very simple lemmas that Noll proved in the appendix of his 1955 paper [Nol55]. The lemma states that subject to some mild conditions, the following identity is satisfied for any antisymmetric tensor-valued function $\boldsymbol{f}(\boldsymbol{x},\boldsymbol{y})$ of two vectors $\boldsymbol{x}$ and $\boldsymbol{y}$,

$$\int_{\boldsymbol{y}\in\mathbb{R}^3} \boldsymbol{f}(\boldsymbol{x},\boldsymbol{y})\,d\boldsymbol{y} = -\frac{1}{2}\operatorname{div}_{\boldsymbol{x}} \int_{\boldsymbol{z}\in\mathbb{R}^3} \left[\int_{s=0}^{1} \boldsymbol{f}\left(\boldsymbol{x}+s\boldsymbol{z},\boldsymbol{x}-(1-s)\boldsymbol{z}\right) ds\right]\otimes\boldsymbol{z}\,d\boldsymbol{z}. \tag{8.127}$$

For example, for $\boldsymbol{f}\in\mathbb{R}^3$ this expression in indicial notation is

$$\int_{\boldsymbol{y}\in\mathbb{R}^3} f_i(\boldsymbol{x},\boldsymbol{y})\,dy = -\frac{1}{2}\frac{\partial}{\partial x_j}\int_{\boldsymbol{z}\in\mathbb{R}^3} g_i(\boldsymbol{x},\boldsymbol{z})z_j\,dz,$$

where we have defined $\boldsymbol{g}(\boldsymbol{x},\boldsymbol{z}) \equiv \int_{s=0}^{1} \boldsymbol{f}\left(\boldsymbol{x}+s\boldsymbol{z},\boldsymbol{x}-(1-s)\boldsymbol{z}\right) ds$. The proof for this lemma is given in [AT10, Appendix C]. Applying Noll's lemma in Eqn. (8.127) to Eqn. (8.126) and using $\boldsymbol{x}-\boldsymbol{y}=(\boldsymbol{x}+s\boldsymbol{z})-(\boldsymbol{x}-(1-s)\boldsymbol{z})=\boldsymbol{z}$ gives

$$\operatorname{div}\boldsymbol{\sigma}^{\mathrm{pt,V}} = \operatorname{div}\frac{1}{2}\int_{\boldsymbol{z}\in\mathbb{R}^3}\left[\frac{\boldsymbol{z}\otimes\boldsymbol{z}}{\|\boldsymbol{z}\|}\int_{s=0}^{1}\sum_{\substack{\alpha,\beta \\ \alpha\neq\beta}}\left\langle\varphi^{\alpha\beta}\delta(\boldsymbol{r}^\alpha-\boldsymbol{x}^+)\delta(\boldsymbol{r}^\beta-\boldsymbol{x}^-);f\right\rangle ds\right] d\boldsymbol{z}, \tag{8.128}$$

where $\boldsymbol{x}^+$ and $\boldsymbol{x}^-$ are defined as (see also Fig. 8.6):

$$\boldsymbol{x}^+(\boldsymbol{z},s)\equiv\boldsymbol{x}+s\boldsymbol{z}, \qquad \boldsymbol{x}^-(\boldsymbol{z},s)\equiv\boldsymbol{x}-(1-s)\boldsymbol{z}. \tag{8.129}$$

[29] Note that this condition is satisfied for any interatomic potential, not just pair potentials.

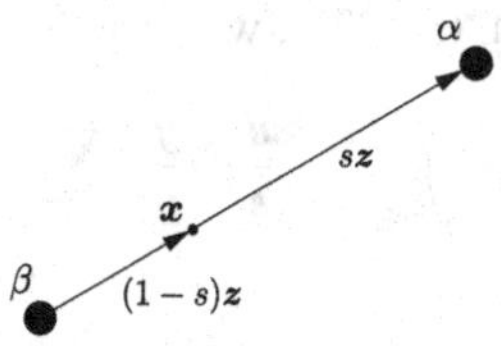

Fig. 8.6 A schematic diagram to help to explain the vectors appearing in Noll's potential stress expression in Eqn. (8.130). The bond α–β is defined by the vector $\boldsymbol{z}$. When $s = 0$, atom α is located at point $\boldsymbol{x}$, and when $s = 1$, atom β is located at $\boldsymbol{x}$.

In Eqn. (8.128), we have that the divergence of $\boldsymbol{\sigma}^{\mathrm{pt,V}}$ is equal to the divergence of the expression on the right-hand side, we can therefore obtain an explicit[30] expression for $\boldsymbol{\sigma}^{\mathrm{pt,V}}$,

$$\boldsymbol{\sigma}^{\mathrm{pt,V}}(\boldsymbol{x};t) = \frac{1}{2}\int_{\boldsymbol{z}\in\mathbb{R}^3}\left[\frac{\boldsymbol{z}\otimes\boldsymbol{z}}{\|\boldsymbol{z}\|}\int_{s=0}^{1}\sum_{\substack{\alpha,\beta\\ \alpha\neq\beta}}\langle\varphi^{\alpha\beta}\delta(\boldsymbol{r}^{\alpha}-\boldsymbol{x}^{+})\delta(\boldsymbol{r}^{\beta}-\boldsymbol{x}^{-});f\rangle\,ds\right]dz. \tag{8.130}$$

We refer to this as the *Noll stress*. In indicial notation this is

$$\sigma_{ij}^{\mathrm{pt,V}} = \frac{1}{2}\iiint_{-\infty}^{+\infty}\left[\frac{z_i z_j}{\|\boldsymbol{z}\|}\int_{s=0}^{1}\sum_{\substack{\alpha,\beta\\ \alpha\neq\beta}}\langle\varphi^{\alpha\beta}\delta(\boldsymbol{r}^{\alpha}-\boldsymbol{x}^{+})\delta(\boldsymbol{r}^{\beta}-\boldsymbol{x}^{-});f\rangle\,ds\right]dz_1 dz_2 dz_3.$$

The Noll stress is a general expression for the potential part of the pointwise Cauchy stress. Its derivation involved no approximations. It has the following important features:

1. Equation (8.130) is correct for any classical interatomic model that depends only on the positions of the atoms (see Section 5.8.1).
2. Although, the expression for $\boldsymbol{\sigma}^{\mathrm{pt,V}}$ appears complex it is actually conceptually quite simple. The potential stress $\boldsymbol{\sigma}^{\mathrm{pt,V}}$ at a point $\boldsymbol{x}$ is the superposition of the expectation values of the forces in all possible bonds passing through $\boldsymbol{x}$. The variable $\boldsymbol{z}$ selects a bond length and direction and the variable s slides the bond through $\boldsymbol{x}$ from end to end (see Fig. 8.6). For a given bond and position, the expression

$$\langle\varphi^{\alpha\beta}\delta(\boldsymbol{r}^{\alpha}-\boldsymbol{x}^{+})\delta(\boldsymbol{r}^{\beta}-\boldsymbol{x}^{-});f\rangle,$$

where $\boldsymbol{x}^{+}$ and $\boldsymbol{x}^{-}$ are defined in Eqn. (8.129), is the expectation value of the force in that bond when it connects atoms α and β, while the rest of the atoms are allowed to fully explore their phase space subject to the weighting of the distribution function.
3. $\boldsymbol{\sigma}^{\mathrm{pt,V}}$ is symmetric. This is clear because the term $\boldsymbol{z}\otimes\boldsymbol{z}$ in Eqn. (8.130) is symmetric. Since the kinetic part of the pointwise stress tensor, $\boldsymbol{\sigma}^{\mathrm{pt,K}}$, in Eqn. (8.119) is also symmetric, this means that *the total pointwise stress tensor is symmetric for all interatomic*

[30] The fact that a divergence-free term can be added to $\boldsymbol{\sigma}^{\mathrm{pt,V}}$ is one of the sources of nonuniqueness of the stress tensor discussed in Section 8.3.4.

potentials. Thus the pointwise stress tensor satisfies the balance of angular momentum (see Section 2.3.3) in addition to the balance of linear momentum.

4. Since $\boldsymbol{\sigma}^{\mathrm{pt,V}}$ depends on the nature of the force decomposition in Eqn. (8.121) and different extensions of a given potential energy can result in different force decompositions, we conclude that the pointwise stress tensor is *nonunique* for all interatomic potentials (including pair potentials). However, we show in Section 8.3.4 that the difference due to any two pointwise stress tensors, resulting from different extensions for the interatomic potential energy, tends to zero as the volume of the domain over which these pointwise quantities are spatially averaged tends to infinity. Therefore, as expected, the macroscopic stress tensor, which is defined in the thermodynamic limit, is always unique and is independent of the potential energy extension.
5. Another source of nonuniqueness is that any expression of the form, $\boldsymbol{\sigma}^{\mathrm{pt,V}} + \widehat{\boldsymbol{\sigma}}$, where $\operatorname{div} \widehat{\boldsymbol{\sigma}} = 0$, is also a solution to Eqn. (8.128). We address this issue in Section 8.3.4, where we show that in the thermodynamic limit under equilibrium conditions the spatially-averaged counterpart to $\boldsymbol{\sigma}^{\mathrm{pt,V}}$ converges to the virial stress derived in Section 8.1.3.

We do not discuss the case of nonsymmetric pointwise stress tensors, and pointwise couple stress tensors which can exist in multipolar media (see, for example, [Tou64, TN65, Jau67, BH82]). In order to obtain a nonsymmetric stress at the microscopic level, some of the assumptions leading to the derivation of the Noll stress have to be relaxed. There are several possibilities:

1. It was assumed that the atoms in the microscopic system are point particles without internal structure. If internal structure is allowed, additional variables can enter into the potential energy (such as polarization, spin, etc.). In this case, a more general formulation of the Noll lemma allowing for curved *paths of interaction* between the particles is possible, which leads to a nonsymmetric stress tensor [SH82, AT10].
2. If the forces between the particles also depend on the velocities of the particles (for example, the forces derived from the Biot–Savart law for moving charges), the resulting stress tensor can be nonsymmetric. This was studied within the Irving–Kirkwood–Noll procedure by Pitteri [Pit90].
3. Murdoch and coworkers [Mur85, MB94, Mur03, Mur07] show that nonsymmetric stress tensors can be obtained by discarding the statistical mechanics foundation of Irving and Kirkwood and adopting an approach based solely on spatial averaging. The implications of this are discussed in [AT10].

8.2.5 Spatial averaging and macroscopic fields

In the previous sections, we have used the Irving–Kirkwood–Noll procedure to construct continuous fields from the underlying discrete microscopic system using statistical mechanics phase averaging. Although the resulting fields resemble the continuum mechanics fields and satisfy the continuum conservation equations, they are not macroscopic continuum fields. For example the pointwise stress field in Eqn. (8.118) will be highly non-uniform, exhibiting a crisscross pattern with higher stresses along bond directions, even when macroscopically the material is nominally under uniform or even zero stress.

A true macroscopic measurement is by necessity an average over some spatial region surrounding the continuum point where it is nominally defined. Thus if $g^{\mathrm{pt}}(\boldsymbol{x};t)$ is an Irving–Kirkwood–Noll pointwise field, such as density, stress or energy, the corresponding macroscopic field $g(\boldsymbol{x};t)$ is given by

$$g(\boldsymbol{x};t) = \int_{\boldsymbol{y}\in\mathbb{R}^3} w(\boldsymbol{y}-\boldsymbol{x}) g^{\mathrm{pt}}(\boldsymbol{y};t)\, d\boldsymbol{y}, \tag{8.131}$$

where $w(\boldsymbol{r})$ is a weighting function.

The important thing to note is that due to the linearity of the phase averaging in the Irving–Kirkwood–Noll procedure, the averaged macroscopic function $g(\boldsymbol{x};t)$ satisfies the same balance equations as does the microscopic measure $g^{\mathrm{pt}}(\boldsymbol{x};t)$.

Weighting function The key to the spatial averaging operation is the weighting function $w(\boldsymbol{r})$. Normally, the weighting function is selected to have compact support, so that $w(\boldsymbol{r}) = 0$ for distances greater than a specified microstructural length scale.[31] The weighting function has units of 1/volume and must satisfy the normalization condition [Mur03],

$$\int_{\boldsymbol{r}\in\mathbb{R}^3} w(\boldsymbol{r})\, d\boldsymbol{r} = 1. \tag{8.132}$$

This condition ensures that the correct macroscopic stress is obtained when the microscopic stress is uniform. For a spherically-symmetric distribution, $w(\boldsymbol{r}) = \hat{w}(r)$, where $r = \|\boldsymbol{r}\|$. The normalization condition in this case is

$$\int_0^\infty \hat{w}(r) 4\pi r^2\, dr = 1. \tag{8.133}$$

The simplest choice for $\hat{w}(r)$ is a spherically-symmetric uniform distribution over a specified radius r_w:

$$\hat{w}(r) = \begin{cases} 1/V_w & \text{if } r \le r_w, \\ 0 & \text{otherwise,} \end{cases} \tag{8.134}$$

where $V_w = \frac{4}{3}\pi r_w^3$ is the volume of the sphere where $\hat{w}(r)$ is nonzero. This function is discontinuous at $r = r_w$. If this is a concern, a "mollifying" function that smoothly takes $w(r)$ to zero at r_w over some desired range can be added [Mur07]. Another possible choice for $\hat{w}(r)$ is a Gaussian function that weights points closer to $\boldsymbol{x}$ more strongly than those further away [Har82]:

$$\hat{w}(r) = \pi^{-3/2} r_w^{-3} \exp\left[-r^2/r_w^2\right]. \tag{8.135}$$

[31] The size of the compact support represents the length scale over which continuum fields are being measured. Consider, for example, the case of a small hole in very large isotropic plate subjected to a uniaxial stress σ_∞ at two of the plate edges. We know from elasticity theory that the maximum stress at the surface of the hole will be $3\sigma_\infty$ (see, for example, [TG51]). However, the stress measured in an experiment will depend on the size of the strain gauge. If the strain gauge is much larger that the grain size of the material (but small relative to the hole size), then we would expect to measure $3\sigma_\infty$. However, if the strain gauge is smaller than a single grain, the measured stress will vary from grain to grain as the gauge is moved. If the strain gauge is even smaller, it will pick up stress concentrations associated with impurities and defects in the material. This example points to the basic fact that a continuum stress field is length-scale-dependent. In the microscopic definition for stress this length scale is set by the range of spatial averaging weighting function.

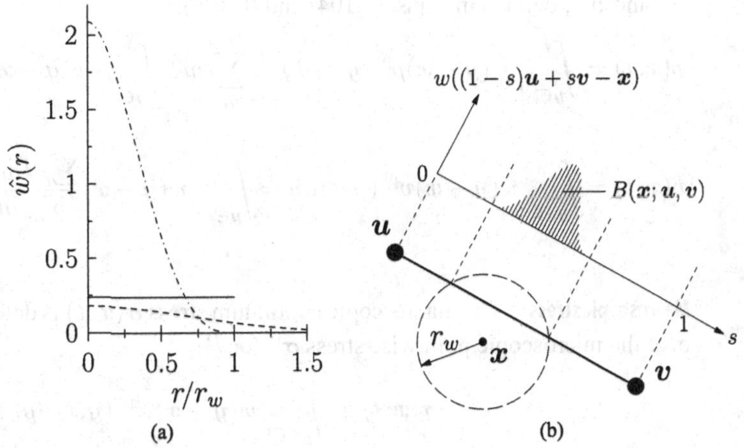

Fig. 8.7 (a) Three weighting functions for spatial averaging: uniform weighting (solid line) in Eqn. (8.134); Gaussian weighting (dashed line) in Eqn. (8.135); quartic spline weighting (dash-dot line) in Eqn. (8.136). The areas under the curves are not equal because the normalization in Eqn. (8.132) is according to volume. (b) The bond function $B(\boldsymbol{x};\boldsymbol{u},\boldsymbol{v})$ in Eqn. (8.146). The graph above the $\boldsymbol{u}$–$\boldsymbol{v}$ bond shows the result for a quartic spline weighting function. B is the area under the curve.

This function does not have compact support; however, it decays rapidly with distance so that a numerical cutoff can be imposed where its value drops below a specified tolerance. Yet another possibility is a quartic spline used in meshless method applications (where it is called a kernel function) [BKO+96],

$$\hat{w}(r) = \begin{cases} \dfrac{105}{16\pi r_w^3}\left(1+3\dfrac{r}{r_w}\right)\left(1-\dfrac{r}{r_w}\right)^3 & \text{if } r \le r_w, \\ 0 & \text{otherwise.} \end{cases} \tag{8.136}$$

This spline has the advantage that it goes smoothly to zero at $r = r_w$, i.e. $\hat{w}(r_w) = 0$, $\hat{w}'(r_w) = 0$, and $\hat{w}''(r_w) = 0$.

The above choices for $w(r)$ (plotted in Fig 8.7(a)) are rather arbitrary. An alternative approach described by Murdoch and Bedeaux in [MB94] is based on the requirement that "repeated spatial averaging should produce nothing new." In other words, spatially averaging a quantity that was already spatially averaged should give the same average. This leads to the following explicit expression for the weighting function:

$$\hat{w}(r) = \frac{1}{2\pi^2 r^3}\left(\sin\frac{r}{r_w} - \frac{r}{r_w}\cos\frac{r}{r_w}\right). \tag{8.137}$$

Note that although predominantly positive, this weighting function also takes on small negative values in its tail.

Macroscopic density and velocity The macroscopic continuum density and velocity fields, ρ and $\boldsymbol{v}$, are defined from Eqn. (8.131) to be the spatial average of the pointwise quantities,

ρ^{pt} and $\boldsymbol{v}^{\mathrm{pt}}$, defined in Eqns. (8.104) and (8.106):

$$\rho(\boldsymbol{x};t) = \int_{\boldsymbol{y}\in\mathbb{R}^3} w(\boldsymbol{y}-\boldsymbol{x})\rho^{\mathrm{pt}}(\boldsymbol{y};t)\,d\boldsymbol{y} = \sum_\alpha m^\alpha \int_{\boldsymbol{y}\in\mathbb{R}^3} w(\boldsymbol{y}-\boldsymbol{x})\,\langle \delta(\boldsymbol{r}^\alpha-\boldsymbol{y});f\rangle\,d\boldsymbol{y}, \tag{8.138}$$

$$\boldsymbol{v}(\boldsymbol{x};t) = \int_{\boldsymbol{y}\in\mathbb{R}^3} w(\boldsymbol{y}-\boldsymbol{x})\boldsymbol{v}^{\mathrm{pt}}(\boldsymbol{y};t)\,d\boldsymbol{y} = \int_{\boldsymbol{y}\in\mathbb{R}^3} w(\boldsymbol{y}-\boldsymbol{x})\frac{\sum_\alpha \langle m^\alpha \boldsymbol{v}^\alpha \delta(\boldsymbol{r}^\alpha-\boldsymbol{y});f\rangle}{\sum_\alpha m^\alpha \langle \delta(\boldsymbol{r}^\alpha-\boldsymbol{y});f\rangle}\,d\boldsymbol{y}. \tag{8.139}$$

Macroscopic stress The macroscopic continuum stress $\boldsymbol{\sigma}(\boldsymbol{x};t)$ is defined as a spatial average over the microscopic pointwise stress $\boldsymbol{\sigma}^{\mathrm{pt}}(\boldsymbol{x};t)$:

$$\boldsymbol{\sigma}(\boldsymbol{x};t) \equiv \int_{\boldsymbol{y}\in\mathbb{R}^3} w(\boldsymbol{y}-\boldsymbol{x})\boldsymbol{\sigma}^{\mathrm{pt}}(\boldsymbol{y};t)\,d\boldsymbol{y}. \tag{8.140}$$

Substituting Eqn. (8.118) into Eqn. (8.140) and using Eqns. (8.119) and (8.130), the macroscopic stress tensor divides into kinetic and potential parts:

$$\boldsymbol{\sigma}(\boldsymbol{x};t) = \boldsymbol{\sigma}^{\mathrm{K}}(\boldsymbol{x};t) + \boldsymbol{\sigma}^{\mathrm{V}}(\boldsymbol{x};t), \tag{8.141}$$

where

$$\boldsymbol{\sigma}^{\mathrm{K}}(\boldsymbol{x};t) = -\sum_\alpha m^\alpha \int_{\boldsymbol{y}\in\mathbb{R}^3} w(\boldsymbol{y}-\boldsymbol{x})\,\langle \boldsymbol{v}^\alpha_{\mathrm{rel}} \otimes \boldsymbol{v}^\alpha_{\mathrm{rel}}\delta(\boldsymbol{r}^\alpha-\boldsymbol{y});f\rangle\,d\boldsymbol{y} \tag{8.142}$$

and

$$\boldsymbol{\sigma}^{\mathrm{V}}(\boldsymbol{x};t) = \frac{1}{2}\int_{\boldsymbol{y}\in\mathbb{R}^3} w(\boldsymbol{y}-\boldsymbol{x}) \int_{\boldsymbol{z}\in\mathbb{R}^3} \left[\frac{\boldsymbol{z}\otimes\boldsymbol{z}}{\|\boldsymbol{z}\|}\int_{s=0}^{1} \sum_{\substack{\alpha,\beta\\ \alpha\neq\beta}} \langle \varphi'^{\alpha\beta}\delta(\boldsymbol{r}^\alpha-\boldsymbol{y}-s\boldsymbol{z})\delta(\boldsymbol{r}^\beta-\boldsymbol{y}+(1-s)\boldsymbol{z});f\rangle\,ds\right] d\boldsymbol{z}\,d\boldsymbol{y}. \tag{8.143}$$

The potential part of the stress can be expressed in a more convenient form by introducing the following change of variables [AT10],

$$\boldsymbol{u} = \boldsymbol{y} + s\boldsymbol{z}, \qquad \boldsymbol{v} = \boldsymbol{y} - (1-s)\boldsymbol{z},$$

or inversely

$$\boldsymbol{y} = (1-s)\boldsymbol{u} + s\boldsymbol{v}, \qquad \boldsymbol{z} = \boldsymbol{u} - \boldsymbol{v}. \tag{8.144}$$

The Jacobian of the transformation is

$$J = \det\begin{bmatrix} \partial\boldsymbol{y}/\partial\boldsymbol{u} & \partial\boldsymbol{y}/\partial\boldsymbol{v} \\ \partial\boldsymbol{z}/\partial\boldsymbol{u} & \partial\boldsymbol{z}/\partial\boldsymbol{v}\end{bmatrix} = \det\begin{bmatrix}(1-s)\boldsymbol{I} & s\boldsymbol{I}\\ \boldsymbol{I} & -\boldsymbol{I}\end{bmatrix} = -1.$$

Substituting Eqn. (8.144) into Eqn. (8.143) and using, $|J| = 1$, to account for the change of variables gives

$$\boldsymbol{\sigma}^{\mathrm{V}}(\boldsymbol{x};t) = \frac{1}{2}\iint_{\mathbb{R}^3\times\mathbb{R}^3} \frac{(\boldsymbol{u}-\boldsymbol{v})\otimes(\boldsymbol{u}-\boldsymbol{v})}{\|\boldsymbol{u}-\boldsymbol{v}\|} \times \sum_{\substack{\alpha,\beta \\ \alpha\neq\beta}} \langle \varphi^{\alpha\beta}\delta(\boldsymbol{r}^\alpha-\boldsymbol{u})\delta(\boldsymbol{r}^\beta-\boldsymbol{v}); f\rangle\, B(\boldsymbol{x};\boldsymbol{u},\boldsymbol{v})\, d\boldsymbol{u}\, d\boldsymbol{v}, \tag{8.145}$$

where

$$B(\boldsymbol{x};\boldsymbol{u},\boldsymbol{v}) \equiv \int_{s=0}^{1} w((1-s)\boldsymbol{u}+s\boldsymbol{v}-\boldsymbol{x})\, ds, \tag{8.146}$$

is called the "bond function." B equals the integrated weight of the bond for a weighting function centered at $\boldsymbol{x}$. See Fig. 8.7(b) for an example of a bond function for a quartic spline weighting function.

8.3 Practical methods: the stress tensor

The expressions for the macroscopic fields derived in the previous section are satisfying since they demonstrate how continuum mechanics and statistical mechanics are mutually consistent. They also help to clarify what the macroscopic measures mean directly in terms of the motion of atoms. The remaining obstacle is the derivation of a practical method for computing the phase averages that appear in these expressions. This can be difficult for a number of reasons. First, the appropriate distribution function may not be known for the problem being considered [MB93]. Second, even if the distribution function is known, calculation of phase averages is a difficult problem. One possibility that addresses both of these concerns, is to estimate the phase average from an MD simulation of the system being studied. In this case, the phase average is replaced with a time average which provides an estimate for the measure over the time scales accessible to the simulation. In performing such an operation, we must note the following:

1. Under conditions of thermodynamic equilibrium, ensemble averages can be replaced by time averages provided that the system is assumed to be ergodic. Strictly speaking this time average should be done for infinite time, but for practical reasons we are restricted to finite time. See discussion in Section 7.2.3.
2. The Hardy stress tensor is valid under nonequilibrium conditions assuming that the system is in local thermodynamic equilibrium at all points at every instant of time. This is plausible only when there is a clear separation of time scales between the microscopic

equilibration time scale τ and macroscopic times. Here, τ is not being defined rigorously. Roughly speaking, τ must be sufficiently small that macroscopic observables do not vary appreciably over it.

Under these assumptions, we may replace ensemble averages with time averages in the spatially-averaged quantities defined earlier. In this section, we focus on the calculation of the stress tensor.

8.3.1 The Hardy stress

In MD simulations (see Chapter 9), the phase averages appearing in the kinetic and potential parts of the stress tensor in Eqns. (8.142) and (8.145) are replaced by time averages over the period $[t, t+\tau]$, so that

$$\boldsymbol{\sigma}^{\mathrm{K}}(\boldsymbol{x};t) = -\frac{1}{N_\tau}\sum_{i=1}^{N_\tau}\left[\sum_\alpha m^\alpha w(\boldsymbol{r}^{\alpha(i)} - \boldsymbol{x})\boldsymbol{v}_{\mathrm{rel}}^{\alpha(i)} \otimes \boldsymbol{v}_{\mathrm{rel}}^{\alpha(i)}\right], \tag{8.147}$$

where N_τ is the number of timesteps during the time interval τ over which the averaging is performed ($\tau = N_\tau \Delta\tau$) and $\boldsymbol{r}^{\alpha(i)}$ and $\boldsymbol{v}_{\mathrm{rel}}^{\alpha(i)}$ are the position and relative velocity of atom α at timestep i. Similarly, the potential part of the stress is

$$\boldsymbol{\sigma}^{\mathrm{V}}(\boldsymbol{x};t) = \frac{1}{2N_\tau}\sum_{i=1}^{N_\tau}\left[\sum_{\substack{\alpha,\beta\\ \alpha\neq\beta}} \frac{\boldsymbol{r}^{\alpha\beta(i)} \otimes \boldsymbol{r}^{\alpha\beta(i)}}{r^{\alpha\beta(i)}} \varphi^{\alpha\beta(i)} B(\boldsymbol{x};\boldsymbol{r}^{\alpha(i)},\boldsymbol{r}^{\beta(i)})\right], \tag{8.148}$$

where $\boldsymbol{r}^{\alpha\beta(i)} = \boldsymbol{r}^{\beta(i)} - \boldsymbol{r}^{\alpha(i)}$ is the vector connecting atoms α and β at timestep i, $r^{\alpha\beta(i)} = \left\|\boldsymbol{r}^{\alpha\beta(i)}\right\|$, $\varphi^{\alpha\beta(i)}$ is the magnitude of the force between atoms α and β at timestep i and B is the bond function given in Eqn. (8.146). It is sometimes convenient to rewrite Eqn. (8.148) as

$$\boldsymbol{\sigma}^{\mathrm{V}}(\boldsymbol{x};t) = \frac{1}{2N_\tau}\sum_{i=1}^{N_\tau}\left[\sum_{\substack{\alpha,\beta\\ \alpha\neq\beta}} \boldsymbol{f}^{\alpha\beta(i)} \otimes \boldsymbol{r}^{\alpha\beta(i)} B(\boldsymbol{x};\boldsymbol{r}^{\alpha(i)},\boldsymbol{r}^{\beta(i)})\right], \tag{8.149}$$

where $\boldsymbol{f}^{\alpha\beta(i)} \equiv \varphi^{\alpha\beta(i)}\boldsymbol{r}^{\alpha\beta(i)}/r^{\alpha\beta(i)}$ is the force term in the central force decomposition at timestep i. The total stress follows as

$$\boldsymbol{\sigma}(\boldsymbol{x};t) = \boldsymbol{\sigma}^{\mathrm{K}}(\boldsymbol{x};t) + \boldsymbol{\sigma}^{\mathrm{V}}(\boldsymbol{x};t). \tag{8.150}$$

Equations (8.147) and (8.148) are the familiar Hardy expressions for the kinetic and potential parts of the stress tensor [Har82].[32] We therefore see that Hardy's stress tensor

[32] Hardy's original derivation was limited to pair potentials. Here, the results are extended to any interatomic potential with $\varphi^{\alpha\beta}$ defined in Eqn. (8.122).

emerges seamlessly from the Irving–Kirkwood–Noll procedure after using Noll's analytical solution for the potential stress, applying spatial averaging to obtain a macroscopic stress measure and finally replacing the statistical mechanics phase averaging with molecular dynamics time averaging. This is presented schematically in Fig. 8.1.

The Hardy stress is reduced to other well-known stress measures by selecting specific weighting functions and taking suitable limits as explained below.

8.3.2 The virial stress tensor and atomic-level stresses

A simple special case is obtained when the weighting function, $w(\boldsymbol{y}-\boldsymbol{x})$, is constant within a given domain, Ω, and zero elsewhere:[33]

$$w(\boldsymbol{y}-\boldsymbol{x}) = \begin{cases} 1/\operatorname{Vol}(\Omega) & \text{if } \boldsymbol{y} \in \Omega, \\ 0 & \text{otherwise,} \end{cases} \tag{8.151}$$

where Ω contains the point $\boldsymbol{x}$. In addition to this, the bond function, B, in Eqn. (8.148) is evaluated approximately using its definition in Eqn. (8.146) by only counting those bonds α–β that lie entirely within the averaging domain. Hence, $B(\boldsymbol{x};\boldsymbol{r}^\alpha,\boldsymbol{r}^\beta)$ is given by

$$B(\boldsymbol{x};\boldsymbol{r}^\alpha,\boldsymbol{r}^\beta) = \begin{cases} 1/\operatorname{Vol}(\Omega) & \text{if bond } \alpha\text{-}\beta \in \Omega, \\ 0 & \text{otherwise.} \end{cases} \tag{8.152}$$

Substituting Eqns. (8.151) and (8.152) into the kinetic and potential parts of the Hardy stress in Eqns. (8.147) and (8.148) gives after combining the expressions:

$$\boldsymbol{\sigma}(\boldsymbol{x};t) = \frac{1}{N_\tau \operatorname{Vol}(\Omega)} \sum_{i=1}^{N_\tau} \left[-\sum_{\alpha\in\Omega} m^\alpha \boldsymbol{v}_{\text{rel}}^{\alpha(i)} \otimes \boldsymbol{v}_{\text{rel}}^{\alpha(i)} + \frac{1}{2} \sum_{\substack{\alpha,\beta\in\Omega \\ \alpha\neq\beta}} \varphi'^{\alpha\beta(i)} \frac{\boldsymbol{r}^{\alpha\beta(i)} \otimes \boldsymbol{r}^{\alpha\beta(i)}}{r^{\alpha\beta(i)}} \right]. \tag{8.153}$$

This is the *virial stress tensor* given earlier in Section 8.1.3 (see Eqn. (8.68)).[34] The difference between this stress and the Hardy stress with a uniform weighing function is that in the Hardy expression, bonds crossing out of Ω to surrounding atoms and bonds between external atoms that cross Ω are included, with the bond function B equal to the fractional length of the bond lying within Ω. This difference becomes negligible as the size of Ω becomes large relative to the cutoff radius of the potential. The reason for this is that the number of bonds connected with external atoms that cross into or through Ω scales with the surface area of Ω, whereas the number of internal bonds scales with the volume of Ω. As Ω grows, the surface-to-volume ratio shrinks and the contribution of external bonds is reduced. Tsai [Tsa79] has discussed this effect and explicitly computed the error in the

[33] This is similar to the uniform weighting function in Eqn. (8.134), however, Ω may be any shape. Typically, a rectangular box is taken for simplicity.

[34] Recall that in the statistical mechanics derivation of the virial stress in Section 8.1, the position and momenta are center-of-mass coordinates. Therefore, the expression $m^\alpha \boldsymbol{v}_{\text{rel}}^\alpha$ of this section, corresponds to the center-of-mass momentum $\boldsymbol{p}^\alpha$ of Section 8.1.

virial expression due to its neglect of external bonds for a one-dimensional system. Since, taking the averaging domain size to infinity is equivalent to taking the thermodynamic limit in this context, the Hardy and virial stress expressions become identical in this limit.[35] This means that the Irving–Kirkwood–Noll procedure is consistent with the results of equilibrium statistical mechanics in the thermodynamic limit.

8.3.3 The Tsai traction: a planar definition for stress

Cauchy's original definition of stress discussed in Section 2.3.2 emerges from the concept of the traction acting across internal surfaces of a solid. It is therefore natural to attempt to define stress at the atomic level in a similar vein in terms of the force in bonds intersecting a given plane. This approach actually goes back to Cauchy himself as part of his efforts in the 1820s to define the stress in crystalline systems and is described in detail in Note B in Love's classical book on the theory of elasticity [Lov27]. Cauchy's derivation appears to have been limited to zero-temperature equilibrium states where the atoms are stationary. The approach was extended by Tsai [Tsa79] to the dynamical setting by also accounting for the momentum flux of atoms moving across the plane. The expression in [Tsa79] appears to be based on intuition. In this section, we show how the Tsai traction can be systematically derived from the Hardy stress tensor, which itself was derived from the spatially-averaged stress tensor defined in Section 8.2.5.

Derivation of the Tsai traction from the Hardy stress The traction definition is presented as an independent definition of stress. Tsai calls it different from "the virial theorem in spirit and details." Subsequent authors have accepted this statement and explored the conditions under which the traction definition and virial definitions agree [CY91, SWSW06]. However, we show below that similar to the virial stress, Tsai's traction definition can be obtained from the Hardy stress as a special case when a uniform weighting function is used and the spatial averaging volume is collapsed onto a plane. Consequently, the virial stress and Tsai's planar definition are equivalent. The difference only lies in the additional approximations associated with the virial definition. Our derivation below follows [AT10], see that reference for more details.

We begin with the Hardy stress in Eqn. (8.150) and select as the averaging domain a generalized cylinder $\mathcal{C}_h$ of height h and cross-sectional area A, centered at point $\boldsymbol{x}$, with its axis oriented along the direction $\boldsymbol{n}$ (see Fig. 8.8). We assume a constant weighting function, w_h, over the domain of the cylinder:

$$w_h(\boldsymbol{y}-\boldsymbol{x}) = \begin{cases} 1/hA & \text{if } \boldsymbol{y} \in \mathcal{C}_h, \\ 0 & \text{otherwise.} \end{cases} \tag{8.154}$$

We then define the traction, $\boldsymbol{t}(\boldsymbol{x}, \boldsymbol{n}; t)$, at point $\boldsymbol{x}$ on a plane with normal $\boldsymbol{n}$ at time t as

$$\boldsymbol{t}(\boldsymbol{x}, \boldsymbol{n}; t) = \lim_{h\to 0} \boldsymbol{\sigma}_{w_h} \boldsymbol{n} = \lim_{h\to 0} \left(\boldsymbol{\sigma}^{\mathrm{K}}_{w_h} \boldsymbol{n} + \boldsymbol{\sigma}^{\mathrm{V}}_{w_h} \boldsymbol{n} \right), \tag{8.155}$$

[35] Note that for a periodic system the Hardy stress with uniform weighting and the virial stress are identical since the "internal" bonds (α–$\beta \in \Omega$) in the virial description also account for bonds exiting the domain and wrapping around to the other side.

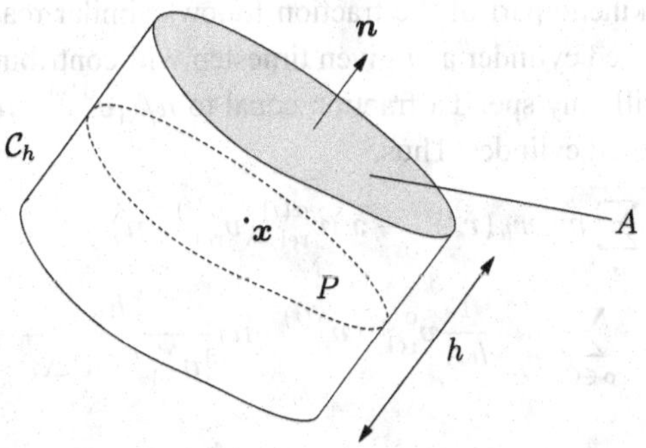

Fig. 8.8 A generalized cylinder $\mathcal{C}_h$ with cross-sectional area A and height h and normal $\boldsymbol{n}$ centered on point $\boldsymbol{x}$. The cross section passing through $\boldsymbol{x}$ is the plane P.

where a subscript "w_h" indicates that the variable is obtained by spatially averaging with respect to this weighting function. In this limit, the generalized cylinder $\mathcal{C}_h$ collapses onto the plane P passing through $\boldsymbol{x}$ (see Fig. 8.8). Substituting in Eqns. (8.147) and (8.148) and using Eqn. (8.122) and the tensor identity in Eqn. (2.25) gives

$$\boldsymbol{t}(\boldsymbol{x},\boldsymbol{n};t) = \frac{1}{N_\tau}\sum_{i=1}^{N_\tau}\lim_{h\to 0}\left[-\sum_\alpha m^\alpha w_h(\boldsymbol{r}^{\alpha(i)}-\boldsymbol{x})\boldsymbol{v}_{\mathrm{rel}}^{\alpha(i)}(\boldsymbol{v}_{\mathrm{rel}}^{\alpha(i)}\cdot\boldsymbol{n}) + \frac{1}{2}\sum_{\substack{\alpha,\beta\\ \alpha\neq\beta}}\boldsymbol{f}^{\alpha\beta(i)}(\boldsymbol{r}^{\alpha\beta(i)}\cdot\boldsymbol{n})B_h(\boldsymbol{x};\boldsymbol{r}^{\alpha(i)},\boldsymbol{r}^{\beta(i)})\right], \tag{8.156}$$

where B_h is the bond function for a generalized cylinder of height h. The rest of the notation is the same as that used in the previous section. Atom α has position $\boldsymbol{r}^\alpha$, relative velocity $\boldsymbol{v}_{\mathrm{rel}}^\alpha$ and mass m^α. The vector connecting atoms α and β is $\boldsymbol{r}^{\alpha\beta} = \boldsymbol{r}^\beta - \boldsymbol{r}^\alpha$. The force on atom α due to atom β is $\boldsymbol{f}^{\alpha\beta}$ as defined in Eqn. (8.122). A superscript "(i)" indicates the quantity is evaluated at timestep i.

Let us consider the potential term first. As h approaches zero, the generalized cylinder will no longer contain complete bonds. In this case, the bond function B_h equals the fraction of the length of the bond lying within the generalized cylinder per unit volume:

$$B_h(\boldsymbol{x};\boldsymbol{r}^\alpha,\boldsymbol{r}^\beta) = \frac{1}{hA}\frac{h}{|\boldsymbol{r}^{\alpha\beta}\cdot\boldsymbol{n}|} = \frac{1}{A|\boldsymbol{r}^{\alpha\beta}\cdot\boldsymbol{n}|}, \tag{8.157}$$

for any bond α–β crossing the cylinder. The potential part of the traction is therefore

$$\lim_{h\to 0}\frac{1}{2}\sum_{\substack{\alpha,\beta\\ \alpha\neq\beta}}\boldsymbol{f}^{\alpha\beta(i)}(\boldsymbol{r}^{\alpha\beta(i)}\cdot\boldsymbol{n})B_h(\boldsymbol{x};\boldsymbol{r}^{\alpha(i)},\boldsymbol{r}^{\beta(i)}) = \frac{1}{A}\sum_{\alpha\beta\cap P}\boldsymbol{f}^{\alpha\beta(i)}\frac{\boldsymbol{r}^{\alpha\beta(i)}\cdot\boldsymbol{n}}{|\boldsymbol{r}^{\alpha\beta(i)}\cdot\boldsymbol{n}|}. \tag{8.158}$$

Note that we have replaced "$\frac{1}{2}\sum_{\substack{\alpha,\beta\\ \alpha\neq\beta}}$" that visits every bond twice and therefore has the one-half factor with "$\sum_{\alpha\beta\cap P}$" that visits once each bond that pierces the generalized cylinder $\mathcal{C}_h$ and (in the limit) the resulting plane P.

The kinetic part of the traction follows similar reasoning. Only atoms lying inside the generalized cylinder at a given timestep will contribute to the kinetic term. As $h \to 0$, an atom will only spend a fraction equal to $h/(|\boldsymbol{v}_{\text{rel}}^{\alpha(i)} \cdot \boldsymbol{n}|\Delta\tau)$ of the full timestep $\Delta\tau$ in the generalized cylinder. Thus,

$$\lim_{h\to 0} -\sum_{\alpha} m^\alpha w_h(\boldsymbol{r}^{\alpha(i)} - \boldsymbol{x})\boldsymbol{v}_{\text{rel}}^{\alpha(i)}(\boldsymbol{v}_{\text{rel}}^{\alpha(i)} \cdot \boldsymbol{n})$$
$$= \lim_{h\to 0} - \sum_{\alpha\in\mathcal{C}_h} m^\alpha \frac{1}{hA}\boldsymbol{v}_{\text{rel}}^{\alpha(i)}(\boldsymbol{v}_{\text{rel}}^{\alpha(i)} \cdot \boldsymbol{n})\frac{h}{|\boldsymbol{v}_{\text{rel}}^{\alpha(i)} \cdot \boldsymbol{n}|\Delta\tau} = -\frac{1}{A\Delta\tau}\sum_{\alpha\leftrightarrow P} m^\alpha \boldsymbol{v}_{\text{rel}}^{\alpha(i)} \frac{\boldsymbol{v}_{\text{rel}}^{\alpha(i)} \cdot \boldsymbol{n}}{|\boldsymbol{v}_{\text{rel}}^{\alpha(i)} \cdot \boldsymbol{n}|}, \tag{8.159}$$

where "$\sum_{\alpha\leftrightarrow P}$" denotes a sum over atoms that cross P in a given timestep i. Note that some care is required when evaluating the velocity of the center of mass in the limit $h \to 0$ relative to which $\boldsymbol{v}_{\text{rel}}^\alpha$ is defined. See [AT10] for details.

Combining Eqns. (8.158) and (8.159) and summing over all timesteps gives

$$\boldsymbol{t}(\boldsymbol{x}, \boldsymbol{n}; t) = \frac{1}{N_\tau A}\sum_{i=1}^{N_\tau}\left[-\frac{1}{\Delta\tau}\sum_{\alpha\leftrightarrow P} m^\alpha \boldsymbol{v}_{\text{rel}}^{\alpha(i)}\frac{\boldsymbol{v}_{\text{rel}}^{\alpha(i)} \cdot \boldsymbol{n}}{|\boldsymbol{v}_{\text{rel}}^{\alpha(i)} \cdot \boldsymbol{n}|} + \sum_{\alpha\beta\cap P} \boldsymbol{f}^{\alpha\beta(i)}\frac{\boldsymbol{r}^{\alpha\beta(i)} \cdot \boldsymbol{n}}{|\boldsymbol{r}^{\alpha\beta(i)} \cdot \boldsymbol{n}|}\right]. \tag{8.160}$$

We refer to this expression as the *Tsai traction*. The first term in the square brackets is the momentum flux across the plane P during the time interval $[t+i\Delta\tau, t+(i+1)\Delta\tau]$, where $\Delta\tau$ is the molecular dynamics time increment (in units of time). The momentum carried across the surface by an atom crossing in timestep i is $m^\alpha \boldsymbol{v}_{\text{rel}}^{\alpha(i)}$, and $(\boldsymbol{v}_{\text{rel}}^{\alpha(i)} \cdot \boldsymbol{n})/|\boldsymbol{v}_{\text{rel}}^{\alpha(i)} \cdot \boldsymbol{n}|$ provides the correct sign. The second term in the square brackets is the average resultant of the forces in all the bonds that cross plane P in the time interval $[t + i\Delta\tau, t + (i + 1)\Delta\tau]$. The expression $(\boldsymbol{r}^{\alpha\beta(i)} \cdot \boldsymbol{n})/|\boldsymbol{r}^{\alpha\beta(i)} \cdot \boldsymbol{n}|$ provides the correct sign so that an attractive force between the atoms translates to a positive contribution to the stress.[36]

Relation between the potential and kinetic stress in the Tsai expression The physical significance of the kinetic part of the stress tensor is discussed in Section 8.1.3. We revisit this issue here, since seeing it in this context may again raise questions. Consider a crystalline solid at a relatively low temperature under uniform stress. The atoms vibrate about their mean positions with an amplitude that is small relative to the nearest-neighbor spacing. Now imagine placing a Tsai plane P between two crystal lattice planes and measuring the traction across it. If P is midway between the lattice planes (see Fig. 8.9(a)), we expect that relatively few atoms will cross P and that consequently the kinetic stress will be small or even zero. Whereas if P is close to a lattice plane there will be many such crossings (for

[36] An attractive force between two atoms means that the bond between them is in a state of tension since if the atoms were released they would move towards each other. This is consistent with the standard continuum mechanics sign convention for stress which is adopted in this book whereby tensile stresses are positive and compressive stresses are negative. Some authors, such as Tsai [Tsa79], adopt the opposite convention where a compressive stress is taken to be positive.

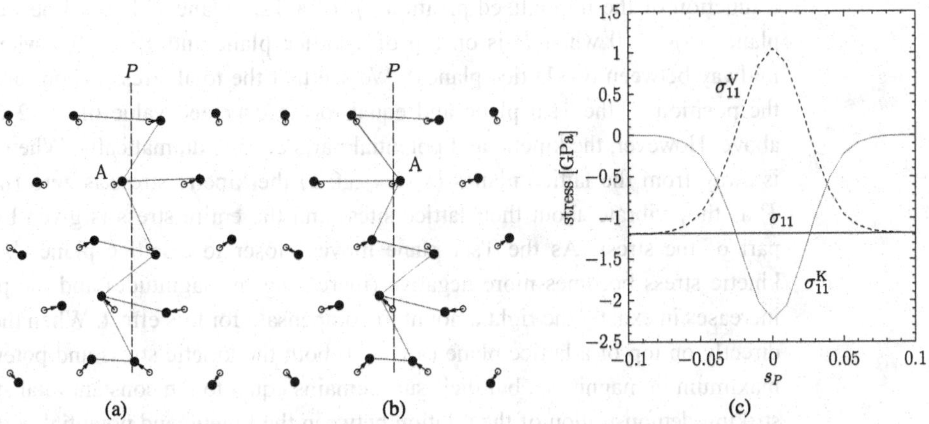

Fig. 8.9 The effect of the position of the Tsai plane P on the resulting traction. (a), (b) Schematic diagrams of a two-dimensional triangular lattice with (a) plane P positioned midway between lattice planes and (b) plane P positioned almost on top of a lattice plane. The open circles correspond to the ideal lattice positions. The black circles are atoms that are shown in midvibration relative to the lattice site as indicated by the arrow. The bonds crossing the Tsai plane appear as dotted lines. Note that the atom labeled A is on opposite sides of the Tsai plane depending on the position of the plane. (c) The results from an MD simulation showing the kinetic stress σ_{11}^{K}, potential stress σ_{11}^{V} and total stress $\sigma_{11} = \sigma_{11}^{K} + \sigma_{11}^{V}$ as a function of the normalized position $s_P = (x_P - x_L)/\Delta x$ of the Tsai plane P, where x_P is the position of P, x_L is the position of a lattice plane, and Δx is the spacing between lattice planes.

example the atom labeled A in Fig. 8.9(b) crosses the Tsai plane) and the kinetic stress will be large in magnitude. This seems to suggest that the traction changes as a function of the position of P, which would be incorrect since the system is under uniform stress. The reason that this does not occur is that every time an atom crosses P the bonds connected with it reverse directions with respect to P changing a positive contribution to the potential stress to a negative one and vice versa (see the bonds connected with atom A in Figs. 8.9(a) and 8.9(b)). This effect on the potential stress exactly compensates for the change in magnitude of the kinetic stress leaving the total stress constant.

The effect described above is demonstrated numerically in Fig. 8.9(c). This graph shows the results obtained from an MD simulation of a periodic block of 4000 aluminum atoms at 310 K [AT10]. The interactions between the atoms are modeled with a pair functional [EA94]. The periodic length of the box is set based on the zero-temperature lattice spacing. As shown in Eqn. (8.76), upon heating a compressive stress is built up in the box equal to $\sigma_{11} = \sigma_{22} = \sigma_{33} = -(c_{11} + 2c_{12})\alpha\Delta T$, where c_{ij} are the entries of the elasticity matrix in Voigt notation (see Section 2.5.5) and α is the coefficient of thermal expansion. Substituting in the appropriate values for Ercolessi–Adams aluminum [EA94] (c_{11} = 118.1 GPa, c_{12} = 62.3 GPa, $\alpha = 1.6 \times 10^{-5}\ \mathrm{K}^{-1}$) and ΔT = 310 K, gives[37] $\sigma = -1.2$ GPa. Figure 8.9(c) shows the potential, kinetic and total stress in the horizontal direction as

[37] Strictly speaking, we should be accounting for the temperature dependence of the elastic constants. See footnote 15 on page 459.

a function of the normalized position s_P of a Tsai plane P located between two lattice planes ($s_P = 0$ when P is on top of a lattice plane and $s_P = 0.5$ when it is exactly midway between two lattice planes). We see that the total stress is constant regardless of the position of the Tsai plane and equal to the expected value of -1.2 GPa computed above. However, the kinetic and potential parts change dramatically. When the Tsai plane is away from the lattice planes ($s_P = \pm 0.1$) the kinetic stress is zero (no atoms cross P as they vibrate about their lattice sites) and the entire stress is given by the potential part of the stress. As the Tsai plane moves closer to a lattice plane ($|s_P| \to 0$), the kinetic stress becomes more negative (increasing in magnitude) and the potential stress increases in exactly the right amount to compensate for this effect. When the Tsai plane is directly on top of a lattice plane ($s_P = 0$) both the kinetic stress and potential stress are maximum in magnitude, but their sum remains equal to the constant total stress. This is a striking demonstration of the relation between the kinetic and potential parts of the stress tensor.

Advantages and disadvantages of the Tsai expression Tsai's traction definition is appealing since it is more closely related to Cauchy's original concept of stress. It also provides a spatially more localized measure since it provides the stress across a plane instead of an average over a volume. However, this latter advantage comes with a price. Since the stresses are computed from an average over a plane rather than over a volume, to obtain comparable accuracy to the volumetric approaches, the plane needs to be much larger than the cross section through the volumetric domain. Thus, while the stress obtained from the traction definition is more localized in the direction normal to the plane than the volume average, it is less localized in the perpendicular directions (see discussion in Section 8.3.5).

An additional downside to the traction definition is that it provides only three of the six independent components of the stress tensor from a single plane measurement. For example if the plane is taken normal to the x_1-axis, i.e. $\boldsymbol{n} = \boldsymbol{e}_1$ so that $\boldsymbol{t} = \boldsymbol{\sigma n} = \sigma_{11}\boldsymbol{e}_1 + \sigma_{21}\boldsymbol{e}_2 + \sigma_{31}\boldsymbol{e}_3$, the three stress components obtained from Eqn. (8.160) are

$$\sigma_{k1} = \frac{1}{N_\tau A}\sum_{i=1}^{N_\tau}\left[-\frac{1}{\Delta\tau}\sum_{\alpha\leftrightarrow P} m^\alpha v_{\mathrm{rel},k}^{\alpha(i)}\frac{v_{\mathrm{rel},1}^{\alpha(i)}}{|v_{\mathrm{rel},1}^{\alpha(i)}|} + \sum_{\alpha\beta\cap P} f_k^{\alpha\beta(i)}\frac{r_1^{\alpha\beta(i)}}{|r_1^{\alpha\beta(i)}|}\right], \qquad k = 1, 2, 3.$$

To obtain the full stress tensor at least two planes must be evaluated at each point, More accurate results are obtained if three planes are evaluated and the off-diagonal terms are averaged to ensure a symmetric stress tensor.

The issue of symmetry is interesting. It is not clear from the perspective put forward by Tsai [Tsa79], whether the stress tensor obtained from the Tsai traction would be symmetric or even well defined, i.e. it is not clear if another choice of planes will give suitably transformed components of the same stress tensor. Our derivation suggests that a stress tensor constructed from the Tsai traction should be well defined and symmetric, at least in a weak sense, since it is a limit of the Hardy stress, which has these properties. A numerical test described in [AT10, Section 6.3] examining the stress concentration about a hole in a plate suggests that the stress tensor associated with the Tsai traction is indeed symmetric. See also the discussion in Section 8.3.5.

8.3.4 Uniqueness of the stress tensor

In the derivation presented so far, we have demonstrated that the microscopically-based Hardy stress, virial stress and Tsai traction can all be obtained from nonequilibrium statistical mechanics in a two-step procedure:

1. The Irving–Kirkwood–Noll procedure [IK50, Nol55] is applied to obtain a *pointwise* stress tensor, $\boldsymbol{\sigma}^{\mathrm{pt}} = \boldsymbol{\sigma}^{\mathrm{pt,K}} + \boldsymbol{\sigma}^{\mathrm{pt,V}}$, consisting of kinetic and potential parts. The potential part of the stress is obtained for arbitrary multibody potentials, which generalizes the original Irving–Kirkwood–Noll approach that was limited to pair potentials. This generalization is based on the important result from Section 5.8.1 that the force on an atom can always be expressed as a sum of *central* forces regardless of the nature of the interatomic potential.
2. The pointwise stress obtained in the first step is spatially averaged to obtain the *macroscopic* stress $\boldsymbol{\sigma}$.

This two-step procedure provides a general unified framework from which various stress definitions can be derived including *all* of the main definitions commonly used in practice (see Fig. 8.1). In particular, it is shown that this procedure leads directly to the stress tensor derived by Hardy in [Har82]. The traction of Cauchy and Tsai [Cau28a, Cau28b, Tsa79] is obtained from the Hardy stress in the limit that the averaging domain is collapsed to a plane. The virial stress of Clausius and Maxwell [Cla70, Max70] is an approximation of the Hardy stress tensor for a uniform weighting function where bonds that cross the averaging domain are neglected. The Hardy stress and virial stress become identical in the thermodynamic limit. In this manner, clear connections are established between all of the major stress definitions in use today. It is demonstrated that the stress tensors obtained in this way are symmetric for all interatomic potentials.

Three possible sources of nonuniqueness for the stress tensor were identified during the procedure described above:

1. Since there are multiple extensions to the interatomic potential energy function (see Section 5.8.2 and Appendix A), different force decompositions are possible and hence different pointwise stress tensors can be obtained. Admal and Tadmor [AT10, Section 5.5] showed that the macroscopic stress tensor, calculated as a spatial average of the pointwise stress tensor with constant weighting function, is always unique in the thermodynamic limit, i.e. the difference between the spatially-averaged pointwise stress tensors resulting from two different extensions, $\Delta\boldsymbol{\sigma}^{\mathrm{pt}}$, tends to zero, as the volume of the averaging domain is increased.
2. For a given pointwise stress tensor, a new pointwise stress, which also satisfies the balance of linear momentum, can be obtained by adding an arbitrary tensor field with zero divergence. This issue is well known (see, for example, [SH82, WAD95, MRT06 footnote 13]). We do not resolve it here, however, we point to two results that strengthen the argument that pointwise stress obtained through the Irving–Kirkwood–Noll procedure, i.e. the Noll stress, is the correct solution. First, Lehoucq and Silling [LS08] have shown that the Noll stress is a minimal solution in a variational sense. The physical significance

of this variational principle still needs to be studied, but the result is suggestive. Second, we showed above in Section 8.3.2 that the stress tensor obtained by spatially averaging the Noll stress converges to the virial stress in the thermodynamic limit. Let us explore the implications of this.

Consider the pointwise stress, $\boldsymbol{\sigma}^{\text{pt}}$, obtained through the Irving–Kirkwood–Noll procedure, which satisfies the balance of linear momentum, and a new pointwise stress, $\hat{\boldsymbol{\sigma}}^{\text{pt}} = \boldsymbol{\sigma}^{\text{pt}} + \tilde{\boldsymbol{\sigma}}^{\text{pt}}$, where $\operatorname{div} \tilde{\boldsymbol{\sigma}}^{\text{pt}} = \mathbf{0}$. Clearly, $\hat{\boldsymbol{\sigma}}^{\text{pt}}$ also satisfies the balance of linear momentum and is therefore also a valid solution. The spatially-averaged stress obtained from the new definition is

$$\hat{\boldsymbol{\sigma}}(\boldsymbol{x};t) = \int_{\boldsymbol{y}\in\mathbb{R}^3} w(\boldsymbol{y}-\boldsymbol{x})\hat{\boldsymbol{\sigma}}^{\text{pt}}(\boldsymbol{y};t)\,d\boldsymbol{y} = \int_{\boldsymbol{y}\in\mathbb{R}^3} w(\boldsymbol{y}-\boldsymbol{x})(\boldsymbol{\sigma}^{\text{pt}}(\boldsymbol{y};t)+\tilde{\boldsymbol{\sigma}}^{\text{pt}}(\boldsymbol{y};t))\,d\boldsymbol{y}. \tag{8.161}$$

We showed in Section 8.3.2 that in the thermodynamic limit the spatially-averaged stress $\boldsymbol{\sigma}$ obtained from $\boldsymbol{\sigma}^{\text{pt}}$ converges to the virial stress. We also expect $\hat{\boldsymbol{\sigma}}$ to equal the virial stress in this limit (since any macroscopic stress must converge to this value under equilibrium conditions). Therefore, Eqn. (8.161) reduces to

$$\lim_{\text{TD}} \int_{\mathbb{R}^3} w(\boldsymbol{y}-\boldsymbol{x})\tilde{\boldsymbol{\sigma}}^{\text{pt}}(\boldsymbol{y};t)\,d\boldsymbol{y} = \mathbf{0}, \tag{8.162}$$

where $\lim_{\text{TD}}$ refers to the thermodynamic limit. Equation (8.162) places a strong constraint on allowable forms for $\tilde{\boldsymbol{\sigma}}^{\text{pt}}$, the implications of which remain an area of open research.

3. The Irving–Kirkwood–Noll procedure can be generalized to arbitrary curved "paths of interaction" which leads to the possibility of nonsymmetric expressions for the pointwise stress tensor. This is discussed in [AT10] by Admal and Tadmor, who show that curved paths of interaction are only possible in systems where the discrete particles making up the system possess internal structure, such as internal polarization or spin. For systems of discrete particles without internal structure only straight bonds are possible due to symmetry arguments.

8.3.5 Hardy, virial and Tsai stress expressions: numerical considerations

Before concluding this chapter, we make a few brief comments regarding the ease of use and accuracy of the various stress expressions derived above. These comments are based on the numerical tests that are described in [AT10, Section 6]. Two mechanical boundary-value problems were considered in these tests: (1) a system under uniform uniaxial tension; and (2) a plate with a hole subjected to a uniaxial tension at the boundaries. In both cases, the stresses computed using the Hardy, Tsai and virial stress expressions were compared with the "exact" anisotropic linear elasticity solution (computed using the material properties predicted by the interatomic model used in the microscopic calculations). The accuracy of the solution as a function of the averaging domain size was considered, as well as other considerations like ease of use and the symmetry of the resulting stress tensor. All of the performed numerical experiments are static and at zero temperature since the kinetic part of the stress tensor is the same for all three expressions.

Overall, we can summarize as follows. Of the three definitions, the Hardy stress is generally preferred. (A similar conclusion was reached by Zimmerman *et al.* [ZWH$^+$04].) The Hardy stress tends to be the smoothest and provides good accuracy away from surfaces as long as the length scale over which the continuum fields vary is large relative to the atomic spacing. In situations where either of those conditions breaks down, the Tsai stress provides a better localized measure of stress. The virial stress is less accurate than both, but it has the advantage of being the easiest to compute. The evaluation of the bond function in the Hardy stress makes it a bit more difficult to compute, but comparable to the virial stress. The Tsai traction is most difficult and time consuming to compute, since it requires the detection of bonds and atoms that cross a given plane during the averaging process. Furthermore, this evaluation must be performed for three separate planes in order to obtain the full stress tensor in three dimensions. For more details see the discussion in [AT10].

Exercises

8.1 [SECTION 8.1] In this problem we investigate the application of a canonical transformation in a simple one-dimensional system. The Hamiltonian of a harmonic oscillator is given by $\mathcal{H} = p^2/(2m) + kq^2/2$, where q and p are the position and momentum variables, m is the mass of the oscillator and k is the spring constant.

1. Write the canonical equations of motion for this system. Combine the equations into a single second-order ordinary differential equation for q.
2. Show that $q(t) = q_0 \cos(\omega t + \phi)$, where $\omega = \sqrt{k/m}$ and q_0 and ϕ are constants, is a solution to the ordinary differential equation derived above.
3. Consider the canonical transformation defined by the generator function, $G_1 = q\widehat{q}$. Obtain an expression for the Hamiltonian and the canonical equations of motion in terms of the new canonical variables $\widehat{q}$ and $\widehat{p}$.
4. Derive a second-order ordinary differential equation for $\widehat{q}$ from the canonical equations derived in the previous part. Solve this equation and show that the solution is equivalent to the one given above for q. Explain the role played by the canonical transformation.

8.2 [SECTION 8.1] Rederive Eqn. (8.37) for a system S consisting of a finite one-dimensional chain of N^{S} atoms. The chain is divided into an internal system A embedded in an outer system B. The atoms interact via a general interatomic potential $\mathcal{V}^{\mathsf{A+B}}(\boldsymbol{r}^{\mathsf{A}}, \boldsymbol{r}^{\mathsf{B}})$. Show that the stress is given by

$$P = -\frac{1}{L_0} \sum_{\alpha \in \mathsf{A}} f^{\text{int},\alpha} \left(R^\alpha + \frac{\partial u^\alpha}{\partial F} \right), \tag{8.163}$$

where L_0 is the length of system A in the reference configuration, $f^{\text{int},\alpha} = -\partial \mathcal{V}^{\mathsf{A}}/\partial r^\alpha$, and F is the one-dimensional deformation gradient. The other symbols have the same meaning as in Section 8.1.2.

8.3 [SECTION 8.1] Show that $\partial u^\alpha/\partial F$ in Eqn. (8.163) can be expressed as

$$\frac{\partial u^\alpha}{\partial F} = \sum_{\gamma \in \mathsf{S}} \left[\frac{\partial^2 \mathcal{V}^{\mathsf{A+B}}}{\partial u^\alpha \partial u^\gamma} \right]^{-1} \frac{\partial^2 \mathcal{V}^{\mathsf{A+B}}}{\partial u^\gamma \partial F}.$$

8.4 [SECTION 8.1] Let S be a finite one-dimensional chain of N^{S} *identical* atoms. The chain is subjected to a stretch defined by the deformation gradient F. Let A be a subsystem of S consisting of N atoms whose ends are far from the ends of S. The stress in system A is given by Eqn. (8.163). Assuming short-range atomic interactions, show that if $N \ll N^{\mathrm{S}}$, then the contribution to the stress coming from terms that include $\partial u^\alpha/\partial F$ are negligible, therefore

$$P = -\frac{1}{L_0}\sum_{\alpha\in\mathrm{A}} f^{\mathrm{int},\alpha} R^\alpha. \tag{8.164}$$

8.5 [Section 8.1] Let A be a one-dimensional chain of N identical atoms embedded in a much larger finite chain of N^{S} atoms. The chain is subjected to a stretch defined by the deformation gradient F. The atoms in the chain interact via a pair potential, $\phi(r)$, with second-neighbor interactions. Thus, the energy of atom α in the chain is

$$E^\alpha = \sum_{\substack{\beta=\alpha-2\\ \beta\neq\alpha}}^{\alpha+2} \phi(|r^{\alpha\beta}|),$$

where $r^{\alpha\beta} = r^\beta - r^\alpha$ and r^α is the position of atom α.

1. Obtain an explicit expression for the stress in A. Assume that $N \ll N^{\mathrm{S}}$ so that Eqn. (8.164) can be used. Write your expression in terms of $\phi(r)$, N, F and the equilibrium lattice parameter a_0 of an infinite chain.
2. Show that in the limit $N \to \infty$, the expression you obtained is physically sensible.

8.6 [Section 8.1] A rectangular three-dimensional block of a simple cubic crystal is oriented so that the [100], [010] and [001] directions coincide with the X_1-, X_2- and X_3-axes of a Cartesian coordinate system. The block occupies the domain in the reference configuration defined by $-n_i a_0 \le X_i \le n_i a_0$, where n_i are positive integers and a_0 is the equilibrium lattice parameter. The block is subjected to a shear deformation by holding m (001) planes at the bottom of the block fixed and rigidly displacing m (001) planes at the top in the [100] direction. (Note that the expressions derived below will depend on m as a parameter.)

1. Use Eqn. (8.38) to obtain an expression for the P_{13} component of the first Piola–Kirchhoff stress tensor in terms of the external forces acting on the constrained planes. Denote by $f^{\mathrm{lay}}(X_3)$ the force in the 1-direction acting on the layer located at height X_3. (Neglect edge and corner effects.) Note that due to symmetry, $f^{\mathrm{lay}}(X_3) = -f^{\mathrm{lay}}(-X_3)$.
2. Compare the expression for P_{13} computed in the previous part with the common sense definition for the engineering shear stress as the total shearing force divided by the reference cross-section area. Show that the two definitions agree only in the limit of $n_3 \to \infty$. Based on this, comment on the validity of Eqn. (8.38) as a definition for a macroscopic stress.

8.7 [Section 8.1] Consider a periodic, nonorthogonal supercell defined by the vectors $\boldsymbol{a}_i = \boldsymbol{F}\boldsymbol{A}_i$, where $\boldsymbol{A}_i$ are reference supercell vectors and $\boldsymbol{F}$ is the deformation gradient. The positions of the N atoms in the cell are $\boldsymbol{r}^\alpha = s_i^\alpha \boldsymbol{a}_i + \boldsymbol{u}^\alpha$, where $\boldsymbol{s}^\alpha$ are fixed scaled coordinates and $\boldsymbol{u}^\alpha$ are displacements. The atoms in the cell interact via a potential energy function $\mathcal{V}^{\mathrm{int}}(\{r^{\alpha\mathring{\beta}}\})$, where

$$r^{\alpha\mathring{\beta}} = \left\| \boldsymbol{r}^\beta_{\mathrm{per}}(\mathring{n}_1,\mathring{n}_2,\mathring{n}_3) - \boldsymbol{r}^\alpha \right\|, \qquad \boldsymbol{r}^\beta_{\mathrm{per}}(n_1,n_2,n_3) = \boldsymbol{r}^\beta + n_i\boldsymbol{a}_i,$$

$$(\mathring{n}_1,\mathring{n}_2,\mathring{n}_3) = \arg\min_{n_1,n_2,n_3} \left\| \boldsymbol{r}^\beta_{\mathrm{per}}(n_1,n_2,n_3) - \boldsymbol{r}^\alpha \right\|.$$

Under equilibrium conditions, where $\partial\mathcal{V}^{\mathrm{int}}/\partial\boldsymbol{u}^\alpha = \boldsymbol{0}$ ($\alpha = 1,\dots,N$), derive an expression for the first Piola–Kirchhoff stress from $\boldsymbol{P} = (1/V_0)\partial\mathcal{V}^{\mathrm{int}}/\partial\boldsymbol{F}$, where V_0 is the volume of the reference supercell. Show that your result is consistent with the one given in Eqn. (8.69), which was adapted from a calculation on a finite system.

8.8 [Section 8.1] An infinite set of atoms is arranged in an fcc crystal structure with lattice parameter a. The atoms interact via a nearest-neighbor pair potential $\phi(r)$.

1. Using Eqn. (8.69), derive an expression for the Cauchy stress tensor at 0 K. Express your result in terms of $\phi(r)$ and a as a matrix of the components of $\boldsymbol{\sigma}$ in the principal crystallographic coordinate system.
2. Using your results above, obtain an equation for determining the 0 K equilibrium lattice parameter a_0.

3. Solve for the equilibrium lattice parameter for the special case of a Morse potential given in Eqn. (5.38). Express your result in terms of the Morse potential parameters, σ, ε and r_e, as necessary.
4. Plot the pressure, $p = -\frac{1}{3}\operatorname{tr}\sigma$ (see Eqn. (2.89)$_1$) versus the lattice parameter, a, for a Morse potential with parameters $\sigma = \varepsilon = r_e = 1$ in the range $a \in [0.5a_0, 5a_0]$. What is the physical significance of the minimum of this function?

8.9 [SECTION 8.1] Derive the result in Eqn. (8.76).

9 Molecular dynamics

In Chapter 6, we discussed molecular statics (MS), where the potential energy landscape is used to understand materials phenomena. But MS has its limitations, of course, because real systems of atoms at finite temperature are in constant motion. Molecular dynamics (MD) simulations follow the motion of all of the atomic nuclei in the system by treating them as classical, Newtonian particles and integrating the equations of motion:

$$m^{\alpha}\ddot{\boldsymbol{r}}^{\alpha} = \boldsymbol{f}^{\alpha} \qquad \alpha = 1, \dots, N. \tag{9.1}$$

Here N is the number of atoms, m^{α} is the mass of atom α, $\boldsymbol{r}^{\alpha}$ is its position and $\boldsymbol{f}^{\alpha}$ is the time-dependent force acting on it due to external effects and the presence of its neighbors. A superposed dot denotes time differentiation, so that $\ddot{\boldsymbol{r}}^{\alpha}$ is the acceleration of atom α. The force $\boldsymbol{f}^{\alpha}$ is generally a function of the positions of all the other atoms, as we have seen in Chapter 5. In some cases (such as when we implement the thermostats of Section 9.4), $\boldsymbol{f}^{\alpha}$ may depend on the particle velocities as well. In this chapter, we discuss the theory behind the different types of MD simulation one might implement: the MD "ensembles" in the language of statistical mechanics.[1]

Treating the atoms as classical particles and replacing the electrons with effective force laws has an important consequence for dynamic simulations: the transfer of heat between the moving atoms and the free electron gas is ignored. For systems with highly ionic or covalent bonding character, this may not be a big problem, but in metals there is often significant heat conduction by the electrons. For some physical phenomena, where heating or cooling is relatively gentle and conduction by the electrons is expected to rapidly carry away any local temperature gradients, this may be approximately corrected using a thermostat (the subject of Section 9.4). For other problems, the localization of temperature and the rate of heat exchange between the phonons and the electrons may be a critical part of the problem (for example during irradiation damage or laser pulse heating). In these cases, additional modeling considerations are required (see, for example, [Che05]). Here, we focus on simulations where electron–phonon coupling can be ignored.

9.1 Brief historical introduction

Before diving into the theory of MD, it is worthwhile to briefly review the history of the method. The interest in simulating the dynamics of molecules extends back to the end of

[1] Historically, statistical mechanics is the branch of science from which MD evolved, and as such it is often described in that context.

the nineteenth century when the discrete structure of matter was at last starting to become widely accepted (see Section 3.1) and the theory of statistical mechanics was emerging. Due to the large number of particles involved, direct calculation of the trajectories of the particles was not possible at that time. Instead researchers resorted to "analog" MD studies, where spheres of different materials were placed in physical containers and agitated to simulate the behavior of liquids.

Perhaps the first study of this type was performed by William Sutherland in 1891 [Sut91a, Sut91b, Sut91c].[2] Sutherland was studying liquids and postulated that freezing occurred when neighboring particles formed a cage trapping a given particle. To understand the physics of this process Sutherland performed what can be called the first "MD" (marble dynamics) experiments by putting 100 close-packed marbles into a box with five colored marbles at the center [Cur86]. Agitation led to slow diffusion when 16 marbles were removed. Rapid diffusion of the five colored ones occurred once 20–25 marbles were removed. Sutherland had a number of other quite insightful propositions about melting which were later borne out more formally and via experiment. This approach was similar to the bubble raft approach of Bragg and Nye [BN47] but preceded it by over 50 years.

Sutherland's work was followed by other studies which employed balls of different materials. For example, Morrell and Hildebrand [MH36] used gelatine balls to measure the radial distribution function of a liquid (the probability of finding two particles at a given distance). In 1960, Turnbull and Cormia [TC60] used a system of glass spheres to study the change in behavior from gaslike to liquidlike to solidlike (crystalline) as the density is increased. Pierański *et al.* [PMKW78] and Pouligny *et al.* [PMRC90] used ball bearings to study the melting transition, and more recently Wu *et al.* [WWN05] used a similar approach to study structural phase transitions.

The invention of the electronic computer in the 1940s meant that an alternative approach to simulating the dynamics of molecules based on calculation rather than observation became feasible. The development of MD emerged from earlier work on the Monte Carlo method, which is attributed to Stanislaw Ulam and John von Neumann,[3] and was later developed (and named) by Metropolis and coworkers [MU49, MRR+53]. The first MD simulations were carried out in the late 1950s by Berni J. Alder[4] in collaboration with

[2] We thank Bill Curtin from Brown University for sharing this information with us. The description that follows of Sutherland's work is based on his communication. He also coined the term "marble dynamics."

[3] Enrico Fermi also deserves credit for this discovery. According to his student and collaborator, Emilio Segre, Fermi was quietly performing Monte Carlo simulations 15 years before Ulam and von Neumann began working with this technique using newly invented computers in the mid 1940s. Metropolis relates the following story told him by Segre about Fermi's time in the early 1930s at the University of Rome, where he was studying neutron diffusion [Met87]: "Fermi took great delight in astonishing his Roman colleagues with his remarkably accurate, 'too-good-to-believe' predictions of experimental results. After indulging himself, he revealed that his 'guesses' were really derived from the statistical sampling techniques that he used to calculate [using his small mechanical adding machine] whenever insomnia struck in the wee morning hours! And so it was that nearly fifteen years earlier, Fermi had independently developed the Monte Carlo method."

[4] Berni Alder was awarded the US National Medal of Science in 2009 for "establishing powerful computer methods useful for molecular dynamics simulations, conceiving and executing experimental shock-wave simulations to obtain properties of fluids and solids at very high pressures, and developing Monte Carlo methods for calculating the properties of matter from first principles, all of which contributed to major achievements in the science of

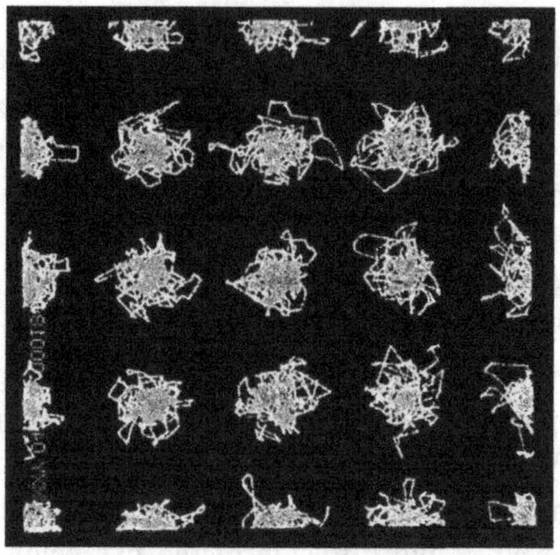

Fig. 9.1 Results from one of the first MD simulations ever performed by the inventors of the method, Alder and Wainwright [AW59]. The figure shows the trace left by 32 hard-sphere particles in a periodic MD simulation of a face-centered cubic (fcc) crystal. Reprinted with permission from [AW59]. Copyright 1959, American Institute of Physics.

Thomas Wainwright at Lawrence Livermore National Laboratory [AW57, AW59]. These were simulations of several hundred "hard-sphere" particles interacting via a square well potential with periodic boundary conditions (PBCs). In a hard-sphere simulation, particles travel freely through space and only interact during the collision phase. An example of the results obtained from these initial MD simulations is shown in Fig. 9.1 for a solid system. Using this new approach, Alder and his coworkers were able to compute the equation of state of the system along with various other equilibrium and nonequilibrium properties. The calculations were run on a state-of-the-art IBM-704 electronic computer. For 108 particles, about 2000 collisions per hour could be simulated. The authors acknowledged the help of Shirley Campbell and Mary Shephard with the computer coding.

The next significant step in the development of MD was taken by Gibson *et al.* [GGMV60] in 1960, who used a somewhat more realistic continuous interatomic potential to study radiation damage in copper.[5] For the first time, the equations of motion were integrated using a finite difference procedure (see Section 9.3.1). Following this there was an explosion in the number of articles using MD techniques as shown in Tab. 9.1.[6] A noteworthy milestone is the work of Aneesur Rahman at Argonne National Laboratory on simulations of liquid argon using the Lennard-Jones potential [Rah64]. What sets this simulation apart is the use of a realistic interatomic potential with attractive and repulsive parts which captured the physics of the system being studied. Later Stillinger and Rahman performed the first realistic simulation of liquid water [SR74]. Rahman continued to be a major figure in the field of molecular simulation for many years.

condensed matter." Dr. Alder is currently a professor emeritus at the University of California, Davis. At 84 he continues to be active in research.

[5] Gibson *et al.* [GGMV60] used a Born–Mayer potential to describe the repulsion between the atoms and mimicked the cohesion between the particles by applying an external pressure.

[6] The table was generated using Web of Science, searching for articles with the term "molecular dynamics" in either the title or the abstract. The numbers for the early years are an underestimate since the term "molecular dynamics" was still not widely used.

Table 9.1. Number of "molecular dynamics" articles published, by year.

Years	Number of articles
1957–1960	4
1961–1970	52
1971–1980	583
1981–1990	2,237
1991–2000	27,359
2001–2009	51,878

From here on the MD method continued to evolve with the development of more sophisticated interatomic potentials, improved integration techniques and methods for simulating different ensembles (constant temperature, constant stress, etc.).[7] In addition, increasing computing power together with more efficient algorithms and parallelization enabled the simulation of larger systems for longer times. The world record in 2010 appears to have been a simulation of one trillion atoms performed at Lawrence Livermore National Laboratory on the BlueGene/L architecture [GK08].

9.2 The essential MD algorithm

As we have mentioned before, MD comes historically from the realm of statistical mechanics, and as such it is typical to speak of the *ensemble* one chooses to simulate with a given MD run. In Chapter 7, we introduced the concept of phase space sampling and discussed the correspondence between time-averaged and phase-averaged predictions of state variables. In MD, the goal is often a computational realization of the time-averaging process, where properties of interest are extracted from a system simulated in equilibrium over some time period. In other instances, atom trajectories are the focus, as simulations are used to reveal possible mechanisms of material deformation.

As in MS, the forces on the atoms are given by

$$\boldsymbol{f}^{\alpha} = -\partial\mathcal{V}/\partial\boldsymbol{r}^{\alpha}, \tag{9.2}$$

where $\mathcal{V}$ is the potential energy of the system. The potential energy consists of an internal part, $\mathcal{V}^{\text{int}}$, reflecting the interactions between the atoms making up the system, and an external part, $\mathcal{V}^{\text{ext}}$, due to external fields and constraints (see Section 5.1). Having prescribed the forces on the atoms, their motion can be obtained by numerically integrating the equations of motion in Eqn. (9.1) using one of the methods discussed below.

The temperature and pressure (or more generally stress) of the simulation box can be monitored during the simulation as output. The Cauchy stress and true pressure can

[7] There is a large number of sophisticated MD codes available, both freely and commercially. The companion website to this book links to several of them.

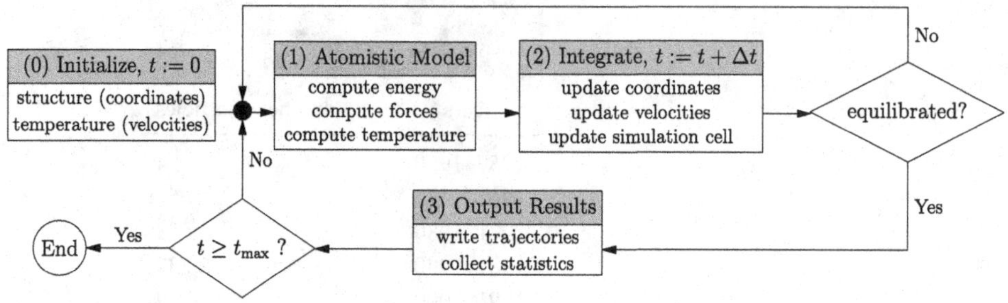

Fig. 9.2 Flow chart of the MD solution process. See text for details.

be obtained from Eqn. (8.69).[8] The temperature is the time-averaged vibrational kinetic energy of the system (Eqn. (7.104)),

$$T = \frac{2}{3Nk_B}\overline{\mathcal{T}^{\text{vib}}} = \frac{2}{3Nk_B}\overline{\sum_{\alpha=1}^{N}\frac{1}{2}m^{\alpha}\left\|\boldsymbol{v}_{\text{rel}}^{\alpha}\right\|^2}, \tag{9.3}$$

where $\boldsymbol{v}_{\text{rel}}^{\alpha}$ is the velocity of atom α relative to the center of mass velocity (see Eqn. (7.11)), and we have invoked the assumption that time averages can be used in place of phase averages (see Chapter 7). It is sometimes convenient to define an "instantaneous temperature," $\mathcal{T}_{\text{inst}}$, based on the instantaneous vibrational energy of the system and Eqn. (9.3):

$$\mathcal{T}_{\text{inst}}(\boldsymbol{v}_{\text{rel}}) = \frac{2}{3Nk_B}\left(\sum_{\alpha=1}^{N}\frac{1}{2}m^{\alpha}\left\|\boldsymbol{v}_{\text{rel}}^{\alpha}\right\|^2\right). \tag{9.4}$$

This is not the true temperature, which can only be computed through a phase (or time) average as in Eqn. (9.3); it will fluctuate around the true temperature.

The flow chart in Fig. 9.2 illustrates the essential steps in an MD simulation. The initialization step labeled (0) involves the definition of the initial structure, the boundary conditions (see Section 6.4.2) and the particle velocities (normally set to an initial temperature, see Section 9.3.3). Subsequently, the main integration loop proceeds until a desired time, t_{max}, has been simulated, typically after thousands or millions of timesteps. The simulation may or may not require a period of time to equilibrate the system before starting the collection of data, and therefore the "equilibrated?" decision in the flow chart depends on the desired results. There are three main steps in the integration loop (although for more sophisticated numerical integrators there can be some overlap between steps (1) and (2)). Step (1) is the evaluation of the potential and the kinetic energy and the forces based on the current positions and velocities. This is a modular component of the simulation, since any of the atomistic models described in Chapter 5 can be used as the force model at this point. For simulations with temperature control (see Section 9.4), the atomic forces are also modified by the thermostat algorithm during step (1). In step (2), the coordinates and velocities of the atoms are updated according to the integrator algorithm (see Section 9.3.1). For

[8] In a periodic system, the volume in Eqn. (8.69) is the volume of the periodic unit cell and the sum on α is over atoms within the primary unit cell. Forces on an atom α must include contributions from all atoms β whether β lies in the primary unit cell or in one of the periodic copies.

constant stress simulations (Section 9.5), the shape and size of the simulation box may also be updated during this step. Finally, step (3) provides the useful output from the simulation. This may involve collection of statistical data (for time averages of temperature, pressure, etc.) or the writing of atomic trajectories for later analysis and visualization. In some cases, analysis of the trajectories may be done during the simulation by specially-written routines. Depending of the details of the simulation, it may be necessary to allow for some equilibration time before including step (3) in the loop (see Section 9.3.4).

The remainder of this chapter discusses the details of each of these essential MD steps for the four most important types of MD simulation: constant energy/strain, constant temperature/strain, constant energy/stress and constant temperature/stress.

9.3 The NVE ensemble: constant energy and constant strain

The easiest MD simulation to implement is one with constant number of atoms, constant volume and constant energy (Thus the use of the notation NVE. The most common MD ensembles are denoted in this way by the three state variables that are prescribed. For more on why exactly three quantities are needed to specify an ensemble, see Section 5.1.4 of [TME12].). This is akin to an isolated system in statistical mechanics, the so-called microcanonical ensemble (see Section 7.3). In modeling solids, we can equally speak of constant volume or *constant strain*. Fixing the volume and shape of a periodic simulation cell is equivalent to holding the macroscopically applied strain at some specified value and watching how the microscopic degrees of freedom respond. In this case we build a simulation cell of fixed size exactly as we described in the context of MS in Chapter 6.[9]

An important property of the NVE ensemble is that the total energy of the system, given by the Hamiltonian $\mathcal{H}$ in Eqn. (7.2) is conserved as the system evolves (see Section 7.1.5 for the proof). We must therefore use an energy conserving numerical integrator as described in the next section.

9.3.1 Integrating the NVE ensemble: the velocity-Verlet (VV) algorithm

Since we cannot analytically solve for the complex motion of N particles obeying Eqn. (9.1), we need to employ a suitable numerical integrator.[10] Ideally, this discrete scheme should behave as closely as possible like the continuous system of equations. Specifically, there

[9] In general, this cell could be nonorthogonal, but still of fixed volume of course.

[10] The use of numerical integration and the finite precision of digital computers leads to an interesting philosophical question at the heart of MD. It is well known that infinitesimal perturbations to the positions and momenta of a Hamiltonian system can cause the trajectory of this system to deviate exponentially from its original trajectory. This is referred to as the *Lyapunov instability* (see footnote 10 on page 389). This means that the discretized trajectory obtained from an MD simulation is never close to a single "real" trajectory that would be obtained from an exact solution of Eqn. (9.1), but is continuously hopping from one real trajectory to another. Despite this we believe that the results of an MD simulation have physical meaning. This is a deep question at the heart of atomistic simulations of materials. See, for example, the discussion regarding this in the small entertaining book by Hoover [Hoo99].

are three features of Eqn. (9.1) that we would like to preserve. First, Eqn. (9.1) conserves the total Hamiltonian (kinetic + potential energy) of a closed system (see Section 7.1.5). Second, Eqn. (9.1) is time-reversible, which is to say that starting from some final state and moving backwards in time will retrace the same trajectory as originally followed from the initial state. Third, the equation is *symplectic*, which means that it obeys Liouville's equation (Eqn. (7.23)). A numerical scheme that has these three features will provide solutions with the best possible accuracy and stability. Although there are several algorithms from which to choose that meet these criteria, we focus on the one that is most commonly used in MD simulations, the so-called velocity-Verlet algorithm.

The *velocity-Verlet (VV) algorithm* [SABW82] can be derived as follows.[11] We start from some time t, where the positions of the particles are $\boldsymbol{r}(t) = (\boldsymbol{r}^1(t), \ldots, \boldsymbol{r}^N(t))$ and the velocities are $\boldsymbol{v}(t) = (\boldsymbol{v}^1(t), \ldots, \boldsymbol{v}^N(t))$. Given the positions,[12] we have at our disposal a suite of atomic models (Section 4.5 through Chapter 5) from which we can find the forces on the atoms, $\widehat{\boldsymbol{f}}(t) = \boldsymbol{f}(\boldsymbol{r}(t))$, where $\boldsymbol{f}(\boldsymbol{r}) = (\boldsymbol{f}^1(\boldsymbol{r}), \ldots, \boldsymbol{f}^N(\boldsymbol{r}))$ and similarly for $\widehat{\boldsymbol{f}}$. We can then approximate the position and velocity vectors at some short time Δt later by a Taylor expansion to second order:

$$\boldsymbol{r}(t + \Delta t) \approx \boldsymbol{r}(t) + \boldsymbol{v}(t)\Delta t + \boldsymbol{a}(t)\frac{(\Delta t)^2}{2}, \tag{9.5}$$

where $\boldsymbol{v}(t) = \dot{\boldsymbol{r}}(t)$ and $\boldsymbol{a}(t) = \ddot{\boldsymbol{r}}(t)$. When discussing time-discretized integration, it will be helpful to adopt a more concise notation. We introduce a new subscript such that the previous equation can be replaced with

$$\boldsymbol{r}_{t+\Delta t} \approx \boldsymbol{r}_t + \boldsymbol{v}_t\Delta t + \boldsymbol{a}_t\frac{(\Delta t)^2}{2}. \tag{9.6}$$

This notation also reminds us of the discrete nature of the equations, since while $\boldsymbol{v}(t)$ suggests a continuous function of time, $\boldsymbol{v}_t$ is meant to suggest a variable evaluated at a particular instance of time, t. We truncate the Taylor series in this equation at second order for the practical reason that the acceleration is directly obtainable from the forces via Eqn. (9.1), whereas higher derivatives are not. We also need a way to propagate the velocity to time $t + \Delta t$, which we can obtain again by recourse to a Taylor series

$$\boldsymbol{v}_{t+\Delta t} \approx \boldsymbol{v}_t + \boldsymbol{a}_t\Delta t. \tag{9.7}$$

However, we note that one could equally well apply the Taylor series in reverse, as an approximation for $\boldsymbol{v}_t$ if we already knew $\boldsymbol{v}_{t+\Delta t}$. This leads to

$$\boldsymbol{v}_t \approx \boldsymbol{v}_{t+\Delta t} - \boldsymbol{a}_{t+\Delta t}\Delta t,$$

[11] The velocity-Verlet algorithm is an extension of the *Verlet method*, named after Loup Verlet who used it in one of the early papers on MD [Ver67]. In the simple Verlet algorithm, the atomic positions at a given timestep are computed from the two prior timesteps, while the evolution of the velocity is not directly addressed. This introduces some minor inconveniences in MD, since the velocity is also an important variable and initial conditions are typically given in velocity terms. Therefore, the VV algorithm explicitly updates both positions and velocities. The VV algorithm also goes by the name "leap-frog algorithm" in some disciplines.

[12] For now, we confine ourselves to simulations where the forces are only a function of atomic *positions*. In Section 9.4.7, we will discuss the situation where forces also depend on atomic velocities.

which we can rearrange as

$$\boldsymbol{v}_{t+\Delta t} \approx \boldsymbol{v}_t + \boldsymbol{a}_{t+\Delta t}\Delta t. \tag{9.8}$$

Blending these two approximations for $\boldsymbol{v}_{t+\Delta t}$, by simply averaging Eqns. (9.7) and (9.8) will ensure the time-reversibility of the algorithm. Thus we take

$$\boldsymbol{v}_{t+\Delta t} = \boldsymbol{v}_t + \frac{\boldsymbol{a}_t + \boldsymbol{a}_{t+\Delta t}}{2}\Delta t, \tag{9.9}$$

and use Eqn. (9.6) to integrate the positions. Clearly, this makes the velocity time-reversible, as Eqn. (9.9) is symmetric with respect to times t and $t + \Delta t$. To see that it also makes the positions time-reversible, we take the reverse timestep $-\Delta t$ from $t + \Delta t$ to t in Eqn. (9.6)

$$\boldsymbol{r}_t = \boldsymbol{r}_{t+\Delta t} + \boldsymbol{v}_{t+\Delta t}(-\Delta t) + \boldsymbol{a}_{t+\Delta t}\frac{(-\Delta t)^2}{2}.$$

Inserting Eqn. (9.9) for $\boldsymbol{v}_{t+\Delta t}$ and simplifying, we find that

$$\boldsymbol{r}_t = \boldsymbol{r}_{t+\Delta t} - \boldsymbol{v}_t\Delta t - \boldsymbol{a}_t\frac{(\Delta t)^2}{2},$$

which is an exact reversal of Eqn. (9.6). Thus an algorithm based on Eqns. (9.6) and (9.9) will be exactly time-reversible.[13] Algorithm 9.1 summarizes the details of the VV integrator for a system containing a single atomic species with mass m.

Algorithm 9.1 The unoptimized VV algorithm (single species system)

$t := 0$
Require: $\boldsymbol{r}_0$, $\boldsymbol{v}_0$ given.
$\widehat{\boldsymbol{f}}_0 := \boldsymbol{f}(\boldsymbol{r}_0)$
$\boldsymbol{a}_0 := \widehat{\boldsymbol{f}}_0/m$
while $t \le t_{\text{final}}$ **do**
  $\boldsymbol{r}_{t+\Delta t} := \boldsymbol{r}_t + \boldsymbol{v}_t\Delta t + \boldsymbol{a}_t\frac{(\Delta t)^2}{2}$
  $\widehat{\boldsymbol{f}}_{t+\Delta t} := \boldsymbol{f}(\boldsymbol{r}_{t+\Delta t})$
  $\boldsymbol{a}_{t+\Delta t} := \widehat{\boldsymbol{f}}_{t+\Delta t}/m$
  $\boldsymbol{v}_{t+\Delta t} := \boldsymbol{v}_t + \Delta t(\boldsymbol{a}_t + \boldsymbol{a}_{t+\Delta t})/2$
  $t := t + \Delta t$
end while

As written in Algorithm 9.1, one would need to store two copies of the acceleration vector (at t and $t + \Delta t$), in addition to the current copies of the positions, velocities and forces (for a total of $5 \times 3N$ real numbers). If storage is a concern this can easily be rewritten to reduce the storage to $3 \times 3N$ by using the forces directly and by breaking the velocity integration into two steps. This is how the VV routine will typically appear in an MD code, so we rewrite it accordingly in Algorithm 9.2.

We built time-reversibility into the VV algorithm in the above derivation. We next want to show that it is both energy-conserving and symplectic.

[13] Note that the practical matter of computing $\boldsymbol{a}_t$ before knowing $\boldsymbol{r}_t$ makes the reverse algorithm hard to implement, but the important point is that the algorithm is, in fact, theoretically reversible.

Algorithm 9.2 The optimized VV algorithm (single species system)

t:=0

Require: $\boldsymbol{r}_0, \boldsymbol{v}_0$ given.

$\widehat{\boldsymbol{f}}_0 := \boldsymbol{f}(\boldsymbol{r}_0)$

while $t \leq t_{\text{final}}$ **do**

$\quad \boldsymbol{r}_{t+\Delta t} := \boldsymbol{r}_t + \boldsymbol{v}_t \Delta t + \widehat{\boldsymbol{f}}_t \frac{(\Delta t)^2}{2m}$

$\quad \boldsymbol{v}_{t+\Delta t/2} := \boldsymbol{v}_t + \frac{\widehat{\boldsymbol{f}}_t}{2m} \Delta t$

$\quad \widehat{\boldsymbol{f}}_{t+\Delta t} := \boldsymbol{f}(\boldsymbol{r}_{t+\Delta t})$

$\quad \boldsymbol{v}_{t+\Delta t} := \boldsymbol{v}_{t+\Delta t/2} + \frac{\widehat{\boldsymbol{f}}_{t+\Delta t}}{2m} \Delta t$

$\quad t := t + \Delta t$

end while

Energy conservation of the VV algorithm The VV algorithm is energy conserving – but only in an approximate way. More precisely, the algorithm does indeed conserve a quantity that can be identified as the Hamiltonian, but this quantity is only a nearby approximation to the exact Hamiltonian of the system being studied. To show this in general is beyond the scope of our discussion (the proof can be found in, for example, [HL00]), but it is easy to show this using the simple system of the *harmonic oscillator* introduced in Chapter 7. For this system, the VV algorithm conserves a Hamiltonian that goes to the exact Hamiltonian in the limit of small time steps $\Delta t \to 0$, at a rate of $O(\Delta t^2)$. We show this in the following example.

Example 9.1 (The shadow Hamiltonian of a one-dimensional harmonic oscillator) The Hamiltonian of a one-dimensional harmonic oscillator is

$$\mathcal{H}(r, v) = \frac{1}{2}mv^2 + \frac{1}{2}kr^2,$$

where r is the position of the particle with mass m, v is its velocity and k is the stiffness of the spring tending to restore the particle to $r = 0$. The force is simply

$$f = -\frac{\partial \mathcal{H}}{\partial r} = -kr.$$

Let us rewrite the steps of the VV algorithm in Algorithm 9.2 that take us from some timestep t, for which we know the position r_t and velocity v_t, to the next timestep $t + \Delta t$ as

$$\begin{aligned}
r_{t+\Delta t} &= r_t + v_t \Delta t - \frac{kr_t}{2m} \Delta t^2, \\
v_{t+\Delta t/2} &= v_t - \frac{kr_t}{m}\frac{\Delta t}{2}, \\
f_{t+\Delta t} &= -kr_{t+\Delta t}, \\
v_{t+\Delta t} &= v_{t+\Delta t/2} - \frac{kr_{t+\Delta t}}{m}\frac{\Delta t}{2} = v_t - \frac{kr_t}{m}\frac{\Delta t}{2} - \frac{kr_{t+\Delta t}}{m}\frac{\Delta t}{2}.
\end{aligned}$$

We can now compute the difference between the Hamiltonian before and after the timestep, $\mathcal{H}_{t+\Delta t} - \mathcal{H}_t$ (where $\mathcal{H}_t = \mathcal{H}(r_t, v_t)$). Through some algebraic stubbornness, we can show

$$\mathcal{H}_{t+\Delta t} - \mathcal{H}_t = \frac{k^2 \Delta t^2}{8m}(r_{t+\Delta t}^2 - r_t^2). \tag{9.10}$$

Clearly, the Hamiltonian is perturbed by the timestep, although by an amount that will be small for small Δt. On the other hand, the following slightly modified Hamiltonian is *identically* conserved by the VV step:

$$\widehat{\mathcal{H}}(r, v) = \mathcal{H}(r, v) - \frac{k^2 \Delta t^2}{8m} r^2.$$

In the numerical integration literature, this is referred to as the "shadow Hamiltonian" conserved by the integrator. That it is exactly conserved can be readily verified by evaluating $\widehat{\mathcal{H}}_{t+\Delta t} - \widehat{\mathcal{H}}_t$ as we did for the original Hamiltonian above. The difference between the additional term at $t + \Delta t$ and at t exactly cancels the error term in Eqn. (9.10).

We can also see that this modified Hamiltonian goes to the exact Hamiltonian for small timesteps. This result is readily generalizable to harmonic oscillators with multiple degrees of freedom, and the essential idea follows for any Hamiltonian: the VV algorithm conserves some nearby approximation to the exact system.

The symplecticity of the VV algorithm[14] It is convenient for our discussion of symplecticity to return to the notation used in Chapter 7. Specifically, we define a vector $\boldsymbol{y} = (y_1, \dots, y^{2n}) = (q_1, \dots, q_n, p_1, \dots, p_n)$, where $n = 3N$ for a system of N atoms. This vector is just the concatenation of the position degrees of freedom and the momenta, $p_i = mv_i$. We will sometimes find it convenient to write this as $\boldsymbol{y} = (\boldsymbol{q}, \boldsymbol{p})$. We take the vector $\boldsymbol{y}$ to be a solution to Hamilton's equations (Eqn. (4.10)), subject to the initial condition $\boldsymbol{y} = \boldsymbol{y}_0 = (\boldsymbol{q}_0, \boldsymbol{p}_0)$ at $t = 0$. It is convenient to rewrite this as

$$\frac{d\boldsymbol{y}(t)}{dt} = \mathbf{A}\frac{\partial \mathcal{H}}{\partial \boldsymbol{y}}, \qquad \boldsymbol{y}(0) = \boldsymbol{y}_0, \tag{9.11}$$

where

$$\mathbf{A} = \begin{bmatrix} \mathbf{0} & \mathbf{I} \\ -\mathbf{I} & \mathbf{0} \end{bmatrix},$$

$\mathbf{0}$ is an $n \times n$ matrix of zeroes and $\mathbf{I}$ is the $n \times n$ identity matrix. The introduction of $\mathbf{A}$ also lets us write very compact expressions for local areas in the $(\boldsymbol{q}, \boldsymbol{p})$ phase space. We define two vectors, $\boldsymbol{x}^{\mathsf{A}} = (\boldsymbol{q}^{\mathsf{A}}, \boldsymbol{p}^{\mathsf{A}})$ and $\boldsymbol{x}^{\mathsf{B}} = (\boldsymbol{q}^{\mathsf{B}}, \boldsymbol{p}^{\mathsf{B}})$. The area spanned by these vectors and projected onto each of the (q_i, p_i) planes is

$$\omega_i = \det \begin{bmatrix} q_i^{\mathsf{A}} & q_i^{\mathsf{B}} \\ p_i^{\mathsf{A}} & p_i^{\mathsf{B}} \end{bmatrix} = q_i^{\mathsf{A}} p_i^{\mathsf{B}} - q_i^{\mathsf{B}} p_i^{\mathsf{A}} \qquad \text{(no sum)}.$$

Next, we note that the volume conservation property of Liouville's theorem is equivalent to the conservation of the sum of these projected areas over all n dimensions. Thus,

$$\Omega \equiv \sum_{i=1}^{n} \omega_i = (\boldsymbol{x}^{\mathsf{A}})^T \mathbf{A} \boldsymbol{x}^{\mathsf{B}} \tag{9.12}$$

must be conserved for any $\boldsymbol{x}^{\mathsf{A}}(t)$ and $\boldsymbol{x}^{\mathsf{B}}(t)$. Next, we consider trajectories of the form

$$\boldsymbol{x}^{\mathsf{A}}(t) = \mathbf{J}(t)\boldsymbol{y}_0^{\mathsf{A}}, \qquad \boldsymbol{x}^{\mathsf{B}}(t) = \mathbf{J}(t)\boldsymbol{y}_0^{\mathsf{B}},$$

[14] This presentation is partly based on [HLW06]. See this reference for a more detailed rigorous discussion.

where

$$\mathbf{J}(t) = \frac{\partial \boldsymbol{y}(t)}{\partial \boldsymbol{y}_0}$$

is a linear mapping, $\boldsymbol{y}(t)$ is the solution to Eqn. (9.11) and $\boldsymbol{y}_0^{\mathsf{A}}$ and $\boldsymbol{y}_0^{\mathsf{B}}$ are two constant vectors. Inserting this into Eqn. (9.12), we see that symplecticity imposes the following condition on the linear mapping:

$$\mathbf{J}^T \mathbf{A} \mathbf{J} = \mathbf{A}. \tag{9.13}$$

The following proof shows that this is exactly satisfied for Hamiltonian systems that satisfy Eqn. (9.11). In other words, Eqn. (9.13) is an alternative expression for Liouville's theorem.

Proof We start with the assumption that Eqn. (9.11) holds. Differentiating Eqn. (9.11) with respect to $\boldsymbol{y}_0$ (and reversing the order of differentiation on the left-hand side) we find

$$\frac{d}{dt}\left(\frac{\partial \boldsymbol{y}}{\partial \boldsymbol{y}_0}\right) = \mathbf{A}\frac{\partial^2 \mathcal{H}}{\partial \boldsymbol{y} \partial \boldsymbol{y}}\frac{\partial \boldsymbol{y}}{\partial \boldsymbol{y}_0},$$

or after simplification

$$\frac{d\mathbf{J}}{dt} = \mathbf{A}\frac{\partial^2 \mathcal{H}}{\partial \boldsymbol{y} \partial \boldsymbol{y}}\mathbf{J}. \tag{9.14}$$

(A result that we will make use of in a moment.) Our goal is to show that Eqn. (9.13) holds for all t. Clearly, Eqn. (9.13) is identically satisfied at $t = 0$ since by definition $\mathbf{J}(0) = \mathbf{I}$. Now we can take the time derivative of the left-hand side of Eqn. (9.13) to find

$$\frac{d}{dt}\left(\mathbf{J}^T \mathbf{A} \mathbf{J}\right) = \left(\frac{d\mathbf{J}}{dt}\right)^T \mathbf{A}\mathbf{J} + \mathbf{J}^T \mathbf{A}\left(\frac{d\mathbf{J}}{dt}\right) = 0.$$

The last step follows from using the result of Eqn. (9.14), noting that $\mathbf{A}^T = \mathbf{A}^{-1}$ and $\mathbf{A}\mathbf{A} = -\mathbf{I}$. In other words, $\mathbf{J}^T \mathbf{A} \mathbf{J}$ is a constant whose initial value is simply $\mathbf{A}$, and Eqn. (9.13) holds for all time provided Eqn. (9.11) holds. □

A numerical integrator such as the VV algorithm is, in effect, a linear mapping from one timestep to the next. To see that such an integrator is in fact symplectic, we need to verify that a timestep satisfies Eqn. (9.13). In the case of the VV algorithm we write

$$\mathbf{J}_{t+\Delta t} = \frac{\partial \boldsymbol{y}_{t+\Delta t}}{\partial \boldsymbol{y}_t} = \begin{bmatrix} \dfrac{\partial \boldsymbol{q}_{t+\Delta t}}{\partial \boldsymbol{q}_t} & \dfrac{1}{m}\dfrac{\partial \boldsymbol{q}_{t+\Delta t}}{\partial \boldsymbol{v}_t} \\ m\dfrac{\partial \boldsymbol{v}_{t+\Delta t}}{\partial \boldsymbol{q}_t} & \dfrac{\partial \boldsymbol{v}_{t+\Delta t}}{\partial \boldsymbol{v}_t} \end{bmatrix},$$

where we have, without loss of generality, defined the initial condition to be the state of the system at time t and the final state to be at time $t + \Delta t$.

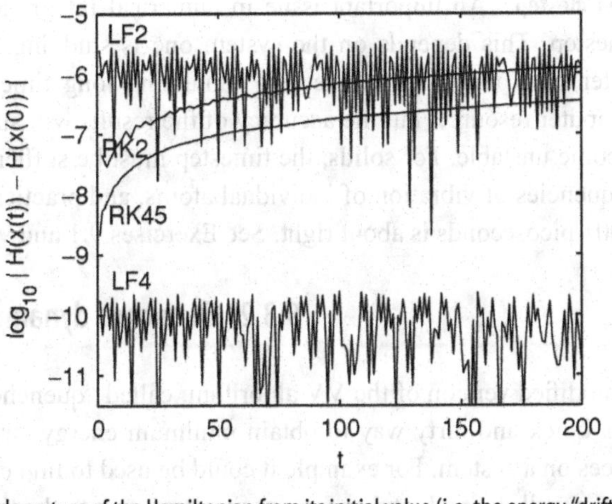

Fig. 9.3 A log–linear plot of the departure of the Hamiltonian from its initial value (i.e. the energy "drift"). Runge–Kutta methods (labeled RK2 and RK45) are compared with the stable energy conservation of the symplectic leap-frog (LF2 and LF4) methods. The VV algorithm is equivalent to the LF2 method. Reproduced from [MQ06], with permission.

From lines 4, 5 and 7 in Algorithm 9.2, it is straightforward to derive[15]

$$\mathbf{J} = \begin{bmatrix} \mathbf{I} - \dfrac{\Delta t^2}{2m}\mathbf{K}_t & \dfrac{\Delta t}{m}\mathbf{I} \\ -\dfrac{\Delta t}{2}\mathbf{K}_t & \mathbf{I} \end{bmatrix},$$

where $\mathbf{K}_t = -\partial\widehat{\boldsymbol{f}}(t)/\partial\boldsymbol{q}(t)$ is the stiffness matrix. One can readily verify that this mapping satisfies Eqn. (9.13). The VV algorithm is therefore symplectic.

The importance of using symplectic integrators for Hamiltonian systems is subtle, but they generally perform better than nonsymplectic integrators in that there is less energy drift. Even if an integrator theoretically conserves energy in exact mathematics, numerical precision errors lead to some drift over time. That drift is minimized for the symplectic integrators. In Fig. 9.3, we show a result from [MQ06], illustrating the departure from exact conservation of the Hamiltonian for a simple system with two degrees of freedom. The curve labeled LF2 is the second-order "leap-frog" method (identical to the VV algorithm) and LF4 is the fourth-order leap-frog scheme, which is more accurate than the VV algorithm but requires six times as many force evaluations per timestep. The other curves, labeled RK2 and RK45, are two examples of Runge–Kutta integrators that are not symplectic. We can see that while the RK schemes initially have better accuracy than the LF2 scheme, their energy continues to drift, while the LF schemes oscillate about a constant energy error. For studies requiring very long simulation times, this can have significant consequences for the results.

[15] Note that the derivative of $\widehat{\boldsymbol{f}}_{t+\Delta t}$ with respect to $\boldsymbol{q}_t$ is zero and we have assumed that the forces are independent of the velocity.

The timestep An important issue in numerical integration is the appropriate size of the timestep. This depends on the system one is studying. Of course, longer timesteps are better from the point of view of simulating long time periods with the least possible computer resource, but the accuracy of the results will suffer, and the integrator may even become unstable. For solids, the timestep must be sufficiently small to resolve the largest frequencies of vibration of individual atoms, and practically speaking a timestep of about 0.001 picoseconds is about right. See Exercises 9.1 and 9.2 where this issue is explored.

9.3.2 Quenched dynamics

A modified version of the VV algorithm, called "quenched dynamics" (QD) is often used as a quick and dirty way to obtain minimum energy structures or to equilibrate a set of forces on a system. For example it could be used to find equilibrium in systems that do not have a well-defined total energy, as an alternative to the the force-based conjugate gradient (CG) method discussed in Section 6.3.1. The QD algorithm is also essential for finding transition paths in the nudged elastic band method of Section 6.3.1. Through a very simple modification, an existing MD code can be converted for this purpose (see Exercise 9.3).

The idea is to follow the dynamics dictated by the forces on the system, but to gradually bleed off energy by occasionally zeroing the velocities. This will force the system to fall into an energy minimum (or into a configuration where all the forces are equilibrated). First, each degree of freedom in the system is given a mass. This mass can be a real atomic mass or a fictitious value that can be treated as a tuning parameter to improve the rate of convergence. At each step, the calculated velocity is projected along the direction of the force, so that only the component of velocity parallel to the force vector is used. Finally, if the force and velocity point in opposite directions (i.e. $\boldsymbol{v} \cdot \boldsymbol{f} < 0$), the velocity is set to zero. This requires the addition of a conditional statement to modify the velocities before line 8 of Algorithm 9.2, as shown in Algorithm 9.3. An improvement on this simple minimization approach is the so-called FIRE algorithm (fast inertial relaxation engine) developed by Bitzek *et al.* [BKG$^+$06]. Based on the QD idea, FIRE heuristically modifies the forces on atoms based on their inertia to improve convergence rates.

9.3.3 Temperature initialization

Often, it is desirable to initialize an NVE simulation to a certain preset average temperature.[16] This can be achieved by imposing random initial velocities on the atoms, although there are some subtleties that must be appreciated in order to understand the result. The simplest approach, as outlined in [FS02] and [BC06], is to work with the instantaneous temperature using Algorithm 9.4. The result of this procedure will be an initial instantaneous temperature exactly equal to the set temperature, and an initial kinetic energy of

$$\mathcal{T}^{\text{vib}} = \frac{3}{2} N k_{\text{B}} T_{\text{set}}.$$

[16] In the NVE ensemble, the energy is fixed and consequently the temperature exhibits small fluctuations about its mean temperature. The larger the system, the smaller the fluctuation. See Fig. 7.4 and the derivation in [RG81].

Algorithm 9.3 The QD algorithm (single species system)

1: t:=0
Require: $\boldsymbol{r}_0$, $\boldsymbol{v}_0$ given.
2: $\widehat{\boldsymbol{f}}_0 := \boldsymbol{f}(\boldsymbol{r}_0)$
3: **while** $t \leq t_{\text{final}}$ **do**
4: $\quad \boldsymbol{r}_{t+\Delta t} := \boldsymbol{r}_t + \boldsymbol{v}_t \Delta t + \widehat{\boldsymbol{f}}_t \frac{(\Delta t)^2}{2m}$
5: $\quad \boldsymbol{v}_{t+\Delta t/2} := \boldsymbol{v}_t + \frac{\widehat{\boldsymbol{f}}_t}{2m} \Delta t$
6: $\quad \widehat{\boldsymbol{f}}_{t+\Delta t} := \boldsymbol{f}(\boldsymbol{r}_{t+\Delta t})$
7: $\quad \boldsymbol{v}_{t+\Delta t} := \boldsymbol{v}_{t+\Delta t/2} + \frac{\widehat{\boldsymbol{f}}_{t+\Delta t}}{2m} \Delta t$
8: $\quad v_p := \boldsymbol{v}_{t+\Delta t} \cdot \widehat{\boldsymbol{f}}_{t+\Delta t}$ {From here to line 13 is the modification of VV}
9: $\quad$ **if** $v_p < 0$ **then**
10: $\quad\quad \boldsymbol{v}_{t+\Delta t} := 0$
11: $\quad$ **else**
12: $\quad\quad \boldsymbol{v}_{t+\Delta t} := v_p \widehat{\boldsymbol{f}}_{t+\Delta t} / \left\| \widehat{\boldsymbol{f}}_{t+\Delta t} \right\|$
13: $\quad$ **end if**
14: $\quad t := t + \Delta t$
15: **end while**

Algorithm 9.4 MD temperature initialization

1: Set each velocity component of each particle to a random value between −0.5 and 0.5.
2: Compute the resulting momentum of the center of mass: $\boldsymbol{P}_{\text{com}} = \sum_{\alpha=1}^{N} m^\alpha \boldsymbol{v}^\alpha$.
3: Adjust all velocities so that the center of mass momentum is zero (this is so that the resulting temperature will only be due to the vibrational part of the velocities):

$$\boldsymbol{v}^\alpha := \boldsymbol{v}^\alpha - \frac{\boldsymbol{P}_{\text{com}}}{M_{\text{tot}}}, \quad \text{where} \quad M_{\text{tot}} = \sum_{\alpha=1}^{N} m^\alpha. \tag{9.15}$$

4: Compute the resulting instantaneous temperature, T_{inst} using Eqn. (9.4).
5: Scale all velocities by $\sqrt{T_{\text{set}}/T_{\text{inst}}}$ to achieve the desired set temperature T_{set}.

However, this will not usually result in the system equilibrating to $T = T_{\text{set}}$. The equilibration temperature is dictated by the equipartition theorem discussed in Section 7.3.6, which states that both

$$\langle \mathcal{T}^{\text{vib}} \rangle = \frac{3}{2} N k_{\text{B}} T \qquad \text{and} \qquad \langle \mathcal{W} \rangle = -3 N k_{\text{B}} T \tag{9.16}$$

are satisfied at equilibrium. If, for example, the atoms are initially in their perfect crystal positions, the initial forces on the atoms will be zero and the instantaneous value of the virial, $\mathcal{W}$, will also be zero. Therefore the system will equilibrate to a temperature lower than T_{set} so that both Eqns. (9.16) can be satisfied while still conserving total energy. Consider a harmonic lattice of identical noninteracting particles, for which we have

$$\mathcal{V}_{\text{HO}}^{\text{int}} = \sum_{\alpha=1}^{N} \frac{1}{2} k \left(\boldsymbol{r}^\alpha - \boldsymbol{R}^\alpha \right) \cdot \left(\boldsymbol{r}^\alpha - \boldsymbol{R}^\alpha \right), \qquad \boldsymbol{f}^\alpha = -k \left(\boldsymbol{r}^\alpha - \boldsymbol{R}^\alpha \right),$$

where $\boldsymbol{R}^\alpha$ is the position of atom α in the perfect lattice. Putting the expression for the force back into $\mathcal{V}_{\mathrm{HO}}^{\mathrm{int}}$, we can evaluate

$$\langle \mathcal{V}_{\mathrm{HO}}^{\mathrm{int}} \rangle = -\frac{1}{2}\left\langle \sum_{\alpha=1}^{N} \boldsymbol{f}^\alpha \cdot (\boldsymbol{r}^\alpha - \boldsymbol{R}^\alpha) \right\rangle = -\frac{1}{2}\left\langle \sum_{\alpha=1}^{N} \boldsymbol{f}^\alpha \cdot \boldsymbol{r}^\alpha \right\rangle + \frac{1}{2}\sum_{\alpha=1}^{N} \langle \boldsymbol{f}^\alpha \rangle \cdot \boldsymbol{R}^\alpha. \tag{9.17}$$

The average value of the force in the second term should be zero as each atom oscillates about its equilibrium position, while from Eqn. (7.108) the remaining term gives

$$\langle \mathcal{V}_{\mathrm{HO}}^{\mathrm{int}} \rangle = -\frac{1}{2}\langle \mathcal{W} \rangle .$$

As such, we expect that

$$\langle \mathcal{V}_{\mathrm{HO}}^{\mathrm{int}} \rangle + \langle \mathcal{T}_{\mathrm{HO}}^{\mathrm{vib}} \rangle = 3Nk_{\mathrm{B}}T,$$

after using Eqns. (9.16). Since energy is conserved and the initial potential energy was zero, this sum must be equal to the initial kinetic energy, so that

$$3Nk_{\mathrm{B}}T = \frac{3}{2}Nk_{\mathrm{B}}T_{\mathrm{set}},$$

meaning that the equilibrium temperature will be exactly $T = T_{\mathrm{set}}/2$.

For more realistic descriptions of the potential energy, this will not be exact, but at moderate temperatures it will be closely approximated since in this case the departures of the atoms from the perfect lattice sites will be small. The important message here is that initializing a perfect crystal to an instantaneous temperature as described above is only initializing the kinetic energy, while the initial potential energy is zero. At equilibrium, about half of the kinetic energy will be transformed into potential energy and the equilibration temperature will be about half the initial instantaneous temperature.

Another common approach to temperature initialization is to set the velocities of the atoms such that they are consistent with the expected distribution of velocities in an equilibrated system. From Eqn. (7.119), we can evaluate the expectation value of the Dirac delta, $\delta(\boldsymbol{p}^\alpha - \boldsymbol{p}_*)$. This gives the probability of atom α having a certain momentum $\boldsymbol{p}_*$:

$$\langle \delta(\boldsymbol{p}^\alpha - \boldsymbol{p}_*) \rangle = \frac{\int_\Gamma \delta(\boldsymbol{p}^\alpha - \boldsymbol{p}_*) e^{-\mathcal{H}(\boldsymbol{r},\boldsymbol{p})/k_{\mathrm{B}}T}\, d\boldsymbol{r} d\boldsymbol{p}}{\int_\Gamma e^{-\mathcal{H}(\boldsymbol{r},\boldsymbol{p})/k_{\mathrm{B}}T}\, d\boldsymbol{r} d\boldsymbol{p}},$$

where $d\boldsymbol{r}$ is shorthand for $d\boldsymbol{r}^1 \cdots d\boldsymbol{r}^N$ and similarly for $d\boldsymbol{p}$. By the separability of the terms in the Hamiltonian, most of the terms in the numerator and denominator cancel, leaving

$$\begin{aligned}\langle \delta(\boldsymbol{p}^\alpha - \boldsymbol{p}_*) \rangle &= \frac{e^{-\|\boldsymbol{p}_*\|^2/2m^\alpha k_{\mathrm{B}}T}}{\int_{-\infty}^{\infty}\int_{-\infty}^{\infty}\int_{-\infty}^{\infty} e^{-\|\boldsymbol{p}^\alpha\|^2/2m^\alpha k_{\mathrm{B}}T}\, dp_1^\alpha dp_2^\alpha dp_3^\alpha} \\ &= \frac{e^{-\|\boldsymbol{p}_*\|^2/2m^\alpha k_{\mathrm{B}}T}}{(2\pi m^\alpha k_{\mathrm{B}}T)^{3/2}} \equiv f_{\mathrm{MB}}(\boldsymbol{p}_*),\end{aligned} \tag{9.18}$$

which is known as the *Maxwell–Boltzmann distribution*. Because of its familiar Gaussian (normal) form, it is straightforward to implement computer code that chooses random velocities from this distribution.[17] Note, however, that the same problem exists as described

[17] Most random number generators provide uniformly distributed random numbers in the range $0 < x \le 1$. One convenient transform from uniformly distributed random numbers x to a normally distributed random number

earlier: equipartition cannot be satisfied without also knowing how to prescribe the initial positions of the atoms, so this remains an approximate thermal initialization procedure.

9.3.4 Equilibration time

Since the initial configuration may be significantly different from an equilibrium state, it is typically necessary to run an MD simulation for some time before recording the configurations used to generate averaged quantities like temperature or stress. It is difficult to make general statements about how long this equilibration time should be, as it depends on many variables, including the desired temperature, the interatomic model being used and the quality of the initialization "guess." Normally, one needs to perform test simulations to determine the required equilibration time for a given problem. The necessary time can be established in such tests by monitoring quantities such as the potential energy, instantaneous temperature and pressure as a function of time. Equilibration is reached once these quantities have stabilized and lost any significant "memory" of the initial configuration.

A more direct measure of this "loss of memory" is to use a *correlation function*, which is a measure of how a variable at one time (usually taken as $t = 0$) relates to another variable at a later time. *Autocorrelation* functions relate a particular variable at $t = 0$ to itself at a later time. For example, the velocity autocorrelation function (VAF) is defined as

$$\mathrm{VAF}(t) = \overline{\sum_{\alpha=1}^{N} \boldsymbol{v}^{\alpha}(0) \cdot \boldsymbol{v}^{\alpha}(t)},$$

where the overbar indicates a time average. To be truly equilibrated, the VAF of a system must have gone to zero, implying that it has lost all memory of its initial state.

In Fig. 9.4, a system of 256 Ar atoms arranged in a periodic fcc lattice was modeled with a Lennard–Jones potential. In the first case, all atoms were initialized to zero velocity, but their positions were perturbed from the perfect lattice randomly. In the second case the initial positions were the perfect lattice sites, but the velocities were initialized to $T_{\mathrm{set}} = 100$ K according to Algorithm 9.4. The displacement-initialized run equilibrates after about 1000 timesteps, whereas the velocity-initialized run takes a little longer. This is more obvious from the VAF plot normalized by the number of atoms (dashed line, right-hand axis), which takes on the order of 2500 steps to adequately decay. Note also that final equilibrated temperature is about half the initialization value, as discussed in Section 9.3.3.

9.4 The NVT ensemble: constant temperature and constant strain

It is not often that one performs experiments on materials within the microcanonical (constant energy) ensemble. A slightly more realistic situation is an experiment at approximately

y with standard deviation 1 is the Box–Muller transform given by $y = \sqrt{-2 \ln x_1} \cos(2\pi x_2)$, where x_1 and x_2 are two uniformly distributed random numbers between 0 and 1 [BM58]. The normally distributed variable y can be used to generate velocity coordinates consistent with a given temperature T by the simple rescaling, $v = y\sqrt{k_B T/m}$.

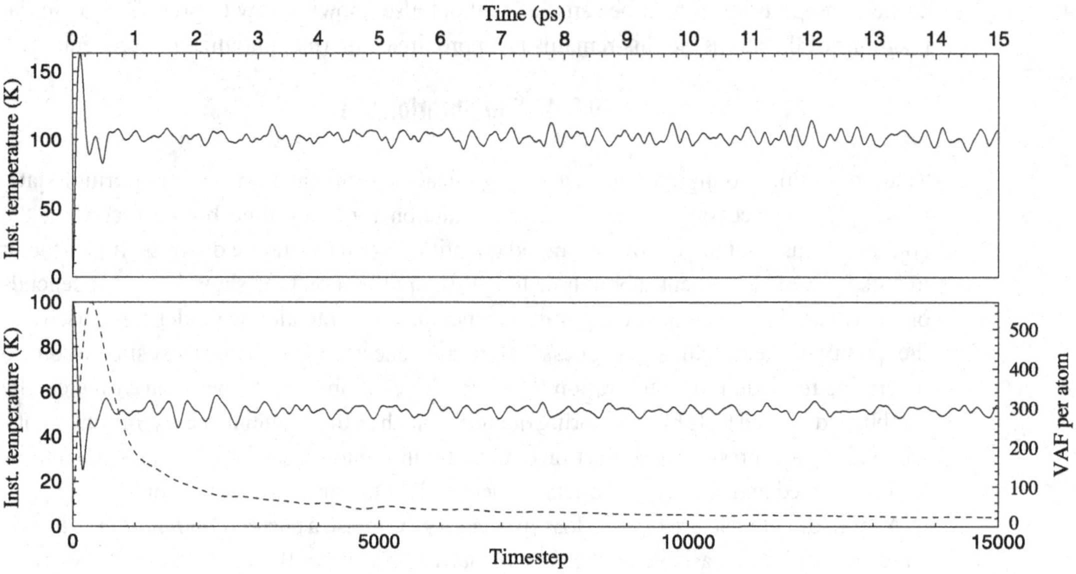

Fig. 9.4 Equilibration of an NVE MD simulation. See text for details.

constant temperature, mimicking the canonical ensemble. The early 1980s seems to have been the heyday of "thermostat" methods aimed at controlling the temperature of MD simulations.[18] While many methods were proposed, over time it became clear that certain properties of the thermostats made some of them more desirable than others. For example, a key feature should be that the system samples a true canonical ensemble. Here, we discuss four approaches. The first two, velocity rescaling and the isokinetic ensemble, very clearly do not sample the canonical ensemble and should therefore be avoided. They serve as simple pedagogical examples. Then, we discuss at length the two most popular canonical thermostats: the stochastic Langevin thermostat and the deterministic Nosé–Hoover (NH) thermostat based on the method of the extended Hamiltonian.

9.4.1 Velocity rescaling

We have already defined temperature in Eqn. (9.3) and instantaneous temperature in Eqn. (9.4). How can we perform a simulation whereby the true temperature remains constant? A naïve approach would be simply to impose that the instantaneous temperature remain constant by scaling the velocities of the atoms at every timestep. For example, we could insert the following steps before line 8 of Algorithm 9.2:

Compute T_{inst} from Eqn. (9.4).
$\boldsymbol{v}^{\mathrm{new}}_{t+\Delta t} := \sqrt{T_{\mathrm{set}}/T_{\mathrm{inst}}}\, \boldsymbol{v}_{t+\Delta t}$.
$\boldsymbol{v}_{t+\Delta t} := \boldsymbol{v}^{\mathrm{new}}_{t+\Delta t}$.
Remove the center of mass motion using Eqn. (9.15).

[18] A comprehensive review of thermostat methods is presented in [Hun05].

This algorithm will keep the instantaneous temperature set to exactly T_{set}, and therefore prescribe the system temperature for all time. The last step is important to avoid a spurious effect cleverly referred to as "the flying ice cube" in the MD literature [HTC98]. If velocity rescaling is performed without correctly maintaining the rigid body motion at zero, the vibrational motion is systematically drained into the rigid body motion. The end result of a simulation of a finite system in vacuum is a frozen block of material traveling at high center of mass velocity. Even with care taken to avoid the flying ice cube effect, velocity rescaling is still inconsistent with the statistical mechanics of the canonical ensemble, where we know that the kinetic energy is not constant even in a system in thermodynamic equilibrium. Rather, the kinetic energy should exhibit well-defined fluctuations as a function of the number of atoms (see Exercise 7.13). Velocity rescaling simulations cannot capture these fluctuations and therefore do not reproduce a canonical ensemble.[19]

9.4.2 Gauss' principle of least constraint and the isokinetic thermostat

Hamiltonian mechanics describes the behavior of a special class of systems, specifically those for which the energy is conserved. But many systems are more general than this, as they may exchange energy with their environment to satisfy some other macroscopic constraints. In the context of this section, the obvious example is the canonical ensemble where the average temperature is kept constant through exchanges of energy with the environment. Modeling such a system using Hamiltonian mechanics would be unwieldy, as the only way to conserve energy would be to model the complex interactions between the system of interest and its entire environment.

A more general mechanics was put forward by Gauss [Lan70, EM90] and is called Gauss' principle of least constraint. This permits the imposition of physical constraints on the system of the form

$$g(\boldsymbol{r}, \dot{\boldsymbol{r}}, t) = 0.$$

The equations of motions are obtained by a minimization over the accelerations:

$$\frac{\partial}{\partial \ddot{\boldsymbol{r}}}(C - \lambda G) = 0, \tag{9.19}$$

where C is defined as

$$C = \frac{1}{2}\sum_{\alpha=1}^{N} m^{\alpha}\left(\ddot{\boldsymbol{r}}^{\alpha} - \frac{\boldsymbol{f}^{\alpha}}{m^{\alpha}}\right)^2, \tag{9.20}$$

[19] Another type of thermostat is the Berendsen thermostat [BPV+84], which is based on the Langevin thermostat (discussed shortly) without the stochastic term. This is sometimes confused with velocity rescaling because it makes use of a damping coefficient that depends of the ratio of the set temperature to the instantaneous temperature. It is not the same as simple rescaling, and is not nearly as problematic. However, the Berendsen thermostat is like velocity rescaling in that it does not reproduce the canonical ensemble either [Hun05], and for this reason we do not include it here.

λ is a Lagrange multiplier and G is the *acceleration-dependent* form of the constraint. (G is the first time derivative of $g(\boldsymbol{r}, \dot{\boldsymbol{r}}, t)$ when g is a function of velocity, or the second derivative if g depends only on position.)

It is easy to see that if $G = 0$, Eqn. (9.19) recovers Newton's equations, corresponding to a Hamiltonian system that conserves energy. Any form for G is permissible, each producing a system which obeys a non-Hamiltonian, constrained mechanics.[20]

Gauss' principle can be used to set up a sort of thermostat, albeit one that suffers from similar problems to velocity rescaling. Consider a system constrained to stay at constant instantaneous temperature, T:

$$\sum_{\alpha=1}^{N} \frac{1}{2} m^{\alpha} \dot{\boldsymbol{r}}^{\alpha} \cdot \dot{\boldsymbol{r}}^{\alpha} = \frac{3}{2} N k_{\mathrm{B}} T.$$

The constraint function is then

$$g(\boldsymbol{r}, \dot{\boldsymbol{r}}, t) = \sum_{\alpha=1}^{N} \frac{1}{2} m^{\alpha} \dot{\boldsymbol{r}}^{\alpha} \cdot \dot{\boldsymbol{r}}^{\alpha} - \frac{3}{2} N k_{\mathrm{B}} T,$$

which we differentiate once to obtain the acceleration-dependent constraint:

$$G(\dot{\boldsymbol{r}}, \ddot{\boldsymbol{r}}, t) = \sum_{\alpha=1}^{N} m^{\alpha} \ddot{\boldsymbol{r}}^{\alpha} \cdot \dot{\boldsymbol{r}}^{\alpha}. \tag{9.21}$$

Substituting Eqn. (9.21) into Eqn. (9.19) gives

$$\frac{\partial}{\partial \ddot{\boldsymbol{r}}} \left(\frac{1}{2} \sum_{\alpha=1}^{N} m^{\alpha} \left(\ddot{\boldsymbol{r}}^{\alpha} - \frac{\boldsymbol{f}^{\alpha}}{m^{\alpha}} \right)^2 - \lambda \sum_{\alpha=1}^{N} m^{\alpha} \ddot{\boldsymbol{r}}^{\alpha} \cdot \dot{\boldsymbol{r}}^{\alpha} \right) = 0,$$

which leads to equations of the form:

$$m^{\alpha} \ddot{\boldsymbol{r}}^{\alpha} = \boldsymbol{f}^{\alpha} - m^{\alpha} \lambda \dot{\boldsymbol{r}}^{\alpha}. \tag{9.22}$$

This expression for $m^{\alpha} \ddot{\boldsymbol{r}}^{\alpha}$ can be inserted into Eqn. (9.21), so that the constraint $G = 0$ is

$$\sum_{\alpha=1}^{N} (\boldsymbol{f}^{\alpha} - m^{\alpha} \lambda \dot{\boldsymbol{r}}^{\alpha}) \cdot \dot{\boldsymbol{r}}^{\alpha} = 0,$$

which can then be solved for λ:

$$\lambda = \frac{\sum_{\alpha=1}^{N} \boldsymbol{f}^{\alpha} \cdot \dot{\boldsymbol{r}}^{\alpha}}{\sum_{\alpha=1}^{N} m^{\alpha} \left\| \dot{\boldsymbol{r}}^{\alpha} \right\|^2}. \tag{9.23}$$

Equations (9.22) and (9.23) form a type of thermostat. The function λ plays the role of a variable damping coefficient by multiplying the atomic velocity to introduce viscous drag

[20] In Hamiltonian systems, the fundamental governing principle is that energy is conserved. By a non-Hamiltonian system, we mean any system for which some external constraints are imposed such that the energy is not conserved. This is common, for example, where the system of interest is a small part of some larger system, such as a heat bath.

forces on the atoms. The equations constitute the *Gaussian isokinetic equations of motion*, first proposed independently by Evans [Eva83] and Hoover *et al.* [HLM82]. A simulation using these equations would strictly fix the kinetic energy (and therefore the temperature) to its initial value. While this simple example serves to illustrate the power of the Gauss principle, it does not lead to a canonical thermostat, since it does not reproduce the expected fluctuations in the instantaneous temperature (see Exercise 7.13).

9.4.3 The Langevin thermostat

The Langevin thermostat is sometimes referred to as a *stochastic dynamics* approach because it incorporates random forces into the equations of motion of the atoms. These random forces are meant to reproduce the effect of a heat bath, which could be due to heat transfer to an (unmodeled) solvent surrounding protein molecules, for example, or equally due to the heat transfer to a free electron gas in a solid metal crystal. The end effect is a method by which the average temperature of the system can be held fixed while permitting the system to sample the phase space of the canonical ensemble, as we outline briefly here.

The Langevin equation of motion for the atoms is

$$m^{\alpha}\ddot{\boldsymbol{r}}^{\alpha} = \boldsymbol{f}^{\alpha} - \gamma^{\alpha} m^{\alpha}\dot{\boldsymbol{r}}^{\alpha} + \boldsymbol{G}^{\alpha}(t), \tag{9.24}$$

where $\boldsymbol{f}^{\alpha}$ is the interatomic force derived from the gradient of the interatomic potential and γ^{α} is a damping constant associated with atom α. The feature of the equation that makes it stochastic is $\boldsymbol{G}^{\alpha}(t)$, a random, time-varying force. In order to ensure that the system traces a canonical distribution of microstates at temperature T, we will see below that the random forces must satisfy

$$\langle \boldsymbol{G}^{\alpha} \rangle = 0, \qquad \left\langle G_i^{\alpha}(t) G_j^{\beta}(t') \right\rangle = 2\gamma^{\alpha} m^{\alpha} k_{\mathrm{B}} T \delta_{ij} \delta^{\alpha\beta} \delta(t - t'). \tag{9.25}$$

These conditions on $\boldsymbol{G}^{\alpha}$ are met in practice if each component of the force is independently selected from a normally distributed random variable with variance $2\gamma^{\alpha} m^{\alpha} k_{\mathrm{B}} T/\Delta t$ at every timestep (where Δt is the size of the timestep).[21]

A related approach, known as the Andersen thermostat [And80], imposes the random forces only for a relatively small number of timesteps, and at random intervals. It can be shown that the Andersen and Langevin thermostats have the same long-time limiting behavior, but the Andersen approach has more tunability, affording the possibility of a gentler influence on the system.

Brownian motion The reason why the Langevin equation successfully reproduces a canonical ensemble at temperature T is not at all obvious from the equation itself, but is due

[21] The inclusion of Δt^{-1} in the variance is subtle, but comes from the discretization of the integral implicit in the time averaging one must do to approximate the phase average of $\left\langle G_i^{\alpha}(t) G_j^{\beta}(t') \right\rangle$. Footnote 17 on page 506 explains how one can generate normally distributed pseudorandom numbers given uniformly distributed pseudo-random numbers between 0 and 1.

to the theoretical basis of the equation in Brownian motion. The physical assumption is that there are two levels of interaction. First, the atoms in the system interact with each other through the interatomic forces. Second, the atoms are in constant collisions with some medium that constitutes the heat bath. The key assumption is that the collisions with the medium are much more frequent and exert smaller forces than the collisions between the atoms themselves, allowing us to decouple the two. The atomic collisions are assumed to lead to smooth, continuously varying forces, whereas the rapidity of the heat bath collisions allows us to treat them as entirely random and discontinuous.

A detailed explanation of Langevin dynamics is presented in [Cha43], while a brief discussion of its application specifically to MD is presented in [BH86]. Here we only present a flavor of why the approach works. For simplicity, let us consider the case where the atoms do not interact with each other but only with the heat bath (i.e. $\boldsymbol{f}^{\alpha} = 0$ in Eqn. (9.24)). Under these circumstances the Langevin equation for any atom can be written in terms of the velocity as

$$m\dot{\boldsymbol{v}} = -\gamma m\boldsymbol{v} + \boldsymbol{G}(t), \tag{9.26}$$

where we have dropped the α superscript for clarity. One can readily verify that this equation can be formally solved for the velocity as

$$\boldsymbol{v}(t) = \boldsymbol{v}_0 e^{-\gamma t} + e^{-\gamma t}\int_0^t e^{\gamma\tau}\frac{\boldsymbol{G}(\tau)}{m}\,d\tau, \tag{9.27}$$

where $\boldsymbol{v}_0$ is the initial velocity. This is not a solution in the usual, deterministic sense of the word, since $\boldsymbol{G}$ is stochastic. Rather, we can only really make sense of this velocity in terms of its statistical properties. For example we would like to see that if we take a time average over a long enough time interval, then the phase averages expected from equipartition (see Section 7.3.6) are recovered, i.e.

$$\overline{\boldsymbol{v}\cdot\boldsymbol{v}} = 3k_{\mathrm{B}}T/m. \tag{9.28}$$

In addition, we would like to ensure that the canonical distribution of velocities (the Maxwell–Boltzmann distribution in Eqn. (9.18)) is satisfied in the limit of long simulation time. Since the term in Eqn. (9.27) involving $\boldsymbol{v}_0$ goes to zero as $t \to \infty$, the probability distribution of the velocity will be the same as the probability distribution of the remaining term:

$$e^{-\gamma t}\int_0^t e^{\gamma\tau}\frac{\boldsymbol{G}(\tau)}{m}\,d\tau.$$

Chandrasekhar [Cha43] showed that this quantity obeys the Maxwell–Boltzmann distribution provided that $\boldsymbol{G}(t)$ obeys the conditions of Eqn. (9.25). From this, it follows that equipartition is satisfied, and so the Langevin approach indeed leads to a canonical ensemble.

The choice of γ^{α} is arbitrary, and in principle any nonzero value will ensure a canonical ensemble.[22] In practice, however, too small a value of γ^{α} will lead to poor temperature

[22] The Langevin equation was originally developed to explain the Brownian motion of particles suspended in a fluid. In that interpretation, the damping comes from viscous losses, and is determined by the size and mass of the particles as well as the fluid properties. For a spherical particle of radius r and mass m, $\gamma = 6\pi r\mu/m$

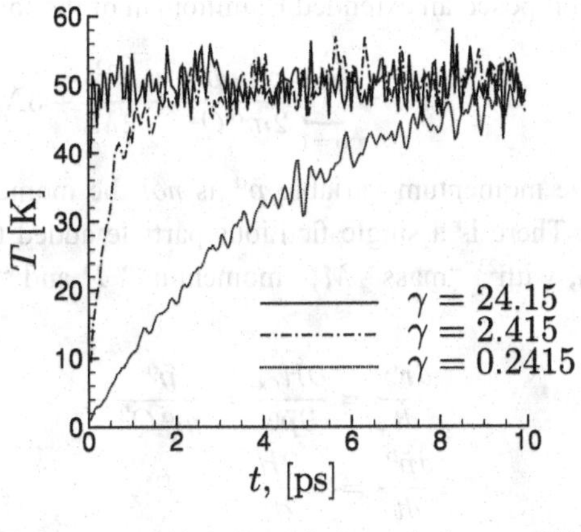

Fig. 9.5 The effect of the damping coefficient, γ, on the behavior of the Langevin thermostat. Units of γ are ps^{-1}. See text for details.

control and slow equilibration. On the other hand, an artificially large value of the damping constant may lead to the stochastic forces overwhelming the interatomic forces that are presumably of the greatest interest. In Fig. 9.5 we show the results of an MD simulation using the Langevin thermostat on an fcc crystal of Lennard-Jones Ar. The system contains 256 atoms, initially at 0 K with atoms randomly displaced from their equilibrium sites to a maximum of 0.1 Å. The thermostat is set to 50 K. We see that the damping coefficient has an effect on the rate of equilibration, but not on the final long-time limit.

9.4.4 The Nosé–Hoover (NH) thermostat

The Langevin approach described above is stochastic in nature, and is therefore not time-reversible. An alternative deterministic approach that does have this property is the so-called *extended system method*, the most common variation being the *NH thermostat*. This method was originally developed by Nosé [Nos84], but in a form that required an inconvenient rescaling of the time variable to interpret the results. Later, Hoover [Hoo85] reformulated it to restore the usual meaning of the time variable.

The essential idea of extended methods is as follows. We start from a system of N atoms with positions and momentum $(\boldsymbol{r}, \boldsymbol{p})$ and introduce additional fictitious particles into the system with positions $\boldsymbol{Q}$ and momenta $\boldsymbol{P}$. We are now at liberty to write any function of $(\boldsymbol{r}, \boldsymbol{p}, \boldsymbol{Q}, \boldsymbol{P})$ as the Hamiltonian of the system and determine the equations of motion for the system using the standard method of Hamiltonian mechanics (Eqn. (4.10)). Whether this extended system behaves according to physical reality is something we must determine after the fact, by studying the properties of the resulting equations of motion.

where μ is the dynamic viscosity of the surrounding fluid [Cha43]. Of course, this interpretation is not valid for atoms moving through an electronic heat bath, but the principle of choosing a different γ^α for each atomic species can be followed to reflect variations in atomic mass. In practice, the thermostat properties usually work well even if a single γ value is used for all the atoms, and so this approach is normally followed.

Nosé proposed an extended Hamiltonian of the form:

$$\mathcal{H}_{\text{ex}} = \sum_{\alpha=1}^{N} \frac{\tilde{\boldsymbol{p}}^{\alpha} \cdot \tilde{\boldsymbol{p}}^{\alpha}}{2m^{\alpha} Q^2} + \frac{P^2}{2M} + 3Nk_{\text{B}}T \ln Q + \mathcal{V}(\boldsymbol{r}), \tag{9.29}$$

where the momentum variable $\tilde{\boldsymbol{p}}^{\alpha}$ is *not* the momentum of particle α (as we shall see shortly). There is a single fictitious particle added to the system, moving along one dimension, with[23] "mass" M, "momentum" P and "position" Q. Applying Eqns. (4.10) yields

$$\begin{aligned}
\frac{d\boldsymbol{r}^{\alpha}}{dt} &= \frac{\partial \mathcal{H}_{\text{ex}}}{\partial \tilde{\boldsymbol{p}}^{\alpha}} = \frac{\tilde{\boldsymbol{p}}^{\alpha}}{m^{\alpha} Q^2}, \\
\frac{d\tilde{\boldsymbol{p}}^{\alpha}}{dt} &= -\frac{\partial \mathcal{H}_{\text{ex}}}{\partial \boldsymbol{r}^{\alpha}} = \boldsymbol{f}^{\alpha}, \\
\frac{dQ}{dt} &= \frac{\partial \mathcal{H}_{\text{ex}}}{\partial P} = \frac{P}{M}, \\
\frac{dP}{dt} &= -\frac{\partial \mathcal{H}_{\text{ex}}}{\partial Q} = \frac{1}{Q}\left(\sum_{\alpha=1}^{N} \frac{\tilde{\boldsymbol{p}}^{\alpha} \cdot \tilde{\boldsymbol{p}}^{\alpha}}{m^{\alpha} Q^2} - 3Nk_{\text{B}}T \right).
\end{aligned} \tag{9.30}$$

A microcanonical simulation of the evolution of the atomic and fictitious positions that will conserve $\mathcal{H}_{\text{ex}}$ is now straightforward using these equations, but it is not clear what physical significance this has until we manipulate the equations further. First, we make a change of time variables $dt = Q dt_{\text{new}}$ and use the superior dot notation to indicate differentiation with respect to the new time:

$$\frac{\dot{\boldsymbol{r}}^{\alpha}}{Q} = \frac{\tilde{\boldsymbol{p}}^{\alpha}}{m^{\alpha} Q^2}, \qquad \frac{\dot{\tilde{\boldsymbol{p}}}^{\alpha}}{Q} = \boldsymbol{f}^{\alpha}, \qquad \frac{\dot{Q}}{Q} = \frac{P}{M}, \qquad \dot{P} = \sum_{\alpha=1}^{N} \frac{\tilde{\boldsymbol{p}}^{\alpha} \cdot \tilde{\boldsymbol{p}}^{\alpha}}{m^{\alpha} Q^2} - 3Nk_{\text{B}}T. \tag{9.31}$$

Using a little algebra, the first three of these equations can be combined to form a single, second-order equation for the evolution of the atomic positions, leading to the final evolution equations for the system

$$m^{\alpha} \ddot{\boldsymbol{r}}^{\alpha} = \boldsymbol{f}^{\alpha} - \gamma m^{\alpha} \dot{\boldsymbol{r}}^{\alpha}, \tag{9.32}$$

and the fourth equation becomes

$$\dot{\gamma} = \frac{1}{M}\left(\sum_{\alpha=1}^{N} \frac{\boldsymbol{p}^{\alpha} \cdot \boldsymbol{p}^{\alpha}}{m^{\alpha}} - 3Nk_{\text{B}}T \right), \tag{9.33}$$

where we have defined $\gamma = P/M$ and distinguished the real particle momentum $\boldsymbol{p}^{\alpha}$ from $\tilde{\boldsymbol{p}}^{\alpha} = Q\boldsymbol{p}^{\alpha}$. These two equations embody the NH thermostat. We see that the first

[23] Note that the units of these variables are not the physical units of mass, momentum or position, but they play similar roles in the final equations.

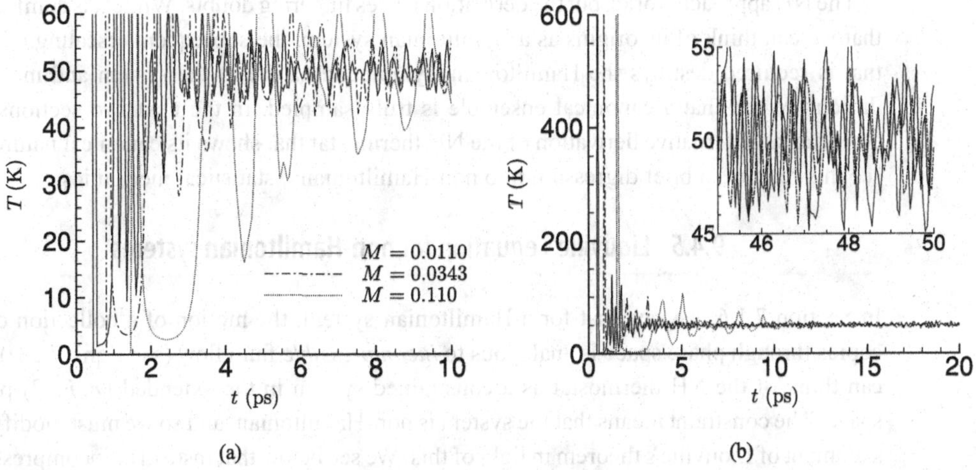

Fig. 9.6 The effect of the fictitious particle mass, M, on the behavior of the NH thermostat. The system and initial conditions are as for Fig. 9.5. Units of M are eV · ps^2: (a) the initial equilibration; (b) the long-time behavior. The inset shows that the magnitude of the temperature fluctuations after equilibration are the same regardless of the choice of M.

is similar to the Langevin approach without the stochastic term. The second determines the evolution of the "damping" coefficient, although since γ can be positive or negative the name "damping" perhaps no longer applies. Since the quantity inside the parentheses in Eqn. (9.33) is the difference between the desired temperature and the instantaneous temperature, we can see how feedback is built into the thermostat. If the instantaneous temperature is below the desired temperature for a particular timestep, γ will decrease (perhaps even becoming negative) to allow the velocities of the atoms to increase, or vice versa. Equation (9.33) is sometimes rewritten by defining a new fictitious mass $\widetilde{M}$ to replace M as

$$\dot{\gamma} = \frac{1}{\widetilde{M}} \left(T_{\text{inst}} - T \right), \quad \text{where} \quad \widetilde{M} = \frac{3k_{\text{B}}}{N} M. \tag{9.34}$$

This form is more convenient because the influence of $\widetilde{M}$ does not change with the number of atoms.

As with the Langevin approach, there is one free parameter in the method; in this case it is the fictitious particle mass, M. A large mass will have a large inertia and will thereby make heat flow between the system and the heat bath sluggish, while too small a mass will not allow good control of the temperature. This is illustrated in Fig. 9.6, where we see that the initial equilibration is affected by the choice of M, while the long-time temperature fluctuations are not. These results can also be compared to the Langevin thermostat results in Fig. 9.5. It is clear that the Langevin thermostat tends to equilibrate more rapidly than the NH thermostat, and without the initial wild fluctuations seen in Fig. 9.6. These transient fluctuations in the NH response can be mitigated to some extent by good temperature initialization (see Section 9.3.3), but in general the Langevin approach will have a more stable and rapid response to changes in set-temperature.

The NH approach works, but the derivation leaves lingering doubts. While it is comforting that we can think of its origins as a Hamiltonian system, the subsequent rescaling of time that is required destroys the Hamiltonian, symplectic nature of the system and makes it difficult to see that a canonical ensemble is truly sampled. In the next two sections, we construct an alternative derivation of the NH thermostat that shows its canonical nature. To see this requires a brief digression into non-Hamiltonian[24] statistical mechanics.

9.4.5 Liouville's equation for non-Hamiltonian systems

In Section 7.1.6, we saw that for a Hamiltonian system, the motion of a collection of its copies through phase space is analogous to *incompressible* fluid flow (see Eqn. (7.23)). We can think of the NH thermostat as a constrained system in the extended $(\boldsymbol{r}, \dot{\boldsymbol{r}}, P)$ phase space. The constraint means that the system is non-Hamiltonian, and so we must modify the statement of Liouville's theorem in light of this. We see below that instead of incompressible fluid flow, the modified form is analogous to the flow of a *compressible* fluid.

Consider an element of phase space volume, $d\boldsymbol{y}$, spanning the hypercube defined by the points $\boldsymbol{y}$ and $\boldsymbol{y} + d\boldsymbol{y}$, where we have returned to the concise notation $\boldsymbol{y} = (y_1, \dots, y_{2n}) = (q_1, \dots, q_n, p_1, \dots, p_n)$. The product $f(\boldsymbol{y}; t)d\boldsymbol{y}$, where $f(\boldsymbol{y}; t)$ is the distribution function, is the probability of finding a system inside $d\boldsymbol{y}$. This can be interpreted as the fractional number of systems, dN, lying within the hypercube $(\boldsymbol{y}, \boldsymbol{y} + d\boldsymbol{y})$, i.e.

$$dN = f(\boldsymbol{y}; t)d\boldsymbol{y}. \tag{9.35}$$

The face of this hypercube with its normal along the y_1 direction has area $dy_2 dy_3 \dots dy_{2n}$, and the flow of systems through this face must be equal to the density of points on the face, times the face area, times the component of "velocity," $v_1 = \partial y_1/\partial t$, that carries systems into the cube. Thus

$$N_1^{\text{in}} = f(y_1, y_2, \dots, t) v_1(y_1, y_2, \dots, t) dy_2 \cdots dy_{2n}.$$

Similarly, the fraction leaving the face at $y_1 + dy_1$ is

$$N_1^{\text{out}} = f(y_1 + dy_1, y_2, \dots, t) v_1(y_1 + dy_1, y_2, \dots, t) dy_2 \cdots dy_{2n}.$$

This second flux can be approximated by a Taylor expansion in terms of f and v_1:

$$N_1^{\text{out}} \approx \left[f(y_1, y_2, \dots, t) + \frac{\partial f}{\partial y_1} dy_1 \right] \left[v_1(y_1, y_2, \dots, t) + \frac{\partial v_1}{\partial y_1} dy_1 \right] dy_2 \cdots dy_{2n}.$$

Equations for the flow into and out of any other pair of faces along the ith direction (N_i^{in} and N_i^{out}) can be written similarly. Conservation of the number of systems in phase space requires that the time rate of change of the number of systems inside the elemental volume must be equal to the net flow of systems into and out of $d\boldsymbol{y}$:

$$\frac{\partial}{\partial t} dN = \sum_{i=1}^{2n} \left(N_i^{\text{in}} - N_i^{\text{out}} \right) = -\sum_{i=1}^{2n} \left[f \frac{\partial v_i}{\partial y_i} + v_i \frac{\partial f}{\partial y_i} \right] d\boldsymbol{y},$$

[24] See footnote 20 on page 510.

where the second equality follows after cancellation of terms and neglecting higher-order terms. Dividing through by dy and using Eqn. (9.35), we have concisely

$$\frac{\partial f}{\partial t} + \sum_{i=1}^{2n} \frac{\partial (f v_i)}{\partial y_i} = 0. \tag{9.36}$$

This is Liouville's equation for the more general case of a non-Hamiltonian system. Note that when the system is Hamiltonian, $\sum_i \partial v_i / \partial y_i = 0$ (see Eqn. (7.22)). Using this in Eqn. (9.36) recovers the familiar Liouville's equation of Eqn. (7.31). We now use Eqn. (9.36) to derive the NH thermostat, showing that it is a canonical thermostat at the same time.

9.4.6 An alternative derivation of the NH thermostat

In his paper, Hoover [Hoo85] presented the following alternative derivation for the NH thermostat. Consider a non-Hamiltonian system that is a function, not only of the atomic positions and momenta $(\boldsymbol{r}, \boldsymbol{p})$, but also of one additional momentum $P(= M\gamma)$ and *impose* the following canonical distribution on the system:

$$f(\boldsymbol{r}, \boldsymbol{p}, \gamma) = C \exp\left(-\frac{1}{k_B T}\left[\mathcal{V}(\boldsymbol{r}) + \frac{M\gamma^2}{2} + \sum_{\alpha=1}^{N} \frac{\boldsymbol{p}^\alpha \cdot \boldsymbol{p}^\alpha}{2m^\alpha}\right]\right),$$

where C is a normalization constant. Next, we ask what form the equations of motion must take in order that the generalized Liouville equation of Eqn. (9.36), which is fundamental, will be satisfied. First, we expand Eqn. (9.36) for our system to[25]

$$\frac{\partial f}{\partial t} + \sum_{\alpha=1}^{N} \operatorname{div}_{\boldsymbol{r}^\alpha}(f \dot{\boldsymbol{r}}^\alpha) + \sum_{\alpha=1}^{N} \operatorname{div}_{\boldsymbol{p}^\alpha}(f \dot{\boldsymbol{p}}^\alpha) + \frac{\partial (f\dot{\gamma})}{\partial \gamma} = 0, \tag{9.37}$$

and then derive the various terms in the equation:

$$\frac{\partial f}{\partial t} = 0, \qquad \sum_{\alpha=1}^{N} \operatorname{div}_{\boldsymbol{r}^\alpha}(f \dot{\boldsymbol{r}}^\alpha) = \frac{f}{k_B T} \sum_{\alpha=1}^{N} \frac{\boldsymbol{p}^\alpha \cdot \boldsymbol{f}^\alpha}{m^\alpha} + f \sum_{\alpha=1}^{N} \operatorname{div}_{\boldsymbol{r}^\alpha} \dot{\boldsymbol{r}}^\alpha,$$

$$\sum_{\alpha=1}^{N} \operatorname{div}_{\boldsymbol{p}^\alpha}(f \dot{\boldsymbol{p}}^\alpha) = -\frac{f}{k_B T} \sum_{\alpha=1}^{N} \frac{\boldsymbol{p}^\alpha \cdot \dot{\boldsymbol{p}}^\alpha}{m^\alpha} + f \sum_{\alpha=1}^{N} \operatorname{div}_{\boldsymbol{p}^\alpha} \dot{\boldsymbol{p}}^\alpha, \qquad \frac{\partial (f\dot{\gamma})}{\partial \gamma} = -\frac{f}{k_B T} M \dot{\gamma} \gamma + f \frac{\partial \dot{\gamma}}{\partial \gamma}.$$

[25] The divergence expressions in the following equations are defined as $\operatorname{div}_{\boldsymbol{z}} \boldsymbol{g}(\boldsymbol{z}) = \sum_{i=1}^{3} \partial g_i / \partial z_i$.

Summing these terms as in Eqn. (9.37), and dividing through by the common term f/k_BT (which we take to be nonzero), we obtain

$$\sum_{\alpha=1}^{N}\left(\frac{\boldsymbol{p}^\alpha\cdot\boldsymbol{f}^\alpha}{m^\alpha}-\frac{\boldsymbol{p}^\alpha\cdot\dot{\boldsymbol{p}}^\alpha}{m^\alpha}\right)-M\dot{\gamma}\gamma+k_BT\left(\sum_{\alpha=1}^{N}(\text{div}_{\boldsymbol{p}^\alpha}\dot{\boldsymbol{p}}^\alpha+\text{div}_{\boldsymbol{r}^\alpha}\dot{\boldsymbol{r}}^\alpha)+\frac{\partial\dot{\gamma}}{\partial\gamma}\right)=0. \tag{9.38}$$

In principle, we might be able to find other thermostat prescriptions that also satisfy this equation, but the NH equations (Eqns. (9.32) and (9.33)) satisfy it directly, as follows. Equation (9.32) gives the expression for $\dot{\boldsymbol{p}}^\alpha$ and explicitly shows its dependence on $\boldsymbol{p}^\alpha$:

$$\dot{\boldsymbol{p}}^\alpha=\boldsymbol{f}^\alpha-\gamma\boldsymbol{p}^\alpha. \tag{9.39}$$

This allows us to evaluate the divergence:

$$\text{div}_{\boldsymbol{p}^\alpha}\dot{\boldsymbol{p}}^\alpha=-\gamma\text{div}_{\boldsymbol{p}^\alpha}\boldsymbol{p}^\alpha=-3\gamma. \tag{9.40}$$

The second of the NH equations (Eqn. (9.33)) prescribes $\dot{\gamma}$, and imposes that it is independent of γ, so that

$$\partial\dot{\gamma}/\partial\gamma=0.$$

Finally, the NH equations tacitly impose that $\dot{\boldsymbol{r}}^\alpha$ is independent of $\boldsymbol{r}^\alpha$, since no relation between these variables is prescribed, making

$$\text{div}_{\boldsymbol{r}^\alpha}\dot{\boldsymbol{r}}^\alpha=0.$$

Putting these simplifications into Eqn. (9.38) leads to

$$\sum_{\alpha=1}^{N}\left(\frac{\boldsymbol{p}^\alpha\cdot\boldsymbol{f}^\alpha}{m^\alpha}-\frac{\boldsymbol{p}^\alpha\cdot\dot{\boldsymbol{p}}^\alpha}{m^\alpha}\right)-M\dot{\gamma}\gamma-3N\gamma k_BT=0. \tag{9.41}$$

Using Eqn. (9.39) and solving this for $\dot{\gamma}$ we recover Eqn. (9.33). In other words, Liouville's equation for the canonical distribution is satisfied if the NH equations describe the motion of the particles. Conversely, the NH equations of motion sample the canonical ensemble.

9.4.7 Integrating the NVT ensemble

Choosing a thermostat There is no unequivocal "best" thermostat, as each needs to be assessed for a given application. For example an issue with the NH thermostat, not discussed here, is its possible lack of ergodicity, i.e. an incomplete sampling of phase space. This can be addressed by introducing a so-called "NH chain," whereby the single thermostat particle is replaced by a chain of such particles [MKT92], each thermostatting the previous one, but this complicates the method. The Langevin approach, on the other hand, is ergodic but its performance can be overly sensitive to the value of the damping constant. Clearly, when using any thermostat it is necessary to carefully assess whether the results are affected by the limitations of the thermostat or the choice of its parameters.

Both the Langevin and NH thermostats are relatively easy to incorporate into an NVE integrator that uses the VV algorithm. The resulting Langevin or NH integrator is no longer symplectic, but since this property is already lost for these thermostats, this is of secondary concern.[26] Here, we look at the practical issues of this implementation in some detail.

The Langevin thermostat integrator The VV algorithm for the Langevin thermostat is straightforward and is shown in Algorithm 9.5. The only differences from Algorithm 9.2 are at lines 2 and 6. In this case the force is redefined to include the damping and stochastic terms from Eqn. (9.24), i.e.

$$\boldsymbol{f}^{\mathrm{L}}(\boldsymbol{r},\boldsymbol{v}) = \boldsymbol{f} - \gamma m \dot{\boldsymbol{r}} + \boldsymbol{G}, \qquad \widehat{\boldsymbol{f}}(t) = \boldsymbol{f}^{\mathrm{L}}[\boldsymbol{r}(t),\boldsymbol{v}(t)].$$

Also, because the velocity $\boldsymbol{v}_{t+\Delta t}$ is not yet computed when we arrive at line 6, it is approximated by $\boldsymbol{v}_{t+\Delta t/2}$ just for the purpose of computing the drag forces. This is done to avoid an integrator that is implicit, which would require an expensive internal iteration to self-consistency within each timestep. Note that there are more efficient ways to integrate the Langevin equations (see, for example, [Sha03]) but the algorithm presented here provides greater versatility in codes that are designed to swap between thermostats easily.

Algorithm 9.5 The VV algorithm modified for the Langevin thermostat

1: t:=0
Require: $\boldsymbol{r}_0$, $\boldsymbol{v}_0$ given.
2: $\widehat{\boldsymbol{f}}_0 := \boldsymbol{f}^{\mathrm{L}}(\boldsymbol{r}_0,\boldsymbol{v}_0)$
3: **while** $t \le t_{\mathrm{final}}$ **do**
4: $\quad \boldsymbol{r}_{t+\Delta t} := \boldsymbol{r}_t + \boldsymbol{v}_t \Delta t + \widehat{\boldsymbol{f}}_t \dfrac{(\Delta t)^2}{2m}$
5: $\quad \boldsymbol{v}_{t+\Delta t/2} := \boldsymbol{v}_t + \dfrac{\widehat{\boldsymbol{f}}_t}{2m}\Delta t$
6: $\quad \widehat{\boldsymbol{f}}_{t+\Delta t} := \boldsymbol{f}^{\mathrm{L}}(\boldsymbol{r}_{t+\Delta t},\boldsymbol{v}_{t+\Delta t/2})$
7: $\quad \boldsymbol{v}_{t+\Delta t} := \boldsymbol{v}_{t+\Delta t/2} + \dfrac{\widehat{\boldsymbol{f}}_{t+\Delta t}}{2m}\Delta t$
8: $\quad t := t + \Delta t$
9: **end while**

The NH thermostat integrator Integrating the NH thermostat is slightly more involved than the Langevin case because there are evolution equations for the damping coefficient itself. These are normally integrated just before the evaluation of the forces. As such, the complete algorithm is as shown in Algorithm 9.6, where lines 6 and 7 are just straightforward discretizations of Eqn. (9.34) and $\dot{\gamma} = d\gamma/dt$, respectively. Again, we have redefined the forces to include the NH damping terms, i.e.

$$\boldsymbol{f}^{\mathrm{NH}}(\boldsymbol{r},\boldsymbol{v}) = \boldsymbol{f} - \gamma m \boldsymbol{v}, \qquad \widehat{\boldsymbol{f}}(t) = \boldsymbol{f}^{\mathrm{NH}}[\boldsymbol{r}(t),\boldsymbol{v}(t)].$$

[26] However, it also proves very difficult to design an integrator for the NH thermostat that adequately prevents drift in the value of the extended Hamiltonian of the system. See [Ezr06] and [LM02] for some studies along this line. For a discussion of a symplectic variant of the NH thermostat, the so-called Nosé–Poincaré thermostat, see [BLL99].

As with the Langevin integrator, the velocity used at line 8 is $\boldsymbol{v}_{t+\Delta t/2}$ to avoid the need for an implicit integrator [Pli09, BLL99]. The instantaneous temperature in line 6 of this algorithm, T_{inst}, is as defined in Eqn. (9.4).

Algorithm 9.6 The VV algorithm modified for the NH thermostat

1: t:=0

Require: $\boldsymbol{r}_0, \boldsymbol{v}_0$ given.

2: $\widehat{\boldsymbol{f}}_0 := \boldsymbol{f}^{\text{NH}}(\boldsymbol{r}_0, \boldsymbol{v}_0)$

3: **while** $t \leq t_{\text{final}}$ **do**

4: $\quad \boldsymbol{r}_{t+\Delta t} := \boldsymbol{r}_t + \boldsymbol{v}_t \Delta t + \widehat{\boldsymbol{f}}_t \dfrac{(\Delta t)^2}{2m}$

5: $\quad \boldsymbol{v}_{t+\Delta t/2} := \boldsymbol{v}_t + \dfrac{\widehat{\boldsymbol{f}}_t}{2m} \Delta t$

6: $\quad \dot{\gamma}_{t+\Delta t} := \dot{\gamma}_t + \Delta t [T_{\text{inst}}(\boldsymbol{v}_{t+\Delta t/2} - T] / \widetilde{M}$

7: $\quad \gamma_{t+\Delta t} := \gamma_t + \dfrac{\Delta t}{2} [\dot{\gamma}_t + \dot{\gamma}_{t+\Delta t}]$

8: $\quad \widehat{\boldsymbol{f}}_{t+\Delta t} := \boldsymbol{f}^{\text{NH}}(\boldsymbol{r}_{t+\Delta t}, \boldsymbol{v}_{t+\Delta t/2})$

9: $\quad \boldsymbol{v}_{t+\Delta t} := \boldsymbol{v}_{t+\Delta t/2} + \dfrac{\widehat{\boldsymbol{f}}_{t+\Delta t}}{2m} \Delta t$

10: $\quad t := t + \Delta t$

11: **end while**

9.5 The finite strain NσH ensemble: applying stress

Here, we discuss how to apply stress in an MD simulation. The origin of constant stress methods is the extended Hamiltonian approach first put forward by Andersen [And80]. Andersen's method was limited to imposing a constant pressure, whereas in solids we are often interested in the full stress tensor.

The most widely used formulation of the constant stress MD ensemble is that of Parrinello–Rahman (PR) [PR81], although there have been alternative formulations (see, for example, [Ray83] and [Wen91]). Perhaps the clearest discussion of the methods is that of Ray and Rahman [RR84], and we take a similar line of reasoning here. We start with a rigorous derivation of the equations of motion of the system from the point of view of a canonical transformation of the Hamiltonian. Because we believe our resulting equations of motion to be distinct from any in the literature, we choose to give the formulation derived here a unique name. We will therefore call this the "finite strain NσH ensemble," based on observations made later in Section 9.5.4 that this approach is correct even in the case of large deformations of the simulation box (something which is not the case for the PR approach). Later, we will look at the approximations made in the PR and related methods, and investigate when these approximations will work in practice.

The goal of all these methods is to augment the equations of motion for the particles with a set of equations of motion for the variables defining the shape (and volume) of the simulation box. In other words, Eqn. (9.1) is replaced with a set of equations of the form

$$m^\alpha \ddot{\boldsymbol{r}}^\alpha = \boldsymbol{f}^\alpha(\boldsymbol{q}, \boldsymbol{p}, \boldsymbol{F}, \dot{\boldsymbol{F}}), \qquad M\ddot{\boldsymbol{F}} = \boldsymbol{f}^{\text{cell}}(\boldsymbol{q}, \boldsymbol{p}, \boldsymbol{F}, \dot{\boldsymbol{F}}), \tag{9.42}$$

where $\boldsymbol{q}$ and $\boldsymbol{p}$ are the concatenated positions and momenta:

$$\boldsymbol{q} = (r_1^1, r_2^1, r_3^1, r_1^2, \dots, r_3^N), \qquad \boldsymbol{p} = (p_1^1, p_2^1, p_3^1, p_1^2, \dots, p_3^N),$$

and M is a constant fictitious "mass" assigned to the box (with units of mass times length-squared). We choose to define the shape of the simulation box using the deformation gradient, $\boldsymbol{F}$, referred to some initial box, although other choices are possible (see more below). The two equations in Eqn. (9.42) are coupled through atomic forces ($\boldsymbol{f}$) and configurational forces on the cell shape ($\boldsymbol{f}^{\text{cell}}$). These forces can generally depend on atomic positions and velocities as well as the current box shape and its time rate of change. We derive these equations using the canonical transformation of variables introduced in Section 8.1.1.

9.5.1 A canonical transformation of variables

Starting from the Hamiltonian of a system of particles, we have

$$\mathcal{H}(\boldsymbol{q}, \boldsymbol{p}) = \mathcal{V}^{\text{int}}(\boldsymbol{r}^1, \dots, \boldsymbol{r}^N) + \sum_{\alpha=1}^{N} \frac{\|\boldsymbol{p}^\alpha\|^2}{2m^\alpha}. \tag{9.43}$$

Using the method of canonical transformations from Section 8.1.1, we would now like to introduce a transformation to a new set of variables

$$\widehat{\boldsymbol{q}} = (R_1^1, R_2^1, R_3^1, R_1^2, \dots, R_3^N), \qquad \widehat{\boldsymbol{p}} = (\mathfrak{P}_1^1, \mathfrak{P}_2^1, \mathfrak{P}_3^1, \mathfrak{P}_1^2, \dots, \mathfrak{P}_3^N),$$

where $\boldsymbol{R}^\alpha$ and $\boldsymbol{\mathfrak{P}}^\alpha$ are the reference position and momentum[27] of atom α. We choose a generating function of the same form used in Section 8.1.3 (see Eqn. (8.46)):

$$G_3(\boldsymbol{p}, \widehat{\boldsymbol{q}}, t) = -\sum_{\alpha=1}^{N} \boldsymbol{p}^\alpha \cdot \boldsymbol{F}(t)\boldsymbol{R}^\alpha, \tag{9.44}$$

where $\boldsymbol{F}(t)$ is the time-dependent deformation gradient relating the reference configuration to an evolving deformed configuration. Applying Eqns. (8.14), we find expressions for the transformed coordinates and momenta in terms of $\boldsymbol{r}^\alpha$ and $\boldsymbol{p}^\alpha$:

$$\boldsymbol{r}^\alpha = -\frac{\partial G_3}{\partial \boldsymbol{p}^\alpha} = \boldsymbol{F}\boldsymbol{R}^\alpha, \qquad \boldsymbol{\mathfrak{P}}^\alpha = -\frac{\partial G_3}{\partial \boldsymbol{R}^\alpha} = \boldsymbol{F}^T \boldsymbol{p}^\alpha, \tag{9.45}$$

[27] We use the fraktur font for the reference momentum to prevent confusion with the first Piola–Kirchhoff stress $\boldsymbol{P}$.

and the new Hamiltonian:

$$\widehat{\mathcal{H}}(\widehat{\boldsymbol{q}},\widehat{\boldsymbol{p}};\boldsymbol{F}) = \mathcal{H}(\boldsymbol{q}(\widehat{\boldsymbol{q}};\boldsymbol{F}),\boldsymbol{p}(\widehat{\boldsymbol{p}};\boldsymbol{F})) + \frac{\partial G_3}{\partial t}(\boldsymbol{p}(\widehat{\boldsymbol{p}};\boldsymbol{F}),\widehat{\boldsymbol{q}},t)$$
$$= \mathcal{V}^{\text{int}}(\boldsymbol{F}\boldsymbol{R}^1,\ldots,\boldsymbol{F}\boldsymbol{R}^N) + \sum_{\alpha=1}^{N} \frac{\left\|\boldsymbol{F}^{-T}\boldsymbol{\mathfrak{P}}^\alpha\right\|^2}{2m^\alpha} - \sum_{\alpha=1}^{N} \boldsymbol{F}^{-T}\boldsymbol{\mathfrak{P}}^\alpha \cdot \dot{\boldsymbol{F}}\boldsymbol{R}^\alpha, \tag{9.46}$$

where we have used Eqn. (9.45) to simplify this last expression.

Other canonical transformations are possible. Instead of using the deformation gradient in Eqn. (9.44), Parrinello and Rahman used the simulation box H as defined in Eqn. (6.30). This leads to a different interpretation for the transformed coordinates $\widehat{\boldsymbol{q}}$ and $\widehat{\boldsymbol{p}}$. We prefer our interpretation because it leads more directly to the connection with continuum mechanics concepts. In any event, the two choices are essentially interchangeable since the two matrices are related through a constant initial box shape, i.e.

$$\mathsf{H} = \boldsymbol{F}\mathsf{H}_0, \qquad \boldsymbol{F} = \mathsf{H}\mathsf{H}_0^{-1}. \tag{9.47}$$

The idea now is to let the box deform in response to an applied stress, so that the equilibrium shape of the box corresponds to the state where the internal stress is equal to the imposed stress. To do so, we associate a potential energy and a kinetic energy with the shape of the cell, such that

$$\mathcal{V}_{\text{cell}} = -V_0\widetilde{\boldsymbol{S}} : \boldsymbol{E}, \qquad \mathcal{K}_{\text{cell}} = \frac{\boldsymbol{\pi} : \boldsymbol{\pi}}{2M}, \tag{9.48}$$

where V_0 is the volume of the simulation box in the reference configuration. We recognize the potential energy as the strain energy associated with the constant applied second Piola–Kirchhoff stress, $\widetilde{\boldsymbol{S}}$. The kinetic energy is an artificial construct that allows us to derive the equations of motion, where $\boldsymbol{\pi}$ is a second-order tensor that is a fictitious momentum conjugate to the deformation gradient, $\boldsymbol{F}$ (i.e. $\boldsymbol{\pi} = M\dot{\boldsymbol{F}}$). We stress that the kinetic energy $\mathcal{K}_{\text{cell}}$ is not a physical energy, but rather a trick that will allow us to derive an equation of motion for the box; recall that our goal is a computational scheme that will let the box find a shape that puts it in equilibrium with the applied stress. The mass M determines how quickly the box responds when it is out of equilibrium, and it is a parameter that we must set by experience. Alternative forms of $\mathcal{K}_{\text{cell}}$ are discussed in Section 9.5.5.

Our new extended Hamiltonian is now

$$\widehat{\mathcal{H}}_{\text{ex}}(\widehat{\boldsymbol{q}},\widehat{\boldsymbol{p}},\boldsymbol{F},\boldsymbol{\pi}) = \widehat{\mathcal{H}}(\widehat{\boldsymbol{q}},\widehat{\boldsymbol{p}};\boldsymbol{F}) + \mathcal{V}_{\text{cell}}(\boldsymbol{F}) + \mathcal{K}_{\text{cell}}(\boldsymbol{\pi}),$$
$$= \mathcal{V}^{\text{int}}(\boldsymbol{F}\boldsymbol{R}^1,\ldots,\boldsymbol{F}\boldsymbol{R}^N) + \sum_{\alpha=1}^{N} \frac{\left\|\boldsymbol{F}^{-T}\boldsymbol{\mathfrak{P}}^\alpha\right\|^2}{2m^\alpha} - \sum_{\alpha=1}^{N} \boldsymbol{F}^{-T}\boldsymbol{\mathfrak{P}}^\alpha \cdot \dot{\boldsymbol{F}}\boldsymbol{R}^\alpha$$
$$-V_0\widetilde{\boldsymbol{S}} : \boldsymbol{E} + \frac{\boldsymbol{\pi} : \boldsymbol{\pi}}{2M}. \tag{9.49}$$

It is expressed in terms of the extended variable set $(\widehat{q}, \widehat{p}, \boldsymbol{F}, \boldsymbol{\pi})$. The equations of motion of the extended system are given by Hamilton's equations ($\alpha = 1, \ldots, N$):

$$\dot{\boldsymbol{R}}^\alpha = \frac{\partial \widehat{\mathcal{H}}_{\text{ex}}}{\partial \mathfrak{P}^\alpha}, \qquad \dot{\mathfrak{P}}^\alpha = -\frac{\partial \widehat{\mathcal{H}}_{\text{ex}}}{\partial \boldsymbol{R}^\alpha}, \qquad \dot{\boldsymbol{F}} = \frac{\partial \widehat{\mathcal{H}}_{\text{ex}}}{\partial \boldsymbol{\pi}}, \qquad \dot{\boldsymbol{\pi}} = -\frac{\partial \widehat{\mathcal{H}}_{\text{ex}}}{\partial \boldsymbol{F}}. \tag{9.50}$$

The first of Hamilton's equations leads to

$$\boldsymbol{F}^{-T} \mathfrak{P}^\alpha = m^\alpha \left(\boldsymbol{F} \dot{\boldsymbol{R}}^\alpha + \dot{\boldsymbol{F}} \boldsymbol{R}^\alpha \right), \tag{9.51}$$

which is equivalent to $\boldsymbol{p}^\alpha = m^\alpha \dot{\boldsymbol{r}}^\alpha$, while the second equation gives

$$\dot{\mathfrak{P}}^\alpha = \boldsymbol{F}^T \boldsymbol{f}^{\text{int},\alpha} + \dot{\boldsymbol{F}}^T \boldsymbol{F}^{-T} \mathfrak{P}^\alpha, \tag{9.52}$$

where $\boldsymbol{f}^{\text{int},\alpha} = -\partial \mathcal{V}^{\text{int}} / \partial \boldsymbol{r}^\alpha$ is the usual force on an atom. Inserting Eqn. $(9.45)_2$ for $\mathfrak{P}^\alpha$, we see that this is equivalent to

$$\dot{\boldsymbol{p}}^\alpha = \boldsymbol{f}^{\text{int},\alpha}. \tag{9.53}$$

It is clear that the canonical transformation leaves the original equations of motion intact (as it should). We can rewrite Eqn. (9.53) in terms of the transformed variables $\boldsymbol{R}^\alpha$ using Eqn. (9.51) and Eqn. $(9.45)_2$:

$$m^\alpha \ddot{\boldsymbol{R}}^\alpha = \boldsymbol{F}^{-1} \left[\boldsymbol{f}^{\text{int},\alpha} - 2m^\alpha \dot{\boldsymbol{F}} \dot{\boldsymbol{R}}^\alpha - m^\alpha \ddot{\boldsymbol{F}} \boldsymbol{R}^\alpha \right]. \tag{9.54}$$

However, in practice it will be more convenient to use Eqn. (9.53).

Equations of motion for the shape of the box come from the third and fourth of Hamilton's equations in Eqn. (9.50). The third leads simply to

$$\dot{\boldsymbol{F}} = \frac{\boldsymbol{\pi}}{M}, \tag{9.55}$$

which establishes the relation between the simulation box shape and its momentum. The fourth of Hamilton's equations involves the derivative of $\widehat{\mathcal{H}}_{\text{ex}}$ with respect to $\boldsymbol{F}$. To compute this, we separately consider the parts of $\widehat{\mathcal{H}}_{\text{ex}}$ that depend on $\boldsymbol{F}$. Referring back to Eqns. (9.49) and (9.46), we see that the terms we need to consider are $\mathcal{H}$, $\partial G_3/\partial t$ and $\mathcal{V}_{\text{cell}}$. Calculating the corresponding derivatives, we have:

$$\frac{\partial \mathcal{H}}{\partial \boldsymbol{F}} = V \boldsymbol{\sigma}^{\text{inst}} \boldsymbol{F}^{-T}, \tag{9.56a}$$

$$\frac{\partial}{\partial \boldsymbol{F}} \left(\frac{\partial G_3}{\partial t} \right) = -\sum_{\alpha=1}^{N} \left(-\boldsymbol{p}^\alpha \otimes \boldsymbol{l} \boldsymbol{r}^\alpha + \boldsymbol{l}^T \boldsymbol{p}^\alpha \otimes \boldsymbol{r}^\alpha \right) \boldsymbol{F}^{-T}, \tag{9.56b}$$

$$\frac{\partial \mathcal{V}_{\text{cell}}}{\partial \boldsymbol{F}} = -V_0 \boldsymbol{F} \widetilde{\boldsymbol{S}}, \tag{9.56c}$$

where in the derivation we have used the definitions in Eqn. (9.45), the expression for $\partial \boldsymbol{F}^{-1} / \partial \boldsymbol{F}$ in Eqn. (8.58), the relation $\dot{\boldsymbol{F}} = \boldsymbol{l} \boldsymbol{F}$ given in Eqn. (2.71) (where $\boldsymbol{l}$ is the velocity gradient defined in Eqn. (2.69)) and the definition of the Lagrangian strain tensor in Eqn. (2.60). In Eqn. (9.56a), $V = J V_0$ is the volume of the simulation box in the deformed configuration (where $J = \det \boldsymbol{F}$ is the Jacobian), and $\boldsymbol{\sigma}^{\text{inst}}$ is the instantaneous Cauchy stress tensor defined in Eqn. (8.73) (see also Eqn. (8.62)):

$$\boldsymbol{\sigma}^{\text{inst}} = -\frac{1}{V}\sum_{\alpha=1}^{N}\left[\frac{\boldsymbol{p}^{\alpha}\otimes\boldsymbol{p}^{\alpha}}{m^{\alpha}} + \boldsymbol{f}^{\text{int},\alpha}\otimes\boldsymbol{r}^{\alpha}\right]. \tag{9.57}$$

Combining Eqns. (9.56a)–(9.56c) and substituting into Eqn. $(9.50)_4$ gives

$$\dot{\boldsymbol{\pi}} = V\left[-\boldsymbol{\sigma}^{\text{inst}} + \frac{\boldsymbol{F}\widetilde{\boldsymbol{S}}\boldsymbol{F}^T}{J} + \frac{1}{V}\sum_{\alpha=1}^{N}\left(-\boldsymbol{p}^{\alpha}\otimes\boldsymbol{l}\boldsymbol{r}^{\alpha} + \boldsymbol{l}^T\boldsymbol{p}^{\alpha}\otimes\boldsymbol{r}^{\alpha}\right)\right]\boldsymbol{F}^{-T}. \tag{9.58}$$

Substituting in Eqn. (9.55), we find the equation of motion for the simulation cell shape:

$$\mathcal{M}\ddot{\boldsymbol{F}} = \boldsymbol{f}^{\text{cell}}, \tag{9.59}$$

where

$$\boldsymbol{f}^{\text{cell}} = V\left[\frac{\boldsymbol{F}\widetilde{\boldsymbol{S}}\boldsymbol{F}^T}{J} - \boldsymbol{\sigma}^{\text{inst}} - \Delta\boldsymbol{\sigma}\right]\boldsymbol{F}^{-T} \tag{9.60}$$

and we have defined

$$\Delta\boldsymbol{\sigma} \equiv \frac{1}{V}\sum_{\alpha=1}^{N}\left[\boldsymbol{p}^{\alpha}\otimes\boldsymbol{l}\boldsymbol{r}^{\alpha} - \boldsymbol{l}^T\boldsymbol{p}^{\alpha}\otimes\boldsymbol{r}^{\alpha}\right]. \tag{9.61}$$

This equation, combined with Eqn. (9.53), defines the new equations of motion for the system. Note that this equation is in terms of an applied *second Piola-Kirchhoff stress* ($\widetilde{\boldsymbol{S}}$), not the Cauchy stress, as otherwise the definition of $\mathcal{V}_{\text{cell}}$ would lose its physical meaning. This can be related to an effective applied Cauchy stress through Eqn. (2.100):

$$\widetilde{\boldsymbol{\sigma}} = \frac{\boldsymbol{F}\widetilde{\boldsymbol{S}}\boldsymbol{F}^T}{J}, \tag{9.62}$$

but we emphasize that this effective Cauchy stress is only known after the simulation cell is allowed to deform to its equilibrium shape.

At equilibrium, $\ddot{\boldsymbol{F}} \approx 0$, and Eqn. (9.59) reduces to

$$\boldsymbol{\sigma}^{\text{inst}} \approx \frac{\boldsymbol{F}\widetilde{\boldsymbol{S}}\boldsymbol{F}^T}{J} - \Delta\boldsymbol{\sigma}. \tag{9.63}$$

This is *almost* the equation we want, since our goal was to find an equilibrium configuration where the internal Cauchy stress is driven to the applied Cauchy stress given by $\widetilde{\boldsymbol{\sigma}} = \boldsymbol{F}\widetilde{\boldsymbol{S}}\boldsymbol{F}^T/J$ (albeit with persistent thermal fluctuations around the applied value). Next we show that the apparent extra term, $\Delta\boldsymbol{\sigma}$, goes to zero at equilibrium.

Stress at equilibrium conditions Let us take a closer look at the extra term $\Delta\boldsymbol{\sigma}$ in Eqn. (9.63). First, using Eqns. (2.71) and (9.45), we have $\boldsymbol{l}\boldsymbol{r}^\alpha = \dot{\boldsymbol{F}}\boldsymbol{R}^\alpha$. Substituting this into Eqn. (9.61) gives

$$\Delta\boldsymbol{\sigma} = \frac{1}{V}\sum_{\alpha=1}^{N}\left[\boldsymbol{p}^\alpha \otimes \dot{\boldsymbol{F}}\boldsymbol{R}^\alpha - \boldsymbol{l}^T\boldsymbol{p}^\alpha \otimes \boldsymbol{r}^\alpha\right]. \tag{9.64}$$

Using the chain rule, $\dot{\boldsymbol{r}}^\alpha = \dot{\boldsymbol{F}}\boldsymbol{R}^\alpha + \boldsymbol{F}\dot{\boldsymbol{R}}^\alpha$, to replace $\dot{\boldsymbol{F}}\boldsymbol{R}^\alpha$ in the first term in the sum of Eqn. (9.64), and Eqns. (2.71) and $(9.45)_2$ in the second term, we have

$$\Delta\boldsymbol{\sigma} = \frac{1}{V}\sum_{\alpha=1}^{N}\left[\frac{\boldsymbol{p}^\alpha \otimes \boldsymbol{p}^\alpha}{m^\alpha} - \boldsymbol{p}^\alpha \otimes \boldsymbol{F}\dot{\boldsymbol{R}}^\alpha - \boldsymbol{F}^{-T}\dot{\boldsymbol{F}}^T\boldsymbol{F}^{-T}\mathfrak{P}^\alpha \otimes \boldsymbol{r}^\alpha\right].$$

Next, we replace the quantity $\dot{\boldsymbol{F}}^T\boldsymbol{F}^{-T}\mathfrak{P}^\alpha$ in the last term using Eqn. (9.52) to get a total of four terms inside the summation

$$\Delta\boldsymbol{\sigma} = \frac{1}{V}\sum_{\alpha=1}^{N}\left[\frac{\boldsymbol{p}^\alpha \otimes \boldsymbol{p}^\alpha}{m^\alpha} - \boldsymbol{p}^\alpha \otimes \boldsymbol{F}\dot{\boldsymbol{R}}^\alpha - \boldsymbol{F}^{-T}\dot{\mathfrak{P}}^\alpha \otimes \boldsymbol{r}^\alpha + \boldsymbol{f}^{\mathrm{int},\alpha} \otimes \boldsymbol{r}^\alpha\right].$$

The first and last of these terms combine to give the negative of the instantaneous Cauchy stress in Eqn. (9.57), so that we now have

$$\Delta\boldsymbol{\sigma} = -\boldsymbol{\sigma}^{\mathrm{inst}} + \boldsymbol{\sigma}^{\mathrm{eq}}, \tag{9.65}$$

where we have defined

$$\boldsymbol{\sigma}^{\mathrm{eq}} \equiv -\frac{1}{V}\sum_{\alpha=1}^{N}\left[\boldsymbol{p}^\alpha \otimes \boldsymbol{F}\dot{\boldsymbol{R}}^\alpha + \boldsymbol{F}^{-T}\dot{\mathfrak{P}}^\alpha \otimes \boldsymbol{r}^\alpha\right]. \tag{9.66}$$

We can now show that this term, $\boldsymbol{\sigma}^{\mathrm{eq}}$, is equal to $\boldsymbol{\sigma}^{\mathrm{inst}}$ in the limit where $\dot{\boldsymbol{F}} \to \boldsymbol{0}$ (which we expect at equilibrium). We see this by differentiating the two Eqns. (9.45) with respect to time. From the first equation we have

$$\dot{\boldsymbol{r}}^\alpha = \dot{\boldsymbol{F}}\boldsymbol{R}^\alpha + \boldsymbol{F}\dot{\boldsymbol{R}}^\alpha \approx \boldsymbol{F}\dot{\boldsymbol{R}}^\alpha,$$

or in terms of momentum simply

$$\boldsymbol{F}\dot{\boldsymbol{R}}^\alpha \approx \frac{\boldsymbol{p}^\alpha}{m^\alpha}.$$

From the second of Eqns. (9.45) we have

$$\dot{\mathfrak{P}}^\alpha = \dot{\boldsymbol{F}}^T\boldsymbol{p}^\alpha + \boldsymbol{F}^T\dot{\boldsymbol{p}}^\alpha \approx \boldsymbol{F}^T\dot{\boldsymbol{p}}^\alpha,$$

which we invert to get

$$\boldsymbol{F}^{-T}\dot{\mathfrak{P}}^\alpha \approx \dot{\boldsymbol{p}}^\alpha = \boldsymbol{f}^{\mathrm{int},\alpha},$$

where the last equality is from Eqn. (9.53). Substituting these into the definition of $\boldsymbol{\sigma}^{\mathrm{eq}}$ we find that when $\dot{\boldsymbol{F}} \approx \boldsymbol{0}$ then

$$\boldsymbol{\sigma}^{\mathrm{eq}} \approx -\frac{1}{V}\sum_{\alpha=1}^{N}\left[\frac{\boldsymbol{p}^\alpha \otimes \boldsymbol{p}^\alpha}{m^\alpha} + \boldsymbol{f}^{\mathrm{int},\alpha} \otimes \boldsymbol{r}^\alpha\right],$$

which is exactly the definition of $\boldsymbol{\sigma}^{\text{inst}}$ in Eqn. (9.57), i.e.

$$\boldsymbol{\sigma}^{\text{eq}} \to \boldsymbol{\sigma}^{\text{inst}} \quad \text{as} \quad \dot{\boldsymbol{F}} \to \boldsymbol{0}. \tag{9.67}$$

Therefore when $\dot{\boldsymbol{F}}$ is small we expect

$$\Delta\boldsymbol{\sigma} \approx -\boldsymbol{\sigma}^{\text{inst}} + \boldsymbol{\sigma}^{\text{inst}} = \boldsymbol{0}. \tag{9.68}$$

Referring back to Eqn. (9.63), we see that this implies that at equilibrium the instantaneous stress will fluctuate around the imposed stress as required. The fluctuations are a well-defined, characteristic property of the NσE ensemble, in direct analogy to the fluctuations of the instantaneous temperature about the set temperature in the NVT (canonical) ensemble. More about these fluctuations can be found in [RG81, RGH81, Ray82, PR82].

Implementation Putting this form of stress control in an existing MD code is fairly simple. The equations of motion of the atoms are unchanged (as shown in Eqn. (9.53)). The equation of motion of the simulation cell in Eqn. (9.59) can be implemented with the expression for $\boldsymbol{f}^{\text{cell}}$ in Eqn. (9.60). However, a more convenient form in terms of $\boldsymbol{\sigma}^{\text{eq}}$ (which is more directly found from the positions and momenta) is obtained using Eqn. (9.65):

$$M\ddot{\boldsymbol{F}} = V\left[\frac{\boldsymbol{F}\widetilde{\boldsymbol{S}}\boldsymbol{F}^T}{J} - \boldsymbol{\sigma}^{\text{eq}}\right]\boldsymbol{F}^{-T}. \tag{9.69}$$

Finally, an ad hoc damping term can be added to the forces deforming the periodic cell to suppress oscillations and hasten convergence to equilibrium, so that

$$M\ddot{\boldsymbol{F}} = \boldsymbol{f}^{\text{cell,damped}}, \tag{9.70}$$

where

$$\boldsymbol{f}^{\text{cell,damped}} = V\left[\frac{\boldsymbol{F}\widetilde{\boldsymbol{S}}\boldsymbol{F}^T}{J} - \boldsymbol{\sigma}^{\text{eq}} - \gamma_{\text{cell}}\dot{\boldsymbol{F}}\right]\boldsymbol{F}^{-T}, \tag{9.71}$$

and γ_{cell} is a constant damping coefficient. At equilibrium, $\dot{\boldsymbol{F}} \approx \boldsymbol{0}$ and $\ddot{\boldsymbol{F}} \approx \boldsymbol{0}$, so we can rearrange Eqn. (9.70) to get

$$\boldsymbol{\sigma}^{\text{inst}} = \frac{\boldsymbol{F}\widetilde{\boldsymbol{S}}\boldsymbol{F}^T}{J},$$

where we have used Eqn. (9.67) to replace $\boldsymbol{\sigma}^{\text{eq}}$.

Integrating Eqn. (9.70) in time requires that $\boldsymbol{F}$ and $\dot{\boldsymbol{F}}$ be stored as additional variables analogous to atomic positions and momentum. A discrete integration of the cell evolution can then be performed alongside the integration of the ordinary atomic motion. This is straightforward to do since the form of Eqn. (9.70) is directly amenable to the VV algorithm. This modification to Algorithm 9.2 is shown in Algorithm 9.7.

Algorithm 9.7 The VV algorithm for the finite strain NσH ensemble

$t := 0$
Require: $\boldsymbol{r}_0, \boldsymbol{v}_0$ given
Require: $\boldsymbol{F}_0 = \boldsymbol{I}$ and $\dot{\boldsymbol{F}}_0 = \boldsymbol{0}$
$\widehat{\boldsymbol{f}}_0 := \boldsymbol{f}(\boldsymbol{r}_0)$
$\widehat{\boldsymbol{f}}_0^{\text{cell,damped}} := \boldsymbol{f}^{\text{cell,damped}}(\boldsymbol{r}_0, \boldsymbol{v}_0, \boldsymbol{F}_0, \dot{\boldsymbol{F}}_0)$
while $t \le t_{\text{final}}$ **do**
  $\boldsymbol{r}_{t+\Delta t} := \boldsymbol{r}_t + \boldsymbol{v}_t \Delta t + \widehat{\boldsymbol{f}}_t (\Delta t)^2 / 2m$
  $\boldsymbol{v}_{t+\Delta t/2} := \boldsymbol{v}_t + \widehat{\boldsymbol{f}}_t \Delta t / 2m$
  $\boldsymbol{F}_{t+\Delta t} := \boldsymbol{F}_t + \dot{\boldsymbol{F}}_t \Delta t + \widehat{\boldsymbol{f}}_t^{\text{cell,damped}} (\Delta t)^2 / 2M$
  $\dot{\boldsymbol{F}}_{t+\Delta t/2} := \dot{\boldsymbol{F}}_t + \widehat{\boldsymbol{f}}_t^{\text{cell,damped}} \Delta t / 2M$
  $\widehat{\boldsymbol{f}}_{t+\Delta t} := \boldsymbol{f}(\boldsymbol{r}_{t+\Delta t})$
  $\widehat{\boldsymbol{f}}_{t+\Delta t}^{\text{cell,damped}} := \boldsymbol{f}^{\text{cell,damped}}(\boldsymbol{r}_{t+\Delta t}, \boldsymbol{v}_{t+\Delta t/2}, \boldsymbol{F}_{t+\Delta t}, \dot{\boldsymbol{F}}_{t+\Delta t/2})$
  $\boldsymbol{v}_{t+\Delta t} := \boldsymbol{v}_{t+\Delta t/2} + \widehat{\boldsymbol{f}}_{t+\Delta t} \Delta t / 2m$
  $\dot{\boldsymbol{F}}_{t+\Delta t} := \dot{\boldsymbol{F}}_{t+\Delta t/2} + \widehat{\boldsymbol{f}}_{t+\Delta t}^{\text{cell,damped}} \Delta t / 2M$
  $t := t + \Delta t$
end while

One additional detail to be aware of is that the nine components of $\boldsymbol{F}$ are more than is really necessary to define the motion of a simulation cell; leaving them all as degrees of freedom will lead to the cell rotating freely in space during the minimization. This means that some simple constraints must be imposed on $\boldsymbol{F}$ to reduce it from nine to six degrees of freedom. This is more readily achieved by constraining some of the components of the simulation box matrix, H, to their initial values. For example, in the MD code LAMMPS [Pli95, Pli09], the elements of the matrix H_{ij} with $i > j$ are constrained to be zero as in Eqn. (6.37). To work in terms of the simulation box matrix instead of $\boldsymbol{F}$ presents no great difficulty as they are related through Eqn. (9.47). For example using this relation changes Eqn. (9.70) to one that can be applied directly to the evolution of H:

$$M\ddot{\mathsf{H}} = V\left[\frac{\mathsf{H}\mathsf{H}_0^{-1}\widetilde{\boldsymbol{S}}\mathsf{H}_0^{-T}\mathsf{H}^T}{J} - \boldsymbol{\sigma}^{\text{eq}} - \gamma_{\text{cell}}\dot{\mathsf{H}}\mathsf{H}_0^{-1}\right]\mathsf{H}^{-T}\mathsf{H}_0^T\mathsf{H}_0.$$

9.5.2 The hydrostatic stress state

As we have already stated, it is not possible to impose a specific Cauchy stress because it depends, through Eqn. (9.62), on an a priori knowledge of the deformation. However, there is one special case where the Cauchy stress can be imposed directly. This is the hydrostatic stress state, $\widetilde{\boldsymbol{\sigma}} = -\widetilde{p}\boldsymbol{I}$, with the usual definition of a positive compressive pressure. In this case, we can use

$$\mathcal{V}_{\text{cell}} = \widetilde{p}V,$$

in Eqn. (9.48), since p and V are a work conjugate pair (see Eqn. (2.126)). Repeating the derivation for this case, it is straightforward to show that Eqn. (9.70) becomes[28]

$$M\ddot{\boldsymbol{F}} = V\left[-\widetilde{p}\boldsymbol{I} - \boldsymbol{\sigma}^{\text{eq}} - \gamma_{\text{cell}}\dot{\boldsymbol{F}}\right]\boldsymbol{F}^{-T}, \tag{9.72}$$

and the equations of motion for the atoms, Eqn. (9.53), remain unchanged. At equilibrium, $\dot{\boldsymbol{F}} \approx \boldsymbol{0}$ and $\ddot{\boldsymbol{F}} \approx \boldsymbol{0}$, and therefore we can rearrange this equation to get

$$\boldsymbol{\sigma}^{\text{inst}} = -\widetilde{p}\boldsymbol{I},$$

where we have used Eqn. (9.67) to replace $\boldsymbol{\sigma}^{\text{eq}}$.

9.5.3 The Parrinello–Rahman (PR) approximation

The derivations of the PR constant stress ensemble [PR81] and related approaches [Ray83, Wen91] do not look quite like the derivation presented above. Ray and Rahman [RR84] have pointed out that at finite temperature, the PR approach amounts to ignoring the term $\partial G_3/\partial t$ in Eqn. (9.46). Referring back to Eqns. (9.56b) and (9.61), we see that this is equivalent to assuming that $\Delta\boldsymbol{\sigma} = \boldsymbol{0}$ identically in Eqn. (9.60). We have shown above that this is strictly correct only in the limit where $\dot{\boldsymbol{F}}$ is small compared with other time derivatives such as the momenta and accelerations of the particles. In practice, this assumption is nearly valid during transient periods if the simulation parameters are tuned so that the response of the box is not too rapid. Of course, $\dot{\boldsymbol{F}} = \boldsymbol{0}$ is identically true at equilibrium, so this should be a good approximation once the system is well equilibrated. The PR approach also makes use of an ad hoc definition of $\mathcal{V}_{\text{cell}}$ which is only correct for infinitesimal strains and must be interpreted differently depending on the type of applied stress state. In this section we show the results of these approximations.

The effect of ignoring $\partial G_3/\partial t$ Neglecting the $\partial G_3/\partial t$ term changes the equations of motion of the particles. The first of Hamilton's equations (given in Eqn. (9.51) for the finite strain NσH formulation) becomes

$$\boldsymbol{F}^{-T}\mathfrak{P}^{\alpha} = m^{\alpha}\boldsymbol{F}\dot{\boldsymbol{R}}^{\alpha}, \tag{9.73}$$

which using Eqn. $(9.45)_1$ is equivalent to

$$\boldsymbol{p}^{\alpha} = m^{\alpha}(\dot{\boldsymbol{r}}^{\alpha} - \dot{\boldsymbol{F}}\boldsymbol{F}^{-1}\boldsymbol{r}^{\alpha}). \tag{9.74}$$

This shows that the fundamental relationship between the velocities and momenta of the particles is not retained by the PR approximation when $\dot{\boldsymbol{F}} \neq \boldsymbol{0}$. The second of Hamilton's equations (given in Eqn. (9.52) for the finite strain NσH formulation) also loses the term involving $\dot{\boldsymbol{F}}$ to become

$$\dot{\mathfrak{P}}^{\alpha} = \boldsymbol{F}^{T}\boldsymbol{f}^{\text{int},\alpha}. \tag{9.75}$$

[28] In this derivation, it helps to recall that $V/V_0 = J = \det\boldsymbol{F}$ and the derivative of a determinant given in Eqn. (2.5).

Using Eqn. $(9.45)_2$, this is equivalent to

$$\dot{\boldsymbol{p}}^\alpha = \boldsymbol{f}^{\text{int},\alpha} - \boldsymbol{F}^{-T}\dot{\boldsymbol{F}}^T\boldsymbol{F}^{-T}\boldsymbol{\mathfrak{P}}^\alpha. \tag{9.76}$$

Thus, the equations of motion of the particles are modified from Eqn. (9.1) unless $\dot{\boldsymbol{F}} = \boldsymbol{0}$.

There is therefore a practical disadvantage to the PR approach, since it requires us to implement these modified equations into our MD code or alternatively to change the code so that it works with the reference coordinates and momenta $\boldsymbol{R}^\alpha$ and $\boldsymbol{\mathfrak{P}}^\alpha$. The equations of motion for the reference coordinates can be obtained from Eqns. (9.73) and (9.75) derived above. Differentiating the first of these equations with respect to time, and using the result to write the second in terms of $\boldsymbol{R}^\alpha$ gives

$$m^\alpha\ddot{\boldsymbol{R}}^\alpha = \boldsymbol{F}^{-1}\boldsymbol{f}^{\text{int},\alpha} - m^\alpha\boldsymbol{C}^{-1}\dot{\boldsymbol{C}}\dot{\boldsymbol{R}}^\alpha, \tag{9.77}$$

where we have used the definition for the right Cauchy–Green deformation tensor, $\boldsymbol{C} = \boldsymbol{F}^T\boldsymbol{F}$ (see Eqn. (2.59)). In contrast, the finite strain NσH approach of Section 9.5.1, while it requires a slightly more complex expression for the box dynamics, requires no modifications to the particle equations since they are preserved by the canonical transformation. In short, Eqn. (9.54) (the analog of Eqn. (9.77) in PR) need not be implemented and the Newtonian equations for the particle motion do not change.

The fact that the PR approximation does not preserve the equations of motion or the relation between the velocities and momenta of the atoms means that the evolution of the atoms in the simulation box that occurs while the box itself is changing will be non-physical. It is not clear what effect this may have in simulations where the applied loading is driving the internal response, for example in simulations where phase transformations occur in response to a changing applied loading. However, the PR approximation can be safely used in equilibrium simulations where the simulation box shape is not expected to change much. In this case, the PR formalism provides a convenient method to find the equilibrium shape and the results obtained will be the same as those from the finite strain NσH formulation, provided that the trajectory of the system during the initial transient period is discarded from any time averages.

Effect of approximating $\mathcal{V}_{\text{cell}}$ The PR approach defines

$$\mathcal{V}_{\text{cell}}^{\text{PR}} = -V_0(\widetilde{S}_{IJ} + \widetilde{p}\delta_{IJ})E_{IJ} + \widetilde{p}(V - V_0),$$

where $\widetilde{p}$ is the applied pressure and $\widetilde{S}$ is the applied stress. This leads to a very different equation of motion for the box (cf. Eqn. (9.59)) as

$$M\ddot{\boldsymbol{F}} = -V(\boldsymbol{\sigma} + \widetilde{p}\boldsymbol{I})\mathsf{H}^{-T} + V_0\mathsf{H}\mathsf{H}_0^{-1}(\widetilde{\boldsymbol{S}} + \widetilde{p}\boldsymbol{I})\mathsf{H}_0^{-T}. \tag{9.78}$$

The reason that this unusual form was chosen is that it recovers the hydrostatic case exactly when $\widetilde{\boldsymbol{S}} \equiv -\widetilde{p}\boldsymbol{I}$. However, the down side to this is that the interpretation of $\widetilde{\boldsymbol{S}}$ is unclear for more general stress states. In the next section, we use the zero-temperature limit to illustrate the difficulties associated with this approximation.

9.5.4 The zero-temperature limit: applying stress in molecular statics

In the limit of zero temperature (0 K), the extended Hamiltonian of Eqn. (9.49) involves only the potential energy terms; by definition zero temperature means all velocities are zero:

$$\widehat{\mathcal{H}}_{\text{ex}}(\widehat{\boldsymbol{q}}, \mathbf{0}, \boldsymbol{F}, \mathbf{0}) = \mathcal{V}^{\text{int}}(\boldsymbol{F}\boldsymbol{R}^1, \dots, \boldsymbol{F}\boldsymbol{R}^N) - V_0\widetilde{\boldsymbol{S}} : \boldsymbol{E}.$$

We can now invoke the principle of minimum potential energy (Section 2.6). We seek the reference atomic positions $\boldsymbol{R}$ and deformation gradient $\boldsymbol{F}$ that minimize the extended Hamiltonian above.[29] Conditions at an extremum are

$$-\frac{\partial \widehat{\mathcal{H}}_{\text{ex}}}{\partial \boldsymbol{R}^\alpha} = \mathbf{0}, \qquad -\frac{\partial \widehat{\mathcal{H}}_{\text{ex}}}{\partial \boldsymbol{F}} = \mathbf{0},$$

which we recognize as Eqns. $(9.50)_2$ and $(9.50)_4$ in the limit of all momenta going to zero.

As such, we can simply take the zero-temperature limit of Eqns. (9.53) and (9.59) to obtain the system of equations to be solved for $\boldsymbol{r}^\alpha$ and $\boldsymbol{F}$ in the static case:

$$\boldsymbol{f}^{\text{int},\alpha}(\boldsymbol{r}, \boldsymbol{F}) = \mathbf{0}, \qquad \boldsymbol{f}^{\text{cell}}(\boldsymbol{r}, \boldsymbol{F}) = \mathbf{0}, \tag{9.79}$$

where as usual $\boldsymbol{f}^{\text{int},\alpha} = -\partial \mathcal{V}^{\text{int}}/\partial \boldsymbol{r}^\alpha$ and

$$\boldsymbol{f}^{\text{cell}} = V\left[\frac{\boldsymbol{F}\widetilde{\boldsymbol{S}}\boldsymbol{F}^T}{J} - \boldsymbol{\sigma}\right]\boldsymbol{F}^{-T}.$$

We have dropped the superscript inst from $\boldsymbol{\sigma}$ since there are no temperature fluctuations anymore and $\boldsymbol{\sigma} = \boldsymbol{\sigma}^{\text{inst}}$. Also note (see Eqn. (9.61)) that $\Delta\boldsymbol{\sigma} = \mathbf{0}$ when $\boldsymbol{p}^\alpha = \mathbf{0}$ for all α. We can see that when Eqn. $(9.79)_2$ is satisfied then the stress on the simulation cell will be

$$\boldsymbol{\sigma} = \frac{\boldsymbol{F}\widetilde{\boldsymbol{S}}\boldsymbol{F}^T}{J}, \tag{9.80}$$

as desired.

Relation to the PR formulation The static, zero-temperature limit of the PR method similarly follows from Eqns. (9.77) and (9.78), where it is more convenient to work in terms of the reference atomic coordinates, $\boldsymbol{R}^\alpha$, instead of $\boldsymbol{r}^\alpha$.

The conditions at the extremum are again

$$-\frac{\partial \widehat{\mathcal{H}}_{\text{ex}}}{\partial \boldsymbol{R}^\alpha} = \mathbf{0}, \qquad -\frac{\partial \widehat{\mathcal{H}}_{\text{ex}}}{\partial \boldsymbol{F}} = \mathbf{0}.$$

and so the equations to be solved at a minimum are

$$\boldsymbol{g}^\alpha(\boldsymbol{R}, \boldsymbol{F}) = \mathbf{0}, \qquad \boldsymbol{f}^{\text{cell}}(\boldsymbol{R}, \boldsymbol{F}) = \mathbf{0}, \tag{9.81}$$

where $\boldsymbol{g}^\alpha$ comes from the right-hand side of Eqn. (9.77):

$$\boldsymbol{g}^\alpha = \boldsymbol{F}^{-1}\boldsymbol{f}^{\text{int},\alpha} - m^\alpha \boldsymbol{C}^{-1}\dot{\boldsymbol{C}}\dot{\boldsymbol{R}}^\alpha,$$

[29] Note that it is important to *simultaneously* minimize with respect to both $\boldsymbol{R}$ and $\boldsymbol{F}$ and not to adopt a sequential minimization approach, where an effective Hamiltonian is defined, $\mathcal{H}^{\text{eff}}(\boldsymbol{F}) \equiv \min_{\widehat{q}} \widehat{\mathcal{H}}_{\text{ex}}(\widehat{\boldsymbol{q}}, \mathbf{0}, \boldsymbol{F}, \mathbf{0})$, which is then minimized with respect to $\boldsymbol{F}$. See the discussion of potential pitfalls with this approach in Section 11.3.4.

and $\boldsymbol{f}^{\mathrm{cell}}$ is the right-hand side of Eqn. (9.78):

$$\boldsymbol{f}^{\mathrm{cell}} = -V(\boldsymbol{\sigma} + \tilde{p}\boldsymbol{I})\mathsf{H}^{-T} + V_0\mathsf{H}\mathsf{H}_0^{-1}(\widetilde{\boldsymbol{S}} + \tilde{p}\boldsymbol{I})\mathsf{H}_0^{-T}.$$

Now, when Eqn. $(9.81)_2$ is satisfied the stress on the simulation cell is

$$\boldsymbol{\sigma}^{\mathrm{PR}} = \frac{1}{J}\boldsymbol{F}\left(\widetilde{\boldsymbol{S}} + \tilde{p}\boldsymbol{I}\right)\boldsymbol{F}^T - \tilde{p}\boldsymbol{I}. \tag{9.82}$$

It is easy to see that if we now impose the special case of $\widetilde{\boldsymbol{S}} = -\tilde{p}\boldsymbol{I}$, we will have $\boldsymbol{\sigma}^{\mathrm{PR}} = -\tilde{p}\boldsymbol{I}$ and the final pressure will be exactly the prescribed pressure $\tilde{p}$. On the other hand, any other choice of $\widetilde{\boldsymbol{S}}$ leads to an ill-defined stress state since it is unclear what stress measure is actually being used. We demonstrate this in the following example.

Example 9.2 (Superposition of uniaxial tension and hydrostatic pressure) To elucidate the approximate nature of the PR approach, it is instructive to compare $\boldsymbol{\sigma}$ with $\boldsymbol{\sigma}^{\mathrm{PR}}$ in the special case where the simulation box is constrained to be orthogonal and the stress state is a superposition of uniaxial tension and hydrostatic pressure. In this case H_0 is defined in Eqn. (6.34) and the deformed box must also be of the form:

$$\mathsf{H} = \begin{bmatrix} l_1 & 0 & 0 \\ 0 & l_2 & 0 \\ 0 & 0 & l_3 \end{bmatrix}.$$

From Eqn. $(9.47)_2$ it is easy to see that the deformation gradient now takes the simple form:

$$[F] = \begin{bmatrix} l_1/L_1 & 0 & 0 \\ 0 & l_2/L_2 & 0 \\ 0 & 0 & l_3/L_3 \end{bmatrix}. \tag{9.83}$$

From Eqn. (9.82), we can now see that the PR formulation predicts the diagonal components of the Cauchy stress tensor as

$$\sigma_{11}^{\mathrm{PR}} = \frac{l_1 L_2 L_3}{L_1 l_2 l_3}\left(\widetilde{S}_{11} + \tilde{p}\right) - \tilde{p},$$
$$\sigma_{22}^{\mathrm{PR}} = \frac{L_1 l_2 L_3}{l_1 L_2 l_3}\left(\widetilde{S}_{22} + \tilde{p}\right) - \tilde{p},$$
$$\sigma_{33}^{\mathrm{PR}} = \frac{L_1 L_2 l_3}{l_1 l_2 L_3}\left(\widetilde{S}_{33} + \tilde{p}\right) - \tilde{p},$$

whereas the correct diagonal components are obtained from Eqn. (9.80) as

$$\sigma_{11} = \frac{l_1 L_2 L_3}{L_1 l_2 l_3}\widetilde{S}_{11}, \quad \sigma_{22} = \frac{L_1 l_2 L_3}{l_1 L_2 l_3}\widetilde{S}_{22}, \quad \sigma_{33} = \frac{L_1 L_2 l_3}{l_1 l_2 L_3}\widetilde{S}_{33}.$$

Now let us choose a specific stress state to impose. Imagine that we want to superimpose two stress states: a uniaxial tensile stress of magnitude $\bar{S}$ in the x_1 direction and a positive hydrostatic pressure of the same magnitude. These are

$$[S_{\mathrm{u}}] = \begin{bmatrix} \bar{S} & 0 & 0 \\ 0 & 0 & 0 \\ 0 & 0 & 0 \end{bmatrix}, \quad [S_{\mathrm{p}}] = \begin{bmatrix} -\bar{S} & 0 & 0 \\ 0 & -\bar{S} & 0 \\ 0 & 0 & -\bar{S} \end{bmatrix}.$$

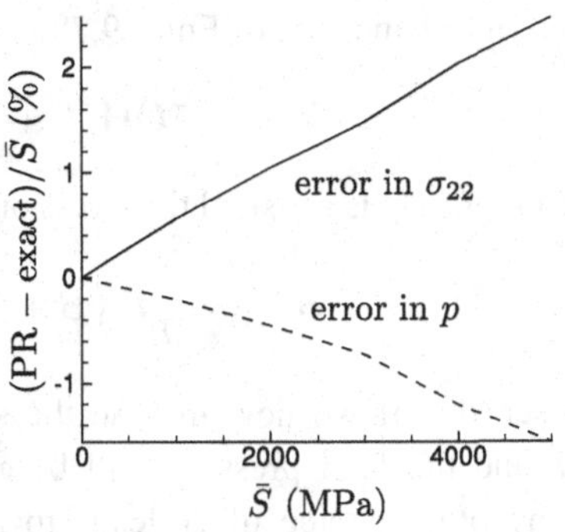

Fig. 9.7 Departure of the PR stresses at equilibrium from the exact finite strain formulation, as a function of the applied stress, $\bar{S}$.

The total applied stress is therefore

$$\left[\widetilde{\boldsymbol{S}}\right] = [\boldsymbol{S}_{\mathrm{u}}] + [\boldsymbol{S}_{\mathrm{p}}] = \begin{bmatrix} 0 & 0 & 0 \\ 0 & -\bar{S} & 0 \\ 0 & 0 & -\bar{S} \end{bmatrix}. \tag{9.84}$$

In the PR formulation, we also need to specify the applied pressure, which is

$$\tilde{p} = -\widetilde{S}_{II}/3 = \frac{2\bar{S}}{3}.$$

We can now examine the differences between the PR and finite strain NσH ensemble formulations. For definiteness, we apply the stress state of Eqn. (9.84) to an fcc crystal of Al, modeled using the pair functional Ercolessi–Adams model of [EA94], with the X_1-axis along [100]. In Fig. 9.7, we plot the discrepancies $(p^{\mathrm{PR}} - p^{\mathrm{exact}})$ and $(\sigma_{22}^{\mathrm{PR}} - \sigma_{22}^{\mathrm{exact}})$, normalized by the applied stress, $\bar{S}$, as the stress is increased. We see that the differences are relatively small, but not insignificant, and that they systematically increase in magnitude as the applied stress (and therefore the resulting deformation) grows.

The above example uses the PR approach as it is commonly implemented, and indeed this seems to be the most appropriate interpretation of the original paper [PR81]. In a follow-up paper, Ray and Rahman [RR84] recognized the problems with the original presentation and clarified them. In essence, they argued that one must treat the PR formulation differently for two different cases. In the first case, when one wishes to apply *only* a hydrostatic pressure, one must interpret $\widetilde{S}_{IJ}$ as the Cauchy stress. Thus one sets $\widetilde{S}_{IJ} = -\tilde{p}\delta_{IJ}$ in Eqn. (9.82) to cancel the first stress term. In the second case where one wishes to apply a general stress, one must set $\tilde{p} = 0$ and interpret $\widetilde{S}_{IJ}$ as the second Piola–Kirchhoff stress. This second case is equivalent to the finite strain NσH implementation presented here.

Clearly, neither the PR nor the finite strain NσH formulation allows us to specify an exact Cauchy stress state to apply to an MS system. However, only the applied stress in the finite strain NσH formulation has a clear physical interpretation as the second Piola–Kirchhoff stress tensor. The resulting Cauchy stress depends on the compliance of the system of atoms in the simulation box.

9.5.5 The kinetic energy of the cell

Recall that it was necessary to introduce a fictitious kinetic energy for the cell, $\mathcal{K}_{\text{cell}}$, in order to obtain dynamical equations to evolve the cell shape (through the deformation gradient, $\boldsymbol{F}$). As we have already discussed, $\dot{\boldsymbol{F}}$ becomes small ($\approx \mathbf{0}$) as the system approaches equilibrium, so that the kinetic energy of the cell will also be small. As a result, the conserved quantity (the Hamiltonian of Eqn. (9.49)) becomes

$$\widehat{\mathcal{H}}_{\text{ex}}(\widehat{\boldsymbol{q}}, \widehat{\boldsymbol{p}}, \boldsymbol{F}, \mathbf{0}) = \mathcal{H}(\boldsymbol{q}(\widehat{\boldsymbol{q}}; \boldsymbol{F}), \boldsymbol{p}(\widehat{\boldsymbol{p}}; \boldsymbol{F})) - V_0 \widetilde{\boldsymbol{S}} : \boldsymbol{E},$$

where $\mathcal{H}$ is the original Hamiltonian of the particles from Eqn. (9.43) and $\widetilde{\boldsymbol{S}}$ is the applied second Piola–Kirchhoff stress.

The requirement that the correct Hamiltonian is recovered in the equilibrium limit can be satisfied by more than one choice for $\mathcal{K}_{\text{cell}}$. Parrinello and Rahman used

$$\mathcal{K}_{\text{cell}}^{\text{PR}} = \frac{M}{2} \dot{\mathsf{H}} : \dot{\mathsf{H}},$$

which differs from our choice in Eqn. (9.48). Other alternatives have been proposed. For example, Wentzcovitch [Wen91] argued that since the choice of the periodic cell shape is not unique for a given problem, the kinetic energy should be independent of this choice. She showed that any form of the kinetic energy of the cell (such as the PR form above) that does not satisfy this invariance can unduly influence the simulation of a phase transformation between crystal structures of radically different Bravais cell shapes. As such, she proposed the form

$$\mathcal{K}_{\text{cell}}^{\text{W}} = \frac{M}{2} \dot{\epsilon} : \dot{\epsilon},$$

where ϵ is the small strain tensor. We note that our form (Eqn. (9.48)) is related to this choice and in fact also satisfies the invariance requirement.

9.6 The NσT ensemble: applying stress at a constant temperature

Simultaneously specifying a constant temperature and applied stress involves no great difficulty once the NVT and NσH ensembles are implemented as discussed above. The extension of the Hamiltonian with $\mathcal{V}_{\text{cell}}$ and $\mathcal{K}_{\text{cell}}$ used to impose the stress is naturally integrated with any of the thermostats previously discussed. Of course, the kinetic energy of the simulation cell should not be considered as a contribution to the temperature (and, in any event, is small in equilibrium). This means that the equations used to determine the drag forces on particles remain decoupled from those for the simulation box, and time integration of the two sets of equations can therefore be simultaneously implemented.

Further reading

- There are many good books on MD. *Computer Simulation of Liquids*, by Allen and Tildesley [AT87] is an especially clear and engaging presentation. Although focused on the application of MD to the study of liquids, many of the topics are relevant to solid phase simulations as well. *Understanding Molecular Simulation* by Frenkel and Smit [FS02] is a thorough and formal account of modern MD simulations, while *The Art of Molecular Dynamics Simulation,* by Rapaport [Rap95], although not as in-depth as the others, is extremely useful as a "recipe book" for implementing MD algorithms.
- Mark E. Tuckerman's *Statistical Mechanics: Theory and Molecular Simulation* [Tuc10] takes the unusual approach of weaving a theoretical discussion of statistical mechanics with the computational aspect of molecular simulation. It provides an interesting perspective from an acknowledged leader in the field.
- *Geometric Numerical Integration*, by Hairer *et al.* [HLW06] and *Integration Methods for Molecular Dynamics* by Leimkuhler *et al.* [LRS96] are good sources for a rigorous mathematical treatment of the subject of numerical integration techniques and their accuracy, convergence and stability. The classic reference on the topic is *Numerical Initial Value Problems in Ordinary Differential Equations,* by Gear [Gea71].
- A good review of thermostat techniques in MD is provided by Hünenberger [Hun05].

Exercises

9.1 [SECTION 9.3] The timestep in an MD simulation should be small enough that the fundamental vibrations in the system are reproduced accurately. If the timestep is too large numerical problems may occur, because atoms come to close to each other. To estimate the maximum timestep, approximate the Lennard–Jones potential (Eqn. (5.36)) by a harmonic one,

$$\phi(r) \approx \frac{k}{2}(r - r_0),$$

by fitting k to the curvature at the minimum, occurring at $r_0 = 2^{1/6}\sigma$. If we assume that there should be at least 20 timesteps in a full period of the motion, what is the maximum timestep? **Hint** The frequency of vibration of a harmonic oscillator is $\omega = \sqrt{k/m}$ where m is the mass.

9.2 [SECTION 9.3] Test the result of Exercise 9.1 using the MiniMol MD program provided on the companion website to this book. Simulate a periodic argon crystal using a Lennard–Jones potential (see full directions on the website). Vary the timestep, keeping the total simulation time roughly the same. Check the algorithm's energy conservation by monitoring the ratio, R, of the fluctuations in total energy and kinetic energy

$$R = \frac{\Delta\mathcal{E}_{\mathrm{tot}}}{\Delta\mathcal{T}^{\mathrm{vib}}}.$$

Ideally, R should be zero, but for a well-behaved system R should be less than about 0.05.

9.3 [SECTION 9.3] Starting from the MD code MiniMol provided on the companion website to this book, modify the code so that it can perform a QD simulation. Compare the energy minimization properties (speed and accuracy) between your QD minimizer and the CG minimizer already provided with the code.

9.4 [Section 9.3] Using the programs provided on the website for this book, study the melting of an argon crystal.[30] Build an fcc crystal and initialize its temperature according to Section 9.3.3. Then run a constant energy MD simulation with free boundary conditions.[31] Run sufficiently long simulations that equilibration is achieved (check this by monitoring the temperature). Repeat this at higher and higher temperatures until you observe melting of the crystal and estimate the melting temperature.

9.5 [Section 9.3] Using the MD program provided on the website for this book, simulate radiation damage in Ar. Initialize an NVE simulation with a periodic, perfect crystal at moderate temperature (well below melting). Determine the equilibration time, and then modify the code to impart a large velocity on a single atom after equilibration. Choose this velocity so that the kinetic energy imparted on the system is on the order of 1000 eV. Monitor the subsequent temperature variation, reequilibration and resulting damage to the crystal.

9.6 [Section 9.4] Repeat Exercise 9.5 using both the Langevin and NH thermostats set to the equilibration temperature found during the previous NVE simulation. Compare the resulting damage to the NVE results.

9.7 [Section 9.4] Using the MD program provided on the book website, simulate the stretching of an Al nanorod. Build a slender crystal of aluminum and use free boundary conditions.[31] Modify the MD code to apply a constant force to four or five planes of atoms at each end of the rod, such that they tend to stretch the rod (apply equal and opposite forces at the two ends). Repeat the simulation at various levels of force and study the resulting deformation.

9.8 [Section 9.5] Derive Eqn. (9.58). See the text after the equation for guidelines on the identities that need to be used in the derivation. **Hint** Use indicial notation to perform the derivation, then switch the final expression to direct notation.

30 Exercises 9.4–9.7 require some means for visualizing MD results. We recommend AtomEye [Li10].

31 Set the periodic lengths significantly larger than the extent of the crystal.

PART IV

MULTISCALE METHODS

10 What is multiscale modeling?

When we say that a problem in science is "multiscale," we broadly mean that it involves phenomena at disparate length and/or time scales spanning several orders of magnitude. More importantly, these phenomena all play key roles in the problem, so that we cannot correctly model the interesting behavior without explicitly accounting for these different scales. In Chapter 1, we looked at a wide range of length and time scales relevant to materials modeling, motivating the case that materials science is filled with multiscale problems. Indeed, the message we have tried to carry throughout this book is that there is a need to model materials at many scales, and to make connections between them. However, when we speak of *multiscale modeling*, we tend to be referring to something more specific, meaning that the problem is tackled with a conscious effort to span multiple scales simultaneously.

In many cases, multiscale methods involve just two scales, a "coarse scale" and "fine scale", each of which plays a role in the problem. Depending on the perspective, multiscale models offer different advantages. For the fine-scale modeler, a multiscale approach allows one to study much larger systems (or longer times) than could be studied using the fine-scale alone. On the other hand, the coarse-scale expert views the multiscale model as a way to establish the constitutive laws of the problem from first principles, or at least from a more fundamental scientific basis than could be realized from the coarse-scale alone. In the former case, the advantage is clearly a practical one. In the latter case, the advantage is often better scientific insight into the *why* of a particular phenomenon, but there is also a practical benefit. For example a constitutive law may be provided in cases where direct experimental determination of the coarse-scale behavior is not possible.

10.1 Multiscale modeling: what is in a name?

Multiscale methods can be classified into two broad categories: "sequential" and "concurrent." Sequential multiscale approaches are methods whereby fine-scale simulations are done as a "pre-processing step" to the coarse-scale model. In modest examples of such approaches, the fine-scale model generates parameters that are needed to complete the coarse-scale model description, after which it is assumed that the fine-scale model is no longer required. In more ambitious sequential methods, the fine-scale model is used to derive the very *form* of the governing equations at the coarse scale, often leading to deep physical insight into the origins of the coarse-scale behavior.

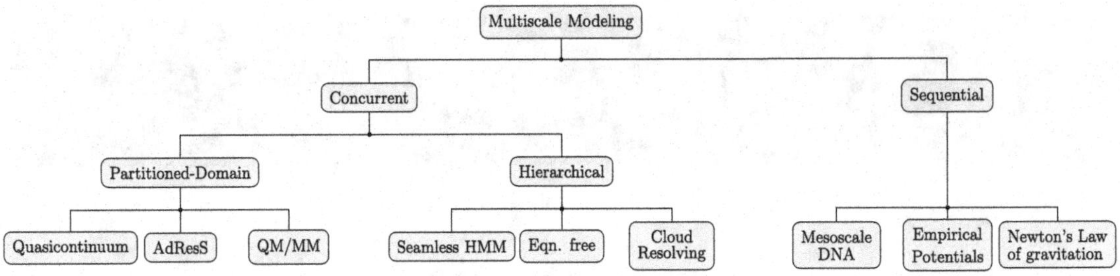

Fig. 10.1 The taxonomy of multiscale modeling. Representative examples on the bottom row are elaborated on during the discussion in this chapter. (AdResS = adaptive resolution scheme; QM/MM = quantum mechanics/molecular mechanics. HMM = heterogeneous multiscale method.)

Concurrent methods, as the name suggests, employ models at two or more scales simultaneously. Within concurrent methods, there is a further classification into two sub-categories: "partitioned-domain" and "hierarchical" methods. In partitioned-domain concurrent approaches, the physical problem is divided into two or more contiguous regions, with a different model scale used in each. Hierarchical methods, on the other hand, make use of both scales everywhere, with the coarse-scale model making regular appeals to the fine-scale model to determine a constitutive law and, conversely, the fine-scale model looking to the coarse-scale model for its boundary conditions.[1] In Fig. 10.1, we summarize this taxonomy and include some of the examples of each type of model that will be discussed presently.

In the next few chapters, our main focus will be on multiscale models of the concurrent, partitioned-domain variety. However, we take the opportunity in this brief chapter to quickly review examples of other kinds of multiscale models from several fields of science.

As Tab. 10.1[2] suggests, the phrase "multiscale modeling"[3] was coined in the early 1980s, and since that time the application of the term in various scientific disciplines has grown rapidly. Initial use was in reference to fluid mechanics modeling of turbulent flows [Sch83a, MPT83]. In turbulent fluids, there are flow eddies at many different length scales, and the transfer of energy between them is at once critical to analyzing the flow

[1] What we call "sequential methods" here are sometimes referred to as "hierarchical methods" in the literature. This is not to be confused with the definition of hierarchical methods that we have introduced above.

[2] The data in Tab. 10.1 were generated using Web of Science, searching for either "multiscale model" or "multiple scale model" in the title or abstract. The reader may recall that we produced a similar table for MD simulations on page 495. If we compare the number of articles in the third decade of MD (1981–1990) with the number of articles in the third decade of multiscale (2001–2010), it suggests (rather vaguely) that multiscale modeling is growing about twice as fast as early MD simulations.

[3] Early use of the term "multiscale modeling" was confined to models of the *concurrent* type defined above; in later years the broadening of the definition to include *sequential* models contributed to the growth in use of the term. In fact, it is really only *concurrent* multiscale models that emerged as a new approach in the 1980s and 1990s. Sequential multiscale approaches have been used in various disciplines for centuries, and while new sequential multiscale models continue to be developed and to contribute to our understanding of science, their description as "multiscale" reflects more a change in our awareness of their multiscale nature than a change in the modeling strategy itself.

Table 10.1. Number of "multiscale modeling" articles published, by year

Years	Number of articles
before 1980	0
1981–1985	2
1986–1990	3
1991–1995	25
1996–2000	106
2001–2005	474
2006–2010	1,174

and difficult to model. Early multiscale turbulence simulations treated eddies at multiple characteristic wavelengths, each with a different evolution equation dependent on the length scale. In addition, coupling terms that described the flow of energy between eddies on neighboring scales were derived. These models were truly *multi*scale in the sense that *many* eddy scales were included. In contrast, most modern multiscale approaches could well be described as merely "dual-scale" models, with only two distinct modeling paradigms. The early multiscale turbulence models were concurrent and hierarchical in the sense that we have defined; models at multiple scales existed on overlapping physical domains and ran simultaneously, with information transferred between the scales.

10.2 Sequential multiscale models

In [Phi01], Phillips makes the argument that examples of sequential multiscale models can be found throughout scientific history and across all scientific disciplines long before the term became commonplace. In some sense, "multiscaling" or "coarse graining" is the essence of most modeling efforts in science, as we try to identify a minimal set of key parameters to describe a complex process. Consider, for example, linear elasticity, or more specifically the generalized form of Hooke's law (see Section 2.5.5)

$$\sigma_{ij} = c_{ijkl}\epsilon_{kl}.$$

This is not normally thought of as a multiscale model, and it was not really developed as such. In fact, the proportionality between stress and strain and the more general anisotropic relations were established by empirical observation and considerations of symmetry. The constants that complete this constitutive relation are also normally determined from experiment. But one can turn this example on its head, and consider it to be a coarse-scale model that emerges from the fine-scale model of atoms interacting on a lattice. In this way, the symmetries of the lattice and the nature of the interatomic forces that we assume for the atoms give rise to the elastic constants directly. Indeed, we have provided explicit examples of this in Section 8.1.4. Consideration of linear elasticity from the coarse-scale perspective

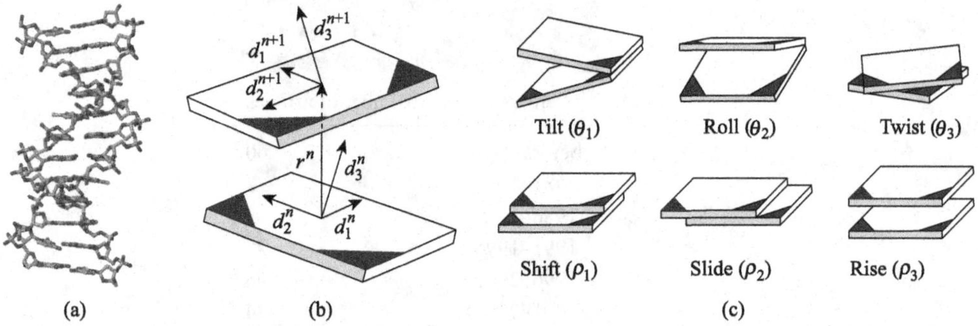

Fig. 10.2 (a) The DNA molecule comprises many individual atoms. (b) The multiscale model of DNA developed by Olson and coworkers. (c) The degrees of freedom in the multiscale model. Reproduced from fig. 1 of [OSC04], with permission.

leads to insights into the form that the *fine-scale* model must adopt. Specifically, it shows us how some forms of atomistic force models lead to incorrect symmetries in c_{ijkl} (see Sections 5.4.2 and 11.5.3).

Other examples of sequential multiscale models have already been discussed throughout this book. Indeed, the entire science of building atomistic models (Chapter 5) can be viewed as an exercise in sequential multiscale modeling, whereby the fine-scale quantum mechanics is replaced by coarse-scale molecular statics (MS) or molecular dynamics (MD) via the classical atomistic model. In Section 4.4.6, we touched briefly on "Car–Parrinello MD," a form of MD whereby both the electronic (quantum mechanics) and nuclear (classical mechanics) dynamics are simultaneously evolved. MD modeling of polymers and complex biological molecules like DNA take this multiscaling a step further (see [PDSK08] for a review of multiscale models of soft materials). In these systems, the number of atoms in a single molecular chain is often prohibitively large, while at the same time groups of these atoms along the chain are relatively rigid and/or inert with respect to conformations of the molecule and interactions between chains. As such, further coarse graining is performed to model the molecule as a string of "beads" that interact elastically with neighboring beads. In the example of Fig. 10.2, "mesoscale DNA" is modeled by treating each base-pair (a single "rung" on the helical ladder) as an oriented brick. Deformation of any of the six degrees of freedom between each pair of bricks leads to elastic energy, with the stiffness of each degree of freedom being estimated empirically. For comparison, we show the full DNA molecule in Fig. 10.2(a) alongside the coarse-grained model in Fig. 10.2(b). This coarse-grained model replaces approximately 75 atoms (representing 225 degrees of freedom) by one rung of the ladder (which has only 6 degrees of freedom), and of course it vastly simplifies the mechanics of bonding by using elastic springs instead of chemical bonds. Such models, whereby classical dynamics simulations are performed on "particles" that represent conglomerations of atoms or molecules are sometimes referred to as "coarse-grained molecular dynamics (CGMD)" simulations. (This is not to be confused with the CGMD method developed by Rudd and Broughton [RB98, RB05], which is more in the spirit of the partitioned-domain approaches discussed later.)

An example of a successful sequential approach that is quite relevant to the theme of this book is the linking of continuum crystal plasticity, discrete dislocation dynamics (DDD) and atomic scale models (MS and MD) [TPO00]. Crystal plasticity is the continuum-based, coarse-scale approach to describing the plastic flow of single crystals (see, for example, [Asa83, Bas94, Gur00] for comprehensive overviews of the subject). It is built on a set of phenomenological hardening rules that determine the resistance to plastic flow along a given slip system based on the accumulated slip on it and the other slip systems. The details of these hardening rules (i.e. constitutive laws) must be developed theoretically and then parameterized experimentally. However, experimental quantification is difficult in this case, because it is almost impossible to design experiments that sufficiently isolate individual causes and effects. A natural connection to finer modeling scales can then be made by determining these hardening rules with DDD simulations.[4] In DDD models, individual dislocations are discretized into segments that are allowed to move, interact and multiply under applied loading. This allows for a direct simulation of some of the phenomenological assumptions of crystal plasticity. For example the hardening of one slip system can be "measured" numerically as a function of dislocation density on other slip systems. However, DDD models have their own inherent phenomenology. For example the relationship between the velocity of a particular dislocation and the applied stress or temperature it experiences must be prescribed. In addition, the model requires a phenomenological set of rules for the strength of junctions which form between dislocations that pass each other at short distances. These details are extremely difficult to determine experimentally, but they are a natural subject of inquiry for MD, MS, and the partitioned-domain methods discussed in this book.

As a final sequential multiscale example, consider the Peierls–Nabarro model of dislocations discussed in Section 6.5.7. In this case, the fine scale is the determination of the γ-surface, whereby a fully atomistic model is used to determine the energy versus slip relationship. Then, continuum mechanics is used as the coarse scale with the γ-surface as part of its constitutive law.[5]

10.3 Concurrent multiscale models

Concurrent multiscale methods are almost exclusively *computational methods*, in the sense that they are intended for implementation on a computer. It is no surprise then that their ascendancy coincides with the development of supercomputers and parallel processing. These methods are usually *designed* for computer algorithms and are well adapted to the parallel environment, with natural ways to divide the labor between processors. Let us take a brief look at the literature on these methods before turning to the details of some of them in the next two chapters.

[4] Some key references on DDD are [KC92, vdGN95, FGC96, VFG98, ZRH98, FC99, WFvdGN02]. Also, the book by Bulatov and Cai [BC06] discusses dislocation dynamics in some detail.

[5] Actually the Peierls–Nabarro model can also be viewed as *a concurrent* (partitioned-domain) approach, if we view the slip plane as the fine-scale domain and the continuum half-spaces as the coarse scale.

10.3.1 Hierarchical methods

As with sequential multiscale models, there are examples of *hierarchical* multiscale models that predate the use of the "multiscale" title. The best example in this book is the Cauchy–Born rule discussed at length in Chapter 11, the ideas for which were laid out in the early part of the twentieth century. We shall see that the finite element method is ideally suited to accept the Cauchy–Born rule as its constitutive model, and it was this realization that led to a renewed interest in the Cauchy–Born framework in the 1990s. Many partitioned-domain multiscale approaches (discussed in Chapters 12 and 13) make use of a hierarchical implementation of the Cauchy–Born concept in their coarse-scale region.

The first "self-identifying" multiscale models in the early 1980s were hierarchical methods applied to the problem of modeling the evolution of a star's atmosphere. Later, similar ideas were extended to the problem of modeling Earth's atmosphere for the purposes of predicting weather and long-time climate trends. In this endeavor, the body of work on "cloud-resolving" atmospheric models by Grabowski and his coworkers looms large as a major contribution to the field (see, for example [GS99, Gra00, GS02, Gra04, Gra08, JGP09, XMG09]). In these models, the coarse scale is the flow of the atmosphere over the surface of the earth, while the fine scale is the behavior of individual clouds. Like the earlier examples, these models aim to predict complex turbulent fluid flow with key characteristics at multiple scales.

In our discussion of MD, we mentioned in passing that one of the great challenges of MD is the modeling of thermal conductivity. This is because empirical atomistic models, by design, remove the electronic degrees of freedom, and therefore electronic thermal transport is not possible. This is especially problematic in metals, whose thermal conductivity is dominated by the electronic component, and makes it almost impossible to study any metallic system that is significantly far from thermal equilibrium. For example simulations of nanoindentation, laser ablation and even rapid fracture processes could be artificially influenced by the inability of the system to rapidly transport heat from local hot spots. Efforts to improve this modeling take the form of hierarchical multiscale models. For example Finnis *et al.* [FAF91] developed a phonon–electron coupling model that they embedded into MD simulations of the irradiation damage process. The model takes the form of an electronically-motivated "thermostat" in the sense that it introduces drag terms on the molecular equations of motion. More recently, Padgett and Brenner [PB05] and Wagner *et al.* [WJTP08] have developed hierarchical methods whereby a continuum grid is superimposed over the MD region in order to provide a framework for heat conduction between the atoms and an electronic "heat bath."

More recent hierarchical methods have evolved into general *frameworks* that are not, in principle, tied to any specific application. For example the so-called "equation-free" method proposed by Kevrekidis, Gear and coworkers [EKO06, EKAE06, CDGK04, KGH04] is a method whereby any fine-scale system can be studied with the goal of *establishing* the coarse-scale governing equations. Consider a system where we can quantify and characterize the macrostate, but we do not know the physics of its evolution. For example imagine that we have a certain dislocation density everywhere in a solid and we want to know how that density will evolve under a nonuniform applied load. Imagine further that we have a reliable fine-scale model that allows us to study dislocation density "point-wise" as a function of a uniformly applied stress (this might be a DDD model, as discussed earlier). Starting from an initial macrostate (the dislocation density field, ρ, and the stress field, σ), the fine scale is initialized on a grid and a "burst" of fine scale simulation is performed for a period of time Δt. The new macrostate is then established from the evolved microstate and numerical estimates of field derivatives (e.g. $\Delta\rho/\Delta t$, $\Delta\sigma/\Delta t$) can be determined. In this way, the macroscale governing equations can emerge from the simulations. Interesting applications of the equation-free method include the study of gene regulatory networks [EKAE06] and disease propagation through a population [CGLK04].

E. *et al.* have proposed a general strategy for hierarchical multiscale methods [ERVE09] that they call the "seamless" method. This is an evolution of their earlier "heterogeneous multiscale method (HMM)" [EEL+07] and it is related to other frameworks like that of Brandt [Bra05]. Unlike the equation-free method, the seamless HMM is a "top-down" approach in that the structure of the equations at the coarse scale is assumed to be known. What is missing is some set of parameters (essentially what we would call the constitutive law) that can be computed from a fine-scale model. Thus we have some incomplete macroscopic model:

$$\partial U = L(U; D),$$

where U is the coarse-scale field and D is the unknown constitutive information. This model is solved using a coarse-scale solver that ensures accuracy and stability. A fine-scale model is run at several grid points throughout the coarse-scale model. The fine scale is denoted symbolically as

$$\partial u = \mathcal{L}(u; U),$$

where u are the atomic coordinates, which we assume can be mapped to the coarse scale through a projection operator, $U = Qu$. Finally, a link is established between the scales that allows the coarse-scale constitutive law to be estimated from the fine scale:

$$D = D(u).$$

The original HMM procedure was to take a coarse-scale solver step, and then at each integration point in the macro model, initialize the fine-scale solver and run it for a while in order to obtain the necessary data from which the macroscopic parameters could be obtained. However, the problems with this approach are that the initialization can be costly, it is difficult to know how long the fine-scale simulation needs to be performed in order to equilibrate and it is also hard to get good statistics unless the fine-scale systems are large and/or the simulation times are long. In the seamless HMM, the coarse- and fine-scale simulations run concurrently and exchange information at every timestep. Let us say that the coarse-scale simulation allows a maximum timestep ΔT to still be stable and accurate, while the largest allowable fine scale timestep is δt. Further, assume we know that M fine-scale timesteps are required for the fine scale to equilibrate. Then the idea is to set the coarse-scale timestep to $\Delta t = \Delta T/M$, which is smaller than necessary. Now both the fine-scale solver and the coarse-scale solver are stepped M steps forward. Note that the clocks at the two scales will not be synchronized, since $\delta t \neq \Delta t$, but at each step we nonetheless update the macroscopic constraint imposed on the fine-scale simulation and the coarse-scale constitutive data based on the fine-scale data. The effect of this procedure is that the rate at which the constraint is changed at the fine scale is greatly accelerated, but since the fine-scale system has sufficient time to equilibrate, the results should be the same provided that two assumptions hold: (1) the fine-scale system is not sensitive to the rate at which the constraint is applied and (2) the simulation time at the fine scale is long compared with the relaxation time.

General frameworks like the seamless HMM just outlined are of interest because they establish a systematic approach to building a model for virtually any multiscale problem. However, it is in the details of a problem that the practical viability of any specific implementation is determined. Whether or not the approach produces a model that is computationally efficient and beneficial depends very much on the specifics of both the coarse- and fine-scale phenomena in question. Another drawback of many hierarchical methods is that it is difficult to control the statistical error in the estimation of the coarse-scale parameters from the fine-scale data. The result is either poor accuracy at the coarse scale or high cost at the fine scale, associated with trying to get larger statistical samples. Partitioned-domain methods, on the other hand, are usually developed with a particular application in mind, and for that reason, pragmatic questions of efficiency are usually considered from the start. Applications of such methods to problems of deformation in crystalline solids have been extensively developed and tested, and for that reason we focus on partitioned-domain models for the remainder of this book.

10.3.2 Partitioned-domain methods

In a sense, partitioned-domain models are much less ambitious than hierarchical methods. Hierarchical models use the fine-scale data to generate constitutive information (or even the governing equations themselves) everywhere in the coarse-scale model. Partitioned-domain models take the more pragmatic approach of using the coarse-scale model simply as a way to "cheat" at doing the fine-scale simulation. As the name suggests, these models run the full fine-scale simulation in only part of the physical domain of the system, while it is assumed that the coarse-scale model is sufficiently accurate for the rest of the problem. The earliest example of a partitioned-domain method applied to crystalline solids was the so-called FEAt (finite-element and atomistic) model of [KGF91]. The well-known quasicontinuum (QC) method first appeared a few years later [TOP96, SMT$^+$98, SMT$^+$99] and continues to be developed today [TM09]. The QC method in particular is described in detail in Section 12.6.

In partitioned-domain methods, the coarse- and fine-scale models are independent and fully defined a priori (both spatially and in terms of their parameters) and the key modeling effort is in defining the communication between the domains across an interface. The nature of this communication is the central topic of Chapter 12. The primary disadvantage with the partitioned-domain approach is that the accuracy in much of the domain is inherently limited to the accuracy of the coarse-scale model. A decision has to be made where to draw the boundary between the "exact" fine-scale model and the more restrictive coarse-scale approach. For some partitioned-domain models, one cannot be completely sure when the limits of the coarse-scale model are being exceeded and the overall results can therefore be doubtful. For other models, efforts have been made to improve this by including adaptive refinement of the model (usually in the form of growing the extent of the fine-scale domain, see Section 12.6.5).

In the following chapters, we provide a detailed discussion of partitioned-domain models developed for studying the deformation of solids, in particular methods that couple empirical atomistic models (like those discussed in Chapter 5) with continuum mechanics. In essence, these are methods that perform MS (Chapter 6) and MD (Chapter 9) simulations at reduced computational cost. We note in passing that there have been efforts in the materials community to extend these ideas to the coupling between MS/MD and quantum calculations like density functional theory (DFT) (for a comprehensive review of these methods applied to the solid state, see [BKC09]). In chemistry and biology, this sort of coupling is more common and falls under the heading of "QM/MM" (quantum mechanics/molecular mechanics) models, where they are applied to develop a detailed quantum mechanical understanding of chemical reactions in complex molecules (see reviews in [SdVG$^+$03, SW03, FG05, ST07]). We do not pursue QM/MM models in this book. Partitioned-domain methods have also been developed extensively for applications outside the modeling of crystalline solids. For example, the so-called "AdResS (adaptive resolution scheme)" method [PDSK07] is a partitioned-domain approach that combines full MD simulation with CGMD simulation (the sequential multiscale method mentioned earlier) to study polymer molecules in solution.

Static versus dynamic methods We divide the discussion of partitioned-domain methods into two categories: static (0 K) and dynamic approaches. In the discussion of static methods in Chapter 12, we provide a detailed overview of 14 different methods that have been proposed in the literature, showing their similarities and differences. Then, we make direct comparisons between their relative speed and accuracy on a simple benchmark problem. Dynamic methods are discussed in Chapter 13, where we shall see that such methods are built on the foundations of the static techniques, but that their dynamic nature introduces additional modeling artifacts that need to be addressed. For example there exists an impedance mismatch between the fine-scale and coarse-scale regions that leads to spurious wave reflections at the interface. Also, the dynamics of the coarse-scale model is artificially affected by an entropylike contribution that comes from the motion of the coarse-scale grid itself. While the implementation of static partitioned-domain models is fairly well established and the associated errors well understood, the picture for dynamic models is still not complete. In Chapter 13, we present an admittedly selective view of these dynamic models that comes mainly from our own reflections on the topic.

10.4 Spanning time scales

Models that span the time scales of crystalline solids typically follow a different tack than those designed to span spatial scales. This is largely because, as noted in Chapter 1, there is generally a wide gap of about seven orders of magnitude in time between processes like atomic vibration (that ultimately limit MD simulations) and diffusion processes (that control many important materials phenomena). This separation of scales enables approaches based on transition state theory (TST, see, for example, [Zwa01]) to work. Although the subject of spanning time scales is beyond the scope of this book, we touch on it briefly here.

One of the main applications of temporal multiscale methods is complex diffusion processes. Diffusion is controlled by random hopping of atomic configurations between energy minima in the potential energy surface (see Fig. 6.1). At low to moderate temperatures, configurations tend to stay near the bottom of an energy basin, only rarely making excursions that take them up over a barrier and into a neighboring basin. Chains of many such rare events lead to diffusion. TST can be used to predict that diffusion rates will be controlled by the Arrhenius form:

$$\text{rate} \propto \exp\left(-\frac{\Delta E}{k_B T}\right), \tag{10.1}$$

where ΔE is the activation energy of a particular process and T is the temperature. Therefore, the key input to TST models is an accurate calculation of energy barriers. This can be done using an approach like the *nudged elastic band* method which is described in Section 6.3.1. One way to accelerate time in rare event systems is to catalog all possible processes and then to use a *kinetic Monte Carlo* (KMC) simulation to evolve the system according to the probabilities dictated by the Arrhenius form (for a description of KMC see, for example, [MS06, BC06]). When all possible processes are not known in advance, an approach like "on-the-fly" KMC can be used where, first, possible pathways are found by searching for saddle points exiting from the current minimum and then KMC is applied [HJ01].

An alternative set of approaches based on directly accelerating MD simulations have been developed by Voter and coworkers (see [VMG02] for a review). The simplest method is *parallel replica dynamics* [Vot98]. The idea is simply to run an MD simulation and wait for basin-hops to occur. However, since these are rare events, multiple realizations of the system are run simultaneously, each on a different one of M processors. Each realization is initialized differently to allow them to sample different parts of phase space. Once a hop is detected at time t, all processors are stopped and reset to start from the hopped state, and the global clock is advanced to Mt. Thus, the speed-up is limited, in principle, only by the number of available processors.

Two other approaches, also due to Voter and his coworkers, aim to speed up the processes of interest by modifying the two key variables in Eqn. (10.1): the energy barrier ΔE and the temperature T. In so-called *hyperdynamics* [Vot97], the potential energy surface is systematically modified by adding a "bias potential," essentially to make the basins less deep without affecting the saddle points between them. Rare events occur more frequently on the modified landscape, and the speed-up can be estimated based on TST assumptions and knowledge of the bias potential. Finally, the method of "temperature-assisted dynamics" [MSV01, MV01, MV02] essentially runs the system at a higher temperature. The problem with doing this is that while you increase the hopping rate, you are likely to encounter hops that would not have taken place at the real (lower) temperature of interest. As such, we again can appeal to TST under some modest assumptions, in order to be able to identify the nonphysical transitions and filter them out of the simulation.

These methods have seen impressive success in modeling problems like surface layer deposition and the formation and growth of surface islands, where simulations have been accelerated by factors of 10^8 or more.

Further reading

- Reviews that aim to give a general overview of multiscale methods, providing a cross-section of the various types, can be found in the work of Ghoniem *et al.* [GBKH03] and Lu and Kaxiras [LK05], while the *Handbook of Materials Modeling* edited by Yip [Yip06] provides encyclopedic entries on many multiscale methods.
- The review by Curtin and Miller [CM03] discusses the details of static, partitioned-domain approaches, with application to the plastic deformation of crystals. A review with an emphasis on *hierarchical methods* can be found in the work of E *et al.* [EEL$^+$07].
- An effort to systematically build a taxonomy for multiscale methods, to help identify common mathematical features between methods developed for different applications, is provided by Gravemeier *et al.* [GLW08].
- A discussion of the state-of-the-art in 2008 for multiscale methods in soft matter science can be found in [PDSK08].
- Reviews of dynamic methods and methods that include temperature can be found in [MT07, Gil10]. Two reviews by W. K. Liu and his coworkers [LKZP04, PL04] are good sources for elucidating the challenges associated with wave reflections from the atomistic/continuum interface in dynamic partitioned-domain models.
- An excellent review of accelerated dynamics methods that bridge time scales is [VMG02].

11 Atomistic constitutive relations for multilattice crystals

In this chapter, we derive expressions for the free energy, stress and elasticity tensors for crystalline systems under equilibrium conditions. These expressions can be used as constitutive relations in continuum mechanics (see Section 2.5) under the assumption that a continuum system is in a state of "local thermodynamic equilibrium" at each point. The advantage of the "atomistic constitutive relations" derived here is that they inherently possess basic properties of the material such as its symmetries and lattice-invariant shears which are difficult to incorporate into standard continuum models for crystals. It is also hoped that atomistic models are more predictive than macroscopic phenomenological models, but of course this depends on the transferability of the interatomic model as discussed in Section 5.7.2. The use of atomistic constitutive relations within a continuum finite element framework will be our first example of a multiscale method in Chapter 12.

Chapter 8 has already dealt with the derivation of microscopic expressions for stress and elasticity, so why are we revisiting this problem again? The reason is that the statistical mechanics expressions in Chapter 8 place no restrictions on the positions of the atoms aside from overall macroscopic constraints. This is an excellent model for fluids, where atoms move freely through space. However, in solid systems, atoms are arranged in energetically-favorable patterns about which they vibrate with an overall magnitude dictated by the temperature of the system. Figures 11.1(a) and 11.1(b) show that an atom in an aluminum crystal remains quite close to its mean position in the lattice in the solid state. Once the crystal melts (Fig. 11.1(c)), the atoms are released from the lattice constraint and a qualitatively different, liquid-like, behavior is observed. It is exactly the definite arrangement of the atoms in the solid state that lends to solids their resistance to shape change.[1]

Our objective is to suitably modify the standard statistical mechanics theory of Chapter 7 to account for the "solidity" of solids. The first thing we must realize is that this solidity is a fiction that depends on the observation time scale. Although the term "solid" suggests permanence, in fact, the solid state is inherently *metastable* (see the discussion of thermodynamic equilibrium on page 62). Over time, the atoms in a solid will move around by exchanging positions with their neighbors, defects will diffuse through the material eventually exiting through surfaces, new defects will form to relax local concentrations of stress, and phase transformations may occur. A solid is therefore a dynamical

[1] It is worth pointing out that the nonequilibrium formulation of Irving and Kirkwood, which appears in Chapter 8, was also derived with fluids in mind. However, the time dependence of the distribution functions means that restrictions on atomic positions, as described in this chapter for solids, could be incorporated within that formulation. Our discussion in this chapter is limited to equilibrium systems.

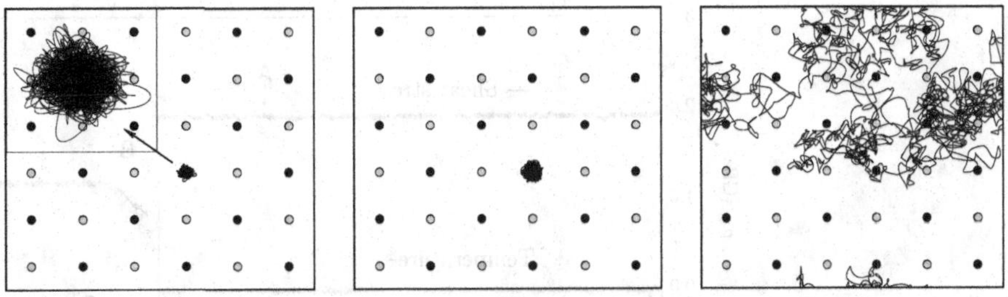

Fig. 11.1 The trajectory of a single atom in an aluminum fcc lattice is shown projected onto a plane for three temperatures: (a) 300 K, (b) 600 K and (c) 2000 K. In (a) and (b), the points correspond to the mean positions of the atoms in the lattice (normalized so that thermal expansion is factored out). The black atoms lie on the projection plane and the gray atoms are off-plane. The trajectory of one atom near the center is shown as a line that overlaps itself to form a small blob. Details of the trajectory at 300 K are shown in the inset. In (c), the lattice has melted, and the points correspond to the mean positions the atoms would have had otherwise. The black line corresponds to the continuous trajectory of a single atom that now moves widely through the simulation domain. In all three frames, the trajectories represent 50 000 timesteps (about 1 ns in real time) in a constant energy MD simulation of 108 atoms with periodic boundary conditions (PBCs) (indicated by the surrounding frame) using a pair functional model for aluminum due to [EA94].

system which is constantly evolving over time. "*Panta rei*" (everything flows) as the mystic Heraclitus believed. There is a time-scale issue, though. A solid may indeed be flowing, but it is doing so so slowly that it appears stable. This is what is meant by the term "metastable."[2]

Figure 11.2 shows a good example of the metastable nature of solids (motivated by Fig. 2 in [Pen02]). A block of aluminum is gripped on the top and bottom and subjected to a constant applied shear (configuration A). The block remains stable in this state for some time until the shear is relaxed by the passage of a dislocation (configuration C), which causes the shear strain in the block to reduce significantly as shown in the graph. Because the block is small and the shear quite large, this loss of stability occurs quickly enough to be accessible in a molecular dynamics (MD) simulation. In a macroscopic system, the waiting time can be very long, since the nucleation of the dislocation involves the coordinated motion of a huge number of atoms. Nevertheless, the loss of stability will eventually occur and hence the solid state is metastable. We therefore see that the ability to resist shape change, the hallmark of solids, exists only on time scales that are short enough to preclude escape from one metastable state to another.

[2] Markus Reiner made this point most eloquently in an amusing after dinner talk, where he coined the concept of the "Deborah number" [Rei64]. Deborah was a prophetess and judge in Israel around 1200 BC. After a successful campaign against the Canaanite general Sisera, she sings in victory "the mountains flowed before the Lord" (Judges 5:5). This inspired Reiner to define the Deborah number, D = time of relaxation/time of observation. "The greater the Deborah number, the more solid the material; the smaller the Deborah number, the more fluid it is." On the time scale of God, a very long observation time, even mountains appear to flow.

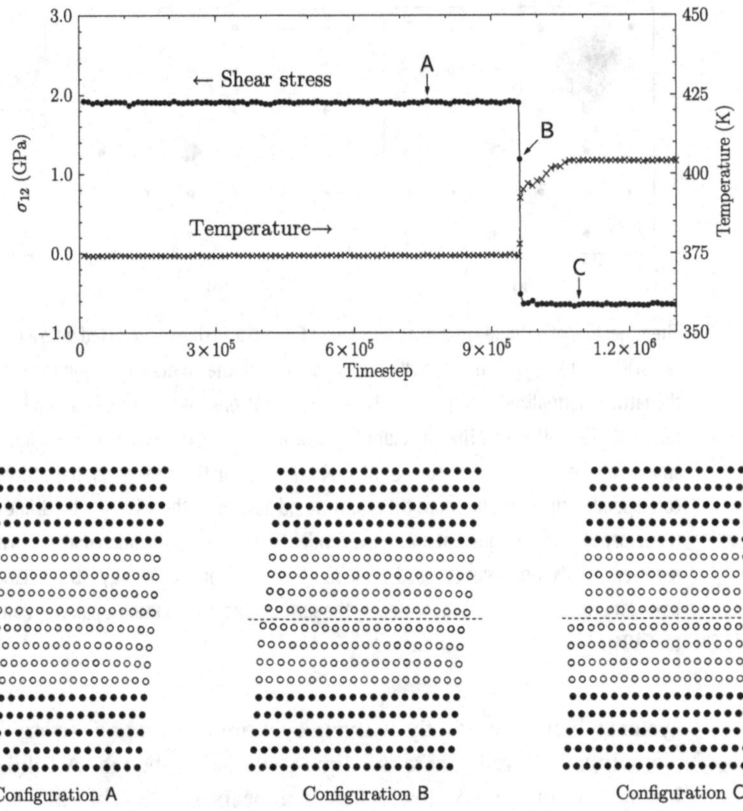

Fig. 11.2 A block of aluminum is gripped on the top and bottom and subjected to a simple shear of magnitude $\gamma = 0.1$ in the horizontal direction. (See Fig. 2.4(c) for the definition of γ.) There are 1080 atoms, modeled by a pair functional [EA94]. The grips are simulated by keeping the layers of atoms (shown as filled circles) within $2r_{\text{cut}}$ of the top and bottom fixed in the sheared position, where r_{cut} is the cutoff radius of the interatomic potential. This eliminates surface effects, although the sides of the crystal remain free surfaces. PBCs are applied in the out-of-plane direction. The aluminum is oriented to facilitate slip, with the horizontal direction along a $\langle 110 \rangle$ direction (Burgers vector direction in fcc) and the vertical direction along a $\langle 111 \rangle$ direction (normal to the slip plane in fcc). After application of the boundary conditions, the unconstrained atoms are randomly perturbed with an amplitude of 0.2 Å to set the initial temperature, and a constant energy (microcanonical) MD simulation is performed. The shear stress σ_{12} (computed using the virial expression with the phase average replaced by an MD time average) and temperature are monitored and plotted above. The uniformly sheared state is shown as configuration A. The system remains in this state for many timesteps at a shear stress of 1.90 GPa and temperature of 374 K, until a sudden nucleation of a dislocation causes a drop of the shear stress to -0.65 GPa; (fcc dislocations are discussed in Section 6.5.5.) The nucleation occurs in two steps indicated as B and C in the graph and configurations below it. First, a partial dislocation nucleates and sweeps through the system along the slip plane indicated by the dashed line in configuration B. A small step is formed on the surface. Almost immediately, a second partial dislocation nucleates and sweeps through on the same plane. The steps on the surface now correspond to one full Burgers vector. Note that the temperature of the system increases by 30 K following the passage of the dislocation as the potential energy in the sheared crystal is released and transformed into kinetic energy.

Recall that a standard statistical mechanics phase average integrates over *all* configurations accessible to the system regardless of the amount of time one would have to wait to reach each configuration. The resulting phase average will therefore provide a smeared view of the liquidlike behavior that a solid system will exhibit over infinite time. Take, for example, a statistical mechanics system of N atoms confined to a box with volume V. We saw in Section 7.4.5 that the free energy of such a system in the thermodynamic limit depends only on its temperature T and the number density N/V. As a consequence, the shear components of the stress and elasticity tensors (which are derivatives of the free energy with respect to shear strain) are identically zero [BCF86]. Thus, a statistical mechanics system behaves like a fluid with no resistance to changes of shape. We saw an example of this in Fig. 11.2, where the shear stress in a solid subjected to simple shear was relaxed by escaping from the sheared metastable state to a new defective metastable state. The statistical mechanics results for such a system correspond to an average over these two states and all other states accessible to it. The question is: how can the properties of a *single* metastable state (like configuration A in Fig. 11.2) be computed? How to do this remains an area of active research [Bal02, Section 4.1].

The method we pursue here is based on the *restricted ensemble* approach used by Penrose and Lebowitz [PL71, LP79] to study the liquid metastable state and more recently extended to solid systems [Pen02]. In our derivation, we assume that the solid is a multilattice crystal, i.e. a crystal with possibly more than one atom per unit cell. This description is quite general, since it also applies to periodic systems that are not necessarily perfect crystals, such as defective crystals and (periodic) amorphous materials. As the name "restricted ensemble" suggests, the basic idea is to restrict the statistical mechanics ensemble by only including configurations that are consistent with the metastable state being considered. One can view the restriction as a set of internal constraints placed on the system in addition to macroscopic constraints, such as the overall volume or deformation. The key is to find a suitable restriction which will keep a system trapped in its metastable state without unduly affecting its behavior. We will consider only the simplest possible restriction where the atoms are constrained to remain close to specified *mean positions*. The imposition of fixed mean positions necessitates the introduction of a kinematics linking the motion of these positions to the applied deformation gradient and internal shifts (i.e. motions of the atoms in the unit cell). This kinematics, referred to as the "Cauchy–Born rule," assumes that the lattice vectors defining the periodic cell deform in an affine manner in agreement with the applied macroscopic deformation, while the atoms in the unit cell are free to displace arbitrarily. Failures and extensions of the Cauchy–Born rule are also briefly discussed.

Following the restricted ensemble approach, we consider the more limited case, where the excursions of the atoms about their mean positions are assumed to be small, and use a harmonic expansion of the potential energy to analytically integrate phase averages. The resulting "quasiharmonic" expressions for the free energy and its derivatives (stress and elasticity) play an important role in the dynamic multiscale methods of Chapter 13. Finally, zero-temperature limits for all expressions are obtained that are important for molecular statics (MS, Chapter 6) and static multiscale methods (Chapter 12). In addition to general expressions, specific forms are obtained for a number of common interatomic potentials.

11.1 Statistical mechanics of systems in metastable equilibrium

The notion of a state of *metastable equilibrium* was introduced in Section 2.4.1. A system in metastable equilibrium is not strictly in thermodynamic equilibrium, since given sufficient time its properties will change; however, a time scale exists on which it exhibits all of the behavior expected of an equilibrium system. According to Penrose and Lebowitz [PL71], a metastable thermodynamic state has the following properties:

1. Only one thermodynamic phase is present.
2. A system that starts in this state is likely to take a long time to get out.
3. Once the system has gotten out, it is unlikely to return.

These properties suggest that it should be possible to use statistical mechanics to compute the properties of a system in metastable equilibrium, such as its temperature, free energy, stress, and so on. Strictly speaking these are not thermodynamic state variables, since the system is not in thermodynamic equilibrium, but they play the same role. Penrose and Lebowitz refer to them as *analogs* for the thermodynamic variables [PL71].

11.1.1 Restricted ensembles

The traditional treatment of the metastable state has been phenomenological. A well-known example is the van der Waals theory of the liquid–gas transition, which leads to the van der Waals equation of state for a fluid (see, for example, [Hua63, Section 2.3]). The van der Waals idea has been applied to other types of phase transformations in liquid and solid systems [CWA83]. It has also led to more sophisticated phenomenological methods which are described in most modern books on statistical mechanics and thermodynamics. We do not discuss these methods here. Instead, we focus on the more rigorous *restricted ensemble* approach,[3] which is closer to basic statistical mechanics ideas and is therefore more consistent with the approach taken elsewhere in this book. It is also completely general, unlike the phenomenological approaches which are often tied to specific applications. The basic idea is quite simple. We confine the system to a part of configuration space R by introducing a characteristic function $\chi(\boldsymbol{q})$, defined as

$$\chi(\boldsymbol{q}) \equiv \begin{cases} 1 & \text{if } \boldsymbol{q} \text{ is in } R, \\ 0 & \text{otherwise,} \end{cases} \tag{11.1}$$

where $\boldsymbol{q} = (q_1, \ldots, q_{3N})$ is a point in configuration space. The statistical mechanics phase average defined in Eqn. (7.28) is restricted to region R in configuration space through the

[3] The restricted ensemble approach can be traced back to early studies in statistical mechanics, see [Yuk91, Section 2.5] for a review.

following definition:

$$\langle A \rangle_{\chi} \equiv \int_{\Gamma} A(\boldsymbol{q}, \boldsymbol{p}) f(\boldsymbol{q}, \boldsymbol{p}) \chi(\boldsymbol{q}) \, d\boldsymbol{q} d\boldsymbol{p}, \tag{11.2}$$

where $A(\boldsymbol{q}, \boldsymbol{p})$ is a phase function of the set of positions $\boldsymbol{q} = (q_1, \dots, q_{3N})$ and momenta $\boldsymbol{p} = (p_1, \dots, p_{3N})$, and Γ is the full (unrestricted) phase space. As usual,

$$d\boldsymbol{q}d\boldsymbol{p} = dq_1 \cdots dq_{3N} \, dp_1 \cdots dp_{3N}$$

represents an element of phase space volume. As an alternative to introducing the characteristic function, one can modify the potential energy of the system to include an additional term that becomes very large (infinite) when the system is not inside R [PL71, CD98].

The remaining question is how to suitably define the region R for a given problem. Very generally, we can say that R should be sufficiently restrictive to confine the system to the metastable state but not so restrictive as to significantly affect its properties. The specifics of a particular choice of R are strongly tied to the physics of the problem being studied. For example, Penrose [Pen02, Pen06] studied a two-dimensional model of a solid system consisting of polarized particles arranged on a square lattice. Region R for this case is defined as the set of configurations in which the angles between the bonds lie in the range 60–120° (to prevent dislocation nucleation) along with additional constraints on the particles near the edges of the domain to impose the boundary conditions. Other examples for liquid systems are [PL71, LP79, ERH85, CD98].

In principle, once the region R is defined, properties of the modified system can be computed as phase averages over the restricted ensemble according to Eqn. (11.2). In practice, obtaining analytical results for the modified system can be quite difficult and such attempts have so far been mostly limited to simple idealized systems [PL71, CO77, Sew80, ERH85, Sti95, CD98, Pen02]. In contrast, it is quite easy to apply restricted ensemble constraints in Monte Carlo and MD computations. An early example is the Monte Carlo calculation of the elastic constants of fcc argon by Squire *et al.* [SHH69]. In these simulations, the particles are constrained to remain inside spherical cells centered on the lattice sites with radius equal to the nearest-neighbor distance. Any Monte Carlo move that would lead to this constraint being violated is automatically rejected. A more recent example of the same approach is the simulations of Corti and Debenedetti [CD94] of vapor–liquid equilibrium of a Lennard-Jones fluid. A similar procedure can be applied to MD simulations of metastable states, although, in a sense, such simulations are de facto carried out in a restricted ensemble simply due to the limitations of the accessible time scales. It is, however, straightforward to explicitly impose the constraints if desired. For example, one could apply the *basin-constrained MD* technique introduced by Sørensen and Voter [SV00] as part of the *temperature-accelerated dynamics* method for accelerating time in MD simulations. A similar idea is described by Stillinger [Sti95] in his discussion of metastable states.

11.1.2 Properties of a metastable state from a restricted canonical ensemble

In our analysis, we adopt the restricted ensemble approach applied to the canonical ensemble, but consider only the simplest case where the positions of atoms are constrained to remain close to specified mean positions – exactly as was done in the Squire *et al.* [SHH69] simulation described above. Thus, R is the set of all configurations where each atom is within a distance r_ϵ of a specified mean position. Let $\boldsymbol{r}^\alpha$ be the position of atom α, $\bar{\boldsymbol{r}}^\alpha$ be the *mean* position of the same atom, and $\boldsymbol{w}^\alpha$ its displacement.[4] Then,

$$\boldsymbol{r}^\alpha = \bar{\boldsymbol{r}}^\alpha + \boldsymbol{w}^\alpha \qquad \alpha = 1, \dots, N, \tag{11.3}$$

where $\|\boldsymbol{w}^\alpha\| \le r_\epsilon$ and N is the number of atoms. An explicit definition for the mean positions will be given later after the Helmholtz free energy is defined.

Clearly the magnitude of r_ϵ plays an important role in the results. If r_ϵ is too small it could affect the thermal vibrations of the atoms and also interfere with large amplitude phonons. If it is too large it could fail to prevent escape from the metastable state. For example, nucleation of a dislocation requires atomic motions of the order of the Burgers vector, which has length $a/\sqrt{6}$ for the common Shockley partial in fcc crystals (see Section 6.5.5). Hence in order to prevent the nucleation of any dislocations in an fcc crystal, we must have $r_\epsilon \le a/\sqrt{6}$. For example, for aluminum this is about 1.646 Å. By comparison, the root-mean-square displacement of aluminum atoms due to thermal vibrations at 300 K is about 0.50 Å [Dov93, Chapter 4]. So for aluminum at 300 K, it appears to be possible to select a value for r_ϵ that will prevent dislocation nucleation without overly interfering with thermal vibration. For other systems, other temperatures and other possible escapes from the metastable state (such as due to phase transformations) this must be reassessed.

As usual, we introduce the concatenated notation (see Section 7.1),

$$\boldsymbol{q} = (r_1^1, r_2^1, r_3^1, \dots, r_3^N), \qquad \bar{\boldsymbol{q}} = (\bar{r}_1^1, \bar{r}_2^1, \bar{r}_3^1, \dots, \bar{r}_3^N), \qquad \boldsymbol{w} = (w_1^1, w_2^1, w_3^1, \dots, w_3^N).$$

Equation (11.3) can thus be rewritten as $\boldsymbol{q} = \bar{\boldsymbol{q}} + \boldsymbol{w}$. In terms of the characteristic function, the constraint defining R is

$$\chi(\boldsymbol{w}) \equiv \begin{cases} 1 & \text{if } \|\boldsymbol{w}^\alpha\| \le r_\epsilon, \quad \forall \alpha \in [1, N], \\ 0 & \text{otherwise.} \end{cases} \tag{11.4}$$

The restricted phase average in Eqn. (11.2) for the canonical ensemble (see Section 7.4) is

$$\mathcal{A}(\bar{\boldsymbol{q}}, T) = \langle A \rangle_\chi = \frac{1}{h^{3N} Z(\bar{\boldsymbol{q}}, T)} \int_{-\infty}^{\infty} \cdots \int_{-\infty}^{\infty} A(\bar{\boldsymbol{q}} + \boldsymbol{w}, \boldsymbol{p}) e^{-\mathcal{H}(\bar{\boldsymbol{q}} + \boldsymbol{w}, \boldsymbol{p})/k_B T} \chi(\boldsymbol{w}) \, d\boldsymbol{w} d\boldsymbol{p}, \tag{11.5}$$

[4] We use the symbol $\boldsymbol{w}$ for displacements due to atomic vibrations about mean positions in the deformed configuration, reserving $\boldsymbol{u}$ for the displacements relative to a given reference configuration as is done in Sections 12.2 and 13.2.2.

where h is Planck's constant, k_B is Boltzmann's constant, T is the temperature of the heat bath and $\mathcal{H}(\boldsymbol{q}, \boldsymbol{p})$ and $Z(\bar{\boldsymbol{q}}, T)$ are the Hamiltonian and partition function of the system:

$$\mathcal{H}(\boldsymbol{q}, \boldsymbol{p}) = \mathcal{T}(\boldsymbol{p}) + \mathcal{V}^{\text{int}}(\boldsymbol{q}) = \sum_{i=1}^{3N} \frac{p_i^2}{2m_i} + \mathcal{V}^{\text{int}}(q_1, \ldots, q_{3N}), \tag{11.6}$$

$$Z(\bar{\boldsymbol{q}}, T) = \frac{1}{h^{3N}} \int_{-\infty}^{\infty} \cdots \int_{-\infty}^{\infty} e^{-\mathcal{H}(\bar{\boldsymbol{q}}+\boldsymbol{w}, \boldsymbol{p})/k_B T} \chi(\boldsymbol{w})\, d\boldsymbol{w} d\boldsymbol{p}. \tag{11.7}$$

Note that for the canonical ensemble, the Hamiltonian depends on the *internal* potential energy (see Eqn. (8.42) along with the accompanying explanation and footnote 7 on page 451). Also, note that the integration in Eqns. (11.5) and (11.7) is over the atomic displacements $\boldsymbol{w}$ and not the atomic positions $\boldsymbol{q}$. For this reason, the $N!$ term included in the definitions in Eqns. (7.116) and (7.118) is missing. This is because χ breaks the permutation symmetry of the system and therefore prevents the double counting without the need for the $N!$ term (see the discussion of "correct Boltzmann counting" after Eqn. (7.88) and also [Pen02] and [Maz00, Section 6.4.5]).

As shown after Eqn. (7.116), the partition function can be decomposed into a kinetic part, which can be integrated analytically, and a potential part called the *configurational integral*. Doing so for the present case, we have, $Z(\bar{\boldsymbol{q}}, T) = Z^{\text{K}}(T) Z^{\text{V}}(\bar{\boldsymbol{q}}, T)$, where

$$Z^{\text{K}}(T) = \prod_{i=1}^{3N} \left(\frac{2\pi m_i k_B T}{h^2} \right)^{1/2}, \qquad Z^{\text{V}}(\bar{\boldsymbol{q}}, T) = \int e^{-\mathcal{V}^{\text{int}}(\bar{\boldsymbol{q}}+\boldsymbol{w})/k_B T} \chi(\boldsymbol{w}) d\boldsymbol{w}. \tag{11.8}$$

Using the decomposition of the partition function, we can obtain convenient expressions for the special cases when the phase function depends only on momenta or positions:

$$\mathcal{A}(T) = \langle A(\boldsymbol{p}) \rangle_\chi = \frac{1}{h^{3N} Z^{\text{K}}} \int A(\boldsymbol{p}) e^{-\mathcal{T}(\boldsymbol{p})/k_B T}\, d\boldsymbol{p}, \tag{11.9}$$

$$\mathcal{A}(\bar{\boldsymbol{q}}, T) = \langle A(\boldsymbol{q}) \rangle_\chi = \frac{1}{Z^{\text{V}}(\bar{\boldsymbol{q}})} \int A(\bar{\boldsymbol{q}} + \boldsymbol{w}) e^{-\mathcal{V}^{\text{int}}(\bar{\boldsymbol{q}}+\boldsymbol{w})/k_B T} \chi(\boldsymbol{w})\, d\boldsymbol{w}. \tag{11.10}$$

As in Eqn. (11.5), the integration for all momentum and displacement degrees of freedom is over $-\infty$ to ∞ but for brevity from now on we use a single integration symbol to refer to integration over all momenta or all positions. Note that momentum properties are not affected by the constraint. So, for example, the temperature of the system will be equal to the temperature of the heat bath just as for the unrestricted ensemble.

The Helmholtz free energy defined in Eqn. (7.129) has the same form for the restricted canonical ensemble with the exception that the free energy now depends on the mean positions of the atoms:

$$\Psi(\bar{\boldsymbol{q}}, T) = -k_B T \ln Z(\bar{\boldsymbol{q}}, T). \tag{11.11}$$

Making use of the decomposition of the partition function, the Helmholtz free energy is

$$\Psi(\bar{\boldsymbol{q}}, T) = -k_B T \ln Z^{\text{K}}(T) - k_B T \ln Z^{\text{V}}(\bar{\boldsymbol{q}}, T). \tag{11.12}$$

The mean positions concatenated in $\bar{q}$ correspond to a local minimum of the Helmholtz free energy (see Section 2.6), and therefore satisfy the equations [Sut92, Section 2])

$$\left.\frac{\partial \Psi(\boldsymbol{q},T)}{\partial q_i}\right|_{\boldsymbol{q}=\bar{\boldsymbol{q}}} = 0, \qquad \left.\frac{\partial^2 \Psi(\boldsymbol{q},T)}{\partial q_i \partial q_j}\right|_{\boldsymbol{q}=\bar{\boldsymbol{q}}} w_i w_j \geq 0 \qquad \forall \boldsymbol{w} \in \mathbb{R}^{3N}, \boldsymbol{w} \neq \boldsymbol{0}. \tag{11.13}$$

The first condition ensures that $\bar{q}$ is an extremum point of the free energy surface. The second condition (positive semi-definiteness of the Hessian of Ψ) ensures that the extremum is a local minimum provided that the only zero eigenvalues are associated with the rigid-body (translation and rotation) modes. For (infinite) periodic systems it also necessary to ensure stability with respect to quasiuniform deformations[5] [Kan95a].

The solution to Eqn. (11.13) is of course not unique. The free energy landscape tends to inherit the structure of the potential energy landscape, which, as explained in Section 6.1, contains a huge number of local minima. For solid systems, many of the local minima of the free energy will be sufficiently deep to ensure that a system with the corresponding mean positions will be in a state of metastable equilibrium as defined in Section 11.1.

Next, we would like to compute the stress and elasticity for the metastable state using the restricted ensemble, which involves taking derivatives of the free energy with respect to the deformation gradient $\boldsymbol{F}$. This requires us to define a relationship between the mean positions and the applied deformation as described in the next section.

11.2 Relating mean positions to applied deformation: the Cauchy–Born rule

In our derivation of the restricted ensemble approach, we have introduced the idea that atoms have well-defined mean positions, but aside from that, we have left the structure of the material undefined. In this section, we become more specific and consider the case of multilattice crystals[6] and then introduce *Cauchy–Born kinematics* that relate the applied deformation to the mean positions of the atoms in the material.

11.2.1 Multilattice crystals and mean positions

We consider a multilattice crystal with a basis of N_B atoms (see Section 3.6). The position in a stress-free reference configuration (see Section 2.2.2) of basis atom λ associated with

[5] A "quasiuniform" deformation is one involving a uniform deformation of the periodic system together with internal "shifts" [ETS06b]. See Section 11.2.2, where quasiuniform deformation is described in the context of Cauchy–Born kinematics.

[6] It is worth noting that although we use the language and notation of multilattice crystals, the derivation applies equally to a general periodic system with a nonorthogonal cell as discussed in Section 6.4.2. The "lattice vectors" of the crystal correspond to the vectors defining the periodicity of the system and the "basis atoms" correspond to the atoms in the periodic cell.

lattice site $\boldsymbol{\ell} = (\ell_1, \ell_2, \ell_3)$, denoted $[\boldsymbol{\ell}\lambda]$, is

$$\boldsymbol{R}^{[\boldsymbol{\ell}\lambda]}(t) = \bar{\boldsymbol{R}}^{[\boldsymbol{\ell}\lambda]} + \boldsymbol{W}^{[\boldsymbol{\ell}\lambda]}(t), \qquad \forall \boldsymbol{\ell} \in \mathbb{Z}^3, \qquad \lambda = 0, \dots, N_{\mathrm{B}} - 1, \tag{11.14}$$

where $\bar{\boldsymbol{R}}^{[\boldsymbol{\ell}\lambda]}$ is the mean position of the atom, and $\boldsymbol{W}^{[\boldsymbol{\ell}\lambda]}(t)$ is its instantaneous displacement due to thermal vibration. (See Eqn. (3.15) for a definition of the notation.) The mean positions of the atoms form a multilattice crystal and are thus given by Eqn. (3.15) as

$$\bar{\boldsymbol{R}}^{[\boldsymbol{\ell}\lambda]} = \ell_i \boldsymbol{A}_i + \bar{\boldsymbol{Z}}^{\lambda}, \tag{11.15}$$

where $\boldsymbol{A}_i$ are the reference lattice vectors, and $\bar{\boldsymbol{Z}}^{\lambda}$ is the mean reference position of basis atom λ relative to a lattice site. As usual, we apply the summation convention to spatial indices (see Section 2.1.1).

The application of external loading to the crystal will cause it to deform to a new arrangement in the deformed configuration, where the positions of the atoms will also satisfy a relationship like Eqn. (11.14), i.e.

$$\boldsymbol{r}^{[\boldsymbol{\ell}\lambda]}(t) = \bar{\boldsymbol{r}}^{[\boldsymbol{\ell}\lambda]} + \boldsymbol{w}^{[\boldsymbol{\ell}\lambda]}(t). \tag{11.16}$$

Here $\bar{\boldsymbol{r}}^{[\boldsymbol{\ell}\lambda]}$ and $\boldsymbol{w}^{[\boldsymbol{\ell}\lambda]}(t)$ are the mean position and displacement of atom $[\boldsymbol{\ell}\lambda]$ in the deformed configuration. We have adopted the standard continuum mechanics convention of using capital letters for the reference configuration and lower-case letters for the deformed configuration (see Section 2.2.2).

11.2.2 Cauchy–Born kinematics

The question we need to address now is how the mean positions of atoms in the periodic cell displace in response to the applied deformation, characterized by the deformation gradient $\boldsymbol{F}$. The answer is not obvious since the mean positions we are referring to are introduced as additional constraints, and are not equal to the expectation values computed from statistical mechanics. It is therefore necessary to introduce an additional postulate.[7]

An early step in this direction was taken by Cauchy in 1828 [Cau28a, Cau28b], who was studying the elasticity of simple crystals and assumed that the atoms behave as material points embedded in the continuum, and simply follow the macroscopic deformation exactly. Cauchy did not consider temperature, and was therefore referring to absolute atom positions and not mean positions. The "Cauchy rule" for a uniform deformation defined by the deformation gradient $\boldsymbol{F}$ is then simply

$$\boldsymbol{r}^{[\boldsymbol{\ell}]} = \boldsymbol{F}\boldsymbol{R}^{[\boldsymbol{\ell}]}. \tag{11.17}$$

Using the small-strain equivalent to this rule and assuming pair potential interactions, Cauchy computed the stress and elastic properties of crystalline solids. These results are reviewed in [Lov27, Note B].

Born [Bor23, BH54] later recognized that the Cauchy rule is overly restrictive and does not apply to most crystals. He extended Cauchy's ideas to finite temperature by postulating

[7] The choice is not arbitrary, of course, since the configuration defined by the mean positions must satisfy force equilibrium as discussed presently.

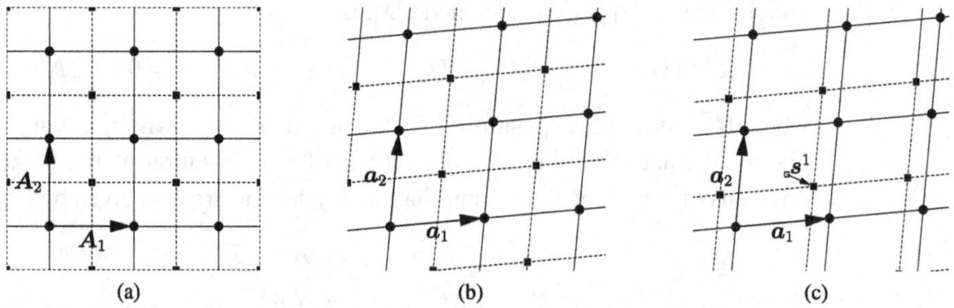

Fig. 11.3 Cauchy–Born kinematics for a two-dimensional multilattice crystal with two basis atoms. The circles and squares correspond to the mean positions of atoms that are assumed to be of different species (otherwise this would be a simple crystal in its essential description). The two sublattices (solid and dashed) in the reference configuration are shown in (a), while (b) and (c) depict the Cauchy–Born kinematics in two stages. In (b), the deformation gradient is applied and the sublattices transform to new sublattices with lattice vectors $\boldsymbol{a}_i = \boldsymbol{F}\boldsymbol{A}_i$. In (c), sublattice 1 is shifted with respect to sublattice 0 by $\bar{\boldsymbol{s}}^1$.

that "Cauchy's rule" applies to mean atomic positions, with the atoms experiencing thermal vibrations about these points. He also considered the case of multilattices, and recognized that in addition to the macroscopic deformation set by $\boldsymbol{F}$, an internal deformation associated with the relative motion of sublattices must be considered. Thus, according to the "Cauchy–Born rule," the mean positions of the atoms transform according to

$$\bar{\boldsymbol{r}}^{[\ell\lambda]} = \boldsymbol{F}\bar{\boldsymbol{R}}^{[\ell\lambda]} + \bar{\boldsymbol{s}}^{\lambda}, \tag{11.18}$$

where $\bar{\boldsymbol{s}}^{\lambda}$ are called the *shift vectors*. A deformation of this type is called a *quasiuniform* deformation [ETS06b]. If we think of a multilattice as a set of N_B interpenetrating sublattices as shown in Fig. 11.3,[8] then Eqn. (11.18) means that each sublattice undergoes an affine transformation defined by $\boldsymbol{F}$ and, in addition, the sublattices can "shift" relative to each other as shown in Fig. 11.3. Without loss of generality, we can take $\bar{\boldsymbol{s}}^0 = \boldsymbol{0}$, so that $\bar{\boldsymbol{s}}^{\lambda}$ $(\lambda > 0)$ represents the shift of sublattice λ relative to the origin $(\lambda = 0)$ sublattice.

An atom's instantaneous position follows by substituting Eqn. (11.18) into Eqn. (11.16):

$$\boldsymbol{r}^{[\ell\lambda]}(t) = \boldsymbol{F}\bar{\boldsymbol{R}}^{[\ell\lambda]} + \bar{\boldsymbol{s}}^{\lambda} + \boldsymbol{w}^{[\ell\lambda]}(t). \tag{11.19}$$

Note that the shift vectors are mean quantities, unlike the displacement $\boldsymbol{w}^{[\ell\lambda]}(t)$, which describes the time-dependent microscopic fluctuations. The shifting of the sublattices, also called "shuffling," is an internal mode of deformation that the crystal has in addition to the macroscopic deformation. We will see in Section 11.3.4 that, under equilibrium conditions, the mean shift positions are obtained from the requirement that that the mean forces acting on the basis atoms are zero. See also [ETS06b, Section 2.3] for a discussion of this issue.

Another useful observation is made by substituting the mean reference positions from Eqn. (11.15) into Eqn. (11.18). We see that the mean positions of the atoms in the deformed

[8] See also Fig. 3.18(c) and accompanying text.

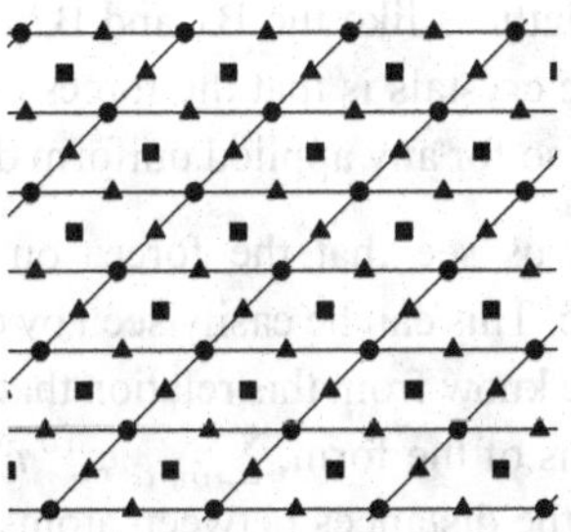

Fig. 11.4 A two-dimensional centrosymmetric multilattice crystal with four atoms per unit cell ($N_B = 4$). The different symbols represent different atomic species. The unit cells are indicated by solid lines.

configuration form a new multilattice:

$$\bar{\boldsymbol{r}}^{[\boldsymbol{\ell}\lambda]} = \ell_i \boldsymbol{a}_i + \bar{\boldsymbol{z}}^{\lambda}, \tag{11.20}$$

where

$$\boldsymbol{a}_i = \boldsymbol{F}\boldsymbol{A}_i, \qquad \bar{\boldsymbol{z}}^{\lambda} = \boldsymbol{F}\bar{\boldsymbol{Z}}^{\lambda} + \bar{\boldsymbol{s}}^{\lambda}. \tag{11.21}$$

Therefore, the Cauchy–Born rule amounts to the statement that *an* N_B*-lattice crystal remains an* N_B*-lattice crystal (with equal or lesser periodicity) under the action of a uniform deformation.*

Before closing this section, we introduce some derivatives of the instantaneous atomic position in Eqn. (11.19) that will be needed in the derivations that follow:

$$\frac{\partial r_k^{[\boldsymbol{\ell}\lambda]}}{\partial F_{iJ}} = \frac{\partial}{\partial F_{iJ}}\left[F_{kL}\bar{R}_L^{[\boldsymbol{\ell}\lambda]} + \bar{s}_k^{\lambda} + w_k^{[\boldsymbol{\ell}\lambda]}\right] = \delta_{ik}\bar{R}_J^{[\boldsymbol{\ell}\lambda]}, \tag{11.22}$$

$$\frac{\partial r_k^{[\boldsymbol{\ell}\lambda]}}{\partial \bar{s}_i^{\mu}} = \frac{\partial}{\partial \bar{s}_i^{\mu}}\left[F_{kL}\bar{R}_L^{[\boldsymbol{\ell}\lambda]} + \bar{s}_k^{\lambda} + w_k^{[\boldsymbol{\ell}\lambda]}\right] = \delta_{ik}\delta_{\mu\lambda}. \tag{11.23}$$

11.2.3 Centrosymmetric crystals and the Cauchy–Born rule

An important special case of the Cauchy–Born rule occurs for centrosymmetric crystals. We define a *centrosymmetric crystal* as a crystal in which *every* atom is at a center of inversion (see Section 3.4.1).[9] An example of a two-dimensional centrosymmetric multilattice crystal is shown in Fig. 11.4. A centrosymmetric crystal has the property that if the origin of a coordinate system is located at any atom α in the crystal, then for any neighbor β of that atom located at a relative position[10] $\bar{\boldsymbol{r}}^{\alpha\beta} = \bar{\boldsymbol{r}}^{\beta} - \bar{\boldsymbol{r}}^{\alpha}$ there is another neighbor β' of the same species located at $\bar{\boldsymbol{r}}^{\alpha\beta'} = -\bar{\boldsymbol{r}}^{\alpha\beta}$. All simple crystals are centrosymmetric as well as many

[9] Note that in crystallography the term "centrosymmetric crystal" is sometimes associated more broadly with any crystal possessing a center of inversion in its space group. Clearly for multilattice crystals our definition and this definition can differ.

[10] Here superscripts like α in $\bar{\boldsymbol{r}}^{\alpha}$ are short-hand for the multilattice notation $[\boldsymbol{\ell}\lambda]$ in $\bar{\boldsymbol{r}}^{[\boldsymbol{\ell}\lambda]}$ used above. The bars remind us that we are dealing with the mean positions of the atoms.

important multilattices like the B1 and B2 crystal structures.[11] An important property of centrosymmetric crystals is that the forces on the atoms in such a structure are identically zero and remain so for any applied uniform deformation without shifting of the sublattices.

Proof First, let us see that the forces on the atoms are indeed zero if the crystal is centrosymmetric. This can be easily seen by considering the central force decomposition in Eqn. (5.107). We know from this relation that the force on any atom α can always be written as a sum of terms of the form, $\sum_{\beta \neq \alpha} \varphi^{\alpha\beta} \bar{\boldsymbol{r}}^{\alpha\beta} / \bar{r}^{\alpha\beta}$, where $\bar{r}^{\alpha\beta} = \|\bar{\boldsymbol{r}}^{\alpha\beta}\|$ and where $\varphi^{\alpha\beta}$ is a function of the distances between atoms in the crystal (not necessarily just $\bar{r}^{\alpha\beta}$). Due to centrosymmetry, this decomposition can be written as a sum over pairs of neighbors β and β' that satisfy $\bar{\boldsymbol{r}}^{\alpha\beta'} = -\bar{\boldsymbol{r}}^{\alpha\beta}$ (as explained above) that add up to zero:

$$\frac{\varphi^{\alpha\beta}}{\bar{r}^{\alpha\beta}} \bar{\boldsymbol{r}}^{\alpha\beta} + \frac{\varphi^{\alpha\beta'}}{\bar{r}^{\alpha\beta'}} \bar{\boldsymbol{r}}^{\alpha\beta'} = \frac{\varphi^{\alpha\beta}}{\bar{r}^{\alpha\beta}} (\bar{\boldsymbol{r}}^{\alpha\beta} - \bar{\boldsymbol{r}}^{\alpha\beta}) = \mathbf{0},$$

where we have used the fact that distances are unaffected by an inversion operation. Thus the forces on all the atoms in a multilattice crystal are identically zero.

In the second step of the proof, we need to show that a centrosymmetric crystal remains centrosymmetric when subject to a uniform deformation without shifts. Consider a crystal whose mean positions form a centrosymmetric crystal in the reference configuration. According to Eqn. (11.15), the mean positions are

$$\bar{\boldsymbol{R}}^{[\boldsymbol{\ell}\lambda]} = \ell_i \boldsymbol{A}_i + \bar{\boldsymbol{Z}}^\lambda = (\ell_i + \bar{\zeta}_i^\lambda) \boldsymbol{A}_i,$$

where in the second expression we have used fractional coordinates (see Section 3.6). If we now apply a uniform deformation without shifts ($\bar{\boldsymbol{s}}^\lambda = \mathbf{0}$) to the crystal, according to the Cauchy–Born rule in Eqn. (11.18) the mean positions in the deformed configuration are

$$\bar{\boldsymbol{r}}^{[\boldsymbol{\ell}\lambda]} = \boldsymbol{F} \bar{\boldsymbol{R}}^{[\boldsymbol{\ell}\lambda]} = \boldsymbol{F} (\ell_i + \bar{\zeta}_i^\lambda) \boldsymbol{A}_i = (\ell_i + \bar{\zeta}_i^\lambda) \boldsymbol{a}_i,$$

where $\boldsymbol{a}_i = \boldsymbol{F}\boldsymbol{A}_i$. Now, clearly since $\bar{\boldsymbol{R}}^{[\boldsymbol{\ell}\lambda]} = (\ell_i + \bar{\zeta}_i^\lambda)\boldsymbol{A}_i$ is a centrosymmetric crystal, so is $(\ell_i + \bar{\zeta}_i^\lambda)\boldsymbol{a}_i$. Therefore the deformed crystal will be centrosymmetric without shifts. □

We have shown that the forces on the atoms in a centrosymmetric crystal are zero. This means that a centrosymmetric crystal structure corresponds to a stationary point in configuration space. Note, however, that it is not guaranteed that this is a *minimum* configuration because the second condition in Eqn. (11.13) may not be satisfied. This occurs when a centrosymmetric crystal undergoes a phase transformation to a different crystal structure (see footnote 15 on page 570). Situations like this are addressed by the *cascading Cauchy–Born approach*, references for which are provided in the discussion of "Failures of the Cauchy–Born rule" in the next section.

11.2.4 Extensions and failures of the Cauchy–Born rule

The Cauchy–Born rule is of central importance when developing theoretical and computational continuum formulations based on atomistic models. It is therefore not surprising that

[11] Of course many important crystal structures are not centrosymmetric. Some examples include the diamond structure that both carbon atoms and silicon atoms adopt, ferroelectric crystals, martensitic phases of shape-memory alloys and many inorganic minerals.

it has been extensively studied. These studies divide into three broad categories: theoretical studies of the rule and its applications, extensions to the rule and identification of situations where it fails. In the interest of brevity, we will not review this body of work here, but briefly point the interested reader to further reading on the subject below.

Theoretical studies of the Cauchy–Born rule The mathematical and physical implications of the Cauchy–Born rule have been explored in depth for a variety of applications [Sta50, BH54, Eri70, Par76, Wei83, Eri84, TSBK99, BBL02, ETS06a, EM07, Eri08].

Extensions to the Cauchy–Born rule The Cauchy–Born rule has been extended in various ways to accommodate noncrystalline structures. Examples include extensions to amorphous structures [GK98, KG98], thin films and shells [FJ00, AB02, YE06], nanotubes [ZHG+02, AB04], objective structures [Jam06] and the incorporation of surface effects [PKW06]. Another type of extension involves the generalization of the rule to nonuniform deformation by including higher-order gradients [LCN03, SS03, GWZ06].

Failures of the Cauchy–Born rule The Cauchy–Born rule can break down for a number of reasons. First, the basic assumption that the atoms in the crystal undergo a quasiuniform deformation can be violated by the formation of defects or other inhomogeneities in the deformation [WYPW93, SES07]. Second, in some cases the energy-minimizing deformation requires an increase in the periodic size of the unit cell [FT02, CDKM06]. This issue has been addressed by a new approach referred to as *cascading Cauchy–Born kinematics* [DELT07, SET11], where the energy-minimizing unit cell is identified through a phonon-stability analysis (see additional comments on this in Section 12.1). Third, in certain cases deformations leading to multiple phases, such as mechanical twinning, can violate the Cauchy–Born rule regardless of the size of the unit cell [Pit85, Zan92, Zan96, Eri97, PZ00].

11.3 Finite temperature constitutive relations for multilattice crystals

In this section, we derive expressions for the Helmholtz free energy, stress tensor and elasticity tensor as statistical mechanics phase averages within the canonical ensemble suitably restricted to maintain the solid structure. The solid being considered is a multilattice crystal represented by a supercell with PBCs.

11.3.1 Periodic supercell of a multilattice crystal

We consider a periodic (see Section 6.4.2), nonorthogonal supercell of a multilattice crystal defined by the vectors $\boldsymbol{l}_i = \nu_i \boldsymbol{a}_i$ (no sum on i), where $\boldsymbol{a}_i$ are the lattice vectors of the crystal and ν_i are positive integers. The supercell contains $N_{\text{lat}} = \nu_1\nu_2\nu_3$ unit cells each with N_{B} basis atoms for a total of $N = N_{\text{lat}}N_{\text{B}}$ atoms.

In order for the periodic expressions for the free energy, stress and elasticity derived below to be equivalent to those of an infinite crystal, the integers ν_i must be taken to infinity. This is of course computationally impossible. The selection of finite values for ν_i means that modes of deformation with periodicity larger than that of the supercell are precluded. The correct choice for ν_i is problem-dependent. Without additional information, one should verify that obtained results are independent of ν_i by increasing (or decreasing) the size of the supercell and verifying that the results are not significantly changed. (We explore this below for a one-dimensional chain in Example 11.1.) Another practical constraint on the magnitude of ν_i is related to the minimum image convention that we adopt in our derivation (see Section 6.4.2). This requires the supercell to be sufficiently large to ensure that given the cutoff radius of the interatomic potential, an atom in the supercell does not interact with multiple copies of other atoms in the cell or with copies of itself.

The atoms within the supercell are identified by lattice point $\boldsymbol{\ell} = (\ell_1, \ell_2, \ell_3)$ and sub-lattice number $\lambda = 0, \dots, N_\mathrm{B} - 1$. Using the notation of Section 11.2.1, the mean position of atom $[\boldsymbol{\ell}\lambda]$ is

$$\bar{\boldsymbol{r}}^{[\boldsymbol{\ell}\lambda]} = \ell_i \boldsymbol{a}_i + \bar{\boldsymbol{z}}^\lambda, \tag{11.24}$$

where the summation convention is observed. To simplify the derivation, rather than using the explicit notation in Eqn. (11.24), we will simply number the atoms in the supercell consecutively, identifying each atom with a scalar $\alpha = 1, \dots, N$. The downside of this approach is that the scalar α provides no information on the lattice point and sublattice number of an atom. We therefore define two functions, $\mathrm{latpt}(\alpha)$ and $\mathrm{sublat}(\alpha)$, which respectively return the lattice point and sublattice of atom α. So, for example, if atom 939 is associated lattice point $\boldsymbol{\ell} = (2, 3, 7)$ and belongs to sublattice 2, then $\mathrm{latpt}(939) = (2, 3, 7)$ and $\mathrm{sublat}(939) = 2$. These functions are a bit cumbersome to write down, so we introduce the following short-hand notation using the superposed symbols $\diamond$ and $\#$:

$$\overset{\diamond}{\alpha} \equiv \mathrm{latpt}(\alpha), \qquad \overset{\#}{\alpha} \equiv \mathrm{sublat}(\alpha). \tag{11.25}$$

Think of the diamond representing the unit cell and the hash symbol suggesting sub-lattices. Using this notation, the mean position of atom α is

$$\bar{\boldsymbol{r}}^\alpha = \overset{\diamond}{\alpha}_i \boldsymbol{a}_i + \bar{\boldsymbol{z}}^{\overset{\#}{\alpha}}. \tag{11.26}$$

This is a bit odd, but in most cases it will greatly simplify the expressions we derive. In some cases, we will also need the set of all atoms in the supercell that belong to sublattice λ. We denote this as $\mathcal{S}^\lambda$, with the definition

$$\mathcal{S}^\lambda = \{\alpha \mid \overset{\#}{\alpha} = \lambda\}. \tag{11.27}$$

So far we have defined the mean positions of the atoms. The atoms vibrate about these positions with displacements $\boldsymbol{w}^\alpha$ so that their instantaneous positions are

$$\boldsymbol{r}^\alpha = \bar{\boldsymbol{r}}^\alpha + \boldsymbol{w}^\alpha. \tag{11.28}$$

As usual, we use concatenated notation, $\boldsymbol{q}$, $\bar{\boldsymbol{q}}$ and $\boldsymbol{w}$, to refer to all atoms at once.

Definition of a reference configuration The definitions given above apply to any arbitrary multilattice crystal structure. We now single out one of these structures as a *reference configuration*. The mean positions of the atoms in the reference configuration are[12]

$$\bar{\boldsymbol{R}}^{\alpha} = \overset{\diamond}{\alpha}_i \boldsymbol{A}_i + \bar{\boldsymbol{Z}}^{\overset{\#}{\alpha}}. \tag{11.29}$$

The lattice vectors of the reference structure $\boldsymbol{A}_i$ and the mean basis atom positions $\{\bar{\boldsymbol{Z}}^{\lambda}\}$ will be determined in the next section from the requirement that this is an equilibrium configuration, i.e. one that minimizes the free energy. Since the free energy depends on the temperature, $\boldsymbol{A}_i$ and $\bar{\boldsymbol{Z}}^{\lambda}$ will also depend on the temperature.

The application of a uniform deformation to the reference configuration, characterized by the deformation gradient $\boldsymbol{F}$, maps the mean positions to new mean positions in the *deformed configuration* that also form a multilattice crystal structure. This hypothesis is referred to as the *Cauchy–Born rule* and was discussed in Section 11.2.2. In our short-hand notation the Cauchy–Born rule given in Eqn. (11.18) states

$$\bar{\boldsymbol{r}}^{\alpha} = \boldsymbol{F}\bar{\boldsymbol{R}}^{\alpha} + \bar{\boldsymbol{s}}^{\overset{\#}{\alpha}}, \tag{11.30}$$

where $\{\bar{\boldsymbol{s}}^{\lambda}\}$ are the mean shifts of the sublattices. If we substitute Eqn. (11.29) into Eqn. (11.30) and compare the result with Eqn. (11.26), we see that the lattice vectors in the deformed configurations are related to the reference lattice vectors by $\boldsymbol{a}_i = \boldsymbol{F}\boldsymbol{A}_i$. As a result the supercell vectors are mapped in the same manner:

$$\boldsymbol{l}_i = \boldsymbol{F}\boldsymbol{L}_i, \tag{11.31}$$

where $\boldsymbol{L}_i = \nu_i \boldsymbol{A}_i$.

Application of PBCs Next we consider PBCs which affect the manner in which distances are computed between atoms. As noted above, we require that the supercell is sufficiently large so that the minimum image convention can be applied. In this case, the distance between two atoms α and β is

$$r^{\alpha\mathring{\beta}} = \left\|\boldsymbol{r}^{\alpha\mathring{\beta}}\right\| = \left[r_n^{\alpha\mathring{\beta}} r_n^{\alpha\mathring{\beta}}\right]^{1/2}, \tag{11.32}$$

where

$$\boldsymbol{r}^{\alpha\mathring{\beta}} = \bar{\boldsymbol{r}}^{\beta}_{\mathrm{per}}(\mathring{n}_1, \mathring{n}_2, \mathring{n}_3) + \boldsymbol{w}^{\beta} - \bar{\boldsymbol{r}}^{\alpha} - \boldsymbol{w}^{\alpha}, \tag{11.33}$$

and

$$\bar{\boldsymbol{r}}^{\beta}_{\mathrm{per}}(n_1, n_2, n_3) = \bar{\boldsymbol{r}}^{\beta} + n_i \boldsymbol{l}_i, \qquad (\mathring{n}_1, \mathring{n}_2, \mathring{n}_3) = \arg\min_{n_1, n_2, n_3} \left\|\bar{\boldsymbol{r}}^{\beta}_{\mathrm{per}}(n_1, n_2, n_3) - \bar{\boldsymbol{r}}^{\alpha}\right\|. \tag{11.34}$$

The values of $\mathring{n}_i$ will be 0 or ± 1 depending on whether atom β or one of its periodic images is closest to α. The superposed circle in $\boldsymbol{r}^{\alpha\mathring{\beta}}$ reminds us that PBCs are applied when constructing the relative position vector connecting atoms α and β. Due to antisymmetry,

[12] We adopt the continuum mechanics convention of denoting variables in the reference configuration with capital letters.

we have that $\boldsymbol{r}^{\alpha\mathring{\beta}} = -\boldsymbol{r}^{\beta\mathring{\alpha}}$, which means that $r^{\alpha\mathring{\beta}} = r^{\beta\mathring{\alpha}}$. Also note that the PBCs imply that the displacement of a periodic copy of an atom is the same as that of the atom itself.

In what follows we will require the derivatives of the relative position vector $\boldsymbol{r}^{\alpha\mathring{\beta}}$ and the distance $r^{\alpha\mathring{\beta}}$ with respect to $\boldsymbol{F}$ and $\{\bar{\boldsymbol{s}}^{\lambda}\}$. In preparation for computing these, we substitute Eqn. (11.30) into Eqn. (11.33) and use Eqns. $(11.34)_1$ and (11.31) to obtain

$$\boldsymbol{r}^{\alpha\mathring{\beta}} = \bar{\boldsymbol{r}}^{\alpha\mathring{\beta}} + \boldsymbol{w}^{\beta} - \boldsymbol{w}^{\alpha} \tag{11.35}$$

where

$$\bar{\boldsymbol{r}}^{\alpha\mathring{\beta}} = \boldsymbol{F}\bar{\boldsymbol{R}}^{\alpha\mathring{\beta}} + \overset{\#}{\bar{\boldsymbol{s}}}{}^{\mathring{\beta}} - \overset{\#}{\bar{\boldsymbol{s}}}{}^{\mathring{\alpha}}, \qquad \bar{\boldsymbol{R}}^{\alpha\mathring{\beta}} = \bar{\boldsymbol{R}}^{\beta} - \bar{\boldsymbol{R}}^{\alpha} + \mathring{n}_i \boldsymbol{L}_i. \tag{11.36}$$

Combining Eqns. $(11.36)_1$ and $(11.36)_2$ and using Eqn. (11.30), we also have that

$$\bar{\boldsymbol{r}}^{\alpha\mathring{\beta}} = \bar{\boldsymbol{r}}^{\beta} - \bar{\boldsymbol{r}}^{\alpha} + \mathring{n}_i \boldsymbol{l}_i. \tag{11.37}$$

It is important to note that $\bar{\boldsymbol{R}}^{\alpha\mathring{\beta}}$ in Eqn. $(11.36)_2$ is *not* necessarily the periodic relative position vector that would be computed for atoms α and β in the reference configuration. This is because the parameters $\mathring{n}_i$ are computed in Eqn. $(11.34)_2$ based on positions in the *deformed* configuration. It is possible that due to the shifting of the sublattices the copy of β closest to α will be different in the reference and deformed configurations.

The explicit dependence of $\boldsymbol{r}^{\alpha\mathring{\beta}}$ on $\boldsymbol{F}$ and $\{\bar{\boldsymbol{s}}^{\lambda}\}$ is now clear and we can compute the desired derivatives. These are

$$\frac{\partial r_n^{\alpha\mathring{\beta}}}{\partial F_{iJ}} = \delta_{in}\bar{R}_J^{\alpha\mathring{\beta}}, \qquad \frac{\partial r_n^{\alpha\mathring{\beta}}}{\partial \bar{s}_i^{\lambda}} = \delta_{in}(\overset{\#}{\delta}{}^{\mathring{\beta}\lambda} - \overset{\#}{\delta}{}^{\mathring{\alpha}\lambda}), \tag{11.38}$$

where δ is the Kronecker delta (δ_{in} is zero unless $i = n$ and $\overset{\#}{\delta}{}^{\mathring{\alpha}\lambda}$ is zero unless the sublattice number of atom α is equal to λ). The derivatives of the distance follow using Eqn. (11.32) as

$$\frac{\partial r^{\alpha\mathring{\beta}}}{\partial F_{iJ}} = \frac{r_i^{\alpha\mathring{\beta}}}{r^{\alpha\mathring{\beta}}}\bar{R}_J^{\alpha\mathring{\beta}}, \qquad \frac{\partial r^{\alpha\mathring{\beta}}}{\partial \bar{s}_i^{\lambda}} = \frac{r_i^{\alpha\mathring{\beta}}}{r^{\alpha\mathring{\beta}}}(\overset{\#}{\delta}{}^{\mathring{\beta}\lambda} - \overset{\#}{\delta}{}^{\mathring{\alpha}\lambda}). \tag{11.39}$$

With the geometric derivatives in place, we can proceed to the calculation of the Helmholtz free energy and then to the determination of the reference configuration in Section 11.3.3.

11.3.2 Helmholtz free energy density of a multilattice crystal

The expression for the Helmholtz free energy Ψ is given in Eqn. (11.12). We see there that the free energy depends on the mean positions $\bar{q}$ and the temperature T. The dependence on the mean positions comes in through the potential energy. However, we know that due to the *principle of interatomic potential invariance* (see Section 5.3.2), the potential energy of the system can only depend on the distances between the atoms, $\mathcal{V}^{\text{int}} = \mathcal{V}^{\text{int}}(\{r^{\alpha\mathring{\beta}}\})$, where $\{r^{\alpha\mathring{\beta}}\}$ is short-hand for the $N(N-1)/2$ distances defined by

$$\{r^{\alpha\mathring{\beta}} \mid \alpha, \beta = 1, \dots, N, \alpha < \beta\}.$$

The superposed circle reminds us that PBCs are applied in the calculation of the distances as defined in Eqn. (11.32).

The above means Helmholtz free energy is a function of the mean distances $\{\bar{r}^{\alpha\mathring{\beta}}\}$:

$$\Psi(\{\bar{r}^{\alpha\mathring{\beta}}\}, T) = -k_B T \ln Z^K(T) - k_B T \ln Z^V(\{\bar{r}^{\alpha\mathring{\beta}}\}, T). \tag{11.40}$$

The kinetic and potential parts of the partition function, given in Eqn. (11.8), are then

$$Z^K(T) = \prod_{\alpha=1}^{N} \left(\frac{2\pi m^{\overset{\#}{\alpha}} k_B T}{h^2} \right)^{3/2}, \qquad Z^V(\{\bar{r}^{\alpha\mathring{\beta}}\}, T) = \int e^{-\mathcal{V}^{\text{int}}(\{r^{\alpha\mathring{\beta}}\})/k_B T} \chi(\boldsymbol{w})\, d\boldsymbol{w}, \tag{11.41}$$

where $m^{\overset{\#}{\alpha}}$ is the mass of an atom belonging to sublattice $\overset{\#}{\alpha}$, $d\boldsymbol{w} = dw_1 \cdots dw_{3N}$, and the integration bounds for each degree of freedom are from $-\infty$ to ∞.

The mean positions $\{\bar{r}^{\alpha\mathring{\beta}}\}$ depend on the deformation gradient $\boldsymbol{F}$ and the mean shifts $\{\bar{\boldsymbol{s}}^{\lambda}\}$ through Eqn. (11.36)$_1$. Therefore, we define the Helmholtz free energy density in terms of these variables:

$$\psi(\boldsymbol{F}, \{\bar{\boldsymbol{s}}^{\lambda}\}, T) \equiv \frac{1}{M} \Psi(\{\bar{r}^{\alpha\mathring{\beta}}\}, T), \tag{11.42}$$

where M is the mass contained in the supercell,

$$M = N_{\text{lat}} M_{\text{cell}}, \qquad M_{\text{cell}} = \sum_{\lambda=0}^{N_B-1} m^{\lambda}, \tag{11.43}$$

and M_{cell} is the mass associated with a single lattice site.

11.3.3 Determination of the reference configuration

The reference configuration of a crystal structure at a given temperature T is defined by the set of lattice vectors $\{\boldsymbol{A}_i(T)\}$ and mean basis atom positions $\{\bar{\boldsymbol{Z}}^{\lambda}(T)\}$ that minimize the free energy of the system (see Section 2.6). The procedure is outlined in Algorithm 11.1.

With the reference lattice vectors known, the volume of the unit cell in the reference configuration can be computed:

$$\Omega_0(T) = |\boldsymbol{A}_1(T) \cdot (\boldsymbol{A}_2(T) \times \boldsymbol{A}_3(T))|. \tag{11.44}$$

Of course since the reference lattice vectors depend on temperature, so does the reference unit cell volume. This is the phenomenon of *thermal expansion*.[13]

Algorithm 11.1 can be performed repeatedly over a range of temperatures, from $T_{\min}$ to $T_{\max}$ in increments of ΔT. In this case the initial guess at each step can be taken as the solution to the previous step. Prior to performing the minimization in step 2, it is important to perturb the shifts slightly in case the initial guess is trapped at a stationary point.

[13] Thermal expansion is explored further in Exercise 11.1 for a simple one-dimensional chain. See also Examples 11.1 and 11.2 in the quasiharmonic approximation.

Algorithm 11.1 Determination of reference configuration: full minimization

1: Select initial guesses for the lattice vectors and mean basis atom positions:

$$\{\boldsymbol{A}_i^{\text{guess}}\} \text{ and } \{\bar{\boldsymbol{Z}}^\lambda_{\text{guess}}\}.$$

2: Minimize the free energy density in Eqn. (11.42) with respect to the deformation gradient and mean shifts at the specified temperature T:

$$\min_{\boldsymbol{F},\{\bar{\mathbf{s}}^\lambda\}} \psi(\boldsymbol{F}, \{\bar{\mathbf{s}}^\lambda\}, T) \quad \rightarrow \quad \text{reference configuration.} \tag{A}$$

The solution to Eqn. (A) gives the minimizing deformation gradient $\boldsymbol{F}_{\text{min}}(T)$ and mean shifts $\{\bar{\mathbf{s}}^\lambda_{\text{min}}(T)\}$.

3: Compute the reference configuration lattice vectors and mean positions from:

$$\boldsymbol{A}_i(T) = \boldsymbol{F}_{\text{min}}(T)\boldsymbol{A}_i^{\text{guess}}, \qquad \bar{\boldsymbol{Z}}^\lambda(T) = \boldsymbol{F}_{\text{min}}(T)\bar{\boldsymbol{Z}}^\lambda_{\text{guess}} + \bar{\mathbf{s}}^\lambda_{\text{min}}(T).$$

Methods for performing the minimization in Eqn. (A) in Algorithm 11.1 are discussed in Section 6.2. Typically, an approach like the conjugate gradient (CG) method will be used which requires the gradient of the function being minimized. Let us compute the gradient of ψ with respect to $\boldsymbol{F}$ and $\{\bar{\mathbf{s}}^\lambda\}$. The kinetic part of the free energy does not contribute to the gradient since it does not depend on the positions of the atoms. The gradient then follows from Eqns. (11.42) and (11.40) as

$$\frac{\partial \psi}{\partial \boldsymbol{F}} = -\frac{1}{M}\frac{k_{\text{B}}T}{Z^{\text{V}}}\frac{\partial Z^{\text{V}}}{\partial \boldsymbol{F}}, \qquad \frac{\partial \psi}{\partial \bar{\mathbf{s}}^\lambda} = -\frac{1}{M}\frac{k_{\text{B}}T}{Z^{\text{V}}}\frac{\partial Z^{\text{V}}}{\partial \bar{\mathbf{s}}^\lambda}. \tag{11.45}$$

We now need to compute the gradient of Z^{V} with respect to $\boldsymbol{F}$ and $\{\bar{\mathbf{s}}^\lambda\}$. Using Eqn. $(11.41)_2$, the derivative of Z^{V} with respect to $\boldsymbol{F}$ is[14]

$$\begin{aligned}
\frac{\partial Z^{\text{V}}}{\partial \boldsymbol{F}} &= -\frac{1}{k_{\text{B}}T}\int \frac{\partial \mathcal{V}^{\text{int}}}{\partial \boldsymbol{F}} e^{-\mathcal{V}^{\text{int}}/k_{\text{B}}T}\chi(\boldsymbol{w})\,d\boldsymbol{w} \\
&= -\frac{1}{k_{\text{B}}T}\sum_{\substack{\alpha,\beta \\ \alpha<\beta}} \int \frac{\partial \mathcal{V}^{\text{int}}}{\partial r^{\alpha\mathring{\beta}}}\frac{\partial r^{\alpha\mathring{\beta}}}{\partial \boldsymbol{F}} e^{-\mathcal{V}^{\text{int}}/k_{\text{B}}T}\chi(\boldsymbol{w})\,d\boldsymbol{w} \\
&= -\frac{1}{k_{\text{B}}T}\sum_{\substack{\alpha,\beta \\ \alpha<\beta}} \int \boldsymbol{f}^{\alpha\mathring{\beta}} \otimes \bar{\boldsymbol{R}}^{\alpha\beta} e^{-\mathcal{V}^{\text{int}}/k_{\text{B}}T}\chi(\boldsymbol{w})\,d\boldsymbol{w},
\end{aligned} \tag{11.46}$$

where Eqn. $(11.39)_1$ was used and $\boldsymbol{f}^{\alpha\mathring{\beta}}$ (originally defined in Eqn. (5.107)) is

$$\boldsymbol{f}^{\alpha\mathring{\beta}} = \varphi^{\alpha\mathring{\beta}}\frac{\boldsymbol{r}^{\alpha\mathring{\beta}}}{r^{\alpha\mathring{\beta}}}, \tag{11.47}$$

[14] The binary operator $\otimes$ in Eqn. (11.46) is the tensor product. If $\boldsymbol{a}$ and $\boldsymbol{b}$ are vectors, then $\boldsymbol{a}\otimes\boldsymbol{b}$ is a second-order tensor with components: $[\boldsymbol{a}\otimes\boldsymbol{b}]_{ij} = a_i b_j$.

where $\varphi^{\alpha\mathring{\beta}} = \partial\mathcal{V}^{\text{int}}/\partial r^{\alpha\mathring{\beta}}$. Following a similar procedure and using Eqn. (11.39)$_2$, the derivative of Z^{V} with respect to $\bar{s}^\lambda$ is

$$\frac{\partial Z^{\text{V}}}{\partial \bar{s}^\lambda} = -\frac{1}{k_{\text{B}}T}\sum_{\substack{\alpha,\beta\\ \alpha<\beta}}\int f^{\alpha\mathring{\beta}}(\delta^{\overset{\#}{\beta}\lambda} - \delta^{\overset{\#}{\alpha}\lambda})e^{-\mathcal{V}^{\text{int}}/k_{\text{B}}T}\chi(w)\,dw. \tag{11.48}$$

The Kronecker deltas limit the summation to pairs of atoms where one or the other (but not both) belong to the λ sublattice. Equation (11.48) can be simplified as follows. Recall that the force on an atom for all interatomic potentials can always be expressed as a sum of central terms (see Eqn. (5.107) in Section 5.8.2). For a periodic system this is

$$f^{\text{int},\alpha} = \sum_{\substack{\beta\\ \beta\neq\alpha}} f^{\alpha\mathring{\beta}}. \tag{11.49}$$

Now returning to Eqn. (11.48), focus on the sum $\sum_{\alpha,\beta,\ \alpha<\beta} f^{\alpha\mathring{\beta}}(\delta^{\overset{\#}{\beta}\lambda} - \delta^{\overset{\#}{\alpha}\lambda})$. Replace $\sum_{\alpha,\beta,\ \alpha<\beta}$ with the symmetric $\frac{1}{2}\sum_{\alpha,\beta,\ \alpha\neq\beta}$ and separate the resulting expression into two parts limiting the sums according to the effect of the Kronecker deltas. This gives

$$\sum_{\substack{\alpha,\beta\\ \alpha<\beta}} f^{\alpha\mathring{\beta}}(\delta^{\overset{\#}{\beta}\lambda} - \delta^{\overset{\#}{\alpha}\lambda}) = \frac{1}{2}\left[\sum_{\beta\in S^\lambda}\sum_{\substack{\alpha\\ \alpha\neq\beta}} f^{\alpha\mathring{\beta}} - \sum_{\alpha\in S^\lambda}\sum_{\substack{\beta\\ \beta\neq\alpha}} f^{\alpha\mathring{\beta}}\right] = -\sum_{\alpha\in S^\lambda} f^{\text{int},\alpha}, \tag{11.50}$$

where S^λ is the set of all atoms belonging to sublattice λ (see Eqn. (11.27)). To obtain the final result on the right, we swapped the dummy indices on the first sum in the square brackets, used the fact that $f^{\alpha\mathring{\beta}} = -f^{\beta\mathring{\alpha}}$ and applied Eqn. (11.49). Now substituting Eqn. (11.50) into Eqn. (11.48), we have that

$$\frac{\partial Z^{\text{V}}}{\partial \bar{s}^\lambda} = \frac{1}{k_{\text{B}}T}\sum_{\alpha\in S^\lambda}\int f^{\text{int},\alpha}e^{-\mathcal{V}^{\text{int}}/k_{\text{B}}T}\chi(w)\,dw. \tag{11.51}$$

The gradient of ψ follows by substituting Eqns. (11.46) and (11.51) into Eqn. (11.45):

$$\frac{\partial\psi}{\partial F} = \frac{1}{M}\sum_{\substack{\alpha,\beta\\ \alpha<\beta}}\left\langle f^{\alpha\mathring{\beta}}\otimes\bar{R}^{\alpha\mathring{\beta}}\right\rangle_\chi, \qquad \frac{\partial\psi}{\partial\bar{s}^\lambda} = -\frac{1}{M}\sum_{\alpha\in S^\lambda}\left\langle f^{\text{int},\alpha}\right\rangle_\chi. \tag{11.52}$$

where $\langle\cdot\rangle_\chi$ is the phase average over the restricted ensemble constrained by χ. Since the arguments do not depend on momenta, the simplified relation in Eqn. (11.10) is used. We will see in the next section that $\partial\psi/\partial F$ is closely related to the stress in the system. The term $\partial\psi/\partial\bar{s}^\lambda$ is related to the mean force acting on an atom belonging to sublattice λ. (Simply multiply by $-M_{\text{cell}}$ to get the force.)

With the gradient expressions in Eqn. (11.52) known, the minimization of Eqn. (A) in Algorithm 11.1 can be carried out and the reference lattice vectors and mean basis atom positions determined. At the minimum, the gradient of ψ is zero, so that

$$\frac{\partial \psi}{\partial \boldsymbol{F}} = \boldsymbol{0}, \qquad \frac{\partial \psi}{\partial \bar{\boldsymbol{s}}^\lambda} = \boldsymbol{0} \quad \lambda = 0, \ldots, N_{\mathrm{B}} - 1. \tag{11.53}$$

These conditions imply that the configuration has zero stress and that the mean force acting on each basis atom is zero. We defer examples of the minimization process to Section 11.4.2 after we have introduced the quasiharmonic approximation.

11.3.4 Uniform deformation and the macroscopic stress tensor

At this stage we know how to compute the free energy density for our system using Eqn. (11.42), and how to compute the reference configuration of the crystal using the procedure of the previous section. We would now like to consider situations where the crystal is subjected to an affine deformation as described in Eqn. (11.30) and to compute the resulting stress tensor (and later the elasticity tensor). There is a problem, though. The free energy density in Eqn. (11.42) is a function of both the deformation gradient $\boldsymbol{F}$ and the mean sublattice shifts $\{\bar{\boldsymbol{s}}^\lambda\}$, while the macroscopic thermodynamics expression for free energy depends only on the deformation gradient. Why do the shifts disappear?

The "standard" approach is to assume that the sublattice shifts represent microscopic processes that cannot be controlled or observed at the macroscopic level – much like the "microstates" of statistical mechanics. It is therefore reasonable to assume that a separation of time scales exists so that for given macroscopic conditions, characterized by $\boldsymbol{F}$ and T, the shifts have sufficient time to relax to their equilibrium configuration:

$$\bar{\boldsymbol{s}}^\lambda(\boldsymbol{F}) = \underset{\{\overset{*}{\boldsymbol{s}}^\lambda\}}{\arg\min}\ \psi(\boldsymbol{F}, \{\overset{*}{\boldsymbol{s}}^\lambda\}, T). \tag{11.54}$$

The minimization in Eqn. (11.54) can be either local or global. Normally local minimization is used for practical reasons, but global minimization has the theoretical advantage of providing a unique result. To perform the minimization the gradient $\partial\psi/\partial\bar{\boldsymbol{s}}^\lambda$ in Eqn. $(11.52)_2$ is used and at the minimum, Eqn. 11.53 is satisfied:[15]

$$\frac{\partial \psi}{\partial \bar{\boldsymbol{s}}^\lambda} = -\frac{1}{M} \sum_{\alpha \in \mathcal{S}^\lambda} \left\langle \boldsymbol{f}^{\mathrm{int},\alpha} \right\rangle_\chi = \boldsymbol{0}. \tag{11.55}$$

[15] An important special case arises for centrosymmetric crystals. As shown in Section 11.2.3, the forces on the atoms in a centrosymmetric crystal are identically zero for any value of $\boldsymbol{F}$ due to the inversion symmetries of the lattice. Given Eqn. $(11.52)_2$, this means that $\partial\psi/\partial\bar{\boldsymbol{s}}^\lambda = \boldsymbol{0}$ when setting $\bar{\boldsymbol{s}}^\lambda = \boldsymbol{0}$ for all basis atoms. However, satisfaction of the condition $\partial\psi/\partial\bar{\boldsymbol{s}}^\lambda = \boldsymbol{0}$ does not ensure that the zero shift configuration corresponds to a minimum as required by Eqn. (11.54). It is quite possible for centrosymmetric crystals to lose stability and undergo a phase transformation to a different crystal structure. (See, for example, [DELT07, SET11].) In this case, the zero shift configuration actually corresponds to a saddle point and not a minimum. For this reason it is very important when dealing with centrosymmetric crystals to slightly perturb the initial guesses for the shifts $\{\overset{*}{\boldsymbol{s}}^\lambda\}$ at the start of the minimization process in Eqn. (11.54).

We can then define an *effective* or *homogenized* free energy density which depends only on the deformation gradient and temperature:

$$\psi^{\text{eff}}(\boldsymbol{F},T) \equiv \psi(\boldsymbol{F},\{\bar{\boldsymbol{s}}^{\lambda}(\boldsymbol{F})\},T). \tag{11.56}$$

This approach appears to be reasonable – and we will adopt it in the rest of the chapter – but it is not always correct. There are situations where shifts can manifest themselves at the macroscopic level as, for example, when an electric field is applied to an ionic crystal causing the charged sublattices to be displaced relative to each other [BGM71]. This realization led to the development in the 1970s of a generalized thermodynamics for multilattice crystals (or "non-Bravais crystals" as they are referred to in this literature). In this new *general regime* of thermodynamics, "external strain" and "internal strain" are treated on an equal footing as opposed to the *macroscopic regime* in which the shifts are minimized out as explained above [BGM71]. The term "internal strain" refers to a set of rotationally-invariant measures introduced to replace shifts. Later the term "inner elasticity" was coined to represent the microscopic aspects of the deformation of the crystal [Cou78a]. A series of papers were written during this period computing various orders of the "inner elastic constants" (i.e. derivatives of ψ with respect to internal strain) for different crystal systems (see, for example, [Mar75a, Mar75b, Mar75c, Cou78b]).

From the above it may seem that at least for purely mechanical loading the effective strain energy approach is safe. However, it can fail in this case as well. For example, suppose that we wish to use ψ^{eff} to compute the reference configuration of the lattice as described in the previous section. It seems possible to replace the full minimization of ψ with respect to $\boldsymbol{F}$ and $\{\bar{\boldsymbol{s}}^{\lambda}(\boldsymbol{F})\}$ which appears in Eqn. (A) in Algorithm 11.1, with a minimization of ψ^{eff} with respect to just $\boldsymbol{F}$, i.e.

$$\min_{\boldsymbol{F}} \psi^{\text{eff}}(\boldsymbol{F},T) \quad \rightarrow \quad \text{reference configuration.} \tag{11.57}$$

But Eqn. (A) in Algorithm 11.1 and Eqn. (11.57) are only equivalent if the minimizing deformation pathway does not involve coupling between the deformation gradient and shifts. This is the case for centrosymmetric crystals (see Section 11.2.3) but it is not true in general. An example demonstrating where the effective strain energy approach fails for a prototypical shape-memory alloy is given in [SET11]. In such cases, the minimum obtained from Algorithm 11.1 with Eqn. (11.57) can actually correspond to a saddle point of the original function ψ.

All of this suggests that a robust continuum theory should include shifts or internal strain fields as variables alongside the macroscopic deformation field. This remains an area of current interest (see, for example, [Eri97]), but is not part of the standard continuum mechanics and thermodynamics in widespread use. Therefore, we adopt the effective free energy in Eqn. (11.56), but consider the reader warned regarding its limitations.

We are now ready to compute the stress tensor. As explained in Section 2.5.2, the derivative of the *effective* free energy density with respect to the deformation gradient gives the first Piola–Kirchhoff stress tensor (see Eqn. $(2.171)_2$). Referring to Eqn. (11.56), we have

$$P_{iJ}(\boldsymbol{F},T) = \rho_0 \frac{\partial \psi^{\text{eff}}(\boldsymbol{F},T)}{\partial F_{iJ}} = \rho_0 \left[\frac{\partial \psi}{\partial F_{iJ}} + \sum_{\lambda=0}^{N_B-1} \frac{\partial \psi}{\partial \bar{s}_m^\lambda} \frac{\partial \bar{s}_m^\lambda}{\partial F_{iJ}} \right], \tag{11.58}$$

where ρ_0 is the reference mass density. For a multilattice crystal, $\rho_0 = M_{\text{cell}}/\Omega_0$, where M_{cell} and Ω_0 (defined in Eqn. (11.43)$_2$ and (11.44)) are the mass and reference volume of a single unit cell. Note that Ω_0 depends on the temperature due to thermal expansion. The second term in the square brackets in Eqn. (11.58) is zero due to Eqn. (11.53)$_2$. Therefore,

$$\boldsymbol{P}(\boldsymbol{F},T) = \rho_0 \frac{\partial \psi(\boldsymbol{F}, \{\bar{\boldsymbol{s}}^\lambda\}, T)}{\partial \boldsymbol{F}}. \tag{11.59}$$

Substituting in $\partial\psi/\partial\boldsymbol{F}$ from Eqn. (11.52)$_1$, we obtain the following expression for the first Piola–Kirchhoff stress tensor:

$$\boldsymbol{P} = \frac{1}{V_0} \sum_{\substack{\alpha,\beta \\ \alpha<\beta}} \left\langle \boldsymbol{f}^{\alpha\mathring{\beta}} \otimes \bar{\boldsymbol{R}}^{\alpha\mathring{\beta}} \right\rangle_\chi = \frac{1}{V_0} \sum_{\substack{\alpha,\beta \\ \alpha<\beta}} \left\langle \varphi^{\alpha\mathring{\beta}} \frac{\boldsymbol{r}^{\alpha\mathring{\beta}} \otimes \bar{\boldsymbol{R}}^{\alpha\mathring{\beta}}}{r^{\alpha\mathring{\beta}}}, \right\rangle_\chi, \tag{11.60}$$

where $\langle \cdot \rangle_\chi$ is defined in Eqn. (11.10) and $V_0 = N_{\text{lat}}\Omega_0$ is the volume of the supercell in the reference configuration. Since Ω_0 depends on the temperature, so does V_0. Equation (11.60) is very similar to Eqn. (8.66) derived earlier in Chapter 8 for the unrestricted ensemble. *We note, however, that the expression derived here lacks the kinetic terms which appear in Eqn. (8.66) and uses* mean *positions in the reference configuration.* We will see below after obtaining an expression for the Cauchy stress that the kinetic terms can be recovered by reverting to instantaneous atom positions.[16]

Difference between periodic and nonperiodic expressions for the stress tensor It is worth pointing out at this stage that the due to the PBCs, Eqn. (11.60) cannot be simply recast in terms of forces as was done in going from Eqn. (8.60) to Eqn. (8.66) in Section 8.1.3. If we follow that procedure for the present case, we find

$$\frac{1}{2} \sum_{\substack{\alpha,\beta \\ \alpha\neq\beta}} \boldsymbol{f}^{\alpha\mathring{\beta}} \otimes \bar{\boldsymbol{R}}^{\alpha\mathring{\beta}} = -\sum_{\alpha} \boldsymbol{f}^{\text{int},\alpha} \otimes \bar{\boldsymbol{R}}^\alpha + \frac{1}{2} \sum_{\substack{\alpha,\beta \\ \alpha\neq\beta}} \boldsymbol{f}^{\alpha\mathring{\beta}} \otimes \mathring{n}_i(\alpha,\beta)\boldsymbol{L}_i, \tag{11.61}$$

where we have stressed the fact that $\mathring{n}_i$, defined in Eqn. (11.34)$_2$, depends on α and β. Using Eqn. (11.61) in a symmetric version of Eqn. (11.60) (i.e. with $\sum_{\substack{\alpha,\beta \\ \alpha<\beta}}$ replaced with $\frac{1}{2}\sum_{\substack{\alpha,\beta \\ \alpha\neq\beta}}$) gives

$$P_{iJ} = -\frac{1}{V_0} \left[\sum_{\alpha} \boldsymbol{f}^{\text{int},\alpha} \otimes \bar{\boldsymbol{R}}^\alpha - \frac{1}{2} \sum_{\substack{\alpha,\beta \\ \alpha\neq\beta}} \boldsymbol{f}^{\alpha\mathring{\beta}} \otimes \mathring{n}_i(\alpha,\beta)\boldsymbol{L}_i \right]. \tag{11.62}$$

[16] The issue of the presence or absence of kinetic terms in the stress tensor has been a controversial one. The matter is clarified in the discussion below.

The second term in the square brackets includes contributions from pairs of atoms connected via the PBC for which at least one of the factors $\mathring{n}_1$, $\mathring{n}_2$ and $\mathring{n}_3$ is nonzero. It accounts for the traction applied to the periodic cell by the surrounding periodic images. This term explains why the stress in a periodic system is not zero even if the force on each of the atoms is zero.[17]

Derivation of an expression for the Cauchy stress The Cauchy stress, $\boldsymbol{\sigma} = \boldsymbol{P}\boldsymbol{F}^T / \det \boldsymbol{F}$ (see Eqn. (2.96)$_2$), follows from Eqn. (11.60) as

$$\begin{aligned}\boldsymbol{\sigma} &= \frac{1}{V}\sum_{\substack{\alpha,\beta\\ \alpha<\beta}} \left\langle \boldsymbol{f}^{\alpha\mathring{\beta}} \otimes \boldsymbol{F}(\bar{\boldsymbol{R}}^{\beta} - \bar{\boldsymbol{R}}^{\alpha} + \mathring{n}_i \boldsymbol{L}_i)\right\rangle_{\chi},\\ &= \frac{1}{V}\sum_{\substack{\alpha,\beta\\ \alpha<\beta}} \left\langle \boldsymbol{f}^{\alpha\mathring{\beta}} \otimes (\bar{\boldsymbol{r}}^{\beta} - \overset{\#}{\boldsymbol{s}}{}^{\beta} - \bar{\boldsymbol{r}}^{\alpha} + \overset{\#}{\boldsymbol{s}}{}^{\alpha} + \mathring{n}_i \boldsymbol{l}_i)\right\rangle_{\chi},\\ &= \frac{1}{V}\sum_{\substack{\alpha,\beta\\ \alpha<\beta}} \left\langle \boldsymbol{f}^{\alpha\mathring{\beta}} \otimes \bar{\boldsymbol{r}}^{\alpha\mathring{\beta}}\right\rangle_{\chi} - \frac{1}{2V}\sum_{\substack{\alpha,\beta\\ \alpha\neq\beta}} \left\langle \boldsymbol{f}^{\alpha\mathring{\beta}} \otimes (\overset{\#}{\boldsymbol{s}}{}^{\beta} - \overset{\#}{\boldsymbol{s}}{}^{\alpha})\right\rangle_{\chi},\end{aligned} \tag{11.63}$$

where $V = (\det \boldsymbol{F})V_0$ is the volume of the supercell in the deformed configuration. In the above equation, Eqn. (11.36)$_2$ was used in the first line, Eqns. (11.30) and (11.31) were used in the second line and Eqn. (11.37) was used in the third line. Note also that the last term has been symmetrized with the change of the sum to $\frac{1}{2}\sum_{\substack{\alpha,\beta\\ \alpha\neq\beta}}$. This last term can be evaluated using the following identity (see Exercise 11.3):

$$\frac{1}{2}\sum_{\substack{\alpha,\beta\\ \alpha\neq\beta}} \boldsymbol{f}^{\alpha\mathring{\beta}} \otimes (\overset{\#}{\boldsymbol{s}}{}^{\beta} - \overset{\#}{\boldsymbol{s}}{}^{\alpha}) = -\sum_{\lambda}\Big(\sum_{\alpha\in\mathcal{S}^{\lambda}} \boldsymbol{f}^{\mathrm{int},\alpha}\Big) \otimes \bar{\boldsymbol{s}}^{\lambda}, \tag{11.64}$$

where $\mathcal{S}^{\lambda}$ is the set of all atoms belonging to sublattice λ defined in Eqn. (11.27). Substituting Eqn. (11.64) into Eqn. (11.63) gives

$$\boldsymbol{\sigma} = \frac{1}{V}\sum_{\substack{\alpha,\beta\\ \alpha<\beta}} \left\langle \boldsymbol{f}^{\alpha\mathring{\beta}} \otimes \bar{\boldsymbol{r}}^{\alpha\mathring{\beta}}\right\rangle_{\chi} + \frac{1}{V}\sum_{\lambda}\left[\sum_{\alpha\in\mathcal{S}^{\lambda}} \left\langle \boldsymbol{f}^{\mathrm{int},\alpha}\right\rangle_{\chi}\right] \otimes \bar{\boldsymbol{s}}^{\lambda}. \tag{11.65}$$

The term in the square brackets is zero due to Eqn. (11.55). Therefore, the Cauchy stress is

$$\boldsymbol{\sigma} = \frac{1}{V}\sum_{\substack{\alpha,\beta\\ \alpha<\beta}} \left\langle \boldsymbol{f}^{\alpha\mathring{\beta}} \otimes \bar{\boldsymbol{r}}^{\alpha\mathring{\beta}}\right\rangle_{\chi} = \frac{1}{V}\sum_{\substack{\alpha,\beta\\ \alpha<\beta}} \left\langle \varphi'_{\alpha\mathring{\beta}} \frac{\boldsymbol{r}^{\alpha\mathring{\beta}} \otimes \bar{\boldsymbol{r}}^{\alpha\mathring{\beta}}}{r^{\alpha\mathring{\beta}}}\right\rangle_{\chi}. \tag{11.66}$$

We expect the Cauchy stress to be symmetric, but the symmetry of $\boldsymbol{\sigma}$ is *not* apparent from Eqn. (11.66) because the instantaneous and mean relative position vectors are not equal

[17] The forces will be zero in any configuration of static equilibrium for which the shifts are minimized out subject to a given value of the deformation gradient.

and therefore $r_i^{\alpha\mathring{\beta}}\bar{r}_j^{\alpha\mathring{\beta}} \neq r_j^{\alpha\mathring{\beta}}\bar{r}_i^{\alpha\mathring{\beta}}$. To demonstrate the symmetry of $\boldsymbol{\sigma}$, we must recover the kinetic part of the stress tensor which is done next.

Recovery of the kinetic terms for the stress tensor It is striking that Eqn. (11.66) lacks the kinetic terms normally associated with the stress tensor (see Eqn. (8.68)). The reason for this is that the positions in Eqn. (11.66) are *mean* positions, whereas the positions in Eqn. (8.68) are instantaneous positions.[18] By reverting to the instantaneous positions, the kinetic terms can be recovered. To see this, substitute Eqn. (11.35) into Eqn. (11.66) and use the symmetric form of the summation:

$$\boldsymbol{\sigma} = \frac{1}{2V}\sum_{\substack{\alpha,\beta \\ \alpha\neq\beta}} \left\langle \boldsymbol{f}^{\alpha\mathring{\beta}} \otimes \boldsymbol{r}^{\alpha\mathring{\beta}} \right\rangle_\chi - \frac{1}{2V}\sum_{\substack{\alpha,\beta \\ \alpha\neq\beta}} \left\langle \boldsymbol{f}^{\alpha\mathring{\beta}} \otimes (\boldsymbol{w}^\beta - \boldsymbol{w}^\alpha) \right\rangle_\chi . \tag{11.67}$$

Using an analogous expression to Eqn. (8.64), the sum in the second term on the right-hand side can be rewritten as

$$\frac{1}{2}\sum_{\substack{\alpha,\beta \\ \alpha\neq\beta}} \boldsymbol{f}^{\alpha\mathring{\beta}} \otimes (\boldsymbol{w}^\beta - \boldsymbol{w}^\alpha) = -\sum_\alpha \boldsymbol{f}^{\text{int},\alpha} \otimes \boldsymbol{w}^\alpha . \tag{11.68}$$

The Cauchy stress is therefore

$$\boldsymbol{\sigma} = \frac{1}{V}\left[\frac{1}{2}\sum_{\substack{\alpha,\beta \\ \alpha\neq\beta}} \left\langle \boldsymbol{f}^{\alpha\mathring{\beta}} \otimes \boldsymbol{r}^{\alpha\mathring{\beta}} \right\rangle_\chi + \sum_\alpha \left\langle \boldsymbol{f}^{\text{int},\alpha} \otimes \boldsymbol{w}^\alpha \right\rangle_\chi \right]. \tag{11.69}$$

Now from Newton's second law we have[19]

$$\boldsymbol{f}^{\text{int},\alpha} = m^{\overset{\#}{\alpha}}\ddot{\boldsymbol{r}}^\alpha = m^{\overset{\#}{\alpha}}\ddot{\boldsymbol{w}}^\alpha , \tag{11.70}$$

where m^λ is the mass of an atom belonging to sublattice λ. The second equality in Eqn. (11.70) follows from Eqn. (11.28) since $\bar{\boldsymbol{r}}^\alpha$ is constant. Assuming that in the thermodynamic limit, phase averages are equal to infinite time averages, the phase average in the

[18] This difference has been a source of some confusion in the literature. In particular, an article by Zhou [Zho03] claiming that the kinetic terms should not appear in the microscopic expression for the stress tensor started a long debate in the literature on this subject. The source of the discrepancy is exactly the difference between continuum velocities that correspond to the motion of mean positions and the instantaneous velocities associated with thermal fluctuations. The existence of a stress expression without kinetic terms was pointed out by Hoover in [Hoo86, Section II.C] and in a more recent article [HHL09]. For more on this, see the discussion of the physical significance of the kinetic stress on page 457.

[19] Note that we are assuming that there are no external fields present. If they exist they would be included as part of a body force term as was done in the Irving and Kirkwood derivation of Section 8.2.4.

second term of Eqn. (11.69) can be replaced by the following expression:

$$\lim_{\tau\to\infty}\frac{1}{\tau}\int_0^\tau \overset{\#}{m^\alpha}\ddot{w}^\alpha\otimes w^\alpha\,dt = \lim_{\tau\to\infty}\frac{\overset{\#}{m^\alpha}}{\tau}\int_0^\tau\left[\frac{d}{dt}(\dot{w}^\alpha\otimes w^\alpha)-\dot{w}^\alpha\otimes\dot{w}^\alpha\right]dt,$$
$$= \lim_{\tau\to\infty}\overset{\#}{m^\alpha}\left[\frac{1}{\tau}(\dot{w}^\alpha\otimes w^\alpha)|_0^\tau - \frac{1}{\tau}\int_0^\tau \dot{w}^\alpha\otimes\dot{w}^\alpha\,dt\right],$$
$$= -\lim_{\tau\to\infty}\frac{\overset{\#}{m^\alpha}}{\tau}\int_0^\tau \dot{w}^\alpha\otimes\dot{w}^\alpha\,dt. \tag{11.71}$$

The final expression follows by dropping the first term in the square brackets of the second line by assuming that $\dot{w}^\alpha\otimes w^\alpha$ is bounded.[20] Reverting back to a phase average and substituting Eqn. (11.71) into Eqn. (11.69), we have

$$\boldsymbol{\sigma} = \frac{1}{V}\left[\frac{1}{2}\sum_{\substack{\alpha,\beta\\ \alpha\neq\beta}}\left\langle \varphi'_{\alpha\beta}\frac{\boldsymbol{r}^{\alpha\mathring{\beta}}\otimes\boldsymbol{r}^{\alpha\mathring{\beta}}}{r^{\alpha\mathring{\beta}}}\right\rangle_\chi - \sum_\alpha\left\langle\frac{\boldsymbol{p}^\alpha\otimes\boldsymbol{p}^\alpha}{\overset{\#}{m^\alpha}}\right\rangle_\chi\right], \tag{11.72}$$

where Eqn. (11.47) was used and $\boldsymbol{p}^\alpha = \overset{\#}{m^\alpha}\dot{\boldsymbol{w}}^\alpha$ is the center-of-mass momentum of atom α. Equation (11.72) includes the expected kinetic terms and is equivalent to the standard form of the virial stress in Eqn. (8.68).

Equation (11.72) also shows that the Cauchy stress under equilibrium conditions in a multilattice crystal is symmetric. Note that the symmetry is conditional on satisfaction of Eqn. (11.55). Otherwise the potential part of the stress tensor would have terms of the form $\boldsymbol{r}^{\alpha\mathring{\beta}}\otimes(\boldsymbol{r}^{\alpha\mathring{\beta}}-(\overset{\#}{\bar{\boldsymbol{s}}^\beta}-\overset{\#}{\bar{\boldsymbol{s}}^\alpha}))$ and would not be symmetric in general! (See Exercise 11.4.)

11.3.5 Elasticity tensor

The mixed elasticity tensor is given by $D_{iJkL} = \rho_0\partial^2\psi^{\rm eff}/\partial F_{iJ}\partial F_{kL}$ (see Eqn. (2.190)). Using Eqn. (11.58), this is

$$D_{iJkL} = \rho_0\left[\frac{\partial^2\psi}{\partial F_{iJ}\partial F_{kL}} + \sum_{\lambda=0}^{N_B-1}\frac{\partial^2\psi}{\partial F_{iJ}\partial\bar{s}^\lambda_m}\frac{\partial\bar{s}^\lambda_m}{\partial F_{kL}}\right.$$
$$\left. + \sum_{\lambda=0}^{N_B-1}\left(\frac{\partial^2\psi}{\partial\bar{s}^\lambda_m\partial F_{kL}} + \sum_{\mu=0}^{N_B-1}\frac{\partial^2\psi}{\partial\bar{s}^\lambda_m\partial\bar{s}^\mu_n}\frac{\partial\bar{s}^\mu_n}{\partial F_{kL}}\right)\frac{\partial\bar{s}^\lambda_m}{\partial F_{iJ}} + \sum_{\lambda=0}^{N_B-1}\frac{\partial\psi}{\partial\bar{s}^\lambda_m}\frac{\partial^2\bar{s}^\lambda_m}{F_{iJ}F_{kL}}\right], \tag{11.73}$$

where as usual the summation convention is applied to spatial indices. This expression can be significantly simplified. To see this, we need to compute the term $\partial\bar{s}^\lambda_m/\partial F_{kL}$ and a

[20] This is similar to the assumption of "stationary motion" in the derivation of the time averaged virial theorem (see, for example, [AT10, Appendix A]) and to the assumption in the Irving–Kirkwood derivation in Section 8.2.1 that phase average arguments remain bounded at infinity.

related identity. The shifts $\bar{s}^{\lambda}_{m}$ depend on $\boldsymbol{F}$ through the minimization in Eqn. (11.54). At the minimum Eqn. $(11.53)_2$ is satisfied. As explained in footnote 15 on page 570, a special case arises for centrosymmetric crystals. In this case (assuming no phase transformation occurs), the minimum corresponds to the configuration where all shifts are zero regardless of the value of $\boldsymbol{F}$. This means that

$$\partial \bar{\boldsymbol{s}}^{\lambda}/\partial \boldsymbol{F} = \boldsymbol{0}. \tag{11.74}$$

If a phase transformation occurs or the crystal is not centrosymmetric, then $\partial \bar{\boldsymbol{s}}^{\lambda}/\partial \boldsymbol{F}$ will be nonzero. To obtain it, we follow the approach of Tadmor *et al.* [TSBK99]. Taking the partial derivative of Eqn. $(11.53)_2$ with respect to $\boldsymbol{F}$, we have in indicial form:

$$\frac{\partial^2 \psi}{\partial \bar{s}^{\lambda}_{m} \partial F_{kL}} + \sum_{\mu=0}^{N_{\mathrm{B}}-1} \frac{\partial^2 \psi}{\partial \bar{s}^{\lambda}_{m} \partial \bar{s}^{\mu}_{n}} \frac{\partial \bar{s}^{\mu}_{n}}{\partial F_{kL}} = 0, \tag{11.75}$$

where as usual the summation convention is applied to repeating spatial indices (n in this case). Inverting Eqn. (11.75), we obtain the desired derivative:

$$\frac{\partial \bar{s}^{\mu}_{n}}{\partial F_{kL}} = -\sum_{\lambda=0}^{N_{\mathrm{B}}-1} \left[\frac{\partial^2 \psi}{\partial \bar{s}^{\lambda}_{m} \partial \bar{s}^{\mu}_{n}} \right]^{-1} \frac{\partial^2 \psi}{\partial \bar{s}^{\lambda}_{m} \partial F_{kL}}. \tag{11.76}$$

We now return to Eqn. (11.73) and see how we can simplify it. The last term in this equation drops out due to Eqn. $(11.53)_2$ and the term in the round parentheses drops out due to the identity in Eqn. (11.75). This leaves

$$D_{iJkL} = \rho_0 \left[\frac{\partial^2 \psi}{\partial F_{iJ} \partial F_{kL}} + \sum_{\lambda=0}^{N_{\mathrm{B}}-1} \frac{\partial^2 \psi}{\partial F_{iJ} \partial \bar{s}^{\lambda}_{m}} \frac{\partial \bar{s}^{\lambda}_{m}}{\partial F_{kL}} \right]. \tag{11.77}$$

For a centrosymmetric crystal the sum drops out due to Eqn. (11.74). Otherwise, a more symmetric expression is obtained by replacing $\partial \bar{\boldsymbol{s}}^{\lambda}/\partial \boldsymbol{F}$ in the above equation with Eqn. (11.76):

$$D_{iJkL} = \rho_0 \left[\frac{\partial^2 \psi}{\partial F_{iJ} \partial F_{kL}} - \sum_{\lambda,\mu=0}^{N_{\mathrm{B}}-1} \left[\frac{\partial^2 \psi}{\partial \bar{s}^{\lambda}_{m} \partial \bar{s}^{\mu}_{n}} \right]^{-1} \frac{\partial^2 \psi}{\partial \bar{s}^{\lambda}_{m} \partial F_{iJ}} \frac{\partial^2 \psi}{\partial \bar{s}^{\mu}_{n} \partial F_{kL}} \right]. \tag{11.78}$$

We see that $\boldsymbol{D}$ has two terms: the first is the standard macroscopic expression, while the second is due the internal shift vectors. Depending on the symmetry of the crystal this term can be nonzero and significant even for an undistorted crystal (i.e. $\boldsymbol{F} = \boldsymbol{I}$) [TSBK99]. For example for silicon at zero temperature computed using a Stillinger–Weber potential, D_{2323} (corresponding to the c_{44} elastic constant in Voigt notation) is reduced by 49% from $0.732 \text{ meV}/\text{Å}^3$ to $0.376 \text{ meV}/\text{Å}^3$ due to internal relaxation.

Our next task is to obtain an explicit expression for $\boldsymbol{D}$ in terms of the potential energy, its derivatives and the geometry of the crystal. We follow the same procedure as for the stress tensor. First, we substitute in Eqns. (11.42) and (11.40) and note that only Z^{V} contributes to the derivatives. We use Eqns. (11.46) and (11.48) as well as the second derivatives that are not given here for reasons of brevity. After some tedious algebra, the terms appearing in Eqn. (11.78) are obtained. The first term is

$$\rho_0 \frac{\partial^2 \psi}{\partial F_{iJ} \partial F_{kL}} = \left\langle D^0_{iJkL} \right\rangle_\chi - \frac{V_0}{k_B T} \mathrm{Cov}_\chi(P^{\mathrm{inst}}_{iJ}, P^{\mathrm{inst}}_{kL}), \tag{11.79}$$

where $\mathrm{Cov}_\chi(A, B)$ is the restricted covariance of A and B,

$$\mathrm{Cov}_\chi(A, B) \equiv \langle AB \rangle_\chi - \langle A \rangle_\chi \langle B \rangle_\chi, \tag{11.80}$$

the instantaneous first Piola–Kirchhoff stress $\boldsymbol{P}^{\mathrm{inst}}$ is given below in Eqn. $(11.85)_1$ and

$$D^0_{iJkL} = \frac{1}{2V_0} \sum_{\substack{\alpha,\beta \\ \alpha \neq \beta}} \left[\frac{1}{2} \sum_{\substack{\gamma,\delta \\ \gamma \neq \delta}} \kappa^{\alpha\mathring{\beta}\gamma\mathring{\delta}} \frac{r_i^{\alpha\mathring{\beta}} r_k^{\gamma\mathring{\delta}}}{r^{\alpha\mathring{\beta}} r^{\gamma\mathring{\delta}}} \bar{R}_J^{\alpha\mathring{\beta}} \bar{R}_L^{\gamma\mathring{\delta}} + \frac{\varphi^{\alpha\mathring{\beta}}}{r^{\alpha\mathring{\beta}}} \left(\delta_{ik} - \frac{r_i^{\alpha\mathring{\beta}} r_k^{\alpha\mathring{\beta}}}{(r^{\alpha\mathring{\beta}})^2} \right) \bar{R}_J^{\alpha\mathring{\beta}} \bar{R}_L^{\alpha\mathring{\beta}} \right], \tag{11.81}$$

is the zero-temperature expression for the mixed elasticity tensor in the absence of shifts. (A similar expression for the spatial elasticity tensor was obtained by Born and is given in Eqn. (8.84).) The term $\kappa^{\alpha\mathring{\beta}\gamma\mathring{\delta}}$ appearing in Eqn. (11.81) is the *bond stiffness* defined in Eqn. (5.117).

The second term in Eqn. (11.78) is

$$\rho_0 \sum_{\lambda,\mu=0}^{N_B - 1} \left[\frac{\partial^2 \psi}{\partial \bar{s}^\lambda_m \partial \bar{s}^\mu_n} \right]^{-1} \frac{\partial^2 \psi}{\partial \bar{s}^\lambda_m \partial F_{iJ}} \frac{\partial^2 \psi}{\partial \bar{s}^\mu_n \partial F_{kL}} = \frac{1}{V_0} \sum_{\lambda,\mu=0}^{N_B - 1} \left[A^{\lambda\mu}_{mn} \right]^{-1} B^\lambda_{iJm} B^\mu_{kLn}, \tag{11.82}$$

where

$$\begin{aligned} A^{\lambda\mu}_{mn} = {} & \frac{1}{4} \sum_{\substack{\alpha,\beta \\ \alpha \neq \beta}} \sum_{\substack{\gamma,\delta \\ \gamma \neq \delta}} \left\langle \kappa^{\alpha\mathring{\beta}\gamma\mathring{\delta}} \frac{r_m^{\alpha\mathring{\beta}} r_n^{\gamma\mathring{\delta}}}{r^{\alpha\mathring{\beta}} r^{\gamma\mathring{\delta}}} \left(\delta^{\overset{\#}{\beta}\lambda} - \delta^{\overset{\#}{\alpha}\lambda} \right) \left(\delta^{\overset{\#}{\delta}\mu} - \delta^{\overset{\#}{\gamma}\mu} \right) \right\rangle_\chi \\ & + \frac{1}{2} \sum_{\substack{\alpha,\beta \\ \alpha \neq \beta}} \left\langle \frac{\varphi^{\alpha\mathring{\beta}}}{r^{\alpha\mathring{\beta}}} \left(\delta_{mn} - \frac{r_m^{\alpha\mathring{\beta}} r_n^{\alpha\mathring{\beta}}}{(r^{\alpha\mathring{\beta}})^2} \right) \left(\delta^{\overset{\#}{\beta}\lambda} - \delta^{\overset{\#}{\alpha}\lambda} \right) \left(\delta^{\overset{\#}{\beta}\mu} - \delta^{\overset{\#}{\alpha}\mu} \right) \right\rangle_\chi \\ & - \frac{(N_{\mathrm{lat}})^2}{k_B T} \mathrm{Cov}_\chi(f_m^{\lambda,\mathrm{inst}}, f_n^{\mu,\mathrm{inst}}) \end{aligned} \tag{11.83}$$

and

$$\begin{aligned} B^\lambda_{iJm} = {} & \frac{1}{4} \sum_{\substack{\alpha,\beta \\ \alpha \neq \beta}} \sum_{\substack{\gamma,\delta \\ \gamma \neq \delta}} \left\langle \kappa^{\alpha\mathring{\beta}\gamma\mathring{\delta}} \frac{r_i^{\alpha\mathring{\beta}} r_m^{\gamma\mathring{\delta}}}{r^{\alpha\mathring{\beta}} r^{\gamma\mathring{\delta}}} \bar{R}_J^{\alpha\mathring{\beta}} \left(\delta^{\overset{\#}{\delta}\lambda} - \delta^{\overset{\#}{\gamma}\lambda} \right) \right\rangle_\chi \\ & + \frac{1}{2} \sum_{\substack{\alpha,\beta \\ \alpha \neq \beta}} \left\langle \frac{\varphi^{\alpha\mathring{\beta}}}{r^{\alpha\mathring{\beta}}} \left(\delta_{im} - \frac{r_i^{\alpha\mathring{\beta}} r_m^{\alpha\mathring{\beta}}}{(r^{\alpha\mathring{\beta}})^2} \right) \bar{R}_J^{\alpha\mathring{\beta}} \left(\delta^{\overset{\#}{\beta}\lambda} - \delta^{\overset{\#}{\alpha}\lambda} \right) \right\rangle_\chi \\ & + \frac{V_0 N_{\mathrm{lat}}}{k_B T} \mathrm{Cov}_\chi(P^{\mathrm{inst}}_{iJ}, f_m^{\lambda,\mathrm{inst}}). \end{aligned} \tag{11.84}$$

The instantaneous first Piola–Kirchhoff stress and mean force terms appearing in the above expressions are defined as

$$P_{iJ}^{\text{inst}} \equiv \frac{1}{2V_0} \sum_{\substack{\alpha,\beta \\ \alpha \neq \beta}} \varphi'^{\alpha\beta} \frac{\mathring{r}_i^{\alpha\beta} \bar{R}_J^{\alpha\beta}}{\mathring{r}^{\alpha\beta}}, \qquad f_i^{\lambda,\text{inst}} \equiv -\frac{1}{2N_{\text{lat}}} \sum_{\substack{\alpha,\beta \\ \alpha \neq \beta}} \varphi'^{\alpha\beta} \frac{\mathring{r}_i^{\alpha\beta}}{\mathring{r}^{\alpha\beta}} \left(\delta^{\overset{\#}{\beta}\lambda} - \delta^{\overset{\#}{\alpha}\lambda} \right). \tag{11.85}$$

For a centrosymmetric (csym) crystal, the shift-related term in Eqn. (11.78) is identically zero, and the elasticity tensor reduces to the expression in Eqn. (11.79),

$$D_{iJKL}^{\text{csym}} = \langle D_{iJkL}^{0} \rangle_{\chi} - \frac{V_0}{k_B T} \operatorname{Cov}_{\chi}(P_{iJ}^{\text{inst}}, P_{kL}^{\text{inst}}), \tag{11.86}$$

where $\boldsymbol{D}^0$ and $\boldsymbol{P}^{\text{inst}}$ are given in Eqns. (11.81) and $(11.85)_1$. An expression for the spatial elasticity tensor can be obtained by using Eqn. $(2.195)_2$. This is done in Section 8.1.4 for the general statistical mechanics expressions which include the kinetic terms. Recall that the relations derived in this section assume a restricted ensemble and are expressed in terms of *mean* atomic positions in the reference configuration. These relations could be adjusted to have the same form as those in Section 8.1.4 by recovering the kinetic terms as was done for the stress tensor in the previous section.

11.4 Quasiharmonic approximation

The restricted ensemble approach adopted so far in this chapter was based on the idea that atoms in a solid remain close to well-defined mean positions. We can further capitalize on this observation by adding the assumption, correct at low to moderate temperatures, that the displacements of the atoms from their mean positions are *small*. Consequently the potential energy appearing in the partition function can be replaced with its Taylor expansion, retaining terms only up to second order in displacement. The resulting expression can be integrated exactly to give us the *quasiharmonic approximation*[21] for the free energy and its derivatives with respect to deformation – the stress and elasticity tensors. In this section we derive the quasiharmonic approximation for the general case of arbitrary multilattice crystals. In the case of noncentrosymmetric crystals, a subtle point related to the force terms in the potential energy expansion is discussed and clarified.

11.4.1 Quasiharmonic Helmholtz free energy

The derivation starts with the Helmholtz free energy Ψ which is given in Eqn. (11.40) and consists of kinetic and potential parts, $\Psi = \Psi^{\text{K}} + \Psi^{\text{V}}$, where

$$\Psi^{\text{K}} = -k_B T \ln Z^{\text{K}}, \qquad \Psi^{\text{V}} = -k_B T \ln Z^{\text{V}} \tag{11.87}$$

[21] The distinction between this approach and the more approximate *strict harmonic approximation* is made in Section 11.4.4.

and where Z^{K} and Z^{V} are the kinetic and potential (configurational) parts of the partition function given in Eqn. (11.41). As described in Section 11.3.1, we consider a periodic system where the periodic cell contains N_{lat} lattice sites with N_{B} basis atoms for a total of $N = N_{\mathrm{lat}}N_{\mathrm{B}}$ atoms. The partition function expressions are

$$Z^{\mathrm{K}} = \left(\frac{2\pi k_{\mathrm{B}}T}{h^2}\right)^{3N/2}\left(\prod_{\lambda=0}^{N_{\mathrm{B}}-1} m^{\lambda}\right)^{3N_{\mathrm{lat}}/2}, \quad Z^{\mathrm{V}} = \int e^{-\mathcal{V}^{\mathrm{int}}(\bar{\boldsymbol{q}}+\boldsymbol{w})/k_{\mathrm{B}}T}\chi(\boldsymbol{w})\,d\boldsymbol{w}. \tag{11.88}$$

We recall that $\bar{\boldsymbol{q}}$ represents the concatenated mean positions and that these depend on the deformation gradient $\boldsymbol{F}$ and shifts $\{\bar{\boldsymbol{s}}^{\lambda}\}$ through Eqn. (11.30). For convenience, we write the $\mathcal{V}^{\mathrm{int}}$ as a function of positions, although as explained in Section 11.3.2, we realize that ultimately the dependence is through interatomic distances.

From Eqns. (11.87) and (11.88), we see that the kinetic free energy is in closed form; however, the configurational free energy involves a complicated multi-dimensional integral which cannot be analytically evaluated for an arbitrary potential energy function. To make progress, we replace $\mathcal{V}^{\mathrm{int}}(\bar{\boldsymbol{q}}+\boldsymbol{w})$ with its Taylor expansion retaining only terms to second order:

$$\begin{aligned}\mathcal{V}^{\mathrm{int}}(\bar{\boldsymbol{q}}+\boldsymbol{w}) &\approx \mathcal{V}^{\mathrm{int}}(\bar{\boldsymbol{q}}) + \left.\frac{\partial\mathcal{V}^{\mathrm{int}}}{\partial q_i}\right|_{\boldsymbol{q}=\bar{\boldsymbol{q}}} w_i + \frac{1}{2}\left.\frac{\partial^2\mathcal{V}^{\mathrm{int}}}{\partial q_i\partial q_j}\right|_{\boldsymbol{q}=\bar{\boldsymbol{q}}} w_i w_j \\ &= \mathcal{V}^{\mathrm{int}}(\bar{\boldsymbol{q}}) - f_i^{\mathrm{int}}(\bar{\boldsymbol{q}})w_i + \frac{1}{2}\Phi_{ij}(\bar{\boldsymbol{q}})w_i w_j,\end{aligned} \tag{11.89}$$

where $\boldsymbol{f}^{\mathrm{int}} \in \mathbb{R}^{3N}$ is the concatenated vector of forces on the atoms and $\boldsymbol{\Phi} \in \mathbb{R}^{3N\times 3N}$ is the force constant matrix:

$$\Phi_{ij} \equiv \frac{\partial^2\mathcal{V}^{\mathrm{int}}}{\partial q_i\partial q_j}. \tag{11.90}$$

(An expression for the force constant matrix in terms of interatomic distances is given in Exercise 11.5.) As usual the summation convention is applied to spatial indices.

Let us divide the potential energy in Eqn. (11.89) into two parts:

$$\mathcal{V}^{\mathrm{int}}(\bar{\boldsymbol{q}}+\boldsymbol{w}) \approx \mathcal{V}^{\mathrm{int}}_{\mathrm{qh}}(\boldsymbol{w};\bar{\boldsymbol{q}}) - \mathcal{W}(\boldsymbol{w};\bar{\boldsymbol{q}}), \tag{11.91}$$

where $\mathcal{V}^{\mathrm{int}}_{\mathrm{qh}}(\boldsymbol{w};\bar{\boldsymbol{q}})$ includes the constant and quadratic terms,

$$\mathcal{V}^{\mathrm{int}}_{\mathrm{qh}}(\boldsymbol{w};\bar{\boldsymbol{q}}) \equiv \mathcal{V}^{\mathrm{int}}(\bar{\boldsymbol{q}}) + \frac{1}{2}\Phi_{ij}(\bar{\boldsymbol{q}})w_i w_j, \tag{11.92}$$

and $\mathcal{W}(\boldsymbol{w};\bar{\boldsymbol{q}})$ includes the force terms,

$$\mathcal{W}(\boldsymbol{w};\bar{\boldsymbol{q}}) \equiv f_i^{\mathrm{int}}(\bar{\boldsymbol{q}})w_i. \tag{11.93}$$

Except for the special case where the mean positions form a centrosymmetric crystal (Section 11.2.3), the force terms in $\mathcal{W}(\boldsymbol{w};\bar{\boldsymbol{q}})$ will *not* be zero in general. This can happen because the mean positions are obtained by minimizing the *free energy*, not the potential energy of the system. However, we argue below that these forces are small and, as a result, we can show (in Eqn. (11.96)) that $\mathcal{W}$ can be neglected when computing phase averages.

Substituting Eqn. (11.91) in Eqn. $(11.88)_2$, the configurational integral is

$$Z^{\mathrm{V}} = \int e^{-\mathcal{V}_{\mathrm{qh}}^{\mathrm{int}}(\boldsymbol{w};\bar{\boldsymbol{q}})/k_{\mathrm{B}}T} e^{\mathcal{W}(\boldsymbol{w};\bar{\boldsymbol{q}})/k_{\mathrm{B}}T} \chi(\boldsymbol{w})\, d\boldsymbol{w}. \tag{11.94}$$

To evaluate this expression, we make two approximations. First, we drop the χ constraint from the integral so that the integration bounds on each degree of freedom are $-\infty$ to ∞. This can be justified by noting that for $k_{\mathrm{B}}T \ll 1$ the term $e^{-\mathcal{V}_{\mathrm{qh}}^{\mathrm{int}}/k_{\mathrm{B}}T}$ effectively bounds the atoms to remain close to $\bar{\boldsymbol{q}}$ since the probability of finding a configuration away from $\bar{\boldsymbol{q}}$ decays like $e^{-w^2/k_{\mathrm{B}}T}$. Second, we assume that $\mathcal{W}(\boldsymbol{w};\bar{\boldsymbol{q}})$ is a small number close to zero for small values of $\boldsymbol{w}$ (which is the portion of configuration space contributing significantly to the phase average). The physical reasoning underlying this assumption is that for $k_{\mathrm{B}}T \ll 1$, $\bar{\boldsymbol{q}}$ is close to some configuration $\boldsymbol{q}^0$ which corresponds to a minimum of the potential energy, where $\mathcal{W}(\boldsymbol{w};\boldsymbol{q}^0) = 0$. We can therefore expand the second exponent about $\mathcal{W} = 0$. The quasiharmonic approximation to the configuration free energy is then

$$\begin{aligned} Z_{\mathrm{qh}}^{\mathrm{V}}(\bar{\boldsymbol{q}},T) &= \int e^{-\mathcal{V}_{\mathrm{qh}}^{\mathrm{int}}(\boldsymbol{w};\bar{\boldsymbol{q}})/k_{\mathrm{B}}T} \left[1 + \frac{\mathcal{W}(\boldsymbol{w};\bar{\boldsymbol{q}})}{k_{\mathrm{B}}T} + \frac{1}{2!}\left(\frac{\mathcal{W}(\boldsymbol{w};\bar{\boldsymbol{q}})}{k_{\mathrm{B}}T}\right)^2 + \cdots\right] d\boldsymbol{w} \\ &= \int e^{-\mathcal{V}_{\mathrm{qh}}^{\mathrm{int}}(\boldsymbol{w};\bar{\boldsymbol{q}})/k_{\mathrm{B}}T}\, d\boldsymbol{w} + \frac{1}{k_{\mathrm{B}}T}\int \mathcal{W}(\boldsymbol{w};\bar{\boldsymbol{q}}) e^{-\mathcal{V}_{\mathrm{qh}}^{\mathrm{int}}(\boldsymbol{w};\bar{\boldsymbol{q}})/k_{\mathrm{B}}T}\, d\boldsymbol{w} + O(\mathcal{W}^2). \end{aligned} \tag{11.95}$$

The second integral is zero because the integrand is an odd function of $\boldsymbol{w}$. Therefore,

$$Z_{\mathrm{qh}}^{\mathrm{V}}(\bar{\boldsymbol{q}},T) = \int e^{-\mathcal{V}_{\mathrm{qh}}^{\mathrm{int}}(\boldsymbol{w};\bar{\boldsymbol{q}})/k_{\mathrm{B}}T}\, d\boldsymbol{w} + O(\mathcal{W}^2). \tag{11.96}$$

We see that up to second order in $\mathcal{W}$ the force terms in Eqn. (11.89) do not contribute to the configurational integral.[22] (The result in Eqn. (11.96) is exact for the special case of centrosymmetric crystals where the force terms are identically zero as shown in Section 11.2.3.)

We must now evaluate the integral in Eqn. (11.96). Substituting in Eqn. (11.92), we have

$$Z_{\mathrm{qh}}^{\mathrm{V}}(\bar{\boldsymbol{q}},T) = e^{-\mathcal{V}^{\mathrm{int}}(\bar{\boldsymbol{q}})/k_{\mathrm{B}}T} \int_{-\infty}^{\infty} \cdots \int_{-\infty}^{\infty} e^{-\Phi_{ij}(\bar{\boldsymbol{q}})w_i w_j/2k_{\mathrm{B}}T}\, dw_1 \ldots dw_{3N}. \tag{11.97}$$

The integration of Eqn. (11.97) is facilitated by a suitable change of basis that renders Φ_{ij} diagonal. We know from linear algebra that the principal basis of a symmetric matrix like Φ_{ij} is obtained by solving the eigenvalue equation (see Section 2.1.4):

$$\boldsymbol{\Phi}\boldsymbol{\Lambda} = \lambda\boldsymbol{\Lambda}. \tag{11.98}$$

A nontrivial solution requires that

$$\det(\boldsymbol{\Phi} - \lambda \boldsymbol{I}) = \boldsymbol{0}. \tag{11.99}$$

[22] This subtlety in the quasiharmonic derivation is often not appreciated. Most books drop the force terms from the start in Eqn. (11.89) even though this is only correct for the *strict* harmonic approximation described in Section 11.4.4. In that case, the harmonic expansion is performed about a minimum of the *potential* energy and so the force terms drop out. But as we saw above, the case is a bit more involved for the quasi-harmonic model.

For a symmetric matrix, Eqn. (11.99) has $3N$ real solutions, called the eigenvalues of $\boldsymbol{\Phi}$, which we denote λ_r $(r = 1, .., 3N)$. Each eigenvalue[23] has a corresponding eigenvector $\boldsymbol{\Lambda}_r$ obtained by solving Eqn. (11.98) together with the additional normalization condition, $\|\boldsymbol{\Lambda}_r\| = 1$. As shown on page 39, for a symmetric matrix the set of eigenvectors form an orthonormal basis, called the *principal basis*, relative to which $\boldsymbol{\Phi}$ is diagonal with components equal to its eigenvalues:[24]

$$[\boldsymbol{\Phi}] = \begin{bmatrix} \lambda_1 & 0 & \cdots\cdots & 0 \\ 0 & \lambda_2 & 0 \ \cdots & 0 \\ \cdots & \cdots & \cdots & \cdots \\ 0 & \cdots\cdots & 0 & \lambda_{3N} \end{bmatrix}. \tag{11.100}$$

The eigenvectors satisfy the orthonormality condition and the completeness relation (see Eqns. (2.42) and (2.43)):

$$\Lambda_{ri}\Lambda_{si} = \delta_{rs}, \qquad \Lambda_{ri}\Lambda_{rj} = \delta_{ij}, \tag{11.101}$$

where Λ_{ri} is the ith component of $\boldsymbol{\Lambda}_r$, and the summation convention applies to all indices. Equations $(11.101)_1$ and $(11.101)_2$ together imply that Λ_{ri} is orthogonal, a fact that will be used below.

We can now perform a change of variables in Eqn. (11.97) to the new basis. Adopting vector notation, we have that

$$\boldsymbol{w} = w_i \boldsymbol{e}_i = \mu_r \boldsymbol{\Lambda}_r, \tag{11.102}$$

where $\boldsymbol{e}_i$ are the original basis vectors and μ_r are the displacement coordinates relative to the principal basis. Dotting Eqn. (11.102) with $\boldsymbol{\Lambda}_s$ and using Eqn. $(11.101)_1$ gives

$$\mu_s = (\boldsymbol{\Lambda}_s \cdot \boldsymbol{e}_i) w_i = \Lambda_{si} w_i, \qquad w_j = \Lambda_{sj}\mu_s, \tag{11.103}$$

where the second inverse relation follows by using the completeness relation in Eqn. $(11.101)_2$. Substituting Eqn. $(11.103)_2$ into Eqn. (11.97) and noting that the Jacobian of the transformation is unity because Λ_{ri} is orthogonal, we have

$$Z^{\mathrm{V}}_{\mathrm{qh}}(\bar{\boldsymbol{q}}, T) = e^{-\mathcal{V}^{\mathrm{int}}(\bar{\boldsymbol{q}})/k_B T} \int_{-\infty}^{\infty} \cdots \int_{-\infty}^{\infty} e^{-\sum_{r=1}^{n_{\mathrm{nz}}} \lambda_r(\bar{\boldsymbol{q}})\mu_r^2/2k_B T} \, d\mu_1 \ldots d\mu_{n_{\mathrm{nz}}}, \tag{11.104}$$

where n_{nz} is the number of nonzero eigenvalues[23]. For a three-dimensional periodic crystal, $n_{\mathrm{nz}} = 3N - 3$. Equation (11.104) can be integrated analytically with the result:

$$Z^{\mathrm{V}}_{\mathrm{qh}}(\bar{\boldsymbol{q}}, T) = e^{-\mathcal{V}^{\mathrm{int}}(\bar{\boldsymbol{q}})/k_B T} (2\pi k_B T)^{n_{\mathrm{nz}}/2} \left[\prod_{r=1}^{n_{\mathrm{nz}}} \lambda_r(\bar{\boldsymbol{q}}) \right]^{-1/2}. \tag{11.105}$$

[23] Due to translational invariance of the interatomic potential, we expect that in a d-dimensional space, d of the eigenvalues will be zero. (This is demonstrated for a one-dimensional case in Example 11.2.) In general, there are also zero modes associated with rigid-body rotation, but these are eliminated in the present case by the application of periodic boundary conditions. We order the eigenvalues so that the zero eigenvalues are last in the list. Later, we remove the zero eigenvalues when computing the partition function.

[24] The eigenvalues and eigenvectors of $\boldsymbol{\Phi}$ have important physical meanings related to the lattice dynamics of a crystal. This need not concern us here, however, the interested reader is referred to the very accessible introductory book on the subject by Dove [Dov93].

The quasiharmonic approximation for Ψ^{V} follows from Eqn. $(11.87)_2$ as

$$\Psi^{\mathrm{V}}_{\mathrm{qh}}(\bar{q},T) = \mathcal{V}^{\mathrm{int}}(\bar{q}) + \frac{k_{\mathrm{B}}T}{2}\ln\left[\frac{\prod_{r=1}^{n_{\mathrm{nz}}}\lambda_r(\bar{q})}{(2\pi k_{\mathrm{B}}T)^{n_{\mathrm{nz}}}}\right]. \tag{11.106}$$

Since the product of the eigenvalues of a matrix is equal to its determinant, it is common to rewrite Eqn. (11.106) as

$$\Psi^{\mathrm{V}}_{\mathrm{qh}}(\bar{q},T) = \mathcal{V}^{\mathrm{int}}(\bar{q}) + \frac{k_{\mathrm{B}}T}{2}\ln\left[\frac{\det \boldsymbol{\Phi}^*(\bar{q})}{(2\pi k_{\mathrm{B}}T)^{n_{\mathrm{nz}}}}\right], \tag{11.107}$$

where

$$\det \boldsymbol{\Phi}^*(\bar{q}) = \prod_{r=1}^{n_{\mathrm{nz}}}\lambda_r(\bar{q}). \tag{11.108}$$

However, it is important to understand that the force constant matrix appearing in this relation must have its rigid-body modes removed, i.e.[25]

$$\boldsymbol{\Phi}^* = \sum_{r=1}^{n_{\mathrm{nz}}}\lambda_r \boldsymbol{\Lambda}_r \otimes \boldsymbol{\Lambda}_r. \tag{11.109}$$

The total quasiharmonic free energy is then $\Psi_{\mathrm{qh}}(\bar{q}) = \Psi^{\mathrm{K}} + \Psi^{\mathrm{V}}_{\mathrm{qh}}(\bar{q})$, where $\Psi^{\mathrm{V}}_{\mathrm{qh}}$ is given in either Eqn. (11.106) or Eqn. (11.107) and Ψ^{K} follows from Eqns. $(11.87)_1$ and $(11.88)_1$.[26]

11.4.2 Determination of the quasiharmonic reference configuration

The reference configuration at a given temperature T corresponds to the minimum of the free energy with respect to the lattice vectors $\{\boldsymbol{A}_i(T)\}$ and mean basis atom positions $\{\bar{\boldsymbol{Z}}^\lambda(T)\}$ as described in Algorithm 11.1. The minimization process requires the gradients of the free energy with respect to the deformation gradient $\boldsymbol{F}$ and mean shifts $\{\bar{\boldsymbol{s}}^\lambda\}$. Since the kinetic part of the free energy does not depend on positions, we need only differentiate $\Psi^{\mathrm{V}}_{\mathrm{qh}}$. First, let us compute the derivative of $\Psi^{\mathrm{V}}_{\mathrm{qh}}$ with respect to the $\mathring{r}^{\alpha\beta}$ which we will require repeatedly:

$$\begin{aligned}\frac{\partial \Psi^{\mathrm{V}}_{\mathrm{qh}}}{\partial \mathring{r}^{\alpha\beta}} &= \frac{\partial \mathcal{V}^{\mathrm{int}}}{\partial \mathring{r}^{\alpha\beta}} + \frac{k_{\mathrm{B}}T}{2}\frac{(2\pi k_{\mathrm{B}}T)^{n_{\mathrm{nz}}}}{\prod_{t=1}^{n_{\mathrm{nz}}}\lambda_t}\frac{\sum_{s=1}^{n_{\mathrm{nz}}}\frac{\partial \lambda_s}{\partial \mathring{r}^{\alpha\beta}}\prod_{t=1,t\neq s}^{n_{\mathrm{nz}}}\lambda_t}{(2\pi k_{\mathrm{B}}T)^{n_{\mathrm{nz}}}} \\ &= \mathring{\varphi}^{\alpha\beta} + \frac{k_{\mathrm{B}}T}{2}\sum_{s=1}^{n_{\mathrm{nz}}}\frac{1}{\lambda_s}\frac{\partial \lambda_s}{\partial \mathring{r}^{\alpha\beta}} \\ &= \mathring{\varphi}^{\alpha\beta} + \frac{k_{\mathrm{B}}T}{2}\sum_{s=1}^{n_{\mathrm{nz}}}\frac{\partial(\ln \lambda_s)}{\partial \mathring{r}^{\alpha\beta}} = \mathring{\varphi}^{\alpha\beta} + \frac{k_{\mathrm{B}}T}{2}\frac{\partial(\ln \boldsymbol{\Phi}^*)}{\partial \mathring{r}^{\alpha\beta}}.\end{aligned} \tag{11.110}$$

[25] The relation in Eqn. (11.109) comes from the spectral decomposition of $\boldsymbol{\Phi}$. See page 39.

[26] The kinetic partition function in $(11.88)_1$ must be modified to remove the terms associated with gross motion of the entire system, which corresponds to the zero eigenmodes of the internal potential energy function. See Exercise 11.6.

Now we compute $\partial\Psi^{\mathrm{V}}_{\mathrm{qh}}/\partial\boldsymbol{F}$ and $\partial\Psi^{\mathrm{V}}_{\mathrm{qh}}/\partial\bar{\boldsymbol{s}}^{\lambda}$ using Eqn. (11.110) and the derivatives in Eqn. (11.39):

$$\frac{\partial\Psi^{\mathrm{V}}_{\mathrm{qh}}}{\partial\boldsymbol{F}}=\sum_{\substack{\alpha,\beta\\ \alpha<\beta}}\frac{\partial\Psi^{\mathrm{V}}_{\mathrm{qh}}}{\partial r^{\alpha\mathring{\beta}}}\frac{\partial r^{\alpha\mathring{\beta}}}{\partial\boldsymbol{F}}=\sum_{\substack{\alpha,\beta\\ \alpha<\beta}}\left[\varphi^{\alpha\mathring{\beta}}+\frac{k_{\mathrm{B}}T}{2}\frac{\partial(\ln\boldsymbol{\Phi}^{*})}{\partial r^{\alpha\mathring{\beta}}}\right]\frac{\boldsymbol{r}^{\alpha\mathring{\beta}}\otimes\bar{\boldsymbol{R}}^{\alpha\beta}}{r^{\alpha\mathring{\beta}}}\tag{11.111}$$

$$\frac{\partial\Psi^{\mathrm{V}}_{\mathrm{qh}}}{\partial\bar{\boldsymbol{s}}^{\lambda}}=\sum_{\substack{\alpha,\beta\\ \alpha<\beta}}\frac{\partial\Psi^{\mathrm{V}}_{\mathrm{qh}}}{\partial r^{\alpha\mathring{\beta}}}\frac{\partial r^{\alpha\mathring{\beta}}}{\partial\bar{\boldsymbol{s}}^{\lambda}}=\sum_{\substack{\alpha,\beta\\ \alpha<\beta}}\left[\varphi^{\alpha\mathring{\beta}}+\frac{k_{\mathrm{B}}T}{2}\frac{\partial(\ln\boldsymbol{\Phi}^{*})}{\partial r^{\alpha\mathring{\beta}}}\right]\frac{\boldsymbol{r}^{\alpha\mathring{\beta}}}{r^{\alpha\mathring{\beta}}}(\delta^{\#}_{\mathring{\beta}\lambda}-\delta^{\#}_{\mathring{\alpha}\lambda}).\tag{11.112}$$

Given the gradients, Algorithm 11.1 can be applied to obtain the reference configuration.

The zero static internal stress approximation (ZSISA) approach An approximation to the full minimization in Algorithm 11.1, which is often applied in quasiharmonic calculations, is to perform the minimization with respect to the shifts using only the potential energy and then to obtain the reference configuration by minimizing the free energy with respect to the deformation gradient. This approach, referred to as the "zero static internal stress approximation" (ZSISA) [ABB96] or the "statically constrained" approach [WYW10], is often applied in first-principles quasiharmonic calculations due to the computational expense of the energy evaluations.

Mathematically, ZSISA works like this. First, the shifts are obtained for a given deformation gradient by minimizing the potential energy (exactly as was done for the free energy in Eqn. (11.54)):

$$\bar{\boldsymbol{s}}^{\lambda}(\boldsymbol{F})=\arg\min_{\{\mathring{\boldsymbol{s}}^{\lambda}\}}\mathcal{V}^{\mathrm{int}}(\bar{\boldsymbol{q}}(\boldsymbol{F},\{\mathring{\boldsymbol{s}}^{\lambda}\})).\tag{11.113}$$

Then an effective potential energy function which depends only on the deformation gradient is defined:

$$\mathcal{V}^{\mathrm{int}}_{\mathrm{eff}}(\boldsymbol{F})\equiv\mathcal{V}^{\mathrm{int}}(\bar{\boldsymbol{q}}(\boldsymbol{F},\{\bar{\boldsymbol{s}}^{\lambda}(\boldsymbol{F})\})).\tag{11.114}$$

The "ZSISA" configurational free energy is obtained by replacing $\mathcal{V}^{\mathrm{int}}$ in Eqn. (11.106) with $\mathcal{V}^{\mathrm{int}}_{\mathrm{eff}}$:

$$\Psi^{\mathrm{V,ZSISA}}_{\mathrm{qh}}(\boldsymbol{F},T)=\mathcal{V}^{\mathrm{int}}_{\mathrm{eff}}(\boldsymbol{F})+\frac{k_{\mathrm{B}}T}{2}\ln\left[\frac{\prod_{r=1}^{n_{\mathrm{nz}}}\lambda^{\mathrm{eff}}_{r}(\boldsymbol{F})}{(2\pi k_{\mathrm{B}}T)^{n_{\mathrm{nz}}}}\right],\tag{11.115}$$

where $\lambda^{\mathrm{eff}}_{r}$ are the eigenvalues of $\partial^{2}\mathcal{V}^{\mathrm{int}}_{\mathrm{eff}}/\partial q_i\partial q_j$. Finally, the reference configuration is obtained by minimizing $\Psi^{\mathrm{V,ZSISA}}_{\mathrm{qh}}$ with respect to $\boldsymbol{F}$. The minimization is usually done with numerical derivatives due to the complexity of the definition of $\Psi^{\mathrm{V,ZSISA}}_{\mathrm{qh}}$, i.e.

$$\frac{\partial\Psi^{\mathrm{V,ZSISA}}_{\mathrm{qh}}}{\partial F_{iJ}}(\boldsymbol{F},T)\approx\frac{\Psi^{\mathrm{V,ZSISA}}_{\mathrm{qh}}(F_{iJ}+\epsilon,T)-\Psi^{\mathrm{V,ZSISA}}_{\mathrm{qh}}(F_{iJ}-\epsilon,T)}{2\epsilon},\tag{11.116}$$

where ϵ is a small number.

Note that the minimization in Eqn. (11.113) can depend on the initial guess used for the shifts, and therefore both $\mathcal{V}^{\mathrm{int}}_{\mathrm{eff}}(\boldsymbol{F})$ and $\Psi^{\mathrm{V,ZSISA}}_{\mathrm{qh}}(\boldsymbol{F},T)$ can be multivalued functions of $\boldsymbol{F}$. In an implementation of ZSISA this can be addressed by performing a preprocessing step and computing the sets of shifts $\{\bar{s}^\lambda\}$ in Eqn. (11.113) for a series of values of F relevant to some physical process (for example a range of volumes for the case of hydrostatic loading). Subsequently, when computing $\mathcal{V}^{\text{int}}_{\text{eff}}(F)$ during the minimization process, the stored set of shifts associated with the closest value of F is used as the initial guess.

The accuracy of the ZSISA approach is discussed by Allan *et al.* [ABB96]. They show analytically that the minimizing deformation gradient ("external strain") is accurate to first order in temperature. The minimizing shifts ("internal strain"), however, are not as accurate and this can have important consequences on the predictions of a quasiharmonic calculations. We show the difference between the full minimization of Algorithm 11.1 and the ZSISA approach for a one-dimensional chain in the following example.

Example 11.1 (Reference configuration of a four-lattice chain of atoms) We consider a periodic system consisting of N_{lat} unit cells (we explore the effect of N_{lat} on the results below) each containing $N_{\text{B}} = 4$ atoms. Basis atoms 0 and 2 are of species "a" and basis atoms 1 and 3 are of species "b." The atoms interact via a modified Lennard-Jones potential:

$$\phi(r) = \begin{cases} \varphi(r) + Ar^2 + B & r \leq r_{\text{cut}}, \\ 0 & r > r_{\text{cut}}, \end{cases}$$

where

$$\varphi(r) = 4\varepsilon\left[\left(\frac{\sigma}{r}\right)^{12} - \left(\frac{\sigma}{r}\right)^{6}\right],$$

$$A = -\frac{1}{2r_{\text{cut}}}\varphi'(r_{\text{cut}}), \qquad B = \frac{r_{\text{cut}}}{2}\varphi'(r_{\text{cut}}) - \varphi(r_{\text{cut}}).$$

The term $Ar^2 + B$ ensures that the potential and its first derivative go smoothly to zero at the cutoff radius r_{cut}. Three sets of parameters are given for a–a, b–b and a–b interactions:[27]

$$\varepsilon_{\text{aa}} = 1.0 \qquad \varepsilon_{\text{bb}} = 1.0 \qquad \varepsilon_{\text{ab}} = 0.1,$$
$$\sigma_{\text{aa}} = 1.25 \qquad \sigma_{\text{bb}} = 0.85 \qquad \sigma_{\text{ab}} = 0.475,$$

where ε and σ have arbitrary units of energy and length respectively. The cutoff radius is $r_{\text{cut}} = 2.0$ for all three potentials. The potential energy of the chain is

$$\mathcal{V}^{\text{int}} = \frac{1}{2}\sum_{\substack{\alpha,\beta \\ \alpha\neq\beta}} \phi(r^{\alpha\mathring{\beta}}; \text{spec}(\alpha), \text{spec}(\beta)),$$

where $\text{spec}(\alpha)$ is the species ("a" or "b") of atom α. The species determines which set of parameters is used in the function $\varphi(r)$. The force constant matrix for this chain is

$$\Phi^{\alpha\beta} = \frac{\partial^2\mathcal{V}^{\text{int}}}{\partial r^\alpha \partial r^\beta} = \begin{cases} \sum_{\substack{\gamma \\ \gamma\neq\alpha}} \phi''(r^{\alpha\mathring{\gamma}}; \text{spec}(\alpha), \text{spec}(\gamma)) & \alpha = \beta, \\ -\phi''(r^{\alpha\mathring{\beta}}; \text{spec}(\alpha), \text{spec}(\beta)) & \alpha \neq \beta. \end{cases}$$

Note that for $\alpha \neq \beta$, $\Phi^{\alpha\beta} = 0$ if $r^{\alpha\mathring{\beta}} > r_{\text{cut}}$.

[27] These parameters constitute a slight modification of the parameters of a potential given in a paper by Dobson *et al.* [DELT07]. The Dobson *et al.* potential leads to a one-dimensional two-lattice crystal which is centrosymmetric in the reference configuration. At a critical strain this crystal undergoes a phase transformation to a noncentrosymmetric four-lattice crystal structure. In contrast, the current potential has a noncentrosymmetric structure from the start in the reference configuration as will be presently seen.

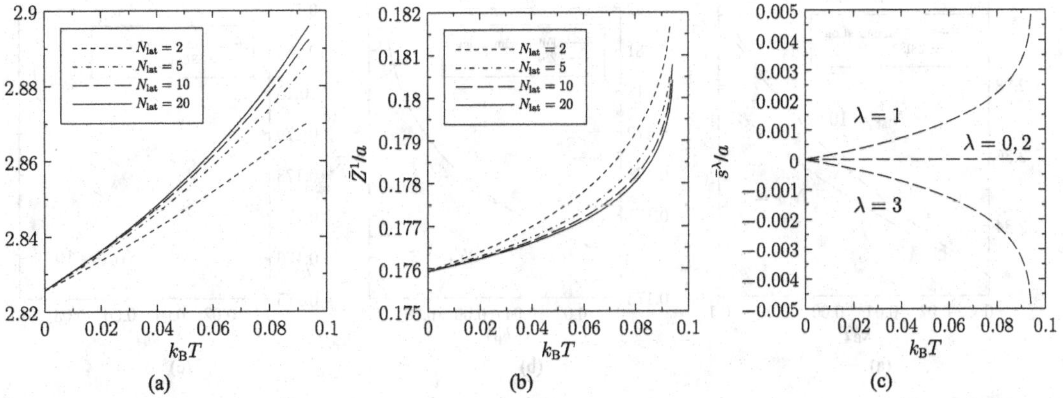

Fig. 11.5 The effect of supercell size on the quasiharmonic reference configuration obtained using full minimization for the four-lattice chain of atoms in Example 11.1. (a) Lattice parameter $a(T)$ as a function of k_BT in arbitrary units of length and energy. The different lines correspond to different supercell sizes as indicated by the key. (b) Normalized position of basis atom 1, $\bar{Z}^1/a$, as a function of k_BT. (c) The normalized shifts of all basis atoms $\{\bar{s}^\lambda/a\}$ for $N_{lat} = 10$. The shifts are measured relative to the positions of the basis atoms at $k_BT = 0$ which are $\{\bar{Z}^\lambda\} = (0, 0.1759, 0.5, 0.8241) \times a(0)$, where $a(0) = 2.826$.

Fig. 11.6 The reference configuration at $k_BT = 0$ of the four-lattice chain in Example 11.1 for $N_{lat} = 10$. Black atoms are of species "a" and white atoms are of species "b". The box indicates the periodic supercell.

Our objective is to compute the reference configuration as a function of temperature, i.e. the lattice parameter $a(T)$ and the positions of the basis atoms $\{\bar{Z}^\lambda(T)\}$. We use the full minimization approach in Algorithm 11.1. The results are plotted in Fig. 11.5. The lattice parameter and one of the shifts are shown in Figs. 11.5(a) and 11.5(b) as a function of temperature for different values of N_{lat} which sets the supercell size. We see that as anticipated in Section 11.3.1 there is a strong dependence on N_{lat}. For the present problem, a choice of $N_{lat} = 10$ is a reasonable compromise relative to larger more expensive calculations, giving close to converged results. The normalized shifts of all the basis atoms relative to their positions at $k_BT = 0$ for this supercell size are shown in Fig. 11.5(c). The four-lattice structure for $k_BT = 0$ is illustrated in Fig. 11.6, clearly showing the noncentrosymmetric structure.

Next, we compare the results from the ZSISA approach with those of the exact full minimization given above. The results are shown in Fig. 11.7. The comparison is made for two of the supercell sizes considered above: $N_{lat} = 2$ and $N_{lat} = 10$. Only the position of basis atom 1 is plotted since basis atoms 0 and 4 are fixed at their zero-temperature positions and basis atom 3 is symmetry related to 1. We see that for both values of N_{lat}, the ZSISA value for the lattice parameter provides a good approximation to the one obtained from the full minimization. However, the ZSISA approximation to the position of basis atom 1 is only good for the smaller supercell size ($N_{lat} = 2$). For the larger supercell the ZSISA prediction is very poor (and of course this is the more relevant comparison since $N_{lat} = 10$ is the converged result). The explanation for this is that in ZSISA the shifts are obtained from Eqn. (11.113) by minimizing the potential energy which is insensitive to the supercell size as all unit cells experience identical deformation in the static limit. These results are in agreement with the analytical results of Allan *et al.* [ABB96] which show that the minimum obtained from ZSISA is

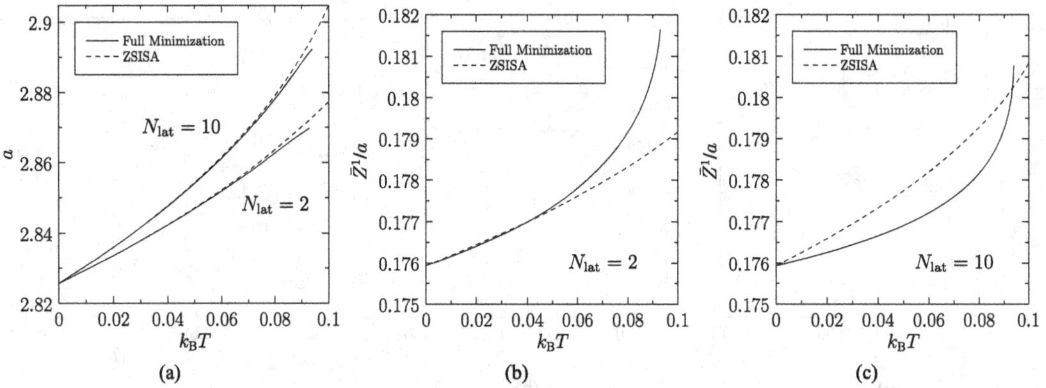

Fig. 11.7 The quasiharmonic reference configuration obtained from a full minimization versus that obtained from ZSISA for the four-lattice chain of atoms in Example 11.1. (a) Lattice parameter $a(T)$ as a function of $k_{\rm B}T$ in arbitrary units of length and energy. The different lines correspond to different supercell size and calculation method as indicated. (b) Normalized position of basis atom 1, $\bar{\boldsymbol{Z}}^1/a$, as a function of $k_{\rm B}T$ for $N_{\rm lat} = 2$. (c) Same as (b) for $N_{\rm lat} = 10$.

accurate to first order in temperature with respect to "external strain" (i.e. the lattice parameter in this case) but that the "internal strain" (i.e. sublattice shifts) do not have the same accuracy. The limitation of the ZSISA approach is clear from this example.

Finally, it is of interest to compare the quasiharmonic approximation results with the exact solution for this problem. We do not have the exact solution, but we can compare with the results from a *self-consistent* calculation which includes higher-order anharmonic terms. The self-consistent approach is described in general in Wallace's book [Wal72] and was applied to a one-dimensional chain of atoms by Guthikonda and Elliott [GE10]. Briefly, the idea is to include higher-order anharmonic terms in $\mathcal{W}$ defined in Eqn. (11.93) and to then obtain an effective harmonic potential by minimizing the contribution of $\mathcal{W}$ to statistical phase averages. In Fig. 11.8, we compare for $N_{\rm lat} = 10$ the quasiharmonic results obtained using full minimization with the self-consistent results kindly provided by Guthikonda and Elliott for our case. The lattice parameters and positions of basis atom 1 are compared. The self-consistent calculations are significantly more time consuming (about 5–10 times slower than the quasiharmonic calculations) and require derivatives up to fourth order of the potential energy. However, this approach dramatically increases the temperature range over which meaningful results are obtained. It is clear from the comparison that the quasiharmonic results become suspect once the lattice parameter begins to significantly deviate from the linear behavior predicted by the self-consistent approach.

11.4.3 Quasiharmonic stress and elasticity tensors

The quasiharmonic approximations for the stress and elasticity tensors follow as first and second derivatives of the Helmholtz free energy with respect to the deformation gradient. As explained in Section 11.3.4, the calculation of stress involves the definition of an effective free energy (see Eqn. (11.56)):

$$\Psi^{\rm eff}_{\rm qh}(\boldsymbol{F}, T) \equiv \min_{\{\overset{*}{\boldsymbol{s}}{}^{\lambda}\}} \Psi_{\rm qh}(\bar{\boldsymbol{q}}(\boldsymbol{F}, \{\overset{*}{\boldsymbol{s}}{}^{\lambda}\}), T). \tag{11.117}$$

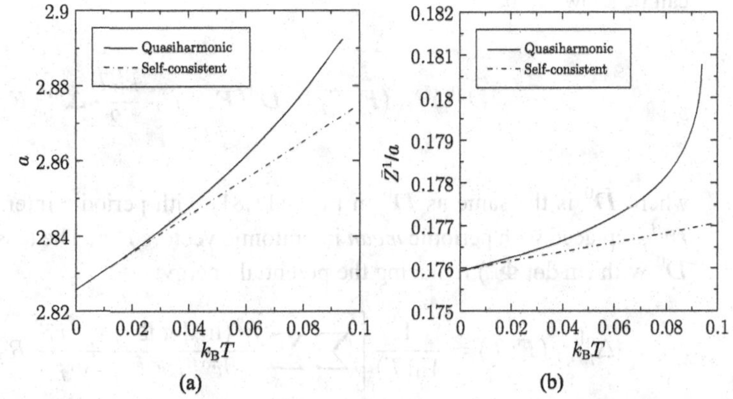

Fig. 11.8 The reference configuration obtained from the quasiharmonic approximation using full minimization versus that obtained from the self-consistent approach for the four-lattice chain of atoms in Example 11.1 for $N_{\text{lat}} = 10$. (a) Lattice parameter $a(T)$ as a function of $k_B T$ in arbitrary units of length and energy. (b) Normalized position of basis atom 1, $\bar{Z}^1/a$, as a function of $k_B T$.

The first Piola–Kirchhoff stress is then given by

$$\boldsymbol{P}^{\text{qh}}(\boldsymbol{F},T) = \frac{1}{V_0(T)} \frac{\partial \Psi^{\text{eff}}_{\text{qh}}(\boldsymbol{F},T)}{\partial \boldsymbol{F}}, \tag{11.118}$$

where $V_0(T) = N_{\text{lat}}\Omega_0(T)$ and $\Omega_0(T)$ is the volume of the crystal unit cell in the reference configuration at temperature T as defined in Eqn. (11.44). Following the procedure leading from Eqns. (11.56) to Eqns. (11.60) and using Eqn. (11.111), the quasiharmonic approximation for the first Piola–Kirchhoff stress is

$$\boldsymbol{P}^{\text{qh}}(\boldsymbol{F},T) = \frac{1}{V_0(T)} \sum_{\substack{\alpha,\beta \\ \alpha<\beta}} \left[\left(\varphi'^{\alpha\beta} + \frac{k_B T}{2} \frac{\partial (\ln \det \boldsymbol{\Phi}^*)}{\partial \bar{r}^{\alpha\beta}} \right) \frac{\bar{\boldsymbol{r}}^{\alpha\beta} \otimes \bar{\boldsymbol{R}}^{\alpha\beta}}{\bar{r}^{\alpha\beta}} \right]. \tag{11.119}$$

Equation (11.119) should be compared with the similar relation in Eqn. (11.60). Note that the phase average is gone and has been replaced with the temperature-dependent term coming from the quasiharmonic approximation. Also, all interatomic distances refer to the mean positions of the atoms. The Cauchy stress and second Piola–Kirchhoff stresses follow immediately from Eqns. $(8.40)_2$ and $(8.40)_3$.

The quasiharmonic approximation for the mixed elasticity tensor is given by $D^{\text{qh}}_{iJkL} = (1/V_0)\partial^2\Psi^{\text{eff}}_{\text{qh}}/\partial F_{iJ}\partial F_{kL}$ (see Eqn. (2.190)). For a centrosymmetric (csym) crystal,[28] this

[28] For reasons of brevity, we do not give the elasticity tensor for noncentrosymmetric crystals, which would include the terms in Eqn. (11.82). See [Kan95b, Kan95a] for a more general derivation of the elasticity tensor for noncentrosymmetric crystals.

can be shown to be

$$\boldsymbol{D}^{\text{qh,csym}}(\boldsymbol{F},T) = \bar{\boldsymbol{D}}^0(\boldsymbol{F},T) + \frac{k_B T}{2}\bar{\boldsymbol{\Delta}}^{\text{qh}}(\boldsymbol{F},T), \tag{11.120}$$

where $\bar{\boldsymbol{D}}^0$ is the same as $\boldsymbol{D}^0$ in Eqn. (11.81) with periodic interatomic vectors, such as $\boldsymbol{r}^{\alpha\mathring{\beta}}$, replaced with periodic *mean* interatomic vectors $\bar{\boldsymbol{r}}^{\alpha\mathring{\beta}}$. The tensor $\bar{\boldsymbol{\Delta}}^{\text{qh}}$ is analogous to $\bar{\boldsymbol{D}}^0$ with $(\ln\det\boldsymbol{\Phi}^*)$ replacing the potential energy:

$$\Delta^{\text{qh}}_{iJkL}(\boldsymbol{F},T) = \frac{1}{V_0(T)}\left[\sum_{\substack{\alpha,\beta\\ \alpha<\beta}}\sum_{\substack{\gamma,\delta\\ \gamma<\delta}} \frac{\partial^2(\ln\det\boldsymbol{\Phi}^*)}{\partial\bar{r}^{\alpha\mathring{\beta}}\partial\bar{r}^{\gamma\mathring{\delta}}}\frac{\bar{r}_i^{\alpha\mathring{\beta}}\bar{r}_k^{\gamma\mathring{\delta}}}{\bar{r}^{\alpha\mathring{\beta}}\bar{r}^{\gamma\mathring{\delta}}}\bar{R}_J^{\alpha\mathring{\beta}}\bar{R}_L^{\gamma\mathring{\delta}} \right.$$
$$\left. + \sum_{\substack{\alpha,\beta\\ \alpha<\beta}} \frac{\partial(\ln\det\boldsymbol{\Phi}^*)}{\partial\bar{r}^{\alpha\mathring{\beta}}}\left(\delta_{ik} - \frac{\bar{r}_i^{\alpha\mathring{\beta}}\bar{r}_k^{\alpha\mathring{\beta}}}{(\bar{r}^{\alpha\mathring{\beta}})^2}\right)\frac{\bar{R}_J^{\alpha\mathring{\beta}}\bar{R}_L^{\alpha\mathring{\beta}}}{\bar{r}^{\alpha\mathring{\beta}}}\right]. \tag{11.121}$$

The temperature dependence of the elasticity tensor in Eqn. (11.120) is nonlinear in general due to the thermal expansion of the reference configuration which enters both terms through $V_0(T)$. In many cases, however, the nonlinearity is weak and a linear response is observed over a broad temperature range. We saw this earlier in the calculation of the elastic constants of argon in Example 8.2 and the corresponding graph in Fig. 8.5. In those results, obtained from MD calculations, linear behavior is observed over the entire temperature range up to melting except for quantum effects near 0 K where the experimental curve turns over. We also see this below for a one-dimensional chain.

Example 11.2 (Quasi-harmonic stress and elasticity for a chain of atoms) To demonstrate the use of Eqn. (11.106) and its derivatives, let us consider the simplest case of a periodic, one-dimensional chain of N identical atoms with mass m and nearest-neighbor, pair potential interactions. For this simple case, we will be able to obtain closed-form analytic expressions. This problem was studied in Example 6.2 and the chain is depicted in Fig. 6.7. The concatenated position vector is $\boldsymbol{q} = (r^1, \dots, r^N)$. Since the chain is a simple lattice the mean positions of the atoms coincide with the positions that minimize the potential energy:

$$\bar{\boldsymbol{q}} = (0, b, 2b, \dots, (N-1)b).$$

Here $b = L/N$ is the spacing between atoms and L is the periodic length of the chain. The potential energy in this state follows from Eqn. (6.38) as

$$\mathcal{V}^{\text{int}}(\bar{\boldsymbol{q}}) = N\phi(b). \tag{11.122}$$

Using Eqn. (6.39), the force constant matrix is

$$\begin{aligned}\Phi_{ij} = \left.\frac{\partial^2\mathcal{V}^{\text{int}}}{\partial q_i\partial q_j}\right|_{\boldsymbol{q}=\bar{\boldsymbol{q}}} &= \frac{\partial}{\partial q_j}\left[-\phi'(q_{i+1}-q_i) + \phi'(q_i - q_{i-1})\right]\Big|_{\boldsymbol{q}=\bar{\boldsymbol{q}}} \\ &= \left[-\phi''(q_{i+1}-q_i)(\delta_{i+1,j}-\delta_{ij}) + \phi''(q_i-q_{i-1})(\delta_{ij}-\delta_{i-1,j})\right]\Big|_{\boldsymbol{q}=\bar{\boldsymbol{q}}} \\ &= -\phi''(b)(\delta_{i+1,j} - 2\delta_{ij} + \delta_{i-1,j}).\end{aligned}$$

In matrix form this is

$$[\Phi] = \phi''(b) \begin{bmatrix} 2 & -1 & 0 & \cdots & \cdots & \cdots & 0 & -1 \\ -1 & 2 & -1 & 0 & \cdots & \cdots & \cdots & 0 \\ 0 & -1 & 2 & -1 & 0 & \cdots & \cdots & 0 \\ \vdots & \ddots & \ddots & \ddots & \ddots & \ddots & \ddots & \vdots \\ 0 & \cdots & \cdots & 0 & -1 & 2 & -1 & 0 \\ 0 & \cdots & \cdots & \cdots & 0 & -1 & 2 & -1 \\ -1 & 0 & \cdots & \cdots & \cdots & 0 & -1 & 2 \end{bmatrix}$$

Matrix Φ is *symmetric* and *circulant*.[29] The eigenvalues of such matrices can be obtained analytically [Dav94]. For our example, the eigenvalues are

$$\lambda_j = 2\phi''(b)\left(1 - \cos\frac{2\pi j}{N}\right) = 4\phi''(b)\sin^2\frac{\pi j}{N} \qquad j = 1, \dots, N.$$

We note that as expected (see footnote 23 on page 581) the last eigenvalue is zero, so $n_{\mathrm{nz}} = N - 1$. The product of the nonzero eigenvalues is

$$\prod_{j=1}^{N-1} \lambda_j = [\phi''(b)]^{N-1} 4^{N-1} \sin^2\frac{\pi}{N} \sin^2\frac{2\pi}{N} \cdots \sin^2\frac{(N-1)\pi}{N} = [\phi''(b)]^{N-1} N^2, \quad (11.123)$$

where we have used the trigonometric identity, $\prod_{j=1}^{N-1} \sin(j\pi/N) = N/2^{N-1}$. Substituting Eqns. (11.122) and (11.123) into Eqn. (11.106), the free energy is

$$\Psi^{\mathrm{V}}_{\mathrm{qh}}(b, T) = N\phi(b) + \frac{k_{\mathrm{B}}T}{2}(N-1)\ln\frac{\phi''(b)}{2\pi k_{\mathrm{B}}T} + k_{\mathrm{B}}T\ln N.$$

In the thermodynamic limit, where $N \to \infty$ while keeping $b = L/N$ constant, we obtain the following configurational free energy per atom:

$$\psi^{\mathrm{V}}_{\mathrm{qh}}(b, T) = \lim_{N\to\infty}\frac{\Psi^{\mathrm{V}}_{\mathrm{qh}}}{N} = \phi(b) + \frac{k_{\mathrm{B}}T}{2}\ln\frac{\phi''(b)}{2\pi k_{\mathrm{B}}T}.$$

The kinetic free energy per atom is

$$\frac{\Psi^{\mathrm{K}}}{N} = -\frac{k_{\mathrm{B}}T}{2}\ln\frac{2\pi m k_{\mathrm{B}}T}{h^2}.$$

Adding together the configurational and kinetic terms gives the final expression for the Helmholtz free energy density:

$$\psi_{\mathrm{qh}}(b, T) = \phi(b) + k_{\mathrm{B}}T\ln\frac{h\omega(b)}{2\pi k_{\mathrm{B}}T}, \quad (11.124)$$

where $\omega = \sqrt{\phi''(b)/m}$ is the natural frequency of the chain. The stress and elastic constant can be computed by setting $b = Fa$, where F is the deformation gradient and $a = a(T)$ is the equilibrium lattice parameter at temperature T, and taking the first and second derivatives of ψ_{qh} with respect to F:

$$P^{\mathrm{qh}}(b, T) = \frac{1}{a}\frac{\partial\psi_{\mathrm{qh}}}{\partial F} = \phi'(b) + \frac{k_{\mathrm{B}}T}{2}\frac{\phi'''(b)}{\phi''(b)}, \quad (11.125)$$

$$D^{\mathrm{qh}}(b, T) = \frac{1}{a}\frac{\partial^2\psi_{\mathrm{qh}}}{\partial F^2} = a(T)\left[\phi''(b) + \frac{k_{\mathrm{B}}T}{2}\frac{\phi^{(4)}(b)\phi''(b) - (\phi'''(b))^2}{(\phi''(b))^2}\right]. \quad (11.126)$$

[29] These are just some of the many special properties of force constant matrices. See, for example, [LL61, Wal72].

The temperature dependence of the equilibrium lattice parameter, $a = a(T)$, is obtained from the condition $P^{\mathrm{qh}}(a(T), T) = 0$:

$$\phi'(a) + \frac{k_B T}{2} \frac{\phi'''(a)}{\phi''(a)} = 0. \tag{11.127}$$

It is clear from this equation that thermal expansion is an anharmonic effect since for a quadratic potential energy, $\phi''' = 0$, and a would be independent of temperature. We can use Eqn. (11.127) to obtain an expression for the coefficient of thermal expansion α, which is defined by

$$a(T) = a_0 + \alpha T, \tag{11.128}$$

where a_0 is the 0 K equilibrium lattice parameter. To do so, we expand $\phi'(a)$, $\phi''(a)$ and $\phi'''(a)$ to first order:

$$\begin{aligned}
\phi'(a) &\approx \phi'(a_0) + \phi''(a_0)(a - a_0) = \phi''(a_0)\alpha T,\\
\phi''(a) &\approx \phi''(a_0) + \phi'''(a_0)(a - a_0) = \phi''(a_0) + \phi'''(a_0)\alpha T,\\
\phi'''(a) &\approx \phi'''(a_0) + \phi^{(4)}(a_0)(a - a_0) = \phi'''(a_0) + \phi^{(4)}(a_0)\alpha T,
\end{aligned}$$

where we have used Eqn. (11.128) and the equilibrium condition, $\phi'(a_0) = 0$. Substituting these relations into Eqn. (11.127) and retaining only linear terms in T gives

$$\alpha = -\frac{k_B}{2} \frac{\phi'''(a_0)}{(\phi''(a_0))^2}. \tag{11.129}$$

The results for a one-dimensional chain of copper atoms are presented in Figs. 11.9 and 11.10. The atomic interactions are modeled using a modified Morse potential [MM81],

$$\phi(r) = \frac{D_0}{2B - 1}\left[e^{-2A\sqrt{B}(r - r_0)} - 2Be^{-A(r - r_0)/\sqrt{B}}\right]. \tag{11.130}$$

The standard Morse potential is recovered for $B = 1$. The modification was introduced to improve agreement with experimental values for thermal expansion of face-centered cubic (fcc) crystals. The parameters for copper are $r_0 = 2.5471$ Å, $A = 1.1857$ Å^{-1}, $D_0 = 0.9403 \times 10^{-19}$ J and $B = 2.265$ [MM81]. The mass of a copper atom is $m = 1.055 \times 10^{-25}$ kg. Figure 11.9 shows the temperature dependence of the lattice parameter, equilibrium free energy and equilibrium elastic constant as a function of temperature. Figure 11.10 shows the dependence of the free energy, stress and the elastic constant on deformation at $T = 300$ K. These are the finite-temperature constitutive relations of the chain at 300 K.

11.4.4 Strict harmonic approximation

The derivation given above is referred to as the "quasiharmonic approximation" to distinguish it from what can be called the *strict harmonic approximation*. In the strict case, the eigenvalues of the force constant matrix, $\lambda_r(\bar{q})$, which actually depend on the deformation gradient $\boldsymbol{F}$ through $\bar{q}$, are treated *as if* they were strictly constants independent of $\boldsymbol{F}$ whenever derivatives of λ_r with respect to $\boldsymbol{F}$ are taken. The strict harmonic approximation also

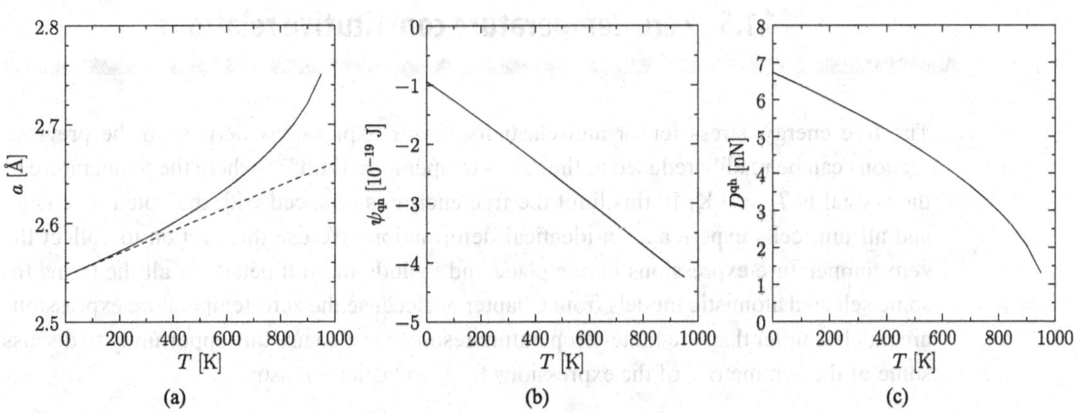

Fig. 11.9 Temperature dependence of thermodynamic properties of a one-dimensional chain of copper atoms modeled via the modified Morse potential in Eqn. (11.130). (a) Lattice parameter $a(T)$ computed from Eqn. (11.127) compared with the linear prediction of Eqn. (11.128) (dashed) using the coefficient of thermal expansion computed from Eqn. (11.129) which is $\alpha = 1.1376 \times 10^{-4}$ Å/K. (b) The equilibrium Helmholtz free energy density $\psi_{\rm qh}(a(T), T)$ computed from Eqn. (11.124). As expected the free energy *decreases* with increasing temperature due to the increase in entropy. (c) The equilibrium elastic constant $D^{\rm qh}(a(T), T)$ computed from Eqn. (11.126). We see, as explained earlier, that the dependence is linear over much of the temperature range.

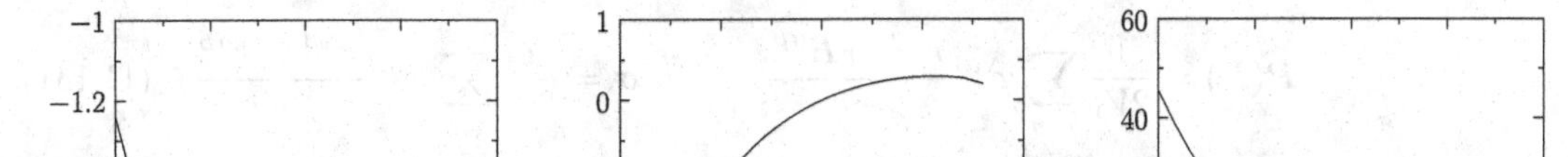

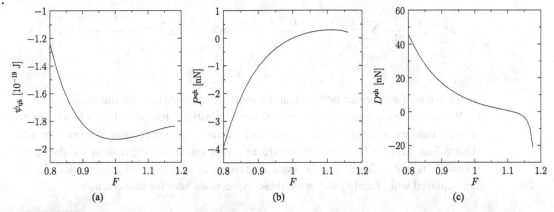

Fig. 11.10 Effect of deformation on a one-dimensional chain of copper atoms modeled via the modified Morse potential in Eqn. (11.130) at 300 K. The equilibrium lattice parameter for this temperature computed from Eqn. (11.127) is $a(300) = 2.5852$ Å. (a) The Helmholtz free energy density $\psi_{\rm qh}(Fa(300), 300)$ computed from Eqn. (11.124). (b) The stress (force) $P^{\rm qh}(Fa(300), 300)$ computed from Eqn. (11.125). The maximum stress the chain can sustain at 300 K is 0.295 nN. (c) The elastic constant $D^{\rm qh}(Fa(300), 300)$ computed from Eqn. (11.126).

implies that the solid has no thermal expansion. All this means that the stress and elasticity tensors computed from the strict harmonic free energy have no temperature dependence. (Simply set $T = 0$ in Eqns. (11.119) and (11.120) to get the corresponding strict harmonic expressions.)

11.5 Zero-temperature constitutive relations

The free energy, stress tensor and elasticity tensor expressions derived in the previous sections can be readily reduced to the "zero-temperature limit,"[30] where the temperature of the crystal is $T = 0$ K. In this limit the free energy is replaced with the potential energy and all unit cells experience an identical deformation. We use this section to collect the zero-temperature expressions in one place and include the full details of all the terms for some selected atomistic models from Chapter 5. Because the zero-temperature expressions are much simpler than the finite-temperature results, we also use this opportunity to discuss some of the symmetries of the expressions for the elasticity tensor.

11.5.1 General expressions for the stress and elasticity tensors

The zero-temperature expression for the first Piola–Kirchhoff stress follows immediately from either Eqn. (11.60) or Eqn. (11.119), while we can use Eqn. (11.72) to obtain the Cauchy stress. At zero temperature, there is no fluctuation in atomic positions or forces and therefore phase averages are trivial and mean positions are absolute positions. The stresses at 0 K are therefore simply

$$\boldsymbol{P}(\boldsymbol{F}) = \frac{1}{2V_0} \sum_{\substack{\alpha,\beta \\ \alpha \neq \beta}} \varphi^{\alpha\mathring{\beta}} \frac{\boldsymbol{r}^{\alpha\mathring{\beta}} \otimes \boldsymbol{R}^{\alpha\mathring{\beta}}}{r^{\alpha\mathring{\beta}}}, \qquad \boldsymbol{\sigma} = \frac{1}{2V} \sum_{\substack{\alpha,\beta \\ \alpha \neq \beta}} \varphi^{\alpha\mathring{\beta}} \frac{\boldsymbol{r}^{\alpha\mathring{\beta}} \otimes \boldsymbol{r}^{\alpha\mathring{\beta}}}{r^{\alpha\mathring{\beta}}} \tag{11.131}$$

(note that we have written both so that the sums are symmetric in the indices α and β).

We have already encountered the 0 K mixed elasticity tensor in Eqn. (11.81) for the special case when there are no shifts (this is always true for centrosymmetric crystals). Using Eqn. (2.195) it is straightforward to recast this as an expression for the spatial elasticity tensor, $\mathbf{c}$, leading to the same expression as in Eqn. (8.84), except that β and δ are replaced with $\mathring{\beta}$ and $\mathring{\delta}$. We repeat these expressions here for convenience:

$$D_{iJkL} = \frac{1}{2V_0} \sum_{\substack{\alpha,\beta \\ \alpha \neq \beta}} \left[\frac{1}{2} \sum_{\substack{\gamma,\delta \\ \gamma \neq \delta}} \kappa^{\alpha\mathring{\beta}\gamma\mathring{\delta}} \frac{r_i^{\alpha\mathring{\beta}} r_k^{\gamma\mathring{\delta}}}{r^{\alpha\mathring{\beta}} r^{\gamma\mathring{\delta}}} R_J^{\alpha\mathring{\beta}} R_L^{\gamma\mathring{\delta}} + \frac{\varphi^{\alpha\mathring{\beta}}}{r^{\alpha\mathring{\beta}}} \left(\delta_{ik} - \frac{r_i^{\alpha\mathring{\beta}} r_k^{\alpha\mathring{\beta}}}{(r^{\alpha\mathring{\beta}})^2} \right) R_J^{\alpha\mathring{\beta}} R_L^{\alpha\mathring{\beta}} \right], \tag{11.132}$$

$$c_{ijkl} = \frac{1}{2V} \sum_{\substack{\alpha,\beta \\ \alpha \neq \beta}} \left[\frac{1}{2} \sum_{\substack{\gamma,\delta \\ \gamma \neq \delta}} \kappa^{\alpha\mathring{\beta}\gamma\mathring{\delta}} \frac{r_i^{\alpha\mathring{\beta}} r_j^{\alpha\mathring{\beta}} r_k^{\gamma\mathring{\delta}} r_l^{\gamma\mathring{\delta}}}{r^{\alpha\mathring{\beta}} r^{\gamma\mathring{\delta}}} - \varphi^{\alpha\mathring{\beta}} \frac{r_i^{\alpha\mathring{\beta}} r_j^{\alpha\mathring{\beta}} r_k^{\alpha\mathring{\beta}} r_l^{\alpha\mathring{\beta}}}{(r^{\alpha\mathring{\beta}})^3} \right], \tag{11.133}$$

[30] We adopt a classical view of atomic behavior and neglect zero-point motion due to quantum effects.

(We drop the "0" superscript that appears in Eqn. (11.81) where it was used to identify the static part of a more general term.)

In Sections 5.8.3 and 5.8.4, we provided the expressions $\varphi^{\alpha\beta}$ and $\kappa^{\alpha\beta\gamma\delta}$ within the framework of several common forms for the atomic interactions (pair functionals and cluster potentials). Here, we note that these same expressions are valid with the second and fourth superscripts replaced with their periodic counterparts, i.e. $\varphi^{\alpha\beta}$ becomes $\varphi^{\alpha\mathring{\beta}}$ and $\kappa^{\alpha\beta\gamma\delta}$ becomes $\kappa^{\alpha\mathring{\beta}\gamma\mathring{\delta}}$. In the next section, we will complete the process of substituting these quantities into the above expressions to obtain final expressions for some of the common interatomic models. These are not necessarily the most efficient forms for implementation. For instance, in the case of a uniformly deformed crystal with a small unit cell, the minimum image convention usually leads to some redundancy in the summations. The forms we present here are convenient, however, because they are modular and general: the results for φ and κ for a new interatomic form can be readily derived and inserted into Eqns. (11.131)–(11.133). When we present the specific forms of $\boldsymbol{\sigma}$ and $\mathbf{c}$ below, we will also include some more optimized forms for comparison.

11.5.2 Stress and elasticity tensors for some specific interatomic models

Pair potentials The simple nature of $\varphi^{\alpha\beta}$ for pair potentials makes the expression for the stress straightforward in this case. Using Eqn. (5.108) in Eqn. $(11.131)_2$ we obtain simply

$$\boldsymbol{\sigma}^{(2)} = \frac{1}{2V} \sum_{\substack{\alpha,\beta \\ \alpha\neq\beta}} \phi_{\alpha\beta}(r^{\alpha\mathring{\beta}}) \frac{\boldsymbol{r}^{\alpha\mathring{\beta}} \otimes \boldsymbol{r}^{\alpha\mathring{\beta}}}{r^{\alpha\mathring{\beta}}}, \tag{11.134}$$

where the superscript "(2)" indicates that this is the pair term in a cluster potential. The elasticity tensor is a little more complicated. Using Eqn. (5.118) in Eqn. (11.133) and simplifying leads to

$$c^{(2)}_{ijkl} = \frac{1}{2V} \sum_{\substack{\alpha,\beta \\ \alpha\neq\beta}} \left[\phi''_{\alpha\beta}(r^{\alpha\mathring{\beta}}) - \frac{\phi'_{\alpha\beta}(r^{\alpha\mathring{\beta}})}{r^{\alpha\mathring{\beta}}} \right] \frac{r^{\alpha\mathring{\beta}}_i r^{\alpha\mathring{\beta}}_j r^{\alpha\mathring{\beta}}_k r^{\alpha\mathring{\beta}}_l}{(r^{\alpha\mathring{\beta}})^2}. \tag{11.135}$$

For uniformly deformed simple crystals, the periodic system can be fully described by a single atom in a primitive unit cell. As such, the above summations involve significant redundancy, because the minimum periodic image convention means that the simulation cell will always be much larger that the primitive cell. In the finite-temperature case, it is *necessary* to have a simulation cell as large as possible in order to get good sampling of phase space, but at zero temperature there is no thermal motion and all atoms are truly identical. We can therefore recast these expressions by looking at a single atom (labeled 1 in the following) in a primitive unit cell and removing the restriction of the minimum periodic image. To build the environment of this atom, we allow as many periodic copies as required to completely embed the atom in a crystal larger than the cutoff radius of the pair potential (we denote these copies of the atom by $\beta = 2, \ldots, N$). This reduces the double summations above to single summations as

$$\sigma_{ij}^{(2)} = \frac{1}{2\widehat{\Omega}} \sum_{\beta=2}^{N} \phi'(r^{1\beta}) \frac{r_i^{1\beta} r_j^{1\beta}}{r^{1\beta}}, \tag{11.136}$$

$$c_{ijkl}^{(2)} = \frac{1}{2\widehat{\Omega}} \sum_{\beta=2}^{N} \left[\phi''(r^{1\beta}) - \frac{\phi'(r^{1\beta})}{r^{1\beta}} \right] \frac{r_i^{1\beta} r_j^{1\beta} r_k^{1\beta} r_l^{1\beta}}{(r^{1\beta})^2}, \tag{11.137}$$

where $\widehat{\Omega}$ is the volume of the primitive unit cell in the deformed configuration and we have dropped the subscripts on ϕ since all atoms in a simple crystal are of the same type. We have also stopped using the superposed $^\circ$ notation, since it is now the case that every index β refers to a periodic copy of the single atom in the simulation cell.

Three-body cluster potentials For three-body cluster potentials, we use Eqn. (5.110) in Eqn. $(11.131)_2$, so that

$$\boldsymbol{\sigma}^{(3)} = \frac{1}{2V} \sum_{\substack{\alpha,\beta,\gamma \\ \alpha \neq \beta \neq \gamma}} \frac{\partial \phi_{\alpha\beta\gamma}}{\partial r^{\alpha\mathring{\beta}}} \frac{\boldsymbol{r}^{\alpha\mathring{\beta}} \otimes \boldsymbol{r}^{\alpha\mathring{\beta}}}{r^{\alpha\mathring{\beta}}}. \tag{11.138}$$

The elasticity tensor is obtained by putting Eqn. (5.119) into Eqn. (11.133) and simplifying:

$$c_{ijkl}^{(3)} = \frac{1}{2V} \sum_{\substack{\alpha,\beta,\gamma \\ \alpha \neq \beta \neq \gamma}} \left[\frac{\partial^2 \phi_{\alpha\beta\gamma}}{(\partial r^{\alpha\mathring{\beta}})^2} - \frac{1}{r^{\alpha\mathring{\beta}}} \frac{\partial \phi_{\alpha\beta\gamma}}{\partial r^{\alpha\mathring{\beta}}} \right] \frac{r_i^{\alpha\mathring{\beta}} r_j^{\alpha\mathring{\beta}} r_k^{\alpha\mathring{\beta}} r_l^{\alpha\mathring{\beta}}}{(r^{\alpha\mathring{\beta}})^2}$$
$$+ \frac{1}{V} \sum_{\substack{\alpha,\beta,\gamma \\ \alpha \neq \beta \neq \gamma}} \frac{\partial^2 \phi_{\alpha\beta\gamma}}{\partial r^{\alpha\mathring{\beta}} \partial r^{\alpha\mathring{\gamma}}} \frac{r_i^{\alpha\mathring{\beta}} r_j^{\alpha\mathring{\beta}} r_k^{\alpha\mathring{\gamma}} r_l^{\alpha\mathring{\gamma}}}{r^{\alpha\mathring{\beta}} r^{\alpha\mathring{\gamma}}}. \tag{11.139}$$

For uniformly deformed simple crystals, the expressions above can be simplified, but not as readily as in the case of the pair potential. This is because any three-body cluster will always involve *two periodic image atoms* in addition to the atom in the primitive unit cell. To some degree, this reduces the advantage of the simplification, and we find little is gained from it in the case of the elasticity tensor. However, we can write the stress as

$$\sigma_{ij}^{(3)} = \frac{1}{\widehat{\Omega}} \sum_{\alpha=2}^{N} \left[\frac{1}{3} \sum_{\substack{\beta=2 \\ \beta \neq \alpha}}^{N} \frac{\partial \phi_{1\alpha\beta}}{\partial r^{1\alpha}} \frac{r_i^{1\alpha} r_j^{1\alpha}}{r^{1\alpha}} + \frac{1}{6} \sum_{\substack{\beta=2 \\ \beta \neq \alpha}}^{N} \frac{\partial \phi_{1\alpha\beta}}{\partial r^{\alpha\beta}} \frac{r_i^{\alpha\beta} r_j^{\alpha\beta}}{r^{\alpha\beta}} \right]. \tag{11.140}$$

Pair functionals (embedded atom method (EAM), Finnis–Sinclair method, etc.) The "embedding" part of the pair functional introduces a term in the stress that is added to the pair potential contribution. Using Eqn. (5.115) in Eqn. $(11.131)_2$, we obtain

$$\boldsymbol{\sigma}^{(\text{embed})} = \frac{1}{V} \sum_{\substack{\alpha,\beta \\ \alpha \neq \beta}} U'_\alpha(\rho^\alpha) g'_\beta(r^{\alpha\mathring{\beta}}) \frac{\boldsymbol{r}^{\alpha\mathring{\beta}} \otimes \boldsymbol{r}^{\alpha\mathring{\beta}}}{r^{\alpha\mathring{\beta}}}. \tag{11.141}$$

The elasticity tensor is considerably more complicated. Using Eqn. (5.120) in Eqn. (11.133) and simplifying leads to

$$c_{ijkl}^{(\mathrm{embed})} = \frac{1}{V}\left(\sum_{\alpha} U_{\alpha}''(\rho^{\alpha})\left[\sum_{\substack{\beta \\ \beta\neq\alpha}} g_{\beta}'(r^{\alpha\mathring{\beta}})\frac{r_i^{\alpha\mathring{\beta}} r_j^{\alpha\mathring{\beta}}}{r^{\alpha\mathring{\beta}}}\right]\left[\sum_{\substack{\gamma \\ \gamma\neq\alpha}} g_{\gamma}'(r^{\alpha\mathring{\gamma}})\frac{r_k^{\alpha\mathring{\gamma}} r_l^{\alpha\mathring{\gamma}}}{r^{\alpha\mathring{\gamma}}}\right] + \sum_{\substack{\alpha,\beta \\ \alpha\neq\beta}} U_{\alpha}'(\rho^{\alpha})\left[g_{\beta}''(r^{\alpha\mathring{\beta}}) - \frac{g_{\beta}'(r^{\alpha\mathring{\beta}})}{r^{\alpha\mathring{\beta}}}\right]\frac{r_i^{\alpha\mathring{\beta}} r_j^{\alpha\mathring{\beta}} r_k^{\alpha\mathring{\beta}} r_l^{\alpha\mathring{\beta}}}{(r^{\alpha\mathring{\beta}})^2}\right). \tag{11.142}$$

As we did for the case of pair potentials, we can simplify these for the case of a uniformly deformed simple crystal as

$$\sigma_{ij}^{(\mathrm{embed})} = \frac{1}{\widehat{\Omega}}\sum_{\beta=2}^{N} U'(\rho) g'(r^{1\beta})\frac{r_i^{1\beta} r_j^{1\beta}}{r^{1\beta}}, \tag{11.143}$$

$$c_{ijkl}^{(\mathrm{embed})} = \frac{1}{\widehat{\Omega}}\left(U''(\rho)\left[\sum_{\beta=2}^{N} g'(r^{1\beta})\frac{r_i^{1\beta} r_j^{1\beta}}{r^{1\beta}}\right]\left[\sum_{\gamma=2}^{N} g'(r^{1\gamma})\frac{r_k^{1\gamma} r_l^{1\gamma}}{r^{1\gamma}}\right] + U'(\rho)\sum_{\beta=2}^{N}\left[g''(r^{1\beta}) - \frac{g'(r^{1\beta})}{r^{1\beta}}\right]\frac{r_i^{1\beta} r_j^{1\beta} r_k^{1\beta} r_l^{1\beta}}{(r^{1\beta})^2}\right), \tag{11.144}$$

where we have again dropped unnecessary subscripts and the periodicity notation.

11.5.3 Crystal symmetries and the Cauchy relations

Examining expressions like Eqns. (11.132) and (11.133), we can see the symmetries that the elasticity tensors possess.[31] Some of these symmetries are independent of the atomistic model and of any assumed crystal structure. For example, we can see that $c_{ijkl} = c_{jikl} = c_{ijlk} = c_{klij}$ regardless of the form of $\varphi^{\alpha\beta}$, the form of $\kappa^{\alpha\beta\gamma\delta}$ or the crystal structure.

Additional symmetries depend on the arrangement of the atoms. For example, let us consider an fcc crystal. Since fcc crystals have cubic symmetry, we expect to find only three independent, nonzero elastic constants (c_{11}, c_{12} and c_{44} in Voigt notation, see Section 2.5.5). This will indeed be the case for any atomic model, but let us use the simple pair potential form of the spatial elasticity tensor in Eqn. (11.137) and restrict our attention to a near-neighbor model just to keep the discussion simple.[32] Now, consider one of the components that must be zero for cubic symmetry, $c_{16} = c_{1112}$. It becomes

$$c_{1112}^{(2)} = \frac{1}{2\widehat{\Omega}}\sum_{\beta=2}^{N}\left[\phi''(r^{1\beta}) - \frac{\phi'(r^{1\beta})}{r^{1\beta}}\right]\frac{r_1^{1\beta} r_1^{1\beta} r_1^{1\beta} r_2^{1\beta}}{(r^{1\beta})^2}, \tag{11.145}$$

[31] These symmetries are equally present in the finite-temperature expressions, but they are easier to see at 0 K.
[32] The result generalizes to any number of neighbors, see the Exercises 11.12 and 11.13.

where $r^{1\beta} = a_0/\sqrt{2}$ for every near neighbor β. There will be 12 contributions to the sum, for which the interatomic vectors $\boldsymbol{r}^{1\beta}$ ($\beta = 1, 2, \ldots, 12$) will be

$$\frac{a_0}{2}(\pm1, \pm1, 0), \qquad \frac{a_0}{2}(\pm1, 0, \pm1), \qquad \frac{a_0}{2}(0, \pm1, \pm1).$$

For all but four of these, there is a zero in either $r_1^{1\beta}$ or $r_2^{1\beta}$, so their product $r_1^{1\beta} r_1^{1\beta} r_1^{1\beta} r_2^{1\beta}$ will be zero. For the remaining four neighbors, it is easy to see that the contributions from $\boldsymbol{r}^{1\beta} = a_0(1, 1, 0)/2$ and $\boldsymbol{r}^{1\beta} = a_0(1, -1, 0)/2$ cancel each other, as do $\boldsymbol{r}^{1\beta} = a_0(-1, 1, 0)/2$ and $\boldsymbol{r}^{1\beta} = a_0(-1, -1, 0)/2$. Similarly, we can verify that any component of c_{ijkl} which must be zero by symmetry is in fact zero by Eqn. (11.137).

By considering other combinations of the $\boldsymbol{r}^{1\beta}$ components in the nearest-neighbor fcc model, one can also see that, for example, $c_{1122} = c_{1133} = c_{2233}$ as required for cubic symmetry. However, there is an addition relationship between the moduli components that emerges in the case of the pair potential. Specifically, we see that c_{1212} and c_{1122} (or c_{44} and c_{12} in Voigt notation) are exactly the same. Indeed, for any simple pair potential

$$P_C \equiv \mathsf{c}_{12} - \mathsf{c}_{44} = 0, \tag{11.146}$$

where P_C is called the *Cauchy pressure*. This is one of the six *Cauchy relations* between the elastic constants that occur for pair potential interactions. The entirety of the Cauchy relations is embodied in $c_{ijkl} = c_{ikjl}$, for which the explicit nontrivial relationships are

$$c_{1122} = c_{1212}, \qquad c_{1133} = c_{3131}, \qquad c_{2233} = c_{2323},$$

$$c_{1123} = c_{1231}, \qquad c_{2231} = c_{2312}, \qquad c_{3312} = c_{3123}.$$

The Cauchy relations reduce the number of independent elastic constants[33] from 21 to 15.

Initially, this observation led to a great deal of heated debate, with supporters of the *rari-constant theory* (Cauchy, Poisson, Kelvin, Lamé) believing that the Cauchy relations applied to real materials, while supporters of the *multi-constant theory* (Green, Stokes, Kirchhoff) did not accept the Cauchy relations. The debate was settled in 1887 by Voigt, who performed experiments that demonstrated conclusively that the Cauchy relations were violated by real materials [DZ99]. For example the Cauchy pressure P_C in Eqn. (11.146) is nonzero for many cubic crystals. We now understand that the Cauchy relations arise from limiting the interatomic model to a pair potential.

As we noted in Section 5.4.2, there are only two classes of material for which a pair potential is truly representative of the physics of bonding: materials governed by van der Waals or by ionic bonding. It is interesting to note that these materials do in fact obey the Cauchy relations – one can experimentally verify that $P_C = 0$ in these crystals.

For bonding that is neither purely van der Waals nor ionic, the environmental dependence of each bond must not be overlooked and a simple pair potential fails; the example of the incorrect elastic moduli is but one clear manifestation of this limitation. Fortunately, we can see that the original expressions we derived for the elasticity tensors (Eqns. (11.81)

[33] Interestingly, it can be shown that the elasticity tensor can be uniquely decomposed into two parts, one with 15 independent parameters and one with 6, which exactly correspond to the Cauchy relations [HI02, Fos09]. The implications of this result from the perspective of pair potentials is still unclear.

and (11.133)) using a generic interatomic potential are not limited to satisfying the Cauchy relations. The problem arose only once the simplifications of the pair potential were made in Eqn. (11.135). Indeed, one can verify that there is a nonzero Cauchy pressure for either the pair functional form or the three-body form.

The Cauchy pressure for pair functional models Consider a cubic crystal described by a pair functional such as EAM. Examining the terms in Eqn. (11.144), we can see that the component indices appear only through the components of the interatomic vectors, and that interchanging indices from 1122 to 1212 has no effect on the second line of the equation. Taking the difference $c_{1122} - c_{1212}$ will therefore cancel these two lines, leaving

$$P_C = c_{1122} - c_{1212} = \frac{U''(\rho)}{\Omega}\left[\Lambda_{11}\Lambda_{22} - \Lambda_{12}^2\right], \tag{11.147}$$

where we have defined for convenience

$$\Lambda_{ij} = \left[\sum_{\beta=2}^{N} g'(r^{1\beta})\frac{r_i^{1\beta} r_j^{1\beta}}{r^{1\beta}}\right]. \tag{11.148}$$

For crystals with cubic symmetry, one can easily show (similar to what was shown for c_{1112} above) that $\Lambda_{11} = \Lambda_{22}$ and $\Lambda_{12} = 0$, leaving

$$P_C = \frac{U''(\rho)}{\Omega}(\Lambda_{11})^2. \tag{11.149}$$

Indeed, the pair functional form restores a nonzero Cauchy pressure. However, the important point is that the sign of P_C is determined by the curvature of U at equilibrium. The "physically motivated" forms that are used for the function U when developing a pair functional model always have a positive curvature U''. This is clear from Eqn. (5.55) in the case of the Finnis–Sinclair form and from the physical picture of what U represents in the EAM description: it is the energy associated with placing a positively charged ion in a uniform electron gas. One would expect this quantity to be negative and to be monotonically decreasing with increased electron density. In other words, both of these forms for U point to a positive value of the Cauchy pressure.

Since it seems that pair functionals cannot fit a negative Cauchy pressure, is this consistent with the type of bonding we expect them to describe? Indeed, we find that experimental negative Cauchy pressures are only found for body-centered cubic (bcc) metals and intermetallic compounds which tend to form due to the covalent nature of their bonding. "Purely" metallic materials that tend to form fcc crystals also tend to have positive Cauchy pressures, and it is in fact for these systems that the pair functional was developed. Conversely, this points to the inappropriateness of using a pair functional for covalent materials.

Further reading

- Most statistical mechanics books focus on gases and liquids. Those that consider solids normally do so by assuming that the atoms in a solid experience small fluctuations about lattice sites for which the harmonic approximation is suitable. Examples are discussed below. We are not aware of other books that adopt a restricted ensemble approach to solids as we do here. To find out more about this topic, the reader is referred to articles on this subject including [PL71, CO77, LP79, Sew80, ERH85, Yuk91, Sti95, CD98, Pen02, Pen06]. Most of these references deal with liquid systems, a notable exception is the work of Penrose on two-dimensional square lattice of polarized material [Pen02, Pen06].

- Some examples of statistical mechanics books that discuss solids within the harmonic approximation are [MM77, Wei83, Maz00]. Weiner's book [Wei83] in particular is clear and grounded in the principles of continuum mechanics which are adopted in this book.

- For readers wishing a more general introduction to the harmonic approximation and lattice dynamics. The following books are recommended:
  1. Dove's book titled *Introduction to Lattice Dynamics* [Dov93] provides a very friendly accessible introduction. It is an excellent place to start when new to the subject.
  2. Although first published in 1954, the classic book by Born and Huang called the *Dynamical Theory of Crystal Lattices* [BH54] continues to be an invaluable reference.
  3. Wallace's book called the *Thermodynamics of Crystals* [Wal72] provides an authoritative comprehensive discussion of lattice dynamics and much more. It is not an easy read, but is the standard text to which many other books refer.
  4. Leibfried and Ludwig's treatise on the "Theory of anharmonic effects in crystals" [LL61] discusses the extension of the harmonic theory to include anharmonic effects. It provides a thorough and generally accessible introduction to the topic.

Exercises

11.1 [Section 11.3] Consider an infinite chain of identical atoms with mass m interacting via a general interatomic potential, $\mathcal{V}^{\text{int}}(\boldsymbol{q})$, where $\boldsymbol{q}$ is the concatenated coordinates of the atoms. The chain is in contact with a heat bath at temperature T. The atoms in the chain are restricted by a characteristic function χ to remain with a specified distance of their mean positions, $\bar{r}^{\alpha}(F) = \alpha F a(T)$, where F is the deformation gradient and $a(T)$ is the equilibrium lattice parameter at temperature T.

1. Obtain an expression for the Helmholtz free energy density $\psi(F, T)$ for the chain.
2. Compute the stress in the chain from $P(F,T) = \rho_0 \partial\psi/\partial F$, where ρ_0 is the reference mass density.
3. Using your expression for $P(F,T)$, write an equation whose solution gives the temperature-dependent equilibrium lattice parameter $a(T)$.

11.2 [Section 11.3] Starting from the definition of $\boldsymbol{r}^{\alpha\mathring{\beta}}$ in Eqns. (11.35) and (11.36) do the following:

1. Derive the results in Eqns. (11.38) and (11.39). Note that $r^{\alpha\mathring{\beta}} = [r_n^{\alpha\mathring{\beta}} r_n^{\alpha\mathring{\beta}}]^{1/2}$.

2. Show that

$$\frac{\partial}{\partial F_{jK}}\left(\frac{r_i^{\alpha\mathring{\beta}}}{r^{\alpha\mathring{\beta}}}\right) = \delta_{ij}\frac{\bar{R}_K^{\alpha\mathring{\beta}}}{r^{\alpha\mathring{\beta}}} - \frac{r_i^{\alpha\mathring{\beta}} r_j^{\alpha\mathring{\beta}} \bar{R}_K^{\alpha\mathring{\beta}}}{(r^{\alpha\mathring{\beta}})^3}.$$

11.3 [SECTION 11.3] Derive Eqn. (11.64).

11.4 [SECTION 11.3] Equation (11.72), which shows that the Cauchy stress is symmetric, was derived under the assumption that Eqn. (11.55) is satisfied. Let us see what happens without this condition. Starting with the final relation in Eqn. (11.63), recover the kinetic part of the stress tensor using the procedure shown on page 574. (There is no need to rederive what is already in the book.) Show that the resulting expression for the Cauchy stress is not symmetric. What are the implications of your result for "general regime" thermodynamics discussed in Section 11.3.4. (See the discussion of multipolar theories in footnote 31 on page 59).

11.5 [SECTION 11.3] Show that the force constant matrix defined in Eqn. (11.90) has the following form when written in terms of interatomic distances for a periodic system:

$$\Phi_{ij}^{\alpha\beta} = \sum_{\substack{\gamma,\delta \\ \gamma\neq\alpha,\ \delta\neq\beta}} \kappa^{\alpha\mathring{\gamma}\beta\mathring{\delta}} \frac{r_i^{\alpha\mathring{\gamma}} r_j^{\beta\mathring{\delta}}}{r^{\alpha\mathring{\gamma}} r^{\beta\mathring{\delta}}} + \begin{cases} -\dfrac{\varphi^{\alpha\mathring{\beta}}}{r^{\alpha\mathring{\beta}}}\left[\delta_{ij} - \dfrac{r_i^{\alpha\mathring{\beta}} r_j^{\alpha\mathring{\beta}}}{(r^{\alpha\mathring{\beta}})^2}\right] & \text{if } \alpha\neq\beta, \\ \sum_{\substack{\gamma \\ \gamma\neq\alpha}} \dfrac{\varphi^{\alpha\mathring{\gamma}}}{r^{\alpha\mathring{\gamma}}}\left[\delta_{ij} - \dfrac{r_i^{\alpha\mathring{\gamma}} r_j^{\alpha\mathring{\gamma}}}{(r^{\alpha\mathring{\gamma}})^2}\right] & \text{if } \alpha=\beta, \end{cases}$$

where κ is defined in Eqn. (5.117). Note that instead of the concatenated notation in Eqn. (11.90), here $\Phi_{ij}^{\alpha\beta} = \partial^2\mathcal{V}^{\text{int}}/\partial r_i^\alpha \partial r_j^\beta$.

11.6 [SECTION 11.4] The calculation of the quasiharmonic configurational integral in Eqn. (11.104) requires the removal of zero eigenmodes associated with the rigid-body motion of the entire system. A similar modification is necessary for the kinetic part of the partition function defined in Eqn. $(7.121)_1$.

1. Show that if the kinetic partition function is not modified (so that Ψ^{K} in Eqn. $(11.87)_1$ is computed using Eqn. $(11.88)_1$ and added to $\Psi_{\text{qh}}^{\text{V}}$ in Eqn. (11.106) to obtain Ψ_{qh}), then Ψ_{qh} will have the wrong units, i.e. not units of energy.
2. To remove the contribution of rigid-body motion to Ψ^{K}, it is first necessary to change the momentum variables in accordance with the change from $\boldsymbol{w}$ to $\boldsymbol{\mu}$ in Eqn. (11.103) using an appropriate canonical transformation. Find a generating function of the following form: $G_2(\boldsymbol{q}, \widehat{\boldsymbol{p}}) = \sum_s f_s(\boldsymbol{q})\widehat{p}_s$, that performs the necessary transformation. (See Section 8.1.1.)
3. Use the canonical transformation determined above to perform the necessary change of variables in the integration of Z^{K}. Write an analytical expression for the kinetic part of the partition function with the same form as Eqn. $(11.88)_1$, where rigid-body modes have been removed. Show that this new expression gives the correct units for the total free energy Ψ_{qh}.

11.7 [SECTION 11.4] Derive Eqns. (11.119) and (11.120).

11.8 [SECTION 11.4] This problem refers to Example 11.2.

1. Use a computer to reproduce the six graphs appearing in Figs. 11.9 and 11.10. As explained in Example 11.2, these curves were obtained from the modified Morse potential in Eqn. (11.130). See the figure captions and text after the equation for details of the computations.
2. Generate a new graph for this potential showing the maximum stress that the chain can sustain as a function of temperature in the range $T \in [0, 1000\text{ K}]$. Discuss your result.

11.9 [SECTION 11.5] Implement a subroutine that computes Eqns. (11.137) and (11.144) in the MD code MiniMol provided on the companion website to this book. Use it to compute the zero-temperature moduli for fcc crystals of the Ar and Al models provided with the code. Verify that cubic symmetry holds for both materials, and that the Cauchy relations are obeyed for Ar but not for Al. (Note that the published elastic constants for these models are $c_{11} = 0.7371\text{ eV/Å}^3$, $c_{12} = 0.3888\text{ eV/Å}^3$ and $c_{44} = 0.2291\text{ eV/Å}^3$ for Al and $c_{11} = 0.0195\text{ eV/Å}^3$ and $c_{12} = c_{44} = 0.0110\text{ eV/Å}^3$ for Ar.)

11.10 [SECTION 11.5] Using the subroutine written for the previous exercise, study the effect of pressure on the elastic moduli. Specifically, plot c_{11}, c_{12}, c_{44} and the bulk modulus B (see definition of B in Exercise 2.15) versus pressure by imposing different lattice constants and recording the resulting pressure and moduli at zero temperature.

11.11 [SECTION 11.5] Compute the elasticity tensor for an equilibrated fcc crystal modeled using the pair potential that you created in Exercise 5.2 and the subroutine written for Exercise 11.9. How do the results compare to those obtained for fcc Ar using the Lennard-Jones potential provided with MiniMol? (See full directions on the companion website to this book.)

11.12 [SECTION 11.5] In the discussion following Eqn. (11.145), we verified that $c_{1112} = 0$ for a near-neighbor pair potential model in an fcc crystal. Verify that this also holds for a second-neighbor model in both fcc and bcc crystals.

11.13 [SECTION 11.5] Show that $\Lambda_{11} = \Lambda_{22}$ and $\Lambda_{12} = 0$ (see Eqn. (11.148)) for both fcc and bcc crystals up to second-neighbor interactions.

12 Atomistic–continuum coupling: static methods

In Chapter 2, we reviewed the essential concepts of continuum mechanics (which are covered in full detail in the companion book to this one [TME12]) and talked at length about atomistic models in Chapter 5 and static solution methods for these models in Chapter 6. The focus of the current chapter is on ways to couple the two approaches – continuum and atomistic – in search of a "best of both worlds" model that combines their strengths.[1]

The models discussed in this chapter achieve this coupling by using a discretized approximation to continuum mechanics called the finite element method (FEM).[2] We provide a brief review of FEM in the next section (see [TME12] for a more detailed discussion).

12.1 Finite elements and the Cauchy–Born rule

An overview of finite elements The problem we wish to solve with FEM is the static boundary value problem embodied in Fig. 12.1(a). A body B_0 in the reference configuration has surface ∂B_0 with surface normal $\boldsymbol{N}$. This surface is divided into a portion (∂B_{0u}) over which the displacements are prescribed as $\bar{\boldsymbol{u}}$ and the remainder (∂B_{0t}) which is either free or subject to a prescribed traction, $\bar{\boldsymbol{T}}$. Our goal is to determine the resulting stress, strain and displacement fields throughout the body due to the applied loads.

This boundary-value problem is conveniently recast as an energy minimization problem using the *principle of minimum potential energy* from Section 2.6. The total potential energy, Π, is

$$\Pi = \int_{B_0} W(\boldsymbol{F}(\boldsymbol{X}))dV_0 - \int_{B_0} \rho_0 \bar{\boldsymbol{b}} \cdot \boldsymbol{u}\, dV_0 - \int_{\partial B_{0t}} \bar{\boldsymbol{T}} \cdot \boldsymbol{u}\, dA_0, \qquad (12.1)$$

where ρ_0 is the reference mass density, $\bar{\boldsymbol{b}}$ is an applied field of body forces and $\bar{\boldsymbol{T}}$ are applied tractions. Next, Π is minimized subject to the constraint that $\boldsymbol{u}(\boldsymbol{X}) = \bar{\boldsymbol{u}}$ for $\boldsymbol{X} \in \partial B_{0u}$. The strain energy density, W, is assumed to be a function of the deformation gradient $\boldsymbol{F}$ and provides the *constitutive behavior* of the material in the body. We will return to $W(\boldsymbol{F})$ shortly.

In FEM, the first step is to replace the continuous variable $\boldsymbol{u}(\boldsymbol{X})$ with a discrete variable, $\mathbf{u}(\mathbf{X})$, stored at a finite number of points (called the nodes) in the body, as illustrated

[1] Parts of this chapter have appeared as a topical review in [MT09b] and are reproduced here with the permission of the publisher.

[2] There are also multiscale methods that use a "meshless method" instead of finite elements to approximate the continuum region. See, for example, the review by Park and Liu [PL04]. We do not discuss such methods here.

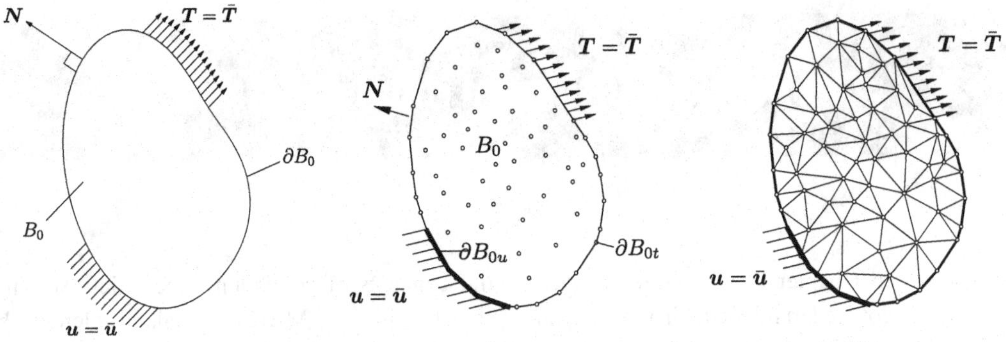

Fig. 12.1 The general continuum mechanics boundary value problem in (a) is approximated by storing the displacements at a set of discrete nodal points as in (b), between which elements as in (c) are constructed to interpolate the fields throughout the body.

schematically in Fig. 12.1(b). Between the nodes, elements are constructed as in Fig. 12.1(c) to provide interpolation functions (or "shape functions") that approximate the continuum displacement field based on the discrete values stored at the nodes. This can be written as a sum over the n_{nodes} nodes,

$$\widetilde{u}_i(\boldsymbol{X}) = \sum_{I=1}^{n_{\text{nodes}}} S^I(\boldsymbol{X})\mathsf{u}_i^I, \tag{12.2}$$

where there is a shape function S^I associated with every node I and u_i^I is the ith component of the displacement stored at node I. The $\widetilde{\cdot}$ notation reminds us that the field is an approximation, dependent on the number and arrangement of the nodes and elements. This equation appears to suggest that the displacement at each point depends on all nodes in the model. In fact, the displacement at a point depends only on nearby nodes since the shape functions have *compact support*, meaning that S^I is nonzero only inside an element in direct contact with node I. As a result, the displacement field inside an element is determined only from the nodes defining that particular element. The deformation gradient, $\boldsymbol{F}$, can be determined from this interpolated displacement field at any point in the body. Recalling the definition in Eqn. (2.62), $\boldsymbol{F} = \boldsymbol{I} + \partial \boldsymbol{u}/\partial \boldsymbol{X}$, we have

$$\widetilde{F}_{iJ} = \delta_{iJ} + \frac{\partial \widetilde{u}_i}{\partial X_J} = \delta_{iJ} + \sum_{I=1}^{n_{\text{nodes}}} \frac{\partial S^I}{\partial X_J}\mathsf{u}_i^I, \tag{12.3}$$

where δ_{iJ} is the Kronecker delta function. Putting Eqns. (12.2) and (12.3) into Eqn. (12.1) provides an approximate form amenable to computer solution:

$$\widetilde{\Pi} = \sum_{e=1}^{n_{\text{elem}}} \int_{B_0^e} W(\widetilde{\boldsymbol{F}}(\mathbf{u}))\,dV_0 - \sum_{I=1}^{n_{\text{nodes}}} \sum_{i=1}^{n_{\text{d}}} \mathsf{f}_i^{\text{ext},I}\mathsf{u}_i^I, \tag{12.4}$$

where there are n_{nodes} nodes and n_{elem} elements in n_{d}-dimensional space. The components of the externally applied force, $\mathsf{f}_i^{\text{ext},I}$, come from the applied tractions and body forces as

$$\mathsf{f}_i^{\mathrm{ext},I} = \int_{\partial B_{0t}} \bar{T}_i S^I \, dA_0 + \int_{B_0} \rho_0 \bar{b}_i S^I \, dV_0. \tag{12.5}$$

Minimization of $\widetilde{\Pi}$ with respect to the vector of nodal displacements $\mathbf{u}$ leads to the equilibrium condition

$$\mathsf{f}_i^{\mathrm{ext},I} + \mathsf{f}_i^{\mathrm{int},I} = 0, \tag{12.6}$$

where we have defined the internal force vector as

$$\mathsf{f}_i^{\mathrm{int},I} \equiv \int_{B_0} \frac{\partial W}{\partial \widetilde{F}_{iJ}} \frac{\partial S^I}{\partial X_J} \, dV_0. \tag{12.7}$$

The minimization process is usually carried out using a Newton–Raphson (NR) approach as described in Section 6.2.7. The integrals in Eqns. (12.4), (12.5) and (12.7) are handled by numerical quadrature (integration) (see Section 9.3.2 of [TME12]). The result is that the integrands need only be evaluated at a handful of points inside each element (called quadrature points or "Gauss points" in honor of Gauss' development of the technique).

The Cauchy–Born rule in finite elements A central component of FEM is the constitutive law $W(\boldsymbol{F})$ and, for the implementation of the NR method, its derivatives

$$P_{iJ} = \frac{\partial W(\boldsymbol{F})}{\partial F_{iJ}}, \qquad D_{iJkL} = \frac{\partial W(\boldsymbol{F})}{\partial F_{iJ} \partial F_{kL}}. \tag{12.8}$$

A natural connection can be made between FEM and atomistics through the Cauchy–Born constitutive rule that was introduced in Section 11.2.2. For a *simple* lattice material, we can imagine a constitutive model subroutine for an FEM code that, given an arbitrary deformation gradient and the known crystal structure of the undeformed material, returns the stress and tangent stiffness. The deformed lattice vectors are determined from Eqn. $(11.21)_1$, and used to construct a periodic unit cell of the deformed crystal. From this cell, one can compute $W(\boldsymbol{F})$ (and its derivatives) to obtain the stress and elasticity tensors as discussed in Section 11.5. These define the constitutive response at a continuum material point.

For a *multilattice* crystal material, the above prescription cannot be directly applied because the strain energy density also depends on the sublattice shifts, $W = W(\boldsymbol{F}, \{\boldsymbol{s}^\lambda\})$, where $\lambda = 1, \ldots, N_{\mathrm{B}}$ and N_{B} is the number of sublattices (or the number of basis atoms). One possibility is to define an *effective* strain energy density by minimizing out the shifts for a given value of the deformation gradient (see Section 11.3.4):

$$W^{\mathrm{eff}}(\boldsymbol{F}) \equiv \min_{\{\boldsymbol{s}^\lambda\}} W(\boldsymbol{F}, \{\boldsymbol{s}^\lambda\}). \tag{12.9}$$

The effective strain energy density can then be used within the FEM exactly as $W(\boldsymbol{F})$ is used for a simple lattice. Although attractive, this approach can fail under certain conditions as discussed in Section 11.3.4 and in the article by Sorkin *et al.* [SET11]. In such cases it is necessary to modify the FEM implementation to include the shifts as global degrees of freedom along with the displacements of the nodes. This is discussed in [SET11].

The use of an atomistic constitutive law in static FEM is both powerful and quite general, since we can in principle use any atomistic model ranging from density functional theory

(DFT) to simple pair potentials. Indeed, such an approach is the basis of the so-called "local" regions in the quasicontinuum method (see Section 12.3.2), where it has been used to study such things as phase transformations in silicon under high pressure due to nanoindentation [STK00, STBK01] and the response of ferroelectric lead titanate ($PbTiO_3$) to electrical and mechanical loading [TWSK02]. It has been adapted to the study of carbon nanotubes by accounting for curvature effects on the otherwise two-dimensional crystal structure of a graphene sheet [JZL⁺03]. With simple pair potentials, this approach has been used to study dislocation nucleation in a two-dimensional bubble raft [vVLZ⁺03, vVS02], while indentation of an Al crystal was studied using full DFT [FHCO04].

Such a constitutive model has a number of desirable properties. It has the full nonlinear elastic behavior of the underlying atomistic model. It also immediately possesses the correct symmetries, lattice invariances and material frame-indifference. Most interestingly, the only input to the model is, if you will, entirely from "first principles." If the atomic model could be made perfectly true to the atomistic nature of the crystal, the continuum formulation requires no additional parameters or assumed functional forms. Of course atomistic constitutive models also have limitations. First, the Cauchy–Born assumption that the deformation is affine at the microscale can fail under certain conditions (see Section 11.2.4). Second, the discussion in this chapter is limited to zero-temperature statics. Atomistic constitutive relations for systems at finite temperature, $T > 0$ K, can be derived from statistical mechanics (see Section 11.3). However, practical implementations of these models require additional approximations like the quasiharmonic approach described in Section 11.4 and applied within the "hot-QC" method in Section 13.2.

12.2 The essential components of a coupled model

A large number of "partitioned-domain" methods (see Section 10.3.2) have been proposed for coupling of atomistics and FEM. Our goal here is to present them under a unified description and terminology. The basic idea of partitioned-domain methods is to divide the body of interest into a region that can be treated using continuum mechanics (usually solved approximately using FEM) and a region that will be treated atomistically. The goal is to model large problems, using the more expensive atomistic model where it is needed and the cheaper FEM where it is possible. As we shall see, there are many similarities between the various methods, but there are key, subtle differences that determine the accuracy and efficiency of each approach.

A generic partitioned-domain model is presented in Fig. 12.2. In Fig. 12.2(a), we schematically illustrate an idealized partitioning, where the body is unambiguously divided into two regions. Region B^{A} is treated atomistically, while region B^{C} is modeled as a continuum. An interface between the two regions is identified, ∂B^{I}, across which compatibility and equilibrium are enforced. However, the finite range of interactions of the atoms typically requires a more elaborate interfacial region with finite width, B^{I}, as illustrated in Fig. 12.2(b). Boundary conditions, in the form of prescribed tractions on ∂B^C_{0t}, forces on ∂B^A_{0t} or displacements on $\partial B_{0u} = \partial B^A_{0u} \cup \partial B^C_{0u}$ are applied to induce deformation.

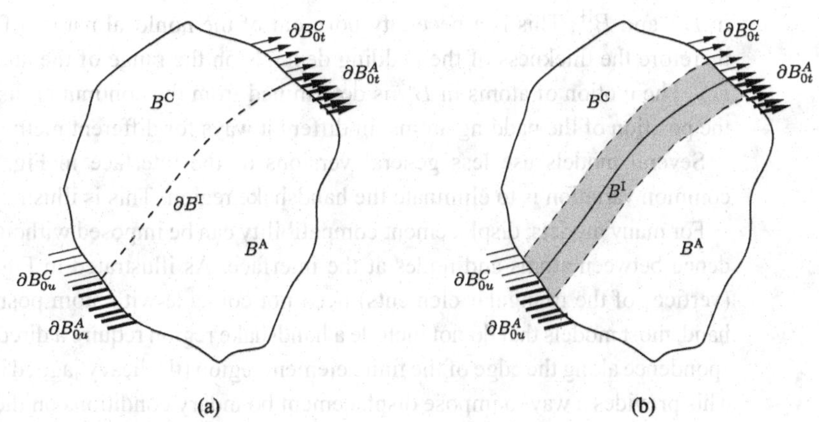

Fig. 12.2 (a) A general partitioned-domain problem. B^{C} is modeled as continuum while all atoms in B^{A} are explicitly treated as discrete degrees of freedom. (b) Some methods require a blended interfacial region, B^{I}, of finite volume.

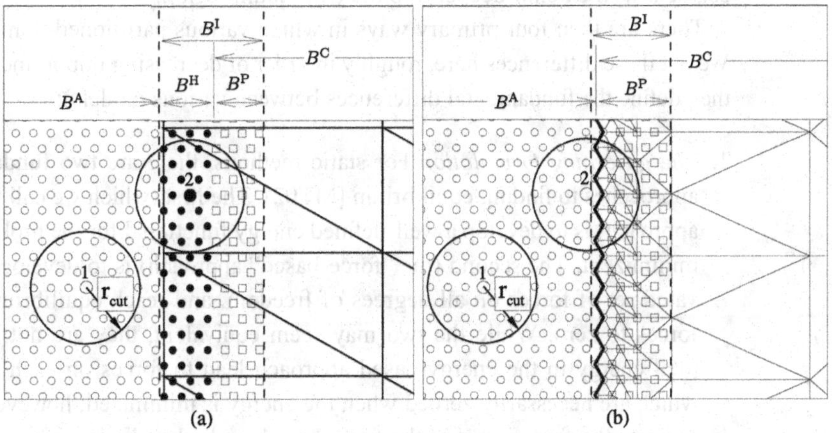

Fig. 12.3 A generic interface in a coupled atomistic–continuum problem. The finite cutoff radius of the atoms means that an atom like 1 cannot "see" into the continuum, while atom 2 can. The model in (a) includes a handshake region, B^{H}, while the model in (b) does not. Padding atoms are shown as open squares, handshake atoms as black circles and regular atoms as white circles. See the text for details.

The key to any multiscale method lies in the way that it handles the interface region. Most methods adopt one of the strategies depicted in Fig. 12.3. Figure 12.3(a) shows perhaps the most generic interface, which in various limiting cases leads to all the models discussed here. The interface region B^{I}, shown by the dashed lines, has been further subdivided into two parts. We refer to these as the "handshake region", B^{H} (filled circles), and the "padding region", B^{P} (open squares). The size and nature of these regions depend on the specifics of the multiscale model, as discussed below. The handshake region is neither fully atomistic nor fully continuum. It is a region where there is some degree of mixing between the two descriptions of the material as we explain later. The padding region is continuum in nature, however it is used to generate atoms that provide the boundary conditions to the atoms in B^{A} and B^{H}. This is a necessity born out of the nonlocal nature of atomic bonds, and therefore the thickness of the padding depends on the range of the atomistic interactions, r_{cut}. The motion of atoms in B^{P} is determined from the continuum displacement fields at the position of the padding atoms, in different ways for different methods.

Several models use less general versions of the interface in Fig. 12.3(a). The most common variation is to eliminate the handshake region. This is illustrated in Fig. 12.3(b).

For many models, displacement compatibility can be imposed without a clear correspondence between atoms and nodes at the interface. As illustrated in Fig. 12.3(a) the nodes (vertices of the triangular elements) need not coincide with atom positions. On the other hand, most models that do not include a handshake region require a direct atom–node correspondence along the edge of the finite element region (the heavy jagged line in Fig. 12.3(b)). This provides a way to impose displacement boundary conditions on the finite elements, as the nodes along this edge move in lock step with the corresponding atoms. The cost of this is a restriction on the finite element mesh, which must be refined down to the atomic scale near the interface (we refer to this as a "fully-refined" mesh). Normally however, the mesh coarsens as it extends deeper into the continuum region.

There are then four primary ways in which various partitioned-domain methods differ. We list these differences here, roughly in order of decreasing importance (to the extent that they define the fundamental differences between various models):

1. *The governing formulation* For static methods, there are two fundamentally different approaches to finding equilibrium [MT02]. The first (which we call the "energy-based" approach) is to develop a well-defined energy functional for the problem and rigorously minimize it. The second (the "force-based" approach) is to develop a physically motivated set of forces on all degrees of freedom, and reach equilibrium by driving these forces to zero. While the two may seem equivalent, they are not. Derivatives of the total energy in the energy-based approach lead to forces on each degree of freedom which are necessarily zeroed when the energy is minimized; however, these are not the same as the forces used in the force-based methods. The force-based methods have *no well-defined total energy*. Instead, they start from a definition of the coupling in terms of a physically-motivated prescription of the forces on unequilibrated atoms or nodes.

   As we shall discuss later, the disadvantage of the energy-based approach is that it is extremely difficult to eliminate the nonphysical side effects of the coupled energy functional, dubbed "ghost forces" in the literature.[3] On the other hand, force-based methods can be slow to equilibrate, can converge to unstable equilibrium states (such as saddle points or maxima), are nonconservative and can be numerically unstable. Later, we will quantitatively compare the energy-based and force-based approaches.
2. *The coupling boundary conditions* The coupling between the continuum and atomistic regions requires compatibility conditions in each direction. This means providing a

[3] "Ghost forces" are nonphysical forces that exist near the continuum–atomistic interface. Their existence was first discussed in the original derivation of the quasicontinuum method [Tad96]. Several authors have since then devoted considerable time to discussing the source and nature of these ghost forces [CM03, SMSJ04, ELY06, BPB+08, DL08b]. The name "ghost forces" was coined by David Rodney and appears for the first time in [SMT+99].

prescription for finding the displacements of atoms in the padding region, $\mathcal{B}^{\mathrm{P}}$, from the nodal displacement information in $\mathcal{B}^{\mathrm{C}}$, as well as a way to determine the displacement boundary conditions for the finite element nodes along the edge of the mesh closest to the atomistic region. As we describe the various methods, we will see that compatibility can be imposed in a "strong" or "weak" sense, as we elaborate here.

"Strong compatibility" imposed by the continuum on the atoms implies that the padding atoms are constrained to move as though they are "glued" to the finite elements in which they reside. Thus the displacement, $\boldsymbol{u}^{\alpha}$, of padding atom α is set to

$$\boldsymbol{u}^{\alpha} := \tilde{\boldsymbol{u}}(\boldsymbol{R}^{\alpha}), \tag{12.10}$$

where $\tilde{\boldsymbol{u}}$ is the interpolated finite element displacement field defined in Eqn. (12.2) and $\boldsymbol{R}^{\alpha}$ is the reference coordinate position of padding atom α.[4]

Strong compatibility is imposed by the atomistic region on the continuum by defining a subset of the nodes (typically a row like the jagged, heavy line in Fig. 12.3(b)) that are coincident with some of the atoms in $\mathcal{B}^{\mathrm{A}}$. These atoms are formally part of $\mathcal{B}^{\mathrm{A}}$. Their contribution to the total energy or forces (depending on whether the method is energy-based or forced-based) is computed as appropriate for the atomistic region. However, their positions impose a displacement boundary condition on the finite elements of $\mathcal{B}^{\mathrm{C}}$.

The main disadvantage with strong compatibility is the added complexity in mesh generation near the interface. The mesh must usually be fully refined and consistent with the underlying atomic configuration (which is normally a crystalline lattice and therefore requires additional lattice-based algorithms [AST09]). To avoid these additional requirements, some methods employ what we call "weak compatibility," where displacement boundary conditions are enforced only in some average sense, or with some type of penalty method. These methods make mesh generation much easier, but we will show later that they are generally less accurate. Examples of weak compatibility will be presented in the context of specific methods later.

3. *The handshake region* The idea of the handshake region is to provide a gradual transition from the atomistic to the continuum model. The handshake region is treated very differently in different coupled models. In some, the atomistic-to-continuum transition is abrupt as in Fig. 12.3(b), and there is no handshake region (e.g., the quasicontinuum (QC) and coupled length scales (CLS) methods of Section 12.3.2 and Section 12.3.3, the finite element atomistic (FEAt) and coupled atomistic and discrete dislocations (CADD) methods of Section 12.5.2). In others, a handshake region exists as in Fig. 12.3(a), but the ways in which it blends the continuum and atomistics descriptions can differ substantially. We discuss these details below.

[4] In some methods, a special case is adopted where the finite element mesh is fully refined to the atomic scale in the entire padding region. That is to say that a node is defined to lie on top of every atom in $\mathcal{B}^{\mathrm{P}}$. This portion of the $\mathcal{B}^{\mathrm{C}}$ region is not *atomistic*, since the region is still treated using the continuum finite element formulation. It is merely a continuum region with very small elements. A fully-refined padding region of this type serves to simplify the implementation, since as far at the atoms in $\mathcal{B}^{\mathrm{A}}$ are concerned the nodes in the continuum are effectively atoms as well. As such, the interpolation of displacements to the padding atoms is no longer required, but this does not otherwise significantly alter the model.

4. *Treatment of the continuum* As we have mentioned, B^{C} is usually modeled with FEM, but the details of the finite element formulation and the constitutive law adopted to describe the material response differ amongst the methods. In some cases, a simple, small-strain, linear elastic, finite element formulation is used with elastic constants fitted to the properties of the atomistic model used in B^{A}. In others, a finite strain (nonlinear) formulation is used together with the Cauchy–Born rule (or some more sophisticated appeal to the atomistic model) to describe the constitutive response in the nonlinear range. One exception is the so-called "coarse-grained molecular dynamics" (CGMD) model developed by Rudd and Broughton [RB98, RB00, RB05], where the continuum treatment is substantially different from conventional finite elements. Unfortunately, we have not been able to include CGMD in the study presented in this chapter.

We now turn to a description of the different methods considered here. In doing so, we divide the methods into either energy-based or force-based and explore the essential differences between these two classes of methods.

12.3 Energy-based formulations

12.3.1 Total energy functional

In an energy-based formulation, it is assumed that the total potential energy Π of a body B can be written as the sum of the potential energies of the three subbodies, B^{C}, B^{A} and B^{H}, from which it is composed. (Recall that the padding region, B^{P}, is really considered a part of the continuum.) Thus

$$\Pi = \Pi^{\mathrm{A}} + \widetilde{\Pi}^{\mathrm{C}} + \widetilde{\Pi}^{\mathrm{H}}, \tag{12.11}$$

where the tilde on Π^{C} and Π^{H} indicates that the potential energy of the continuum is approximated by a finite element formulation. For this approach to work, a prescription must be provided for computing the energy of each of these regions. This is straightforward in the continuum, where an energy density is postulated to exist at every point. In the atomistic region, the analogy of an energy density is a well-defined energy per atom, which most empirical atomistic models permit.[5] Thus the contributions from B^{A} and B^{C} are

$$\Pi^{\mathrm{A}} = \sum_{\alpha \in B^{\mathrm{A}}} E^{\alpha} - \sum_{\alpha \in B^{\mathrm{A}}} \boldsymbol{f}^{\mathrm{ext},\alpha} \cdot \boldsymbol{u}^{\alpha}, \tag{12.12}$$

$$\Pi^{\mathrm{C}} = \int_{B^{\mathrm{C}}} W(\boldsymbol{F})\, dV - \int_{\partial B^{C}_{0t}} \bar{\boldsymbol{T}} \cdot \boldsymbol{u}\, dA, \tag{12.13}$$

[5] It is worth noting, however, that a quantum mechanical DFT description cannot be uniquely decomposed into an energy-per-atom description. Even for empirical models, the issue of the uniqueness of partitioning energy to atoms remains an open question. See Admal and Tadmor [AT11] for more on this.

where $\boldsymbol{u}^\alpha$ is the displacement of atom α, $\boldsymbol{u}(\boldsymbol{X})$ is the continuum displacement field, the energy of atom α is represented by E^α and applied forces and tractions are denoted by $\boldsymbol{f}^{\text{ext},\alpha}$ and $\bar{\boldsymbol{T}}$, respectively. Once discretized by finite elements, the continuum contribution is approximated by numerical quadrature as

$$\widetilde{\Pi}^{\text{C}} = \sum_{e=1}^{n_{\text{elem}}} \sum_{q=1}^{n_{\text{q}}} w_q V_0^e W(\boldsymbol{F}(\boldsymbol{X}_q^e)) - \bar{\mathbf{f}}^T \mathbf{u}, \tag{12.14}$$

where n_{elem} is the number of elements, V_0^e is the volume[6] of element e, n_{q} is the number of quadrature points, $\boldsymbol{X}_q^e$ is the position of quadrature point q of element e in the reference configuration, w_q are the associated quadrature weights and $\mathbf{f}$ and $\mathbf{u}$ represent the vector of applied forces and nodal displacements, respectively, in the finite element region.

The energy of the handshake region is a partition-of-unity blending of the atomistic and continuum energy descriptions. For example, one energy-based method which includes a handshake region is the bridging domain (BD) method described in [XB04]. Within the BD handshake region, both the continuum and atomistic energies are used, but their contributions are weighted according to a function Θ that varies linearly from 1 at the edge of B^{H} closest to B^{C} to 0 at the edge closest to B^{A}. Assuming no externally applied forces on this region to simplify the notation, the energy of the handshake region is

$$\Pi^{\text{H}} = \sum_{\alpha \in B^{\text{H}}} (1 - \Theta(\boldsymbol{R}^\alpha))E^\alpha + \int_{B^{\text{H}}} \Theta(\boldsymbol{X}) W(\boldsymbol{F}(\boldsymbol{X}))\, dV. \tag{12.15}$$

If we take the specific case of constant strain triangular elements, and make the relatively minor restriction that the handshake region starts and ends along a contiguous line of element edges (in other words, every element is either entirely "in" or entirely "out" of region B^{H}), it is simple to evaluate the integrals above. Since W is constant within each constant strain element in this case, the energy becomes

$$\widetilde{\Pi}^{\text{H}} = \sum_{\alpha \in B^{\text{H}}} (1 - \Theta(\boldsymbol{R}^\alpha))E^\alpha + \sum_{e \in B^{\text{H}}} \Theta(\boldsymbol{X}^e_{\text{cent}}) W(\boldsymbol{F}(\boldsymbol{X}^e_{\text{cent}}))V_0^e, \tag{12.16}$$

where $\boldsymbol{X}^e_{\text{cent}}$ is the coordinate of the quadrature point of element e (the centroid of the triangle in this specific case).

We emphasize that the padding atoms in region B^{P}, described previously, are distinct from the atoms contained in region B^{H}; B^{H} and B^{P} do not overlap. The padding region corresponds to a set of atoms whose energy is not explicitly included in the energy functional but whose positions are required to provide the full neighbor environment to other atoms. In fact, a padding region is still needed in the BD model (in addition to the atoms in the handshake region). The padding region lies inside B^{C} along the edge of B^{H}.

For energy-based methods with no handshake, the energy reduces to

$$\Pi = \Pi^{\text{A}} + \widetilde{\Pi}^{\text{C}}. \tag{12.17}$$

[6] For a two-dimensional model, V_0^e is equal to the area of the element multiplied by the periodic length in the out-of-plane direction.

Given the positions of the atoms in the atomistic region, the displacements of the nodes in the finite element region and a prescription for generating the padding atoms, we can compute the energy of Eqn. (12.11) or Eqn. (12.17), and its derivatives as needed. The total potential energy is then minimized subject to the imposed displacement boundary conditions to obtain the equilibrium configuration of the system.

We now review several energy-based partitioned-domain methods from the literature, focusing on their static limit. Our starting point will be the QC method because this is the method with which we are most familiar. Where a model is similar to QC or some other model already described, we will simply state this similarity without rehashing the details. This is not meant to imply, necessarily, that one method was a direct off-shoot from another. It shows only that different methods, developed independently, often share common traits.

12.3.2 The quasi-continuum (QC) method

The two-dimensional version of the QC method [TOP96, SMT$^+$99] is freely available as open source software at [TM09], complete with example input files, a tutorial guide and a reference manual. The QC method is an energy-based method with no handshake regions, so the energy functional of Eqn. (12.17) is used. However, there is a conceptual advantage in developing this equation from a point of view that makes no distinction between atoms and nodes, for reasons that will become clear later when we introduce cluster-based methods. In the QC literature, any node or atom that is retained in the model as part of the set of degrees of freedom is identified as a "representative atom" or simply "repatom." The terminology "nonlocal repatom" and "local repatom" is used to distinguish between repatoms in the atomistic region where the atomic bonding is nonlocal and repatoms in the continuum region where a local constitutive relations is applied.[7]

The idea is that we seek to compute a good approximation to the total energy of the fully-atomistic system, which is a function of the positions of all the atoms in the body, $\boldsymbol{r}^\alpha$ ($\alpha = 1, \ldots, N$). As usual we can assume, without loss of generality, that we know some reference configuration of the atoms, $\boldsymbol{R}^\alpha$, and then work in terms of displacements, $\boldsymbol{u}^\alpha = \boldsymbol{r}^\alpha - \boldsymbol{R}^\alpha$. The potential energy of the atomistic system is then

$$\Pi = \mathcal{V}^{\text{int}}(\boldsymbol{u}^1, \ldots, \boldsymbol{u}^N) - \sum_{\alpha=1}^{N} \boldsymbol{f}^{\text{ext},\alpha} \cdot \boldsymbol{u}^\alpha, \tag{12.18}$$

where $\mathcal{V}^{\text{int}}$ is the interaction energy of the atoms, and we have included the possibility of external forces applied to the atoms.

We reduce generality somewhat by assuming a class of atomistic models that permits the identification of a site energy, E^α, for each atom, and thus

$$\mathcal{V}^{\text{int}}(\boldsymbol{u}) = \sum_{\alpha=1}^{N} E^\alpha(\boldsymbol{u}), \tag{12.19}$$

where $\boldsymbol{u} = (\boldsymbol{u}^1, \ldots, \boldsymbol{u}^N)$.

[7] See Section 2.5.1 for more on the local versus nonlocal terminology of continuum mechanics.

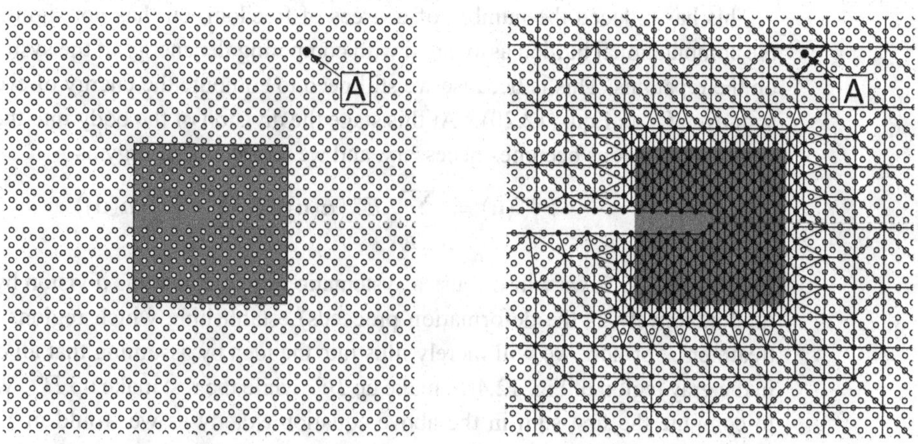

Fig. 12.4 (a) The atoms in the vicinity of a crack tip are divided into the atomistic region (shaded) and the continuum region (unshaded). (b) Filled circles show the selected repatoms, which are then meshed by linear triangular elements.

We start by reducing the number of degrees of freedom in the problem. To achieve this, we constrain the motion of most of the atoms to follow the motion of the repatoms. How we choose these repatoms will be discussed in Section 12.6, but we can imagine something similar to the illustration in Fig. 12.4, where the density of repatoms varies according to the expected severity of the deformation. In some critical regions (as around a crack tip in this example) all atoms are selected as repatoms, while very few are selected in regions that are experiencing a close to uniform deformation. For now, we assume that the displacements of some n_R repatoms, $n_R \ll N$, are chosen as the degrees of freedom. A finite element mesh with constant strain triangles (in two dimensions) or constant strain tetrahedra (in three dimensions) is defined by the repatoms as shown in Fig. 12.4(b), and the repatoms behave as the finite element nodes. The displacements of all other atoms can then be determined from the displacement of these node/repatoms through interpolation. In this way, the potential energy of the atoms becomes a function of *only* the node/repatom displacement vector, $\mathbf{u}$:

$$\widetilde{\mathcal{V}}^{\text{int}}(\mathbf{u}) \equiv \mathcal{V}^{\text{int}}(\boldsymbol{u}(\mathbf{u})) = \sum_{\alpha=1}^{N} E^{\alpha}(\boldsymbol{u}(\mathbf{u})), \tag{12.20}$$

where the dependence of the atom displacements on the repatom displacements is through the finite element shape functions. By the compact support of these functions, we know, for example, that the displacement of atom A in Fig. 12.4(b) is a linear interpolation of the displacements of the three repatoms forming the nodes of the highlighted element.

Clearly, the constraints introduced by the interpolation of the displacements will mean some level of approximation in the model, but we can presumably control the error by recognizing regions where the element size needs to be refined. In the fully-refined limit where every atom is chosen as a repatom we recover ordinary lattice statics, since in this limit there are no constrained atoms between the repatoms.

This has reduced the number of degrees of freedom, but has not significantly lowered the computational burden associated with computing the energy of the system or the forces on the repatoms. This is because we still need to compute the energy of every atom in the sum over E^α in Eqn. (12.20).[8] At this stage the division of the body into the atomistic and continuum regions becomes necessary, and we rewrite the energy as

$$\widetilde{\mathcal{V}}^{\text{int}}(\mathbf{u}) = \sum_{\alpha \in \mathcal{B}^{\text{A}}} E^\alpha(\boldsymbol{u}(\mathbf{u})) + \sum_{\alpha \in \mathcal{B}^{\text{C}}} E^\alpha(\boldsymbol{u}(\mathbf{u})). \tag{12.21}$$

In Section 12.6.5, we will discuss a systematic way to make this division by automatically (and adaptively as the deformation proceeds) establishing which regions need atomistic treatment. For now, we will merely illustrate the idea by assuming that all of the repatoms in the gray region of Fig. 12.4(b) make up $\mathcal{B}^{\text{A}}$, while the rest make up $\mathcal{B}^{\text{C}}$. We then propose to compute the first sum in the above equation exactly as we would in a fully-atomistic model; the only difference from the exact result comes from the fact that the padding atoms in neighboring regions are undergoing a constrained deformation. The second sum, on the other hand, we replace by finite elements and the Cauchy–Born constitutive rule, as discussed in Section 12.1. Thus for an atom in $\mathcal{B}^{\text{C}}$ we assume that it is in a region of approximately uniform deformation, characterized by $\boldsymbol{F}$, so that

$$E^\alpha \approx \Omega_0 W(\boldsymbol{F}(\boldsymbol{R}^\alpha)), \tag{12.22}$$

where Ω_0 is the Wigner–Seitz volume of a single atom in the reference configuration. The second sum over atoms in Eqn. (12.21) can now be replaced by a sum over *elements* as

$$\widetilde{\mathcal{V}}^{\text{int}}(\mathbf{u}) \approx \sum_{\alpha \in \mathcal{B}^{\text{A}}} E^\alpha(\boldsymbol{u}(\mathbf{u})) + \sum_{e \in \mathcal{B}^{\text{C}}} \nu_e \Omega_0 W(\boldsymbol{F}^e(\mathbf{u})), \tag{12.23}$$

where ν_e is the number of atoms associated with element e and $\boldsymbol{F}^e$ is the deformation gradient there, which depends on $\mathbf{u}$ through Eqn. (12.3).

A practical point concerns the determination of ν_e, as it would seem difficult to unambiguously assign atoms to specific elements if they lie at element corners or edges. In fact, $\nu_e \Omega_0$ can simply be taken as the total volume of the element, eliminating the need for explicitly counting atoms. This leaves one subtlety at the atomistic–continuum interface, as shown in Fig. 12.5, where the darkly shaded element has a node/atom that belongs to $\mathcal{B}^{\text{A}}$ and is therefore already accounted for in the atomistic contributions. For this element, the volume used should be the total volume less the volume overlapping the Voronoi cell of the atom in $\mathcal{B}^{\text{A}}$. In this way, $\nu_e \Omega_0$ values for elements touching $\mathcal{B}^{\text{A}}$ are reduced to avoid the double counting of energy. In practice, approximations to the Voronoi cell can be used to expedite calculations without too much consequence for accuracy [SMT$^+$99].

We have introduced constant strain elements because they provide the most natural justification for the Cauchy–Born rule that was used: within each element there is indeed a uniform strain applied to the crystal. It is also clear that Eqn. (12.17) is formally equivalent to the more traditional implementation of finite elements when Eqn. (12.17) is written for the case of constant strain triangles with a single quadrature point. The only difference

[8] It is worth pointing out that up to this point there are still no ghost forces in the model. These are introduced next as the continuum approximation is imposed.

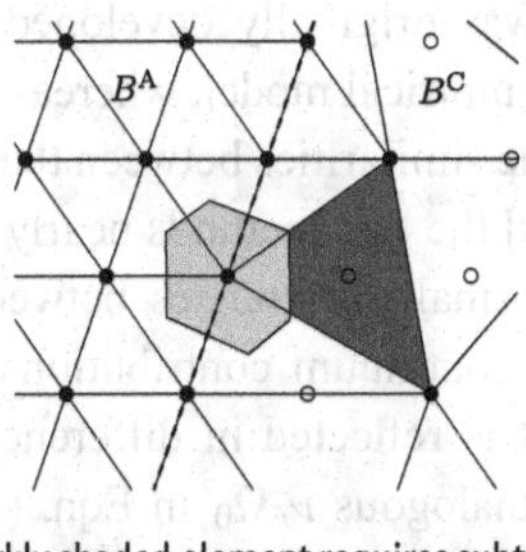

Fig. 12.5 Computing the effective volume of the darkly shaded element requires subtraction of the part overlapping the Voronoi cells of atoms inside B^A (one is lightly shaded). The dashed line shows the boundary between regions B^A and B^C.

is the modified weights near the atomistic–continuum interface to avoid double counting. The QC method can easily be formulated for higher-order finite elements as well, although the benefit is questionable: regions requiring higher-order elements are likely to require an atomistic treatment instead of the Cauchy–Born approximation. More details about the implementation and use of the QC method are given in Section 12.6. See also the one-dimensional example in Section 12.4.

12.3.3 The coupling of length scales (CLS) method

The CLS method was developed in a series a papers [ABBK98, BABK99, ABBH00, Rud01] focused initially on the problem of fracture in silicon. It was later applied to the vibrational response of nanoscale resonators [RB99].

In addition to the atomistic–continuum coupling that is our focus here, CLS also includes a coupling between molecular dynamics (MD) and the tight-binding approximation to quantum mechanics. As such, there are three concurrent regimes in the CLS approach: continuum coupled to classical atomistic coupled to quantum mechanics.[9] For this particular discussion our focus will be only on the atomistic/continuum coupling.

CLS is an energy-based method like QC. Compared with the QC energy in Eqn. (12.23), the energy of the CLS method takes the form[10]

$$\widetilde{\mathcal{V}}^{\rm int}(\mathbf{u}) = \mathcal{V}_A^{\rm int}(\boldsymbol{u}(\mathbf{u})) + \sum_{e=1}^{n_{\rm elem}} W(\boldsymbol{F}^e(\mathbf{u}))w^e, \tag{12.24}$$

where $\mathcal{V}_A^{\rm int}$ is the energy contained in region B^A, and the second sum is over all the elements in the continuum region. For most elements, the weight w^e is simply the volume of the element. The exceptions are elements directly adjacent to the atomistic–continuum interface, where the weights are reduced to avoid double counting.

[9] More recently, MD–QM coupling (or more precisely MS–QM coupling, since the focus has been on the static case) of this type has been developed by a number of researchers [FHCO04, LTK06].

[10] The original CLS method used a small-strain approximation to describe the continuum region rather than the Cauchy–Born rule used in QC. However, conceptually the methods are similar. To facilitate the comparison, we generalize the original CLS approach here by adopting a nonlinear CB model in the continuum region.

Because CLS was originally developed for silicon,[11] it was crafted in terms of the Stillinger–Weber empirical model, whereas QC was focused on pair functionals more suitable for metals. The similarities between the models are largely disguised by these different starting points, and the two methods nearly converge given the same underlying atomistic model. There are small differences between QC and CLS in the method of weighting the atomistic and continuum contributions at the interface to avoid double counting of energy. The effect is reflected in differences between the values of w^e in Eqn. (12.24) for CLS and the analogous $\nu_e\Omega_0$ in Eqn. (12.23) for QC. These differences lead to only slight changes in the error and the rate of convergence of the models. It is not clear that either the QC or the CLS weighting scheme is superior; it seems that the relative error is problem-dependent.

12.3.4 The bridging domain (BD) method

One way to try to mitigate the interfacial error is to make the transition from nonlocal atomistics to local continuum mechanics less abrupt. Although somewhat ad hoc, this can be achieved by the simple blending of energy in the handshake region between the two descriptions as described earlier in Section 12.3. This approach is related to the so-called Arlequin method [BDE$^+$08], which is a general approach to coupling different models based on a blended energy functional. The BD method is perhaps the first application of the blending region within an energy-based scheme [XB04], and is certainly representative of such models. Briefly, the BD model makes use of Eqn. (12.11) to define the energy, including a handshake region with a specifically-defined blend of the atomistic and continuum energy functionals via Eqn. (12.15).

In the limit where the thickness of the handshake region goes to zero and the size of the finite elements at the interface approaches the atomic spacing, we recover, more or less, the QC or CLS formulation. This is not precisely so, since the weights of the atomistic and continuum energy contributions in QC, CLS and BD would still be slightly different at the interface. However, these differences will not substantially change the strength of the ghost forces that will occur in the interfacial region.

An important practical difference between the methods discussed previously and the BD method is the way in which compatibility is imposed between the atomistic and continuum regions. In the BD method, displacement constraints are introduced into the handshake region to force the displacement $\boldsymbol{u}^\alpha$ for every atom α in B^{H} to follow the displacements dictated by the interpolated displacement field at the atoms reference position, $\boldsymbol{u}(\boldsymbol{R}^\alpha)$. This is done using the augmented Lagrangian method, as follows. First, we introduce the vector

$$\boldsymbol{h}^\alpha \equiv \boldsymbol{u}(\boldsymbol{R}^\alpha) - \boldsymbol{u}^\alpha = \sum_{I=1}^{n_{\mathrm{nodes}}} S^I(\boldsymbol{R}^\alpha)\mathbf{u}^I - \boldsymbol{u}^\alpha, \tag{12.25}$$

[11] Note that for a multilattice crystal, such as silicon, the strain energy density in Eqn. (12.24) should also depend on the "shifts" of the sublattices. See, for example, [TSBK99]. We do not include this since here we focus on simple lattice crystals.

which we want to constrain to be equal to zero. This can be achieved by redefining the handshake region energy in Eqn. (12.16) to be

$$\widetilde{\Pi}^{\mathrm{H}} = \sum_{\alpha \in B^{\mathrm{H}}} (1 - \Theta(\boldsymbol{R}^\alpha)) E^\alpha(\boldsymbol{u}) + \sum_{e \in B^{\mathrm{H}}} \Theta(\boldsymbol{X}^e_{\mathrm{cent}}) W(\boldsymbol{F}(\boldsymbol{X}^e_{\mathrm{cent}})) V^e_0 + \sum_{\alpha \in B^{\mathrm{H}}} \left[\beta_1 \boldsymbol{\lambda}^\alpha \cdot \boldsymbol{h}^\alpha + \frac{\beta_2}{2} \boldsymbol{h}^\alpha \cdot \boldsymbol{h}^\alpha \right], \tag{12.26}$$

where $\boldsymbol{\lambda}^\alpha = (\lambda^\alpha_1, \lambda^\alpha_2, \lambda^\alpha_3)$ are multipliers for the degrees of freedom of atom α, and β_1 and β_2 are penalty functions that can be chosen to optimize computational efficiency. Energy minimization can then proceed with $\boldsymbol{\lambda}$ as additional degrees of freedom. Note that the original BD paper was presented as a dynamic method. As such, forces on atoms and nodes come from the differentiation of this energy functional and include effects due to the constraint term. Since this term constitutes no more than a mathematical "trick" without clear physical significance (for example what is the right choice of the penalties β_1 and β_2?), it is not clear how to interpret the resulting dynamics near the interface. In the static case, on the other hand, this is a convenient tool for finding a constrained minimum energy configuration, and the magnitude of the ghost forces can be examined. The use of a finite handshake region has the effect of smearing out the ghost forces, reducing the ghost force on a given atom or node, but also introduces ghost forces on more atoms and nodes. This will be more clearly evident in the discussion of Section 12.7.

Relaxing the condition of "strong compatibility" In the QC and CLS methods, the "last row" of atoms is identified and made coincident with the "first row" of finite element nodes. The positions of these atoms provide the displacement boundary condition for the finite element region, which in turn imposes its displacement field on the "padding" atoms. This is what we have called the "strong compatibility" approach.

Using the multipliers of the BD method is not a strong compatibility approach. Although they do, in the static limit, lead to the atoms in the handshake region identically conforming to the continuum displacement fields, they relax the compatibility constraint in two important ways. The first is that there is no longer a need to identify a one-to-one correspondence between atoms and nodes along an interface. The second is that in dynamic systems, departures from strong compatibility are possible and lead to additional forces on the atoms and nodes in the handshake region.

The most useful feature of relaxing strong compatibility is that this eliminates the need for one-to-one correspondence between atoms and nodes at the interface. This makes mesh generation easier, since elements can be made in any convenient way without regard for compatibility with the underlying atomic lattice.[12] On the other hand, we will show later that the trade-off is a reduced accuracy in the solution. An additional downside of weak compatibility is that extending the atomistic region at the expense of the continuum region, or vice versa, through automatic mesh adaption, becomes more difficult.

12 When strong compatibility is enforced, additional numerical algorithms for moving nodes onto lattice sites are required. See, for example, [AST09].

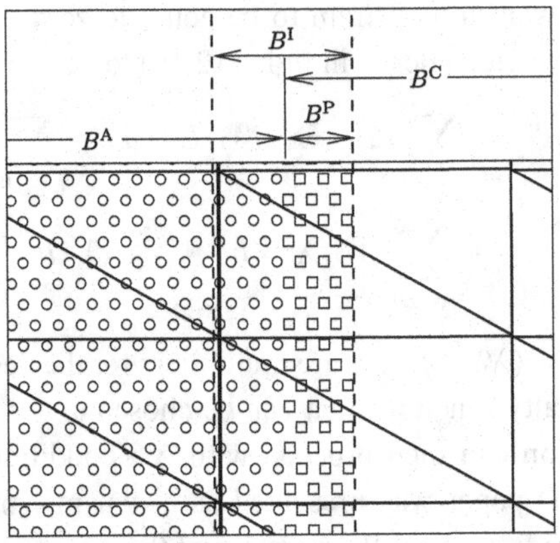

Fig. 12.6 The BSM interface region. There is no handshake region and the finite elements exist throughout the body in order to store the coarse-scale displacement field.

12.3.5 The bridging scale method (BSM)

The original formulation of the BSM [WL03] focused on dynamic simulations, while a later paper [QWL04] presented a static version of the approach. We discuss the static method here, leaving the dynamical implementation to Chapter 13.

BSM is an energy-based method with no handshake region (like QC and CLS), but it is conceptually very different from other methods. This is because in the BSM there are (in principle) no separate continuum and atomistic regions. Instead, the BSM introduces the notion of *coarse-scale* and *fine-scale* displacement fields that exist *everywhere* in the body. As such, the picture is one where atoms exist throughout the body with a displacement

$$\boldsymbol{u} = \boldsymbol{u}' + \boldsymbol{u}'', \tag{12.27}$$

where $\boldsymbol{u}'$ and $\boldsymbol{u}''$ are the coarse- and fine-scale displacements, respectively. The former is stored on a finite element mesh everywhere (overlapping both B^{C} and B^{A}). The fine-scale displacements are assumed to exist in B^{C} as the manifestation of heat, but they are only tracked and stored inside B^{A}. The BSM interface region is illustrated in Fig. 12.6.

The coarse-scale displacement field is stored on the finite element grid, and is defined as a least-squares fit to the underlying atomic displacements. In the atomistic region, this fit is explicitly determined from the atoms, while in the continuum region the inverse thinking is applied: it is imagined that the untracked atoms are vibrating around equilibrium positions that are known from the finite element displacements. In the zero-temperature static limit, the fine-scale displacements are assumed to be zero everywhere in B^{C}, so the static BSM is similar to the QC or CLS approach with two additional features: the partitioning of atomic displacements using Eqn. (12.27) and a different compatibility condition. This is a mix of weak and strong compatibility, which we describe next by referring to Fig. 12.6.

Compatibility is first imposed on the atomistic region by constraining the square atoms in the padding region (B^{P}) to follow the coarse-scale displacements of the finite elements using Eqn. (12.10) (strong compatibility). Since it is assumed that there is no fine-scale displacement in B^{C} at zero temperature, this completely determines their positions. On the other hand, compatibility of the continuum region is imposed at all nodes lying inside B^{A}. Their displacements are constrained to follow the least-squares fit of the atomic displacements in neighboring elements (i.e. $\boldsymbol{u}'$). This fully constrains elements deep inside B^{A}, whose energy is not included in the model, and constrains some of the nodes of elements straddling B^{I}. As we will see, this amounts to the same weak compatibility as described in more detail for the composite grid atomistic continuum method (CACM) model in the next section, with a particular choice of the weighting function ω^I in Eqn. (12.28). In this way, the BSM mixes elements of weak and strong compatibility: strong compatibility for the motion of the padding atoms and weak compatibility for the enforcement of the finite element boundary condition.

The full energy of the atoms shown as open circles in Fig. 12.6 is included in the BSM, while the energy of any finite element overlapping the filled circles is reduced by an amount that reflects the volume of the overlap. So, for instance, the energy of an element completely inside B^{A} is not included at all, the energy an element overlapping B^{I} is partially included, and that of an element deep inside B^{C} is included as usual. Formally, then, the energy is identical to the QC energy of Eqn. (12.23), but with different weighting factors ν_e and a relaxation of the constraint that interfacial elements must be fully refined.

The introduction of the coarse-scale displacement with Eqn. (12.27) within the atomistic region leads to a larger set of dependent variables for which the equations must be solved; instead of satisfying equations for $\boldsymbol{u}$, we must now solve equations for two fields $\boldsymbol{u}'$ and $\boldsymbol{u}''$. As outlined in [QWL04], this means that the static BSM requires an iterative approach to the solution, alternatively holding fixed the fine-scale variables while solving for the coarse-scale and vice versa. As we will discuss more quantitatively in Section 12.7, iterative approaches like this tend to be quite slow.

12.3.6 CACM: iterative minimization of two energy functionals

The problem of ghost forces stems from trying to combine two energy functionals from different models (an atomistic and a continuum model) into a single, coupled energy expression. A way to avoid ghost forces while retaining an energy formulation was proposed by Datta *et al.* [DPS04] in the so-called CACM. In CACM, no attempt is made to combine the atomistic and continuum energy functionals. Instead, they work with the two energy functionals separately.

The CACM method can be described using our generic interface pictured in Fig. 12.3(a) without a handshake region.[13] First, the continuum problem is defined by setting the boundary condition on nodes that lie inside the atomistic region. This is done in an average

[13] For convenience Datta *et al.* [DPS04] chose to fill the entire body with a finite element mesh that overlaps the atomistic region similar to Fig. 12.6. Because of the local nature of the finite element energy functional, elements deep inside B^{A} that do not overlap the padding region do not affect the solution and so the formulation is equivalent to Fig. 12.3(a). The overlapping elements are convenient, however, to simplify the mesh building process, since there is neither the need to mesh a nonconvex region containing a "hole" in which the atoms reside, nor the need for a complicated fine-scale mesh around the atomistic/continuum interface.

sense, using a form of weak compatibility. In general

$$\bar{\mathbf{u}}^I = \sum_{\alpha=1}^{N} \omega^I(\boldsymbol{R}^\alpha)\boldsymbol{u}^\alpha, \tag{12.28}$$

where the $\bar{\cdot}$ indicates a prescribed displacement and ω^I is some general weighting function for node I. The CACM developers [DPS04] chose to define ω^I as a cubic function over a sphere of radius r_{cut} surrounding node I:

$$\omega^I(\boldsymbol{R}^\alpha) = 1 - 3\left(\frac{r}{r_{\text{cut}}}\right)^2 + 2\left(\frac{r}{r_{\text{cut}}}\right)^3,$$

where $r = \|\boldsymbol{X}^I - \boldsymbol{R}^\alpha\|$. Choosing

$$\omega^I(\boldsymbol{R}^\alpha) = \delta(\boldsymbol{X}^I - \boldsymbol{R}^\alpha), \tag{12.29}$$

introduces the special case of constraining the node I to move exactly in step with an atom α at the same position. Other choices will see the displacement of node I take on some weighted average of the displacements of atoms in its immediate vicinity.

Given this scheme to find the finite element nodal displacements within $\mathcal{B}^{\text{A}}$ from the atomic displacements, the nodes within $\mathcal{B}^{\text{A}}$ are held fixed and the energy of the continuum region is minimized without any consideration of the atomistic energy. Once a minimum is found, the atoms lying in the padding region are positioned according to the current displacement field in the finite elements (in this instance, CACM uses strong compatibility through Eqn. (12.10), so that CACM combines characteristics of strong and weak compatibility). These padding atoms are then fixed, and the energy of the atomistic system is minimized, again without regard for the continuum energy. This leads to a new set of atomic positions, which are used to find new displacements for the nodes inside $\mathcal{B}^{\text{A}}$, and the process repeats until the change in atomistic energy during an iteration is below a specified tolerance. This type of an approach is often referred to as an "alternating Schwarz" method in the mathematical literature.

The main advantage of CACM is its modularity. In principle, it can be implemented with a simple interface between two fully independent codes: a lattice statics code and a finite element code. The coupling can be treated as a postprocessing step on the solution from each iteration, and since most of the time will be spent in minimizing the two energy functionals, the implementation of the coupling need not be especially efficient. However, the main disadvantage of this method is that it will be extremely slow to converge for nonlinear problems. In Section 12.7, we will clearly illustrate this slow convergence.

12.3.7 Cluster-based quasicontinuum (CQC-E)

If we return to the discussion of the QC method from Section 12.3.2, we can follow a different path starting from Eqn. (12.20). Recall that in the QC method, the step following Eqn. (12.20) was to divide the body into two regions that would ultimately be treated differently: one where the energy per atom is computed using the fully-atomistic approach

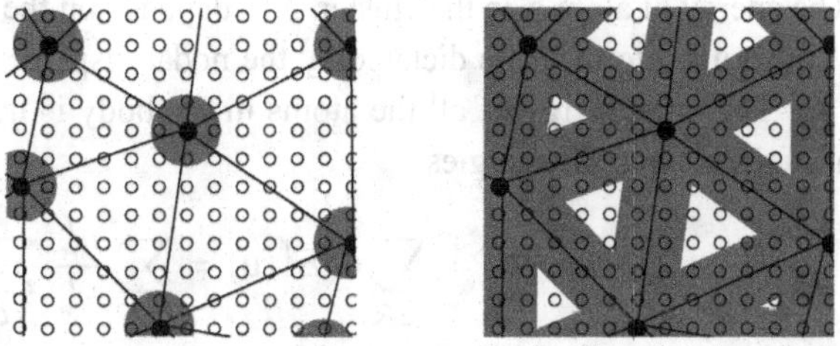
(a) (b)

Fig. 12.7 (a) Clusters of repatoms around each node used to compute the energy in CQC-E. (b) Atoms that need to be included in the "clusters" in CQC-E for an exact elimination of the ghost forces (see Section 12.4).

(the "nonlocal region") and one where the energy was found using the Cauchy–Born rule (the "local region"). We will next describe a different approach that was originally posited as a force-based method in [KO01] (see Section 12.5.5), and later recast as an energy-based method [ES09].

The idea is illustrated in Fig. 12.7(a). As outlined before, we select a handful of "repatoms" to also act as nodes, and these are connected by a finite element mesh. The atoms between the nodes are constrained to move according to the interpolated finite element displacement field. The CQC-E method[14] now posits that the energy of *all* of the atoms can be estimated by only computing the energy of a handful of atoms around each node. These are the atom "clusters" highlighted in the figure. The number of atoms in the cluster, ν, is a parameter of the model, with larger clusters improving accuracy while increasing the computational effort. Where it is necessary to indicate the size of the cluster used for a given model, we will indicate it in parenthesis. For example, CQC(13)-E indicates an energy-based cluster method with each cluster containing 13 atoms.

The energy of Eqn. (12.20) can then approximated as follows. The energy of a single node/repatom I is found by computing the average energy of all the atoms in its cluster, the set of which are denoted $\mathcal{C}^I$:

$$\bar{E}^I(\mathbf{u}) \approx \frac{1}{\nu^I} \sum_{\alpha \in \mathcal{C}^I} E^\alpha(\boldsymbol{u}[\mathbf{u}]), \tag{12.30}$$

where ν^I is the number of atoms in cluster I. Based on the discussion so far, the superscript on ν seems unnecessary since we have suggested that all clusters are the same size. In fact, in highly refined regions (and especially the atomistic region) there will be overlap of the clusters for any $\nu > 1$. This is resolved by assigning atoms only to their nearest repatom, so that no atom is part of more than one cluster (ambiguous cases that are equidistant from several repatoms are resolved randomly). As such, some repatoms near the atomistic/continuum interface will have $\nu^I < \nu$, and all of the atomistic region will have $\nu^I = 1$ regardless of the nominal choice of ν for the model.

[14] CQC-E refers to the energy-based cluster quasicontinuum method. In Section 12.5.5, we discuss the original force-based CQC-F method.

The energy of atom α in the cluster, E^α, depends on the positions of all atoms, but these are constrained to move as dictated by the nodal displacements $\mathbf{u}$ as in Eqns. (12.10) and (12.2). The total energy of all the atoms in the body is now approximated by a weighted sum of these repatom energies:

$$\mathcal{E}_{\text{tot}}(\mathbf{u}) = \sum_{I\in\mathcal{R}} n^I \bar{E}^I(\mathbf{u}) = \sum_{I\in\mathcal{R}} \frac{n^I}{\nu^I} \sum_{\alpha\in C^I} E^\alpha(\boldsymbol{u}[\mathbf{u}]), \tag{12.31}$$

where $\mathcal{R}$ denotes the set of atoms that are selected to be repatom/nodes and n^I is a weight assigned to each repatom. The weights n^I can be chosen in a variety of ways, and can be thought of as the number of atoms whose energy is *represented* by repatom I. For example one might draw the Voronoi diagram of the repatoms and then count or otherwise approximate the number of atoms falling within the cell of repatom I. Other schemes for computing n^I are possible, as well as different weighted-averaging techniques to compute $\bar{E}^I$ from the cluster atom energies (Eqn. (12.30)). These differences do not strongly affect the behavior, speed or accuracy of the method, although a sensible restriction on the selection of the n^I values is that

$$\sum_{I\in\mathcal{R}} n^I = N, \tag{12.32}$$

where N is the total number of atoms in the entire body.

In regions where the distance between repatoms becomes small, the method naturally goes over to the fully-atomistic limit if we insist that clusters cannot overlap. Instead, they shrink to $\nu^I = 1$ in the fully-refined limit, where the repatom weights become $n^I = 1$ as well. In that case, it is clear that Eqn. (12.31) reduces to the lattice statics energy expression.

Although it is no longer as apparent, there is still a division between "atomistic" and "continuum" regions in this model. In essence, repatoms with $n^I = 1$ constitute $\mathcal{B}^{\mathrm{A}}$, whereas $\mathcal{B}^{\mathrm{C}}$ is spanned by elements touching repatoms with $n^I > 1$. In $\mathcal{B}^{\mathrm{C}}$, the continuum assumption is not in the form of a new constitutive law, since the energy $\bar{E}^I$ for repatoms in this region is still obtained from a (cluster-averaged) fully-atomistic energy calculation subject to the constrained deformation $\boldsymbol{u}(\mathbf{u})$ (Eqn. (12.2)). But there is still a continuum assumption made here, in that the method assumes a smooth variation of the energy from one atom to the next.

For clusters of size $\nu^I = 1$ in coarsely meshed regions, the accuracy of CQC(1)-E is quite poor, and it is necessary to use larger clusters or a ghost-force correction technique as shown in [ES09]. We will discuss this at more length in subsequent sections.

12.4 Ghost forces in energy-based methods

The approximations inherent in energy-based coupling schemes lead to errors known as "ghost forces."[15] *All energy-based coupling methods suffer from ghost forces to various*

[15] See footnote 3 on page 606.

degrees.[16] To define the ghost forces, we build a model in which the atoms are on their equilibrium crystal lattice sites and the finite elements are unstressed and undeformed. We avoid surface effects by embedding the atomistic region inside the continuum region. In the absence of coupling effects, this should be an equilibrium configuration, where the forces on all atoms and nodes are zero. Any nonzero forces are unphysical and will lead to spurious distortions of the body upon relaxation. These unphysical forces are the ghost forces. This definition is related to the *patch test* for finite elements,[17] but note that the patch test requires application of a uniform deformation to remain uniform. The definition of ghost forces is more limited requiring only that the *undeformed* reference state remains undeformed if it is in fact an equilibrium configuration of the underlying atoms.

Ghost forces arise from the inherent mismatch between the *nonlocal* atomistic region and the *local* continuum region. Thus while the forces on an atom can be affected by atoms a finite distance away, a basic tenet of *local* continuum mechanics is that the state variables at a material point are fully determined by the deformation *at that point*.[18]

Since ghost forces arise from nonlocality, it is not surprising that they identically vanish for atomistic models that involve only *nearest-neighbor* interactions in the equilibrium configuration. Clearly, such models are "local" in the sense that they only depend on the immediate environment of an atom. This is true regardless of the type of model. Thus, a nearest-neighbor pair potential has no ghost forces and neither does the three-body Stillinger–Weber potential for Si (which only involves nearest-neighbor at equilibrium).[19] Another special case where ghost forces identically vanish is that of harmonic lattices. In this case, instead of a collection of contributions to the force on an atom summing to zero, each contribution is itself identically zero. Ghost forces are therefore avoided without the the need for strict balancing of forces. There are also special cases where for certain coupling methods, ghost forces can be made to identically vanish in a one-dimensional model, but they exist in higher dimensions. This happens for example for the CQC-E method of Section 12.3.7 if sufficiently large clusters are used (see Fig. 12.7(b)).

The above discussion means that when testing the impact of ghost forces on the accuracy of coupled methods it is vital to use a sufficiently sophisticated test where ghost force effects exist. In general, the test should be at least two-dimensional and use an anharmonic, next-to-nearest neighbor interaction model in $\mathcal{B}^{\mathrm{A}}$. This is likely to make analytical analysis of the method difficult, but a numerical test for ghost forces is a useful alternative.

16 This statement may seem controversial, but we believe it to be accurate, as we have not yet seen a method that avoids the problem. It should be noted that there are special cases where for *all the methods* ghost forces disappear (see discussion of this below). But for the most general potentials in two and three dimensions all energy-based methods suffer from this effect.

17 The patch test is described in Section 9.3.6 of the companion volume to this book [TME12]. Briefly, it says that boundary conditions associated with uniform strain should produce exactly uniform strain throughout a traditional FEM mesh, regardless of the element arrangement.

18 This is the continuum principle of "local action." Nonlocal continuum formulations and so-called "gradient corrected" continuum mechanics are possible but are not considered here. See Section 2.5.1.

19 For this reason, the original implementation of CLS which used the Stillinger–Weber potential did not suffer from ghost forces, but an implementation of CLS that uses a potential with long-range interactions would.

To demonstrate the ghost force effect clearly, we follow the approach of [MT02, CM03, DL08a, DL08b] and study a one-dimensional chain of atoms, modeled using an energy-based coupled method.

12.4.1 A one-dimensional Lennard-Jones chain of atoms

Consider a one-dimensional chain of atoms that interact via a truncated Lennard-Jones potential. Specifically, we will assume that each atom on the chain interacts only with its nearest and second-nearest neighbors, according to the potential

$$\phi(r) = 4\varepsilon \left[\left(\frac{\sigma}{r}\right)^{12} - \left(\frac{\sigma}{r}\right)^{6} \right]. \tag{12.33}$$

Later in the discussion it will be useful to also have the first and second derivatives of ϕ with respect to r, so we include them both here:

$$\phi'(r) = \frac{24\varepsilon}{r}\left[-2\left(\frac{\sigma}{r}\right)^{12} + \left(\frac{\sigma}{r}\right)^{6} \right], \qquad \phi''(r) = \frac{24\varepsilon}{r^2}\left[26\left(\frac{\sigma}{r}\right)^{12} - 7\left(\frac{\sigma}{r}\right)^{6} \right]. \tag{12.34}$$

The energy of the chain can be written as $\mathcal{V}^{\text{int}} = \sum_{\alpha}^{N} E^{\alpha}$, where

$$E^{\alpha} = \frac{1}{2}\sum_{\beta \neq \alpha} \phi(r^{\alpha\beta}) = \phi(2a) + \phi(a), \tag{12.35}$$

and a is the distance between atoms along the chain. It is straightforward to minimize this energy with respect to the lattice constant a to find the equilibrium lattice constant

$$a_0 = \left(\sigma \frac{2080}{4097}\right)^{-1/6} = 1.11961\sigma, \tag{12.36}$$

and the resulting equilibrium cohesive energy per atom in the chain

$$E^0_{\text{coh}} = -(\phi(2a_0) + \phi(a_0)) = 1.03124\varepsilon. \tag{12.37}$$

Note that the negative sign is introduced to reflect the convention that a cohesive energy is a positive quantity, even though it represents a reduction in energy from the atoms in isolation (zero energy) to the crystal ($-E^0_{\text{coh}}$). An important point is that this value of a_0 is not the same as the value of r at which $\phi(r)$ is minimum, which occurs at $r = 2^{1/6}\sigma = 1.12246\sigma$. The equilibrium lattice constant is slightly smaller because of the inclusion of second-neighbor interactions.[20]

Now we imagine the infinite chain of atoms to be in equilibrium, such that the lattice constant is a_0. Clearly, the energy of any subset of N atoms from the chain is just

$$\mathcal{V}^{\text{int}}_{\text{chain}} = -NE^0_{\text{coh}} \tag{12.38}$$

and the force on every atom is identically zero.

[20] For our discussion here, we can avoid the numerical issues surrounding the truncation of the potential (see Section 5.3.3) by assuming that an atom always interacts with exactly two neighbors on either side, regardless of the length of their bonds.

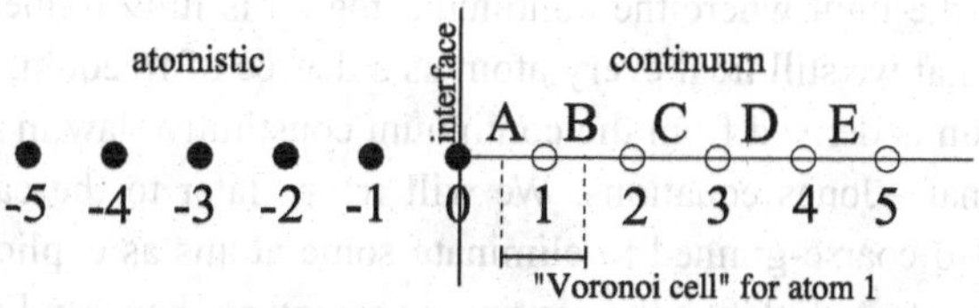

Fig. 12.8 A one-dimensional chain of atoms modeled with a partitioned-domain approach. Atom 0 lies on the atomistic–continuum interface. On the continuum side, elements A–E define the energy of the system.

12.4.2 A continuum constitutive law for the Lennard-Jones chain

The strain energy density W must be defined as a function of the deformation in order to formulate the continuum constitutive law. In the simplest form, we can assume that the continuum elements are linear elastic and that the strains will remain small. Thus,

$$W = W_0 + \frac{1}{2} D \hat{\epsilon}^2, \tag{12.39}$$

where W_0 is the energy density of the undeformed crystal and $\hat{\epsilon}$ is the small strain measure for this one-dimensional example.[21] Normally W_0 is taken to be zero for convenience in continuum models, but here it is more natural to set it to the cohesive energy per unit volume (length in one dimension) of the crystal. Thus, from Eqns. (12.36) and (12.37),

$$W_0 = \frac{-E^0_{\rm coh}}{a_0} = -0.92107 \frac{\varepsilon}{\sigma}. \tag{12.40}$$

We need to make the continuum model consistent with the Lennard-Jones atoms by fitting the constant D to the limit of an infinitesimal uniform stretching of the atomic chain. Returning to Eqn. (12.35) for the energy of a single atom in the chain with arbitrary a, we divide by the volume per atom in the undeformed chain to get the strain energy density. The Lennard-Jones expression for W is then

$$W(\hat{\epsilon}) = \frac{\phi(2a) + \phi(a)}{a_0}, \qquad a = F a_0 = (1 + \hat{\epsilon}) a_0. \tag{12.41}$$

We see from Eqn. (12.39) that the second derivative of W with respect to $\hat{\epsilon}$ gives us D directly. Thus, we take the same derivative of Eqn. (12.41) and find

$$D = \left. \frac{\partial^2 W}{\partial \hat{\epsilon}^2} \right|_{a=a_0} = a_0 \left[4\phi''(2a_0) + \phi''(a_0) \right] = 66.317 \varepsilon / \sigma. \tag{12.42}$$

12.4.3 Ghost forces in a generic energy-based model of the chain

Now consider a coupled model where the interface between the atomistic and continuum region is at atom number 0 as shown in Fig. 12.8.[22] To the left of 0 ($\alpha \leq 0$) is the atomistic region, and to the right is the region to be approximated by the continuum. For now, let us

[21] We introduce the hat on $\hat{\epsilon}$ to clearly distinguish it from the Lennard-Jones energy parameter, ε.
[22] Note that we will take the limit of no handshake region for this discussion.

take the limit where the continuum region is fully refined to the atomic scale. That is to say that we still treat every atom as a degree of freedom, but the energy in the continuum region is derived from the continuum constitutive law in Eqn. (12.39), rather than the full Lennard–Jones equations. We will return later to the case where the continuum region is also coarse-grained to eliminate some atoms as explicit degrees of freedom. Thus the elements A–E define the continuum region as shown in Fig. 12.8.

We use the QC formulation to illustrate the ghost force phenomenon, but we note that it is representative of the same problem in the other energy-based methods. Starting from Eqn. (12.23) we have for the one-dimensional chain

$$\Pi(\mathbf{u}) = \sum_{\alpha \le 0}^{\text{atoms}} E^\alpha(\mathbf{u}) + \sum_{e}^{\text{elements}} \nu_e \Omega_0 W(\hat{\epsilon}^e), \tag{12.43}$$

where $\hat{\epsilon}^e$ is a function of $\mathbf{u}$ as clarified momentarily. Here, we have ignored any externally applied loads or tractions for simplicity. The quantity $\nu_e \Omega_0$ is identically the "volume" (length in one dimension) of element e for elements far from the atomistic–continuum interface, and is reduced somewhat adjacent to the interface to avoid double counting of energy (see page 612). In this simple model, this affects only element "A" in the figure, half of whose volume is associated with atom 0. Since this atom's energy will be explicitly added to the atomistic energy, the number of atoms remaining in element A is $\nu_A = 1/2$. Because the elements are fully refined in this simple model, $\nu_e = 1$ for all other elements $B, C, \dots$ Finally, the energy of atom α is

$$E^\alpha = \frac{1}{2} \sum_{\substack{\beta=\alpha-2 \\ \beta \ne \alpha}}^{\alpha+2} \phi(r^{\alpha\beta}), \tag{12.44}$$

where the interatomic distances $r^{\alpha\beta}$ are functions of the displacements of the nodes, $\mathbf{u}$.

We confine our attention to constant strain (linear interpolation) elements, so that for an element spanning nodes I and J, we have the simple deformation mapping

$$x = (1 + \hat{\epsilon})X, \qquad \hat{\epsilon} = \frac{\mathsf{u}^J - \mathsf{u}^I}{\mathsf{X}^J - \mathsf{X}^I}. \tag{12.45}$$

Now we are in a position to compare the exact energy of this atomic chain with the approximation made by the coupled model, first in the limit of no deformation ($\mathbf{u} = \mathbf{0}$). Let us write the energy of the small subset of the atoms near the interface; atoms -4 to 4 will suffice. In the fully atomistic description of the chain, this will simply be $-9E^0_{\text{coh}}$.

For the energy of the coupled model, we need to evaluate Eqn. (12.43) for the subregion between atoms -4 and 4. Atoms -4, -3, -2 are treated atomistically, and therefore their contributions are the same as in the exact model. Atoms -1 and 0 have bonds which extend into the continuum, so we need to have "pad atoms" attached to nodes 1 and 2 to correctly coordinate the atoms. On the continuum side we have to determine how to partition the energy so that we are truly computing the energy of atoms -4 through 4, and not some of the contribution from atoms 5 and beyond. This is sensibly resolved by identifying a Voronoi cell around every atom in the continuum region, and assigning the energy contained in that Voronoi cell to the atom it surrounds. In the simple one-dimensional chain, this amounts to assigning half of an element's energy to the atom on its left end and the other half to the atom on its right, so that we want to include only half the energy of element E. Hence, the energy of atoms -4 to 4 in the coupled model can be written as

$$\Pi(\mathbf{0}) = -3E^0_{\mathrm{coh}} + [\phi(2a_0) + \phi(a_0)] + [\phi(2a_0) + \phi(a_0)] + \left(\nu_A + \nu_B + \nu_C + \nu_D + \frac{1}{2}\nu_E\right)\Omega_0\left(\frac{\phi(2a_0) + \phi(a_0)}{a_0}\right). \quad (12.46)$$

The first term is the energy of the three atoms, $-4, -3, -2$, that are treated atomistically. The term in the first set of square brackets is the energy of atom -1, which has one bond spanning the interface to atom 1. The second set of square brackets encloses the energy of atom 0 and the final terms come from the energy contained in the continuum elements, where each has the same value of W with $a = a_0$ in Eqn. (12.41). Note that the factor of $1/2$ in front of ν_E is due to the Voronoi partitioning we just discussed, while all $\nu_e = 1$ for this fully-refined model, except for $\nu_A = 1/2$ as mentioned earlier. Comparing these terms with Eqn. (12.37) we see that this reduces to exactly

$$\Pi(\mathbf{0}) = -9E^0_{\mathrm{coh}} \quad (12.47)$$

for the undeformed model, as expected for a chain of nine atoms.

The important question we wish to address, however, is: what are the forces on these atoms? Since they are in their equilibrium positions for $a = a_0$, we want the forces to be zero. In fact, we will see that it is apparently not possible to achieve this in general within the energy-based framework.[23]

The force on any atom β follows from Eqn. (12.43) as

$$f^\beta = -\frac{\partial \Pi}{\partial \mathrm{u}^\beta}. \quad (12.48)$$

Just for convenience, we can look at the derivatives themselves and drop the negative sign required to define physical forces. Thus

$$\frac{\partial \Pi}{\partial \mathrm{u}^\beta} = \sum_{\alpha \le 0}^{\mathrm{atoms}} \frac{\partial E^\alpha}{\partial \mathrm{u}^\beta} + \sum_{e}^{\mathrm{elements}} \nu_e \Omega_0 \frac{\partial W_e}{\partial \mathrm{u}^\beta}, \quad (12.49)$$

where

$$\frac{\partial W_e}{\partial \mathrm{u}^\beta} = \begin{cases} D\hat{\epsilon}/L_e, & \text{if } \beta \text{ is the right-hand node of element } e, \\ -D\hat{\epsilon}/L_e, & \text{if } \beta \text{ is the left-hand node of } e, \\ 0 & \text{otherwise} \end{cases} \quad (12.50)$$

and L_e is the undeformed length of element e. The derivatives of any one atomic site energy E^α with respect to the position of another atom, u^β also need to be computed. Rather than defining a general notation it is easiest in this case to just catalog the short list

[23] This can be achieved for certain special configurations. See the discussion in Section 12.4.5.

of possibilities:

$$\frac{\partial E^\alpha}{\partial \mathrm{u}^\beta} = \begin{cases} -\phi'(2a_0)/2 & \text{for } \alpha - \beta = 2, \\ -\phi'(a_0)/2 & \text{for } \alpha - \beta = 1, \\ 0 & \text{for } \alpha - \beta = 0, \\ \phi'(a_0)/2 & \text{for } \alpha - \beta = -1, \\ \phi'(2a_0)/2 & \text{for } \alpha - \beta = -2. \end{cases} \tag{12.51}$$

We can now look carefully at all of the contributions to the forces on each atom. We will start with atom -2, since all atoms $\alpha \leq -2$ are equivalent. We can see by inspection of Eqn. (12.49) and Fig. 12.8 that there will be the following contributions to $\partial\Pi/\partial\mathrm{u}^{-2}$:

$$\left\{\frac{\partial E^{-4}}{\partial \mathrm{u}^{-2}}, \frac{\partial E^{-3}}{\partial \mathrm{u}^{-2}}, \frac{\partial E^{-2}}{\partial \mathrm{u}^{-2}}, \frac{\partial E^{-1}}{\partial \mathrm{u}^{-2}}, \frac{\partial E^{0}}{\partial \mathrm{u}^{-2}}\right\} = \frac{1}{2}\left\{\phi'(2a_0), \phi'(a_0), 0, -\phi'(a_0), -\phi'(2a_0)\right\}.$$

Clearly, the sum of all these terms is identically zero as we expect.

As we move to atom -1, the situation changes slightly. In the fully atomistic chain, we would find all the same terms as we did for the force on atom -2, with the indices shifted by one to the right. However, the energy of atom 1 is no longer explicitly included in the sum of energies: it has been replaced by the continuum model. Thus $\partial E^1/\partial\mathrm{u}^{-1}$ is not included, and none of the continuum element energies depend on atom -1 to make up for this "missing" contribution to the forces. The forces on atom -1 are therefore

$$\left\{\frac{\partial E^{-3}}{\partial \mathrm{u}^{-1}}, \frac{\partial E^{-2}}{\partial \mathrm{u}^{-1}}, \frac{\partial E^{-1}}{\partial \mathrm{u}^{-1}}, \frac{\partial E^{0}}{\partial \mathrm{u}^{-1}}\right\} = \frac{1}{2}\left\{\phi'(2a_0), \phi'(a_0), 0, -\phi'(a_0)\right\},$$

and the sum of these terms leaves the nonzero ghost force

$$\frac{\partial\Pi}{\partial\mathrm{u}^{-1}} = \frac{1}{2}\phi'(2a_0). \tag{12.52}$$

Moving to atom 0 at the interface, we will find a similar problem to that which occurred with atom -1, but it will now happen twice. Specifically, the fact that the energies of atoms 1 and 2 are not directly computed but are indirectly treated using the continuum elements means that two ghost force contributions $\phi'(2a_0)/2$ and $\phi'(a_0)/2$ will appear in the forces on atom 0. In addition, we note that the energy of element A also depends on the displacement of node 0. Thus there is a term

$$\frac{\partial E^A}{\partial\mathrm{u}^0} = -\frac{D\hat{\epsilon}}{L_A}\nu_A\Omega_0. \tag{12.53}$$

In the undeformed state that we are currently considering, this contribution is still zero since $\hat{\epsilon}$ is zero. However, note that in the case of a uniform strain applied to the entire crystal (the usual finite element patch test), this term starts to contribute a ghost force. At zero strain, though, we have the ghost force on atom 0 equal to

$$\frac{\partial\Pi}{\partial\mathrm{u}^0} = \frac{\phi'(a_0) + \phi'(2a_0)}{2}. \tag{12.54}$$

It is also interesting to note that in the case of a *nearest- neighbor* pair potential model, the ghost forces on atoms -1 and 0 vanish. This is because in order to equilibrate the crystal at zero strain with only nearest-neighbor interactions, a_0 is modified so that $\phi'(a_0) = 0$, and of course the contribution from $\phi(2a_0)$ is not included. This suggests that a nearest-neighbor model is, in a sense, no longer really nonlocal, as an atom is only influenced by its *immediate* neighborhood defined by its *nearest* neighbors.

Atoms 1 and 2 have no forces on them due to the finite element energies (since $\hat{\epsilon} = 0$ in terms like Eqn. (12.53)), but there are terms due to the dependence of the energy of atoms 0 and -1 on the positions of atoms 1 and 2. These are

$$\frac{\partial \Pi}{\partial \mathrm{u}^1} = \frac{\phi'(a_0) + \phi'(2a_0)}{2}, \qquad \frac{\partial \Pi}{\partial \mathrm{u}^2} = \frac{1}{2}\phi'(2a_0). \tag{12.55}$$

A general property of ghost forces is that they are self-equilibrating, i.e. they sum to zero. For atomistic models with relatively short range, this means that the error due to the ghost forces is confined to a small region about the atomistic–continuum interface. To see that these forces self-equilibrate, we must invoke the equilibrium of the underlying crystal structure. In this one-dimensional case, this is to say that the distance between atoms, a_0, corresponds to an energy minimum. For the second-neighbor pair potential, the energy per atom of the perfect, infinite chain as a function of the atomic spacing, a, is

$$E^0(a) = \phi(a) + \phi(2a).$$

The condition of equilibrium is that $dE^0/da = 0$ when $a = a_0$, which implies that

$$\phi'(a_0) = -2\phi'(2a_0).$$

Using this result in Eqns. (12.52), (12.54) and (12.55), one can readily verify that the sum of the four ghost forces is indeed zero as expected.

The discussion here has precluded methods like the BD approach, where the handshake region permits the use of weak compatibility at the interface and does not require full refinement of the mesh along ∂B^{I} to the atomic scale. We shall see in our quantitative comparison of a specific example in Section 12.7 that this does not eliminate ghost forces. Rather, it has the effect of spreading the ghost forces out over a wider region depending on the width of B^{H} (also, see Exercise 12.2).

It therefore appears that regardless of the coupling method used in an energy-based method, ghost forces cannot be eliminated in a general way. These unphysical forces are an unavoidable consequence of a well-defined total energy functional for the coupled system. In Section 12.4.5, we present a few ways to correct these forces approximately, while in Section 12.7 we will see quantitative examples of the errors that ghost forces introduce.

12.4.4 Ghost forces in the cluster-based quasicontinuum (CQC-E)

As we have said, the source of the ghost forces is the incompatibility between the nonlocal atomistic model and the local continuum. Therefore, one may ask whether it is possible to formulate an efficient coarse-grained model that does not depend on a local continuum model, and whether such an approach will eliminate the problem of the ghost forces.

One option [KGF91] is to reformulate the continuum side of the problem in a nonlocal way. For example, the strain energy of the continuum can be represented as an integral of

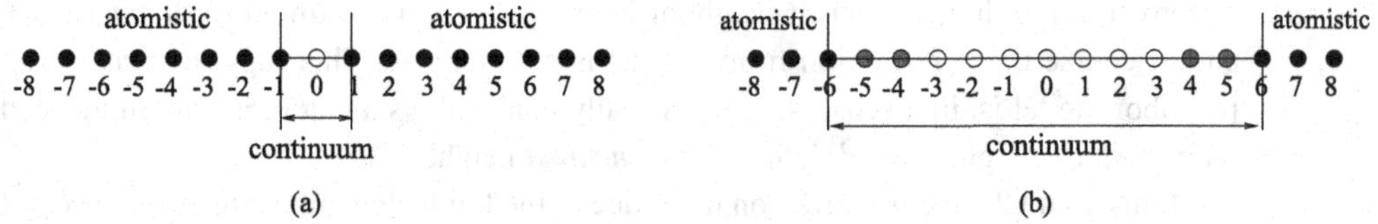

Fig. 12.9 A one-dimensional chain of atoms modeled with a partitioned-domain approach and a fully nonlocal constitutive law: (a) only atom 0 is "coarsened-out;" (b) atoms −5 through 5 are constrained to move according to atoms −6 and 6, while atoms −3 through 3 are coarsened-out.

the form (see Section 2.5):

$$\Pi^{\mathrm{C}}_{\mathrm{NL}} = \int_{B_0} W(\boldsymbol{F}(\boldsymbol{X}))dV + \int_{B_0}\int_{B_0'} \kappa(\boldsymbol{F}(\boldsymbol{X}), \boldsymbol{F}(\boldsymbol{X}'))\, dV dV', \tag{12.56}$$

where the nonlocal kernel function κ determines the contributions from deformation at a distance. However, any determination of κ is necessarily an approximation to the real constitutive law, which we already know to be the sum over atomic neighbors, and some level of ghost force error will likely remain.

Alternatively, the CQC-E method, discussed in Section 12.3.7, uses the full nonlocal atomistic model *everywhere*, and uses the finite elements only as a means to remove degrees of freedom from the problem by constraining the motion of certain atoms. However, as illustrated by [ES09], this approach is still not free of ghost forces. Here, we discuss why this is the case by examining the simplest possible implementation of the energy-based CQC-E model.

Consider an infinite chain of the same Lennard-Jones atoms as previously introduced, but where we make the single, rather minimal, step towards model reduction (shown in Fig. 12.9(a)): atom 0 is not free to move on its own but instead is forced to displace as dictated by atoms −1 and 1. Specifically, we imagine a single finite element spanning atoms −1 and 1, and use it to interpolate the position of atom 0 via

$$u^0 = \frac{\mathsf{u}^{-1} + \mathsf{u}^1}{2}. \tag{12.57}$$

First, imagine that we compute the energy of the chain *exactly as for the fully-atomistic case*. That is to say, we sum the energies of all atoms, including atom 0:

$$\widetilde{\mathcal{V}}^{\mathrm{int}} = \sum_{\alpha}^{N} E^{\alpha}. \tag{12.58}$$

One can readily verify that there will be no ghost forces in this case (see the exercises). At zero deformation, the interpolation of atom 0 from atoms −1 and 1 has no effect and the energy is exact. For uniform stretching of the chain as well, Eqn. (12.57) puts atom 0 in exactly the right place for a uniform deformation and all the correct atomistic forces are present. However, the model is not especially useful as a way to save computations. While we have fewer degrees of freedom with atom 0 enslaved to atoms −1 and 1, we still have to explicitly compute the energy of atom 0.

To make the model more computationally attractive, we assume that we can approximate the *energy* of atom 0 as the average of the energies of atoms −1 and 1, $E^0 = (E^{-1}+E^1)/2$. This is essentially the approach taken in the CQC-E method to estimate the energy of atoms not explicitly considered in the clusters. The approximation of the energy of the atoms in the chain running from atom −3 to 3 then becomes

$$\widetilde{\mathcal{V}}^{\text{int}} = E^{-3} + E^{-2} + \frac{3}{2}(E^{-1} + E^1) + E^2 + E^3. \tag{12.59}$$

In this way we are never explicitly referring to a local, continuum-based constitutive rule, but we are still making a continuum assumption that the atomic site energy is linearly varying through the finite element.

In the undeformed state, we can check, for instance, the force on atom 1. We find that

$$\frac{\partial \widetilde{\mathcal{V}}^{\text{int}}}{\partial u^1} = \frac{\partial E^{-2}}{\partial u^1} + \frac{3}{2}\left(\frac{\partial E^{-1}}{\partial u^1} + \frac{\partial E^1}{\partial u^1}\right) + \frac{\partial E^2}{\partial u^1} + \frac{\partial E^3}{\partial u^1}. \tag{12.60}$$

These terms are straightforward, except for the dependence of the position of atom 0 on atom 1, though Eqn. (12.57), which leads to an additional dependence of the total energy on atom 1 not found in the atomistic model. We can use Eqn. (12.51) to see that these terms are specifically

$$\left\{\frac{\partial E^{-2}}{\partial u^1}, \frac{\partial E^{-1}}{\partial u^1}, \frac{\partial E^1}{\partial u^1}, \frac{\partial E^2}{\partial u^1}, \frac{\partial E^3}{\partial u^1}\right\} = \frac{1}{2}\left\{\phi'(2a_0), \left(\phi'(2a_0) + \phi'(a_0)\frac{\partial u^0}{\partial u^1}\right),\right.$$
$$\left.\left(\phi'(2a_0) + \phi'(a_0) - \phi'(a_0)\frac{\partial u^0}{\partial u^1} - \phi'(2a_0)\right), -\phi'(a_0), -\phi'(2a_0)\right\},$$

where the terms involving $\partial u^0/\partial u^1$ are from chain rule differentiation. When added together according to Eqn. (12.60) these terms lead to the nonzero ghost force

$$\frac{\partial \widetilde{\mathcal{V}}^{\text{int}}}{\partial u^1} = \frac{1}{4}\left(\frac{3}{2}\phi'(a_0) - \phi'(2a_0)\right). \tag{12.61}$$

We see, then, that ghost forces can appear even when we have made no explicit use of a local (or continuum) model in any region. The energy in this simple example was computed everywhere using the fully nonlocal atomistic description, and the only model reduction came from two features: (1) the kinematic constraint of atom 0 and (2) the approximation of the energy of atom 0 as the average of that from atoms 1 and −1.

It is the second of the two features that leads to the ghost forces in the coarse-grained system. In essence, the coarse-grained system adopts a configuration that lowers the energy of atoms 1 and −1, at the cost of raising the energy of atom 0. However, since the energy of atom 0 is only approximated as the average energy of atoms 1 and −1 and is not explicitly calculated, this cost is not considered by the model. Thus, although the fully-atomistic energy of the system is raised by any distortion of the perfect crystal, the coarse-grained approximate system wants to adopt a distorted crystal as the equilibrium state.

12.4.5 Ghost force correction methods

Several methods have been proposed to either eliminate or at least mitigate ghost forces. The first method proposed to eliminate them in a fundamental sense from an energy-based formulation was the creation of so-called "quasi-nonlocal atoms" in [SMSJ04]. These atoms occupy a transition region between B^{A} and B^{C}, and are designed to interact differently with the two regions. The approach is limited to certain crystal structures and certain orientations of the atomistic–continuum interface within the crystal. Later, a more general reconstruction scheme was proposed in [ELY06], but it is still not clear how to generalize this approach beyond a planar interface within a crystal. Furthermore, the goal in many coupled methods – to include automatic remeshing and adaption of the model to respond to error measures or changes in the deformation field – makes these approaches expensive.

Klein and Zimmerman [KZ06] have proposed a generalization of the coupling used in the static BSM, including a careful accounting of the causes of ghost forces and ways to minimize them (but not completely eliminate them). However, that method is limited to pair potential atomistic models.

Dead-load ghost force correction In the implementation of the QC method in [SMT+99], an approximate ghost force correction is proposed. It turns out that this approach is very general; it will work within the framework of any method for which the ghost forces can be analytically derived or computationally estimated. As we shall see in our quantitative example, this correction seems to go a long way in improving accuracy with almost negligible extra computational effort. The proposed ghost force correction is to explicitly compute the ghost forces in some suitable configuration, and then add the negative of these forces as dead loads on the affected atoms or nodes. We define $\boldsymbol{g}^I$ as the ghost force experienced by atom or node I (which could be zero if the atom/node is far from the continuum–atomistic interface). Then the potential due to these ghost forces is added to the potential energy from Eqn. (12.23) to obtain the total potential to be minimized

$$\Pi = \sum_{\alpha \in B^{\mathrm{A}}} E^{\alpha}(\boldsymbol{u}(\mathbf{u})) + \sum_{e \in B^{\mathrm{C}}} \nu_e \Omega_0 W(\boldsymbol{F}^e(\mathbf{u})) - \sum_{I=1}^{n_R} \boldsymbol{g}^I \cdot \mathbf{u}^I. \tag{12.62}$$

For an undeformed model this exactly eliminates the ghost forces by construction. It becomes less clear how to proceed when the model is nonuniformly strained, for two reasons: (1) it may not be possible to clearly identify which internal forces are "ghost" and which are "real", and (2) the ghost forces will not remain constant as the deformation changes, and so the dead load assumption is only an approximation (with an indeterminate error associated with it).

To deal with the first difficulty, the QC developers *defined* the ghost forces in an intuitive way that ensured that the method will have no ghost forces in the undeformed configuration while still providing a clear prescription for ghost force computation under general deformation. This "ghost force" is defined as follows. For an atom whose energy is explicitly included in the total energy (atoms included in the first sum of Eqn. (12.23)), the ghost force is defined as any force the atom would not feel if its environment was truly atomistic everywhere. At the same time, ghost forces on the nodes defining the continuum are those forces that the nodes would not feel if their environment were truly just the continuum. When implementing the forces, it then becomes clear which contributions are the ghost forces.[24] Using this same guiding principle, it is possible to incorporate a ghost force correction within any of the other existing methods. Indeed, we shall discuss next how this was adopted within the CQC method in [ES09] and how we propose a similar correction to the atomistic-to-continuum (AtC) force-based method in Section 12.5.6.

The second difficulty can be treated by occasionally recomputing the ghost forces, effectively changing the energy functional that is being minimized from time to time. In the limit of continuously updating the ghost forces (say for every evaluation of the forces in an energy minimization or dynamic simulation), one is no longer minimizing any energy functional at all, since the energy functional is continuously changing. Instead, one has moved to the realm of the "force-based" methods that we will discuss in Section 12.5.

Dead-load ghost force correction in CQC-E It has been demonstrated [ES09] that ghost force correction via constant dead loads, proposed for the original QC method, can also be efficiently applied to the CQC-E method. Normally, in order to obtain good accuracy in CQC-E, large cluster sizes have to be used, but this significantly slows down the method. It turns out that good accuracy can be obtained with small clusters (even "clusters" of only one atom at each repatom site) if a ghost force correction is used. For more details on how this can be done efficiently, see [ES09] and [MT09b, Section 4.7.2].

12.5 Force-based formulations

12.5.1 Forces without an energy functional

The existence of ghost forces is, it seems, a necessary consequence of having a well-defined energy functional. An alternative approach is to abandon the energy-based approach and instead start from forces directly. As we show in this section, methods of this type can indeed eliminate the ghost forces. However, the lack of a well-defined energy can be problematic for a number of reasons:

1. If the forces are not constructed carefully, it may not be possible to find a solution at all. Assuming a stable model is built, one may still find "equilibrium" solutions that are not physical because they correspond to saddle points or maxima (rather than minima) on some unknown energy surface.
2. Methods for equilibrating nonconservative forces do not tend to be as efficient as energy minimization algorithms.

[24] This prescription is closely related to the "force-based" CADD/FEAt method to be discussed in Section 12.5. Essentially, the differences between the forces used in CADD/FEAt and the actual derivative of the energy functional in the QC/CLS method are defined as the ghost forces.

3. The lack of an energy functional also means that force-based methods cannot be directly used to compute the difference in energy between different equilibrium states or the activation energies along transition paths that are often the goal of static methods.
4. Another difficulty that is not often mentioned, but can be cause for practical concern, is that debugging force-based schemes can be more difficult since the workhorse of energy-based method debugging, the numerical derivative test, may not be applicable in the absence of an energy functional.
5. In dynamic systems, an unstable, nonconservative model is grounds for considerable caution. For example, a force-based dynamical method cannot simulate a system in the microcanonical ensemble since the energy of the system will not be conserved.

In essence, a force-based method is founded on the following philosophy: to eliminate ghost forces, *design* the method so that the forces are identically zero when the perfect crystal is in its correct equilibrium state. Let us consider, again, the coupled problem shown in Fig. 12.3(b), with the left-hand region modeled atomistically and the right-hand region treated using finite elements. We now assume that two independent potential energy functionals exist: one that treats the entire body atomistically, Π^{atom}, and one that models the entire body as a continuum using finite elements, $\widetilde{\Pi}^{\text{FEM}}$. For Π^{atom}, we imagine that the atoms underlying the continuum region are in positions determined by their reference crystal structure and the displacement field of the finite elements. We could then, in principle, compute the potential energy of this atomic configuration. Differentiation of this potential energy with respect to the position of any one of the real atoms gives us the force on this atom. In practice, the finite range of interaction used to define an interatomic model means that we only need to worry about a padding of atoms in the continuum near the atomistic–continuum interface, as discussed previously.

At the same time, forces on the finite element nodes are computed by starting from the finite element energy functional, $\widetilde{\Pi}^{\text{FEM}}$. To this end, we can now imagine that the entire body is modeled by finite elements, with the actual continuum region using the same elements as defined for the coupled model. Assuming a strictly local finite element formulation, the details of the finite element mesh inside the atomistic region are not important, since the force on a node is computed entirely from elements in direct contact with it. In other words, we only need to know the positions of the handful of atoms defining the displacement boundary condition for the elements near the atom–continuum interface.

Thus, the force $\boldsymbol{f}^\alpha$ on atom α and the force $\mathbf{f}^I$ on node I are

$$\boldsymbol{f}^\alpha = -\frac{\partial \Pi^{\text{atom}}}{\partial \boldsymbol{u}^\alpha}, \qquad \mathbf{f}^I = -\frac{\partial \widetilde{\Pi}^{\text{FEM}}}{\partial \mathbf{u}^I}. \tag{12.63}$$

Note that this is not the same as minimizing the combined energy functional $\Pi^{\text{atom}} + \widetilde{\Pi}^{\text{FEM}}$. The forces computed in Eqn. (12.63) cannot be derived from a unified energy functional. But the advantage is that the ghost forces will be zero by design.

In the following sections, we look at the details of several existing force-based methods. We will see that even some force-based methods, with what seem to be very reasonable prescriptions for the forces, can have their own spurious effects that are similar to the ghost force artifacts that plague energy-based implementations.

12.5.2 FEAt and CADD

The earliest force-based method (and, indeed, the earliest of all the methods presented here) was the FEAt method of [KGF91]. In this approach, there is no handshake region and strong compatibility is enforced, making it comparable to the QC and CLS methods but with a force-based rather than energy-based governing formulation. The FEAt method took the additional step of introducing a nonlocal elasticity formulation in the finite elements (see Eqn. (12.56)) in an effort to mitigate the abrupt transition from a local continuum to nonlocal atoms.

More recently, the same force-based coupling was used in the development of the CADD method [SMC02, SMC04]. The focus of this development was the connection to discrete dislocation methods in the continuum region. Incorporating elastic dislocations necessitates the use of linear elasticity (as opposed to the Cauchy–Born rule) in the continuum region, but in the limit where there are no dislocations in the continuum, CADD can use the Cauchy–Born rule or any other nonlinear constitutive law in the finite elements. In this limit (no dislocations and the Cauchy–Born rule) CADD shares the features of the QC method, but with a force-based rather than energy-based coupling scheme. In fact, CADD can be described as a force-based QC formulation.

The CADD–FEAt coupling is therefore simple to describe in relation to the QC method as presented in Section 12.3.2. There is no handshake region and strong compatibility sets the positions of the padding atoms and the nodes along the heavy, jagged line in Fig. 12.3(b). The *forces* on every atom in B^{A} are computed as if the continuum did not exist, from the derivative with respect to atom positions of an energy functional:

$$\Pi^{\mathrm{A}\cup\mathrm{P}} = \sum_{\alpha\in\{B^{\mathrm{A}}\cup B^{\mathrm{P}}\}} E^{\alpha} - \sum_{\alpha\in\{B^{\mathrm{A}}\cup B^{\mathrm{P}}\}} \boldsymbol{f}^{\mathrm{ext},\alpha}\cdot\boldsymbol{u}^{\alpha}. \tag{12.64}$$

This is, in effect, the same as Π^{atom} introduced in the last section. Note that this energy functional is fundamentally different from Eqn. (12.13) since it contains the padding atoms as well.[25] Similarly, the *forces* on the nodes are obtained from the derivative with respect to nodal positions of an energy functional:

$$\widetilde{\Pi}^{\mathrm{C}} = \sum_{e=1}^{n_{\mathrm{elem}}}\sum_{q=1}^{n_{\mathrm{q}}} w_q V_0^e W(\boldsymbol{F}(\boldsymbol{X}_q^e)) - \bar{\mathbf{f}}^T\mathbf{u}, \tag{12.65}$$

but without regard for the energy of the atoms. These forces are then used to move the atoms and nodes (either dynamically or towards equilibrium in a static solution using a suitably modified conjugate gradient (CG) method or quasi-Newton scheme) and the forces are recomputed for the new atom and node positions.

The strong compatibility employed in CADD–FEAt means that there is effectively a displacement boundary condition on the atoms in the form of a hard constraint on the

25 Note also that force-based approaches require a padding region which is twice as thick as that for a comparable energy-based method. This is because the forces are derived from an energy functional that involves more atoms, as evidenced in Eqn. (12.64). These extra atoms must be properly coordinated by additional padding, which means that force-based methods will be slightly slower than energy-based methods, all other things being equal.

motion of the padding atoms. Conversely, it means that there is a displacement boundary condition on the finite elements, as the last row of atoms impose their displacements on the first row of nodes. Note that this is not an iterative, alternating Schwarz type of an approach, even though the presentation may suggest it is. The equations for the forces (i.e. the selective derivatives of Eqns. (12.64) and (12.65)) are solved simultaneously with the strong compatibility conditions as a constraint. Of all the methods presented here, only the BSM and CACM methods of Sections 12.3.5 and 12.3.6, respectively, require repeated iterations between the atomistic and continuum domains.

12.5.3 The hybrid simulation method (HSM)

Like QC and CLS, both FEAt and CADD require strong compatibility to be enforced in the interface region. Another force-based method that we will call the hybrid simulation method (HSM) was proposed in [LHM+07]. It partially removes the strong compatibility condition of the CADD–FEAt method by generalizing the coupling to include a handshake region of finite width. There is still a padding region where strong compatibility between the atoms and the continuum displacement field is used, but atoms in the handshake region are free to move as they like. In fact, the purpose of the handshake region in this model is to provide a prescription for a weak compatibility scheme for the boundary condition on the finite element region. As in CACM, this method relaxes strong compatibility by an averaging method using Eqn. (12.28). In the original paper [LHM+07], the developers used a function ω^I defined on a sphere of radius r_{av} surrounding node I and linearly decaying to zero at $r = r_{\text{av}}$. The size of the sphere could be varied to study the effects of this averaging, and of course other weighting functions could be used. The main advantage of this approach is that it makes mesh generation easier. However, in Section 12.7, we will see that it tends to be less accurate than strong compatibility for a comparable mesh design.

We discussed earlier how the BD method could be shown to reduce to the QC or CLS method in certain limiting cases. Similarly, the HSM method reduces to the CADD–FEAt coupling method if we choose Eqn. (12.29) for ω^I, refine the elements to atomic spacing at the interface and shrink the width of the handshake region to zero. In addition, the original HSM method used nonlinear elasticity instead of the simpler linear version or the more expensive Cauchy–Born rule. It is straightforward, however, to exchange the constitutive model in any of these methods.

12.5.4 The atomistic-to-continuum (AtC) method

The AtC method was presented and analyzed in a series of papers [FNS+07, PBL08, BPB+08, BBL+07]. This method is, in essence, a force-based version of the BD method discussed earlier. Recall that the BD method used Eqn. (12.11), with the energy of the handshake region coming from the blending of a continuum and atomistic energy via Eqn. (12.15). The AtC method achieves its coupling by blending at the level of forces, as follows.

The derivation starts by assuming that the atomistic and continuum regions are completely uncoupled, even though they physically overlap in the handshake region. The forces on the atoms and nodes are

$$\boldsymbol{f}^{\alpha}=\sum_{\beta\neq\alpha}\boldsymbol{f}^{\alpha\beta},\qquad \mathbf{f}^{I}=-\sum_{e=1}^{n_{\mathrm{elem}}}\int_{B_e}\boldsymbol{P}(\widetilde{\boldsymbol{F}}(\mathbf{u}))\frac{\partial S^{I}}{\partial \boldsymbol{X}}\,dV. \tag{12.66}$$

Here $\boldsymbol{f}^{\alpha}$ is the force on atom α, $\boldsymbol{f}^{\alpha\beta}$ is the force on atom α due to the presence of atom β, $\mathbf{f}^{I}$ is the force residual on node I, $\boldsymbol{P}$ is the first Piola–Kirchhoff stress tensor obtained from the finite element constitutive law and S^{I} is the shape function of node I. The numerical approximant to the deformation gradient $\widetilde{\boldsymbol{F}}$ is obtained from the finite element displacement field. We have not included externally applied forces, to simplify the discussion somewhat.

The forces between atoms are gradually weakened across the handshake region from the atomistic to the continuum side, using a weight function, η, that linearly decreases from 1 to 0. The weight for atom α is

$$\eta^{\alpha}=\eta(\boldsymbol{R}^{\alpha}). \tag{12.67}$$

Then, the force between two atoms α and β is weakened by a factor

$$\eta^{\alpha,\beta}=\frac{\eta^{\alpha}+\eta^{\beta}}{2}, \tag{12.68}$$

so that the atomic forces become

$$\boldsymbol{f}^{\alpha}=\sum_{\beta\neq\alpha}\eta^{\alpha,\beta}\boldsymbol{f}^{\alpha\beta}. \tag{12.69}$$

The symmetric definition of $\eta^{\alpha,\beta}$ ensures that Newton's third law is satisfied, i.e. the weakened force exerted on atom α due to the presence of atom β is equal to the weakened force due to α exerted on β.

A complementary weight function,[26] $\Theta=1-\eta$, is used to gradually weaken the finite element nodal forces across the handshake region from the continuum to the atomistic side:

$$\mathbf{f}^{I}=-\sum_{e=1}^{n_{\mathrm{elem}}}\int_{B_e}\Theta(\boldsymbol{X})\boldsymbol{P}(\widetilde{\boldsymbol{F}}(\mathbf{u}))\frac{\partial S^{I}}{\partial \boldsymbol{X}}\,dV, \tag{12.70}$$

which can be evaluated using numerical quadrature:

$$\mathbf{f}^{I}=\sum_{e=1}^{n_{\mathrm{elem}}}\sum_{q=1}^{n_{\mathrm{q}}}\Theta(\boldsymbol{X}_q^{e})w_q V_0^{e}\left(-\boldsymbol{P}\frac{\partial S^{I}}{\partial \boldsymbol{X}}\right). \tag{12.71}$$

Up to this point, there has been no coupling; all that has been done is to systematically weaken the forces through the handshake region using two complementary functions η and Θ. Two additional steps are required to effect the coupling. First, atoms *inside* the handshake region are constrained to follow the displacements of the finite elements (strong compatibility) via Eqns. (12.10) and (12.2).[27] Second, the forces computed on the handshake

[26] Note that Θ is the same as the weight function introduced in the BD method earlier.

[27] In the atomistic region, the atoms are free to move according to the dictates of the modified forces in Eqn. (12.69). Since $\eta=1$ everywhere inside B^{A}, the introduction of η will only have a relatively small effect, confined to atoms with neighbors in the handshake region where η is less than 1.

region atoms are projected to the finite element nodes. Thus, the forces on the nodes are

$$\mathbf{f}^I = \sum_{e=1}^{n_{\rm elem}} \sum_{q=1}^{n_q} \Theta(\boldsymbol{\xi}_q^e) w_q V_0^e \left(-\boldsymbol{P} \frac{\partial S^I}{\partial \boldsymbol{X}} \right) + \sum_{\alpha \in \mathcal{S}^I} \sum_{\beta \neq \alpha} \eta^{\alpha,\beta} \boldsymbol{f}^{\alpha\beta} S^I(\boldsymbol{R}^\alpha). \qquad (12.72)$$

By the compact support of the finite element shape functions, a node only receives force contributions from atoms inside the elements contacting the node. We emphasize this by defining the set of atoms $\alpha \in \mathcal{S}^I$ as all atoms within the compact support of node I. We see below in Section 12.5.6 that the AtC scheme leads to spurious forces at the interface.

12.5.5 Cluster-based quasicontinuum (CQC-F)

In Section 12.3.7, we discussed the energy-based variant of the cluster-based quasicontinuum method (CQC-E). This approach can also be couched as a force-based method, and in fact the original formulation in [KO01] was force-based.

In the force-based CQC-F method, clusters are chosen around each node, exactly as in the energy-based version. However, instead of approximating the energy of the missing atoms via the energy of the cluster atoms, the *forces* on the nodes are determined from the forces on atoms in the cluster. For a given configuration of the nodes, the interpolated displacement fields between the nodes determine the deformed positions of the cluster atoms and any adjacent atoms required to build neighbor lists. Then, the force $\boldsymbol{f}^\alpha$ on any cluster atom, α, can be determined, and the force on node I follows as

$$\mathbf{f}^I = \sum_{J=1}^{n_{\rm nodes}} n^J \sum_{\alpha \in \mathcal{C}^J} \boldsymbol{f}^\alpha S^I(\boldsymbol{R}^\alpha). \qquad (12.73)$$

In words, this means the following. Each node J represents n^J atoms. The value of n^J can be determined in a number of ways [KO01], but one physically sensible approach is to assign all atoms within the Voronoi cell of node J to n^J as discussed earlier. Then a cluster of atoms, denoted $\mathcal{C}^J$, is selected around each node J. The force on each cluster atom α is computed and is multiplied by the weight factor n^J. Finally, this weighted force $n^J \boldsymbol{f}^\alpha$ is distributed amongst the nodes defining the element in which α lies, using the partition-of-unity shape functions $S^I(\boldsymbol{R}^\alpha)$. Writing this as a sum over all the nodes as in Eqn. (12.73) is convenient notation, but we must remember that the compact support of the shape functions greatly reduces the computational effort this entails.

The developers of CQC-F observed that taking "clusters" of a single atom at each node leads to extremely poor results; the method is unstable. Making the clusters larger restores stability, but still leaves a certain level of error in the analysis. This can be controlled by making larger and larger clusters, but at the expense of increased computational cost. We revisit this more quantitatively in Section 12.7.

12.5.6 Spurious forces in force-based methods

Although the prescriptions of forces for the various force-based methods all appear reasonable, they are not all equally good. As an example, it is worth illustrating how the AtC method leads to spurious forces that need to be corrected to achieve reasonable accuracy.

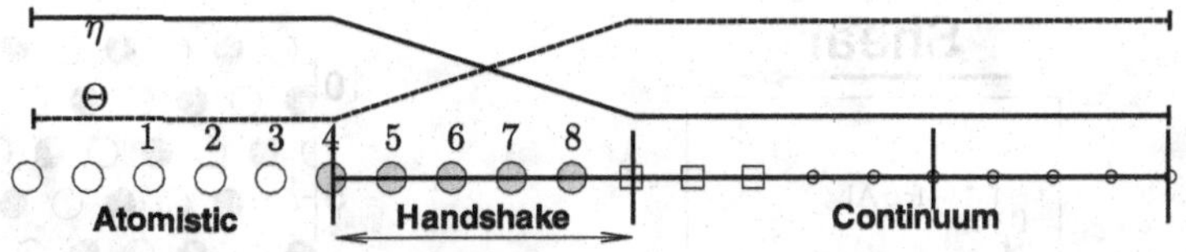

Fig. 12.10 A one-dimensional implementation of the AtC method. As in Fig. 12.3, handshake region atoms appear as filled circles and padding atoms are open squares. Elements and nodes are indicated by heavy lines, while the graph above the model indicates the variation of the weighting functions.

Consider the one-dimensional chain illustrated in Fig. 12.10 and modeled using the AtC prescription of forces described in Section 12.5.4. The atoms are shown at their equilibrium spacing, and the strain in the elements is zero everywhere, so there should be no forces on this initial configuration. For definiteness, imagine that the atoms interact via the second-neighbor Lennard-Jones potential used in Section 12.4.1. Now consider an atom like the one labeled "3" in this figure. It interacts with atoms 1, 2, 4 and 5, so that the total force on atom 3 is zero in the equilibrium configuration:

$$f^3 = f^{13} + f^{23} + f^{43} + f^{53} = 0. \tag{12.74}$$

Note that the sum of these interatomic forces is zero, but the four contributions are not zero on their own. In fact, the symmetry of the chain implies that

$$f^{13} = -f^{53}, \qquad f^{23} = -f^{43}. \tag{12.75}$$

When we now weight the AtC atomic forces according to Eqn. (12.69), we have

$$f^3 = \eta^{1,3} f^{13} + \eta^{2,3} f^{23} + \eta^{4,3} f^{43} + \eta^{5,3} f^{53}. \tag{12.76}$$

All of these weights will be equal to 1 with the exception of $\eta^{5,3}$, which is somewhat less than 1 due to the value of the blend function η at the position of atom 5. Using this result and Eqn. (12.75), we are left with a nonzero force on atom 3:

$$f^3 = (\eta^{5,3} - 1) f^{53} \neq 0. \tag{12.77}$$

Similar nonzero forces exist on atoms 4–8 in the handshake region, which are projected onto the nodes at either end. As such, there are nonzero ghost forces throughout the interfacial region in this force-based method.

As we shall see in Section 12.7, these forces seriously damage the accuracy of the AtC method. Fortunately, they can be approximately corrected using exactly the same dead-load correction approach for ghost forces described in Section 12.4.5. Any force present in the undeformed model is spurious, and so once these are computed they can be subtracted from the forces obtained in deformed configurations.

It is worth noting that the need for this correction is only obvious if a sufficiently complex interatomic model is studied. A harmonic model, even with second-neighbor interactions, is a special case where all forces $f^{\alpha\beta}$ are identically zero for an undeformed perfect lattice. Since this is the model that was used in the development of the AtC method, the existence of these spurious forces was initially overlooked.

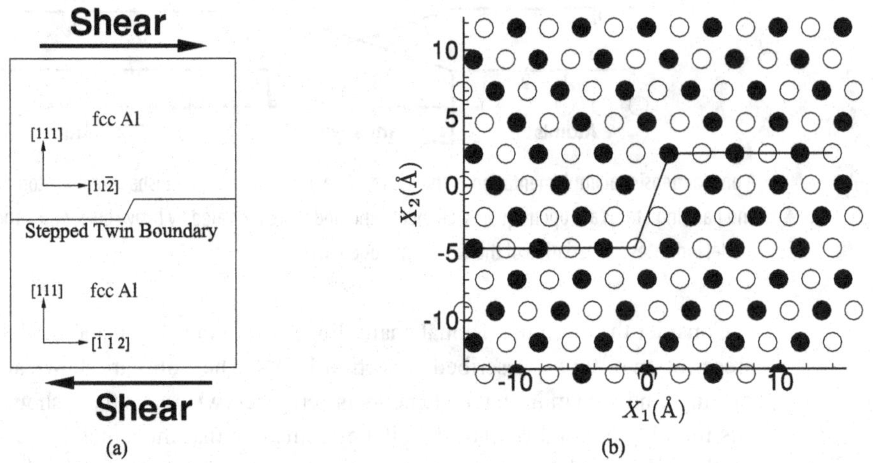

Fig. 12.11 (a) Shear of a stepped twin boundary in an fcc Al bi-crystal. (b) Close-up of the twin boundary in the vicinity of the step, with the solid line showing the boundary.

12.6 Implementation and use of the static QC method

Here, we present a practical example showing how a multiscale method is implemented and how a multiscale simulation is performed. We focus on the QC method which is described in Section 12.3.2. We discuss such issues as how to set up a QC model, how to apply a solution algorithm and how to control any error that may arise. A two-dimensional implementation of the QC method is provided at the website associated with this book, complete with a reference manual, tutorial and example input files. Here, we make use of one of those examples to discuss the simulation details.

12.6.1 A simple example: shearing a twin boundary

Consider a twin boundary in face-centered cubic (fcc) aluminum, perfect but for a small step, as illustrated in Fig. 12.11. A twin boundary in the fcc structure is a special, high-symmetry grain boundary formed when two crystals are brought together in the orientations indicated in Fig. 12.11(a). It is referred to as a twin boundary since the two crystals are related by a simple mirror symmetry across the boundary plane. In Fig. 12.11(b), we can see this symmetry, where the atoms are colored according to the AB stacking in the X_3 direction.

A question of interest may be, "how does this stepped boundary respond to an applied shear load?" We will use this relatively simple example as a vehicle to illustrate the steps of a typical QC simulation. We use a two-dimensional model for simplicity, but we note that all of this discussion is readily extended to a three-dimensional problem.

Algorithm 12.1 A typical solution procedure for the static QC method.

1: Define the grains
2: Define the initial mesh
3: Define the initial atomistic region
4: $\gamma := 0$
5: **while** $\gamma < \gamma_{\max}$ **do**
6:     increment the boundary condition, $\mathcal{B}(\gamma)$
7:     compute the ghost force correction
8:     **while** $\|\mathbf{f}\| > $ tol **do**
9:         compute the tangent stiffness matrix (Hessian) and forces
10:        take a Newton–Raphson step
11:        **if** refinement criteria are met **then**
12:            refine mesh
13:            recompute atomistic region
14:            recompute ghost force correction
15:        **end if**
16:        **if** displacements have changed significantly **then**
17:            recompute Verlet lists
18:            store current displacements for comparison
19:        **end if**
20:     **end while**
21:     $\gamma := \gamma + \Delta\gamma$
22: **end while**

Algorithm 12.1 gives the steps one might typically employ for such a simulation. In the following sections, we provide further details. Note that we have defined a variable, γ, that parameterizes the loading. In general, γ may represent the depth of penetration of an indenter, or the force applied to a sample. In this example, it corresponds to the shear parameter which defines the simple shear deformation (see Section 2.2.2):

$$u_1 = \gamma X_2. \tag{12.78}$$

Here u_1 is the X_1-component of the displacement and X_2 is the reference coordinate. For the first step, $\gamma = 0$ and any relaxation is only due to rearrangements of the atoms near the twin boundary. Subsequent values of γ are then introduced by superimposing a uniform shear strain increment on the relaxed configuration of the previous load step. For example after load step n, the atomic displacements are relaxed to $\boldsymbol{u}_{(n)}$. As the boundary positions are incremented from $\gamma_{(n)}$ to $\gamma_{(n+1)} = \gamma_{(n)} + \Delta\gamma$, atoms and nodes are displaced in the X_1 direction according to

$$u^0_{1(n+1)} = u_{1(n)} + \Delta\gamma R_2, \tag{12.79}$$

while $u^0_{i(n+1)} = u_{i(n)}$ for $i = 2, 3$. From this initial guess, $\boldsymbol{u}^0_{(n+1)}$, all atoms and nodes except those on the outer boundary (which are held fixed) are allowed to relax to the solution, $\boldsymbol{u}_{(n+1)}$ for step $n + 1$.

12.6.2 Setting up the model

The model is set up in lines 1–3 of Algorithm 12.1.

Defining the grains The starting point for a QC simulation is a crystal lattice, defined by an origin atom and a set of lattice vectors as discussed in Section 3.6. To allow the QC method to model polycrystals, it is necessary to divide space into regions, each of which will be considered a single grain with a unique crystal structure. The shape of each grain is defined by a polygon in two dimensions. Physically, it makes sense that the polygons defining each grain do not overlap, although it may be possible to have holes between the grains. In our example, it is easy to see how the shape of the two grains could be defined to include the grain boundary step. Mathematically, the line defining the boundary should be shared identically by the two grains, but this can lead to numerical complications; for example in checking whether the two grains overlap. Fortunately, realistic atomistic models are unlikely to encounter atoms that are less than an ångstrom or so apart, and so there exists a natural "tolerance" in the definition of these polygons. For example a gap between grains of 0.1 Å will usually provide sufficient numerical resolution between the grains without any atoms falling "in the gap" (and therefore being omitting from the model).

In the QC implementation, the definition of the grains is separate from the definition of the actual volume of material to be simulated. This simulation volume is defined by a finite element mesh between an initial set of repatoms. Each element in this mesh must lie within one or more of the grain polygons described above, but the finite element mesh need not fill the entire volume of the defined grains. It is useful to think of the actual model (the mesh) being "cut-out" from the previously defined grain structure. For our problem, a sensible choice for the initial mesh is shown in Fig. 12.12(a), where the twin boundary lies approximately (to within the height of the step) along the line $X_2 = 0$. Elements whose centroids lie above or below the grain boundary are assumed to contain material oriented according to the lattice of the upper or lower grain, respectively.

Since our interest here is atomic-scale processes along the grain boundary, it is clear that the model shown in Fig. 12.12(a), with elements approximately 50 Å in width, will not provide the necessary resolution. We need to define an atomistic region $\mathcal{B}^A$ inside this continuum mesh. The QC code comes equipped with tools that help automate this process to some degree, whereby we can identify a region we wish to designate as $\mathcal{B}^A$ and the code will automatically find and add all the repatoms within this region as outlined in Algorithm 12.2. At line 7, an element is subdivided differently depending on the elements used. For two-dimensional, constant strain triangles, the midside point $\boldsymbol{X}_s$ of each side is determined. Next, the grain occupied by $\boldsymbol{X}_s$ is established, and then the nearest lattice site in the underlying crystal structure of the grain is found.[28] If this lattice site is not *already*

[28] See [AST09] for efficient algorithms for finding the nearest lattice site to a point and other lattice-related tasks.

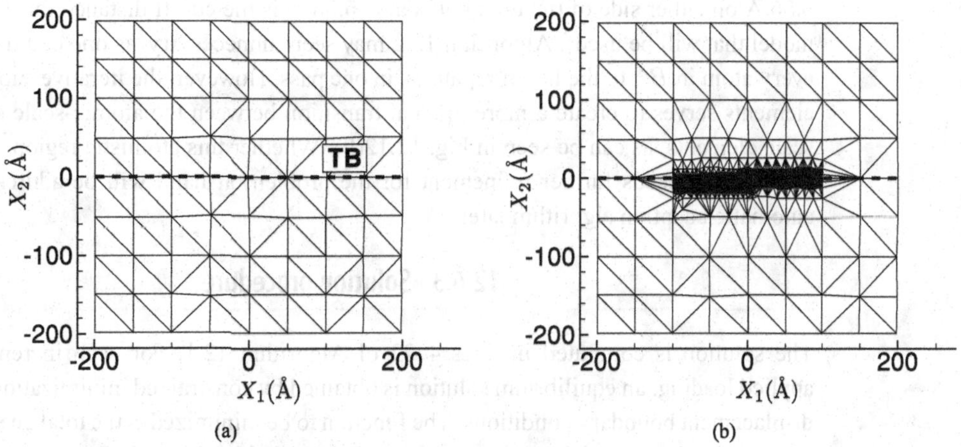

Fig. 12.12 (a) The initial coarse mesh defining the geometry of the problem and (b) the mesh after the atomistic region has been added. Note that in the QC method, elements are stored between the atoms as a convenient book-keeping and visualization tool, but these vacuum elements do not contribute to the calculation.

Algorithm 12.2 Inserting the atomistic region into an existing continuum mesh.

1: Define a region of the body as the atomistic region, B^{A}
2: numel := total number of elements in the mesh
3: numel$_0$:= 0
4: **while** numel$_0 \neq$ numel **do**
5:     numel$_0$:= numel
6:     **for all** elements i overlapping B^{A} **do**
7:         add repatoms on sides of element i by subdivision
8:     **end for**
9:     rebuild mesh (constrained Delaunay triangulation)
10:     numel := number of elements in the mesh
11: **end while**

---

a repatom of the model it is added to the array of repatoms. This is repeated for all three sides of the triangle.[29] At line 9, the mesh is regenerated using Delaunay triangulation[30] to incorporate the new repatoms, and the process is repeated. When the number of elements stops growing, it signifies that every element inside B^{A} is as small as it can be, as there are no atoms left unrepresented in these elements.

We have applied the above process to our example to produce the mesh shown in Fig. 12.12(b). The resulting atomistic region extends from -100 Å to 100 Å along X_1 and

[29] Of course we can optimize this by noting which element shares a side with the element being processed and later avoiding repeating the calculation. A similar process in three dimensions with tetrahedral elements involves adding a point at the centroid of each of the four sides of the element.

[30] The details of Delaunay triangulation in two or three dimensions are discussed in the book by O'Rourke [O'Ro98].

5.55 Å on either side of the twin boundary (5.55 Å is the cutoff distance for the atomistic model that will be used). Algorithm 12.2 may seem unnecessary, as one could simply add every atom in B^{A} to the list of repatoms in one pass. However, the iterative subdividing of elements serves to create a more gradual transition between the atomic-scale and coarse-scale elements, as can be seen in Fig. 12.12(b). Whether this atomistic region is sufficient or the mesh needs further refinement for the problem at hand will be addressed by the automatic adaption algorithm later.

12.6.3 Solution procedure

The solution is computed in lines 4–22 of Algorithm 12.1. For each increment of the applied loading, an equilibrium solution is obtained by constrained minimization subject to displacement boundary conditions. The function to be minimized is the total energy (elastic energy plus the potential energy of the external applied loads), as in Eqn. (12.18), with the QC approximation for $\widetilde{\mathcal{V}}^{\mathrm{int}}$ from Eqn. (12.23):

$$\Pi(\mathbf{u}) = \sum_{\alpha\in B^{\mathrm{A}}} E^{\alpha}(\boldsymbol{u}(\mathbf{u})) + \sum_{e\in B^{\mathrm{C}}} \nu_e \Omega_0 W(\boldsymbol{F}^e) - \sum_{I=1}^{n_R} \mathbf{f}^{\mathrm{ext},I} \cdot \mathbf{u}^I,$$

where n_R is the number of representative atoms. We can obtain the solution using any of the minimization techniques described in this or other books (see Section 6.2), so long as we can find the forces and possibly the stiffness (Hessian) of the system. The forces are determined from the derivatives

$$\mathrm{f}_i^I = -\frac{\partial \Pi}{\partial \mathrm{u}_i^I} = -\sum_{\alpha\in B^{\mathrm{A}}} \sum_{\beta=1}^{N} \frac{\partial E^{\alpha}(\boldsymbol{u})}{\partial u_j^{\beta}} \frac{\partial u_j^{\beta}}{\partial \mathrm{u}_i^I} - \sum_{e\in B^{\mathrm{C}}} \nu_e \Omega_0 \frac{\partial W(\boldsymbol{F}^e)}{\partial F_{kL}^e} \frac{\partial F_{kL}^e}{\partial \mathrm{u}_i^I} + \mathrm{f}_i^{\mathrm{ext},I}, \qquad (12.80)$$

where N is the number of atoms in B^{A} and in the adjacent padding region, and the summation convention is applied to spatial indices. Recall that u_i^I refers to the ith displacement component for repatom I (an atom that is explicitly included either inside B^{A} or as a node in B^{C}), while u_j^{β} is the displacement of any atom β. Thus from Eqns. (12.10) and (12.2) we have

$$\frac{\partial u_j^{\beta}}{\partial \mathrm{u}_i^I} = S^I(\boldsymbol{R}^{\beta})\delta_{ij},$$

where S^I is the shape function for node/repatom I. Note that much of the time this term will be zero. Due to the compact support of the shape functions (see page 602), this term only contributes when an atom β is either coincident with node I or lying inside an element that touches node I. This only occurs if a local element that is not fully refined is within the cutoff distance of a nonlocal repatom. The dependence of the deformation gradient on the repatom displacements is known via Eqn. (12.3), and therefore

$$\frac{\partial F_{kL}^e}{\partial \mathrm{u}_i^I} = \frac{\partial S^I}{\partial X_L}\delta_{ik}.$$

This is only nonzero if repatom I is a node on element e. Also appearing in Eqn. (12.80) are $\partial W/\partial \boldsymbol{F}$ (which we recognize as the first Piola–Kirchhoff stress $\boldsymbol{P}$ to be obtained in the QC method from the Cauchy–Born rule, see Section 11.5) and $\partial E^\alpha/\partial u^\beta$ (which is the contribution to the force on atom β due to the energy of atom α, see Chapter 5). In the case of a pair functional (see Section 5.5, Eqn. (5.54)), we can write

$$\frac{\partial E^\alpha}{\partial u_k^\beta} = \sum_{\substack{\gamma=1\\ \gamma\neq\alpha}}^{N} \left[U'_\alpha(\rho^\alpha) g'_\gamma(r^{\alpha\gamma}) + \frac{1}{2}\phi'_{\alpha\gamma}(r^{\alpha\gamma}) \right] \frac{\partial r^{\alpha\gamma}}{\partial u_k^\beta}. \tag{12.81}$$

The partial derivative follows from Eqn. (5.13) as

$$\frac{\partial r^{\alpha\gamma}}{\partial u_k^\beta} = -(\delta^{\alpha\beta} - \delta^{\gamma\beta}) \frac{r_k^{\alpha\gamma}}{r^{\alpha\gamma}}. \tag{12.82}$$

where δ is the Kronecker delta, so that

$$\frac{\partial E^\alpha}{\partial u_k^\beta} = \begin{cases} \sum_{\substack{\gamma=1\\ \gamma\neq\alpha}}^{N} \left[U'_\alpha(\rho^\alpha) g'_\gamma(r^{\alpha\gamma}) + \frac{1}{2}\phi'_{\alpha\gamma}(r^{\alpha\gamma}) \right] \frac{-r_k^{\alpha\gamma}}{r^{\alpha\gamma}} & \text{for } \alpha = \beta, \\ \left[U'_\alpha(\rho^\alpha) g'_\beta(r^{\alpha\beta}) + \frac{1}{2}\phi'_{\alpha\beta}(r^{\alpha\beta}) \right] \frac{r_k^{\alpha\beta}}{r^{\alpha\beta}} & \text{for } \alpha \neq \beta. \end{cases} \tag{12.83}$$

For minimization methods that require the Hessian, we can write

$$\frac{\partial^2 \Pi}{\partial \mathrm{u}_i^I \partial \mathrm{u}_j^J} = \sum_{\alpha\in\mathcal{B}^{\mathrm{A}}} \sum_{\beta=1}^{N} \sum_{\epsilon=1}^{N} \frac{\partial^2 E^\alpha(\boldsymbol{u})}{\partial u_k^\beta \partial u_l^\epsilon} \frac{\partial u_k^\beta}{\partial \mathrm{u}_i^I} \frac{\partial u_l^\epsilon}{\partial \mathrm{u}_j^J} + \sum_{e\in\mathcal{B}^{\mathrm{C}}} \nu_e \Omega_0 \frac{\partial^2 W(\boldsymbol{F}^e)}{\partial F_{kL}^e \partial F_{mN}^e} \frac{\partial F_{kL}^e}{\partial \mathrm{u}_i^I} \frac{\partial F_{mN}^e}{\partial \mathrm{u}_j^J},$$

where the second derivative of the strain energy density W is the mixed elasticity tensor $\boldsymbol{D}$ in Eqn. (2.190) which can be found using the Cauchy–Born rule (Section 11.5). The additional new term is the second derivative of the site energy with respect to atomic positions, which appears in the first sum. This is a quantity that can be tedious to derive for many interatomic models, and we have avoided presenting such details in the book up to now. For a pair functional this follows from Eqn. (12.81) as

$$\begin{aligned} \frac{\partial^2 E^\alpha}{\partial u_k^\beta \partial u_l^\epsilon} = {} & U''_\alpha(\rho^\alpha) \left[\sum_{\substack{\gamma=1\\ \gamma\neq\alpha}}^{N} g'_\gamma(r^{\alpha\gamma}) \frac{\partial r^{\alpha\gamma}}{\partial u_k^\beta} \right] \left[\sum_{\substack{\gamma=1\\ \gamma\neq\alpha}}^{N} g'_\gamma(r^{\alpha\gamma}) \frac{\partial r^{\alpha\gamma}}{\partial u_l^\epsilon} \right] \\ & + \sum_{\substack{\gamma=1\\ \gamma\neq\alpha}}^{N} \left[\left(U'_\alpha(\rho^\alpha) g''_\gamma(r^{\alpha\gamma}) + \frac{1}{2}\phi''_{\alpha\gamma}(r^{\alpha\gamma}) \right) \frac{\partial r^{\alpha\gamma}}{\partial u_k^\beta} \frac{\partial r^{\alpha\gamma}}{\partial u_l^\epsilon} \right. \\ & \left. + \left(U'_\alpha(\rho^\alpha) g'_\gamma(r^{\alpha\gamma}) + \frac{1}{2}\phi'_{\alpha\gamma}(r^{\alpha\gamma}) \right) \frac{\partial^2 r^{\alpha\gamma}}{\partial u_k^\beta \partial u_l^\epsilon} \right], \end{aligned} \tag{12.84}$$

where the partial derivatives are in Eqn. (12.82) and below:

$$\frac{\partial^2 r^{\alpha\gamma}}{\partial u_k^\beta \partial u_l^\epsilon} = (\delta^{\alpha\beta} - \delta^{\gamma\beta})(\delta^{\alpha\epsilon} - \delta^{\gamma\epsilon}) \left[\frac{\delta_{kl}}{r^{\alpha\gamma}} - \frac{r_k^{\alpha\gamma} r_l^{\alpha\gamma}}{(r^{\alpha\gamma})^3} \right]. \tag{12.85}$$

In lines 8–20 of Algorithm 12.1, we illustrate the use of the Newton–Raphson method (see Section 6.2.7) in QC. After each Newton–Raphson iteration, a check is made to see if either of the refinement criteria (to be discussed shortly) are met, in which case the mesh is

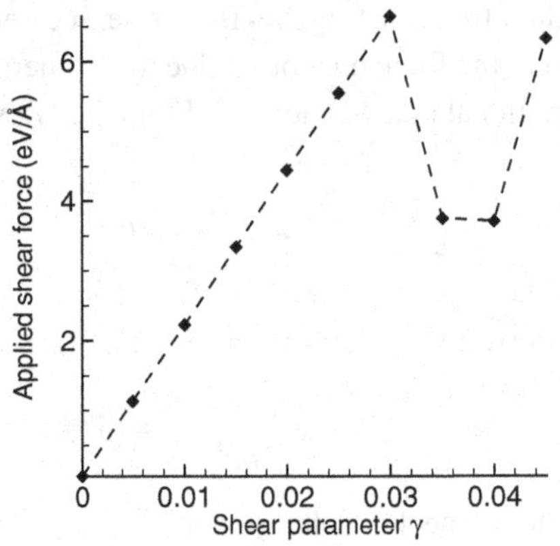

Fig. 12.13 Applied load versus shear strain on the bi-crystal.

adapted and the size of the atomistic region is adjusted. In addition, we must monitor the displacements so that the neighbor lists for repatoms are kept current (using the method of Verlet lists, see Section 6.4.1). The Newton–Raphson iterations repeat until the norm of the residual force vector falls below some user-defined tolerance, as in line 8 of Algorithm 12.1.

12.6.4 Twin boundary migration

For this example, the outermost nodes along the top, bottom and sides of the mesh are held fixed to an applied displacement, and we can therefore monitor the effective applied force on these nodes.[31] This load equals the sum of the X_1-components of the resultant forces on the boundary repatoms on the upper half of the model. The plot in Fig. 12.13 of the applied load versus the shear parameter reveals an essentially linear response for the first six load steps, and then a sudden load drop from step 6 to step 7. Beyond this point, the load starts to rise again due to limitations of the finite size of the model as explained below. The initial load drop corresponds to the first inelastic behavior of the boundary as shown in Fig. 12.14. In Fig. 12.14(a), a close-up of the relaxed step at an applied strain of $\gamma = 0.03$ is shown, while Fig. 12.14(b) shows the relaxed configuration after the next strain increment at $\gamma = 0.035$. The boundary has migrated down by one atomic spacing. This occurs due to the nucleation and motion of two Shockley partial dislocations, one in each direction, from the corners of the step along the boundary. This can be seen clearly by comparing the finite element mesh between the repatoms in Figs. 12.14(a) and (b). Because the mesh is triangulated in the reference configuration, the effect of plastic slip is the shearing of a row of elements in the wake of the moving dislocations. The dislocations move until they run up against the rigid boundaries of the model at which point the load begins to rise again.

A global view of the mesh at different levels of strain is shown in Fig. 12.15. Here, we see the effect of mesh adaption, which extends the atomistic region along the twin boundary

[31] In FEM, the force residual can be computed on a constrained node in the same way as any other node. At equilibrium, unconstrained nodes will have zero force, and the constrained nodes will have a force equal and opposite to an applied force that would lead to the same equilibrium as is created by the constraints.

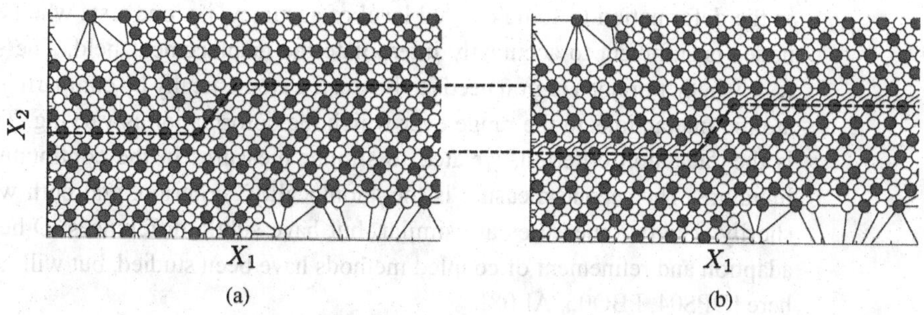

Fig. 12.14 Comparison of the twin boundary before (a) and after (b) the motion of the two Shockley partials. The effect is that the boundary has migrated one atomic plane in the $-X_2$ direction.

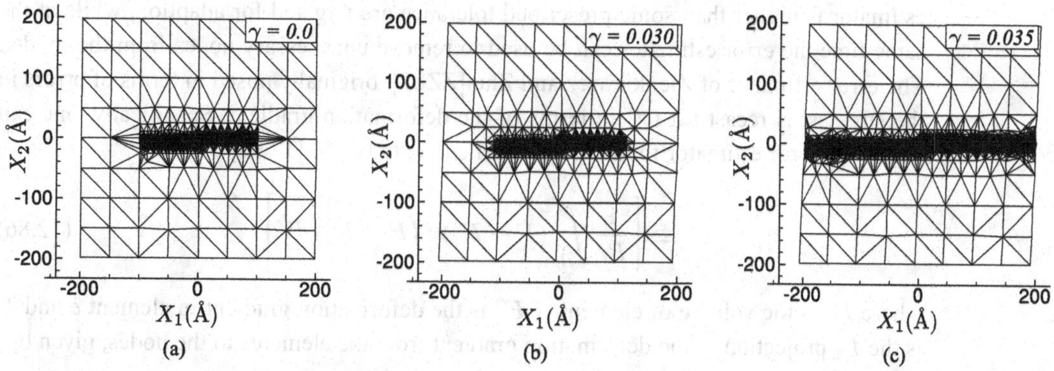

Fig. 12.15 (a)–(c) Evolution of the mesh as the shear increases on the twin boundary.

as the Shockley partials move into the previously continuum region. One challenge in modeling dislocation motion in crystals at the atomic scale is evident in this simulation. In crystals with a low Peierls resistance like the fcc crystal modeled here, dislocations move long distances under small applied stresses. In order to rigorously compute the equilibrium position of these dislocations, it was necessary to refine the model all the way to the outer boundaries. If the model did not have the capability to adapt, the final positions of the dislocations would be incorrect as they would be stopped by the atomistic–continuum interface. We discuss the QC adaption strategy next.

12.6.5 Automatic model adaption

An important feature of a multiscale method is the ability to adapt the model automatically in response to the deformation. In multiscale models, it is necessary to insert an artificial boundary between the atomistic and continuum regions and to build a finite element mesh. Since the location of the boundary and the mesh details are usually initialized in an intuitive, ad hoc manner, it is desirable to have quantitative measures built into the model that can assess the accuracy of these choices and refine them as necessary.

In the QC method this is achieved using a combination of two criteria. First, the model is tested to see whether the refinement of the FEM mesh is sufficient to capture variations in the deformation to some desired level of accuracy, i.e. one tests whether the density of nodes is sufficient to accurately interpolate the displacement field. The second criterion tests whether or not a region needs to be treated atomistically by comparing the range of the atomic interactions to the range over which the deformation is varying. The first measure is the "adaption criterion," related to similar measures in the continuum finite element literature. The second measure is the "nonlocality" criterion, and both will be discussed shortly. The two measures are similar, but have subtle differences. Other approaches to adaption and refinement of coupled methods have been studied, but will not be elaborated here [DPS04, PBO06, AL08].

The QC adaption criterion Typically in FEM, a scalar measure is defined to quantify the error introduced into the solution by the current density of nodes. Elements in which this error estimator is higher than some prescribed tolerance are targeted for adaption, while at the same time the error estimator can be used to remove unnecessary nodes from the model. The error estimator of Zienkiewicz and Zhu [ZZ87], originally posed in terms of errors in the stresses, is recast for QC in terms of the deformation gradient. Specifically, one can define the error estimator for a specific element, e, to be [SMT+99]

$$\varepsilon_e \equiv \left[\frac{1}{B_e} \int_{B_e} (\bar{\boldsymbol{F}} - \boldsymbol{F}^e) : (\bar{\boldsymbol{F}} - \boldsymbol{F}^e)\, dV \right]^{1/2}, \qquad (12.86)$$

where B_e is the volume of element e, $\boldsymbol{F}^e$ is the deformation gradient in element e and $\bar{\boldsymbol{F}}$ is the L_2 projection of the deformation gradient from the elements to the nodes, given by

$$\bar{F}_{iJ}(\boldsymbol{X}) = \sum_{\alpha=1}^{n_{\text{nodes}}} S^\alpha(\boldsymbol{X}) F_{iJ}^{\text{avg},\alpha}. \qquad (12.87)$$

Here, S^α is the shape function of repatom α, and $\boldsymbol{F}^{\text{avg},\alpha}$ is the deformation gradient averaged over all elements touching repatom α. This has the effect of comparing the piecewise constant deformation gradient, $\boldsymbol{F}^e$, with the interpolated nodal average of $\bar{\boldsymbol{F}}$, the latter of which serves as an estimate for the field if higher order (quadratic) elements were used. The error is defined as the difference between the actual solution and this estimate of the higher-order solution. If this error is small, it implies that the higher-order solution is well represented by the lower-order elements in the region, and thus no refinement is required. The integral in Eqn. (12.86) can be computed quickly and accurately using numerical quadrature (see Chapter 9 of [TME12]). Elements for which the error ε_e is greater than some prescribed error tolerance are targeted for refinement that then proceeds by the subdivision of elements already discussed in the context of Algorithm 12.2, line 7.

The same error estimator is used in the QC method to coarsen the mesh by removing unnecessary nodes. In this process, a node is temporarily removed from the mesh and the surrounding region is locally remeshed. If all of the elements produced by this remeshing process have a value of the error estimator below the threshold, the node can be eliminated.

The local–nonlocal criterion To decide whether a given repatom should be treated as part of the continuum region B^C or the atomistic region B^A, the following *nonlocality criterion* is used.[32] Nonlocal treatment is triggered by a significant *variation* in the deformation gradient on the atomic scale. Thus, a cutoff distance (r_{nl}) is empirically chosen to be between 2 and 3 times the cutoff radius of the atomic interactions. The deformation gradients in every element within r_{nl} of a given repatom are compared by looking at the differences between their eigenvalues.[33] The criterion is then [SMT+99]:

$$\max_{a,b;k} |\lambda_k^a - \lambda_k^b| < \epsilon, \tag{12.88}$$

where λ_k^a is the kth eigenvalue of the right stretch tensor $\boldsymbol{U}_a = \sqrt{\boldsymbol{F}_a^T \boldsymbol{F}_a}$ in element a, $k = 1, \ldots, 3$ and the indices a and b run over all elements within r_{nl} of a given node.[34] The node will be considered "local" (and thus treatable using the continuum assumptions) if this inequality is satisfied, and nonlocal otherwise. In practice, the tolerance ϵ is determined empirically. Many of the adaptive results obtained with the QC method have used a value of 0.1, which seems to give largely satisfactory results.

Ideally, such a criterion would automatically generate clusters of atoms around points of discontinuity or abrupt transition in the deformation. In practice, this process is made more efficient by building such clusters as soon as any one nonlocal atom is found. Once a repatom is designated "nonlocal" by this criterion, all atoms within a sphere of radius r_{nl} around this repatom are added to the model and explicitly treated as part of $\mathcal{B}^{\text{A}}$ (using Algorithm 12.2). The effect of this is clusters of atoms in regions of rapidly varying deformation that become the "atomistic regions." This creates a small number of well-defined interfacial regions like the one illustrated in Fig. 12.3(b).

Adaption in the twin boundary example In this example, we can see some of the adaption which takes place between load steps in Fig. 12.15. The main adaption takes place during the motion of the dislocations between $\gamma = 0.03$ and $\gamma = 0.035$, and Fig. 12.16 shows a close-up of the mesh around the Shockley partial moving to the right during the minimization process. We can see in Fig. 12.16(a)–(c) how the relaxation moves the dislocation up against the atomistic/continuum interface, which triggers adaption around the defect core. This creates new out-of-balance degrees of freedom and therefore leads to further minimization, which moves the dislocation further along. The process repeats until the dislocation reaches its true equilibrium position near the boundary of the model.

12.7 Quantitative comparison between the methods

Table 12.1 summarizes the methods discussed in this chapter and their key features. The acronyms are meant to help keep the various methods straight. For example, we distinguish between the energy-based cluster method (CQC-E) of Section 12.3.7, the same method

[32] In the QC literature, $\mathcal{B}^{\text{C}}$ is referred to as the "local" region while $\mathcal{B}^{\text{A}}$ is called "nonlocal" to reflect the nature of the different constitutive laws.

[33] There are other possibilities. See, for example, [TOP96].

[34] See Section 2.2.4 for a discussion of the right stretch tensor.

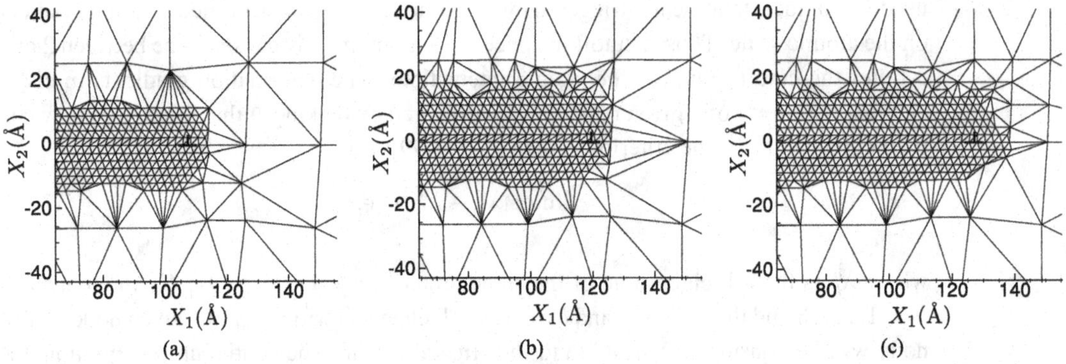

Fig. 12.16 (a)–(c) Close-up of the adaption taking place during the solution at a strain level of $\gamma = 0.03$. The partial dislocation core (denoted by $\perp$) moving along the twin boundary triggers adaption ahead of it, allowing it to continue to move.

with a ghost-force correction (CQC-GFC) of Section 12.4.5 and the force-based version (CQC-F) of Section 12.5.5. Any method with a ghost-force correction includes "GFC" in its acronym. Some acronyms are taken from the original papers, others are coined here.

The implementation of all of the methods in a single computer code [MT09a] and their performance on a nontrivial system were discussed in [MT09b].[35] The result of the study was a comprehensive comparison of the relative accuracy and speed of the different approaches, which we summarize here. (We invite the reader to turn to the original reference for all the details.) The code is written so that all methods are treated on as equal a footing as possible. The main exception is the solution method: the CG algorithm (see Section 6.2.5) is used for the energy-based methods and the CG-FB algorithm (see Section 6.3.1) for the force-based approaches.[36] All methods use the identical atomistic model and the identical continuum constitutive law (the Cauchy–Born rule).

12.7.1 The test problem

An important issue in comparing multiscale methods is the determination of a suitable test problem. Making the problem too simple might hide problems or mask differences between methods. For example, we have already seen how using linear springs as the atomistic model hides spurious forces in the AtC method. As another example, it is possible to completely eliminate ghost forces in the CQC-E method in one dimension, but not in higher dimensions. On the other hand, making the problem too complex will make it difficult to analyze the results. It is also important to choose a problem that is robust and has a well-defined unique

[35] We have made the code itself and all data files available at the QC website [TM09].

[36] In Section 6.3.1, we showed (in the context of the NEB method) that CG-FB suffers from instabilities if the system of equations for the forces has a Hessian which is not positive definite. See also detailed discussions of this in [DLO10a, DLO10b]. We make use of CG-FB with this caveat in mind, and it seems that these instabilities did not occur for the systems studied here. In Section 12.6, we used the Newton–Raphson method, but here this was avoided because it would have required difficult coding of the Hessian for all the different methods.

Table 12.1. Summary of the methods discussed in this chapter.

Method	Acronym	Section	Continuum model	Handshake	Coupling boundary condition	Governing formulation
Quasicontinuum	QC	Section 12.3.2	Cauchy–Born	None	Strong compatibility	Energy-based
Coupling of length scales	CLS	Section 12.3.3	Linear elasticity	None	Strong compatibility	Energy-based
Bridging domain	BD	Section 12.3.4	Cauchy–Born	Linear mixing of energy	Weak compatibility (penalty)	Energy-based
Bridging scale method	BSM	Section 12.3.5	Cauchy–Born	None	Weak/strong Mix (least-squares fit)	Energy-based
Composite grid atomistic continuum method	CACM	Section 12.3.6	Linear elasticity	None	Weak compatibility (average atomic positions)	Iterative energy-based (two energy functionals)
Cluster-energy quasicontinuum	CQC-E	Section 12.3.7	Averaging of atomic clusters	None	Strong compatibility	Energy-based
Ghost force corrected quasicontinuum	QC-GFC	Section 12.4.5	Cauchy–Born	None	Strong compatibility	Energy-based with dead load GFC
Ghost force corrected cluster-energy QC	CQC-GFC	Section 12.4.5	Averaging of atomic clusters	None	Strong compatibility	Energy-based with dead load GFC
Finite-element/atomistics method	FEAt	Section 12.5.2	Nonlinear, nonlocal elasticity	None	Strong compatibility	Force-based
Coupled atomistics and discrete dislocations	CADD	Section 12.5.2	Linear elasticity	None	Strong compatibility	Force-based
Hybrid simulation method	HSM	Section 12.5.3	Nonlinear elasticity	Atomic averaging for nodal B.C.	Weak compatibility (average atomic positions)	Force-based
Concurrent AtC coupling	AtC	Section 12.5.4	Linear elasticity	Linear mixing of stress and atomic force	Strong compatibility	Force-based
Ghost force corrected concurrent AtC coupling	AtC-GFC	Section 12.5.6	Linear elasticity	Linear mixing of stress and atomic force	Strong compatibility	Force-based
Cluster-force quasicontinuum	CQC-F	Section 12.5.5	Averaging of atomic clusters	None	Strong compatibility	Force-based

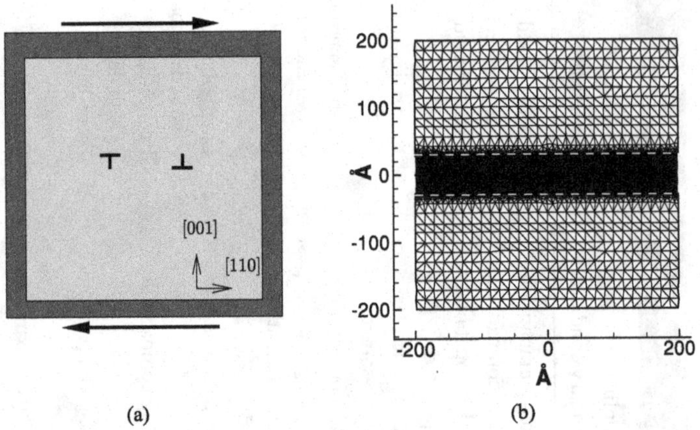

Fig. 12.17 (a) Test problem used to compare the various multiscale techniques. A Lomer dipole, 40 Å wide, is centered in the model. The darker region around the edges of the model is held fixed to various levels of applied shear. (b) A typical mesh for the problem, with the atomistic region extending from −30 Å to 30 Å.

solution, otherwise the comparison between methods becomes meaningless. The problem described below satisfies these criteria.

We consider a block of single crystal aluminum[37] containing a dipole of Lomer dislocations [HL92] near the center, as shown in Fig. 12.17(a). The model is two-dimensional, but with PBCs in the third dimension so that the crystal is semi-infinite and the dislocations are straight. This problem is studied both "fully atomistically" using molecular statics (MS) (we call this the "exact" solution), as well as with the various multiscale methods using different meshes that systematically increase the size of the atomistic region. The meshes will be denoted by an extension "10," "20" or "30" as appropriate. The numbers indicate the extent of the fully-refined atomistic region along the middle of the model. For example, "10" means that the atomistic region extends from $X_2 = -10$ Å to $X_2 = +10$ Å. In Fig. 12.17(b), an example "30" mesh is shown. In all cases, the Lomer dipole lies on the $X_2 = 0$ plane, with the two cores initially at $X_1 = \pm 20$ Å.

The accuracy and speed of convergence of the methods were tested on the initial configuration and after applying incremental shear strains exactly as described in Section 12.6 in the context of a twin boundary example. The shear induces deformation of the dislocation cores that eventually causes them to move left and right towards the outer edges.

12.7.2 Comparing the accuracy of multiscale methods

The displacement error We first quantify the error in the results by comparing the displacement fields from the multiscale models to the exact atomic displacements. In the multiscale approaches, there is always a fully-atomistic region, where ideally we desire perfect agreement between the multiscale and exact results, and a coarse-grained or continuum region where the positions of most of the atoms are not explicitly tracked. However, given the

[37] Modeled using the pair functional of [EA94]; a sufficiently "multibodied" model to ensure a rigorous test of the methods.

reference position of an atom, we can obtain the displacement field values at that position and compare these to the true atomic displacements. We define an atom-by-atom error measure of the displacement error as follows. Denoting the exact displacement vector of atom α as $\boldsymbol{u}^{\alpha}_{\text{exact}}$ and the displacement obtained from a multiscale model as $\boldsymbol{u}^{\alpha}$, we define the error as the L_2-norm of the difference between these two vectors:

$$e^{\alpha} \equiv \left\| \boldsymbol{u}^{\alpha} - \boldsymbol{u}^{\alpha}_{\text{exact}} \right\|. \tag{12.89}$$

The global displacement error for a given model is similarly defined as the L_2-norm of the difference between the global displacement vectors ($\boldsymbol{u}$ and $\boldsymbol{u}_{\text{exact}}$ are of length $3N$ for N atoms), normalized by the number of atoms (in this case, $N = 27\,760$):

$$e \equiv \sqrt{\frac{\left\| \boldsymbol{u} - \boldsymbol{u}_{\text{exact}} \right\|^2}{N}} = \sqrt{\frac{\sum_{\alpha=1}^{N} (e^{\alpha})^2}{N}}. \tag{12.90}$$

As defined here, both e^{α} and e have units of ångstroms. To turn the global error into a percent error, $e_{\%}$, we divide by the average of the atomic displacement norm in the exact solution:

$$e_{\%} = \frac{e}{u_{\text{avg}}} \times 100, \tag{12.91}$$

where

$$u_{\text{avg}} = \frac{1}{N} \sum_{\alpha=1}^{N} \left\| \boldsymbol{u}^{\alpha}_{\text{exact}} \right\| = 0.05755 \text{ Å}. \tag{12.92}$$

Plots showing the magnitudes of the atomic displacement error, e^{α}, are presented in Figs. 12.18 and 12.19. All plots are for the crystal under no applied shear. In both figures, each column shows the results for a specific method, while moving down the column shows the effect of refining the model used for a given method by expanding the atomistic region. Figure 12.18 shows the results for the force-based methods, going from the most accurate to the least accurate as we move from left to right. The scale bar on the right applies not only to this figure, but to Fig. 12.19 as well. Figure 12.19 presents the results for the energy-based methods, again going from most accurate on the left to least accurate on the right. Note that once we made the decision to use the same Cauchy–Born constitutive law for all models in this comparison, some of the methods became indistinguishable, hence the combined results for "QC/CLS" and "CADD/FEAt."

We clearly see the effects of ghost forces along the atomistic–continuum interface in the BD, QC/CLS, BSM, CQC(13)-E and AtC models. The band of error at ± 10 Å, ± 20 Å and ± 30 Å in each of the models respectively is the effect of spurious ghost force relaxations. The correction of the ghost forces with dead loads, as in QC-GFC, CQC(1)-GFC and AtC-GFC, can be seen to almost entirely remove this component of the error (for example, we can directly compare the contours for QC/CLS-30 with QC-GFC-30).

The energy error The displacement error presented above is useful for two principal reasons: it serves as a convenient relative measure of the error of the various methods and it provides a good way to visualize the distribution of the error throughout the system. However, it is not easy to assess the significance of this error; it is not easy to say what accuracy is "good enough" in these terms. A better measure is the error in the total energy.

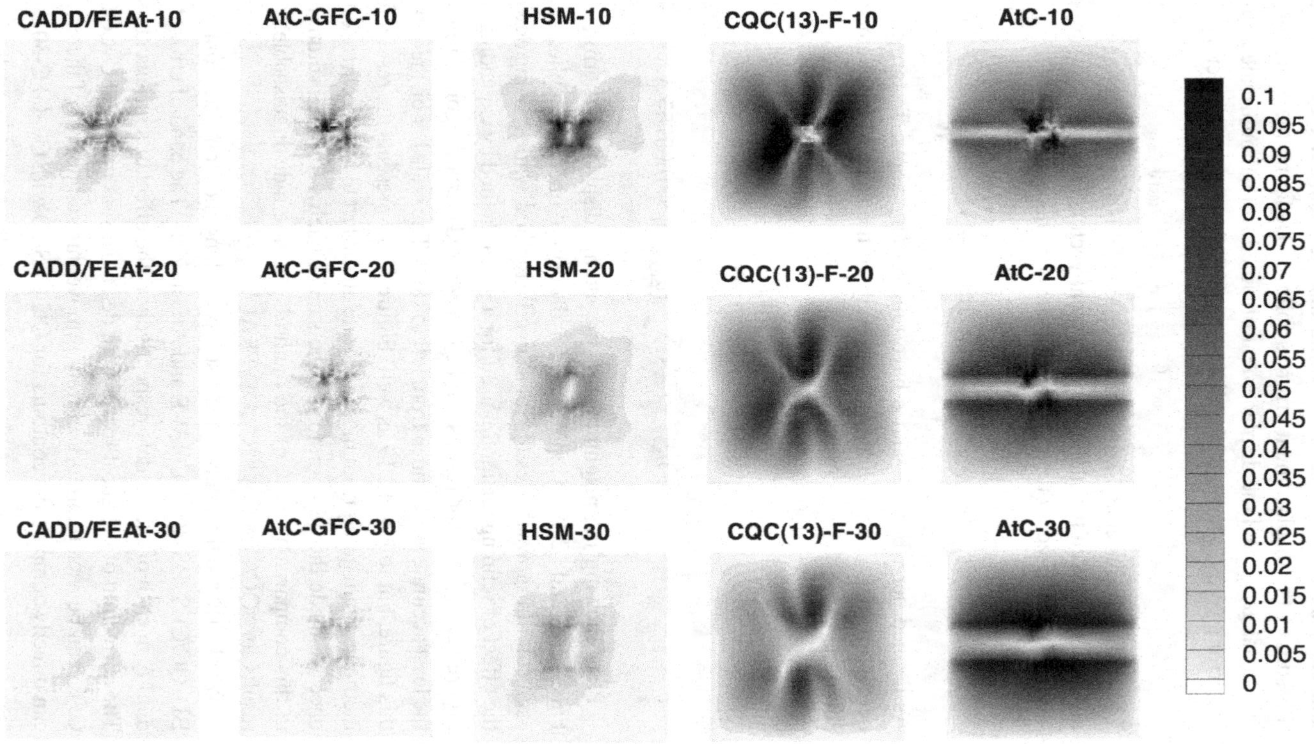

Fig. 12.18 Plots of the atomic displacement error, e^{α}, in ångstroms, for the force-based multiscale models. Each image is of the entire model, which is shown in Fig. 12.17.

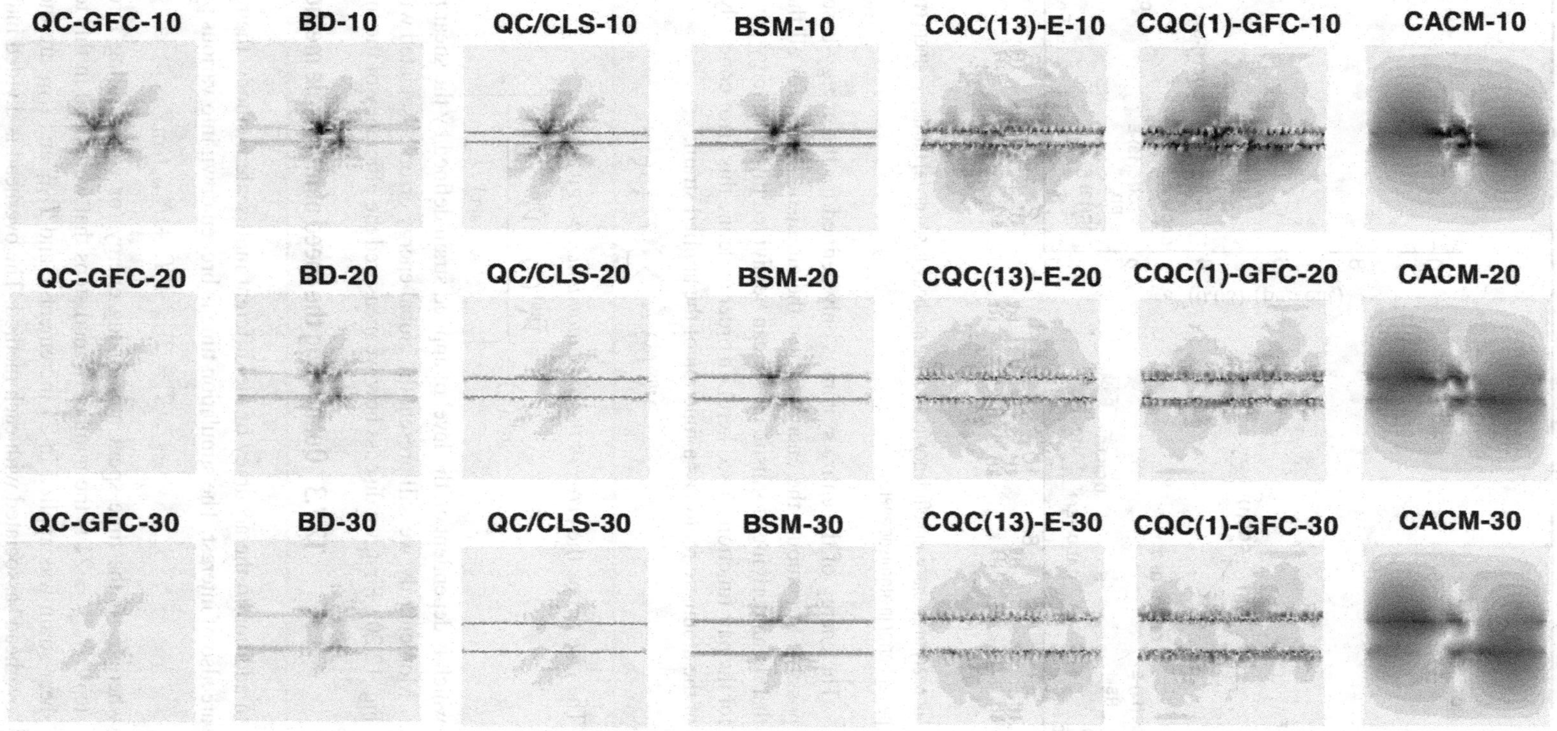

Fig. 12.19 Plots of the atomic displacement error, e^{α} in ångstroms, for the energy-based multiscale models. The contour key from Fig. 12.18 applies to this figure as well. Each image is of the entire model, which is shown in Fig. 12.17.

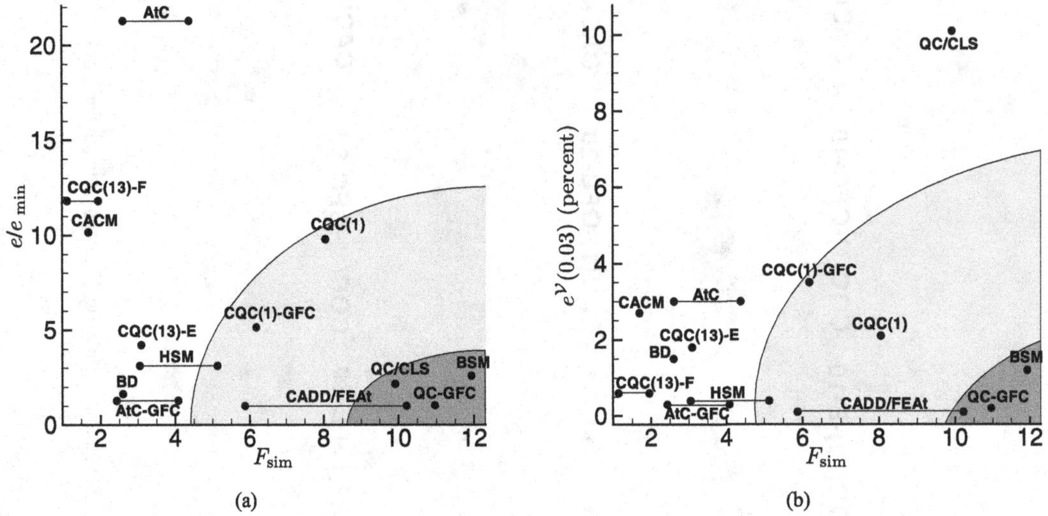

Fig. 12.20 Summary of accuracy and efficiency for all of the methods, based on (a) the displacement error norm and (b) the percent error in atomistic energy.

The energy of the entire system is only defined for the energy-based methods, but it is possible to compute the energy of just the atomistic region for all the methods provided that the underlying atomistic model can be divided into an energy per atom (as is the case for the pair functional used here). In order to define this error concisely we treat the energy of the atomistic region as a function of the applied strain:

$$\widehat{\mathcal{V}}_{\mathrm{A}}^{\mathrm{int}}(\gamma) = \sum_{\alpha \in B^{\mathrm{A}}} E^{\alpha}(\boldsymbol{u}(\gamma)). \tag{12.93}$$

Then we define the *error in the total atomistic energy* as

$$e^{\mathcal{V}}(\gamma) \equiv \left| \frac{\widehat{\mathcal{V}}_{\mathrm{A}}^{\mathrm{int}}(\gamma) - \widehat{\mathcal{V}}_{\mathrm{A,exact}}^{\mathrm{int}}(\gamma)}{\widehat{\mathcal{V}}_{\mathrm{A,exact}}^{\mathrm{int}}(\gamma)} \right|, \tag{12.94}$$

which is dependent on the level of applied strain defined by the shear parameter γ.

Momentarily, we will present the global energy errors associated with each method (see Fig. 12.20). First, we discuss how we measured the efficiency of the methods.

12.7.3 Quantifying the speed of multiscale methods

In addition to the accuracy of the different multiscale methods, their relative efficiencies are also of interest. The simulation time is broken down into various parts as

$$T = T_{\mathrm{OH}} + T_{\mathrm{s}} + N_{\mathrm{step}} t_{\mathrm{OH}}, \tag{12.95}$$

where T_{s} is the time spent minimizing energy (or equilibrating forces) within all load steps, $N_{\mathrm{step}} = 24$ is the number of load steps that were used in this test problem (i.e. the shear strain was applied in 24 increments), and T_{OH} and t_{OH} are the time consumed in "overhead" associated with each method. The overhead is divided into "global" overhead, T_{OH}, that has to be carried out once at the start of the simulation and "timestep" overhead, t_{OH}, that must be repeated each timestep. The former, for example, may include building the model mesh and related data structures, while the latter may include such things as updating the ghost forces for the deformed configuration. To be fair to all the methods, we have only compared the total solve times, T_s, after subtracting all overhead costs. To make the comparison easier, we define an efficiency factor:

$$F_{\text{sim}} = \frac{T_s^{\text{Exact}}}{T_s^{\text{Model}}}. \tag{12.96}$$

This is the factor by which the simulation time is improved by the multiscale models, defined such that bigger is better. We next compare the efficiency factors for all the methods.

12.7.4 Summary of the relative accuracy and speed of multiscale methods

The full numerical details of all the accuracy and speed studies presented in [MT09b] are too lengthy to reproduce here. Instead, we provide only an essential summary of the results.

The accuracy and efficiency results are combined in two graphs in Fig. 12.20, where each method is represented by a point in "error–speed" space for the "model-30" results. In Fig. 12.20(a), the accuracy is represented by the displacement error, while Fig. 12.20(b) shows the accuracy in terms of the total energy error at an applied shear[38] of $\gamma = 0.03$, $e^{\mathcal{V}}(0.03)$. The force-based methods are indicated by two points joined by a line because the speed-up depends on how we define the efficiency factor.[39] The ideal model would be in the lower right-hand corner, with high efficiency and high accuracy (low error). These graphs allow us to identify, somewhat subjectively, the best performing models in the dark gray region, intermediate performance in light gray and poor performance in white.

The force-based methods are slower, to a large extent because of the rather unsophisticated CG-FB solver we have used. Normalizing for the differences between the CG and CG-FB solver, the force-based CADD/FEAt method is essentially as efficient as the comparable energy-based method (QC-GFC) in improving the speed of the calculation. However, to see this efficiency in absolute terms, it is necessary to develop a force-based solver that is as fast, or faster, than a comparable energy-based one (from the results shown here CG-FB is about 1.7 times slower than CG). Finding the optimal solver for the force-based methods is certainly an area worthy of further study.

Intuitively, it might seem reasonable to expect that using a handshake region would improve accuracy, by providing a more gradual transition from the atomistic to the continuum description. In fact, this seems not to be the case, as handshake methods tended to be both slower and less accurate. This is also true of methods using weak compatibility. Handshake regions and weak compatibility are appealing features because they make mesh generation easier, but at least for the current test problem, this comes at a rather high price in terms of ultimate performance.

38 The relative error of the methods is similar for all levels of γ until the dislocations start to move, at which time differences in the distance traveled by the dislocations in a given load step make comparison difficult. See [MT09b] for a comparison of the predictions of the methods for dislocation motion.

39 We define the efficiency factor for the force-based methods by comparing their speed to that of the exact MS simulation, which is an energy-based method. However, we can solve the MS problem using *either* CG or CG-FB. The latter leads to a more favorable efficiency conclusion for the force-based methods by putting them on a more equal footing.

Cluster-based methods are generally slower and less accurate than methods which appeal to a traditional continuum constitutive law such as linear elasticity or the Cauchy–Born rule. This finding is consistent with, and analytically explained by, a mathematical study [LO09] comparing cluster-based and element-based models. Our results shown in Fig. 12.20 also indicate that errors in the continuum region significantly contribute to the overall error of cluster methods.

The CACM model was implemented here as a representative of the class of coupled methods that use an iterative solver. In these approaches, one part (either the atomistic or continuum region) of the problem is held fixed to provide the boundary conditions during energy minimization of the other part, and then the roles are switched back and forth until the change in energy during an iteration is small. Figure 12.20 shows that such methods can be extremely slow, to the point of potentially eliminating any benefit of the coupled method over the full atomistic solution if reasonable accuracy is expected.

Multiscale methods like the ones discussed here show much promise to improve the efficiency of atomistic calculations, but they have not yet fully realized this potential. This is in part because the focus has mainly been on development of the methodology as opposed to the large-scale application to materials problems. One of the disadvantages of multiscale methods is that, relative to straightforward atomistics, they tend to be more difficult to implement. In order for multiscale methods to compete with (or eventually replace) atomistics, it is necessary that they be implemented in three-dimensional, parallel codes optimized to the same degree as modern atomistic packages. One of the barriers to this is the wide variety of existing multiscale approaches from which to choose; it is not clear where to invest the time required to create a highly optimized code. This chapter has taken a step towards removing this barrier by demonstrating that, to a large extent, there is a great deal of similarity between the methods and that they can be described within a common framework. We have also taken a first step towards identifying certain "best practice" strategies for getting the most efficient and accurate implementation possible.

Once an optimized multiscale implementation is developed, its potential for significantly improving the efficiency of atomistic calculations is enormous. Here, we have demonstrated 12-fold improvements in efficiency for an extremely modest test problem, using a very poorly written code and comparing with an atomistic model (EAM) that is amongst the least expensive available. For materials problems where deformation tends to localize and long-range effects are important, an optimized code could easily yield 100- to 1000-fold gains in efficiency. This efficiency would be further increased if we considered more expensive interatomic models like the Brenner–Tersoff potentials [Ter88a, Bre90], MGPT potentials [Mor88, Mor94], or the ReaxFF model [vDDLG01].

Exercises

12.1 [Section 12.4] Derive expressions for the forces on atoms 1 and 2 in the Lennard-Jones chain shown in Fig. 12.9(a) when its energy is computed as in Eqn. (12.58), subject to a linear displacement field of the form $u(X) = AX$. Verify that the forces are zero for all A (i.e. there are no ghost forces).

12.2 [SECTION 12.4] Derive expressions for the ghost forces for a one-dimensional BD model. Use the chain shown in Fig. 12.10 and the energy expression of Eqn. (12.26) to compute the ghost forces on atoms 3, 4 and 5 when the chain is undeformed.

12.3 [SECTION 12.6] In this rather lengthy exercise, you will become familiar with the QC method by exploring the accuracy of the grain boundary (GB) shear example described in Section 12.6 and in the tutorial that accompanies the QC code. You will also learn how to modify this example to create an entirely new simulation.

1. Run the GB shear example as explained in the QC tutorial that is provided at the companion website to this book. Reduce the time step to obtain the critical shear strain that causes the grain boundary to migrate with a tolerance of ± 0.001. Plot the force–displacement (pde) curve and indicate the critical strain on the plot. **Hint**: To save computation time you can take large steps until approaching the critical strain and then switch to smaller steps. Remember, though, that this is a nonlinear problem so your large steps cannot be too large or the simulation will become unstable.
2. In this part we will test how different parameters in the QC simulation affect the critical shear strain. These changes should always be applied to the standard simulation you performed in part 1 of this question, i.e. the changes are not cumulative. In all cases discuss the effect on the critical strain and show the resulting pde curve.
   - Turn off the ghost force correction.
   - Increase the proximity factor (PROXFACT) to 3.0. What does this factor control in the QC simulation? Plot the mesh used for the simulation in this case.
   - Remove the no-adaption zones.
   - Increase PROXFACT to 3.0 *and* remove the no-adaption zones. Based on this result and the previous two results, discuss whether PROXFACT or the constraint imposed by the no-adaption zone is more important.
3. In this part you will modify the GB shear example in some way to explore a different mechanical problem. The modification you introduce is entirely up to you. Whatever your modification, the objective is still to measure the critical shear strain required to move the twin boundary. Here are some possibilities:
   - Study the effect of the twin boundary step height on the critical shear strain.
   - Introduce a void or inclusion into one of the grains and study the effect of this defect on the critical shear strain. Modify the distance of this new defect from the twin step to see how this affects the results. It will also make a difference whether the defect is above or below the twin boundary since the boundary moves in a specific direction.
   - Turn one of the sides of the model (left or right) into a free surface and see how this affects the critical shear strain. Change the distance of the step from the surface (by making the model smaller) to see how this affects the results.
   - Create another stepped twin boundary close to the existing one and see if you can get the two boundaries to annihilate themselves by shearing the crystal. Compute the critical shear strain for annihilation. Explore the effect of the distance between the boundaries.

   In any simulation you perform select the simulation parameters (dtime, PROXFACT and no-adaption zone size) based on your previous results to ensure an acceptable accuracy. Depending on what you choose to do, you may need to modify the user routines (`user_mesh.f`, `user_bcon.f`, `user_pdel.f` and/or the the input file for the simulation. Discuss the results of your simulations in detail and draw any conclusions that you can.

13 Atomistic–continuum coupling: finite temperature and dynamics

In Chapter 12, we discussed a suite of static atomistic–continuum coupling methods, all with the goal of being an approximate, more efficient alternative to molecular statics (MS) (Chapter 6). While there are many such methods, we demonstrated that they can all be described within a common framework, and summarized the various options that one can select within that framework to define a specific variant of the coupling procedure. We deliberately avoided the question of *dynamic* multiscale methods in that discussion.

Here, we look at methods whose goal is to be an approximate, low-cost alternative to *molecular dynamics* (MD) (Chapter 9). The general idea is still embodied in Fig. 12.2, as we envision partitioning the body into two regions: one ($\mathcal{B}^{\mathrm{A}}$) that will be treated atomistically using full MD and another ($\mathcal{B}^{\mathrm{C}}$) that can be approximated using a continuum approach (commonly discretized using finite elements). Instead of seeking an equilibrium configuration for the body, we now have time-dependent boundary conditions (including prescribed displacements, atomic forces and continuous tractions) and a set of initial conditions from which we want to follow the evolution of the system. In the atomistic region, this means the trajectories of all the atoms. In the continuum region, this may include a discretized approximation to the mean displacement, velocity and temperature fields.

Superficially, this sounds like a straightforward extension of static methods, but in fact there are additional difficulties that make the problem much more complex. For example there is only one "goal" for MS and static multiscale methods: one seeks an equilibrium configuration of atoms. Questions of optimal minimization tools and concerns about multiple local minima are important problems, but they do not fundamentally change the methodology of coarse-graining a static atomistic system into a multiscale one. MD, on the other hand, embodies a wider variety of goals, and it is important to use the right formulation for each. In this case, there are two main types of questions we can ask:

1. We might be interested in *equilibrium* properties of a system at finite temperature.
2. We may want to study rapid, *nonequilibrium* processes that are an important part of the phenomenon of interest.

Both cases are *dynamic* processes in that the atoms will be at finite temperature and therefore in constant motion. However, in the first (equilibrium) case, the boundary conditions are changing slowly (in principle, infinitely slowly) on the atomic scale. In the second (nonequilibrium) case, both the atomic positions and the boundary conditions are time-dependent. The nonequilibrium situation is more challenging in a way because it requires the propagation of waves and the transfer of heat between the partitioned domains. The very nature of the concepts of temperature and heat introduce fundamental questions about nonequilibrium multiscale methods. At the level of individual atoms there is no well-defined temperature or heat flow; there are only atomic velocities and collisions. On the other hand, the continuum theories that we hope to use to effect our coarse-graining are often predicated on the notion of a temperature field and properties that depend, pointwise, on that temperature.

A practical problem that arises in nonequilibrium multiscale methods is the tendency for waves to be reflected at the atomistic–continuum interface, ∂B^{I} (see Fig. 12.2). By design, the continuum region (B^{C}) will have a reduced density of finite element nodes compared with the density of atoms, so that there will be a range of short-wavelength phonons present in the atomistic region (B^{A}) that will not be supportable on B^{C}. When these phonons reach the interface, it will appear to them as a rigid boundary, and they will be reflected back into B^{A}. It is worth noting that the same problem arises in conventional dynamic finite element calculations, where the discreteness of the mesh leads to anisotropic wave dispersion and reflection in nonuniform grids [Baž78, MB82]. Reflections can also arise if the constitutive behavior is significantly different in the two regions: for example a linear elastic B^{C} coupled to a nonlinear B^{A} also creates an artificial impedance to wave transmission at the interface. The wave reflection problem is made worse by our explicit goal in multiscale models: to make B^{A} as small as possible and to represent B^{C} as coarsely as we can. If even a small part of the heat generated in the atomistic region is artificially confined, it can ultimately destroy the usefulness of the simulation.

Following a brief summary of dynamic finite elements, we discuss in detail the finite-temperature implementation of the quasicontinuum (QC) method – so-called "hot-QC" – described in [DTMP05, DTLMK11]. This is a method designed to answer questions of the first type (equilibrium) posed above. We end the chapter with a brief review of some of the methods aimed at the second (nonequilibrium) type of multiscale problem.

13.1 Dynamic finite elements

The finite element method[1] (FEM) is readily adapted to dynamic systems by employing d'Alembert's principle. Essentially, this states that we can replace a dynamic system obeying Newton's second law by an equivalent static system if we apply an inertial force, $-\rho_0 \dot{\boldsymbol{v}} dV_0$ (where $\boldsymbol{v}$ is the velocity and ρ_0 is the mass density in the reference configuration), to every differential element in the continuum. In the context of FEM, the appropriate equation of static equilibrium was given in Eqn. (12.6),[2] where we see that this inertial term enters as a body force, $\boldsymbol{b}^{\text{inert}} = -\dot{\boldsymbol{v}}$. Thus there is an additional term in the external residual force of the form

$$f_i^{\text{inert},I} = \int_{B_0} \rho_0(-\dot{v}_i) S^I \, dV_0. \tag{13.1}$$

[1] Finite elements are discussed at length in the companion volume to this one [TME12], and are reviewed briefly in Section 12.1 of this book.

[2] Note that we derived Eqn. (12.6) using the principle of minimum potential energy. Strictly speaking, the dynamic equations cannot be derived from this perspective, but the equilibrium condition in Eqn. (12.6) is still applicable and can be obtained using other variational techniques. See, for example, [ZT05].

Next, we discretize both the displacement field *and* the velocity field, storing a vector of each at every node. Just as we did for the displacements in Eqn. (12.2), we interpolate the velocity field as

$$v_i(\mathbf{X}) = \sum_{I=1}^{n_{\mathrm{nodes}}} S^I(\mathbf{X})\mathsf{v}_i^I, \tag{13.2}$$

where v_i^I is the nodal velocity and S^I is the shape function associated with node I. The acceleration field, $\dot{\boldsymbol{v}}$, can be formally determined from the time derivative of the nodal velocities as[3]

$$\dot{v}_i(\mathbf{X}) = \sum_{I=1}^{n_{\mathrm{nodes}}} S^I(\mathbf{X})\dot{\mathsf{v}}_i^I. \tag{13.3}$$

Using this in Eqn. (13.1) and substituting back into Eqn. (12.6) we can obtain the equations of motion for the nodes:

$$\mathsf{f}_i^I = \sum_J \mathsf{M}^{IJ}\dot{\mathsf{v}}_i^J. \tag{13.4}$$

Here, we have defined

$$\mathsf{f}_i^I \equiv \int_{B_0} \frac{\partial W}{\partial \widetilde{F}_{iJ}}\frac{\partial S^I}{\partial X_J}\,dV_0 - \int_{\partial B_{0t}} \bar{T}_i S^I\,dA_0$$

as the nodal force vector, comprising internal forces due to stresses (the first term on the right-hand side) and external forces due to the applied tractions (the second term). Note that we used Eqn. (12.3) to simplify this expression and we have neglected any body forces other than the inertial term. We have also defined the *consistent mass matrix* as

$$\mathsf{M}^{IJ} \equiv \int_{B_0} \rho_0(\mathbf{X})S^I(\mathbf{X})S^J(\mathbf{X})\,dV_0. \tag{13.5}$$

The contribution to the mass matrix from each finite element can be evaluated efficiently using numerical quadrature, and as such Eqn. (13.4) provides an efficient set of dynamic equations of motion for the nodes in the mesh.

Often, an FEM implementation replaces the consistent mass matrix of Eqn. (13.5) with a so-called *lumped mass matrix*, which is a diagonal matrix derived from M in some ad hoc way.[4] When the mass matrix is diagonal, it becomes trivial to invert Eqn. (13.4) and write an explicit time-integration scheme for the evolution of the nodal displacements. Although this introduces an additional approximation into the equations, there is a substantial saving in the calculation of M^{-1}, and lumped mass schemes still yield surprisingly good accuracy. In fact, in some cases, a lumped mass matrix can actually improve accuracy over the use of the consistent mass matrix through a fortuitous cancellation of errors [ZT05, page 475].

[3] The nodal accelerations $\dot{\mathbf{v}}^I$ appearing in this relation are not additional degrees of freedom. Rather they will be computed as part of a numerical integration scheme to evolve the nodal velocities and displacements with time.

[4] Most commonly, all entries in a row of M are added and the sum becomes the diagonal element. The off-diagonal elements are made zero.

Referring to Eqn. (13.4), and recalling the details of MD discussed in Chapter 9, we see that with a diagonal mass matrix, the FEM system of nodes becomes computationally indistinguishable from a system of atoms once we identify the residual vector in FEM as the force applied to each node. The only differences between a system of atoms and a system of nodes are that: (1) the mass of each node is different (and can vary widely depending on the relative density of the nodes in different regions and the details of the mass-lumping) and (2) the potential energy function, instead of coming from an interatomic potential, comes from the constitutive law defining the stress–strain relationship. As such, any of the time integration schemes introduced for an atomistic system (such as the velocity-Verlet algorithm of Section 9.3.1) can be equally applied to the time integration of the positions and velocities of the nodes. This forms the basis for the coupled dynamic methods discussed in this chapter.

13.2 Equilibrium finite temperature multiscale methods

One approach to dynamic multiscale methods is to extend the concept of an MD *thermostat* (see Section 9.4) to the multiscale setting. This naturally limits the multiscale model to problems that are in or near thermodynamic equilibrium, since a thermostat is designed to maintain a uniform average temperature (consistent with the canonical ensemble) throughout the body. In MD, we often push the limits of where to apply such thermostats, using them to study problems like fracture, nanoindentation or dislocations under stress. Fundamentally, these *are not* equilibrium problems; there will be localized heating as bonds break or dislocations form or move and therefore a thermostat is not strictly correct. However, we can often justify the action of the thermostat if the forces it exerts are small. This is usually the case if the amount of energy released by the bond-breaking process is small relative to the total kinetic energy of the system, since this only raises the average system temperature by a small amount. Qualitatively, we can believe that this circumstance mimics the gentle effect of the "heat bath," which may take the form of collisions with the conduction electrons in a metal, for example. However, our goal in multiscale methods is usually to make the atomistic region very small, and to use a minimal number of degrees of freedom elsewhere. In this case, even a small localized heating may lead to large average temperature rise in the atomistic region, and we may start to worry that the physics of the problem is being adversely affected by the computational "trick" of the thermostat.

For this reason, the simple approach of applying a thermostat only to the atomistic region needs to be treated carefully. Instead, it seems more fundamentally sound to attempt to derive a coarse-scale model that takes into account the missing entropy of the removed atoms at finite temperature and to apply the thermostat globally (to the entire atomistic and continuum region simultaneously). One such approach is the "coarse-grained molecular dynamics (CGMD)" model developed by Rudd and Broughton [RB00, RB05]. The CGMD approach is developed in such a way that, formally at least, one can exactly reproduce MD results by retaining a sufficient number of terms in the approximations to the free energy. In practice, this becomes computationally impractical and instead a smaller, more approximate number of terms need to be treated. Although not explicitly implemented within the framework of a thermostat, an effective Langevin thermostat arises naturally from the formulation. An alternative approach is our focus here, as we describe the work of Dupuy *et al.* [DTMP05, DTLMK11] within the framework of the QC method.

13.2.1 Effective Hamiltonian for the atomistic region

We consider a system consisting of a large number of N atoms whose positions and momenta are given by $\boldsymbol{q} = (\boldsymbol{q}^1, \ldots, \boldsymbol{q}^N)$ and $\boldsymbol{p} = (\boldsymbol{p}^1, \ldots, \boldsymbol{p}^N)$, where $\boldsymbol{q}^\alpha$ and $\boldsymbol{p}^\alpha$ are the position and momentum of the atom α. The Hamiltonian is

$$\mathcal{H}(\boldsymbol{q}, \boldsymbol{p}) = \mathcal{V}(\boldsymbol{q}) + \mathcal{T}(\boldsymbol{p}), \tag{13.6}$$

where $\mathcal{V}(\boldsymbol{q})$ is the potential energy and $\mathcal{T}(\boldsymbol{p})$ is the kinetic energy of the system, which is

$$\mathcal{T}(\boldsymbol{p}) = \sum_{\alpha=1}^{N} \frac{\|\boldsymbol{p}^\alpha\|^2}{2m^\alpha}, \tag{13.7}$$

where m^α is the mass of atom α.

The equilibrium statistical properties at temperature T of such systems are given by phase space averages in the canonical ensemble. We are considering solid systems and therefore adopt a *restricted ensemble* view (see Section 11.1.1) where the phase averages are restricted to the portion of phase space accessible to the solid system. We assume that the particles remain close to their mean positions, $\bar{\boldsymbol{q}} = (\bar{\boldsymbol{q}}^1, \ldots, \bar{\boldsymbol{q}}^N)$, so that

$$\boldsymbol{q} = \bar{\boldsymbol{q}} + \boldsymbol{w}, \tag{13.8}$$

where $\boldsymbol{w} = (\boldsymbol{w}^1, \ldots, \boldsymbol{w}^N)$ are the vibration displacements of the atoms relative to their mean positions.[5] We are envisioning a system where the mean atomic positions define a crystal lattice, although these positions will be determined later from the condition that they minimize the free energy of the system (see Eqn. (13.25)).

In this description, observables are restricted phase averages, which can be computed using Eqn. (11.5) (repeated here for convenience):

$$\mathcal{A}(\bar{\boldsymbol{q}}, T) = \langle A \rangle_\chi = \frac{1}{h^{3N} Z(\bar{\boldsymbol{q}})} \iint A(\bar{\boldsymbol{q}} + \boldsymbol{w}, \boldsymbol{p}) e^{-\mathcal{H}(\bar{\boldsymbol{q}}+\boldsymbol{w},\boldsymbol{p})/k_{\mathrm{B}}T} \chi(\boldsymbol{w})\, d\boldsymbol{w} d\boldsymbol{p}, \tag{13.9}$$

where k_{B} is Boltzmann's constant, h is Planck's constant, χ is the characteristic function defining the restricted ensemble and $Z(\bar{\boldsymbol{q}})$ is the partition function of the restricted canonical ensemble:

$$Z(\bar{\boldsymbol{q}}) = \frac{1}{h^{3N}} \iint e^{-\mathcal{H}(\bar{\boldsymbol{q}}+\boldsymbol{w},\boldsymbol{p})/k_{\mathrm{B}}T} \chi(\boldsymbol{w})\, d\boldsymbol{w} d\boldsymbol{p}.$$

As usual in partitioned-domain methods, we divide the problem into an atomistic region ($\mathcal{B}^{\mathrm{A}}$) and a continuum region ($\mathcal{B}^{\mathrm{C}}$), as shown in Fig. 12.4. For definiteness, the qualifiers

[5] We denote vibrational displacements from the *evolving* mean positions ($\boldsymbol{w} = \boldsymbol{q} - \bar{\boldsymbol{q}}$) differently from absolute displacements from reference positions ($\boldsymbol{u} = \boldsymbol{q} - \boldsymbol{Q}$) for reasons that will become clear later.

"A" and "C" refer to the variables in the atomistic and continuum regions respectively throughout this chapter. We renumber the atoms so that the N_A atoms lying within the atomistic region appear first followed by the N_C atoms in the continuum region, where $N_A + N_C = N$. The coordinates of the atoms are then given by

$$\begin{aligned} \boldsymbol{q} &= (\boldsymbol{q}^1, \dots, \boldsymbol{q}^{N_A}, \boldsymbol{q}^{N_A+1}, \dots, \boldsymbol{q}^N) \\ &\equiv (\boldsymbol{q}_A^1, \dots, \boldsymbol{q}_A^{N_A}, \boldsymbol{q}_C^1, \dots, \boldsymbol{q}_C^{N_C}) = (\boldsymbol{q}_A, \boldsymbol{q}_C), \end{aligned}$$

and the momenta are similarly stored as

$$\boldsymbol{p} = (\boldsymbol{p}_A^1, \dots, \boldsymbol{p}_A^{N_A}, \boldsymbol{p}_C^1, \dots, \boldsymbol{p}_C^{N_C}) = (\boldsymbol{p}_A, \boldsymbol{p}_C).$$

We continue to employ a restricted ensemble, so that Eqn. (13.8) holds with

$$\bar{\boldsymbol{q}} = (\bar{\boldsymbol{q}}_A, \bar{\boldsymbol{q}}_C), \qquad \boldsymbol{w} = (\boldsymbol{w}_A, \boldsymbol{w}_C).$$

We are interested in phase functions that depend only on the positions and momenta of atoms in the atomistic region, $A = A(\boldsymbol{q}_A, \boldsymbol{p}_A)$. Hence, our aim is to compute

$$\mathcal{A}(\bar{\boldsymbol{q}}, T) = \frac{1}{h^{3N} Z(\bar{\boldsymbol{q}})} \iint A(\bar{\boldsymbol{q}}_A + \boldsymbol{w}_A, \boldsymbol{p}_A) e^{-\mathcal{H}(\bar{\boldsymbol{q}}+\boldsymbol{w}, \boldsymbol{p})/k_B T} \chi(\boldsymbol{w})\, d\boldsymbol{w} d\boldsymbol{p} \qquad (13.10)$$

by a method that is computationally more efficient than running an MD simulation on the full set of atoms. We show below that this can be done by constructing an effective potential for the atomistic region where the continuum degrees of freedom are minimized out.

An effective potential for the atomistic region We follow the procedure outlined in [SSP99] and [DTMP05, DTLMK11] to construct an effective Hamiltonian, $\mathcal{H}_A$, for the atomistic region, which will depend on the system's temperature T.

Consider the integral in the numerator of Eqn. (13.10). This integral can be divided into atomistic and continuum parts:

$$\begin{aligned} \iint \left[\iint e^{-\mathcal{H}(\bar{\boldsymbol{q}}_A + \boldsymbol{w}_A, \bar{\boldsymbol{q}}_C + \boldsymbol{w}_C, \boldsymbol{p}_A, \boldsymbol{p}_C)/k_B T} \chi_C(\boldsymbol{w}_C)\, d\boldsymbol{w}_C d\boldsymbol{p}_C \right] \\ \times A(\bar{\boldsymbol{q}}_A + \boldsymbol{w}_A, \boldsymbol{p}_A) \chi_A(\boldsymbol{w}_A)\, d\boldsymbol{w}_A d\boldsymbol{p}_A, \end{aligned} \qquad (13.11)$$

where $\chi_A(\boldsymbol{w}_A)$ and $\chi_C(\boldsymbol{w}_C)$ are the characteristic functions for the atomistic and continuum regions separately, and we have used the fact that $\chi(\boldsymbol{w}) = \chi_A(\boldsymbol{w}_A)\chi_C(\boldsymbol{w}_C)$, which is correct since the system is confined to its restricted configuration space only if both the atomistic *and* continuum regions satisfy their respective constraints. Referring to the expression in the square brackets in Eqn. (13.11), we now define the effective Hamiltonian $\widehat{\mathcal{H}}_A$ for the atomistic region through the following relation:

$$e^{-\widehat{\mathcal{H}}_A(\boldsymbol{q}_A, \boldsymbol{p}_A, \bar{\boldsymbol{q}}_C, T)/k_B T} \equiv \frac{1}{h^{3N_C}} \iint e^{-\mathcal{H}(\boldsymbol{q}_A, \bar{\boldsymbol{q}}_C + \boldsymbol{w}_C, \boldsymbol{p}_A, \boldsymbol{p}_C)/k_B T} \chi_C(\boldsymbol{w}_C)\, d\boldsymbol{w}_C d\boldsymbol{p}_C, \qquad (13.12)$$

from which we obtain the desired definition:

$$\widehat{\mathcal{H}}_{\mathrm{A}}(\boldsymbol{q}_{\mathrm{A}},\boldsymbol{p}_{\mathrm{A}},\bar{\boldsymbol{q}}_{\mathrm{C}},T)=-k_{\mathrm{B}}T\ln\frac{1}{h^{3N_{\mathrm{C}}}}\iint e^{-\mathcal{H}(\boldsymbol{q}_{\mathrm{A}},\bar{\boldsymbol{q}}_{\mathrm{C}}+\boldsymbol{w}_{\mathrm{C}},\boldsymbol{p}_{\mathrm{A}},\boldsymbol{p}_{\mathrm{C}})/k_{\mathrm{B}}T}\chi_{\mathrm{C}}(\boldsymbol{w}_{\mathrm{C}})d\boldsymbol{w}_{\mathrm{C}}d\boldsymbol{p}_{\mathrm{C}}. \tag{13.13}$$

Here $\mathcal{H}$ is the Hamiltonian of the entire system in Eqn. (13.6). The positions $\bar{\boldsymbol{q}}_{\mathrm{C}}$ will be determined later by minimizing the free energy of the system at a given temperature T (as explained in Section 11.1.2) with $\boldsymbol{q}_{\mathrm{A}}$ as fixed parameters (see Eqn. (13.25)), hence $\bar{\boldsymbol{q}}_{\mathrm{C}} = \bar{\boldsymbol{q}}_{\mathrm{C}}(\boldsymbol{q}_{\mathrm{A}},T)$, and so[6]

$$\mathcal{H}_{\mathrm{A}}(\boldsymbol{q}_{\mathrm{A}},\boldsymbol{p}_{\mathrm{A}},T) \equiv \widehat{\mathcal{H}}_{\mathrm{A}}(\boldsymbol{q}_{\mathrm{A}},\boldsymbol{p}_{\mathrm{A}},\bar{\boldsymbol{q}}_{\mathrm{C}}(\boldsymbol{q}_{\mathrm{A}},T),T). \tag{13.14}$$

Substituting Eqn. (13.12) into Eqn. (13.11) and then back into Eqn. (13.10) and applying a similar procedure to the denominator, we have

$$\begin{aligned}\mathcal{A}(\bar{\boldsymbol{q}}_{\mathrm{A}},T) &= \frac{1}{h^{3N_{\mathrm{A}}}Z_{\mathrm{A}}(\bar{\boldsymbol{q}}_{\mathrm{A}},T)}\\ &\times\iint A(\bar{\boldsymbol{q}}_{\mathrm{A}}+\boldsymbol{w}_{\mathrm{A}},\boldsymbol{p}_{\mathrm{A}})e^{-\mathcal{H}_{\mathrm{A}}(\bar{\boldsymbol{q}}_{\mathrm{A}}+\boldsymbol{w}_{\mathrm{A}},\boldsymbol{p}_{\mathrm{A}},T)/k_{\mathrm{B}}T}\chi_{\mathrm{A}}(\boldsymbol{w}_{\mathrm{A}})\,d\boldsymbol{w}_{\mathrm{A}}d\boldsymbol{p}_{\mathrm{A}},\end{aligned} \tag{13.15}$$

with the reduced partition function

$$Z_{\mathrm{A}}(\bar{\boldsymbol{q}}_{\mathrm{A}},T) = \frac{1}{h^{3N_{\mathrm{A}}}}\iint e^{-\mathcal{H}_{\mathrm{A}}(\bar{\boldsymbol{q}}_{\mathrm{A}}+\boldsymbol{w}_{\mathrm{A}},\boldsymbol{p}_{\mathrm{A}},T)/k_{\mathrm{B}}T}\chi_{\mathrm{A}}(\boldsymbol{w}_{\mathrm{A}})\,d\boldsymbol{w}_{\mathrm{A}}d\boldsymbol{p}_{\mathrm{A}}. \tag{13.16}$$

In Eqn. (13.15), we now have an expression for the phase average of our phase function $A(\boldsymbol{q}_{\mathrm{A}},\boldsymbol{p}_{\mathrm{A}})$ at a given temperature T directly in terms of the atomistic variables (although it depends indirectly on the continuum region through the effective Hamiltonian $\mathcal{H}_{\mathrm{A}}$). To explore how this expression can be computed efficiently, we note that since the Hamiltonian can be separated into potential energy and kinetic energy, the effective Hamiltonian has a similar separation:

$$\mathcal{H}_{\mathrm{A}}(\boldsymbol{q}_{\mathrm{A}},\boldsymbol{p}_{\mathrm{A}},T) = \mathcal{V}_{\mathrm{A}}(\boldsymbol{q}_{\mathrm{A}},T) + \mathcal{T}_{\mathrm{A}}(\boldsymbol{p}_{\mathrm{A}},T), \tag{13.17}$$

with

$$\mathcal{V}_{\mathrm{A}}(\boldsymbol{q}_{\mathrm{A}},T) = -k_{\mathrm{B}}T\ln\frac{1}{h_q^{3N_{\mathrm{C}}}}\int e^{-\mathcal{V}(\boldsymbol{q}_{\mathrm{A}},\bar{\boldsymbol{q}}_{\mathrm{C}}+\boldsymbol{w}_{\mathrm{C}})/k_{\mathrm{B}}T}\chi_{\mathrm{C}}(\boldsymbol{w}_{\mathrm{C}})\,d\boldsymbol{w}_{\mathrm{C}}, \tag{13.18}$$

where we recall that $\bar{\boldsymbol{q}}_{\mathrm{C}} = \bar{\boldsymbol{q}}_{\mathrm{C}}(\boldsymbol{q}_{\mathrm{A}},T)$ and

$$\mathcal{T}_{\mathrm{A}}(\boldsymbol{p}_{\mathrm{A}},T) = -k_{\mathrm{B}}T\ln\frac{1}{h_p^{3N_{\mathrm{C}}}}\int e^{-\mathcal{T}(\boldsymbol{p}_{\mathrm{A}},\boldsymbol{p}_{\mathrm{C}})/k_{\mathrm{B}}T}\,d\boldsymbol{p}_{\mathrm{C}}. \tag{13.19}$$

In Eqns. (13.18) and (13.19), h_q and h_p are constants with units of position and momentum, respectively, that satisfy the relation, $h_q h_p = h$, where h is Planck's constant. The

[6] In this equation and below, we will often use the presence or absence of a $\widehat{\ }$ to emphasize that a function has been modified only through a change of variables.

particular choice of h_q and $h_p = h/h_q$ is arbitrary since these constants cancel out in calculations of phase averages. The effective kinetic energy $\mathcal{T}_A(\boldsymbol{p}_A, T)$ in Eqn. (13.19) can be evaluated analytically by substituting in Eqn. (13.7) and integrating each momentum degree of freedom from $-\infty$ to ∞. The result is

$$\mathcal{T}_A(\boldsymbol{p}_A, T) = \sum_{\alpha=1}^{N_A} \frac{\|\boldsymbol{p}_A^\alpha\|^2}{2m_A^\alpha} - \frac{3k_B T}{2} \sum_{\beta=1}^{N_C} \ln \frac{2\pi k_B T m_C^\beta}{h_p^2}, \tag{13.20}$$

where m_A^α is the mass of atom α in the atomistic region, and m_C^β is the mass of atom β in the continuum region.

While the effective kinetic energy is analytically computable, the calculation of the effective potential energy, $\mathcal{V}_A(\boldsymbol{q}_A, T)$, is more challenging. It involves the integration of a complex, generally anharmonic, potential energy function over all degrees of freedom in the continuum region. As such, it is similar to the concept of the *potential of mean force* introduced by Kirkwood [Kir35]. Direct calculation of this term would actually be *more demanding* than the calculation of the complete potential energy $\mathcal{V}(\boldsymbol{q})$ and the forces on all atoms – clearly this is not an effective coarse-graining method without further simplification. Therefore, the purpose of the next section is to demonstrate how it is possible to expedite this calculation using a quasiharmonic approximation for the free energy.

Efficient calculation of the effective potential energy In the next step of this model, we recognize that the effective potential energy $\mathcal{V}_A$ is the free energy of a system where the atoms in the continuum region are at thermodynamic equilibrium at the specified temperature T. We begin by computing this free energy term using the *quasiharmonic approximation* as discussed in Section 11.4. Specifically, the lattice dynamics of the system associated with the vibrations of the atoms in $\mathcal{B}^C$ are treated within the harmonic approximation. This means that the potential energy of the system is expanded about a given equilibrium structure of the crystal defined by $\bar{\boldsymbol{q}}_C$, which is determined as part of this process by ensuring that the corresponding Helmholtz free energy, Ψ, is minimal with respect to these coordinates. More precisely, we first estimate the free energy of the system by expanding the potential energy function to second order around a candidate crystal structure defined by the coordinates $\mathring{\boldsymbol{q}}_C$:

$$\mathcal{V}(\boldsymbol{q}_A, \mathring{\boldsymbol{q}}_C + \boldsymbol{w}_C) \approx \mathcal{V}(\boldsymbol{q}_A, \mathring{\boldsymbol{q}}_C) + \sum_{\alpha=1}^{N_C} \frac{\partial \mathcal{V}}{\partial \boldsymbol{q}_C^\alpha}(\boldsymbol{q}_A, \mathring{\boldsymbol{q}}_C) \cdot \boldsymbol{w}_C^\alpha + \frac{1}{2} \sum_{\alpha,\beta=1}^{N_C} \boldsymbol{w}_C^\alpha \cdot \boldsymbol{\Phi}_C^{\alpha\beta}(\boldsymbol{q}_A, \mathring{\boldsymbol{q}}_C) \boldsymbol{w}_C^\beta, \tag{13.21}$$

where $\boldsymbol{w}_C^\alpha = \boldsymbol{q}_C^\alpha - \mathring{\boldsymbol{q}}_C^\alpha$ is the displacement of atom α relative to $\mathring{\boldsymbol{q}}_C^\alpha$ and $\boldsymbol{\Phi}_C^{\alpha\beta}$ is the 3×3 force constant matrix associated with atoms α and β in the continuum region, defined as

$$\boldsymbol{\Phi}_C^{\alpha\beta}(\boldsymbol{q}_A, \boldsymbol{q}_C) = \frac{\partial^2 \mathcal{V}}{\partial \boldsymbol{q}_C^\alpha \partial \boldsymbol{q}_C^\beta}(\boldsymbol{q}_A, \boldsymbol{q}_C). \tag{13.22}$$

The quasiharmonic approximation for the Helmholtz free energy associated with the atoms in the continuum region follows from Eqn. (11.107) as

$$\Psi_{\mathrm{C}}^{\mathrm{QH}}(\boldsymbol{q}_{\mathrm{A}}, \mathring{\boldsymbol{q}}_{\mathrm{C}}, T) = \mathcal{V}(\boldsymbol{q}_{\mathrm{A}}, \mathring{\boldsymbol{q}}_{\mathrm{C}}) + \frac{k_{\mathrm{B}}T}{2} \ln \left[\frac{h_q^{6N_{\mathrm{C}}} \det \boldsymbol{\Phi}_{\mathrm{C}}(\boldsymbol{q}_{\mathrm{A}}, \mathring{\boldsymbol{q}}_{\mathrm{C}})}{(2\pi k_{\mathrm{B}} T)^{3N_{\mathrm{C}}}} \right]. \tag{13.23}$$

Note that the constant term $h_q^{6N_{\mathrm{C}}}$ has been added to the numerator of the log function to obtain the correct units for $\Psi_{\mathrm{C}}^{\mathrm{QH}}$. Also, $\boldsymbol{\Phi}_{\mathrm{C}}$ is the $3N_{\mathrm{C}} \times 3N_{\mathrm{C}}$ matrix formed from the two-atom $\boldsymbol{\Phi}_{\mathrm{C}}^{\alpha\beta}$ force constant matrices defined in Eqn. (13.22):

$$\boldsymbol{\Phi}_{\mathrm{C}} = \begin{bmatrix} \boldsymbol{\Phi}_{\mathrm{C}}^{11} & \cdots & \boldsymbol{\Phi}_{\mathrm{C}}^{1N_{\mathrm{C}}} \\ \boldsymbol{\Phi}_{\mathrm{C}}^{21} & \cdots & \boldsymbol{\Phi}_{\mathrm{C}}^{2N_{\mathrm{C}}} \\ \vdots & & \vdots \\ \boldsymbol{\Phi}_{\mathrm{C}}^{N_{\mathrm{C}}1} & \cdots & \boldsymbol{\Phi}_{\mathrm{C}}^{N_{\mathrm{C}}N_{\mathrm{C}}} \end{bmatrix}. \tag{13.24}$$

Note that $\boldsymbol{\Phi}_{\mathrm{C}}$ must have rigid-body modes removed as explained after Eqn. (11.108).

The mean positions, $\bar{\boldsymbol{q}}_{\mathrm{C}}$, are now determined by locally minimizing $\Psi_{\mathrm{C}}^{\mathrm{QH}}$ with respect to the trial positions:

$$\bar{\boldsymbol{q}}_{\mathrm{C}}(\boldsymbol{q}_{\mathrm{A}}, T) = \arg\min_{\mathring{\boldsymbol{q}}_{\mathrm{C}}} \Psi_{\mathrm{C}}^{\mathrm{QH}}(\boldsymbol{q}_{\mathrm{A}}, \mathring{\boldsymbol{q}}_{\mathrm{C}}, T). \tag{13.25}$$

In the low-temperature regime, which is consistent with the approximation in Eqn. (13.21), the quasiharmonic approximation for the effective potential energy follows as

$$\begin{aligned} \mathcal{V}_{\mathrm{A}}(\boldsymbol{q}_{\mathrm{A}}, T) &= \Psi_{\mathrm{C}}^{\mathrm{QH}}(\boldsymbol{q}_{\mathrm{A}}, \bar{\boldsymbol{q}}_{\mathrm{C}}(\boldsymbol{q}_{\mathrm{A}}, T), T) + O(T^2) \\ &\equiv \widetilde{\mathcal{V}}_{\mathrm{A}}(\boldsymbol{q}_{\mathrm{A}}, T) + O(T^2), \end{aligned} \tag{13.26}$$

where $\widetilde{\mathcal{V}}_{\mathrm{A}}$, defined on the second line, is the approximation to $\mathcal{V}_{\mathrm{A}}$ and $O(T^2)$ indicates that the expression is accurate to second order in temperature (see Exercise 13.3). Thus, we are using the low-temperature regime to obtain a tractable approximation of the effective potential $\mathcal{V}_{\mathrm{A}}(\boldsymbol{q}_{\mathrm{A}}, T)$ (see [BBLP10] for the study of a related question, in the one-dimensional setting, under a different regime, namely when $N_{\mathrm{C}} \gg N_{\mathrm{A}}$).

The quasiharmonic approximation in Eqn. (13.26) provides a good estimate for $\mathcal{V}_{\mathrm{A}}$. Unfortunately, this expression is computationally intractable for most systems, since $\Psi_{\mathrm{C}}^{\mathrm{QH}}$ includes the determinant of the full $3N_{\mathrm{C}} \times 3N_{\mathrm{C}}$ force constant matrix. We therefore go one step further and adopt the *local harmonic model* proposed in [LNS89] in which all of the off-diagonal coupling terms in Eqn. (13.24) are set to zero, i.e. $\boldsymbol{\Phi}_{\mathrm{C}}^{\alpha\beta} := \boldsymbol{0}$, for $\alpha \neq \beta$. Although this is a drastic simplification of the total force constant matrix (for example it precludes the existence of phonons), the local harmonic assumption can be shown to be valid for the calculation of complex structure at finite temperature from 0 K up to half of the melting temperature of the crystal [RL02]. In addition, we note that this assumption is invoked only in the continuum region, where it should hold with reasonable accuracy, since the atoms are expected to be located close to their mean positions in a defect-free crystal.

Using the local harmonic approximation, the determinant of the force constant matrix can be written as the product of the determinants of the local 3×3 force constant matrices, $\mathbf{\Phi}_{\mathrm{C}}^{\alpha\alpha}$ (no sum on α), of the atoms $\alpha = 1, \ldots, N_{\mathrm{C}}$ in the continuum region:

$$\det \mathbf{\Phi}_{\mathrm{C}}^{\mathrm{LH}} = \prod_{\alpha=1}^{N_{\mathrm{C}}} \det \mathbf{\Phi}_{\mathrm{C}}^{\alpha\alpha}. \tag{13.27}$$

The Helmholtz free energy can therefore be written as

$$\Psi_{\mathrm{C}}^{\mathrm{LH}}(\boldsymbol{q}_{\mathrm{A}}, \mathring{\boldsymbol{q}}_{\mathrm{C}}, T) = \mathcal{V}(\boldsymbol{q}_{\mathrm{A}}, \mathring{\boldsymbol{q}}_{\mathrm{C}}) + \frac{k_{\mathrm{B}}T}{2} \sum_{\alpha=1}^{N_{\mathrm{C}}} \ln \left[\frac{h_q^6 \det \mathbf{\Phi}_{\mathrm{C}}^{\alpha\alpha}(\boldsymbol{q}_{\mathrm{A}}, \mathring{\boldsymbol{q}}_{\mathrm{C}})}{(2\pi k_{\mathrm{B}}T)^3} \right]. \tag{13.28}$$

As was done in Eqns. (13.25) and (13.26), the effective potential energy is approximated as the minimum of the free energy in Eqn. (13.28) with respect to the trial positions $\mathring{\boldsymbol{q}}_{\mathrm{C}}$:

$$\widetilde{\mathcal{V}}_{\mathrm{A}}(\boldsymbol{q}_{\mathrm{A}}, T) = \min_{\mathring{\boldsymbol{q}}_{\mathrm{C}}} \Psi_{\mathrm{C}}^{\mathrm{LH}}(\boldsymbol{q}_{\mathrm{A}}, \mathring{\boldsymbol{q}}_{\mathrm{C}}, T). \tag{13.29}$$

The approximate effective Hamiltonian for the atomistic region is then

$$\widetilde{\mathcal{H}}_{\mathrm{A}}(\boldsymbol{q}_{\mathrm{A}}, \boldsymbol{p}_{\mathrm{A}}, T) = \widetilde{\mathcal{V}}_{\mathrm{A}}(\boldsymbol{q}_{\mathrm{A}}, T) + \mathcal{T}_{\mathrm{A}}(\boldsymbol{p}_{\mathrm{A}}, T) \tag{13.30}$$

where $\widetilde{\mathcal{V}}_{\mathrm{A}}$ is given above and $\mathcal{T}_{\mathrm{A}}$ is given in Eqn. (13.20).

13.2.2 Finite temperature QC framework

The effective Hamiltonian in Eqn. (13.30) is well defined. However, it requires the calculation of the free energy in Eqn. (13.28), which is difficult since $\Psi_{\mathrm{C}}^{\mathrm{LH}}$ depends on a large number of degrees of freedom (as many as the original potential energy), and also involves a sum over all the atoms which have been integrated out. This computation is made tractable by adopting the QC strategy of representing the deformation in the continuum region in terms of a discretized displacement field and then applying the Cauchy–Born rule to obtain an approximation for $\Psi_{\mathrm{C}}^{\mathrm{LH}}$.

The first step in constructing the finite-temperature QC ("hot-QC") framework is the introduction of a *reference configuration* relative to which deformations are measured. We denote the *mean* positions of the atoms in this reference configuration as $\bar{\boldsymbol{Q}} = (\bar{\boldsymbol{Q}}^1, \ldots, \bar{\boldsymbol{Q}}^N)$. The choice of a reference configuration is problem specific. For example for a single crystal the reference configuration can be taken as the unstressed crystal with all atoms at their ideal lattice positions.

We relate the mean positions of the atoms at time t, $\bar{\boldsymbol{q}}(t) = (\bar{\boldsymbol{q}}^1(t), \ldots, \bar{\boldsymbol{q}}^N(t))$, to their mean reference positions by

$$\bar{\boldsymbol{q}}^\alpha(t) = \bar{\boldsymbol{Q}}^\alpha + \bar{\boldsymbol{u}}^\alpha(t) \qquad \alpha = 1, \ldots, N. \tag{13.31}$$

Here $\bar{\boldsymbol{u}}^\alpha(t)$ is the *mean displacement* of atom α at time t defined as

$$\bar{\boldsymbol{u}}^\alpha(t) \equiv \lim_{\Delta t \to \infty} \frac{1}{\Delta t} \int_0^{\Delta t} \boldsymbol{u}^\alpha(t+\tau)\, d\tau, \tag{13.32}$$

where $\boldsymbol{u}^\alpha = \boldsymbol{q}^\alpha - \boldsymbol{Q}^\alpha$ is the instantaneous displacement of atom α. Note that implicit in what follows is the assumption that a separation of time scales exist between the macroscopic time t and the thermal vibrations of atoms. Basically, it is assumed that the limit in Eqn. (13.32) exists and that it converges to its final value for $\Delta t \ll t_{\mathrm{char}}$, where t_{char} is a characteristic time scale over which the mean displacement field varies appreciably.

The second step in constructing the hot-QC framework is to make the assumption that the mean displacement field in the continuum region is spatially slowly varying and can be accurately approximated by an interpolation based on a discretized representation. As in 0 K QC, a subset of all atoms in the model is selected to be *representative atoms* or "repatoms" for short. The set of repatoms includes all atoms in the atomistic region and a small number of atoms in the continuum region, as shown schematically in Fig. 12.4. The criteria for repatom selection are discussed in Section 12.6.

The repatoms serve as the nodes in a finite element mesh, as shown in Fig. 12.4(b). In two dimensions, linear triangular elements are used and, in three dimensions, linear tetrahedral elements are used. We assume that the mesh is "fully refined" near the atomistic–continuum interface, i.e. that the mesh has been refined down to the atomic scale there. This means that no element with a node in the atomistic region contains any internal atoms. This is not a requirement of the method, but is done to improve accuracy and to simplify the implementation. The number of nodes in the atomistic and continuum regions, respectively, are $n^{\mathrm{A}}_{\mathrm{nodes}}$ (which is equal to N_{A}) and $n^{\mathrm{C}}_{\mathrm{nodes}}$, such that $n^{\mathrm{A}}_{\mathrm{nodes}} + n^{\mathrm{C}}_{\mathrm{nodes}} = n_{\mathrm{nodes}}$, where n_{nodes} is the total number of nodes (which is equal to the number of repatoms, n_R).

The mean position of any atom in the continuum region is now approximated using finite element interpolation:[7]

$$\bar{\boldsymbol{q}}^\alpha_{\mathrm{C}} = \bar{\boldsymbol{Q}}^\alpha_{\mathrm{C}} + \sum_{I=1}^{n^{\mathrm{C}}_{\mathrm{nodes}}} S^I(\bar{\boldsymbol{Q}}^\alpha_{\mathrm{C}})\bar{\mathbf{u}}^I_{\mathrm{C}} \qquad \alpha = 1, \ldots, N_{\mathrm{C}}, \tag{13.33}$$

where $S^I(\boldsymbol{X})$ and $\bar{\mathbf{u}}^\alpha_{\mathrm{C}}$ are the finite element shape function and mean displacement associated with node I in the continuum region (see Section 12.1). We use upper-case letters to distinguish nodal indices from atom indices (Greek letters). Finite element shape functions possess the "Kronecker delta property," whereby $S^I(\bar{\boldsymbol{Q}}^\alpha_{\mathrm{C}}) = 1$ if atom α is colocated with node I, and zero if atom α is colocated with node $J \neq I$. Therefore, if the sum in Eqn. (13.33) is evaluated at a nodal position, it returns the mean displacement of that node.

We are now ready to revisit the calculation of $\widetilde{\mathcal{V}}_{\mathrm{A}}$ in Eqn. (13.29). Rather than minimizing $\Psi^{\mathrm{LH}}_{\mathrm{C}}$ with respect to the full set of positions $\mathring{\boldsymbol{q}}_{\mathrm{C}} = (\mathring{\boldsymbol{q}}^\alpha_{\mathrm{C}})_{1 \le \alpha \le N_{\mathrm{C}}}$, we will minimize this function over equilibrium configurations that can be obtained from the smooth displacement

[7] There is no need to use interpolation to obtain the positions of atoms in the atomistic region, since all atoms are represented there. In fact, the finite element mesh is not needed in the atomistic region (except for the elements spanning the interface with the continuum region). It is retained to facilitate meshing and to enable easy expansion and shrinking of the atomistic region.

field defined by the nodal displacements, $\mathbf{u}_C = (\mathbf{u}_C^I)_{1\le I\le n_{\rm nodes}^C}$. As in Eqn. (13.33), we have

$$\mathring{q}_C^\alpha = \bar{Q}_C^\alpha + \sum_{I=1}^{n_{\rm nodes}^C} S^I(\bar{Q}_C^\alpha)\mathbf{u}_C^I \qquad \alpha = 1,\ldots,N_C. \tag{13.34}$$

In this way the number of degrees of freedom in the continuum region is reduced from $3N_C$ to $3n_{\rm nodes}^C$. To make this explicit, let us recast the Helmholtz free energy in Eqn. (13.28) as a function of the nodal displacements:

$$\widehat{\Psi}_C^{\rm LH}(\boldsymbol{q}_A, \mathbf{u}_C, T) = \mathcal{V}(\boldsymbol{q}_A, \mathring{\boldsymbol{q}}_C(\mathbf{u}_C)) + \frac{k_B T}{2}\sum_{\alpha=1}^{N_C} \ln\left[\frac{h_q^6 \det \boldsymbol{\Phi}_C^{\alpha\alpha}(\boldsymbol{q}_A, \mathring{\boldsymbol{q}}_C(\mathbf{u}_C))}{(2\pi k_B T)^3}\right]. \tag{13.35}$$

To proceed, we write the total potential energy as a sum over the energies of the atoms:

$$\mathcal{V}(\boldsymbol{q}_A, \mathring{\boldsymbol{q}}_C) = \sum_{\alpha=1}^{N_A} E_A^\alpha(\boldsymbol{q}_A, \mathring{\boldsymbol{q}}_C) + \sum_{\beta=1}^{N_C} E_C^\beta(\boldsymbol{q}_A, \mathring{\boldsymbol{q}}_C).$$

Here E_A^α is the energy of atom α in the atomistic region and E_C^β is the energy of atom β in the continuum region. Both functions depend on $\boldsymbol{q}_A$ and $\mathring{\boldsymbol{q}}_C$ since atoms in one region can have neighbors that lie in the other. We also define the atom energy functions in terms of the nodal displacements:

$$\widehat{E}_A^\alpha(\boldsymbol{q}_A, \mathbf{u}_C) \equiv E_A^\alpha(\boldsymbol{q}_A, \mathring{\boldsymbol{q}}_C(\mathbf{u}_C)), \qquad \widehat{E}_C^\alpha(\boldsymbol{q}_A, \mathbf{u}_C) \equiv E_C^\alpha(\boldsymbol{q}_A, \mathring{\boldsymbol{q}}_C(\mathbf{u}_C)).$$

Next, let us define ψ_C^α as the free energy of atom α in the continuum region:

$$\psi_C^\alpha(\boldsymbol{q}_A, \mathbf{u}_C, T) \equiv \widehat{E}_C^\alpha(\boldsymbol{q}_A, \mathbf{u}_C) + \frac{k_B T}{2} \ln\left[\frac{h_q^6 \det \boldsymbol{\Phi}_C^{\alpha\alpha}(\boldsymbol{q}_A, \mathring{\boldsymbol{q}}_C(\mathbf{u}_C))}{(2\pi k_B T)^3}\right]. \tag{13.36}$$

We can then write Eqn. (13.35) as

$$\widehat{\Psi}_C^{\rm LH}(\boldsymbol{q}_A, \mathbf{u}_C, T) = \sum_{\alpha=1}^{N_A} \widehat{E}_A^\alpha(\boldsymbol{q}_A, \mathbf{u}_C) + \sum_{\beta=1}^{N_C} \psi_C^\beta(\boldsymbol{q}_A, \mathbf{u}_C, T). \tag{13.37}$$

The free energy now depends on a small number of degrees of freedom. However, we still require an efficient method for computing $\widehat{\Psi}_C^{\rm LH}$ and its gradient without having to visit every atom in the problem as implied by the sum in the second term of Eqn. (13.37). We invoke again the slow variation of the displacement field and make use of the Cauchy–Born approximation. We assume that the atomic neighborhood of each atom in the continuum region is uniform, and consequently its free energy contribution can be computed as if it were in an infinite crystal homogeneously strained by the deformation gradient $\boldsymbol{F}$ at its position. Recall (see Eqn. (12.3)) that we can compute the deformation gradient at a position $\boldsymbol{X}$ in an element from the shape functions, and that the deformation gradient is constant within each element since we have opted to use linear shape functions. We therefore replace the sum in Eqn. (13.37) over all atoms in the continuum region with a sum over the elements spanning this region with appropriate weighting:

$$\sum_{\beta=1}^{N_{\mathrm{C}}} \psi_{\mathrm{C}}^{\beta} \approx \sum_{e=1}^{n_{\mathrm{elem}}^{\mathrm{C}}} \nu_e \psi^{\mathrm{CB}}(\boldsymbol{F}_e, T). \tag{13.38}$$

Here $n_{\mathrm{elem}}^{\mathrm{C}}$ is the number of elements that have at least one node in the continuum region, $\boldsymbol{F}_e$ is the deformation gradient in element e and $\psi^{\mathrm{CB}}(\boldsymbol{F}, T)$ is the "Cauchy–Born" free energy density, i.e. the free energy per atom in an infinite crystal subjected to a uniform deformation gradient $\boldsymbol{F}$. The weighting factor ν_e is the number of atoms "associated" with element e. If all of the nodes of element e are inside the continuum region, then $\nu_e = V_e/\Omega_0$, where V_e is the volume[8] of element e and Ω_0 is the crystal unit cell volume in the reference configuration. For elements crossing into the atomistic region, ν_e must be proportionally reduced to prevent double counting of atoms in the atomistic region (see the discussion surrounding Fig. 12.5 for more details).

Now the Helmholtz free energy in Eqn. (13.37) is simply the sum of element free energies for the continuum region plus the energy of the atoms in the atomistic region:

$$\widehat{\Psi}_{\mathrm{C}}^{\mathrm{LH}}(\boldsymbol{q}_{\mathrm{A}}, \mathbf{u}_{\mathrm{C}}, T) = \sum_{\alpha=1}^{N_{\mathrm{A}}} \widehat{E}_{\mathrm{A}}^{\alpha}(\boldsymbol{q}_{\mathrm{A}}, \mathbf{u}_{\mathrm{C}}) + \sum_{e=1}^{n_{\mathrm{elem}}^{\mathrm{C}}} \nu_e \psi^{\mathrm{CB}}(\boldsymbol{F}_e, T). \tag{13.39}$$

We note the similarity between this expression and the calculation of the potential energy at 0 K using the standard QC method (see Eqn. (12.23)). The only difference is that the free energy density function $\psi^{\mathrm{CB}}(\boldsymbol{F}, T)$ now replaces the 0 K *strain energy density* function $W(\boldsymbol{F})$.

Using this framework, the approximate effective potential in Eqn. (13.29) is estimated by minimizing the free energy $\widehat{\Psi}_{\mathrm{C}}^{\mathrm{LH}}(\boldsymbol{q}_{\mathrm{A}}, \mathbf{u}_{\mathrm{C}}, T)$ with respect to $\mathbf{u}_{\mathrm{C}}$ subject to the appropriate continuum boundary conditions:

$$\widetilde{\widetilde{\mathcal{V}}}_{\mathrm{A}}(\boldsymbol{q}_{\mathrm{A}}, T) = \min_{\mathbf{u}_{\mathrm{C}}} \widehat{\Psi}_{\mathrm{C}}^{\mathrm{LH}}(\boldsymbol{q}_{\mathrm{A}}, \mathbf{u}_{\mathrm{C}}, T). \tag{13.40}$$

The mean displacement $\bar{\mathbf{u}}_{\mathrm{C}}$ that appears in Eqn. (13.33) is the value of $\mathbf{u}_{\mathrm{C}}$ that minimizes Eqn. (13.40). We now describe how Eqn. (13.40) can be used in hot-QC simulations.

13.2.3 Hot-QC-static: atomistic dynamics embedded in a static continuum

The approximate effective potential energy in Eqn. (13.40) and the effective kinetic energy in Eqn. (13.20) can be combined to give an approximation to the effective Hamiltonian,

[8] For a two-dimensional QC model, the volume of an element is equal to its area multiplied by the length of the periodic cell in the out-of-plane direction.

defined in Eqn. (13.17):

$$\overset{\approx}{\mathcal{H}}_A(\boldsymbol{q}_A, \boldsymbol{p}_A, T) = \overset{\approx}{\mathcal{V}}_A(\boldsymbol{q}_A, T) + \mathcal{T}_A(\boldsymbol{p}_A, T). \tag{13.41}$$

This function can be used to perform an equilibrium MD simulation of the atomistic region at a given temperature T with the continuum region maintained at its free energy minimum. This is the *hot-QC-static* approach, where we note that only the continuum is "static."

In practice, a hot-QC-static simulation would proceed by numerically integrating the equations of motion for the atoms in the atomistic region,

$$m_A^\alpha \ddot{\boldsymbol{q}}_A^\alpha = \boldsymbol{f}_A^\alpha \qquad \alpha = 1, \ldots, N_A, \tag{13.42}$$

with the temperature imposed using a suitable thermostat. In Eqn. (13.42), $\boldsymbol{f}_A^\alpha$ is the time-dependent, temperature-dependent force acting on atom α given by

$$\boldsymbol{f}_A^\alpha = -\frac{\partial \overset{\approx}{\mathcal{V}}_A}{\partial \boldsymbol{q}_A^\alpha} + \boldsymbol{f}_A^{\text{ext},\alpha}. \tag{13.43}$$

In this relation, $\boldsymbol{f}_A^{\text{ext},\alpha}$ is the external force acting on atom α due to external fields and due to the thermostat used to maintain the temperature T (see Section 9.4).

The equilibrium MD simulation described above can be used to compute time average analogs to phase averages of the form

$$\langle A \rangle_\chi^{\text{approx}} \equiv \frac{1}{h^{3N_A} \overset{\approx}{Z}_A(\bar{\boldsymbol{q}}_A, T)} \times \iint A(\bar{\boldsymbol{q}}_A + \boldsymbol{w}_A, \boldsymbol{p}_A) e^{-\overset{\approx}{\mathcal{H}}_A(\bar{\boldsymbol{q}}_A + \boldsymbol{w}_A, \boldsymbol{p}_A, T)/k_B T} \chi_A(\boldsymbol{w}_A)\, d\boldsymbol{w}_A\, d\boldsymbol{p}_A, \tag{13.44}$$

with the approximate reduced partition function

$$\overset{\approx}{Z}_A(\bar{\boldsymbol{q}}_A, T) = \frac{1}{h^{3N_A}} \iint e^{-\overset{\approx}{\mathcal{H}}_A(\bar{\boldsymbol{q}}_A + \boldsymbol{w}_A, \boldsymbol{p}_A, T)/k_B T} \chi_A(\boldsymbol{w}_A)\, d\boldsymbol{w}_A\, d\boldsymbol{p}_A.$$

In view of Eqn. (13.15), the approximate phase averages computed from MD should satisfy

$$\langle A \rangle_\chi^{\text{approx}} (\bar{\boldsymbol{q}}_A, T) \approx \langle A \rangle_\chi (\bar{\boldsymbol{q}}_A, T), \tag{13.45}$$

where A is a property of the system which depends *only* on the positions and momenta of the atoms in the atomistic region, and where $\langle A \rangle_\chi (\bar{\boldsymbol{q}}_A, T)$ is the phase average of Eqn. (13.10) computed with the full system (at temperature T). However, the numerous simplifications that were performed in the construction of the coarse-grained potential, $\overset{\approx}{\mathcal{V}}_A$, obviously limit the accuracy of the agreement in Eqn. (13.45). The errors due to finite element interpolations have been described in the context of the QC method at 0 K and will persist here. In addition, we have used the *quasiharmonic approximation* and ultimately its local variant, the *local harmonic approximation*. The quasiharmonic assumption limits the accuracy of the calculation of properties such as the partition function and phase averages of $A(\boldsymbol{q}_{\mathrm{A}}, \boldsymbol{p}_{\mathrm{A}})$ to second order in temperature:

$$\approx{Z}_{\mathrm{A}}(T) = Z_{\mathrm{A}}(T) + O(T^2), \tag{13.46}$$

$$\langle A(\boldsymbol{q}_{\mathrm{A}}, \boldsymbol{p}_{\mathrm{A}})\rangle_{\chi}^{\mathrm{approx}}(T) = \langle A(\boldsymbol{q}_{\mathrm{A}}, \boldsymbol{p}_{\mathrm{A}})\rangle_{\chi}(T) + O(T^2). \tag{13.47}$$

This is a consequence of Eqn. (13.26). Several studies have shown that the quasiharmonic approximation remains valid for temperatures up to the half the melting temperature of the crystal (see, for example, [Foi94]). Strictly speaking, the *local harmonic* model does not guarantee the same accuracy, but numerical tests in [RL02] have shown that it provides accuracy comparable to that of the quasiharmonic approximation.

13.2.4 Hot-QC-dynamic: atomistic and continuum dynamics

The hot-QC-static approach described in the previous section is efficient in the sense that it involves a limited number of degrees of freedom $\boldsymbol{q}_{\mathrm{A}}$ and $\boldsymbol{p}_{\mathrm{A}}$. However, at each timestep, this approach requires the calculation of the force in Eqn. (13.43), which involves the minimization of the free energy in the continuum region according to Eqn. (13.40). While the QC formulation significantly reduces the number of degrees of freedom (from $3N_{\mathrm{C}}$ to $3n^{\mathrm{C}}_{\mathrm{nodes}}$), this may still be computationally expensive.

Building on the effective Hamiltonian of Eqn. (13.41), we now discuss a method that can be more efficient than hot-QC-static, by replacing the minimization in the continuum region with dynamic sampling. The resulting formulation is referred to as *hot-QC-dynamic*. The idea is to introduce equations of motion for the macroscopic system, so that the system essentially samples configurations close to the one that minimizes the Helmholtz free energy. This modification has several advantages. First, there is no longer a need to perform a free energy minimization at each timestep. Second, the atomistic and continuum regions evolve simultaneously in time reducing in part (although not canceling) spurious wave reflection effects that may exist at the interface between the atomistic and the continuum region. The outcome of this coarse-graining procedure is the definition of a global Hamiltonian that allows us to use the same integration scheme for the entire system.

The method is as follows. Rather than calculating the displacement field $\bar{\mathbf{u}}_{\mathrm{C}}$ that minimizes the Helmholtz free energy, $\widehat{\Psi}^{\mathrm{LH}}_{\mathrm{C}}(\boldsymbol{q}_{\mathrm{A}}, \mathbf{u}_{\mathrm{C}}, T)$, at every timestep, and deducing the effective potential energy $\approx{\mathcal{V}}_{\mathrm{A}}$ (see Eqn. (13.40)), we let the entire system evolve in time based on the Helmholtz free energy $\widehat{\Psi}^{\mathrm{LH}}_{\mathrm{C}}(\boldsymbol{q}_{\mathrm{A}}, \mathbf{u}_{\mathrm{C}}, T)$. This strategy leads to an effective finite-temperature QC Hamiltonian, $\mathcal{H}^{\mathrm{QC}}$, which depends on both the positions of atoms in the atomistic region and the displacement field in the continuum region. For definiteness, we use the "QC" subscript to refer to this final formulation. We discuss the derivation of $\mathcal{H}^{\mathrm{QC}}$ in the remainder of this section.

Our first task is to introduce a kinetic energy for the continuum region. Following our strategy of adopting a macroscopic formulation, we use the following definition:

$$\mathcal{T}_{\mathrm{C}}(\dot{\mathbf{u}}_{\mathrm{C}}) \equiv \frac{1}{2}(\dot{\mathbf{u}}_{\mathrm{C}})^T \mathbf{M}\, \dot{\mathbf{u}}_{\mathrm{C}}, \tag{13.48}$$

where $\mathbf{M}$ is the mass matrix. In order to unify this continuum approach with MD, we adopt a *lumped mass approach* (see Section 13.1) so that

$$\mathcal{T}_{\mathrm{C}}(\mathbf{\Pi}_{\mathrm{C}}) = \sum_{I=1}^{n^{\mathrm{C}}_{\mathrm{nodes}}} \frac{\left\|\mathbf{\Pi}^I_{\mathrm{C}}\right\|^2}{2M^I_{\mathrm{C}}}, \tag{13.49}$$

where $\mathbf{\Pi}^I_{\mathrm{C}} = M^I_{\mathrm{C}}\dot{\mathbf{u}}^I_{\mathrm{C}}$ is the momentum of node I and M^I_{C} is its mass. Measures sampled in terms of the atomistic degrees of freedom, q_{A} and p_{A}, do not depend on the choice of masses of the nodes in the coarse-grained region. However, the numerical efficiency of the approach can depend on how M^I_{C} are chosen. A particular choice that ensures that the coarse-grained system has the same total mass and the same kinetic free energy as the full system is given by Dupuy *et al.* in [DTMP05]. Next, we turn to the definition of $\mathcal{H}^{\mathrm{QC}}$.

A "naïve" choice for $\mathcal{H}^{\mathrm{QC}}$ A natural first guess for $\mathcal{H}^{\mathrm{QC}}$ would be to combine the kinetic energy of the atomistic and continuum regions with the effective free energy expression derived earlier. We refer to this as a "naïve" choice for reasons that will be discussed momentarily:

$$\mathcal{H}^{\mathrm{QC}}_{\mathrm{naïve}}(q_{\mathrm{A}}, \mathbf{u}_{\mathrm{C}}, p_{\mathrm{A}}, \mathbf{\Pi}_{\mathrm{C}}, T) = \widehat{\Psi}^{\mathrm{LH}}_{\mathrm{C}}(q_{\mathrm{A}}, \mathbf{u}_{\mathrm{C}}, T) + \mathcal{T}_{\mathrm{A}}(p_{\mathrm{A}}, T) + \mathcal{T}_{\mathrm{C}}(\mathbf{\Pi}_{\mathrm{C}}), \tag{13.50}$$

where $\mathcal{T}_{\mathrm{A}}$ and $\mathcal{T}_{\mathrm{C}}$ are respectively the kinetic energies in Eqns. (13.20) and (13.49) for the atomistic and the continuum regions, and $\widehat{\Psi}^{\mathrm{LH}}_{\mathrm{C}}(q_{\mathrm{A}}, \mathbf{u}_{\mathrm{C}}, T)$ is the local harmonic free energy of Eqn. (13.39).

This seems a reasonable approach, as we have already laid out an efficient framework for computing Eqn. (13.39) and the kinetic energy so defined is a natural "coarsening" of the atomistic kinetic energy. However, as shown in [DTLMK11], this definition of the Hamiltonian loses the second-order accuracy in temperature of the hot-QC-static method. Physically, we can interpret this loss of accuracy as follows. The coupling between the atomistic and continuum regions introduced by $\mathcal{H}^{\mathrm{QC}}_{\mathrm{naïve}}$ causes the nodes in the continuum region to vibrate, leading to an additional, unphysical contribution to the entropy of the system we refer to as *mesh entropy*, S^{mesh}. This ultimately modifies both the kinetic and configurational parts of the free energy of the system leading to spurious effects in its equilibrium properties. For example, each node will have an average kinetic energy equal to

$$\left\langle \frac{\left\|\mathbf{\Pi}^I_{\mathrm{C}}\right\|^2}{2M^I_{\mathrm{C}}} \right\rangle^{\mathrm{QC,naïve}}_{\chi} = \frac{3}{2}k_{\mathrm{B}}T, \qquad I = 1, \ldots, n^{\mathrm{C}}_{\mathrm{nodes}}, \tag{13.51}$$

whereas the effective potential $\mathcal{H}_{\mathrm{A}}$ is built on the premise that the nodes do not carry any kinetic energy. Another example is seen in Fig. 13.1, where the thermal expansion of a perfect single crystal is calculated using the effective Hamiltonian $\mathcal{H}_{\mathrm{A}}$, the naïve Hamiltonian $\mathcal{H}^{\mathrm{QC}}_{\mathrm{naïve}}$ and the final finite-temperature Hamiltonian $\mathcal{H}^{\mathrm{QC}}$ described next.

An accurate choice for $\mathcal{H}^{\mathrm{QC}}$ A final strategy which retains dynamical sampling in the continuum region and reasonably accurate equilibrium properties requires a correction for the spurious mesh entropy introduced by the continuum dynamics. If we knew how to compute this mesh entropy exactly, we could simply correct the naïve Hamiltonian by subtracting it, as

$$\mathcal{H}^{\mathrm{QC}}(\boldsymbol{q}_{\mathrm{A}},\mathbf{u}_{\mathrm{C}},\boldsymbol{p}_{\mathrm{A}},\boldsymbol{\Pi}_{\mathrm{C}},T)=\mathcal{H}^{\mathrm{QC}}_{\text{naïve}}(\boldsymbol{q}_{\mathrm{A}},\mathbf{u}_{\mathrm{C}},\boldsymbol{p}_{\mathrm{A}},\boldsymbol{\Pi}_{\mathrm{C}},T)-\Psi^{\mathrm{mesh}}(\boldsymbol{q}_{\mathrm{A}},\mathbf{u}_{\mathrm{C}},T), \tag{13.52}$$

where $\Psi^{\mathrm{mesh}}=TS^{\mathrm{mesh}}$ is the *mesh free energy* associated with the mesh entropy.

Since we cannot exactly calculate Ψ^{mesh}, we make use of a reasonable approximation as shown in [DTLMK11] to restore second-order accuracy with respect to temperature. Writing Ψ^{mesh} as the sum of a kinetic and potential part, $\Psi^{\mathrm{mesh}}=\Psi^{\mathrm{mesh}}_p+\Psi^{\mathrm{mesh}}_q$, we can obtain the kinetic part analytically from the diagonal mass matrix of the continuum region. This is analogous to the result for a collection of atoms obtained from Eqn. $(11.87)_1$ together with Eqn. $(11.8)_1$, with the atoms replaced by nodes with masses M^I_{C}:

$$\Psi^{\mathrm{mesh}}_p(T)\equiv-\frac{3k_{\mathrm{B}}T}{2}\sum_{I=1}^{n^{\mathrm{C}}_{\mathrm{nodes}}}\ln\frac{2\pi k_{\mathrm{B}}TM^I_{\mathrm{C}}}{h_p^2}. \tag{13.53}$$

The potential part, Ψ^{mesh}_q, can be estimated by applying the local harmonic approximation to the *mesh*, which can be viewed as a set of particles (nodes) interacting through a force law dictated by the finite element stiffness matrix. Thus, in analogy to Eqn. (13.28) we can make an estimate for Ψ^{mesh}_q as

$$\Psi^{\mathrm{mesh}}_q(\boldsymbol{q}_{\mathrm{A}},\mathbf{u}_{\mathrm{C}},T)\approx\frac{k_{\mathrm{B}}T}{2}\ln\left[\frac{h_q^{6n^{\mathrm{C}}_{\mathrm{nodes}}}\det\mathbf{K}^{\mathrm{LH}}_{\mathrm{C},0}(\boldsymbol{q}_{\mathrm{A}},\mathbf{u}_{\mathrm{C}})}{(2\pi k_{\mathrm{B}}T)^{3n^{\mathrm{C}}_{\mathrm{nodes}}}}\right], \tag{13.54}$$

where $\mathbf{K}^{\mathrm{LH}}_{\mathrm{C},0}$ is the local harmonic approximation to the zero-temperature stiffness matrix of the continuum region:

$$\left[\mathbf{K}^{\mathrm{LH}}_{\mathrm{C},0}\right]_{IJ}(\boldsymbol{q}_{\mathrm{A}},\mathbf{u}_{\mathrm{C}})=\begin{cases}[\mathbf{K}_{\mathrm{C},0}]_{IJ}(\boldsymbol{q}_{\mathrm{A}},\mathbf{u}_{\mathrm{C}}) & \text{if } I=J,\\ \mathbf{0} & \text{otherwise,}\end{cases} \tag{13.55}$$

where

$$[\mathbf{K}_{\mathrm{C},0}]_{IJ}(\boldsymbol{q}_{\mathrm{A}},\mathbf{u}_{\mathrm{C}})\equiv\frac{\partial^2\mathcal{V}(\boldsymbol{q}_{\mathrm{A}},\boldsymbol{q}_{\mathrm{C}}(\mathbf{u}_{\mathrm{C}}))}{\partial\mathbf{u}^I_{\mathrm{C}}\partial\mathbf{u}^J_{\mathrm{C}}}. \tag{13.56}$$

This completes the formulation, which was shown in [DTLMK11] to retain second-order accuracy in temperature.[9] To summarize, the final hot-QC-dynamic Hamiltonian consists

[9] Actually, the second-order accuracy can only be shown if the *quasiharmonic* approximation is used for both the effective free energy and the correction to the mesh entropy. The local harmonic assumption is a less controlled approximation, but it is reasonable to expect that it would work as well here as in other applications, as discussed on page 666.

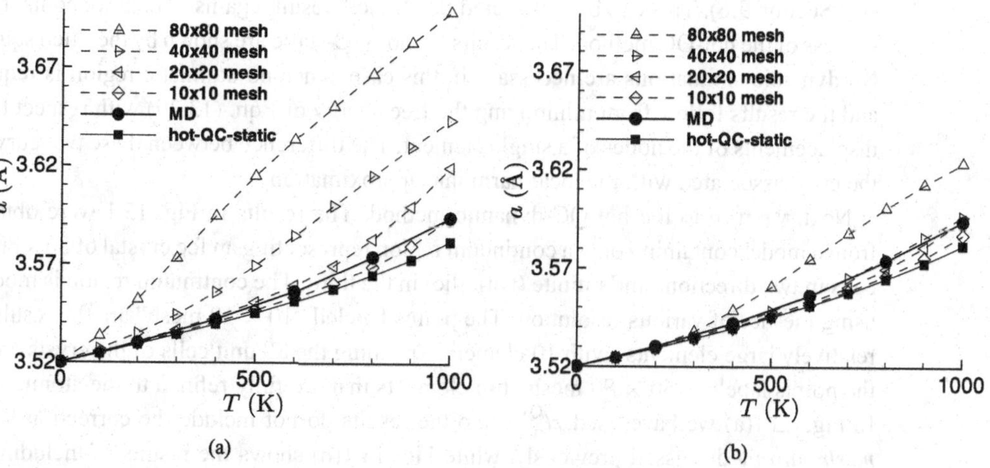

Fig. 13.1 Predicted dependence of the lattice constant on temperature for fcc Ni using the exact MD, hot-QC-static and hot-QC-dynamic models with various degrees of mesh coarsening. In (a), there is no correction for the mesh entropy in the hot-QC-dynamic model ($\mathcal{H}^{\rm QC}_{\rm naive}$ was used), while (b) shows the full, corrected formulation ($\mathcal{H}^{\rm QC}$ was used). See text for details.

of the following potential and kinetic energy parts:

$$\mathcal{V}^{\rm QC}(\boldsymbol{q}_{\rm A}, \mathbf{u}_{\rm C}, T) = \widehat{\Psi}^{\rm LH}_{\rm C}(\boldsymbol{q}_{\rm A}, \mathbf{u}_{\rm C}, T) - \frac{k_{\rm B}T}{2} \ln \left[\frac{h_q^{6n^{\rm C}_{\rm nodes}} \det \mathbf{K}^{\rm LH}_{{\rm C},0}(\boldsymbol{q}_{\rm A}, \mathbf{u}_{\rm C})}{(2\pi k_{\rm B}T)^{3n^{\rm C}_{\rm nodes}}} \right], \tag{13.57}$$

$$\mathcal{T}^{\rm QC}(\boldsymbol{p}_{\rm A}, \boldsymbol{\Pi}_{\rm C}, T) = \mathcal{T}_{\rm A}(\boldsymbol{p}_{\rm A}, T) + \mathcal{T}_{\rm C}(\boldsymbol{\Pi}_{\rm C}) + \frac{3k_{\rm B}T}{2} \sum_{I=1}^{n^{\rm C}_{\rm nodes}} \ln \frac{2\pi k_{\rm B} T M^I_{\rm C}}{h_p^2}. \tag{13.58}$$

Next, we briefly examine two proof-of-principle example applications.

13.2.5 Demonstrative examples: thermal expansion and nanoindentation

Thermal expansion A simple test of the performance of hot-QC method is to calculate the thermal expansion of a lattice.[10] In Fig. 13.1, we show the results of a straightforward simulation whereby a perfect crystal is left to equilibrate at a specified temperature and the resulting average lattice constant is recorded. Several curves are shown, all are based on a face-centered cubic (fcc) crystal of Ni modeled using the pair functional (embedded atom method (EAM)) potential of [AMB95]. The filled circles are from a conventional MD simulation containing 4000 atoms in a periodic, cubic simulation box and equilibrated using the Nosé–Hoover thermostat (see Section 9.4.4) and with the pressure set to zero

[10] The authors thank Dr. Woo Kyun Kim for performing the thermal expansion simulations.

(see Section 9.6). This can be considered the "exact" result, against which we compare the success of the hot-QC method. The results for hot-QC-static are shown by the filled squares. No dynamic simulations are necessary in this case, since no atomistic region is required and the results follow from minimizing the free energy of Eqn. (13.39) with respect to the displacements of the nodes of a single element. The difference between these two curves is the error associated with the local harmonic approximation.

Next, we turn to the hot-QC-dynamic method. The results in Fig. 13.1 were obtained from a model containing only a continuum region representing an fcc crystal of 80×80 unit cells in two directions and infinite (periodic) in the third. The continuum region is modeled using meshes of various resolution. The points labeled "10×10 mesh" are the results for relatively large elements (with 10 elements spanning the 80 unit cells of the crystal) while the points labeled "80×80 mesh" use elements that are fully refined to the atomic scale. In Fig. 13.1(a) we have used $\mathcal{H}^{\text{QC}}_{\text{naïve}}$, so the results do not include the correction for the *mesh entropy* discussed previously, while Fig. 13.1(b) shows the results of including the approximate correction for the mesh entropy, as in the complete formulation embodied by Eqn. (13.57). As we can see, there is a strong mesh dependence and large error when the elements are small, but this is largely corrected by the full formulation. In the limit of large elements, the hot-QC-dynamic results converge to those of hot-QC-static.

Nanoindentation An example that more exhaustively tests the capabilities of the hot-QC approach is that of nanoindentation into a single crystal.[11] In Fig. 13.2(a) we show a model of a Ni crystal, modeled using the pair functional potential of [AMB95]. The fcc crystal is oriented with the $[11\bar{1}]$ direction along X_1 and $[1\bar{1}0]$ along X_2. A cylindrical indenter is simulated by a repulsive potential that forces atoms to stay outside of the indenter [KPH98]. Specifically, we impose an external potential energy of the form

$$\mathcal{V}^{\text{ext}} = \sum_{\alpha=1}^{N_{\text{A}}} P(d^{\alpha}),$$

where d^{α} is the distance between atom α and the center of the indenter,

$$P(d) = \begin{cases} C\,(d-R)^2 & \text{for } d < R, \\ 0 & \text{for } d \geq R \end{cases}$$

and $R = 7\,\text{nm}$ is the indenter radius. The constant C sets the "softness" of the indenter surface and has little effect on the results, except that too large a value can lead to very high instantaneous forces and instabilities. For this simulation, C was set to $100\ \text{eV}/\text{Å}^2$. After an initial equilibration time of 100 ps, the indenter was pressed into the crystal at a rate of 5 m/s. The MD timestep was 1 fs and the temperature was controlled throughout the simulation using the Nosé–Poincaré thermostat proposed in [BLL99]. This is similar to the Nosé–Hoover thermostat discussed at length in Chapter 9, but it has an additional advantage of being symplectic. These details are not critical for the examples presented here, and so they are left to the literature in the interest of brevity.

[11] A "nanoindentation" test is an experiment where a nanoscale indenter is driven into the surface of a material – usually a single crystal. The simulations described below are based on an earlier study in [DTMP05].

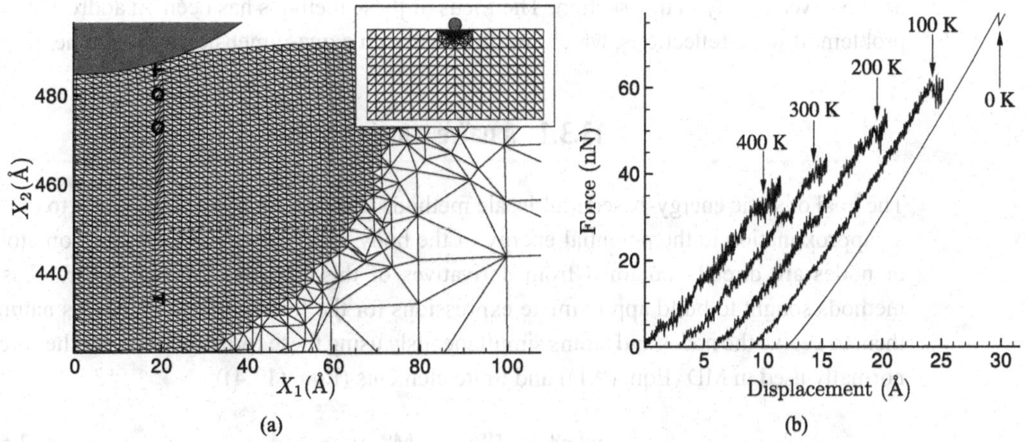

Fig. 13.2 Nanoindentation simulation using hot-QC-dynamic method: (a) dislocations nucleate below the surface in the atomistic region, which is embedded in a larger continuum region as shown in the inset; (b) force versus indentation depth at different temperatures, shifted horizontally to make each curve visible. Arrows indicate the point on each curve where nucleation first occurs.

Figure 13.2(a) shows a typical deformation that results from this simulation, where dislocations nucleate below the surface of the indented crystal after some amount of elastic compression. In this case, the dislocations nucleate two at a time, as a dipole. The two small circles indicate the approximate nucleation site in this particular case (at $T = 100\,\text{K}$), and the two $\perp$ symbols indicate the location of the defects after a few timesteps. As these are partial dislocations, they leave a stacking fault between them. A short time later (not shown), the trailing Shockley partials form at about the same nucleation site, but because there is no model adaption in this example the defect traveling downwards quickly reaches the end of the atomistic region and only the initial nucleation can be taken as an artifact-free result. In Fig. 13.2(b), we show the resulting force versus indentation depth curves as a function of temperature. The forces are averaged over a period of 0.5 ps to reduce thermal noise. Two important features captured by the model are the slight increase in slope with decreasing temperature (due to the temperature dependence of the elastic stiffness) and the decreasing force required for dislocation nucleation as the temperature is increased.

This nanoindentation simulation was performed with about 4200 repatoms, whereas the volume of simulated material contains over 790 000 atoms. As such, even in the ideal case of $O(N)$ scaling, the saving in simulation time is about a 200-fold increase in speed.

13.3 Nonequilibrium multiscale methods

The multiscale methods discussed so far are limited to studying *equilibrium* properties. Methods to model nonequilibrium behavior have been developed by several authors and are reviewed briefly in this section. The focus of these methods has been on addressing the problem of wave reflections, which we illustrate with a one-dimensional system next.

13.3.1 A naïve starting point

The goal of static energy-based multiscale methods (Section 12.3) was essentially to obtain an approximation to the potential energy of the body, as in Eqn. (12.11). Forces on atoms or nodes are directly obtained from derivatives of this function. Likewise, force-based methods sought to build approximate expressions for the forces directly. It seems natural, then, to evolve the nodes and atoms simultaneously using these forces in place for the forces normally used in MD (Eqn. (9.1)) and finite elements (Eqn. (13.4)):

$$m^{\alpha}\ddot{\boldsymbol{r}}^{\alpha} = \boldsymbol{f}^{\alpha}, \qquad \mathsf{M}^{\alpha\beta}\dot{\mathsf{v}}_i^{\beta} = \mathsf{f}_i^{\alpha}. \tag{13.59}$$

Given a method to compute the atomic and nodal forces, these equations can be simultaneously integrated in time using methods like those described in Chapter 9.[12] Indeed, this idea serves as the basis for nonequilibrium multiscale methods, but it is not the complete story. In the next section, we use this simple idea in a one-dimensional model as motivation for the additional steps taken in these methods. The main problem that this will illustrate is the reflection of waves at the atomistic–continuum interface.

Nonequilibrium multiscale methods can be broadly divided into two categories which we will call: (1) kernel function absorbing boundary conditions and (2) damping bands. The first category is theoretically motivated from the generalized Langevin equations (the subject of Section 13.3.3), while the second is more ad hoc, but both are aimed primarily at keeping the behavior of the atomistic region as close as possible to that of the same region embedded in a much larger constant energy MD simulation. Their goal is mainly to remove the wave reflection problem, although damping bands can have the effect of imposing a gentle kind of a thermostat on the atomistic region as we shall discuss. Whether such a thermostat can rigorously maintain an NVT ensemble when the boundary conditions are not changing remains an open question.

13.3.2 Wave reflections

The problem of wave reflections is easily illustrated with a one-dimensional example. We use a nearest-neighbor chain of N atoms connected by linear springs.[13] The reference positions of the atoms are

$$R^{\alpha} = \alpha a_0 \qquad \alpha \in \mathbb{Z},$$

[12] Using the forces from an energy-based approach in this way could be considered an approximation to the NVE ensemble. If forces from a force-based method are used, then it becomes more difficult to control the error. Care must be taken with force-based methods because there is no longer a conserved quantity in the system.

[13] Although we have simplified the model to linear springs to make it amenable to analytic treatment, the essence of the problem persists for a nonlinear system.

and the energy of the chain depends on the displacements from these positions as

$$\mathcal{V}^{\text{int}} = \frac{1}{2}\sum_{\alpha=1}^{N-1} \kappa(u^{\alpha+1} - u^{\alpha})^2, \tag{13.60}$$

where for definiteness, we choose the spring constant and interatomic distances to match a near-neighbor Lennard Jones model with $\varepsilon = 1$ eV and $\sigma = 1$ Å (see Section 12.4.1):

$$a_0 = 1.1225 \text{ Å}, \quad \kappa = 57.14 \text{ eV/Å}^2.$$

We choose the atomic mass[14] to be $m = 40$ amu so that a timestep of $\Delta t = 1$ fs is stable.

We build a finite chain containing 1001 atoms spanning the range from 0 to $1000a_0$, but since we will only consider the early stages of wave propagation we can think of the chain as semi-infinite with the atom at 0 held fixed. Initially, all the atoms are at rest at their equilibrium lattice spacing, until we introduce a velocity distributed as half a sine wave:

$$\dot{u}(0) = \begin{cases} -2\sin\dfrac{2\pi(R-4a_0)}{20a_0} \text{ Å/ps} & 4a_0 \leq R \leq 14a_0, \\ 0 & \text{otherwise.} \end{cases}$$

The exact form of this initial condition is not of great importance; it is chosen as a convenient way to demonstrate wave reflections. The severity of the reflection problem depends on the relation between the nodal spacing in the coarse-scale region and the wavelengths of the phonons comprising the imposed wave.[15] Next, we integrate Eqn. $(13.59)_1$ in time using the velocity-Verlet algorithm (Algorithm 9.2) with a timestep of $\Delta t = 1$ fs. Figure 13.3(a) shows the shape of the resulting displacement wave at $t = 180\Delta t$. This wave remains essentially self-similar (but for a small amount of dispersion[16]) as it moves from left to right along the chain. In Fig. 13.3(b), we see the same wave at $t = 530\Delta t$.

Now, we introduce this wave into a multiscale model of the same harmonic chain. We simply implement the static QC method without ghost force corrections[17] to describe the atomic and nodal forces (see Section 12.3.2), use a lumped mass matrix (as described in footnote 4 on page 660) and simultaneously evolve atoms and nodes by Eqn. (13.59). The QC model consists of an atomistic region extending from 0 to $50a_0$ in the reference configuration and a coarse-grained continuum region where only every fifth atom is represented (this can be seen by examining the horizontal spacing between the points on the curve in Fig. 13.3(c)). Because the atomistic model is a simple nearest-neighbor harmonic chain the Cauchy–Born constitutive model employed in the continuum region reduces to linear elasticity. We therefore fit the continuum constitutive model to the atomistic model in the limit of uniform strain. Because it is uniform, this strain can be obtained from the displacements of any two

[14] Recall 1 amu $= 1.0365 \times 10^{-4}$eV $\cdot$ ps^2/Å^2.

[15] Any wave can be decomposed into a superposition of plane waves (phonons) with different wave vectors, exactly as we did in our discussion of electronic waves in Section 4.3.2. Phonons with wavelengths much longer than the nodal spacing pass almost unaffected by the coarse scale. However, phonons with wavelengths comparable to or smaller than the nodal spacing are partially or even entirely reflected.

[16] Dispersion refers to the fact that the wave speed depends on the wave length (this will be elaborated shortly). For a wave comprised of many different phonons, dispersion means that the wave will "spread out" over time instead of remaining exactly self-similar.

[17] There are actually no ghost forces because this is a nearest-neighbor model as explained in Section 12.4.

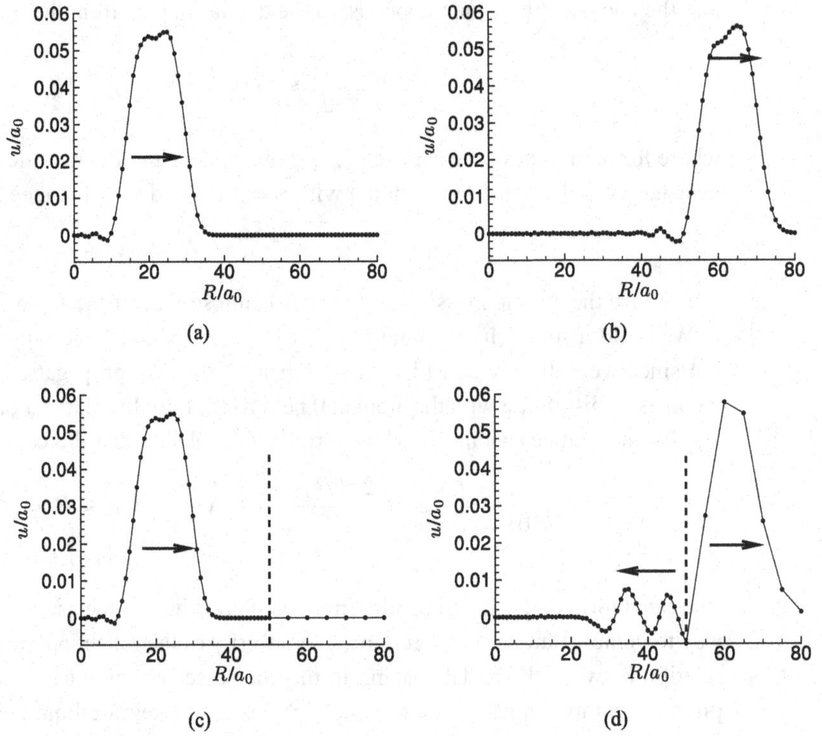

Fig. 13.3 The wave reflection problem in multiscale methods, illustrated in one dimension: (a) and (b) the propagation of a wave in a full MD simulation; (c) and (d) the consequences of the atomistic–continuum coupling, with the atomistic–continuum interface indicated by the dashed line.

neighboring atoms, $\epsilon = (u^{\alpha+1} - u^{\alpha})/a_0 = \Delta u/a_0$, and then the strain energy density (energy per unit length in one dimension) is simply:

$$W = \frac{1}{2}D\epsilon^2 = \frac{1}{2}D\left(\frac{\Delta u}{a_0}\right)^2. \tag{13.61}$$

At the same time, we can write an expression for the energy per unit length of a uniformly stretched chain from Eqn. (13.60):[18]

$$W = \frac{1}{2a_0}\kappa(\Delta u)^2. \tag{13.62}$$

Equating these two expressions for W will give us the elastic modulus, D, in terms of the spring constant, κ, as

$$D = \kappa a_0, \tag{13.63}$$

which completes the continuum constitutive law and allows us to run the multiscale simulation to compare with the fully atomistic results.

[18] This strictly only works for a periodic chain, or in the limit where $N \to \infty$.

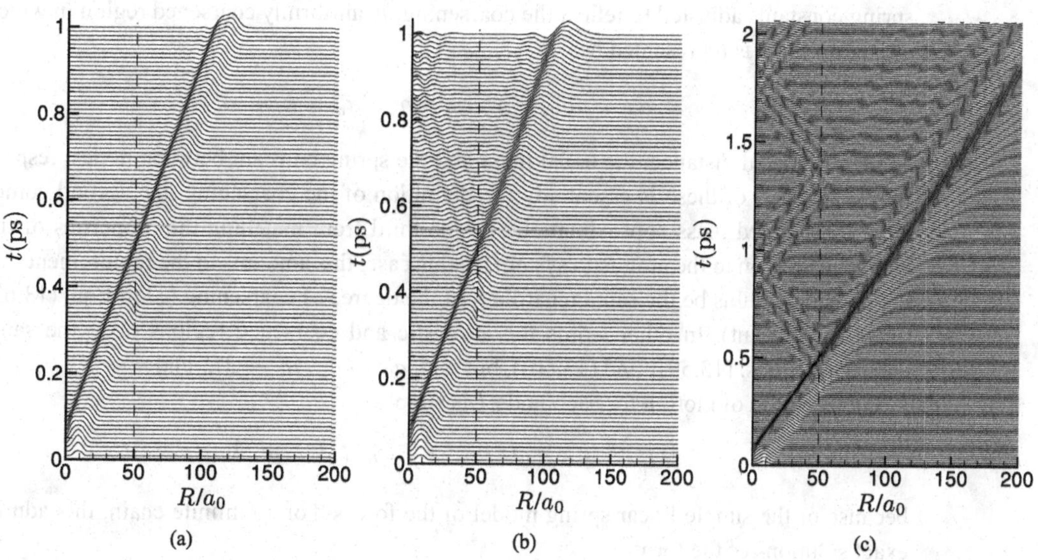

Fig. 13.4 Wave propagation through an atomic chain in (a) is contrasted with wave reflections from the coarsened region in (b). In (c) the time axis is extended to show multiple wave reflections. The position of the atomistic–continuum interface is shown by the dashed line (see text for full details).

In Figs. 13.3(c) and 13.3(d), we see the wave moving through the coupled model. In Fig. 13.3(c), the wave has not yet reached the atomistic–continuum interface and we can see that the displacement field is essentially identical to that of the exact result in Fig. 13.3(a). However, comparing Figs. 13.3(b) and (d) shows a clear consequence of the atomistic–continuum interface, as part of the wave is reflected (indicated by the arrow pointing left). The transmitted wave has lost amplitude and changed its shape, indicating the root cause of the problem: the coarsened continuum region cannot support wavelengths shorter than the spacing between nodes.

Figure 13.4 illustrates the propagation of the wave by lining up multiple snapshots of the chain at different times (one image every $10\Delta t$). In Fig. 13.4(a), the exact MD result shows the more or less self-similar propagation of the wave after an initial transient period.[19] The figure also shows the boundary at $50a_0$ (as a dashed line) which will be the interface between the atomistic and continuum regions in the coarse-grained model. It is clear that in the fully atomistic case shown in Fig. 13.4(a), the atoms to the left of $50a_0$ come to rest once the wave has passed out of this region. The multiscale simulation of the same wave is shown in Fig. 13.4(b), where it is clear that part of the wave is reflected back into the atomistic region. By extending the simulation as shown in Fig. 13.4(c), we see that the reflected waves remain trapped in the atomistic region throughout the simulation.

We can understand this reflection problem by recognizing that the coarsened chain is essentially still just a chain of "atoms" joined by harmonic springs, but with the mass and

[19] Note that we have introduced a *velocity* initial condition, while the displacements are everywhere zero at $t = 0$. One can see that after about five snapshots in Fig. 13.4(a) ($t = 0.05$ ps), the displacement wave is established.

spring constants adjusted to reflect the coarsening. A uniformly coarsened region in which every nth atom is represented has

$$a_n = na_0, \quad M_n = nm, \quad \kappa_n = \kappa/n,$$

as the internodal distance, the nodal mass and the spring constant between nodes, respectively. The first of these is essentially the definition of the coarsening, the second comes from the lumped mass approximation and the third from matching the properties of the coarsened region to the atomistic system. In this case, this amounts to the requirement that the elastic modulus be the same regardless of the degree of coarsening (D in Eqn. (13.63) must be constant). In other words the atomistic and coarsened regions obey the same equations (Eqns. $(13.59)_1$ and (13.60)), but with $a_0 \to a_n$, $m \to M_n$ and $\kappa \to \kappa_n$.

The equation of motion for each node I is simply

$$M_n^I \ddot{u}^I = \kappa_n (u^{I+1} - u^I) - \kappa_n (u^I - u^{I-1})$$

because of the simple linear spring model of the forces. For an infinite chain, this admits exact solutions of the form[20]

$$u^I(R,t) \propto e^{i(kR-\omega t)},$$

where $i = \sqrt{-1}$, k is the wave number and ω is the frequency of vibration. Using this in the equation of motion, one can readily verify that the frequency depends on the wave number through

$$\omega(k) = 2\sqrt{\frac{\kappa_n}{M_n}}\left|\sin\frac{ka_n}{2}\right| = \frac{2}{n}\sqrt{\frac{\kappa}{m}}\left|\sin\frac{kna_0}{2}\right|, \tag{13.64}$$

where the last equality explicitly shows the dependence on the degree of coarsening, n. Equation (13.64) is the *dispersion relation* for the one-dimensional "crystal," and it tells us how waves of a given wave number propagate, since the velocity at which a single phonon of wave number k travels (called the *phase velocity*) is $c = \omega/k$. Any general waveform can be written, via a Fourier series, as a sum of waves over this basis of plane waves, and since each component plane wave has a different phase velocity, a general wave tends to spread out as it propagates.

In Fig. 13.5, we compare the dispersion relations for $n = 1$ (the fully-refined chain) and $n = 5$ (the coarse region in our example). In the long-wavelength limit at $k = 0$, the two regions behave identically. This is as expected, and means that only higher-frequency wavelengths will be reflected. As k grows (going to shorter wavelengths), it is clear that there are values of k for which the phase velocity is zero, and that these $\omega = 0$ points occurs more frequently for $n = 5$. This means that certain wavelengths that can travel freely through a fully-refined ($n = 1$) region cannot travel in the $n = 5$ region and they are therefore reflected back. It may seem strange that waves with a very high magnitude of k have any velocity at all since these should correspond to wavelengths that are much shorter that the distances between the atoms. However, the periodic nature of the dispersion relation is a result of the fact that short wavelength plane waves can always be referred to

[20] See any solid state physics book for a detailed derivation, such as [Dov93] and [AM76].

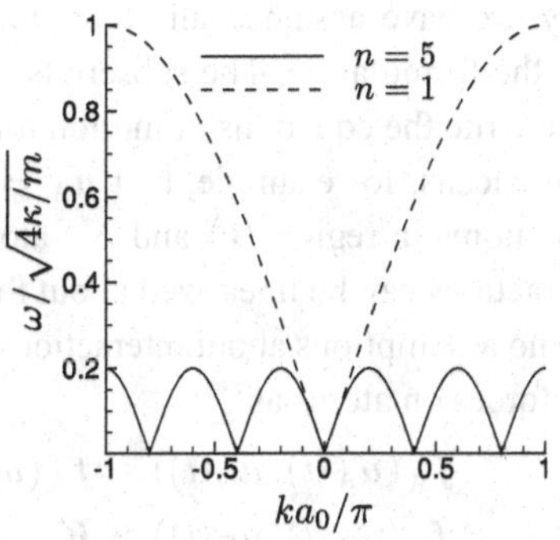

Fig. 13.5 Dispersion relations for the fully-refined ($n = 1$) and coarsened ($n = 5$) harmonic chains.

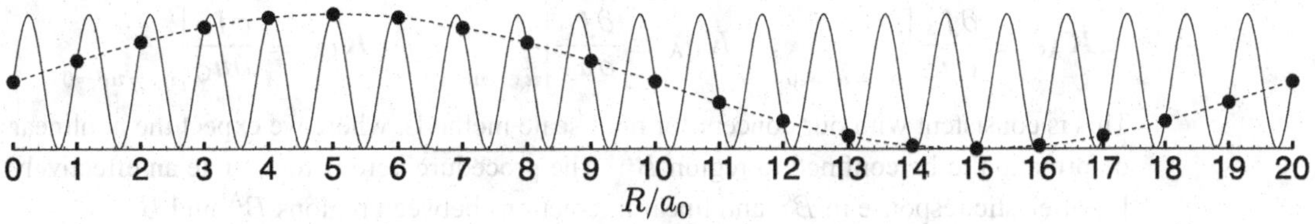

Fig. 13.6 Two very different wave numbers (dashed, $ka_0/\pi = 0.1$ and solid, $ka_0/\pi = 2.1$) produce waves that have the same amplitude at the discrete lattice sites, producing the identical lattice displacements.

an equivalent wave inside the first Brillouin zone (see Section 3.7.2). This is illustrated for two sine waves in Fig. 13.6.

We see, then, that even though we have carefully matched the continuum constitutive law to the atomistic behavior and used a simple, linear material in both regions, there is still a dynamic mismatch when we try to coarsen the model. This mismatch is inherent to the approach, since removing degrees of freedom is both our intention and the cause of the problem. It is for this reason that dynamic multiscale methods must employ some technique to mitigate these effects for nonequilibrium simulations.

13.3.3 Generalized Langevin equations

In a series of papers in the 1970s, Doll, Adelman and Myers [AD74, DMA75, AD76] laid the groundwork for a formal coarse-graining procedure that has since been one focus of attention in dynamic multiscale methods. Although the formal procedure is too computationally intensive to be practical, it provides a rigorous justification and guiding principles for the approximations that have been made, which we discuss briefly below. First, let us review the Adelman–Doll procedure.

As usual, we have a solid composed of N atoms with reference positions $\boldsymbol{R}^\alpha$, displacements $\boldsymbol{u}^\alpha$ and velocities $\dot{\boldsymbol{u}}^\alpha$. Divide these into a primary region B^{A} (this will be the atomistic region) and a secondary region B^{C} (this will become the continuum region), and write equations of motion as usual for atoms in the two regions:

$$m\ddot{\boldsymbol{u}}_{\mathrm{A}}(t) = \boldsymbol{f}_{\mathrm{A}}(\boldsymbol{u}_{\mathrm{A}}(t), \boldsymbol{u}_{\mathrm{C}}(t)), \qquad m\ddot{\boldsymbol{u}}_{\mathrm{C}}(t) = \boldsymbol{f}_{\mathrm{C}}(\boldsymbol{u}_{\mathrm{A}}(t), \boldsymbol{u}_{\mathrm{C}}(t)). \tag{13.65}$$

For simplicity, we have assumed all atoms to have the same mass, although this is not necessary for the formulation. The subscripts identify whether an atom lies in region $\mathcal{B}^{\mathrm{A}}$ or $\mathcal{B}^{\mathrm{C}}$, and we write the equations of motion for all the atoms in a region as a single vector equation. This means, for example, that $\ddot{\boldsymbol{u}}_{\mathrm{A}} = (\ddot{\boldsymbol{u}}^1, \ldots, \ddot{\boldsymbol{u}}^{N_{\mathrm{A}}})$ is a vector of length $3N_{\mathrm{A}}$ (there are N_{A} atoms in region $\mathcal{B}^{\mathrm{A}}$ and N_{C} atoms in $\mathcal{B}^{\mathrm{C}}$). Now, we assume that the C–C and A–C interactions can be linearized about the reference configuration of the atoms, but we will make no assumptions about interactions between atoms within $\mathcal{B}^{\mathrm{A}}$. This means we can write the forces on atoms as

$$\begin{aligned} \boldsymbol{f}_{\mathrm{A}}(\boldsymbol{u}_{\mathrm{A}}(t), \boldsymbol{u}_{\mathrm{C}}(t)) &\approx \boldsymbol{f}_{\mathrm{A}}(\boldsymbol{u}_{\mathrm{A}}(t), 0) + \boldsymbol{K}_{\mathrm{AC}}\boldsymbol{u}_{\mathrm{C}}(t), \\ \boldsymbol{f}_{\mathrm{C}}(\boldsymbol{u}_{\mathrm{A}}(t), \boldsymbol{u}_{\mathrm{C}}(t)) &\approx \boldsymbol{K}_{\mathrm{CA}}\boldsymbol{u}_{\mathrm{A}}(t) + \boldsymbol{K}_{\mathrm{CC}}\boldsymbol{u}_{\mathrm{C}}(t), \end{aligned} \tag{13.66}$$

where we have defined

$$\boldsymbol{K}_{\mathrm{AC}} \equiv \left.\frac{\partial \boldsymbol{f}_{\mathrm{A}}}{\partial \boldsymbol{u}_{\mathrm{C}}}\right|_{\boldsymbol{u}_{\mathrm{A}}=\boldsymbol{u}_{\mathrm{C}}=0}, \qquad \boldsymbol{K}_{\mathrm{CA}} \equiv \left.\frac{\partial \boldsymbol{f}_{\mathrm{C}}}{\partial \boldsymbol{u}_{\mathrm{A}}}\right|_{\boldsymbol{u}_{\mathrm{A}}=\boldsymbol{u}_{\mathrm{C}}=0}, \qquad \boldsymbol{K}_{\mathrm{CC}} \equiv \left.\frac{\partial \boldsymbol{f}_{\mathrm{C}}}{\partial \boldsymbol{u}_{\mathrm{C}}}\right|_{\boldsymbol{u}_{\mathrm{A}}=\boldsymbol{u}_{\mathrm{C}}=0}.$$

This is consistent with our concept for multiscale methods, where we expect the nonlinear deformation to be confined to region $\mathcal{B}^{\mathrm{A}}$. The procedure here is to assume an effectively linear elastic response in $\mathcal{B}^{\mathrm{C}}$ and linear interactions between regions $\mathcal{B}^{\mathrm{A}}$ and $\mathcal{B}^{\mathrm{C}}$.

Our goal is to rewrite the equation of motion for $\boldsymbol{u}_{\mathrm{A}}$ so that it is independent of $\boldsymbol{u}_{\mathrm{C}}$, replacing $\boldsymbol{u}_{\mathrm{C}}$ with its effect on interactions between atoms within $\mathcal{B}^{\mathrm{A}}$. To do this, we substitute Eqns. (13.66) into Eqns. (13.65) and take the Laplace transforms to facilitate a solution to the resulting differential equations, leading to

$$m\left[s^2 \underset{\sim}{\boldsymbol{u}}_{\mathrm{A}}(s) - \boldsymbol{Z}_{\mathrm{A}}(s)\right] = \underset{\sim}{\boldsymbol{f}}_{\mathrm{A}}(s) + \boldsymbol{K}_{\mathrm{AC}}\underset{\sim}{\boldsymbol{u}}_{\mathrm{C}}(s) \tag{13.67a}$$

$$m\left[s^2 \underset{\sim}{\boldsymbol{u}}_{\mathrm{C}}(s) - \boldsymbol{Z}_{\mathrm{C}}(s)\right] = \boldsymbol{K}_{\mathrm{CA}}\underset{\sim}{\boldsymbol{u}}_{\mathrm{A}} + \boldsymbol{K}_{\mathrm{CC}}\underset{\sim}{\boldsymbol{u}}_{\mathrm{C}}(s). \tag{13.67b}$$

In these equations, the $\underset{\sim}{\square}$ notation implies the Laplace transform of a function, which we recall is defined as

$$\underset{\sim}{f}(s) = \mathcal{L}\{f(t)\} = \int_0^\infty e^{-st} f(t)dt.$$

The vector function $\boldsymbol{Z}$ is defined as

$$\boldsymbol{Z} = s\boldsymbol{u}(0) + \dot{\boldsymbol{u}}(0),$$

and comes from the Laplace transform of the second derivative. Properties of the Laplace transform that we have used (or will use soon) are

$$\mathcal{L}\{af(t)\} = a\underset{\sim}{f}(s), \quad \mathcal{L}\left\{\dot{f}(t)\right\} = s\underset{\sim}{f}(s) - f(0), \quad \mathcal{L}\left\{\ddot{f}(t)\right\} = s^2\underset{\sim}{f}(s) - sf(0) - \dot{f}(0)$$

and

$$\mathcal{L}\left\{\int_0^t g(t-\tau)f(\tau)d\tau\right\} = \underset{\sim}{g}\,\underset{\sim}{f}. \tag{13.68}$$

Equation (13.67b) can be solved for $\underset{\sim}{\boldsymbol{u}}_{\mathrm{C}}$, and the result used to eliminate $\underset{\sim}{\boldsymbol{u}}_{\mathrm{C}}$ from Eqn. (13.67a), leading to

$$m\left[s^2 \underset{\sim}{\boldsymbol{u}}_{\mathrm{A}}(s) - \boldsymbol{Z}_{\mathrm{A}}(s)\right] = \underset{\sim}{\boldsymbol{f}}_{\mathrm{A}}(s) + \frac{\boldsymbol{K}_{\mathrm{AC}}\underset{\sim}{\boldsymbol{Y}}(s)\boldsymbol{K}_{\mathrm{CA}}}{m}\underset{\sim}{\boldsymbol{u}}_{\mathrm{A}}(s) + \boldsymbol{K}_{\mathrm{AC}}\underset{\sim}{\boldsymbol{Y}}(s)\boldsymbol{Z}_{\mathrm{C}}, \tag{13.69}$$

where we have defined

$$\underset{\sim}{\boldsymbol{Y}}(s) = \left(s^2 \boldsymbol{I} - \frac{1}{m}\boldsymbol{K}_{\mathrm{CC}}\right)^{-1}.$$

We can now apply the inverse Laplace transform, $\mathcal{L}^{-1}$, recognizing that the product of Laplace transforms in the second term on the right-hand side of Eqn. (13.69) transforms to a convolution integral like Eqn. (13.68):

$$m\ddot{\boldsymbol{u}}_{\mathrm{A}} = \boldsymbol{f}_{\mathrm{A}} + \int_0^t \boldsymbol{\theta}(t-\tau)\boldsymbol{u}_{\mathrm{A}}(\tau)d\tau + \boldsymbol{G}(t), \tag{13.70}$$

where

$$\boldsymbol{\theta}(t) = \mathcal{L}^{-1}\left\{\frac{\boldsymbol{K}_{\mathrm{AC}}\underset{\sim}{\boldsymbol{Y}}(s)\boldsymbol{K}_{\mathrm{CA}}}{m}\right\}$$

and we have defined

$$\boldsymbol{G}(t) = m\left[\dot{\boldsymbol{\theta}}(t)\boldsymbol{K}_{\mathrm{CA}}^{-1}\boldsymbol{u}_{\mathrm{C}}(0) + \boldsymbol{\theta}(t)\boldsymbol{K}_{\mathrm{CA}}^{-1}\dot{\boldsymbol{u}}_{\mathrm{C}}(0)\right]. \tag{13.71}$$

It is sometimes more convenient to express the integral in terms of velocity rather than displacement. We can manipulate the second term on the right-hand side of Eqn. (13.69) as follows:

$$\frac{\boldsymbol{K}_{\mathrm{AC}}\underset{\sim}{\boldsymbol{Y}}(s)\boldsymbol{K}_{\mathrm{CA}}}{m}\underset{\sim}{\boldsymbol{u}}_{\mathrm{A}}(s) = \frac{\boldsymbol{K}_{\mathrm{AC}}\underset{\sim}{\boldsymbol{Y}}(s)\boldsymbol{K}_{\mathrm{CA}}}{sm}\left(s\underset{\sim}{\boldsymbol{u}}_{\mathrm{A}}(s) - \boldsymbol{u}_{\mathrm{A}}(0)\right) + \frac{\boldsymbol{K}_{\mathrm{AC}}\underset{\sim}{\boldsymbol{Y}}(s)\boldsymbol{K}_{\mathrm{CA}}}{sm}\boldsymbol{u}_{\mathrm{A}}(0),$$

where we have multiplied and divided by s, added a term in $\boldsymbol{u}_{\mathrm{A}}(0)$ and subtracted the same term. The first of these two terms is a new convolution integral involving the velocities instead of the displacements, whereas the second term depends on the initial displacements. Taking inverse Laplace transforms of these new terms leads to

$$m\ddot{\boldsymbol{u}}_{\mathrm{A}} = \boldsymbol{f}_{\mathrm{A}} + \int_0^t \boldsymbol{\beta}(t-\tau)\dot{\boldsymbol{u}}_{\mathrm{A}}(\tau)d\tau + \boldsymbol{\beta}(t)\boldsymbol{u}_{\mathrm{A}}(0) + \boldsymbol{G}(t), \tag{13.72}$$

where

$$\boldsymbol{\beta}(t) = \mathcal{L}^{-1}\left\{\frac{\boldsymbol{K}_{\mathrm{AC}}\underset{\sim}{\boldsymbol{Y}}(s)\boldsymbol{K}_{\mathrm{CA}}}{sm}\right\}$$

and we have assumed that $\boldsymbol{u}_{\mathrm{A}}(0) = 0$.

Equations (13.70)–(13.72) represent the main point of the derivation, and are referred to as the *generalized Langevin equations (GLE)* for reasons that become clear if we compare them to the Langevin equation of Section 9.4.3. Up to the assumption of linearity in B^{C}, the results are exact. We see that eliminating the degrees of freedom from B^{C} has the effect of introducing two types of additional forces on the atoms in B^{A}. The first involves a *memory kernel* (either $\boldsymbol{\theta}$ or $\boldsymbol{\beta}$) and an integral over the entire time history of the motion of the atoms in B^{A}. Direct evaluation of these integrals is all but impossible for anything but the simplest example (the one-dimensional harmonic chain). We can see this by noting that the evaluation of the memory kernel requires inversion of a matrix as large as the number of degrees of freedom we have coarsened out. Also, the integral is (in principle at least) over *all time* and involves interactions between every pair of atoms in the atomistic region. Clearly, some approximation to the memory kernels and these integrals will be necessary to make progress.

The second type of additional forces in the GLE, $\boldsymbol{G}(t)$ and $\boldsymbol{\beta}(t)\boldsymbol{u}_{\mathrm{A}}(0)$, are dependent on the initial conditions (positions and velocities) of all the eliminated degrees of freedom in B^{C}. These forces evolve with time due to their dependence on $\boldsymbol{\theta}(t)$. If we imagine that the continuum region is in thermodynamic equilibrium at $t = 0$, the effect of these initial conditions on B^{A} is to impose random forces that sample the canonical distribution. We see in this term then an alternative justification for random forces imposed by the Langevin thermostat (see Section 9.4.3). See [AD76] for a more rigorous discussion of this point.

How does this equation solve the wave reflection problem? In principle there are two ways. The first is through the rather indirect means of providing justification for a Langevin thermostat, which does not so much prevent wave reflections as remove their after-effects by removing spurious heat. However, the memory kernel can, in principle, completely eliminate the wave reflections. This is because the GLE is exact; we have not arbitrarily thrown away degrees of freedom from B^{C}. Rather, we have simply recast their exact equations of motion in terms of the positions of the other atoms (those in B^{A}). The challenge, however, is the substantial computational effort associated with this integral. Is it possible to approximate it in some way that will remain effective in reducing wave reflections?

Calculating the kernel functions Cai *et al.* [CdKBY00] presented a means by which $\boldsymbol{\beta}$ can be computed exactly (but numerically) using Eqn. (13.72). First, it is helpful to recall that Eqn. (13.72) is a *matrix* equation. In indicial form, it becomes

$$m\ddot{u}_i = f_i + \int_0^t \beta_{ij}(t-\tau)\dot{u}_j(\tau)d\tau + \beta_{ij}(t)u_j(0) + G_i(t), \tag{13.73}$$

where we have dropped the general subscript "A" in favor of indices running from 1 to $3N_{\mathrm{A}}$ (with an implied summation on j). To numerically evaluate β_{ij}, Cai *et al.* fixed one component of the displacement of one atom in B^{A} to a constant value ε while simultaneously holding all other displacements in B^{A} fixed to zero. In other words, for a specific index k, they set

$$u_i(t) = \varepsilon\delta_{ik}, \qquad i = 1, \ldots, 3N_{\mathrm{A}}.$$

This eliminates the convolution term (since the velocities are zero) and isolates $\beta_{ik}(t)$ in the third term on the right-hand side of Eqn. (13.73). Next, a full MD simulation is run on the *eliminated degrees of freedom in* B^{C} with the initial conditions $\boldsymbol{u}_{\mathrm{C}} = \dot{\boldsymbol{u}}_{\mathrm{C}} = 0$. (This initial condition eliminates the stochastic force, $G(t)$, from the equation.) During the simulation, one can track the forces on all the fixed atoms in B^{A}, which must be equal to the resulting simplified version of the right-hand side of Eqn. (13.73):

$$F_i(t) = f_i(t) + \varepsilon\beta_{ik}(t), \qquad i = 1, \ldots, 3N_A.$$

Since f_i is known from the underlying interatomic potentials and $F_i(t)$ can be tracked during the simulation, a numerical approximant to β_{ik} can be determined for all i.

Clearly, this is a large computational effort, as it requires $3N_A$ independent MD simulations over infinite time. Practically speaking, however, Cai *et al.* showed that the memory kernel approaches zero over a relatively short time period, t_c, that can be determined for a given atomic system. In addition, the speed of sound in the body, c, can be used to determine a cutoff distance. For a given atom k, elements of β_{ik} can be assumed zero if atom i is further than ct_c away from k. This reduces the computational burden, but still leaves the method too expensive for simulations of large numbers of atoms.

Kernel function absorbing boundary conditions Although they are often cast in a different framework, absorbing boundary conditions methods are essentially dedicated to finding the best way to approximate the kernel function integrals in Eqns. (13.70) and (13.72). E and Huang [EH01, EH02] showed that they could make this approximation with only a handful of discrete contributions to the forces on atoms very near the interface between B^A and B^C. Like Cai *et al.*, they also argued that the memory kernel is fairly short-lived, so that wave reflections can be largely eliminated by considering only a few previous timesteps in the evaluation. Wagner and coworkers [WL03, WKL04] studied a simple one-dimensional case where β_{ij} can be computed analytically, and then showed how Laplace transforms can be used to efficiently obtain approximations to β_{ij} for a given crystal structure in higher dimensions. The absorbing boundary condition is still a subject of active research without consensus on the best way to proceed.

To date (2011) these methods are also effectively "zero-temperature" dynamic methods, in that the stochastic term in the GLE has not been fully exploited as a means of temperature control.[21] It is also the case that the focus of these methods is on the atomistic region. By this, we mean that the goal of these methods is to avoid spurious reflections back into B^A, as opposed to correctly *transmitting* that wave energy into B^C as heat. The correct physical and computational prescription for how to do this remains an area of active research.

13.3.4 Damping bands

Another approach to the wave reflection problem is that of Qu *et al.* [QSCM05], who developed a dynamic multiscale coupling by introducing a band of atoms inside B^A and along ∂B^I where a form of thermostat is applied locally. The damping band of Qu *et al.* is based on a combination of the so-called "stadium boundary conditions" proposed by Holian and Ravelo [HR95] and the Langevin thermostat as implemented by Holland and Marder [HM99] and is shown schematically in Fig. 13.7. The method is more ad hoc than

21 Because of the connection between the GLE and the Langevin thermostat, a multiscale method employing the latter is essentially an implementation of the GLE approach with $\beta_{ij}(t) = \gamma\delta_{ij}\delta(t)$. The hot-QC method of Section 13.2 has been implemented with the Langevin thermostat [DTMP05, DTLMK11], as has an alternative form of the QC method in [MVH+10].

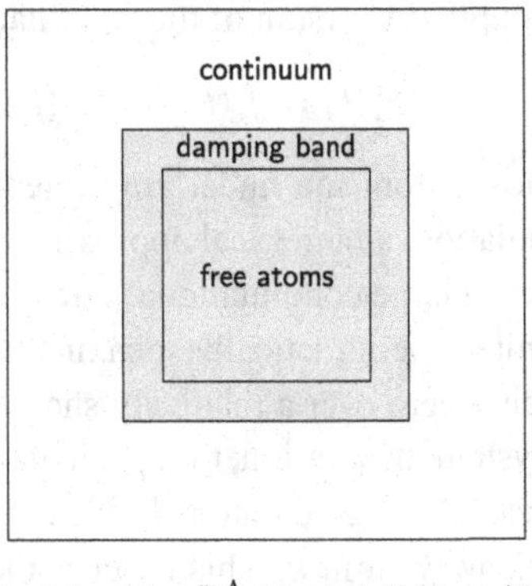

Fig. 13.7 The method of damping bands. The atomistic region B^{A} (shown in light gray) is divided into two subregions. Free atoms evolve with no direct thermostat applied, while atoms in the damping band are subjected to Langevin temperature control.

the other approaches, and requires a bit of experimentation to fit the optimal parameters for a given application. However, it is capable of preserving the correct Hamiltonian dynamics of the free atoms in B^{A}, while effectively coupling to a reservoir that maintains a constant temperature under equilibrium conditions. It also largely eliminates spurious wave reflections at ∂B^{I}. The basic idea is that the dynamics of atoms in the damping band are modified according to the Langevin thermostat (see Eqn. (9.24)), but with the strength of the damping coefficient γ^{α} increasing linearly from the inner to the outer edge of the damping band (hence the term "stadium" boundary condition).

This model mimics the situation where the continuum is at a constant temperature, in the sense that the undamped part of the atomistic region will reach exactly the set temperature of the thermostat given enough time to equilibrate. At the same time, however, nonequilibrium disturbances in atomic velocities can propagate through the free atomistic region without being artificially suppressed by a global thermostat. The damping of the Langevin force also has the additional advantage of adequately eliminating the problem of wave reflections, and the variable damping coefficient provides some degree of tunability in terms of the range of wavelengths that are effectively absorbed by the damped atoms. Shiari and coworkers [SMC05, SMK07] have proposed a simpler type of damping band whereby the Nosé–Hoover thermostat (Section 9.4.4) is applied in the damping region. It has the advantage of straightforward implementation and is efficient at wave absorption, but is apparently not as robust at controlling the temperature as the Langevin approach.

The missing electrons One disadvantage of damping band methods (and for that matter, kernel function methods) is that they are really only applicable to insulators. This is because metals conduct most of their heat through the electrons that are not explicitly accounted for in empirical interatomic models. Of course this is a disadvantage inherited from the underlying atomistics (see discussions on pages 492 and 544), as opposed to something emerging from the multiscale approach, but it must nonetheless be kept in mind. By contrast, global thermostats like those discussed in Section 13.2 are more amenable to metals if we justify the thermostat as effectively mimicking the electronic heat bath, but their use in studying nonequilibrium processes is dubious.

13.4 Concluding remarks

Dynamic multiscale methods like those described in this chapter are in many ways a culmination of all the topics addressed in this book. They depend on continuum mechanics and thermodynamics for the description of the continuum regions, while the atomistic region is of course described by an empirical potential within the framework of molecular dynamics. Methods to make direct links between the atomic scale and the macroscale, such as the Cauchy–Born rule and the constitutive laws derived therefrom, are the natural choice to define the response of the continuum region and ensure a good coupling to the atomistic region. Clearly, statistical mechanics plays an important role in developing the approximations for the free energy expressions derived in Section 13.2, and they will no doubt play a future role as these methods are extended to correctly describe heat flux between the atomistic and continuum domains in the nonequilibrium methods of Section 13.3.

Exercises

13.1 [SECTION 13.2] Derive the result in Eqn. (13.20) by analytically integrating Eqn. (13.19).

13.2 [SECTION 13.2] Derive the result in Eqn. (13.51) for the average kinetic energy of a node in the continuum region using the "naïve" hot-QC Hamiltonian $\mathcal{H}^{\mathrm{QC}}_{\mathrm{naïve}}$.

13.3 [SECTION 13.2] Show that $\widetilde{\mathcal{V}}_{\mathrm{A}}(q_{\mathrm{A}}, T)$ is accurate to second order in $k_{\mathrm{B}}T$ as indicated in Eqn. (13.26). **Hint:** Retain terms above second order in the expansion of $\mathcal{V}$ in Eqn. (13.21), and then use $e^z = \sum_{m=0}^{\infty}(1/m!)z^m$ in the calculation of $\mathcal{V}_{\mathrm{A}}$ in Eqn. (13.18), where z contains the anharmonic terms in the expansion.

A Mathematical representation of interatomic potentials

In Chapter 5 we saw that, as a consequence of the Born–Oppenheimer approximation, the classical potential energy can be represented as a function of the atoms. In particular, the interatomic potential energy (see Eqn. (5.10)) is

$$\mathcal{V}^{\text{int}} = \widehat{\mathcal{V}}^{\text{int}}(\boldsymbol{r}^1, \boldsymbol{r}^2, \dots, \boldsymbol{r}^N), \tag{A.1}$$

where $\boldsymbol{r}^\alpha$ is the position vector of atom α. The "hat" over $\mathcal{V}^{\text{int}}$ indicates that the functional dependence is on absolute particle positions (as opposed to distances later). It is common to assume that[1] $\widehat{\mathcal{V}}^{\text{int}} : \mathbb{R}^{3N} \to \mathbb{R}$ is a continuously differentiable function, and we do the same here.[2] We then argued in Section 5.3.2 that due to material frame-indifference and parity symmetry of the Hamiltonian the invariance principle in Eqn. (5.17) must be satisfied. It is repeated here for convenience:[3]

Principle of interatomic potential invariance The internal potential energy of a system of particles is invariant with respect to the Euclidean group $G \equiv \{\boldsymbol{x} \mapsto \boldsymbol{Q}\boldsymbol{x} + \boldsymbol{c} \mid \boldsymbol{x} \in \mathbb{R}^3, \boldsymbol{Q} \in O(3), \boldsymbol{c} \in \mathbb{R}^3\}$, where $O(3)$ denotes the full orthogonal group. (A.2)

In Section 5.3.2, we showed that *if* the interatomic potential energy function is written as a function of interatomic distances, the function will satisfy Eqn. (A.2). Thus,

$$\mathcal{V}^{\text{int}} = \mathcal{V}^{\text{int}}(r^{12}, r^{13}, \dots, r^{1N}, r^{23}, \dots, r^{N-1,N}) = \mathcal{V}^{\text{int}}(\{r^{\alpha\beta}\}), \tag{A.3}$$

where $r^{\alpha\beta} = \left\|\boldsymbol{r}^\beta - \boldsymbol{r}^\alpha\right\|$ and $\{r^{\alpha\beta}\}$ denotes the set of all distances with $\alpha < \beta$ as defined in Eqn. (5.19). However, one may ask whether interatomic distances are a sufficiently general set of coordinates to describe all energetically distinct atomic configurations. In other words, we know that the representation in terms of interatomic distances in Eqn. (A.3) is a sufficient condition for principle A.2 to be satisfied, but is it also a necessary condition?

A considerable body of work exists in the literature to address the issue of the selection of a suitable set of invariant coordinates for describing atomic configurations. The interested reader may start with [Kea66, Mar75a, LR95, FW99] and the extensive review written

[1] The notation "$f : X \to Y$" used in this section states that f is a function that maps elements from the domain X to the codomain Y.

[2] Note that this assumption may fail in systems undergoing first-order magnetic or electronic phase transformations. This issue is related to the dependence of the ground state energy on the initial guess for the electron density discussed in Section 4.3.6. See, for example, [CvdV99] for a discussion of electronic and magnetic phase transformations in quantum mechanical systems.

[3] See footnote 5.3.2 on page 244 for a definition of the notation.

by Littlejohn and Reinsch [LR97]. In this appendix, we present a rigorous mathematical derivation based on the work of Admal and Tadmor [AT10], which proves the results quoted in Section 5.3.2 and deals with the important issue of obtaining a continuously differentiable extension of the internal potential energy function to $\mathbb{R}^{N(N-1)/2}$. We will see that this leads us to the conclusion that the representation of the interatomic potential in terms of distances is in fact *not* unique. This has important implications for the uniqueness of the atomistic stress tensor as explained in Chapter 8 (although the force acting on a given atom, computed from equivalent potentials, will be the same).

A.1 Interatomic distances and invariance

The invariance requirements for the potential energy function defined in Eqn. (A.1) are stated above in Eqn. (A.2). To exploit this invariance, let us consider the action of the Euclidean group G on $\mathbb{R}^{3N}$, i.e. the action of any combination of translation and rotation (proper or improper), which is represented by an element $g \in G$, on any configuration of N particles represented by a vector $(\boldsymbol{r}^1, \dots, \boldsymbol{r}^N) \in \mathbb{R}^{3N}$:

$$g \cdot (\boldsymbol{r}^1, \dots, \boldsymbol{r}^N) = (\boldsymbol{Q}\boldsymbol{r}^1 + \boldsymbol{c}, \dots, \boldsymbol{Q}\boldsymbol{r}^N + \boldsymbol{c}). \tag{A.4}$$

This action splits $\mathbb{R}^{3N}$ into disjoint sets of equivalence classes [DF04], which we now describe. For any $\boldsymbol{u} = (\boldsymbol{r}^1, \dots, \boldsymbol{r}^N) \in \mathbb{R}^{3N}$, let $O_{\boldsymbol{u}} \subset \mathbb{R}^{3N}$ denote an equivalence class which is defined as[4]

$$O_{\boldsymbol{u}} \equiv \{g \cdot \boldsymbol{u} \mid g \in G\}, \tag{A.5}$$

where $g \cdot \boldsymbol{u}$ denotes the action of g on $\boldsymbol{u}$ defined in Eqn. (A.4). In words, $O_{\boldsymbol{u}}$ represents the set of all configurations which are related to the configuration $\boldsymbol{u}$ by a rigid-body motion and/or reflection. Due to the invariance of the interatomic potential energy, we can view the function $\mathcal{V}^{\text{int}}$ as a function on the set of equivalence classes, i.e.

$$\widehat{\mathcal{V}}^{\text{int}}(O_{\boldsymbol{u}}) = \widehat{\mathcal{V}}^{\text{int}}(\boldsymbol{u}), \tag{A.6}$$

because

$$\widehat{\mathcal{V}}^{\text{int}}(\boldsymbol{v}) = \widehat{\mathcal{V}}^{\text{int}}(\boldsymbol{u}) \quad \forall \boldsymbol{v} \in O_{\boldsymbol{u}}. \tag{A.7}$$

Now, consider a set $S \subset \mathbb{R}^{N(N-1)/2}$, defined as

$$S \equiv \{(r^{12}, r^{13}, \dots, r^{1N}, r^{23}, \dots, r^{(N-1)N}) \mid r^{\alpha\beta} = \|\boldsymbol{r}^\alpha - \boldsymbol{r}^\beta\|, (\boldsymbol{r}^1, \dots, \boldsymbol{r}^N) \in \mathbb{R}^{3N}\}. \tag{A.8}$$

In words, the set S consists of all possible $N(N-1)/2$-tuples of real numbers which correspond to the distances between N particles in $\mathbb{R}^3$. The key here is that not all $N(N-1)/2$ combinations of real numbers constitute a valid set of physical distances. The distances must satisfy certain geometric constraints in order to be physically meaningful, which

[4] The notation "$\{g \cdot \boldsymbol{u} \mid g \in G\}$" should be read as "the set of all $g \cdot \boldsymbol{u}$, such that g is in the Euclidean group G."

reduces the number of independent degrees of freedom from $N(N-1)/2$ to $3N-6$ as explained in Section A.2. In technical terms, the coordinates of any point in S are said to be *embeddable* in $\mathbb{R}^3$ and S represents a $(3N-6)$-dimensional manifold in $\mathbb{R}^{N(N-1)/2}$, commonly referred to as the *shape space* [LR97].

Let ϕ be the mapping taking a point in configuration space to the corresponding set of distances in the shape space S, i.e. $\phi : \mathbb{R}^{3N} \to S : (\boldsymbol{r}^1, \dots, \boldsymbol{r}^N) \mapsto (r^{12}, \dots, r^{(N-1)N})$. Since the Euclidean group preserves distances, it immediately follows that the map

$$\bar{\phi} : \{\text{equivalence classes}\} \to S, \tag{A.9}$$

defined as $\bar{\phi}(O_{\boldsymbol{u}}) = \phi(\boldsymbol{u})$, is a bijection (i.e. a one-to-one and onto mapping) from the set of equivalence classes to the set S. The proof that $\bar{\phi}$ is bijective is similar to the proof of Cauchy's basic representation theorem [Cau50] which is discussed in [TN65, Section 11].[5]

Proof It is clear that $\bar{\phi}$ is surjective (i.e. an "onto" mapping) by the definition of S. We now show that $\bar{\phi}$ is also injective (one-to-one). Consider two equivalence classes $O_{\boldsymbol{u}}$ and $O_{\boldsymbol{v}}$, where $\boldsymbol{u} = (\boldsymbol{x}^1, \dots, \boldsymbol{x}^N)$ and $\boldsymbol{v} = (\boldsymbol{y}^1, \dots, \boldsymbol{y}^N)$ in $\mathbb{R}^{3N}$, such that $\bar{\phi}(O_{\boldsymbol{u}}) = \bar{\phi}(O_{\boldsymbol{v}})$. (Here $\boldsymbol{x}^\alpha$ and $\boldsymbol{y}^\alpha$ represent positions of particles.) By the definition of $\bar{\phi}$ this implies $\phi(\boldsymbol{u}) = \phi(\boldsymbol{v})$. It is enough to show that $O_{\boldsymbol{u}} = O_{\boldsymbol{v}}$ to complete the proof.

Since an equivalence class consists of all configurations that are related by a rigid-body motion and/or reflection, we may take $\boldsymbol{x}^1 = \boldsymbol{y}^1 = \boldsymbol{0}$. Thus, since $\phi(\boldsymbol{u}) = \phi(\boldsymbol{v})$, we have

$$\|\boldsymbol{x}^\alpha\| = \|\boldsymbol{y}^\alpha\| \qquad \forall \alpha = 2, 3, \dots, N. \tag{A.10}$$

Moreover, since

$$\boldsymbol{x}^\alpha \cdot \boldsymbol{x}^\beta = \frac{1}{2}\left(\|\boldsymbol{x}^\alpha\|^2 + \left\|\boldsymbol{x}^\beta\right\|^2 - \left\|\boldsymbol{x}^\alpha - \boldsymbol{x}^\beta\right\|^2\right), \tag{A.11}$$

it follows from Eqn. (A.10) and the fact that $\phi(\boldsymbol{u}) = \phi(\boldsymbol{v})$ that

$$\boldsymbol{x}^\alpha \cdot \boldsymbol{x}^\beta = \boldsymbol{y}^\alpha \cdot \boldsymbol{y}^\beta \qquad \forall \alpha, \beta = 2, 3, \dots, N. \tag{A.12}$$

Let U be a subspace spanned by $\boldsymbol{x}^2, \dots, \boldsymbol{x}^N$. After a suitable rearrangement, we may assume that U is spanned by the linearly-independent set $\boldsymbol{x}^2, \dots, \boldsymbol{x}^p$, where $p \le N$ and $p \le 3$. Since

$$\det\left[\boldsymbol{x}^\alpha \cdot \boldsymbol{x}^\beta\right]_{\alpha,\beta=2,\dots,p} \neq 0 \tag{A.13}$$

is a necessary and sufficient condition for the linear independence of $\boldsymbol{x}^2, \dots, \boldsymbol{x}^p$, it follows from Eqn. (A.12) that the subspace V, spanned by $\boldsymbol{y}^2, \dots, \boldsymbol{y}^N$, is spanned by the linearly independent set $\boldsymbol{y}^2, \dots, \boldsymbol{y}^p$. Now there exists a unique linear transformation $\boldsymbol{Q}$, such that

$$\boldsymbol{Q}\boldsymbol{x}^\alpha = \boldsymbol{y}^\alpha \qquad \forall \alpha = 2, \dots, p, \tag{A.14}$$

and such that $\boldsymbol{Q}$ reduces to the identity on the orthogonal complement of U. Using Eqns. (A.14) and (A.12), we have

$$\boldsymbol{Q}\boldsymbol{x}^\alpha \cdot \boldsymbol{Q}\boldsymbol{x}^\beta = \boldsymbol{y}^\alpha \cdot \boldsymbol{y}^\beta = \boldsymbol{x}^\alpha \cdot \boldsymbol{x}^\beta \qquad \forall \alpha, \beta = 2, 3, \dots, p, \tag{A.15}$$

from which we conclude that $\boldsymbol{Q}$ is orthogonal.

[5] This proof is due to Nikhil C. Admal.

Since $\boldsymbol{x}^2, \dots, \boldsymbol{x}^p$ are linearly independent, we have for the remaining positions that

$$\boldsymbol{x}^\gamma = \sum_{\alpha=2}^{p} c^{\gamma\alpha} \boldsymbol{x}^\alpha \qquad \gamma = p+1, \dots, N, \tag{A.16}$$

where $c^{\gamma 2}, \dots, c^{\gamma p}$ are coefficients for particle γ. Taking the inner product of Eqn. (A.16) with $\boldsymbol{x}^\beta$, we obtain

$$\boldsymbol{x}^\beta \cdot \boldsymbol{x}^\gamma = \sum_{\alpha=2}^{p} c^{\gamma\alpha} (\boldsymbol{x}^\beta \cdot \boldsymbol{x}^\alpha) \qquad \beta = 2, \dots, p, \tag{A.17}$$

which is a system of $p-1$ equations containing the $p-1$ unknowns, $c^{\gamma 2}, \dots, c^{\gamma p}$, for particle γ. The coefficients $c^{\gamma\alpha}$ are uniquely determined because Eqn. (A.13) holds. Substituting Eqn. (A.12) into Eqn. (A.17), we find that

$$\boldsymbol{y}^\gamma = \sum_{\alpha=2}^{p} c^{\gamma\alpha} \boldsymbol{y}^\alpha \qquad \gamma = p+1, \dots, N. \tag{A.18}$$

From Eqns. (A.14), (A.16) and Eqn. (A.18), it immediately follows that

$$\boldsymbol{Q}\boldsymbol{x}^\gamma = \boldsymbol{y}^\gamma \qquad \gamma = p+1, \dots, N. \tag{A.19}$$

Equations (A.14) and (A.19) together imply that $\boldsymbol{Q}\boldsymbol{x}^\alpha = \boldsymbol{y}^\alpha$ for $\alpha = 2, \dots, N$. Therefore, $\boldsymbol{u}$ and $\boldsymbol{v}$ are related to each other by a Euclidean transformation given by $\boldsymbol{Q}$, which implies that $O_{\boldsymbol{u}} = O_{\boldsymbol{v}}$ as required. □

This result means that for every set of equivalent configurations, i.e. configurations related to each other by a rigid-body motion and/or reflection, there exists a unique $N(N-1)/2$-tuple of distances and vice versa. From Eqns. (A.6) and (A.9), it immediately follows that the interatomic potential energy of the system can be completely described by a function $\breve{\mathcal{V}}^{\text{int}} : S \to \mathbb{R}$, defined as

$$\breve{\mathcal{V}}^{\text{int}}(\boldsymbol{s}) \equiv \bar{\mathcal{V}}^{\text{int}}(\bar{\phi}^{-1}(\boldsymbol{s})) \qquad \forall \boldsymbol{s} \in S. \tag{A.20}$$

It is important to note that $\breve{\mathcal{V}}^{\text{int}}(\boldsymbol{s})$ is defined only for points $\boldsymbol{s} = (r^{12}, \dots, r^{N(N-1)})$ that belong to the shape space S. This is because $\breve{\mathcal{V}}^{\text{int}}(\boldsymbol{s})$ is defined from $\bar{\mathcal{V}}^{\text{int}}$ which has no information on nonphysical configurations. We will see later that it will be necessary to extend $\breve{\mathcal{V}}^{\text{int}}(\boldsymbol{s})$ to the full space $\mathbb{R}^{N(N-1)/2}$. But first we must gain a better understanding of the nature of the geometric constraints that define the shape space manifold.

A.2 Distance geometry: constraints between interatomic distances

Once we accept the notion that distances can be used to describe atomic configurations, we are faced with the following fact. If we eliminate rigid-body motion of a system of N

atoms, the number of independent degrees of freedom in three-dimensional space is $3N-6$ for $N \geq 3$ (for $N=2$ there is only one independent degree of freedom).[6] At the same time, a simple combinatorial analysis shows us that the number of interatomic distances is $N(N-1)/2$. For systems of two, three or four atoms, the number of interatomic distances is equal to the number of independent degrees of freedom, but for systems of five or more, there are

$$N(N-1)/2-(3N-6)=N(N-7)/2+6$$

dependencies between the interatomic distances. For example, for $N=5$, there are $(5 \times 4)/2=10$ distances, and only $3 \times 5-6=9$ degrees of freedom.

An entire field, referred to as *distance geometry*, has emerged to describe the geometry of sets of points in terms of the distances between them. The subject was first treated systematically by Karl Menger in the early twentieth century and is summarized in the book by Blumenthal [Blu70]. More recent references include [Cri77, HKC83, SS86, Bra87, CH88]. Distance geometry has been applied extensively in the determination of the structures of proteins and other complex molecules as well as in areas such as the application of geometric constraints in computer-aided design and robotics. Our discussion below is partly based on [PRTT05], which provides a particularly clear explanation of the subject.

The key function in distance geometry is the *Cayley–Menger determinant*[7], χ, which is defined as

$$\chi(\varsigma^{12},\ldots,\varsigma^{1N},\varsigma^{23},\ldots,\varsigma^{(N-1)N})=\det\begin{bmatrix} 0 & s^{12} & s^{13} & \cdots & s^{1N} & 1 \\ s^{21} & 0 & s^{23} & \cdots & s^{2N} & 1 \\ s^{31} & s^{32} & 0 & \cdots & s^{3N} & 1 \\ \vdots & \vdots & \vdots & & \vdots & \vdots \\ s^{N1} & s^{N2} & s^{N3} & \cdots & 0 & 1 \\ 1 & 1 & 1 & \cdots & 1 & 0 \end{bmatrix}, \tag{A.21}$$

where $s^{\alpha\beta}=(\varsigma^{\alpha\beta})^2$. For a system of N points, $(\boldsymbol{r}^1,\ldots,\boldsymbol{r}^N)$, embedded in $\mathbb{R}^3$, the Cayley–Menger determinant evaluated at the point, $(r^{12},\ldots,r^{(N-1)N})$, where $r^{\alpha\beta}=\left\|\boldsymbol{r}^\beta-\boldsymbol{r}^\alpha\right\|$, is related to the volume V_{N-1} of a simplex of N points in an $(N-1)$-dimensional space through the relation:

$$\chi(r^{12},\ldots,r^{(N-1)N})=\frac{2^{N-1}((N-1)!)^2}{(-1)^N}V_{N-1}^2(\boldsymbol{r}^1,\ldots,\boldsymbol{r}^N). \tag{A.22}$$

For $N=2$,

$$\chi(r^{12})=2L^2, \tag{A.23}$$

[6] This is easy to see. Consider a system of N particles with positions $\boldsymbol{r}^1,\ldots,\boldsymbol{r}^N$. The position of each atom in three-dimensional space has three coordinates. Each coordinate is a *degree of freedom*, so the total number of degrees of freedom is $3N$. To eliminate rigid-body translation of the system, three coordinates must be specified to set the center of mass of the system. Next, to eliminate rigid-body rotation of the system, three additional coordinates (representing rotation about three noncollinear and nonplanar axes) must be specified to fix the orientation in space. Thus, the remaining degrees of freedom are $3N-6$.

[7] The Cayley–Menger determinant was first published in 1841 by the British mathematician Arthur Cayley (the same Cayley of the Cayley–Hamilton theorem) while he was still an undergraduate [Cri05]. However, as noted above, it was the Austrian mathematician Karl Menger who systematized the field in the 1920s and 1930s.

where $L = \sqrt{s^{12}}$ is the length of the segment defined by the two points. For $N = 3$,

$$\chi(r^{12}, r^{13}, r^{23}) = -16A^2, \tag{A.24}$$

where A is the area of the triangle defined by the three points. For $N = 4$,

$$\chi(r^{12}, \ldots, r^{34}) = 288V^2, \tag{A.25}$$

where V is the volume of the tetrahedron defined by the four points. For $N \geq 5$, we must have

$$\chi(r^{12}, \ldots, r^{(N-1)N}) = 0, \tag{A.26}$$

for points in $\mathbb{R}^3$ since any simplex with five or more points has zero volume in three-dimensional space.[8]

We now seek to go in the opposite direction. Rather than computing the squared distances $\{s^{\alpha\beta}\}$ from a set of points and using the Cayley–Menger determinants to compute volumes, we seek to verify that a set of distances actually corresponds to a set of points in three-dimensional space. In technical terms, we want the points associated with the distances to be *embeddable* in $\mathbb{R}^3$. In order for this to be the case the following conditions must be satisfied:[9]

1. $\chi(r^{\alpha\beta}, r^{\alpha\gamma}, r^{\beta\gamma}) \leq 0, \quad \forall \alpha < \beta < \gamma,$
2. $\chi(r^{\alpha\beta}, r^{\alpha\gamma}, r^{\alpha\delta}, \ldots, r^{\gamma\delta}) \geq 0, \quad \forall \alpha < \beta < \gamma < \delta,$
3. $\chi(r^{\alpha\beta}, r^{\alpha\gamma}, r^{\alpha\delta}, r^{\alpha\epsilon}, \ldots, r^{\delta\epsilon}) = 0, \quad \forall \alpha < \beta < \gamma < \delta < \epsilon,$
4. $\chi(r^{\alpha\beta}, r^{\alpha\gamma}, r^{\alpha\delta}, r^{\alpha\epsilon}, r^{\alpha\zeta}, \ldots, r^{\epsilon\zeta}) = 0, \quad \forall \alpha < \beta < \gamma < \delta < \epsilon < \zeta.$

For example, condition 1 states that the Cayley–Menger determinants computed for all distinct triplets of points must be negative or zero. Conditions 2–4 apply similarly to sets of four, five and six points. The above conditions are easy to understand in terms of Eqns. (A.23)–(A.26). They enforce the correct sign of areas and volumes in three-dimensional space and the degeneracy condition in Eqn. (A.26).

Example A.1 (A five-atom cluster) As an example, let us see how the Cayley–Menger formalism is applied to a cluster of $N = 5$ atoms. There are $(5 \times 4)/2 = 10$ distances, and only $3 \times 5 - 6 = 9$ degrees of freedom. Therefore, there is one more interatomic distance than is needed to describe the configuration and indeed there is one Cayley–Menger determinant coming from condition 3:

$$\chi(r^{12}, \ldots, r^{45}) = \det \begin{bmatrix} 0 & s^{12} & s^{13} & s^{14} & s^{15} & 1 \\ s^{12} & 0 & s^{23} & s^{24} & s^{25} & 1 \\ s^{13} & s^{23} & 0 & s^{34} & s^{35} & 1 \\ s^{14} & s^{24} & s^{34} & 0 & s^{45} & 1 \\ s^{15} & s^{25} & s^{35} & s^{45} & 0 & 1 \\ 1 & 1 & 1 & 1 & 1 & 0 \end{bmatrix} = 0.$$

[8] This is easier to visualize in two-dimensional space. In that case, a simplex with four vertices (a tetrahedron) would have zero volume since its vertices would be confined to a plane. The same applies to higher-order simplexes and the corresponding higher-order volumes.

[9] Actually, a somewhat stronger theorem can be proved. If the points are numbered in such a way that the first four points satisfy conditions 1 and 2, then conditions 3 and 4 need only be applied to groups of points that include these four points. See [Blu70] for details.

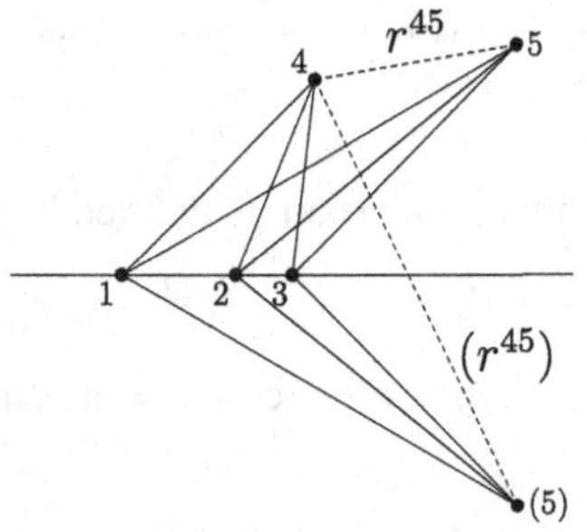

Fig. A.1 A cluster of five atoms. The cluster is shown projected onto the plane normal to the plane defined by atoms 1, 2 and 3 shown as a horizontal line. Atom 4 lies above this plane. There are two possible positions for atom 5 (at the same height above or below the 1–2–3 plane), where the distances $(r^{12}, \dots, r^{35})$ are all the same and only distance r^{45} differs. The alternative position of atom 5 and corresponding distance r^{45} are shown in parentheses. Based on Fig. 2 in [Mar75a].

This expression can be expanded out leading to the following explicit expression [Mar75a]:[10]

$$-\frac{1}{24}\sum_{\alpha=1}^{5}\sum_{\beta=1}^{5}\sum_{\gamma=1}^{5}\sum_{\delta=1}^{5}\sum_{\epsilon=1}^{5}\Big[24s^{\alpha\beta}s^{\beta\gamma}s^{\gamma\delta}s^{\delta\epsilon} - 6s^{\alpha\beta}s^{\beta\gamma}s^{\gamma\delta}s^{\delta\alpha}$$
$$- 8s^{\alpha\beta}s^{\gamma\delta}s^{\delta\epsilon}s^{\epsilon\gamma} - 12(s^{\alpha\beta})^2 s^{\gamma\delta}s^{\delta\epsilon} + 3(s^{\alpha\beta})^2(s^{\gamma\delta})^2\Big] = 0, \quad \text{(A.27)}$$

where in the above expression $s^{\alpha\beta} = s^{\beta\alpha}$ whenever $\alpha > \beta$. For a given set of nine squared distances, say $(s^{11}, \dots, s^{35})$, Eqn. (A.27) provides a quadratic equation for the tenth squared distance, s^{45}:

$$A(s^{45})^2 + Bs^{45} + C = 0, \quad \text{(A.28)}$$

where A, B and C are functions of the other squared distances. (See [KD80] for explicit expressions for A, B and C.) This means that there are two possible solutions for s^{45} (and hence also for r^{45}) when the other distances are set, as demonstrated in Fig. A.1.

The above example shows that that although only nine degrees of freedom are necessary to characterize a cluster of five atoms, selecting a subset of nine distances is not sufficient. For similar reasons, the interatomic potentials function $\breve{\mathcal{V}}^{\text{int}}(\boldsymbol{s})$ defined in Eqn. (A.20) must be expressed as a function of all $N(N-1)/2$ distances and not just an arbitrary subset of $3N-6$ of them. Also, the existence of geometric constraints between the distances means that $\breve{\mathcal{V}}^{\text{int}}(\boldsymbol{s})$ is defined only over a subset of $\mathbb{R}^{N(N-1)/2}$, the so-called *shape space manifold* introduced earlier. To obtain a more general function that can be evaluated at any set of arguments, it is necessary to extend $\breve{\mathcal{V}}^{\text{int}}(\boldsymbol{s})$ to the full space $\mathbb{R}^{N(N-1)/2}$.

A.3 Continuously differentiable extensions of $\breve{\mathcal{V}}^{\text{int}}(\boldsymbol{s})$

The domain of the function $\breve{\mathcal{V}}^{\text{int}}(\boldsymbol{s})$ defined in Eqn. (A.20) is the shape space S, i.e. it is defined for all $\boldsymbol{s} = (r^{12}, \dots, r^{N(N-1)}) \in S$. This means that the $N(N-1)/2$ arguments

[10] Note that Martin [Mar75a] has a small typographical error in his relation.

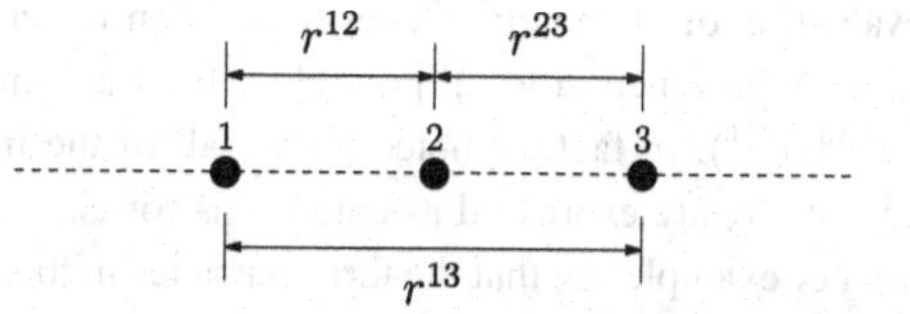

Fig. A.2 Three atoms constrained to move along a line. The distances between the atoms are indicated.

that this function takes must be a set of physical[11] distances computed between the positions of N particles. The function $\breve{\mathcal{V}}^{\text{int}}(s)$ is undefined for a set of arguments, $\zeta^{12}, \ldots, \zeta^{N(N-1)}$, that do not belong to S. In three-dimensional space, this becomes an issue for systems containing more than four particles. As explained in the previous section, for $N \leq 4$ the number of degrees of freedom equals the number of distances, so there are no constraints on the distances, and any set of $N(N-1)/2$ real numbers also corresponds to a set of physical distances. In mathematical terms, $S = \mathbb{R}^{N(N-1)/2}$ for $N = 2, 3, 4$. However, for $N > 4$, $\breve{\mathcal{V}}^{\text{int}}(s)$ is defined only over part of $\mathbb{R}^{N(N-1)/2}$.

We now consider the case where there are more than four particles and restrict our discussion to those systems for which there exists a continuously differentiable extension of $\breve{\mathcal{V}}^{\text{int}}$ to $\mathbb{R}^{N(N-1)/2}$. We will see that this extension is necessary to be able to compute derivatives as explained later. Therefore, we assume that there exists a continuously differentiable function $\mathcal{V}^{\text{int}} : \mathbb{R}^{N(N-1)/2} \to \mathbb{R}$, such that the restriction of $\mathcal{V}^{\text{int}}$ to S is equal to $\breve{\mathcal{V}}^{\text{int}}$:

$$\mathcal{V}^{\text{int}}(s) = \breve{\mathcal{V}}^{\text{int}}(s) \qquad \forall s = (r^{12}, \ldots, r^{(N-1)N}) \in S. \tag{A.29}$$

This is justifiable due to the fact that all interatomic potentials used in practice[12] for a system of N particles are either continuously differentiable functions on $\mathbb{R}^{N(N-1)/2}$, or can easily be extended to one. For example, pair potentials and pair functionals are continuously differentiable functions on $\mathbb{R}^{N(N-1)/2}$, while cluster potentials and cluster functionals can be easily extended to $\mathbb{R}^{N(N-1)/2}$ by expressing the angles appearing in them as a function of distances between particles. To help clarify the idea of the extension of interatomic potential functions, consider the following simple one-dimensional example.

Example A.2 (A one-dimensional chain of three atoms) Consider a chain of three atoms constrained to lie along a line and arranged so that atom 2 lies between atoms 1 and 3 as shown in Fig. A.2. The interatomic potential energy function for this chain is

$$\mathcal{V}^{\text{int}} = \widehat{\mathcal{V}}^{\text{int}}(r^1, r^2, r^3) = \breve{\mathcal{V}}^{\text{int}}(r^{12}, r^{13}, r^{23}), \tag{A.30}$$

where the second equality follows from application of the principle of interatomic potential invariance. The function $\widehat{\mathcal{V}}^{\text{int}}(r^1, r^2, r^3)$ can be evaluated for any position of the atoms along the line (except for the divergent case where the atoms overlap), but the second function, $\breve{\mathcal{V}}^{\text{int}}(r^{12}, r^{13}, r^{23})$ can only be evaluated for distances satisfying the geometric constraint $r^{12} + r^{23} = r^{13}$. Other sets of arguments are simply not physically possible and the original function $\widehat{\mathcal{V}}^{\text{int}}$ can provide no information on them. So, for example, $\breve{\mathcal{V}}^{\text{int}}(1, 2, 1)$ is valid, but $\breve{\mathcal{V}}^{\text{int}}(1, 1, 1)$ lies outside the domain of the function.

[11] A "physical" set of distances is a set of real numbers that satisfy the Cayley–Menger constraints described in Section A.2.

[12] See Chapter 5 for a comprehensive discussion of interatomic potentials.

The evaluation of the potential energy function for nonphysical arguments requires us to extend the domain of the function to all possible sets of arguments. We denote this extended function as $\mathcal{V}^{\text{int}}(\zeta^{12}, \zeta^{13}, \zeta^{23})$. In fact, as noted above, all of the interatomic potential functions discussed in Chapter 5 are already expressed as extensions (or can easily be extended by replacing angles with distances). For example say that the three particles in this example interact via a pair potential model $\phi_{\alpha\beta}(r^{\alpha\beta})$. In this case the extended potential energy is

$$\mathcal{V}^{\text{int}}(\zeta^{12}, \zeta^{13}, \zeta^{23}) = \phi_{12}(\zeta^{12}) + \phi_{13}(\zeta^{13}) + \phi_{23}(\zeta^{23}). \tag{A.31}$$

Clearly, this function can be evaluated for any set $(\zeta^{12}, \zeta^{13}, \zeta^{23})$, whether $\zeta^{\alpha\beta}$ satisfies the geometric constraint between the distances or not. Presumably, as required in Eqn. (A.29), the function has been constructed in a way that it gives the same result as $\widehat{\mathcal{V}}^{\text{int}}(r^1, r^2, r^3)$ when $(\zeta^{12}, \zeta^{13}, \zeta^{23}) = (r^{12}, r^{13}, r^{23})$.

Now, it may seem that the distinction between $\breve{\mathcal{V}}^{\text{int}}(\{r^{\alpha\beta}\})$ and $\mathcal{V}^{\text{int}}(\{\zeta^{\alpha\beta}\})$ is unimportant since in practice $\mathcal{V}^{\text{int}}$ is only evaluated at physical sets of arguments. This is indeed the case if one only wishes to compute the energy of atomic configurations. However, when calculating the force on an atom it is necessary to compute the *derivatives* of $\mathcal{V}^{\text{int}}$ with respect to its arguments. This is done in Section 5.8 to obtain an expression for the force on an atom in terms of the contributions from its neighbors – information which is used in Chapter 8 to obtain an expression for the atomistic (pointwise) stress tensor. For example the partial derivative of $\mathcal{V}^{\text{int}}(\zeta^{12}, \ldots, \zeta^{(N-1)N})$ with respect to ζ_{12} at any point $s = (r^{12}, \ldots, r^{(N-1)N}) \in S$, defined as

$$\frac{\partial \mathcal{V}^{\text{int}}}{\partial \zeta^{12}}(s) \equiv \lim_{\epsilon \to 0} \frac{\mathcal{V}^{\text{int}}(r^{12} + \epsilon, \ldots, r^{N(N-1)/2}) - \mathcal{V}^{\text{int}}(r^{12}, \ldots, r^{N(N-1)/2})}{\epsilon}, \tag{A.32}$$

requires us to evaluate the function at $(r^{12} + \epsilon, \ldots, r^{N(N-1)/2})$ which is *not* a physical configuration. The manner in which $\breve{\mathcal{V}}^{\text{int}}(\{r^{\alpha\beta}\})$ is extended to $\mathcal{V}^{\text{int}}(\{\zeta^{\alpha\beta}\})$ could influence the result obtained from Eqn. (A.32). We show below that this is indeed the case and that this has implications for the uniqueness of the force decomposition mentioned above and the corresponding stress tensor.

A.4 Alternative potential energy extensions and the effect on atomic forces

In this section, we show that multiple extensions of $\breve{\mathcal{V}}^{\text{int}}$ are possible and that this can lead to a nonunique decomposition of the force on a particle, which in turn leads to a nonunique pointwise stress tensor. We start by noting that the internal force on atom α,

$$\boldsymbol{f}^{\text{int},\alpha}(\boldsymbol{u}) = -\frac{\partial \widehat{\mathcal{V}}^{\text{int}}(\boldsymbol{u})}{\partial \boldsymbol{r}^{\alpha}}, \tag{A.33}$$

where $\boldsymbol{u} = (\boldsymbol{r}^1, \ldots, \boldsymbol{r}^N) \in \mathbb{R}^{3N}$, is uniquely defined for any extension. This is because

$$\frac{\partial \widehat{\mathcal{V}}^{\text{int}}(\boldsymbol{u})}{\partial \boldsymbol{r}^{\alpha}} = \frac{\partial \breve{\mathcal{V}}^{\text{int}}(\bar{\phi}^{-1}(\boldsymbol{s}))}{\partial \boldsymbol{r}^{\alpha}} = \frac{\partial \breve{\mathcal{V}}^{\text{int}}(\boldsymbol{s})}{\partial \boldsymbol{r}^{\alpha}} = \frac{\partial \mathcal{V}^{\text{int}}(\boldsymbol{s})}{\partial \boldsymbol{r}^{\alpha}}, \tag{A.34}$$

where $\bar{\phi}^{-1}(\boldsymbol{s}) = O_{\boldsymbol{u}}$, which implies[13] $\phi(\boldsymbol{u}) = \boldsymbol{s}$, and we have used Eqns. (A.6), (A.20) and (A.29) in the first, the second and the last equality respectively.

Next, we repeat the derivation given in Section 5.8.2, which shows that the force on a particle can always be decomposed as a sum of central forces. Using Eqn. (A.34), Eqn. (A.33) takes the form

$$\boldsymbol{f}^{\text{int},\alpha}(\boldsymbol{u}) = - \left. \frac{\partial \mathcal{V}^{\text{int}}(\boldsymbol{s})}{\partial \boldsymbol{r}^\alpha} \right|_{\boldsymbol{s}=\phi(\boldsymbol{u})} = \sum_{\substack{\beta \\ \beta \neq \alpha}} \boldsymbol{f}^{\alpha\beta}(\boldsymbol{u}), \tag{A.35}$$

where $\boldsymbol{s} = \phi(\boldsymbol{u}) = (r^{12}, \dots, r^{(N-1)N})$ and

$$\boldsymbol{f}^{\alpha\beta}(\boldsymbol{u}) \equiv \begin{cases} \dfrac{\partial \mathcal{V}^{\text{int}}}{\partial \zeta^{\alpha\beta}}(\phi(\boldsymbol{u})) \dfrac{\boldsymbol{r}^\beta - \boldsymbol{r}^\alpha}{r^{\alpha\beta}} & \text{if } \alpha < \beta, \\ \dfrac{\partial \mathcal{V}^{\text{int}}}{\partial \zeta^{\beta\alpha}}(\phi(\boldsymbol{u})) \dfrac{\boldsymbol{r}^\beta - \boldsymbol{r}^\alpha}{r^{\alpha\beta}} & \text{if } \alpha > \beta, \end{cases} \tag{A.36}$$

is the contribution to the force on particle α due to the presence of particle β.

Note that $\boldsymbol{f}^{\alpha\beta}$ is parallel to the direction $\boldsymbol{r}^\beta - \boldsymbol{r}^\alpha$ and satisfies $\boldsymbol{f}^{\alpha\beta} = -\boldsymbol{f}^{\beta\alpha}$. We therefore note that the *internal force on a particle, for any interatomic potential that has a continuously differentiable extension can always be decomposed as a sum of central forces, i.e., forces parallel to directions connecting the particle to its neighbors*.

The remaining question is how different potential energy extensions affect the force decomposition in Eqn. (A.35). We have already established in Eqns. (A.33) and (A.34) that the force $\boldsymbol{f}^{\text{int},\alpha}$ is independent of the particular extension used. However, we show below that the individual terms in the decomposition, $\boldsymbol{f}^{\alpha\beta}$, are *not* unique. These terms depend on the manner in which the potential energy, defined on the shape space, is extended to its neighborhood in the higher-dimensional Euclidean space.

In order to construct different extensions, we use the geometric constraints expressed in terms of Cayley–Menger determinants that the distances have to satisfy in order for them to be embeddable in $\mathbb{R}^3$ as outlined in Section A.2. For simplicity let us restrict our discussion to one dimension. It is easy to see that in this case the number of independent coordinates is $N - 1$ and for $N > 2$ the number of interatomic distances exceeds the number of independent coordinates. We return to the case of three points interacting in one dimension described in Example A.2. The standard pair potential representation for this system, which is an extension of the potential energy to the higher-dimensional Euclidean space, is given in Eqn. (A.31) and reproduced here:

$$\mathcal{V}^{\text{int}}(\zeta^{12}, \zeta^{13}, \zeta^{23}) = \phi_{12}(\zeta^{12}) + \phi_{13}(\zeta^{13}) + \phi_{23}(\zeta^{23}). \tag{A.37}$$

As in Example A.2 to keep things simple, let us consider the special case in which the particles are arranged to satisfy $r^1 < r^2 < r^3$, for which $r^{13} = r^{12} + r^{23}$. Using Eqn. (A.35), the internal force, $f^{\text{int},1}$, evaluated at this configuration, is decomposed as

$$f^{\text{int},1}(r^{12}, r^{13}, r^{23}) = -\frac{\partial \mathcal{V}^{\text{int}}}{\partial r^1} = -\frac{\partial \phi_{12}}{\partial r^1} - \frac{\partial \phi_{13}}{\partial r^1} = \underbrace{\phi'_{12}(r^{12})}_{f^{12}} + \underbrace{\phi'_{13}(r^{13})}_{f^{13}}. \tag{A.38}$$

[13] Note that the vector $\boldsymbol{u}$ appearing in Eqn. (A.34) can be replaced by any $\boldsymbol{v} \in O_{\boldsymbol{u}}$.

We now provide an alternative extension to the standard pair potential representation given in Eqn. (A.37). The Cayley–Menger determinant corresponding to a cluster of three points (see Eqn. (A.25)) is identically equal to zero at every point on the shape space. This is because the shape space corresponds to a configuration of three collinear points, and the area of the triangle formed by three collinear points is zero. Thus, we have

$$\begin{aligned}\chi(r^{12}, r^{13}, r^{23}) &= (r^{12} - r^{13} - r^{23})(r^{23} - r^{12} - r^{13})(r^{13} - r^{23} - r^{12})(r^{12} + r^{13} + r^{23}) \\ &= 0. \end{aligned} \tag{A.39}$$

Using the identity in Eqn. (A.39), an alternative extension $\mathcal{V}_*^{\text{int}}$ is constructed:

$$\mathcal{V}_*^{\text{int}}(\zeta^{12}, \zeta^{13}, \zeta^{23}) = \mathcal{V}^{\text{int}}(\zeta^{12}, \zeta^{13}, \zeta^{23}) + \chi(\zeta^{12}, \zeta^{13}, \zeta^{23}). \tag{A.40}$$

Note that $\mathcal{V}_*^{\text{int}}$ is indeed an extension because from Eqn. (A.39) it is clear that $\mathcal{V}_*^{\text{int}}$ is equal to $\mathcal{V}^{\text{int}}$ at every point on the shape space of the system and it is continuously differentiable because $\chi(\zeta^{12}, \zeta^{13}, \zeta^{23})$, being a polynomial, is infinitely differentiable. Let us now see how the internal force, $f^{\text{int},1}$, for the special configuration considered in this example, is decomposed using the new extension:

$$\begin{aligned} f^{\text{int},1} &= -\frac{\partial \mathcal{V}_*^{\text{int}}}{\partial r^1} = -\frac{\partial \mathcal{V}^{\text{int}}}{\partial r^1} - \frac{\partial \chi}{\partial r^1} \\ &= \left(\phi'_{12} - \frac{\partial \chi}{\partial \zeta^{12}}(s)\frac{\partial \zeta^{12}}{\partial r^1}(s)\right) + \left(\phi'_{13} - \frac{\partial \chi}{\partial \zeta^{13}}(s)\frac{\partial \zeta^{13}}{\partial r^1}(s)\right) \\ &= \underbrace{\left(f^{12} - 8r^{12}r^{23}(r^{12} + r^{23})\right)}_{f_*^{12}} + \underbrace{\left(f^{13} + 8r^{12}r^{23}(r^{12} + r^{23})\right)}_{f_*^{13}}. \end{aligned} \tag{A.41}$$

It is clear from Eqns. (A.38) and (A.41) that the central-force decomposition is not the same for the two representations, i.e. $f^{12} \neq f_*^{12}$ and $f^{13} \neq f_*^{13}$; however, the force on particle 1, $f^{\text{int},1}$, is the same in both cases as expected.

It is very interesting to note that $\mathcal{V}_*^{\text{int}}$ is *not* a pair potential but is equivalent to one, i.e. it agrees with a pair potential on the shape space. Thus, the set of continuously differentiable extensions of a given interatomic potential function forms an equivalence class. It is not clear at this stage if these equivalence classes can be fully expressed in terms of the Cayley–Menger determinant constraints.

Although the above example is quite elementary, this process can be extended to any arbitrary number of particles in three dimensions. Any given potential can be altered to an equivalent potential by adding a function of the Cayley–Menger determinants corresponding to any cluster of five or six particles (see Section A.2). This function must be continuously differentiable and equal to zero when all of its arguments are zero. For example, a new representation in three dimensions can be constructed by adding a linear combination of the Cayley–Menger determinants:

$$\mathcal{V}_*^{\text{int}} = \mathcal{V}^{\text{int}}(\zeta^{12}, \ldots, \zeta^{(N-1)N}) + \sum_{k=1}^{m} \lambda_k \chi_k, \tag{A.42}$$

where there are m constraints defined by the Cayley–Menger determinants χ_k, and λ_k are constants.[14]

It is clear from the discussion in this appendix that an interatomic potential energy function, $\mathcal{V}^{\rm int}$, like those discussed in Chapter 5 and used in the rest of the book, is strictly speaking an extension of a potential energy function defined over shape space, $\breve{\mathcal{V}}^{\rm int}(r^{12}, \ldots, r^{(N-1)N})$, which in turn is derived from a function of the positions of the atoms, $\widehat{\mathcal{V}}^{\rm int}(\boldsymbol{r}^1, \ldots, \boldsymbol{r}^N)$. Given this, a derivative of $\mathcal{V}^{\rm int}$ with respect to a distance argument evaluated at an atomic configuration should be written as

$$\left.\frac{\partial \mathcal{V}^{\rm int}(\zeta^{12}, \ldots, \zeta^{(N-1)N})}{\partial \zeta^{\alpha\beta}}\right|_{(\zeta^{12},\ldots,\zeta^{(N-1)N})=(r^{12},\ldots,r^{(N-1)N})}. \tag{A.43}$$

However, in the interest of brevity and to conform to common practices in the field, in the rest of the book we abuse our notation and write $\partial \mathcal{V}^{\rm int}/\partial r^{\alpha\beta}$ in place of Eqn. (A.43). Also, we assume that there exists a continuously differentiable extension whenever we make use of derivatives of $\mathcal{V}^{\rm int}$ as in the calculation of $\varphi^{\alpha\beta}$ in Section 5.8.2 or $\kappa^{\alpha\beta\gamma\delta}$ in Section 8.1.4.

[14] Note that Eqn. (A.42) has the same form as a Lagrangian with the λ terms playing the role of Lagrange multipliers. For a static minimization problem, we seek to minimize $\mathcal{V}^{\rm int}$, without violating the physical constraints relating the distances to each other. (This is equivalent to minimizing $\widehat{\mathcal{V}}^{\rm int}$ with respect to the positions of particles.) Thus, the original constrained minimization of $\mathcal{V}^{\rm int}$ is replaced by the problem of finding the saddle points of $\mathcal{V}^{\rm int}_*$.

References

[AB96] J. E. Angelo and M. I. Baskes. Interfacial studies using the EAM and MEAM. *Interface Sci.*, **4(1–2)**:47–63, 1996.

[AB02] M. Arroyo and T. Belytschko. An atomistic-based finite deformation membrane for single layer crystalline films. *J. Mech. Phys. Solids*, **50**:1941–1977, 2002.

[AB04] M. Arroyo and T. Belytschko. Finite crystal elasticity of carbon nanotubes based on the exponential Cauchy–Born rule. *Phys. Rev. B*, **69**:115415, 2004.

[ABB96] N. L. Allan, T. H. K. Barron, and J. A. O. Bruno. The zero static internal stress approximation in lattice dynamics, and the calculation of isotope effects on molar volumes. *J. Chem. Phys.*, **105**:8300–8303, 1996.

[ABBH00] F. F. Abraham, N. Bernstein, J. Q. Broughton, and D. Hess. Dynamic fracture of silicon: Concurrent simulation of quantum electrons, classical atoms, and the continuum solid. *MRS Bull.*, **25(5)**:27–32, 2000.

[ABBK98] F. F. Abraham, J. Q. Broughton, N. Bernstein, and E. Kaxiras. Spanning the length scales in dynamic simulation. *Comput. Phys.*, **12**:538, 1998.

[ABI09] ABINIT. Abinit website. http://www.abinit.org, 2009.

[ABRT65] N. Aslund, R. F. Barrow, W. G. Richards, and D. N. Travis. Rotational analysis of bands of b-x system of Cu2 and of a-x system of Bi2. *Ark. Fys.*, **30(2)**:171, 1965.

[AD74] S. A. Adelman and J. D. Doll. Generalized Langevin equation approach for alom/solid-surface scattering – collinear atom/harmonic chain model. *J. Chem. Phys.*, **61(10)**:4242–4245, 1974.

[AD76] S. A. Adelman and J. D. Doll. Generalized Langevin equation approach for atom–solid-surface scattering – general formulation for classical scattering off harmonic solids. *J. Chem. Phys.*, **64(6)**:2375–2388, 1976.

[Adk83] C. J. Adkins. *Equilibrium Thermodynamics*. Cambridge: Cambridge University Press, third edition, 1983.

[AGM$^+$90] S. F. Altschul, W. Gish, W. Miller, E. W. Myers, and D. J. Lipman. Basic local alignment search tool. *J. Mol. Biol.*, **215**:403–410, 1990.

[AJ05] M. F. Ashby and D. R. H. Jones. *Engineering Materials 1*. Oxford: Butterworth-Heinemann, third edition, 2005.

[AK07] M. Aoki and T. Kurokawa. A simple environment-dependent overlap potential and cauchy violation in solid argon. *J. Phys. Condens. Matter*, **19**:236228, 2007.

[AL08] M. Arndt and M. Luskin. Error estimation and atomistic-continuum adaptivity for the quasicontinuum approximation of a Frenkel–Kontorova model. *Multiscale Model. Simul.*, **7(1)**:147–170, 2008.

[Alb00] D. Z. Albert. *Time and Chance*. Cambridge: Harvard University Press, 2000.

[AM76] N. W. Ashcroft and N. D. Mermin. *Solid State Physics*. Philadelphia: Saunders College, 1976.

[AMB95] J. E. Angelo, N. R. Moody, and M. I. Baskes. Trapping of hydrogen to lattice-defects in nickel. *Modell. Simul. Mater. Sci. Eng.*, **3(3)**:289–307, 1995.

[And80] H. C. Andersen. Molecular dynamics simulations at constant pressure and/or temperature. *J. Chem. Phys.*, **72(4)**:2384–2393, 1980.

[And94] G. Andrews. *Number Theory*. New York: Dover Publications, 1994.

[ANMPV07] M. Aoki, D. Nguyen-Manh, D. G. Pettifor, and V. Vitek. Atom-based bond-order potentials for modelling mechanical properties of metals. *Prog. Mater. Sci.*, **52(2–3)**:154–195, 2007.

[Apo69] T. M. Apostol. *Calculus*, Volume II. *Multi-variable Calculus and Linear Algebra with Applications*. New York: Wiley, second edition, 1969.

[AR03] G. J. Ackland and S. K. Reed. Two-band second moment model and an interatomic potential for caesium. *Phys. Rev. B*, **67(17)**:174108, 2003.

[ARK+94] J. B. Adams, A. Rockett, J. Kieffer, *et al.* Atomic-level computer simulation. *J. Nucl. Mater.*, **216**:265–74, 1994.

[Arn63] V. I. Arnold. Proof of a theorem by A. N. Kolmogorov on the invariance of quasi-periodic motions under small perturbations of the Hamiltonian. *Russ. Math. Surv.*, **18**:9–36, 1963.

[Arn89] V. I. Arnold. *Mathematical Methods of Classical Mechanics*. New York: Springer-Verlag, second edition, 1989.

[Asa83] R. J. Asaro. Crystal plasticity. *J. Appl. Mech.*, **50(4b)**:921–934, 1983.

[Ash66] N. W. Ashcroft. Electron-ion pseudopotentials in metals. *Phys. Lett.*, **23**:48–50, 1966.

[AST09] M. Arndt, V. Sorkin, and E. B. Tadmor. Efficient algorithms for discrete lattice calculations. *J. Comput. Phys.*, **228**:4858–4880, 2009.

[AT87] M. P. Allen and D. J. Tildesley. *Computer Simulation of Liquids*. Oxford: Clarendon Press, 1987.

[AT10] N. C. Admal and E. B. Tadmor. A unified interpretation of stress in molecular systems. *J. Elast.*, **100**:63–143, 2010.

[AT11] N. C. Admal and E. B. Tadmor. Stress and heat flux for arbitrary multi-body potentials: A unified framework. *J. Chem. Phys*, **134**:184106, 2011.

[AW57] B. J. Alder and T. E. Wainright. Phase transitions for a hard sphere system. *J. Chem. Phys.*, **27**:1208–1209, 1957.

[AW59] B. J. Alder and T. E. Wainright. Studies in molecular dynamics: I. General method. *J. Chem. Phys.*, **31**:459–466, 1959.

[AW95] G. B. Arfken and H. J. Weber. *Mathematical Methods of Physicists*. San Diego: Academic Press, fourth edition, 1995.

[BABK99] J. Q. Broughton, F. F. Abraham, N. Bernstein, and E. Kaxiras. Concurrent coupling of length scales: Methodology and application. *Phys. Rev. B*, **60(4)**:2391–2403, 1999.

[Bak39] C. M. Bakewell. *Source Book in Ancient Philosophy*. New York: Charles Scribner's Sons, revised edition, 1939.

[Bal02] J. M. Ball. Some open problems in elasticity. In P. Holmes, P. Newton and A. Weinstein, editors, *Geometry, Mechanics, and Dynamics*, Chapter 1, pages 3–59. New York: Springer-Verlag, 2002.

[Bas92] M. I. Baskes. Modified embedded-atom potentials for cubic materials and impurities. *Phys. Rev. B*, **46(5)**:2727–2742, 1992.

[Bas94] J. L. Bassani. Plastic-flow of crystals. *Adv. Appl. Mech.*, **30**:191–258, 1994.

[BAT10] B. Bar On, E. Altus, and E. B. Tadmor. Surface effects in non-uniform nanobeams: continuum and atomistic modeling. *Int. J. Solids Struct.*, **47**: 1243–1252, 2010.

[Baž78] Z. P. Bažant. Spurious reflection of elastic waves in nonuniform finite element grids. *Comput. Meth. Appl. Mech. Eng.*, **16**:91–100, 1978.

[BBL02] X. Blanc, C. Le Bris, and P.-L. Lions. From molecular models to continuum mechanics. *Arch. Ration. Mech. Anal.*, **164**:341–381, 2002.

[BBL+07] S. Badia, P. Bochev, R. Lehoucq, *et al.* A force-based blending model for atomistic-to-continuum coupling. *Int. J. Multiscale Comput. Eng.*, **5(5)**:387–406, 2007.

[BBLP10] X. Blanc, C. Le Bris, F. Legoll, and C. Patz. Finite-temperature coarse-graining of one-dimensional models: mathematical analysis and computational approaches. *J. Nonlinear Sci.*, **20(2)**:241–275, 2010.

[BBO+83] B. R. Brooks, R. E. Bruccoleri, B. D. Olafson, *et al.* CHARMM – a program for macromolecular energy, minimization, and dynamics calculations. *J. Comput. Chem.*, **4(2)**:187–217, 1983.

[BC06] V. V. Bulatov and W. Cai. *Computer Simulations of Dislocations*. Oxford Series on Materials Modelling. Oxford: Oxford University Press, 2006.

[BCF86] F. Bavaud, Ph. Choquard, and J.-R. Fontaine. Statistical mechanics of elastic moduli. *J. Stat. Phys.*, **42**:621–646, 1986.

[BDE+08] P. T. Bauman, H. B. Dhia, N. Elkhodja, J. T. Oden, and S. Prudhomme. On the application of the Arlequin method to the coupling of particle and continuum models. *Comput. Mech.*, **42(4)**:511–530, 2008.

[Ber04] S. Berryman. Democritus. In E. N. Zalta, editor, *The Stanford Encyclopedia of Philosophy*. http://plato.stanford.edu/archives/fall2004/entries/democrites/, 2004.

[BF94] G. E. Beltz and L. B. Freund. Analysis of the strained-layer critical thickness concept based on a Peierls–Nabarro model of a threading dislocation. *Philos. Mag. A*, **69**:183–202, 1994.

[BFS91] G. Bozzolo, J. Ferrante, and J. R. Smith. Universal behaviour in ideal slip. *Scr. Metall. Mater.*, **25**:1927–1931, 1991.

[BG90] G. Burns and A. M. Glazer. *Space Groups for Solid State Scientists*. San Diego: Academic Press, second edition, 1990.

[BGCB94] T. S. Bush, J. D. Gale, C. R. A. Catlow, and P. D. Battle. Self-consistent interatomic potentials for the simulation of binary and ternary oxides. *J. Mater. Chem.*, **4(6)**:831–837, 1994.

[BGM71] T. H. K. Barron, T. G. Gibbons, and R. W. Munn. Thermodynamics of internal strain in perfect crystals. *J. Phys. C: Solid State Phys.*, **4**:2805–2821, 1971.

[BH54] M. Born and K. Huang. *Dynamical Theory of Crystal Lattices*. Oxford: Clarendon, 1954.

[BH82] O. Brulin and R. K. T. Hsieh, editors. *Mechanics of Micropolar Media*. Singapore: World Scientific, 1982.

[BH86] R. Biswas and D. R. Hamann. Simulated annealing of silicon atom clusters in Langevin molecular-dynamics. *Phys. Rev. B*, **34(2)**:895–901, 1986.

[BH01] M. I. Baskes and R. G. Hoagland. Dislocation core structures and mobilities in MoSi2. *Acta Mater.*, **49(13)**:2357–2364, 2001.

[BH03] S. G. Brush and N. S. Hall. *The Kinetic Theory of Gasses: An Anthology of Classic Papers with Historical Commentary*. London: Imperial College Press, 2003.

[BHT92] H. Ballamane, T. Halicioglu, and W. A. Tiller. Comparative study of silicon empirical interatomic potentials. *Phys. Rev. B*, **46(4)**:2250–2279, 1992.

[Bir31] G. D. Birkhoff. Proof of the ergodic theorem. *Proc. Nat. Acad. Sci. USA*, **17**:656–660, 1931.

[BK97] V. Bulatov and E. Kaxiras. Semidiscrete variational Peierls framework for dislocation core properties. *Phys. Rev. Lett.*, **78(22)**:4221–4224, 1997.

[BKC09] N. Bernstein, J. R. Kermode, and G. Csanyi. Hybrid atomistic simulation methods for materials systems. *Rep. Prog. Phys.*, **72(2)**, 2009.

[BKG⁺06] E. Bitzek, P. Koskinen, F. Gähler, M. Moseler and P. Gumbsch. Structural relaxation made simple. *Phys. Rev. Lett.*, **97(17)**:170201, 2006.

[BKJ97] M. Z. Bazant, E. Kaxiras and J. F. Justo. Environment-dependent interatomic potential for bulk silicon. *Phys. Rev. B*, **56**:8542–8552, 1997.

[BKO⁺96] T. Belytschko, Y. Krongauz, D. Organ, M. Fleming and P. Krysl. Meshless methods: An overview and recent developments. *Comput. Meth. Appl. Mech. Eng.*, **139**:3–47, 1996.

[BL05] C. Le Bris and P.-L. Lions. From atoms to crystals: a mathematical journey. *Bull. Am. Math. Soc.*, **42**:291–363, 2005.

[BLL99] S. D. Bond, B. J. Leimkuhler, and B. B. Laird. The Nosé–Poincaré method for constant temperature molecular dynamics. *J. Comput. Phys.*, **151**:114–134, 1999.

[Blu70] L. M. Blumenthal. *Theory and Applications of Distance Geometry*. New York: Chelsea, second edition, 1970.

[BM32] M. Born and J. E. Mayer. Zur gittertheorie der ionenkristalle. *Z. Phys.*, **75**:1–18, 1932.

[BM58] G. E. P. Box and M. E. Muller. A note on the generation of random normal deviates. *Ann. Math. Stat.*, **29(2)**:610–611, 1958.

[BM96] G. T. Barkema and N. Mousseau. Event-based relaxation of continuous disordered systems. *Phys. Rev. Lett.*, **77(21)**:4358–4361, 1996.

[BMW03] C. M. Bender, P. N. Meisinger and Q. Wang. All Hermitian hamiltonians have parity. *J. Phys. A: Math. Gen.*, **36**:1029–1031, 2003.

[BN47] L. Bragg and J. F. Nye. A dynamical model of a crystal structure. *Proc. R. Soc. London, Ser. A*, **190(1023)**:474–481, 1947.

[BNW89] M. I. Baskes, J. S. Nelson, and A. F. Wright. Semiempirical modified embedded-atom potentials for silicon and germanium. *Phys. Rev. B*, **40(9)**:6085–6100, 1989.

[Bor23] M. Born. *Atomtheorie des Festen Zustandes*. Leipzig: B. G. Teubner, second edition, 1923.

[BPB+08] S. Badia, M. Parks, P. Bochev, M. Gunzburger and R. Lehoucq. On atomistic-to-continuum coupling by blending. *Multiscale Model. Simul.*, **7(1)**:381–406, 2008.

[BPV+84] H. J. C. Berendsen, J. P. M. Postma, W. F. Vangunsteren, A. Dinola, and J. R. Haak. Molecular-dynamics with coupling to an external bath. *J. Chem. Phys.*, **81(8)**:3684–3690, 1984.

[Bra87] W. Braun. Distance geometry and related methods for protein structure determination from NMR data. *Q. Rev. Biophys.*, **19(3/4)**:115–157, 1987.

[Bra05] A. Brandt. Multiscale solvers and systematic upscaling in computational physics. *Comput. Phys. Commun.*, **169(1–3)**:438–441, 2005. Europhysics Conference on Computational Physics, Genova, Italy, Sep 01–04, 2004.

[Bre90] D. W. Brenner. Empirical potential for hydrocarbons for use in simulating the chemical vapor deposition of diamond films. *Phys. Rev. B*, **42(15)**:9458–9471, 1990.

[Bre92] D. W. Brenner. Erratum: Empirical potential for hydrocarbons for use in simulating the chemical vapor deposition of diamond films. *Phys. Rev. B*, **46(3)**:1948, 1992.

[Bre00] D. W. Brenner. The art and science of an analytic potential. *Phys. Stat. Sol. B*, **217**:23–40, 2000.

[Bri95] J. Bricmont. Science of chaos or chaos in science? *Ann. NY Acad. Sci.*, **775**:131–175, 1995.

[Bro04] H. W. Broer. KAM theory: The legacy of Kolmogorov's 1954 paper. *Bull. Am. Math. Soc.*, **41**:507–521, 2004.

[BS88] A. Banerjea and J. R. Smith. Origins of the universal binding-energy relation. *Phys. Rev. B*, **37(12)**:6632–6645, 1988.

[BSH+02] D. W. Brenner, O. A. Shenderova, J. A. Harrison, *et al.* A second-generation reactive empirical bond order (REBO) potential energy expression for hydrocarbons. *J. Phys. Condens. Matter*, **14(4)**:783–802, 2002.

[BvdSvD95] H. J. C. Berendsen, D. van der Spoel, and R. van Drunen. GROMACS: A message-passing parallel molecular dynamics implementation. *Comput. Phys. Commun.*, **91(1–3)**:43 – 56, 1995.

[BvK12] M. Born and T. von Karman. On fluctuations in spatial grids. *Phys. Z.*, **13**:297–309, 1912.

[CA80] D. M. Ceperley and B. J. Alder. Ground state of the electron gas by a stochastic method. *Phys. Rev. Lett.*, **45(7)**:566–569, 1980.

[Cah99] R. W. Cahn. Slaying the crystal homunculus. *Nature*, **400(6745)**:625, 1999.

[Cal85] H. B. Callen. *Thermodynamics and an Introduction to Thermostatics*. New York: John Wiley and Sons, second edition, 1985.

[Cal01] C. Callender. Taking thermodynamics too seriously. *Stud. Hist. Phil. Mod. Phys.*, **32(4)**:539–553, 2001.

[Cam94] R. C. Cammarata. Surface and interface stress effects in thin films. *Prog. Surf. Sci.*, **46(1)**:1–38, 1994.

[Car90] A. E. Carlsson. Beyond pair potentials in elemental transition metals and semiconductors. In H. Ehrenreich and D. Turnbull, editors, *Solid State Physics*, Volume 43, pages 1–91. Academic Press Inc., 1990.

[CAS10] CASTEP. CASTEP website, http://www.castep.org, 2010.

[Cau50] A. Cauchy. Mémoire sur les systèmes isotropes de points matériels. *Mém. Acad. Sci. Paris*, **22**:615–654, 1850.

[Cau28a] A. Cauchy. *Exercises du Mathématique*, Volume 3, chapter Sur l'équilibre et le mouvement d'un système du points matériels sollicités par des forces d'attraction ou de répulsion mutuelle, pages 227–252. Paris: Chez de Bure Frères, 1828.

[Cau28b] A. Cauchy. *Exercises de Mathématique*, Volume 3, Chapter: De la pression ou tension dans un système de points matériels, pages 253–277. Paris: Chez de Bure Frères, 1828.

[CBA03] M. Cafiero, S. Bubin, and L. Adamowicz. Non-Born–Oppenheimer calculations of atoms and molecules. *Phys. Chem. Chem. Phys.*, **5(8)**:1491–1501, 2003.

[CBC$^+$03] W. Cai, V. V. Bulatov, J. Chang, J. Li, and S. Yip. Periodic image effects in dislocation modelling. *Philos. Mag.*, **83(5)**:539–567, 2003.

[CBM99] G. H. Campbell, J. Belak and J. A. Moriarty. Atomic structure of the sigma 5 (310)/[001] symmetric tilt grain boundary in molybdenum. *Acta Mater.*, **47(15–16)**:3977–3985, 1999.

[CBS03] F. J. Cherne, M. I. Baskes and R. B. Schwarz. Atomistic simulations of the phase stability and elastic properties of nickel-zirconium alloys. *J. Non-Cryst. Solids*, **317(1–2)**:45–51, 2003.

[CC93] S. J. Cook and P. Clancey. Comparison of semi-empirical potential functions for silicon and germanium. *Phys. Rev. B*, **47**:7686–7699, 1993.

[CD94] D. S. Corti and P. G. Debenedetti. A computational study of metastability in vapor–liquid equilibrium. *Chem. Eng. Sci.*, **49(17)**:2717–2734, 1994.

[CD97] C. Creemers and P. Deurinck. Platinum segregation to the (111) surface of ordered Pt80Fe20: LEIS results and model simulations. *Surf. Interface Anal.*, **25(3)**:177–190, 1997.

[CD98] D. S. Corti and P. G. Debenedetti. Statistical mechanics of fluids under internal constraints: Rigorous results for the one-dimensional hard rod fluid. *Phys. Rev. E*, **57(4)**:4211–4226, 1998.

[CDC$^+$08] D. A. Case, T. A. Darden, T. E. Cheatham, III, *et al.* AMBER 10. http://www.ambermd.org, 2008.

[CDGK04] L. Chen, P. G. Debenedetti, C. W. Gear, and I. G. Kevrekidis. From molecular dynamics to coarse self-similar solutions: a simple example using equation-free computation. *J. Non-Newtonian Fluid Mech.*, **120(1–3)**:215–223, 2004. 3rd International Workshop on Nonequilibrium Thermodynamics and Complex Fluids, Princeton, NJ, Aug 14–17, 2003.

[CdKBY00] W. Cai, M. de Koning, V. V. Bulatov, and S. Yip. Minimizing boundary reflections in coupled-domain simulations. *Phys. Rev. Lett.*, **85(15)**:3213–3216, 2000.

[CDKM06] S. Conti, G. Doltzmann, B. Kirchheim, and S. Müller. Sufficient conditions for the validity of the Cauchy–Born rule close to $so(n)$. *J. Eur. Math. Soc.*, **8**:515–530, 2006.

[CvdV99] G. Ceder and A. Van der Ven. Phase diagrams of lithium transition metal oxides: investigations from first principles. *Electrochim. Acta*, **45(1–2)**:131–150, 1999.

[CDvDG05] S. Cheung, W. Q. Deng, A. C. T. van Duin and W. A. Goddard, III. ReaxFF(MgH) reactive force field for magnesium hydride systems. *J. Phys. Chem. A*, **109(5)**:851–859, 2005.

[CG01] P. Cermelli and M. E. Gurtin. On the characterization of geometrically necessary dislocations in finite plasticity. *J. Mech. Phys. Solids*, **49**:1539–1568, 2001.

[CGE80] A. E. Carlsson, C. D. Gelatt, and H. Ehrenreich. Ab-initio pair potential applied to metals. *Philos. Mag. A*, **41(2)**:241–250, 1980.

[CGL$^+$07] C. Cui, Z. Guo, Y. Liu, *et al.* Characteristics of cobalt-based alloy coating on tool steel prepared by powder feeding laser cladding. *Opt. Laser Technol.*, **39(8)**:1544–1550, 2007.

[CGLK04] J. Cisternas, C. W. Gear, S. Levin, and I. G. Kevrekidis. Equation-free modelling of evolving diseases: Coarse-grained computations with individual-based models. *Proc. R. Soc. London, Ser. A*, **460(2050)**:2761–2779, 2004.

[CH88] G. M. Crippen and T. F. Havel. *Distance Geometry and Molecular Conformation.* New York: Wiley, 1988.

[CH09] A. J. Chorin and O. H. Hald. *Stochastic Tools in Mathematics and Science.* New York: Springer, second edition, 2009.

[Cha43] S. Chandrasekhar. Stochastic problems in physics and astronomy. *Rev. Mod. Phys.*, **15(1)**:1–89, 1943.

[Cha99] P. Chadwick. *Continuum Mechanics: Concise Theory and Problems.* Mineola: Dover, second edition, 1999.

[Che05] G. Chen. *Nanoscale Energy Transport and Conversion: A Parallel Treatment of Electrons, Molecules, Phonons, and Photons.* New York: Oxford University Press, 2005.

[CL68] F. Cyrot-Lackmann. Calculation of cohesion and surface tension of transition metals by a tight binding method. *J. Phys. Chem. Solids*, **29(7)**:1235–&, 1968.

[Cla70] R. Clausius. On a mechanical theorem applicable to heat. *Philos. Mag.*, **40**:122–127, 1870.

[CM03] W. A. Curtin and R. E. Miller. Atomistic/continuum coupling methods in multi-scale materials modeling. *Modell. Simul. Mater. Sci. Eng.*, **11(3)**:R33–R68, 2003.

[CN63] B. D. Coleman and W. Noll. The thermodynamics of elastic materials with heat conduction and viscosity. *Arch. Ration. Mech. Anal.*, **13**:167–178, 1963.

[CO77] M. Cassandro and E. Olivieri. A rigorous study of metastability in a continuous model. *J. Stat. Phys.*, **17(4)**:229–244, 1977.

[Col05] L. Colombo. Tight-binding molecular dynamics: A primer. *Riv. Nuovo Cimento*, **28(10)**:1–59, 2005.

[Cop10] Copper Development Association. Low temperature properties of copper. http://www.copper.org/resources/properties/, 2010.

[Cot08] R. Cotterill. *The Material World*. Cambridge: Cambridge University Press, second edition, 2008.

[Cou78a] C. S. G. Cousins. Inner elasticity. *J. Phys. C: Solid State Phys.*, **11**:4867–4879, 1978.

[Cou78b] C. S. G. Cousins. Symmetry of the inner elastic constants. *J. Phys. C: Solid State Phys.*, **11**:4881–4900, 1978.

[CP74] R. D. Crowninshield and M. H. Pope. Response of compact bone in tension at various strain rates. *Ann. Biomed. Eng.*, **2(2)**:217–225, 1974.

[CP85] R. Car and M. Parrinello. Unified approach for molecular dynamics and density-functional theory. *Phys. Rev. Lett.*, **55(22)**:2471–2474, 1985.

[Cri77] G. M. Crippen. A novel approach to calculation of conformation: distance geometry. *J. Comput. Phys.*, **24**:96–107, 1977.

[Cri05] T. Crilly. *Arthur Cayley: Mathematician Laureate of the Victorian Age*. Baltimore: The Johns Hopkins University Press, 2005.

[Cur86] W. A. Curtin. Theories of inhomogeneous media: far-infrared absorption in composites and a density functional theory of freezing. PhD thesis, Cornell University, 1986.

[CvDDG09] K. Chenoweth, A. C. T. van Duin, S. Dasgupta and W. A. Goddard, III. Initiation mechanisms and kinetics of pyrolysis and combustion of JP-10 hydrocarbon jet fuel. *J. Phys. Chem. A*, **113(9)**:1740–1746, 2009.

[CWA83] D. Chandler, J. D. Weeks and H. C. Andersen. Van der Waals picture of liquids, solids, and phase transformations. *Science*, **220(4599)**:787–794, 1983.

[CY91] K. S. Cheung and S. Yip. Atomic-level stress in an inhomogeneous system. *J. Appl. Phys.*, **70(10)**:5688–5690, 1991.

[Dav94] P. J. Davis. *Circulant Matrices*. New York: Chelsea Publishing, second edition, 1994.

[Daw89] M. S. Daw. Model of metallic cohesion – the embedded-atom method. *Phys. Rev. B*, **39(11)**:7441–7452, 1989.

[DB83] M. S. Daw and M. I. Baskes. Semiempirical, quantum mechanical calculation of hydrogen embrittlement in metals. *Phys. Rev. Lett.*, **50(17)**:1285–1288, 1983.

[DB84] M. S. Daw and M. I. Baskes. Embedded-atom method: Derivation and application to impurities, surfaces, and other defects in metals. *Phys. Rev. B*, **29**:6443–6453, 1984.

[DBBW86] M. S. Daw, M. I. Baskes, C. L. Bisson and W. G. Wolfer. Application of the embedded atom method to fracture, dislocation dynamics and hydrogen embrittlement. In R. H. Jones and W. W. Gerberich, editors, *Modeling Environmental Effects on Crack Growth Processes*, pages 99–124. Warrendale: The Materials Society, 1986.

[DCL70] F. DuCastel and F. Cyrot-Lackmann. Moments developments and their application to electronic charge distribution of d-bands. *J. Phys. Chem. Solids*, **31(6)**:1295–1306, 1970.

[DCL71] F. DuCastel and F. Cyrot-Lackmann. Moments developments .2. Application to crystalline structures and stacking fault energies of transition metals. *J. Phys. Chem. Solids*, **32(1)**:285–301, 1971.

[DELT07] M. Dobson, R. S. Elliott, M. Luskin and E. B. Tadmor. A multilattice quasicontinuum for phase transforming materials: Cascading Cauchy Born kinematics. *J. Comput.-Aided Mater. Des.*, **14**:219–237, 2007.

[Deu04] P. Deuflhard. *Newton Methods for Nonlinear Problems*. Berlin: Springer-Verlag, 2004.

[DF04] D. S. Dummit and R. M. Foote. *Abstract Algebra*. Hoboken: John Wiley and Sons, third edition, 2004.

[DFB93] M. S. Daw, S. M. Foiles and M. I. Baskes. The embedded-atom method – a review of theory and applications. *Mater. Sci. Rep.*, **9(7–8)**:251–310, 1993.

[DFS04] R. Drautz, M. Fahnle and J. M. Sanchez. General relations between many-body potentials and cluster expansions in multicomponent systems. *J. Phys. Condens. Matter*, **16(23)**:3843–3852, 2004.

[dGM62] S. R. de Groot and P. Mazur. *Non-Equilibrium Thermodynamics*. Amsterdam: North-Holland Publishing Company, 1962.

[DJ07] T. Dumitrica and R. D. James. Objective molecular dynamics. *J. Mech. Phys. Solids*, **55(10)**:2206–2236, 2007.

[DL08a] M. Dobson and M. Luskin. Analysis of a force-based quasicontinuum approximation. *Math. Modell. Numer. Anal. (ESAIM:M2AN)*, **42(1)**:113–119, 2008.

[DL08b] M. Dobson and M. Luskin. An analysis of the effect of ghost force oscillation on the quasicontinuum error. *Math. Modell. Numer. Anal. (ESAIM:M2AN)*, **43(3)**:591–604, 2008.

[DLO10a] M. Dobson, M. Luskin and C. Ortner. Sharp stability estimates for the force-based quasicontinuum approximation of homogeneous tensile deformation. *Multiscale Model. Simul.*, **8(3)**:782–802, 2010.

[DLO10b] M. Dobson, M. Luskin and C. Ortner. Stability, instability, and error of the force-based quasicontinuum approximation. *Arch. Ration. Mech. Anal.*, **197(1)**:179–202, 2010.

[DMA75] J. D. Doll, L. E. Myers and S. A. Adelman. Generalized Langevin equation approach for atom-solid-surface scattering – inelastic studies. *J. Chem. Phys.*, **63(11)**:4908–4914, 1975.

[Dov93] M. T. Dove. *Introduction to Lattice Dynamics*. Cambridge: Cambridge University Press, 1993.

[DPS04] D. K. Datta, R. C. Picu and M. S. Shephard. Composite grid atomistic continuum method: An adaptive approach to bridge continuum with atomistic analysis. *Int. J. Multiscale Comput. Eng.*, **2(3)**:71–90, 2004.

[DTLMK11] L. M. Dupuy, E. B. Tadmor, F. Legoll, R. E. Miller and W. K. Kim. Finite temperature quasicontinuum. 2011. In preparation.

[DTMP05] L. M. Dupuy, E. B. Tadmor, R. E. Miller and R. Phillips. Finite temperature quasicontinuum: Molecular dynamics without all the atoms. *Phys. Rev. Lett.*, **95**:060202, 2005.

[DZ99] D. X. Du and Q.-S. Zheng. A note on Cauchy–Voigt dispute of independent constants in anisotropic elastic Hooke's law. *Mech. Res. Commun.*, **26**:295–300, 1999.

[DZM+07] R. Drautz, Z. W. Zhou, D. A. Murdick, B. Gillespie, H. N. G. Wadley and D. G. Pettifor. Analytic bond-order potentials for modelling the growth of semiconductor thin films. *Prog. Mater. Sci.*, **52(2–3)**:196–229, 2007.

[EA94] F. Ercolessi and J. B. Adams. Interatomic potentials from first-principles calculations – the force-matching method. *Europhys. Lett.*, **26**:583, 1994.

[EE11] P. Ehrenfest and T. Ehrenfest. *Encyklopädie der mathematische Wissenschaften*, Volume 4 part 32, chapter Begriffliche Grundagen der statistischen Auffassung in der Mechanik. 1911. English translation by M. J. Moravcsik, *The Conceptual Foundations of the Statistical Approach in Mechanics*. Ithaca: Cornell University Press, 1959.

[EEL+07] W. E, B. Engquist, X. Li, W. Ren and E. Vanden-Eijnden. Heterogeneous multiscale methods: A review. *Commun. Comput. Phys.*, **2(3)**:367–450, 2007.

[EH01] W. E and Z. Huang. Matching conditions in atomistic-continuum modeling of materials. *Phys. Rev. Lett.*, **87(13)**:135501, 2001.

[EH02] W. E and Z. Y. Huang. A dynamic atomistic-continuum method for the simulation of crystalline materials. *J. Comput. Phys.*, **182(1)**:234–261, 2002.

[Ein16] A. Einstein. Die grundlage der allgemeinen relativitätstheorie. *Ann. der Phys.*, **49**:769–822, 1916.

[EK87] R. Elber and M. Karplus. A method for determining reaction paths in large molecules: Application to myoglobin. *Chem. Phys. Lett.*, **139(5)**:375–380, 1987.

[EKAE06] R. Erban, I. G. Kevrekidis, D. Adalsteinsson, and T. C. Elston. Gene regulatory networks: A coarse-grained, equation-free approach to multiscale computation. *J. Chem. Phys.*, **124(8)**:084106, 2006.

[EKO06] R. Erban, I. G. Kevrekidis and H. G. Othmer. An equation-free computational approach for extracting population-level behavior from individual-based models of biological dispersal. *Physica D*, **215(1)**:1–24, 2006.

[ELY06] W. E, J. Lu, and J. Z. Yang. Uniform accuracy of the quasicontinuum method. *Phys. Rev. B*, **74(21)**:214115/1–12, 2006.

[EM90] D. J. Evans and G. P. Morriss. *Statistical Mechanics of Nonequilibrium Liquids*. London: Academic Press, 1990.

[EM07] W. E and P. Ming. Cauchy–Born rule and the stability of crystalline solids: Static problems. *Arch. Ration. Mech. Anal.*, **183**:241–297, 2007.

[EMM+08] K. Endo, C. Masumoto, D. Matsumoto, T. Ida, M. Mizuno and N. Kato. Fragment distribution of thermal decomposition for PS and PET with QMD calculations by considering the excited and charged model molecules. *Appl. Surf. Sci.*, **255(4)**:856–859, 2008.

[Erc08a] F. Ercolessi. Lecture notes on tight-binding molecular dynamics, and tight-binding justification of classical potentials. http://www.fisica.uniud.it/~ercolessi/SA/tb.pdf, 2008.

[Erc08b] F. Ercolessi. A Molecular Dynamics Primer, http://fisica.uniud.it/~ercolessi/md/md, 2008.

[ERH85] Z. Elkoshi, H. Reiss, and D. Hammerich. One-dimensional rigorous hole theory for fluids: internally constrained ensembles. *J. Stat. Phys.*, **41(3/4)**:685–708, 1985.

[Eri70] J. L. Ericksen. Nonlinear elasticity of diatomic crystals. *Int. J. Solids Struct.*, **6(7)**:951–957, 1970.

[Eri84] J. L. Ericksen, The Cauchy and Born hypothesis for crystals. In M. Gurtin editor, Phase Transformations and Material Instabilities in Solids, pages 61–77. New York, Academic Press, 1984.

[Eri97] J. L. Ericksen. Equilibrium theory for X-ray observations of crystals. *Arch. Ration. Mech. Anal.*, **139**:181–200, 1997.

[Eri02] A. C. Eringen. *Nonlocal Continuum Field Theories*. New York: Springer, 2002.

[Eri08] J. L. Ericksen. On the Cauchy–Born rule. *Math. Mech. Solids*, **13**:199–220, 2008.

[ERVE02] W. E, W. Ren and E. Vanden-Eijnden. String method for the study of rare events. *Phys. Rev. B*, **66(5)**:052301, 2002.

[ERVE09] W. E, W. Ren and E. Vanden-Eijnden. A general strategy for designing seamless multiscale methods. *J. Comput. Phys.*, **228(15)**:5437–5453, 2009.

[ES79] D. M. Esterling and A. Swaroop. Inter-atomic potentials from experimental phonon-spectra .1. prototypes. *Phys. Status Solidi B*, **96(1)**:401–411, 1979.

[ES09] B. Eidel and A. Stukowski. A variational formulation of the quasicontinuum method based on energy sampling of clusters. *J. Mech. Phys. Solids*, **57(1)**:87–108, 2009.

[ETP86] F. Ercolessi, E. Tosatti and M. Parrinello. Au (100) surface reconstruction. *Phys. Rev. Lett.*, **57(6)**:719–722, 1986.

[ETS06a] R. S. Elliott, N. Triantafyllidis and J. A. Shaw. Stability of crystalline solids – I: Continuum and atomic lattice considerations. *J. Mech. Phys. Solids*, **54**:161–192, 2006.

[ETS06b] R. S. Elliott, J. A. Shaw and N. Triantafyllidis. Stability of crystalline solids – II: Application to temperature-induced martensitic phase transformations in a bi-atomic crystal. *J. Mech. Phys. Solids*, **54(1)**:193–232, 2006.

[Eva83] D. J. Evans. Computer "experiment" for nonlinear thermodynamics of Couette flow. *J. Chem. Phys.*, **78(6)**:3297–3302, 1983.

[Eva98] L. C. Evans. *Partial Differential Equations*, Volume 19 of Graduate Studies in Mathematics. Providence: American Mathematical Society, 1998.

[EW77] J. J. Erpenbeck and W. W. Wood. Molecular dynamics techniques for hard-core systems. In J. Berne, editor, *Statistical Mechanics, Part B: Time-dependent processes*, Volume 6 of Modern Theoretical Chemistry, Chapter 1, pages 1–40. New York: Plenum Press, 1977.

[Ezr06] G. S. Ezra. Reversible measure-preserving integrators for non-Hamiltonian systems. *J. Chem. Phys.*, **125(3)**:034104, 2006.

[FAF91] M. W. Finnis, P. Agnew and A. J. E. Foreman. Thermal excitation of electrons in energetic displacement cascades. *Phys. Rev. B*, **44(2)**:567–574, 1991.

[Far94] D. Farkas. Interatomic potentials for Ti–Al with and without angular forces. *Modell. Simul. Mater. Sci. Eng.*, **2(5)**:975–984, 1994.

[FBD86] S. M. Foiles, M. I. Baskes, and M. S. Daw. Embedded-atom-method functions for the fcc metals Cu, Ag, Au, Ni, Pd, Pt, and their alloys. *Phys. Rev. B*, **33(12)**:7983–7991, 1986.

[FC99] M. C. Fivel and G. R. Canova. Developing rigorous boundary conditions to simulations of discrete dislocation dynamics. *Modell. Simul. Mater. Sci. Eng.*, **7(5)**:753–768, 1999.

[Fey85] R. P. Feynman. *QED: The Strange Theory of Light and Matter*. Princeton: Princeton University Press, 1985.

[FF77] R. Fletcher and T. L. Freeman. A modified Newton method for minimization. *J. Optimiz. Theory App.*, **23**:357–372, 1977.

[FG05] R. A. Friesner and V. Guallar. Ab initio quantum chemical and mixed quantum mechanics/molecular mechanics (QM/MM) methods for studying enzymatic catalysis. *Annu. Rev. Phys. Chem.*, **56**:389–427, 2005.

[FGC96] M. C. Fivel, T. J. Gosling and G. R. Canova. Implementing image stresses in a 3D dislocation simulation. *Modell. Simul. Mater. Sci. Eng.*, **4(6)**:581–596, 1996.

[FGU96] C. Filippi, X. Gonze and C. J. Umrigar. Generalized gradient approximations to density functional theory. In J. M. Seminario, editor, *Recent Developments and Applications of Modern Density Functional Theory*, Volume 4 of Theoretical and Computational Chemistry. Amsterdam: Elsevier, 1996.

[FH65] R. P. Feynman and A. R. Hibbs. *Quantum Mechanics and Path Integrals*. New York: McGraw-Hill, 1965.

[FHCO04] M. Fago, R. L. Hayes, E. A. Carter and M. Ortiz. Density-functional-theory-based local quasicontinuum method: Prediction of dislocation nucleation. *Phys. Rev. B*, **70(10)**:100102, 2004.

[Fin03] M. Finnis. *Interatomic Forces in Condensed Matter*. Oxford: Oxford University Press, 2003.

[Fis64] M. E. Fisher. The free energy of a macroscopic system. *Arch. Ration. Mech. Anal.*, **17**:377–410, 1964.

[FJ00] G. Friesecke and R. D. James. A scheme for the passage from atomic to continuum theory for thin films, nanotubes and nanorods. *J. Mech. Phys. Solids*, **48(6–7)**:1519 – 1540, 2000.

[FL70] M. E. Fisher and J. L. Lebowitz. Asymptotic free energy of a system with periodic boundary conditions. *Commun. Math. Phys.*, **19**:251–272, 1970.

[FLS06] R. P. Feynman, R. B. Leighton and M. Sands. *The Feynman Lectures on Physics, The Definitive and Extended Edition*. Boston: Addison-Wesley, second edition, 2006.

[FMAH94] N. A. Fleck, G. M. Muller, M. F. Ashby and J. W. Hutchinson. Strain gradient plasticity: Theory and experiment. *Acta Metall. Mater.*, **42**:475–487, 1994.

[FNS$^+$07] J. Fish, M. A. Nuggehally, M. S. Shephard, *et al.* Concurrent AtC coupling based on a blend of the continuum stress and the atomistic force. *Comput. Meth. Appl. Mech. Eng.*, **196(45–48)**:4548–4560, 2007.

[FOH08] S. M. Foiles, D. L. Olmsted, and E. A. Holm. Using atomistic simulations to inform mesoscale simulations of microstructural evolution. In A. El-Azab, editor, *Proc. 4th International Conference of Multiscale Materials Modeling (MMM)*, pages 362–368, Tallahassee: Florida State University, 2008.

[Foi85] S. M. Foiles. Calculation of the surface segregation of Ni–Cu alloys with the use of the embedded-atom method. *Phys. Rev. B*, **32(12)**:7685–7693, 1985.

[Foi94] S. M. Foiles. Evaluation of harmonic methods for calculating the free energy of defects in solids. *Phys. Rev. B*, **49**:14930–14938, 1994.

[Fos09] R. L. Fosdick, 2009. Private communications.

[Fow36] R. H. Fowler. *Statistical Mechanics: The Theory of the Properties of Matter in Equilibrium*. Cambridge: Cambridge University Press, second edition, 1936.

[FPA$^+$95] C. Fiolhais, J. P. Perdew, S. Q. Armster, J. M. MacLaren and M. Brajczewska. Dominant density parameters and local pseudopotentials for simple metals. *Phys. Rev. B*, **51**:14001–14011, 1995.

[Fre26] J. Frenkel. Zur theorie der elastizitätsgrenze und der festigkeit kristallinischer körper. *Z. Phys.*, **37(7/8)**:572–609, 1926.

[FS84] M. W. Finnis and J. E. Sinclair. A simple empirical n-body potential for transition-metals. *Philos. Mag. A*, **50(1)**:45–55, 1984.

[FS02] D. Frenkel and B. Smit. *Understanding Molecular Simulation: From Algorithms to Applications*. San Diego: Academic Press, second edition, 2002.

[FSR83] J. Ferrante, J. R. Smith and J. H. Rose. Diatomic molecules and metallic adhesion, cohesion, and chemisorption: A single binding-energy relation. *Phys. Rev. Lett.*, **50(18)**:1385–1386, 1983.

[FT02] G. Friesecke and F. Theil. Validity and failure of the Cauchy–Born hypothesis in a two dimensional mass-spring lattice. *J. Nonlinear Sci.*, **12**:445–478, 2002.

[FW99] J. H. Frederick and C. Woywod. General formulation of the vibrational kinetic energy operator in internal bond-angle coordinates. *J. Chem. Phys.*, **111(16)**:7255–7271, 1999.

[Gau10] Gaussian. Gaussian website. http://www.gaussian.com, 2010.

[GB95] P. Gumbsch and G. E. Beltz. On the continuum versus atomistic descriptions of dislocation nucleation and cleavage in nickel. *Modell. Simul. Mater. Sci. Eng.*, **3(5)**:597–613, 1995.

[GBKH03] N. M. Ghoniem, E. P. Busso, N. Kioussis and H. C. Huang. Multiscale modelling of nanomechanics and micromechanics: an overview. *Philos. Mag.*, **83**(31–34, Sp. Iss. SI):3475–3528, 2003. 1st International Conference on Multiscale Materials Modelling (MMM), London, England, June 17–22, 2002.

[GCL73] J. P. Gaspard and F. Cyrot-Lackmann. Density of states from moments – application to impurity band. *J. Phys. C: Solid State Phys.*, **6(21)**:3077–3096, 1973.

[GE10] V. S. Guthikonda and R. S. Elliott. Modeling martensitic phase transformation in shape memory alloys with the self-consistent lattice dynamics approach. *J. Mech. Phys. Solids*, 2010, submitted.

[Gea71] C. W. Gear. *Numerical Initial Value Problems in Ordinary Differential Equations*. Englewood Cliffs: Prentice-Hall, 1971.

[GF00] I. M. Gelfand and S. V. Fomin. *Calculus of Variations*. Mineola: Dover, 2000. Translated by Richard A. Silverman.

[GFA10] M. E. Gurtin, E. Fried and L. Anand. *The Mechanics and Thermodynamics of Continua*. Cambridge: Cambridge University Press, 2010.

[GFS89] J. J. Gracio, J. V. Fernandes and J. H. Schmitt. Effect of grain-size on substructural evolution and plastic behavior of copper. *Mater. Sci. Eng. A*, **118**:97–105, 1989.

[GGMV60] J. B. Gibson, A. N. Goland, M. Milgram and G. H. Vineyard. Dynamics of radiation damage. *Phys. Rev.*, **120(4)**:1229–1253, 1960.

[GHBF08] P. M. Gullett, M. F. Horstemeyer, M. I. Baskes and H. Fang. A deformation gradient tensor and strain tensors for atomistic simulations. *Modell. Simul. Mater. Sci. Eng.*, **16**:015001, 2008.

[Gib02] J. W. Gibbs. *Elementary Principles in Statistical Mechanics*. New York: C. Scribner and Sons, 1902.

[GIKI98] M. Giesen, G. S. Icking-Konert and H. Ibach. Fast decay of adatom islands and mounds on Cu(111): A new effective channel for interlayer mass transport. *Phys. Rev. Lett.*, **80(3)**:552–555, 1998.

[Gil10] S. P. A. Gill. Nonequilibrium molecular dynamics and multiscale modeling of heat conduction in solids. In T. Dumitrica, editor, *Trends in Computational Nanomechanics*, Volume 9, Chapter 4, pages 83–132. Dordrecht: Springer Science and Business Media, 2010.

[GK98] H. Gao and P. Klein. Numerical simulation of crack growth in an isotropic solid with randomized internal cohesive bonds. *J. Mech. Phys. Solids*, **46**:187–218, 1998.

[GK08] T. C. Germann and K. Kadau. Trillion-atom molecular dynamics becomes a reality. *Int. J. Mod. Phys. C*, **19**:1315–1319, 2008.

[GLW08] V. Gravemeier, S. Lenz and W. A. Wall. Towards a taxonomy for multiscale methods in computational mechanics: Building blocks of existing methods. *Comput. Mech.*, **41(2)**:279–291, 2008.

[GM75] M. E. Gurtin and A. I. Murdoch. A continuum theory of elastic material surfaces. *Arch. Ration. Mech. Anal.*, **57(4)**:291–323, 1975.

[Gol80] H. Goldstein. *Classical Mechanics*. Reading: Addison-Wesley, second edition, 1980.

[Gor76] J. E. Gordon. *The New Science of Strong Materials or Why You Don't Fall Through the Floor*. Princeton: Princeton University Press, second edition, 1976.

[GR87] L. Greengard and V. Rokhlin. A fast algorithm for particle simulations. *J. Comput. Phys.*, **73(2)**:325–348, 1987.

[Gra87] W. T. Grandy Jr. *Foundations of Statistical Mechanics*. Volume I: *Equilibrium Theory*. Dordrecht: D. Reidel Publishing, 1987.

[Gra00] W. W. Grabowski. Cloud microphysics and the tropical climate: Cloud-resolving model perspective. *J. Clim.*, **13(13)**:2306–2322, 2000.

[Gra04] W. W. Grabowski. An improved framework for superparameterization. *J. Atmos. Sci.*, **61(15)**:1940–1952, 2004.

[Gra08] H. Morrison W. W. Grabowski. Modeling supersaturation and subgrid-scale mixing with two-moment bulk warm microphysics. *J. Atmos. Sci.*, **65(3)**:792–812, 2008.

[Gri21] A. A. Griffith. The phenomena of rupture and flow in solids. *Philos. Trans. R. Soc. London, Ser. A*, **221**:163–198, 1921.

[Gri05] D. J. Grifffiths. *Introduction to Quantum Mechanics*. Upper Saddle River: Pearson Prentice-Hall, second edition, 2005.

[GS99] W. W. Grabowski and P. K. Smolarkiewicz. CRCP: A cloud resolving convection parameterization for modeling the tropical convecting atmosphere. *Physica D*, **133(1–4)**:171–178, 1999. 18th Annual International Conference of the Center-for-Nonlinear-Studies, Los Alamos, New Mexico, May 11–15, 1998.

[GS02] W. W. Grabowski and P. K. Smolarkiewicz. A multiscale anelastic model for meteorological research. *Mon. Weather Rev.*, **130(4)**:939–956, 2002.

[GStVB98] C. F. Guerra, J. G. Snijders, G. te Velde and E. J. Baerends. Towards an order-*N* DFT method. *Theor. Chem. Acc.*, **99(6)**:391–403, 1998.

[Gui02] B. Guillot. A reappraisal of what we have learnt during three decades of computer simulations on water. *J. Mol. Liq.*, **101**:219–260, 2002.

[Gur65] M. E. Gurtin. Thermodynamics and the possibility of spatial interaction in elastic materials. *Arch. Ration. Mech. Anal.*, **19**:339–352, 1965.

[Gur95] M. E. Gurtin. The nature of configurational forces. *Arch. Ration. Mech. Anal.*, **131**:67–100, 1995.

[Gur00] M. E. Gurtin. On the plasticity of single crystals: free energy, microforces, plastic-strain gradients. *J. Mech. Phys. Solids*, **48(5)**:989–1036, 2000.

[GW66] M. E. Gurtin and W. O. Williams. On the Clausius–Duhem inequality. *J. Appl. Math. Phys. (ZAMP)*, **17**:626–633, 1966.

[GWZ06] X. Guo, J. B. Wang and H. W. Zhang. Mechanical properties of single-walled carbon nanotubes based on higher order Cauchy–Born rule. *Int. J. Solids Struct.*, **43**:1276–1290, 2006.

[HA87] J. D. Honeycutt and H. C. Andersen. Molecular-dyanmics study of melting and freezing of small Lennard-Jones clusters. *J. Phys. Chem.*, **91**:4950–4963, 1987.

[Haf87] J. Hafner. *From Hamiltonians to Phase Diagrams*. Berlin: Springer-Verlag, 1987.

[Hal05] T. C. Hales. A proof of the Kepler conjecture. *Ann. Math.*, **162(3)**:1065–1185, 2005.

[Har66] W. A. Harrison. *Pseudopotentials in the Theory of Metals*. New York: PWA Benjamin, Inc., 1966.

[Har82] R. J. Hardy. Formulas for determining local properties in molecular-dynamics simulations: Shock waves. *J. Chem. Phys.*, **76**:622–628, 1982.

[Har00] W. A. Harrison. *Applied Quantum Mechanics*. River Edge: World Scientific, 2000.

[HB01] D. Hull and D. Bacon. *Introduction to Dislocations*. Oxford: Oxford University Press, fourth edition, 2001.

[HBF$^+$96] A. P. Horsfield, A. M. Bratkovsky, M. Fearn, D. G. Pettifor and M. Aoki. Bond-order potentials: Theory and implementation. *Phys. Rev. B*, **53**:12694–12712, 1996.

[HCCK02] J. R. Hu, S. C. Chang, F. R. Chen and J. J. Kai. HRTEM investigation of the multiplicity of sigma = 9 [01$\bar{1}$]/(122) symmetric tilt grain boundary in Cu. *Mater. Chem. Phys.*, **74(3)**:313–319, 2002.

[HH70] F. J. Humphreys and P. B. Hirsch. The deformation of single crystals of copper and copper-zinc alloys containing alumina particles. II. Microstructure and dislocation-particle interactions. *Proc. R. Soc. London, Ser. A*, **318(1532)**:73–92, 1970.

[HH99] Y. Huang and F. J. Humphreys. Measurements of grain boundary mobility during recrystallization of a single-phase aluminium alloy. *Acta Mater.*, **47(7)**:2259–2268, 1999.

[HH03] D. Heggie and P. Hut. *The Gravitational Million-Body Problem: A Multidisciplinary Approach to Star Cluster Dynamics*. Cambridge: Cambridge University Press, 2003.

[HHL09] W. G. Hoover, C. G. Hoover and J. F. Lutsko. Microscopic and macroscopic stress with gravitation and rotational forces. *Phys. Rev. E*, **79**:036709, 2009.

[HI02] F. W. Hehl and Y. Itin. The Cauchy relations in linear elasticity theory. *J. Elast.*, **66**:185–192, 2002.

[Hil56] T. L. Hill. *Statistical Mechanics: Principles and Selected Applications*. New York: McGraw-Hill, 1956.

[HJ99] G. Henkelman and H. Jónsson. A dimer method for finding saddle points on high dimensional potential surfaces using only first derivatives. *J. Chem. Phys.*, **111(15)**:7010–7022, 1999.

[HJ00] G. Henkelman and H. Jónsson. Improved tangent estimate in the nudged elastic band method for finding minimum energy paths and saddle points. *J. Chem. Phys.*, **113(22)**:9978–9985, 2000.

[HJ01] G. Henkelman and H. Jónsson. Long time scale kinetic Monte Carlo simulations without lattice approximation and predefined event table. *J. Chem. Phys.*, **115(21)**:9657–9666, 2001.

[HJJ98] G. Mills H. Jónsson and K. W. Jacobsen. Nudged elastic band method for finding minimum energy paths of transitions. In B. J. Berne, G. Ciccoti, and D. F. Coker, editors, *Classical and Quantum Dynamics in Condensed Phase Simulations*, Chapter 16, pages 385–404. Singapore: World Scientific, 1998.

[HJJ00] G. Henkelman, G. Jóhannesson and H. Jónsson. *Progress on Theoretical Chemistry and Physic*, chapter Methods for Finding Saddle Points and Minimum Energy Paths, pages 269–300. Dordrecht: Kluwer Academic Publishers, 2000.

[HK64] P. Hohenberg and W. Kohn. Inhomogeneous electron gas. *Phys. Rev.*, **136(3B)**:B864–B871, 1964.

[HKC83] T. F. Havel, I. D. Kuntz, and G. M. Crippen. The theory and practice of distance geometry. *Bull. Math. Biol.*, **45**:665–720, 1983.

[HL63] G. K. Horton and J. W. Leech. On the statistical mechanics of the ideal inert gas solids. *Proc. Phys. Soc.*, **82**:816–854, 1963.

[HL92] J. P. Hirth and J. Lothe. *Theory of Dislocations*. Malabar: Krieger, 1992.

[HL00] E. Hairer and C. Lubich. Energy conservation by Stormer-type numerical integrators. In D. F. Griffiths and G. A.Watson editors, *Numerical Analysis 1999*, volume 420, pages 169–189, 2000. 18th Dundee Biennial Conference on Numerical Analysis, Dundee, Scotland, Jun 29–Jul 02, CRC Press, 1999.

[HLM82] W. G. Hoover, A. J. C. Ladd and B. Moran. High-strain-rate plastic-flow studied via non-equilibrium molecular-dynamics. *Phys. Rev. Lett.*, **48(26)**:1818–1820, 1982.

[HLW06] E. Hairer, C. Lubich, and G. Wanner. *Geometric Numerical Integration: Structure-Preserving Algorithms for Ordinary Differential Equations*, Volume 31 of Computational Mathematics. Berlin: Springer, second edition, 2006.

[HM86] J. P. Hansen and I. R. McDonald. *Theory of Simple Liquids*. London: Academic Press, second edition, 1986.

[HM99] D. Holland and M. Marder. Cracks and atoms. *Adv. Mater.*, **11**:793–806, 1999.

[HMP91] J. Huang, M. Meyer and V. Pontikis. Core structure of a dissociated edge dislocation and pipe diffusion in copper investigated by molecular-dynamics. *J. de Phys. III*, **1(6)**:867–883, 1991.

[Hol00] G. A. Holzapfel. *Nonlinear Solid Mechanics*. Chichester: Wiley, 2000.

[Hoo85] W. G. Hoover. Canonical dynamics: Equilibrium phase-space distributions. *Phys. Rev. A*, **31(3)**:1695–1697, 1985.

[Hoo86] W. G. Hoover. *Molecular Dynamics*, Volume 258 of Lecture Notes in Physics. Berlin: Springer, 1986. Available at www.williamhoover.info/MD.pdf.

[Hoo87] R. Hooke. *Micrographia* [A facsimile edition]. Lincolnwood: Science Heritage Ltd., 1987.

[Hoo99] W. G. Hoover. *Time Reversibility, Computer Simulation, and Chaos*. Singapore: World Scientific, 1999.

[HPM99] C. Herzig, T. Przeorski and Y. Mishin. Self-diffusion in gamma-TiAl: An experimental study and atomistic calculations. *Intermetallics*, **7(3–4)**:389–404, 1999.

[HR95] B. L. Holian and R. Ravelo. Fracture simulations using large-scale molecular dynamics. *Phys. Rev. B*, **51(17)**:11275–11288, 1995.

[HSMP06] P. D. Haynes, C.-K. Skylaris, A. A. Mostofi and M. C. Payne. ONETEP: linear-scaling density-functional theory with local orbitals and plane waves. *Phys. Stat. Sol. B*, **243(11)**:2489–2499, 2006.

[HTC98] S. C. Harvey, R. K. Z. Tan and T. E. Cheatham. The flying ice cube: Velocity rescaling in molecular dynamics leads to violation of energy equipartition. *J. Comput. Chem.*, **19(7)**:726–740, 1998.

[Hua63] K. Huang. *Statistical Mechanics*. New York: Wiley, 1963.

[Hun05] P. Hunenberger. Thermostat algorithms for molecular dynamics simulations. In *Advanced Computer Simulation Approaches for Soft Matter Sciences I*, volume 173 of Advances in Polymer Science, pages 105–147. Berlin: Springer-Verlag, 2005.

[HWF+98] H. Haas, C. Z. Wang, M. Fahnle, C. Elsasser and K. M. Ho. Environment-dependent tight-binding model for molybdenum. *Phys. Rev. B*, **57(3)**:1461–1470, 1998.

[HXS+00] W. Y. Hu, H. D. Xu, X. L. Shu, *et al.* Calculation of thermodynamic properties of Mg-RE alloys by an analytic modified embedded atom method. *J. Phys. D: Appl. Phys.*, **33(6)**:711–718, 2000.

[HXW+09] K. Han, Y. Xin, R. Walsh, S. Downey, II and P. N. Kalu. The effects of grain boundary precipitates on cryogenic properties of aged 316-type stainless steels. *Mater. Sci. Eng. A*, **516(1–2)**:169–179, 2009.

[IBA00] S. Ismail-Beigi and T. A. Arias. Ab initio study of screw dislocations in Mo and Ta: A new picture of plasticity in bcc transition metals. *Phys. Rev. Lett.*, **63**:1499–1502, 2000.

[IK50] J. H. Irving and J. G. Kirkwood. The statistical mechanical theory of transport processes. IV. the equations of hydrodynamics. *J. Chem. Phys.*, **18**:817–829, 1950.

[Irv08] Irving. Jack Howard Irving [obituary]. *Los Angeles Times*, 2008.

[Jac88] K. W. Jacobsen. Bonding in metallic systems: An effective medium approach. *Comments Condens. Matter Phys.*, **14**:129–161, 1988.

[Jam06] R. D. James. Objective structures. *J. Mech. Phys. Solids*, **54**:2354–2390, 2006.

[Jau67] W. Jaunzemis. *Continuum Mechanics*. New York: Macmillan, 1967.

[JBA02] A. N. Jackson, A. D. Bruce and G. J. Ackland. Lattice-switch monte carlo method: application to soft potentials. *Phys. Rev. E*, **65**:036710, 2002.

[JD97] B. Joós and M. S. Duesbery. The Peierls stress of dislocations: An analytic formula. *Phys. Rev. Lett.*, **78(2)**:266–269, 1997.

[Jen06] F. Jensen. *Introduction to Computational Chemistry*. Chichester: John Wiley & Sons, 2006.

[JG89] R. O. Jones and O. Gunnarsson. The density functional formalism, its applications and prospects. *Rev. Mod. Phys.*, **61(3)**:689–746, 1989.

[JGP09] D. Jarecka, W. W. Grabowski and H. Pawlowska. Modeling of subgrid-scale mixing in large-eddy simulation of shallow convection. *J. Atmos. Sci.*, **66(7)**:2125–2133, 2009.

[JHK$^+$07] B. Jelinek, J. Houze, S. Kim, M. F. Horstemeyer, M. I. Baskes and S.-G. Kim. Modified embedded-atom method interatomic potentials for the Mg–Al alloy system. *Phys. Rev. B*, **75(5)**:054106, 2007.

[JL89] J. Jellinek and D. H. Li. Separation of the energy of overall rotation in any n-body system. *Phys. Rev. Lett.*, **62(3)**:241–244, 1989.

[JNP87] K. W. Jacobsen, J. K. Norskov and M. J. Puska. Interatomic interactions in the effective-medium theory. *Phys. Rev. B*, **35(14)**:7423–7442, 1987.

[Joh64] R. A. Johnson. Interstitials and vacancies in α iron. *Phys. Rev.*, **134(5A)**:A1329–A1336, 1964.

[Joh72] R. A. Johnson. Relationship between two-body interatomic potentials in a lattice model and elastic constants. *Phys. Rev. B*, **6(6)**:2094–2100, 1972.

[Joh88] R. A. Johnson. Analytic nearest-neighbor model for fcc metals. *Phys. Rev. B*, **37(8)**:3924–3931, 1988.

[Jon24a] J. E. Jones. On the determination of molecular fields. I. From the variation of the viscosity of a gas with temperature. *Proc. R. Soc. London, Ser. A*, **106(738)**:441–462, 1924.

[Jon24b] J. E. Jones. On the determination of molecular fields. II. From the equation of state of a gas. *Proc. R. Soc. London, Ser. A*, **106(738)**:463–477, 1924.

[JRD94] B. Joos, Q. Ren, and M. S. Duesbery. Peierls–Nabarro model of dislocations in silicon with generalized stacking fault restoring forces. *Phys. Rev. B*, **50**:5890–5898, 1994.

[JvDGD09] D. Jiang, A. C. T. van Duin, W. A. Goddard, III, and S. Dai. Simulating the initial stage of phenolic resin carbonization via the ReaxFF reactive force field. *J. Phys. Chem. A*, **113(25)**:6891–6894, 2009.

[JZL+03] H. Jiang, P. Zhang, B. Liu, *et al.* The effect of nanotube radius on the constitutive model for carbon nanotubes. *Comput. Mater. Sci.*, **28(3–4)**:429–442, 2003.

[Kah10] A. Kahn. Surface and interface science laboratory website. http://www.princeton.edu/~kahnlab/, 2010.

[Kan95a] L. N. Kantorovich. Thermoelastic properties of perfect crystals with non-primitive lattices. I. General theory. *Phys. Rev. B*, **51(6)**:3520–3534, 1995.

[Kan95b] L. N. Kantorovich. Thermoelastic properties of perfect crystals with nonprimitive lattices. II. Application to KCl and NaCl. *Phys. Rev. B*, **51(6)**:3535–3548, 1995.

[Kax03] E. Kaxiras. *Atomic and Electronic Structure of Solids*. Cambridge: Cambridge University Press, 2003.

[KC92] L. P. Kubin and G. Canova. The modelling of dislocation patterns. *Scr. Metall.*, **27**:957–962, 1992.

[KCDM02] T. Kruml, D. Caillard, C. Dupas and J. L. Martin. A transmission electron microscopy in situ study of dislocation mobility in Ge. *J. Phys. Condens. Matter*, **14(48)**:12897–12902, 2002. Conference on Extended Defects in Semiconductors (EDS 2002), Bologna, Italy, Jun 01–06, 2002.

[KCF+05] W. E. King, G. H. Campbell, A. Frank, *et al.* Ultrafast electron microscopy in materials science, biology and chemistry. *J. Appl. Phys.*, **97**:111101, 2005.

[KD80] M. H. Klapper and D. DeBrota. Use of Caley–Menger determinants in the calculation of molecular structures. *J. Comput. Phys.*, **37**:56–69, 1980.

[Kea66] P. N. Keating. Effect of invariance requirements on the elastic strain energy of crystals with application to the diamond structure. *Phys. Rev.*, **145(2)**:637–645, 1966.

[Kep66] J. Kepler. *On the Six-cornered Snowflake* [edited and translated from the Latin by Colin Hardie, with essays by L. L. Whyte and B. F. J. Mason]. Oxford: Clarendon Press, 1966.

[KF96a] G. Kresse and J. Furthmuller. Efficiency of ab-initio total energy calculations for metals and semiconductors using a plane-wave basis set. *Comput. Mater. Sci.*, **6(1)**:15–50, 1996.

[KF96b] G. Kresse and J. Furthmuller. Efficient iterative schemes for ab initio total-energy calculations using a plane-wave basis set. *Phys. Rev. B*, **54(16)**:11169–11186, 1996.

[KG98] P. Klein and H. Gao. Crack nucleation and growth as strain localization in a virtual-bond continuum. *Eng. Fract. Mech.*, **61**:21–48, 1998.

[KGF91] S. Kohlhoff, P. Gumbsch and H. F. Fischmeister. Crack propagation in bcc crystals studied with a combined finite-element and atomistic model. *Philos. Mag. A*, **64(4)**:851–878, 1991.

[KGH04] I. G. Kevrekidis, C. W. Gear, and G. Hummer. Equation-free: The computer-aided analysis of complex multiscale systems. *AIChE J.*, **50(7)**:1346–1355, 2004.

[KGK00] A. Kelly, G. W. Groves and P. Kidd. *Crystallography and Crystal Defects*. Chichester: Wiley, revised edition, 2000.

[Khi49] A. I. Khinchin. *Mathematical Foundations of Statistical Mechanics*. New York: Dover Publications, 1949.

[Kir35] J. G. Kirkwood. Statistical mechanics of fluid mixtures. *J. Chem. Phys.*, **3(5)**:300–313, 1935.

[Kir46] J. G. Kirkwood. The statistical mechanical theory of transport processes. I. General theory. *J. Chem. Phys.*, **14**:180–201, 1946.

[Kit96] C. Kittel. *Introduction to Solid State Physics*. Hoboken: Wiley, seventh edition, 1996.

[KO01] J. Knap and M. Ortiz. An analysis of the quasicontinuum method. *J. Mech. Phys. Solids*, **49(9)**:1899–1923, 2001.

[Koi65a] W. T. Koiter. The energy criterion of stability for continuous elastic bodies. – I. *Proc. of the Koninklijke Nederlandse Akademie Van Wetenschappen, Ser. B*, **68(4)**:178–189, 1965.

[Koi65b] W. T. Koiter. The energy criterion of stability for continuous elastic bodies.—II. *Proc. of the Koninklijke Nederlandse Akademie Van Wetenschappen, Ser. B*, **68(4)**:190–202, 1965.

[Koi65c] W. T. Koiter. On the instability of equilibrium in the absence of a minimum of the potential energy. *Proc. of the Koninklijke Nederlandse Akademie Van Wetenschappen, Ser. B*, **68(3)**:107–113, 1965.

[Kol54] A. N. Kolmogorov. On the conservation of conditionally periodic motions for a small change in Hamilton's functions [in Russian]. *Dokl. Akad. Nauk SSSR*, **98**:525–530, 1954. English translation in *LNP*, **93**:51–56, 1979.

[KPH98] C. L. Kelchner, S. J. Plimpton and J. C. Hamilton. Dislocation nucleation and defect structure during surface indentation. *Phys. Rev. B*, **58(17)**:11085–11088, 1998.

[Kra08] H. Kragh. Max Planck and quantum theory: The reluctant revolutionary. *Sci. Culture Rev.*, **5(6)**:23–31, 2008.

[Kro94] Herbert Kroemer. *Quantum Mechanics: For Engineering, Materials Science, and Applied Physics*. Englewood Cliffs: Prentice Hall, 1994.

[KS65] W. Kohn and L. J. Sham. Self-consistent equations including exchange and correlation effects. *Phys. Rev.*, **140(4A)**:A1133–A1138, 1965.

[KSB+99] L. Kalé, R. Skeel, M. Bhandarkar, *et al.* NAMD2: Greater scalability for parallel molecular dynamics. *J. Comput. Phys.*, **151(1)**:283–312, 1999.

[Kuh78] T. S. Kuhn. *Black Body Theory and the Quantum Discontinuity: 1894–1912*. Oxford: Oxford University Press, 1978.

[KV95] J. D. Kress and A. F. Voter. Low-order moment expansions to tight binding for interatomic potentials: Successes and failures. *Phys. Rev. B*, **52**:8766–8775, 1995.

[KZ06] P. A. Klein and J. A. Zimmerman. Coupled atomistic-continuum simulations using arbitrary overlapping domains. *J. Comput. Phys.*, **213(1)**:86–116, 2006.

[Lal06] J. N. Lalena. From quartz to quasicrystals: probing nature's geometric patterns in crystalline substances. *Crystallogr. Rev.*, **12(2)**:125–180, 2006.

[Lan70] C. Lanczos. *The Variational Principles of Mechanics*. Mineola: Dover, fourth edition, 1970.

[Lan73] O. E. Lanford. Entropy and equilbrium states in classical statistical mechanics. In A. Lenard, editor, *Statistical Mechanics and Mathematical Problems*, pages 1–113. Berlin: Springer-Verlag, 1973.

[Law93] B. Lawn. *Fracture of Brittle Solids*. Cambridge: Cambridge University Press, 1993.

[LB06] M. J. Louwerse and E. J. Baerends. Calculation of pressure in case of periodic boundary conditions. *Chem. Phys. Lett.*, **421**:138–141, 2006.

[LBK02] G. Lu, V. V. Bulatov, and N. Kioussis. Dislocation constriction and cross-slip: An ab initio study. *Phys. Rev. B*, **66(14)**:144103, 2002.

[LC85] G. V. Lewis and C. R. A. Catlow. Potential models for ionic oxides. *J. Phys. C: Solid State Phys.*, **18(6)**:1149–1161, 1985.

[LCN03] M. J. Leamy, P. W. Chung and R. Namburu. On an exact mapping and a higher order Born rule for use in analyzing graphene carbon nanotubes. Technical Report ARL-TR-3117, US Army Research Laboratory, Aberdeen Proving Ground, MD, 2003.

[Leb93] J. L. Lebowitz. Boltzmann's entropy and time's arrow. *Phys. Today*, **46(9)**:32–38, 1993.

[Leb99] J. L. Lebowitz. Statistical mechanics: A selective review of two central issues. *Rev. Mod. Phys.*, **71**:S346–S357, 1999.

[LHM$^+$07] B. Q. Luan, S. Hyun, J. F. Molinari, N. Bernstein and M. O. Robbins. Multiscale modeling of two-dimensional contacts. *Phys. Rev. E*, **74**:046710, 2007.

[Li10] J. Li. AtomEye website. http://mt.seas.upenn.edu/Archive/Graphics/A/, 2010.

[Lib05] K. G. Libbrecht. The physics of snow crystals. *Rep. Prog. Phys.*, **68(4)**:855–895, 2005.

[Lio38] J. Liouville. Sur la théorie de la variation des constantes arbitraires. *Journal de Mathémathiques Pures et Appliquées*, **3**:342–349, 1838.

[LJ25] J. E. Lennard-Jones. On the forces between atoms and ions. *Proc. R. Soc. London, Ser. A*, **109(752)**:584–597, 1925.

[LJ07] B.-J. Lee and J.-W. Jang. A modified embedded-atom method interatomic potential for the Fe–H system. *Acta Mater.*, **55(20)**:6779–6788, 2007.

[LJCV08] G. Lebon, D. Jou and J. Casas-Vázquez. *Understanding Non-equilitrium Thermodynamics: Foundations, Applications, Frontiers*. Berlin: Springer-Verlag, 2008.

[LK05] G. Lu and E. Kaxiras. Overview of multiscale simulations of materials. In M. Rieth and W. Schommers, editors, *Handbook of Theoretical and Computational Nanotechnology*, Volume 4, Chapter 22. Stevenson Ranch: American Scientific Publishers, 2005.

[LKBK00a] G. Lu, N. Kioussis, V. V. Bulatov and E. Kaxiras. Generalized-stacking-fault energy surface and dislocation properties of aluminum. *Phys. Rev. B*, **62(5)**:3099–3108, 2000.

[LKBK00b] G. Lu, N. Kioussis, V. V. Bulatov, and E. Kaxiras. The Peierls–Nabarro model revisited. *Philos. Mag. Lett.*, **80(10)**:675–682, 2000.

[LKZP04] W. K. Liu, E. G. Karpov, S. Zhang and H. S. Park. An introduction to computational nano mechanics and materials. *Comput. Meth. Appl. Mech. Eng.*, **193**:1529–1578, 2004.

[LL61] G. Leibfried and W. Ludwig. Theory of anharmonic effects in crystals. In F. Seitz and D. Turnbull, editors, *Solid State Physics: Advances in Research and Applications*, Volume 12, pages 275–459. New York: Academic Press, 1961.

[LL80] L. D. Landau and E. M. Lifshitz. *Statistical Physics, Part I.* Oxford: Pergamon, third edition, 1980.

[LM02] F. Legoll and R. Monneau. Designing reversible measure invariant algorithms with applications to molecular dynamics. *J. Chem. Phys.*, **117(23)**:10452–10464, 2002.

[LNS89] R. LeSar, R. Najafabadi, and D. J. Srolovitz. Finite-temperature defect properties from free-energy minimization. *Phys. Rev. Lett.*, **63**:624–627, 1989.

[LO09] M. Luskin and C. Ortner. An analysis of node-based cluster summation rules in the quasticontinuum method. *SIAM J. Numer. Anal.*, **47(4)**:3070–3086, 2009.

[Lov27] A. E. H. Love. *A Treatise on the Mathematical Theory of Elasticity*. Cambridge: Cambridge University Press, 1927.

[LP79] J. L. Lebowitz and O. Penrose. Towards a rigorous molecular theory of metastability. In E. W. Montroll and J. L. Lebowitz, editors, *Fluctuation Phenomena*, pages 293–340. New York: North-Holland, 1979.

[LR95] R. G. Littlejohn and M. Reinsch. Internal or shape coordinates in the *n*-body problem. *Phys. Rev. A*, **52(3)**:2035–2051, 1995.

[LR97] R. G. Littlejohn and M. Reinsch. Gauge fields in the separation of rotations and internal motions in the *n*-body problem. *Rev. Mod. Phys.*, **69(1)**:213–275, 1997.

[LRS96] B. Leimkuhler, S. Reich and R. D. Skeel. *Integration Methods for Molecular Dynamics*, Volume 82 of IMA Volumes in Mathematics and its Applications, pages 161–186. New York: Springer-Verlag, 1996.

[LS97] V. F. Lotrich and K. Szalewicz. Three-body contribution to binding energy of solid argon and analysis of crystal structure. *Phys. Rev. Lett.*, **79(7)**:1301–1304, 1997.

[LS08] R. B. Lehoucq and S. A. Silling. Force flux and the peridynamic stress tensor. *J. Mech. Phys. Solids*, **56**:1566–1577, 2008.

[LSAL03] Y. Li, D. J. Siegel, J. B. Adams and X.-Y. Liu. Embedded-atom-method tantalum potential developed by the force-matching method. *Phys. Rev. B*, **67(12)**:125101, 2003.

[LSB03] B. J. Lee, J. H. Shim and M. I. Baskes. Semiempirical atomic potentials for the fcc metals Cu, Ag, Au, Ni, Pd, Pt, Al, and Pb based on first and second nearest-neighbor modified embedded atom method. *Phys. Rev. B*, **68(14)**, 2003.

[LTK06] G. Lu, E. B. Tadmor, and E. Kaxiras. From electrons to finite elements: A concurrent multiscale approach for metals. *Phys. Rev. B*, **73(2)**:024108, 2006.

[Lut89] J. F. Lutsko. Generalized expressions for the calculation of elastic constants by computer simulation. *J Appl. Phys.*, **65**:2991–2997, 1989.

[Mac04] A. D. MacKerell. Empirical force fields for biological macromolecules: Overview and issues. *J. Comput. Chem.*, **25**:1584–1604, 2004.

[Mal69] L. E. Malvern. *Introduction to the Mechanics of a Continuous Medium*. Englewood Cliffs: Prentice-Hall, 1969.

[Mar75a] J. W. Martin. Many-body forces in metals and the Brugger elastic constants. *J. Phys. C: Solid State Phys.*, **8**:2837–2857, 1975.

[Mar75b] J. W. Martin. Many-body forces in solids and the Brugger elastic constants: II. Inner elastic constants. *J. Phys. C: Solid State Phys.*, **8**:2858–2868, 1975.

[Mar75c] J. W. Martin. Many-body forces in solids: Elastic constants of diamond-type crystals. *J. Phys. C: Solid State Phys.*, **8**:2869–2888, 1975.

[Mar90] E. Marquit. A plea for a correct translation of Newton's law of inertia. *Am. J. Phys.*, **58**:867–870, 1990.

[Mar04] R. M. Martin. *Electronic Structure: Basic Theory and Practical Methods*. Cambridge: Cambridge University Press, 2004.

[Max70] J. C. Maxwell. On reciprocal figures, frames, and diagrams of forces. *Trans. R. Soc. Edin.*, **xxvi**:1–43, 1870.

[Maz00] G. F. Mazenko. *Equilibrium Statistical Mechanics*. New York: John Wiley and Sons, 2000.

[MB82] R. Mullen and T. Belytschko. Dispersion analysis of finite element semidiscretizations of the two-dimensional wave equation. *Int. J. Numer. Methods Eng.*, **18**:11–29, 1982.

[MB93] A. I. Murdoch and D. Bedeaux. On the physical interpretation of fields in continuum mechanics. *Int. J. Eng. Sci.*, **31(10)**:1345–1373, 1993.

[MB94] A. I. Murdoch and D. Bedeaux. Continuum equations of balance via weighted averages of microscopic quantities. *Proc. Math. Phys. Sci.*, **445(1923)**:157–179, 1994.

[MBG$^+$06] J. A. Moriarty, L. X. Benedict, J. N. Glosli, *et al.* Robust quantum-based interatomic potentials for multiscale modeling in transition metals. *J. Mater. Res.*, **21(3)**:563–573, 2006.

[MBR$^+$02] J. A. Moriarty, J. F. Belak, R. E. Rudd, P. Soderlind, F. H. Streitz and L. H. Yang. Quantum-based atomistic simulation of materials properties in transition metals. *J. Phys. Condens. Matter*, **14(11)**:2825–2857, 2002.

[MDF94] M. J. Mills, M. S. Daw and S. M. Foiles. High-resolution transmission electron-microscopy studies of dislocation cores in metals and intermetallic compounds. *Ultramicroscopy*, **56**:79–93, 1994.

[Men00] K. Menger. What is the calculus of variations and what are its applications? In J. Newman, editor, *The World of Mathematics*, Volume 2, Part V, Chapter 8, pages 886–890. Mineola: Dover, 2000.

[Met87] N. Metropolis. The beginning of the Monte Carlo method. *Los Alamos Science*, **15**:125–130, 1987.

[Mey85] H.-D. Meyer. Theory of the Liapunov exponents of Hamiltonian systems and a numerical study on the transition from regular to irregular classical motion. *J. Chem. Phys.*, **84(8)**:3147–3161, 1985.

[MH36] W. E. Morrell and J. H. Hildebrand. The distribution of molecules in a model liquid. *J. Chem. Phys.*, **4(3)**:224–227, 1936.

[MH94] J. E. Marsden and T. J. R. Hughes. *Mathematical Foundations of Elasticity*. New York: Dover, 1994.

[MH00] D. Marx and J. Hutter. Ab initio molecular dynamics: Theory and implementation. In J. Grotendorst, editor, *Modern Methods and Algorithms of Quantum Chemistry*, Volume 1, pages 301–449. Jülich: John von Neumann Institute for Computing, 2000.

[Mis04] Y. Mishin. Atomistic modeling of the γ and γ'-phases of the Ni–Al system. *Acta Mater.*, **52(6)**:1451–1467, 2004.

[MJ94] G. Mills and H. Jónsson. Quantum and thermal effects in H_2 dissociative adsorption – evaluation of free-energy barriers in multidimensional quantum-systems. *Phys. Rev. Lett.*, **72(7)**:1124–1127, 1994.

[MJS95] G. Mills, H. Jónsson, and G. K. Schenter. Reversible work transition-state theory – application to dissociative adsorption of hydrogen. *Surf. Sci.*, **324(2–3)**:305–337, 1995.

[MKT92] G. J. Martyna, M. L. Klein and M. Tuckerman. Nosé–Hoover chains – the canonical ensemble via continuous dynamics. *J. Chem. Phys.*, **97(4)**:2635–2643, 1992.

[MM77] J. E. Mayer and M. G. Mayer. *Statistical Mechanics*. New York: Wiley, second edition, 1977.

[MM81] R. A. MacDonald and W. M. MacDonald. Thermodynamic properties of fcc metals at high temperatures. *Phys. Rev. B*, **24**:1715–1724, 1981.

[MM00] R. Malek and N. Mousseau. Dynamics of Lennard-Jones clusters: A characterization of the activation–relaxation technique. *Phys. Rev. E*, **62(6, Part A)**:7723–7728, 2000.

[MMEG07] M. Mrovec, M. Moseler, C. Elsaesser and P. Gumbsch. Atomistic modeling of hydrocarbon systems using analytic bond-order potentials. *Prog. Mater. Sci.*, **52(2–3)**:230–254, 2007.

[MMP$^+$01] Y. Mishin, M. J. Mehl, D. A. Papaconstantopoulos, A. F. Voter and J. D. Kress. Structural stability and lattice defects in copper: Ab initio, tight-binding, and embedded-atom calculations. *Phys. Rev. B*, **63(22)**:224106, 2001.

[MMP02] Y. Mishin, M. J. Mehl and D. A. Papaconstantopoulos. Embedded-atom potential for B2 – NiAl. *Phys. Rev. B*, **65(22)**, 2002.

[Moo90] D. M. Moody. Unsteady expansion of an ideal gas into a vacuum. *J. Fluid Mech.*, **214**:455–468, 1990.

[Mor29] P. M. Morse. Diatomic molecules according to the wave mechanics. II. Vibrational levels. *Phys. Rev.*, **34(1)**:57–64, 1929.

[Mor88] J. A. Moriarty. Density-functional formulation of the generalized pseudopotential theory: Transition-metal interatomic potentials. *Phys. Rev. B*, **38(5)**:3199–3231, 1988.

[Mor90] J. A. Moriarty. Analytic representation of multiion interatomic potentials in transition-metals. *Phys. Rev. B*, **42(3)**:1609–1628, 1990.

[Mor94] J. A. Moriarty. Angular forces and melting in bcc transition metals: A case study of molybdenum. *Phys. Rev. B*, **49(18)**:12431–12445, 1994.

[Mos62] J. K. Moser. On invariant curves of area-preserving mappings of an annulus. *Nachr. Akad. Wiss. Göttingen, Math.-Phys. Kl. II.*, **1**:1–20, 1962.

[MP76] H. J. Monkhorst and J. D. Pack. Special points for Brillouin-zone integrations. *Phys. Rev. B*, **13(12)**:5188–5192, 1976.

[MP96] R. Miller and R. Phillips. Critical analysis of local constitutive models for slip and decohesion. *Philos. Mag. A*, **73(4)**:803, 1996.

[MPBO96] R. Miller, R. Phillips, G. Beltz and M. Ortiz. A non-local formulation of the Peierls dislocation model. *J. Mech. Phys. Solids*, **46(10)**:1845–68, 1996.

[MPDJ98] M. C. Michelini, R. Pis Diez, and A. H. Jubert. A density functional study of small nickel clusters. *Int. J. Quantum Chem.*, **70(4–5)**:693–701, 1998.

[MPT83] P. S. Marcus, W. H. Press and S. A. Teukolsky. Multiscale model-equations for turbulent convection and convective overshoot. *Astrophys. J.*, **267(2)**:795–821, 1983.

[MQ06] R. I. McLachlan and G. R. W. Quispel. Geometric integrators for ODEs. *J. Phys. A: Math. Gen.*, **39(19)**:5251–5285, 2006.

[MRR$^+$53] N. Metropolis, A. W. Rosenbluth, M. N. Rosenbluth, A. H. Teller and E. Teller. Equation of state calculations by fast computing machines. *J. Chem. Phys.*, **21**:1087–1092, 1953.

[MRT06] S. Morante, G. C. Rossi and M. Testa. The stress tensor of a molecular system: An exercise in statistical mechanics. *J. Chem. Phys.*, **125**:034101, 2006.

[MS00] R. E. Miller and V. B. Shenoy. Size-dependent elastic properties of nanosized structural elements. *Nanotechnology*, **11**:139–147, 2000.

[MS06] G. Martin and F. Soisson. Kinetic Monte Carlo method to model diffusion controlled phase transformations in the solid state. In S. Yip, editor, *Handbook of Materials Modeling*, Part A, *Methods*, Chapter 7.9, pages 2223–2248. New York: Springer Science and Business Media, 2006.

[MS07] R. Maranganti and P. Sharma. Length scales at which classical elasticity breaks down for various materials. *Phys. Rev. Lett.*, **98(19)**, 2007.

[MSV01] F. Montalenti, M. R. Sorensen and A. R. Voter. Closing the gap between experiment and theory: Crystal growth by temperature accelerated dynamics. *Phys. Rev. Lett.*, **87(12)**, 2001.

[MT02] R. E. Miller and E. B. Tadmor. The quasicontinuum method: Overview, applications and current directions. *J. Comput.-Aided Mater. Des.*, **9**:203–239, 2002.

[MT07] R. E. Miller and E. B. Tadmor. Hybrid continuum mechanics and atomistic methods for simulating materials deformation and failure. *MRS Bull.*, **32**:920–926, 2007.

[MT09a] R. E. Miller and E. B. Tadmor. Multiscale benchmark code, `multibench`, 2009. Available at www.qcmethod.org.

[MT09b] R. E. Miller and E. B. Tadmor. A unified framework and performance benchmark of fourteen multiscale atomistic/continuum coupling methods. *Modell. Simul. Mater. Sci. Eng.*, **17**:053001, 2009.

[MTP02] K. L. Merkle, L. J. Thompson and F. Phillipp. Collective effects in grain boundary migration. *Phys. Rev. Lett.*, **88(22)**, 2002.

[MU49] N. Metropolis and S. Ulam. The Monte Carlo method. *J. Am. Stat. Assoc.*, **44**:335–341, 1949.

[Mur83] A. I. Murdoch. The motivation of continuum concepts and relations from discrete considerations. *Q. J. Mech. Appl. Math.*, **36**:163–187, 1983.

[Mur85] A. I. Murdoch. A corpuscular approach to continuum-mechanics: Basic considerations. *Arch. Ration. Mech. Anal.*, **88**:291–321, 1985.

[Mur03] A. I. Murdoch. On the microscopic interpretation of stress and couple stress. *J. Elast.*, **71**:105–131, 2003.

[Mur07] A. I. Murdoch. A critique of atomistic definitions of the stress tensor. *J. Elast.*, **88**:113–140, 2007.

[MV01] F. Montalenti and A. F. Voter. Applying accelerated molecular dynamics to crystal growth. *Phys. Stat. Sol. B*, **226(1)**:21–27, 2001. 2nd Motorola Workshop on Computational Materials and Electronics, Tempe, AZ, Nov. 09–10, 2000.

[MV02] F. Montalenti and A. F. Voter. Exploiting past visits or minimum-barrier knowledge to gain further boost in the temperature-accelerated dynamics method. *J. Chem. Phys.*, **116(12)**:4819–4828, 2002.

[MVH$^+$10] J. Marian, G. Venturini, B. L. Hansen, *et al.* Finite-temperature extension of the quasicontinuum method using Langevin dynamics: Entropy losses and analysis of errors. *Modell. Simul. Mater. Sci. Eng.*, **18(1)**, 2010.

[MW97] J. A. Moriarty and M. Widom. First-principles interatomic potentials for transition-metal aluminides: Theory and trends across the 3d series. *Phys. Rev. B*, **56(13)**:7905–7917, 1997.

[MW99] L. J. Munro and D. J. Wales. Defect migration in crystalline silicon. *Phys. Rev. B*, **59(6)**:3969–3980, 1999.

[NA95] F. R. N. Nabarro and A. S. Argon. Egon Orowan. 2 August 1902–3 August 1989. *Biographical Memoirs of Fellows of the Royal Society*, **41**:317–340, 1995.

[Nab47] F. R. N. Nabarro. Dislocations in a simple cubic lattice. *Proc. Phys. Soc. London*, **59**:256, 1947.

[Nab87] F. R. N. Nabarro. *Theory of Crystal Dislocations*. New York: Dover Books on Physics and Chemistry, 1987.

[New30] I. Newton. *Opticks: or a Treatise on the Reflexions, Refractions, Inflexions and Colours of Light*. London, fourth english edition. William Inngs, 1730.

[NKN+07] A. Nakano, R. K. Kalia, K. Nomura, *et al.* A divide-and-conquer/cellular-decomposition framework for million-to-billion atom simulations of chemical reactions. *Comput. Mater. Sci.*, **38(4)**:642–652, 2007.

[NL80] J. K. Norskøv and N. D. Lang. Effective-medium theory of chemical binding: Application to chemisorption. *Phys. Rev. B*, **21(6)**:2131–2136, 1980.

[NM83] O. H. Nielsen and R. M. Martin. 1st-principles calculation of stress. *Phys. Rev. Lett.*, **50(9)**:697–700, 1983.

[NM85] O. H. Nielsen and R. M. Martin. Quantum-mechanical theory of stress and force. *Phys. Rev. B*, **32(6)**:3780–3791, 1985.

[NM02] R. J. Needs and A. Mujica. Theoretical description of high-pressure phases of semiconductors. *High Pressure Res.*, **22**(2, Sp. Iss. SI):421–427, 2002.

[NMVH07] D. Nguyen-Manh, V. Vitek, and A. P. Horsfield. Environmental dependence of bonding: A challenge for modelling of intermetallics and fusion materials. *Prog. Mater. Sci.*, **52(2–3)**:255–298, 2007.

[Nol55] W. Noll. Die herleitung der grundgleichungen der thermomechanik der kontinua aus der statischen mechanik. *J. Ration. Mech. Anal.*, **4**:627–646, 1955.

[Nos84] S. Nosé. A molecular-dynamics method for simulations in the canonical ensemble. *Mol. Phys.*, **52(2)**:255–268, 1984.

[NPM90] R. M. Nieminen, M. J. Puska, and M. J. Manninen, editors. *Many-Atom Interactions in Solids*, Volume 48 of Proceedings in Physics. Berlin: Springer-Verlag, 1990.

[NRL09] NRL. Crystal lattice structures website. http://cst-www.nrl.navy.mil/lattice/index.html, 2009.

[NW99] J. Nocedal and S. J. Wright. *Numerical Optimization*. New York: Springer Verlag, 1999.

[Ogd84] R. W. Ogden. *Non-linear Elastic Deformations*. Chichester: Ellis Horwood, 1984.

[Omn99] R. Omnès. *Understanding Quantum Mechanics*. Princeton: Princeton University Press, 1999.

[O'Ro98] J. O'Rourke. *Computational Geometry in C*. Cambridge: Cambridge University Press, second edition, 1998.

[Oro34a] E. Orowan. Zur kristallplastizität. I Tieftemperaturplastizitt und beckersche formel. *Z. Phys.*, **89(9–10)**:605–613, 1934.

[Oro34b] E. Orowan. Zur kristallplastizität. II Die dynamische auffassung der kristallplastizität. *Z. Phys.*, **89(9–10)**:614–633, 1934.

[Oro34c] E. Orowan. Zur kristallplastizität. III Über den mechanismus des gleitvorganges. *Z. Phys.*, **89(9–10)**:634–659, 1934.

[Oro44] E. Orowan. Discussion of the significance of tensile and other mechanical test properties of metals. In *Proceedings of the Institute of Mechanical Engineers*, Volume 151, pages 131–146, London: Institute of Mechanical Engineers, 1944.

[OSC04] W. K. Olson, D. Swigon, and B. D. Coleman. Implications of the dependence of the elastic properties of DNA on nucleotide sequence. *Philos. Trans. R. Soc. London, Ser. A*, **362(1820)**:1403–1422, 2004.

[Par76] G. P. Parry. On the elasticity of monatomic crystals. *Math. Proc. Camb. Phil. Soc.*, **80**:189–211, 1976.

[Par04] G. P. Parry. On essential and non-essential descriptions of multilattices. *Math. Mech. Solids*, **9**:411–418, 2004.

[PB05] C. W. Padgett and D. W. Brenner. A continuum-atomistic method for incorporating Joule heating into classical molecular dynamics simulations. *Mol. Simul.*, **31(11)**:749–757, 2005.

[PBL08] M. L. Parks, P. B. Bochev and R. B. Lehoucq. Connecting atomistic-to-continuum coupling and domain decomposition. *Multiscale Model. Simul.*, **7(1)**:362–380, 2008.

[PBO06] S. Prudhomme, P. T. Bauman and J. T. Oden. Error control for molecular statics problems. *Int. J. Multiscale Comput. Eng.*, **4(5–6)**:647–662, 2006.

[PBS66] O. G. Peterson, D. N. Batchelder and R. O. Simmons. Measurements of x-ray lattice constant, thermal expansivity, and isothermal compressibility of argon crystals. *Phys. Rev.*, **150(2)**:703–711, 1966.

[PC03] J. W. Ponder and D. A. Case. Force fields for protein simulations. In Protein Simulations, volume 66 of *Advances in protein chemistry*, pages 27–85. San Diego: Academic Press Inc 2003.

[PDSK07] M. Praprotnik, L. Delle Site and K. Kremer. A macromolecule in a solvent: Adaptive resolution molecular dynamics simulation. *J. Chem. Phys.*, **126(13)**, 2007.

[PDSK08] M. Praprotnik, L. Delle Site and K. Kremer. Multiscale simulation of soft matter: From scale bridging to adaptive resolution. *Annu. Rev. Phys. Chem.*, **59**:545–571, 2008.

[Pei40] R. E. Peierls. The size of a dislocation. *Proc. Phys. Soc. London*, **52**:34, 1940.

[Pei64] R. E. Peierls. *Quantum Theory of Solids*. Oxford: Oxford University Press, second edition, 1964.

[Pen79] O. Penrose. Foundations of statistical mechanics. *Rep. Prog. Phys.*, **42**:1937–2006, 1979.

[Pen99] R. Penrose. *The Emperor's New Mind*. Oxford: Oxford University Press, 1999.

[Pen02] O. Penrose. Statistical mechanics of nonlinear elasticity. *Markov Processes and Related Fields*, **8**:351–364, 2002. Available online at www.ma.hw.ac.uk/∼oliver/.

[Pen05] O. Penrose. *Foundations of Statistical Mechanics: A Deductive Treatment*. Mineola: Dover Publications, 2005.

[Pen06] O. Penrose. Correction to 'statistical mechanics of nonlinear elasticity'. *Markov Processes and Related Fields*, **12**:169, 2006. Available online at www.ma.hw.ac.uk/~oliver/.

[Pet89] D. G. Pettifor. New many-body potential for the bond order. *Phys. Rev. Lett.*, **63(22)**:2480–2483, 1989.

[Pet95] D. G. Pettifor. *Bonding and Structure of Molecules and Solids*. Oxford: Oxford University Press, 1995.

[Phi01] R. Phillips. *Crystals, Defects and Microstructures*. Cambridge: Cambridge University Press, 2001.

[Pit85] M. Pitteri. On $\nu + 1$-lattices. *J. Elast.*, **15**:3–25, 1985.

[Pit86] M. Pitteri. Continuum equuations of balance in classical statistical-mechanics. *Arch. Ration. Mech. Anal.*, **94**:291–305, 1986.

[Pit90] M. Pitteri. On a statistical-kinetic model for generalized continua. *Arch. Ration. Mech. Anal.*, **111**:99–120, 1990.

[Pit98] M. Pitteri. Geometry and symmetry of multilattices. *Int. J. Plast.*, **14**:139–157, 1998.

[PK07] H. S. Park and P. A. Klein. Surface Cauchy–Born analysis of surface stress effects on metallic nanowires. *Phys. Rev. B*, **75(8)**, 085408, 2007.

[PK08] H. S. Park and P. A. Klein. Surface stress effects on the resonant properties of metal nanowires: The importance of finite deformation kinematics and the impact of the residual surface stress. *J. Mech. Phys. Solids*, **56(11)**:3144–3166, 2008.

[PKW06] H. S. Park, P. A. Klein and G. J. Wagner. A surface Cauchy–Born model for nanoscale materials. *Int. J. Numer. Methods Eng.*, **68(10)**:1072–1095, 2006.

[PL71] O. Penrose and J. L. Lebowitz. Rigorous treatment of metastable states in the van der Waals–Maxwell theory. *J. Stat. Phys.*, **3(2)**:211–236, 1971.

[PL04] H. S. Park and W. K. Liu. An introduction and tutorial on multiple-scale analysis in solids. *Comput. Meth. Appl. Mech. Eng.*, **193**:1733–1772, 2004.

[Pla01] M. Planck. Ueber das Gesetz der Energieverteilung im Normalspectrum. *Ann. der Phys.*, **309**:553–563, 1901.

[Pla13] M. Plancherel. Beweis der Unmöglichkeit ergödischer mechanischer Systeme. *Ann. der Phys.*, **42**:1061–1163, 1913.

[Pla20] M. Planck. The genesis and present state of development of the quantum theory (Nobel lecture). http://nobelprize.org/nobel_prizes/physics/laureates/1918/planck-lecture.html, June 2, 1920.

[PLBK01] R. Plass, J. A. Last, N. C. Bartelt and G. L. Kellogg. Nanostructures – self-assembled domain patterns. *Nature*, **412(6850)**:875, 2001.

[Pli95] S. J. Plimpton. Fast parallel algorithms for short-range molecular dynamics. *J. Comput. Phys.*, **117**:1–19, 1995.

[Pli09] S. J. Plimpton. LAMMPS website. http://lammps.sandia.gov, 2009.

[PMC$^+$95] R. O. Piltz, J. R. Maclean, S. J. Clark, *et al.* Structure and properties of silicon XII: A complex tetrahedrally bonded phase. *Phys. Rev. B*, **52(6)**:4072–4085, 1995.

[PMKW78] P. Pierański, J. Malecki, W. Kuczyński and K. Wojciechowski. A hard-disc system, an experimental model. *Philos. Mag. A*, **37**:107–115, 1978.

[PMP01] D. N. Pawaskar, R. Miller, and R. Phillips. Structure and energetics of long-period tilt grain boundaries using an effective hamiltonian. *Phys. Rev. B*, **63**:214105–214118, 2001.

[PMRC90] B. Pouligny, R. Malzbender, P. Ryan and N. A. Clark. Analog simulation of melting in two dimensions. *Phys. Rev. B*, **42(1)**:988–991, 1990.

[PNN10] PNNL. NWChem website. http://www.emsl.pnl.gov/capabilities/computing/nwchem, 2010.

[PO04] D. G. Pettifor and I. I. Oleynik. Interatomic bond-order potentials and structural prediction. *Prog. Mater. Sci.*, **49(3–4)**:285–312, 2004.

[Pol34] M. Polanyi. Über eine Art Gitterstörung, die einen Kristall plastisch machen könnte. *Z. Phys.*, **89(9–10)**:660–664, 1934.

[Pol71] E. Polak. *Computational Methods in Optimization: A Unified Approach*, volume 77 of *Mathematics in Science and Engineering*. Academic Press, New York, 1971.

[PR81] M. Parrinello and A. Rahman. Polymorphic transitions in single crystals: A new molecular dynamics method. *J. Appl. Phys.*, **52(12)**:7182–7190, 1981.

[PR82] M. Parrinello and A. Rahman. Strain fluctuations and elastic-constants. *J. Chem. Phys.*, **76(5)**:2662–2666, 1982.

[PRTT05] J. M. Porta, L. Ros, F. Thomas and C. Torras. A branch-and-prune solver for distance constraints. *IEEE Trans. Rob.*, **21(2)**:176–187, 2005.

[PTA$^+$92] M. C. Payne, M. P. Teter, D. C. Allan, T. A. Arias and J. D. Joannopoulos. Iterative minimization techniques for ab initio total energy calculations: Molecular dyanmics and conjugate gradients. *Rev. Mod. Phys.*, **64**:1045–1097, 1992.

[PTVF92] W. H. Press, S. A. Teukolsky, W. T. Vetterling and B. P. Flannery. *Numerical Recipes in FORTRAN: The Art of Scientific Computing*. Cambridge: Cambridge University Press, second edition, 1992.

[PTVF08] W. H. Press, S. A. Teukolsky, W. T. Vetterling and B. P. Flannery. Numerical recipes: The art of scientific computing. http://www.nr.com, 2008.

[PW92] J. P. Perdew and Y. Wang. Accurate and simple analytic representation of the electron-gas correlation energy. *Phys. Rev. B*, **45(23)**:13244–13249, 1992.

[PZ81] J. P. Perdew and A. Zunger. Self-interaction correction to density-functional approximations for many-electron systems. *Phys. Rev. B*, **23(10)**:5048–5079, 1981.

[PZ00] M. Pitteri and G. Zanzotto. *Continuum Models for Phase Transitions and Twinning in Crystals*. London: CRC/Chapman and Hall, 2000.

[QSCM05] S. Qu, V. Shastry, W. A. Curtin and R. E. Miller. A finite temperature, dynamic, coupled atomistic/discrete dislocation method. *Modell. Simul. Mater. Sci. Eng.*, **13(7)**:1101–1118, 2005.

[QWL04] D. Qian, G. J. Wagner and W. K. Liu. A multiscale projection method for the analysis of carbon nanotubes. *Comput. Meth. Appl. Mech. Eng.*, **193**:1603–32, 2004.

[Rad87] C. Radin. Low temperature and the origin of crystalline symmetry. *Int. J. Mod. Phys. B*, **1(5 & 6)**:1157–1191, 1987.

[Rah64] A. Rahman. Correlations in the motion of atoms in liquid argon. *Phys. Rev.*, **136(2A)**:A405–A411, 1964.

[Rap95] D. C. Rapaport. *The Art of Molecular Dynamics Simulation*. New York: Cambridge University Press, 1995.

[Ray82] J. R. Ray. Fluctuations and thermodynamics properties of anisotropic solids. *J. Appl. Phys.*, **53(9)**:6441–6443, 1982.

[Ray83] J. R. Ray. Molecular-dynamics equations of motion for systems varying in shape and size. *J. Chem. Phys.*, **79(10)**:5128–5130, 1983.

[RB94] J. R. Rice and G. E. Beltz. The activation energy for dislocation nucleation at a crack. *J. Mech. Phys. Solids*, **42**:333–360, 1994.

[RB98] R. E. Rudd and J. Q. Broughton. Coarse-grained molecular dynamics and the atomic limit of finite elements. *Phys. Rev. B*, **58(10)**:R5893–R5896, 1998.

[RB99] R. E. Rudd and J. Q. Broughton. Atomistic simulation of MEMS resonators through the coupling of length scales. *J. Model. Simul. Microsys.*, **1(1)**:29–38, 1999.

[RB00] R. E. Rudd and J. Q. Broughton. Concurrent coupling of length scales in solid state systems. *Phys. Stat. Sol. B*, **217**:251–291, 2000.

[RB05] R. E. Rudd and J. Q. Broughton. Coarse-grained molecular dynamics: Nonlinear finite elements and finite temperature. *Phys. Rev. B*, **72(14)**:144104, 2005.

[RD91] T. J. Raeker and A. E. DePristo. Theory of chemical bonding based on the atom-homogeneous electron-gas system. *Int. Rev. Phys. Chem.*, **10(1)**:1–54, 1991.

[Ref00] K. Refson. MOLDY: a portable molecular dynamics simulation program for serial and parallel computers. *Comput. Phys. Commun.*, **126(3)**:310–329, 2000.

[Rei64] M. Reiner. The Deborah number. *Phys. Today*, **17**:62, 1964.

[Rei85] F. Reif. *Fundamentals of Statistical and Thermal Physics*. Singapore: McGraw-Hill, international edition, 1985.

[RFA02] M. Ruda, D. Farkas and J. Abriata. Interatomic potentials for carbon interstitials in metals and intermetallics. *Sci. Mater.*, **46(5)**:349–355, 2002.

[RFG09] M. Ruda, D. Farkas and G. Garcia. Atomistic simulations in the Fe–C system. *Comput. Mater. Sci.*, **45(2)**:550–560, 2009.

[RFS81] J. H. Rose, J. Ferrante and J. R. Smith. Universal binding energy curves for metals and bimetallic interfaces. *Phys. Rev. Lett.*, **47**:675–678, 1981.

[RG81] J. R. Ray and H. W. Graben. Direct calculation of fluctuation formulae in the microcanonical ensemble. *Mol. Phys.*, **43(6)**:1293–1297, 1981.

[RGH81] J. R. Ray, H. W. Graben, and J. M. Haile. A new adiabatic ensemble with particle fluctuations. *J. Chem. Phys.*, **75(8)**:4077–4079, 1981.

[Ric92] J. R. Rice. Dislocation nucleation from a crack tip: An analysis based on the Peierls concept. *J. Mech. Phys. Solids*, **40**:239, 1992.

[RL02] J. M. Rickman and R. LeSar. Free-energy calculations in materials research. *Annu. Rev. Mater. Res.*, **32**:195–217, 2002.

[Ros13] A. Rosenthal. Beweis der unmöglichkeit ergödischer gasssystemse. *Ann. der Phys.*, **42**:796–806, 1913.

[RPFS00] K. Rościszewski, B. Paulus, P. Fulde, and H. Stoll. Ab initio coupled-cluster calculations for the fcc and hcp structures of rare-gas solids. *Phys. Rev. B*, **62(9)**:5482–5488, 2000.

[RPW$^+$91] J. M. Rickman, S. R. Phillpot, D. Wolf, D. L. Woodraska and S. Yip. On the mechanism of grain-boundary migration in metals – a molecular-dynamics study. *J. Mater. Res.*, **6(11)**:2291–2304, 1991.

[RR84] J. R. Ray and A. Rahman. Statistical ensembles and molecular-dynamics studies of anisotropic solids. *J. Chem. Phys.*, **80(9)**:4423–4428, 1984.

[RS96] J. D. Rittner and D. N. Seidman. $< 110 >$ symmetric tilt grain-boundary structures in fcc metals with low stacking-fault energies. *Phys. Rev. B*, **54(10)**:6999–7015, 1996.

[RSF70] R. A. Rege, E. S. Szekeres and W. D. Forgeng. 3-dimensional view of alumina clusters in aluminum-killed low-carbon steel. *Metall. Trans.*, **1(9)**:2652–&, 1970.

[RSGF84] J. H. Rose, J. R. Smith, F. Guinea and J. Ferrante. Universal features of the equation of state of metals. *Phys. Rev. B*, **29(6)**:2963–2969, 1984.

[Rub00] M. B. Rubin. *Cosserat Theories: Shells, Rods and Points*, Volume 79 of Solid Mechanics and its Applications. Dordrecht: Kluwer, 2000.

[Rud01] R. E. Rudd. Concurrent multiscale modeling of embedded nanomechanics. In V. Bulatov, F. Cleri, L. Colombo, L. Lewis, and N. Mousseau, editors, *Advances in Materials Theory and Modeling – Bridging Over Multiple-Length and Time Scales, Mater. Res. Soc. Symp. Proc.*, Vol. 677, pages AA1.6.1–AA1.6.12. Warrendale: Materials Research Society, 2001.

[Rue63] D. Ruelle. Classical statistical mechanics of a system of particles. *Helv. Phys. Acta*, **36**:183–197, 1963.

[Rue69] D. Ruelle. *Statistical Mechanics: Rigorous Results*. Reading: Benjamin, 1969.

[Rue99] D. Ruelle. Smooth dynamics and new theoretical ideas in nonequilibrium statistical mechanics. *J. Stat. Phys.*, **95**:393–468, 1999.

[Rus06] A. Ruszczyński. *Nonlinear Optimization*. Princeton: Princeton University Press, 2006.

[SABW82] W. C. Swope, H. C. Andersen, P. H. Berens and K. R. Wilson. A computer simulation method for the calculation of equilibrium constants for the formation of physical clusters of molecules: application to small water clusters. *J. Chem. Phys.*, **76**:637–649, 1982.

[SAG$^+$02] J. M. Soler, E. Artacho, J. D. Gale, *et al.* The SIESTA method for ab initio order-*N* materials simulation. *J. Phys. Condens. Matter*, **14(11)**:2745–2779, 2002.

[Sak94] J. J. Sakurai. *Modern Quantum Mechanics*. Reading: Addison-Wesley, revised edition, 1994.

[Sal01] J. Salençon. *Handbook of Continuum Mechanics: General Concepts, Thermoelasticity*. Berlin: Springer, 2001.

[San93] D. E. Sands. *Introdcution to Crystallography*. Mineola: Dover, 1993.

[SB06] A. P. Sutton and R. W. Balluffi. *Interfaces in Crystalline Materials*. Oxford: Oxford University Press, 2006.

[SBR93] Y. Sun, G. E. Beltz and J. R. Rice. Estimates from atomic models of tension-shear coupling in dislocation nucleation from a crack tip. *Mater. Sci. Eng. A*, **170**:67–85, 1993.

[Sch26] E. Schrödinger. An undulatory theory of the mechanics of atoms and molecules. *Phys. Rev.*, **28(6)**:1049–1070, 1926.

[Sch44] E. Schrödinger. *What is Life. The Physical Aspect of the Living Cell.* Cambridge: Cambridge University Press, 1944.

[Sch76] L. A. Schwalbe. Equilibrium vacancy concentration measurements in solid argon. *Phys. Rev. B*, **14(4)**:1722–1732, 1976.

[Sch83a] R. Schiestel. Multiple scale concept in turbulence modeling. 1. Multiple-scale model for turbulence kinetic-energy and mean-square of passive scalar fluctuations. *J. Mec. Theor. Appl.*, **2(3)**:417–449, 1983.

[Sch83b] E. Schrödinger. *My View of the World*. Woodbridge: Ox Bow Press, 1983.

[SDG84] J. M. Sanchez, F. Ducastelle and D. Gratias. Generalized cluster description of multicomponent systems. *Physica A*, **128(1–2)**:334–350, 1984.

[SDKF08] R. Stote, A. Dejaegere, D. Kuznetsov and L. Falquet. CHARMM molecular dynamics simulation tutorial. http://www.ch.embnet.org/MD-tutorial, 2008.

[SdVG$^+$03] P. Sherwood, A. H. de Vries, M. F. Guest, *et al.* QUASI: A general purpose implementation of the QM/MM approach and its application to problems in catalysis. *J. Mol. Struct.*, **632**(Sp. Iss. SI):1–28, 2003.

[Seq09] SeqQuest. SeqQuest website. http://dft.sandia.gov/Quest, 2009.

[SES07] P. Steinmann, A. Elizondo and R. Sunyk. Studies of validity of the Cauchy Born rule by direct comparison of continuum and atomistic modelling. *Modell. Simul. Mater. Sci. Eng.*, **15**:5271–5281, 2007.

[Set06] J. P. Sethna. *Statistical Mechanics: Entropy, Order Parameters, and Complexity*. Oxford: Oxford University Press, 2006.

[SET11] V. Sorkin, R. S. Elliott, and E. B. Tadmor. A local quasicontinuum for 3D multilattice crystalline materials: Application to shape-memory alloys. Phys. Rev. B, 2011, submitted.

[Sew80] G. L. Sewell. Stability, equilibrium and metastability in statistical mechanics. *Phys. Rep.*, **57(5)**:307–342, 1980.

[SGK$^+$06] P. Schwerdtfeger, N. Gaston, R. P. Krawczyk, R. Tonner and G. E. Moyano. Extension of the Lennard–Jones potential: Theoretical investigations into rare-gas clusters and crystal lattices of He, Ne, Ar, and Kr using many-body interaction expansions. *Phys. Rev. B*, **73(6)**:064112, 2006.

[SH82] P. Schofield and J. R. Henderson. Statistical mechanics of inhomogeneous fluids. *Proc. R. Soc. London, Ser. A*, **379(1776)**:231–246, 1982.

[SH86] A. M. Stoneham and J. H. Harding. Interatomic potentials in solid state chemistry. *Annu. Rev. Phys. Chem.*, **37**:52–80, 1986.

[Sha03] T. Shardlow. Splitting for dissipative particle dynamics. *SIAM J. Sci. Comput.*, **24(4)**:1267–1282, 2003.

[She94] J. R. Shewchuk. An introduction to the conjugate gradient method without the agonizing pain. www.cs.cmu.edu/~quake-papers/painless-conjugate-gradient.pdf, 1994.

[SHH69] D. R. Squire, A. C. Holt and W. G. Hoover. Isothermal elastic constants for argon. Theory and Monte Carlo calculations. *Physica*, **42**:388–397, 1969.

[Shi96] T. W. Shield. An experimental study of the plastic strain fields near a notch tip in a copper single crystal during loading. *Acta Mater.*, **44(4)**:1547–1561, 1996.

[Sho48] W. Shockley. Minutes of the meeting at Chicago, Illinois December 29-31, 1947. *Phys. Rev.*, **73(10)**:1217–1236, 1948. ("Half Dislocations", page 1232).

[SIE10] SIESTA. SIESTA website. http://www.icmab.es/siesta, 2010.

[Sil02] S. A. Silling. The reformulation of elasiticity theory for discontinuities and long-range forces. *J. Mech. Phys. Solids*, **48**:175–209, 2002.

[Sin63] Ya. G. Sinai. On the foundations of the ergodic hypothesis for a dynamical system of statistical mechanics. *Dokl. Akad. Nauk SSSR*, **153**:1261–1264, 1963. English translation in *Sov. Math.-Dokl*, **4**:1818–1822, 1964.

[SK54] J.C. Slater and G. F. Koster. Simplified LCAO method for the periodic potential problem. *Phys. Rev.*, **94(6)**:1498–1524, 1954.

[SK08] E. Salomon and A. Kahn. One-dimensional organic nanostructures: A novel approach based on the selective adsorption of organic molecules on silicon nanowires. *Surf. Sci.*, **602(13)**:L79–L83, 2008.

[Skl93] L. Sklar. *Physics and Chance: Philosophical Issues in the Foundations of Statistical Mechanics*. Cambridge: Cambridge University Press, 1993.

[SL08] S. Suresh and J. Li. Materials science: deformation of the ultra-strong. *Nature*, **456**:716–717, 2008.

[SM66] A. J. F. Siegert and E. Meeron. Generalizations of the virial and Wall theorems in classical statistical mechanics. *J. Math. Phys*, **7**:741–750, 1966.

[SM03] A. Suzuki and Y. Mishin. Interaction of point defects with grain boundaries in fcc metals. *Interface Sci.*, **11(4)**:425–437, 2003.

[SM05] F. Sansoz and J. F. Molinari. Mechanical behavior of sigma tilt grain boundaries in nanoscale Cu and Al: A quasicontinuum study. *Acta Mater.*, **53**:1931–1944, 2005.

[SMC02] L. E. Shilkrot, R. E. Miller, and W. A. Curtin. Coupled atomistic and discrete dislocation plasticity. *Phys. Rev. Lett.*, **89(2)**:025501, 2002.

[SMC04] L. E. Shilkrot, R. E. Miller, and W. A. Curtin. Multiscale plasticity modeling: Coupled atomistic and discrete dislocation mechanics. *J. Mech. Phys. Solids*, **52(4)**:755–787, 2004.

[SMC05] B. Shiari, R. E. Miller, and W. A. Curtin. Coupled atomistic/discrete dislocation simulations of nanoindentation at finite temperature. *J. Eng. Mater. Technol., Trans. ASME*, **127(4)**:358–368, 2005.

[SMK07] B. Shiari, R. E. Miller and D. D. Klug. Multiscale simulation of material removal processes at the nanoscale. *J. Mech. Phys. Solids*, **55(11)**:2384–2405, 2007.

[SMSJ04] T. Shimokawa, J. J. Mortensen, J. Schiøtz and K. W. Jacobsen. Matching conditions in the quasicontinuum method: Removal of the error introduced at the interface between the coarse-grained and fully atomistic region. *Phys. Rev. B*, **69**:214104/1–10, 2004.

[SMT+98] V. B. Shenoy, R. Miller, E. B. Tadmor, R. Phillips and M. Ortiz. Quasicontinuum models of interfacial structure and deformation. *Phys. Rev. Lett.*, **80(4)**:742–745, 1998.

[SMT+99] V. B. Shenoy, R. Miller, E. B. Tadmor, D. Rodney, R. Phillips and M. Ortiz. An adaptive methodology for atomic scale mechanics: The quasicontinuum method. *J. Mech. Phys. Solids*, **47**:611–642, 1999.

[SMV00] M. R. Sorensen, Y. Mishin and A. F. Voter. Diffusion mechanisms in Cu grain boundaries. *Phys. Rev. B*, **62(6)**:3658–3673, 2000.

[SOL07] F. Shimizu, S. Ogata and J. Li. Theory of shear banding in metallic glasses and molecular dynamics calculations. *Mater. Trans.*, **48**:2923–2927, 2007.

[SPT00] V. B. Shenoy, R. Phillips, and E. B. Tadmor. Nucleation of dislocations beneath a plane strain indenter. *J. Mech. Phys. Solids*, **48**:649–673, 2000.

[SR74] F. H. Stillinger and A. Rahman. Improved simulation of liquid water by molecular dynamics. *J. Chem. Phys.*, **60(4)**:1545–1557, 1974.

[SS86] M. J. Sippl and H. A. Scheraga. Cayley–Menger coordinates. *Proc. Nat. Acad. Sci. USA*, **83**:2283–2287, 1986.

[SS92] J. R. Smith and D. J. Srolovitz. Developing potentials for atomistic simulations. *Modell. Simul. Mater. Sci. Eng.*, **1**:101–109, 1992.

[SS03] R. Sunyk and P. Steinmann. On higher gradients in continuum-atomistic modelling. *Int. J. Solids Struct.*, **40**:6877–96, 2003.

[SSH+03] E. J. Shin, B. S. Seong, Y. S. Han, K. P. Hong, C. H. Lee and H. J. Kang. Effect of precipitate size and dispersion on recrystallization behavior in Ti-added ultra low carbon steels. *J. Appl. Crystallogr.*, **36(3 Part 1)**:624–628, 2003.

[SSP99] V. Shenoy, V. Shenoy and R. Phillips. Finite temperature quasicontinuum methods. In T. Diaz de la Rubia, E. Kaxiras, V. Bulatov, N. M. Ghoniem and R. Phillips, editors, *Multiscale Modelling of Materials*, Mater. Res. Soc. Symp. Proc., Vol. 538, pages 465–471. Warrendale: Materials Research Society, 1999.

[ST07] H. M. Senn and W. Thiel. QM/MM methods for biological systems. In *Atomistic Approaches in Modern Biology: from Quantum Chemistry to Molecular Simulations*, Volume 268 of Topics in Current Chemistry, pages 173–290. Berlin: Springer, 2007.

[Sta50] I. Stackgold. The Cauchy relations in molecular theory of elasticity. *Q. Appl. Math.*, **8**:169–186, 1950.

[STBK01] G. S. Smith, E. B. Tadmor, N. Bernstein and E. Kaxiras. Multiscale simulations of silicon nanoindentation. *Acta Mater.*, **49(19)**:4089–4101, 2001.

[Sti95] F. H. Stillinger. Statistical mechanics of metastable matter: superheated and stretched liquids. *Phys. Rev. E*, **52(5)**:4685–4690, 1995.

[STK00] G. S. Smith, E. B. Tadmor, and E. Kaxiras. Multiscale simulation of loading and electrical resistance in silicon nanoindentation. *Phys. Rev. Lett.*, **84(6)**:1260–1263, 2000.

[Sut91a] W. Sutherland. A kinetic theory of solids, with an experimental introduction. *Philos. Mag.*, **32(194)**:31–43, 1891.

[Sut91b] W. Sutherland. A kinetic theory of solids, with an experimental introduction. *Philos. Mag.*, **32(195)**:215–225, 1891.

[Sut91c] W. Sutherland. A kinetic theory of solids, with an experimental introduction. *Philos. Mag.*, **32(199)**:524–553, 1891.

[Sut92] A. P. Sutton, Direct free energy minimization methods: application to grain boundaries, *Phil. Trans. R. Soc. London A*, **341**:233–245, 1992.

[SV00] M. R. Sørensen and A. F. Voter. Temperature-accelerated dynamics for simulation of infrequent events. *J. Chem. Phys.*, **112**:9599–9606, 2000.

[SvDL$^+$09] E. Salmon, A. C. T. van Duin, F. Lorant, P.-M. Marquaire and W. A. Goddard, III. Thermal decomposition process in algaenan of *Botryococcus braunii* race l. part 2: Molecular dynamics simulations using the ReaxFF reactive force field. *Org. Geochem.*, **40(3)**:416–427, 2009.

[SW82] F. H. Stillinger and T. A. Weber. Hidden structure in liquids. *Phys. Rev. A*, **25(2)**:978–989, 1982.

[SW83] F. H. Stillinger and T. A. Weber. Dynamics of structural transitions in liquids. *Phys. Rev. A*, **28(4)**:2408–2416, 1983.

[SW84] F. H. Stillinger and T. A. Weber. Packing structures and transitions in liquids and solids. *Science*, **225(4666)**:983–989, 1984.

[SW85] F. H. Stillinger and T. A. Weber. Computer-simulation of local order in condensed phases of silicon. *Phys. Rev. B*, **31(8)**:5262–5271, 1985.

[SW03] A. Shurki and A. Warshel. Structure/function correlations of proteins using MM, QM/MM, and related approaches: Methods, concepts, pitfalls, and current progress. In *Protein Simulations*, Volume 66 of Advances in Protein Chemistry, pages 249–313. San Diego: Elsevier, 2003.

[SWPF97] B. Schonfelder, D. Wolf, S. R. Phillpot and M. Furtkamp. Molecular-dynamics method for the simulation of grain-boundary migration. *Interface Sci.*, **5(4)**:245–262, 1997.

[SWSW06] Z. H. Sun, X. X. Wang, A. K. Soh and H. A. Wu. On stress calculations in atomistic simulations. *Modell. Simul. Mater. Sci. Eng.*, **14**:423–431, 2006.

[SZ03] D. Shilo and E. Zolotoyabko. Stroboscopic x-ray imaging of vibrating dislocations excited by 0.58 GHz phonons. *Phys. Rev. Lett.*, **91(11)**, 2003.

[Sza93] D. Szasz. Ergodicity of classical billiard balls. *Physica A*, **194**:86–92, 1993.

[TA86] N. Triantafyllidis and E. C. Aifantis. A gradient approach to localization of deformation. 1. Hyperelastic materials. *J. Elast.*, **16**:225–237, 1986.

[Tad96] E. B. Tadmor. The quasicontinuum method. PhD thesis, Brown University, 1996.

[Tay34a] G. I. Taylor. The mechanism of plastic deformation of crystals. Part I. Theoretical. *Proc. R. Soc. London, Ser. A*, **145(855)**:362–387, 1934.

[Tay34b] G. I. Taylor. The mechanism of plastic deformation of crystals. part II. comparison with observations. *Proc. R. Soc. London, Ser. A*, **145(855)**:388–404, 1934.

[Tay99] C. C. W. Taylor. *The Atomists: Leucippus and Democritus: Fragments, a Text and Translation with a Commentary*. Toronto: University of Toronto Press, 1999.

[TC60] D. Turnbull and R. L. Cormia. A dynamic hard sphere model. *J. Appl. Phys.*, **31(4)**:674–678, 1960.

[Ter86] J. Tersoff. New empirical model for the structural properties of silicon. *Phys. Rev. Lett.*, **56(6)**:632–635, 1986.

[Ter88a] J. Tersoff. Empirical interatomic potential for silicon with improved elastic properties. *Phys. Rev. B*, **38(14)**:9902–9905, 1988.

[Ter88b] J. Tersoff. New empirical approach for the structure and energy of covalent systems. *Phys. Rev. B*, **37(12)**:6991–7000, 1988.

[Ter89] J. Tersoff. Modeling solid-state chemistry: Interatomic potentials for multicomponent systems. *Phys. Rev. B*, **39(8)**:5566–5568, 1989.

[TESM09] E. B Tadmor, R. S. Elliott, J. P. Sethna and R. E. Miller. Knowledgebase of interatomic models (KIM). http://openkim.org, 2009.

[TF00] K. Tillmann and A. Forster. Critical dimensions for the formation of interfacial misfit dislocations of In0.6Ga0.4As islands on GaAs(001). *Thin Solid Films*, **368(1)**:93–104, 2000.

[TG51] S. P. Timoshenko and J. N. Goodier. *Theory of Elasticity*. New York: McGraw-Hill, 1951.

[TH03] E. B. Tadmor and S. Hai. A Peierls criterion for the onset of deformation twinning at a crack tip. *J. Mech. Phys. Solids*, **51(5)**:765–793, 2003.

[TIP+05] S. Tejima, M. Iizuka, N. Park, S. Berber, H. Nakamura and D. Tomanek. Large scale nanocarbon simulations. In *Proceedings of the 2005 NSTI Nanotechnology Conference and Trade Show*, Volume 2, pages 181–184. Danville: Nano Science and Technology Institute, 2005.

[TKH91] M. Toda, R. Kubo and N. Hashitsume. *Statistical Physics II: Nonequilibrium Statistical Mechanics*, Volume 31 of Springer Series in Solid-State Sciences. Berlin: Springer-Verlag, second edition, 1991. Third corrected printing 1998.

[TKS92] M. Toda, R. Kubo and N. Saitô. *Statistical Physics I: Equilibrium Statistical Mechanics*, Volume 30 of Springer Series in Solid-State Sciences. Berlin: Springer-Verlag, second edition, 1992. Third corrected printing 1998.

[TM77] W. R. Tyson and W. A. Miller. Surface free energies of solid metals: Estimation from liquid surface tension measurements. *Surf. Sci.*, **62(1)**:267–276, 1977.

[TM04] P. A. Tipler and G. Mosca. *Physics for Scientists and Engineers*, Volume 2. New York: W. H. Freeman, fifth edition, 2004.

[TM09] E. B. Tadmor and R. E. Miller. Quasicontinuum method website. http://www.qcmethod.org, 2009.

[TME12] E. B. Tadmor, R. E. Miller, and R. S. Elliott. *Continuum Mechanics and Thermodynamics: From Fundamental Concepts to Governing Equations*. Cambridge: Cambridge University Press, 2012.

[TN65] C. Truesdell and W. Noll. The non-linear field theories of mechanics. In S. Flügge, editor, *Handbuch der Physik*, Volume III/3, pages 1–603. Berlin: Springer, 1965.

[TN04] C. Truesdell and W. Noll. In S. S. Antman, editor, *The Non-Linear Field Theories of Mechanics*. Berlin: Springer-Verlag, third edition, 2004.

[TOP96] E. B. Tadmor, M. Ortiz and R. Phillips. Quasicontinuum analysis of defects in solids. *Philos. Mag. A*, **73(6)**:1529–1563, 1996.

[Tor72] I. McC. Torrens. *Interatomic Potentials*. New York: Academic Press, 1972.

[Tou64] R. A. Toupin. Theories of elasticity with couple-stress. *Arch. Ration. Mech. Anal.*, **17**:85–112, 1964.

[TPO96] E. B. Tadmor, R. Phillips, and M. Ortiz. Mixed atomistic and continuum models of deformation in solids. *Langmuir*, **12(19)**:4529–4534, 1996.

[TPO00] E. B. Tadmor, R. Phillips, and M. Ortiz. Hierarchical modeling in the mechanics of materials. *Int. J. Solids Struct.*, **37**:379–389, 2000.

[TPT93] C. Tserbak, H. M. Polatoglou, and G. Theodorou. Unified approach to the electronic structure of strained Si/Ge superlattices. *Phys. Rev. B*, **47(12)**:7104–7124, 1993.

[Tru68] C. Truesdell. *Essays in the History of Mechanics*. New York: Springer-Verlag, 1968.

[Tru77] C. Truesdell. *A First Course in Rational Continuum Mechanics*. New York: Academic Press, 1977.

[Tsa79] D. H. Tsai. The virial theorem and stress calculation in molecular dynamics. *J. Chem. Phys.*, **70**:1375–1382, 1979.

[TSBK99] E. B. Tadmor, G. S. Smith, N. Bernstein and E. Kaxiras. Mixed finite element and atomistic formulation for complex crystals. *Phys. Rev. B*, **59(1)**:235–245, 1999.

[TT60] C. Truesdell and R. Toupin. The classical field theories. In S. Flügge, editor, *Handbuch der Physik*, Volume III/1, pages 226–793. Berlin: Springer, 1960.

[Tuc10] M. E. Tuckerman. *Statistical Mechanics: Theory and Molecular Simulations*. Oxford: Oxford University Press, 2010.

[Tuc11] M. E. Tuckerman. Statistical mechanics course (G25.2651) lecture notes. http://www.nyu.edu/classes/tuckerman/stat.mech/lectures.html, 2011.

[TWSK02] E. B. Tadmor, U. V. Waghmare, G. S. Smith and E. Kaxiras. Polarization switching in $PbTiO_3$: An *ab initio* finite element simulation. *Acta Mater.*, **50**:2989–3002, 2002.

[Uff07] J. Uffink. Compendium of the foundations of classical statistical physics. In J. Butterfield and J. Earman, editors, *Handbook for Philsophy of Physics*, pages 923–1074. Amsterdam: Elsevier, 2007.

[UoC09] Department of Chemistry University of Cambridge. History of the theory sector of the chemistry department. www-theor.ch.cam.ac.uk/history.html, 2009.

[Var57] Y. P. Varshni. Comparative study of potential energy functions for diatomic molecules. *Rev. Mod. Phys.*, **29(4)**:664–682, 1957.

[VAR$^+$03] M. Veleva, S. Arsene, M. C. Record, J. L. Bechade and J. Bai. Hydride embrittlement and irradiation effects on the hoop mechanical properties of pressurized water reactor (PWR) and boiling-water reactor (BWR) ZIRCALOY cladding tubes. Part II. Morphology of hydrides investigated at different magnifications and their interaction with the processes of plastic deformation. *Metall. Mater. Trans. A*, **34(3)**:567–578, 2003.

[VAS09] VASP. VASP website. http://cms.mpi.univie.ac.at/vasp, 2009.

[VC87] A. F. Voter and S. P. Chen. High temperature ordered intermetallic alloys. In R. W. Siegal, J. R. Weertmong, and R. Sinclair, editors, *Characterization of Defects in Materials, Mater. Res. Soc. Symp. Proc.*, Volume 82, page 175, Pittsburgh: Materials Research Society, 1987.

[vDDLG01] A. C. T. van Duin, S. Dasgupta, F. Lorant and W. A. Goddard, III. ReaxFF: A reactive force field for hydrocarbons. *J. Phys. Chem. A*, **105**:9396–9409, 2001.

[vdGN95] E. van der Giessen and A. Needleman. Discrete dislocation plasticity: A simple planar model. *Modell. Simul. Mater. Sci. Eng.*, **3**:689–735, 1995.

[vDMH$^+$08] A. C. T. van Duin, B. V. Merinov, S. S. Han, C. O. Dorso and W. A. Goddard, III. ReaxFF reactive force field for the Y-doped BaZrO3 proton conductor with applications to diffusion rates for multigranular systems. *J. Phys. Chem. A*, **112(45)**:11414–11422, 2008.

[vDSS$^+$03] A. C. T. van Duin, A. Strachan, S. Stewman, Q. S. Zhang, X. Xu, and W. A. Goddard, III. ReaxFF(SiO) reactive force field for silicon and silicon oxide systems. *J. Phys. Chem. A*, **107(19)**:3803–3811, 2003.

[vdZ07] H. van der Zant. Molecular electronics and devices website. http://www.med.tn.tudelft.nl, 2007.

[Ver67] L. Verlet. Computer "experiments" on classical fluids. I. thermodynamical properties of Lennard-Jones molecules. *Phys. Rev.*, **159(1)**:98, 1967.

[VFG98] M. Verdier, M. Fivel, and I. Groma. Mesoscopic scale simulation of dislocation dynamics in fcc metals: Principles and applications. *Modell. Simul. Mater. Sci. Eng.*, **6(6)**:755–770, 1998.

[vH49] L. van Hove. Quelques propriétés générales de l'intégrale de configuration d'un système de particules avec interaction. *Physica*, **15(11–12)**:951 – 961, 1949.

[Vit68] V. Vitek. Intrinsic stacking faults in body-centered cubic crystals. *Philos. Mag.*, **18**:773–786, 1968.

[VKS96] A. F. Voter, J. D. Kress and R. N. Silver. Linear-scaling tight binding from a truncated-moment approach. *Phys. Rev. B*, **53(19)**:12733–12741, 1996.

[vL01] J. H. van Lith. Stir in stillness: A study in the foundations of equilibrium statistical mechanics. PhD thesis, Universiteit Utrecht, 2001. Full text available at http://igitur-archive.library.uu.nl/dissertations/1957294/UUindex.html.

[vVLZ$^+$03] K. J. van Vliet, J. Li, T. Zhu, S. Yip, and S. Suresh. Quantifying the early stages of plasticity through nanoscale experiments and simulations. *Phys. Rev. B*, **67**:104105, 2003.

[VMG02] A. F. Voter, F. Montalenti, and T. C. Germann. Extending the time scale in atomistic simulation of materials. *Annu. Rev. Mater. Res.*, **32**:321–346, 2002.

[vN32] J. von Neaumann. Physical applications of the ergodic hypothesis. *Proc. Nat. Acad. Sci. USA*, **18**:263–266, 1932.

[Vot94] A. F. Voter. The embedded atom method. In J. H. Westbrook and R. L. Fleischer, editors, *Intermetallic Compounds: Principles and Practice*, pages 77–90. London: John Wiley and Sons, Ltd, 1994.

[Vot97] A. F. Voter. Hyperdynamics: Accelerated molecular dynamics of infrequent events. *Phys. Rev. Lett.*, **78**:3908–3911, 1997.

[Vot98] A. F. Voter. Parallel replica method for dynamics of infrequent events. *Phys. Rev. B*, **57(22)**:13985–13988, 1998.

[VRSK98] L. Vitos, A. V. Ruban, H. L. Skriver, and J. Kollár. The surface energy of metals. *Surf. Sci.*, **411(1–2)**:186 – 202, 1998.

[VS82] A. R. Verma and O. N. Srivastava. *Crystallography for Solid State Physics*. New Delhi: Wiley Eastern Ltd., 1982.

[vVS02] K. J. van Vliet and S. Suresh. Simulations of cyclic normal indentation of crystal surfaces using the bubble-raft model. *Philos. Mag. A*, **82(10)**:1993–2001, 2002.

[WAD95] E. Wajnryb, A. R. Altenberger and J. S. Dahler. Uniqueness of the microscopic stress tensor. *J. Chem. Phys.*, **103(22)**:9782–9787, 1995.

[Wal72] D. C. Wallace. *Thermodynamics of Crystals*. Mineola: Dover, 1972.

[Wal94] D. J. Wales. Rearrangements of 55-atom Lennard-Jones and $(C_{60})_{55}$ clusters. *J. Chem. Phys.*, **101(5)**:3750–3762, 1994.

[Wal03] D. J. Wales. *Energy Landscapes*. Cambridge: Cambridge University Press, 2003.

[Wei83] J. H. Weiner. *Statistical Mechanics of Elasticity*. New York: John Wiley and Sons, 1983.

[Wen91] R. M. Wentzcovitch. Invariant molecular-dynamics approach to structural phase transitions. *Phys. Rev. B*, **44(5)**:2358–2361, 1991.

[WFvdGN02] D. Weygand, L. H. Friedman, E. van der Giessen and A. Needleman. Aspects of boundary-value problem solutions with three-dimensional dislocation dynamics. *Modell. Simul. Mater. Sci. Eng.*, **10(4)**:437–468, 2002.

[WGC98] Y. A. Wang, N. Govind, and E. A. Carter. Orbital-free kinetic-energy functionals for the nearly free electron gas. *Phys. Rev. B*, **58(20)**:13465–13471, 1998.

[Wik05] Wikipedia. Photograph of Boltzmann's grave in central cemetery of Vienna – Wikipedia, the free encyclopedia. http://en.wikipedia.org/wiki/File:Zentralfriedhof_Vienna_-_Boltzmann.JPG, 2005. [Online; accessed March 05, 2009].

[Wik08] Wikipedia. Macrostructure of rolled and annealed brass; magnification 400x – Wikipedia, the free encyclopedia. http://en.wikipedia.org/wiki/File:SDC10257.JPG, 2008. [Online; accessed 11-Aug-2010].

[Wim96] E. Wimmer. Computational materials design and processing: perspectives for atomistic approaches. *Mater. Sci. Eng.*, **B37**:72–82, 1996.

[WJTP08] G. J. Wagner, R. E. Jones, J. A. Templeton and M. L. Parks. An atomistic-to-continuum coupling method for heat transfer in solids. *Comput. Meth. Appl. Mech. Eng.*, **197(41–42)**:3351–3365, 2008.

[WKL04] G. J. Wagner, E. G. Karpov and W. K. Liu. Molecular dynamics boundary conditions for regular crystal lattices. *Comput. Meth. Appl. Mech. Eng.*, **193**:1579–1601, 2004.

[WL03] G. J. Wagner and W. K. Liu. Coupling of atomistic and continuum simulations using a bridging scale decomposition. *J. Comput. Phys.*, **190**:249–274, 2003.

[WM92] D. Wolf and K. L. Merkle. Correlation between the structure and energy of grain boundaries in metals. In D. Wolf and S. Yip, editors, *Materials Interfaces: Atomic Level Structure and Properties*, Chapter 3, pages 87–150. London: Chapman and Hall, 1992.

[WW96a] D. J. Wales and T. R. Walsh. Theoretical study of the water pentamer. *J. Chem. Phys.*, **105(16)**:6957–6971, 1996.

[WW96b] T. R. Walsh and D. J. Wales. Rearrangements of the water trimer. *J. Chem. Soc., Faraday Trans.*, **92(14)**:2505–2517, 1996.

[WWN05] S. C. Wu, D. T. Wasan, and A. D. Nikolov. Structural transitions in two-dimensional hard-sphere systems. *Phys. Rev. E*, **71(5)**:056112, 2005.

[WYPW93] J. Wang, S. Yip, S. R. Phillpot and D. Wolf. Crystal instabilities at finite strain. *Phys. Rev. Lett.*, **71(25)**:4182–4185, 1993.

[WYW10] R. M. Wentzcovitch, Y. G. Yu and Z. Wu. Thermodynamic properties and phase relations in mantle minerals investigated by first principles quasiharmonic theory. *Rev. Mineral. Geochem.*, **71**:59–98, 2010.

[XAO95] X.-P. Xu, A. S. Argon, and M. Ortiz. Nucleation of dislocations from crack tips under mixed modes of loading: Implications for brittle versus ductile behaviour of crystals. *Philos. Mag. A*, **72**:415, 1995.

[XB04] S. P. Xiao and T. Belytschko. A bridging domain method for coupling continua with molecular dynamics. *Comput. Meth. Appl. Mech. Eng.*, **193**:1645–69, 2004.

[XM96] W. Xu and J. A. Moriarty. Atomistic simulation of ideal shear strength, point defects and screw dislocations in bcc transition metals: Molybdenum as a prototype. *Phys. Rev. B*, **54(10)**:6941–6951, 1996.

[XM98] W. Xu and J. A. Moriarty. Accurate atomistic simulations of the Peierls barrier and kink-pair formation energy for $\langle 111 \rangle$ screw dislocations in bcc Mo. *Comput. Mater. Sci.*, **9(3–4)**:348–356, 1998.

[XMG09] Y. Xing, A. J. Majda, and W. W. Grabowski. New efficient sparse space-time algorithms for superparameterization on mesoscales. *Mon. Weather Rev.*, **137(12)**:4307–4324, 2009.

[YC82] M. T. Yin and M. L. Cohen. Theory of lattice-dynamical properties of solids: Application to Si and Ge. *Phys. Rev. B*, **26(6)**:3259–3272, 1982.

[YE06] J. Z. Yang and W. E. Generalized Cauchy–Born rules for elastic deformation of sheets, plates and rods. *Phys. Rev. B*, **74**:184110, 2006.

[Yip06] S. Yip, editor. *Handbook of Materials Modeling*, Part A. *Methods*. New York: Springer Science and Business Media, 2006.

[YL52] C. N. Yang and T. D. Lee. Statistical theory of equations of state and phase transitions. I. theory of condensation. *Phys. Rev.*, **87(3)**:404–409, 1952.

[YSM01a] L. H. Yang, P. Soderlind, and J. A. Moriarty. Accurate atomistic simulation of $(a/2)\langle 111 \rangle$ screw dislocations and other defects in bcc tantalum. *Philos. Mag. A*, **81**:1355–1385, 2001.

[YSM01b] L. H. Yang, P. Soderlind and J. A. Moriarty. Atomistic simulation of pressure-dependent screw dislocation properties in bcc tantalum. *Mater. Sci. Eng. A*, **309(Sp. Iss. SI)**:102–107, 2001.

[Yuk91] V. I. Yukalov. Phase transitions and heterophase fluctuations. *Phys. Rep.*, **208(6)**:395–489, 1991.

[Zan92] G. Zanzotto. On the material symmetry group of elastic crystals and the Born rule. *Arch. Ration. Mech. Anal.*, **121**:1–36, 1992.

[Zan96] G. Zanzotto. The Cauchy–Born hypothesis, nonlinear elasticity and mechanical twinning in crystals. *Acta Crystallogr., Sect. A*, **52**:839–849, 1996.

[ZBG09] J. A. Zimmerman, D. J. Bammann and H. Gao. Deformation gradients for continuum mechanical analysis of atomistic simulations. *Int. J. Solids Struct.*, **46**:238–253, 2009.

[ZCvD+04] Q. Zhang, T. Cagin, A. van Duin, W. A. Goddard, III, Y. Qi and L. G. Hector. Adhesion and nonwetting-wetting transition in the Al/α-Al2O3 interface. *Phys. Rev. B*, **69(4)**, 2004.

[ZHG+02] P. Zhang, Y. Huang, P. H. Geubelle, P. A. Klein and K. C. Hwang. The elastic modulus of single-wall carbon nanotubes: A continuum analysis incorporating interatomic potentials. *Int. J. Solids Struct.*, **39**:3893–3906, 2002.

[Zho03] M. Zhou. A new look at the atomic level virial stress: On continuum-molecular system equivalence. *Proc. R. Soc. London, Ser. A*, **459**:2347–2392, 2003.

[ZJ05] J. Zinn-Justin. *Path Integrals in Quantum Mechanics*. Oxford: Oxford University Press, 2005.

[ZJZ+08] R. Zhu, F. Janetzko, Y. Zhang, A. C. T. van Duin, W. A. Goddard, III and D. R. Salahub. Characterization of the active site of yeast RNA polymerase II by DFT and ReaxFF calculations. *Theor. Chem. Acc.*, **120(4–6)**:479–489, 2008.

[ZLC05] B. J. Zhou, V. L. Ligneres and E. A. Carter. Improving the orbital-free density functional theory description of covalent materials. *J. Chem. Phys.*, **122(4)**, 2005.

[ZM67] H. Ziegler and D. McVean. On the notion of an elastic solids. In B. Broberg, J. Hult, and F. Niordson, editors, *Recent Progress in Applied Mechanics (The Folke Odquist Volume)*, pages 561–572. Stockholm: Almquist and Wiksell, 1967.

[ZM03] R. R. Zope and Y. Mishin. Interatomic potentials for atomistic simulations of the Ti–Al system. *Phys. Rev. B*, **68(2)**, 2003.

[ZRH98] H. M. Zbib, M. Rhee, and J. P. Hirth. On plastic deformation and the dynamics of 3D dislocations. *Int. J. Mech. Sci.*, **40**:113–127, 1998.

[ZT05] O. C. Zienkiewicz and R. L. Taylor. *The Finite Element Method*. London: McGraw-Hill, sixth edition, 2005.

[Zun02] J. D. Zund. George David Birkhoff and John von Neumann: A question of priority and the ergodic theorems, 1931–1932. *Historia Mathematica*, **29**:138–156, 2002.

[Zwa01] R. Zwanzig. *Nonequilibrium Statistical Mechanics*. Oxford: Oxford Univ. Press, 2001.

[ZWH+04] J. A. Zimmerman, E. B. Webb, III, J. J. Hoyt, R. E. Johnson and D. J. Bammann. Calculation of stress in atomistic simulation. *Modell. Simul. Mater. Sci. Eng.*, **12**:S319–S332, 2004.

[ZZ87] O. C. Zienkiewicz and J. Z. Zhu. A simple error estimator and adaptive procedure for practical engineering analysis. *Int. J. Numer. Methods Eng.*, **24**:337–357, 1987.

Index

Printed in the United States
by Baker & Taylor Publisher Services